GUIDE TO STABILITY DESIGN CRITERIA FOR METAL STRUCTURES

GUIDE TO STABILITY DESIGN CRITERIA FOR METAL STRUCTURES

Sixth Edition

Edited by
RONALD D. ZIEMIAN

WILEY

JOHN WILEY & SONS, INC.

This book is printed on acid-free paper.

Published by John Wiley & Sons, Inc., Hoboken, New Jersey

Published simultaneously in Canada

For general information about our other products and services, please contact our Customer Care Department within the United States at (800) 762-2974, outside the United States at (317) 572-3993 or fax (317) 572-4002.

Wiley also publishes its books in a variety of electronic formats. Some content that appears in print may not be available in electronic books. For more information about Wiley products, visit our web site at www.wiley.com.

Library of Congress Cataloging-in-Publication Data:

Guide to stability design criteria for metal structures / edited by Ronald D. Ziemian. – 6th ed.
p. cm.
Includes index.
ISBN 978-0-470-08525-7 (cloth)
1. Columns. 2. Girders. I. Ziemian, Ronald D.
TA660.C6G85 2010
624.1′821—dc22

2009020807

Printed in the United States of America

10 9 8 7 6 5 4 3

CONTENTS

PREFACE

Since its founding in 1944, the principal objectives of the Structural Stability Research Council (SSRC, formerly the Column Research Council) have been to foster research on the behavior of compressive components of metal structures and of structural systems and to assist in the development of improved design procedures. The Council provides guidance to practicing engineers and writers of design specifications, codes, and standards in both offering simplified and refined procedures applicable to design and assessing their limitations.

The initial outline of this guide was prepared in 1956 by Lynn S. Beedle and Jonathan Jones. The first edition, published in 1960, was dedicated to the Council's first chairman with these words: "As first Chairman of Column Research Council, Shortridge Hardesty gave freely for twelve years his time, devotion, and material assistance. His mind grasped both the practical problems of engineering application and the fundamental knowledge necessary to research. His influence was a personal inspiration to all who worked in Column Research Council."

This, the sixth edition of the guide, is essentially an evolutionary extension of previous editions, which were published about every decade since the first edition in 1960. In an effort to permit many of the chapters in this edition to be expanded to reflect significant recent growth in knowledge of several topics related to structural stability, the total number of chapters has been reduced by one (Chapter 19—*Selected Topics in Dynamic Stability* has been removed). Many of the chapters that appear under the same title have been extensively revised or entirely rewritten, either for clarity and/or to present and evaluate new research results. These chapters include: *Centrally Loaded Columns* (Chapter 3), *Plates* (Chapter 4), *Beams* (Chapter 5), *Box Girders* (Chapter 7), *Beam-Columns* (Chapter 8), *Horizontally Curved Steel I-Girders* (Chapter 9), *Composite Columns and Structural*

Systems (Chapter 10), *Thin-Walled Metal Construction* (Chapter 13), *Frame Stability* (Chapter 16), *Arches* (Chapter 17), *Stability under Seismic Loading* (Chapter 19), and *Stability Analysis by the Finite Element Method* (Chapter 20). In addition, most of the remaining chapters have been updated and expanded to include recent research and changes to code provisions.

The Council was fortunate in having excellent previous editors to the guide, including Bruce G. Johnston (editions 1 through 3) and Theodore V. Galambos (editions 4 and 5). The Council is indebted to both men for the time and effort that each has devoted to this work. As a result of Dr. Johnston's and Dr. Galambos's guidance, inspiration, and example, the Task Groups of the SSRC have produced the world's foremost collection of ideas and information on the subject of the stability of metal structures in the first five editions of the guide. It is with great esteem that this, the sixth edition of the guide, was shepherded through the publication process by the current editor.

The concept that the Task Groups of the SSRC would have the responsibility for most individual chapters was retained also in this edition. Additional substantial contributions were provided by others, who either wrote or reviewed drafts of chapters. The editor sincerely thanks all those who had a hand in this effort. A list of those primarily responsible for each chapter is presented at the end of Chapter 1.

The first edition of the guide received special financial support from the Engineering Foundation and the Association of American Railroads. Costs in preparing the second edition were borne jointly by the American Institute of Steel Construction and the Column Research Council. Preparation of the third edition was supported by the National Science Foundation and the American Institute of Steel Construction. The preparation of both the fourth and fifth editions was supported by a grant from the National Science Foundation. The sixth edition of the guide was supported financially through generous grants from the American Institute of Steel Construction, the Aluminum Association, John Wiley & Sons, and the Structural Stability Research Council. The Council further enjoys the financial support of many organizations, listed in Appendix C, which provide for its continuity and make it possible to sponsor several publications related to structural stability, including SSRC's annual conference proceedings. Through the members of these organizations the Council also maintains the vital creative interaction between structural engineering practice and research.

The Structural Stability Research Council and the editor deeply thank all those who have provided intellectual and financial support for encouraging this dissemination of research results on the stability of metal structures to the professional and research community. On a personal note, the editor wishes to recognize William McGuire, Professor Emeritus at Cornell University and longtime member of SSRC, for being an excellent mentor and friend and always emphasizing the need to deliver rather sophisticated topics in structural stability in a form useful to the practicing engineer.

R.D. Ziemian, Editor
Lewisburg, PA
December 2009

NOTATION

A	Cross-section area
$[A]$	Stiffness matrix
A_1, A_2	Internal and external cross-section areas of cylinder
A_b	Cross-section area of beam; area of all batten elements within one batten spacing
A_c	Cross-section area of column; area of a single longitudinal element; area of compression flange; area of concrete infill; area of core of buckling restrained brace
A_d	Cross-section area of diaphragm normal to column axis and contributing to support of member
A_e	Effective or elastic cross-section area; required area of end plate; effective area of diaphragm and flanges
A_f	Cross-section area of flange, compression flange, or stiffened flange; cross-section area per unit width of box-girder flange
A_{fc}, A_{ft}	Cross-section area of compression and tension flanges
A_{he}	Net area of horizontal cross section
$\overline{A}_{mn}, \overline{B}_{mn}, \overline{C}_{mn}$	Amplitudes of shell buckling waves
A_r	Area of longitudinal reinforcing bars
A_{ri}	Area of an individual rebar
A_s	Area of steel; cross-section area of steel tube, stiffener, or steel shape; area of small stiffener plus total area of shell between small stiffeners

A_S	Area of large stiffener plus total area of shell between small stiffeners
A_{si}	Area of stiffener i
A_{sr}	Area of longitudinal reinforcing bars
A_{st}	Area of intermediate-stiffener
A_T	Preliminary total area
A_u	Area of web bearing stiffener
A_{vea}	Minimum net area of vertical cross section of diaphragm plating
A_w	Cross-section area of web; area of welded affected zone
A_x	Cross-section area of shell stringer (meridional stiffener)
A_y	Cross-section area of ring stiffener
a	Center-to-center distance between battens; cross-beam spacing; longitudinal spacing of transverse flange stiffeners; depth of compression block; distance between restrained axis and centroid; fastener spacing; total length of column; acceleration; initial angle of extensometer gage point
$a_{\max}$	Spacing of vertical stiffeners
a_t	Length of column end tie plate
$\overline{B}$	Coefficient in postbuckling plate formula
B	Spacing of webs in main beam; average width of diaphragm
$[B]$	Strain-displacement matrix
B_1	Moment amplification factor assuming no lateral translational
B_{1x}, B_{1y}	Major and minor axis moment amplification factors
B_2	Moment amplification factor assuming lateral translation permitted
B_c	Buckling constant intercept for compression in aluminum columns and beam flanges
$[B_{GNL}]$	Geometric nonlinear strain-displacement matrix
$[B_L]$	Linear strain-displacement matrix
$[B_{NL}]$	Nonlinear strain-displacement matrix
B_p	Buckling constant intercept for axial compression in aluminum flat plates
B_s	Buckling constant intercept for shear in aluminum flat elements
B_t	Buckling constant intercept for axial compression in aluminum curved plates
B_{tb}	Buckling constant intercept for bending compression in aluminum curved plates
b	Width; effective width; flange width; width between centroids of chords; width of plate; overall width of longitudinally stiffened panel; distance from cross-section centroid to tension flange; spacing between webs; spacing of stiffeners; web depth
$\{b\}$	Force vector
b_1, b_2	Depths of subpanels
b_e	Effective width

b_f	Width of flange
b_{fc}, b_{ft}	Width of compression and tension flanges
b_o	Width of plate
b_s	Width of angle
b_w	Depth of member between flange centroids
b_x, b_y	Spacing of ringer and stringer (meridional) stiffeners
$\overline{b}_x, \overline{b}_y, \overline{b}_z$	Effective length in x-, y-, and z-directions
C	Elastic-buckling coefficient; nondimensional buckling load parameter; elastic transverse frame stiffness or equivalent spring constant; material constant; compressive force
C, C_R, C_N, C_h	Empirical web crippling coefficients that are cross-section and loading dependent
C'_u	Reduced post-buckling strength
C^*	Effective end restraint
C_1, C_2, C_3, C_4	Coefficients for lateral-torsional buckling of a beam
C_a	Membrane stiffness of a stiffened cylinder
C_b	Equivalent uniform moment factor; bending stiffness of a stiffened cylinder
C_b^*	Modified equivalent uniform moment factor
C_{bb}	Moment modification factor for full bracing condition
C_B	Coefficient that accounts for effect of bracing
C_c	Constant representing transition from elastic to inelastic buckling for compression in aluminum columns and beam flanges
C_d	Factor to account for single or reverse curvature; amplification factor to account for inelastic deformations
C_f	Maximum axial design load
C_H	Coefficient that accounts for load height effect
C_L	Coefficient to account for moment distribution along length of beam
C_m	Equivalent uniform moment factor
C_{mx}, C_{my}	Moment modification factors for major and minor axis bending
C_r	Factored buckling strength for axial compression
C_{req}	Required elastic transverse pony-truss frame stiffness or equivalent spring constant
$C_{ry,T}$	Factored out-of-plane flexural or torsional buckling strength for axial compression
C_s	Effective slenderness coefficient; seismic design coefficient
C_{TF}	Moment modification factor
C_u	Expected brace buckling strength or compressive resistance
C'_u	Reduced postbuckling compressive strength
C_w	Cross-section warping torsion constant
C_x	Coefficient describing linear elastic buckling load

c	Distance from cross-section centroid to compression flange; distance to plastic hinge in girder flange; spring stiffness
c_c	Reduction factor for column buckling
c_f	Unsupported width of flange element
c_i	Edge distance for stiffener i
c_u	Material parameter or constant
D	Diameter; depth of web between flanges, measured along web; plate coefficient; discretization factor; flexural rigidity of a plate per unit width; dead load
D_c	Buckling constant slope for compression in aluminum columns and beam flanges; depth of web in compression
D_f	Cumulated axial deformation in tension and compression up to brace fracture
D_p	Buckling constant slope for axial compression in aluminum flat plates
D_s	Buckling constant slope for shear in aluminum flat elements
D_t	Buckling constant slope for axial compression in aluminum curved plates
D_{tb}	Buckling constant slope for bending compression in aluminum curved plates
D_x, D_z	Flexural stiffness about minor and major axis
D_1, D_2, D_3	Material parameters in orthotropic plates
d	Depth of shape; width of panel; diameter of contact edge in gage hole; distance between initial and final middle ordinates
d_b	Moment amplification factor
d_o	Transverse stiffener separation
d_r	Symmetrical cross-section depth of restraining system
d_{ri}	Distance of rebar to plastic neutral axis
d_s	Depth of stiffener
d_z	Depth of panel zone between continuity plates
$\{dF\}$, $\{dF_{\text{applied}}\}$	Vector of incremental applied nodal forces
$\{dF_{\text{int}}\}$	Vector of internal member forces
$\{dR\}$	Vector of unbalanced nodal forces
$\{d\Delta\}$	Vector of incremental nodal displacements
E	Modulus of elasticity; Young's modulus
E^*	Average or apparent modulus of elasticity (ratio of average stress resisted by plate to average strain)
E_c	Modulus of elasticity of concrete
E_{cm}	Secant modulus of concrete
E_{eff}	Effective modulus
E_{ij}	Tensorial component of Green-Lagrange strain
E_m	Errors in measurement

E_o	Original or initial modulus of elasticity of concrete
E_r	Modulus of elasticity of reinforcing steel
E_R, E_r	Reduced modulus
E_s	Modulus of elasticity of steel; modulus of elasticity of shell plate; secant modulus
E_{st}	Strain hardening modulus
E_t	Effective modulus
E_T, E_t	Tangent modulus
E_x, E_y	Modulus of elasticity of x- and y-stiffeners
e	Eccentricity of load; distance from point of load application to remote edge of plate; distance between bearing stiffener and end plate; eccentricity of bearing reaction along span; radial eccentricity from a true circle; length of shear link
e_0	Out-of-roundness
e_f	Distance from mid-thickness of shell to extreme fiber of stiffener
e_{oh}	Maximum permitted nonverticality
e_u, e_x, e_y, e_z	Eccentricities in x-, y- and z-directions of applied axial load; eccentricities with respect to center of gravity
F	Applied force, restraining force
$\{F\}$	Applied or prescribed load vector; force vector
F_{br}	Brace force
$\overline{F}_{br}$	Brace strength requirement
F_c	Critical column stress; buckling load for arch
F_{cr}	Critical stress in column; critical load of frame; compressive flange stress at elastic LTB of beam
$\{F_{cr,i}\}$	Critical load vector
$F_{cr.mid}$	Elastic LTB stress for nonprismatic member calculated from elastic LTB equations for a prismatic member using cross-section properties at middle of unbraced length
$F_{cr.2}$	Elastic LTB stress in tapered beam that corresponds to largest end moment within unbraced length
F_e	Euler flexural buckling stress; least of elastic flexural, torsional, and flexural-torsional buckling stresses of section; elastic lateral-torsional buckling stress
F'_e	Euler flexural buckling stress divided by factor of safety
F_E	Elastic buckling stress
F_{ec}	Elastic global buckling stress
F_{EXP}	Elastic critical load of arch determined by experiment
F_{GMNIA}	Elastic critical load of arch determined by a geometric and material nonlinear analysis that includes initial imperfections and residual stresses
F_{LBA}	Elastic critical load of arch determined by a linear buckling analysis
F_l	Difference between yield stress and residual stress $(F_y - F_r)$

F_{MNA}	Plastic strength of arch determined by a material nonlinear analysis
F_n	Design stress; global elastic or inelastic buckling stress; flexural resistance in terms of flange major axis bending stress
F_{ni}	Notional load applied at story i
F_o	Limiting stress
F_{ob}	Critical maximum bending stress of equal-leg angles
F_r, F_{rc}	Residual compressive stress
F_s	Resisting force developed by transverse stiffener; force resisted by seismic system
F_{sy}	Yield strength of structural system resisting only lateral loads
F'_{sy}	Yield strength of structural system resisting both lateral and gravity loads
F_T	Tension-field stress
F_u	Ultimate tensile stress; factored design load on frame; ultimate strength of stiffener strut
F'_u	Modified ultimate strength of stiffener strut under combined compression and shear
$F^0_{vcr}, F^0_{bcr}, F^0_{ccr}$	Critical buckling shear stress assuming only shear, bending, or compressive stresses exist
F_y	Yield stress; specified minimum yield stress of steel section
F_{yb}	Yield strength of bottom flange
F_{yc}	Yield strength of compression flange; yield strength of core in buckling restrained brace
$F_{yc.s}$	Yield strength of compression flange at the smaller section at the cross-section transition
$F_{yc.2}$	Yield strength of compression flange at the cross-section corresponding to the largest end moment within the unbraced length
F_{yf}	Yield strength of flange
F_{yr}	Yield strength of longitudinal reinforcement
F_{vcr}	Critical buckling shear stress
$F_{y,\mathrm{web}}$	Yield stress of web
f_b	Global bending stress
f_{bu}	Major axis bending stress in flange
$f_{bu.s}$	Factored stress in compression flange at the smaller section at the cross-section transition
$f_{bu.2}$	Factored stress in compression flange at the cross-section corresponding to the largest end moment within the unbraced length
f_c	Concrete compressive strength accounting for scale effects
f'_c	Compressive strength of concrete
$f_{cr,l}$	Local plate or cross-section buckling stress
$f_{cr,l\text{-flange}}, f_{cr,l\text{-lip}}, f_{cr,l\text{-web}}$	Local plate buckling stresses in flange, lip, and web

f_l	Lateral bending stress in flange
$f_{\max}$	Maximum factored compressive stress
f_{mid}	Flange stress at mid-span
f_r	Required flange stress
f_s	Largest longitudinal flange stress in compression flange
f_v	Governing shear stress; average shear stress at service loads
f_1, f_2	Stress in compression and tension flanges assuming fully participating webs
f_{1R}, f_{2R}	Stress assuming reduced moment of inertia of cross section
f_{1w}, f_{2w}	Governing axial compressive stresses at longitudinal edge of web panel and at opposite edge of panel
f_{2s}	Average maximum bond strength
G	Shear modulus of material; gage length; ratio of column to beam stiffness
G_{eff}	Effective shear modulus of diaphragm
G_r	End-restraint relative stiffness distribution factor
G_s	Shear modulus of elasticity of shell plate
G_x, G_y	Shear modulus of elasticity of x- and y- stiffeners
g	Length modification factor
H	Total frame height; story shear produced by lateral forces; horizontal reaction for arch
H_o^m	Sobolev space of order m
h	Depth between flange centroids; clear web depth between flanges; story height; height of frame; rise of an arch
h_c	Clear depth of web under compression; height of arch crown at initiation of buckling
h_{eff}	Most efficient depth for plate girder
h_i	Initial rise of arch
h_o	Distance between flange centroids; depth of girder
h_s	Depth of stiffener; story height; location of stiffener relative to compression flange
I	Moment of inertia (second moment of area) of cross section; weighting factor for determining critical moment in overhanging beam
$\overline{I}$	Bending stiffness ratio of arch
I_b	Moment of inertia of beam
I_{be}	Mimimum required moment of inertia of transverse stiffener and its effective width of attached flange
I_c	Moment of inertia of individual longitudinal element; minimum moment of inertia of column; moment of inertia of uncracked concrete; moment of inertia at crown of arch; flexural stiffness of column; moment of inertia of core in buckling restrained brace

I_d	Moment of inertia of diagonal member
I_e	Effective moment of inertia about centroid of a section comprising one stiffener plus effective width of shell
I_E	Moment of inertia of large stiffener plus effective width of shell
I_{eff}	Effective moment of inertia
I_f	Second moment of area per unit width of box-girder flange
I_{FE}	Value of I_E that makes large stiffener fully effective
I_g	Gross moment of inertia
I_i	Moment of inertia of stiffener i
I_ℓ	Moment of inertia of stiffener about axis parallel to flange and taken at base of stiffener
I_0	Polar moment of inertia about shear center
I_o	Original moment of inertia
I_p	Polar moment of inertia of cross section about centroid
I_{ps}	Polar moment of inertia of cross section about shear center
I_r	Moment of inertia of reinforcing steel; moment of inertia of restraining system for buckling restrained brace
I_{req}	Required moment of inertia
I_s	Moment of inertia of transverse or longitudinal stiffener; moment of inertia of steel section
I_{sr}	Moment of inertia of longitudinal reinforcing bars
I_{s1}	Moment of inertia of fictitious strut
I_t	Moment of inertia of transverse member about an axis though its centroid
I_w	Moment of inertia of web
I_x, I_y	Major and minor axis moment of inertia
$I_{y.\text{top}}$	Moment of inertia of top flange about minor axis of section
I_{yc}	Moment of inertia of compression portion of section or flange about centroidal axis parallel to web
I_{yd}	Moment of inertia of a width of diaphragm plate, excluding cutouts
I_{yt}	Out-of-plane moment of inertia of tension flange; moment of inertia of tension portion of a section about centroidal axis parallel to web
I_{1E}, I_{2E}	Equivalent moments of inertia of orthotropic shell
J	St.-Venant torsional constant for cross section
J_x, J_y	St.-Venant torsional constant about x- and y-axes
j	Effective width of contact of bearing pad
K	Effective-length factor dependent on member-end conditions; bar stiffness; stiffness (moment per unit length per radian of rotation) of restraining medium; required spring constant of elastic supports for a column; Plate buckling coefficient; thinness factor; spring constant; slenderness factor
$[K]$	Stiffness matrix (typically a tangent stiffness)
$[K_o]$	Linear elastic stiffness matrix

K'	Modified effective-length factor
K_b, K_q, K_1	Coefficients used in calculating ultimate strength of a plate
$[K_{\text{baseline}}]$	Stiffness corresponding to baseline loading condition
K_c	Effective-width factor
$[K_{\text{characteristic}}]$	Stiffness corresponding to characteristic loading condition
K_e, K_l, K_β	Effective-length factors used in arch design
K_{el}	Effective-length factor determined from an elastic buckling analysis
$[K_g]$	Geometric stiffness matrix
K_i	In-plane effective-length factor
K_m	Average effective-length factor of all panel length compression chords in a pony truss
K_o	Out-of-plane effective length factor
K_s	Shear buckling coefficient for cylindrical shells
$[K_s]$	Initial stress stiffness matrix
K_t	Effective-length factor for twisting
$[K_T]$	Tangent stiffness matrix
K_u, K_x	Effective-length factors for bending about u- and x-axes
K_y	Effective-length factor corresponding to lateral bending restraint provided to full cross-section at ends of unbraced length; effective length factor for bending about y-axis
K_z	Effective-length factor corresponding to warping restraint at ends of unbraced length; effective-length factor for bending about z-axis
K_γ	Effective-length factor for tapered (nonprismatic) members
$[K_\sigma]$	Initial stress stiffness matrix
k	Plate buckling coefficient; moment amplification factor; ratio of moments at column face including and excluding panel zone deformations; material constant
k_a, k_{ax}	Factors to account for out-of-plane and in-plane buckling of an arch
k_c	Plate buckling coefficient for long plates
k_{cr}	Plate buckling coefficient that includes effect of stiffeners
k_s, k_v	Plate buckling coefficient for shear buckling
k_1	Coefficient for determining slenderness limits for plates with postbuckling strength
k_1, k_2	Concrete compression block parameters
k_∞	Traditional plate buckling coefficient as plate length tends to infinity
L	Total length of member; unbraced length; width of plate girder wall or infill plate; axial distance between circumferential stiffeners; span of arch; live load
L_b	Unbraced length; length of backspan; length of shell between bulkheads
L_c	Length of cantilever segment
L_d	Length of diagonal member

L_e	Effective width of shell acting as part of a stiffener
L_f	Center-to-center spacing of stiffening rings or large stiffeners
L_F	Center-to-center spacing of large shell stiffeners
L_g	Gage length; length of girder
L_H	Length between brace end hinges
L_{HE}	Longer of two brace segments measured between gusset hinges and brace intersection point
L_i	Height of story i
L_p	Limiting laterally unbraced length for full cross-section yielding
L_{pd}	Limiting laterally unbraced length for full cross-section yielding and adequate amount of rotation capacity
L_r	Limiting laterally unbraced length for inelastic LTB
L_{rbs}	Length of reduced beam section
L_s	Circumferential distance between longitudinal stringers; curved length of arch axis
L_t	Unbraced length for twisting
L_y	Unbraced length for bending about y-axis
l	Length of member; length of column between end tie plates; spacing of cross beams; panel length in a pony-truss bridge
l_f	Horizontal distance to nearest edge of bottom flange
l_s	Length of stiffener between points of effective restraint; total curved length of arch rib
M	Bending moment
M^*	Maximum design moment; maximum bending moment; second-order moment
M_A, M_B, M_C	Absolute value of moments at quarter, center, and three-quarter points of unbraced segment
M_{amys}	Out-of-plane strength for uniform bending of an arch
M_{ays}	Critical bending moment in arch
M_b	Buckling resistance of back-span for over-hanging beam system
$M_{b,Rd}$	Factored lateral-torsional buckling strength
M_{bal}	Balance moment
M_{bottom}	Maximum moment from applied loading causing compression in bottom flange
M_{bx}	Lateral-torsional buckling strength
M_c	Critical moment of cantilever segment assuming no warping restraint; factored cross-section plastic flexural strength
$M_{c,Rd}$	Factored cross-section plastic flexural strength
M_{cr}	Critical moment strength; theoretical elastic lateral-torsional buckling moment
M_{cr+}, M_{cr-}	Positive and negative critical moments in the absence of axial load

$M_{cr,d}$	Critical elastic distortional buckling moment
$M_{cr,e}$	Critical elastic lateral-torsional buckling moment
$M_{cr,g}$	Critical moment for twin-girder system
$M_{cr,l}$	Critical elastic local buckling moment
M_{cx}	Factored lateral-torsional buckling strength
M_D	Design bending resistance
M_E	Elastic critical moment for lateral-torsional buckling of beam
M_{Ed}	Bending moment design load
$M_{\exp}$	Moment capacity determined experimentally
M_f	Plastic moment resisted by flanges alone; maximum design moment
M_L	Smallest meridional half-wave number
$M_{\max}$	Absolute value of maximum moment in unbraced segment; second-order moment; maximum moment imposed at column face
M_n	Nominal moment capacity; factored flexural resistance
$M_{n(L=0)}$	Cross-section moment resistance corresponding to a theoretical "zero-length" member
M_{nd}	Nominal flexural strength for distortional buckling
M_{ne}	Nominal flexural strength for lateral-torsional buckling
M_{nl}	Nominal flexural strength for local buckling
M_{nx}, M_{ny}	Nominal major and minor axis moment resistances
$M_{ocr}, M_{ocr.\text{top}}, M_{ocr.\text{bottom}}$	Elastic lateral-torsional buckling moment capacity under uniform bending calculated by assuming top or bottom flange in compression
M_{ox}, M_{oz}	Critical flexural and lateral-torsional buckling moments for equal-leg angle.
$M_{pb,e}$	Plastic moment capacity using expected yield strength and accounting for strain hardening effects
M_p	Plastic-bending moment
M_{pb}	Plastic strength of beam
M_{pc}	Plastic-hinge moment modified to account for effect of axial compression
M_{pc}^*	Nominal flexural strength reduced for presence of axial force
M_{pe}	Effective plastic moment capacity
M_{pld}	Plastic design resistance irrespective of cross-section class
M_{px}, M_{py}	Plastic-resisting moment about x- and y-axes
M_{py}	Minor axis plastic moment capacity
M_{pz}	Major axis plastic moment capacity
M_r	Maximum design moment; factored cross-section plastic flexural strength
M_{rc}	Initial yield moment in presence of axial force
M_{rx}	Factored lateral-torsional buckling strength
M_s	Cross-section moment capacity of a fully braced member
M_{test}	Moment corresponding to experimental data

M_{th} Theoretical moment capacity
M_{top} Maximum moment from applied loading causing compression in top flange
M_u Ultimate bending strength of beam; major axis bending moment in member; largest meridional half-wave number
M_{ucx}, M_{ucy} Ultimate bending moments about x- and y-axes for a given axial force
M_{ux}, M_{uy} Maximum factored major and minor axis moments
M_{uY}, M_{zY} Moment required to initiate yield in extreme fiber in absence of axial force
M_x, M_y, M_z Moments about x-, y-, and z-axes
M_y, M_{yield} Moment to initiate yielding
M_{yc} Moment to initiate yielding in compression flange
M_{yf} Yield moment corresponding to flange under consideration
M_{yz} Lateral-torsional buckling moment for a beam in uniform bending
M_z Major axis bending moment; z-axis bending moment in equivalent orthotropic shell
M_0 Uniform bending moment; elastic lateral–torsional buckling moment
M_0, M_1 First-order bending moment
$M_{0.03}$ Moment capacity corresponding to a plastic rotation of 0.03 radians
m Number of panels; number of buckled half-waves along length of plate; number of columns in story; buckling coefficient for an arch; meridional half-wave number; material constant
$m_{cl}, m_{cr}, m_{tl}, m_{tr}$ Plastic moments in compression and tension flanges
m_{pb}, m_{pt} Resisting moments of bottom and top flanges
m_{pw} Plastic moment capacity of web
m_r Buckling coefficient for an arch
m_y, m_z Minor and major axis bending moments normalized by plastic moment capacity
N Nominal axial load; number of panels into which longitudinal stiffeners divide plate; width of patch; compressive force per unit length
$\{N\}$ Vector of shape functions
$N(x)$ Shape function
N^* Maximum axial compression based on first-order elastic analysis
N_A In-plane buckling load of an arch in uniform compression
N_{acx} Axial compression capacity of an arch in uniform compression
N_{acys} Out-of-plane strength for uniform compression of an arch
N_{ays} Elastic flexural-torsional buckling load of an arch in uniform compression
$N_{b,Rd}$ Factored buckling strength for axial compression

$N_{b,z,Rd}$	Factored out-of-plane buckling strength for axial compression
N_c	Buckling strength for axial compression
N_{cr}	Critical load of fictitious strut or orthotropic plate
$\overline{N}_{cr}$	Critical stress resultant
N_{cy}	Out-of-plane buckling strength for axial compression
N_E	Critical load for column buckling
N_{Ed}	Maximum axial design load
N_{Ex}	Elastic in-plane buckling load of an arch in uniform compression
N_{Ey}	Minor axis buckling load of a column
N_f	Number of equivalent cycles up to fracture; final middle ordinate
$N_i(x)$	shape function corresponding to approximate solution
N_i	Notional lateral load at level i; initial middle ordinate; ideal non-dimensional stiffness factor
N_L	Smallest circumferential wave length number
N_o	Axial yield load of arch
$N_{pl,rd}$	Squash load for encased shape or concrete filled tube
N_T	Torsional buckling load of a column
N_u	Tangential thrust at arch support that produced inelastic lateral buckling of arch; largest circumferential wave-length number
N_x, N_y, N_z, N_{xy}	x-, y- and z-direction tangential thrust in equivalent orthotropic shell; uniform in-plane force per unit length loadings
$\overline{N}_x$, $\overline{N}_y$	Membrane forces in x- and y-direction
N_y	Yield (squash) load of cross section
n	Ramburg-Osgood coefficient that reflects shape in knee location of a stress–strain curve; number of parallel planes of battens; number of plate panels; Poisson's ratio; number of longitudinal stiffeners; number of braces; number of circumferential lobes existing at collapse; circumferential half-wave number; number of bays
P	Axial load; allowable compressive axial load; critical load; preexisting axial compression in column; axial force; total gravity load above level under consideration; ultimate load
P_{baseline}, $\{P_{\text{baseline}}\}$	Baseline loading condition
P_c	Critical load; factored buckling strength for axial compression
$P_{\text{characteristic}}$, $\{P_{\text{characteristic}}\}$	Characteristic loading condition
P_{co}	Factored out-of-plane buckling strength for axial compression
P_{conc}	Axial load given by concrete section alone under a uniform stress of $0.85f'_c$
P_{cr}, $\{P_{cr}\}$, $\{P_{\text{critical}}\}$	Elastic buckling load of structure; elastic critical load of an open-web or open-flange column

$P_{cr,d}$ — Elastic distortional buckling axial load
$P_{cr,e}$ — Elastic global buckling axial load; minimum of critical elastic flexural, torsional, and torsional-flexural buckling loads
$P_{cr,l}$ — Elastic local buckling axial load
P_e — Euler buckling load; elastic critical load for buckling in plane of applied moments; effective-buckling force
P_E — Euler buckling load; critical load of column with stiff bracing
P_{er} — Overall buckling strength of buckling restrained bracing system
P_{ex}, P_{ey} — Elastic critical loads for major and minor axis flexural buckling
P_f — Failure load
P_F — Flexural–torsional buckling load; critical pressure determined
$P_m, P_{\max}$ — Maximum load
P_m, P_{rL}, P_{rR} — Critical loads for buckling assuming simply supported ends for weakest segment and two adjacent segments
P_{MD}, P_{MS} — Maximum dynamic and static loads
P_n — Nominal compressive strength; web crippling capacity
$P_{n(KL)}$ — Nominal axial compressive strength based on an effective length
$P_{n(L)}$ — Nominal axial compressive strength based on actual member length ($K = 1$)
P_{nd} — Column distortional buckling strength
P_{ne} — Column global buckling strength; nominal axial strength for flexural, Torsional, and torsional-flexural buckling
P_{nl} — Column local buckling strength
P_o — Axial load capacity; nominal axial strength of square CFT column
P_r — Reduced-modulus buckling load; maximum axial design load
P_s — Gravity load supported during an earthquake
P_S — Smaller of yield load P_{yd} or buckling load P_{xe} applied at shear center
P_{se} — Fictional load accounting for destabilizing effects of transverse plate stresses
P_t — Inelastic buckling load based on tangent modulus; torsional buckling load
P_T — Torsional buckling load
P_{test} — Experimental failure load
P_{TF} — Flexural–torsional buckling load
P_{TFO} — Flexural–torsional buckling load for concentric loading
P_u — Factored axial load; gravity load on structure; Euler buckling load about u-axis; ultimate patch load on plate-girder web; gravity load that must be resisted after an earthquake
$\overline{P}_{\text{ult}}$ — Ultimate load of plate under eccentric loading
P_{uy} — Axial load producing failure in absence of bending moment
P_{xe} — Axial flexural buckling load about x-axis; load for buckling in direction of y-axis if displacements ϕ and u are prevented

P_y	Axial yield (squash) load; axial load that initiates yielding
P_{yc}	Yield load of core of buckling restrained brace
P_{ye}	Critical buckling load in direction of symmetry axis
P_{yy}	Minor axis buckling load of unbraced column
P_z	Euler buckling load about z-axis; axial load in curved girder
P_ϕ	Elastic critical load for pure torsional buckling
$P_{\phi e}, P'_{\phi e}$	Concentric torsional buckling load if lateral displacements are prevented
p	Applied pressure; axial load normalized by yield load; live load
p^*	Theoretical critical end pressure for a cylinder compressed only at its ends
p_b	Critical pressure for cylinder with large stiffeners
p_B	Critical pressure assuming large stiffeners are same size as small stiffeners
p_c	Critical pressure under hydrostatic or uniform external conditions; critical pressure for a perfect ring-stiffened cylinder
p_{cr}	Critical buckling pressure
p_p	Plastic collapse pressure when circumferential stress is at yield
p_y	Hydrostatic pressure to initiate yielding
p_1, p_2	Internal and external pressures
Q	Form factor; buckling parameter; shear force required by specification plus shear due to any transverse loading; shear rigidity of diaphragm contributing to support of column
Q_{fv}	Vertical force transmitted to diaphragm by change of flange shape
Q_h	Shear force due to transverse horizontal loads transmitted from top flange
Q_i	Unfactored load effect
Q_w	Applied shear force
q	Web angle; dead load; critical intensity of distributed load; behavior (ductility) factor
R	Radius of cylinder, arch, or spherical shell; hole radius; radius of gyration; rotation capacity; girder web capacity reduction factor; ratio of stresses; radius of curvature; ratio of concentric torsional buckling load to eccentric torsional buckling load; radius of arch; response modification coefficient
$\{R\}$	residual vector representing imbalance between internal and external forces
R'_p	Reduced rotation capacity
R_c	Ratio of compressive stress when buckling occurs in combined shear and direct stress to compressive stress when buckling occurs in pure compression; radius to centroidal axis of combined stiffeners and shell of effective width
R_{cr}	Plastic rotation capacity at M_{cr}

R_d	Radius to centroidal axis of large stiffener plus effective width of shell; ductility force modification factor
R_m	Cross-section monosymmetry parameter
R_M	Factor to account for P-δ effect
R_n	Nominal resistance of structural component
R_o	Outside radius of shell; overstrength force modification factor
R_s	Ratio of shear stress when buckling occurs in combined shear and direct stress to shear stress when buckling occurs in pure shear
R_v	Vertical force transmitted to bearing(s)
R_x, R_y	Radius of shell in x- and y-directions
r	Radius of gyration of section; ratio of minimum to maximum end moments; stress gradient ($\sigma_{\min}/\sigma_{\max}$)
r_b	Radius of gyration of all batten elements
r_c	Radius of gyration of a single longitudinal element
r_e	Radius of gyration of section consisting of a stiffener plus a width of plate
r_i	Minimum radius of gyration of an individual shape in a built-up member
$r_{\min}$	Minimum radius of gyration
r_{ps}	Polar radius of gyration about shear center
r_{se}	Radius of gyration of effective section about axis parallel to plate
r_t	Radius of gyration of an equivalent tee section
r_{tf}	Equivalent radius of gyration for torsional buckling
$r_{t.2}$	Radius of gyration of compression flange plus one-sixth of depth of web in compression
r_u	Radius of gyration of cross section about u-axis
r_x, r_y	Radius of gyration of cross section about centroidal principal x- or y-axis
r_y	Radius of gyration about minor axis; radius of gyration of stiffener (without plate) about axis normal to plate
r_{yb}	Radius of gyration of girder bottom flange
$r_{y,rbs}$	Minor axis radius of gyration of beam at reduced cross section
S	Elastic section modulus; length of compressed bar; girder spacing; distance between exterior girders; section modulus of unreduced cross section; reduction factor; length of curved centroidal axis of arch rib from support to crown; developed length of arch; shear stiffness provided by bracing bars or battens
S_a	Spectral acceleration
S_e, S_{eff}	Effective section modulus
S_f, S_i	Correction for final or initial middle ordinates
S_g	Gross elastic section modulus
S_L	Required section modulus of longitudinal stiffener
$S_{\max}$	Design spacing between cross frames

S_T	Required section modulus of transverse stiffener
S_x	Elastic section modulus about major axis of section to flange under consideration
S_{xc}	Elastic section modulus to compression flange; compressive section modulus of entire section about major axis
S_y, S_z	Minor or major axis elastic section modulus
S_2	Slenderness parameter separating elastic and inelastic buckling
s	Length of arch subjected to live load; critical stress ratio; spacing of intermediate stiffeners; parameter to classify beam slenderness
T	Torque transmitted to diaphragm; natural period; tension force
T'	Natural period including second-order (P–Δ) effects
T_b	Torsion reaction at single central bearing
T_u	Expected brace axial strength in tension
t	Thickness of plate, shell, tube, strip, wall, or specimen; ratio of applied shear stress to material shear yield stress
t_d	Thickness of diaphragm
t_f	Thickness of flange
t_{fc}	Thickness of compression flange
t_{fi}	Equivalent flange thickness
t_g	Thickness of gusset plate perpendicular to brace axis
t_r	Depth of ring lip used to reinforce plate with hole
t_w	Thickness of web; thickness of infill plate
t_0, t', t_s	Thickness of stiffener
t_1, t_2	Equivalent orthotropic shell thicknesses
U_o	Distance between centroid and shear center
u	Lateral displacement; displacement function in x-direction
u_x, u_y	Horizontal and vertical translations
$u(x), \hat{u}(x)$	exact and approximate solutions
V	Normal shearing force; seismic design story shear force; design base shear force
V_B	Shear strength of web; buckling shear capacity of web
V_c	Capacity in shear provided by only concrete
V_{crc}	Beam-action shear force
V_D	Coefficient of variation associated with resistance supplied to that required
V_{DC}	Coefficient of variation associated with design charts
V_E	Coefficient of variation associated with material elastic modulus
V_{ec}	Tension-field shear force
V_{Ed}	Shear design load
V_{Em}	Coefficient of variation associated with errors in measurement
$V_{\exp}, V_{ex}$	Shear strength of girder as determined experimentally
V_{fc}	Frame-action shear force

V_{FEM}	Coefficient of variation associated with finite element results
V_{Fy}	Coefficient of variation associated with material yield strength
V_G	Coefficient of variation associated with cross-section geometry
V_M	Factored shear force acting coincident with maximum moment; coefficient of variation associated with material
$V_{\max}$	Maximum base shear force
V_p	Plastic shear capacity; coefficient of variation associated with fit of design equation to test results (test-to-predicted strength ratio)
V_{p1}, V_{p2}	Tension field shears for a stiffened girder
V_R	Coefficient of variation associated with resistance
V_{Rd}	Design resistance to shear buckling
V_s	Capacity in shear given by steel alone
V_{s+c}	Combined capacity in shear given by steel and concrete
V_S	Coefficient of variation associated with elastic section modulus
V_T	Shear strength defined by tension-field action
V_Z	Coefficient of variation associated with plastic section modulus
V_u	Ultimate shear strength
$\overline{V}_u$	Estimate of maximum shear strength
V_w	Shear strength of web
V_{wc}	Total shear resistance
V_y	Post-buckling resistance of shear wall; lateral load to initiate system yielding in absence of gravity loads; story shear strength
V_y'	Lateral load to initiate system yielding at a reduced lateral load
v	Brace lateral deformation; buckled displacement function in y-direction
W	Effective or total seismic weight of structure; parameter representing ratio of warping stiffness to St. Venant torsion stiffness; magnitude of concentrated load; wind load
w	Width of plate; deflection; thickness of infill plate; larger of width of flange between longitudinal flange stiffeners or distance from a web to nearest longitudinal flange stiffener; spacing of stiffeners; width of sub-panel; minimum total load intensity; uniform vertical load; buckled displacement function in z-direction
w_c	Unit weight of concrete; uniformly distributed load at buckling
w_{cr}	Critical uniformly distributed load
w_d	Uniform load applied to top of diaphragm
w_h	Width of cutout for stiffener
w_i	Weighting function
$w_{\max}$	Maximum uniform load resisted by arch
w_y	Uniform load required to produce yielding in arch
w_z	Width of panel zone between column flanges
x	Displacement; subtended angle in radians
x_0	Distance from shear center to centroid along principal x-axis

Y_i Total factored gravity load at level i

y Location of applied load relative to mid-height of cross-section; height of loading relative shear center; distance from centroid to compressive extreme fiber; critical moment ratio

y_{Bs} Distance from centroid of effective stiffener section to neutral axis of entire cross section of beam

y_c Distance from neutral axis to compression edge of web

y_0 Distance from shear center to centroid along principal y-axis; portion of web in compression

Z Plastic section modulus; curvature parameter

Z_e Effective section modulus of diaphragm and flanges

Z_x, Z_y Plastic section moduli about x- and y-axes

Z_{xc}, Z_{xt} Elastic moduli of effective section for extreme compression and tension fibers

z_i Distance from fiber to centroidal bending axis; distance from centroid of stiffening element to center of plate

α Buckling coefficient; material-specific constant; aspect ratio of length of sides of a plate or panel; ratio of transverse stiffener separation to depth of web between flanges; capacity reduction factor; angle of inclination of tension field with respect to vertical; internal angle of extensometer gage point; non-dimensional local buckling parameter; applied load ratio; smallest eigenvalue

α_{acx} Arch slenderness reduction factor

α_{acys} Arch slenderness reduction factor for out-of-plane buckling

α_{amx} Moment modification factor

$\alpha_{anx}, \alpha_{any}$ In-plane and out-of-plane axial compression modification factors

α_i Empirical coefficient dependent on plate boundary condition along unloaded longitudinal edge

α_{LT} Coefficient

α_m Flexural stiffness ratio; equivalent moment factor

α_R Coefficient of separation

α_s Normalized slenderness ratio of wall; slenderness of tube; moment modification factor

α_{says} Arch slenderness reduction factor for uniform bending

α_x Capacity reduction factor for initial imperfections

β Material-specific constant; plate aspect ratio of length of sides; bending coefficient; brace stiffness; story stiffness; story shear demand-to-capacity ratio; ratio of member end moments; relative rigidity parameter; equivalent uniform moment factor reliability index; distributed body force

$\overline{\beta}$ Brace stiffness per unit length

β_{act} Actual brace stiffness provided

β_{brace}	Stiffness of brace
β_{conn}	Stiffness of connection
β_i	Ideal stiffness of brace; stiffness of story at level *i*
β_L	Stiffness of lateral brace
β_o	Initial twist
$\beta_{\text{req'd}}$	Required stiffness of brace
β_{sec}	Stiffness of web plus any other stiffeners
β_{sys}	Stiffness of braced system
β_T	Torsional stiffness for discrete bracing system
$\overline{\beta}_T$	Torsional stiffness for continuous bracing system
β_x	Coefficient of monosymmetry
γ	Buckling pressure coefficient
γ_c	Partial safety factor for concrete
γ_e	Ratio of elastic buckling load level to required load level
$\gamma_{e.\text{LTB}}$	Ratio of elastic LTB load level to factored load level
γ_f	Normalizing yield strength factor for flange
γ_i	Load factor
$\gamma_{\min}$	Minimum stiffness per unit length of encasing material for buckling restrained brace
γ_p	Plastic rotation demand
γ_r	Partial safety factor for reinforcing steel
γ_s	Partial safety factor for structural steel; stiffener rigidity ratio
γ_u	Strength reduction factor; scale factor
γ_w	Normalizing yield strength factor for web
δ	Lateral deflection along column length (out-of-straightness); difference between initial and final middle ordinates
δ_0	Initial out-of-straightness of member
δ_b, δ_{by}	Moment amplification factor
δ_y	Brace axial deformation at yield
Δ	Maximum equivalent initial imperfection of strut; lateral drift of frame or inter-story (out-of-plumbness); elastic deflection under design seismic loads
$\Delta_{0,\max}$	Maximum lack of verticality
$\Delta_{1\text{st}}, \Delta_{2\text{nd}}$	first- and second-order drifts
Δ_c	Displacement at structural collapse
Δ_f	Cumulative axial deformation in tension and compression up to brace failure
Δ_H	First-order inter-story drift due to lateral forces
$\{\Delta_i\}$	Eigenvector (buckled mode shape)
Δ_{o}	Initial column out-of-straightness; inter-story out-of-plumbness
Δ^{o}	First-order lateral deflection
Δ_{res}	Residual permanent drifts

Δ_T	Total deflection
$\{\Delta u\}$	Incremental nodal displacements
$\Delta x, \Delta y, \Delta z, \theta x, \theta y, \theta z$	Degrees of freedom
Δ_y	System yield deformation
ϵ	Normal strain; lateral displacement normalized by length
$\dot{\epsilon}$	Strain rate
ϵ_B	Strain corresponding to onset of local buckling
ϵ_{cco}	Strain at confined concrete strength σ_{cco}
ϵ_{co}	Strain at f_c
ϵ_0	Original lateral displacement normalized by length
ϵ_R	Residual post-local buckling strain
ϵ_{st}	Strain at onset of strain-hardening
$\epsilon_x, \epsilon_y, \epsilon_z$	Middle surface strain in x-, y- and z-directions
ϵ_y	Strain at initial yielding
η	Factor based on test results to account for nonlinear behavior; imperfection parameter; ratio of tangent modulus to elastic modulus; plasticity reduction factor
η_{ax}	Imperfection parameter
η_b	Shear shape factor of all batten elements
η_c	Shear shape factor of a single longitudinal element
η_i	Concrete confinement coefficient
θ	Stability coefficient; plastic hinge rotation; total rotation capacity $(\theta_y + \theta_p)$; shell geometry parameter; angle of opening of circular arch
$\theta_{\text{ASCE}}, \theta_{\text{EC8}}, \theta_{\text{NBCC}}, \theta_{\text{NZS}}$	Stability coefficient according to ASCE 7, Eurocode, National Building Code of Canada, New Zealand Standard NZ 1170
θ_{cr}	Rotation at onset of buckling
θ_d	Angle of panel diagonal with respect to flange or horizontal
θ_f	Brace plastic hinge rotation causing fracture
$\theta_{\max}$	Total end rotation at M_p during unloading beam response
θ_p	Elastic end rotation at M_p; rotation at full yielding; inelastic story drift
θ_y	Rotation at initial yielding
θ_z	Rotation about z-axis
κ	Moment ratio; reduction factor that accounts for column slenderness; factor that accounts for rotational flexibility of beam-to-column connection contributed by shear deformation in the column web panel zone
κ_n	Reduction factor to account for variable end moments and column slenderness
$\kappa_x, \kappa_y, \kappa_z$	Curvatures at middle surface in x-, y- and z-directions

λ	Slenderness parameter of member; correction factor; ratio of circumference to length of a cylinder; load index or applied load ratio
$\overline{\lambda}$	Slenderness of column
λ_0, λ_1	Material-specific constants
λ_a	Equivalent slenderness parameter
λ_{amys}	Modified slenderness for arch in uniform bending
λ_c	Global slenderness used in column strength curve or associated with global buckling
λ_{cr}	Elastic critical load
λ_d	Slenderness parameter associated with distortional buckling
λ_e	Equivalent slenderness parameter
$\overline{\lambda}_{eff,x}, \overline{\lambda}_{eff,y}, \overline{\lambda}_{eff,z}$	Effective slenderness parameter for buckling about x-, y- and z-axes (length-axis)
λ_{eq}	Equivalent slenderness ratio
λ_f	Slenderness ratio of flange
$\overline{\lambda}_{\mathrm{GM}}$	Non-dimensional slenderness of arch by general method of Eurocode 3
$\lambda_h, \lambda_n, \lambda_s, \lambda_x, \lambda_t$	Correction factors
$\overline{\lambda}_{LT}$	Non-dimensional slenderness parameter
$\overline{\lambda}_{LT,0}$	Bending coefficient
λ_s	Slenderness of arch
λ_T	Slenderness ratio of stiffened arch structure
λ_w	Slenderness ratio of web
$\overline{\lambda}_x, \overline{\lambda}_y, \overline{\lambda}_z$	Slenderness parameters for x-, y- and z-axes
μ	Shear flexibility parameter; brace Ductility of brace; structural ductility factor
μ_c	Ductility of brace in compression
μ_f	Sum of maximum ductility in tension and compression reached before occurrence of fracture
μ_i	Eigenvalue (critical load index)
μ_{Scr}	Critical value for ratio of shear rigidity of a rib and bracing system in-plane containing two ribs to flexural rigidity of two arch ribs
μ_t	Ductility of brace in tension
ν	Poisson's ratio
ν_x, ν_y	Poisson's ratios for flexure in x- and y-directions
ρ	Stress gradient ($\sigma_{\min}/\sigma_{\max}$); ratio of outer-band tension to inner-band tension
ρ_c	Factor to reduce effective area of plate
ρ_D	Bias coefficient associated with resistance supplied to that required
ρ_{DC}	Bias coefficient associated with design charts
ρ_E	Bias coefficient associated with material elastic modulus

ρ_{Em}	Bias coefficient associated with errors in measurement
ρ_{FEM}	Bias coefficient associated with finite element results
ρ_{Fy}	Bias coefficient associated with material yield strength
ρ_G	Bias coefficient associated with cross-section geometry
ρ_i	In-plane slenderness ratio of an arch rib
ρ_M	Bias coefficient associated with material property
ρ_o	Out-of-plane slenderness ratio of an arch rib
$\rho_{o,\max}$	Largest ratio of factored moment (or compression flange stress) to section yield strength
ρ_P	Bias coefficient associated with fit of design equation to test results (test-to-predicted strength ratio)
ρ_R	Bias coefficient associated with resistance
ρ_S	Bias coefficient associated with elastic section modulus
ρ_Z	Bias coefficient associated with plastic section modulus
σ	Normal stress
σ_a	Longitudinal stress at centroid of effective stiffener section; average stress
$\sigma_{\text{ave}}, \sigma_{\text{av}}$	Average stress
σ_b	Bending stress; maximum longitudinal stress due to in-plane bending
σ_B	Stress corresponding to onset of local buckling
σ_{bc}	Maximum bending stress in web; bending stress in panel
σ_{br}	Maximum bending stress due to initial imperfection in restraining system within buckling restrained brace
σ_c	Elastic critical stress; elastic or inelastic compressive stress in column or plate; compression-flange buckling stress
$\overline{\sigma}_c$	Compressive failure stress of long stiffened panel
σ_{cb}	Compressive stress in plate under bending; actual compressive stress at junction of flange and web
σ_{cb}^*	Critical plate buckling stress due to bending only
σ_{cb}	Critical stress due to only bending in plate with hole perforation
σ_{cc}	Axial stress in panel
σ_{cco}	Confined concrete strength
σ_{cp}	Critical buckling stress in web or panel due to bending only
σ_{cr}	Longitudinal stress capacity of square steel tube; linear elastic buckling stress
$\sigma_{cr,c}$	Critical buckling stress in panel due to axial loading only
σ_e	Maximum edge stress; Euler stress of panel
$\sigma_{e(\text{all})}$	Allowable edge stress
σ_{E}	Euler stress of effective strut
σ_{eG}	Stress for general shell instability
σ_f	Longitudinal compressive stress in flange
$\overline{\sigma}_f$	Failure stress of short stiffened panel

σ_{f0}	Stress due to local bending at a point on stiffener farthest from flange plate
σ_{fz}	Stress at midplane of flange plate due to local bending
σ_h	Hoop stress
σ_i	Stress intensity
σ_{lc}	Ultimate stress of compression flange
σ_{ly}	Lower yield point
$\sigma_{\max}, \sigma_{\min}$	Maximum and minimum stresses
σ_n	Critical stress for columns with no part of cross section affected by welding
σ_{Nu}	Ultimate normal stress of arch
σ_p, σ_{PL}	Proportional limit stress
σ_{pw}	Critical stress for columns with part of cross section affected by welding
σ_q	Horizontal stress in middle third of length
σ_r	Residual tension stress
σ_R	Residual post-local buckling stress
σ_{rc}	Maximum compressive residual stress
σ_{sc}	Critical buckling stress of sheet panel with small curvature
σ_{su}	Limiting applied axial stress on effective strut section
σ_t	Tension-field stress
σ_{t1}, σ_{t2}	Diagonal tension stresses in subpanels
σ_{tb}	Bending stress in tube
σ_{ty}	Tension-field yield stress
σ_u	Ultimate stress; ultimate unit strength of arch
σ_{uy}	Upper yield point
σ_w	Critical stress for columns with entire cross section affected by welding
σ_x	Longitudinal stress; total axial stress resulting from any combination of load, bending, and hdyrostatic pressure
σ_{xc}	Critical stress for axial compression
σ_{xe}	Local buckling stress of doubly curved shell
σ_y	Yield stress
$\overline{\sigma}_y$	Modified yield stress
σ_{yd}	Dynamic yield stress
σ_{ye}	Effective yield stress of flange allowing for presence of shear
σ_{yf}	Yield strength of flange material
σ_{yr}	Effective hoop yield stress
σ_{ys}	Yield stress of stiffener material; static yield stress
σ_{yt}	Yield stress of tension flange material
σ_{yw}	Yield strength of web material
σ_z	Stress due to axial load in curved girder

$\sigma_{0.1}, \sigma_{0.2}$	Stresses at 0.1% and 0.2% (yield) offsets
σ_1	Maximum compressive stress; mean longitudinal stress; compressive yield strength at 0.1% offset
σ_2	Maximum compressive stress; compressive yield strength at 0.2% offset
σ_{20}	Euler stress evaluated at *L*/*r* equaling 20
σ_ϕ	Circumferential (hoop) stress
$\sigma_{\phi c}$	Critical circumferential (hoop) stress
$\sigma'_{\phi c}$	Reduced critical circumferential (hoop) stress or external pressure stress
$\sigma^*_{\phi c}$	Critical plate buckling stress under only compression
τ	Shear stress; inelastic stiffness reduction factor
τ_b	Average shear stress
τ_c	Critical shear stress; shear stress in plate; beam-action shear strength; actual average shear stress; elastic buckling stress due to torsion in cylinders; critical stress for torsion or transverse shear
τ^*	Critical plate buckling stress for shear only
τ_{cp}	Critical shear stress for perforated plate
τ_{cr}	Critical shear stress
$\tau_{cr,i}$	Inelastic shear buckling stress
τ_f	Ultimate shear stress from finite element analysis
τ_u	Shear stress for optimal angle of panel diagonal with respect to flange
τ_w	Shear stress in web
τ_{yw}	Shear yield stress of web material
$\tau_{0.1}, \tau_{0.2}$	Shear stresses at 0.1% and 0.2% (yield) offsets
τ_1	In-plane shear stress in flange plate due to torsion; shear yield strength at 0.1% offset
$\overline{v}_o$	Amplitude of initial out-of-plane deflection
ϕ	Cross-section twist; resistance factor; angle between tangent to arch axis and horizontal; curvature
$\{\phi_i\}$	eigenvector of mode *i*
ϕ_b	Resistance factor for bending
ϕ_c	Resistance factor for compression
ϕ_e	Elastic curvature
ϕ_f	Resistance factor for flexure
ϕ_p	Plastic curvature
ϕ_t	Resistance factor for tension
Φ_{LT}	Coefficient
φ	Dynamic strength amplification factor
χ	Slenderness or column reduction factor

χ_{EXP}	Non-dimensional reduction factor (F_{EXP}/F_{MNA})
χ_{GM}	Non-dimensional reduction factor (F_{GMNIA}/F_{MNA})
Ψ	ratio of maximum to minimum compressive stresses in a plate; ratio of wave length to plate width; parameter used to characterize $P-\Delta$ effects at a level
ω	Factor to account for strain hardening
ω_i	Eigenvalue for mode i
ω_L	Boundary member flexibility parameter
Ω_0	System overstrength factor
$\mathcal{L}$	General linear differential operator
Ω	Open set defining problem domain
$\mathcal{R}$	Residual for a given problem

ABBREVIATIONS

AA	Aluminum Association
AASHTO	American Association of State Highway and Transportation Officials
ACI	American Concrete Institute
AIJ	Architectural Institute of Japan
AISC	American Institute of Steel Construction
AISE	Association of Iron and Steel Engineers
AISI	American Iron and Steel Institute
API	American Petroleum Institute
AREA	American Railway Engineering Association
AREMA	American Railway Engineering and Maintenance-of-Way Association
AS	Australian Standards
ASCE	American Society of Civil Engineers
ASME	American Society of Mechanical Engineers
ASTM	American Society for Testing and Materials
ATC	Applied Technology Council
AWS	American Welding Society
AWWA	American Water Works Association
BSI	British Standards Institute
CEN	Comité Européen de Normalisation (European Committee for Standardization)
CISC	Canadian Institute of Steel Construction
CRC	Column Research Council
CRCJ	Column Research Committee of Japan
CSA	Canadian Standards Association

ECCS	European Convention for Constructional Steelwork
FEMA	Federal Emergency Management Agency
FHWA	Federal Highway Administration
GSA	General Services Administration
IASS	International Association for Shell and Spatial Structures
ICC	International Code Council
ISE	Institution of Structural Engineers
ISO	International Organization for Standardization
NACA	National Advisory Committee for Aeronautics
NIST	National Institute of Standards and Technology
NRCC	National Research Council of Canada
NZS	New Zealand Standards
RMI	Rack Manufacturers Institute
SAA	Standards Association of Australia
SJI	Steel Joist Institute
SSRC	Structural Stability Research Council
TIA	Telecommunications Industry Association
WRC	Welding Research Council

CHAPTER 1

INTRODUCTION

1.1 FROM THE METAL COLUMN TO THE STRUCTURAL SYSTEM

This guide contains a summary of modern knowledge on the behavior of metal columns, a knowledge that over time has been expanded and generalized to include any member, component, or structural system with significant portions in compression. Advances in the application of the metal structural column have involved the interrelated development of theory, materials, testing machines, test instruments, design procedures, and design standardization. The history of column theory goes back to the work of the Swiss mathematician Leonard Euler, who in 1744 published his famous column formula (Euler, 1744). The theoretical developments since then represent some of the finest achievements in the discipline of applied mechanics. Bruce G. Johnston, editor of the first three editions of this guide, has given a clear review of this history in the paper "Column Buckling Theory: Historic Highlights" (Johnston, 1983).

Since 1944, when the Column Research Council (CRC) was founded, much of the theoretical and practical work related to metal column design was performed under the auspices of the Council. In 1976 the name of the Council was changed to Structural Stability Research Council (SSRC) to reflect the broadened scope of the research. The history of the CRC-SSRC—its accomplishments and the personalities involved—has been recounted by Bruce G. Johnston (1981) from the firsthand point of view of an active and creative initial and continuous participant.

Ever since its inception the Council has played a leading role in developing rational design criteria based on research not only for metal columns but also for all types of structures and structural elements in which stability can impact behavior. This accumulated knowledge has been disseminated by the SSRC in many forms, but the chief vehicles for presenting the sum of it have been the five previous editions of this guide (1960, 93 pages; 1966, 217 pages; 1975, 616 pages; 1988, 786 pages; 1998, 911 pages). The present sixth edition aims to continue this tradition.

1.2 SCOPE AND SUMMARY OF THE GUIDE

The continued importance and vitality of the research on stability problems is due to technical and economic developments that demand the use of ever-stronger and ever-lighter structures in an increasingly wider range of applications. Such an expansion of use is made possible by developments in (1) manufacturing, such as metallurgy, cold forming, extruding, and welding; (2) theory and understanding of behavior under load; (3) fabrication technology, such as the automated assembly of structural members; (4) computer-aided design; (5) economic competition from nonmetallic materials; and (6) construction efficiency. These developments continually not only change the way in which traditional structures are designed and built, but they also make possible the economical use of material in other areas of application, such as offshore structures, transportation vehicles, and structures for outer space. In all these applications the demands of higher strength and lighter weight inexorably lead to structures in which a consideration of stability must play a crucial role in design. Increased strength and increased slenderness invariably lead to problems with instability.

The third edition of this guide (published in 1975) was a substantial expansion over the second edition (published in 1966), introducing a number of new chapters that reflected the expanded scope of the council. The fourth edition (published in 1988) added three new chapters, two on the fundamental topics of stability theory and finite element analysis of stability problems and one on box girders which dealt with the special stability problems of these structures having very slender plate elements. The fifth edition (published in 1998) included not only several new chapters but also a significant amount of reorganization to make room for these chapters. New topics included the stability of horizontally curved beams, stability of angle members, bracing, and stability under seismic loading.

Although no new chapters have been added in this sixth edition, many of the chapters have been significantly revised to reflect recent developments in various areas of stability research and recently adopted design criteria. Such areas not only include fundamental structural components such as columns, beams, and beam-columns, but also box girders, curved girders, bracing, composite systems, thin walled metal construction, frame stability, arches, stability under seismic loading, and stability analysis by finite element analysis. A few of the chapters, which deal with topics that did not receive a great amount of research interest from the task groups of the SSRC, were left relatively unchanged except for some updating of the literature and introducing one or two new topics. The previously appearing chapter entitled "Selected Topics in Dynamic Stability" was deemed to be somewhat extraneous to present SSRC concerns and in an effort to provide space for additional material has been removed from this edition.

1.3 MECHANICAL PROPERTIES OF STRUCTURAL METALS

Knowledge of the material stress–strain relationship during the elastic and initial inelastic ranges of behavior is an essential requisite to compression member analysis. In the elastic range there are accepted average values of the modulus of elasticity, and test values vary within reasonably small limits. Specified values of the yield point or yield strength[1] (depending on whether the initiation of yielding is a sudden or gradual process) are provided by the various specifications of the American Society for Testing and Materials (ASTM) and by product information from manufacturers.

The initial portions of the typical stress–strain curves for structural metals in tension or compression are shown in Fig. 1.1. The strengths of beams and columns are determined largely by stress–strain characteristics in the range shown. It should be noted that plotting complete curves to the same scale as Fig. 1.1 would take up a horizontal space between 20 and 30 times that available on the page, which is an indication of the inherent ductility of the metals shown.

The structurally significant aspects of a stress–strain curve for carbon or high-strength low-alloy structural steels can be characterized by the following five properties (see Fig. 1.1):

E = modulus of elasticity (slope of stress–strain curve in the elastic range)

σ_{uy} = upper yield point (maximum stress prior to yield stress level)

σ_y = yield-stress level (stress at a constant strain rate in the flat portion of the stress–strain curve after initial yield)

ϵ_{st} = strain at initial strain hardening

$E_{st} = (d\sigma/d\epsilon)_{\epsilon=\epsilon_{st}}$ (initial strain-hardening modulus)

These properties are generally sufficient for calculation of the inelastic strength and plastic deformation of structural steel members.

Structurally significant properties of the aluminum alloys, quenched and tempered steels, cold-worked steels, and stainless steels include the modulus of elasticity E; the yield strength σ_y, preferably determined by the offset method (ASTM Designation A370); and the tangent modulus $E_t = d\sigma/d\epsilon$, which varies with stress for strains greater than the elastic limit. For all steels and aluminum alloys the maximum tensile (ultimate) strength, based on original area, is also a part of the mill test report, although of no direct relevance to compression member behavior.

[1] In this guide the term *yield stress* generally is used to denote either the yield point or yield strength, whichever is applicable.

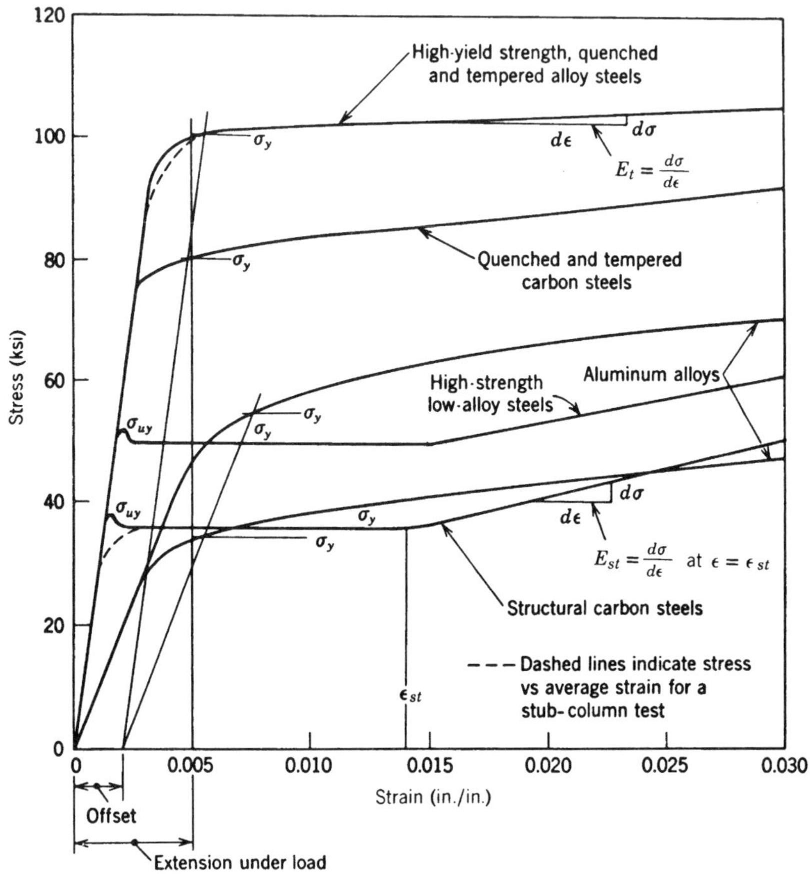

FIGURE 1.1 Initial stress–strain relationships for structural metals in tension or compression.

The yield stress of both steel and aluminum alloy varies with temperature, rate of strain, and the surface characteristics of the test specimen, as well as with the testing machine and test method. The yield stress is a function of the rate of strain, becoming lower as the testing speed is lowered. *Zero strain rate* defines a lower limit of the testing speed corresponding to the lowest yield-stress level for structural steels. ASTM specifications establish a maximum allowable strain rate. Tests made according to these specifications may be suitable for quality control but indicate yield-stress values as much as 15% greater than those from tests at low rates of strain. The influence of strain rate is less, percentagewise, for higher-strength steels.

For as-rolled structural steels, the yield-stress level in a tension or compression test can be regarded as the level of stress, after initial yield, that is sufficient at a given temperature and rate of strain to develop successively new planes of slip in the portions of the test specimen that remain in the elastic state. After initial yielding has proceeded discontinuously from point to point throughout the specimen, general strain hardening begins, and the stress rises with further increase in strain. The sharp yield point may disappear with cold work or heat treatment.

The yield-stress level is structurally more significant than the upper yield point, and its existence for relatively large average strains with no appreciable change in stress is taken advantage of in design methods based on inelastic analyses, which often make the assumption that the stress is constant and equal to the yield stress across yielded portions of the cross section.

The plot of average stress versus strain as determined by a stub-column test of an actual structural cross section is often somewhat different than that resulting from the test of a much smaller tension or compression specimen. Residual stresses that can result from the manufacturing process are one cause of these differences (indicated qualitatively as dashed lines in Fig. 1.1); other factors are the lack of uniformity of yield stress over the cross section and varying degrees of working during the rolling process. Similarly, strain hardening caused by the forming processes in cold-formed members may result in changes in yield stress which tend to shift the curve of average stress versus strain toward higher values of stress and more gradual yield development. Cold-forming effects are particularly pronounced for the stainless steels.

1.4 DEFINITIONS

The following list of terms defines their use in this guide. These terms are supplementary to the list of symbols provided in the notation and include primarily those for which variations in meaning are prevalent in the technical literature.

Beam: a straight or curved structural member, primarily supporting loads applied at right angles to the longitudinal axis. The internal stresses on a transverse cross section may be resolved into one or more of three resultant components: a transverse shear, a bending moment, and a torsional moment.

Beam-Column: a beam that also functions to transmit compressive axial force.

Bifurcation: a term relating to the load–deflection behavior of a perfectly straight and perfectly centered compression element at critical load. Bifurcation can occur in the inelastic range only if the pattern of postyield properties and/or residual stresses is symmetrically distributed so that no bending moment is developed at subcritical loads. At the critical load, a member can be in equilibrium in either a straight or a slightly deflected configuration, and bifurcation results at a branch point in the plot of axial load versus lateral deflection from which two alternative load–deflection plots are valid.

Braced Frame: a frame in which the resistance to both lateral load and frame instability is provided by the combined action of floor diaphragms and a structural core, shear walls, and/or a diagonal, K-brace, or other auxiliary system of bracing.

Buckle: to kink, wrinkle, bulge, or otherwise lose original shape as a result of elastic or inelastic strain.

Buckled: descriptive of the final shape after buckling.

Buckling Load: the load at which a compressed element, member, or frame collapses in service or buckles in an experimental loading test.

Critical Load: the load at which bifurcation (*see* Bifurcation) occurs as determined by a theoretical stability analysis.

Effective Length: the equivalent or effective length (KL) which, in the buckling formula for a pin-ended column, results in the same elastic critical load as for the framed member or other compression element under consideration at its theoretical critical load. Use of the effective-length concept in the inelastic range implies that the ratio between elastic and inelastic critical loads for an equivalent pin-ended column is the same as the ratio between elastic and inelastic critical loads in the beam, frame, plate, or other structural element for which buckling equivalence has been assumed.

Effective Width: a reduced width of plate, slab, or flat segment of a cross section which, assuming uniform stress distribution, leads to the same behavior of a structural member as the actual section of plate and the actual nonuniform stress distribution.

First Yield: a limiting stress level above which a permanent deformation results upon removal of a load.

Initial Imperfection: an unavoidable deviation from perfect geometry which is within the accepted practical tolerance of the particular applicable fabrication technology: for example, initial out-of-straightness (crookedness) of a member, initial out-of-plumb of a story, initial out-of-flatness of a plate, or initial denting or bulging of a shell.

Instability: a condition reached during buckling under increasing load in a compressive member, element, or frame at which the capacity for resistance to additional load is exhausted and continued deformation results in a decrease in load-resisting capacity.

Proportional Limit: the load or stress beyond which there is a significant amount of deviation from a prior linear load–deformation or stress–strain relationship. The term is usually used in connection with a tensile or compressive test, and the sensitivity of the strain or deformation measuring device is a determining factor in the evaluation.

Residual Stress: the stresses that exist in an unloaded member after it has been formed into a finished product. Such stresses can be caused by cold bending, finishing, straightening, flame cambering, oxygen cutting, welding, cooling after rolling, or quenching during heat treatment.

Restraint: deviation from the ideal articulated boundary condition or unbraced condition of an element, a member, or a structure.

Stability: the capacity of a compression member or element to remain in position and support load, even if perturbed slightly out of line or position by an added lateral force. In the elastic range, removal of the added lateral force would result in a return to the prior loaded position, unless the disturbance causes yielding to commence.

Strain-Hardening Modulus: for structural steels that have a flat (plastic) region in the stress–strain relationship, the strain-hardening modulus is the initial slope of the stress–strain curve just beyond the terminus of the flat region. It depends on prior strain and thermal history and exhibits a much greater range of variation than does the elastic modulus of the material.

Stub Column: a short compression test specimen utilizing the complete cross section, sufficiently long to provide a valid measure of the stress–strain relationship as averaged over the cross section, but short enough so that it will not experience flexural or torsional buckling in the elastic or plastic range.

Tangent Modulus: the slope of the stress–strain curve of material in the inelastic range, at any given stress level, as determined by the compression test of a small specimen under controlled conditions. The *effective tangent modulus* (as determined by a stub-column test) is modified by nonhomogeneity of material properties and by residual stresses.

Tangent-Modulus Load: the critical column load obtained by substituting E_t, the tangent modulus, for E in the Euler formula.

Tension-Field Action: a description of the postbuckling behavior of a plate girder panel under shear force, during which compressive stresses cause the web to form diagonal waves and tension stresses develop that are parallel to the wave troughs. These diagonal tensile stresses induce compressive stresses in the transverse (vertical) stiffeners.

Unbraced Frame: a frame in which the resistance to lateral load is provided primarily by the bending resistance of the frame members and their connections.

Yield Point: the maximum stress recorded in a tensile or compressive test of steel specimen prior to entering the plastic range.

Yield Strength: in a tension or compression test, the stress at which there is a specified amount of measured deviation from an extension of the initial linear stress–strain plot, commonly taken as the intersection of the stress–strain curve and a line parallel with the linear portion of the curve but offset by a strain of 0.002.

Yield Stress: a general term, denoting either yield strength, yield-stress level, or yield point, as defined herein.

Yield-Stress Level: for carbon- and low-alloy structural steels, the stress immediately beyond the elastic strain range, within which range the strain appears to increase without change in stress. It may be defined arbitrarily as the stress determined at a strain of 0.005.

1.5 POSTBUCKLING BEHAVIOR

Load-deflection relationships in the postbuckling range have an important bearing on the structural design significance of the critical load. For the idealized "perfect" compression element—one that is linearly elastic, devoid of imperfection, and within which the load-induced stress is perfectly uniform— there exists three different types of postbuckling behavior. As shown by the load–deflection curves in Fig. 1.2, these three cases may be demonstrated by the behavior of (1) a slender column, (2) a stiffened thin plate, and (3) a thin-walled cylinder. For each case, "perfect" elements will result in the behavior illustrated in Fig. 1.2 by solid lines, which for a given situation and corresponding buckling mode are unique and can be determined by a theoretical analysis. The dashed lines in Fig. 1.2 indicate the theoretical behavior for the same elements when a given degree of imperfection is assumed. The response of the imperfect elements (dashed lines) will approach that of the perfect elements (solid lines) as the degree of imperfection is assumed to diminish toward zero. The dashed lines are also indicative of what may be expected in a laboratory test or within an actual structure.

For the case of the elastic behavior of a slender column, the critical load for the perfect member and the maximum load carried by an imperfect one are in reasonable agreement; thus the critical load often provides a satisfactory basis for computing the design strength of the column. For the stiffened thin plate, an added postbuckling strength can be achieved with acceptably small lateral deflections, and the use of a greater strength in relation to the critical load may be justified

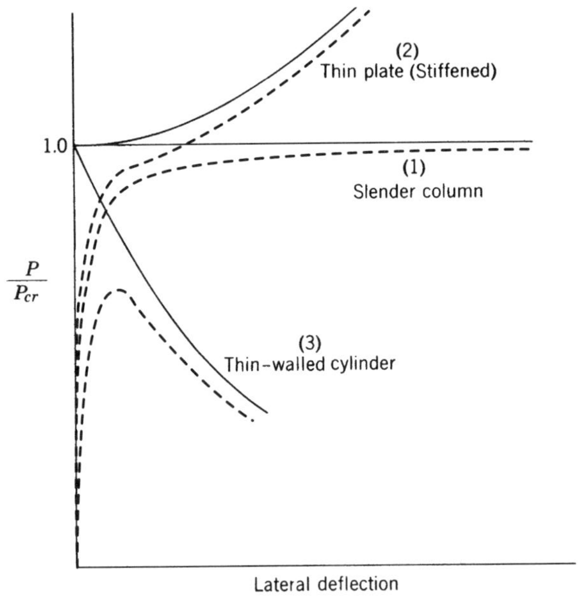

FIGURE 1.2 Elastic postbuckling curves for compressed elements.

in design. In contrast, the load-carrying capacity of a thin-walled cylinder is to an uncertain degree dependent on the amount of imperfection, and even with minimal imperfection can be drastically reduced with respect to the theoretical critical load. It is for this reason that the critical load would not be a suitable criterion on which to base the design strength of a thin-walled cylinder.

Inelastic material behavior may alter the elastic relationships depicted in Fig. 1.2. There are several conditions that may produce yielding. Although the critical load or buckling load may occur when the section is fully elastic (as indicated), subsequent postbuckling bending may produce additional stresses that when combined with the stress due to the axial load exceed the elastic stress range. On the other hand, both the critical and buckling loads may occur in the inelastic range as a result of the presence of residual stresses or the inherent nonlinearity of the material's stress–strain relationship.

When the yielding occurs after the elastic critical or buckling loads are reached, the curves shown in Fig. 1.2 will simply branch into new paths below those shown. For the cases of critical and buckling loads occurring in the inelastic range, the solid lines emanating from the critical-load bifurcation point will each take on a different initial increment of slope. In the case of the slender column, the horizontal solid line indicative of Euler buckling will be replaced by a curved line, initially sloping upward and reaching a maximum (instability) at a point somewhat greater than the critical load. A slender column with very small imperfections tends to approach this behavior as the imperfections lessen in size, with the result that with small imperfections a column under test may reach or slightly exceed the critical (tangent-modulus) load. Most important, the relevance of the critical load, or lack of relevance, is not altered from that pertinent to the completely elastic behavior, as discussed previously. Large initial imperfections, however, may cause the maximum buckling strength to be significantly lower than the critical load. In general, this behavior occurs for all types of structural systems and members and is covered in debt throughout this guide.

1.6 CREDITS FOR THE CHAPTERS IN THE SIXTH EDITION OF THE SSRC GUIDE

This book is the product of the many people who have given generously of their time and talent. This effort is gratefully acknowledged. Following is a recognition of those individuals and groups who have made major contributions arranged by chapters:

Chapter 1, "Introduction," was revised by the editor of this edition of the guide.

Chapter 2, "Stability Theory," was originally written by Alex Chajes for the fourth edition and only slightly modified by the editor of this edition.

Chapter 3, "Centrally Loaded Columns," was revised extensively by the members of SSRC Task Group 1 under the leadership of Robert Driver, with major contributions provided by Reidar Bjorhovde, Murty Madugula, Kim Rasmussen, Lip Teh, Yoon Duk Kim, and Donald White.

Chapter 4, "Plates," was originally written by Shien T. Wang and completely reorganized and updated for this edition by Benjamin Schafer of SSRC Task Group 13.

Chapter 5, "Beams," was fully revised by Donald White, with assistance from Bo Dowswell, Robert Driver, Yuhsi Fukumoto, Yoon Duk Kim, Dagowin la Poutré, and Jennifer Righman as well as advice from members of SSRC Task Group 15. The original text was written by the Theodore Galambos, with assistance from Sriramulu Vinnakota, Nicholas Trahair, Yuhsi Fukumoto, Joachim Lindner, David Nethercot, and Kit Kitipornchai.

Chapter 6, "Plate Girders," was expanded with a new section by Todd Helwig on preliminary sizing. The remainder of the chapter was revised and updated for this edition by Todd Helwig and Daniel Linzell of SSRC Task Group 27. The section on steel plate shear walls was updated by Robert Driver.

Chapter 7, "Box Girders," was originally written by Patrick Dowling and his colleagues at Imperial College in London. The chapter has been updated and revised for this edition by Domenic Coletti, Brandon Chavel, Anthony Flint, and Bernt Johannson of SSRC Task Group 27.

Chapter 8, "Beam-Columns," was originally written by David Nethercot and revised, updated, and reorganized for this edition by Dinar Camotim and Rodrigo Gonçalves.

Chapter 9, "Horizontally Curved Steel Girders," was rewritten for this edition by James Davidson with input from SSRC Task Group 14. The box girder section was provided by Chai Yoo and several figures were contributed by Joseph Hartmann.

Chapter 10, "Composite Columns and Structural Systems," was originally written by Roberto Leon and revised significantly for this edition by Amit Varma of SSRC Task Group 20.

Chapter 11, "Stability of Angle Members," was updated for this edition by Iraj Mamaghani of SSRC Task Group 26 with assistance from Theodore Galambos and David Dinehart.

Chapter 12, "Bracing," was originally written by Todd Helwig and Joseph Yura. This chapter was revised by SSRC Task Group 30 under the direction of Joseph Yura.

Chapter 13, "Thin-Walled Metal Construction," has been completely rewritten by Benjamin Schafer with input from SSRC Task Group 13.

Chapter 14, "Circular Tubes and Shells," was originally written for the fourth edition by former SSRC Task Group 18 Unstiffened Tubular Members and Task Group 22 Stiffened Cylindrical Member. It was updated for the fifth edition by former Task Group 18 Tubular Members and by Donald Sherman, who also updated the chapter for this edition with the assistance of the editor for this edition.

Chapter 15, "Members with Elastic Lateral Restraints," was originally written by Mo Elgaaly and Bruce Johnston and was revised minimally by the editor.

Chapter 16, "Frame Stability," was originally written for the fifth edition by Gregory Deierlein and Donald White. The chapter was updated for this edition by Andrea Surovek, Christopher Foley, the editor, and Ian MacPhedran of SSRC Task Group 4. A new section on structural integrity and disproportionate collapse resistance was prepared by Christopher Foley.

Chapter 17, "Arches," was significantly revised by Dagowin la Poutré.

Chapter 18, "Doubly Curved Shells and Shell-Like Structures," was written by Nicholas Morris and received some revision by the editor.

Chapter 19, "Stability under Seismic Loading," was written originally for the fifth edition by Subhash Goel with input from members of SSRC Task Group 24. The chapter was significantly revised and updated for this edition by Robert Tremblay of SSRC Task Group 24, with the assistance of Chia-Ming Uang for the section on steel moment-resisting frames.

Chapter 20, "Stability Analysis by Finite Element Methods," was completely rewritten by Christopher Earls.

Many of the above chapters include sections on aluminum design, all of which were reviewed and revised by J. Randolph Kissell. The appendixes were updated and slightly expanded by the editor with the assistance of Perry Green and SSRC Task Group 6.

REFERENCES

Euler, L. (1744), "Methodus Inveniendi Lineas Curvas ... ", Bousquet, Lausanne.

Johnston, B. G. (1981), "History of Structural Stability Research Council," *ASCE J. Struct. Div.*, Vol. 107, No. ST8, pp. 1529–1550.

Johnston, B. G. (1983), "Column Buckling Theory: Historic Highlights," *ASCE J. Struct. Eng.*, Vol. 109, No. 9, pp. 2086–2096.

CHAPTER 2

STABILITY THEORY

2.1 INTRODUCTION

The principal subject matter of this guide is the stability[1] of metal structures. This chapter introduces the various types of instability encountered in the other chapters by presenting the solutions to several simple illustrative problems.

Instability is a condition wherein a compression member or structural system loses the ability to resist increasing loads and exhibits instead a decrease in load-carrying capacity. In other words, instability occurs at the maximum point on the load–deflection curve.

Problems in instability of compression members can be subdivided into two categories: those associated with the phenomenon *bifurcation of equilibrium* and those in which instability occurs when the system reaches a maximum, or limit, load without previous bifurcation. In the first case, a perfect member when subjected to increasing load initially deforms in one mode (e.g., axial deformation) and then, at a load referred to as the *critical load*, the deformation suddenly changes into a different mode (e.g., bending deformation). Axially compressed columns, plates, and cylindrical shells experience this type of instability. By comparison, members belonging to the latter category deform in a single mode from the beginning of loading until the maximum load is reached. Shallow arches and spherical caps subjected to uniform external pressure are examples of the second type of instability.

[1]The scope—and complexity—of the subject of stability may be illustrated by two extreme ways in which the elementary verb "buckle" has been used in classical literature:

Melville's *Moby Dick* on cruising in a schooner: "Sideways leaning, we sideways darted: every ropeyarn tingling like a wire: the two tall masts buckling like Indian canes in land tornados." This translation of buckling implies elastic stability under large displacements.

Shakespeare's *King Henry the Fourth* on Northumberland's receipt of devastating news: "And as the wretch, whose fever-weaken'd joints, like strengthless hinges buckle under life." In this case, buckling may be translated as an unstable inelastic mechanism.

2.2 BIFURCATION BUCKLING

2.2.1 Initially Perfect Systems

The critical load of a compression member, obtained from the linear analysis of an idealized perfect member, does not necessarily coincide with the load at which collapse of a real imperfect member occurs. To determine the failure load of an actual member it is necessary to take initial imperfections into account and to consider the entire nonlinear load–deflection curve of the member. Unfortunately, the process of obtaining such a curve is often too difficult and time consuming to be used in routine engineering design. Instead, the maximum load of a compression member is generally calculated by semiempirical means, that is, using curves fitted to numerically obtained maximum-strength curves or combining test results with a qualitative understanding of the nonlinear load–deflection behavior of the imperfect member.

A general understanding of the basic characteristics of the elastic buckling and postbuckling behavior of members that become unstable as a result of bifurcation can be obtained by considering the simple model in Fig. 2.1. The model consists of two rigid bars hinged to one another and to the supports and restrained laterally by a nonlinear elastic spring. A similar model has been used by many, including Budiansky and Hutchinson (1964) and Hoff (1966).

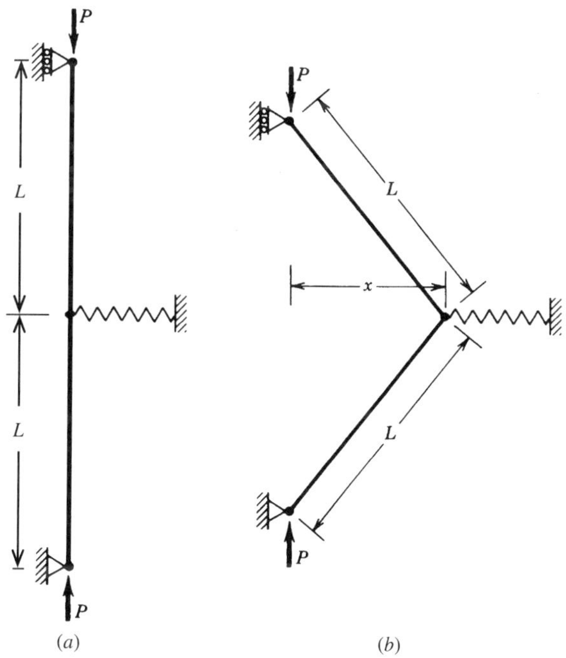

FIGURE 2.1 Bifurcation-buckling model of initially perfect system: (*a*) prior to buckling; (*b*) postbuckling.

The restraining force F exerted by the spring at the juncture of the bars is assumed to be related to the lateral displacement x by an arbitrary nonlinear function

$$F = k_1\epsilon - k_2\epsilon^2 + k_3\epsilon^3 \tag{2.1}$$

where $\epsilon = x/L$. If the model is initially straight, equilibrium in a deformed configuration requires that

$$P\epsilon = \tfrac{1}{2}F(1-\epsilon^2)^{1/2} = \tfrac{1}{2}(k_1\epsilon - k_2\epsilon^2 + k_3\epsilon^3)(1-\epsilon^2)^{1/2} \tag{2.2}$$

By letting ϵ become infinitesimally small in Eq. 2.2, the critical load obtained is

$$P_c = \frac{k_1}{2} \tag{2.3}$$

Based on the work of Koiter (1970), it has been demonstrated that the essential characteristics of the postbuckling behavior of a member can be determined by considering the initial stages of the postbuckling curve in the vicinity of the critical load. Thus ϵ is assumed to be small but finite, $\epsilon^2 \ll 1$, which reduces Eq. 2.2 to

$$P\epsilon = \tfrac{1}{2}(k_1\epsilon - k_2\epsilon^2 + k_3\epsilon^3) \tag{2.4}$$

In view of Eq. 2.3, the foregoing expression can be rewritten in the form

$$P = P_c(1 - a\epsilon + b\epsilon^2) \tag{2.5}$$

where $a = k_2/k_1$ and $b = k_3/k_1$.

Certain structures behave in a symmetrical manner; that is, the buckling characteristics are the same regardless of whether the deformation is positive or negative. To simulate the behavior of such structures, we let $a = 0$. Equation 2.5 then reduces to

$$P = P_c(1 + b\epsilon^2) \tag{2.6}$$

The load–deflection curves corresponding to Eq. 2.6 are shown in Fig. 2.2. The type of behavior depicted by these curves is referred to as *bifurcation buckling*. The member initially deforms in one mode, the prebuckling deformation, and then at the critical load, due to a branch in the load–deflection curve, the deformation suddenly changes into a different pattern, the buckling mode. For example and as stated earlier, axially loaded columns initially shorten due to axial compression. Then at the critical load the member suddenly begins to bend.

The curve in Fig. 2.2*a* results if $b > 0$, and the curve in Fig. 2.2*b* if $b < 0$. These two cases correspond to models with "springs" that become either stiffer or more flexible with increasing lateral deflection. In a similar manner, the stiffness of an actual structure may either increase or decrease subsequent to the onset of buckling.

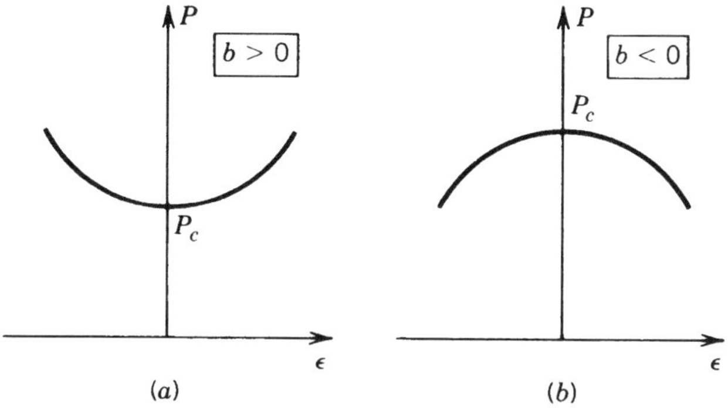

FIGURE 2.2 Symmetric buckling of a bifurcation model: (*a*) stable postbuckling curve; (*b*) unstable postbuckling curve.

In other words, the load required to keep the structure in a deformed configuration may either increase or decrease as the deformation increases in magnitude. If the load that the structure can support subsequent to the onset of buckling increases with increasing deformation, as shown in Fig. 2.2*a*, the structure is said to have a *stable postbuckling curve.* By comparison, if the load decreases, as indicated in Fig. 2.2*b*, the member has an *unstable postbuckling curve.*

An axially compressed plate with restrained edges is an example of a structure with a stable postbuckling curve. As the plate buckles, the buckling deformations give rise to tensile membrane stresses normal to the direction of loading which increase the stiffness of the plate and give it the capacity to resist additional load. By comparison, the guyed tower in Fig. 2.3 has an unstable postbuckling curve. As the top of the tower deflects laterally, some of the cables are stretched, causing them to push down on the post. As a consequence, the external load required to maintain equilibrium decreases with the magnitude of the lateral deflection. The

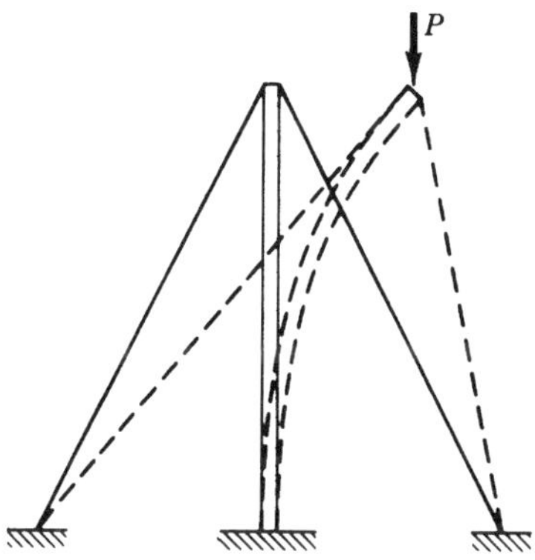

FIGURE 2.3 Guyed tower.

most notorious example of a structure with an unstable postbuckling curve is the axially compressed cylindrical shell. This system, however, does not buckle in a symmetric manner and its behavior is therefore not fully described by Fig. 2.2*b*.

To simulate the behavior of structures that behave in an asymmetric manner, return to Eq. 2.5 and let $b = 0$. Thus

$$P = P_c(1 - a\epsilon) \tag{2.7}$$

The load–deflection curve corresponding to Eq. 2.7 is shown in Fig. 2.4. Unlike the symmetric system, the unsymmetric one becomes stiffer if it deflects in one direction and more flexible if it deflects in the opposite way.

The simple frame in Fig. 2.5*a* is an example of a structure that has an asymmetric postbuckling curve. After the frame buckles as shown in Fig. 2.5*b*, a secondary tension force V is induced in the vertical member. As a consequence the external load P that the structure can support increases with increasing deformations ($\theta < 0$). By comparison, after the frame buckles as indicated in Fig. 2.5*c*, a secondary compression force is induced in the vertical member and the resistance of the system to applied loads decreases with increasing deformations ($\theta > 0$). The foregoing analytically predicted behavior of the frame in Fig. 2.5 has been verified experimentally by Roorda (1965).

2.2.2 Initially Imperfect Systems

The postbuckling curve of an initially perfect system does not by itself give sufficient information to allow one to determine when failure takes place. To obtain that information, one must also consider the initial geometrical imperfections and eccentricities of loading that are present in all real structures.

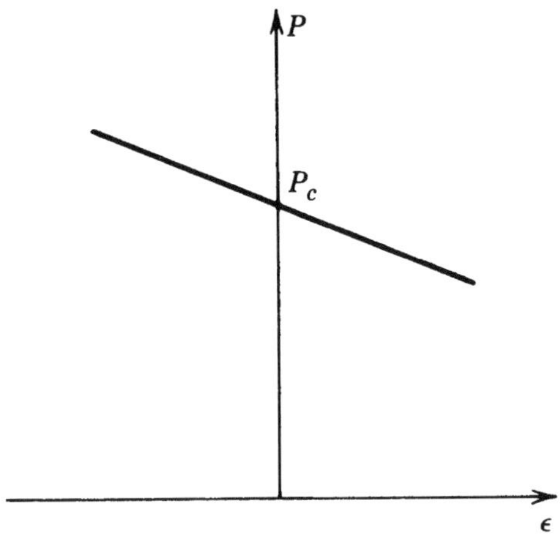

FIGURE 2.4 Asymmetric buckling of a bifurcation model.

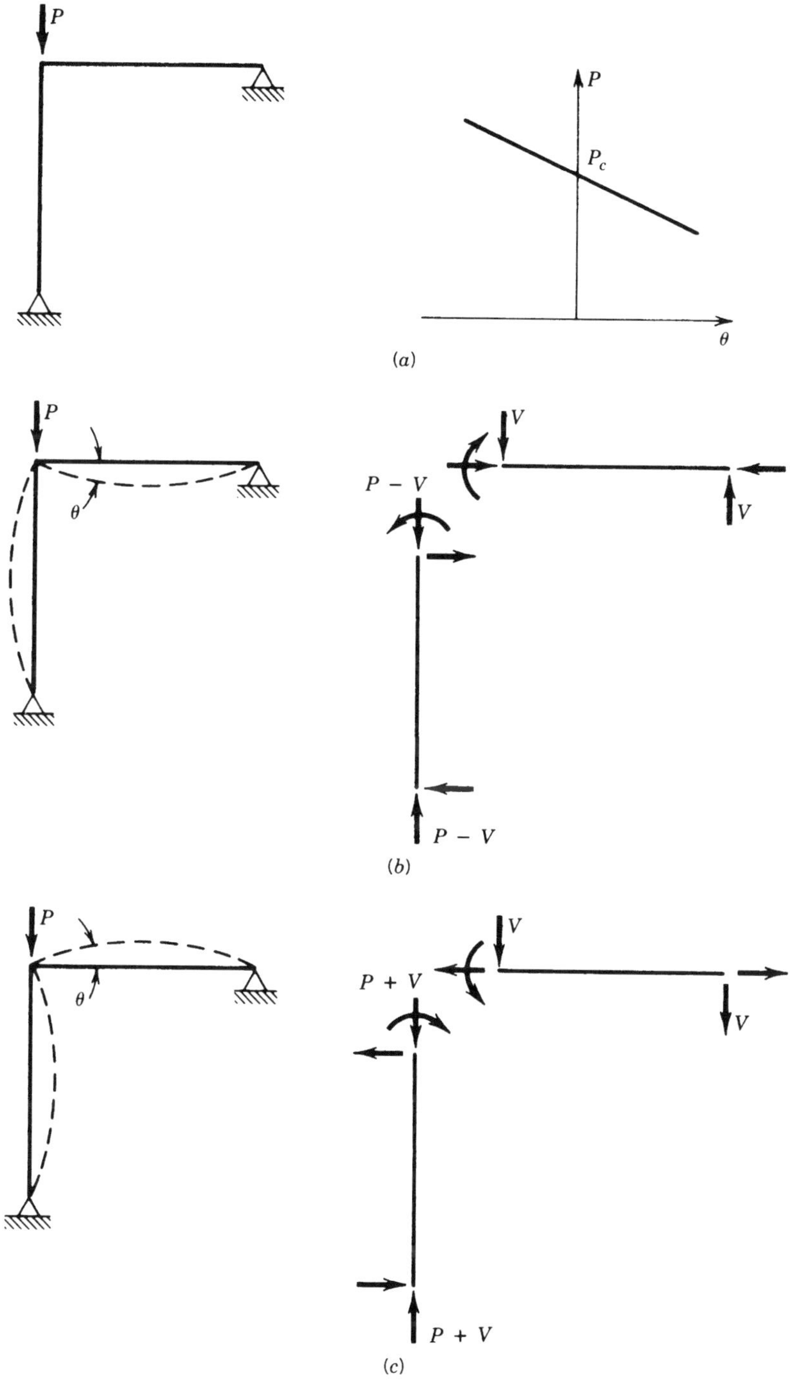

FIGURE 2.5 Buckling of an L-shaped frame.

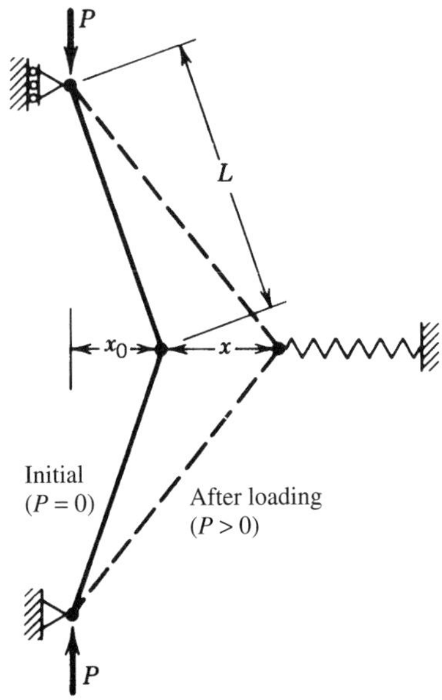

FIGURE 2.6 Model of an initially imperfect system.

Assuming that our model now has an initial deformation x_0, as indicated in Fig. 2.6, Eq. 2.4 takes the form

$$P(\epsilon + \epsilon_0) = \tfrac{1}{2}(k_1\epsilon - k_2\epsilon^2 + k_3\epsilon^3) \tag{2.8}$$

where $\epsilon_0 = x_0/L$. In view of Eq. 2.3, the relation above can be rewritten as

$$P = \frac{P_c(\epsilon - a\epsilon^2 + b\epsilon^3)}{\epsilon + \epsilon_0} \tag{2.9}$$

For symmetric behavior $a = 0$ and

$$P = \frac{P_c(\epsilon + b\epsilon^3)}{\epsilon + \epsilon_0} \tag{2.10}$$

and for asymmetric behavior $b = 0$ and

$$P = \frac{P_c(\epsilon - a\epsilon^2)}{\epsilon + \epsilon_0} \tag{2.11}$$

The load–deflection curves corresponding to Eqs. 2.10 and 2.11 are shown as dashed lines in Fig. 2.7. It is evident from these curves that small initial imperfections do not significantly affect the behavior of systems with stable postbuckling

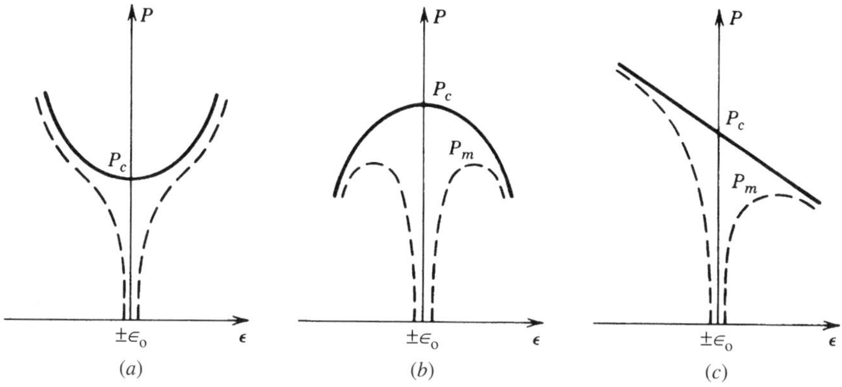

FIGURE 2.7 Postbuckling curves of initially imperfect systems (dashed lines): (*a*) symmetric stable; (*b*) symmetric unstable; (*c*) unsymmetric.

curves. These members can continue to resist increasing loads above the critical load, and failure takes place only after yielding of the material has occurred.

The amount of postbuckling strength that a system with a stable postbuckling curve possesses depends on two factors: the steepness of the postbuckling curve and the relative magnitude of the critical load and the load at which yielding begins. For example, axially compressed plates possess a relatively steep postbuckling curve and as a consequence often exhibit sizable postbuckling strength. Failure loads three or four times as large as the critical load have been obtained (Gerard, 1957). By comparison, the slope of the postbuckling curve of an axially loaded column is extremely small and the failure load of such a member therefore coincides, very nearly, with the critical load.

In addition to possessing a relatively steep postbuckling curve, a system must have a yield load that is considerably in excess of its critical load if the system is to exhibit significant postbuckling strength. A very rough estimate of the postbuckling strength of an axially compressed plate is given by the expression

$$\frac{P_c}{P_f} = \left(\frac{P_c}{P_y}\right)^{1/2} \tag{2.12}$$

where P_c is the critical load, P_f the failure load, and P_y the load when yielding commences. According to Eq. 2.12, a plate possesses significant postbuckling strength when P_c/P_y is considerably less than unity. Hence, only thin plates can be expected to display sizable postbuckling strength.

Whereas small initial imperfections have only a negligible effect on the behavior of systems with stable postbuckling curves, they have a very marked effect on systems with unstable postbuckling curves. As indicated by the curves in Fig. 2.7, the presence of small initial imperfections will cause systems that have unstable postbuckling curves to fail at loads below the critical load. These structures are accordingly referred to as being *imperfection sensitive*.

By setting $dP/d\epsilon = 0$ for Eqs. 2.10 and 2.11, the following approximations of the maximum load P_m can be obtained. For the symmetric system with $b < 0$,

$$\frac{P_m}{P_c} = 1 - 3\left(-\frac{b}{4}\right)^{1/3} \epsilon_0^{2/3} \tag{2.13}$$

and for the asymmetric system,

$$\frac{P_m}{P_c} = 1 - 2\left(a\epsilon_0\right)^{1/2} \tag{2.14}$$

Equations 2.13 and 2.14 indicate that the larger the initial imperfection x_0 and the steeper the postbuckling curve (i.e., the larger a or b), the smaller will be the ratio of P_m to P_c. Axially compressed cylindrical shells that have a very steep post-buckling curve have been found to fail at loads significantly below the critical load (Brush and Almroth, 1975). Using both theory and tests it has been demonstrated that initial imperfections whose magnitude is only 10% of the shell thickness can result in maximum loads whose magnitude is 60% of the critical load (Hutchinson and Koiter, 1970). Conversely, by manufacturing and testing nearly perfect shell specimens, failure loads only slightly below the critical load have been obtained (Tennyson, 1964).

In conclusion, it is evident that the behavior of real imperfect members can be predicted from the shape of the postbuckling curve for perfect systems. Members with stable postbuckling curves will fail at loads equal to or above the critical load, whereas members with unstable postbuckling curves will fail at loads below the critical load.

2.3 LIMIT-LOAD BUCKLING

Buckling that is associated with a bifurcation of equilibrium is not the only form of instability that can occur. A second type of instability that can take place is illustrated by the model in Fig. 2.8. The model consists of a simple arch formed by

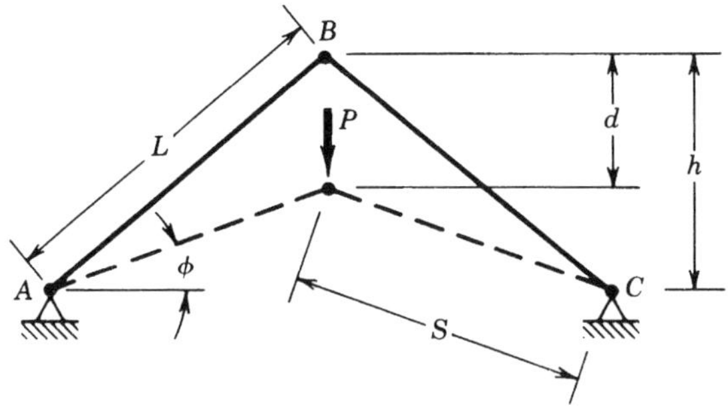

FIGURE 2.8 Limit-load buckling model.

two elastic bars hinged to each other and to the supports. As the load P acting on the model increases, legs AB and BC shorten (due to compressive axial strain) by an amount Δ, and point B moves down a distance d. The axial force F developed in the bars by the applied load P is equal to

$$F = \frac{P}{2\sin\phi} = \frac{PS}{2(h-d)} \tag{2.15}$$

and the axial shortening Δ of each bar is given by

$$\Delta = \frac{F}{K} = \frac{PS}{2K(h-d)} \tag{2.16}$$

in which $S = \sqrt{L^2 + d^2 - 2dh}$ is the length of the compressed bars and $K = AE/L$ is the stiffness of the bars. Substitution of $\Delta = L - S$ in Eq. 2.16 leads to

$$L - S = \frac{PS}{2K(h-d)}$$

or

$$L - \sqrt{L^2 + d^2 - 2dh} = P\frac{\sqrt{L^2 + d^2 - 2dh}}{2K(h-d)} \tag{2.17}$$

If the rise h of the arch is assumed to be small compared to L, Eq. 2.17 reduces to

$$P = \frac{Kh^3}{L^2}(2\delta - 3\delta^2 + \delta^3) \tag{2.18}$$

in which $\delta = d/h$.

The load-deflection relation corresponding to Eq. 2.18 is depicted by the solid curve in Fig. 2.9. It is evident that no bifurcation of equilibrium exists. Instead,

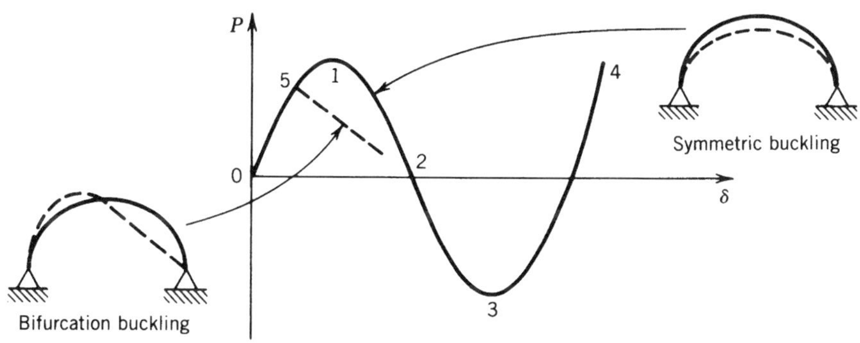

FIGURE 2.9 Load–deflection curve of a limit-load model.

the load and deformation increase simultaneously until a maximum or limit load is reached (point 1) beyond which the system becomes unstable.

If the rise h of the model is large enough compared to L, the axial forces in the legs may reach their critical loads, causing the legs to buckle as hinged–hinged columns before the entire system reaches its limit load at point 1. In that case buckling occurs as a result of a bifurcation of equilibrium at point 5 on the curve.

The behavior of arches and spherical shells subject to uniform external pressure is similar to that described by the curves in Fig. 2.9. Arches and spherical caps with a large rise-to-span ratio fail in an asymmetric mode as a result of bifurcation buckling, whereas shallow arches and spherical caps fail in a symmetric mode due to limit-load buckling.

An extensive treatment of similar and more complex elastic stability phenomena is presented by Thompson and Hunt (1984).

REFERENCES

Brush, D. O., and Almroth, B. O. (1975), *Buckling of Bars, Plates and Shells*, McGraw-Hill, New York.

Budiansky, B., and Hutchinson, J. W. (1964), "Dynamic Buckling of Imperfection-Sensitive Structures," *Proc. 11th Int. Congr. Appl. Mech.*, Munich, Germany.

Gerard, G. (1957), "Handbook of Structural Stability: Part IV. Failure of Plates and Composite Elements," NACA Tech. Note No. 3784, Aug.

Hoff, N. J. (1966), "The Perplexing Behavior of Thin Circular Cylindrical Shells in Axial Compression," *Isr. J. Technol.*, Vol. 4, No. 1, pp 1–28.

Hutchinson, J. W., and Koiter, W. T. (1970), "Postbuckling Theory," *Appl. Mech. Rev.*, Vol. 23, No. 12, pp. 1353–1363.

Koiter, W. T. (1970), "On the Stability of Elastic Equilibrium," Tech. Rep. No. AFFDLTR-70-25, Air Force Flight Dynamics Laboratory, Wright Patterson Air Force Base, OH, Feb.

Roorda, J. (1965), "Stability of Structures with Small Imperfections," *ASCE J. Eng. Mech. Div.*, Vol. 91, No. EM1, pp. 87–106.

Tennyson, R. C. (1964), "An Experimental Investigation of the Buckling of Circular Cylindrical Shells in Axial Compression Using the Photoelastic Technique," Rep. No. 102, Institute of Aerospace Sciences, University of Toronto, Toronto, Ontario, Canada, Nov.

Thompson, J. M. T., and Hunt, G. W. (1984), *Elastic Instability Phenomena*, Wiley, New York.

CHAPTER 3

CENTRALLY LOADED COLUMNS

3.1 INTRODUCTION

The cornerstone of column theory is the Euler column, a mathematically straight, prismatic, pin-ended, centrally loaded[1] strut that is slender enough to buckle without the stress at any point in the cross section exceeding the proportional limit of the material. The *buckling load* or *critical load* or *bifurcation load* (see Chapter 2 for a discussion of the significance of these terms) is defined as

$$P_E = \frac{\pi^2 EI}{L^2} \tag{3.1}$$

where E is the modulus of elasticity of the material, I is the second moment of area of the cross section about which buckling takes place, and L is the length of the column. The Euler load, P_E, is a reference value to which the strengths of actual columns are often compared.

If end conditions other than perfectly frictionless pins can be defined mathematically, the critical load can be expressed by

$$P_E = \frac{\pi^2 EI}{(KL)^2} \tag{3.2}$$

where KL is an *effective length* defining the portion of the deflected shape between points of zero curvature (inflection points). In other words, KL is the length of an equivalent pin-ended column buckling at the same load as the end-restrained

[1]Centrally loaded implies that the axial load is applied through the centroidal axis of the member, thus producing no bending or twisting.

column. For example, for columns in which one end of the member is prevented from translating with respect to the other end, K can take on values ranging from 0.5 to 1.0, depending on the end restraint.

The isolated column can be considered a theoretical concept; it rarely exists in practice. Usually, a column forms part of a structural frame and its stability is interrelated with the stability, stiffness, and strength of the surrounding structure. The structure imposes not only axial forces on the column, but also flexural and torsional forces as well as end restraints. This interrelationship is treated elsewhere in many parts of this guide. This chapter considers only the isolated column because (1) many structural design situations are idealized such that elements can be thought of as centrally loaded columns (e.g., truss members) and (2) the centrally loaded column is a limiting point in the mathematical space defining the interaction between axial and flexural forces in a member of a structure. Thus, an understanding of the behavior of individual centrally loaded columns is essential to the development of design criteria for compression members in general.

Columns are made in a variety of cross sections and by several processes, depending on their size and shape. Most steel columns are prismatic (i.e., the cross section is the same from end to end), although tapered columns are also used in certain circumstances. Virtually all rolled shapes can be used as columns, but some are much more efficient than others because of such factors as the ratio of the governing second moment of area to the weight per unit length, the ratio of the radii of gyration about perpendicular axes, double or single symmetry or asymmetry of the cross section, and the propensity toward torsional or flexural–torsional buckling or local buckling of elements.

Section 3.2 discusses general concepts related to column stability and design, whereas Sections 3.3 and 3.4 focus on key properties that have specific influences on column capacity, namely, imperfections related to residual stresses and out-of-straightness (Section 3.3) and end restraint (Section 3.4). Section 3.5 discusses specific concepts related to the development of column design curves and design criteria.

Although the majority of this chapter deals with either general concepts or the specific behavior of steel columns, other metals are also used for columns in practice. Sections 3.6 and 3.7 cover aspects particular to aluminum and stainless steel columns, respectively.

Section 3.8 deals with tapered columns, generally fabricated by welding flange plates to a tapered web plate. These are common in rigid moment-resisting steel frames, where they act, of course, as beam-columns. For frames with hinged bases, the cross-section depth increases from the base to the knee, where the bending moments are large. A component of the understanding of the behavior of these beam-columns comes from an appreciation of their behavior under axial loads.

Section 3.9 covers issues related to built-up columns. These find use in bridges and mill buildings where either the loads are relatively large or particular circumstances suggest their use. Analysis or rehabilitation of historical structures may also require an understanding of these special structural elements. Section 3.10 provides a discussion pertaining to stepped columns, commonly used to support crane runway girders in industrial structures. Section 3.11 addresses the unique problem of guyed towers.

Although this chapter provides a broad overview of the behavior and design of centrally loaded columns, it is impossible to address all related aspects of such a diverse subject. A comprehensive list of additional reference material on topics related to centrally loaded columns can be found in the annotated bibliography by Driver et al. (2003).

3.2 COLUMN STRENGTH

3.2.1 Critical-Load Theory

The strength of a perfectly straight prismatic column with central loading and well-defined end restraints that buckles elastically in a flexural mode is the Euler load, P_E (Eq. 3.2). When the axial load attains the value P_E, a stable equilibrium configuration is possible even in the presence of lateral deflection (Fig. 3.1*a*), while the load remains essentially constant (Fig. 3.1*b*, lines *OAB*). Even if an initial deflection and/or an initial load eccentricity is present, the maximum load will approach the Euler load asymptotically as long as the material remains elastic (curve *C* in Fig. 3.1*b*).

Many practical columns are in a range of slenderness where at buckling portions of the columns are no longer linearly elastic, and thus one of the key assumptions underlying Euler column theory is violated due to a reduction in the stiffness of the column. This degradation of the stiffness may be the result of a nonlinearity

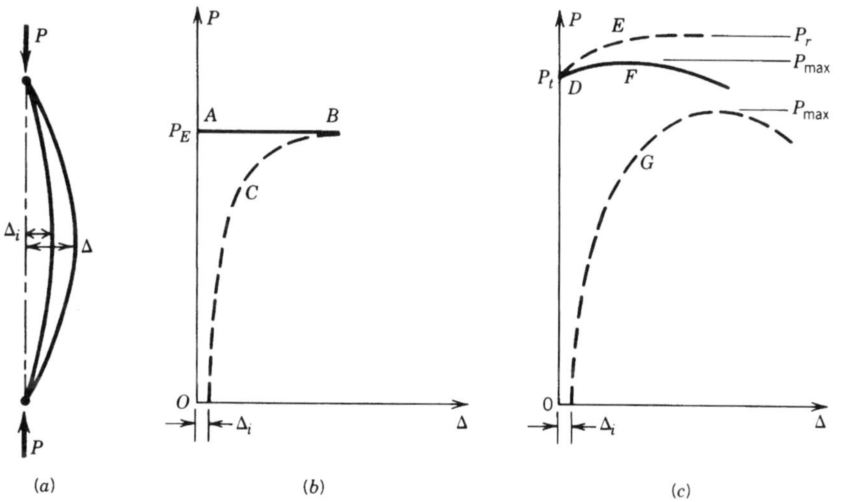

FIGURE 3.1 Behavior of perfect and imperfect columns.

in the material behavior itself (e.g., aluminum, which has a nonlinear stress–strain curve), or it may be due to partial yielding of the cross section at points of compressive residual stress (e.g., steel shapes). The postbuckling behavior of such a column is fundamentally different from the perfect elastic column: bifurcation buckling occurs for an initially straight column at the tangent-modulus load (point D in Fig. 3.1c) defined as

$$P_t = \frac{\pi^2 E_t I}{L^2} \tag{3.3}$$

but further lateral deflection is possible only if the load increases. If there were no further changes in stiffness due to yielding, the load would asymptotically approach the *reduced-modulus load* (point E in Fig. 3.1c)

$$P_r = \frac{\pi^2 E_r I}{L^2} \tag{3.4}$$

as the deflection tends to large values. The increase in load is due to the elastic unloading of some fibers in the cross section, which results in an increase in stiffness. The tangent modulus, E_t, is the slope of the stress–strain curve (Fig. 3.2) when the material is nonlinear, but E_r and E_t when residual stresses are present also depend on the shape of the cross section. Because increased loading beyond the tangent-modulus load results in further yielding, stiffness continues to be reduced and the load–deflection curve achieves a peak ($P_{\max}$, point F in Fig. 3.1c) beyond which it falls off.

The improved understanding of the post-buckling behavior of inelastic columns made possible by Shanley (1947) represented the single most significant step in understanding column behavior since Euler's original development of elastic buckling theory in 1744. Thus, a perfect inelastic column will begin to deflect laterally when $P = P_t$ and $P_t < P_{\max} < P_r$.

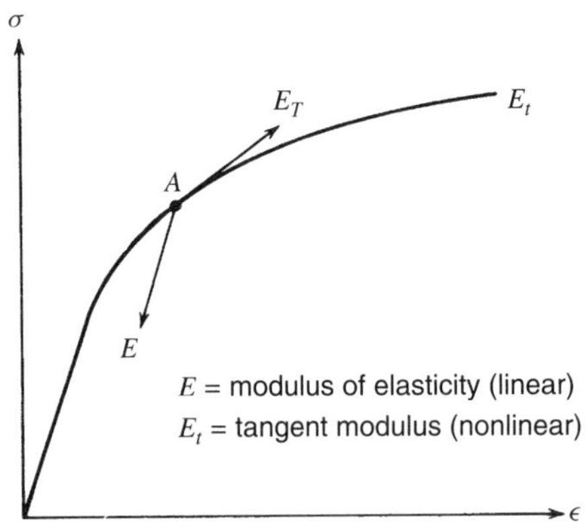

FIGURE 3.2 General stress–strain relationship.

The Euler buckling behavior described above pertains to the flexural (in-plane) mode, which is the dominant mode experienced by most standard hot-rolled shapes with doubly symmetric cross sections in typical applications. It must be recognized, however, that centrally loaded columns may potentially experience a torsional buckling mode or, in the case of singly symmetric or asymmetric sections, a combination mode, generally referred to as flexural–torsional buckling. Because these behaviors are more commonly associated with angle struts and thin-walled compression members, these concepts are covered in Chapters 11 and 13.

3.2.2 Imperfect Column Theory

Geometric imperfections, in the form of tolerable but unavoidable out-of-straightness of the column and/or eccentricity of the applied axial load, introduce bending from the onset of loading, and curve G in Fig. 3.1c characterizes the behavior of an out-of-straight column. Lateral deflection exists from the start of loading, and the maximum load is reached when the internal moment capacity (in the presence of axial load) at the critical section is equal to the external moment caused by the product of the load and the deflection. The maximum load is thus a function of the imperfection. For some types of columns, the nature of the problem is such that the maximum capacity of the imperfect column is closely approximated by the tangent-modulus load of the perfect column, but for many types of columns the imperfections must be included to give a realistic maximum load. In general, the strength of columns must be determined by including both the imperfections and material nonlinearity and/or the residual stress effects.

3.2.3 Approaches to the Design of Metal Columns

Accurate determination of the maximum strength of metal columns is a complicated process involving numerical integration that may use various solution procedures for nonlinear problems. The nonlinear approach is essential when initial imperfections, material nonlinearities, residual stresses, and other column strength parameters have to be considered.

Simplified column formulas are usually provided for design office practice. These formulas incorporate the major strength parameters, such as the yield stress, the column length, and the cross-sectional properties, and resistance factors are prescribed to arrive at acceptable levels of reliability. Many column formulas have been used throughout the history of structural engineering, and the reader can consult standard textbooks, including the previous editions of this Guide, to find the equations that have been used and the rationales behind the various models. A brief description of these models is provided below, strictly to offer the historical background of the most important approaches.

1. ***Empirical formulas based on the results of column tests***. Such formulas are applicable only to the material and geometry for which the tests were performed. The earliest column formulas (from the 1840s) are of this type.

Some subsequent studies (Hall, 1981; Fukumoto et al., 1983), however, have utilized the availability of computerized databases that contain a number of the column tests reported in the literature. The reader is referred to the paper by Hall (1981) for numerous plots that include accumulated test data from the literature for a variety of column types. Empirical factors can account approximately for initial imperfections of geometry and loading, but the formulas do not consider the inelastic basis of general column behavior, nor can they rationally account for end restraint.

2. ***Formulas based on the yield limit state***. These formulas define the strength of a column as the axial load that gives an elastic stress for an initially imperfect column equal to the yield stress. Such column formulas have a long history, also dating back to the middle of the nineteenth century, and they continue to enjoy popularity to the present, for example, the use of the Perry–Robertson (Robertson, 1925) formula.
3. ***Formulas based on the tangent-modulus theory***. Such formulas can account rationally for the bifurcation load, but not the maximum strength, of perfectly straight columns. If the effects of imperfections are such that they just reduce the maximum strength to the tangent-modulus strength, these formulas have empirical justification. On the other hand, if the perfect column can be thought of as an anchor point in an interaction surface, initial imperfections of geometry and loading can be represented as flexural effects in the interaction equation.

   The “CRC Column Strength Curve,” named after the acronym of the former name of the Structural Stability Research Council (i.e, Column Research Council), was recommended in the first edition of this guide (1960) and has been used for many steel design specifications in North America and elsewhere. It is based on the average critical stress for small- and medium-sized hot-rolled wide-flange shapes of mild structural steel, with a symmetrical residual stress distribution typical of such members. The column curves based on the tangent-modulus theory can also accurately account for end restraints (Yura, 1971).
4. ***Formulas based on maximum strength***. State-of-the-art column design formulas are based on extensive studies of the maximum strength of representative geometrically imperfect columns containing residual stresses. The analyses have incorporated comprehensive numerical data, as well as evaluations of test results and how well these compare. Reliability analyses have been performed, leading to the resistance factors that are given in state-of-the-art design standards. The third edition of this guide presented new column curves based on this principle (Bjorhovde, 1972). Subsequently, SSRC published Technical Memorandum No. 5, stating the principle that design of metal structures should be based on the maximum strength, including the effects of geometric imperfections.

   It was also suggested that the strength of columns might be represented better by more than one column curve, thus introducing the concept of multiple column curves (Bjorhovde and Tall, 1971, Bjorhovde, 1972). SSRC curves 1, 2, and 3 and curves 1P, 2P, and 3P are two sets of such curves; another example is the set of five curves in Eurocode 3 [European Committee for Standardization (CEN, 2005)]. The Canadian Standards Association (CSA, 2009) provides two column curves that are based on SSRC curves 1 and 2. The column curve of the American Institute of Steel Construction (AISC) specifications (AISC, 2005a) is the same as SSRC curve 2P, although the equation takes a different form. Finally, end-restraint effects are readily incorporated with the maximum-strength approach.

3.2.4 Local Buckling

When structural members composed of slender elements, such as the flanges and webs of many steel shapes, are loaded axially, the overall column capacity can be limited by the capacity of the individual cross-sectional elements. This phenomenon is known as local buckling and is closely related to classical plate-bucking theory. This topic is covered in Chapter 4.

3.2.5 Bracing

The strength of a compression member can be influenced greatly by the method with which it is braced. Although brace locations between the member ends influence the effective length of the member, as discussed in Section 3.4, the type, strength, and stiffness of the braces, as well as the means of connecting them to the column, can affect the behavior significantly. Torsional buckling modes can only be restrained using braces that restrain twisting deformations. Bracing of members is a complex topic that is largely beyond the scope of this chapter. Column bracing topics are covered in Section 3.4.2 and Chapter 12.

3.3 INFLUENCE OF IMPERFECTIONS

3.3.1 Residual Stresses

Structural steel shapes and plates contain residual stresses that result primarily from nonuniform cooling after rolling. Welded built-up members also exhibit tensile residual stresses in the vicinity of the welds due to the cooling of the weld metal. These are generally equal to the yield stress of the weld metal, which will normally be somewhat greater than the yield stress of the base metal (Tall, 1966; Alpsten and Tall, 1970; Bjorhovde et al., 1972). Flame cutting (also called oxygen cutting) introduces intense heat in a narrow region close to the flame-cut edge. As a result, the material in this region acquires properties that are significantly different from those of the base metal, and residual stresses develop that are often much higher than the yield stress of the parent material (McFalls and Tall, 1970; Alpsten and Tall, 1970; Bjorhovde et al., 1972). Finally, cold forming and cold straightening introduce residual stresses, especially in regions with the most severe bending effects, such as in corners of cold-formed shapes (Alpsten, 1972b; Sherman, 1976; Yu, 1992).

In 1908, in a discussion of the results of column tests at the Watertown Arsenal, residual stresses due to the cooling of hot-rolled steel shapes were cited as the probable cause of the reduced column strength in the intermediate slenderness range (Howard, 1908) The possible influence of residual stresses on the buckling strength of both rolled members and welded plates in girders was subsequently noted by others (Salmon, 1921; Madsen, 1941). Systematic research on the effect of residual stress on column strength was initiated in the late 1940s under the guidance of Research Committee A of the Column Research Council (Osgood, 1951; Yang et al., 1952; Beedle and Tall, 1960). This work continued through the early 1970s in extensive research projects, primarily at Lehigh University (Kishima et al., 1969; McFalls and Tall, 1970; Alpsten and Tall, 1970; Brozzetti et al., 1970a; Bjorhovde et al., 1972). Work in Europe and Canada on these effects must also be noted to appreciate fully the magnitude and complexity of the problem (Sfintesco, 1970; Beer and Schultz, 1970; Alpsten, 1972a; Chernenko and Kennedy, 1991).

At the time of the first edition of this guide (1960), the tangent-modulus curve appeared to be the proper basis for the determination of allowable column design stresses. This curve was based on the effect of typical residual stress distributions in hot-rolled steel shapes. The CRC column curve, based on computed column curves for hot-rolled wide-flange shapes and taken as an approximate average of the major and minor axis buckling curves, served as a basis for the column design provisions of the AISC and CSA specifications. The second edition of this guide (1966) mentioned the increasing use of columns made of (1) high-strength steels with yield stresses up to 70 ksi (480 Mpa) and (2) heat-treated steels with yield stresses up to 100 ksi (690 Mpa) or more. It noted the importance of initial imperfections as well as of residual stresses in determining the strengths of pin-ended columns made of higher strength steels.

One of the possible ways of differentiating between categories of column strength is by the use of the concept of multiple column curves, such as those that were developed through research at Lehigh University (Bjorhovde and Tall, 1971; Bjorhovde, 1972) and those that have been provided by the studies in Europe (Beer and Schultz, 1970). In addition, large numbers of column tests have also been performed, in some cases on a systematic basis, to provide further assurance of the theoretical results obtained by computer studies. The single largest group of such column tests is probably the more than 1000 tests that were conducted at a number of European universities and laboratories, as well as a number of tests on heavy shapes at Lehigh University, under the auspices of the European Convention for Constructional Steelwork (ECCS) (Sfintesco, 1970). Over the years, a great many other tests have also been performed, and these have been summarized by Fukumoto et al. (1983).

Residual Stresses in Hot-Rolled Shapes The magnitude and distribution of residual stresses in hot-rolled shapes depend on the type of cross section, rolling temperature, cooling conditions, straightening procedures, and the material properties of the steel (Beedle and Tall, 1960). Examples of residual stress distributions resulting from cooling without straightening of wide-flange shapes are shown in

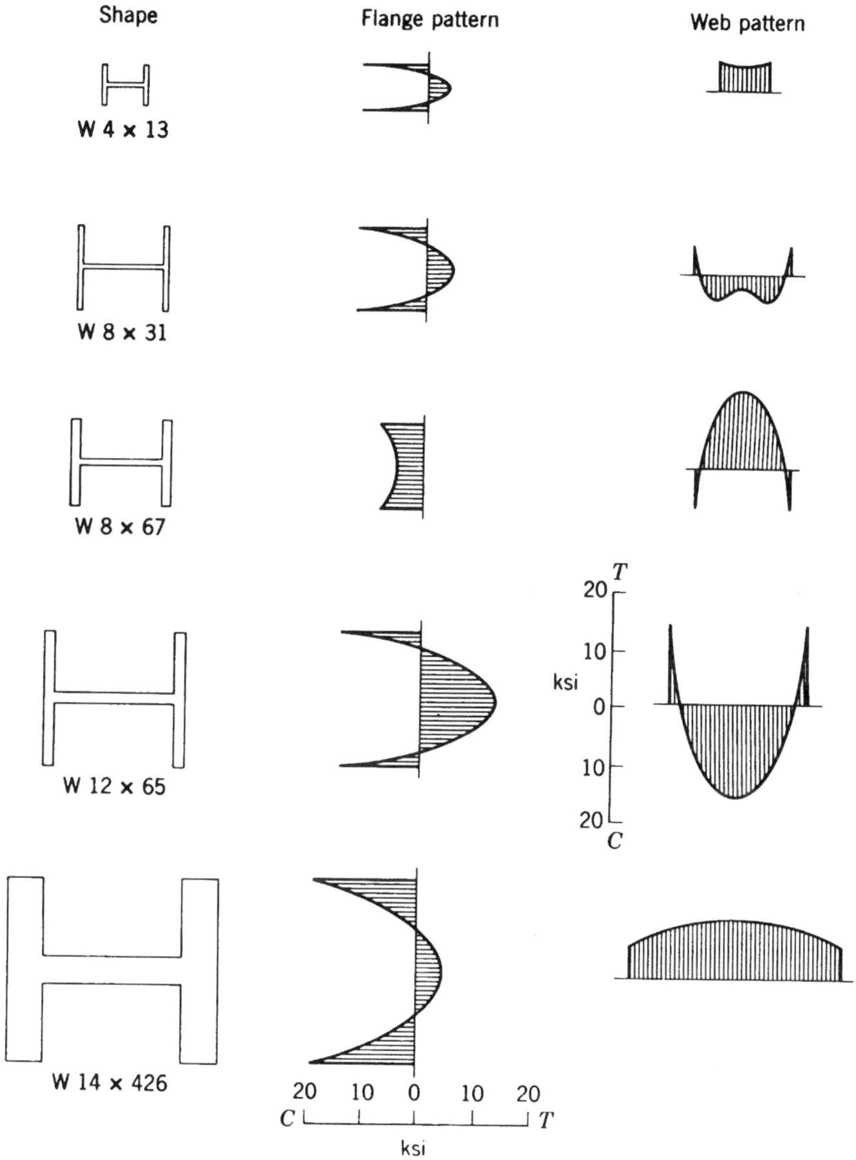

FIGURE 3.3 Residual-stress distribution in rolled wide-flange shapes.

Fig. 3.3 (Tall, 1964). For heavier shapes, residual stresses vary significantly through the thickness. Figure 3.4 shows the measured residual stresses in one of the heaviest rolled shapes that is currently produced (Brozzetti et al., 1970a).

The effect of the strength of the steel on the residual stress distribution is not as great as the effect of geometry (Tall, 1964). Residual stress measurements in the

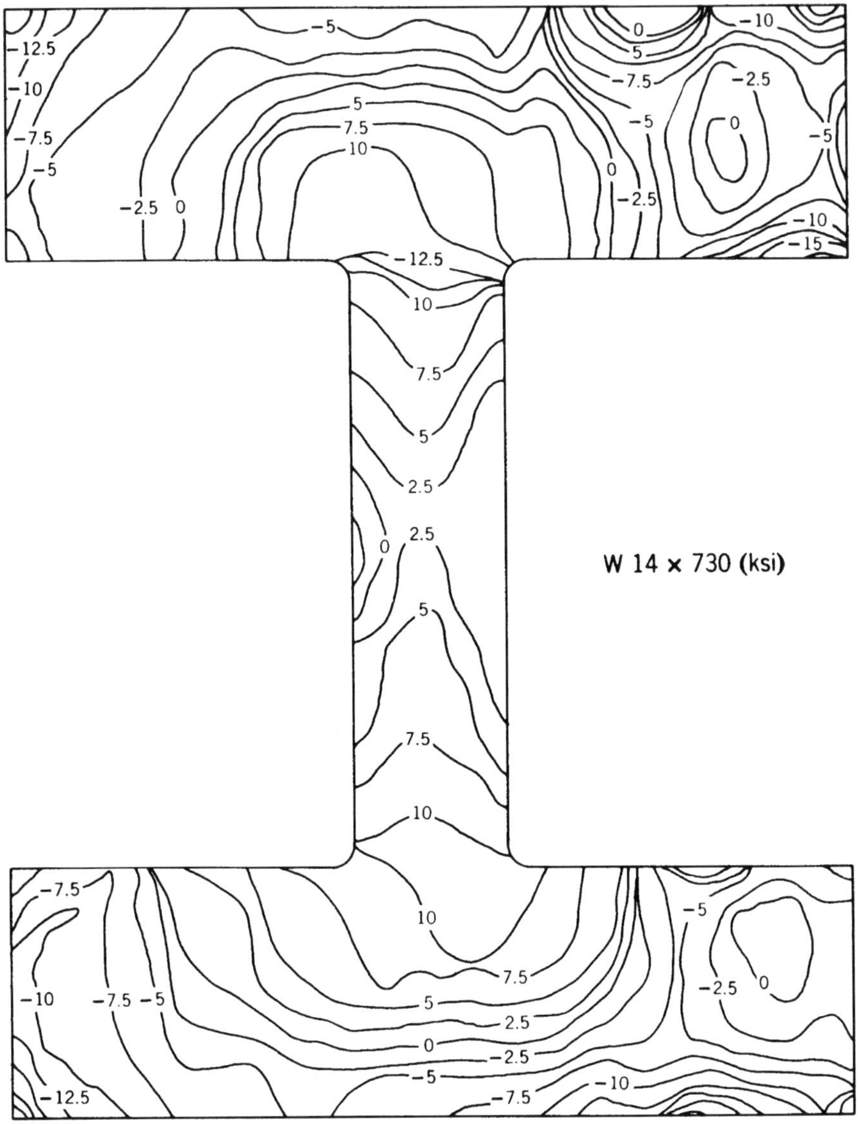

FIGURE 3.4 Residual-stress distribution in W14X730 shape.

flanges of similar shapes made of different steel grades show that the distributions and magnitudes of the residual stresses are very similar. For H-shaped columns, it is residual stresses in the flanges that have the most significant effect on the column strength.

Computed column curves based on the residual stresses in the five shapes of Fig. 3.3 are shown in Fig. 3.5 for buckling about the minor axis. The figure

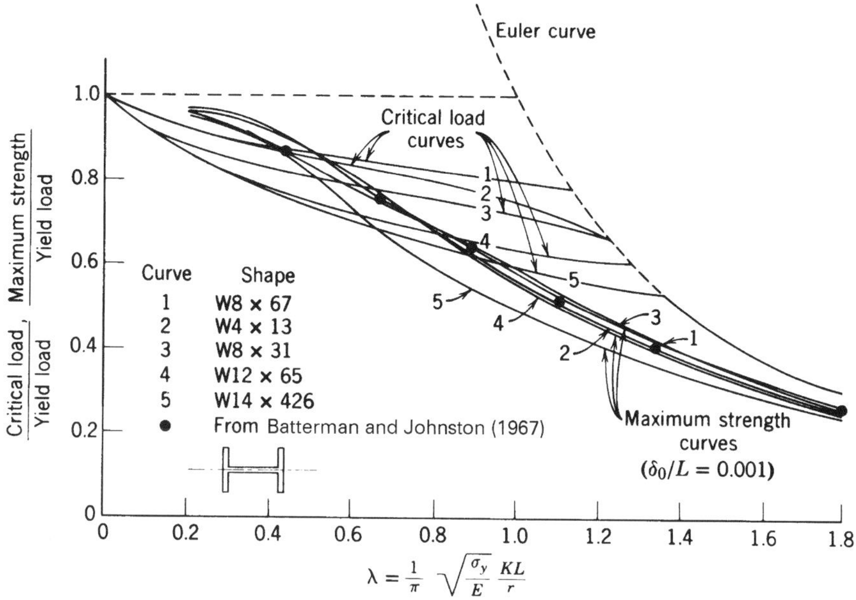

FIGURE 3.5 Critical-load curves for straight columns compared with maximum-strength curves for initially curved rolled steel W-shapes.

also shows the computed maximum strength curves for these shapes, using a combination of the measured residual stresses and an initial out-of-straightness equal to the approximate maximum of $L/1000$ that is permitted by the ASTM standard for delivery of structural shapes, ASTM A6 (the actual maximum is $L/960$, as defined by ASTM A6).

To provide a systematic examination of the separate and the combined effects of the residual stresses and the initial out-of-straightness, extensive column strength analyses were carried out at the University of Michigan (Batterman and Johnston, 1967). The studies included the following parameters:

1. Yield stresses of 36, 60, and 100 ksi (approximately 250, 415, and 700 Mpa)
2. Maximum compressive residual stresses of 0, 10, and 20 ksi (0, 70, and 140 Mpa)
3. Five initial out-of-straightnesses ranging from 0 to $0.004L$
4. Slenderness ratios ranging from 20 to 240.

The mode of failure was flexural buckling about the minor axis. The results of this condition are presented graphically in the work by Batterman and Johnston (1967), and permit the maximum-strength evaluation within the range of the parameters cited. On the basis of a maximum residual stress of 13 ksi (90 Mpa), which is the scaled average maximum for the five sections shown in Fig. 3.3, together with a yield stress of 36 ksi (250 Mpa), the maximum column strength predicted by Batterman and Johnston is shown by the solid circles in Fig. 3.5. The solid curves are from an analysis neglecting the webs. Although the shapes and the residual stress distributions are different, there is good correlation between the two independently developed analysis procedures.

The findings of Batterman and Johnston are corroborated by those of a wide-ranging investigation of column strength, which examined the behavior and strength of a large and diverse number of structural shapes, grades of steel, and manufacturing methods (Bjorhovde, 1972). Bjorhovde's computational procedure is very accurate but requires knowledge of the residual stresses and the out-of-straightness. The work was performed at Lehigh University and included the full range of practical structural steel grades and shapes. A number of welded built-up box and H-shapes were also examined.

The results provided by the studies of Batterman and Johnston (1967) and Bjorhovde (1972) show clearly that:

1. The separate effects of residual stress and initial out-of-straightness cannot be added to give a good approximation of the combined effect on the maximum column strength. In some cases, and for some slenderness ratios, the combined effect is less than the sum of the parts (intermediate slenderness ratios, low residual stresses). In other cases the combined effect is more than the sum of the parts. The latter applies to the intermediate slenderness ratio range for heavy hot-rolled shapes in all steel grades and for welded built-up H-shapes. It is emphasized that the magnitudes of the maximum compressive residual stresses in a large number of these shapes were 50% or more of the yield stress of the steel.
2. As would be expected, residual stresses had little effect on the maximum strength of very slender columns, either straight or initially crooked. Such members have strengths approaching the Euler load. Very slender higher strength steel columns, however, can tolerate much greater lateral deflection before yielding or otherwise becoming unstable.
3. Strengths are slightly underestimated in a computer analysis that is based on the assumption that the initial out-of-straightness will remain in the shape of a half-sine wave during further loading.
4. Differences in column strength caused by variations in the shape of the residual stress pattern are smaller for initially crooked columns than for initially straight columns. This is a result of the early flexural behavior of the initially curved members.

Additional data on the residual stresses and column strengths of very heavy hot-rolled shapes confirmed the findings of Brozzetti et al. (1970a). The relative maximum column strength (i.e., computed maximum strength divided by the yield load) reaches a minimum for flange thicknesses around 3 to 4 in. (75 to 100 mm). The relative strength increases as the flange thickness exceeds this magnitude (Bjorhovde, 1988; 1991).

Residual Stresses in Welded Built-Up Shapes Residual stresses resulting either from welding or from the manufacturing of the component plates have a significant influence on the strength of welded H- and box-section columns. The maximum tensile residual stress at a weld or in a narrow zone adjacent to a flame-cut edge is generally equal to or greater than the yield stress of the plates (Alpsten and Tall, 1970; McFalls and Tall, 1970; Alpsten, 1972a; Brozzetti et al., 1970b; Bjorhovde et al., 1972). Welding modifies the prior residual stresses due either to flame cutting or cooling after rolling.

Figure 3.6 shows that the strengths of welded columns made of higher-strength steels appear to be influenced relatively less by residual stresses than are the strengths of similar columns made of lower-strength steels (Kishima et al., 1969; Bjorhovde, 1972). It is also evident that the differences in strengths of columns with the maximum permissible initial out-of-straightness are less than the differences in critical loads of initially straight columns (see Fig. 3.5).

As shown in Fig. 3.7, plates with mill-rolled edges (often referred to as universal mill plates) have compressive residual stresses at the plate edges, whereas flame-cut plates have tensile residual stresses at the edges. In built-up H-shapes made of universal mill plates, the welding increases the compressive stress at the flange tips, enlarging the region of compressive residual stress and adversely affecting the column strength. Conversely, as illustrated in Fig. 3.8, an H-shaped column made from flame-cut plates will have favorable tensile residual stresses at the flange tips and will therefore have greater strength than a column of the same section with flanges consisting of universal mill plates. It is also seen that for short welded

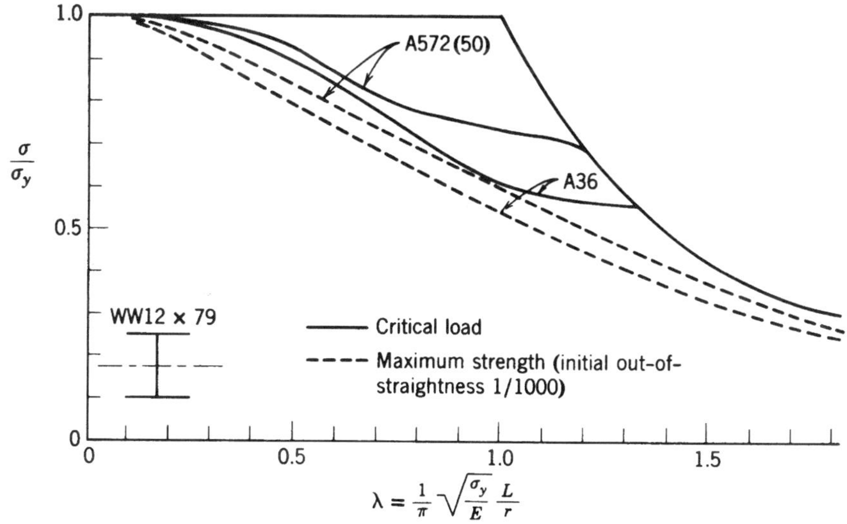

FIGURE 3.6 Critical-load curves for welded WW12X79 of flame-cut plates compared with maximum-strength curves for initially curved members (Kishima et al., 1969; Bjorhovde, 1972).

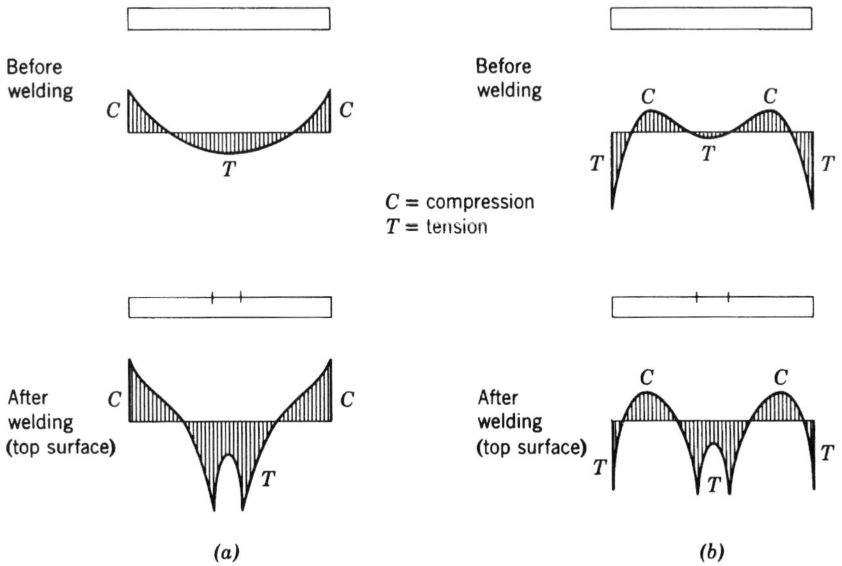

FIGURE 3.7 Qualitative comparison of residual stresses in as-received and center-welded (*a*) universal mill plate; (*b*) oxygen-cut plate.

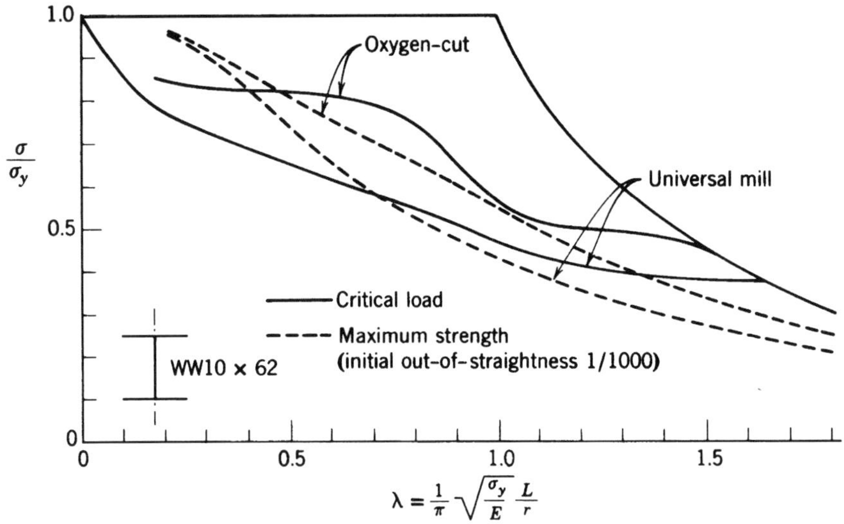

FIGURE 3.8 Comparison of column curves for WW10X62 (A7 steel) with universal mill versus oxygen-cut plates (Bjorhovde, 1972).

columns the maximum strength of an initially curved column may in some cases be greater than the critical load of a straight column. Obviously, the maximum strength of an initially straight column will always be greater than the critical load of the same column with an initial imperfection.

Strength differences between boxsection columns made of universal mill and flame-cut plates tend to be very small because the edge welds override the residual stresses in the component plates (Bjorhovde and Tall, 1971). The sequence of welding can be a significant factor for such columns, particularly for those with large welds (Beer and Tall, 1970).

Several investigations have considered the effects of column size. It has been shown conclusively that welding has a greater influence on the overall distribution of residual stress in small- and medium-sized shapes than in heavy shapes (Kishima et al., 1969; Alpsten and Tall, 1970; Brozzetti et al., 1970b; Bjorhovde et al., 1972).

The distribution of residual stress in heavy plates and shapes is not uniform through the thickness (Brozzetti et al., 1970a; Alpsten and Tall, 1970). As the thickness increases, the difference between surface and interior residual stresses may be as high as 10 ksi (70 Mpa). As an example, Fig. 3.9 shows an isostress diagram for a heavy welded shape made from flame-cut plates. It has been found, however, that calculated critical loads and maximum column strengths are only

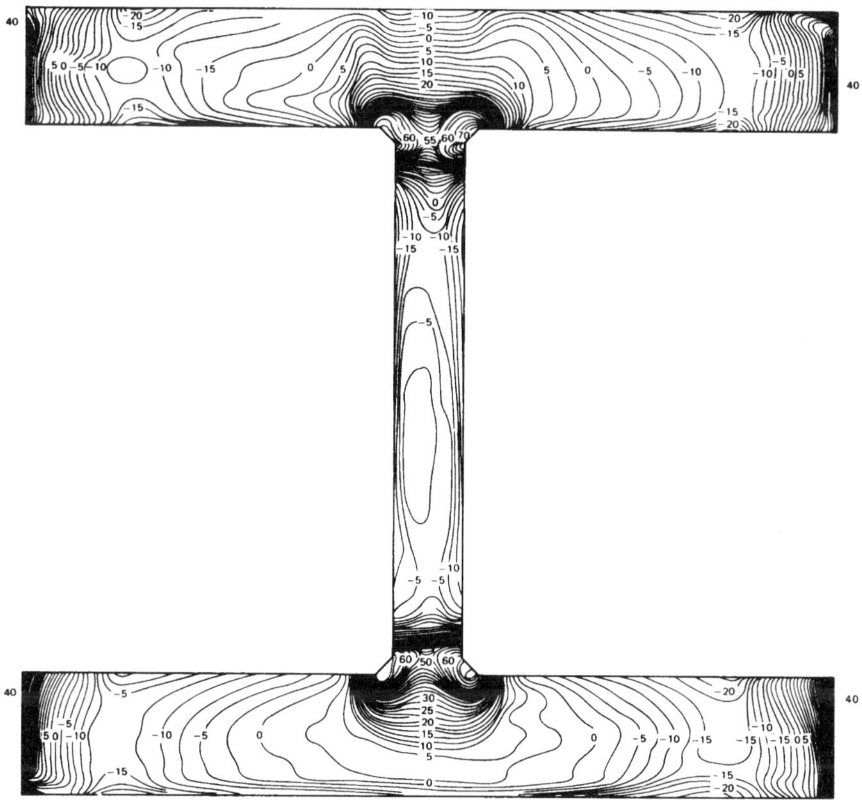

FIGURE 3.9 Isostress diagrams for WW23X681 welded built-up shape (stresses in kips per square inch) (Alpsten and Tall, 1970).

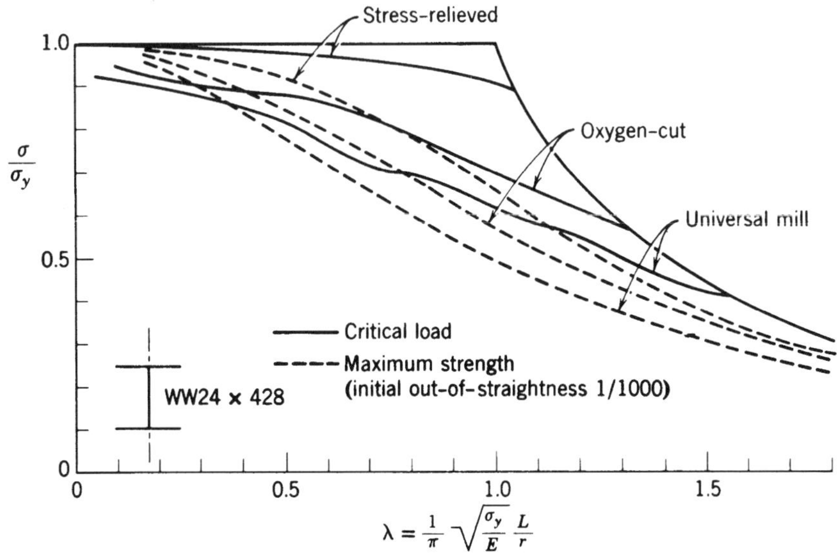

FIGURE 3.10 Column curves for heavy and light welded wide-flange shapes (Bjorhovde, 1972).

a few percent less when based on the complete residual stress distributions, as compared with analyses that assume the stress to be constant through the thickness and equal to the surface-measured residual stress.

In general, shapes made from flame-cut plates exhibit higher strength than shapes that are made from universal mill plates. This is demonstrated by the curves in Fig. 3.10. Similarly, flame-cut shapes tend to have strengths that are comparable with those of similar rolled shapes, whereas universal mill shapes tend to be comparatively weaker.

Figure 3.11 compares the strengths of two typical welded columns with flame-cut flange plates, and one being distinctly heavier than the other. It is seen that the heavier shape tends to be relatively stronger than the lighter one. This is even more accentuated for shapes that are welded from universal mill plates, for which the strength of the lighter shape will be significantly lower than the heavy one (Bjorhovde and Tall, 1971; Bjorhovde, 1972).

In a major study, Chernenko and Kennedy (1991) examined an extensive range of welded built-up H-shapes. In addition to performing maximum-strength computations for columns with a variety of residual stress distributions and out-of-straightnesses, the work also examined statistical data on material and other properties. Resistance factors for use with limit states criteria for welded columns were developed. It is shown that current approaches are conservative. As a consequence, since 1994 these shapes (made from plates with flame-cut edges) have been assigned to the higher of the two column curves of the CSA steel design standard (CSA, 2009).

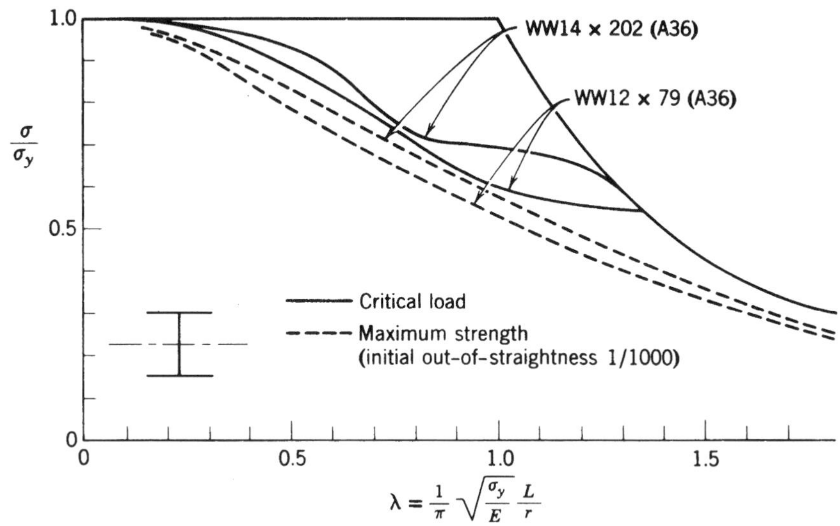

FIGURE 3.11 Comparison of column curves for WW24X428 (A36 steel) with stress-relieved, oxygen-cut, and universal mill plates (Bjorhovde, 1972).

The sequence of welding and the number of welding passes are factors that influence the distribution of residual stresses. Other welding parameters, such as voltage, speed of welding, and temperature and areas of preheating, have less influence (Brozzetti et al., 1970b). Stress-relief annealing of the component plates prior to welding of the shape raises column strength very significantly by reducing the magnitude of the residual stresses, even though it lowers the yield stress of the steel. Figure 3.10 compares the column curves for shapes made from flame-cut and universal mill plates, along with curves for the same shapes made from stress-relieved plates.

Residual Stresses in Cold-Straightened Columns Cold straightening of structural sections to meet tolerances for camber and sweep induces a redistribution and reduction of the residual stresses that were caused by earlier rolling and cooling. In current practice for all steel mills around the world, shapes are cold straightened as a matter of course, either by rotary or gag straightening (Brockenbrough, 1992; Bjorhovde, 2006). In rotary straightening the shape is passed through a series of rolls that bend the member back and forth with progressively diminishing deformation. In gag straightening, concentrated forces are applied locally along the length of the member to bend it to the required straightness. Rotary straightening is applied for small- and medium-size shapes; gag straightening is typically used for heavy shapes.

The rotary straightening process redistributes and reduces the initial residual stresses in the flanges, as shown in Fig. 3.12. In gag straightening, moments that approximate the full plastic value, M_p, are produced at the points where the forces

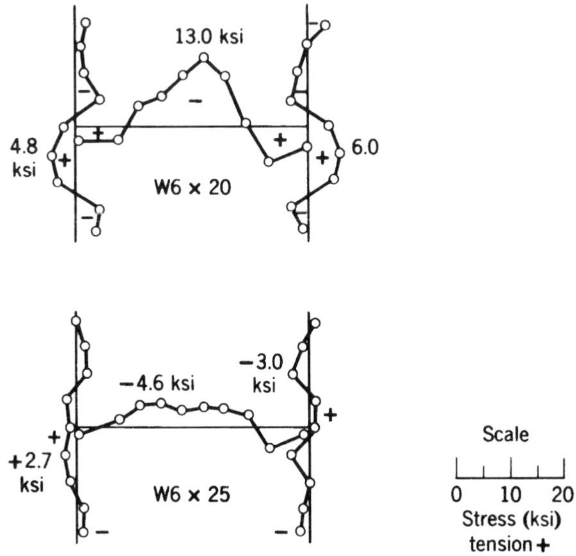

FIGURE 3.12 Residual stresses in rotary straightened shapes.

are applied, and the cooling residual stresses are therefore redistributed only at or near the points of gag pressing. In the usual case of gag straightening, to remove sweep (curvature about the minor axis of a wide-flange shape), the change in residual stress from compression to tension takes place locally at the edges on the side of the flanges where the load is applied. Figure 3.13 shows the residual stresses measured in a W8X31 shape after gag straightening about the minor axis (Huber, 1956). It should be noted that a W8X31 would most likely not be gag straightened today, and instead it would be run through the rotary procedure.

The strength of a cold-straightened column is generally greater than that of the corresponding as-rolled member because of the improved straightness and the redistribution of residual stress (Frey, 1969; Alpsten, 1970). Rotary straightening produces a greater improvement than gag straightening, and according to theoretical analyses and experimental results the column strength may be increased by as much as 20% when compared at the same slenderness ratio and initial out-of-straightness (Alpsten, 1970, 1972b).

The strength and behavior of cold-straightened columns still have not been documented satisfactorily, and research should be undertaken to detail all of the individual influences and effects. This is particularly important in view of the fact that almost all hot-rolled wide-flange shapes are straightened in the mill to meet straightness requirements.

For tubular shapes the situation is somewhat different. The final mill process in most cases is cold forming or rolling, producing a very small initial out-of-straightness, which is then followed in some mills by partial stress relieving (Birkemoe, 1977a; Bjorhovde and Birkemoe, 1979). This is also the case for welded built-up wide-flange columns (Chernenko and Kennedy, 1991).

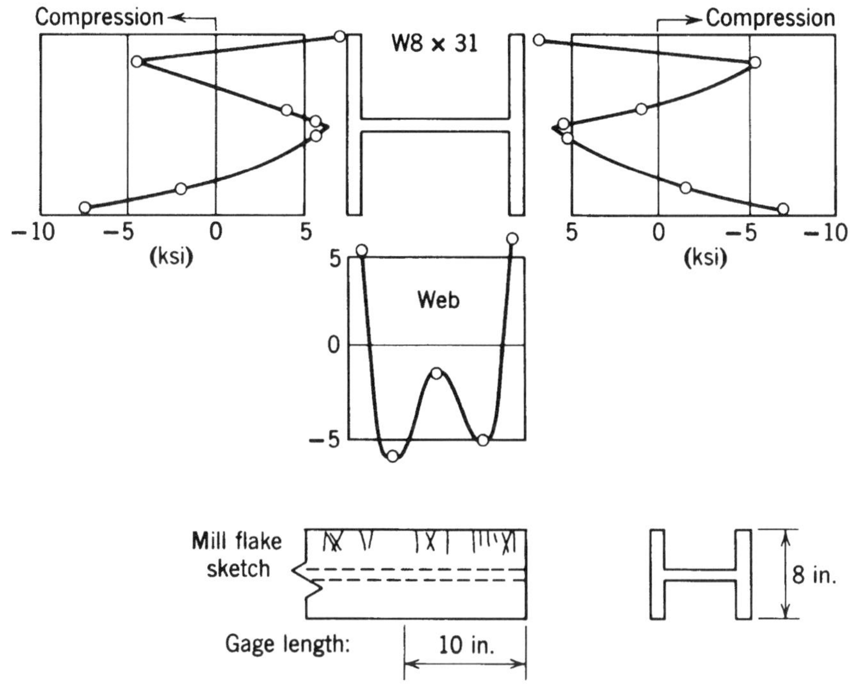

FIGURE 3.13 Residual stresses in a gag-straightened shape, bent about the weak axis to remove sweep (Huber, 1956).

3.3.2 Out-of-Straightness

Another major factor influencing the behavior of columns is the initial out-of-straightness (also referred to as initial crookedness or initial curvature). Some of the characteristics of the behavior and strength of inelastic, initially curved columns have been discussed in the previous evaluation of residual stress influences, and the two parameters interact in many ways. The explanation of the ranges of slenderness ratios and column types, for which the combined effect of residual stress and initial crookedness is more than the sum of the parts, emphasizes the complexity of the phenomenon.

The analyses that have been made of the strength of inelastic, initially curved columns either have made use of assumed values and shapes of the initial out-of-straightness, or have used measured data. The former is by far the most common, mostly because the measurements that are available for columns are scarce. This applies in particular to the magnitude of the maximum out-of-straightness, normally assumed to occur at midlength of the member, as well as the shape of the bent member. The latter is usually thought to be that of a half sine wave (Batterman and Johnston, 1967; Bjorhovde, 1972). The real configuration of the initial out-of-straightness of a column may be very complicated, often expressed as a simultaneous crookedness about both principal axes of the cross section. Systematic measurements have been made in some laboratories in conjunction with testing programs (Beer and Schultz, 1970; Bjorhovde, 1972, 1977; Bjorhovde and Birkemoe, 1979; Fukumoto et al., 1983, Essa and Kennedy, 1993), but very few data are available for columns in actual structures (Tomonaga, 1971; Lindner and Gietzelt, 1984; Beaulieu and Adams, 1980). Chernenko and Kennedy (1991) measured the out-of-straightness of welded wide-flange shapes at the steel mill.

Magnitudes and Limitations The magnitude of the maximum initial out-of-straightness is limited by the structural steel delivery specifications (e.g., ASTM A6 in the United States; CSA G40.20 in Canada) and is normally expressed as a fraction of the length of the member. Hot-rolled wide-flange shapes are required to have a maximum initial crookedness of $L/960$ [measured as 1/8 in. (3 mm) in 10 ft (3 m) of length], which is usually given as $L/1000$ for convenience. Tubular shapes are required to meet a straightness tolerance of $L/480$, commonly given as $L/500$.

The measurements that are available show that most hot-rolled W-shapes tend to have values toward the maximum permissible, with an average of $L/1470$ (Bjorhovde, 1972), although Dux and Kitipornchai (1981) and Essa and Kennedy (1993) give a mean value for the maximum initial imperfection of $L/3300$ and $L/2000$, respectively, for wide-flange shapes of lengths of 20 to 33 feet (6 to 10 m). Tubular members typically exhibit values significantly smaller than the specification limitations, with out-of-straightnesses on the order of $L/3000$ to $L/8000$, with an average of $L/6300$ (Bjorhovde, 1977; Bjorhovde and Birkemoe, 1979). The data for welded wide-flange shapes indicate a relatively small initial crookedness, with a mean of approximately $L/3300$ (Chernenko and Kennedy, 1991). With this in mind, it is rare to encounter columns with out-of-straightnesses larger than the maximum permitted.

In the development of column strength criteria such as the SSRC curves (Bjorhovde, 1972) and the ECCS curves (Beer and Schultz, 1970), the maximum permissible values of the initial out-of-straightness were utilized. This was done for several reasons, the primary one being that $L/1000$ constituted the upper limit of what is acceptable for actually delivered members and therefore could be regarded as a conservative measure. On the other hand, because mean characteristics were used for the other strength parameters, it can be rationally argued that the mean values for out-of-straightness also should be utilized. This was done by Bjorhovde (1972) in parallel with his development of the original SSRC curves, using the mean of $L/1470$ that was determined through statistical evaluations. The resulting multiple column curves are shown in Fig. 3.14, where the curves labeled as 1P, 2P, and 3P have used $L/1470$. For comparison, the SSRC curves have been included in the figure; these have used an initial out-of-straightness of $L/1000$. The mathematical equations for both sets of curves are given in Section 3.5.

Variations in the magnitude of the initial crookedness were considered in the study by Bjorhovde (1972). The strengths of the 112 columns that were included in the investigation were examined using maximum initial out-of-straightnesses of

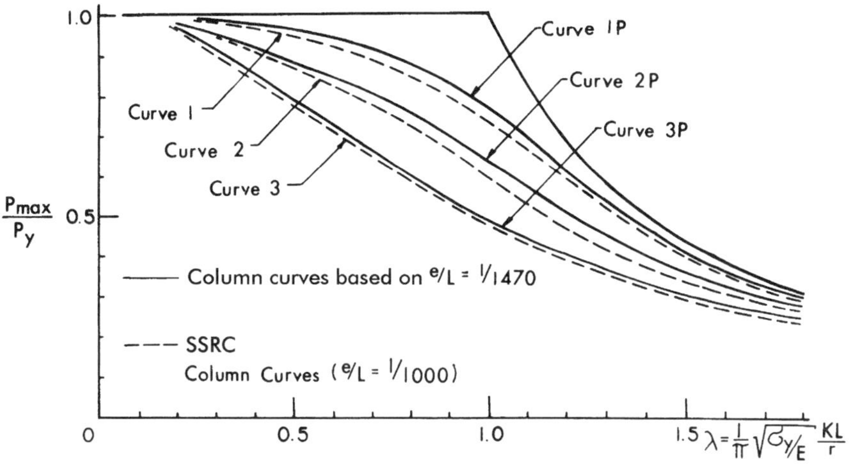

FIGURE 3.14 Comparison of multiple column curves developed on the basis of mean out-of-straightness ($L/1470$) and maximum permissible out-of-straightness ($L/1000$) (Bjorhovde, 1972).

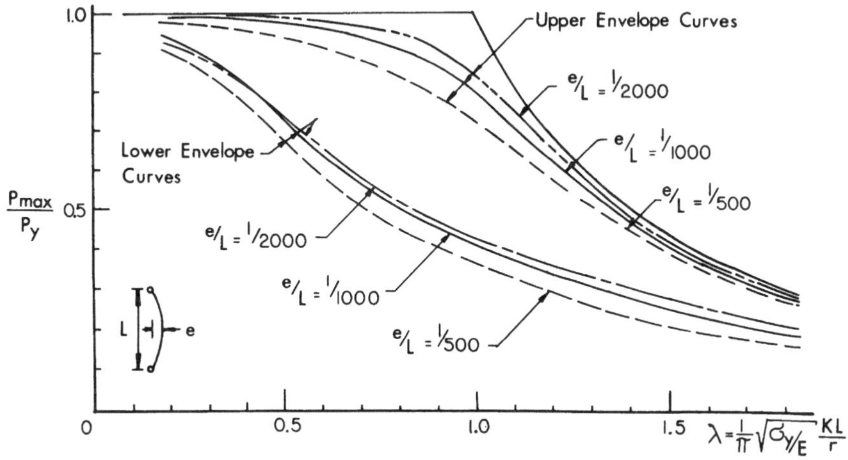

FIGURE 3.15 Column curve bands for 112 columns, based on initial out-of-straightnesses of $L/500$, $L/1000$, and $L/2000$ (Bjorhovde, 1972).

$L/500$, $L/2000$, and $L/10{,}000$. The results for the band of column strength curves are given in Fig. 3.15 (curves that are shown include only the data for $L/500$ to $L/2000$). The results of the studies on the maximum strength of columns emphasize the need for incorporation of the initial out-of-straightness into column strength models that form the basis for design criteria.

3.4 INFLUENCE OF END RESTRAINT

Extensive studies on the influence of end restraint on the strength and behavior of columns have been conducted by Chen (1980), Jones et al. (1980, 1982), Razzaq and Chang (1981), Chapuis and Galambos (1982), Vinnakota (1982, 1983, 1984), and Shen and Lu (1983), among others. In addition, the analysis of frames with semi-rigid connections has been dealt with in several studies (DeFalco and Marino, 1966; Romstad and Subramanian, 1970; Frye and Morris, 1975; Ackroyd, 1979; Lui and Chen, 1988; Nethercot and Chen, 1988; Goto et al, 1993; King and Chen, 1993, 1994; Kishi et al., 1993a,b).

A series of important international workshops on connections in steel structures has provided a large number of references related to the behavior and strength of connections, the influence of connections on column and frame stability, detailed evaluations of methods of frame analysis, and the development of design criteria that take these effects into account. Called semi-rigid or partially restrained (PR) connections, the state-of-the-art of their impact on column stability is very advanced, although design criteria offer only limited practical suggestions. Eurocode 3 (CEN, 2005) provides the most specific criteria, including a classification system for connections. Such a system has also been developed by Bjorhovde et al. (1990). Finally, the Commentary to the AISC specification (AISC, 2005a) offers an extensive assessment of column and frame stability as influenced by connection behavior and strength.

The publications of the international connections workshops are given in the following books, which are listed here because of the large number of papers that are provided by these works:

1. Bjorhovde, Reidar; Brozzetti, Jacques; and Colson, Andre (1988), *Connections in Steel Structure—Behaviour, Strength and Design*, Elsevier Applied Science, London, England.
2. Bjorhovde, Reidar; Colson, Andre; Haaijer, Geerhard; and Stark, Jan W. B. (1992), *Connections in Steel Structures II—Behavior, Strength and Design*, AISC, Chicago, Illinois.
3. Bjorhovde, Reidar; Colson, Andre; and Zandonini, Riccardo (1996), *Connections in Steel Structures III—Behaviour, Strength and Design*, Pergamon/ Elsevier Science, London, England.
4. Leon, Roberto; and Easterling, W. Samuel (2002), *Connections in Steel Structures IV—Behavior, Strength and Design*, AISC, Chicago, Illinois.
5. Bijlaard, F. S. K.; Gresnigt, A. M.; and van der Vegte, G. J. (2005), *Connections in Steel Structures V—Behaviour, Strength and Design*, Bouwen met Staal, The Netherlands.
6. Bjorhovde, Reidar; Bijlaard, F. S. K; and Geschwindner, L. F. (2008), *Connections in Steel Structures VI—Behavior, Stremgth amf Design*, AISC, Chicago, Illinois.

3.4.1 Column Stability as Influenced by End Restraint

Column investigations have examined different aspects of restrained-member behavior, specifically determining the influence of:

1. Type of beam-to-column connection
2. Length of column
3. Magnitude and distribution of residual stress
4. Initial out-of-straightness

The frame analysis studies have focused on evaluations of the drift characteristics of frames with less than fully rigid connections, in part prompted by a study by Disque (1975). Frame-related subjects of this kind, however, are beyond the scope of this chapter. In fact, connection flexibility and member instability are closely related, and their interaction effects can have a significant influence on the overall performance of the frame.

As would be expected, the stiffness of the restraining connection is a major factor. One illustration of the influence is given by the moment–rotation curves in Fig. 3.16 and another by the load–deflection curves for columns with different end restraint that are shown in Fig. 3.17. A British wide-flange shape was used for the data generated for Fig. 3.17, incorporating an initial out-of-straightness of $L/1000$. The curves that are shown apply for a slenderness ratio of $L/r = 120$ ($\lambda = 1.31$), but similar data were developed for longer as well as shorter columns. Other investigators have provided additional load–deflection curves and the primary differences between the individual studies are the methods of column analysis and end-restraint modeling. The resulting load–deflection curves are very

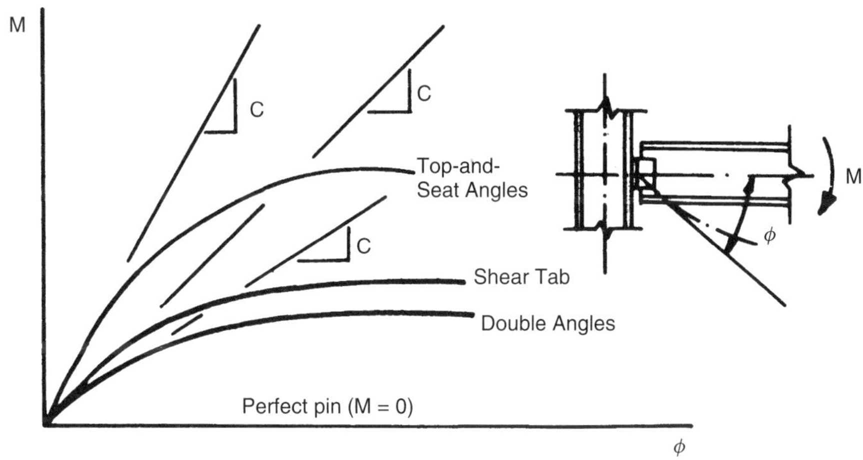

FIGURE 3.16 Moment–rotation curves for some typical simple connections (schematic).

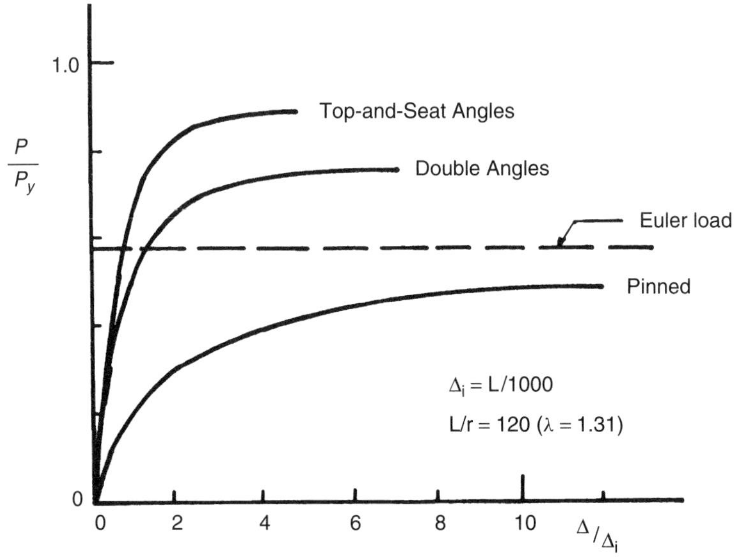

FIGURE 3.17 Typical load–deflection curves for columns (Jones et al., 1982).

similar (Jones et al., 1980, 1982; Razzaq and Chang, 1981; Sugimoto and Chen, 1982).

Figure 3.17 also includes the load–deflection curve for a pin-ended column. As is evident from the figure, the greater the connection restraint, the stiffer will be the initial response of the column, and the greater the maximum load that can be carried as compared to a pin-ended column. The same relative picture emerges for all slenderness ratios, although the magnitude of the increase becomes small for values of L/r less than 50.

A further illustration of the influence of end restraint is given by the data in Fig. 3.18, which shows column strength curves for members with a variety of end conditions (Jones et al., 1982; Lui and Chen, 1983a). The effect of the connection type is again evident, as is the fact that the influence diminishes for shorter columns. Also included in the figure is the Euler curve as well as SSRC curve 2.

It is emphasized that the connections that were used to develop the column curves in Fig. 3.18 are all of the "simple" or partially restrained type. The potential for the structural economies that may be gained by incorporating the end restraint into the column design procedure is clear, although the realistic ranges for the values of λ must be borne in mind and the possible use of bracing to reduce frame drift in designing semirigid frames with flexible base joints must be considered. The ranges for λ have been delineated in Fig. 3.18 for steels with yield stresses of 36 and 50 ksi (250 and 345 Mpa). Consequently, the very large column strength increases that have been reported by several researchers are real (Chen, 1980; Jones et al., 1980, 1982; Razzaq and Chang, 1981; Sugimoto and Chen, 1982; Lui and Chen, 1983b), but they occur for slenderness ratios that are in excess of practical

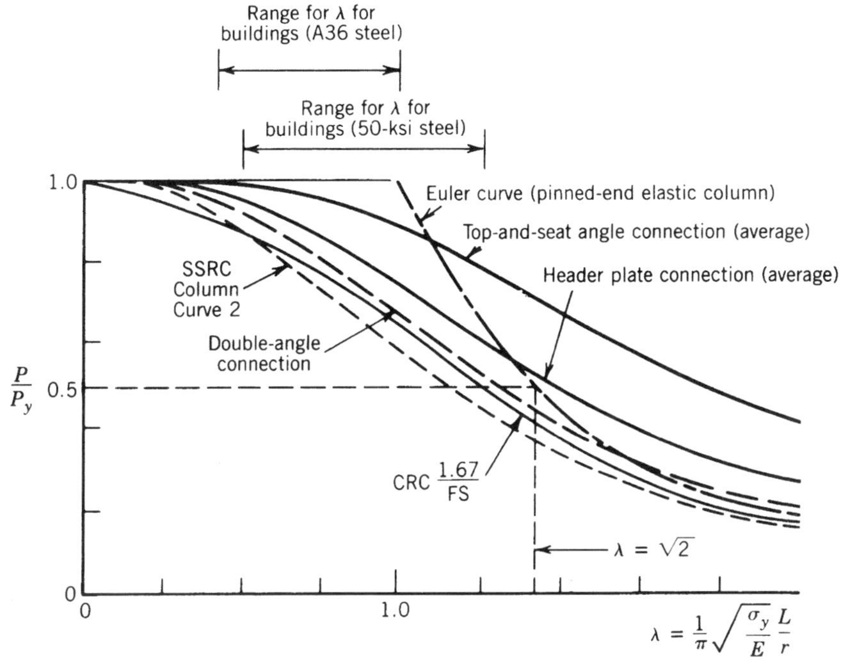

FIGURE 3.18 Column curves for members with different types of end restraint.

values (Bjorhovde, 1981; Ackroyd and Bjorhovde, 1981). In general, end restraint clearly increases strength.

Using the individual column strength studies as the basis, much research has demonstrated the application of end restraint to the design of columns in frames (Bjorhovde, 1984; Chen and Lui, 1991; Chen and Sohal, 1995; Chen et al., 1996). Taking into account actual connection stiffness and the influence of beams, effective-length factors for columns in frames have been developed. The method incorporates the use of alignment charts for the effective length of framed columns and recognizes that buckling of a column in a frame is influenced by the end-restraint relative stiffness distribution factor, G_r, given as

$$G_r = \frac{\sum (EI/L)_{columns}}{C^*} \tag{3.5}$$

where C^* is the effective end restraint that is afforded to a column in a beam and column subassemblage, using connections whose initial stiffness is C, the initial slope of the moment–rotation curve. Barakat and Chen (1990) state that the initial connection stiffness should be used for the design of columns in braced frames, but a secant connection stiffness based on the beam-line concept should be used for columns in unbraced frames. The key findings of this study have been corroborated by major projects aimed at developing practical design methods for semirigid frames (Christopher and Bjorhovde, 1999; Surovek et al., 2005). In particular, the studies of Surovek et al. (2005) provide significant advances in practical applications, and include close correlation with the direct analysis approach of the AISC specification (AISC, 2005a).

The G_r procedure can incorporate design recommendations as well as applications of inelastic K-factor principles, as developed by Yura (1971), expanded by Disque (1973), refined by Barakat and Chen (1990), and implemented in Chen and Lui (1991) and Chen et al. (1996). Practical design examples illustrate the potential for significant structural economies. It is emphasized, however, that in all these methods of analysis the data for the actual end-restraint characteristics of the connection are required. Specifically, the C value must be known. This is a significant drawback, but connection classification schemes similar to those provided by Bjorhovde et al. (1990) and the approaches of Eurocode 3 (CEN, 2005) are useful.

Numerous studies have been aimed at developing methods of accounting for the connection flexibility in providing effective end restraint to framed columns. An extensive review of research on the behavior and modeling of connections is provided in Chen (1987, 1988), Chen and Lui (1991), Beedle (1993), and Chen et al. (1996). Based on the evaluation of different connection models available in the literature, a three-parameter connection power model proposed by Kishi and Chen (1990), together with its large database (Kishi and Chen, 1986; Chen and Kishi, 1989) and design aids (Kishi et al., 1993a), can be recommended for general use.

3.4.2 Effective-Length Factors

The effective-length factor, K, was discussed briefly when introduced in Eq. 3.2. This concept has seen extremely wide acceptance and use for stability assessments of columns in various types of structures since its introduction in the AISC specification in 1961. Despite its wide acceptance, however, it is recognized that the K-factor approach involves a number of major assumptions. Refinements for effective-length computations continue to be made by researchers (e.g., Hellesland and Bjorhovde, 1996a,b), but at the same time it is also recognized that the analytical and computational tools that are available to engineers today clearly necessitate more comprehensive procedures. The direct analysis method (Surovek et al., 2005) that is presented in the most recent AISC specification (AISC, 2005a) is clearly the preferred approach. Detailed examination of this and other procedures is presented in Chapter 16.

The coverage of effective column length in this chapter is limited to certain idealized cases and to certain special situations that occur in compression members of trusses. The effective-length concept has also been applied to members of nonprismatic cross section, whereby they are converted to an equivalent pin-ended member with an effective second moment of area that refers to a particular location of the member (Jimenez and Galambos, 2001).

	(*a*)	(*b*)	(*c*)	(*d*)	(*e*)	(*f*)
Buckled shape of column is shown by dashed line	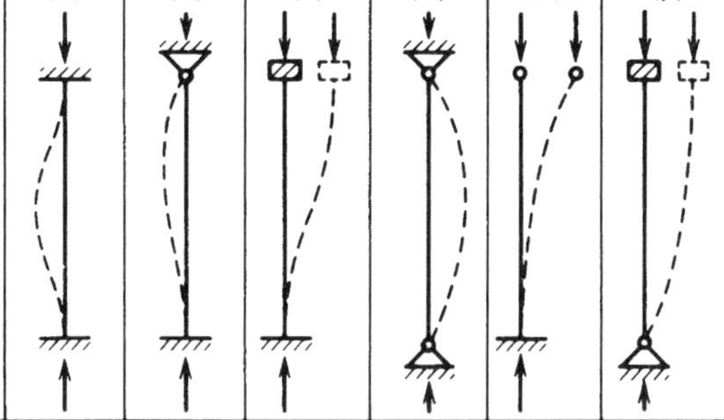					
Theoretical K value	0.5	0.7	1.0	1.0	2.0	2.0
Recommended K value when ideal conditions are approximated	0.65	0.80	1.2	1.0	2.10	2.0
End condition code		Rotation fixed, Translation fixed Rotation free, Translation fixed Rotation fixed, Translation free Rotation free, Translation free				

FIGURE 3.19 Effective-length factors K for centrally loaded columns with various end condition.

Figure 3.19 gives theoretical K values for idealized conditions in which the rotational and/or translational restraints at the ends of the column are either full or nonexistent. At the base, shown fixed for conditions *a*, *b*, *c*, and *e* in Fig. 3.19, the condition of full fixity is approached only when the column is anchored securely to a footing for which the rotation is negligible. Restraint conditions *a*, *c*, and *f* at the top are approached when the top of the column is framed integrally to a girder many times more rigid than itself. Column condition *c* is the same as *a* except that translational restraint is either absent or minimal at the top. Condition *f* is the same as *c* except that there is no rotational restraint at the bottom. The recommended design values of K (see Fig. 3.19) are modifications of the ideal values that reflect the fact that neither perfect fixity nor perfect flexibility is attained in practice. The notion of the perfect pin, however, is retained simply for conservatism (e.g., condition *d*).

The more general determination of K for a compression member as part of any framework requires the application of methods of indeterminate structural analysis, modified to take into account the effects of axial load and inelastic behavior on the rigidity of the members. Gusset plate effects can be included, and for this case extensive charts for modified slope–deflection equations and for moment–distribution stiffness and carryover factors, respectively, have been developed (Goldberg, 1954; Michalos and Louw, 1957). These procedures are not suitable for routine design, but they can be used to determine end restraints and result in modified effective lengths of the component members of a framework.

3.4.3 Effective-Length Factors in Trusses

In triangulated frameworks (trusses), the loads are usually applied at the joints. Thus, if the joints are truly pinned, the members are axially loaded. Deflections of the joints and the truss as a whole are caused by the axial deformations of the members. The angles between members meeting at a joint also change because of these deformations. If the members are connected together at the joints by welds or bolts, the angle changes produce secondary bending stresses. These have little effect on the buckling strength (and tensile strength) of the truss members. Because of local yielding of extreme fibers of the members near the joints, the secondary moments gradually dissipate as the truss is loaded to its ultimate strength. They can therefore be neglected in the buckling analysis (Korol et al., 1986).

When a truss is designed and loaded such that all members reach their factored resistances simultaneously, no member restrains any other. Therefore, the effective-length factors for compression chord members and compression diagonals would be 1.0 for in-plane buckling. In a roof truss of constant or nearly constant depth, and where the compression chord has the same cross section for the full length of the truss, this does not occur, and K may be taken as 0.9. In a continuous truss, K may be taken as 0.85 for the compression chord connected to the joint where the force in the chord changes to compression.

When the magnitude of the force in the compression chord changes at a subpanel point that is not braced normal to the plane on the main truss (Fig. 3.20a), the effective-length factor for chord buckling normal to the plane of the main truss can be approximated from the two compressive forces P_2 and P_1, as follows:

$$K = 0.75 + 0.25\left(\frac{P_1}{P_2}\right) \tag{3.6}$$

where $P_2 < P_1$.

Web members in trusses designed for moving live loads may be designed with $K = 0.85$. This is because the position of the live load that produces maximum force in the web member being designed will result in less than the maximum forces in members framing into it, so that rotational restraints are developed.

The design of vertical web members, U_iL_i, of a K-braced truss (Fig. 3.20b) should be based on the effective length *KL*. Web-member buckling occurs normal to the plane of the truss, and Eq. 3.6 again applies. Also, P_2 is negative in Eq. 3.6 because it is a tensile force. When P_1 and P_2 are equal but opposite in sign, Eq. 3.6 yields a value of $K = 0.5$.

For buckling normal to the plane of a main truss, the web compression members should be designed for $K = 1$ unless detailed knowledge of the makeup of the cross frames (perpendicular to the truss) is available.

In the case of redundant trusses, there is reserve strength above the load at initial buckling of any compression member. Masur (1954) has reviewed developments

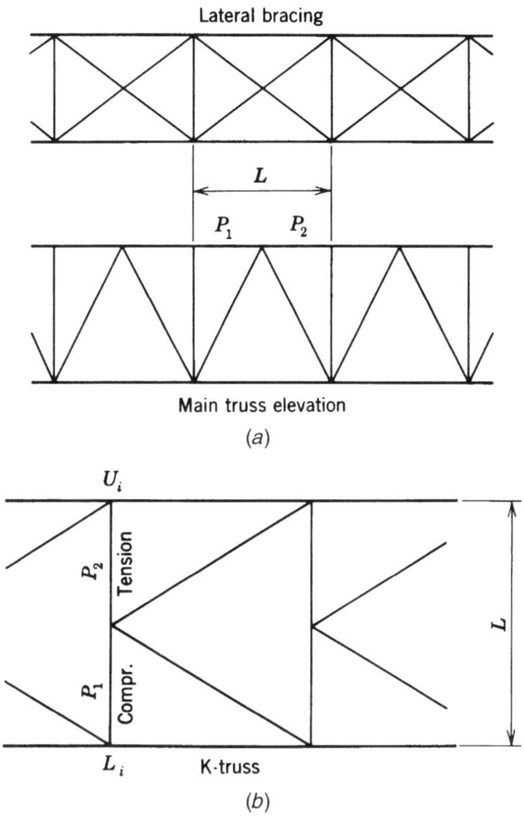

FIGURE 3.20 Effective-length factors in trusses: (*a*) compression chords; (*b*) compression verticals.

on this subject and established upper and lower bounds for the ultimate load of the buckled members of elastic redundant trusses.

3.4.4 Faulty Column Bracing Detail

Numerous structural failures have occurred because of a misunderstanding of the end restraint provided by the structural arrangement shown in Fig. 3.21. The beam (or truss) is continuous over the top of the column. The critical components are the column in compression, compression in the bottom flange of the beam or chord of the truss, and no bottom-flange bracing at point *a* and possibly other points, *b*. The sway at the top of the column shown in section B-B can result in a K factor much greater than 2.0. The bottom flange of the beam can possibly provide bracing to the top of the column if there are braces at points *b* and consideration is given to the compression in the flange when evaluating its stiffness. In general, a brace, such as a bottom chord extension from the joist, should be used at point *a*. Beam

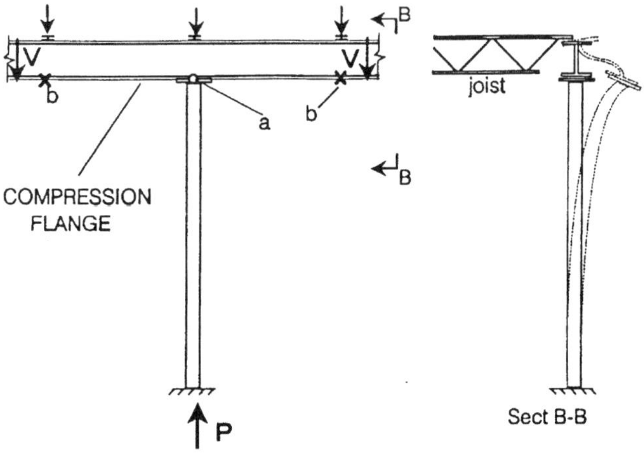

FIGURE 3.21 Structural detail with probable instability.

web stiffeners at the column location will also be effective unless bottom flange lateral buckling is critical.

3.5 STRENGTH CRITERIA FOR STEEL COLUMNS

The position of the SSRC on the basis for the design of columns is stated in Technical Memorandum[2] No. 5 and can be summarized with the following quote:

> Maximum strength, determined by the evaluation of those effects that influence significantly the maximum load-resisting capacity of a frame, member, or element, is the proper basis for the establishment of strength design criteria.

The proper column strength model is therefore one that incorporates both residual stresses and initial out-of-straightness.

3.5.1 Column Design Curves

Multiple Column Curves In a wide-ranging, landmark study, Bjorhovde (1972) examined the deterministic and probabilistic characteristics of column strength in general and developed an extensive database for the maximum strengths of centrally loaded compression members. Covering the full practical range of shapes, steel grades, and manufacturing methods, the study demonstrated the wide variability of column strength. Figure 3.22 illustrates this variability through a collection of 112 maximum-strength column curves. Subsequent and parallel investigations

[2]All SSRC technical memorandums are provided in Appendix B.

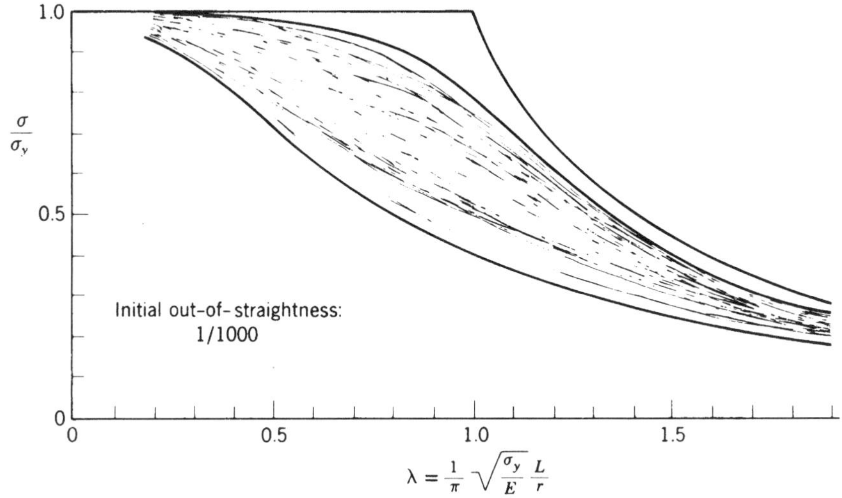

FIGURE 3.22 Maximum-strength column curves for a number of different column types (Bjorhovde, 1972).

of other researchers have added to and confirmed this relatively wide band of column strength (Birkemoe, 1977a; Bjorhovde, 1988, 1991; Bjorhovde and Birkemoe, 1979; Fukumoto et al., 1983; Jacquet, 1970; Kato, 1977; Sherman, 1976).

The key problem in developing a rational, representative, and sufficiently reliable column strength criterion is how to take this large variability into account. This may be achieved by using a mean or other central curve of the band of strength variation of Fig. 3.22, or it may be done by subdividing the band into groups of curves with a mean or similar curve for each group. The latter defines the *multiple column curve concept* (Bjorhovde and Tall, 1971; Bjorhovde, 1972).

Research and development leading to the use of multiple column curves were actively pursued from the late 1950s to the early 1980s. In 1959, the German standard DIN 4114 introduced a special curve for tubes and another curve for all other shapes. Subsequently, the work under the auspices of the ECCS (Beer and Schultz, 1970; Jacquet, 1970; Sfintesco, 1970) resulted in recommended design application and code adoption in several countries (Sfintesco, 1976). The ECCS curves in somewhat modified form are now part of the column design criteria of Eurocode 3 (CEN, 2005). In 1984, the CSA adopted SSRC curve 1 for use with heat-treated tubes; CSA had earlier (1974) adopted SSRC curve 2 as the basic column strength criterion and in 1994 assigned welded wide-flange columns made from a flame-cut plate to SSRC column curve 1 (Chernenko and Kennedy,1991).

Research basic to the development of multiple column curves in North America was conducted at Lehigh University starting in the 1960s (Bjorhovde and Tall, 1971; Bjorhovde, 1972, 1988), and elsewhere (Birkemoe, 1977a,b; Bjorhovde, 1977; Bjorhovde and Birkemoe 1979; Kato, 1977; Sherman, 1976). In addition to length, cross-section dimensions, and the material properties, the maximum strength of steel columns depends on (1) the residual stress magnitude and distribution, (2) the shape and magnitude of the initial out-of-straightness, and (3) the end restraint. The effects of these three variables were discussed in Sections 3.3 and 3.4. Unless special procedures are utilized in the manufacture of steel columns, such as stress relieving or providing actual pins at each end, all three of these effects are present and should be accounted for. The state-of-the-art is such that if the following information is known, accurate calculation of the maximum strength is possible (Bjorhovde, 1972, 1978, 1988; Chen and Lui, 1985):

1. *Material properties* (i.e., the yield stress, σ_y, and the modulus of elasticity, E). In some cases it is necessary to know the variation of the yield stress across the cross section (e.g., welded built-up shapes) or the entire stress–strain curve (e.g., cold-formed shapes).
2. *Cross-section dimensions*. For nonprismatic columns the dimensions along the column length must be known.
3. *Distribution of the residual stresses* in the cross section, including variations through the plate thickness if the shape is tubular or if the plate elements are thick.
4. *Shape and magnitude of the initial out-of-straightness.*
5. *Moment–rotation relationship of the end restraint.*

Maximum strength may be calculated by postulating suitable but realistic idealizations so that closed-form algebraic expressions may be derived or one of many available numerical techniques may be used. In numerical calculations it is usually assumed that deformations are small and that plane sections remain plane. The literature on determining the maximum strength of columns is rich and diverse, but the major methods are described in various textbooks (e.g., Chen and Atsuta, 1976; Chen and Han, 1985).

Residual stresses, initial out-of-straightness, and end restraint vary randomly, and complete statistical information is lacking for most shapes and design situations. In particular, data on end restraint in terms of beam-to-column moment–rotation curves are limited; this is a result of the great variety of connections that are used in steel construction practice. Techniques for evaluating end-restraint effects have been discussed in Section 3.4, and much research has been conducted over the past 25 years (Ackroyd, 1979; Chen, 1980; Jones et al., 1980; 1982; Ackroyd and Bjorhovde, 1981; Bjorhovde, 1981, 1984, 1988; Bjorhovde et al., 1990; Chapuis and Galambos, 1982; Christopher and Bjorhovde, 1999; Lui and Chen, 1983a,b; Sugimoto and Chen, 1982; Shen and Lu, 1983; Surovek et al., 2005). The six books that were published after the international workshops on connections in steel structures provide a large database for analytical and design approaches. The publication data for these books are given in Section 3.4; they are also listed among the references for this chapter. Although the procedures that have been developed are applicable to a range of problems, additional work needs to be done to make such concepts suitable for design specifications.

The reference compression element continues to be the pin-ended, centrally loaded column. The key research work that focused on this issue and the studies that were done for multiple column curves used this basic column concept. An answer to the problem of a suitable specification format was provided by Bjorhovde (1972), who proceeded as detailed in the following.

A computerized maximum-strength analysis was performed first on basic data available from column tests performed at Lehigh University, and it was demonstrated that the method of numerical analysis gave accurate predictions of the test strengths. Next, a set of 112 column curves was generated for members for which measured residual stress distributions were available, assuming that the initial out-of-straightness was of a sinusoidal shape having a maximum amplitude of 1/1000 of the column length and that the end restraint was zero. These shapes encompassed the major shapes used for columns, including rolled and welded shapes from light to heavy dimensions. The column curves thus obtained represent essentially the complete spectrum of steel column behavior. The resulting curves are shown in Fig. 3.22.

Bjorhovde then observed that there were definite groupings among the curves, and from these, three subgroups were identified, each giving a single "average" curve for the subgroup (Bjorhovde and Tall, 1971; Bjorhovde, 1972). The resulting three curves are known as *SSRC column strength curves* 1, 2, *and* 3, and they are reproduced in Figs. 3.23 through 3.25. These figures contain:

1. The number of column curves used as a basis for the statistical analysis and the width of their scatter band
2. The calculated 2.5 and 97.5 percentile lower and upper probability limits for the particular set of curves
3. The column types to which each of the three curves is related

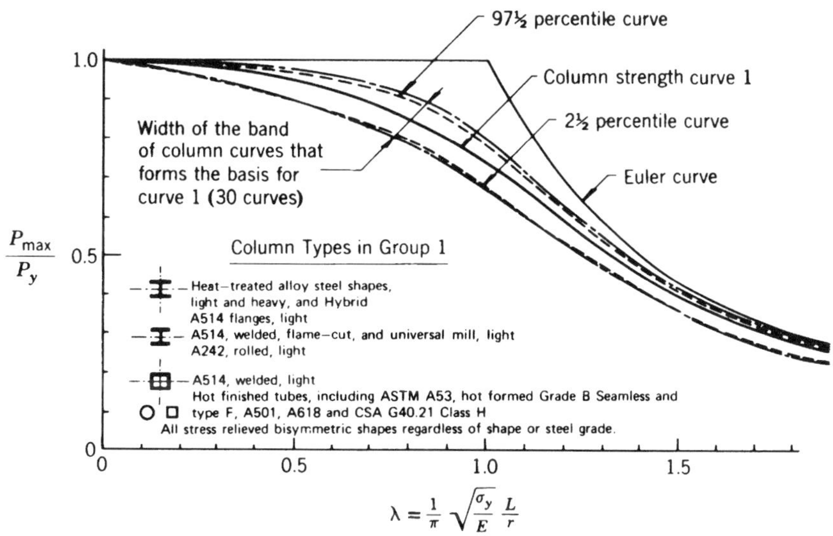

FIGURE 3.23 SSRC Column strength curve 1 for structural steel (Bjorhovde, 1972) (based on maximum strength and initial out-of-straightness of $\delta_0 = 0.001L$).

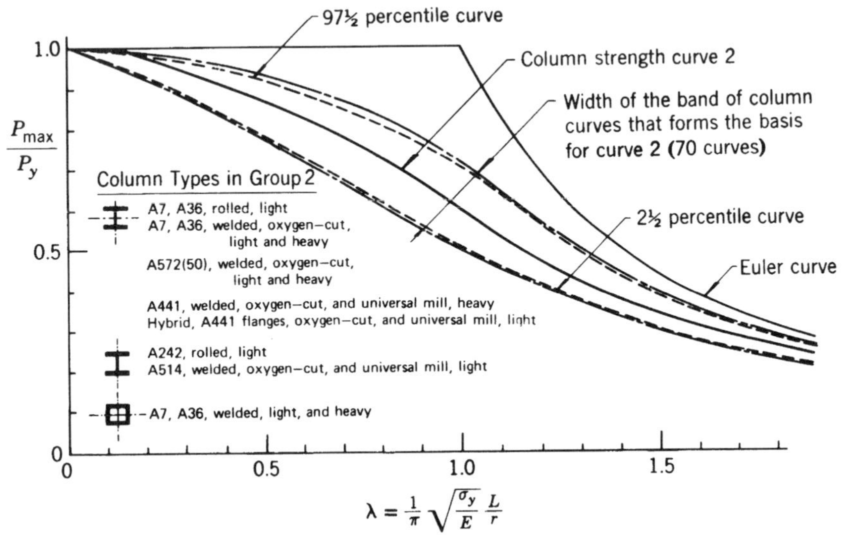

FIGURE 3.24 SSRC column strength curve 2 for structural steel (Bjorhovde, 1972) (based on maximum strength and initial out-of-straightness of straightness of $\delta_0 = 0.001L$).

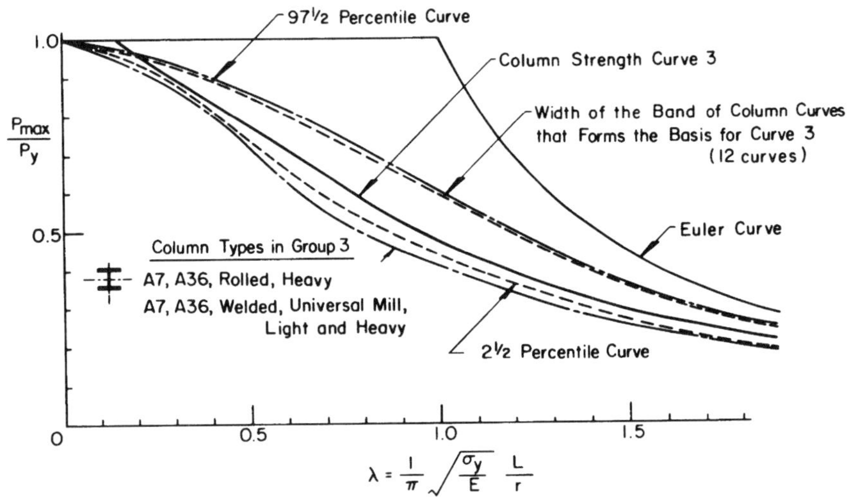

FIGURE 3.25 SSRC column strength curve 2 for structural steel (Bjorhovde, 1972) (based on maximum strength and initial out-of-straightness of straightness of $\delta_0 = 0.001L$).

Algebraic representations of the three column strength curves were obtained by curve fitting, and the resulting equations are as follows:

SSRC curve 1:

$$\begin{array}{lll} 1. & 0 \leq \lambda \leq 0.15 & \sigma_u = \sigma_y \\ 2. & 0.15 \leq \lambda \leq 1.2 & \sigma_u = \sigma_y (0.990 + 0.112\lambda - 0.367\lambda^2) \\ 3. & 1.2 \leq \lambda \leq 1.8 & \sigma_u = \sigma_y (0.051 + 0.801\lambda^{-2}) \\ 4. & 1.8 \leq \lambda \leq 2.8 & \sigma_u = \sigma_y (0.008 + 0.942\lambda^{-2}) \\ 5. & \lambda \geq 2.8 & \sigma_u = \sigma_y \lambda^{-2} (\text{Euler curve}) \end{array} \tag{3.7}$$

SSRC curve 2:

$$\begin{array}{lll} 1. & 0 \leq \lambda \leq 0.15 & \sigma_u = \sigma_y \\ 2. & 0.15 \leq \lambda \leq 1.0 & \sigma_u = \sigma_y (1.035 - 0.202\lambda - 0.222\lambda^2) \\ 3. & 1.0 \leq \lambda \leq 2.0 & \sigma_u = \sigma_y (-0.111 + 0.636\lambda^{-1} + 0.087\lambda^{-2}) \\ 4. & 2.0 \leq \lambda \leq 3.6 & \sigma_u = \sigma_y (0.009 + 0.877\lambda^{-2}) \\ 5. & \lambda \geq 3.6 & \sigma_u = \sigma_y \lambda^{-2} (\text{Euler curve}) \end{array} \tag{3.8}$$

SSRC curve 3:

$$\begin{array}{lll} 1. & 0 \leq \lambda \leq 0.15 & \sigma_u = \sigma_y \\ 2. & 0.15 \leq \lambda \leq 0.8 & \sigma_u = \sigma_y (1.093 - 0.622\lambda) \\ 3. & 0.8 \leq \lambda \leq 2.2 & \sigma_u = \sigma_y (-0.128 + 0.707\lambda^{-1} - 0.102\lambda^{-2}) \\ 4. & 2.2 \leq \lambda \leq 5.0 & \sigma_u = \sigma_y (0.008 + 0.792\lambda^{-2}) \\ 5. & \lambda \geq 5.0 & \sigma_u = \sigma_y \lambda^{-2} (\text{Euler curve}) \end{array} \tag{3.9}$$

These expressions can also be represented accurately by a single equation (Rondal and Maquoi, 1979; Lui and Chen, 1984), as shown below. The maximum deviations from the SSRC curves are −2.1 to +3.6%.

$$\sigma_u = \frac{\sigma_y}{2\lambda^2} \left(Q - \sqrt{Q^2 - 4\lambda^2} \right) \leq \sigma_y \tag{3.10}$$

where

$$Q = 1 + \alpha (\lambda - 0.15) + \lambda^2 \tag{3.11}$$

and

$$\alpha = \begin{cases} 0.103 & \text{for curve 1} \\ 0.293 & \text{for curve 2} \\ 0.622 & \text{for curve 3} \end{cases}$$

Another expression with a single parameter n in a double exponential representation is used in the Canadian steel design standard CSA S16–09 (CSA, 2009; Loov, 1996)

$$\sigma_u = F_y \left(1 + \lambda^{2n}\right)^{-1/n} \tag{3.12}$$

where n equals 2.24 for curve 1 and 1.34 for curve 2. Loov (1996) also provided the value 0.96 for curve 3, although this curve was not adopted by the CSA standard, primarily because welded built-up shapes using universal mill plates are not representative of Canadian practice. These expressions give strengths generally within 1% of the polynomials of Eqs. 3.7 through 3.9 and are never more than 3% unconservative.

Bjorhovde (1972) also developed multiple column curves where the initial out-of-straightness was equal to its mean value of 1/1470 of the column length (Fig. 3.14). The mathematical equations describing these curves are as follows:

SSRC curve 1P:

$$\begin{array}{lll} 1. & 0 \le \lambda \le 0.15 & \sigma_u = \sigma_y \\ 2. & 0.15 \le \lambda \le 1.2 & \sigma_u = \sigma_y(0.979 + 0.205\lambda - 0.423\lambda^2) \\ 3. & 1.2 \le \lambda \le 1.8 & \sigma_u = \sigma_y(0.030 + 0.842\lambda^{-2}) \\ 4. & 1.8 \le \lambda \le 2.6 & \sigma_u = \sigma_y(0.018 + 0.881\lambda^{-2}) \\ 5. & \lambda \ge 2.6 & \sigma_u = \sigma_y\lambda^{-2}\text{(Euler curve)} \end{array} \tag{3.13}$$

SSRC curve 2P:

$$\begin{array}{lll} 1. & 0 \le \lambda \le 0.15 & \sigma_u = \sigma_y \\ 2. & 0.15 \le \lambda \le 1.0 & \sigma_u = \sigma_y(1.030 - 0.158\lambda - 0.206\lambda^2) \\ 3. & 1.0 \le \lambda \le 1.8 & \sigma_u = \sigma_y(-0.193 + 0.803\lambda^{-1} + 0.056\lambda^{-2}) \\ 4. & 1.8 \le \lambda \le 3.2 & \sigma_u = \sigma_y(0.018 + 0.815\lambda^{-2}) \\ 5. & \lambda \ge 3.2 & \sigma_u = \sigma_y\lambda^{-2}\text{(Euler curve)} \end{array} \tag{3.14}$$

SSRC curve 3P:

$$\begin{array}{lll} 1. & 0 \le \lambda \le 0.15 & \sigma_u = \sigma_y \\ 2. & 0.15 \le \lambda \le 0.8 & \sigma_u = \sigma_y(1.091 - 0.608\lambda) \\ 3. & 0.8 \le \lambda \le 2.0 & \sigma_u = \sigma_y(0.021 + 0.385\lambda^{-1} - 0.066\lambda^{-2}) \\ 4. & 2.2 \le \lambda \le 4.5 & \sigma_u = \sigma_y(0.005 + 0.900\lambda^{-2}) \\ 5. & \lambda \ge 4.5 & \sigma_u = \sigma_y\lambda^{-2}\text{(Euler curve)} \end{array} \tag{3.15}$$

A single expression for all of the SSRC-P curves has not been developed; however, this can be achieved relatively easily using the approaches of Rondal and Maquoi (1979), Loov (1996), or Rotter (1982). The single curve that is used in

Chapter E of the previous AISC (1999) Load and Resistance Factor Design (LRFD) specification and the unified AISC (2005a) specification is identical to SSRC 2P. Two equations are used to describe this curve for two regions of slenderness, λ_c (employed in the 1999 specification), or stress ratio F_y/F_e (employed in the 2005 specification). The equations give the same results. For the 1999 specification:

$$F_{cr} = \begin{cases} (0.658^{\lambda_c^2})F_y & \text{for } \lambda_c \leq 1.5 \\ \left(\dfrac{0.877}{\lambda_c^2}\right)F_y & \text{for } \lambda_c > 1.5 \end{cases} \tag{3.16a}$$

where the first equation applies for inelastic buckling and the second equation applies for elastic buckling. For the 2005 specification:

$$F_{cr} = \begin{cases} [0.658^{F_y/F_e}]F_y & \text{for } \dfrac{KL}{r} \leq 4.71\sqrt{\dfrac{E}{F_y}} \quad (\text{or } F_e \geq 0.44F_y) \\ 0.877F_e & \text{for } \dfrac{KL}{r} > 4.71\sqrt{\dfrac{E}{F_y}} \quad (\text{or } F_e < 0.44F_y) \end{cases} \tag{3.16b}$$

where F_e is the Euler flexural buckling stress for the slenderness ratio *KL/r* and F_y $(= \sigma_y)$ is the specified minimum yield stress.

Design Procedure Alternatives It was demonstrated in the preceding section that it is possible to develop multiple column curves into which column types can be grouped for convenience. In developing column design criteria, the following questions should be considered:

1. *What should be the shape and the amplitude of the initial out-of-straightness?* As to the shape, there is general agreement that a sinusoidal shape with the maximum amplitude at midlength is a conservative and reasonable assumption. The maximum amplitude is a crucial quantity, because changes affect the strength, especially in the intermediate slenderness range. Knowledge about initial out-of-straightness is available from measurements of laboratory specimens used for column tests, but there is a lack of field data. Initial out-of-straightness is a function of the manufacturing process, and some column types, such as manufactured tubes, tend to be very straight. In the development of the SSRC and ECCS multiple column curves the position was taken that an initial amplitude of 1/1000 of the length, essentially the mill tolerance, is a reasonable and conservative value for the basis of developing column curves.

In opposition, it can be argued that all geometric imperfections are small enough so that their effect can be relegated to be accounted for by the resistance factor. This was the underlying philosophy of the use of the CRC column curve, which has its basis in tangent-modulus theory, with a factor of safety that depends on the column slenderness ratio. This design approach was entirely sensible when it was initially formulated in the 1950s, but a large body of research work has since shown that the maximum strength can be determined from a knowledge of initial imperfections and the proper design and use of steel columns must take the out-of-straightness explicitly into account.

A task force of SSRC (1985) took an intermediate position, recommending that the basis for the development of design curves for steel columns should be an initial out-of-straightness of 1/1500 of the length. This is close to the average measured in laboratory columns (Bjorhovde, 1972; Fukumoto et al., 1983) and reflects the position of the SSRC in this matter. For all practical purposes, Eqs. 3.13 through 3.15 represent this condition. By its adoption of Eqs. 3.16a and 3.16b, AISC is effectively using *L*/1500 as the governing out-of-straightness criterion.

2. *Should design be based on the concept of multiple column curves or should one composite column curve be used for the design of all steel columns?* The European answer to this question has been to recommend multiple column curves, as shown in Fig. 3.26. As a first step in North America, in 1974 the CSA in 1974 adopted the use of SSRC curve 2 as the basic design curve. In 1984, the CSA also adopted SSRC curve 1 for hollow structural sections, either hot formed or cold formed to final shape and then heat treated. This recommendation was based in part on research on such columns (Bjorhovde, 1977; Birkemoe, 1977b; Bjorhovde and Birkemoe, 1979; Kennedy and Gad Aly, 1980). The original reluctance to adopt multiple column curves for the American structural steel design specification was founded on the belief that the design criteria would become too complicated. Another reason was that it was felt necessary to complete certain additional studies, to ensure that all conceivable column types and materials would be properly assigned to one of the three strength categories (Bjorhovde, 1980). The results of

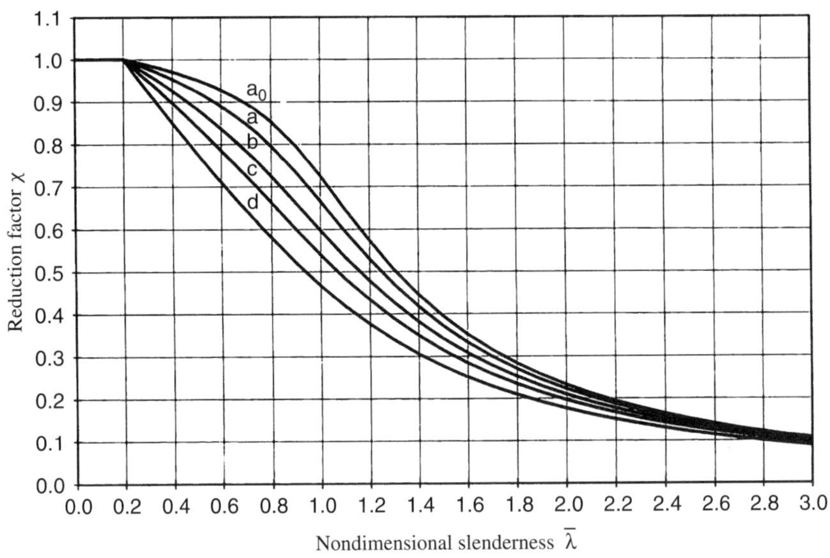

FIGURE 3.26 Eurocode 3 multiple column curves (CEN, 2005).

Fabrication Details		Axis	Specified Minimum Yield Stress of Steel (ksi)				
			≤ 36	37 to 49	50 to 59	60 to 89	≥ 90
Hot-rolled W-shapes	Light and medium W-shapes	Major	2	2	1	1	1
		Minor	2	2	2	1	1
	Heavy W-shapes (flange over 2 in.)	Major	3	2	2	2	2
		Minor	3	3	2	2	2
Welded Built-up H-shapes	Flame-cut plates	Major	2	2	2	1	1
		Minor	2	2	2	2	1
	Universal mill plates	Major	3	3	2	2	2
		Minor	3	3	3	2	2
Welded Box Shapes	Flame-cut and universal mill plates	Major	2	2	2	1	1
		Minor	2	2	2	1	1
Square and Rectangular Tubes	Cold-formed	Major	N/A	2	2	2	2
		Minor	N/A	2	2	2	2
	Hot-formed and cold-formed heat-treated	Major	1	1	1	1	1
		Minor	1	1	1	1	1
Circular Tubes	Cold-formed	N/A	2	2	2	2	2
	Hot-formed	N/A	1	1	1	1	1
All stress-relieved Shapes		Major and Minor	1	1	1	1	1

FIGURE 3.27 Column curve selection table (Bjorhovde, 1972, 1988).

this study are summarized in the column curve selection table (CCS table) shown in Fig. 3.27 (Bjorhovde, 1972, 1988). The CCS table will facilitate the column curve selection process and is also suited for a decision-table format for use with computer-based design.

3. *What end restraint should be assumed for nominally pin-ended columns?* As indicated in Section 3.4, any practical framing scheme or column base condition will increase the column strength. There are really no truly pin-ended columns in existence. Methods have been developed to use this end restraint in determining the maximum strength of columns, but the question of how to use the available information in design is still unresolved. Should explicit restraint factors for different kinds of end conditions be tabulated for use with effective-length-factor alignment charts, or should the design curves implicitly contain minimal end restraints? The latter approach was used in the development of the AISC column curve, which is based on an implicit end restraint producing an elastic effective-length factor of 0.96 ($G = 10$), as well as an initial out-of-straightness of 1/1500 of the length.

Summary In the previous discussion on the strength of steel columns, a number of alternatives were presented. Specification-writing groups need to make decisions to select the column curve or curves satisfying their needs. The necessary theory is available to do so, and much information is on hand. It is the SSRC's opinion that design criteria for steel columns should be based on a column with an initial out-of-straightness column and residual stresses. With this concept as a basis, intelligent choices for column design can be made, resulting in a rational method of design.

3.5.2 Development of Strength Design Criteria

The following gives a description of data and computational techniques that are needed for the development of maximum-strength column curves and other results for pin-ended, centrally loaded columns. Further details can be found in the references of Batterman and Johnston (1967), Bjorhovde (1972, 1988, 1992), Chen and Atsuta (1976), Chen and Han (1985), Chernenko and Kennedy (1991), Kennedy and Gad Aly (1980), and Albert et al. (1995).

When detailed strength and performance data are not available for a specific column shape, it is possible to develop column curves of types that are similar to those that have been presented in this guide. The following gives a brief outline of the assumptions that should be used, the types of data that are required, and the computational technique that is suitable for these types of problems.

Required Data The following data are needed for the computation of the maximum strength of columns:

1. Type of material and its material properties (i.e., yield stress, yield strain, modulus of elasticity).
2. Distribution of the residual stresses in the cross section, including variation through the thickness, if the shape is large or it is tubular.
3. Variation of the yield stress throughout the cross section. This is in most cases needed only for welded built-up shapes and cold-formed shapes, where the yield stress at a weld or a cold-formed corner, for example, may differ significantly from the nominal properties.
4. When the material is of a type or grade that exhibits nonlinear stress–strain characteristics (e.g., stainless steel), a complete, typical stress–strain curve is required.
5. Maximum value of the initial out-of-straightness.

Assumptions for the Analysis The following assumptions are normally conservative in nature, with the result that computed column strengths are usually somewhat less than those obtained in actual tests:

1. Material is linearly elastic, perfectly plastic.
2. The initial and all subsequent deflection shapes of the column can be described by a half sine wave.
3. The residual stresses are constant in an element of the cross section along the full length of the column.

4. Sections that are originally plane remain so for the range of deflections that is suitable for column studies.
5. Yielded fibers in the cross section will unload elastically.
6. The yield stress may vary across the width and through the thickness of the component plates of the cross section but does not vary along the length of the column.
7. In line with assumption 2, only stresses and strains at midlength of the column are considered in the analysis.

It should be pointed out that if detailed yield-stress and other material data are not available for the elements in the cross section, the results of a stub-column test (see Technical Memorandum No. 3 in Appendix B) can be used. If this is not available, tension test results for various parts of the shape can be used; the properties to utilize in the computations should then be based on a weighted average.

Computational Technique Maximum column strength requires the solution of a nonlinear problem. It is best achieved through an incremental, iterative computation algorithm. Internal force and moment equilibrium are established for every load and deflection level, requiring iteration over the cross section to determine when individual fibers yield, unload, or continue to load. The computations are carried to a level where the column cannot take any additional load when an additional amount of deflection is imposed; this constitutes the maximum strength level. It is recommended that deflection increments be used rather than load increments, as convergence problems may be encountered as the maximum load is approached when load increments are used. The procedure above leads to the development of a load–deflection curve for a column of given slenderness ratio or length. To obtain the complete column curve, the process must be repeated for a range of lengths. As an illustration of the basic steps in the column strength computations, Fig. 3.28 gives a flowchart that indicates the necessary major parts of the solution.

3.6 ALUMINUM COLUMNS

3.6.1 Material Properties

The stress–strain curves of aluminum alloys are rounded, or nonlinear, as distinct from the bilinear elastic–perfectly plastic curve used for some structural steels. In the absence of a yield plateau it is standard practice to define the yield stress as the 0.2% offset stress, $\sigma_{0.2}$. Alloying elements, heat treatment, and cold working influence the stress–strain curve but have little effect on the elastic modulus, which for tension falls within the range of 9900 to 10,200 ksi (68 to 70 Gpa) for other than aircraft alloys. A value of 10,000 ksi (68.3 Gpa) is often used.

Yield strength is more strongly influenced by heat treatment and cold working than is the ultimate strength. For precipitation heat-treated alloys there is also an

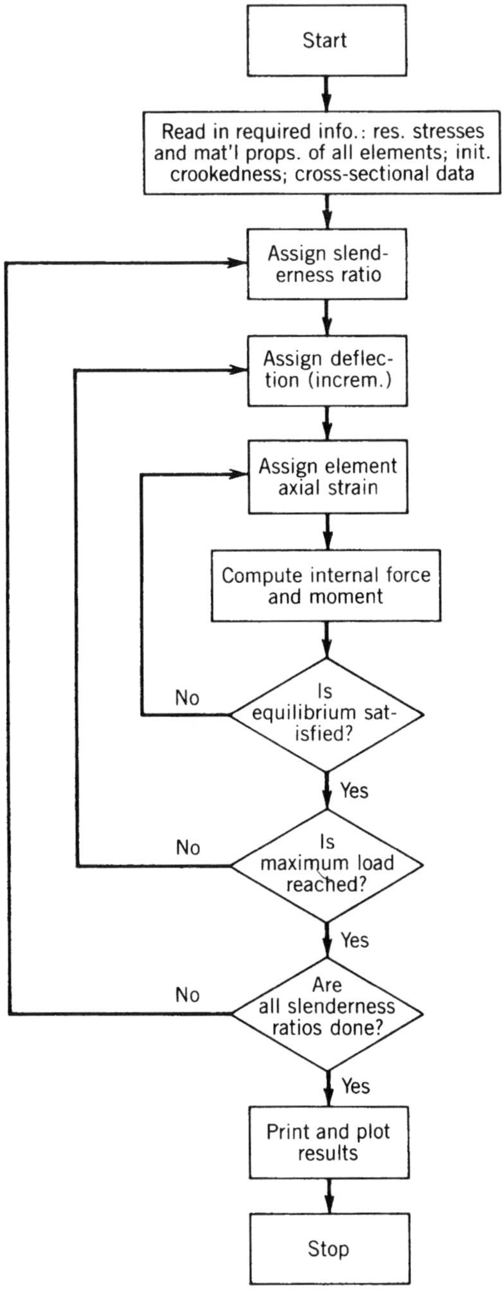

FIGURE 3.28 Flowchart for preparation of column strength curve.

increased sharpness of the "knee" between the elastic and plastic ranges, which is significant for columns in the lower range of slenderness ratios. For this reason, column formulas for aluminum alloys are divided into two groups, precipitation heat-treated tempers and all other tempers, which reflect the differing ratio $\sigma_{0.2}/\sigma_{0.1}$. Guaranteed values for the yield strength, defined by the 0.2% offset, and the ultimate strength are established at levels at which 99% of the material is expected to conform at a 0.95 confidence level. In practice, typical values are about 15% above the guaranteed value; thus, the use of the guaranteed value in a design formula which has been formulated on the basis of measured values provides an additional factor to be considered when selecting resistance factors for columns with short and medium slenderness ratios.

3.6.2 Imperfections

Deviations of real columns from the perfect geometric and material idealizations are of two essential forms: geometric imperfections such as eccentricity and initial out-of-straightness and built-in residual stresses such as those arising from welding. Residual stresses in aluminum extruded members are small because of the method of production and the straightening of the finished member by stretching. The fact that residual stress effects on column strength of aluminum extruded members are insignificant has been confirmed in European studies (Mazzolani and Frey, 1977). The value of the yield strength does not vary significantly across a profile (Bernard et al., 1973).

Welding introduces residual tensile stresses in the weld bead on the order of the yield strength for the annealed material and compressive stresses elsewhere. There is also a local reduction in mechanical properties affected by welding that is significant in precipitation heat-treated or cold-worked material. Residual stresses created by cold forming are related to the yield strength in the same manner as in steel; longitudinal bends, however, are considered to have little influence on column strength, as a result of either the residual stresses or any strain hardening. Geometric imperfections fall into two groups: those that are length dependent, such as initial out-of-straightness, and those independent of length, such as imperfections introduced by tolerances on cross-section dimensions.

Commercial tolerances on initial out-of-straightness are *L*/960 for most extruded structural members. If a member with such an initial out-of-straightness forms two or more bays of the chord of a truss, the final out-of-straightness will be negligible in comparison to the inaccuracies in the assembly. Even in laboratory tests, the initial out-of-straightness of the member as supplied has usually been less than the error in centering the specimen. End moments, due to frame action or eccentricities in joints, will in most cases dominate any moments due to initial crookedness. For large assembled columns, such as latticed masts, an initial out-of-straightness of *L*/1000 has been found to be representative, and design as a beam-column using this value has been adopted.

3.6.3 Strength of Aluminum Alloy Columns

In design applications, aluminum alloy column strength has generally been based on the tangent-modulus theory because of the good agreement with column test results. Evaluation of the maximum strength of both straight and initially crooked columns is practicable with computers. A systematic study of the effects of important parameters that affect column strength has been made by Batterman and Johnston (1967) and Hariri (1967).

Initially Straight Columns Such members were studied by Duberg and Wilder (1950), and Johnston (1963, 1964). In studying the behavior above the critical load, Batterman and Johnston (1967) assumed the stress–strain curve of the material to be represented by the average of a large number of tests of aluminum alloy 6061-T6. By considering both major and minor axis buckling of an H-type section, the practical range of the shape effect was approximately covered. A section having a depth equal to the width was chosen, with flange thickness approximately one-tenth of the depth, and with a web having a thickness two-thirds that of the flange. The maximum increase in strength above the tangent-modulus load was found to be about 2% for minor axis bending. This small difference further justifies the use of the tangent-modulus load as a reasonable basis for estimating the strength of initially straight aluminum alloy columns.

Columns with Initial Out-of-Straightness The maximum strength of initially curved pin-ended aluminum alloy columns can be evaluated by use of a computer, as described previously (Batterman and Johnston, 1967). For a typical H-shape of alloy 6061-T6 with buckling about the strong axis, Fig. 3.29 shows typical plots of load versus midlength lateral deflection for an L/r value of 40 and for initial midlength out-of-straightness ranging from zero to $0.004L$. Figure 3.30 illustrates the strength reduction factor, referenced to the critical load and plotted as a function of L/r for both major and minor axis bending.

The effects of initial out-of-straightness are accentuated in unsymmetrical sections, such as the T-section, as illustrated by the computer-generated load–deflection curves in Fig. 3.31. An initial crookedness of $0.001L$ (with the flange on the convex side) reduced the ultimate strength in comparison with the tangent-modulus load by about 18% at $L/r = 40$, which is about twice the reduction shown in Fig. 3.30 for the doubly symmetric section. Buckling of axially loaded straight members (as confirmed by tests) will occur so as to put the flange of the T-section on the convex side of the column. The shaded lines at the top of Fig. 3.31 indicate the upper bounds of theoretical strength of a straight column, namely the inelastic buckling gradients and the reduced-modulus strengths for buckling in either of the two possible directions (Johnston, 1964). Tests have also indicated that the effect of end eccentricity is somewhat more deleterious than the effect of the same magnitude of initial out-of-straightness (Hariri, 1967).

The effect of end restraint on aluminum columns with an initial out-of-straightness has been studied by Chapuis and Galambos (1982). They conclude: “It

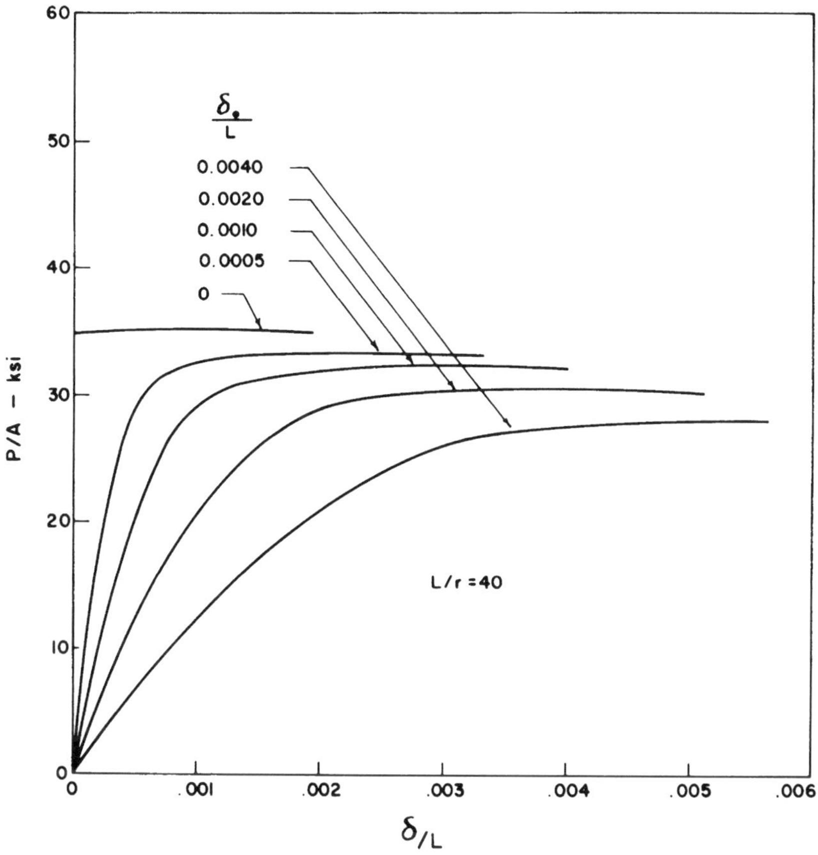

FIGURE 3.29 Effect of initial out-of-straightness on load–deflection relationships of an aluminum alloy column (Batterman and Johnston, 1967).

is conservative to base a design on K_{el} (the effective-length factor determined from elastic buckling analysis) and a column curve derived for pinned crooked columns."

3.6.4 Effects of Welding

For most alloys used in structural applications, the heat of welding reduces the strength of the metal in a narrow zone around the weld, thereby diminishing the capacity of columns of low and intermediate slenderness ratios. Welding can also introduce residual stresses and an initial out-of-straightness in the column. Rectangular section columns with longitudinal and transverse welds have been tested by Brungraber and Clark (1962). Wide-flange and tubular box section columns with longitudinal welds were tested by Mazzolani (1985), and experiments of tubular box section columns with transverse welds were performed by Hong (1991).

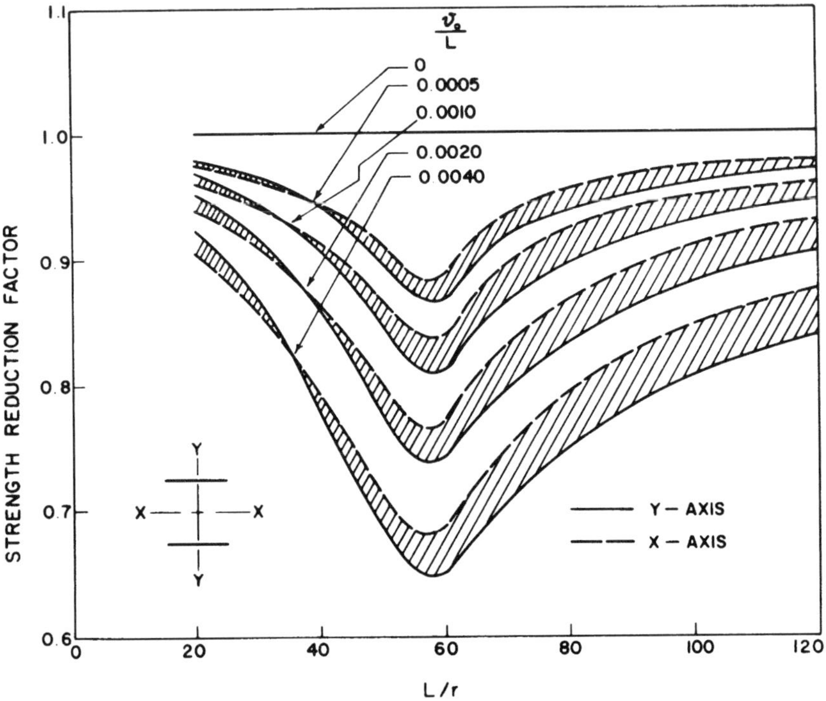

FIGURE 3.30 Ratio of maximum strength/tangent-modulus load of aluminum alloy columns for different amounts of initial out-of-straigthness (Batterman and Johnston, 1967).

Lai and Nethercot (1992) provided analytical results for aluminum wide-flange columns with transverse welds.

For columns with longitudinal welds or with transverse welds that affect only a portion of the cross section, the test values are reasonably predicted by

$$\sigma_{pw} = \sigma_n - \frac{A_w}{A}(\sigma_n - \sigma_w) \tag{3.17}$$

where

σ_{pw} = critical stress for columns with part of the cross section affected by welding

σ_n = critical stress for the same column if there were no welds

σ_w = critical stress for the same column if the entire cross section were affected by welding

A_w = area of affected zone

A = total area of cross section

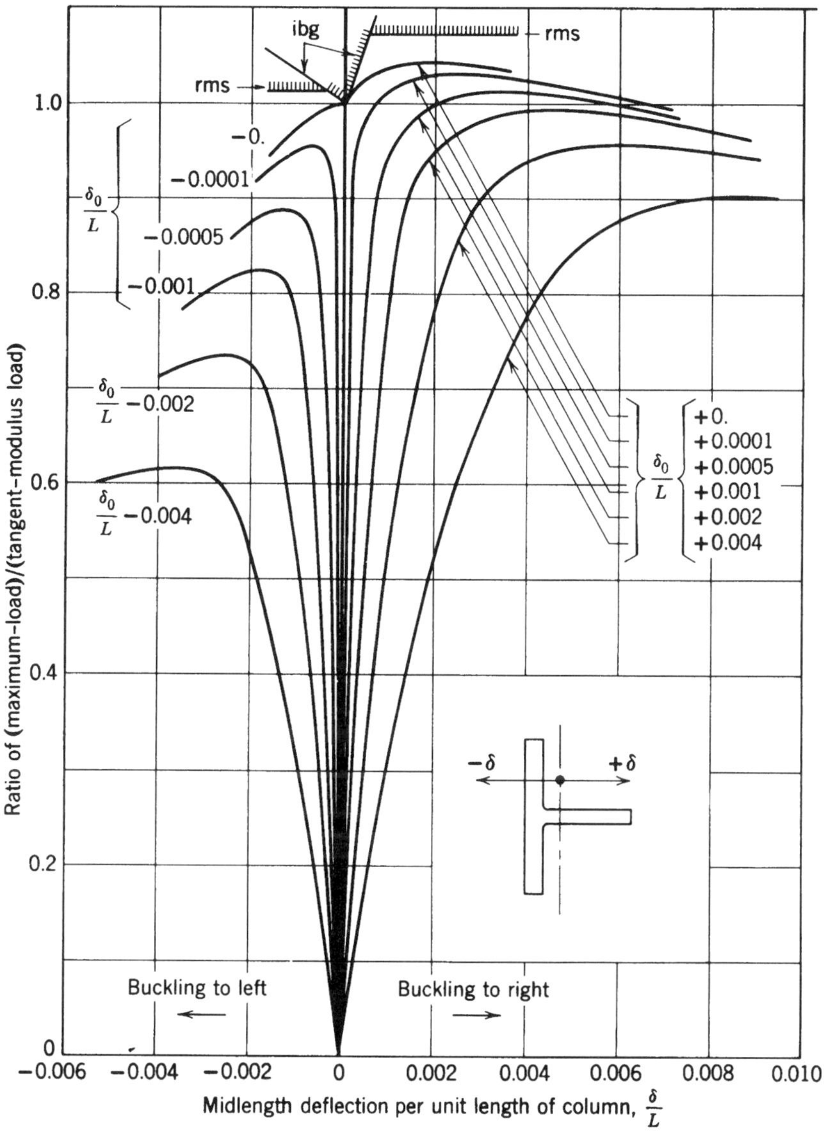

FIGURE 3.31 Theoretical behavior of straight and initially crooked T-section columns of aluminum alloy (Hariri, 1967).

Test data from Brungraber and Clark (1962) have been summarized by Sharp (1993) and are compared to straight-line approximations of predictions based on the tangent-modulus theory in Fig. 3.32 for columns with longitudinal welds and in Fig. 3.33 for columns with transverse welds. The measured values of the

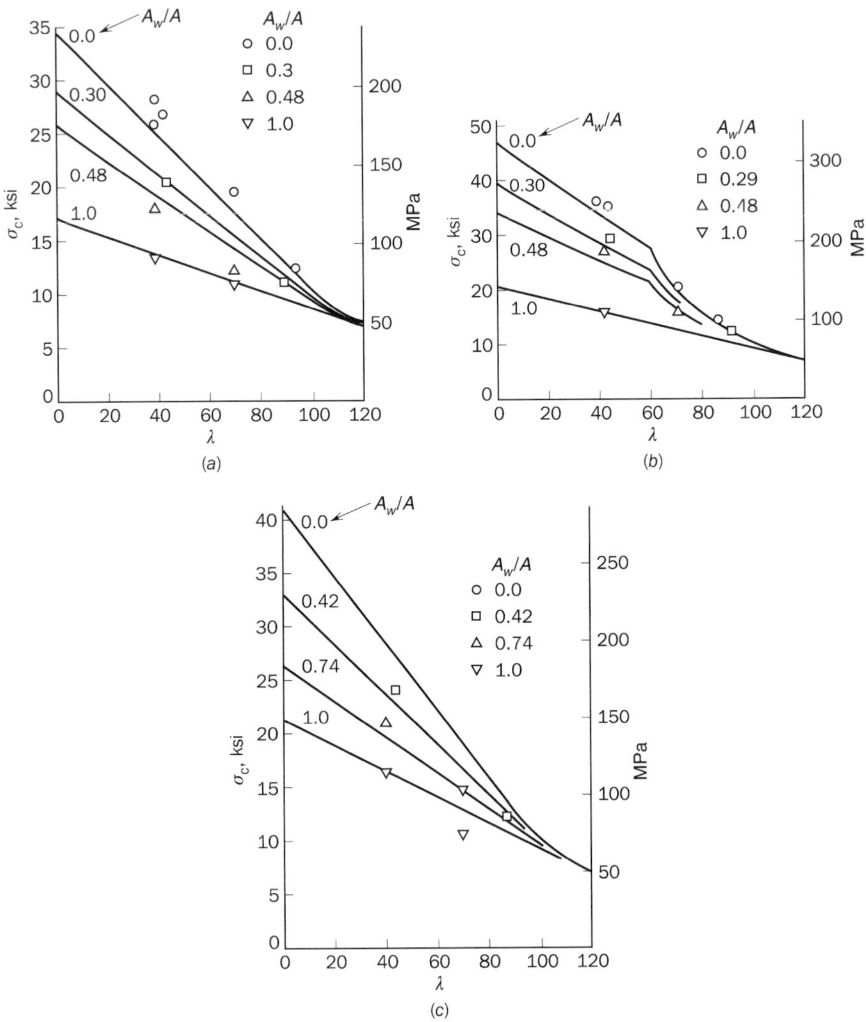

FIGURE 3.32 Columns with longitudinal welds: (*a*) 5154-H32 rectangular sections; (*b*) 6061-T6 rectangular sections; (*c*) 5456-H321 rectangular sections (Sharp, 1993).

heat-affected areas were used in the calculations for the data in these figures. In the absence of measured values, the heat-affected zone may be assumed to be a width of 1 in. (25 mm) on either side of the center of a groove weld or the root of a fillet weld. For columns with transverse welds that affect the entire cross section, the strength reduction is dependent on the location of the welds and the amount of material of reduced strength (Lai and Nethercot, 1992; Fig. 3.33). In general, heat-treatable alloys have a greater loss of strength due to welding than do non-heat-treatable alloys.

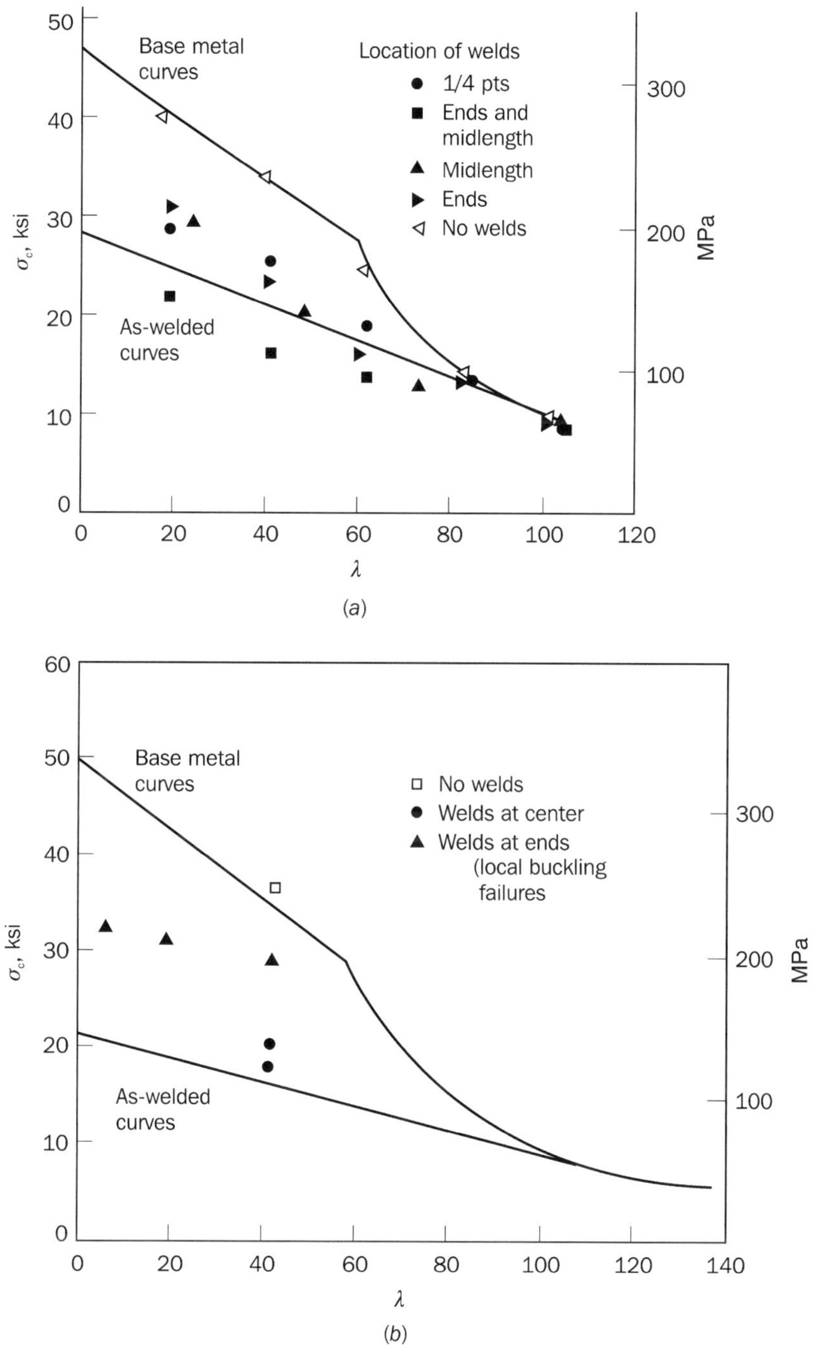

FIGURE 3.33 Columns with transverse welds: 6061-T6 rectangular sections [0.5 × 2.0 in. and 0.25 x 2.0 in. (12.7 x 50.8 mm and 6.3 x 50.8 mm]; (*b*) 6061-T6 tubes 3 in. (76 mm) outside diameter, 0.25 in. (6.4 mm) wall (Sharp, 1993).

3.6.5 Design of Aluminum Alloy Columns

U.S. and Australian Design Practice [Aluminum Association (AA), 2005; Standards Association of Australia (SAA), 1997]. The design formulas are based on the tangent-modulus formula, simplified to a straight line in the inelastic range, which can be expressed as

$$\sigma_c = B_c - D_c\,\lambda \quad \text{for } \lambda < C_c \tag{3.18}$$

in which $\lambda = KL/r$ and for artificially aged tempers,

$$B_c = \sigma_y[1 + (\sigma_y/2250)^{1/2}] \tag{3.19a}$$

$$D_c = (B_c/10)(B_c/E)^{1/2} \tag{3.19b}$$

$$C_c = 0.41B_c/D_c \tag{3.19c}$$

and for other tempers,

$$B_c = \sigma_y[1 + (\sigma_y/1000)^{1/2}] \tag{3.20a}$$

$$D_c = (B_c/20)(6B_c/E)^{1/2} \tag{3.20b}$$

$$C_c = 0.67B_c/D_c \tag{3.20c}$$

In no case can the stress exceed the yield strength. This classification is based on values of the ratios $\sigma_{0.2}/\sigma_{0.1}$ of 1.04 for artificially aged tempers and 1.06 for other tempers. For $\lambda > C_c$ the Euler elastic buckling formula is used.

Figures 3.34 and 3.35 compare the foregoing formulas with the results of tests on aluminum columns (Clark and Rolf, 1966). The test specimens were considered to have fixed ends (flat ends on rigid platens), and the deviations from straightness were less than $0.001L$. The effective-length factor K was assumed to be 0.5 in plotting the test results.

Allowable stresses for building design in the AA specification are obtained by applying a constant factor of safety of 1.95 to the straight-line and Euler formulas. Thus, the specifications do not directly consider the initial crookedness, which is specified by ASTM material specifications as $L/960$ for most extruded shapes. Batterman and Johnston (1967) showed that a small initial out-of-straightness can appreciably reduce the factor of safety, especially in the transition region between elastic and inelastic buckling. Figure 3.36 illustrates the results of calculations for columns of 6061-T6 alloy with $\delta_0 = 0.001L$. In a discussion of the specification, Hartmann and Clark (1963) noted that the effects of small amounts of initial out-of-straightness or eccentricity may be offset by the use of conservative values of the equivalent-length factor as a basis for the specification formulas. To illustrate this, Batterman and Johnston (1967) considered the hypothetical case of a column

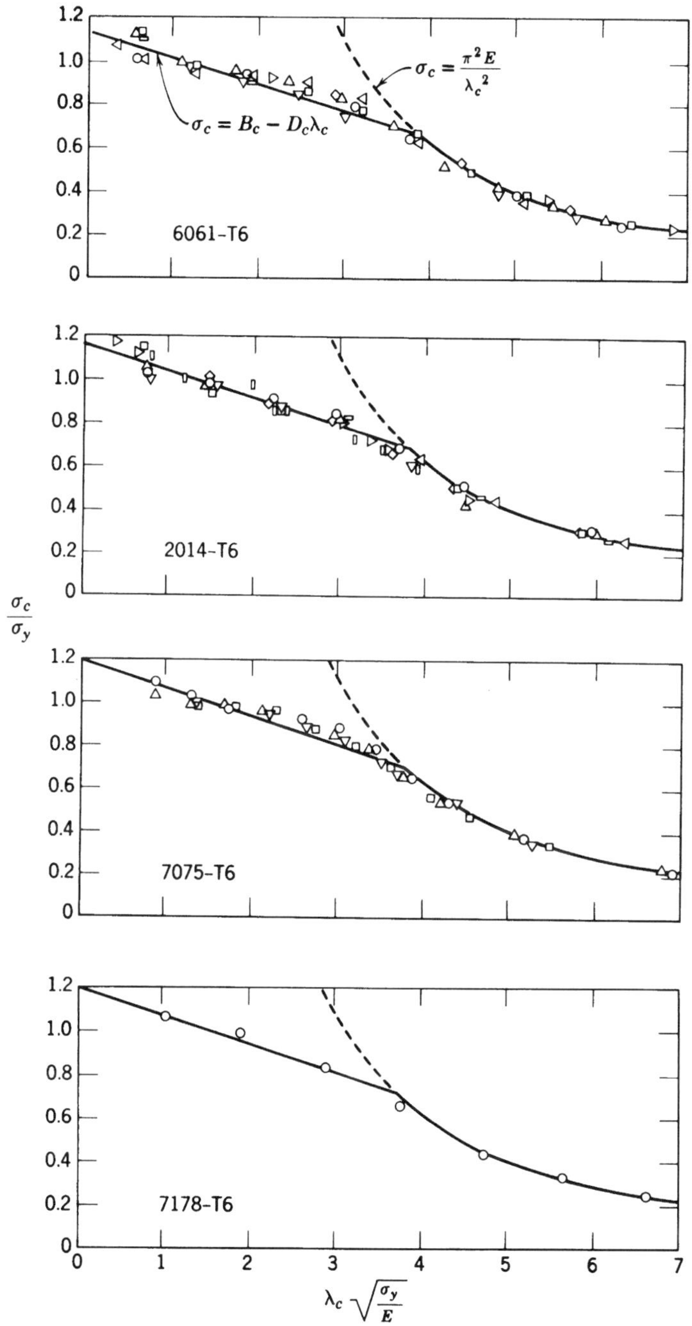

FIGURE 3.34 Column strength of aluminum alloys (artificially aged) (Clark and Rolf. 1966).

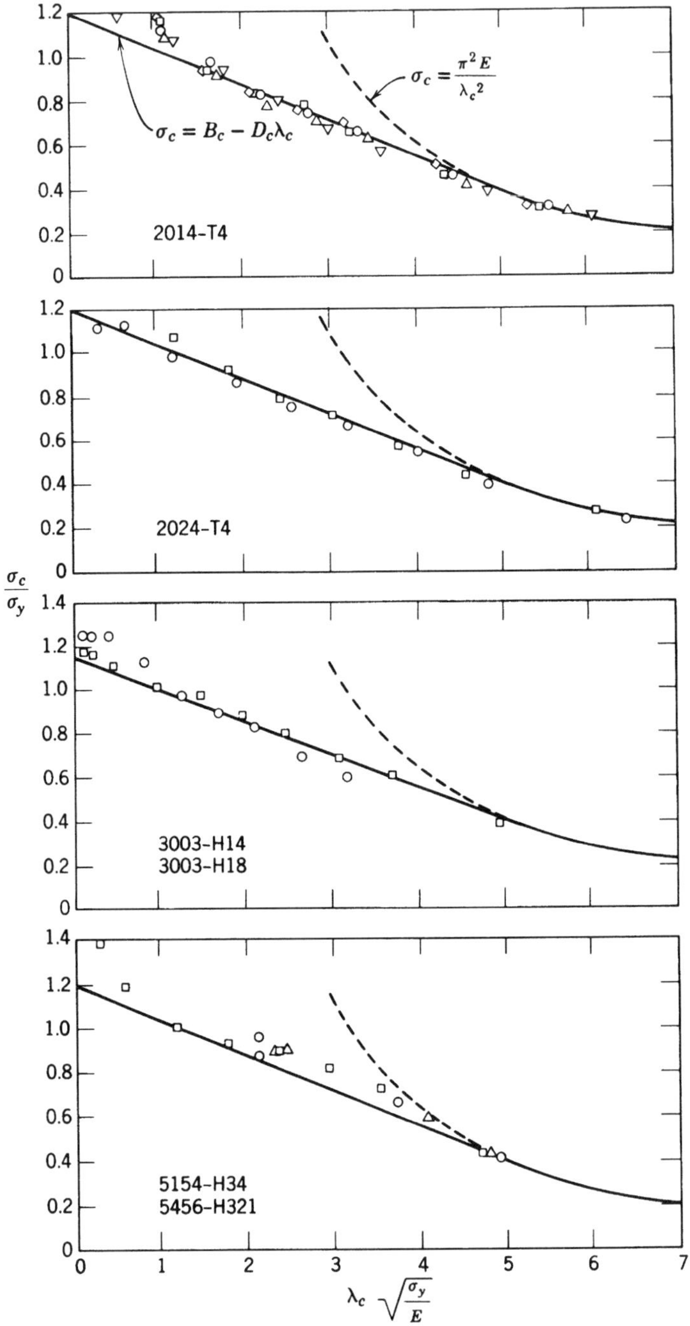

FIGURE 3.35 Column strength of aluminum alloys (not artificially aged) (Clark and Rolf. 1966).

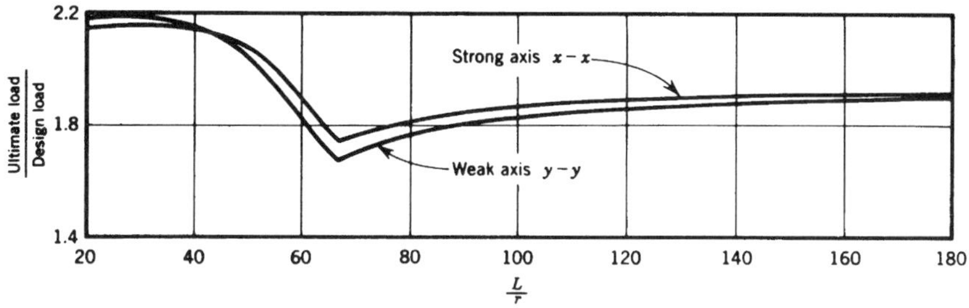

FIGURE 3.36 Design load factor for wide-flange shapes of aluminum alloy columns with $\delta_0 = 0.001L$ (Batterman and Johnston, 1967).

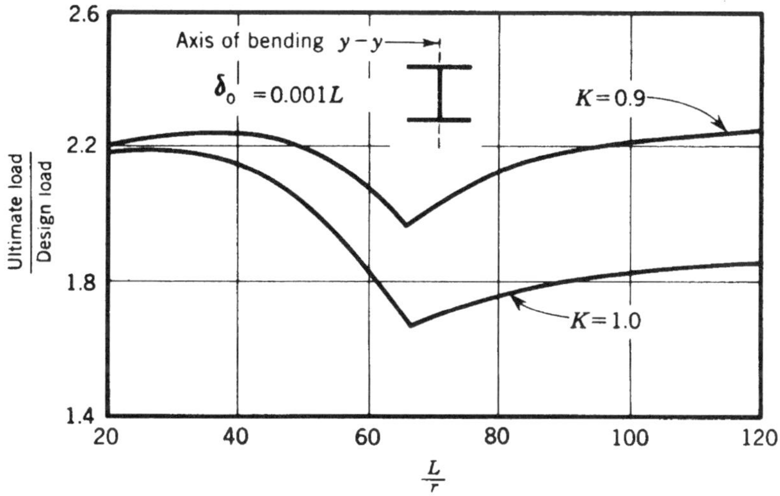

FIGURE 3.37 Comparison of safety factors for columns with and without end restraint (Batterman and Johnston, 1967).

with typical material properties and ends restrained such that throughout loading to maximum strength the column segment between inflection points is always $0.9L$. For bending about the minor axis *(y–y)* and an initial out-of-straightness of $0.001L$, the theoretical safety factor is plotted in Fig. 3.37, both for this case and for the case of pinned ends. Thus, for a relatively slight amount of end restraint, the safety factor varies from values slightly above 2.2 in the short-column range to a minimum of 1.97 at $L/r = 67$. It then increases to slightly above 2.2 at $L/r = 120$. The effect of the initial out-of-straightness in reducing the safety factor from 1.95 to 1.67 is thus offset if the column is restrained at the ends such that an actual K value of 0.9 is produced. In the load and resistance factor design version of the AA (2005) specification, a variable resistance factor is used to account for the effects of initial crookedness.

European Design Practice An earlier British code [Institution of Structural Engineers (ISE), 1962] adopted the Perry–Robertson formula, which can be expressed as

$$\frac{\sigma_c}{\sigma_y} = \frac{1}{2\lambda^2} \left\{(1 + \eta + \lambda^2) - [(1 + \eta + \lambda^2)^2 - 4\lambda^2]^{1/2}\right\} \tag{3.21}$$

The factor η is based on test results (Baker and Roderick, 1948) and thus incorporates all sources of imperfection and nonlinearity present in the tests. Two values are used, $0.0015L/r$ for heat-treated and $0.003L/r$ for non-heat-treated alloys. In a later code, straight-line formulas of the type used in North America were adopted. The British limit-states design specification, BS 8118 [British Standards Institute (BSI), 1991], provides curves relating the critical stress σ_{cr} to the slenderness ratio KL/r for three representative aluminum alloys for use by designers.

Research by Mazzolani and Frey (1980) and Bernard et al. (1973) provided the background information on imperfections, residual stresses, and the influence of welds, which led to the preparation of the ECCS (1978) recommendations and eventually the Eurocode rules. The Ramberg-Osgood formula was used to model the stress–strain relationship in the form

$$\epsilon = \frac{\sigma}{E} + 0.002\left(\frac{\sigma}{\sigma_{0.2}}\right)^n \tag{3.22}$$

Steinhardt's suggestion (1971) that n is equal to the value of the yield strength in kN/cm^2 gives close agreement with test results (n is a coefficient that reflects the shape in the knee location of the stress–strain curve).

Three curves have been adopted using values of n of 20, 15, and 10 to represent the various alloy types. Included in the computer evaluation were initial imperfections due to crookedness and eccentricity as well as unsymmetrical sections. The curves are shown in Fig. 3.38. The uppermost curve is for heat-treated, symmetrical, open or solid sections, with the lowest curve being for all hollow sections and all non-heat-treated sections other than symmetric open or solid sections. This lowest curve lies substantially below those in other standards, as it represents an extreme combination of adverse influences.

The ECCS column curves were presented in tabular form and were therefore not readily adaptable for design purposes. Rasmussen and Rondal (2000b) obtained analytic expressions that were in close agreement with the a-, b- and c-column curves shown in Fig. 3.39. The analytic column curves were based on the Perry–Robertson formula (Eq. 3.21) and a nonlinear expression for the imperfection parameter (η) as

$$\eta = \alpha\left[(\lambda - \lambda_1)^\beta - \lambda_0\right] \tag{3.23}$$

where the values of α, β, λ_0, and λ_1 defining the a-, b- and c-curves are shown in Table 3.1.

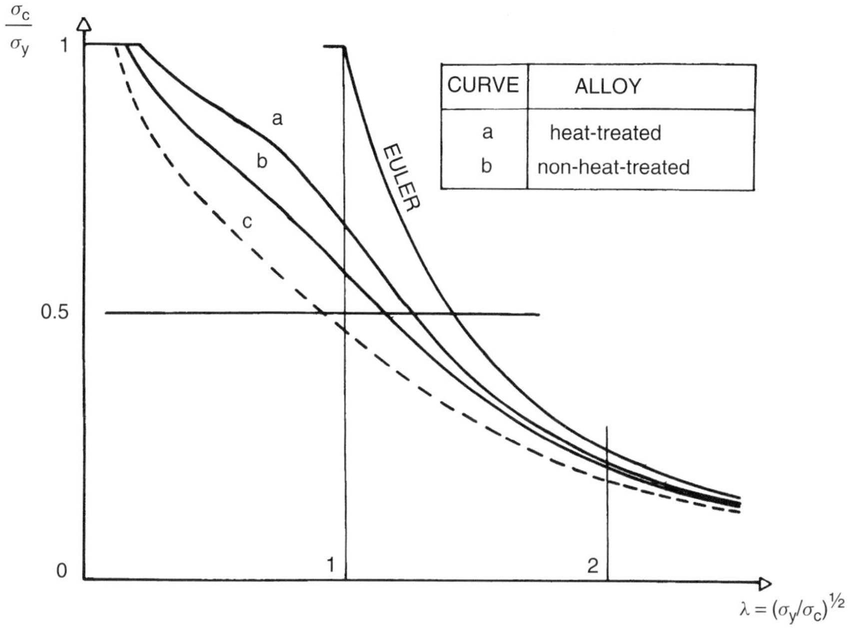

FIGURE 3.38 Basic column-buckling curves (ECCS aluminum structures.)

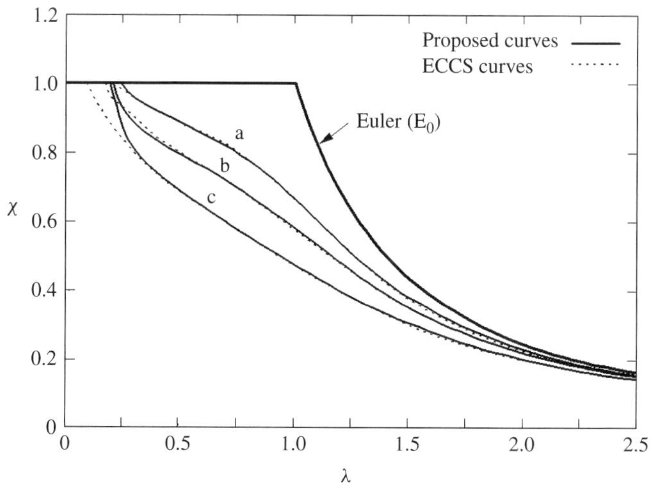

FIGURE 3.39 ECCS *a, b,* and *c* curves for aluminum columns and analytical fits (Rasmussen and Rondal, 2000b).

TABLE 3.1 Values of α, β, λ_0, and λ_1 for Analytic Approximations to ECCS Column Curves

Column Curve	α	β	λ_0	λ_1
a	0.4	0.2	0.55	0.2
b	0.7	0.15	0.55	0.2
c	0.95	0.25	0.35	0.2

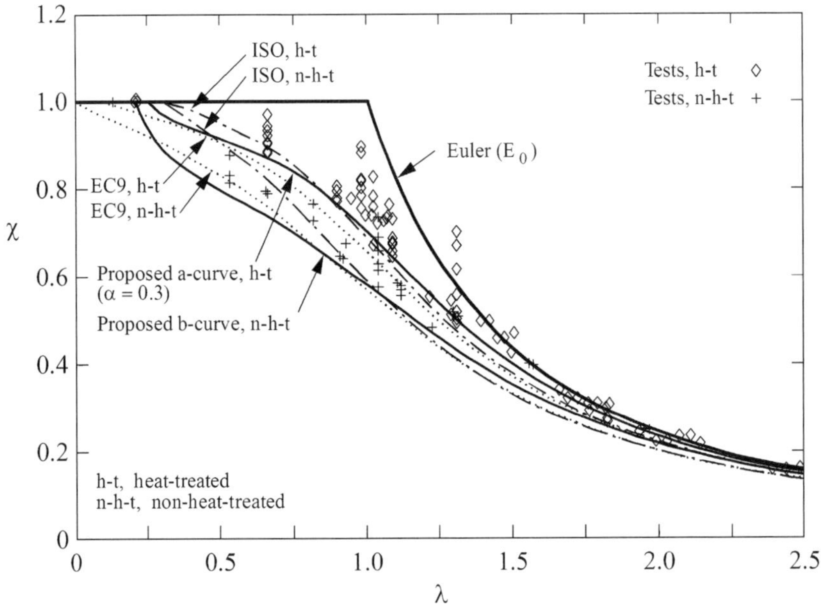

FIGURE 3.40 Comparison of ECCS column curves with tests (Rasmussen and Rondal, 2000b).

Figure 3.40 compares the ECCS *a* and *b* curves with tests on aluminum alloy columns. It can be seen that the *a* curve follows closely the trend of the test points for heat-treated alloys, while the *b* curve follows the trend of the test points for non-heat-treated alloys. The figure also includes the column curves of Eurocode 9 (CEN, 2007) and the International Organization for Standardization (ISO, 1992) specification for aluminum structures.

Canadian Design Practice The Canadian aluminum design standard S157 (CSA, 2005) uses two strength curves based on the Perry–Robertson relationship. The concept of a "limiting stress" F_o is used, which can be the yield stress or a lower stress that accounts for local buckling or welding effects. The slenderness, $\overline{\lambda}$, is a generalized term, as the expressions are used for all member instabilities.

The critical column stress, F_c, is defined as follows:

$$F_c = \overline{F} F_o \tag{3.24}$$

$$\overline{F} = \beta - \left(\beta^2 - \frac{1}{\overline{\lambda}^2}\right)^{1/2} \tag{3.25}$$

$$\beta = \frac{1 + \alpha(\overline{\lambda} - \overline{\lambda}_0) + \overline{\lambda}^2}{2\overline{\lambda}^2} \tag{3.26}$$

$$\overline{\lambda} = \left(\frac{F_o}{F_e}\right)^{1/2} = \frac{KL}{\pi r}\left(\frac{F_o}{E}\right)^{1/2} \tag{3.27}$$

For columns, $\overline{\lambda}_0$ is 0.3. The two curves are distinguished by the α parameter, which is 0.2 for heat-treated members and 0.4 for members that are not heat-treated.

3.7 STAINLESS STEEL COLUMNS

The past two decades have seen unprecedented attention paid to the potential use of stainless steel for structural purposes. Several large-scale European research programs have been conducted under the auspices of the European Coal and Steel Community (ECSC) to investigate the strength of cold-formed and fabricated columns and beams at ambient and elevated temperatures, among many other aspects of the behavior and design of stainless steel structures [Burgan et al., 2000; Steel Construction Institute (SCI), 2000]. Research in South Africa (van den Berg, 2000) has included tests on cold-formed, hot-rolled, and fabricated stainless steel structural members with an emphasis on ferritic stainless steels and a chromium-weldable steel type 3Cr12 developed by Columbus Stainless Steel. In the past decade, research on tubular stainless steel structural members and connections has been undertaken in Australia (Rasmussen, 2000; Rasmussen and Hasham, 2001; Rasmussen and Young, 2001) and in Finland (Talja and Salmi, 1995). Recent attention has been paid to the use of high-strength stainless steel tubular columns (Young and Lui, 2006; Gardner et al., 2006).

The new findings reported over the last 20 years have triggered several revisions of the European standard for stainless steel structures, Part 1.4 of Eurocode 3 (CEN, 2006a). This standard complements the American Society of Civil Engineers ASCE-8 standard (ASCE, 2003) for cold-formed stainless steel structural members, which was first published by AISI in the late 1960s and has been revised several times since, including a conversion in 1991 to the LRFD format.

3.7.1 Stainless Steel Materials

Stainless steel is the common name applied to a range of iron-based alloys whose prime corrosion-resisting element is chromium. The minimum chromium content is approximately 11% and the maximum around 30%. The carbon content ranges from 0.02 to 0.12%, which is generally lower than that of ordinary structural carbon steels where the carbon content lies around 0.15 to 0.25%. Most stainless steels contain additional alloying elements, notably nickel, manganese, and titanium. The stainless steel alloys used for structural applications pertain to the following groups:

- Austenitic alloys
- Ferritic alloys
- Duplex alloys

The names of the austenitic and ferritic groups reflect the metallurgical (crystalline) structure of the alloys, being austenite and ferrite. The austenitic group is characterized by high contents of chromium and nickel, and includes the most commonly used alloys, AISI 304 and AISI 316. The ferritic steels contain medium to low contents of chromium and have no nickel. The duplex alloys contain a mixture of austenite and ferrite, which leads to high strength and good corrosion and fatigue properties.

A typical stress–strain curve for stainless steel is shown in Fig. 3.41. It is noted that there is no yield plateau and the proportionality stress, σ_p, is low compared to the equivalent yield stress, σ_y, which for structural applications is defined as the 0.2% proof stress. The gradual yielding taking place at stresses beyond the proportionality stress leads to a softening of the material and consequently a reduction in resistance to buckling. For this reason, the design strength curves for stainless steel

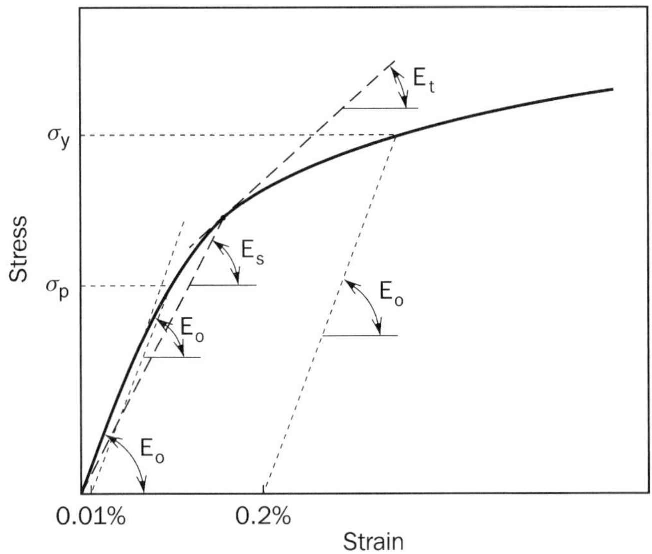

FIGURE 3.41 Typical stress–strain behavior of stainless steel (van den Merwe and van der Berg, 1992).

columns and unbraced beams differ from those for structural carbon steel columns and beams. Furthermore, unlike structural carbon steels, there are clear differences in the mechanical properties associated with the longitudinal direction of rolling and transverse direction, and the material responds differently in compression and tension.

The equivalent yield stress of austenitic stainless steel alloys can be enhanced greatly by cold working. This property is utilized in the production of cold-formed stainless steel tubes, which can be designed to and marketed at an enhanced equivalent yield stress. European tube manufacturers now exploit the strain-hardening capacity to raise the equivalent yield stress of cold-formed rectangular and circular hollow sections in AISI 304 and AISI 316 austenitic stainless steel from a nominal annealed value of 32 ksi (240 Mpa) to a nominal value of 50 ksi (350 Mpa).

While the equivalent yield stress and ultimate tensile strength vary significantly from alloy to alloy and depend on the degree of cold working, the initial elastic modulus lies in the narrow range from 28,000 ksi (195,000 Mpa) to 29,500 ksi (205,000 Mpa) and is only slightly lower than the elastic modulus for carbon structural steels. Representative values of the equivalent yield stress of austenitic alloys in their annealed states lie in the range from 32 ksi (220 Mpa) to 39 ksi (270 Mpa), while equivalent yield-stress values in excess of 50 ksi (350 Mpa) are achievable in the cold-worked condition. For ferritic alloys in the annealed state, the yield stress typically ranges from 38 ksi (260 Mpa) to 50 ksi (350 Mpa). Duplex alloys have equivalent yield-stress values in excess of 65 ksi (450 Mpa) in the annealed condition.

3.7.2 Residual Stresses

The coefficients of expansion of austenitic stainless steel alloys are generally larger than those of structural carbon steel, while the thermal conductivity is lower. The combination of a larger coefficient of expansion and a lower thermal conductivity has the potential of inducing higher welding residual stresses than those experienced in fabricating carbon structural steel sections.

Data on residual stresses in hot-rolled and fabricated (welded) stainless steel structural shapes are limited. Bredenkamp et al. (1992) found that the magnitudes and distribution of the residual stresses are comparable to those found in similar-sized carbon steel welded shapes, as shown in Fig. 3.42, while Lagerquist and Olsson (2001) observed considerably higher residual stresses in fabricated stainless steel sections compared to their carbon steel counterparts.

Rasmussen and Hancock (1993) reported residual stress measurements in cold-formed stainless steel square and circular hollow sections. The measurements showed that the membrane residual stresses were negligible, while the maximum bending residual stresses were on the order of the equivalent yield stress.

3.7.3 Column Strength

Research by Johnson and Winter (1967) and van den Berg and co-workers (van den Berg, 2000) supported using the tangent-modulus approach to account for gradual

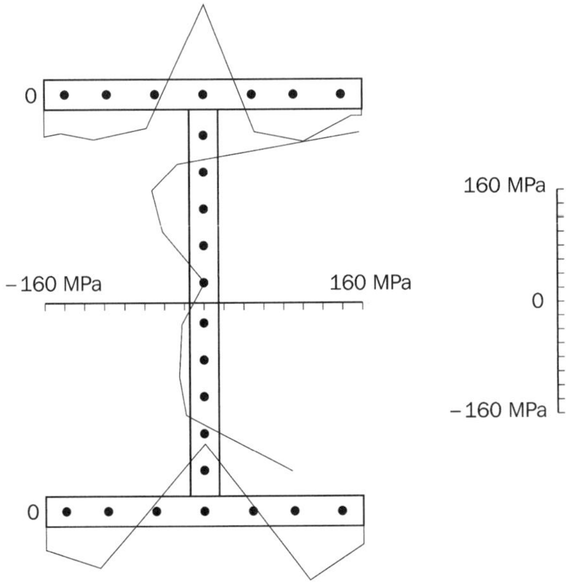

FIGURE 3.42 Residual-stress pattern for a welded stainless steel I-shape (Bredenkamp et al., 1992).

yielding in designing stainless steel columns. The ASCE standard for cold-formed stainless steel structural members (ASCE, 2003) is still based on this approach whereby the column strength is determined by replacing the elastic modulus by the tangent modulus, E_t, in the Euler formula, as described in Section 3.2.1. The tangent-modulus approach is iterative because the tangent modulus depends on the stress level (P_t/A) at the point of bifurcation, where the inelastic buckling load, P_t, is the object of the calculation.

Because the tangent modulus varies differently with stress level for various alloys, reflecting differences in their mechanical response, different column curves are derived for various alloys, as shown in Fig. 3.43. Unlike carbon structural steels, the nondimensional column strength curve for nonlinear materials, such as stainless steel, aluminum, and titanium, raises as the equivalent yield stress increases (Rasmussen and Rondal, 1997b). For this reason, a relatively high nondimensional column curve is derived for the duplex alloy S31803, popularly known as 2205, as shown in Fig. 3.43.

The Australian standard for cold-formed stainless steel structures (AS/NZS, 2001) is based on the ASCE standard. It features supplementary design rules based on research conducted at the University of Sydney, notably on tubular members. As an alternative to the tangent-modulus approach, the Australian standard also allows the column strength to be determined using the explicit method developed by Rasmussen and Rondal (1997a,b). According to this method, the column strength is

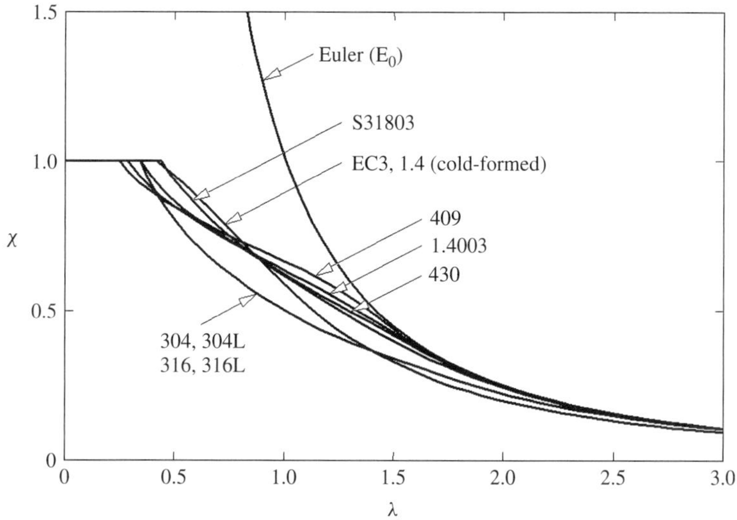

FIGURE 3.43 American (ASCE, 2003) and Australian (AS/NZS, 2001) column curves for cold-formed stainless steel sections based on the tangent-modulus approach.

given by

$$P_n = A_e \chi F_y \tag{3.28}$$

where the slenderness reduction factor, χ, is determined from the Perry–Robertson formulas,

$$\chi = \frac{1}{\phi + \sqrt{\phi^2 - \lambda^2}} \leq 1 \tag{3.29}$$

$$\phi = \tfrac{1}{2}(1 + \eta + \lambda^2) \tag{3.30}$$

$$\lambda = \frac{l_e}{r}\sqrt{\frac{F_y A_e}{\pi^2 A E_0}} \tag{3.31}$$

The imperfection parameter, η, appearing in Eq. 3.30 is tailored to produce strength curves for columns made from nonlinear materials (Rasmussen and Rondal, 1997b) and is given by Eq. 3.23. It involves the material-specific constants α, β, λ_0, and λ_1.

The Australian standard provides values of α, β, λ_0, and λ_1 for common structural stainless steel alloys. Figure 3.44 shows strength curves obtained using the explicit approach. They are generally slightly lower than those obtained using the tangent modulus (Fig. 3.43) because they incorporate geometric imperfections. Accordingly, a resistance factor of 0.9 is specified in the Australian standard

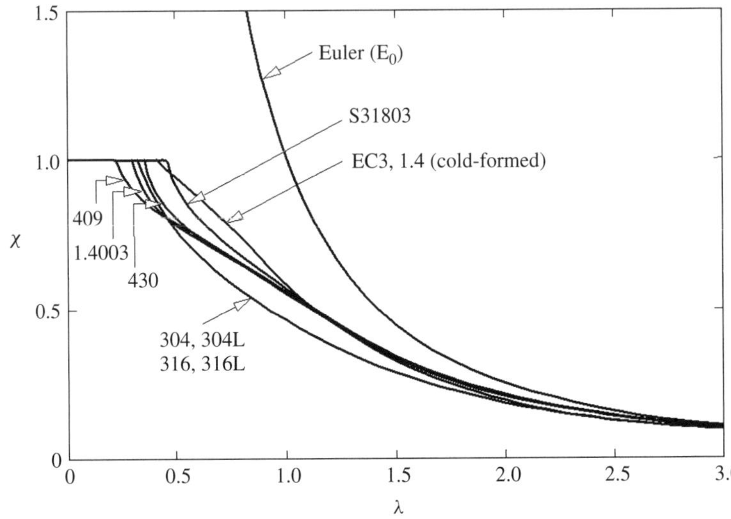

FIGURE 3.44 Australian column curves (AS/NZS, 2001) for cold-formed stainless steel sections based on explicit approach.

when using the explicit approach, whereas a resistance factor of 0.85 is specified when using the iterative tangent-modulus approach.

Part 1.4 of Eurocode 3 (CEN, 2006a) provides supplementary rules for stainless steel structural members and connections. In contrast to the American and Australian standards, Eurocode 3, Part 1.4, is applicable to both cold-formed and fabricated sections. For column design, it uses the Perry–Robertson equations (Eqs. 3.28 to 3.31) in conjunction with the standard linear expression for the imperfection parameter also used for structural carbon steels:

$$\eta = \alpha\,(\lambda - \lambda_0) \tag{3.32}$$

Two column curves are specified in Eurocode 3, Part 1.4, one for cold-formed sections, $(\alpha, \lambda_0) = (0.49, 0.4)$, and one for fabricated (welded) sections, $(\alpha, \lambda_0) = (0.76, 0.2)$. The curve for cold-formed sections is compared with the Australian and American column curves in Figs. 3.43 and 3.44. The European curve for cold-formed sections is relatively high because it is calibrated (Burgan et al., 2000) against tests with high equivalent yield-stress values (Rasmussen and Rondal, 2000a). The column curve for welded sections is based on European tests on AISI 304, AISI 316, and S31803 alloys (Burgan et al., 2000). Further tests on welded and hot-rolled sections have been conducted in South Africa (Bredenkamp et al., 1994; Bredenkamp and van den Berg, 1995).

3.8 TAPERED COLUMNS

Tapered structural members are beams, columns, or beam-columns that have a continuously varying cross section along their length. Such structural components are used in many applications, such as gable frames, towers, and architecturally exposed steel columns in stores, halls, or airports. These types of members provide several advantages over prismatic members. Tapered members can be made more efficient structurally by the use of larger cross sections at locations where the demands are higher and smaller cross sections elsewhere. Furthermore, fabricators equipped to produce web-tapered members can create a wide range of optimized profiles from a minimal stock of different plates and coil. This can result in time and cost savings compared to the alternative ordering or stocking of an array of rolled shapes. It is important to note that tapered members are often used in applications where the members are dominated by bending. Metal building manufacturers commonly taper the webs of their primary members to match the section modulus or capacity to the moment envelope determined from the various design loadings. There are no structural advantages of tapering concentrically loaded columns subjected to a constant axial force.

Due to the variation of the cross section along the member length, and the corresponding variation in the flexural, axial, and torsional stiffness, the stability analysis of tapered members is more complicated than that of prismatic members. Therefore, exact analytical solutions for the buckling of tapered columns are available only for relatively simple cases involving elastic flexural buckling. For more general cases, numerical methods and energy methods may be used for calculating the buckling loads of tapered and of general nonprismatic columns. Also, a number of useful procedures have been recommended for simplified approximate calculation of elastic buckling strengths for I-section members with linear web taper and assumed simply supported end conditions.

Although there are a large number of recommended procedures for calculating elastic flexural buckling loads, a relatively small number of studies have been conducted to evaluate the inelastic buckling capacity and the design strength of tapered columns. Timoshenko and Gere (1961) provided one of the earliest discussions of the calculation of inelastic buckling loads of non-prismatic bars. They suggested an approximate calculation of the inelastic strength of variable cross-section bars using column curves based on the tangent modulus E_t at the cross section with the largest axial stress. In more recent years, a number of researchers have studied the effect of residual stress, initial out-of-straightness, and end restraints from adjacent unbraced lengths on the inelastic flexural buckling of tapered members (Kim et al, 1995; Jimenez-Lopez, 1998; and Jimenez and Galambos, 2001).

In 1966, the CRC and the Welding Research Council (WRC) initiated the first concerted effort to address the complete strength behavior of metal building frames using tapered I-section members. This work was a continuation of earlier work at Columbia University under the direction of Butler (Butler and Anderson, 1963; Butler, 1966). From this effort, Lee and his associates at the University of New York at Buffalo developed an overall design approach for tapered steel members documented in Lee et al. (1972, 1981). Two of the main characteristics of this approach are the concept of equivalent prismatic members and the mapping of the elastic buckling strength of tapered members to the elastic or inelastic design strength of equivalent prismatic members. Recently, these concepts have been refined further by White and Kim (2006) and published by the Metal Buildings Manufacturers Association in an MBMA/AISC design guide (Kaehler et al., 2008). The MBMA/AISC design guide extends the analysis and design procedures provided in the 2005 AISC specification to the design of frames using web-tapered and general non-prismatic members.

In this section, various procedures are reviewed for calculating the elastic buckling strength of tapered columns and methods for calculating the column axial strength of tapered members are summarized. The concept of an equivalent prismatic member, originally suggested by Timoshenko and Gere (1961) and subsequently applied and developed by Lee et al. (1972), is addressed in detail. Furthermore, recent updates by White and Kim (2006) and Kaehler et al. (2008) are discussed. These updates simplify and generalize the application of this concept.

3.8.1 Elastic Buckling Strength of Tapered Columns

The elastic buckling of tapered columns can be determined by solving the governing differential equation of equilibrium for the case of linear buckling. For tapered columns, the second moment of area is a function of the longitudinal coordinate of the member. This analytically exact solution is possible only for special cases of the variation of the second moment of area and involves substantial mathematical manipulation. Timoshenko (1936) solved the elastic flexural buckling of a simply supported column with a step in the cross section and a cantilever column with a second moment of area that varies according to the power of the distance along the member length. Bleich (1952) provided analytical solutions for simply supported I-section columns with linear or parabolic tapers. For these solutions, Bleich assumed an approximate variation of the second moment of area as a power function along the unbraced length.

Various numerical and energy method solutions can be used for calculating the elastic flexural buckling load of general tapered columns. Timoshenko (1936) discussed energy method solutions for the elastic flexural buckling of nonprismatic columns. He also discussed a procedure called the method of successive approximations, which allows the estimation of elastic flexural buckling loads for any variation of the geometry and/or axial loading along the member length. Other discussions of the method of successive approximations are provided by Salvadori (1951), Bleich (1952), Chen and Lui (1987), and Bažant and Cedolin (1991). Newmark (1943) showed that the method of successive approximations could be used with finite difference expressions to provide effective practical solutions for the elastic flexural buckling strengths of columns. This numerical method, which is also known as the *numerical integration method,* or *Newmark's method*, can be applied to members with any variation in cross section and applied loads. This method is particularly powerful in that it provides both upper and lower bounds on the exact solution and can be iterated to obtain any desired degree of accuracy. Furthermore, the use of the deflected shape associated with a uniform transverse load often results in accurate solutions without iteration. Timoshenko and Gere (1961) provided an example elastic buckling solution for a stepped column subject to an end compression force using Newmark's method. Kaehler et al. (2008) summarize the general procedures and the implementation details of the method.

The finite element method has been employed extensively for more general cases involving general boundary conditions and modes other than just flexural buckling. Recent solutions of this type include Ronagh et al. (2000a,b) and Boissonnade and Maquoi (2005). It should be noted that these and other authors have documented the fact that the use of prismatic beam finite elements for the analysis of tapered members (by subdividing the members into small segments) can lead to significant errors when the behavior is influenced by torsion. Kaehler et al. (2008) provide an extensive annotated bibliography outlining various numerical procedures that have been employed historically for the calculation of column elastic buckling loads as well as other theoretical and design calculations for frames using web-tapered members.

For tapered I-section columns with a linearly tapered web and prismatic flanges, Kaehler et al. (2008) also provide simple procedures for calculating the elastic buckling strengths with assumed simply supported end conditions. The limit states of in-plane and out-of-plane flexural buckling, torsional buckling, flexural–torsional buckling, and constrained-axis torsional buckling are considered. These approaches involve the use of an average or weighted average cross section along the length, along with the analytical linear buckling equations for prismatic members.

3.8.2 Design Strength of Tapered Columns

As noted above, a joint task committee of the CRC and the WRC was formed in 1966 to address the strength behavior of tapered members. Prawel et al. (1974) provided the first set of experimental tests conducted under the guidance of this joint task committee. In this research, the inelastic stability of tapered I-section beam-columns was studied. In addition, residual stresses were measured in representative welded tapered I-sections. These tests and other analytical studies served as the basis of the overall design procedures developed by Lee et al. (1972, 1981).

One of the key characteristics of the design approaches by Lee et al. (1972, 1981) is the concept of an equivalent prismatic member. These researchers developed member length modification factors, which mapped a given linearly tapered member to a hypothetical prismatic member composed of the cross section at the shallow end of the tapered member. The modified length of the equivalent prismatic member was determined such that this hypothetical member would buckle elastically at the same total load or maximum stress as the given linearly tapered member. Figure 3.45 shows an equivalent prismatic column having the smaller end cross section and the same elastic in-plane flexural buckling load as the given tapered column.

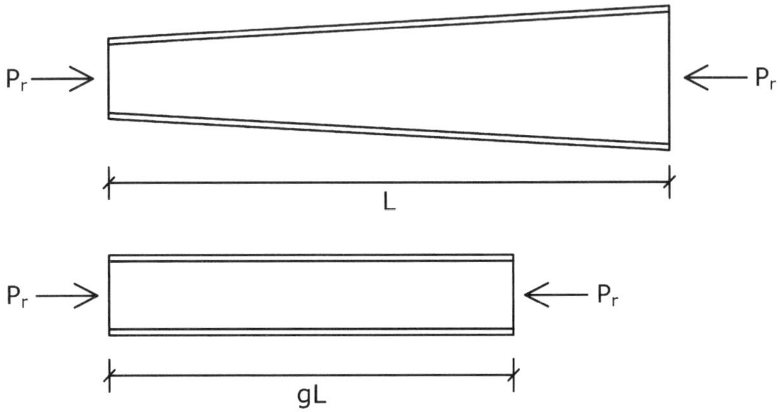

FIGURE 3.45 Equivalent prismatic column with modified length gL (Lee et al., 1972, 1981).

An equation for a length modification factor, g, was determined based on curve fitting to representative results from members with five different cross sections. This length modification factor only addressed the in-plane flexural buckling strength of columns with simply supported end conditions. For the rotational restraint effects from the adjacent members, design charts for an effective-length factor K_γ were derived. Once the equivalent prismatic column was determined with the smallest cross section of the given tapered column and the modified length $K_\gamma gL$ (L being the length of the physical tapered column), the AISC Allowable Stress Design (ASD) equations were used to determine the column elastic or inelastic design strengths. It is important to note that this mapping of the elastic buckling strength to the elastic or inelastic design strength is the same mapping of the theoretical elastic buckling strength to the design resistance employed for prismatic members. This approach is conservative because it assumes the same extent of yielding in all the nominal cross sections along the length, even though the actual cross sections are of larger area than the smallest reference section. The above design procedures were adopted in the AISC (1978) specification provisions. A detailed summary of the design procedures by Lee et al. (1972, 1981) is provided in Chapter 9 of the fourth edition of this guide (Galambos, 1988) and in Section 3.3 of the fifth edition of this guide (Galambos, 1998).

The provisions within the AISC specifications from AISC (1978) through AISC (1999) were limited only to I-section members with equal-size flanges and linearly varying web depths. This, combined with the unpopularity of design charts without underlying equations for calculation of the corresponding parameters, has led to limited acceptance of the AISC provisions. Metal building manufacturers have tended to develop their own specific mappings of the AISC prismatic member equations for design of the wide range of general nonprismatic member geometries encountered in practice. A number of the metal building manufacturers have made substantial investments of their own resources into research to validate their design approaches. To complicate matters, the AISC provisions for design of prismatic I-section members have been greatly improved over the past 40 years relative to the 1963 specification procedures upon which the tapered-web member provisions of AISC (1978) were based. This has led to awkward differences in the design equations for prismatic and linearly tapered I-section members in the AISC (1986, 1993, 1999) provisions. As a result, the AISC Specification Committee decided to drop the explicit consideration of non-prismatic I-section members entirely from the unified 2005 AISC provisions (AISC, 2005a) in favor of subsequent development of separate updated guidelines for these member types. It was anticipated that the subsequent developments could take significant advantage of the many advances that have been implemented for member and frame stability design in the time since the seminal work by Lee et al. (1981).

White and Kim (2006) conducted a pilot study for the application of the AISC (2005a) stability analysis and design provisions to metal building frames with web-tapered members. General procedures developed in this research were later adopted in the MBMA/AISC design guide for frame design using web-tapered members (Kaehler et al., 2008). The following basic concepts employed by White and Kim (2006) and in the MBMA/AISC design guide are essentially the same as those used by Lee et al. (1972, 1981): (1) the concept of an equivalent prismatic member and (2) the mapping of the elastic buckling strength of tapered members to the elastic or inelastic design strength of equivalent prismatic members. Instead of focusing on length modification factors, however, the updated procedures require two different quantities to determine an equivalent prismatic member: (1) the ratio of the elastic buckling load level to the required load level $\gamma_e (= P_e/P_r = F_e/f_r)$ and (2) the maximum value of the ratio of the required flange stress to the yield stress $(f_r/F_y)_{\max}$ within the unbraced length. The equivalent prismatic column is a hypothetical member that has the same γ_e and the same $(f_r/F_y)_{\max}$ as the physical tapered column under consideration. Figure 3.46 shows the equivalent prismatic column concept used in the MBMA/AISC design guide. The product of these two parameters, $\gamma_e (f_r/F_y)_{\max}$, is the ratio of the elastic buckling load level to the yield load level. By using this ratio and the design equations for prismatic members, one can calculate the inelastic buckling strength of tapered members. In the MBMA/AISC design guide, the AISC (2005a) design equations are used for mapping the elastic buckling strength of tapered members to the inelastic buckling strength of equivalent prismatic members.

The explicit use of γ_e has several advantages compared to the use of the length modification factors. One advantage is that if designers obtain the appropriate value of γ_e, the mapping of elastic buckling strength to inelastic buckling strength is possible as a design approximation for any type of nonprismatic member and buckling limit state. Furthermore, if an eigenvalue buckling analysis is performed, the eigenvalue obtained from the analysis is the value of γ_e. Therefore, complex considerations involving general nonprismatic geometries, different end-restraint effects, and nonuniform axial forces can be directly accounted for by determining the appropriate γ_e. Furthermore, by focusing on the elastic buckling load level γ_e

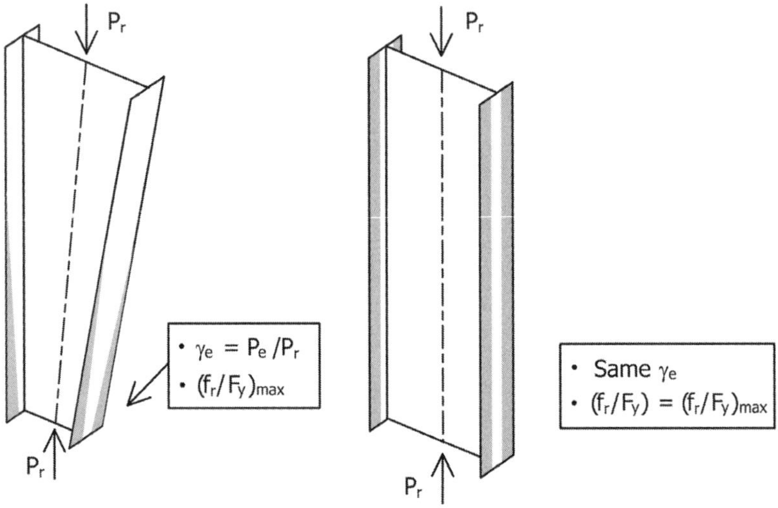

FIGURE 3.46 Equivalent prismatic column concept in MBMA/AISC design guide (Kaehler et al., 2008; White and Kim, 2006).

along with the flange stress ratio f_r/F_y, designers can use any applicable design standard for calculating the design strength of tapered members based on the above concept in the MBMA/AISC design guide.

3.9 BUILT-UP COLUMNS

The effect of shear in built-up columns sets apart the design of these members from that of other columns. The importance of designing the elements connecting the main longitudinal components for shear was tragically demonstrated by the failure of the first Quebec Bridge during construction over the St. Lawrence River, Canada, in 1907. Bridge design practice in North America today reflects the lessons learned from the extensive research that followed that failure. Wyly (1940) concluded that about three-fourths of the failures of laced columns resulted from local, rather than general, column failure. Moreover, the critical load of a built-up column is less than that of a comparable solid column because the effect of longitudinal shear on flexural deflections is much greater for the former. Thus, the longitudinal shear in built-up columns needs to be evaluated in order to (1) determine the possible reduction in the buckling load and (2) design the lacing bars, battens, and their connections.

Three common types of built-up columns are illustrated in Fig. 3.47. They are used when the loads to be carried are large or when a least-weight member or a member with similar radii of gyration in orthogonal directions is desired. Laced or latticed columns (Fig. 3.47*a*) are frequently used in guyed antenna towers, in derrick booms, and in space exploration vehicles. In modern bridge construction,

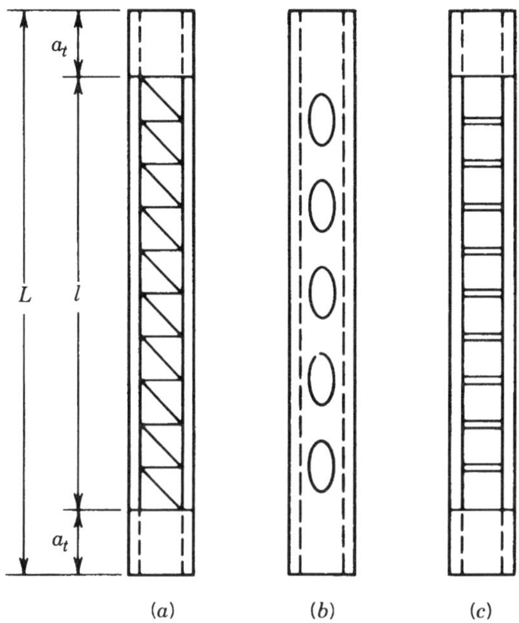

FIGURE 3.47 Common types of built-up columns: (*a*) laced; (*b*) perforated cover plated; (*c*) battened.

perforated cover-plated columns (Fig. 3.47*b*) are likely to be used rather than laced columns. Of the three, battened columns (Fig. 3.47*c*) are the least resistant to shear and they are not generally used for bridge or building construction. Box columns with perforated cover plates designed to specification rules require no special considerations for shear effects.

Engesser (1891) considered the effect of shear on the Euler load of an axially loaded column. His buckling formulas have been reevaluated by Ziegler (1982). Bleich (1952) and Timoshenko and Gere (1961) discuss the shear effects in axially loaded columns and laced columns. Additional references are given in the third edition of this guide (Johnston, 1976) and the SSRC Centrally Loaded Column Research Inventory (Driver et al., 2003).

The shear in a column may be caused by:

1. Lateral loads from wind, earthquake, gravity, or other sources
2. The slope of the column with respect to the line of thrust due both to unintentional initial curvature and the increased curvature during buckling
3. The eccentricity of the load due to either end connections or fabrication imperfections

The slope effect is most important for slender columns and the eccentricity effect for stocky columns.

Worldwide, the design requirements for shear in built-up columns vary widely (Beedle, 1991). Eurocode 3 (CEN, 2005) recommends evaluating shear on the basis of the end slope due to a specified initial out-of-straightness, magnified by the effect of axial load and added to the transverse shear due to the applied loads. The American Association of State and Highway Transportation Officials (AASHTO, 2007) and the American Railway Engineering and Maintenance-of-Way (AREMA, 2008) provide an empirical formula for shear to be added to that due to the weight of the member or the external forces

$$V = \frac{P}{100}\left[\frac{100}{(l/r)+10} + \frac{(l/r)\,F_y}{c_u}\right] \tag{3.33}$$

where

V = normal shearing force, lb (N)
P = allowable compressive axial load on members, lb (N)
l = length of member, in. (m)
r = radius of gyration of section about the axis perpendicular to plane of lacing or perforated plate, in. (m)
F_y = specified minimum yield point of type of steel being used, psi (MPa)
C_u = 3,300,000 when the unit of psi is used and 22,750 when Mpa is used.

The AISC specification for buildings (2005a) calls for the calculation of an increased slenderness ratio for use in the column formula to account for the shear effect. The design formula is based on the research by Zandonini (1985) and Aslani and Goel (1991).

3.9.1 Effect of Shear Distortion on Critical Load

Shear distortion reduces the compression capacity of built-up columns. The shear flexibility of battened or laced structural members can be characterized by a shear flexibility parameter, μ (Lin et al., 1970). The parameter μ takes account of the added distortion due to axial force or bending in the web elements. It is assumed that end stay plates do not undergo shear deformations. The effect of shear on the elastic critical load is depicted in Fig. 3.48 for three basic end conditions:

1. Both ends hinged (subscript *h*)
2. One end fixed, one hinged (subscript *f-h*)
3. Both ends fixed (subscript *f*)

One enters Fig. 3.48 with a chosen value of a_t/L and a calculated shear flexibility factor of μ. The load ratio P_{cr}/P_e can then be read for the appropriate end conditions, after which the effective-length factor may be calculated:

$$K = \sqrt{\frac{P_e}{P_{cr}}} \tag{3.34}$$

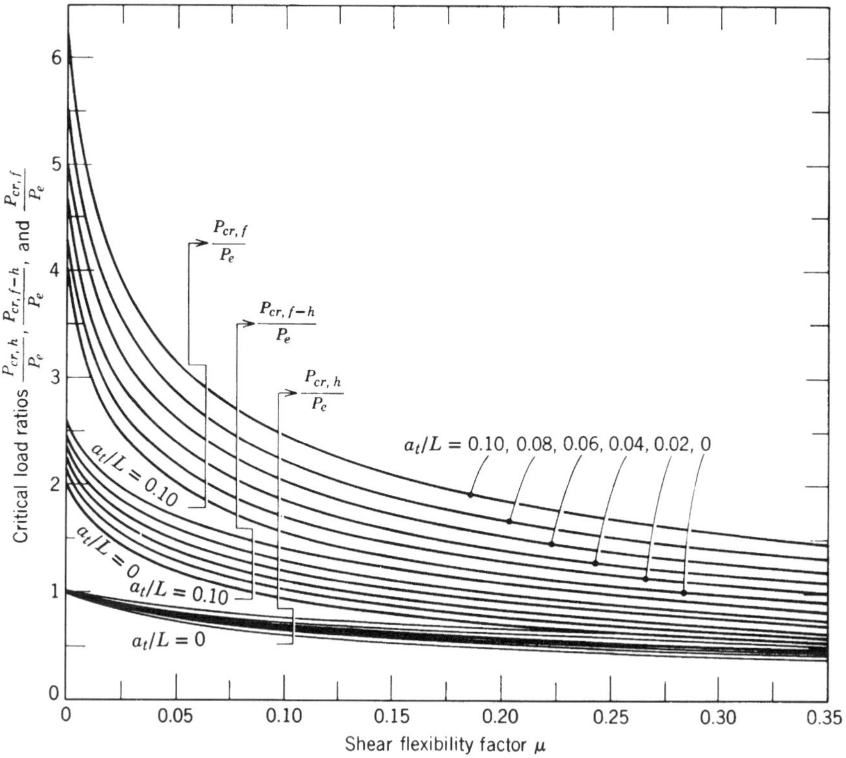

FIGURE 3.48 Critical loads of columns with various end conditions, shear flexibility factor μ, and rigid stay plates.

In Eq. 3.34 and Fig. 3.48, P_{cr} is the elastic critical load of the given open-web or open-flange column, and $P_e = \pi^2 EI/L^2$, the Euler load for a solid column with both ends hinged. Compressive resistance is then obtained for a given specification as a function of KL/r.

For a laced column, the elastic critical load, P_{cr}, may be determined using a method proposed by Razdolsky (2005).

3.9.2 Laced Columns

For a typical laced member (Fig. 3.47*a*), consisting of two main longitudinal elements and two planes of diagonal lacing and transverse struts, Lin et al. (1970) provide the following formula for the shear flexibility factor, μ:

$$\mu = \frac{\xi_b}{1+\xi_a}\left(\frac{b}{l}\right)^2 \frac{A_c}{A_d}\left\{\frac{b}{\xi_a a}\left[1+\left(\frac{\xi_a a}{b}\right)^2\right]^{3/2} + \frac{b}{\xi_a a}\frac{A_d}{A_c}\right\} \tag{3.35}$$

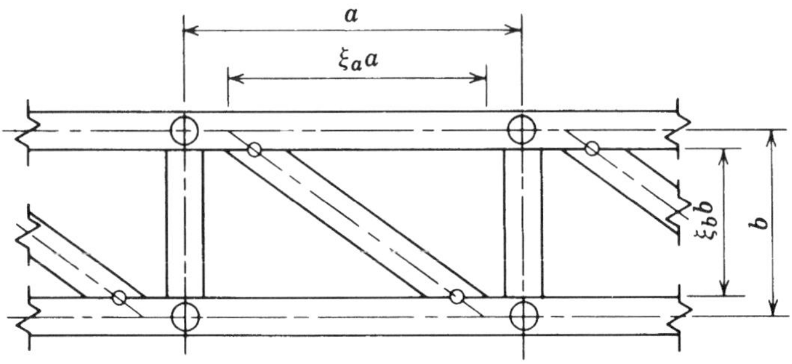

FIGURE 3.49 Typical panel of a laced structural member. Circles indicate where hinges were assumed in the analysis.

The notation used in Eq. 3.35 is defined in Figs. 3.47*a* and 3.49. The third edition of this guide (Johnston, 1976) provides an illustrative example of the application of Eq. 3.35 to a typical design.

In view of the usual small effect of shear in laced columns, Bleich (1952, p. 174) has suggested that a conservative estimate of the influence of 60° or 45° lacing, as generally specified in bridge design practice, can be made by modifying the effective-length factor, K (determined by end-restraint conditions), to a new factor, K', as follows:

$$K' = \begin{cases} K\sqrt{1 + \dfrac{300}{(KL/r)^2}} & \text{for } KL/r > 40 \\ 1.1K & \text{for } KL/r \le 40 \end{cases} \tag{3.36}$$

The change in K is significant only for small KL/r values, in which case there is little change in the compressive design strength.

Bridge design practice in the United States (AASHTO, 2007; AREMA, 2008) requires that the slenderness ratio of the portion of the flange between lacing bar connections have a slenderness ratio of no more than 40, or two-thirds that of the entire member. Canadian bridge specifications (CSA, 2006) change the foregoing limits to 60 and three-fourths, respectively.

The lacing bars and their connections must be designed to act either in tension or in compression, and the rules for general column design apply to them as well. In exceptional cases, such as very large members, double diagonals can be designed as tension members and the truss system completed by compression struts. The importance of adequate and tight-lacing bar connections has been demonstrated experimentally (Hartmann et al., 1938).

Crane booms frequently consist of latticed columns. Vroonland (1971) analyzed nonuniform booms under combined lateral and axial loads and included the effects of deflections and intermediate lateral supports. Brolin et al. (1972) tested four booms, varying from 60 ft (18.2 m) to 200 ft (61.0 m) in length, to destruction. The failure loads were in good agreement with the analyses of Vroonland. Failures

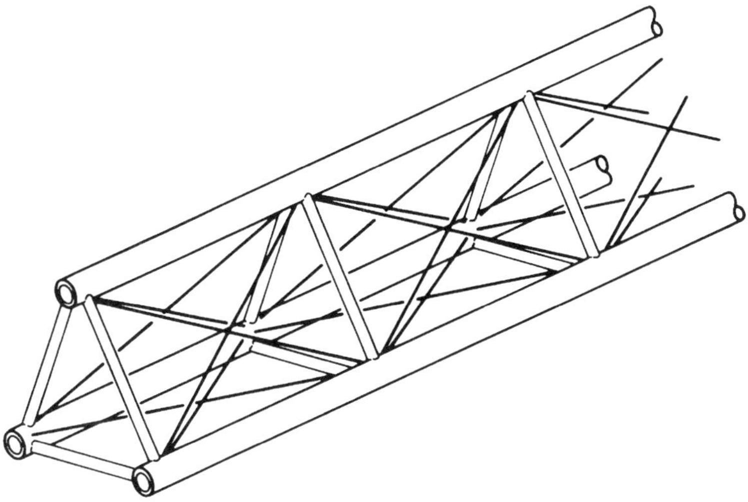

FIGURE 3.50 Triangular latticed column.

occurred when the compressive force in an individual chord member reached the failure load predicted for the unsupported length between brace points.

Interest in laced members has recently been renewed by the needs of the space industry for minimum-weight members to carry very small loads. Figure 3.50 illustrates a member with three longitudinal tubular chords (longerons) with tension diagonals and transverse compression struts. The overall slenderness and chord slenderness are approximately equal and are very large.

Miller and Hedgepeth (1979) studied the harmful effects of initial out-of-straightness of the column as a whole combined with that of the chords between struts. They provide charts that predict the maximum buckling load as a fraction of the Euler bifurcation load of a perfectly straight member. For slenderness ratios of both the member and chord of about 279, the maximum load is less than 50% of the Euler load. Crawford and Benton (1980) studied similar members and obtained comparable results. These studies indicate that for latticed members with very large slenderness ratios the interaction of local and overall out-of-straightness is the dominating factor, not the effect of shear.

3.9.3 Battened Columns

Figure 3.51 shows the basic elements of a battened column consisting of two main longitudinal chords rigidly connected by battens in one, two, or more planes. The battens act as the web and transmit shear from one chord to the other by flexural action in combination with local bending of the chords.

Because a battened column acts as a Vierendeel truss, it is more flexible in shear than either a laced column or even a column with perforated cover plates. The effect of shear distortions can be significant and should be considered in calculating the

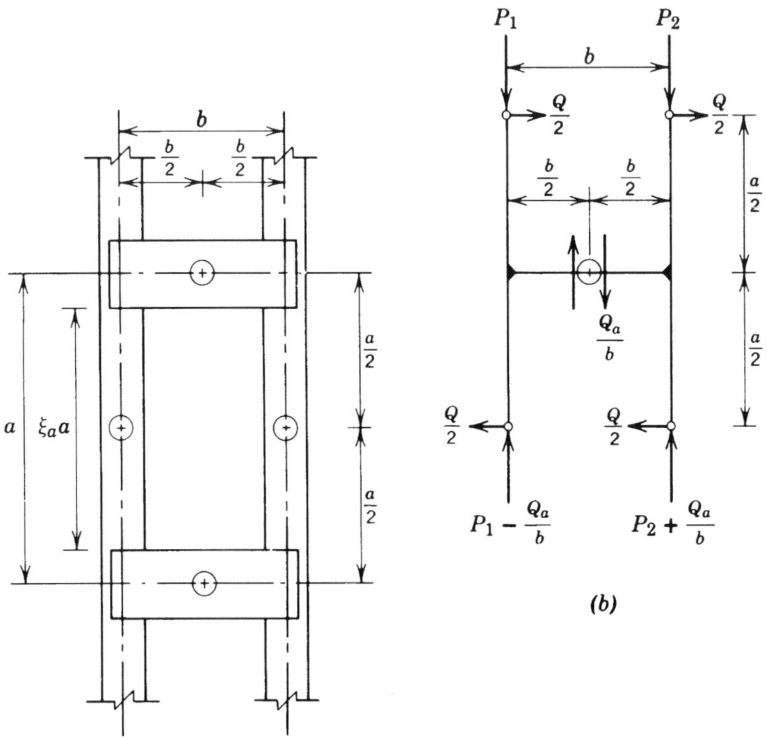

FIGURE 3.51 Typical panel of battened structural member: (*a*) geometrical configuration; (*b*) force equilibrium. Circles indicate assumed points of inflection.

compressive strength of the column. Battened columns are generally not allowed in current U.S. design specifications but are in some circumstances used in Canada. Battened columns are used for antenna towers and on occasion for secondary members. It should be noted that, while the lower portion of columns in mill buildings that support crane runway girders may look like battened columns, they really are spaced columns because the battens are not rigidly connected to the chords, as discussed in Section 3.9.4.

The typical unit of a battened column has a length a equal to the center-to-center distance between two battens, and a width b between centroids of the chords. The properties of the battened column are characterized by shear shape factors and the second moments of area of the battens and chords.

In developing the shear flexibility effect for the highly redundant battened member, Lin et al. (1970) assumed points of inflection for symmetric members at the midpoints of the battens and midway between the battens for the chords. The analysis is conservative because the overall continuity of the longitudinal members is neglected. The shear flexibility parameter is then given by

$$\mu = \left[\frac{1}{(l/r_c)^2} + \left(\frac{b}{2l} \right)^2 \right] \left[\frac{A_c}{A_b} \left(\frac{ab}{6r_b^2} + 5.2\frac{a}{b}\eta_b \right) + 2.6\xi_a \eta_c + \frac{\xi_a^3}{12} \left(\frac{a}{r_c} \right)^2 \right] \tag{3.37}$$

The nomenclature for Eq. 3.37 is shown in Fig. 3.51 and (additionally) as follows:

A_c, A_b = areas of a single longitudinal and of all batten elements within a unit, respectively

r_c, r_b = radii of gyration of the longitudinal and of the batten elements, respectively

η_c, η_b = shear shape factors of the longitudinal and of the batten elements, respectively, where the shear shape factor is the ratio of the total cross-section area to the shear area (Timoshenko and Gere, 1961)

l = length of column between end tie plates

Equation 3.37 accounts for the amplification of deflection in the column segments between battens. These may be neglected if the slenderness ratio of the chord segment between battens, $\xi_a\ a/r_c$, does not exceed $80\sqrt{\sigma_a}$, where σ_a is the average stress in ksi at the specified load level. This limit is considerably more stringent than the batten spacing requirements in some specifications. Because it is likely to be impractical to determine the partial rigidity of semirigid batten connections experimentally, it is recommended that whenever semirigid connections exist, the column be considered as a system of spaced columns.

Example 3.1: Battened Column As shown in Fig. 3.52, batten flange stiffeners have been included to eliminate local web distortion and to ensure that the

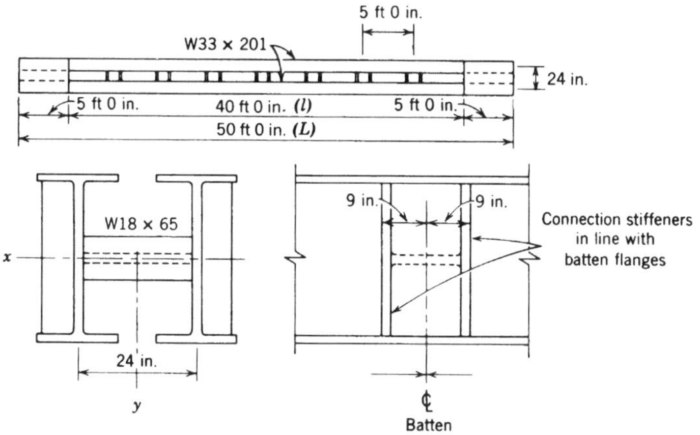

FIGURE 3.52 Example details.

(continued)

battens are fully effective. If such stiffeners were not provided, the column could be treated as a spaced column. From the AISC manual,

$$
\begin{aligned}
&W33 \times 201 \text{ (longitudinal)}\\
&A_c = 59.1\text{ in.}^2\\
&r_c = r_y = 3.56\text{ in.}\\
&r_x = 14.0\text{ in.}\\
&I_y = 749\text{ in.}^4\\
&W18 \times 65 \text{ (batten)}\\
&A_b = 19.1\text{ in.}^2\\
&r_b = r_x = 7.49\text{ in.}\\
&a = 60\text{ in.}\\
&b = 24\text{ in.}\\
&\xi_a = \frac{60-18}{60} = \frac{42}{60} = 0.70\\
&l = 480\text{ in.}\\
&L = 600\text{ in.}\\
&\eta_c = 1.6\\
&a/L = 0.10\\
&\eta_b = 2.6
\end{aligned}
$$

For the combined cross section,

$$
I_y = \frac{59.1 \times 24^2}{2} + 2 \times 749 = 18{,}519\text{ in.}^4
$$

$$
r = \sqrt{\frac{18{,}519}{2 \times 59.1}} = 12.5\text{ in.} < r_x
$$

Substituting Eq. 3.37 yields

$$
\mu = \left[\frac{1}{(480/3.56)^2} + \left(\frac{24}{2 \times 480}\right)^2\right]\left[\frac{59.1}{19.1}\left(\frac{60 \times 24}{6 \times 7.49^2} + 5.2 \times \frac{60}{24} \times 2.6\right)\right.
$$
$$
\left. + 2.6 \times 0.7 \times 1.6 + \frac{0.70^3}{12}\left(\frac{60}{3.56}\right)^2\right] = 0.088
$$

The critical load ratio for three sets of end conditions is obtained from Fig. 3.48 and the effective-length factor is evaluated from Eq. 3.34. The results are given in Table 3.2. If no reduction were made, the value of KL/r for the pinned–pinned case would be $600/12.5 = 48$.

TABLE 3.2 Results for Example 3.1

				$\phi_c F_{cr}$ (AISC) (ksi)	
End Conditions	P_{cr}/P_c (Fig. 3.48)	K	KL/r	$F_y = 36$ ksi	$F_y = 50$ ksi
Hinged-hinged	0.80	1.12	53.8	27.8	36.4
Hinged-fixed	1.45	0.83	39.9	29.8	40.0
Fixed-fixed	2.52	0.63	30.3	30.9	42.2

Bleich (1952) gives the following approximate formula for the effective length of a battened column with both ends pinned

$$\frac{KL}{r} = \sqrt{\left(\frac{L}{r}\right)^2 + \frac{\pi^2}{12}\left(\frac{a}{r_c}\right)^2} \tag{3.38}$$

where L/r is the slenderness ratio of the column as a whole and a/r_c is the slenderness ratio of one chord center-to-center of battens. Bleich (1952) estimates that the buckling strength of a steel column having an L/r ratio of 110 is reduced by about 10% when $a/r_c = 40$, and by greater amounts for larger values of a/r_c.

For Example 3.1, $a/r_c = 60/3.56 = 16.9$ and Eq. 3.38 gives

$$\frac{KL}{r} = \sqrt{\left(\frac{600}{12.5}\right)^2 + \frac{\pi^2}{12}(16.9)^2} = 50.4$$

which is lower than the 53.8 determined by Eqs. 3.34 and 3.37.

An unconservative assumption has been made that the addition of two stay plates gives the end regions full rigidity with respect to shear. It has also been assumed, however, that the effective length of a battened column is not decreased by the bending resistance due to the stay plates in the end regions. The two assumptions offset each other and their net effects may be neglected. Alternatively, more refined methodologies that consider the effects of stay plates have been proposed by Gjelsvik (1990) and Paul (1995).

The design of the chords and the batten plates and their connections should take into account the local bending resulting from specified shear forces. The batten plates and their connections to the chords are designed for the combination of shear, Qa/nb, and moment at the connection of

$$M_b = \frac{Qa}{2n} \tag{3.39}$$

(continued)

where Q is the shear required by specification plus shear due to any transverse loading, a is the center-to-center distance between battens, and n is the number of parallel planes of battens.

The section capacity of the chords should be checked for the combination of axial load and the bending moment:

$$M_c = \frac{Q\xi_a a}{4} \tag{3.40}$$

3.9.4 Stay Plates and Spaced Columns

End stay plates in battened columns may contribute significantly to the buckling strength. Their importance is revealed by the study of a spaced column, defined herein as the limiting case of a battened column in which the battens are attached to the longitudinal column elements by pinned connections. The battens then act simply as spacers, with no shear transmitted between the longitudinal elements. Without end stay plates, the buckling strength of such a spaced column is no greater than the sum of the critical loads of the individa1 longitudinal components of the built-up member. The strengthening effect of the end stay plates arises from two sources: (1) a shortening of the length within which the column components can bend about their own axes and (2) enforcing of the longitudinal components to buckle in a modified second-mode shape and thus have an elastic buckling coefficient that may approximate four times that of the first mode. The buckling load of a spaced column with end tie plates is a lower bound to the buckling load of the battened column (also with tie plates) but with low or uncertain moment resistance in the connections between battens and the longitudinal components. Such columns are sometimes used in mill building construction. In addition to their contribution to column strength, end tie plates perform their usual role of distributing the applied forces or moment to the component elements of either laced or battened columns. They also provide a means of transmitting load to another member or to a footing. With regard to the distance along the column between spacer elements, the same rule as for battens should provide a conservative basis for design.

Example 3.1 involved a battened column. If the stiffener plates to provide rigidity of the batten connections were omitted, the behavior would approach the conditions assumed for the spaced column because the attachment of the battens to the column webs would not transmit moment effectively.

Using the notation adopted for battened columns, the second moment of area of two longitudinals about the y–y axis would be, as in Example 3.1,

$$I = \frac{A_c b^2}{2} + 2I_c \tag{3.41}$$

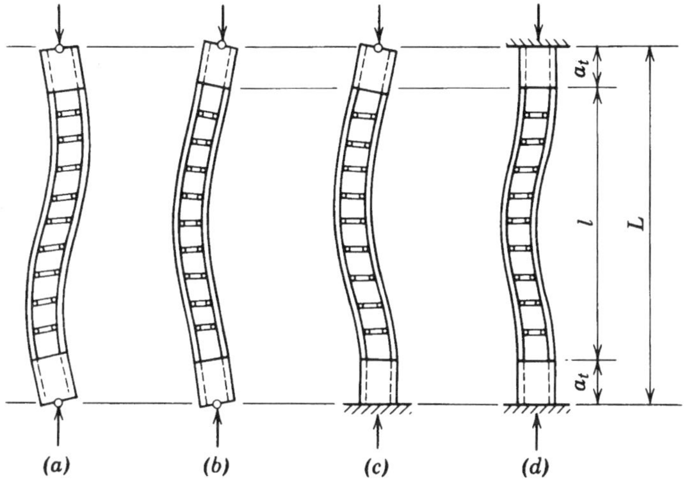

FIGURE 3.53 Spaced column buckling modes: (*a*) mode A, hinged–hinged; (*b*) mode E, hinged–hinged; (*c*) mode C, hinged–fixed; (*d*) mode D, fixed-fixed.

where I_c is the second moment of area of one of the individual longitudinal elements.

The ratio I/I_c in mill building columns is usually at least 40 and could be greater than 100. This ratio is used as a parameter for the determination of the buckling load of a spaced column with end tie plates.

Four modes of buckling for spaced columns without sidesway are treated in Johnston (1971), as illustrated in Fig. 3.53. Spaced columns with sidesway permitted also are considered briefly.

For the pinned-end condition, the spaced column with end stay plates buckles either in S-curvature (mode A in Fig. 3.53*a*) or in mode B curvature (Fig. 3.53*b*), depending on the values of I/I_c and a_t/L. It is noted that in S-curvature there is no differential change of length of the two longitudinal column components between the end tie plates; thus, they buckle under identical loads $P/2$, and the critical load is independent of the ratio I/I_c. When the column buckles in mode B curvature (Fig. 3.53*b*), the component on the concave side (at the center) shortens more than on the convex; thus, there arises overall moment resistance due to the difference in the axial forces. This resistance is in addition to the moments induced in the components as a result of their own curvature. The critical loads for S-curvature buckling could be less than those for mode B curvature when the ratio I/I_c is relatively small and a_t/L is large.

In practice, the base of a column will often be fixed to a footing, and S-curvature buckling cannot take place. Buckling will be in mode C, as illustrated in Fig. 3.53*c*, but as I/I_c gets large it will tend toward the shape with both ends fixed, as illustrated in Fig. 3.53*d*. In fixed-end buckling (Fig. 3.53*d*), as in mode A, the moment resistance is simply the sum of the moments in the component parts, with no contribution due to differential axial forces as in Figs. 3.53*b* or *c*.

The fixed-end case is the simplest to evaluate, because the critical load is simply twice the critical load of a longitudinal component with both ends fixed, that is, the Euler load with an effective length of $0.5L$ multiplied by 2.

Added moment resistance in a spaced column due to differential changes in component length occurs only when the end rotations are different in magnitude and/or sense, as in Figs. 3.53*b* and *c*. Within length l between the tie plates there is no shear transfer between the longitudinals; hence, the differential axial forces and the resisting moment that they contribute must remain constant within l.

Although Johnston (1971) gives critical-load information for a variety of end conditions, including the four shown in Fig. 3.53, the fixed-base and pinned-top case (Fig. 3.53*c*) is possibly of greatest practical application. In terms of the overall length $L = l + 2a_t$, the equation for elastic critical load is written

$$P_{cr} = \frac{CEI_c}{L^2} \tag{3.42}$$

in which C is termed the elastic buckling coefficient. For the determination of approximate critical loads in the inelastic range or for evaluation of compressive strengths by column design formulas, it is convenient to determine the effective-length factor, K, for use in the equation

$$P_{cr} = \frac{2\pi^2 EI_c}{(KL)^2} \tag{3.43}$$

where

$$K = \pi\sqrt{\frac{2}{C}} \tag{3.44}$$

Elastic-buckling coefficients, C, for the pinned–fixed case are plotted in Fig. 3.54, the use of which is illustrated in the following example.

Example 3.2: Spaced Column The column is identical to the one in Example 3.1, but omitting the batten-flange stiffeners and considering the design as a spaced column,

$$\frac{a_t}{L} = \frac{5}{50} = 0.10 \qquad \frac{I}{I_c} = \frac{18{,}519}{749} = 24.7$$

From Fig. 3.54, $C = 100$. By Eq. 3.44,

$$K = \pi\sqrt{\frac{2}{100}} = 0.445$$

$$\frac{KL}{r_c} = \frac{0.445 \times 600}{3.56} = 75.0$$

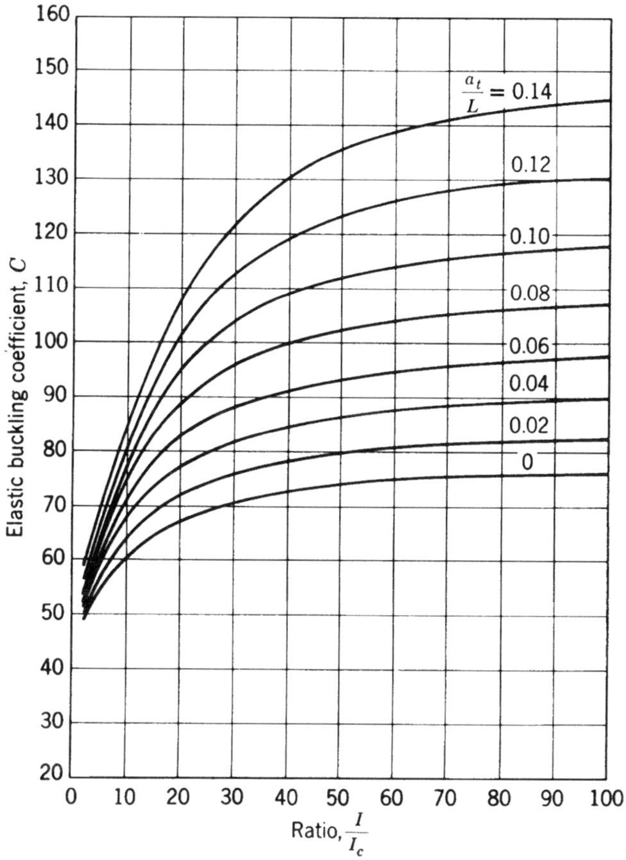

FIGURE 3.54 Elastic buckling coefficients, one end hinged, one end fixed (mode C of Fig. 3.53).

(Note that K is now referenced to L/r_c, not L/r.) Results using the AISC specification (2005a) are presented in Table 3.3.

TABLE 3.3 Results for Example 3.2

	$\phi_c F_{cr}$ ksi	
F_y	Spaced Column	Battened (Example 3.1)
36	24.1	29.8
50	29.9	40.0

In a limited number of tests on spaced columns (Freeman, 1973), the maximum failure loads fell short of the predicted, due in part to open holes in the bolted specimens and to deformation of the end stay plates. Pending further tests,

(continued)

it is recommended that only half of the length of the end stay plates be considered effective and that 90% of the theoretical failure loads be used as a basis for design.

3.9.5 Columns with Perforated Plates

White and Thűrlimann (1956) provide (in addition to the results of their own research) a digest of investigations at the National Bureau of Standards (Stang and Greenspan, 1948) and give recommendations for the design of columns with perforated cover plates. The following design suggestions for such columns are derived from the White–Thűrlimann study and from AASHTO specifications (2007).

When perforated cover plates are used, the following provisions govern their design:

1. The ratio of length, in the direction of stress, to width of perforation should not exceed 2.
2. The clear distance between perforations in the direction of stress should not be less than the distance between points of support [i.e., $(c - a) \geq d$ in Fig. 3.55*a*].
3. The clear distance between the end perforation and the end of the cover plate should not be less than 1.25 times the distance between points of support.
4. The point of support should be taken as the inner line of fasteners or fillet welds connecting the perforated plate to the flanges. For plates butt welded to the flange edge of rolled segments, the point of support may be taken as the weld whenever the ratio of the outstanding flange width to the flange thickness of the rolled segment is less than 7. Otherwise, the point of support should be taken as the root of the flange of the rolled segment.
5. The periphery of the perforation at all points should have a minimum radius of $1\frac{1}{2}$ in. (38 mm).
6. The transverse distance from the edge of a perforation to the nearest line of longitudinal fasteners, divided by the plate thickness, that is, the *b/t* ratio of the plate adjacent to a perforation (see Fig. 3.55), should conform to minimum specification requirements for plates in main compression members.

3.10 STEPPED COLUMNS

Stepped columns are often used in mill buildings to support roof or upper wall loads and runway girders on which cranes travel, as illustrated in Fig. 3.56.

A notional load approach to the design of mill building columns has been presented by Schmidt (2001). In general, however, the effective length of a uniform or prismatic column having the same buckling characteristics as that of the stepped

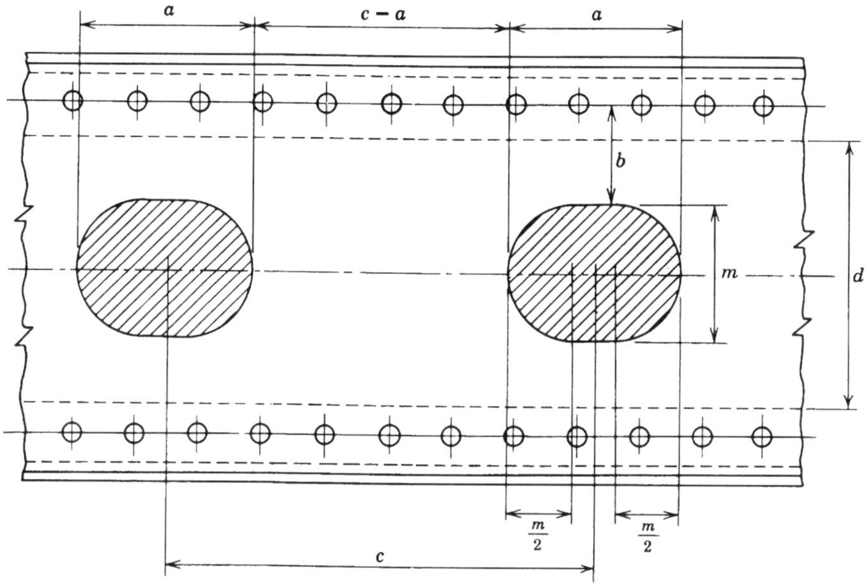

(a)

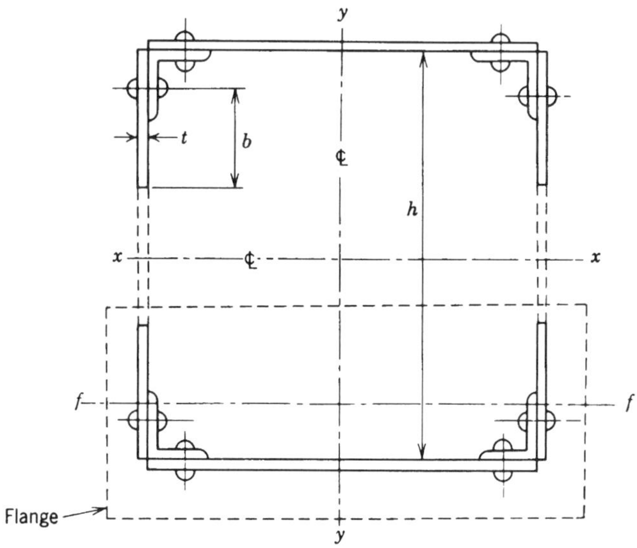

(b)

FIGURE 3.55 Column with perforated web plates.

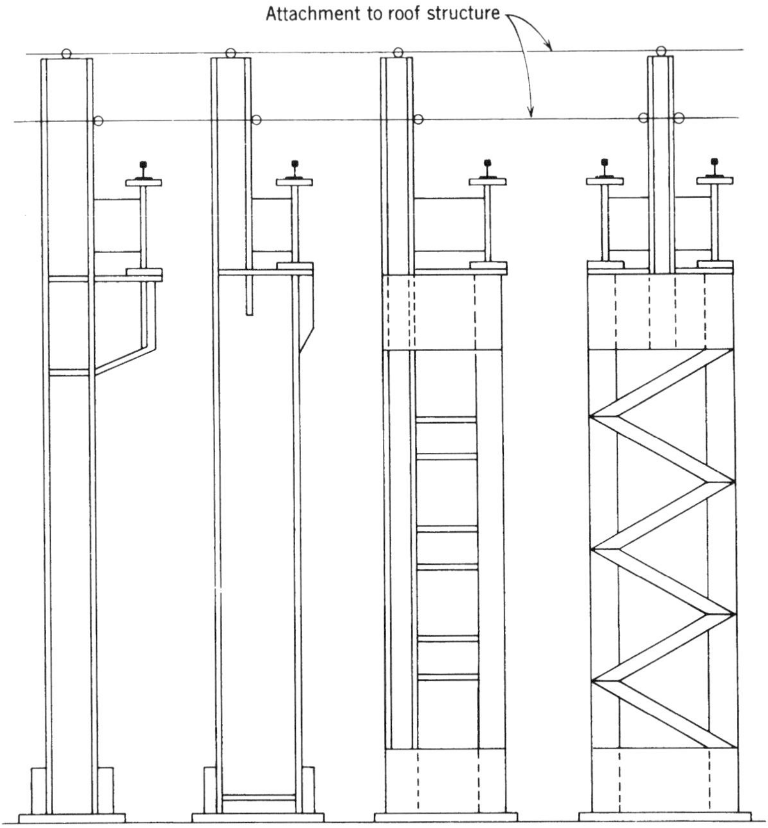

FIGURE 3.56 Type of columns in mill buildings.

column is needed to determine the axial compressive resistance and the Euler buckling load for potential modes of failure by buckling or bending about the x–x axis (see Fig. 3.57). Tables to determine effective-length factors are provided in the Association of Iron and Steel Engineers guide (AISE, 1979) for a range of the three parameters defined in Fig. 3.57 and for the cases when the column base is either fixed or pinned and the column top is pinned. The parameter a is the ratio of the length of the upper (reduced) segment to the total length. The parameter B is the ratio of the second moment of area (about the centroidal x–x axes) of the combined (lower) column cross section to that of the upper section. The parameter P_1/P_2 is the ratio of the axial force acting in the upper segment (roof and upper wall loads) to that applied to the lower segment (crane girder reactions with an allowance for lower wall loads and the column weight). Other notations are also given in Fig. 3.57. The AISE tables give ranges for a from 0.10 to 0.50, for B from 1.0 to 100, and for P_1/P_2 from 0.0 to 0.25. Huang (1968) provides values of the

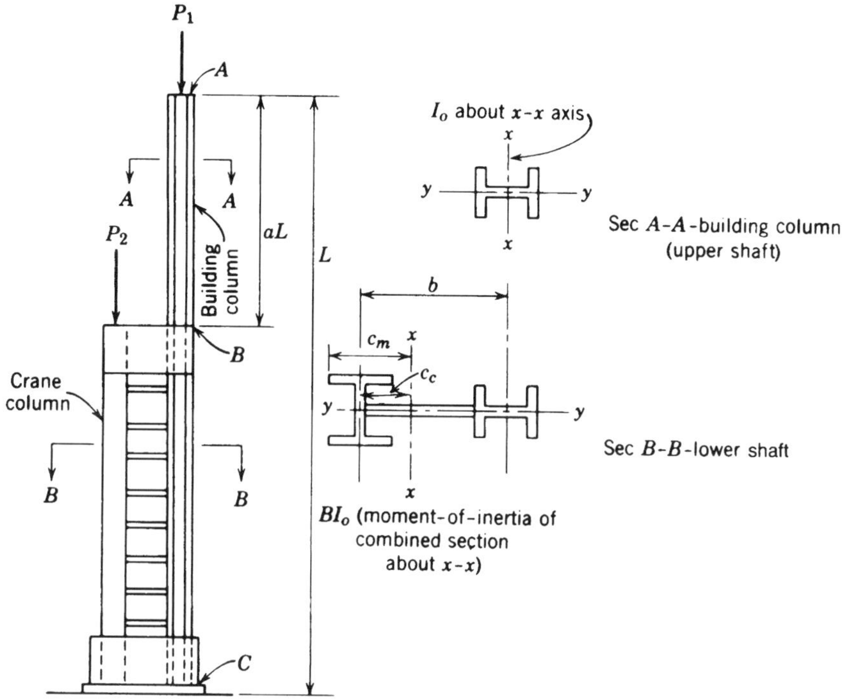

FIGURE 3.57 Notation for stepped columns (AISE, 1979).

effective-length factors in graphs over a somewhat different range of parameters for stepped columns with fixed bases and pinned tops and for values of P_1/P_2 up to 1.0. Further values tabulated for the effective-length factors for stepped columns are given in Timoshenko and Gere (1961), the Column Research Committee of Japan handbook (CRCJ, 1971, English edition), and Young (1989). A method for determining the effective-length factor of a stepped column without using charts was proposed by Lui and Sun (1995). A hand calculation method for determining the buckling load of a stepped column was also proposed by Fraser and Bridge (1990).

The interaction equations used for the design of stepped columns depend on the potential modes of failure and therefore on how the columns are braced. Stepped columns are usually laterally unsupported over the entire length for buckling or bending about the x–x axis. For buckling about the y–y axis, lateral support is usually provided at the level of the crane runway girder seat, location B in Fig. 3.57. Therefore, the following potential failure modes exist:

1. Bending of the overall column in-plane about the x–x axis. It is for this case that the equivalent length for the stepped column is needed in order that the compressive resistance and the Euler buckling load may be determined.

2. Buckling about the y–y axis of the lower segment.
3. Yielding of the cross section of the lower segment.
4. Buckling about the y–y axis of the upper (reduced) segment of the column.
5. Yielding of the cross section of the upper segment.

The interaction equations for axial compression and biaxial bending are

$$\frac{P_u}{\phi P_n} + \frac{8}{9}\left(\frac{M_{ux}}{\phi_b M_{nx}} + \frac{M_{uy}}{\phi_b M_{ny}}\right) \leq 1.0 \quad \text{for } P/\phi P_n \geq 0.2 \tag{3.45a}$$

$$\frac{P_u}{2\phi P_n} + \left(\frac{M_{ux}}{\phi_b M_{nx}} + \frac{M_{uy}}{\phi_b M_{ny}}\right) \leq 1.0 \quad \text{for } P/\phi P_n < 0.2 \tag{3.45b}$$

where M_{ux} would include second-order effects, which could be modeled by

$$B_{1x} = \frac{C_{mx}}{1 - P_u/P_{ex}} \geq 1.0 \tag{3.45c}$$

with a similar term for B_{1y} related to bending about the y–y axis, M_{uy}.

It is conservatively assumed that the moments M_{uy} due to eccentric crane girder reactions are all taken by the lower segment and therefore the moments about the y–y axis in the upper segment are zero.

Following the AISC LRFD procedures, failure modes 1, 2, and 3 are considered together, as are modes 4 and 5. When using the AISC LRFD equations for failure modes 1, 2, and 3 combined, the following values are suggested. The factored axial load, P_u, is $P_1 + P_2$, and M_{ux} and M_{uy} are the maximum modified factored moments within the length. Based on AISE (1979), C_{mx} is taken as 0.85 when all bents are under simultaneous wind load with sidesway. For crane load combinations, with only one bent under consideration, take C_{mx} equal to 0.95. Because it has been assumed that the lower segment takes all the moments about the y–y axis, C_{my} equals 0.6 when the base is pinned and 0.4 when the base is fixed about the y–y axis. The compressive resistance is the least of ϕP_{nx} of the equivalent column and ϕP_{ny} of the bottom segment, taking, in this case, $K = 0.8$ if the base is considered to be fully fixed, and $K = 1.0$ if the base is considered to be pinned. The factored moment resistance $\phi_b M_{nx}$ is based on the possibility of lateral–torsional buckling and $\phi_b M_{ny}$ is the full cross-section strength. This being the case, both factors B_{1x} and B_{1y} as given previously are limited to a value of' 1.0 or greater. The value of P_{ex} is as determined for the equivalent column and P_{ey} is based on 0.8 of the lower segment for a fixed base and 1.0 of the lower segment if the base is pinned. When using the AISC LRFD equations for failure modes 4 and 5 combined for the upper segment, the values suggested parallel those for the lower segment with the exceptions that there are moments only about the x–x axis and C_{mx} depends directly on the shape of the moment diagram.

The five failure modes could also be checked independently when the values for checking are as follows: For bending failure about the x–x axis, failure mode 1 (the predominant moments are about the x–x axis), the factored axial load is $P_1 + P_2$, and M_{ux} and M_{uy} are the maximum modified factored moments within the length. C_{mx} is taken as 0.85 when all bents are under simultaneous wind load with sidesway. For crane load combinations, with only one bent under consideration, take C_{mx} equal to 0.95. Because it has been assumed that the lower segment resists all the moments about the y–y axis, $C_{my} = 0.6$ when the base is pinned and 0.4 when the base is fixed about the y–y axis. The value of B_1 is not restricted. The compressive resistance is the least of ϕP_{nx} of the equivalent column and ϕP_{ny} of the bottom segment with $K = 0.8$ if the base is taken to be fully fixed and with $K = 1.0$ if the base is taken as pinned. The factored moment resistances, $\phi_b M_{nx}$ and $\phi_b M_{ny}$, are the full cross-section strengths. The value of P_{ex} is as determined for the equivalent column and P_{ey} is based on 0.8 of the lower segment for a fixed base and 1.0 of the lower segment if the base is pinned.

The values for checking the last four potential failure modes, modes 2 to 5, independently (for y–y axis buckling and for cross-section strength) are as follows. The factored force effects P_u, M_{ux}, and M_{uy} are those for the appropriate segment. When checking for cross-section strength, the resistances $\phi P_n, \phi_b M_{nx}$, and $\phi_b M_{ny}$ are the full factored cross-section strengths for the appropriate segment. When checking for y–y axis buckling, ϕP_{ny} is based on y–y axis end conditions for the segment being considered. The effective length is the full length AB for the upper segment and the length BC for the lower segment if full base fixity does not exist and 0.80 of the length BC if fully fixed at the base. In Eq. 3.45, M_{nx} is the appropriate lateral–torsional buckling strength taking into account the shape of the moment diagram, and M_{ny} is the full cross-section strength. C_{mx} and C_{my} have minimum values of 1.0 because the variation of moment has been accounted for in finding the moment resistance. The value of P_{ex} is that determined for the equivalent column and P_{ey} is taken as that for the appropriate segment.

A demonstration of the LRFD approach to the design of stepped crane columns has been presented by MacCrimmon and Kennedy (1997). The design of a column with constant cross section subject to lateral loads and moments applied at a bracket has been considered by Adams (1970). The upper and lower segments are treated as separate beam-columns. Horne and Ajmani (1971) and Albert et al. (1995) take into account the lateral support provided by the girts attached to one flange of the column.

3.11 GUYED TOWERS

Guyed towers are used for communication structures, for electrical transmission structures (ASCE, 1972), and, more recently, for compliant offshore platforms and wind energy conversion systems. They consist of a mast supported at one or more levels by guy cables that are anchored at the base. The masts are generally three-dimensional steel trusses of either triangular or square cross section, but occasionally may be of circular cross section. The loading on a guyed tower may include:

1. Self-weight
2. Initial guy tensions (also called guy pretensions)

3. Wind, ice, and earthquake loads
4. Loads due to falling ice, sudden rupture of guy cables, conductors, ground wire
5. For communication towers, loads on antennas
6. For electrical transmission towers, loads on conductors and ground wire
7. For compliant offshore platforms, operational loads and wave loading
8. Transportation and erection loads

The annual failure rate of existing towers was estimated by Magued et al. (1989) to be 55/100,000. This is considered to be high when compared to other structures. Contrary to most other structures for which the dead load is the major load, wind loads and combined wind and ice loads are the major loads on towers (Wahba et al., 1993).

The complete design of a guyed tower includes consideration of vibration and fatigue phenomena in which ice-covered cables tend to vibrate with large amplitudes perpendicular to the direction of wind. Galloping, which generally results from some form of icing on the cables when there is sufficient energy to produce large oscillations, may often lead to failure of the tower. Saxena et al. (1989) reported that there were many incidents where some form of ice storm accompanied by moderate wind speed led to the collapse of transmission and communication towers. Considerable work has been done on the prediction of dynamic behavior and the effect of wind and ice loads on such towers (McCaffrey and Hartman, 1972; Novak et al., 1978; Nakamoto and Chiu, 1985; Saxena et al., 1989). Simplified methods for estimating the dynamic response using a series of static patch loads have been developed (Gerstoft and Davenport, 1986; Davenport and Sparling, 1992). McClure et al. (1993) identified sudden ice shedding and earthquakes as some of the other causes of concern for the safety of these towers.

The following deals only with buckling aspects of guyed tower design. Although instability may not be the failure mode of all guyed towers, the possibility of general instability of a guyed tower should be investigated by the designer (Hull, 1962; Goldberg and Gaunt, 1973; Williamson, 1973; Chajes and Chen, 1979; Chajes and Ling, 1981; Costello and Phillips, 1983).

Because the governing equilibrium equations are not linear homogeneous equations, instability is not a bifurcation problem, but rather occurs as a large deformation for a small increase in applied load. The stability of a guyed tower is influenced principally by (1) the cross-section area of guy cables, (2) the second moment of area of the cross section of the mast, and (3) the initial tension in the guy cables. For the case of buckling of a guyed tower, the axial stiffness of the guy cables is, perhaps, more important than the bending stiffness of the mast. The buckling load can be increased effectively by increasing the cross-section area of guy cables up to the point where the mast would buckle in a number of half-waves equal to the number of guy levels. To increase the buckling load further, the second moment of area (i.e., the bending stiffness) of the mast must be increased. An increase in guy pretensions has both beneficial and detrimental effects on the stability of guyed towers. An increase in guy pretension is beneficial as it results in reduced deflections of the mast. On the other hand, an increase in guy pretension is detrimental as it increases the compressive force in the mast, thus reducing its resistance to buckling. Initially, the increase in guy pretension results in an increase in buckling load because of reduced tower deflections. After a certain stage, however, the increase in the buckling load produced by increasing guy pretension will be more than offset by the decrease in the buckling load due to increasing compressive force on the mast.

The analysis of a guyed tower is complicated because of its geometrically nonlinear behavior even at low-load levels. This nonlinearity is due to the increase in axial stiffness of guy cables with increasing tension and the decrease in the bending stiffness of the mast due to the increase in compressive forces. A guyed tower is usually analyzed by approximating it as an equivalent continuous beam-column on nonlinear elastic supports (Cohen and Perrin, 1957; Rowe, 1958; Dean 1961; Livesley and Poskitt, 1963; Goldberg and Meyers, 1965; Odley, 1966; Williamson and Margolin, 1966; Rosenthal and Skop, 1980; Ekhande and Madugula, 1988). Issa and Avent (1991) presented a method to obtain the forces in all individual members of the mast directly by the use of discrete field mechanics techniques. Sometimes the analytical model for the guyed tower analysis may be a finite element representation using beam-column elements for the mast and cable elements for the guys [International Association for Shell and Spatial Structures (IASS), 1981].

Two basic approaches are used for the second-order elastic analysis (geometric nonlinear analysis): the stability function approach and the geometric stiffness approach. This type of analysis includes both $P - \Delta$ (chord rotation) and $P - \delta$ (member curvature) effects, thus eliminating the need for moment amplification factors. The use of member effective-length factors (K-factors) is also not required when the design is carried out in accordance with CSA S16-09 (CSA, 2009). As all the force resultants acting on the ends of the members are established from the second-order analysis, the member design is based on its actual length, that is, the member is designed assuming that $K = 1$. The tower legs can be designed as beam-columns in the usual manner.

3.11.1 Strengthening of Existing Towers

Solid round and pipe steel members are widely used as leg and bracing members in lattice guyed communication towers because they are subjected to less wind load than angle members and have the same compressive strength about all axes leading to economy in design. The increase in instant voice and data communication and the advent of new technology such as digital television make it necessary for owners to re-evaluate the capacity of existing antenna towers. If there are additional loads (e.g., antennas) added to an antenna tower and/or if there is an increase in the strength requirements, the existing antenna tower has to be reinforced to bring it into compliance with the latest antenna tower standards. The method generally used to increase the compressive strength of the members consists of attaching an angle/rod/split pipe section to the members. The reinforcing members are connected to the main members at intervals by means of clamps (e.g., U-bolts) which are sometimes welded at the ends.

Work was initiated at the University of Windsor in 2004 to determine the strength of reinforced members (Bhuiyan, 2005; Kumalasari et al., 2004, 2005a, 2006b; Siddiqui, 2005; Tickle, 2004; Zeineddine, 2004). Tests and preliminary analyses have been carried on steel members that include:

(i) Solid round sections strengthened with rods
(ii) Solid round sections strengthened with angles
(iii) Solid round sections strengthened with split pipe
(iv) Pipe sections strengthened with split pipe

Because there are very few facilities available in North America for testing large-size rods up to 300 mm diameter which are used in practice, the only realistic method of design has to be based on numerical modeling of the strengthened member, validated by experimental data on small-size specimens.

3.11.2 Strength of Bolted Ring-Type and Flange-Type Connections

Guyed lattice communication tower sections are fabricated using welded splices, and these welded sections are bolted together in the field. There are two types of bolted splices, namely, bolted ring-type splices for leg diameters up to 65 mm (2.5 in.) and bolted flange-type splices for leg diameters greater than 65 mm (2.5 in.). The leg members of a steel tower are subjected to compressive loads due to dead load and guy tension and tensile–compressive loads due to bending moments caused by wind/seismic loads. The strength of splices between the leg sections is vital because it affects the overall performance of the tower.

Bolted Ring-Type Splices The Canadian Institute of Steel Construction *Handbook of Steel Construction* (CISC, 2006) and the AISC *Steel Construction Manual* (AISC, 2005b) do not provide any guidance for analyzing bolted ring-type splices subject to tension. Ignoring the eccentricity of the splice and using the axial tensile capacity of the bolt to design the splice is unsafe, as shown by Kumalasari et al. (2005b) and Shen (2002). The splice should be designed for combined stresses due to axial tension and bending, with a moment equal to the axial load times half the distance between the center of the leg and the center of the bolt (to take into account the fixity of the splice).

Bolted Flange-Type Splices One common type of splice for solid round leg members of guyed lattice communication towers consists of circular flange plates welded to the members and bolted together. These are subjected to tension due to the applied lateral loads (wind or earthquake). Both the *Handbook of Steel Construction* (CISC, 2006) and the *Steel Construction Manual* (AISC, 2005b) discuss prying action only in tee-type and angle-type connections subjected to tensile force and no guidance is provided to determine the prying force in bolted steel circular flange connections. In order to use the formulas given in those publications, Kumalasari et al. (2006a) and Hussaini (2004) suggest that the value of "p", that is, the length of flange tributary to each bolt (bolt pitch), be taken as the distance between the centers of bolts measured along the bolt circle (which is equal to the circumference of the bolt circle divided by the number of bolts).

3.11.3 Dynamic Impact Factor due to Sudden Guy Ruptures and Guy Slippage

There are several instances of collapse of guyed towers due to the failure of the guy. The most significant failure was the collapse of a 646 m mast (the tallest guy mast in the world) in Gabin, Poland. Guys could be ruptured either intentionally due to sabotage or accidentally (e.g., an airplane or farming equipment hitting one of the guy wires). Also, during tower erection, a guy wire can slip at the anchor location before it is set firmly in place. Several different factors, such as guy damping and mast damping, affect the behavior of the mast after guy rupture or slippage. The dynamic impact factor can be very high for guy wires located immediately below the ruptured guy wire.

The effects of sudden guy rupture at the mast attachment point were investigated numerically by Kahla (1997, 2000). Eurocode 3.1 (CEN, 2006b) suggests the dynamic effects caused by guy rupture can be conservatively estimated by (1) the static method and (2) a simplified energy approach. Nielsen (1999) compared the Eurocode procedures with full dynamic analysis for a 244-m mast in Finland and found that both the static method and the simplified energy approach are conservative.

3.11.4 Aeolian Vibration and Galloping of Guy Wires

The mast guys can be subjected to high-frequency, low-amplitude vortex-excited vibrations (Aeolian vibrations) and to low-frequency, large-amplitude vibrations (guy galloping). Guys may be subject to low-amplitude resonant-type vibrations at low wind speeds caused by vortex excitation at high frequency. Guys may be subject to galloping excitation when coated with ice or thick grease. The accretion of ice or grease can form aerodynamic shapes, which provide lift and drag instabilities. These result in low-frequency, high-amplitude vibration. Similar vibrations can also occur under conditions of rain.

Guyed structures are characterized by a certain vibration frequency spectrum (Ciesielski and Koziol, 1995). To suppress guy vibrations that occur under wind, Eurocode prescribes that dampers should be mounted on guys in all cases where initial tension is greater than 10% of the rated breaking strength of the guy.

The dynamic response of guyed towers differs considerably from self-supporting towers, primarily due to the nonlinear behavior of the guys attached to them. To avoid failures, such as fatigue in connections, insulators, anchorage system, and the individual wires of the guys, high and low frequency guyed dampers have been developed to control vibrations of tall guyed towers. Various dampers used for vibration control of guyed towers are discussed in the ASCE Task Committee Report on Dynamic Response of Lattice Towers (Madugula, 2002) and these are briefly described below.

Dampers Used for Control of Aeolian Vibrations Appropriate dampers should be installed in all cases where Aeolian vibrations are predicted or have been observed. Various types of dampers are available, such as:

1. Tuned inertial-type (Stockbridge) dampers consisting of two masses on the ends of short cantilever lengths of spiral strand tuned to the same frequency as the guy
2. Rope loops clamped to the guys
3. Hydraulic dashpots

The tuned inertial type dampers have a long and successful history. There is evidence that partial control of galloping may be obtained by the attachment of a hemp rope from guy to guy, connecting the points of maximum amplitude of two or more guys. The dashpot damper is one of the most common types of guy dampers that have proven effective.

Dampers Used for Control of Guy Galloping Dampers in use to suppress or reduce galloping excitations are the sandamper, the snubber damper, and the hanging chain damper. The sandamper consists of a closed cylinder partially filled with dry sand such that when the cylinder rotates, energy is dissipated. The size, diameter, and mass of the cylinder and the amount of sand must be determined by the modes to be damped and the amount of damping required for controlling the guy vibrations. The snubber damper consists of a tie-down attached to the lower portion of a guy by a pulley. The tie-down has a standard automotive shock absorber installed. The hanging chain damper consists of a heavy chain suspended from a guy rope. Hanging chains, however, constitute a major obstacle to guy inspection and regreasing—both operations involve running a man-riding chair up the full length of each guy. Also, there is a long-term possibility of mechanical damage to the guy wire where the chain is suspended from the guy wire. In addition, the dampers themselves represent a long-term maintenance liability and would require extensive fencing to protect them from grazing farm animals. Guy galloping can probably be controlled with Stockbridge-type dampers if they can be obtained in the appropriate sizes. These are not readily available in the large sizes encountered on guys for tall masts (Wiskin 1995).

REFERENCES

AA (2005), *Specifications for Aluminum Structures*, Aluminum Association, Arlington, VA.

AASHTO (1996), *Standard Specification for Highway Bridges*, 16th ed., American Association of State Highway and Transportation Officials, Washington, DC.

AASHTO (2007), *Standard Specification for Highway Bridges*, 4th ed., American Association of State Highway and Transportation Officials, Washington, DC.

Ackroyd, M. H. (1979), "Nonlinear Inelastic Stability of Flexibly Connected Plane Steel Frames," PhD. dissertation, University of Colorado, Boulder, CO.

Ackroyd, M. H., and Bjorhovde, R. (1981), "Effect of Semi-rigid Connections on Steel Column Strength" (discussion), *J. Const. Steel Res. (London)*, Vol. 1, No. 3, pp. 48–51.

Adams, P. F. (1970), "Segmental Design of Steel Beam-Columns," *Proc. Can. Struct. Eng. Conf.*, Toronto, Ontario, Canada.

AISC (1978), *Specification for the Design, Fabrication and Erection of Structural Steel Buildings*, American Institute of Steel Construction, Chicago, IL.

AISC (1986), *Load and Resistance Factor Design Specification for Structural Steel Buildings*, American Institute of Steel Construction, Chicago, Il.

AISC (1993), *Load and Resistance Factor Design Specification for Structural Steel Buildings*, American Institute of Steel Construction, Chicago, IL.

AISC (1999), *Load and Resistance Factor Design Specification for Structural Steel Buildings*, American Institute of Steel Construction, Chicago, IL.

AISC (2005a), "Specification for Structural Steel Buildings," ANSI/AISC Standard 360-05, American Institute of Steel Construction, Chicago, IL.

AISC (2005b), *Steel Construction Manual*, 13th ed., American Institute of Steel Construction, Chicago, IL.

AISE (1979), "Guide for the Design and Construction of Mill Buildings," AISE Tech. Rep. No. 13, Association of Iron and Steel Engineers, Pittsburgh, PA.

Albert, C, Srna, P., Kennedy, S. J., and Kennedy, D. J. L. (1995), "Inelastic Distortional Strength of Steel Beam-Columns," *Int. Conf. Struct. Stab. Design*, Oct. 30–Nov. 1, Australian Institute of Steel Construction, Sydney, Australia.

Alpsten, G. A. (1970), "Residual Stresses and Strength of Cold-Straightened Wide Flange Shapes," *Jernkontorets Ann.*, Vol. 154 (in Swedish), pp. 1–9.

Alpsten, G. A. (1972a), "Prediction of Thermal Residual Stresses in Hot Rolled Plates and Shapes," 9th Congr. IABSE, Final Rep., Amsterdam, May, pp. 1–13.

Alpsten, G. A. (1972b), "Residual Stresses, Yield Stress and Column Strength of Hotrolled and Roller-Straightened Steel Shapes," Colloq. Column Strength, Paris.

Alpsten, G. A., and Tall, L. (1970) "Residual Stresses in Heavy Welded Shapes," *Weld. J.*, Vol. 49, No. 3, pp. 93–105.

AREMA (2008), *Manual for Railway Engineering*, American Railway Engineering and Maintenance-of-Way Association, Chicago, IL.

ASCE (1972), "Guide for the Design of Aluminum Transmission Towers," Task Committee on Lightweight Alloys of the Committee on Metals of the Structural Division, American Society of Civil Engineers, *Proc. ASCE*., Vol. 98, No. ST12.

ASCE (2003). *Specification for the Design of Cold-formed Stainless Steel Structural Members*, SEI/ASCE-8, American Society of Civil Engineers, New York.

Aslani, F. and Goel, S. C. (1991), "An Analytical Criterion for Buckling Strength of Built-up Compression Members," *AISC Eng. J.*, Vol. 28, No. 4, pp. 159–168.

AS/NZS (2001). *Cold-formed Stainless Steel Structures*, AS/NZS 4673, Standards Australia, Sydney.

Baker, J. F., and Roderick, J. W. (1948), "The Strength of Light Alloy Struts," Res. Rep. No. 3, Aluminum Development Association, London.

Barakat, M. A., and Chen, W. F. (1990), "Practical Analysis of Semi-rigid Frames," *AISC Eng. J.*, Vol. 27, No. 2, pp. 54–68.

Batterman, R. H., and Johnston, B. G. (1967), "Behavior and Maximum Strength of Metal Columns," *ASCE J. Struct. Div.*, Vol. 93, No. ST2, pp. 205–230.

Bažant, Z. P., and Cedolin, L. (1991), *Stability of Structure—Elastic, Inelastic, Fracture and Damage Theories*, Oxford University Press, New York.

Beaulieu, D. and Adams, P. F. (1980), "Significance of Structural Out-of-Plumb Forces and Recommendations for Design," *Can. J. Civil Eng.*, Vol. 7, No. 1, pp. 105–113.

Beedle, L. S. (ed.) (1991), *Stability of Metal Structures: A World View*, Structural Stability Research Council, Bethlehem, PA.

Beedle, L. S. (ed.) (1993), *Semi-rigid Connections in Steel Frames*, Council on Tall Buildings and Urban Habitat, McGraw-Hill, New York.

Beedle, L. S., and Tall, L. (1960), "Basic Column Strength," *ASCE J. Struct. Div.*, Vol. 86, No. ST5, pp. 139–173.

Beer, H., and Schultz, G. (1970), "Theoretical Basis for the European Column Curves," *Constr. Met.*, No. 3, p. 58.

Beer, G., and Tall, L. (1970), "The Strength of Heavy Welded Box Columns," Fritz Eng. Lab. Rep. No. 337.27, Lehigh University, Bethlehem, PA, Dec.

Bernard, A., Frey, F., Janss, J., and Massonnet, C. (1973), "Research on the Buckling Behavior of Aluminum Columns," *IABSE Mem.* (in French), Vol. 33, No. 1, pp. 1–33.

Bhuiyan, P. A. (2005), "Effect of Spacers on the Compressive Strength of Solid Round Steel Members Reinforced with Split Pipe,' M.A.Sc. Major Paper, University of Windsor, Windsor, Ontario, Canada.

Bijlaard, F. S. K., Gresnigt, A. M., and van der Vegte, G. J. (2005), *Connections in Steel Structures V—Behaviour, Strength and Design*, Bouwen met Staal, The Netherlands.

Birkemoe, P. C. (1977a), "Column Behavior of Heat-Treated Cold-Formed Hollow Structural Shapes," *Stability of Structures Under Static and Dynamic Loads, Proc. 2nd Int. Colloq.*, Washington, DC.

Birkemoe, P. C. (1977b), "Development of Column Curves for H.S.S.," *Int. Symp. Hollow Struct. Sec.*, CIDECT, Toronto, Ontario, Canada.

Bjorhovde, R. (1972), "Deterministic and Probabilistic Approaches to the Strength of Steel Columns," Ph.D. dissertation, Lehigh University, Bethlehem, PA.

Bjorhovde, R. (1977), "Strength and Behavior of Cold-Formed H.S.S. Columns," Struct. Eng. Rep. No. 65, University of Alberta, Edmonton, Alberta, Canada, Dec.

Bjorhovde, R. (1978), "The Safety of Steel Columns," *ASCE J. Struct. Div.*, Vol. 104, No. ST3, pp. 463–477.

Bjorhovde, R. (1980), "Research Needs in Stability of Metal Structures," *ASCE J. Struct. Div.*, Vol. 106, No. ST12, pp. 2425–2442.

Bjorhovde, R. (1981), "End Restraint and Column Stability" (discussion), *ASCE J. Struct. Div.*, Vol. 107, No. ST8, pp. 1696–1700

Bjorhovde, R. (1984), "Effect of End Restraint on Column Strength: Practical Applications," *AISC Eng. J.*, Vol. 21, No. 1, pp. 1–13.

Bjorhovde, R. (1988), "Columns: From Theory to Practice," *AISC Eng. J.*, Vol. 25, No. 1, pp. 21–34.

Bjorhovde, R. (1991), "The Strength of Heavy Columns," *J. Constr. Steel Res.*, Vol. 19, No. 4, pp. 313–320.

Bjorhovde, R. (1992), "Compression Members," in *Constructional Steel Design: An International Guide* (P. J. Dowling, J. E. Harding, and R. Bjorhovde, eds.), Elsevier Applied Science, London, Chap. 2.3.

Bjorhovde, R. (2006), "Cold Bending of Wide-Flange Shapes for Construction," *Eng. J. AISC*, Vol. 43, Fourth Quarter, pp. 271–286.

Bjorhovde, R., and Birkemoe, P. O. (1979), "Limit States Design of H.S.S. Columns," *Can. J. Civ. Eng.*, Vol. 8, No. 2, pp. 276–291.

Bjorhovde, R., and Tall, L. (1971), "Maximum Column Strength and the Multiple Column Curve Concept," Fritz Eng. Lab. Rep. No. 338.29, Lehigh University, Bethlehem, PA, Oct.

Bjorhovde, R., Brozzetti, J., Alpsten, G. A., and Tall, L. (1972), "Residual Stresses in Thick Welded Plates," *Weld J.*, Vol. 51, No. 8, pp. 392–405.

Bjorhovde, R., Brozzetti, J., and Colson, A. (1988), *Connections in Steel Structures—Behaviour, Strength and Design*, Elsevier Applied Science, London, England.

Bjorhovde, R., Colson, A. and Brozzetti, J. (1990), "Classification System for Beam-to-Column Connections," *J. Struct. Eng.*, ASCE, Vol. 116, No. 11, pp. 3063–3080.

Bjorhovde, R., Colson, A., Haaijer, G., and Stark, J. W. B. (1992), *Connections in Steel Structures II—Behavior, Strength and Design*, AISC, Chicago, IL.

Bjorhovde, R., Colson, A., and Zandonini, R. (1996), *Connections in Steel Structures III—Behaviour, Strength and Design*, Pergamon/Elsevier Science, London.

Bjorhovde, R., Bijlaard, F. S. K., and Geschwindner, L. F. (2008), *Connections in Steel Structures VI—Behaviour, Strength and Design*, AISC, Chicago, IL.

Bleich, F. (1952), *Buckling Strength of Metal Structures*, McGraw-Hill, New York.

Boissonnade, N., and Maquoi, R. (2005), "A Geometrically and Materially Non-linear 3-D Beam Finite Element for the Analysis of Tapered Steel Members," *Steel Struct.*, Vol. 5, pp. 413–419.

Bredenkamp, P. J. and van den Berg, G. J. (1995), "The Strength of Stainless Steel Built-up I-Section Columns," *J. Constr. Steel Res.*, Vol. 34, No. 2–3, pp. 131–144.

Bredenkamp, P. J., van den Berg, G. J., and van der Merwe, P. (1992), "Residual Stresses and the Strength of Stainless Steel I-Section Columns," *Proc. SSRC 1992 Annu. Tech. Session*, Pittsburgh, PA, Apr.

Bredenkamp, P. J., Human, J. J., and van den Berg, G. J. (1994), "The Strength of Hot-Rolled Stainless Steel Columns," *Proc. SSRC 1994 Annu. Tech. Session*, Apr., Milwaukee, WI.

Brockenbrough, R. L. (1992), "Material Behavior," in *Constructional Steel Design: An International Guide* (P. J. Dowling, J. E. Harding, and R. Bjorhovde, eds.), Elsevier Applied Science, London, Chaps. 1.1 and 1.2.

Brolin, C. A., Durscher, H. E., and Serentha, G. (1972), "Destructive Testing of Crane Booms," *SAE Trans.*, Vol. 81, Pap. 720784.

Brozzetti, J., Alpsten, G. A., and Tall, L. (1970a) "Residual Stresses in a Heavy Rolled Shape 14WF730," Fritz Eng. Lab. Rep. No. 337.1, Lehigh University, Bethlehem, PA, Jan.

Brozzetti, J., Alpsten, G. A., and Tall, L. (1970b), "Welding Parameters, Thick Plates and Column Strength," Fritz Eng. Lab. Rep. No. 337.21, Lehigh University, Bethlehem, PA, Feb.

Brungraber, R. J., and Clark, J. W. (1962), "Strength of Welded Aluminum Columns," *Trans. Am. Soc. Civ. Eng.*, Vol. 127, Part 11, pp. 202–226.

BSI (1991), *Structural Use of Aluminium: Code of Practice for Design*, BS 81181, British Standards Institution, London

Burgan B. A., Baddoo, N. R., and Gilsenan, K. A. (2000), "Structural Design of Stainless Steel Members—Comparison between Eurocode 3, Part 1.4 and Test Results," *J. Constr. Steel Res.*, Vol. 54, No. 1, pp. 51–73.

Butler, D. J. (1966), "Elastic Buckling Tests on Laterally and Torsionally Braced Tapered I-Beams," *Weld. J. Res. Suppl.*, Vol. 44, No. 1, pp. 41–48.

Butler, D. J., and Anderson, G. C. (1963), "The Elastic Buckling of Tapered Beam Columns," *Weld J. Res. Suppl.*, Vol. 42, No. 1, pp. 29–36.

CEN (2005), Eurocode 3: *Design of Steel Structures*, Part 1.1: General Rules and Rules for Buildings, EN 1993-1-1, Comité Européen de Normalisation (CEN), European Committee for Standardization Brussels, Belgium.

CEN (2006a), Eurocode 3: *Design of Steel Structures*, Part 1.4: Supplementary Rules for Stainless Steel, EN-1993-1-4, Comité Européen de Normalisation (CEN), European Committee for Standardization, Brussels, Belgium.

CEN (2006b), *Design of Steel Structures*, Part 3.2: Supplementary Rules for Towers, Masts and Chimneys, EN 1993-3-2, Comité Européen de Normalisation (CEN), European Committee for Standardization, Brussels, Belgium.

CEN (2007), Eurocode 9: *Design of Aluminium Alloy Structures: General Rules*, EN 1999-9, Comité Européen de Normalisation (CEN), European Committee for Standardization, Brussels, Belgium.

Chajes, A., and Chen, W. S. (1979), "Stability of Guyed Towers," *ASCE J. Struct. Div.*, Vol. 105, No. ST1, pp. 163–174.

Chajes, A., and Ling, D. (1981), "Post-buckling Analysis of Guyed Towers," *ASCE, J. Struct. Eng.*, Vol. 107, No. 12, pp. 2313–2323.

Chapuis, J., and Galambos, T. V. (1982), "Restrained Crooked Aluminum Columns," *ASCE J. Struct. Div.*, Vol. 108, No. ST3, pp. 511–524.

Chen, W. F. (1980), "End Restraint and Column Stability," *ASCE J. Struct. Div.*, Vol. 105, No. ST11, pp. 2279–2295.

Chen, W. F. (ed.) (1987), "Joint Flexibility in Steel Frames," Special Issue, *J. Constr. Steel Res.*, Vol. 8.

Chen, W. F. (ed.) (1988), *Steel Beam-to-Column Building Connections*, Elsevier Applied Science, London.

Chen, W. F., and Atsuta, T. (1976), *Theory of Beam-Columns*, Vol. 1, McGraw-Hill, New York.

Chen, W. F., and Han, D. C. (1985), *Tubular Members in Offshore Structures*, Pitman, London.

Chen, W. F., and Kishi, N. (1989), "Semi-rigid Steel Beam-to-Column Connections: Data Base and Modeling," *ASCE J. Struct. Eng.*, Vol. 115, No. 1, pp. 105–119.

Chen, W. F., and Lui, E. M. (1985), "Stability Design Criteria for Steel Members and Frames in the United States," *J. Constr. Steel Res.*, Vol. 5, No. 1, pp. 51–94.

Chen, W. F., and Lui, E. M. (1987), *Structural Stability, Theory and Implementation*, Elsevier, New York.

Chen, W. F., and Lui, E. M. (1991), *Stability Design of Steel Frames*, CRC Press, Boca Raton, FL.

Chen, W. F., and Sohal, I. S. (1995), *Plastic Design and Second-Order Analysis of Steel Frame*, Springer-Verlag, New York.

Chen, W. F., Goto, Y., and Liew, J. Y. R. (1996), *Stability Design of Semi-rigid Frames*, Wiley, New York.

Chernenko, D. E., and Kennedy, D. J. L. (1991), "An Analysis of the Performance of Welded Wide Flange Columns," *Can. J. Civil Eng.*, Vol. 18, pp. 537–555.

Christopher, J. E. and Bjorhovde, R. (1999), "Semi-Rigid Frame Design Methods for Practicing Engineers," *Eng. J. AISC*, Vol. 36, First Quarter, pp. 12–28.

Ciesielski, R., and Koziol, K. (1995), "Dynamic Investigations of a Guyed Mast Model," *Proceedings of IASS Working Group no. 4 for Masts and Towers*, Winchester, England, International Association for Shell and Spatial Structures, Madrid, Spain.

CISC (2008), *Handbook of Steel Construction*, 9th ed., Canadian Institute of Steel Construction, Markham, Ontario, Canada.

Clark, J. W., and Rolf, R. L. (1966), "Buckling of Aluminum Columns, Plates and Beams," *ASCE J. Struct. Div.*, Vol. 92, No. ST3, pp. 17–38.

Cohen, E., and Perrin, H. (1957), "Design of Multi-level Guyed Towers: Structural Analysis," *ASCE J. Struct. Div.*, Vol. 83, No. ST5, pp. 1356–1 to 1356–29.

Column Research Committee of Japan (CRCJ) (1971), *Handbook of Structural Stability*, English ed., Corona Publishing Company, Tokyo.

Costello, G. A., and Phillips, J. W. (1983), "Post-buckling Behavior of Guyed Towers," *ASCE J. Struct. Eng.*, Vol. 109, No. 6, pp. 1450–1459.

Crawford, R. F., and Benton, M. D. (1980), "Strength of Initially Wavy Lattice Columns," *AIAA J.*, Vol. 18, No. 5, p. 581.

CSA (2001), *Antennas, Towers, and Antenna-supporting Structures*. CSA S37-01, Canadian Standards Association (CSA), Toronto, Ontario, Canada.

CSA (2005), *Strength Design in Aluminum*. CSA S157-05, Canadian Standards Association (CSA), Toronto, Ontario, Canada.

CSA (2006), *Canadian Highway Bridge Design Code*, CSA-S6-06, Canadian Standards Association (CSA), Toronto, Ontario, Canada.

CSA (2009), *Design of Steel Structures*, CSA-S16-09, Canadian Standards Association (CSA), Toronto, Ontario, Canada.

Davenport, A. G., and Sparling, B. F. (1992), "Dynamic Gust Response Factors for Guyed Towers," *J. Wind Eng. Ind. Aerodyn*, Vol. 41–44, pp. 2237–2248.

Dean, D. L. (1961), "Static and Dynamic Analysis of Guy Cables," *ASCE J. Struct. Div.*, Vol. 87, No. ST1, pp. 1–21.

DeFalco, F., and Marino, F. J. (1966), "Column Stability in Type 2 Construction," *AISC Eng. J.*, Vol. 3, No. 2, pp. 67–71.

Disque, R. O. (1973), "Inelastic K-Factor in Column Design," *AISC Eng. J.*, Vol. 10, No. 2, pp. 33–35.

Disque, R. O. (1975), "Directional Moment Connections: A Proposed Design Method for Unbraced Steel Frames," *AISC Eng. J.*, Vol. 12, No. 1, pp. 14–18.

Driver, R. G., Grondin, G. Y., and Teh, L. (2003), *Centrally Loaded Column Research Inventory*, Structural Stability Research Council, Task Group 1, University of Missouri-Rolla.

Duberg, J. E., and Wilder, T. W. (1950), "Column Behavior in the Plastic Strength Range," *J. Aeronaut. Sci.*, Vol. 17, No. 6, p. 323.

Dux, P. F., and Kitipornchai, S. (1981), "Inelastic Beam-Buckling Experiments," Res. Rep. No. CE24, Department of Civil Engineering, University of Queensland, St. Lucia, Australia.

ECCS (1978), *European Recommendation for Aluminium Structures*, Committee T2, European Convention for Constructional Steelwork, Brussels, Belgium.

Ekhande, S. G., and Madugula, M. K. S. (1988), "Geometric Nonlinear Analysis of Three-Dimensional Guyed Towers," *Comput. Struct.*, Vol. 29, No. 5, pp. 801–806.

Engesser, F. (1891), "Die Knickfestigkeit gerader Stabe," *Zentralbl. Bauverwaltung*, Vol. 11, pp. 483–486.

Essa, H. S., and Kennedy, D. J. L. (1993), "Distortional Buckling of Steel Beams," Struct. Eng. Rep. No. 185, Department of Civil Engineering, University of Alberta, Edmonton, Alberta, Canada.

Fraser, D. J., and Bridge, R. Q. (1990), "Buckling of Stepped Crane Columns," *J. Constr. Steel Res.*, Vol. 16, No. 1, pp. 23–38.

Freeman, B. G. (1973), "Tie Plate Effects in Weakly Battened Columns," Ph.D. dissertation, Department of Civil Engineering, University of Arizona, Tucson, AZ.

Frey, F. (1969), "Effect of the Cold-Bending of H-Shaped Rolled Steel Sections on Column Strength," *IABSE Mem.* (in French), Vol. 291, pp. 101–124.

Frye, M. J., and Morris, G. A. (1975), "Analysis of Flexibly Connected Steel Frames," *Can. J. Civ. Eng.*, Vol. 2, No. 3, pp. 280–291.

Fukumoto, Y., Nethercot, D. A., and Galambos, T. V. (1983), "Experimental Data for the Buckling of Steel Structures–NDSS Stability of Metal Structures," *Proc. 3rd Int. Colloq. SSRC*, Toronto, Ontario, Canada, May, pp. 609–630.

Galambos, T. V. (ed.) (1988), *Guide to Stability Design Criteria for Metal Structures*, 4th ed., Structural Stability Research Council, Wiley Interscience, New York, 786 pp.

Galambos, T. V. (ed.) (1998), *Guide to Stability Design Criteria for Metal* Structures, 5th ed., Structural Stability Research Council, Wiley Interscience, New York, 911 pp.

Gardner, L., Talja, A., and Baddoo, N. R. (2006), "Structural Design of High-strength Austenitic Stainless Steel." *Thin-walled Struct.* Vol. 44, pp. 517–528.

Gere, J. M., and Carter, W. O. (1962), "Critical Buckling Loads for Tapered Columns," *ASCE J. Struct. Div.*, Vol. 88, No. ST1, Proc. Pap. 3045, pp. 1–11.

Gerstoft, P., and Davenport, A. G. (1986), "A Simplified Method for Dynamic Analysis of a Guyed Mast," *J. Wind Eng. Ind. Aerodyn.*, Vol. 23, pp. 487–499.

Gjelsvik, A. (1990), "Buckling of built-up columns, with or without Stay Plate," *ASCE J. Eng. Mech.*, Vol. 116, No. 5, pp. 1142–1159.

Goldberg, J. E. (1954), "Stiffness Charts for Gusseted Members Under Axial Load," *Trans. ASCE*, Vol. 119, p. 43.

Goldberg, J. E., and Gaunt, J. T. (1973), "Stability of Guyed Towers," *ASCE J. Struct. Div*, Vol. 99, No. ST4, pp. 741–756.

Goldberg, J. E., and Meyers, V. J. (1965), "A Study of Guyed Towers," *ASCE J. Struct. Div.*, Vol. 91, No. ST4, pp. 57–76.

Goto, Y., Suzuki, S., and Chen, W. F. (1993), "Stability Behavior of Semi-rigid Sway Frames," *Eng. Struct.*, Vol. 15, No. 3, pp. 209–219.

Hall, D. H. (1981), "Proposed Steel Column Strength Criteria," *ASCE J. Struct. Div.*, Vol. 107, No. ST4, pp. 649–670.

Hariri, R. (1967), "Post-buckling Behavior of Tee-Shaped Aluminum Columns," Ph.D. dissertation, University of Michigan, Ann Arbor, MI, June.

Hartmann, E. C., and Clark, J. W. (1963), "The U.S. Code," *Proc. Symp. Alum. Struct. Eng.*, London, June 11–12, The Aluminum Federation.

Hartmann, E. C., Moore, R. L., and Holt, M. (1938), "Model Tests of Latticed Structural Frames," Alcoa Res. Lab. Tech. Pap. No. 2, Pittsburgh, PA.

Hellesland, J. and Bjorhovde, R. (1996a), "Restraint Demand Factors and Effective Length of Braced Columns," *J. Struct. Eng., ASCE*, Vol. 122, No. 10, pp. 1216–1224.

Hellesland, J., and Bjorhovde, R. (1996b), "Improved Frame Stability Analysis with Effective Lengths," *J. Struct. Eng., ASCE*, Vol. 122, No. 11, pp. 1275–1283.

Hong, G. M. (1991), "Effects of Non-central Transverse Welds on Aluminum Columns," in *Aluminum Structures: Recent Research and Development* (S. I. Lee, ed.), Elsevier Applied Science, London.

Horne, M. R., and Ajmani, J. L. (1971), "Design of Columns Restrained by Side-Rails and the Post-bucking Behavior of Laterally Restrained Columns," *Struct. Eng.*, Vol. 49, No. 8, pp. 339–352.

Howard, J. E. (1908), "Some Results of the Tests of Steel Columns in Progress at the Watertown Arsenal," *Proc. A.S.T.M.*, Vol. 8, p. 336.

Huang, H. C. (1968), "Determination of Slenderness Ratios for Design of Heavy Mill Building Stepped Columns," *Iron Steel Eng.*, Vol. 45, No. 11, p. 123.

Huber, A. W. (1956), "The Influence of Residual Stress on the Instability of Columns," Ph.D. dissertation, Lehigh University, Bethlehem, PA.

Hull, F. H. (1962), "Stability Analysis of Multi-level Guyed Towers," *ASCE J. Struct. Div.*, Vol. 88, No. ST2, pp. 61–80.

Hussaini, A. (2004), "Tensile Strength of Bolted Flange Connections." M.A.Sc. Major Paper, University of Windsor, Windsor, Ontario, Canada.

IASS (1981), *Recommendations for Guyed Masts*, Working Group 4, International Association for Shell and Spatial Structures, Madrid, Spain.

ISO (1992), "Aluminium Structures: Material and Design, Part 1: Ultimate Limit State - Static Loading," Technical Report, Doc. No 188, ISO Committee TC 167/SC3, International Organization for Standardization, Geneva.

ISE (1962), *Report on the Structural Use of Aluminum*, Institution of Structural Engineers, London.

Issa, R. R. A., and Avent, R. R. (1991), "Microcomputer Analysis of Guyed Towers as Lattices," *ASCE J. Struct. Div.*, Vol. 117, No. 4, pp. 1238–1256.

Jacquet, J. (1970), "Column Tests and Analysis of Their Results," *Constr. Met.*, No. 3, pp. 13–36.

Jimenez-Lopez, G. A. (1998), "Inelastic Stability of Tapered Structural Members," Ph. D. dissertation, University of Minnesota, Minneapolis-St. Paul, MN.

Jimenez, G., and Galambos, T. V. (2001), "Inelastic Stability of Pinned Tapered Columns," Proceedings of SSRC Annual Technical Session, Ft. Lauderdale, FL, pp. 143–158

Johnson A. L. and Winter, G. (1966), "The Structural Performance of Austenitic Stainless Steel Members," Report No. 327, Department of Structural Engineering, Cornell University, Ithaca, NY.

Johnston, B. G. (1963), "Buckling Behavior Above the Tangent Modulus Load," *Trans. ASCE*, Vol. 128, Part 1, pp. 819–848.

Johnston, B. G. (1964), "Inelastic Buckling Gradient," *ASCE J. Eng. Mech. Div.*, Vol. 90, No. EM5, pp. 31–48.

Johnston, B. G. (1971), "Spaced Steel Columns," *ASCE J. Struct. Div.*, Vol. 97, No. ST5, p. 1465.

Johnston, B. G. (ed.) (1976), *Guide to Stability Design Criteria for Metal Structures*, 3rd ed., Wiley, New York.

Jones, S. W., Kirby, P. A., and Nethercot, D. A. (1980), "Effect of Semi-rigid Connections on Steel Column Strength," *J. Constr. Steel Res.*, Vol. 1, No. 1, pp. 35–46.

Jones, S. W., Kirby, P. A., and Nethercot, D. A. (1982), "Columns with Semi-rigid Joints," *ASCE J. Struct. Div.*, Vol. 108, No. ST2, pp. 361–372.

Kaehler, R. C., White, D. W., and Kim, Y. D. (2008), *Frame Design Using Web-Tapered Members*, Metal Building Manufacturers Association, Cleveland, OH, and American Institute of Steel Construction, Chicago, IL.

Kahla, N. B. (1997), "Nonlinear Dynamic Response of a Guyed Tower to a Sudden Guy Rupture," *Eng. Struct.*, Vol. 19, pp. 879–890.

Kahla, N. B. (2000), "Response of a Guyed Tower to a Guy Rupture under No Wind Pressure," *Eng. Struct.*, Vol. 22, pp. 699–706.

Kato, B. (1977), "Column Curves for Cold-Formed and Welded Tubular Members," *2nd Int. Colloq. Stab. Steel Struct.*, Liege, Belgium, pp. 53–60.

Kennedy, D. J. L. and Gad Aly, M. M. (1980), "Limit States Design of Steel Structures Performance Factors," *Can. J. Civil Eng.*, Vol. 7, No. 1, pp. 45–77.

Kim, M. C., Lee, G. C., and Chang, K. C. (1995), "Inelastic Buckling of Tapered Members with Accumulated Strain," *Int. J. Struct. Eng. Mech.*, Vol. 3, No. 6, pp. 611–622.

King, W. S., and Chen, W. F. (1993), "LRFD Analysis for Semi-rigid Frame Design," *AISC Eng. J.*, Vol. 30, No. 4, pp. 130–140.

King, W. S., and Chen, W. F. (1994), "Practical Second-Order Inelastic Analysis of Semi-rigid Frames," *ASCE J. Struct. Eng.*, Vol. 120, No. 7, pp. 2156–2175.

Kishi, N., and Chen, W. F. (1986), "Data Base of Steel Beam-to-Column Connections," Struct. Eng. Rep. No. CE-STR-86-26, School of Civil Engineering, Purdue University, West Lafayette, Ind., 2 vols.

Kishi, N., and Chen, W. F. (1990), "Moment-Rotation Relations of Semi-rigid Connections with Angles," *ASCE J. Struct. Eng.*, Vol. 116, No. 7, pp. 1813–1834.

Kishi, N., Chen, W. F., Goto, Y., and Matsuoka, K. G. (1993a), "Design Aid of Semi-rigid Connections for Frame Analysis," *AISC Eng. J.*, Vol. 30, No. 3, pp. 90–103.

Kishi, N., Goto, Y., Chen, W. F., and Matsuoka, K. (1993b), "Analysis Program for the Design of Flexibly Jointed Frames," *Comput. Struct.*, Vol. 49, No. 4, pp. 705–713.

Kishima, V., Alpsten, G. A., and Tall, L. (1969), "Residual Stresses in Welded Shapes of Flame-Cut Plates in ASTM A572(50) Steel," Fritz Eng. Lab. Rep. No. 321.2, Lehigh University, Bethlehem, PA, June.

Korol, R. M., Rutenberg, A., and Bagnariol, D. (1986), "On Primary and Secondary Stresses in Triangulated Trusses," *J. Constr. Steel Res.*, Vol. 6, pp. 123–142.

Kumalasari, C. (2004), "Compressive Strength of Solid Round Steel Leg and Bracing members of Lattice Communication Towers Reinforced with Rods or Angles," M.A.Sc. Thesis, University of Windsor, Windsor, Ontario, Canada.

Kumalasari, C., Ding, Y., and Madugula, M. K. S. (2006a), "Prying Action in Bolted Steel Circular Flange Connections," *Can. J. Civil Eng.*, April 2006.

Kumalasari, C., Ding, Y., Madugula, M. K. S., and Ghrib, F. (2006b), "Compressive strength of Solid Round Steel Members Strengthened with Rods or Angles," *Can. J. Civil Eng.*, April 2006.

Kumalasari, C., Madugula, M. K. S., and Ghrib, F. (2005a), "Reinforcement of Lattice Communication towers with Rods and Angles," Proc., Annual General Conference of the Canadian Society for Civil Engineering, June 2-4, Toronto, Ontario, Canada, p. GC–255.

Kumalasari, C., Shen, L., Madugula, M. K. S., and Ghrib, F. (2005b), "Tensile Strength of Bolted Ring-type Splices of Solid Round Leg Members of Guyed Communication Towers," *Can. J. Civil Eng.*, Vol. 32, No. 3, pp. 595–600.

Lai, Y. F. W., and Nethercot, D. A. (1992), "Strength of Aluminium Members Containing Local Transverse Welds," *Eng. Struct.*, Vol. 14, No. 4, pp. 241–254.

Lagerquist, O., and Olsson, A. (2001), "Residual Stresses in Welded I-girders Made from Stainless Steel and Structural Steel," in *Proceedings of the 9th Nordic Steel Construction Conference* (P. Makelainen et al., eds.), Helsinki, Finland, pp. 737–744.

Lee, G. C., Morrell, M. L., and Ketter, R. L. (1972), "Design of Tapered Members," *Weld. Res. Counc. Bull. No. 173*, June, pp. 1–32.

Lee, G. C., Ketter, R. L., and Hsu, T. L. (1981), *The Design of Single Story Rigid Frames*, Metal Building Manufacturer's Association, Cleveland, OH.

Leon, R. T. and Easterling, W. S. (2002), *Connections in Steel Structures IV—Behavior, Strength and Design*, AISC, Chicago, IL.

Lin, F. J., Glauser, E. C., and Johnston, B. G. (1970), "Behavior of Laced and Battened Structural Members," *ASCE J. Struct. Div.*, Vol. 96, No. ST7, p. 1377.

Lindner, J., and Gietzelt, R. (1984), "Imperfectionsannahmen für Stützenschiefstellungen," *Stahlbau*, Vol. 53, No. 4, pp. 97–101.

Livesley, R. K., and Poskitt, T. J. (1963), "Structural Analysis of Guyed Masts," *Proc. Inst. Civil Eng. (London)*, Vol. 24, pp. 373–386.

Loov, R. (1996), "A Simple Equation for Axially Loaded Steel Column Design Curves," *Can. J. Civil Eng.*, Vol. 23, No. 11, pp. 272–276.

Lui, E. M., and Chen, W. F. (1983a), "Strength of H-Columns with Small End-Restraints," *Struct. Eng.*, Vol. 61B, No. 1, Part B, pp. 17–26.

Lui, E. M., and Chen, W. F. (1983b), "End Restraint and Column Design Using LRFD," *Eng. J. Am. Inst. Steel Constr.*, Vol. 20, No. 1, pp. 29–39.

Lui, E. M., and Chen, W. F. (1984), "Simplified Approach to the Analysis and Design of Columns with Imperfections," *AISC Eng. J.*, Vol. 21, No. 2, pp. 99–117.

Lui, E. M., and Chen, W. F. (1988), "Behavior of Braced and Unbraced Semi-rigid Frames," *Solids Struct.*, Vol. 24, No. 9, pp. 893–913.

Lui, E. M., and Sun, M. (1995), "Effective Lengths of Uniform and Stepped Crane Columns," *AISC Eng. J.*, Vol. 32, No. 3, pp. 98–106.

MacCrimmon, R. A., and Kennedy D. J. L. (1997), "Load and Resistance Factor Design and Analysis of Stepped Crane Columns in Industrial Buildings," *AISC Eng. J.*, Vol. 34, No. 1, pp. 26–36.

Madsen, I. (1941), "Report of Crane Girder Tests." *Iron Steel Eng.*, Vol. 18, No. 11, p. 47.

Madugula, M. K. S. (ed.) (2002), "Dynamic Response of Lattice Towers and Guyed Masts," Task Committee Report, Structural Engineering Institute of the American Society of Civil Engineers, Reston, VA.

Magued, M. H., Bruneau, M., and Dryburgh, R. B. (1989), "Evolution of Design Standards and Recorded Failures of Guyed Towers in Canada," *Can. J. Civil Eng.*, Vol. 16, No. 5, pp. 725–732.

Masur, E. F. (1954), "Lower and Upper Bounds to the Ultimate Loads of Buckled Trusses," *Q. Appl. Math.*, Vol. 11, No. 4, p. 385.

Maugh, L. C. (1963), *Statically Indeterminate Structures*, 2nd ed., Wiley, New York.

Mazzolani, F. M. (1985), *Aluminium Alloy Structures*, Pitman, Marshfield, MA.

Mazzolani, F. M., and Frey, F. (1977), "Buckling Behavior of Aluminium Alloy Extruded Members," *2nd Int. Colloq. Stabil. Steel Struct.*, Liege, Belgium, pp. 85–94.

Mazzolani, F. M., and Frey, F. (1980), "The Bases of the European Recommendations for the Design of Aluminum Alloy Structures," *Alluminio*, No. 2, pp. 77–94.

McCaffrey, R. J., and Hartmann, A. J. (1972), "Dynamics of Guyed Towers," *ASCE J. Struct. Div.*, Vol. 98, No. ST6, pp. 1309–1323.

McClure, G., Guevara, E., and Lin, N. (1993), "Dynamic Analysis of Antenna-Supporting Towers," *Proc. Annu. Conf. Can. Soc. Civ. Eng.*, June 8–11, Fredericton, New Brunswick, Canada, pp. 335–344.

McFalls, R. K., and Tall, L. (1970), "A Study of Welded Columns Manufactured from Flame-Cut Plates," *AWS Weld. J.*, Vol. 49, No. 4, pp. 141s–153s.

Michalos, J., and Louw, J. M. (1957), "Properties for Numerical Analysis of Gusseted Frameworks," *Proc. Am. Railw. Eng. Assoc.*, Vol. 58, p. 1.

Miller, R. K., and Hedgepeth, J. M. (1979), "The Buckling of Latticed Columns with Stochastic Imperfections," *Int. J. Solids Struct.*, Vol. 15, pp. 71–84.

Nakamoto, R. T., and Chiu, A. N. L. (1985), "Investigation of Wind Effects on Guyed Towers," *ASCE J. Struct. Div.*, Vol. 111, No. 11, pp. 2320–2332.

Nethercot, D. A., and Chen, W. F. (1988), "Effects of Connections on Columns," *J. Constr. Steel Res.*, Vol. 10, pp. 201–239.

Newmark, N. M. (1943), "A Numerical Procedure for Computing Deflections, Moments and Buckling Loads," *Trans. ASCE*, Vol. 108, pp. 1161–1188.

Nielsen, M. G. (1999), "Analysis of Guy Failure," *Proceedings of IASS Working Group No. 4 for Masts and Towers, Krakow, Poland*, International Association for Shell and Spatial Structures, Madrid, Spain.

Novak, M., Davenport, A. G., and Tanaka, H. (1978), "Vibration of Towers Due to Galloping of Iced Cables," *ASCE J. Eng. Mech. Div.*, Vol. 104, No. EM2, pp. 457–473.

Odley, E. G. (1966), "Analysis of High Guyed Towers," *ASCE J. Struct. Div.*, Vol. 92, No. ST1, pp. 169–197.

Osgood, W. R. (1951), "The Effect of Residual Stress in Column Strength," *Proc. First U.S. Natl. Congn. Appl. Mech.*, June, p. 415.

Paul, M. (1995), "Buckling Loads for Built-up Columns with Stay Plates," *ASCE J. Eng. Mech.*, Vol. 121, No. 11, pp. 1200–1208.

Prawel, S. P., Morrell, M. L., and Lee, G. C. (1974), "Bending and Buckling Strength of Tapered Structural Members," *Weld. Res. Suppl.*, Vol. 53, February, pp. 75–84.

Rasmussen K. J. R. (2000), "Recent Research on Stainless Steel Tubular Structures," *J. Constr. Steel Res.*, Vol. 54, No. 1, pp. 75–88.

Rasmussen K. J. R. and Hancock, G. J. (1993), "Stainless Steel Tubular Members. I: Columns," *J. Struct. Eng. ASCE*, Vol. 119, No. 8, pp. 2349–2367.

Rasmussen K. J. R. and Hasham, A. S. (2001), "Tests of X- and K-joints in CHS Stainless Steel Tubes," *J. Struct. Eng. ASCE*, Vol. 127, No. 10, pp. 1183–1189.

Rasmussen K. J. R. and Rondal, J. (1997a), "Explicit Approach to Design of Stainless Steel Columns," *J. Struct. Eng. ASCE*, Vol. 123, No. 7, pp. 857–863.

Rasmussen K. J. R. and Rondal, J. (1997b), "Strength Curves for Metal Columns," *J. Struct. Eng. ASCE*, Vol. 123, No. 6, pp. 721–728.

Rasmussen K. J. R. and Rondal, J. (2000a), "Column Curves for Stainless Steel Alloys," *J. Struct. Eng. ASCE*, Vol. 54, No. 1, pp. 89–107.

Rasmussen K. J. R. and Rondal, J. (2000b), "Strength Curves for Aluminium Alloy Columns," *Eng. Struct.*, Vol. 22, No. 11, pp. 1505–1517.

Rasmussen K. J. R. and Young, B. (2001), "Tests of X- and K-joints in SHS Stainless Steel Tubes," *J. Struct. Eng. ASCE*, Vol. 127, No. 10, pp. 1173–1182.

Razdolsky, A. G. (2005), "Euler Critical Force Calculation for Laced columns," *ASCE J. Engrg. Mech.*, Vol. 131, No. 10, pp. 997–1003.

Razzaq, Z., and Chang, J. G. (1981), "Partially Restrained Imperfect Columns," *Proc. Conf. Joints Struct. Steelwork*, Teesside, England. Apr., Pentech Press, Chichester, West Sussex, England.

Robertson, A. 1925. "The Strength of Struts." *Selected Engineering Papers*, No. 28, The Institution of Civil Engineers, London.

Romstad, K. M., and Subramanian, C. V. (1970), "Analysis of Frames with Partial Connection Rigidity," *ASCE J. Struct. Div.*, Vol. 96, No. ST11, pp. 2283–2300.

Ronagh, H. R., Bradford, M. A., and Attard, M. M. (2000a), "Nonlinear Analysis of Thin-walled Members of Variable Cross-section. Part I: Theory," *Comput. Struct.*, Vol. 77, pp. 285–299.

Ronagh, H. R., Bradford, M. A., and Attard, M. M. (2000b), "Nonlinear Analysis of Thin-walled Members of Variable Cross-section. Part II: Application," *Comput. Struct.*, Vol. 77, pp. 301–313.

Rondal, J., and Maquoi, R. (1979), "Single Equation for SSRC Column Strength Curves," *ASCE J. Struct. Div.*, Vol. 105, No. ST1, pp. 247–250.

Rosenthal, F., and Skop, R. A. (1980), "Guyed Towers Under Arbitrary Loads," *ASCE J. Struct. Div.*, Vol. 106, No. ST3, pp. 679–692.

Rotter, J. M. (1982), "Multiple Column Curves by Modifying Factors," *ASCE J. Struct. Div.*, Vol. 108, No. ST7, pp. 1665–1669.

Rowe, R. S. (1958), "Amplification of Stress and Displacement in Guyed Towers," *ASCE J. Struct. Div.*, Vol. 84, No. ST6, pp. 1821.1–1821.20.

SAA (1997), *Aluminum Structures—Limit State: Design*, AS 1664.1, Standards Association of Australia, North Sydney, New South Wales, Australia.

Salmon, E. H. (1921), *Columns*, Oxford Technical Publishers, London.

Salvadori, M. G. (1951), "Numerical Computation of Buckling Loads by Finite Differences," *Trans. ASCE*, Vol. 116, pp. 590–625.

Saxena, R., Popplewell, N., Trainor, P. G. S., and Shah, A. H. (1989), "Vibration of Complex Guyed Towers," *Proc. ASME 12th Bienn. Conf. Mech. Vib. Noise*, Montreal, Quebec, Canada, Sept. 17–21, pp. 1–7.

Schmidt, J. A. (2001), "Design of Mill Building Columns Using Notional Loads," *AISC Eng. J.*, Vol. 38, No. 2, pp. 90–99.

SCI (2000), "Development of the Use of Stainless Steel in Construction," Main Work Package Reports–Vols. 1 and 2, Document RT810, Ver. 01, Steel Construction Institute, London.

Sfintesco, D. (1970), "Experimental Basis for the European Column Curves," *Constr. Met.*, No. 3, p. 5.

Sfintesco, D. (ed.) (1976), *ECCS Manual on the Stability of Steel Structures*, 2nd ed., European Convention for Constructional Steelwork, Brussels, Belgium.

Shanley, F. R. (1947), "Inelastic Column Theory," *J. Aeronaut. Sci.*, Vol. 14, No. 5, p. 261.

Sharp, M. L. (1993), *Behavior and Design of Aluminum Structures*, McGraw-Hill, New York.

Shen, L. (2002), "Strength of Bolted Ring-type Connections of Solid Round Leg Members of Guyed Communication Towers," M.A.Sc. Thesis, University of Windsor, Windsor, Ontario, Canada.

Shen, Z.-Y., and Lu, L.-W. (1983), "Analysis of Initially Crooked, End Restrained Columns," *J. Constr. Steel Res.*, Vol. 3, No. 1, pp. 10–18.

Sherman, D. R. (1976), *Tentative Criteria for Structural Applications of Steel Tubes*, American Iron and Steel Institute, Washington, DC.

Siddiqui, M. M. R. (2005), "Additional Tests on Compressive Strength of Solid Round Steel Members Reinforced with Split Pipe," M.A.Sc. Major Paper, University of Windsor, Windsor, Ontario, Canada.

Stang, A. H., and Greenspan, M. (1948), "Perforated Cover Plates for Steel Columns: Summary of Compression Properties," *U.S. Nat. Bur. Stand. J. Res.*, Vol. 40, No. 5 RP.

Steinhardt, G. (1971), "Aluminum in Engineered Construction," *Aluminum*, No. 47 (in German).

Sugimoto, H., and Chen, W. F. (1982), "Small End Restraint Effects in Strength of H-Columns," *ASCE J. Struct. Div.*, Vol. 108, No. ST3, pp. 661–681.

Surovek, A. E., White, D. W., and Leon, R. T. (2005), "Direct Analysis for Design Evaluation of Partially Restrained Steel Framing Systems," *J. Struct. Eng. ASCE*, Vol. 131, No. 9, pp. 1376–1389.

Talja, A. and Salmi, P. (1995), "Design of Stainless Steel RHS Beams, Columns and Beam-columns." Research Notes 1619, Technical Research Centre of Finland, VTT, Espoo.

Tall, L. (1964), "Recent Developments in the Study of Column Behavior," *J. Inst. Eng. Aust.*, Vol. 36, No. 12, pp. 319–333.

Tall, L. (1966), "Welded Built-up Columns," Fritz Eng. Lab. Rep. No. 249.29, Lehigh University, Bethlehem, PA, Apr.

Tickle, V. (2004), "Compressive Strength of Solid Round Steel Leg Members of Lattice Communication Towers Strengthened with Split Pipe," M.A.Sc. Major Paper, University of Windsor, Windsor, Ontario, Canada.

Timoshenko, S. P. (1936), *Theory of Elastic Stability*, McGraw-Hill, New York.

Timoshenko, S. P., and Gere, J. M. (1961), *Theory of Elastic Stability*, 2nd ed., McGraw-Hill, New York.

Tomonaga, K. (1971), "Actually Measured Errors in Fabrication of Kasumigaseki Building," *Proc. 3rd Reg. Conf. Plan. Des. Tall Build.*, Tokyo, Sept.

Trahair, N. S. (1988), *The Behavior and Design of Steel Structures*, Chapman and Hall, London.

van den Berg, G. J. (2000), "The Effect of the Non-linear Stress-strain Behaviour of Stainless Steels on Member Capacity," *J. Constr. Steel Res.*, Vol. 54, No. 1, pp. 135–160.

van der Merwe, P., and van den Berg, G. J. (1992), "Design Criteria for Stainless Steel Structural Members," *Proc. Conf. Appl. Stainless Steel 92*, Stockholm, Sweden, June.

Vinnakota, S. (1982), "Planar Strength of Restrained Beam-Columns," *ASCE J. Struct. Div.*, Vol. 108, No. ST11, pp. 2496–2516.

Vinnakota, S. (1983), "Planar Strength of Directionally and Rotationally Restrained Steel Columns," *Proc. 3rd Int. Colloq. Stab. Met. Struct.*, Toronto, May, pp. 315–325.

Vinnakota, S. (1984), "Closure," *ASCE J. Struct. Div.*, Vol. 110, No. ST2, pp. 43–435.

Vroonland, E. J. (1971), "Analysis of Pendant-Supported Latticed Crane Booms," *SAE Trans.*, Vol. 80, Pap. 710697.

Wahba, Y. M. F., Madugula, M. K. S., and Monforton, G. R. (1993), "Combined Wind and Ice Loading on Antenna Towers," *Can. J. Civil Eng.*, Vol. 20, No. 6, pp. 1047–1056.

Wang, C.-K. (1967), "Stability of Rigid Frames with Non-uniform Members," *ASCE J. Struct. Div.*, Vol. 93, No. ST1, pp. 275–294.

White, D., W., and Kim, Y. D. (2006), "A Prototype Application of the AISC (2005) Stability Analysis and Design Provisions to Metal Building Structural Systems," Report prepared for Metal Building Manufacturers Association, School of Civil and Environmental Engineering, Georgia Institute of Technology, Jan.

White, M. W., and Thürlimann, B. (1956), "Study of Columns with Perforated Cover Plates," *AREA Bull. No. 531*.

Williamson, R. A. (1973), "Stability Study of Guyed Towers Under Ice Loads," *ASCE J. Struct. Div.*, Vol. 99, No. ST12, pp. 2391–2408.

Williamson, R. A., and Margolin, M. N. (1966), "Shear Effects in Design of Guyed Towers," *ASCE J. Struct. Div.*, Vol. 92, No. ST5, pp. 213–233.

Wiskin, G. (1995), "Recent Guyed Mast Vibrations," *Proceedings of IASS Working Group No. 4 for Masts and Towers, Winchester, England*, International Association for Shell and Spatial Structures, Madrid, Spain.

Wyly, L. T. (1940), "Brief Review of Steel Column Tests," *J. West. Soc. Eng.*, Vol. 45, No. 3, p. 99.

Yang, H., Beedle, L. S., and Johnston, B. G. (1952), "Residual Stress and the Yield Strength of Steel Beams," *Weld. J. Res. Suppl.*, Vol. 31, pp. 224–225.

Young, W. C. (1989), *Roark's Formulas for Stress and Strain*, McGraw-Hill, New York.

Young, B., and Lui, W.-M. (2006), "Tests of Cold-formed High Strength Stainless Steel Compression Members," *Thin-walled Struct.*, Vol. 44, pp. 224–234.

Yu, W. W. (1992), *Cold-Formed Steel Design*, 2nd ed., Wiley, New York.

Yura, J. A. (1971), "The Effective Length of Columns in Unbraced Frames," *AISC Eng. J.*, Vol. 8, No. 2, pp. 37–42.

Zandonini, R. (1985), "Stability of Compact Built-up Struts: Experimental Investigation and Numerical Simulation," *Costruzioni Metalliche*, No. 4.

Zeineddine, M. (2004), "Compressive Strength of Steel Pipe Leg Members Reinforced with Split Pipe(s)," M.A.Sc. Major Paper, University of Windsor, Windsor, Ontario, Canada.

Ziegler, H. (1982), "Arguments for and Against Engesser's Buckling Formulas," *Ing. Arch.*, Vol. 52, pp. 105–113.

CHAPTER 4

PLATES

4.1 INTRODUCTION

Metal compression or flexural members typically employ cross-sectional shapes which may be idealized as a composition of slender elements, or flat thin[1] plates (Fig. 4.1, except Fig. 4.1*a*). When a plate element is subjected to direct compression, bending, shear, or a combination of these stresses in its plane, the plate may buckle locally before the member as a whole becomes unstable or before the yield stress of the material is reached. Such local buckling behavior is characterized by distortion of the cross section of the member. Contrary to the notion of buckling being a sudden or discontinuous phenomenan, the almost inevitable presence of initial out-of-planeness (imperfections) results in a gradual growth of this cross-sectional distortion with no sudden discontinuity in real behavior at the theoretical critical load.

The theoretical or elastic critical local bucking load is not on its own a satisfactory basis for design. Ultimate strength of plates may be less than the critical local buckling load due to yielding or may be in excess of the critical local buckling load due to beneficial postbuckling reserve. For example, a plate loaded in uniaxial compression, with both longitudinal edges supported, will undergo stress redistribution as well as develop transverse tensile membrane stresses after buckling that provide postbuckling reserve. Thus, additional load may often be applied without structural damage. Initial imperfections in such a plate may cause deformations to begin below the buckling load, yet unlike an initially imperfect column, the plate may sustain loads greater than the theoretical buckling load. Despite its limitations, the critical local buckling load typically forms the basis for an initial evaluation of plates and is the focus of Section 4.2.

[1]A plate may be categorized as thick, thin, or membrane. In contrast to a thick plate, a thin plate assumes that transverse shear deformations are negligible when compared to bending deflections. A membrane assumes that the thickness is so small that the bending stiffness tends to vanish. This chapter deals primarily with thin plates.

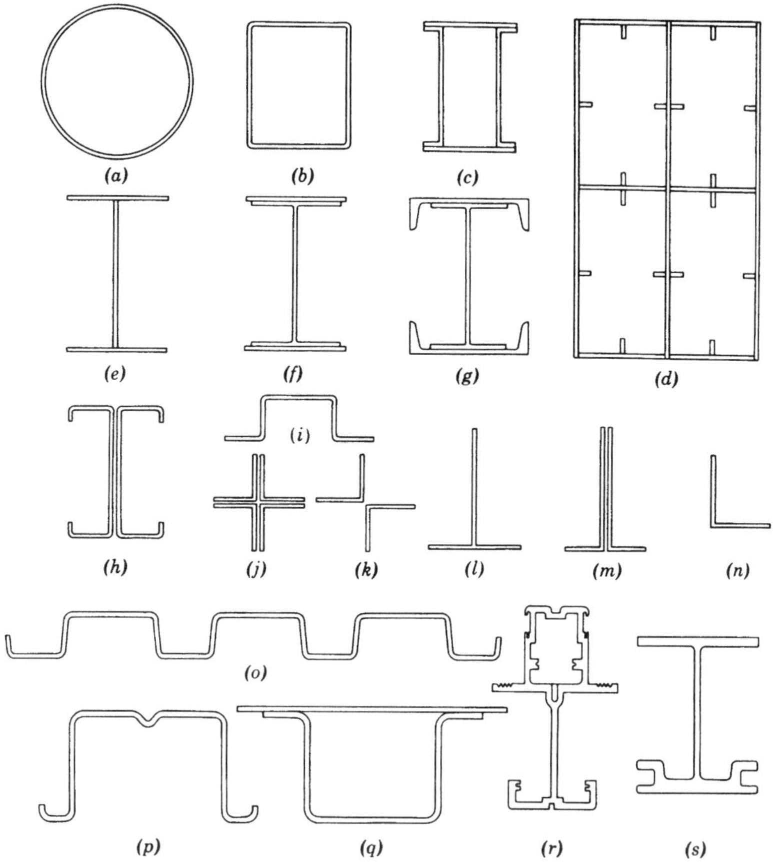

FIGURE 4.1 Compression or flexural members.

After considering elastic local bucking of flat plates in Section 4.2, the inelastic buckling, postbuckling, and ultimate strength of flat plates are discussed in Section 4.3. Particular attention is focused on the effective-width method; an approximate method for ultimate-strength determination which is still used extensively in design today. Flat plates are often augmented with stiffeners, both longitudinally and transversally, and these stiffeners improve and complicate the response as detailed in Section 4.4. One common approximation for stiffened plates is to treat the stiffened plate as a flat plate with orthotropic properties. This approximation is detailed in Section 4.5. Finally, in Section 4.6, the concluding section of this chapter, the interaction of plate elements that make up cross sections such as those of Fig. 4.1 are discussed along with the three modes of cross-sectional instability: local, distortional, and global bucking. Additional materials related to the topics presented in this chapter are provided in Chapters 6, 7, and 13 of this guide.

4.2 ELASTIC LOCAL BUCKLING OF FLAT PLATES

An examination of the buckling behavior of a single plate supported along its edges is an essential preliminary step toward the understanding of local buckling behavior of plate assemblies. The buckling stresses are obtained from the concept of bifurcation of an initially perfect structure. In practice, however, the response of the structure is continuous due to the inevitable presence of initial imperfections. Thus the critical stress is best viewed as a useful index to the behavior, as thin plates can carry additional loads well after initial buckling. Postbuckling and strength of plates are discussed in Section 4.3.

When the member cross section is composed of various connected elements (see Fig. 4.1), a lower bound of the critical stress can be determined by assuming, for each plate element, a simple support condition for each edge attached to another plate element or a free condition for any edge not so attached. The smallest value of the critical stress found for any of the elements is a lower bound of the critical stress for the cross section. Because this lower bound approximation may be excessively conservative for many practical design situations, more thorough methods are discussed in Section 4.6.

4.2.1 Uniform Compression

Long Rectangular Plates In 1891, Bryan presented the analysis of the elastic critical stress for a rectangular plate simply supported along all edges and subjected to a uniform longitudinal compressive stress. The elastic critical stress of a long plate segment is determined by the plate width-to-thickness ratio b/t, by the restraint conditions along the longitudinal boundaries, and by the elastic material properties (elastic modulus E and Poisson's ratio ν). The elastic critical stress, σ_c, is expressed as

$$\sigma_c = k\frac{\pi^2 E}{12(1-\nu^2)(b/t)^2} \tag{4.1}$$

in which k is a *plate buckling coefficient* determined by a theoretical critical-load analysis; k is a function of plate geometry and boundary conditions such as those shown in Fig. 4.2.

Short Plates When a plate element is relatively short in the direction of the compressive stress, there may exist an influence in the elastic buckling stress due to the fact that the buckled half-waves which take integer values are forced into a finite-length plate. Figure 4.3 demonstrates how k varies as a function of normalized plate length; the variation is a function of the plate boundary conditions and the loading. Full analytical solutions for k as a function of a/b and m may be found in Timoshenko and Gere (1961), Allen and Bulson (1980), and others. When a plate

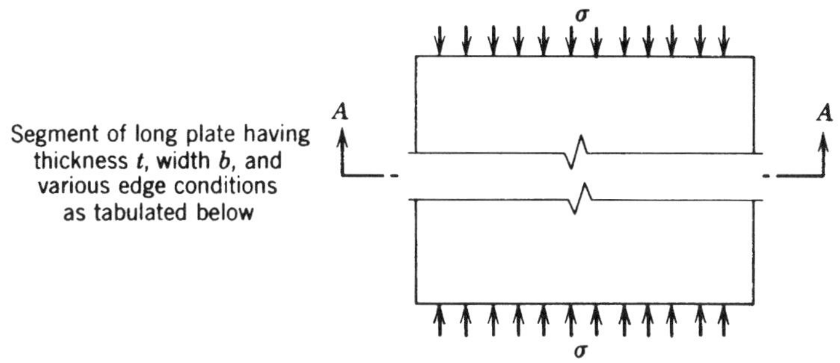

Case	Description of edge support	k
1	Both edges simply supported	4.00
2	One edge simply supported, the other fixed	5.42
3	Both edges fixed	6.97
4	One edge simply supported, the other free	0.425
5	One edge fixed, the other free	1.277

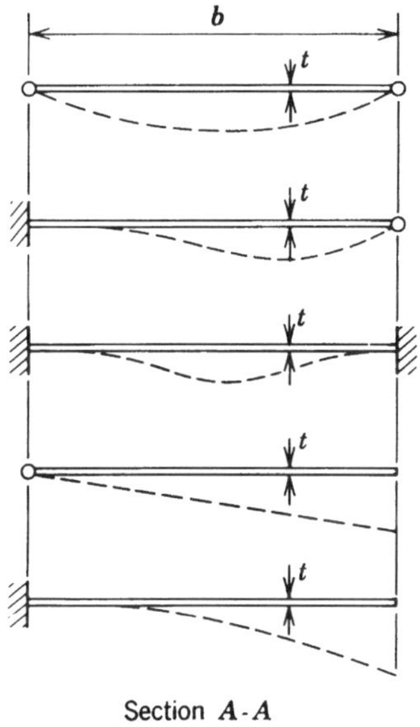

FIGURE 4.2 Local plate buckling coefficient, k of Eq. 4.1, for plates in compression with varied boundary conditions.

element is very short in the direction of the compressive stress (i.e., $a/b \ll 1$), the critical stress may be conservatively estimated by assuming that a unit width of plate behaves like a column.

4.2.2 Compression and Bending

When compression plus bending loads are applied to a structural member, plate elements of the member can be subjected to in-plane stresses which vary along the loaded edges of the plate, from a maximum compressive stress, σ_1, to a minimum stress, σ_2, as shown in Fig. 4.4. For this situation, elastic critical plate stresses are

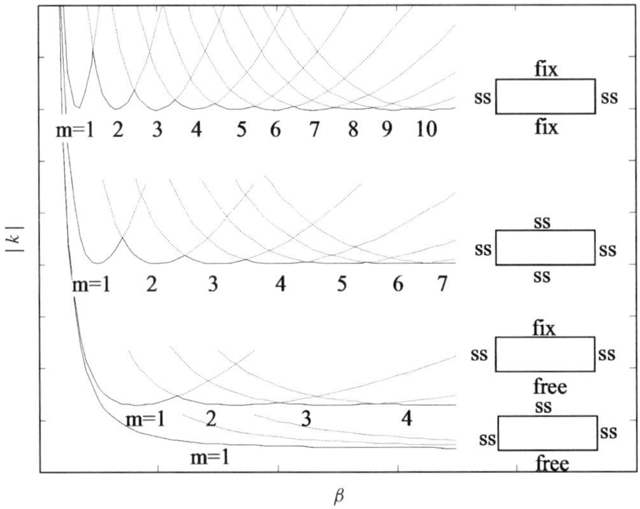

FIGURE 4.3 Plate buckling coefficient, k, as a function of normalized plate length ($\beta = a/b$) for different boundary conditions, m = number of buckled half-waves along the length of the plate (Yu and Schafer, 2007).

Loading	Ratio of Bending Stress to Uniform Compresion Stress σ_{cb}/σ_c	Minimum Buckling Coefficient, $^*k_c$					
		Unloaded Edges Simply Supported	Unloaded Edges Fixed	Top Edge Free		Bottom Edge Free	
				Bottom Edge Simply Supported	Bottom Edge Fixed	Top Edge Simply Supported	Top Edge Fixed
$\sigma_2 = -\sigma_1$	∞ (pure bending)	23.9	39.6	0.85	2.15		
$\sigma_2 = -2/3\sigma_1$	5.00	15.7					
$\sigma_2 = -1/3\sigma_1$	2.00	11.0					
$\sigma_2 = 0$	1.00	7.8	13.6	0.57	1.61	1.70	5.93
$\sigma_2 = 1/3\sigma_1$	0.50	5.8					
$\sigma_2 = \sigma_1$	0.0 (pure compression)	4.0	6.97	0.42	1.33	0.42	1.33

*Values given are based on plates having loaded edges simply supported and are conservative for plates having loaded edges fixed.

FIGURE 4.4 Plate buckling coefficients for long plates under compression and bending (Brockenbrough and Johnston, 1974; Bijlaard, 1957).

dependent on the edge support conditions and the ratio of bending stress to uniform compression stress. Long plate values of k_c that can be substituted for k in Eq. 4.1 are tabulated in Fig. 4.4 for several cases. For plates with a free edge the k_c values vary slightly with Poisson's ratio.

Beyond the k values provided in Fig. 4.4 closed-form expressions also exist for the compression and bending case. For plates simply supported on all four sides, k may be found based on the work of Peköz (1987) to be

$$k_c = 4 + 2(1 - \psi)^3 + 2(1 - \psi) \tag{4.2}$$

where $\psi = \sigma_1/\sigma_2 = f_1/f_2$. For plates with one longitudinal edge free, Bambach and Rasmussen (2004a) provide a series of solutions summarized in Fig. 4.5. These closed-form expressions for k are employed in the AISI (2007) specification. Diagrams for buckling coefficients for rectangular plates under combined bending and compressive stresses in two perpendicular directions are given by Yoshizuka and Naruoka (1971).

4.2.3 Shear

When a plate is subjected to edge shear stresses as shown in Fig. 4.6, it is said to be in a state of *pure shear.* The critical shear buckling stress can be obtained by

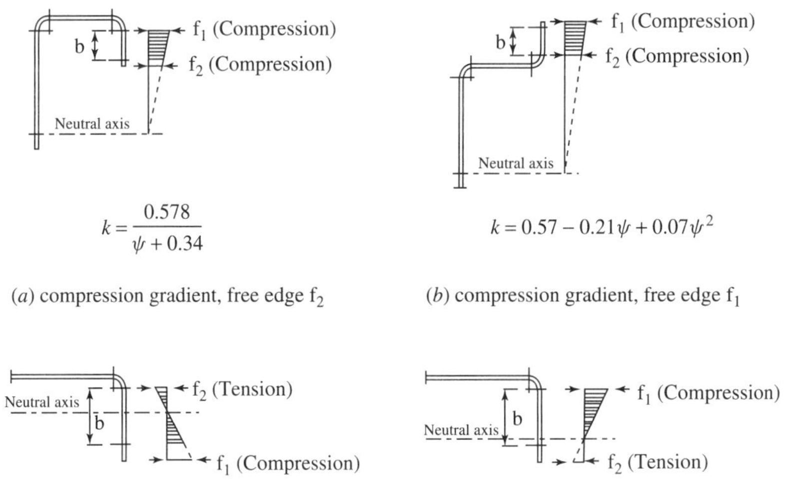

$k = \dfrac{0.578}{\psi + 0.34}$

(*a*) compression gradient, free edge f_2

$k = 0.57 - 0.21\psi + 0.07\psi^2$

(*b*) compression gradient, free edge f_1

$k = 0.57 - 0.21\psi + 0.07\psi^2$

(*c*) bending, free edge in compression

$k = 1.70 - 5\psi + 17.1\psi^2$

(*d*) bending, free edge in tension

notes : $\psi = |f_2/f_1|,\ f_1 \geq f_2$

FIGURE 4.5 Plate buckling coefficients for unstiffened elements in compression and bending (Bambach and Rasmussen, 2004a).

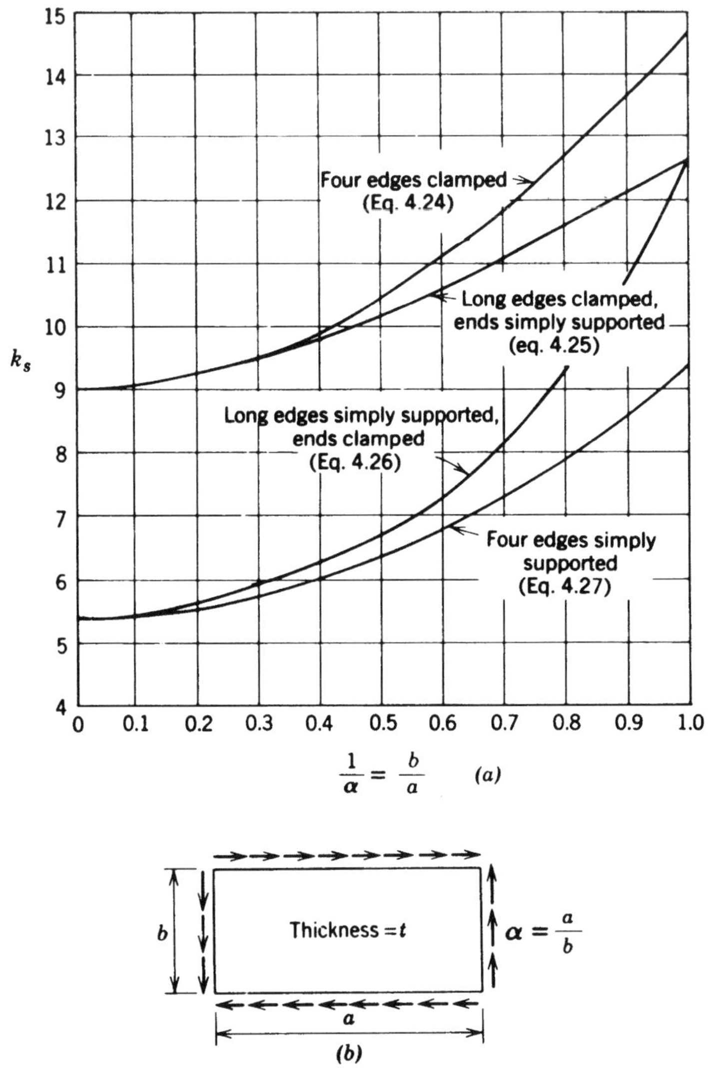

FIGURE 4.6 Plate buckling coefficients for plates in pure shear (side b is shorter side).

substituting τ_c and k_s for σ_c and k in Eq. 4.1, in which k_s is the buckling coefficient for shear buckling stress. Critical stress coefficients k_s for plates subjected to pure shear have been evaluated for three conditions of edge support. In Fig. 4.6 these are plotted with the side b, as used in Eq. 4.1, always assumed shorter in length than side a. Thus α is always greater than 1 and by plotting k_s in terms of $1/\alpha$, the complete range of k_s can be shown and the magnitude of k_s remains manageable for small values of α. For application to plate-girder design, however, it is convenient to define b (or h in plate-girder applications) as the vertical dimension of the plate-girder web for a horizontal girder. Then α may be greater or less than unity and empirical formulas for k_s together with source data are as follows.

Plate Simply Supported on Four Edges Solutions developed by Timoshenko (1910), Bergmann and Reissner (1932), and Seydel (1933) are approximated by Eqs. 4.3a and 4.3b, in which $\alpha = a/b$:

$$k_s = \begin{cases} 4.00 + \dfrac{5.34}{\alpha^2} & \text{for } \alpha \le 1 \quad (4.3a) \\ 5.34 + \dfrac{4.00}{\alpha^2} & \text{for } \alpha \ge 1 \quad (4.3b) \end{cases}$$

Plate Clamped on Four Edges In 1924, Southwell and Skan obtained $k_s = 8.98$ for the case of the infinitely long rectangular plate with clamped edges (edges restrained from out-of-plane rotation). For the finite-length rectangular plate with clamped edges, Moheit (1939) obtained

$$k_s = \begin{cases} 5.6 + \dfrac{8.98}{\alpha^2} & \text{for } \alpha \le 1 \quad (4.4a) \\ 8.98 + \dfrac{5.6}{\alpha^2} & \text{for } \alpha \ge 1 \quad (4.4b) \end{cases}$$

Plate Clamped on Two Opposite Edges and Simply Supported on Remaining Two Edges A solution for this problem has been given by Iguchi (1938) for the general case and by Leggett (1941) for the specific case of the square plate. Cook and Rockey (1963) later obtained solutions considering the antisymmetric buckling mode, which was not considered by Iguchi. The expressions below were obtained by fitting a polynomial equation to the Cook and Rockey results as shown in Bulson (1970).

For long edges clamped

$$k_s = \begin{cases} \dfrac{8.98}{\alpha^2} + 5.61 - 1.99\alpha & \text{for } \alpha \le 1 \quad (4.5a) \\ 8.98 + \dfrac{5.61}{\alpha^2} - \dfrac{1.99}{\alpha^3} & \text{for } \alpha \ge 1 \quad (4.5b) \end{cases}$$

and for short edges clamped

$$k_s = \begin{cases} \dfrac{5.34}{\alpha^2} + \dfrac{2.31}{\alpha} - 3.44 + 8.39\alpha & \text{for } \alpha \le 1 \quad (4.6a) \\ 5.34 + \dfrac{2.31}{\alpha} - \dfrac{3.44}{\alpha^2} + \dfrac{8.39}{\alpha^3} & \text{for } \alpha \ge 1 \quad (4.6b) \end{cases}$$

Curves for $\alpha \ge 1$ are plotted in Fig. 4.6. Tension and compression stresses exist in the plate, equal in magnitude to the shear stress and inclined at 45°. The destabilizing influence of compressive stresses is resisted by tensile stresses in the perpendicular direction, often referred to as "tension-field action." Unlike the case of edge compression, the buckling mode is composed of a combination of several waveforms, complicating the buckling analysis for shear.

4.2.4 Shear and Compression

The case of shear combined with longitudinal compression with all sides simply supported was treated by Iguchi (1938). His results are approximated by the interaction equation

$$\frac{\sigma_c}{\sigma_c^*} + \left(\frac{\tau_c}{\tau_c^*}\right)^2 = 1 \tag{4.7}$$

where σ_c^* and τ_c^* denote critical stress, respectively, under compression or shear alone.

Equation 4.7 is shown graphically in Fig. 4.7 and only applies for ratios of a/b greater than unity. Batdorf and associates (Batdorf and Houbolt, 1946; Batdorf and Stein, 1947) have shown that when the loaded side b is more than twice as long as a, Eq. 4.7 becomes overly conservative. This situation is the exception in actual practice, and Eq. 4.7 may be accepted for engineering design purposes.

4.2.5 Shear and Bending

For a plate simply supported on four sides, under combined bending and pure shear, Timoshenko (1934) obtained a reduced k_c as a function of τ_c/τ_c^* for α values of 0.5, 0.8, and 1.0, where τ_c is the actual shearing stress and τ_c^* is the buckling stress for pure shear. This problem was also solved by Stein (1936) and Way (1936) whose results for four values of α are plotted in Fig. 4.8. Chwalla (1936) suggested the following approximate interaction formula, which agrees well with the graphs of Fig. 4.8:

$$\left(\frac{\sigma_{cb}}{\sigma_{cb}^*}\right)^2 + \left(\frac{\tau_c}{\tau_c^*}\right)^2 = 1 \tag{4.8}$$

For a plate simply supported on four sides, under combined bending and direct stress at the ends (of dimension b), combined with shear, an approximate evaluation of the critical combined load is obtained by use of the three-part interaction formula (Gerard and Becker, 1957/1958)

$$\frac{\sigma_c}{\sigma_c^*} + \left(\frac{\sigma_{cb}}{\sigma_{cb}^*}\right)^2 + \left(\frac{\tau_c}{\tau_c^*}\right)^2 = 1 \tag{4.9}$$

The foregoing problem, with the further addition of vertical compressive force along the top and bottom edges of length a, has been treated by McKenzie (1964) with results given in the form of interaction graphs. The results are in good

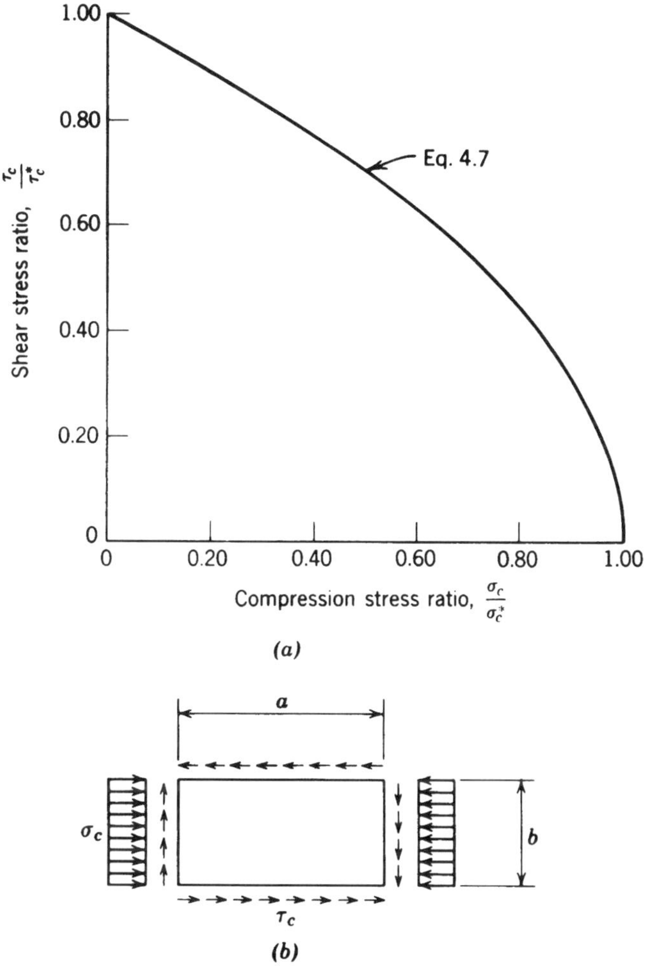

FIGURE 4.7 Interaction curve for buckling of flat plates under shear and uniform compression.

agreement with the special case of Eq. 4.9. Interaction equation 4.9, valid when a/b is greater than unity, is shown graphically in Fig. 4.9 as presented in Brockenbrough and Johnston (1974).

4.2.6 Biaxial Compression

Pavlovic and Baker (1989) presented an exact solution for the stability of a rectangular plate under biaxial compression. The case when both longitudinal and transverse stresses were uniform was used as a starting point to investigate the much more complex problem of partial loading on two opposite edges. A parametric

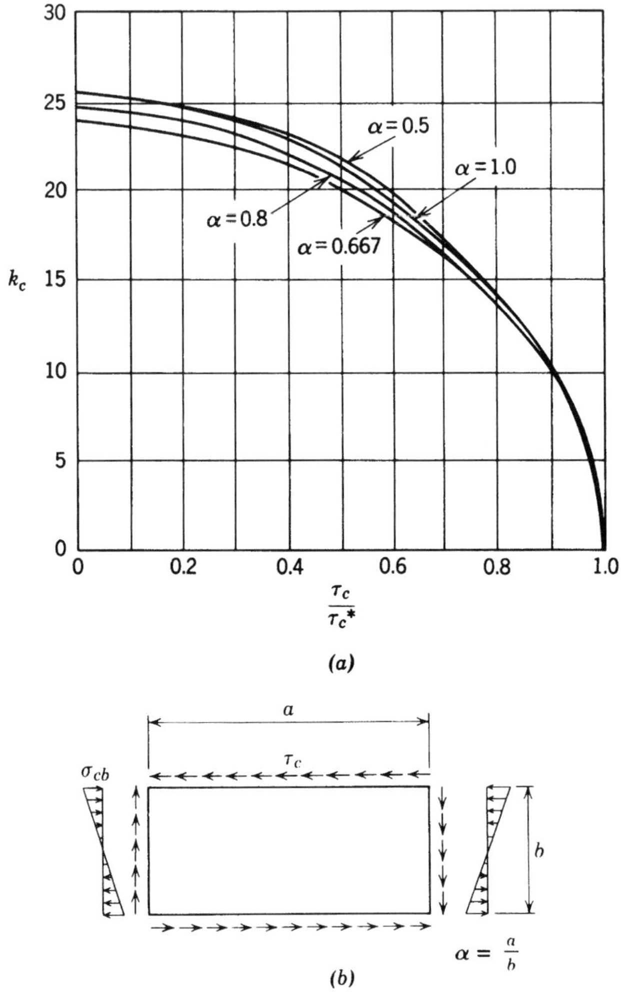

FIGURE 4.8 Buckling coefficients for plates in combined bending and shear.

study was carried out covering different plate geometrics and load ratios and varying edge lengths over which the applied load acted. The limiting cases of very long and very wide plates were considered in the study, as was the problem in which two opposite edges were subjected to concentrated forces.

4.2.7 Longitudinally Varying Compression

When shapes such as those of Fig. 4.1 are used in beams, they are typically exposed to a moment gradient. Under a moment gradient, the plate which compromises the

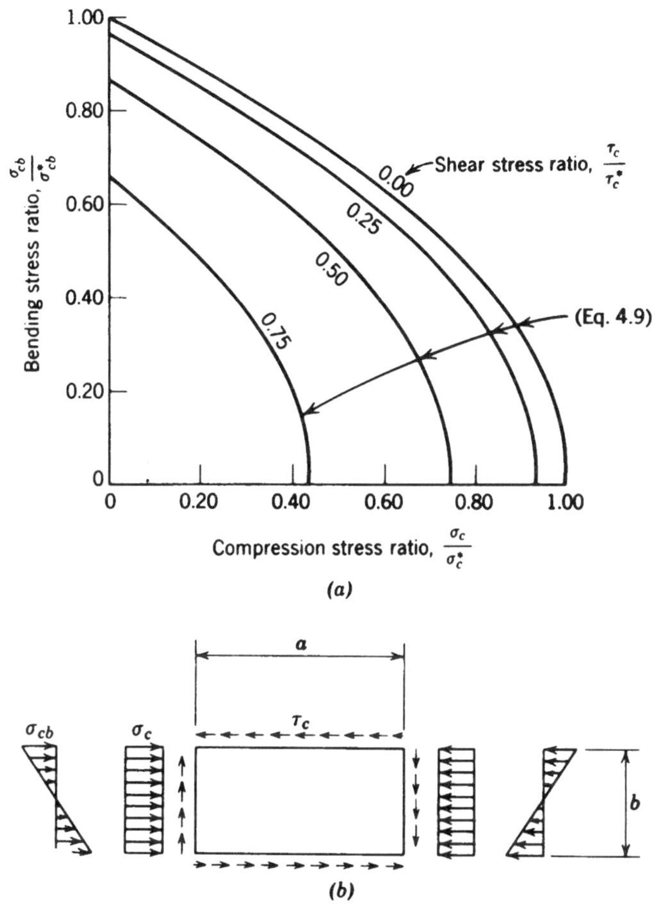

FIGURE 4.9 Interaction curve for buckling of flat plates under shear, compression, and bending.

compression flange is subject to a longitudinal variation in the compression force, as shown in Fig. 4.10.

This case was first treated by Libove et al. (1949), and recently Yu and Schafer (2007) provided a closed-form expression for k,

$$k = k_\infty + \frac{\alpha_1 r + \alpha_2}{\alpha_3 r + \alpha_4 + \beta^{\alpha_5}} \tag{4.10}$$

where k_∞ is the traditional plate-buckling coefficient (i.e., k for pure compression as the plate length tends to infinity), r is the stress gradient $(\sigma_{\min}/\sigma_{\max})$, β is the plate aspect ratio (a/b), and α_1 through α_5 are empirical coefficients dependent on the plate boundary conditions along the unloaded longitudinal edges. Table 4.1 provides these coefficients for several support conditions.

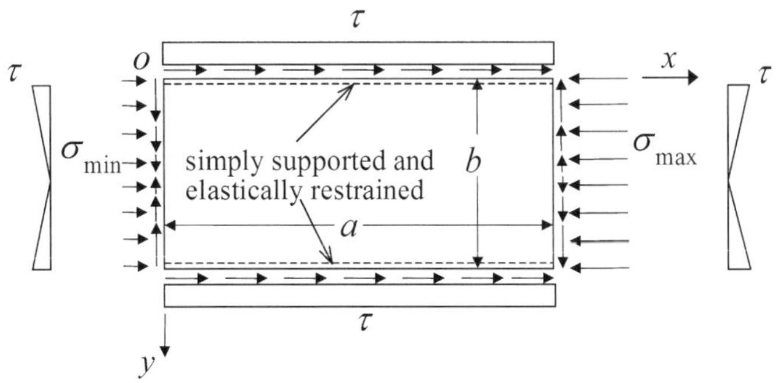

FIGURE 4.10 Simply supported plate subject to a longitudinal stress gradient; note that shear stresses are required for equilibrium (Yu and Schafer, 2007).

TABLE 4.1 Coefficients for Plate Buckling under Longitudinal Stress Gradients

	k_∞	α_1	α_2	α_3	α_4	α_5
ss–ss	4.000	−1.70	1.70	0.20	−0.20	0.75
fix–fix	6.970	−2.20	2.20	0.20	−0.20	0.65
ss–free	0.425	−0.80	1.00	0.00	−0.60	0.95
fix–free	1.277	−0.60	0.60	0.00	−0.65	0.60

Note: applicable for $1 \leq \beta \leq 30$ and $-1 \leq r \leq 1$, ss = simply supported.

4.2.8 Plates on Foundations

One method often used to constrain local buckling in metal plates is to position the thin plate adjacent to a foundation of some sort; for example, a thin steel plate may be adjacent to a concrete beam or the bottom flange of a beam may rest against a floor, foundation slab, or the ground. If the foundation is two way (i.e., works in tension and compression), then the buckling load essentially follows the foundation stiffness, and traditional methods such as those of Timoshenko and Gere (1961) may readily be used with the energy from the foundation included in the formulation. The more common case is a foundation that is only engaged when the thin plate buckles into the foundation; this is known as a tensionless or one-way foundation. The elastic critical local buckling load for this case has received some attention. In particular, the work of Smith et al. (1999) provide plate-buckling coefficients, k, for one-way foundations in compression, bending, shear, and combinations thereof. An important practical finding that may be obtained from Shahwan and Waas (1998) is that the increase in the elastic critical local buckling load, even for a rigid one-way foundation, is limited. For instance, in uniform compression the increase in k due to a rigid one-way foundation is only to 5.33 (compared to 4.0). This suggests that one-way foundations provide only a limited increase in local buckling capacity.

4.2.9 Singly Curved Plates

It is not uncommon that a member with one of the shapes of Fig. 4.1 may be curved either in plan or profile. In such a case the plates making up the cross section are singly curved. Featherston and Ruiz (1998) provide a summary of the fundamental work in this area as well as additional testing to verify the available plate-buckling coefficients. A common application for this situation is the horizontally curved steel I-girder treated in Chapter 9 of this guide. Recent work on elastic buckling of plates in I-girders (Davidson et al., 1999) is detailed in Chapter 9.

4.2.10 Square Plate with a Central Hole

Designers frequently find it necessary to introduce openings in the webs of girders and other large plate structures. The introduction of an opening changes the stress distribution within the member and will, in many instances, also change the mode of failure. Buckling is a key aspect in the behavior of thin perforated plates. This section covers the case of a square plate with a single hole, while the following section covers the case of a rectangular plate with multiple holes.

Compression Though limited in many respects the first problem to see significant study in this area was that of a square plate with a central hole and either simply supported or clamped-edge conditions (Levy et al., 1947; Kumai, 1952; Schlack, 1964; Kawai and Ohtsubo, 1968; Fujita et al., 1970). These studies indicate reductions in the plate-buckling coefficient of 25% from the unperforated k for holes on the order of 50% of the width of the square plate. Square holes are shown to result in greater reductions than round holes of the same diameter (Yang, 1969). Uniform stress loading is shown to be more critical than uniform displacement loading on the member edges. Also, it has been demonstrated that by suitably reinforcing the hole, it is possible to increase the critical stress beyond that of the unperforated plate (Levy et al., 1947).

Shear The buckling of a square plate with a central circular cutout has been examined by Rockey et al. (1969) using the finite element method. The relationship between the buckling stress of the plate and the relative size of the hole (d/b) was obtained for both simply supported and clamped-edge conditions. Rockey's work suggests a simple linear relationship between the critical stress and the d/b ratio in the form

$$\tau_{cp} = \tau_c \left(1 - \frac{d}{b}\right) \tag{4.11}$$

where τ_{cp} and τ_c are the critical stresses for the perforated plate and unperforated plate, respectively. The relationship holds for both clamped and simply supported end conditions (Fig. 4.11). Shear buckling of square perforated plates was also investigated by Grosskurth et al. (1976) using the finite element approach. They considered the case of uniform shear deformation instead of uniform shear stress

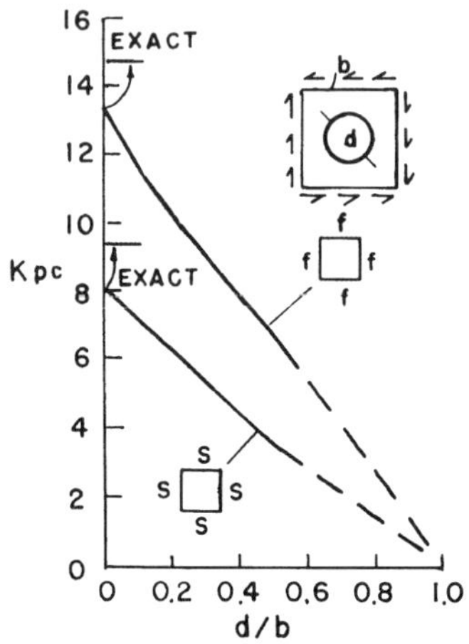

FIGURE 4.11 Simply supported and clamped plates with hole under shear loading (Rockey et al., 1969).

and obtained critical stresses that were in closer agreement with (although larger in magnitude than) the experimental values.

The behavior of plates with cutouts reinforced by a ring formed by a pressing process was also studied by Rockey, both analytically and experimentally (Rockey, 1980). It was found that the buckling stress increases with t_r/t, the ratio of the depth of the lip to the plate thickness, and the larger the hole, the greater must be the t_r/t ratio to achieve a buckling strength equal to that of the unperforated plate.

Combined Loads Elastic buckling of square plates with holes under combined loads of compression, bending, and shear are considered by Brown and Yettram (1986).

4.2.11 Rectangular Plate with Multiple Holes

Compression Recent research on elastic buckling of rectangular plates with holes under compression loading provide a more nuanced picture than the results on square plates presented in the previous section. In a square plate with a hole the buckling mode shape is constricted to the width of the plate; in rectangular plates longer buckling mode shapes are possible and likely. Research employing thin-shell finite element analysis shows that the presence of holes can either increase or decrease the critical elastic buckling stress as well as change the length and quantity of the buckled half-waves, depending upon the quantity of hole material removed relative to the size of the plate as well as the hole spacing (May and Ganaba, 1988; Brown, 1990; Brown and Yettram, 2000; El-Sawy and Nazmy, 2001; Moen and Schafer, 2006).

However, the research on square plates with holes, in particular Kawai and Ohtsubo (1968), did lead to a useful approximation of elastic buckling stress for

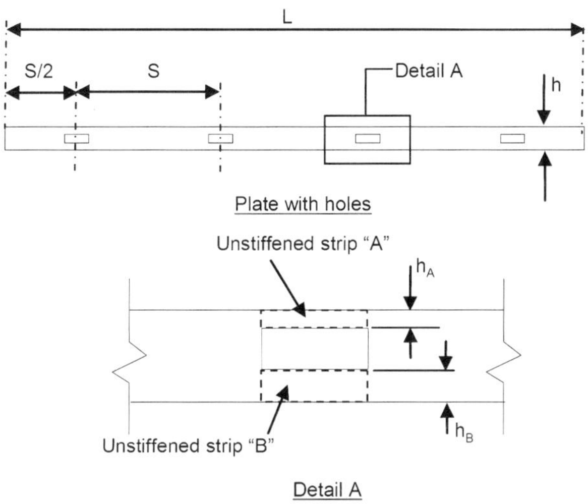

FIGURE 4.12 Illustration of unstiffened strips adjacent to a hole.

rectangular plates with holes, based on assuming the strips adjacent to the hole act as unstiffened elements (i.e., plates with one free edge) in compression (see Fig. 4.12). For a plate in uniform compression the "unstiffened strip" buckling stress uses Eq. 4.1 with h_A or h_B replacing b and $k = 0.425$ per Fig. 4.2, case 4.

The unstiffened strip approximation is useful, but one must also consider the possibility of buckling away from the hole. In addition the length of the hole, as well as its width, can influence the results. Moen and Schafer (2008) provide expressions for the plate-buckling coefficient, k, that account for these effects. Further, their work also covers the case of a rectangular plate with holes in which one longitudinal edge is supported and the other free.

Bending The problem of a rectangular plate with holes in bending has seen little attention, in part due to the fact that shear is often more critical in this situation. Moen and Schafer (2008) provide plate-buckling coefficients for stiffened and unstiffened elements under bending with holes present along the length of the plate.

Shear The problem of a long shear web with holes has been examined by Michael (1960); he suggested semiempirical expressions for the critical stress in terms of *d/a* and *a/b* (notation indicated in Fig. 4.13). These expressions are plotted in Fig. 4.13 and are applicable for the web fixed along the top and bottom edges.

Combined Loading Redwood and Uenoya (1979) have investigated the problem of webs with holes subjected to combined bending and shear. They studied the problem of shear webs with aspect ratios from 1.5 to 2.5 with circular or rectangular holes. They suggested a classic circular interaction formula for τ_c and σ_{cb}

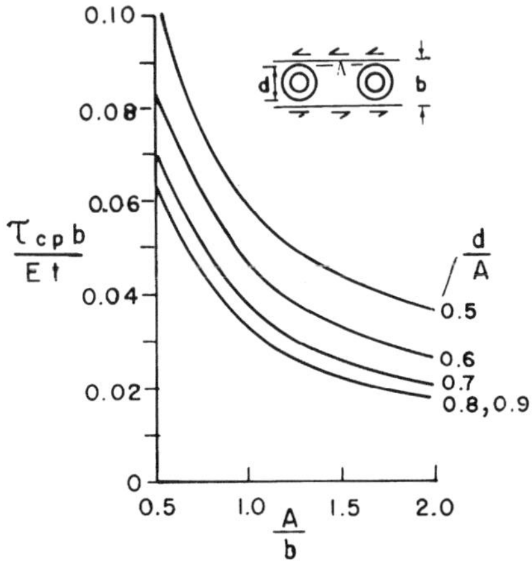

FIGURE 4.13 Critical shear stress for webs with holes (Michael, 1960).

(critical values of the maximum shear and bending stresses, respectively) in the form

$$\left(\frac{\tau_c}{\tau_{cp}}\right)^2 + \left(\frac{\sigma_{cb}}{\sigma_{cbp}}\right)^2 = 1.0 \tag{4.12}$$

in which τ_{cp} and σ_{cbp} are the pure shear and pure bending critical stresses of the plate with the hole. These, in turn, can be expressed in terms of the corresponding critical stresses of plates without holes $(\tau_c^*, \sigma_{cb}^*)$ and the relative sizes of the holes with respect to the plate dimensions. With the notation indicated in Fig. 4.14, the

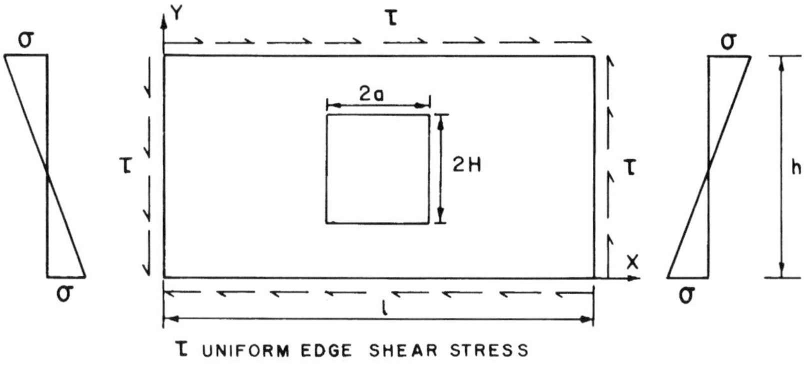

FIGURE 4.14 Perforated rectangular plate under combined action of shear and bending.

expressions for plates with rectangular holes take the form

$$\sigma_{cbp} = \left[1.02 - 0.04\left(\frac{a}{H}\right)\right]\sigma^*_{cb} \qquad \text{but} \leq \sigma^*_{cb} \tag{4.13a}$$

$$\tau_{cp} = \left[1.24 - 1.16\left(\frac{2H}{h}\right) - 0.17\left(\frac{a}{H}\right)\right]\tau^*_c \qquad \text{but} \leq \tau^*_c \tag{4.13b}$$

and for circular holes,

$$\sigma_{cbp} = \sigma^*_{cb} \tag{4.14a}$$

$$\tau_{cp} = \left[1.15 - 1.05\left(\frac{2R}{h}\right)\right]\tau^*_c \qquad \text{but} \leq \tau^*_c \tag{4.14b}$$

where R is the radius of the hole.

The values of σ^*_{cb} and τ^*_c can be obtained from a knowledge of the aspect ratio and boundary conditions of the plate. For the example of a simply supported plate with an aspect ratio of 2, these stresses are given with sufficient accuracy by the expressions

$$\sigma^*_{cb} = 23.90\ \sigma^*_c \tag{4.15a}$$

$$\tau^* = 6.59\ \sigma^*_c \tag{4.15b}$$

in which

$$\sigma^*_c = \frac{\pi^2 E}{12(1-\nu^2)}\left(\frac{t}{h}\right)^2 \tag{4.15c}$$

4.3 INELASTIC BUCKLING, POSTBUCKLING, AND STRENGTH OF FLAT PLATES

The elastic critical plate-buckling stresses, or corresponding plate-buckling coefficients (k's), provided in the previous section represent an important benchmark for understanding the behavior of thin plates. Elastic critical buckling stresses, however, do not directly indicate the actual behavior that may occur in such plates; material and geometric effects often complicate the response.

It is common, though artificial, to use the elastic critical buckling stress as a benchmark for delineating different forms of plate buckling. When material yielding occurs prior to the elastic critical buckling stress, this is known as inelastic buckling. Strengths at magnitudes greater than the elastic critical buckling stress, and the associated deformations that occur under such loading, are referred to as postbuckling and may be either elastic or inelastic. Finally, ultimate strength refers to the maximum load the plate may carry, typically independent of deformation, which may be quite large.

Actual plate response under load is more complicated than the simple notions of inelastic buckling and postbuckling, due in part to unavoidable imperfections. In an imperfect plate, out-of-plane deformations begin to amplify immediately upon loading. Such deformations lead to second-order (geometrically nonlinear) forces and strains that must be accounted for throughout the loading/deformation, and thus the notions of buckling and postbuckling are not definitively distinct. Under load the resulting stress field is complicated and varies along the length, across the width, and through the thickness of the plate. Residual stresses that may exist in the plate further complicate the response. A plate with an applied stress well below the elastic critical plate-buckling stress may still have portions of the plate yielding, and as result a definitive regime where a plate enters inelastic buckling is difficult. For materials without a distinct yield point (e.g., aluminum, stainless steel) the distinction between elastic and inelastic buckling becomes even more difficult.

Currently, inelastic buckling, postbuckling, and the strength of thin plates (and plate assemblages such as Fig. 4.1) are most robustly examined through the use of numerical methods such as finite element analysis. Finite element models for stability critical structures are discussed further in Chapter 21, but key considerations for plates include: the manner in which shear in the plate is handled (namely Kirchhoff vs. Mindlin plate theory); the material stress–strain relation including residual stresses and strains; the yield criterion employed (von Mises is by far the most common in metals); the hardening law (isotropic hardening is the most common for static loading but is inadequate if large strain reversals are present); magnitude and distribution of geometric imperfections; inclusion of higher order strain terms in the development of the plate stiffness; enforcement of equilibrium on the deformed geometry; details of the boundary conditions; and the order of the elements and the discretization of the plate in terms of both element density and element aspect ratio. Finite element analysis is not the only method able to provide postbuckling and collapse analysis of plates. The finite strip (Bradford and Azhari, 1995; Kwon and Hancock, 1991; Lau and Hancock, 1986, 1989) and more recently generalized beam theory (Goncalves and Camotim, 2007; Silvestre and Camotim, 2002) have proven to be able to provide reliable solutions.

For routine design, fully nonlinear numerical collapse analysis of thin plates remains too involved of a task; in this situation one typically turns to classical and semiempirical approaches. These design approximations are the focus of this section. In particular, the effective-width method, which is discussed in detail, has wide use as an approximate technique for determining the ultimate strength of plates that accounts for inelastic buckling and postbuckling.

4.3.1 Inelastic Buckling

The concept of inelastic buckling is an attempt to extend the elastic critical buckling approximations of Section 4.2 to situations where material yielding has already occurred. Bleich (1952) generalized the expression for the critical stress of a flat plate under uniform compressive stress in either the elastic or inelastic range as

$$\sigma_c = k \frac{\pi^2 E \sqrt{\eta}}{12(1 - \nu^2)(b/t)^2} \tag{4.16}$$

where $\eta = E_t/E$ and E_t is the tangent modulus. This modification of Eq. 4.1 to adapt it to a stress higher than the proportional limit is a conservative approximation to the solution of a complex problem that involves a continuous updating of the constitutive relations depending on the axial stress carried (Stowell, 1948; Bijlaard, 1949, 1950).

In combined loading, the work of Stowell (1949) and Peters (1954) suggests that the inelastic buckling interaction is not the same as the elastic buckling interaction. Under combined compressive and shear stress for loads applied in constant ratio, Peters found that a circular stress-ratio interaction formula as expressed by Eq. 4.17 was conservative and agreed better with experimental results than the elastic buckling interaction of Eq. 4.7:

$$\left(\frac{\sigma_c}{\sigma_c^*}\right)^2 + \left(\frac{\tau_c}{\tau_c^*}\right)^2 = 1 \tag{4.17}$$

4.3.2 Postbuckling

Postbuckling of plates may be understood through an analogy to a simple grillage model, as shown in Fig. 4.15. In the grillage model the continuous plate is replaced by vertical columns and horizontal ties. Under edge loading the vertical columns will buckle. If these columns were not connected to the ties, they would buckle at the same load with no postbuckling reserve. The ties, however, are stretched as the columns buckle outward, and thus they tend to restrain motion and in turn provide a postbuckling reserve.

In an actual plate, the tension in the transverse ties is represented by membrane tension and shear. Note also that the columns nearer to the supported edge are restrained by the ties more than those in the middle. This too occurs in an actual plate, as more of the longitudinal membrane compression is carried near the edges of the plate than in the center. Thus, the grillage model provides a working analogy for both the source of the postbuckling reserve and, more importantly, the redistribution of longitudinal membrane stresses.

The elastic postbuckling stiffness may be measured in terms of the apparent modulus of elasticity E^* (the ratio of the average stress carried by the plate to average strain, see Fig. 4.16). The following values of E^* for long plates ($a \gg b$) with several typical longitudinal edge conditions are given by Allen and Bulson (1980) and are considered sufficiently accurate up to twice the critical stress:

Simply supported longitudinal edges

Sides straight but free to move laterally

$$E^* = 0.5E \tag{4.18a}$$

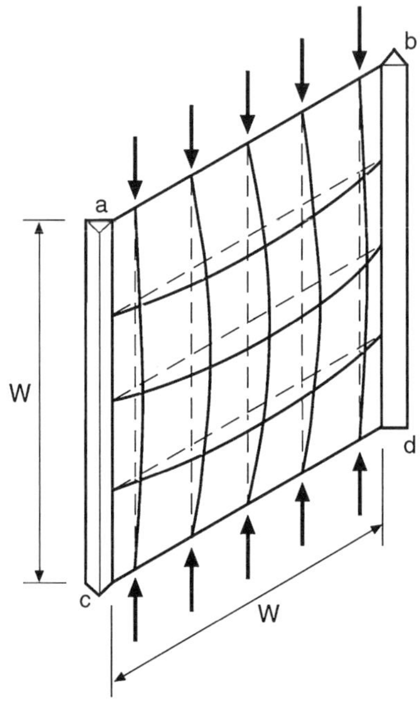

FIGURE 4.15 Simple model for postbuckling of a flat plate in uniform compression (AISI, 2007).

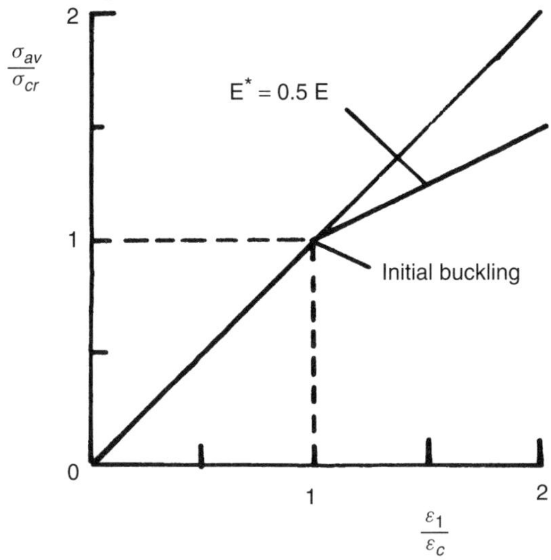

FIGURE 4.16 Postbuckling stiffness of plates having simply supported edges (Allen and Bulson, 1980).

Sides free to move

$$E^* = 0.408E \tag{4.18b}$$

Clamped (fixed) longitudinal edges

Sides straight but free to move laterally

$$E^* = 0.497E \tag{4.18c}$$

One longitudinal edge simply supported, the other free

$$E^* = 0.444E \tag{4.18d}$$

Another approach to accounting for the loss in stiffness in the postbuckling phase is to assume a correlation between the strength loss of the plate and the stiffness loss. The effective width of the plate (as defined in the following section) is used in place of the actual width for determining cross-sectional moment of inertia. This method is employed in the cold-formed steel specification (AISI, 2007) and discussed further in Chapter 13.

Nemeth and Michael (1990) presented an experimental study of the buckling and postbuckling behavior of square and rectangular compression-loaded aluminum plates with centrally located circular, square, and elliptical cutouts. The results indicated that the plates exhibit overall trends of increasing buckling strain and decreasing initial postbuckling stiffness with increasing cutout width. It was further observed that the reduction in initial postbuckling stiffness due to a cutout generally decreases as the plate aspect ratio increases. Specifically, square plates with elliptical cutouts having large cutout width–plate width ratios generally lose prebuckling and initial postbuckling stiffness as the cutout height increases.

4.3.3 Strength and Effective Width

Uniform Compression As indicated in the previous section, local buckling causes a loss of stiffness and a redistribution of stresses. Uniform edge compression in the longitudinal direction results in a nonuniform stress distribution after buckling (Fig. 4.17) and the buckled plate derives almost all of its stiffness from the longitudinal edge supports. It is worth noting that the longitudinal stress distribution of Fig. 4.17 is an idealization of the membrane stress at the cross section where the buckling wave is at maximum deformation; the variation in longitudinal membrane stress along the length of the plate and the bending stress at the face of the plate provide an entirely different stress distribution than that illustrated in Fig. 4.17.

An important semiempirical method of estimating the maximum strength of plates is by the use of the *effective-width* concept. The fact that much of the load

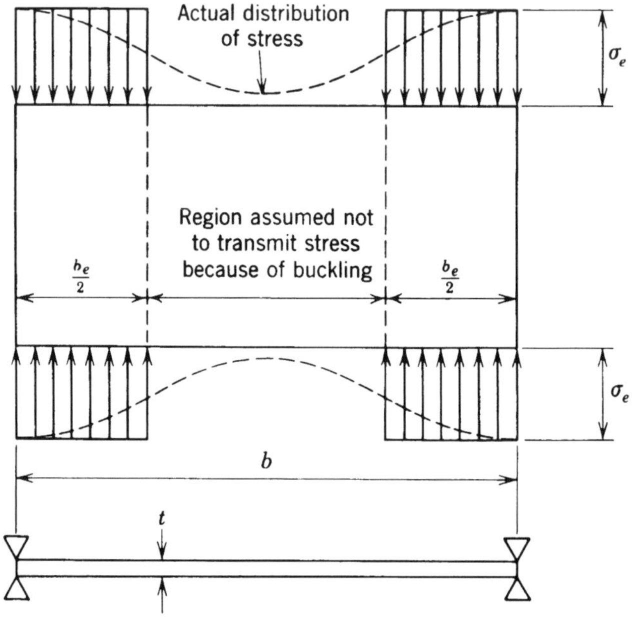

FIGURE 4.17 Longitudinal stress distribution after buckling and definition of effective width.

is carried by the region of the plate in the close vicinity of the edges suggests the simplifying assumption that the maximum edge stress acts uniformly over two "strips" of plate and the central region is unstressed (Fig. 4.17). Thus, only a fraction of the width is considered effective in resisting applied compression. It is important to note that the concept of effective width is not confined to the calculation of postbuckling strength of uniformly compressed plates. It has become the means of allowing for local buckling effects in columns, panels, or flexural members that have the dual function of supporting loads and acting as walls, partitions, bulkheads, floors, or roof decking. In a plate structure, use of the effective width leads to an effective cross section consisting of portions of members meeting along a junction. It is near these junctions that the plates will begin to yield preceding failure.

The effective-width concept has been used in design specifications for many years. Specifications of the AISI (2007), the AA (2005), and the AISC (2005a) all permit the use of an effective width in the design of members having plate elements with b/t ratios greater than the limits for full effectiveness. Implementations of the effective-width method, however, vary greatly for these three design specifications.

The effective-width concept seems to have had its origin in the design of ship plating (Murray, 1946). It had been found that longitudinal bending moments in ships caused greater deflections than those calculated using section properties based on the gross area of the longitudinal members. Deflections could be calculated more accurately by considering only a strip of plate over each stiffener having a width of 40 or 50 plate thicknesses as effective in acting with the stiffeners in resisting longitudinal bending.

The advent of all-metal aircraft construction provided another opportunity for the use of the effective-width concept, because it was advantageous to consider some of the metal skin adjacent to stiffeners as being part of the stiffener in calculating the strength of aircraft components. Cold-formed members used in steel buildings also provide useful applications of stiffened-sheet construction. A thorough discussion of the effective-width concept as applied to cold-formed steel design has been prepared by Winter (1983).

Tests by Schuman and Back (1930) of plates supported in V-notches along their unloaded edges demonstrated that, for plates of the same thickness, increasing the plate width beyond a certain value did not increase the ultimate load that the plate could support. It was observed that wider plates acted as though narrow side portions or effective load-carrying areas took most of the load. Newell (1930) and others were prompted by these tests to develop expressions for the ultimate strength of such plates. The first to use the effective-width concept in handling this problem was von Kármán et al. (1932). He derived an approximate formula for the effective width of simply supported plates, and in an appendix to his paper, Sechler and Donnell derived another formula based on slightly different assumptions. Subsequently, many other effective-width formulas have been derived, some empirical that are based on approximate analyses and some based on the large-deflection plate-bending theory that employ varying degrees of rigor.

For plates under uniform compression, stiffened along both edges parallel to the direction of the applied compression stress, von Kármán et al. (1932) developed the following approximate formula for effective width, based on the assumption that two strips along the sides, each on the verge of buckling, carry the entire load:

$$b_e = \left[\frac{\pi}{\sqrt{3(1-\nu^2)}}\sqrt{\frac{E}{\sigma_e}}\right]t \tag{4.19}$$

Combining Eqs. 4.19 and 4.1, for $k = 4$ (simple edge supports), the formula suggested by Ramberg et al. (1939) is obtained (see Fig. 4.17 for notation):

$$\frac{b_e}{b} = \sqrt{\frac{\sigma_c}{\sigma_e}} \tag{4.20}$$

From Fig. 4.17, the average stress is

$$\sigma_{\text{av}} = \frac{b_e}{b}\sigma_e \tag{4.21}$$

Substituting Eq. 4.20 into Eq. 4.21 with the edge stress equal to the yield stress ($\sigma_e = \sigma_y$) gives

$$\sigma_{\text{av}} = \sqrt{\sigma_c \sigma_y} \tag{4.22}$$

As a result of many tests and studies of postbuckling strength, Winter (1947) and Winter et al. (1950) suggested the formula for effective width that was adopted in the 1946 through 1962 editions of the AISI specifications for cold-formed steel members,

$$\frac{b_e}{t} = 1.9\sqrt{\frac{E}{\sigma_e}}\left(1 - 0.475\sqrt{\frac{E}{\sigma_e}}\frac{t}{b}\right) \tag{4.23}$$

or, alternatively, in the form of Eq. 4.5,

$$\frac{b_e}{b} = \sqrt{\frac{\sigma_c}{\sigma_e}}\left(1 - 0.25\sqrt{\frac{\sigma_c}{\sigma_e}}\right) \tag{4.24}$$

Equations 4.23 and 4.24 are basically the same as Eqs. 4.19 and 4.20, respectively, but include a correction coefficient determined from tests and reflecting the total effect of various imperfections, including initial out-of-planeness. Equation 4.23 was found to be satisfactory also for austenitic stainless steel in the annealed and flattened condition (Johnson and Winter, 1966) and for quarter- and half-hard type 301 stainless steel (Wang, 1969; Wang et al., 1975).

Introducing the coefficient $B = b/t\sqrt{\sigma_e/E}$, Winter's formula, Eq. 4.23, can be written as

$$\frac{b_e}{b} = \frac{1.90}{B} - \frac{0.90}{B^2} \tag{4.25}$$

A formula proposed by Conley et al. (1963) is nearly the same as that proposed by Winter and can be expressed as

$$\frac{b_e}{b} = \frac{1.82}{B} - \frac{0.82}{B^2} \tag{4.26}$$

A useful form of Eq. 4.25 or 4.26 is obtained by introducing the material yield strength into the dimensionless parameter B. If $\overline{B}$ is defined as

$$\overline{B} = B\sqrt{\frac{\sigma_y}{\sigma_e}} = \frac{b}{t}\sqrt{\frac{\sigma_y}{E}} \tag{4.27}$$

and if $\sigma_{\text{av}}/\sigma_e$ is substituted for b_e/b and both sides of the equation are multiplied by σ_e/σ_y, Eq. 4.26 can be written as

$$\frac{\sigma_{\text{av}}}{\sigma_y} = \frac{1.82}{\overline{B}}\sqrt{\frac{\sigma_e}{\sigma_y}} - \frac{0.82}{\overline{B}^2} \tag{4.28}$$

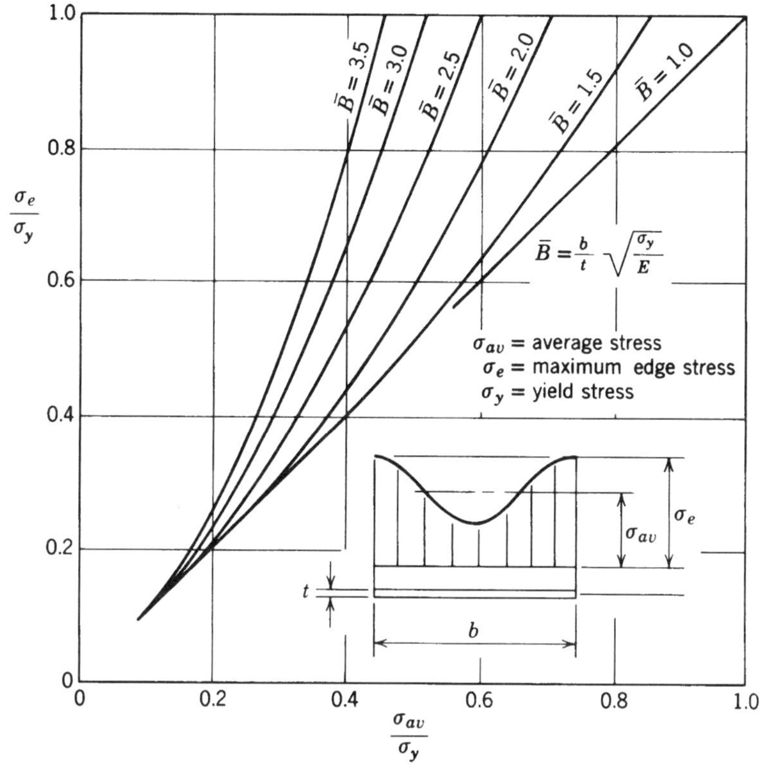

FIGURE 4.18 Chart for determining σ_e/σ_y.

By introducing discrete values of $\overline{B}$ into Eq. 4.28, the relationships shown in Fig. 4.18 between σ_{av}/σ_y and σ_e/σ_y for B values greater than 1.0 were determined. In doing so, it was assumed that there would be no loss of plate effectiveness for values of $\overline{B} \leq 1.0$ and thus the straight line from (0, 0) to (1.0, 1.0) was drawn for $\overline{B} = 1.0$. Lines of constant $\overline{B}$ when plotted fully are tangent to the $\overline{B} = 1.0$ line, and only their upper portions are shown. Thus for any given strength level of plate steel, a relationship between average stress after buckling and the maximum or edge stress of the plate panel is established as a function of the actual *b/t* ratio. This relationship is valid for stiffened plates in which the ratio of stiffener cross-sectional area to plate panel cross-sectional area is small. If the cross section of a structural member includes a buckled plate, the effective-width approach should be used in computing deflections, in determining the location of the neutral axis, or in other calculations where the effective moment of inertia or radius of gyration of the member is important.

In the 1968 and later editions of the AISI specification for cold-formed steel members (e.g., AISI, 2007), the coefficients in Eqs. 4.23 and 4.24 were reduced slightly giving the following expressions for effective width:

$$\frac{b_e}{t} = 1.9\sqrt{\frac{E}{\sigma_e}}\left(1.0 - 0.415\sqrt{\frac{E}{\sigma_e}}\frac{t}{b}\right) \tag{4.29}$$

or

$$\frac{b_e}{b} = \sqrt{\frac{\sigma_c}{\sigma_e}}\left(1.0 - 0.22\sqrt{\frac{\sigma_c}{\sigma_e}}\right) \tag{4.30}$$

The limiting value of *b/t* when all of the width is considered to be effective is obtained by setting b_e equal to b. The AISI value thus obtained from Eq. 4.14 is $(b/t)_{\text{lim}} = 221/\sqrt{\sigma}$. AISC values of effective width (2005) remain slightly more liberal than those of AISI. (See section below titled *Width-to-thickness limits* for further information.)

In the calculation of the ultimate compression load for plates supported along the two unloaded edges, σ_e is taken equal to the compressive yield stress for steel. For aluminum alloys and magnesium alloys, σ_e is taken as 0.7 times the yield strength, as determined by the offset method. If the buckling stress σ_c exceeds 0.7 times the yield strength, however, the load capacity as determined by inelastic plate-buckling analysis may be taken as $bt\sigma_c$ in which σ_c is determined by Eq. 4.16 (inelastic buckling) and the effective width need not be calculated. The use of the ultimate compressive buckling load in specifications for aluminum structures is discussed later in this section.

Equation 4.30 can be used to determine a nondimensional ultimate-strength curve for steel plates in the postbuckling range. The average stress on the plate at ultimate load, σ_{av}, is the ultimate load divided by the total area. From Eq. 4.30,

$$\sigma_{\text{av}} = \frac{P_{\text{ult}}}{bt} = \sqrt{\sigma_c\sigma_y}\left(1.0 - 0.22\sqrt{\frac{\sigma_c}{\sigma_y}}\right) \tag{4.31}$$

In Fig. 4.19 the average stress at ultimate load, by Eq. 4.31, is compared with the uniform-edge compression stresses to cause buckling. A method for predicting the strength of simply supported plates, taking into account initial out-of-flatness, is given by Abdel-Sayed (1969) and Dawson and Walker (1972). Out-of-flatness, residual stress, and strain hardening are considered by Dwight and Ratcliffe (1968).

For a plate supported along only one longitudinal edge, the effective width has been experimentally determined by Winter as

$$\frac{b_c}{b} = 1.19\sqrt{\frac{\sigma_c}{\sigma_e}}\left(1 - 0.30\sqrt{\frac{\sigma_c}{\sigma_e}}\right) \tag{4.32}$$

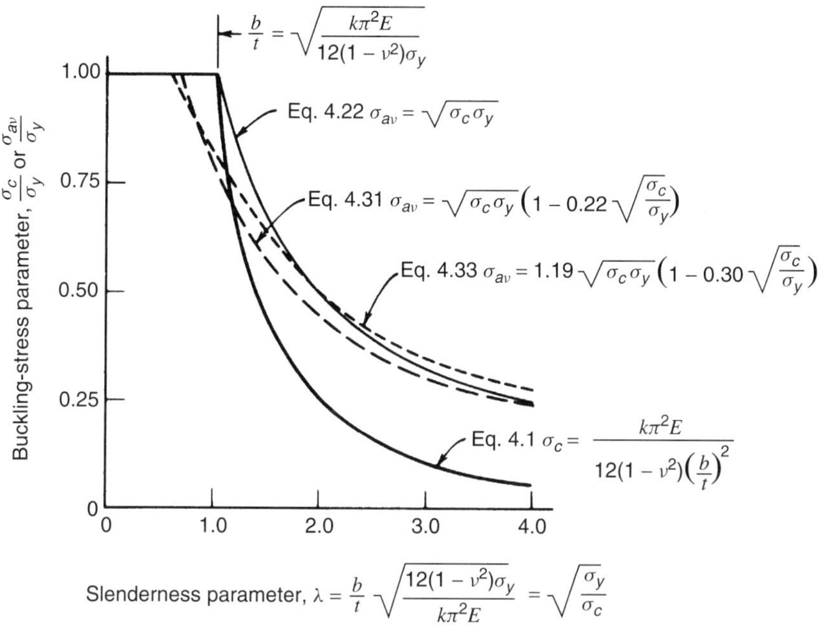

FIGURE 4.19 Nondimensional buckling curves for plates under uniform-edge compression (Brockenbrough and Johnston, 1974).

This equation has also been confirmed by an analysis carried out by Kalyanaraman and co-workers (Kalyanaraman et al., 1977; Kalyanaraman and Peköz, 1978). The average stress at ultimate load can then be expressed as

$$\sigma_{\text{av}} = 1.19\sqrt{\sigma_c\sigma_y}\left(1 - 0.30\sqrt{\frac{\sigma_c}{\sigma_y}}\right) \tag{4.33}$$

Equation 4.33 is also shown in Fig. 4.19. This curve actually falls above and to the right of Eq. 4.31, but it should be pointed out that in the elastic range σ_c for a plate supported on both longitudinal edges is about eight times as large as that for the same plate supported along only one longitudinal edge.

Considering that Eq. 4.29 is an appropriate formula for determining the effective design width of stiffened compression elements with a k value of 4.0, a generalized formula for different stiffened compression elements with various rotational edge restraints can be written as

$$\frac{b_e}{t} = 0.95\sqrt{\frac{kE}{\sigma_e}}\left(1 - 0.209\sqrt{\frac{kE}{\sigma_e}}\frac{t}{b}\right) \tag{4.34}$$

It is worth noting that Eq. 4.34 has also been extended to unstiffened elements. For example, in determining the effective width of a plate simply supported on only one longitudinal edge, $k = 0.425$ of Fig. 4.2 could be used in Eq. 4.15. This basic approach when an appropriate k value is used to determine the strength regardless of the boundary conditions (or loading) is known as the "unified method." It has been shown to provide adequate strength predictions for a variety of conditions (Peköz, 1987) and forms the basis for the cold-formed steel specification discussed further in Chapter 13.

Jombock and Clark (1962) list 14 effective-width formulas, along with their sources, and discuss the assumptions on which they are based. Because the effective-width concept is also well developed in current specifications and commentaries, it is suggested that this work be referenced for further information on this topic.

Certain box-girder-bridge structural failures in the early 1970s should alert designers of plate structures that many aspects of plate behavior need further investigation. For example, the concepts of effective width and average stress at failure consider plate strains only up to the maximum plate capacity—that is, the ultimate load as characterized by reaching the yield stress at the edge. Research has shown that inelastic strains beyond this point may lead to a sudden and substantial reduction of the plate's load-carrying capacity (Dwight and Moxham, 1969; Dwight, 1971).

Compression and Bending One approach to determining the ultimate strength of a plate in bending, or compression and bending, is to substitute the plate-buckling coefficients from Section 4.2.2 into Winter's effective-width formula (Eq. 4.34). One difficulty that remains after determination of the effective width b_e is how to distribute b_e to the plate edges. For example, for a simply supported plate in bending and designating b_e/b as ρ, Fig. 4.20 provides a rational method for distributing the effective width based on maintaining force and moment equilibrium in the effective plate (Schafer and Peköz, 1999). Other alternative methods exist, for example, LaBoube and Yu (1982), Peköz (1987) and Usami (1982). For plates with one longitudinal edge free the plate-buckling coefficients of Fig. 4.5 may be used with Eq. 4.34 to provide effective-width predictions. Distribution of the effective width is provided by Bambach and Rasmussen (2004b) and has been adopted in AISI (2007).

A direct solution for the ultimate strength of plates under compression and bending without recourse to effective width was studied and proposed by Rhodes and Harvey (1971, 1976, 1977). In the case of simply supported plates under eccentric loading P, the failure loads for plates with various loading eccentricities can be accurately predicted by a simple expression of the form (Rhodes and Harvey, 1977)

$$\overline{P}_{\text{ult}} = \frac{pb}{\pi^2 D} = \frac{\overline{\sigma}_y + 11.4}{b(e/b) + 0.85} \tag{4.35}$$

where $D = Et^3/12(1 - \nu^2)$, $\overline{\sigma}_y = \sigma_y(b^2 t/\pi^2 D)$, and e is the distance from the point of load application to the remote edge of the plate.

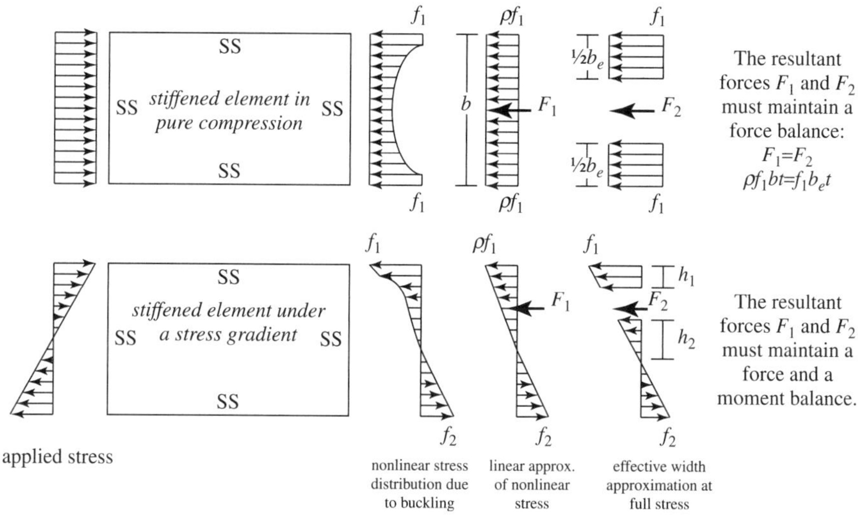

FIGURE 4.20 Distributing effective width in plates under a stress gradient (Schafer and Peköz 1999).

Longitudinally Varying Compression It has been demonstrated that the use of the plate-buckling coefficients of Section 4.2.7 in Winter's effective-width formula leads to reliable and conservative predictions of the strength of plates under longitudinally varying compression, a case that occurs in the compression flange of members subjected to moment gradient (Yu and Schafer, 2006).

Plates with Holes The simplest method for determining the ultimate strength and effective width of plates with holes is to adopt the "unstiffened strip" approach described in Section 4.2.11, in particular Fig. 4.12. The strips of plate material adjacent to the hole are treated as plates with one longitudinal edge free, that is, unstiffened strips. The plate-buckling coefficient may then be determined as a plate with one edge free (e.g., Fig. 4.2), and the effective width is calculated using Winter's formula. This approach, which was developed by Vann (1971), Yu and Davis (1973), and Miller and Peköz (1994), is essentially that adopted in AISI (2007).

Width-to-Thickness Limits In the inelastic design of steel structures, it is necessary that the moment capacity of the member not be impaired by local buckling until the required plastic rotation is achieved. This can be accomplished by limiting the width-to-thickness ratios of elements that are vulnerable to local buckling in the inelastic range. Lay (1965) has proposed such limiting width-to-thickness ratios for flanges of I-beams. Further studies of this topic can be found in the following references: Dawe and Kulak (1984, 1986), Kuhlmann (1989), and Kemp, (1996). Such provisions are also available in design specifications (AISC, 2005a). Additional width-to-thickness limits have also been adopted for seismic applications where cyclic inelasticity must be considered (AISC, 2005b). Earls (1999) shows that care must be taken in the use of these limits, particularly as new high-strength materials are introduced and the buckling limit states may not follow the simple empirical criteria previously employed.

4.3.4 Strength in Shear and Combined Loadings with Shear

The initial mode of buckling in pure shear, which takes the form of a half-wave in the tension direction and at least one full wave in the compression direction (Fig. 4.21*a*), undergoes a change in the advanced postbuckling range and eventually takes on the form of a family of diagonal folds (Fig. 4.21*b*). These folds carry significant tensile stresses developed in the postbuckling range and the displacement pattern is called a *tension field.* The dominance of this tension-field behavior has led to approaches other than the effective-width method for determining the strength of thin plates in shear.

The maximum shear load that can be applied before failure occurs is due to a loss of material strength in the tension field. This load is also influenced by the rigidity of the edge members supporting the plate. This problem is dealt with in greater detail in Chapter 6 for bridge applications and is discussed further in Section 4.3.8 within the context of steel plate shear walls for buildings.

For a plate with infinitely stiff edge members, the maximum shear strength can be estimated by the formula (Allen and Bulson, 1980)

$$\overline{V}_u = \tau_c bt + \tfrac{1}{2}\sigma_{ty} bt \tag{4.36}$$

where

$$\sigma_{ty} = \sqrt{\sigma_y^2 - 0.75\tau_c^2} - 1.5\tau_c \qquad \text{provided that } \tau_c \ll \sigma_y \tag{4.37}$$

Stein (1989) presented buckling and postbuckling results for plates loaded by in-plane shear. The buckling results had been plotted to show the effects of thickness on the stress coefficient for aluminum plates. Results were given for various length-to-width ratios. Postbuckling results for thin plates with transverse shearing flexibility were compared to results from classical theory. The plates were considered to be long with side edges simply supported, with various in-plane edge conditions, and the plates were subjected to a constant shearing displacement along the side edges.

Elangovan and Prinsze (1992) carried out a numerical investigation using a finite element buckling analysis to determine the critical shear stress of flat rectangular plates with two opposite edges free. The parameters considered in their study were plate size and the boundary conditions at the two edges loaded in shear. Results showed a considerable difference in the buckling strength of plates if the in-plane displacement normal to the loaded edges were restrained either at one or at both of those edges.

Information on the postbuckling strength of plate elements subjected to the combined action of shear and compression is limited. A semiempirical method

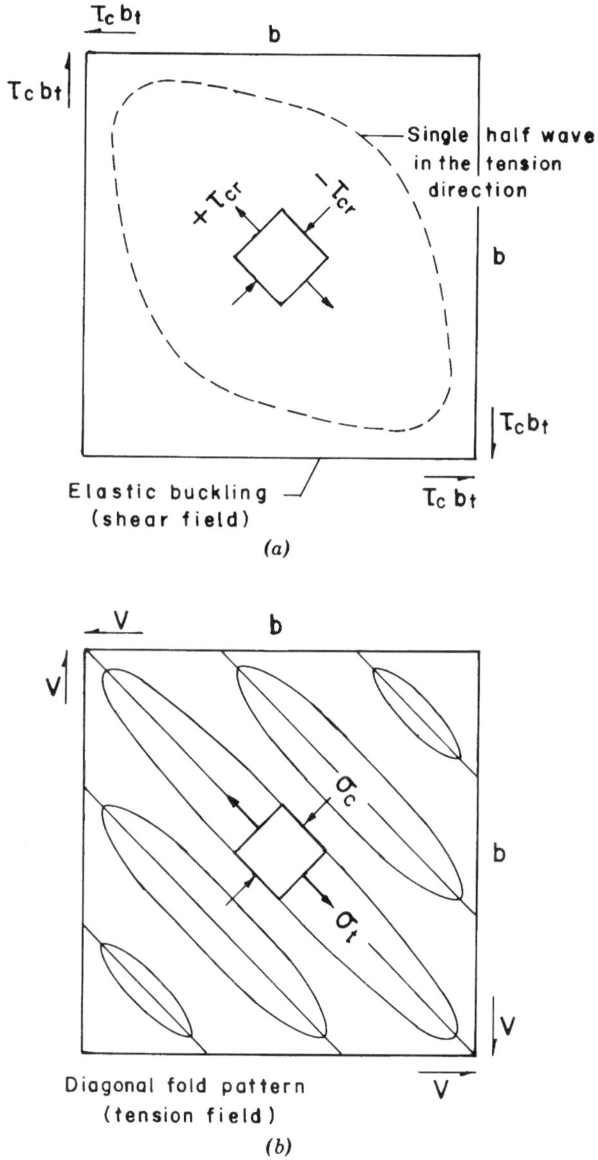

FIGURE 4.21 Shear and tension fields in square plate.

for the determination of stress levels at which permanent buckles occur in a long plate with simply supported edges under the combined action of uniform axial compression and shear has been suggested by Zender and Hall (1960). Additional information on postbuckling strength of plates subjected to the combined action of shear and bending can be found in Chapters 6 and 7.

4.3.5 Strength of Biaxially Compressed Plates

Few simplified design methods exist for postbuckling and strength prediction of biaxially compressed plates. For example, the effective-width method for plates under biaxial compression has seen little study. Isami and Hidenori (1990) employed mechanism analysis methods (see Section 4.3.7) to provide predictions for biaxially compressed plates. Postbuckling and strength prediction relies primarily on geometric and material nonlinear finite element analysis. For example, biaxial compression of ship plates with particular attention paid to the role of initial imperfections in determining the ultimate strength is studied by Paik et al. (2004). In addition, they show that the collapse behavior of biaxially compressed plates can vary significantly from those under only longitudinal compression.

4.3.6 Average Stress and Strength of Aluminum Plates

As an alternative to the effective-width concept for thin plates, another approach is to use the average stress at failure and the actual (unreduced) plate width. This is the basis for slender elements in the Aluminum Association's *Specification for Aluminum Structures* (AA, 2005). In applying this specification, the designer does not, in general, calculate an effective width but uses instead the average stress at failure for plate elements. For plates that buckle in the inelastic stress range, the average stress at failure is considered to be the same as the local buckling stress, because plates of these proportions have little postbuckling strength (Jombock and Clark, 1968). Inelastic local buckling strength for aluminum plates is represented in AA (2005) by the following straight-line approximation to Eq. 4.1 (Clark and Rolf, 1966):

$$\sigma_c = B_p - D_p k_1 \frac{b}{t} \tag{4.38}$$

where

$$B_p = \sigma_y \left[1 + \frac{(\sigma_y)^{1/3}}{k_2} \right] \tag{4.39}$$

$$D_p = \frac{(B_p)^{3/2}}{k_3 (E)^{1/2}} \tag{4.40}$$

$$k_1 = \sqrt{\frac{12(1 - \nu^2)}{k}} \tag{4.41}$$

For aluminum products that are artificially aged (temper designations beginning with T5, T6, T7, T8, or T9), $k_2 = 11.4\,\text{ksi}^{1/3}$ $(78.6\,\text{MPa}^{1/3})$ and $k_3 = 10$. For other aluminum products (temper designations beginning with 0, H, T1, T2, T3, or T4), $k_2 = 7.6\,\text{ksi}^{1/3}$ $(52.4\,\text{MPa}^{1/3})$ and $k_3 = 10\sqrt{2/3}$. Equation 4.20 has been shown to agree well with the results of tests on aluminum plate elements (Clark and Rolf,

1966; Jombock and Clark, 1968). For plates that buckle elastically, the average stress at failure is represented for purposes of the aluminum specification as

$$\sigma_{\text{av}} = \sqrt{\sigma_e \sigma_c} \tag{4.42}$$

Note that Eq. 4.42 corresponds to Eq. 4.22. Jombock and Clark (1968) demonstrated that the edge stress at failure σ_e for aluminum plates could be represented by a function of the intercept B_p in Eq. 4.38. This results in a simple relationship between the ultimate-strength curves corresponding to elastic and inelastic buckling. Generally, this edge stress at failure for aluminum alloys is about $0.7\sigma_y$.

The formulas used in the Aluminum Association Specification (AA, 2005) are illustrated in Fig. 4.22. Comparisons with test results were published by Jombock and Clark (1968). The use of the average stress at failure for slender sections results in some simplification, because the designer does not have to calculate an effective width. However, it sacrifices some of the flexibility of the effective-width approach. For example, the average stress-at-failure method does not treat the change in moment of inertia of a member when its compression elements are in the postbuckling range and hence does not readily lend itself to calculation of deflections. Therefore, the Aluminum Association Specification (AA, 2005) includes an effective-width formula to be used in calculating deflections in the postbuckling range. Chapter 13 provides further discussion on this topic.

The Aluminum Association Specification (AA, 2005) treats postbuckling strength of webs in bending by means of an average stress approach similar to

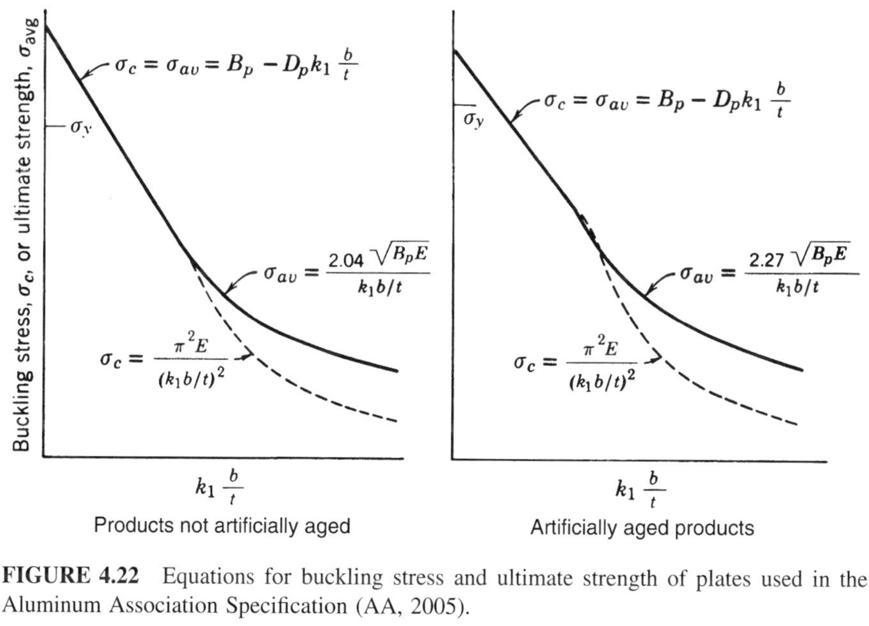

FIGURE 4.22 Equations for buckling stress and ultimate strength of plates used in the Aluminum Association Specification (AA, 2005).

that used for plates in compression. This approach is compared with test results by Jombock and Clark (1968).

4.3.7 Yield-Line Analysis and Plastic Buckling of Plates

Yield-line analysis represents an extension of a concentrated plasticity approach, such as the two dimensional plastic hinge analysis of beams. For out-of-plane loading (e.g., applied loads on floor slabs) yield-line analysis has a long history. However, for in-plane loading that has been the focus of this chapter, its use is less widespread. One of the primary difficulties with the extension to in-plane loads is that yielding mechanisms for in-plane loaded plates typically involve yielded zones (membrane extension) as well as yield lines (where bending and twist) occur. Successful applications include connection design, particularly for tubular and hollow structural shape (HSS) sections (Kosteski et al., 2003); determination of the limiting rotation for beams (Gioncu and Petcu, 1997; Shi et al., 1996); and patch loading or crippling of webs. Although Bambach (2008) provides a recent summary of the limited use of yield-line analysis for generalized collapse analysis of plates, the most complete treatments still remain those of Murray and Khoo (1981) and Murray (1984).

Inoue and Kato (1993) studied an analytical evaluation of the effective plastic shear modulus of fully yielded steel plates at the instant of buckling. It was assumed that yielding of steel was to follow the Tresca yield criterion and that plastic deformation of a steel plate was to be caused by slip planes. The Tresca yield criterion provides lower bending stiffnesses than those obtained from the von Mises yield criterion, but it does not lower the plastic shear modulus of the material at any point on the yield plateau. They proposed a new theory that assumes a nonuniform distribution of slips depending on the orientation of an infinite number of possible slip planes at each point in the plate. The twisting of the plate is then accompanied by distortion of its sectional shape, and this mode of buckling is shown to provide a considerable reduction in the effective plastic shear modulus. Applying these sectional stiffnesses and solving differential equilibrium equations lead to a lower bifurcation strength, which provides much better correlations with experimental results than those of previous predictions.

4.3.8 Energy Dissipation and Steel Plate Shear Walls

The collapse of metal plate structures generally involve either fracture or stability limit states. Stability limit states such as inelastic local plate buckling are typically preferred over fracture limit states, because they can provide some postbuckling resistance and hence offer greater ductility and the potential for energy dissipation. For seismic design, energy dissipation is a fundamental need for most structures. Rather than rely on inherent or assumed energy dissipation, modern designs often attempt to explicitly use specific element(s) for energy dissipation. A properly designed and detailed thin plate undergoing postbuckling behavior may provide reliable energy dissipation.

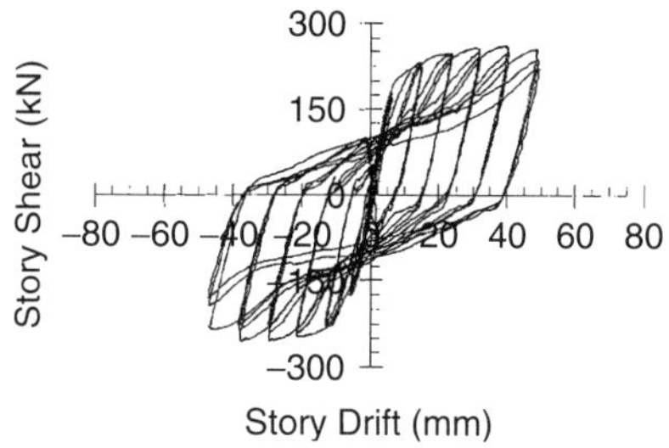

FIGURE 4.23 Hysteretic response of a steel plate shear wall under cyclic loading (Lubell et al., 2000).

Steel plate shear walls are an important example of the application of metal plate structures for the purposes of energy dissipation. Utilizing the postbuckling response of plates in shear, experimental tests under cyclic load have been used to verify the potential energy dissipation of unstiffened steel plate shear walls (Caccese et al., 1993, Elgaaly, 1998, Lubell et al., 2000). A typical hysteretic response of a steel plate shear wall is shown in Fig. 4.23. Currently phenomenological models are favored for characterizing the response of steel plate shear walls. While such models have limitations, they are readily implemented in nonlinear dynamic analysis, which allows the influence of steel plate shear walls on full structural systems to be studied (Roberts and Sabouri-Ghomi, 1992; Sabouri-Ghomi and Roberts, 1992). Recent and ongoing research indicates that perforated steel plates may offer further advantages for energy dissipation, as shown in Fig. 4.24 (Hitaka and Matsui, 2003).

4.4 BUCKLING, POSTBUCKLING, AND STRENGTH OF STIFFENED PLATES

This section deals with the buckling and ultimate strength of stiffened flat plates under various combinations of loadings. The behavior of the stiffened plate as a unit is emphasized rather than the stability of its individual elements. In design, however, the structural properties of all of the components must be considered.

Stiffened plates can fail through instability in essentially two different ways: overall buckling (also referred to as stiffener buckling or distortional buckling in the literature), which occurs when the stiffeners buckle along with the plate, and local buckling with the stiffeners forming nodal lines and the plate panels buckling between the stiffeners. In either case, the stiffness of the combination may be such that initial buckling takes place at fairly low stress levels. Nevertheless, a significant amount of postbuckling strength may remain in the stiffened plate, provided that proper attention is given to the structural design and fabrication details. A great deal of information on this subject can be found in the *Handbook of Structural Stability* (CRCJ, 1971) and in a book by Troitsky (1976).

Finally, as discussed in the introduction to Section 4.3, it is becoming increasingly common to study buckling, postbuckling, and strength through geometric

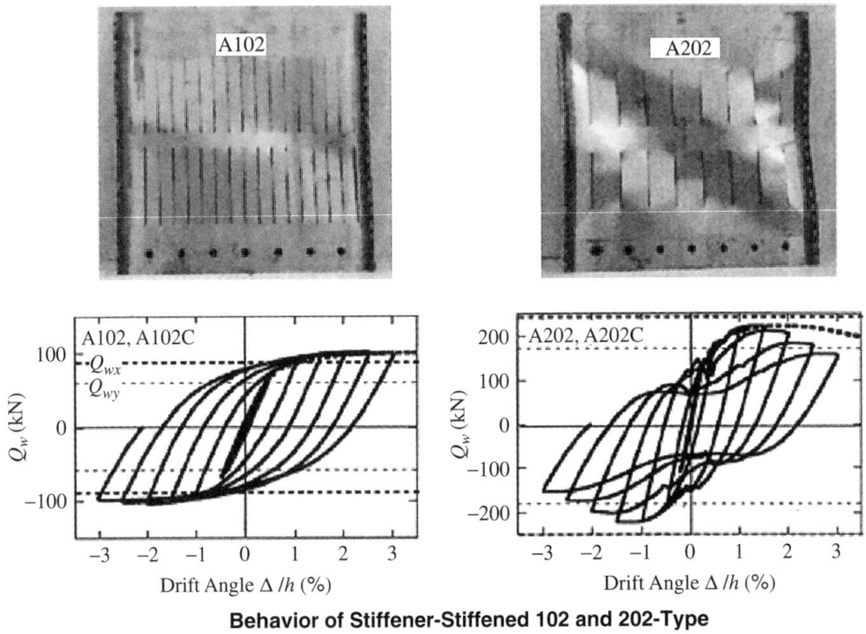

FIGURE 4.24 Hysteretic response of steel plate shear walls with perforations (Hitaka and Matsui, 2003).

and material nonlinear finite element analysis using plate or shell elements. Such models provide the capability to accurately include boundary conditions, imperfections, residual stresses, and so on, and hence are a powerful tool in the study of stiffened plates. For the most part, the discussion provided herein focuses on analytical methods and tools that can be used with far greater expediency than a complete nonlinear finite element analysis. For further discussion on stiffened plates see Chapters 6, 7, and 13, and for information on finite element modeling for stability problems see Chapter 21.

4.4.1 Compression: Buckling

It is often economical to increase the compressive strength of a plate element by introducing longitudinal and/or transverse stiffeners (Fig. 4.25). In this section, methods are presented for determining the compressive strength of stiffened-plate panels. The edges of the plate are assumed to be simply supported in all cases, and it is also assumed that individual elements of the panel and stiffeners are not subject to instability.

Longitudinal Stiffeners Seide and Stein (1949), Bleich and Ramsey (1951), and Timoshenko and Gere (1961) present charts and tables for determining the critical stress of plates simply supported on all edges and having one, two, or

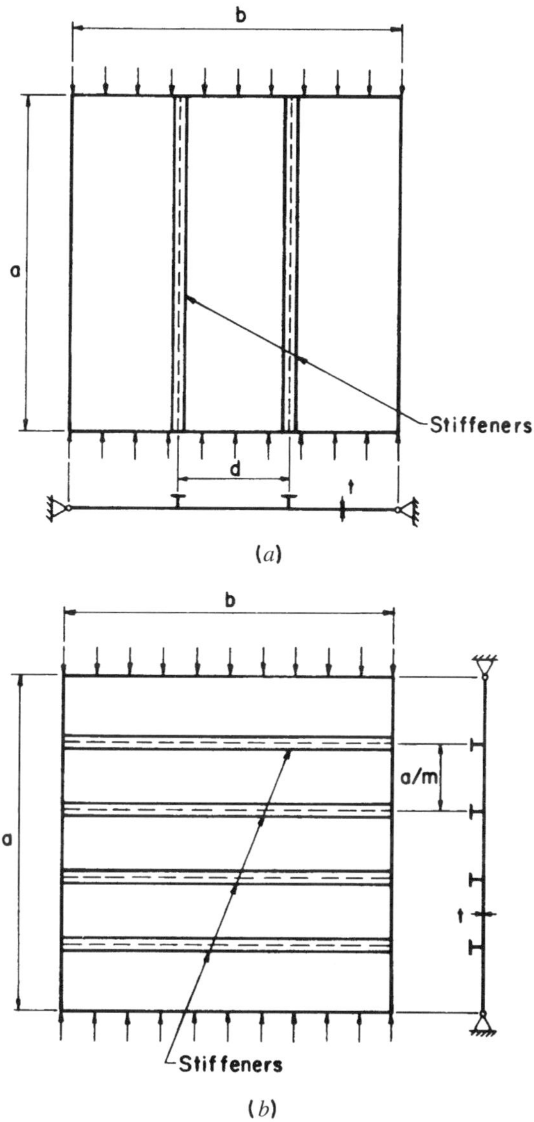

FIGURE 4.25 Stiffened-plate panels: (*a*) panel with longitudinal stiffeners; (*b*) panel with transverse stiffeners.

three equally spaced longitudinal stiffeners parallel to the direction of the applied compressive load. The solutions of Seide and Stein are also useful for other numbers of equally spaced stiffeners. In all of these solutions the stiffeners are assumed to have zero torsional rigidity. If we consider only overall buckling with a single half sine wave in the longitudinal and transverse direction, then this classical solution may be provided in a compact form (Schafer and Peköz, 1998),

$$\sigma_{cr} = k_{cr} \frac{\pi^2 D}{b_o^2 t} \tag{4.43}$$

$$k_{cr} = \frac{\left(1 + \beta_{cr}^2\right)^2 + 2\sum_i \gamma_i \sin^2(\pi\alpha_i)}{\beta_{cr}^2 \left(1 + 2\sum \delta_i \sin^2(\pi\alpha_i)\right)} \tag{4.44}$$

$$\beta_{cr} = \left(2\sum_i \gamma_i \sin^2(\pi\alpha_i) + 1\right)^{1/4} \tag{4.45}$$

$$\gamma_i = \frac{EI_i}{b_o D} \quad \delta_i = \frac{(A_s)_i}{b_o t} \quad \alpha_{\mathrm{i}} = \frac{c_i}{b_o} \quad D = \frac{Et^3}{12\left(1 - \upsilon^2\right)} \tag{4.46}$$

and the summations are over the i stiffeners, I_i is the moment of inertia of stiffener i, b_o is the plate width, A_{si} is the area of stiffener i, and c_i is the distance stiffener i is from the edge. This form of the expressions is adopted in AISI (2007).

A conservative method of analysis proposed by Sharp (1966) divides the analysis of the stiffened plate into two parts: one applying to short panels in which the buckled configuration takes the form of a single half-wave in both the longitudinal and transverse directions and another applying to long panels in which several longitudinal waves may occur along with a single half-wave in the transverse direction. In very short panels, the stiffener and a width of plate equal to the stiffener spacing d are analyzed as a column of length a with a slenderness ratio

$$\left(\frac{L}{r}\right)_{\mathrm{eq}} = \frac{\alpha}{r_e} \tag{4.47}$$

where r_e is the radius of gyration of the section consisting of a stiffener plus a width of plate equal to d.

In long panels the critical stress is larger than that calculated by the use of Eq. 4.47. In this case an equivalent slenderness ratio is defined for use in column strength formulas:

$$\frac{L}{r_{\mathrm{eq}}} = \sqrt{6(1 - \nu^2)}\frac{b}{t}\sqrt{\frac{1 + (A_s/bt)}{1 + \sqrt{(EI_e/bD) + 1}}} \tag{4.48}$$

where $b = Nd$ = overall width of longitudinally stiffened panel

N = number of panels into which the longitudinal stiffeners divide the plate

I_e = moment of inertia of section consisting of the stiffener plus a width of plate equal to d

A_s = cross-sectional area of stiffener
$D = Et^3/[12(1 - \nu^2)]$

The smaller of the slenderness values taken from Eqs. 4.47 and 4.48 is then used in the analysis, and it is assumed that the plate is fully effective over the panel width d. For greater values of d, buckling of the stiffeners and of the plate between the stiffeners would need to be considered.

A flat aluminum sheet with multiple longitudinal stiffeners or a formed stiffened sheet subjected to a uniform longitudinal compression (Sherbourne et al., 1971) will buckle into waves of length Ψb, in which b is the plate width and $\Psi = 1.8(I_x/t^3)^{1/4}$, where I_x is the moment of inertia in the strong direction and t is the plate thickness. For a formed aluminum stiffened sheet this becomes $\Psi = 1.8(\rho r_x/t)^{1/2}$, in which ρ is the ratio of the developed sheet width to the net width and r_x is the radius of gyration in the strong direction. If the spacing, a, of transverse supports is less than Ψb, the elastic critical stress may be approximated as

$$\sigma_c = \frac{\pi^2 E[1 + (a/\Psi b)^4]}{(a/r_x)^2} \tag{4.49a}$$

If the spacing exceeds Ψb, then

$$\sigma_c = \frac{2\pi^2 E}{(\Psi b/r_x)^2} \tag{4.49b}$$

Using this method, the Canadian standard (CSA, 1984) reduces the design procedure to determine the equivalent slenderness ratios, which are, for the two above cases,

$$\lambda = \frac{a/r_x}{[1 + (a/\Psi b)^4]^{1/2}} \tag{4.50a}$$

$$\lambda = \frac{0.7\Psi b}{r_x} \tag{4.50b}$$

Transverse Stiffeners The required size of transverse stiffeners for plates loaded in uniaxial compression is defined by Timoshenko and Gere (1961) for one, two, or three equally spaced stiffeners and by Klitchieff (1949) for any number of stiffeners. The stiffeners as sized provide a nodal line for the buckled plate and thus prohibit overall buckling of the stiffened panel. The strength of the stiffened panel would be limited to the buckling strength of the plate between stiffeners. These references also provide formulas for calculating the buckling strength for smaller stiffeners. With $\gamma = EI_s/bD$, the required minimum value of γ given by Klitchieff is

$$\gamma_{\min} = \frac{(4m^2 - 1)[(m^2 - 1)^2 - 2(m^2 + 1)\beta^2 + \beta^4]}{2m\left[5m^2 + 1 - \beta^2\right]\alpha^3} \tag{4.51}$$

where

$$\beta = \frac{\alpha^2}{m} \quad \text{and} \quad \alpha = \frac{a}{b} \tag{4.52}$$

and m is the number of panels, $m - 1$ the number of stiffeners, and EI_s the flexural rigidity of one transverse stiffener. These stiffener sizes should be understood as ideal stiffeners; unavoidable imperfections and the desire to avoid coupled instabilities will often require the use of stiffeners greater than those of Eq. 4.51.

An approximate analysis that errs on the conservative side but gives estimates of the required stiffness of transverse stiffeners for plates, either with or without longitudinal stiffeners, may be developed from a consideration of the buckling of columns with elastic supports. Timoshenko and Gere (1961) show that the required spring constant K of the elastic supports for a column (for the supports to behave as if absolutely rigid) is given by

$$K = \frac{mP}{Ca} \tag{4.53}$$

where

$$P = \frac{m^2\pi^2(EI)_c}{\alpha^2} \tag{4.54}$$

and C is the constant which depends on m and decreases from 0.5 for $m = 2$ to 0.25 for infinitely large m, $(EI)_c$ the flexural stiffness of column, m the number of spans, and a the total length of the column. In the case of a transversely stiffened plate (Fig. 4.25*b*), a longitudinal strip is assumed to act as a column which is elastically restrained by the transverse stiffeners. Assuming also that the loading from the strip to the stiffener is proportional to the deflection of the stiffener, the spring constant for each column support can be estimated. For a deflected shape of a half sine wave the spring constant is

$$K = \frac{\pi^4(EI)_s}{b^4} \tag{4.55}$$

Equating Eqs. 4.53 and 4.55 and inserting the value given for P result in

$$\frac{(EI)_s}{b(EI)_c} = \frac{m^3}{\pi^2 C(a/b)^3} \tag{4.56}$$

In the case of panels without longitudinal stiffeners $(EI)_c = D$ and the left side of Eq. 4.56 is γ. Values of C are tabulated by Timoshenko and Gere (1961) for $m \leq 11$. As shown by Fig. 4.26, these values are given approximately by

$$C = 0.25 + \frac{2}{m^3} \tag{4.57}$$

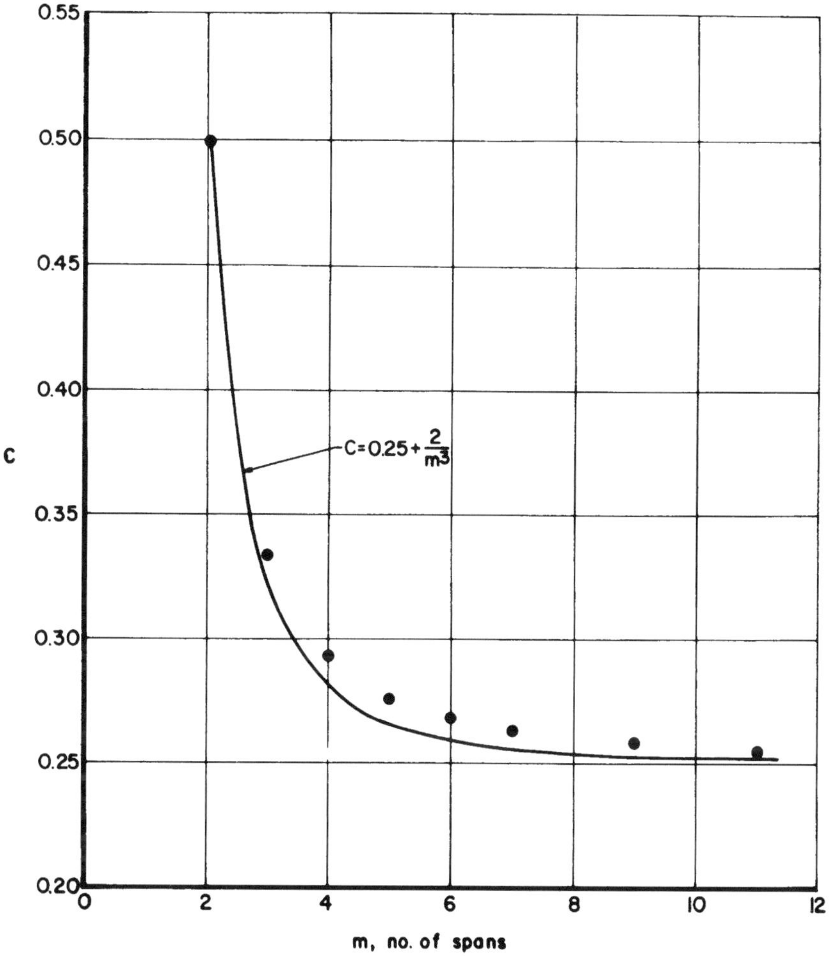

FIGURE 4.26 Buckling of columns with elastic supports.

In Table 4.2 values calculated using Eq. 4.56 are compared to corresponding values tabulated by Timoshenko and Gere (1961) for one, two, and three stiffeners and to those calculated by Eq. 4.41 for 10 stiffeners. Equation 4.56 is always conservative and is highly accurate for cases in which several stiffeners subdivide the panel.

Longitudinal and Transverse Stiffeners For a combination of longitudinal and transverse stiffeners, Gerard and Becker (1957/1958) provide figures showing the minimum value of γ as a function of α for various combinations of equally stiff longitudinal and transverse stiffeners. The same procedure used to establish Eq. 4.56 can also be applied to this case. The difference in the development is that

TABLE 4.2 Limiting Values of γ for Transverse Stiffeners

	One Stiffener		Two Stiffeners		Three Stiffeners		Ten Stiffeners	
a/b Ratio	Titnoshenko and Gere (1961)	Eq. 4.56	Timoshenko and Gere (1961)	Eq. 4.56	Timoshenko and Gere (1961)	Eq. 4.56	Eq. 4.51	Eq. 4.56
0.5	12.8	1.30	65.5	65.5	177.0	177.0	4170	4220
0.6	7.25	7.5	37.8	38.0	102.0	122.0	2420	2440
0.8	2.82	3.2	15.8	16.0	43.1	43.2	1020	1030
1.0	1.19	1.6	7.94	8.20	21.9	22.1	522	528
1.2	0.435	0.94	4.43	4.73	12.6	12.8	301	304
$\sqrt{2}$	0	0.57	2.53	2.90	7.44	7.85	185	187

TABLE 4.3 Dependence of K on Number of Longitudinal Stiffeners

Number and Spacing of Longitudinal Stiffeners	Spring Constant, K	$\gamma = \dfrac{(EI)_s}{b(EI)_c}$
One centrally located	$\dfrac{48(EI)_s}{b^3}$	$\dfrac{0.206m^3}{C(a/b)^3}$
Two equally spaced	$\dfrac{162}{5}\dfrac{(EI)_s}{b^3}$	$\dfrac{0.152m^3}{C(a/b)^3}$
Four equally spaced	$\dfrac{18.6(EI)_s}{b^3}$	$\dfrac{0.133m^3}{C(a/b)^3}$
Infinite number equally spaced	$\dfrac{\pi^4(EI)_s}{b^3}$	$\dfrac{m^3}{\pi^2 C(a/b)^3} = \dfrac{0.1013m^3}{C(a/b)^3}$

the spring constant K of the support is dependent on the number of longitudinal stiffeners, as shown in Table 4.3. The flexural stiffness $(EI)_c$ in these cases is equal to the average stiffness per unit width of the plate–longitudinal stiffener combination.

The required size of transverse stiffeners in a panel also containing longitudinal stiffeners is thus approximately

$$\frac{(EI)_s}{b(EI)_c} = \frac{m^3}{\pi^2 C(a/b)^3}\left(1 + \frac{1}{N-1}\right) \tag{4.58}$$

in which N and m are the number of panels into which the longitudinal and transverse stiffeners divide the plate, respectively. This formula yields essentially the same values as does the formula presented for aluminum panels in the *Alcoa Structural Handbook* (Aluminum Company of America, 1960). With this size of transverse stiffener, the strength of the panel is limited to the buckling strength of the longitudinally stiffened panel between transverse stiffeners.

Stiffener Type The methods of analysis described above are directly applicable to open-section stiffeners having negligible torsional stiffness and hence are conservative when applied to stiffeners with appreciable torsional stiffness. The influence of torsional stiffness on overall panel buckling has been studied by Kusuda (1959) for the case of one longitudinal or one transverse stiffener. Stiffeners with large torsional rigidity also provide partial or complete fixity of the edges of subpanel plating, thereby increasing their critical stresses.

It has been shown by Lind (1973) and Fukumoto et al. (1977) that the stiffener type affects the buckling mode as well as the ultimate carrying capacity of the stiffened plate. It has also been shown by Tvergaard (1973) and Fok et al. (1977) that local imperfections of the stiffeners can influence significantly the overall buckling behavior of stiffened-plate panels.

4.4.2 Compression: Postbuckling and Strength

Buckling of a stiffened-plate panel may occur by primary instability with a half-wavelength which is on the order of the panel length or by local instability with a half-wavelength which is on the order of the width of plate elements of the plate and stiffeners. As plate panel length increases, the ultimate stress decreases, until at large slenderness ratio the panel fails in flexure as a long column. The ultimate strength in the long-column range can be predicted by the Euler column formula, where the radius of gyration is computed for the combined section of the stiffener and the effective width of the plate. At an intermediate slenderness ratio there is a transition in the mode of failure from the purely local mode to one dominated by overall panel failure. In this transition zone the panel fails through a combination of the primary buckling and flexural modes and may involve twisting of the stiffeners.

It is well known that plates supported at their edges are often able to sustain compressive load far in excess of their buckling loads. The margin between the buckling load and the ultimate load in plates, known as the postbuckling strength, depends on whether the critical stress is reached below or above the proportional limit of the material. If the buckling stress is well below the proportional limit, the ultimate load may be many times greater than the buckling strength, depending on the aspect ratio, a/b, of the plate element. The reserve strength in the postbuckling range approaches zero as the buckling stress approaches the yield strength of the material. Section 4.3 provides further discussion of the distinctions between elastic buckling, inelastic buckling, and the postbuckling strength of plates.

Initial buckling modes of stiffened flat panels vary with the slenderness ratio of the panel and whether the type of construction is monolithic or built-up. As a basis for evaluating postbuckling strength, the following sections provide a brief discussion regarding predictions of initial buckling stresses and ultimate strength.

Local Instability Mode The initial buckled form has one transverse sinusoidal half-wave, with perhaps a number of longitudinal half-waves. As the compressive load is increased, the central portion of the transverse half-wave becomes flattened, and as shown in Fig. 4.27 the transverse deformation is no longer a simple sinusoidal curve. The well-known effective-width approach to the analysis of the postbuckling strength of a flat plate is based on the stress distribution associated with this buckled form and was discussed in detail in Section 4.3.

Another change of buckled form is possible when a rectangular flat plate, simply supported on all sides, is subjected to uniform-edge compression and is free to expand laterally. A dynamic snap from one buckled form to another may occur by a sudden change of wavelength of buckles along the direction of compression. The behavior of the flat plate in this sense is analogous to the elastic postbuckling of a column supported laterally by a nonlinear elastic medium (Tsien, 1942; Stein, 1959; Koiter, 1963). A column supported laterally by a finite number of nonlinear elastic restraints buckles initially into m sinusoidal half-waves over its length; then subsequently the buckled form may become unstable, and the column may snap

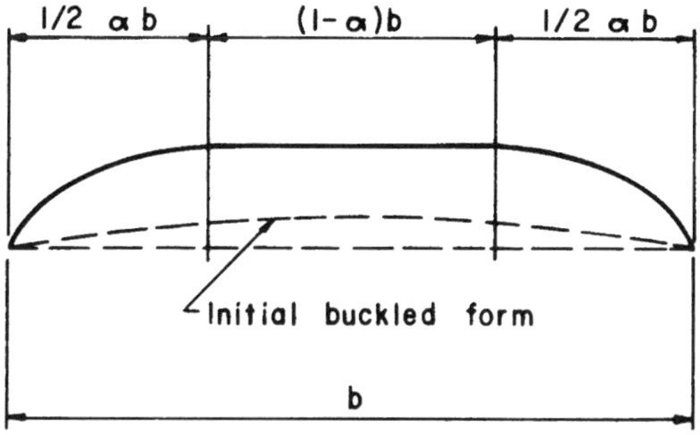

FIGURE 4.27 Initial and final buckled shapes.

into n half-waves, with n being greater than m. The exact analysis of transition between the two modes of buckling is not known at present. Sherbourne et al. (1971) determined the terminal wavelength for flat plates at the ultimate capacity.

Failure Strength of Very Short Stiffened Panels For a short stiffened plate with a slenderness ratio smaller than 20, Gerard and Becker (1957/1958) note that the failure stress is independent of the panel length. The average stress at failure in this slenderness range is known as the crippling, crushing, or local-failure stress and will be represented by $\overline{\sigma}_f$. Although Gerard and Becker (1957/1958) present a method for determining the crippling stress of short longitudinally stiffened panels, the yield stress of the material in most cases can be considered the failure stress for short stiffened panels.

Ultimate Strength of Intermediate and Long Stiffened Panels Gerard and Becker (1957/1958) describe a method to predict the buckling failure of stiffened panels based on a curve analogous to the Johnson parabola shown in Fig. 4.28. At stresses lower than the local buckling stress σ_c and the proportional limit σ_{pl},

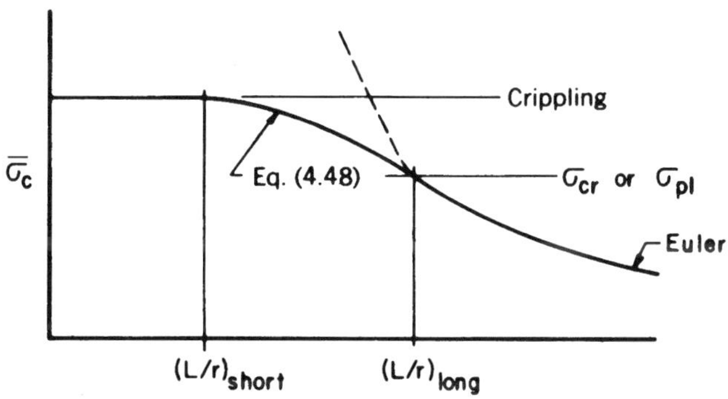

FIGURE 4.28 Column curve for stiffened panels.

the Euler column equation is used. In the transition range between $L/r = 20$ and the long column, a parabola of the form

$$\overline{\sigma}_c = \overline{\sigma}_f \left[1 - \frac{\sigma_c}{\sigma_e} \left(1 - \frac{\sigma_c}{\overline{\sigma}_f} \right) \left(\frac{\sigma_{20}^{1/2} - \sigma_e^{1/2}}{\sigma_{20}^{1/2} - \sigma_c^{1/2}} \right)^2 \right] \tag{4.59}$$

is used, where $\overline{\sigma}_c$ is the failure stress, $\sigma_e = \pi^2 E/(L/r)^2$ the Euler stress for the panel, $\overline{\sigma}_f$ the failure stress for a short stiffened panel, and σ_{20} the Euler stress evaluated at $L/r = 20$. Many direct-reading column charts have been prepared for predicting panel ultimate strength. An example of this type of plot is shown in Fig. 4.29. Gerard and Becker (1957/1958) provide references and examples of this work.

In determining panel strength, it is necessary to estimate the effective column length of the plate–stiffener combination as well as the effective width of plating that acts in conjunction with the stiffener. When the critical stress for the individual panel of plating between stiffeners is greater than the critical stress for the stiffened panel, the plating may be assumed completely effective, and the effective column length of the panel is determined by the end conditions. When the critical stress for the individual plating panel is significantly less than that for the stiffened panel as a unit, the ultimate strength of the stiffened plate is considered to be the lesser of (1) the load that causes the stress at the juncture of plate and stiffener to reach the yield strength of the material or (2) the column strength of the stiffener in

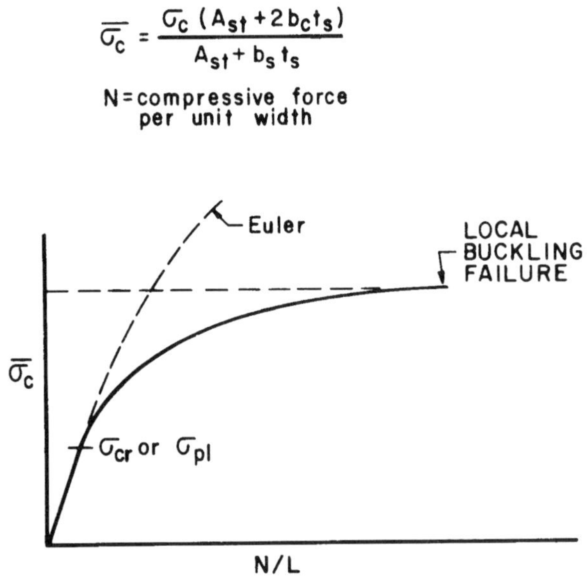

FIGURE 4.29 Column chart by Gerard and Becker (1957/1958).

conjunction with an effective width of plating that is less than the actual width of plating between stiffeners.

The application of the first criterion for ultimate strength assumes that the stiffener is stiff enough to allow the plate-edge stress to reach the yield stress before the stiffener buckles as a column and that the stress distribution across the buckled plate can be determined. The second criterion assumes that the effective column length of the stiffener and the effective width (see Section 4.2) of the associated plating can be properly determined and that residual stresses and initial imperfections due to fabrication are properly accounted for with regard to column buckling.

To provide a better understanding of the behavior of stiffened plates, a theoretical study was carried out by Wittrick (1968). The collapse of box-girder bridges in the early 1970s generated a great interest in research on various aspects of box girders, especially the interactive buckling of an assembly of plates and the ultimate strength of a stiffened plate. The research efforts in many ways were centered around the Merrison Committee (1973). Murray (1975) reported on an analysis-and-design procedure for the collapse load for the stiffened plates. Crisfield (1975) presented a finite element formulation for the large-deflection elastic–plastic full-range analysis of stiffened plating. A simple approach for the design of stiffened steel compression flanges was proposed by Dwight and Little (1976).

A theoretical and experimental study on the inelastic buckling strength of stiffened plates was reported by Fukumoto et al. (1977). Residual stresses were considered and the stiffened plates treated had relatively low width-to-thickness ratios and with relatively rigid stiffeners. It was found that partial yielding in the flat stiffeners considerably reduced the buckling strength of a stiffened plate. In the case of plates stiffened by T-type stiffeners, the strength reduction due to partial yielding in the stiffeners was much less pronounced. This is probably due to the fact that the T-stiffeners were more rigid and less susceptible to initial imperfections than were the flat stiffeners. Murray (1973) has indicated that based on experimental results there is often little margin of strength above the load at which yielding first occurs in a stiffener. This phenomenon can lead to a triggering effect which results in a sudden failure of the stiffened plate. Such a viewpoint is also shared by Horne and Narayanan (1977).

Horne and Narayanan (1975, 1977) have proposed a design method for the prediction of collapse loads of stiffened plates subjected to axial compression. They have compared the results obtained from their method, methods proposed by other researchers, and experiments. The results are all close to the observed strength. Horne and Narayanan's method is based on the British Perry–Robertson column formula and consists of analyzing the stiffened plates as a series of isolated columns comprising a stiffener and the associated effective width of plating. The criterion for plate failure is the attainment of yielding stress at the plate–stiffener boundary. The criterion for the stiffener-initiated failure is by yielding or by instability of the stiffener.

Little (1976) and Elsharkawi and Walker (1980) have studied the effects of continuity of longitudinal stiffeners on the failure mode and strength of stiffened plates consisting of several bays between cross frames. Little has indicated that there is a tendency for longitudinal continuity to be strengthened where failure occurs in the plate and weakened where failure occurs in the stiffener. A simplified design method to account for the effects of continuity has been proposed by Elsharkawi and Walker (1980).

Nonlinear finite element analysis has been compared to tests and shown to provide reliable predictions of the ultimate strength of stiffened panels in a variety of limit states (Grondin et al., 1998). The same researchers went on to perform a wide-ranging parametric study that resulted in design guidance for stiffened plates based on relevant plate and stiffener dimensions (Grondin et al., 1999). The specific role of the stiffener geometry, particularly its torsional stiffness, has been the focus of tests (Ghavami, 1994) as well as further nonlinear finite element analysis (Sheikh et al., 2003).

Desmond et al. (1981a,b) have presented an experimental study of edge-stiffened compression flanges and intermediately stiffened compression flanges. This work has been extended for edge stiffeners (Schafer and Peköz, 1999) and intermediate stiffeners (Schafer and Peköz, 1998). Effective-width procedures for local and overall (distortional) buckling of the stiffened panels were proposed and have been adopted by AISI (2007). Nguyen and Yu (1982) have reported a study on longitudinally reinforced cold-formed steel beam webs. Based on the experimental results, an effective-width procedure is proposed to predict the ultimate strength of such members subjected to bending. Additional information on this subject matter is presented in Chapters 6, 7, and 13.

4.4.3 Compression and Shear: Buckling

Initial Buckling Analytical and experimental results on the buckling behavior of stiffened plates under combined compression and shear are relatively scarce. Recourse is therefore usually made to data for unstiffened plates supplemented with whatever data are available for the type of longitudinally stiffened construction considered to be most important in practice, typically that shown in Fig. 4.30. The case of unstiffened rectangular plates under combined compressive and shear stresses has been presented in Section 4.2.4, which showed that a simple parabolic relationship of the form of Eq. 4.60 is satisfactory for engineering purposes for all ranges of elastic restraint from free rotation to complete fixity. The relationship takes the form

$$R_c + R_s^2 = 1 \tag{4.60}$$

where R_c is the ratio of compressive stress when buckling occurs in combined shear and direct stress to compressive stress when buckling occurs in pure compression and R_s is the ratio of shear stress when buckling occurs in combined shear and direct stress to shear stress when buckling occurs in pure shear.

Johnson (1957) treated the problem of long plates with one and two stiffeners acted on by axial compression and shearing stresses. The stiffeners were assumed

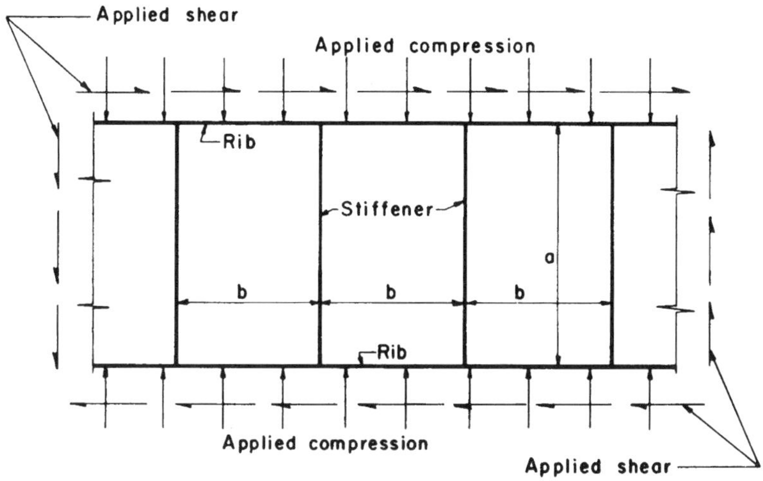

FIGURE 4.30 Longitudinally stiffened plate under combined compression and shear stresses.

to have bending stiffness only and the resulting interaction curves show discontinuities which reflect the mode into which the plate buckles. When the bending stiffness ratio of the stiffener to that of the plate (EI/bD) is low, the buckle goes through the stiffener. When the stiffness ratio, however, is increased so that nodal lines occur along the stiffener lengths, buckling takes place without deflection along the stiffeners. In this case another interaction relation exists between compressive and shearing stresses. Another somewhat more limited study of the interaction relationship of infinitely wide stiffened plates under compression and shearing stresses was reported by Harris and Pifko (1969). In this study the finite element method was used. The stiffeners were assumed to have both bending and torsional stiffness, and the grid refinement used was judged to be adequate to ensure accurate results. The results shown in Fig. 4.31 were compared to the parabolic expression (Eq. 4.49). Except for the case of assumed large torsional stiffness ratio ($GJ/bD = 10^6$) the analytical points follow the parabolic relationship very well.

Buckling in the Inelastic Range Analytic prediction of the interaction relationship for stiffened panels which buckle in the inelastic range under combined compression and shear is practically nonexistent in the literature. One such computation reported by Harris and Pifko (1969) was made for an integrally stiffened panel made of aluminum 2024-T351 and having the dimensions shown in Fig. 4.32. The predicted interaction curves are shown in this figure for both the elastic case, which agrees very well with the parabolic relationship (Eq. 4.60), and the inelastic buckling case. Because of the limited nature of the data, no general relationship for the inelastic buckling case can be derived. However, it should be noted from Fig. 4.32 that the circular relationship lies above the analytical curve of the inelastic buckling case.

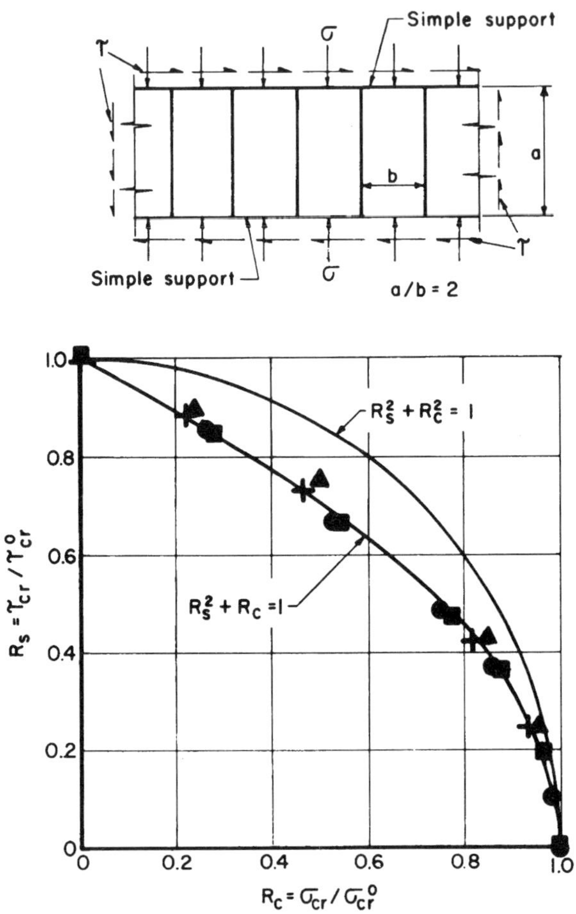

A/bt	EI/bD	CJ/bD	Symbol
0.4	100	0	●
0.4	100	1	+
0.4	100	10^6	▲
0.4	1000	0	■

FIGURE 4.31 Analytical interaction relations for infinitely wide stiffened plate under combined compression and shearing stresses.

4.4.4 Compresion and Shear: Postbuckling and Strength

Postbuckling behavior for the combination of compression and shear is complex, and an exact treatment of the problem has not yet been achieved. Because of the importance of predicting the strength of stiffened panels loaded beyond initial buckling in the aircraft industry, semiempirical methods have been in existence

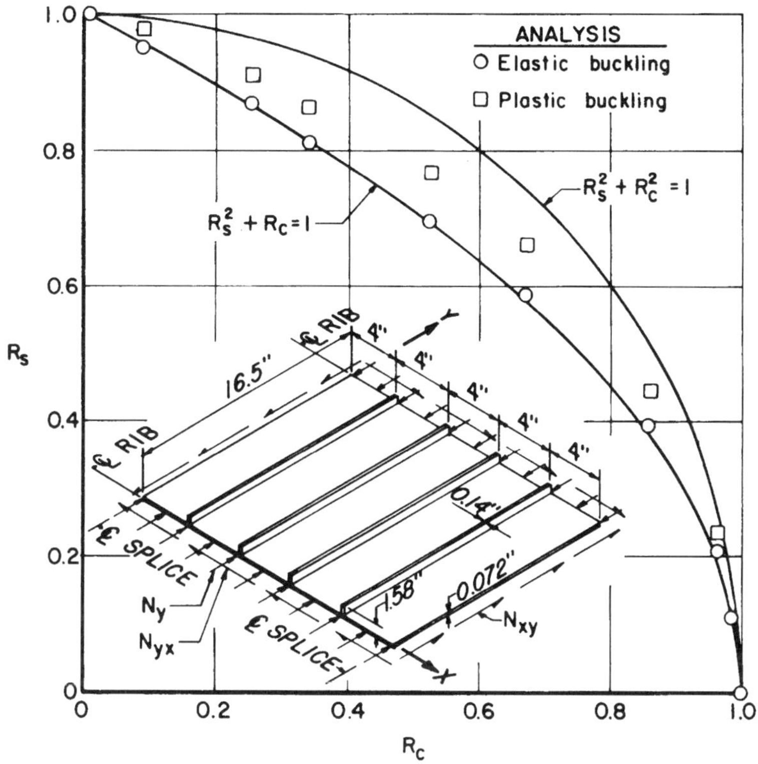

FIGURE 4.32 Analytically predicted elastic and inelastic interaction curves for an integrally stiffened panel.

for many years. In the case of light stiffened plates under combined compression and shear, the structure will usually take considerable load in excess of the initial buckling load of the plate. Chapter 6 covers the design aspects of using these postbuckling or tension-field theories; hence no further discussion of these methods is made here.

4.4.5 Laterally Loaded Plates in Compression

Plates with stiffeners in the direction of the axial load that are also subjected to a distributed lateral load are commonly encountered as the bottom plates in ships. A study based on a large-deflection postbuckling theory (Supple, 1980) shows that lateral pressure produces effects similar to initial geometric imperfections with a long buckling wavelength. It is further shown that a sufficiently high pressure induces stable postbuckling in the long-wave mode.

The stiffened plating is supported on heavy transverse structural members that can be assumed rigid. The panels have width-to-length ratios of about 2.5 to 4.0, and because there is essentially no interaction in the transverse direction, the panel

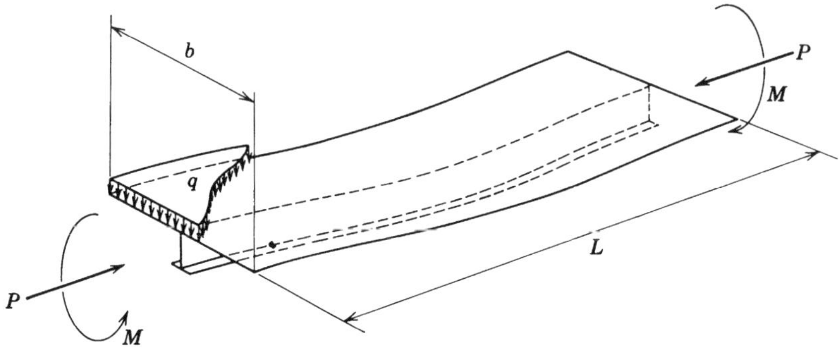

FIGURE 4.33 Idealized beam-column.

essentially behaves as if it were a wide beam-column under axial and lateral loads. It is convenient to isolate one longitudinal stiffener together with a plate of the width equal to the spacing of the stiffeners b and to consider that all other stiffeners behave in a similar manner. Such a beam-column is shown in Fig. 4.33.

Determination of the maximum values of the lateral and axial loads that can be sustained by the member (i.e., analysis for the ultimate strength) requires consideration of the yield strength of plate and stiffener materials; large-deflection theory; partial yielding of the section; and the postbuckling and post-ultimate-strength behavior of the plate. The complexity of the problem suggests that numerical methods be used. Some solutions have been obtained for standard wide-flange shapes by Ketter (1962) and Lu and Kamalvand (1968) in the form of interaction curves between the axial and lateral loads. Unfortunately, these results cannot be used for plating because the yield strength of the material and the postbuckling and post-ultimate-strengths were not considered. Another complication is that the wide-flange sections are doubly symmetrical, whereas the section shown in Fig. 4.33 is singly symmetrical with a very wide top flange.

This means, for example, that although it was possible to nondimensionalize the results for wide-flange sections in terms of L/r and make them applicable with a very small error to any wide-flange section, the unsymmetrical section of the stiffened panel must be treated as a special case for every combination of relative proportions. Nomographs for the ultimate strengths of stiffened plates with a yield point of 47 ksi and the ranges of geometrical and loading parameters commonly encountered in ship structures are given by Vojta and Ostapenko (1967).

4.5 BUCKLING OF ORTHOTROPIC PLATES

Problems related to rectangular plates with stiffeners parallel to one or both pairs of sides can be solved approximately by methods applicable to orthotropic plate theory. An orthotropic plate is one whose material properties are orthogonally anisotropic; a uniformly stiffened plate is reduced to this case by effectively "smearing" the stiffness characteristics of its stiffeners over the domain of the plate. Clearly, the theory is best applicable when the spacing of the stiffeners is small.

The calculation of buckling strength of orthotropic plates is based on the solution of the following differential equation governing the small deflection $w(x, y)$ of the buckled plate:

$$D_1 \frac{\partial^4 w}{\partial x^4} + 2D_3 \frac{\partial^4 w}{\partial x^2\, \partial y^2} + D_2 \frac{\partial^4 w}{\partial y^4} + N_x \frac{\partial^2 w}{\partial x^2} + N_y \frac{\partial^2 w}{\partial y^2} + 2N_{xy} \frac{\partial^2 w}{\partial x\, \partial y} = 0 \tag{4.61}$$

where

$$D_1 = \frac{(EI)_x}{1 - \nu_x \nu_y}$$

$$D_2 = \frac{(EI)_y}{1 - \nu_x \nu_y}$$

$$D_3 = \tfrac{1}{2}(\nu_y D_1 + \nu_x D_2) + 2(GI)_{xy}$$

in which N_x, N_y, and N_{xy} are in-plane forces per unit width (Fig. 4.34); $(EI)_x$ and $(EI)_y$ are flexural stiffnesses per unit width of beam strips in the x and y directions, respectively; ν_x and ν_y are flexural Poisson ratios; and $2(GI)_{xy}$ is a measure of torsional stiffness.

Theoretical buckling data for several cases of rectangular plates with supported edges ($w = 0$) under uniform in-plane loadings, N_x, N_y, and N_{xy}, applied singly or

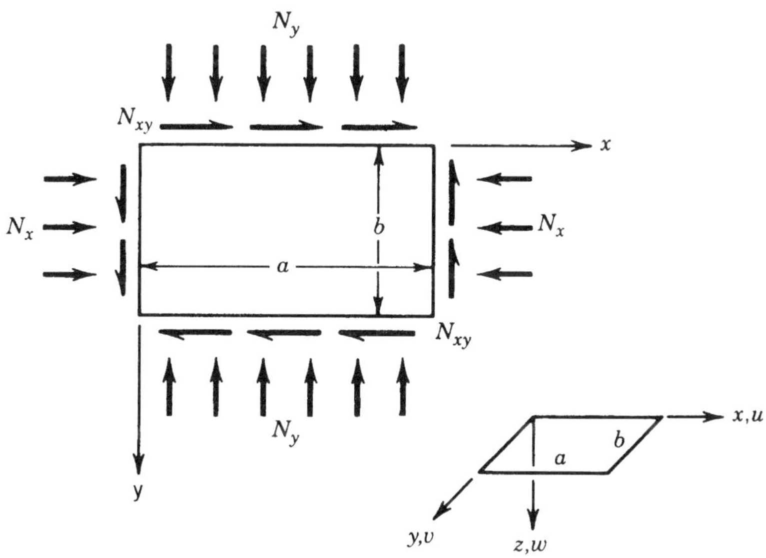

FIGURE 4.34 Plate subjected to axial and shear stress.

in certain combinations, are presented. The following are some of the additional notations that will be employed: a and b are the lengths of the plate in the x and y directions, respectively (see Fig. 4.34), and m and n are integers representing the number of buckles or half-waves for sinusoidal buckling patterns in the x and y directions, respectively.

4.5.1 Compression

Uniaxial Compression in x Direction, Loaded Edges Simply Supported, Unloaded Edges Elastically Restrained against Rotation For these conditions, referring to Fig. 4.34, N_x is the only loading, the edges at $x = 0$ and $x = a$ are simply supported, and the edges $y = 0$ and $y = b$ are each elastically restrained against rotation by a restraining medium whose stiffness (moment per unit length per radian of rotation) is K. The quantity Kb/D_2, to be denoted by ϵ, will be used as a dimensionless measure of this stiffness. An exact solution leads to the following formula defining the value of N_x that can sustain a buckle pattern containing m sinusoidal half-waves in the x direction,

$$\frac{N_x b^2}{\pi^2 D_3} = \left(\frac{b}{a/m}\right)^2 \frac{D_1}{D_3} + 2 + f_1\left(\epsilon, \left(\frac{a/m}{b}\right)^2 \frac{D_2}{D_3}\right) \tag{4.62}$$

where f_1 is the function plotted in Fig. 4.35. If $(a/mb)^2(D_2/D_3) > 0.4$, Eq. 4.62 can be very closely approximated by the formula

$$\frac{N_x b^2}{\pi^2 D_3} = \left(\frac{b}{a/m}\right)^2 \frac{D_1}{D_3} + 2 + f_2(\epsilon) + \left(\frac{a/m}{b}\right)^2 \frac{D_2}{D_3} f_3(\epsilon) \tag{4.63}$$

where f_2 and f_3 are the functions plotted in Fig. 4.36. The buckling load is the smallest N_x obtained by substituting different integer values of $m (m = 1, 2, 3, \ldots)$ into Eq. 4.62 or, as applicable, Eq. 4.63. In performing this minimization, one should take into account the fact that for most practical restraining media the stiffness K is not fixed but is a function of the half-wavelength a/m of the edge rotation. K may also depend on the axial load in the restraining medium and therefore on N_x, necessitating a trial-and-error calculation to determine N_x for any selected m. Negative values of K are physically possible but are excluded from consideration in Fig. 4.35 and 4.36. Unequal restraints along the edges $y = 0$ and $y = b$ can be handled approximately by first assuming the $y = 0$ constraint to be present at both edges, then the $y = b$ constraint, and averaging the two values of N_x thus obtained.

Uniaxial Compression, Loaded Edges Simply Supported, Unloaded Edges Clamped Here N_x is the only loading, the edges $x = 0$ and $x = a$ are simply supported, and the other two edges are clamped. The exact solution is contained in the condition $\epsilon = \infty$ of the previous example. An approximate solution

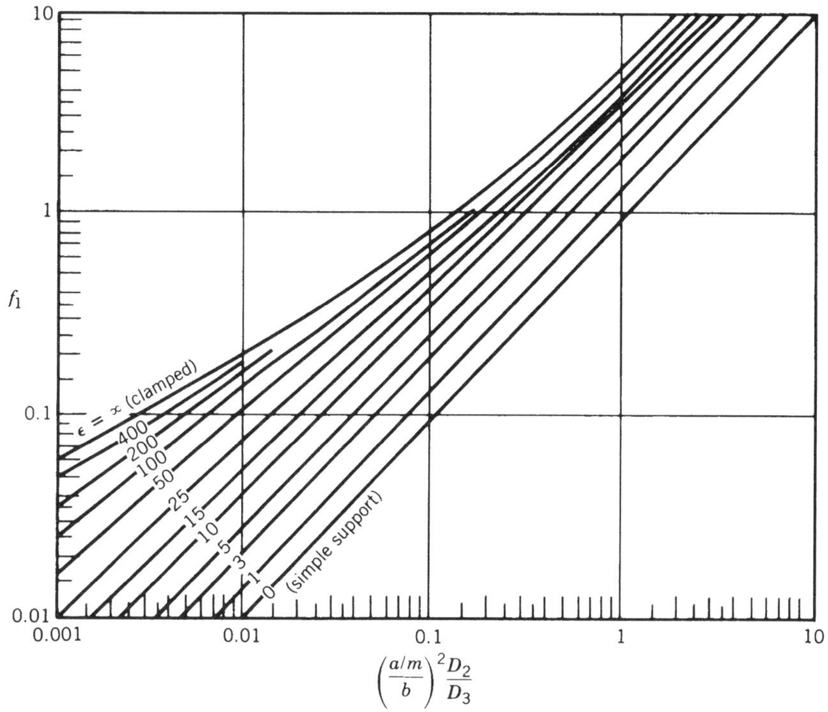

FIGURE 4.35 Function f_1 in Eq. 4.62.

is given by Wittrick (1952) in the form

$$\frac{N_x b^2}{\pi^2 \sqrt{D_1 D_2}} = k - c\left(1 - \frac{D_3}{\sqrt{D_1 D_2}}\right) \tag{4.64}$$

where $c = 2.4$ and k is taken from curve (c) of Fig. 4.37. Equation 4.64 is virtually equivalent to Eq. 4.63 and must therefore be subject to the same restriction [i.e., $(a/mb)^2(D_2/D_3) > 0.4$].

Uniaxial Compression, All Edges Clamped Wittrick (1952) gives an approximate solution for these conditions in the form of Eq. 4.64 with $c = 2.46$ and k taken from curve (d) of Fig. 4.37.

4.5.2 Biaxial Compression

Biaxial Compression, All Edges Simply Supported When all four edges are simply supported, combinations of N_x and N_y that can sustain a buckle pattern of m sinusoidal half-waves in the x direction and n sinusoidal half-waves in the y

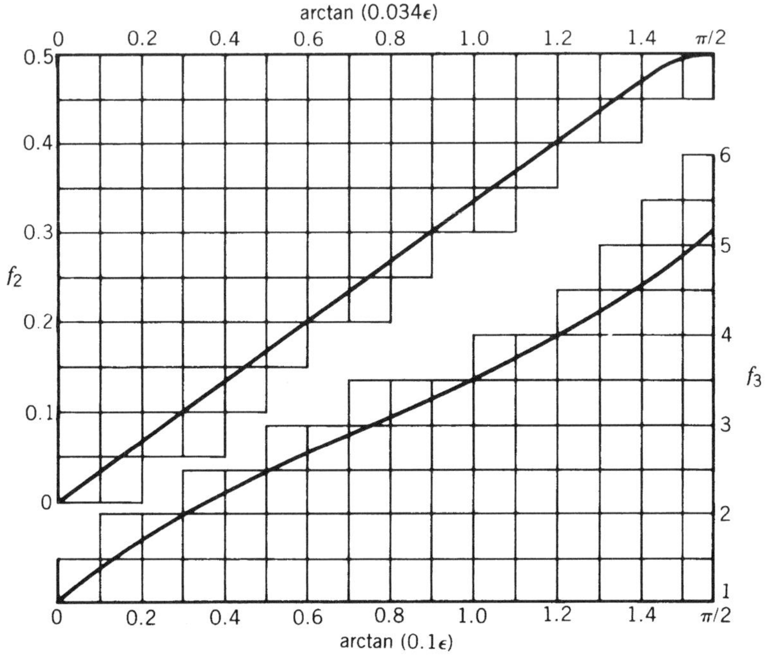

FIGURE 4.36 Functions f_2 and f_3 in Eq. 4.63.

direction are defined exactly by the interaction equation

$$\frac{1}{n^2}\frac{N_x b^2}{\pi^2 D_3} + \frac{1}{m^2}\frac{N_y a^2}{\pi^2 D_3} = \left(\frac{b/n}{a/m}\right)^2 \frac{D_1}{D_3} + 2 + \left(\frac{a/m}{b/n}\right)^2 \frac{D_2}{D_3} \tag{4.65}$$

In using Eq. 4.65, one must substitute different combinations of m and n until that combination is found which minimizes N_x for a given N_y, N_y for a given N_x or N_x and N_y simultaneously for a given ratio between them. It can be shown (Libove, 1983) that these minima will occur with at least one of the two integers equal to unity. Therefore, only the combinations $m = 1$, $n = 1, 2, 3, \ldots$ and $n = 1$, $m = 2, 3, \ldots$ need to be tried. The condition

$$\frac{N_y b^2}{\pi^2 D_2} < 1 \tag{4.66}$$

is sufficient to ensure that $n = 1$ will govern. In that case, with N_y regarded as given, Wittrick (1952) has shown that the N_x required for buckling is defined by

$$\frac{N_x b^2}{\pi^2 D_3} = 2 + (k - 2)\left[\frac{D_1 D_2}{D_3^2}\left(1 - \frac{N_y b^2}{\pi^2 D_2}\right)\right]^{1/2} \tag{4.67}$$

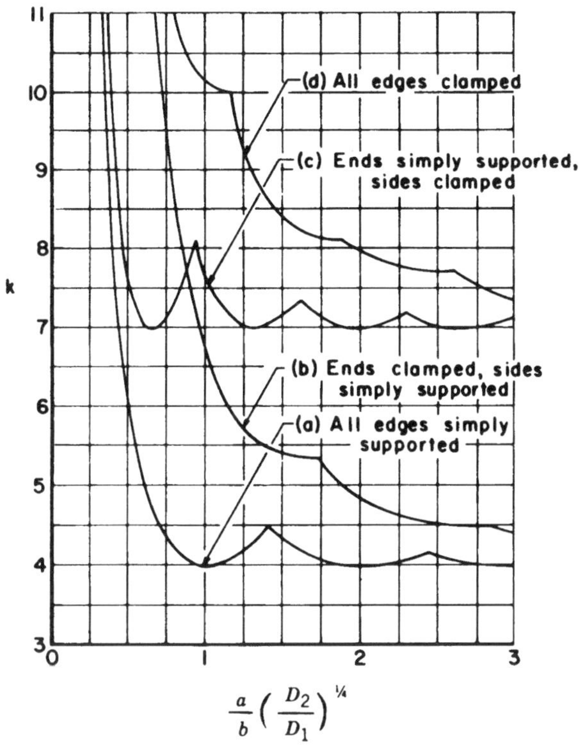

FIGURE 4.37 Buckling coefficients for orthotropic plates.

where k is the function plotted as curve (a) in Fig. 4.37. The special case of a very long plate may be treated as the limit of $a/b = \infty$. For this case it can be deduced from Eq. 4.65 that if $N_x b^2/\pi^2 D_3 \leq 2$, the longitudinal compression N_x is too small to sustain a multilobed buckle pattern, and the plate will buckle in a cylindrical mode (i.e., as a wide plate-column of length b) when $N_y b^2/\pi^2 D_2 = 1$. On the other hand, if $N_x b^2 D_3 > 2$, the buckle pattern will be sinusoidally lobed in the x direction with a half-wave length-to-width ratio of

$$\frac{a/m}{b} = \left[\frac{D_2}{D_1}\left(1 - \frac{N_y b^2}{\pi^2 D_2}\right)\right]^{-1/4} \tag{4.68}$$

and the buckling will occur when N_x and N_y satisfy Eq. 4.67 with k set equal to 4. The interaction curve of Fig. 4.38 summarizes the results just given for the case $a/b = \infty$.

Compression in x Direction Applied to Clamped Edges, Limited Compression in y Direction Applied to Simply Supported Edges Referring to Fig. 4.34, this section considers the case in which the edges $x = 0$ and $x = a$ are

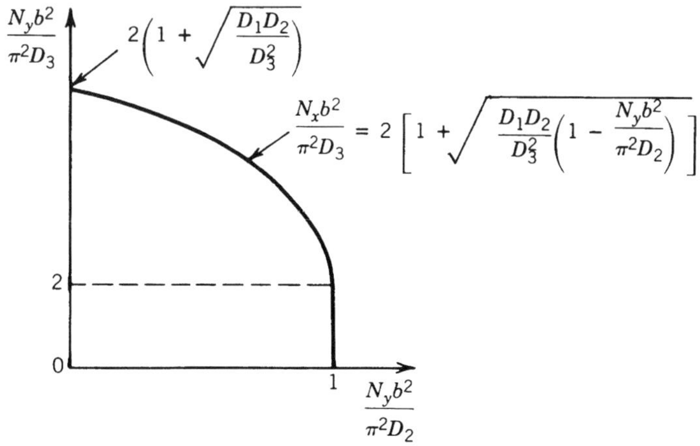

FIGURE 4.38 Interaction curve for orthotropic plates.

clamped, the edges $y = 0$ and $y = b$ are simply supported, N_{xy} is absent, and N_y satisfies the inequality given by Eq. 4.66. With N_y regarded as given, the value of N_x required to cause buckling is defined by Eq. 4.67 with k taken from curve (b) of Fig. 4.37 (Wittrick, 1952).

4.5.3 Shear

Various Boundary Conditions Theoretical data for the shear flow N_{xy} required to cause buckling of rectangular orthotropic plates have been collected by Johns (1971). Three of his graphs are reproduced in Fig. 4.39 with $k_s = N_{xy}b^2/\pi^2 D_1^{1/4} D_2^{3/4}$. They apply, respectively, to the boundary conditions of (a) all edges simply supported; (b) edges $y = 0$ and $y = b$ simply supported, the other two edges clamped; and (c) all edges clamped.

To make use of the buckling data presented, one must know the values of the elastic constants appearing in Eq. 4.61. These constants are best determined experimentally, by tests such as those described by Libove and Batdorf (1948) and Becker and Gerard (1963). If the plates are of a simple enough construction, the constants can also be evaluated theoretically. For example, for a sheet of thickness t and shear modulus G orthogonally stiffened by x-direction stiffeners of torsional stiffness C_1 spaced a distance b_1 apart and y-direction stiffeners of torsional stiffness C_2 spaced a distance a_1 apart, $(EI)_x$ may be taken as the flexural stiffness of the composite beam consisting of one x-direction stiffener and its associated width b_1 of sheet, divided by b_1, while $(EI)_y$ is computed in an analogous manner using a y-direction stiffener and its associated width a_1 of the sheet. For such plates, υ_x and υ_y may usually be taken as zero with little error, which results in

$$D_3 = 2(GI)_{xy} = \frac{Gt^3}{6} + \frac{1}{2}\left(\frac{C_1}{b_1} + \frac{C_2}{a_1}\right) \tag{4.69}$$

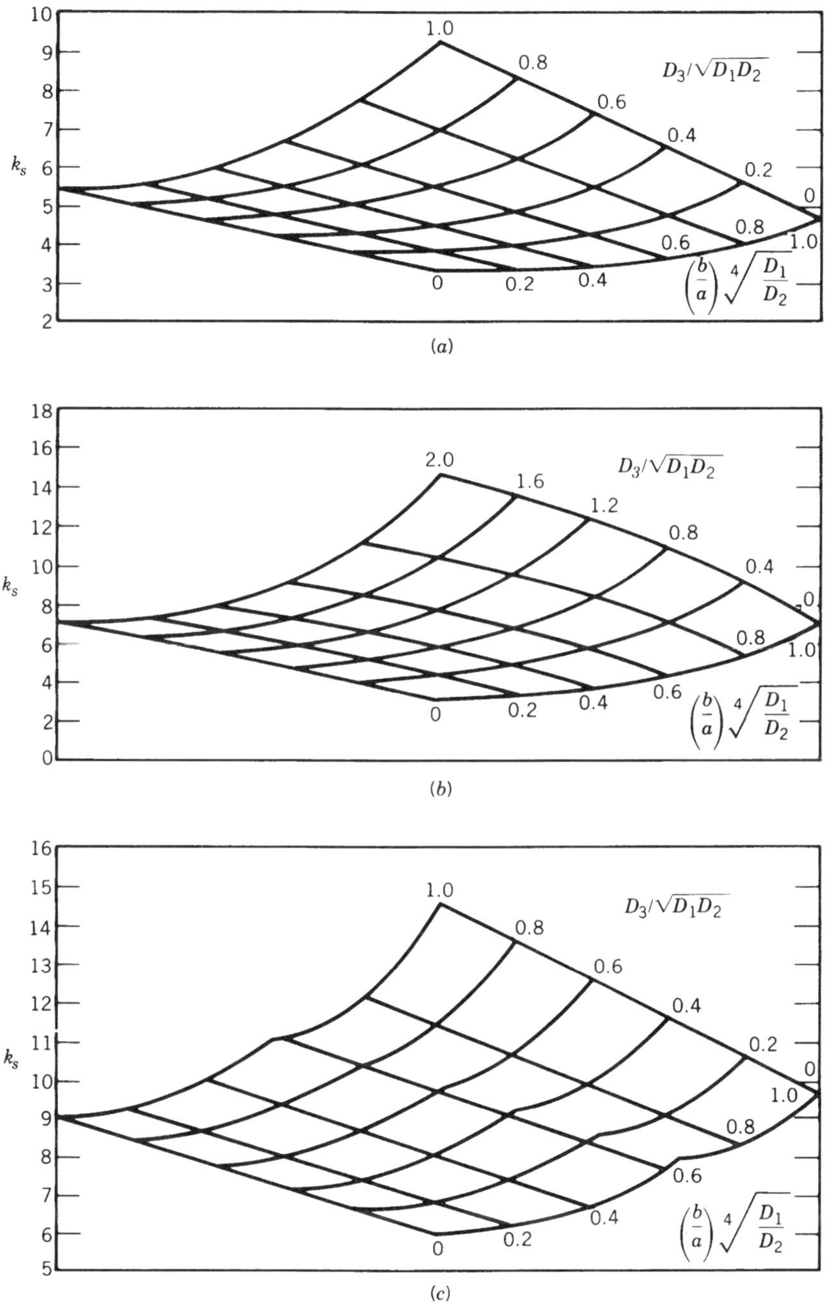

FIGURE 4.39 Shear buckling coefficients for orthotropic plates. [Adapted from Johns (1971), reprinted with the permission of Her Majesty's Stationery Office.]

Plates with integral waffle-like stiffening may also be modeled as orthotropic plates, provided that the ribbings are so oriented as to create axes of elastic symmetry parallel to the plate edges; formulas for estimating the elastic constants of such plates are derived by Dow et al. (1954). Corrugated-core sandwich plates with the corrugations parallel to the x or y axis can similarly be treated as orthotropic plates, and formulas for their elastic constants are developed by Libove and Hubka (1951); however, for such sandwich plates, deflections due to transvcrsc shcar, neglected in the present discussion, may sometimes be significant. It is common practice to treat a corrugated plate also as an orthotropic plate, and the appropriate elastic constants when the profile of the plate is sinusoidal are discussed by Lau (1981). There are indications (Perel and Libove, 1978), however, that modeling the corrugated plate as an orthotropic plate may lead to an underestimate of its shear buckling strength.

The orthotropic plate model has an additional shortcoming when applied to stiffened plates, namely its neglect of any coupling between in-plane forces and out-of-plane deflections. That is, underlying Eq. 4.61 is the tacit assumption that there exists a reference plane in which the forces N_x, N_y, and N_{xy} can be applied without producing any curvatures or twist. In the case of a sheet with identical stiffening on both sides, there does of course exist such a plane—it is the middle surface of the sheet. If the stiffening is one-sided, however, it is usually not possible to find a reference plane that will eliminate completely the coupling between in-plane forces and out-of-plane deflections. It has been shown (Whitney and Leissa, 1969; Jones, 1975) that such coupling, occurring in the context of composite laminated plates, can have a marked effect on the buckling loads. Therefore, it is very likely that in metal plates with one-sided stiffening it can also have a marked effect on the buckling loads. A thorough investigation of this effect, using an appropriately generalized orthotropic plate theory, would be a worthwhile subject for future research, noting that some work has preceded in this direction (Mikami and Niwa, 1996).

Finally, orthotropic plate theory is incapable of modeling local buckling, that is, buckling in which the buckle wavelengths are of the same order as the stiffener spacings or the widths of the plate elements of which the stiffeners are composed. Wittrick and Horsington (1984) have developed more refined approaches that can account for local buckling and modes of buckling in which local and overall deformations appear in conjunction. Their methods are applicable to plates with unidirectional stiffening possessing certain boundary conditions and subjected to combinations of shear and biaxial compression.

4.6 INTERACTION BETWEEN PLATE ELEMENTS

The preceding sections have focused on the behavior of a single plate element supported along one or both of its longitudinal edges with or without stiffeners. The structural sections employed in practice (Fig. 4.1) are composed of plate elements arranged in a variety of configurations. It is clear that the behavior of an assembly of plates would be governed by an interaction between the plate components. In this section the mechanics of such an interaction and its implication in design are discussed briefly.

4.6.1 Buckling Modes of a Plate Assembly

Unlike a single plate element supported along the unloaded edges, a plate assembly can buckle in one of several possible modes. For the case of axial compression, instability can take on one of the following modes:

Local Buckling (Mode I): This is the purely local buckling mode discussed earlier. The mode involves out-of-plane deformation of the component plates with the junctions remaining essentially straight, and it has a half-wavelength of the same order of magnitude as the widths of the plate elements.

Distortional Buckling (Mode II): The buckling process may involve in-plane bending of one or more of the constituent plates as well as out-of-plane bending of all the elements, as in a purely local mode. Such a buckling mode is referred to as a *distortional buckling mode*, *stiffener buckling mode, local torsional mode*, or *orthotropic mode*, depending on the context. The associated wavelengths are considerably greater than those of local buckling (mode I), but there is a half-wavelength at which the critical stress is a minimum.

Global Buckling (Mode III): The plate structure may buckle as a column in flexural or flexural–torsional mode with or without interaction of local buckling.

The three modes are illustrated for an I-beam in major axis bending in Fig. 4.40, based on the finite strip analysis results of Hancock (1978). Finite strip analysis has become a popular alternative to shell finite element analysis for this problem with several programs available for this analysis (Papangelis and Hancock, 1995; Schafer and Adany, 2006). Another analysis option seeing significant recent attention is generalized beam theory (Silvestre and Camotim, 2002). Attention in this section is primarily on local buckling (mode I). Global buckling (mode III) and the interaction between local and global modes as well as distortional buckling (mode II) and its interaction with the other modes are treated in detail in Chapter 13 and elsewhere in this guide.

4.6.2 Local Buckling of a Plate Assembly

A prismatic plate structure is often viewed simply as consisting of *stiffened* and *unstiffened* plate elements. The former are plate elements supported on both of their longitudinal edges by virtue of their connection to adjacent elements, while the latter are those supported only along one of their longitudinal edges. Thus, the critical local buckling stress of a plate assembly may be taken as the smallest of the critical stresses of the plate elements, each treated as simply supported along its junctions with other plates. This stress will usually be conservative because the element providing the lower bound will be restrained by the more stable adjoining plate elements as long as the longitudinal edge joints provide effective continuity.

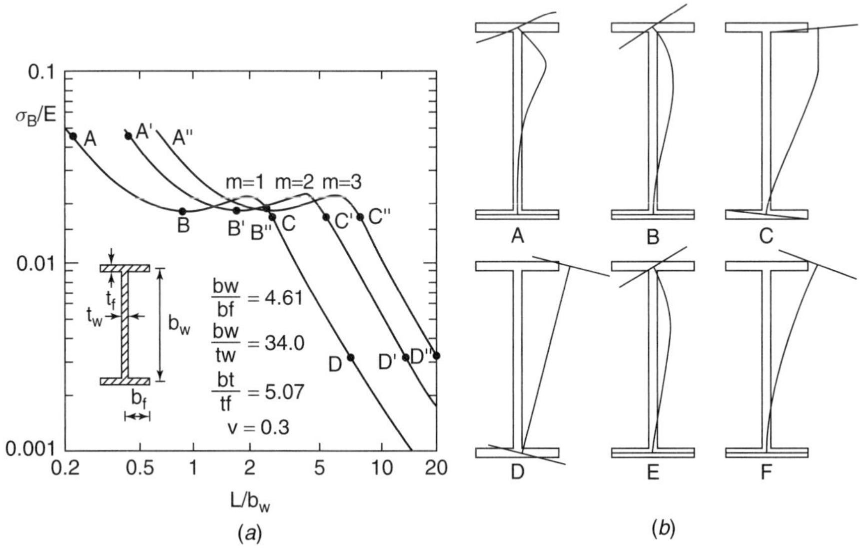

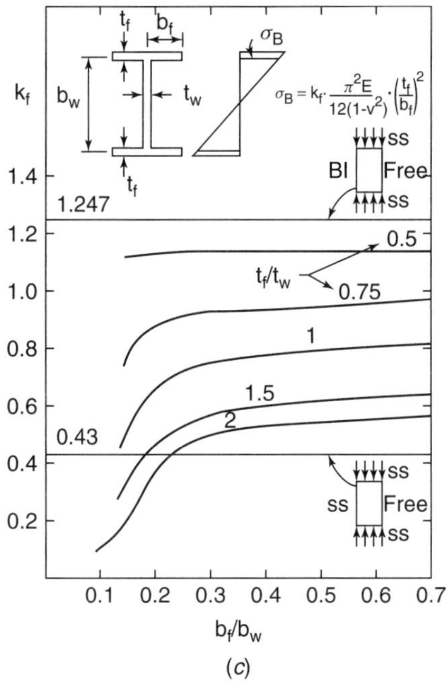

FIGURE 4.40 Stability response of an I-beam in major axis bending using the finite strip method: (*a*) half-wavelength vs. buckling stress, (*b*) buckling mode shapes, (*c*) plate-buckling coefficient (Hancock, 1978).

More complete information on k-factors as influenced by the interaction between plate components can be found in several references (Bleich, 1952; Stowell et al., 1952; Gerard and Becker, 1957/1958; Timoshenko and Gere, 1961). Such calculation, however, must be used with caution for the following reasons:

1. The results can be unduly conservative when the plate structure consists of elements with widely varying slenderness. This is the result of neglecting the rotational restraints at the junctions.
2. The results are inapplicable unless it is ensured that all the plate elements buckle locally (i.e., the junctions remain essentially straight). If, on the other hand, distortional or global (modes II or III) is critical, the result of such a simplified calculation would overpredict strength.

The intervention of distortional or stiffener buckling (mode II) may be averted in practice by designing "out" the stiffener buckling mode by providing stiffeners (edge or intermediate) of adequate rigidity. This may be advantageous because of the limited postbuckling resistance associated with the mode II type of buckling. However, it is not always practical or economical, and hence designs in which distortional buckling must be considered are discussed further in Chapter 13.

Figure 4.41 gives the variations of the local buckling coefficients k_w for a wide-flange I-section, a box section, and a Z- or channel section, all with respect to the geometrical properties of the shape. Each of these charts is divided into two portions by a dashed line curve (Kroll et al., 1943). In each portion, buckling of one of the elements dominates over the other, but the proportions exactly on the dashed line represent the case when the buckling stresses of the elements are equal. Additional charts and related information may be found in the references, such as the Japanese *Handbook of Structural Stability* (CRCJ, 1971). In addition, free or low-cost computational tools (Papangelis and Hancock, 1995; Schafer and Adany, 2006; Silvestre and Camotim, 2002) are now readily available and able to provide similar results.

4.6.3 Postbuckling of a Plate Assembly

Interaction between the elements of a plate assembly exists because of equilibrium and compatibility conditions that must be satisfied at the junction. In the case of local buckling it is possible to simplify these conditions considerably, as has been shown by Benthem (1959). Because the in-plane displacements are much smaller than the out-of-plane displacements, it is possible to assume that normal displacements are zero for each plate element meeting at a corner. Because the bending rigidity is significantly less than the extensional rigidity, it is also possible to assume that the in-plane membrane meets another plate element at an angle (Walker, 1964; Graves-Smith, 1968; Rhodes and Harvey, 1971; Tien and Wang, 1979).

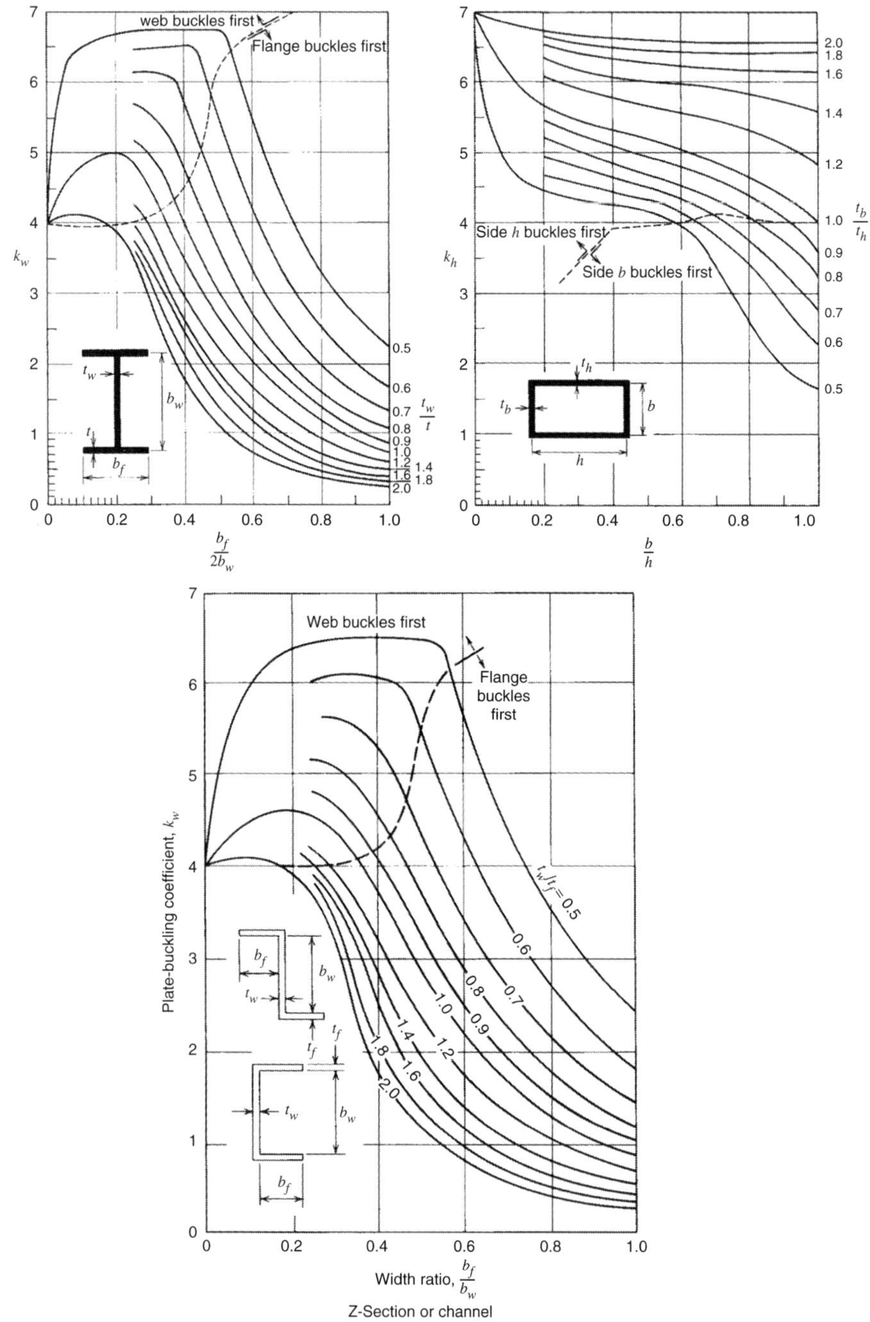

FIGURE 4.41 Plate-buckling coefficients for plate assemblies (Kroll et al., 1943).

In the postbuckling range, in-plane displacements and membrane stresses dominate the behavior of the buckled plates. The interactions between plate elements along the junctions become very complex. The problem is compounded when interaction between global and local buckling is considered or when interaction between distortional and the other modes is considered. It is for these reasons that postbuckling and strength of plate assemblies considering all possible instabilities (local, distortional, and global) remains a topic of considerable interest in research.

The earlier references on interaction of local and overall instability in the postbuckling range include those by Bijlaard and Fisher (1952), Cherry (1960), Graves-Smith (1969), Klöppel et al. (1969), Sharp (1970), and Škaloud and Zornerova (1970). Research in this field has also been carried out by Rhodes and Harvey (1976), Graves-Smith and Sridharan (1978, 1980), Kalyanaraman and Peköz (1978), Thomasson (1978), Little (1979), Rhodes and Loughlan (1980), Hancock (1980, 1981), Sridharan and Graves-Smith (1981), Fukumoto and Kubo (1982), Sridharan (1982), and Bradford and Hancock (1984). Using the concept of effective width, the problem has also been investigated by DeWolf et al. (1974), Wang and Pao, (1981), and Desmond et al. (1981a,b).

As design methods for the combination of local and global buckling mature, recent research attention has focused on distortional buckling. Lau and Hancock (1987) provided analytical formula for predicting elastic distortional buckling and later Hancock et al. (1994) provided empirical formulas for predicting the ultimate strength in distortional buckling for a variety of different cold-formed steel cross sections. Bradford (1992) has examined the distortional mode in the context of hot-rolled steel I-beams, while Schafer (2002) and Schafer and Peköz (1999) examined local, distortional, and global buckling of cold-formed steel beams and columns in developing the methods currently used in the AISI (2007) specification. The application aspect of this subject is discussed further for thin-walled members in Chapter 13.

REFERENCES

Abdel-Sayed, G. (1969), "Effective Width of Thin Plates in Compression," *ASCE J. Struct. Div.*, Vol. 95, No. ST10, pp. 2183–2204.

American Institute of Steel Construction (AISC) (2005a), *Specification for Structural Steel Buildings*, AISC 360-05, AISC, Chicago, IL.

American Institute of Steel Construction (AISC) (2005b), *Seismic Provisions for Structural Steel Buildings*, AISC 341-05, AISC, Chicago, IL.

American Iron and Steel Institute (AISI) (2007), *Specification for the Design of Cold-Formed Steel Structural Members*, AISI-S100-07, AISI, Washington, DC.

Allen, H. G., and Bulson, P. S. (1980), *Background to Buckling*, McGraw-Hill, Maidenhead, Berkshire, England.

Aluminum Association (AA) (2005), *Specification for Aluminum Structures*, AA, Arlington, VA.

Aluminum Company of America (1960), *Alcoa Structural Handbook*, Alcoa, Pittsburgh, PA.

Bambach, M. R. (2008), "Inclined Yield Lines in Flange Outstands," *Struct. Eng. Mech.*, Vol. 29, No. 6, pp. 623–642.

Bambach, M. R., and Rasmussen, K. J. R. (2004a), "Design Provisions for Sections Containing Unstiffened Elements with Stress Gradient," *J. Struct. Eng.*, Vol. 130, No. 10, pp. 1620–1628.

Bambach, M. R., and Rasmussen, K. J. R. (2004b), "Effective Widths of Unstiffened Elements with Stress Gradient," *J. Struct. Eng.*, Vol. 130, No. 10, pp. 1611–1619.

Batdorf, S. B., and Houbolt, J. C. (1946), "Critical Combinations of Shear and Transverse Direct Stress for an Infinitely Long Flat Plate with Edges Elastically Restrained Against Rotation," NACA Rep. No. 847.

Batdorf, S. B., and Stein, M. (1947), "Critical Combinations of Shear and Direct Stress for Simply Supported Rectangular Flat Plates," NACA Tech. Note No. 1223.

Becker, H., and Gerard, G. (1963), "Measurement of Torsional Rigidity of Stiffened Plates," NASA Tech. Note No. D-2007.

Benthem, J. P. (1959), "The Reduction in Stiffness of Combinations of Rectangular Plates in Compression After Exceeding the Buckling Load," NLL-TR S-539.

Bergmann, S., and Reissner, H. (1932), "Über die Knickung von rechteckigen Platten bei Schubbeanspruchung," *Z. Flugtech Motorluftschiffahrt*, Vol. 23, p. 6.

Bijlaard, P. P. (1949), "Theory and Tests on Plastic Stability of Plates and Shells," *J. Aeronaut. Sci.*, Vol. 16, pp. 529–541.

Bijlaard, P. P. (1950), "On the Plastic Buckling of Plates," *J. Aeronaut. Sci.*, Vol. 17, p. 493.

Bijlaard, P. P. (1957), "Buckling of Plates Under Nonhomogeneous Stress," *ASCE J. Eng. Mech. Div.*, Vol. 83, No. EM3, Proc. Paper 1293.

Bijlaard, P. P., and Fisher, G. P. (1952), "Column Strength of H Sections and Square Tubes, in Post Buckling Range of Component Plates," NACA Tech. Note No. 2994.

Bleich, F. (1952), *Buckling Strength of Metal Structures*, McGraw-Hill, New York.

Bleich, F., and Ramsey, L. B. (1951), *A Design Manual on the Buckling Strength of Metal Structures*, Society of Naval Architects, Washington, DC.

Bradford, M. A. (1992), "Lateral-Distortional Buckling of Steel I-Section Members," *J. Constr. Steel Res.*, Vol. 23, Nos. 1–3, pp. 97–116.

Bradford, M. A., and Azhari, M. (1995), "Inelastic Local Buckling of Plates and Plate Assemblies Using Bubble Functions," *Eng. Struct.*, Vol. 17, No. 2, pp. 95–103.

Bradford, M. A., and Hancock, G. J. (1984), "Elastic Interaction of Local and Lateral Buckling in Beams," *Thin-Wall. Struct.*, Vol. 2, No. 1, pp. 1–25.

Brockenbrough, R. L., and Johnston, B. G. (1974), *U.S. Steel Design Manual*, 2nd ed., U.S. Steel Corporation, Pittsburgh, PA.

Brown, C. J. (1990), "Elastic Buckling of Perforated Plates Subjected to Concentrated Loads," *J. Comput. Struct.*, Vol. 36, No. 6, pp. 1103–1109.

Brown, C. J., and Yettram, A. L. (1986), "The Elastic Stability of Square Perforated Plates Under Combinations of Bending, Shear and Direct load," *Thin-Wall. Struct.*, Vol. 4, No. 3, pp. 239–246.

Brown, C. J., and Yettram, A. L. (2000), "Factors Influencing the Elastic Stability of Orthotropic Plates Containing a Rectangular Cut-Out," *J. Strain Anal. Eng. Design*, Vol. 35, No. 6, pp. 445–458.

Bryan, G. H. (1891), "On the Stability of a Plane Plate Under Thrusts in Its Own Plane, with Applications to the 'Buckling' of the Sides of a Ship," *Proc. London Math. Soc.*, Vol. 22, pp. 54–67.

Bulson, P. S. (1970), *Stability of Flat Plates*, American Elsevier, New York.

Caccese, V., Elgaaly, M., and Chen, R. (1993), "Experimental Study of Thin Steel-Plate Shear Walls Under Cyclic Load," *J. Struct. Eng. N.Y.*, Vol. 119, No. 2, pp. 573–587.

Canadian Standards Association (CSA) (1984), *Cold Formed Steel Structural Members*, CSA Standard S136, CSA, Rexdale, Ontario, Canada.

Cherry, S. (1960), "The Stability of Beams with Buckled Compression Flanges," *Struct. Eng.*, Vol. 38, No. 9, pp. 277–285.

Chwalla, E. (1936), "Beitrag zur Stabilitätstheorie des Stegbleches vollwandiger Träger," *Stahlbau*, Vol. 9, p. 81.

Clark, J. W., and Rolf, R. L. (1966), "Buckling of Aluminum Columns, Plates and Beams," *ASCE J. Struct. Div.*, Vol. 92, No. ST3, pp. 17–38.

Column Research Committee of Japan (CRCJ) (1971), *Handbook of Structural Stability*, Corona Publishing Co., Tokyo.

Conley, W. F., Becker, L. A., and Allnutt, R. B. (1963), "Buckling and Ultimate Strength of Plating Loaded in Edge Compression: Progress Report 2. Unstiffened Panels," David Taylor Model Basin Rep. No. 1682.

Cook, I. T., and Rockey, K. C. (1963), "Shear Buckling of Rectangular Plates with Mixed Boundary Conditions," *Aeronaut. Q.*, Vol. 14, No. 4, pp. 349–355.

Crisfield, M. A. (1975), "Full Range Analysis of Steel Plates and Stiffened Plating Under Uniaxial Compression," *Proc. Inst. Civ. Eng.*, Part 2, pp. 1249–1255.

Davidson, J. S., Ballance, S. R., and Yoo, C. H. (1999), "Analytical Model of Curved I-Girder Web Panels Subjected to Bending," *J. Bridge Eng.*, Vol. 4, No. 3, pp. 204–212.

Dawe, J. L., and Kulak, G. L. (1984), *"Local Buckling of W-Shaped Columns and Beams," ASCE J. Struct. Eng.*, Vol. 110, No. 6, pp. 1292–1304.

Dawe, J. L., and Kulak, G. L. (1986), "Local Buckling of Beam-Columns," *ASCE J. Struct. Eng.*, Vol. 112, No. 11, pp. 2447–2461.

Dawson, R. G., and Walker, A. C. (1972), "Post-Buckling of Geometrically Imperfect Plates," *ASCE J. Struct. Div.*, Vol. 98, No. ST1, pp. 75–94.

Desmond, T. P., Peköz, T., and Winter, G. (1981a), "Edge Stiffeners for Thin-Walled Members," *ASCE J. Struct. Div.*, Vol. 107, No. ST2, Proc. Pap. 16056, pp. 329–354.

Desmond, T. P., Peköz, T., and Winter, G. (1981b), "Intermediate Stiffeners for Thin Walled Members," *ASCE J. Struct. Div.*, Vol. 107, No. ST4, Proc. Pap. 16194, pp. 627–648.

DeWolf, J. T., Peköz, T., and Winter, G. (1974), "Local and Overall Buckling of Cold Formed Members," *ASCE J. Struct. Div.*, Vol. 100, No. ST10, pp. 2017–2306.

Dow, N. F., Libove, C., and Hubka, R. E. (1954), "Formulas for the Elastic Constants of Plates with Integral Waffle-Like Stiffening," NACA Rep. No. 1195.

Dwight, J. B. (1971), "Collapse of Steel Compression Panels," *Proc. Conf. Dev. Bridge Des. Constr.*, Crosby Lockwood, London.

Dwight, J. B., and Little, G. H. (1976), "Stiffened Steel Compression Flanges, a Simpler Approach," *Struct. Eng.*, Vol. 54, No. 12, pp. 501–509.

Dwight, J. B., and Moxham, K. E. (1969), "Welded Steel Plates in Compression," *Struct. Eng.*, Vol. 47, No. 2, pp. 49–66.

Dwight, J. B., and Ratcliffe, A. T. (1968), "The Strength of Thin Plates in Compression," in *Thin-Walled Steel Structures* (Eds. K. C. Rockey and H. V. Hill), Crosby Lockwood, London.

Earls, C. J. (1999), "On the Inelastic Failure of High Strength Steel I-Shaped Beams," *J. Constr. Steel Res.*, Vol. 49, No. 1, pp. 1–24.

Elangovan, P. T., and Prinsze, M. (1992), "Critical Shear Stress of Flat Rectangular Plates with Free Edges," *J. Thin-Wall. Struct.*, Vol. 13, No. 5, pp. 409–425.

Elgaaly, M. (1998), "Thin Steel Plate Shear Walls Behavior and Analysis," *Thin-Wall. Struct.*, Vol. 32, Nos. 1–3, pp. 151–180.

El-Sawy, K. M., and Nazmy, A. S. (2001), "Effect of Aspect Ratio on the Elastic Buckling of Uniaxially Loaded Plates with Eccentric Holes," *Thin-Wall. Struct.*, Vol. 39, No. 12, pp. 983–998.

Elsharkawi, K., and Walker, A. C. (1980), "Buckling of Multibay Stiffened Plate Panels," *ASCE J. Struct. Div.*, Vol. 106, No. ST8, pp. 1695–1716.

Featherston, C. A., and Ruiz, C. (1998), "Buckling of Curved Panels Under Combined Shear and Compression," *Proc. Inst. Mech. Eng. Part C: J. Mech. Eng. Sci.*, Vol, 212, No. 3, pp. 183–196.

Fok, W. C., Walker, A. C., and Rhodes, J. (1977), "Buckling of Locally Imperfect Stiffeners in Plates," *ASCE J. Eng. Mech. Div.*, Vol. 103, No. EM5, pp. 895–911.

Fujita, Y., Yoshida, K., and Arai, H. (1970), "Instability of Plates with Holes," 3rd Rep., *J. Soc. Nav. Archit. Jpn.*, Vol. 127, pp. 161–169.

Fukomoto, Y., and Kubo, M. (1982), "Buckling in Steel U Shaped Beams," *ASCE J. Struct. Div.*, Vol. 108, No. ST5, pp. 1174–1190.

Fukomoto, Y., Usami, T., and Yamaguchi, K. (1977), "Inelastic Buckling Strength of Stiffened Plates in Compression," *IABSE Period.* 3/1977, *IABSE Proc.* p-8/77, pp. 1–15.

Gerard, G., and Becker, H. (1957/1958), *Handbook of Structural Stability*, six parts, NACA Tech. Notes Nos. 3781–3786.

Ghavami, K. (1994), "Experimental Study of Stiffened Plates in Compression Up to Collapse," *J. Constr. Steel Res.*, Vol. 28, No. 2, pp. 197–221.

Gioncu, V., and Petcu, D. (1997), "Available Rotation Capacity of Wide-Flange Beams and Beam-Columns: Part 1. Theoretical Approaches," *J. Constr. Steel Res.*, Vol. 43, Nos. 1–3, pp. 161–217.

Goncalves, R., and Camotim, D. (2007), "Thin-Walled Member Plastic Bifurcation Analysis Using Generalised Beam Theory," *Adv. Eng. Software*, Vol. 38, Nos. 8–9, pp. 637–646.

Graves-Smith, T. R. (1968), "The Postbuckled Behavior of Thin Walled Columns," paper presented at the 8th Cong. IABSE, New York, pp. 311–320.

Graves-Smith, T. R. (1969), "The Ultimate Strength of Locally Buckled Columns of Arbitrary Length," in *Thin-Walled Steel Structures* (Eds. K. C. Rockey and H. V. Hill), Crosby Lockwood, London.

Graves-Smith, T. R., and Sridharan, S. (1978), "A Finite Strip Method for the Buckling of Plate Structures Under Arbitrary Loading," *Int. J. Mech. Sci.*, Vol. 20, pp. 685–693.

Graves-Smith, T. R., and Sridharan, S. (1980), "The Local Collapse of Elastic Thin-Walled Columns," *J. Struct. Mech.*, Vol. 13, No. 4, pp. 471–489.

Grondin, G. Y., Chen, Q., Elwi, A. E., and Cheng, J. J. (1998), "Stiffened Steel Plates Under Compression and Bending," *J. Constr. Steel Res.*, Vol. 45, No. 2, pp. 125–148.

Grondin, G. Y., Elwi, A. E., and Cheng, J. J. (1999), "Buckling of Stiffened Steel Plates—A Parametric Study," *J. Constr. Steel Res.*, Vol. 50, No. 2, pp. 151–175.

Grosskurth, J. F., White, R. N., and Gallagher, R. H. (1976), "Shear Buckling of Square Perforated Plates," *ASCE J. Eng. Mech. Div.*, Vol. 102, No. EM6, Proc. Pap. 12641, pp. 1025–1040.

Hancock, G. J. (1978), "Local, Distortional, and Lateral Buckling of I-Beams," *ASCE J. Struct. Div.*, Vol. 104, No. 11, pp. 1787–1798.

Hancock, G. J. (1980), "Design Methods for Thin-Walled Box Columns," Rep. No. 359, University of Sydney, Sydney, Australia.

Hancock, G. J. (1981), "Interaction Buckling of I-Section Columns," *ASCE J. Struct. Div.*, Vol. 107, No. ST1, pp. 165–181.

Hancock, G. J., Kwon, Y. B., and Stefan Bernard, E. (1994), "Strength Design Curves for Thin-Walled Sections Undergoing Distortional Buckling," *J. Constr. Steel Res.*, Vol. 31, Nos. 2–3, pp. 169–186.

Harris, H. G., and Pifko, A. B. (1969), "Elastic-Plastic Buckling of Stiffened Rectangular Plates," *Proc. Symp. Appl. Finite Elem. Methods Div. Eng.*, Vanderbilt University, Nashville, TN, Nov.

Hitaka, T., and Matsui, C. (2003), "Experimental Study on Steel Shear Wall with Slits," *J. Struct. Eng.*, Vol. 129, No. 5, pp. 586–595.

Horne, M. R., and Narayanan, R. (1975), "An Approximate Method for the Design of Stiffened Steel Compression Panels," *Proc. Inst. Civ. Eng.*, Part 2, Vol. 59, pp. 501–514.

Horne, M. R., and Narayanan, R. (1977), "Design of Axially Loaded Stiffened Plates," *ASCE J. Struct. Div.*, Vol. 103, No. ST11, pp. 2243–2257.

Iguchi, S. (1938), "Die Kninckung der rechteckigen Platte durch Schubkräfte," *Ing. Arch.*, Vol. 9, p. 1.

Inoue, T., and Kato, B. (1993), "Analysis of Plastic Buckling of Steel Plates," *Int. J. Solids Struct.*, Vol. 30, No. 6, pp. 835–856.

Isami, T., and Hidenori, K. (1990), "New Strength Prediction Method for Biaxially Compressed Plates," *Proc. 39th Jpn. Nat. Congr. Appl. Mech. Theor. Appl. Mech.*, Vol. 39, University of Tokyo Press, Tokyo, pp. 109–118.

Johns, D. J. (1971), "Shear Buckling of Isotropic and Orthotropic Plates: A Review," Tech. Rep. R&M No. 3677, British Aeronautical Research Council, London.

Johnson, A. E., Jr. (1957), "Charts Relating the Compressive and Shear Buckling Stress of Longitudinally Supported Plates to the Effective Deflectional Stiffness," NACA Tech. Note No. 4188.

Johnson, A. L., and Winter, G. (1966), "Behavior of Stainless Steel Columns and Beams," *ASCE J. Struct. Div.*, Vol. 92, No. ST5, p. 97.

Jombock, J. R., and Clark, J. W. (1962), "Post-Buckling Behavior of Flat Plates," *Trans. Am. Soc. Civ. Eng.*, Vol. 127, Part I, pp. 227–240.

Jombock, J. R., and Clark, J. W. (1968), "Bending Strength of Aluminum Formed Sheet Members," *ASCE J. Struct. Div.*, Vol. 94, No. ST2, pp. 511–528.

Jones, R. M. (1975), *Mechanics of Composite Materials*, Scripta, Silver Spring, MD, pp. 264–267.

Kalyanaraman, V., and Peköz, T. (1978), "Analytical Study of Unstiffened Elements," *ASCE J. Struct. Div.*, Vol. 104, No. ST9, pp. 1507–1524.

Kalyanaraman, V., Peköz, T., and Winter, G. (1977), "Unstiffened Compression Elements," *ASCE J. Struct. Div.*, Vol. 103, No. ST9, pp. 1833–1848.

Kawai, T., and Ohtsubo, H. (1968), "A Method of Solution for the Complicated Buckling Problems of Elastic Plates with Combined Use of Raleigh–Ritz Procedure in the Finite Element Method," *Proc. 2nd Air Force Conf. Matrix Methods Struct. Mech.*, Oct.

Kemp, A. R. (1996), "Inelastic Local and Lateral Buckling in Design Codes," *ASCE J. Struct. Eng.*, Vol. 122, No. 4, pp. 374–382.

Ketter, R. L. (1962), "Further Studies on the Strength of Beam-Columns," *Trans. ASCE*, Vol. 127, No. 2, p. 224.

Klitchieff, J. M. (1949), "On the Stability of Plates Reinforced by Ribs," *J. Appl. Mech.*, Vol. 16, No. 1.

Klöppel, K., Schmied, R., and Schubert, J. (1969), "Ultimate Strength of Thin Walled Box Shaped Columns Under Concentric and Eccentric Loads: Post Buckling Behavior of Plate Elements and Large Deformations Are Considered," *Stahlbau*, Vol. 38, pp. 9–19, 73–83.

Koiter, W. T. (1963), "Introduction to the Post-Buckling Behavior of Flat Plates," paper presented at the Int. Colloq. Comportement Postcritiques Plaques Utilisée Constr. Metall., Institute of Civil Engineering, Liege, Belgium.

Kosteski, N., Packer, J. A., and Puthli, R. S. (2003), "A Finite Element Method Based Yield Load Determination Procedure for Hollow Sturctural Section Connections," *J. Constr. Steel Res.*, Vol. 59, No. 4, pp. 453–471.

Kroll, W. D., Fisher, G. P., and Heimerl, G. J. (1943), "Charts for the Calculation of the Critical Stress for Local Instability of Columns with I, Z, Channel and Rectangular Tube Sections," NACA Wartime Rep. No. L-429.

Kuhlmann, U. (1989), "Definition of Flange Slenderness Limits on the Basis of Rotation Capacity Values," *J. Constr. Steel Res.*, Vol. 14, No. 1, pp. 21–40.

Kumai, T. (1952), "Elastic Stability of a Square Plate with a Central Circular Hole Under Edge Thrust," *Rep. Inst. Appl. Mech. Kyusyn Univ.*, Vol. 1, No. 2, p. 1.

Kusuda, T. (1959), "Buckling of Stiffened Panels in Elastic and Strain-Hardening Range," Rep. No. 39, Transportation Technical Research Institute (Unyn-Gijutsu Kenkyrijo Mejiro, Toshima-Ku), Tokyo.

Kwon, Y. B., and Hancock, G. J. (1991), "A Nonlinear Elastic Spline Finite Strip Analysis for Thin-Walled Sections," *Thin-Wall. Struct.*, Vol. 12, No. 4, pp. 295–319.

LaBoube, R. A., and Yu, W. W. (1982), "Bending Strength of Webs of Cold-Formed Steel Beams," *ASCE J. Struct. Div.*, Vol. 108, No. ST7, pp. 1589–1604.

Lau, J. H. (1981), "Stiffness of Corrugated Plate," *ASCE J. Eng. Mech. Div.*, Vol. 17, No. EM1, pp. 271–275.

Lau, S. C. W., and Hancock, G. J. (1986), "Buckling of Thin Flat-Walled Structures by a Spline Finite Strip Method," *Thin-Wall. Struct.*, Vol. 4, No. 4, pp. 269–294.

Lau, S. C. W., and Hancock, G. J. (1987), "Distortional Buckling Formulas for Channel Columns," *J. Struct. Eng. N.Y.*, Vol. 113, No. 5, pp. 1063–1078.

Lau, S. C. W., and Hancock, G. J. (1989), "Inelastic Buckling Analyses of Beams, Columns and Plates Using the Spline Finite Strip Method," *Thin-Wall. Struct.*, 7, Nos. 3–4, pp. 213–238.

Lay, M. G. (1965), "Flange Local Buckling in Wide-Flange Shapes," *ASCE J. Struct. Div.*, Vol. 91, No. ST6, pp. 49–66.

Legget, D. M. A. (1941), "The Buckling of a Square Panel in Shear When One Pair of Opposite Edges Is Clamped and the Other Pair Simply Supported," Tech. Rep. R&M No. 1991, British Aeronautical Research Council, London.

Levy, S., Wolley, R. M., and Knoll, W. D. (1947), "Instability of Simply Supported Square Plate with Reinforced Circular Hole in Edge Compression,' *J. Res. Natl. Bur. Stand.*, Vol. 39, pp. 571–577.

Libove, C. (1983), "Buckle Pattern of Biaxially Compressed Simply Supported Orthotropic Rectangular Plates," *J. Compos. Mater.*, Vol. 17, No. 1, pp. 45–48.

Libove, C., and Batdorf, S. B. (1948), "A General Small-Deflection Theory for Flat Sandwich Plates," NACA Rep. No. 899.

Libove, C., Ferdman, S., and Reusch, J. J. (1949), "Elastic Buckling of a Simply Supported Plate Under a Compressive Stress That Varies Linearly in the Direction of Loading," NACA Technical Note No. 1891.

Libove, C., and Hubka, R. E. (1951), "Elastic Constants for Corrugated-Core Sandwich Plates," NACA Tech. Note No. 2289.

Lind, N. C. (1973), "Buckling of Longitudinally Stiffened Sheets," *ASCE J. Struct. Div.*, Vol. 99, No. ST7, pp. 1686–1691.

Little, G. H. (1976), "Stiffened Steel Compression Panels: Theoretical Failure Analysis," *Struct. Eng.*, Vol. 54, No. 12, pp. 489–500.

Little, G. H. (1979), "The Strength of Square Steel Box-Columns: Design Curves and Their Theoretical Basis," *Struct. Eng.*, Vol. 57 . No. 2, pp. 49–61.

Lu, L. W., and Kamalvand, H. (1968), "Ultimate Strength of Laterally Loaded Columns," *ASCE J. Struct. Div.*, Vol. 94, No. ST6, pp. 1505–1524.

Lubell, A. S., Prion, H. G. L., Ventura, C. E., and Rezai, M. (2000), "Unstiffened Steel Plate Shear Wall Performance Under Cyclic Loading," *J. Struct. Eng. N.Y.*, Vol. 126, No. 4, pp. 453–460.

May, I. M., and Ganaba, T. H. (1988), "Elastic Stability of Plates with and Without Openings," *J. Eng. Comput.*, Vol. 5, No. 1, pp. 50–52.

McKenzie, K. I. (1964), "The Buckling of Rectangular Plate Under Combined Biaxial Compression, Bending and Shear," *Aeronaut. Q.*, Vol. 15, No. 3, pp. 239–246.

Merrison Committee (1973), "Inquiry into the Basis of Design and Method of Erection of Steel Box Girder Bridges (Interior Design and Workmanship Rules)," Merrison Committee, London.

Michael, M. E. (1960), "Critical Shear Stress for Webs with Holes," *J. R. Aeronaut. Soc.*, Vol. 64, p. 268.

Mikami, I., and Niwa, K. (1996), "Ultimate Compressive Strength of Orthogonally Stiffened Steel Plates," *J. Struct. Eng.*, Vol. 122, No. 6, pp. 674–681.

Miller, T. H., and Peköz, T. (1994), "Unstiffened Strip Approach for Perforated Wall Studs," *ASCE J. Struct. Eng.*, Vol. 120, No. 2, pp. 410–421.

Moen, C. D., and Schafer, B. W. (2008), "Simplified Methods for Predicting Elastic Buckling of Cold-Formed Steel Structural Members with Holes," *Proceedings of the 19th International Specialty Conference on Cold-Formed Steel*, St. Louis, MO, pp. 17–32.

Moen, C. D., and Schafer, B. W. (2006), "Impact of Holes on the Elastic Buckling of Cold-Formed Steel Columns with Applications to the Direct Strength Method," paper presented at the Eighteenth International Specialty Conference on Cold-Formed Steel Structures, Orlando, FL, pp. 269–283.

Moheit, W. (1939), "Schubbeulung rechteckiger Platten mit eingespannten Rändern," Thesis, Technische Hochschule Darmstadt, Leipzig, Germany.

Murray, J. M. (1946), "Pietzker's Effective Breadth of Flange Reexamined," *Engineering*, Vol. 161, p. 364.

Murray, N. W. (1973), "Buckling of Stiffened Panel Loaded Axially and in Bending," *Struct. Eng.*, Vol. 51, No. 8, pp. 285–301.

Murray, N. W. (1975), "Analysis and Design of Stiffened Plates for Collapse Load," *Struct. Eng.*, Vol. 53, No. 3, pp. 153–158.

Murray, N. W. (1984). *Introduction to the Theory of Thin-Walled Structures*, Oxford University Press.

Murray, N. W., Khoo, P. S. (1981), "Some Basic Plastic Mechanisms in Thin-Walled Steel Structures," *Int. J. Mech. Sci.*, 23 (12), pp. 703–713.

Nemeth, J., and Michael, P. (1990), "Buckling and Post-Buckling Behavior of Compression-Loaded Isotropic Plates with Cutouts," *Proc. 31st Struct. Dyn. Mater. Conf.*, April 2–4, Part 2, American Institute of Aeronautics and Astronautics, New York, pp. 862–876.

Newell, J. S. (1930), "The Strength of Aluminum Alloy Sheets," *Airway Age*, Vol. 11, p. 1420 ; Vol. 12, p. 1548.

Nguyen, R. P., and Yu, W. W. (1982), "Longitudinally Reinforced Cold-Formed Steel Beam Webs," *ASCE J. Struct. Div.*, Vol. 108, No. ST11, pp. 2423–2442.

Paik, J. K., Thayamballi, A. K., and Lee, J. M. (2004), "Effect of Initial Deflection Shape on the Ultimate Strength Behavior of Welded Steel Plates Under Biaxial Compressive Loads," *J. Ship Res.*, Vol. 48, No. 1, pp. 45–60.

Papangelis, J. P., and Hancock, G. J. (1995), "Computer Analysis of Thin-Walled Structural Members," *Comput. Struct.*, Vol. 56, No. 1, pp. 157–176.

Pavlovic, M. N., and Baker, G. (1989), "Effect of Non-Uniform Stress Distribution on the Stability of Biaxially Compressed Plates," *Int. J. Struct. Mech. Mater. Sci.*, Vol. 28, Nos. 1–4, pp. 69–89.

Peköz, T. (1987), "Development of a Unified Approach to the Design of Cold-Formed Steel Members," American Iron and Steel Institute, Washington, DC.

Perel, D., and Libove, C. (1978), "Elastic Buckling of Infinitely Long Trapezoidally Corrugated Plates in Shear," *Trans. ASME J. Appl. Mech.*, Vol. 45, pp. 579–582.

Peters, R. W. (1954), "Buckling of Long Square Tubes in Combined Compression and Torsion and Comparison with Flat-Plate Buckling Theories," NACA Tech. Note No. 3184.

Ramberg, W., McPherson, A. E., and Levy, S. (1939), "Experiments on Study of Deformation and of Effective Width in Axially Loaded Sheet-Stringer Panels," NACA Tech. Note No. 684.

Redwood, R. G., and Uenoya, M. (1979), "Critical Loads for Webs with Holes," *ASCE J. Struct. Div.*, Vol. 105, No. ST10, pp. 2053–2068.

Rhodes, J., and Harvey, J. M. (1971), "Effects of Eccentricity of Load or Compression on the Buckling and Post-Buckling Behavior of Flat Plates," *Int. J. Mech. Sci.*, Vol. 13, pp. 867–879.

Rhodes, J., and Harvey, J. M. (1976), "Plain Channel Section Struts in Compression and Bending Beyond the Local Buckling Load," *Int. J. Mech. Sci.*, Vol. 8, pp. 511–519.

Rhodes, J., and Harvey, J. M. (1977), "Examination of Plate Post Buckling Behavior," *ASCE J. Eng. Mech. Div.*, Vol. 103, No. EM3, pp. 461–480.

Rhodes, J., and Loughlan, J. (1980), "Simple Design Analysis of Lipped Simple Columns," *Proc. 5th Int. Spec. Conf. Cold-Formed Steel Struct.*, St. Louis, MO.

Roberts, T. M., and Sabouri-Ghomi, S. (1992), "Hysteretic Characteristics of Unstiffened Perforated Steel Plate Shear Panels," *Thin-Wall. Struct.*, Vol. 14, No. 2, pp. 139–151.

Rockey, K. C. (1980), "The Buckling and Post-Buckling Behavior of Shear Panels Which Have a Central Circular Cut Out," in *Thin Walled Structures* (Eds. J. Rhodes and A. C. Walker), Granada, London.

Rockey, K. C., Anderson, R. G., and Cheung, Y. K. (1969), "The Behavior of Square Shear Webs Having a Circular Hole," in *Thin-Walled Steel Structures* (Eds. K. C. Rockey and H. V. Hill), Crosby Lockwood, London.

Sabouri-Ghomi, S., and Roberts, T. M. (1992), "Nonlinear Dynamic Analysis of Steel Plate Shear Walls Including Shear and Bending Deformations," *Eng. Struct.*, Vol. 14, No. 5, pp. 309–317.

Schafer, B. W. (2002), "Local, Distortional, and Euler Buckling of Thin-Walled Columns," *J. Struct. Eng.*, Vol. 128, No. 3, pp. 289–299.

Schafer, B. W., and Adany, S. (2006), "Buckling Analysis of Cold-Formed Steel Members Using CUFSM: Conventional and Constrained Finite Strip Methods," paper presented at the Eighteenth International Specialty Conference on Cold-Formed Steel Structures, Orlando, FL, pp. 39–54.

Schafer, B. W., and Peköz, T. (1998), "Cold-Formed Steel Members with Multiple Longitudinal Intermediate Stiffeners," *J. Struct. Eng.*, Vol. 124, No. 10, pp. 1175–1181.

Schafer, B. W., and Peköz, T. (1999), "Laterally Braced Cold-Formed Steel Flexural Members with Edge Stiffened Flanges," *J. Struct. Eng.*, Vol, 125, No. 2, pp. 118–126.

Schlack, A. L. (1964), "Elastic Stability of Pierced Square Plates," *Exp. Mech.*, p. Vol. 4, No. 6, pp. 167–172.

Schuman, L., and Back, G. (1930), "Strength of Rectangular Flat Plates Under Edge Compression," NACA Tech. Rep. No. 356.

Seide, P., and Stein, M. (1949), "Compressive Buckling of Simply Supported Plates with Longitudinal Stiffeners," NACA Tech. Note No. 1825.

Seydel, E. (1933), Über das Ausbeulen von rechteckigen isotropen oder orthogonalanisotropen Platten bei Schubbeanspruchung," *Ing. Arch.*, Vol. 4, p. 169.

Shahwan, K. W., and Waas, A. M. (1998), "Buckling of Unilaterally Constrained Infinite Plates," *J. Eng. Mech.*, Vol. 124, No. 2, pp. 127–136.

Sharp, M. L. (1966), "Longitudinal Stiffeners for Compression Members," *ASCE J. Struct. Div.*, Vol. 92, No. ST5, pp. 187–212.

Sharp, M. L. (1970), "Strength of Beams or Columns with Buckled Elements," *ASCE J. Struct. Div.*, Vol. 96, No. ST5, pp. 1011–1015.

Sheikh, I. A., Elwi, A. E., and Grondin, G. Y. (2003), "Stiffened Steel Plates Under Combined Compression and Bending," *J. Constr. Steel Res.*, Vol. 59, No. 7, pp. 911–930.

Sherbourne, A. N., Marsh, C., and Lian, C. Y. (1971), "Stiffened Plates in Uniaxial Compression," *IABSE Publ.*, Vol. 31-I, Zurich, pp. 145–178.

Shi, Y. J., Chan, S. L., and Wong, Y. L. (1996), "Modeling for Moment-Rotation Characteristics for End-Plate Connections," *J. Struct. Eng.*, Vol. 122, No. 11, pp. 1300–1306.

Silvestre, N., and Camotim, D. (2002), "Second-Order Generalised Beam Theory for Arbitrary Orthotropic Materials," *Thin-Wall. Struct.*, Vol. 40, No. 9, pp. 791–820.

Škaloud, M., and Zornerova, M. (1970), "Experimental Investigation into the Interaction of the Buckling of Compressed Thin-Walled Columns with the Buckling of Their Plate Elements," Acta Tecnica CSAV No. 4.

Smith, S. T., Bradford, M. A., and Oehlers, D. J. (1999), "Local Buckling of Side-Plated Reinforced-Concrete Beams. I: Theoretical Study," *J. Struct. Eng.*, Vol. 125, No. 6, pp. 622–633.

Southwell, R. V., and Skan, S. (1924), "On the Stability Under Shearing Force of a Flat Elastic Strip," *Proc. R. Soc.*, Vol. A105 .

Sridharan, S. (1982), "A Semi-Analytical Method for the Post-Local-Torsional Buckling Analysis of Prismatic Plate Structures," *Int. J. Numer. Methods Eng.*, Vol. 18, pp. 1685–1607.

Sridharan, S., and Graves-Smith, T. R. (1981), "Post-Buckling Analysis with Finite Strips," *ASCE J. Eng. Mech. Div.*, Vol. 107, No, EM5, pp. 869–888.

Stein, M. (1959), "The Phenomenon of Change in Buckle Pattern in Elastic Structures," NACA Tech. Note No. R-39.

Stein, M. (1989), "Effect of Transverse Shearing Flexibility on Post-Buckling on Plates in Shear," *AIAA J.*, Vol. 27, No. 5, pp. 652–655.

Stein, O. (1936), "Stabilität ebener Rechteckbleche unter Biegung and Schub," *Bauingenieur*, Vol. 17, p. 308.

Stowell, E. Z. (1948), "A Unified Theory of Plastic Buckling of Columns and Plates," NACA Tech. Note No. 1556.

Stowell, E. Z. (1949), "Plastic Buckling of a Long Flat Plate Under Combined Shear and Longitudinal Compression," NACA Tech. Note No. 1990.

Stowell, E. Z., Heimerl, G. J., Libove, C., and Lundquist, E. E. (1952), "Buckling Stresses for Flat Plates and Sections," *Trans. Am. Soc. Civ. Eng.*, Vol. 117, pp. 545–578.

Supple, W. J. (1980), "Buckling of Plates Under Axial Load and Lateral Pressures," in *Thin Walled Structures* (Eds. J. Rhodes and A. C. Walker), Granada, London.

Thomasson, P.-O. (1978), "Thin-Walled C-Shaped Panels in Axial Compression," Document D.I. 1978, Swedish Council for Building Research, Stockholm.

Tien, Y. L., and Wang, S. T. (1979), "Local Buckling of Beams Under Stress Gradient," *ASCE J. Struct. Div.*, Vol. 105, No. ST8, pp. 1571–1588.

Timoshenko, S. P. (1910), "Einge Stabilitätsprobleme der Elastizitätstheorie," *Z. Math. Phys.*, Vol. 58, p. 337.

Timoshenko, S. P. (1934), "Stability of the Webs of Plate Girders," *Engineering*, Vol. 138, p. 207.

Timoshenko, S. P., and Gere, J. M. (1961), *Theory of Elastic Stability*, 2nd ed., McGraw-Hill, New York.

Troitsky, D. S. C. (1976), *Stiffened Plates, Bending, Stability and Vibrations*, Elsevier, New York.

Tsien, H. S. (1942), "Buckling of a Column with Nonlinear Lateral Supports," *J. Aeronaut. Sci.*, Vol. 9, No. 4, pp. 119–132.

Tvergaard, V. (1973), "Imperfection-Sensitivity of a Wide Integrally Stiffened Panel Under Compression," *Int. J. Solids Struct.*, Vol. 9, pp. 177–192.

Usami, T. (1982), "Post Buckling of Plates in Compression and Bending," *ASCE J. Struct. Div.*, Vol. 108, No. ST3, pp. 591–609.

Vann, P. W. (1971), "Compressive Buckling of Perforated Plate Elements," paper presented at the First Specialty Conference on Cold-formed Structures, Rolla, MO, pp. 58–64.

Vojta, J. F., and Ostapenko, A. (1967), "Ultimate Strength Design Curves for Longitudinally Stiffened Plate Panels with Large *b/t*," Fritz Eng. Lab. Rep. No. 248.19, Lehigh University, Bethlehem, PA.

von Kármán, T., Sechler, E. E., and Donnell, L. H. (1932), "Strength of Thin Plates in Compression," *Trans. A.S.M.E.*, Vol. 54, No. APM-54-5, p. 53.

Walker, A. C. (1964), "Thin Walled Structural Forms Under Eccentric Compressive Load Actions," Ph.D. thesis, University of Glasgow, Glasgow, Scotland.

Wang, S. T. (1969), "Cold-Rolled Austenitic Stainless Steel: Material Properties and Structural Performance," Cornell Univ. Dept. Struct. Eng. Rep. No. 334.

Wang, S. T., and Pao, H. Y. (1981), "Torsional-Flexural Buckling of Locally Buckled Columns," *Int. J. Comput. Struct.*, Vol. 11, pp. 127–136.

Wang, S. T., Errera, S. J., and Winter, G. (1975), "Behavior of Cold Rolled Stainless Steel Members," *ASCE J. Struct. Div.*, Vol. 101, No. ST11, pp. 2337–2357.

Way, S. (1936), "Stability of Rectangular Plates Under Shear and Bending Forces," *J. Appl. Mech. ASME*, Vol. 3, No. 4, pp. A131–A135.

Whitney, J. M., and Leissa, A. W. (1969), "Analysis of Heterogeneous Anisotropic Plates," *Trans. ASME J. Appl. Mech.*, Vol., 36, pp. 261–266.

Winter, G. (1947), "Strength of Thin Steel Compression Flanges," *Trans. ASCE*, Vol. 112, p. 527.

Winter, G. (1983), "Commentary on the 1968 Edition of the Specification for the Design of Steel Structural Members," in *Cold-Formed Steel Design Manual*, American Iron and Steel Institute, Washington, DC.

Winter, G., Lansing, W., and McCalley, R. B. (1950), "Four Papers on the Performance of Thin Walled Steel Structures," Eng. Exp. Strn. Rep. No. 33, Cornell University, Ithaca, NY, pp. 27–32, 51–57.

Wittrick, W. H. (1952), "Correlation Between Some Stability Problems for Orthotropic and Isotropic Plates Under Bi-Axial and Uniaxial Direct Stress," *Aeronaut. Q.*, Vol. 4, pp. 83–99.

Wittrick, W. H. (1968), "A Unified Approach to the Initial Buckling of Stiffened Panels in Compression," *Aeronaut. Q.*, Vol. 19, p. 265.

Wittrick, W. H., and Horsington, R. W. (1984), "Buckling and Vibration of Composite Folded Plate Structures of Finite Length in Combined Shear and Compression," *Proc. R. Soc. London*, Vol. A392, pp. 107–144.

Yang, H. T. Y. (1969), "A Finite Element Formulation for Stability Analysis of Doubly Curved Thin Shell Structures," Ph.D. thesis, Cornell University, Ithaca, NY.

Yoshizuka, J., and Naruoka, M. (1971), "Buckling Coefficient of Simply Supported Rectangular Plates Under Combined Bending and Compressive Stresses in Two Perpendicular Directions," *Stahlbau*, Vol. 40, p. 217.

Yu, C., and Schafer, B. W. (2006), "Effect of Longitudinal Stress Gradient on the Ultimate Strength of Thin Plates," *Thin-Wall. Struct.*, Vol. 44, No. 7, pp. 787–799.

Yu, C., and Schafer, B. W. (2007), "Effect of Longitudinal Stress Gradients on Elastic Buckling of Thin Plates," *J. Eng. Mech.*, Vol. 133, No. 4, pp. 452–463.

Yu, W. W., and Davis, C. S. (1973), "Cold Formed Steel Members with Perforated Elements," *ASCE J. Struct. Div.*, Vol. 99, No. ST10, p. 2061.

Zender, W., and Hall, J. B., Jr. (1960), "Combinations of Shear Compressive Thermal and Compressive Load Stresses for the Onset of Permanent Buckles in Plates," NASA Tech. Note No. D-384.

CHAPTER 5

BEAMS

5.1 INTRODUCTION

Beams, girders, joists, and trusses subjected to flexure typically have much greater strength and stiffness in the plane in which the loads are applied (in the plane associated with bending about their major principal axis) than in the plane associated with bending about their minor principal axis. Unless these members are properly braced against lateral deflection and twisting, they are subject to failure by lateral–torsional buckling (LTB) prior to the attainment of their full in-plane capacity. They are especially prone to this type of buckling during the construction phase, when braces are either absent or different in type from the permanent ones.

Lateral–torsional buckling is a limit state of structural usefulness where the deformation changes from predominantly in-plane bending to combined lateral deflection and twisting. The final failure pattern involves lateral deflection and twisting in combination with various extents of yielding and flange and/or web local buckling depending on the specific member characteristics. Lateral–torsional buckling can be avoided by:

1. Providing properly spaced and designed lateral and/or torsional bracing
2. Connecting open-section beam groups intermittently by triangulated lacing or diaphragms
3. Using cross sections that are torsionally stiff, such as box-shaped sections
4. Ensuring that the required design moment does not exceed the LTB capacity

The principal variable affecting LTB strength is the distance between lateral and/or torsional braces. Other variables include the type and position of the loads, the restraints at the ends and at intermediate positions along the beam axis, the type of cross section, continuity at supports, the presence or absence of stiffening devices that restrain warping at critical locations, the material properties, the magnitude and distribution of residual stresses, prestressing forces, initial imperfections of geometry and loading, changes in the cross section (e.g., steps or taper in the cross section, holes, and copes), cross-section distortion, and interaction between local and overall buckling. The design of flexural members generally must consider (i) the influence of potential local buckling (e.g., compression flange local buckling and/or web local buckling in flexure, shear, or combined flexure and shear); (ii) the postbuckling resistance of the cross-section elements; and (iii) the yield and/or rupture strength of cross-scction tension elements.

This chapter focuses predominantly on the LTB resistance of straight flexural members loaded in bending about their major principal axis. Several chapters in this guide further explore the behavior and design of specific types of flexural members, including: Chapter 6, which addresses the broader considerations associated with the design of built-up members having relatively thin webs, such as typical welded I-section bridge girders and steel plate shear walls; Chapter 7, which deals with box section members built-up from thin-plate components; Chapter 9, which focuses on horizontally curved steel I-girders; and Chapter 13, which concentrates on thin-walled members commonly employed in cold-formed steel and aluminum applications. It should be also noted that Chapter 12 provides valuable information for the design of beam and girder bracing.

The analytical aspects of determining LTB strength are complex, and closed-form solutions exist only for the simplest cases. Much of the seminal research in this field was conducted intensively during the mid-nineteenth century and the corresponding literature is extensive. The third and fifth editions of this guide (Johnston, 1976; Galambos, 1998) review the early work and more recent developments, while Lee (1960), Nethercot (1983), and Trahair and Bradford (1988) also provide excellent background. Lateral–torsional buckling theory and its design applications are covered extensively in textbooks in various languages. Some of the English-language texts are those of Bleich (1952), Timoshenko and Gere (1961), Vlasov (1961), Chajes (1974), Brush and Almroth (1975), Heins (1975), Galambos (1968a), Kirby and Nethercot (1979), Allen and Bulson (1980), Chen and Lui (1987), Bažant and Cedolin (1991), Trahair (1993), Simitses and Hodges (2006), and Galambos and Surovek (2008).

As shown in Fig. 5.1, the LTB strength may be represented either as a plot of bending moment capacity versus unbraced length L or as a plot of the bending moment capacity versus a nondimensional slenderness parameter $\overline{\lambda}_{LT}$ taken as

$$\overline{\lambda}_{LT} = \sqrt{M_{n(L=0)}/M_{cr}} \tag{5.1}$$

where $M_{n(L=0)}$ is the cross-section-based resistance corresponding to a theoretical "zero-length" member and M_{cr} is the theoretical elastic LTB moment corresponding to a bifurcation from in-plane bending response to out-of-plane lateral bending and twisting. The dashed curve in Fig. 5.1 shows a typical variation in M_{cr} versus L or $\overline{\lambda}_{LT}$. The solid curve shows a representative flexural resistance of physical members having a common cross section and different lengths. The response can be subdivided into three ranges for purposes of discussion: (1) elastic buckling, which governs for long beams (usually of importance mostly during construction);

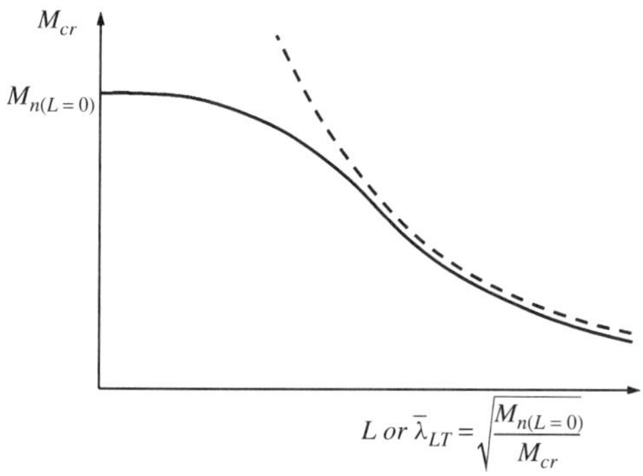

FIGURE 5.1 Beam strength and elastic buckling curves corresponding to the lateral–torsional buckling limit state.

(2) inelastic buckling, when instability occurs after some portions of the beam have yielded; and (3) the cross-section limit of resistance, where the unbraced length is short enough so that LTB occurs after the cross-section resistance is achieved. Typically, the latter two ranges are the only ones of importance for the completed structure.

Different design standards vary substantially from one another in the ways that they characterize the physical bending resistance in Fig. 5.1 via algebraic equations. Section 5.3 provides an overview of the procedures appearing in several prominent (2008) standards. In a broad sense, however, all of the standards use the same approach. They are either explicitly or implicitly based on the calculation of a member elastic LTB resistance M_{cr} followed by a "mapping" of this theoretical resistance to a nominal resistance of the physical member. This "mapping" accounts for reductions in the resistance from the theoretical elastic buckling value due to various factors, including geometric imperfections, local and/or distorsional buckling, and yielding, which can be accentuated by residual stress effects. Both the mean or average effects from the above factors as well as the dispersion in member resistances associated with the variability and uncertainty in these factors must be addressed in arriving at nominal resistance equations.

Because the calculation of the theoretical elastic critical moment M_{cr} is central to the determination of nominal flexural resistances, much of this chapter is devoted to determining this fundamental quantity. In this regard, Section 5.2 provides an overview of procedures for obtaining the elastic LTB moment of prismatic I-section members. The procedures range from the most basic calculations assuming uniform bending and simply supported end conditions on an isolated unbraced length to more refined calculations that account for various detailed load and/or displacement boundary conditions, including:

- Nonuniform bending (moment gradient) along the member length
- Load height effects, including tipping restraint

- Restraint of lateral translation and/or twisting at the ends of the unbraced lengths due to attachment to lateral or torsional bracing components
- Restraint of lateral bending and/or warping at the ends of the unbraced lengths due to attachment to other structural components or continuity with adjacent unbraced lengths
- Unequal bracing on separate flanges, or continuous lateral and/or torsional restraint from a slab, decking, or sheeting attached to one side of the member
- Overhangs or cantilevers having only partially braced or free end conditions
- Lifting, suspending, or holding of members during construction
- System LTB of groups of flexural members

Of course, providing adequate bracing strength and stiffness is an essential part of ensuring the integrity of beam flexural resistance calculations and this is discussed in Chapter 12.

The chapter is organized as follows. Section 5.2 provides an overview of procedures for calculating the elastic LTB resistance of beams for a broad range of conditions. Section 5.3 then presents the mapping from the elastic LTB resistance to the nominal design resistance from various current and emerging standards. Section 5.4 addresses the calculation of the LTB resistance for nonprismatic I-section flexural members, including both variable web-depth members and/or members with steps or transitions in the cross section along the unbraced length. Section 5.5 discusses the current state-of-the-art with respect to calculation of the lateral buckling resistance of composite I-section members in negative bending. An overview of LTB concepts and calculations for other types of prismatic members is provided in Section 5.6. The chapter closes with a brief discussion of design for inelastic deformation capacity in Section 5.7, and overarching considerations for current and future design practice in Section 5.8.

5.2 ELASTIC LATERAL–TORSIONAL BUCKLING, PRISMATIC I-SECTION MEMBERS

5.2.1 Simply Supported Doubly Symmetric Members, Uniform Moment

This is the simplest of all the cases, and closed-form equations for its buckling solution are well established (Timoshenko and Gere, 1961). The ends of the beam are assumed to be prevented from lateral deflection ($u = 0$) and from twisting ($\phi = 0$), but they are free to rotate laterally ($u'' = 0$) and the end cross section is free to warp ($\phi'' = 0$). Solution to the governing differential equations (Galambos and Surovek, 2008) gives the critical buckling moment

$$M_{ocr} = \frac{\pi}{L}\sqrt{EI_y GJ}\sqrt{1 + W^2} \tag{5.2}$$

with

$$W = \frac{\pi}{L}\sqrt{\frac{EC_w}{GJ}} \tag{5.3}$$

in which L is the span length, E and G are the elastic and shear moduli, and I_y, J, and C_w are the minor axis moment of inertia, St.-Venant torsion constant, and warping constant, respectively. The second square-root term in Eq. 5.2 represents the effect of warping torsional stiffness. The cross-sectional constants I_y, J, and C_w are listed in handbooks or they can be derived [see, e.g., Bleich (1952) or Galambos (1968a)]. Equation 5.2 also may be written as

$$M_{ocr} = \left(\frac{\pi}{L}\right)^2 E\sqrt{I_y C_w}\sqrt{1 + \frac{1}{W^2}} \tag{5.4}$$

which, using $C_w = I_y h_o^2/4$ for doubly symmetric I-shaped members, may be expressed as (Johnston, 1976)

$$M_{ocr} = \frac{\pi^2 E I_y}{L^2}\frac{h_o}{2}\sqrt{1 + \frac{1}{W^2}} \tag{5.5}$$

where h_o is the distance between the flange centroids. Equation 5.5 is useful for consideration of the behavior of thin-web members with relatively narrow flanges compared to the section depth. In these cases, $1/W^2$ is typically quite small relative to 1.0 even for relatively long unbraced lengths that fail by elastic LTB. In addition, the influence of GJ on the LTB resistance is typically reduced due to distortion of the web into an S-shape in the buckling mode of these types of members. Therefore, Eq. 5.5 is used in some standards without the square-root term as a practical approximation for the elastic LTB resistance of thin (slender) web members. Conversely, for I-section members with stockier webs and flange width–section depth ratios d/b_f closer to 1.0, W^2 can be quite smaller relative to 1.0. In these cases, the square-root term in Eq. 5.5 is a substantial part of the elastic LTB resistance.

The AISC (2005) and AASHTO (2007) specifications present Eq. 5.5 in the form

$$M_{ocr} = \frac{\pi^2 E}{(L/r_t)^2} S_{xc}\sqrt{1 + \frac{0.078}{X^2}\left(\frac{L}{r_t}\right)^2} \tag{5.6}$$

where S_{xc} is the elastic section modulus to the compression flange, which is equal to the elastic section modulus to the tension flange for a doubly symmetric I-section. The term r_t may be calculated from

$$r_t^2 = \frac{\sqrt{I_y C_w}}{S_x} \tag{5.7}$$

for a doubly symmetric section, which can be approximated closely by the radius of gyration of an equivalent tee section composed of one-third of the depth of the web in compression D_c (equal to one-sixth of the total depth of the web for a doubly symmetric section) resulting in

$$r_t^2 \cong \frac{b_{fc}^2}{12\left(1 + \frac{1}{3}\frac{D_c t_w}{b_{fc} t_{fc}}\right)} \tag{5.8}$$

Finally,

$$X^2 = \frac{S_{xc} h_o}{J} \tag{5.9}$$

which, for a doubly symmetric I-section, is proportional to the ratio of the cross-section major axis bending and St. Venant torsional rigidities, that is, $X^2 \cong 2I_x/J$, or $X^2 \cong \alpha EI_x/GJ$. It is important to note that over the full range of ASTM A6 wide-flange shapes, X^2 varies from 13 for the heavy-column type (low d/b_f) W14x808 section to 2500 for the relatively thin-walled beam-type (high d/b_f) W30x90 section.

5.2.2 Simply Supported Singly Symmetric Members, Uniform Moment

The above equations apply only to doubly symmetric members. For singly symmetric beams (Fig. 5.2), the shear center S and the centroid C do not coincide. In this case, the critical moment for bending in the plane of symmetry is given by (Galambos, 1968a)

$$M_{ocr} = \frac{\pi^2 EI_y}{L^2}\left\{\frac{\beta_x}{2} + \sqrt{\left(\frac{\beta_x}{2}\right)^2 + \left[\frac{C_w}{I_y} + \frac{GJ}{EI_y}\frac{L^2}{\pi^2}\right]}\right\} \tag{5.10}$$

The only term not defined previously is the coefficient of monosymmetry β_x. The general expression for β_x is

$$\beta_x = \frac{1}{I_x}\int_A y(x^2 + y^2)\, dA - 2y_o \tag{5.11}$$

where I_x is the major axis moment of inertia, y_o is the distance of the shear center from the cross-section centroid (Fig. 5.2), which is negative when the larger flange is in compression, x and y are centroidal coordinates, and integration is over the cross-section area A. The parameter β_x is equal to zero for doubly symmetric shapes. The calculation of β_x and other cross-sectional properties for a general singly symmetric I section is shown in Fig. 5.2.

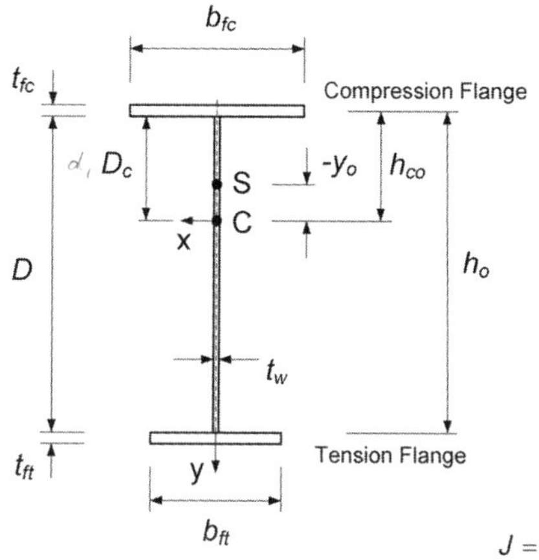

$$\alpha = \frac{1}{1+\left(\frac{b_{fc}}{b_{ft}}\right)^3\left(\frac{t_{fc}}{t_{ft}}\right)} = \frac{1}{\frac{I_{yc}}{I_{yt}}+1}$$

$$I_{yc} = \frac{t_{fc}b_{fc}^3}{12} \qquad I_{yt} = \frac{t_{ft}b_{ft}^3}{12}$$

$$C_w = h_o^2 b_{fc}^3 t_{fc}\alpha/12 = h_o^2 I_{yc}\alpha$$

$$y_o = -h_{co} + \alpha h_o$$

$$J = \frac{Dt_w^3}{3} + \frac{b_{fc}t_{fc}^3}{3}\left(1-0.63\frac{t_{fc}}{b_{fc}}\right) + \frac{b_{ft}t_{ft}^3}{3}\left(1-0.63\frac{t_{ft}}{b_{ft}}\right)$$

$$\beta_x = \frac{1}{I_x}\left[\left(\frac{b_{ft}^3}{12}(D-D_c)t_{ft} + \frac{b_{ft}^3}{24}t_{ft}^2 + b_{ft}t_{ft}(D-D_c)^3 + 1.5b_{ft}(D-D_c)^2t_{ft}^2 + b_{ft}(D-D_c)t_{ft}^3 + \frac{b_{ft}}{4}t_{ft}^4\right)\right.$$
$$-\left(\frac{b_{fc}^3}{12}D_ct_{fc} + \frac{b_{fc}^3}{24}t_{fc}^2 + b_{fc}t_{fc}D_c^3 + 1.5b_{fc}D_c^2t_{fc}^2 + b_{fc}D_ct_{fc}^3 + \frac{b_{fc}}{4}t_{fc}^4\right)$$
$$\left.+\left(\frac{(D-D_c)^4}{4}t_w + \frac{t_w^3}{24}(D-D_c)^2 - \frac{D_c^4}{4}t_w - \frac{t_w^3}{24}D_c^2\right)\right] - 2y_o$$

FIGURE 5.2 Torsional properties for singly symmetric I-beams.

For practical purposes, β_x of the section shown in Fig. 5.2 can be approximated by (Kitipornchai and Trahair, 1980)

$$\beta_x = 0.9h_o\left(\frac{2I_{yc}}{I_y}-1\right)\left[1-\left(\frac{I_y}{I_x}\right)^2\right] \tag{5.12}$$

where I_{yc} is the minor axis moment of inertia of the compression flange and I_y the minor axis moment of inertia of the full cross section. For typical beam-type I-sections (d/b_{fc} and d/b_{ft} both greater than about 1.7), the last bracketed term in Eq. 5.12 is for all practical purposes equal to 1.0. Based on this assumption and substituting $I_y = I_{yc} + I_{yt}$, Eq. 5.12 can be written as

$$\beta_x = 0.9h_o\alpha\left(\frac{I_{yc}}{I_{yt}}-1\right) \tag{5.13}$$

where α is defined in Fig. 5.2. White and Jung (2004) compare the buckling predictions from the combination of Eqs. 5.10 and 5.13 for a comprehensive range of beam- and column-type I-sections and show that the error in the corresponding solutions relative to the exact analytical calculation of β_x ranges from 6% unconservative to 5% conservative.

The AISC 2005 and AASHTO 2007 Specifications provide Eq. 5.6 as an approximation for the critical buckling moment of singly symmetric I-section members. White and Jung (2004) show that Eq. 5.6, with r_t defined by Eq. 5.8 and X^2 defined by Eq. 5.9, provides improved accuracy relative to other similar approximations. Over a reasonably comprehensive range of cases, the combination of Eqs. 5.6, 5.8, and 5.9 gives errors ranging from −35% conservative for sections with the smaller flange in compression to +9% unconservative for sections with the larger flange in compression. The larger errors are for rather extreme singly symmetric geometries. For more common sections, the errors are much smaller. White and Jung (2004, 2007) indicate that the conservatism of these equations for small I_{yc}/I_y is appropriate and acceptable, given the tendency of web distortion to reduce the LTB strength of members with the smaller flange in compression. It should be noted that Eq. 5.7 is applicable only for doubly symmetric I-sections. For singly symmetric sections, Eq. 5.8 must be used to calculate r_t.

5.2.3 Moment Gradient Effects

For beams subjected to unequal end moments (Fig. 5.3), numerical or approximate solutions are required to obtain the buckling load. The most extensive early work for this case was performed by Massonnet (1947), Horne (1954), and Salvadori (1955, 1956), and their results have been verified many times by other researchers using various numerical techniques. Salvadori (1955) found that a simple modification to Eq. 5.2,

$$M_{cr} = C_b M_{ocr} \tag{5.14}$$

can account for the effect of moment gradient in this case, where M_{ocr} is obtained from Eq. 5.2 and C_b is commonly referred to as the *equivalent uniform moment factor.* Various lower-bound formulas have been proposed for C_b, but the most commonly accepted are

$$C_b = 1.75 + 1.05\kappa + 0.3\kappa^2 \leq 2.56 \tag{5.15a}$$

$$C_b = [0.6 - 0.4\kappa]^{-1} \leq 2.5 \tag{5.15b}$$

The moment ratio κ is defined as positive for double-curvature bending (see Fig. 5.3). The value of C_b is mildly dependent on W (Nethercot, 1983; Galambos, 1998), especially when κ approaches +1; however, either Eq. 5.15a or 5.15b provides acceptable accuracy for most purposes.

Equations 5.15a and 5.15b are applicable for use with linearly varying moment diagrams between brace points. Kirby and Nethercot (1979) presented an alternative equation for C_b that applies to more general shapes of nonlinear moment

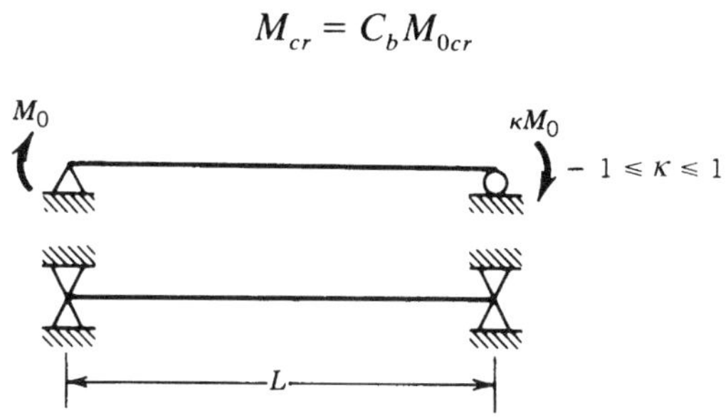

FIGURE 5.3 Nonuniform moment.

diagrams between brace points. Their equation, which appears in the AISC (2005) Specification, is

$$C_b = \frac{12.5M_{\text{max}}}{2.5M_{\text{max}} + 3M_A + 4M_B + 3M_C} \tag{5.15c}$$

where M_{max} is the absolute value of the maximum moment in the unbraced segment and M_A, M_B, and M_C are the absolute values of the moments at the quarter, center, and three-quarter points of the unbraced segment, respectively. Wong and Driver (2010) compare Eq. 5.15c and several other "quarter-point" C_b equations from various standards and the literature for doubly symmetric I-section members. They explain that the quarter-point equations generally have the following shortcomings:

1. They are independent of the actual sign of the moments within the span.
2. They are not capable of capturing the effect of abrupt changes in the moment diagram for segments loaded by concentrated moments (such as when a cantilever post affixed to the beam flange is loaded parallel to the beam axis).

Wong and Driver (2010) show that Eq. 5.15c provides good results for the majority of the cases considered. It gives results, however, that are from zero to 32% unconservative compared to numerical solutions by Austin (1961), Nethercot and Trahair (1976), and Serna et al. (2006) for the cases illustrated in Fig. 5.4. Wong and Driver recommend the following equation as an improved quarter-point formula:

$$C_b = \frac{4M_{\text{max}}}{\sqrt{M_{\text{max}}^2 + 4M_A^2 + 7M_B^2 + 4M_C^2}} \leq 2.5 \tag{5.15d}$$

For many practical cases, Eq. 5.15d gives slightly larger C_b values compared to Eq. 5.15c.

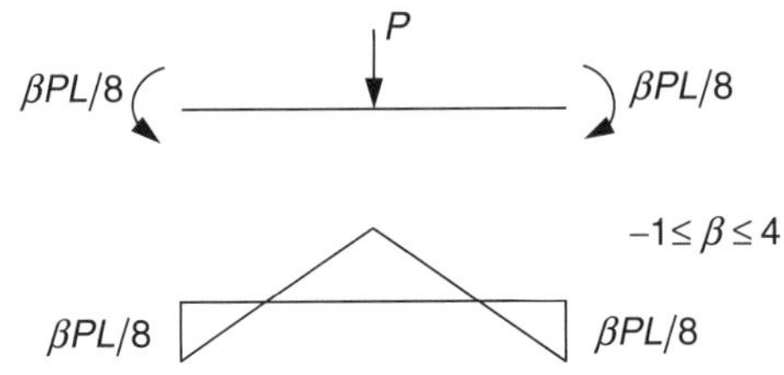

FIGURE 5.4 Cases for which Eq. 5.15c gives unconservative results.

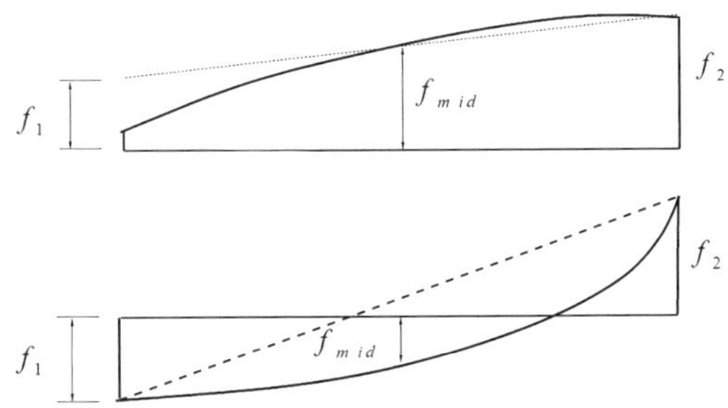

FIGURE 5.5 AASHTO (2007) application of Eq. 5.15a.

Wong and Driver (2010), Kim and White (2007), and White (2006) also discuss the application of a form of Eq. 5.15a to cases with nonlinear moment diagrams. This form is employed by the AASHTO (2007) specifications and uses the corresponding flange stresses rather than the moments (although for prismatic members, the results are the same when moments are used). The nonlinearity in the flange stress diagram is captured by fitting a line though the maximum flange compressive stress at the end of the unbraced length and through the stress in the same flange at the middle of the unbraced length. This is illustrated in Fig. 5.5. In cases where the use of the smaller f_1 would tend to be unconservative this approach tends to give more accurate results than Eqs. 5.15c and 5.15d for typical members having multiple brace points along the span and only slightly nonlinear moment diagrams between the brace points. Kim and White (2007) discuss this modified application of Eq. 5.15a as well as a similar modification to Eq. 5.15c for assessment of the LTB resistance of nonprismatic doubly and singly symmetric I-section members with linearly varying web depths. White (2006) discusses the application of the modified form of Eq. 5.15a in the context of moving loads and the moment envelopes.

If the flange compressive stress is larger at the midpoint of the unbraced length compared to the maximum compressive end stress, the above approach uses $C_b = 1.0$. In these situations, Eqs. 5.15c and 5.15d give more accurate (less conservative) results.

5.2.4 Adjustments for Load Height Effects and for Double-Curvature Bending of Singly Symmetric I-Section Members

Equations 5.15c and 5.15d and the modified version of Eq. 5.15a are all based strictly on elastic LTB of members with transverse loads applied theoretically at the shear center. Furthermore, although the AASHTO (2007) procedure addresses LTB of singly symmetric members loaded in single- or double-curvature bending and the other equations are applicable to singly symmetric members in single-curvature bending based on the findings by Helwig et al. (1997), the other equations do not address double-curvature bending on singly symmetric I-section members. Helwig et al. (1997) provide several modifications and recommendations for the application of Eq. 5.15c to doubly and singly symmetric I-section members with $0.1 \leq I_{yc}/I_y \leq 0.9$ (or $0.11 < I_{yc}/I_{yt} \leq 9$), including consideration of the influence of the height of the applied transverse loads on the cross section. They recommend a modified form of Eq. 5.15c,

$$C_b^* = C_b \left(1.4^{2y/h_o}\right) R_m \leq 3.0 \qquad (5.16a)$$

where y is taken as the location of the applied load relative to the midheight of the cross section, negative for loading above the midheight and positive for loading below the midheight. In addition, the parameter R_m is defined as

$$R_m = \begin{cases} 1.0 & \text{for unbraced lengths subjected to single-curvature bending} \\ 0.5 + 2\left(\dfrac{I_{y.\text{top}}}{I_y}\right)^2 & \text{for unbraced lengths subjected to double-curvature bending} \end{cases} \qquad (5.16b)$$

where $I_{y.\text{top}}$ is the moment of inertia of the top flange about the minor axis of the section. For transverse loadings on nonhorizontal members, the up direction should be taken as the direction opposite to that of the applied transverse loading, and the top flange should be taken as the flange on the opposite side of the midheight from the *direction* of the loading. For double-curvature bending cases, one value is calculated for C_b^* for each unbraced length in this approach; the elastic buckling strength of the unbraced segment, however, must be checked using two different values for M_{ocr}. That is, the load level corresponding to elastic LTB is determined as the smallest load level satisfying the two equations

$$C_b^* M_{ocr.\text{top}} = M_{\text{top}} \quad \text{and} \quad C_b^* M_{ocr.\text{bottom}} = M_{\text{bottom}} \qquad (5.16c)$$

where $M_{ocr.\text{top}}$ and $M_{ocr.\text{bottom}}$ are the respective elastic LTB capacities under uniform bending calculated by treating the top flange as the compression flange and the bottom flange as the compression flange and M_{top} and M_{bottom} are the maximum moments from the applied loading causing compression in the top and bottom flanges, respectively. Helwig et al. (1997) show that Eqs. 5.16 can give results that are unconservative by as much as 30% when there is more than one change of curvature between the brace points. They also show unconservative errors as large as 24% for single-curvature cases with bottom-flange loading. The authors, however, explain that these unconservative results may be mitigated by other factors discussed below.

Regarding the estimates determined using Eqs 5.16, it can be observed that for top-flange loading, the elastic LTB resistance is reduced by a factor of $1/1.4 = 0.71$, whereas for bottom-flange loading, it is increased by a factor of 1.4, with the exception that C_b^* is limited to a maximum value of 3.0. The coefficient 1.4 is an estimate of values obtained from various numerical solutions [see Section 5.2.7 of this chapter and the previous edition of this guide (Galambos, 1998)]. Furthermore, for $I_{y.top}/I_y = 0.1$, the elastic LTB resistance is reduced by a factor of $R_m = 0.52$, whereas for $I_{y.top}/I_y = 0.9$, it is increased by a factor of $R_m = 2.12$, with C_b^* limited to a maximum value of 3.0.

In addition to the development of Eqs. 5.16, Helwig et al. (1997) provide the following recommendations for calculating the elastic LTB resistance of singly symmetric I-section members subjected to top-flange loading:

1. The distance y should be measured from the cross-section midheight, and not from the location of the shear center. The authors show that use of the midheight as the y origin gives the best correlation with benchmark solutions.
2. Due to the beneficial effects of tipping restraint and the warping restraint associated with continuity across the brace points:

   a. $C_b^* = 1.0$ is generally conservative for single-curvature bending cases with intermediate bracing in the span between the supports resisting the transverse loads and in-plane bending.

   b. Eqs. 5.16 are generally conservative for double-curvature bending cases with intermediate bracing within the span between the vertical supports.

Tipping restraint is illustrated in Fig. 5.6 and can be defined as the development of a beneficial restoring torque due to a shift in the contact point between the top flange and the components through which the loads are applied, for example, a slab, as the member starts to buckle. Helwig et al. (1997) note that important factors to consider when evaluating tipping restraint are that cross-sectional distortion

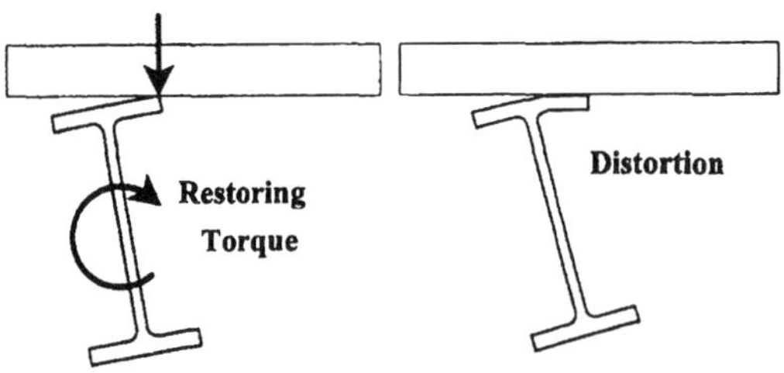

FIGURE 5.6 Beneficial restoring torque from tipping restraint.

significantly reduces the benefits from tipping restraint and no tipping restraint should be assumed from wet concrete.

The results of the above studies may be synthesized for design application as follows:

1. For ordinary situations having intermediate bracing within the span, the influence of load height may be neglected.
2. For situations where there is no intermediate bracing within the span between the supports in the plane of loading:

   a. The influence of load height should be considered for top flange loading. Equations 5.16 may be applied for any of these cases involving single-curvature bending or for double-curvature bending with only one change in sign of the moment between the brace points.
   b. Equations 5.16 may be applied for consideration of bottom- flange loading when L_b/h_o is less than or equal to 10. For $L_b/h_o > 10$, the unconservative error in Eqs. 5.16 may be larger than 10%. Helwig et al. (1997) provide numerical solutions for a number of loadings that can be used as estimates for C_b^* in these cases. It is always conservative to neglect the influence of bottom-flange loading.

It should be noted that the above equations do not apply to cantilever segments with free end conditions. Section 5.2.9 addresses this case.

5.2.5 Inflection Points Are Not Brace Points

It is important to address a common misconception that often has been applied inappropriately in stability design. An inflection point *cannot* be considered as a brace point. In many cases where this issue is raised, the top flange is laterally braced by a slab or joists along the entire span while the bottom flange is unbraced. Figure 5.7 illustrates the fallicy of assuming that an inflection point is a brace

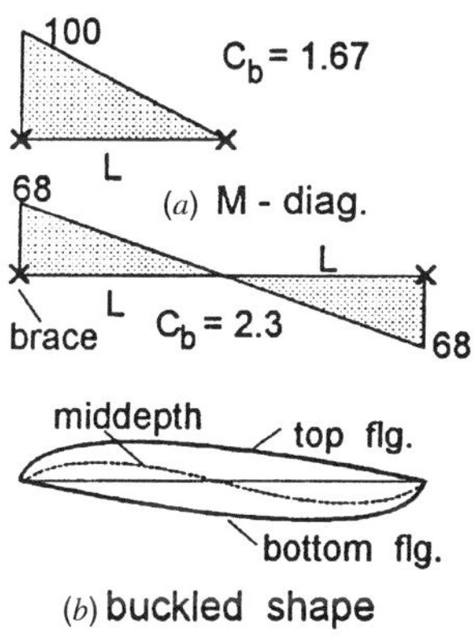

FIGURE 5.7 Fallacy of assuming that an inflection point is a brace point.

point for a hypothetical doubly symmetric I-section beam. The unbraced length L in Fig. 5.7a has a moment only at one end, and thus its C_b is 1.75 by Eqs. 5.15a and 5.15d or 1.67 by Eqs. 5.15b and 5.15c. Conversely, the unbraced length $2L$ in Fig. 5.7b has equal end moments and double-curvature bending. Thus its C_b is 2.56 (by Eq. 5.15a), 2.5 (by Eq. 5.15b), and 2.3 (by Eqs. 5.15c and 5.15d). The $2L$ span with the inflection point will buckle at a load that is as low as 58% of the beam with span L, depending on the value of W (Eq. 5.3). If the inflection point were a brace point, the critical moment of both beams would be the same. The buckled shape of the $2L$ beam shows that the top flange and bottom flange move laterally in opposite directions at the midspan inflection point. In fact, it is not sufficient to provide a lateral brace only on one flange at the midspan (Yura, 2001). The movement of both flanges must be restrained either by lateral or torsional bracing for the midspan of the $2L$ beam to be considered as a brace point.

5.2.6 Influence of Continuous Bracing on One Flange

Figure 5.8 summarizes a number of approximate equations for C_b recommended by Yura and Helwig (2009) for several common cases of I-section beams with continuous top-flange lateral bracing, neglecting torsional restraint of the top flange. These C_b values are used in design with L_b taken as the span length between the supports, assuming that twist is prevented at the supports. Two cases are shown, in the plane of the loading: (1) the top flange laterally braced with gravity loading applied to the top flange and (2) the top flange braced with uplift loading applied to this flange. Similar to the previous discussion in Section 5.2.3, the equations presented may be used for situations involving nonhorizontal members by taking the top flange as the flange opposite to the direction of the loading in case 1 and by taking the top flange as the flange in the direction of the loading in case 2. This, for example, would allow consideration of wind pressure or suction loading on vertical members braced by an outside wall diaphragm or sheeting. Essa and Kennedy (1995) present design charts for suspended construction that consider the torsional restraint provided by joists attached to the top flange in addition to the lateral restraint. In these cases, the distortion of the I-section web must be addressed in determining the buckling resistance. The influence of cross-section distortion is not explicitly addressed in the equations provided in Fig. 5.8.

5.2.7 Prismatic End-Restrained Beams

End restraint has a pronounced effect on the elastic lateral–torsional buckling strength of I-section members. Several special cases of idealized end boundary conditions are shown in Fig. 5.9, where u is the member lateral displacement and ϕ is the cross-section twist. Nethercot and Rockey (1972) give the following equation for the loading cases shown in Fig. 5.10:

$$M_{cr} = CM_{ocr} \tag{5.17}$$

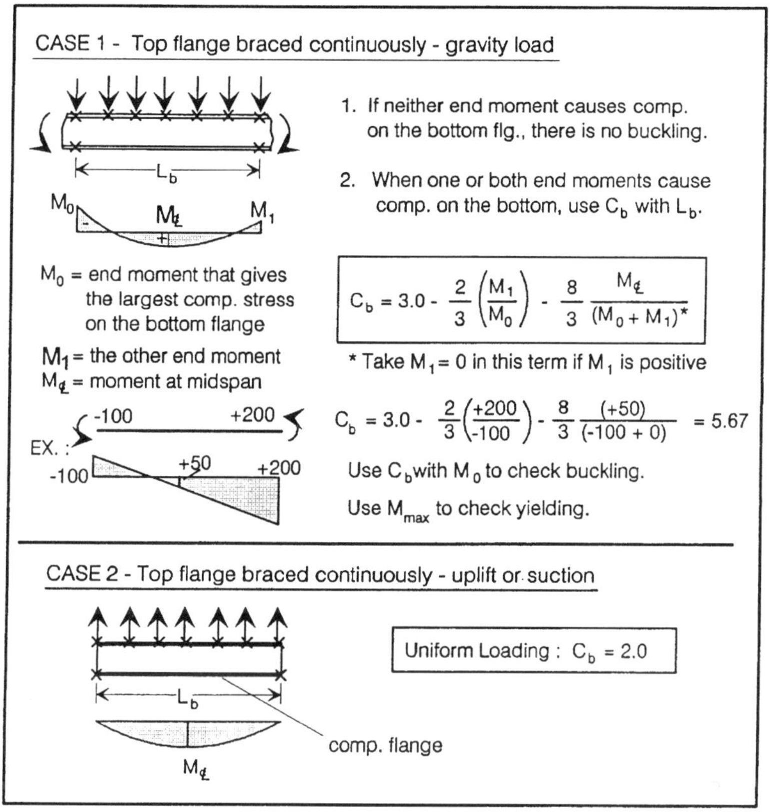

FIGURE 5.8 Other equations for *Cb* addressing special cases of continuous restraint at one flange.

where $C = A/B$ for top-flange loading, $C = A$ for loading through the midheight, and $C = AB$ for bottom-flange loading, with A and B determined by the equations given in Fig. 5.10, which are applicable for doubly symmetric I-section members.

In restraint condition II of Fig. 5.9, the end is free to rotate laterally but is prevented from warping. This condition can be achieved most effectively by welding boxed stiffeners (*tube-type warping restraints* at or near the end supports, as illustrated by Fig. 5.11) (Ojalvo and Chambers, 1977). Other warping-restraint cases have been recommended and their effectiveness has been analyzed (Vacharajittiphan and Trahair, 1974; Ojalvo and Chambers, 1977; Heins and Potocko, 1979; Szewczak et al., 1983; Lindner and Gietzelt, 1983; Lindner, 1995). The boxed stiffener detail shown in Fig. 5.11, however, is clearly the most efficient and the most effective.

The influence of preventing warping at the member ends can be demonstrated by considering a 48-ft (14.63-m) long simply supported W24×55 steel beam [$W = 0.503$; $M_{ocr} = 644$ in.-kips (731 kN-m) of length; $A = 1.507$; $B = 1.230$] subjected

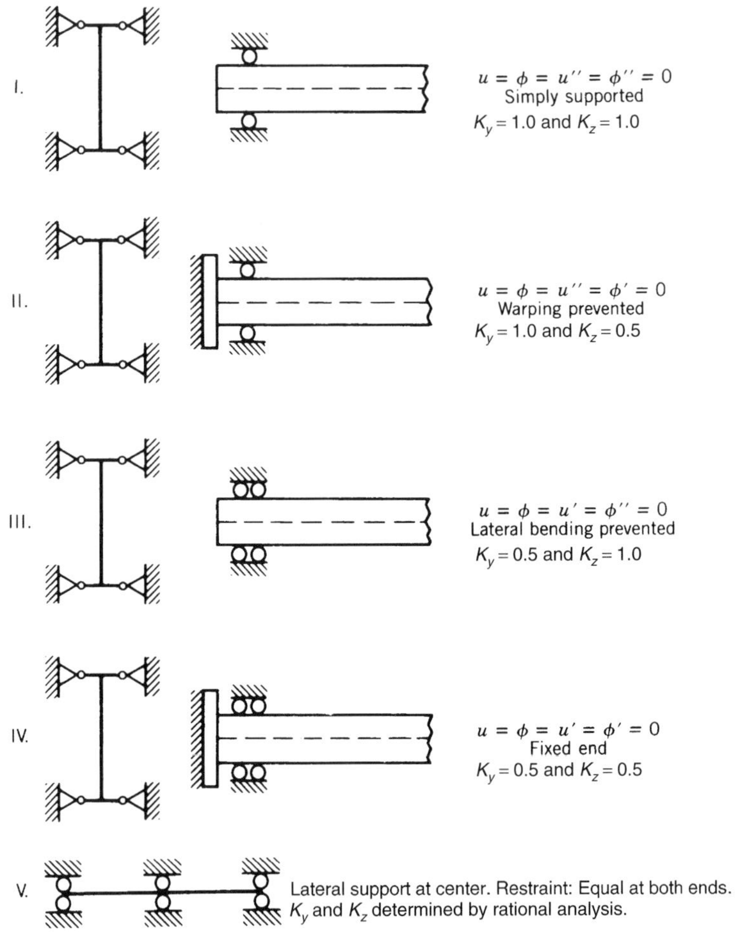

FIGURE 5.9 Idealized end restraints.

to a uniformly distributed load. The critical loads are a function of load height with values of:

For top flange loading: $w_{cr} = 0.228$ kip/ft (3.33 kN/m)

For midheight loading: $w_{cr} = 0.281$ kip/ft (4.10 kN/m)

For bottom flange loading: $w_{cr} = 0.346$ kip/ft (5.05 kN/m)

One can show that the elastic buckling load is increased by 33% for the above warping prevented boundary condition compared to the case of free warping (idealized restraint condition I in Fig. 5.9) in all three cases. For other lengths and sections this ratio differs, but the important point is that the buckling load can be increased substantially with a modest expenditure.

Loading	Restraint	A	B
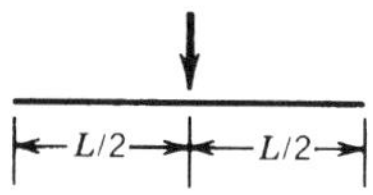	I	1.35	$1 - 0.180W^2 + 0.649W$
	II	$1.43 + 0.485W^2 + 0.463W$	$1 - 0.317W^2 + 0.619W$
	III	$2.0 - 0.074W^2 + 0.304W$	$1 - 0.207W^2 + 1.047W$
	IV	$1.916 - 0.424W^2 + 1.851W$	$1 - 0.466W^2 + 0.923W$
	V	$2.95 - 1.143W^2 + 4.070W$	1
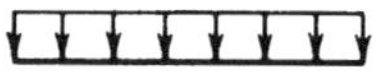	I	1.13	$1 - 0.154W^2 + 0.535W$
	II	$1.2 + 0.416W^2 + 0.402W$	$1 - 0.225W^2 + 0.571W$
	III	$1.9 - 0.120W^2 + 0.006W$	$1 - 0.100W^2 + 0.806W$
	IV	$1.643 - 0.405W^2 + 1.771W$	$1 - 0.339W^2 + 0.625W$
	V	$2.093 - 0.947W^2 + 3.117W$	$1.073 + 0.044W$

FIGURE 5.10 Restraint categories.

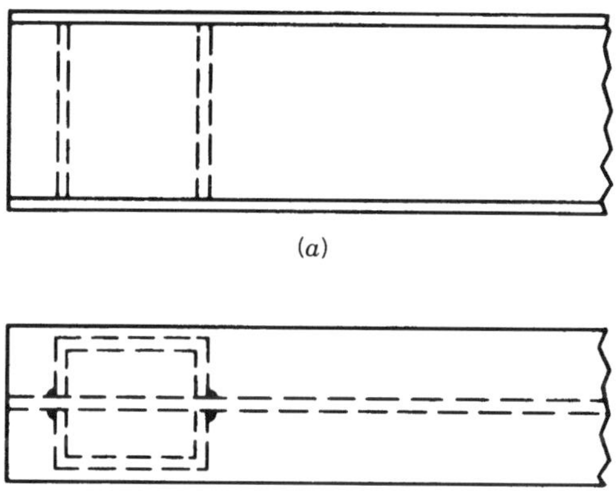

(*a*)

(*b*)

FIGURE 5.11 Tube-type warping restraints.

In the context of the equations presented in Sections 5.2.1 through 5.2.6, the influence of end restraint may be determined by using LTB effective lengths. In general, however, there are two LTB effective-length factors:

1. The effective-length factor corresponding to the lateral bending restraint provided to the full cross section at the ends of the unbraced length, denoted by K_y
2. The effective-length factor corresponding to the warping restraint at the ends of the unbraced length, denoted by K_z

Approximate values of these effective-length factors are shown in Fig. 5.9. More exact values are provided for a number of end-restraint combinations by Vlasov (1961). Solutions obtained by numerical methods for special cases of end restraint are tabulated by Austin et al. (1957), Clark and Hill (1960), and others. Clark and Hill (1960) give approximate solutions with coefficients.

Given the equivalent uniform moment factor C_b and the effective-length factors K_y and K_z, the elastic LTB resistance of doubly symmetric I-section members may be determined from the modified form of Eq. 5.2:

$$M_{cr} = C_b \frac{\pi}{K_y L_b} \sqrt{EI_y GJ} \sqrt{1 + \frac{EC_w}{GJ} \frac{\pi^2}{(K_z L_b)^2}} \tag{5.18a}$$

where L_b is the actual length between the brace points. For both doubly and singly symmetric I-section members, the equivalent modified form of Eq. 5.6 is

$$M_{cr} = C_b \frac{\pi^2 E}{(K_y L_b / r_t)^2} S_{xc} \sqrt{\left(\frac{K_y}{K_z}\right)^2 + \frac{0.078}{X^2} \left(\frac{K_y L_b}{r_t}\right)^2} \tag{5.18b}$$

Alternately, for singly symmetric I-section members, the exact elastic LTB expression from open-section thin-walled beam theory (Eq. 5.10) may be modified to

$$M_{cr} = C_b \frac{\pi^2 EI_y}{(K_y L_b)^2} \left\{ \frac{\beta_x}{2} + \sqrt{\left(\frac{\beta_x}{2}\right)^2 + \left[\frac{C_w}{I_y}\left(\frac{K_y}{K_z}\right)^2 + \frac{GJ}{EI_y}\frac{(K_y L_b)^2}{\pi^2}\right]} \right\} \tag{5.19a}$$

The use of the C_b values from Sections 5.2.3 through 5.2.6 with the above K_z and K_y values gives acceptable results compared to benchmark solutions. The bracing strength and stiffness demands, however, can be increased substantially in members designed based on LTB effective-length factors less than 1. Therefore, caution must be exercised in the design of the bracing system for these types of members.

A form similar to Eq. 5.19a, derived by Clark and Hill (1960) and utilized in Eurocode 3 developments (CEN, 2002; Andrade et al., 2007) is

$$M_{cr} = C_1 \frac{\pi^2 EI_y}{(K_y L_b)^2} \left\{ \left(C_2 y + C_3 \frac{\beta_x}{2}\right) + \sqrt{\left(C_2 y + C_3 \frac{\beta_x}{2}\right)^2 + \left[\frac{C_w}{I_y}\left(\frac{K_y}{K_z}\right)^2 + \frac{GJ}{EI_y}\frac{(K_y L_b)^2}{\pi^2}\right]} \right\} \tag{5.19b}$$

where y is the height of the loading relative to the shear center, positive for loads applied above the shear center (i.e., loads acting toward the shear center), and C_1, C_2 and C_3 are factors that depend predominantly on the moment diagram and end-restraint conditions. In general, these parameters have some dependency on the other variables in this equation (Andrade et al., 2007).

5.2.8 Consideration of Continuity with Adjacent Unbraced Segments

When lateral and/or torsional bracing is provided at the supports and also at intermediate points, the elastic LTB resistance still may be estimated using any of the applicable equations from Section 5.2.1 through 5.2.6. The smallest load corresponding to the applied moment reaching the critical moment M_{cr} in the different unbraced segments is then a conservative lower bound to the elastic buckling load for the entire member. This is because each unbraced segment is assumed to be laterally and torsionally simply supported (Fig. 5.9 case I) when using the prior equations.

In some situations, the member LTB capacity can be increased substantially by the end restraint from adjacent segments on the most critical unbraced segment (Nethercot and Trahair, 1976; Nethercot, 1983). Considerable research has been performed in the past on analyzing the effects of member continuity across brace locations (Vacharajittiphan and Trahair, 1975; Nethercot, 1973a; Powell and Klingner, 1970; Trahair, 1969; Hartmann, 1967; Salvadori, 1955). Nethercot and Trahair (1976), Trahair (1977), and Trahair and Bradford (1988) have recommended the following simple method, which is based on the analogy that the buckling behavior of continuous doubly symmetric beams is the same as the behavior of end-restrained nonsway columns. The method gives conservative solutions except in some extreme cases of segments with high moment gradients, large values of W (Eq. 5.3), and high restraint stiffnesses from the adjacent segments, in which slight overestimates are obtained.

By using the analogy of a nonsway column, the nonsway column nomograph (Johnston, 1976; AISC, 2005) can be used to obtain an effective length of the critical beam segment. This method assumes that the lateral and warping restraint are identical and that in-plane restraint is accounted for by the bending moment diagram. The steps needed for the analysis are:

1. Compute the in-plane bending moment diagram (Fig. 5.12*a*).
2. Determine C_b and M_{cr} for each unbraced segment, using the actual unbraced length as the effective length in Eq. 5.14, and identify the segment with the lowest critical load. The overall critical applied loads for buckling assuming simply supported ends for the weakest segment and the two adjacent segments are denoted by P_m, P_{rL}, and P_{rR}, respectively (Fig. 5.12*b*).
3. Compute the stiffness ratios for the three segments as

$$\alpha_m = \frac{2EI_y}{L_{bcr}} \tag{5.20a}$$

for the critical segment and as

$$\alpha_r = n\frac{EI_y}{L_b}\left(1 - \frac{P_m}{P_r}\right) \tag{5.20b}$$

for each adjacent segment, where $n = 2$ if the far end of the adjacent segment is continuous, $n = 3$ if it is pinned, and $n = 4$ if it is fixed.
4. Determine the stiffness ratios $G = \alpha_m/\alpha_r$ and obtain the effective-length factor $K = K_y = K_z$ from the nonsway column nomographs.

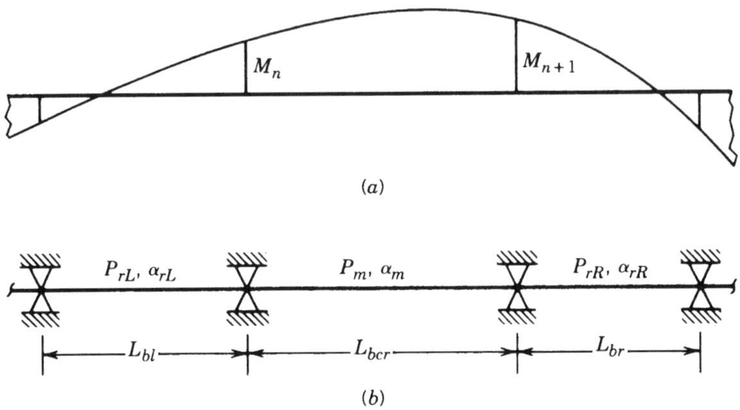

FIGURE 5.12 Computation of the effect of end restraint: (*a*) in-plane BMD; (*b*) braced segments.

5. Compute the critical moment and the buckling load of the critical segment from either of Eqs. 5.18.

The above method is illustrated by an example in Fig. 5.13. When the end restraint of the critical segment is accounted for, the buckling load is increased by 34%. This method may be extended to singly symmetric I-section members by utilizing the tee section composed of the compression flange and one-third of the depth of the web in compression for the controlling direction of bending as the equivalent column (White, 2006). As noted in the previous section, the bracing strength and stiffness demands can be increased substantially in members designed based on LTB effective-length factors less than 1. Therefore, caution must be exercised in the design of the bracing system for these types of members.

5.2.9 Built-In Cantilevers and Overhanging Beams

Solutions in the form of tables, charts, and approximate formulas for the elastic lateral–torsional buckling load of cantilever beams are available for a variety of loading and boundary conditions (Poley, 1956; Timoshenko and Gere, 1961; Nethercot and Rockey, 1972; Nethercot, 1973b; Trahair, 1983b, 1993; Essa and Kennedy, 1994; Andrade et al., 2007). Buckled shapes for unbraced cantilever beams show that the tension flange translates farther than the compression flange. Nethercot and Al-Shankyty (1979), Kitipornchai et al. (1984), and Assadi and Roeder (1985) showed that lateral bracing for cantilever beams is most efficient when it is located at the tension flange and that compression-flange lateral bracing is only slightly beneficial. The critical load is reduced as the point of load application is moved relative to the shear center in the opposite direction from the loading. Singly symmetric cantilever beams with symmetry about the vertical axis have been studied by Anderson and Trahair (1972), Roberts and Burt (1985), and

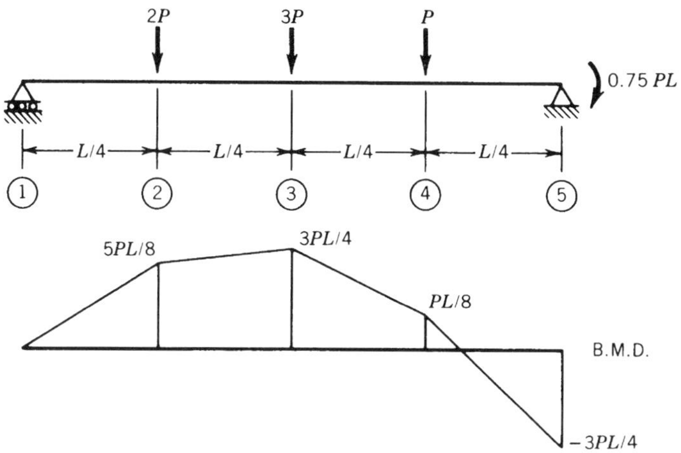

Segment	M_{max}	C_b (Eq. 5.15a)	M_{cr} (Eq. 5.14) [in.-kips (kN-m)]	P_{cr} [kips (kN)]
1-2	$0.625PL$	1.75	8411 (950)	22.43 (99.8)
2-3	$0.75PL$	1.083	5205 (588)	11.57 (51.5)
3-4	$0.75PL$	1.583	7608 (860)	16.91 (72.2)
4-5	$0.75PL$	1.933	9291 (1050)	20.65 (91.9)

$L = 600$ in. (15.24 m); W24 × 55; $I_y = 29.1$ in^4 (12.1×10^6 mm^4)
$J = 1.18$ in^4 (0.491×10^6 mm^4); $C_w = 3870$ in^6 (1.039×10^{12} mm^6). Critical segment: 2–3.

(*a*)

$$\alpha_{12} = \frac{3EI_y}{L/4}\left(1 - \frac{11.57}{22.43}\right) = 1.453\left(\frac{4EI_y}{L}\right)$$

$$\alpha_{23} = \frac{2EI_y}{L/4} = 2\left(\frac{4EI_y}{L}\right)$$

$$\alpha_{34} = \frac{2EI_y}{L/4}\left(1 - \frac{11.57}{16.91}\right) = 0.632\left(\frac{4EI_y}{L}\right)$$

$$G_2 = 2/1.453 = 1.4$$

$$G_3 = 2/0.632 = 3.2$$

From nonsway nomograph: $K = 0.85$.

$$M_{cr} = 6990 \text{ in.-kips } (790 \text{ kN-m})$$

$$P_{cr} = 15.53 \text{ kips } (69.1 \text{ kN})$$

(*b*)

FIGURE 5.13 Example of end restraint.

Wang et al. (1987). Andrade et al. (2007) and Wang and Kitipornchai (1986) have provided approximate equations for calculation of the elastic LTB strength of singly symmetric cantilevers with several loading and boundary conditions. The following section presents selected approaches for calculating the elastic LTB resistance of prismatic doubly symmetric I-beams.

Built-In Cantilevers For built-in cantilever beams where one end is fixed against warping, twist, and deflection in all directions, Dowswell (2004) developed equations to predict the lateral–torsional buckling capacity by curve fitting finite element data. The following equations are valid for $0.5 \le W \le 2.5$, where W is defined by Eq. 5.3. Essa and Kennedy (1994) provide a valid solution for other W values. In Dowswell's approach, the critical moment may be written as

$$M_{cr} = C_L C_H C_B \frac{\sqrt{EI_y GJ}}{L} \tag{5.21}$$

The coefficient C_L accounts for the moment distribution along the length of the beam.

For beams with a point load at the free end

$$C_L = 3.95 + 3.52W \tag{5.22a}$$

For beams with a uniformly distributed load

$$C_L = 5.83 + 8.71W \tag{5.22b}$$

The coefficient C_H accounts for the effect of load height. If the load is applied at or below the level of the shear center, $C_H = 1.0$. The following load height coefficients apply when the load is applied at the top flange (loads applied above the shear center may be taken conservatively as top-flange loads):

For beams with a concentrated load at the free end

$$C_H = 0.76 - 0.51W + 0.13W^2 \tag{5.23a}$$

For beams with a uniform load

$$C_H = 0.49 - 0.27W + 0.06W^2 \tag{5.23b}$$

The coefficient C_B accounts for the effect of bracing. If the beam is not braced, $C_B = 1.0$. For beams with continuous tension-flange lateral bracing or discrete tension-flange lateral bracing at the tip, C_B is obtained as:

For beams with the load at the level of the shear center

$$C_B = 1.42 + 0.88W - 0.26W^2 \tag{5.24a}$$

For beams with the load at the top flange, or conservatively for loads above the shear center,

$$C_B = 1.48 + 0.16W \tag{5.24b}$$

Overhanging Beams An overhanging beam is a cantilever beam that is continuous over a support. It consists of a cantilever segment on at least one end and a backspan. Interaction between the cantilever and backspan causes the segment with the highest buckling capacity to restrain the segment with the lowest buckling capacity. Essa and Kennedy (1994) have developed equations for the buckling capacity of a cantilever segment with a concentrated load at the tip. At the support, they assume lateral restraint at the top and bottom flange. They also assume the backspan to be unrestrained. When the backspan is more critical, they recommend that the overall buckling resistance of the overhanging beam may be calculated with good accuracy as the buckling resistance of the backspan M_b assuming no warping restraint at the support. This moment can be calculated using Eq. 5.2, assuming uniform bending over the backspan. When the overhanging span is more critical, the buckling moment may be written as

$$M_{cr} = M_c + I\,(M_b - M_c) \tag{5.25}$$

where M_c is the critical moment of the cantilever segment assuming no warping restraint. It may be taken as:

For beams loaded at the top flange

$$M_c = 1.5\frac{GJ}{d} \tag{5.26a}$$

For beams loaded at the shear center

$$M_c = \frac{4}{L_c}\sqrt{EI_y GJ} \tag{5.26b}$$

The interaction factor is determined by:

For cantilevers with no bracing

$$I = -0.08 + 0.18\frac{L_b}{L_c} - 0.009\left(\frac{L_b}{L_c}\right)^2 \tag{5.27a}$$

For cantilevers with top-flange lateral bracing at the tip

$$I = 0.064 + 0.162\frac{L_b}{L_c} - 0.009\left(\frac{L_b}{L_c}\right)^2 \tag{5.27b}$$

where L_c is the length of the cantilever segment and L_b is the length of the backspan.

Equations for the capacity of overhanging beams with other loading and restraint conditions are given in Trahair (1983b, 1993). Essa and Kennedy (1995) have developed a design method for beams in cantilever-suspended span construction, where secondary framing members provide both lateral and torsional restraints to the top flange of the beam. Tanner (1985), Essa and Kennedy (1994), Ozdemir and Topkaya (2006), and Trahair (2008) provide expressions for calculating the buckling resistance of overhanging monorail beams, where the compression flange is unbraced at the support.

The 2005 AISC specification indicates that Eq. 5.14 is to be used with $C_b = 1$ for cantilevers or overhangs where the free end is unbraced. Using Dowswell's equations for built-in cantilevers, it can be observed that for free-end cantilevers with a top-flange tip load, the AISC approach dramatically overestimates the LTB resistance. More accurate equations are available in Dowswell (2004) and Trahair (1983b). The AISC predictions can still overpredict these solutions by more than a factor of 4 for practical short cantilever lengths. For tip loads applied at the centroid and for all cases with uniformly distributed loads, however, the AISC specification solution is conservative.

For unbraced cantilever overhangs, the backspan flexibility exacerbates the above unconservative errors for the case of top-flange tip loading. Furthermore, the above equations indicate significant unconservative error of the AISC equations for centroidal tip loading on overhangs. These solutions are corroborated by Trahair (1983b), although Trahair's equations indicate that the maximum overprediction by the AISC equations is not as severe, with a maximum unconservative error of approximately 60% over a range of practical overhang and backspan lengths. The AISC solution is generally accurate to conservative for uniformly distributed loading on overhangs, assuming adequate bracing at the support and the same member cross section for the backspan.

5.2.10 Stability of I-Beams under Self-Weight during Lifting

During construction, the need often arises to lift beams into place. After bracing the beam or connecting it to other members, the beam is stabilized for imposition of further loading. It is often the case that the beam is most susceptible to buckling during the lifting operation. Due to the unusual boundary conditions involved, the prior equations are not strictly applicable for assessing the LTB resistance while the beam is being lifted. For instance, a beam with vertical cables at its ends does not have the buckling capacity of an equivalent simply supported beam. In addition, a midpoint lift does not exhibit cantilever behavior from the perspective of buckling.

Essa and Kennedy (1993) have recommended the following approach to assess the stability of prismatic doubly symmetric I-section members lifted under their self-weight for the case of two vertical lifting cables attached to the top flange (Fig. 5.14). This situation occurs when a spreader beam is used. The recommendations are based on the finite element verification and simplification of solutions originally developed by Dux and Kitipornchai (1989). The uniformly distributed

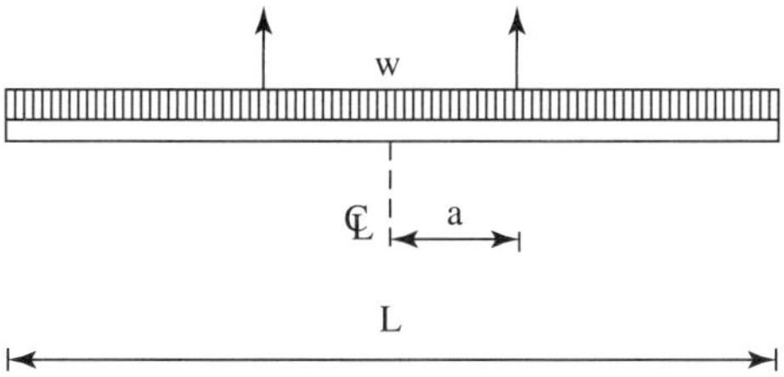

FIGURE 5.14 Doubly symmetric I-section member suspended from two vertical lifting cables.

buckling weight per unit length is approximated by the equation

$$w_{cr} = \frac{C\sqrt{EI_y GJ}}{L^3} \tag{5.28}$$

where C is a nondimensional buckling load parameter and L is the total member length. The value of C is dependent on the location of the cables along the lifted member, a, and on the beam torsional parameter W. Figure 5.15 shows results reported by Essa and Kennedy for W values of 0.2, 0.4, 0.6, and 0.8, which indicate that the buckling resistance is largest when the cables are placed near the quarter points. In the vicinity of this optimum location, however, the buckling capacity is very sensitive to the position of the cables. Therefore, great care should be exercised to ensure the correct position of the cables if they are placed in this region. As the cable attachment positions move from the quarter points toward either the middle or the ends of the beam, the buckling strength decreases. Essa and Kennedy provide the following equation for the case when the cable attachments are between the midspan and the quarter points:

$$C = \frac{1000W}{9.91 - 5.47\,(a/L) - 324.61\,(a/L)^2 + 794\,(a/L)^3} \tag{5.29}$$

Essa and Kennedy were not able to provide a simple curve fit for the right-hand branch in Fig. 5.15, and thus they recommend interpolating between the curves in Fig. 5.15 for this case. Attachment of the cables at the ends yields the most critical conditions for buckling. Dux and Kitipornchai (1989) provide numerous graphical solutions for the member buckling under its self-weight and the induced axial loading when the cables in this problem are inclined.

For the case of the spreader beam configuration shown in Fig. 5.16, Essa and Kennedy (1994) express the total lift weight at buckling of the spreader beam as

$$P_{cr} = \frac{(0.5 + 6W)\sqrt{EI_y GJ}}{L^2} \tag{5.30}$$

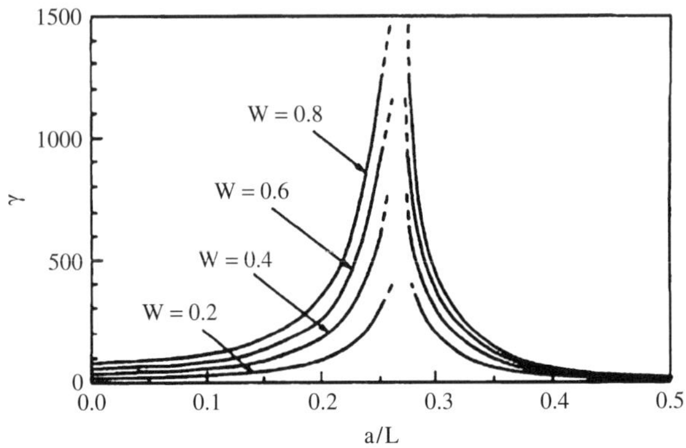

FIGURE 5.15 Normalized buckling strength for different load points.

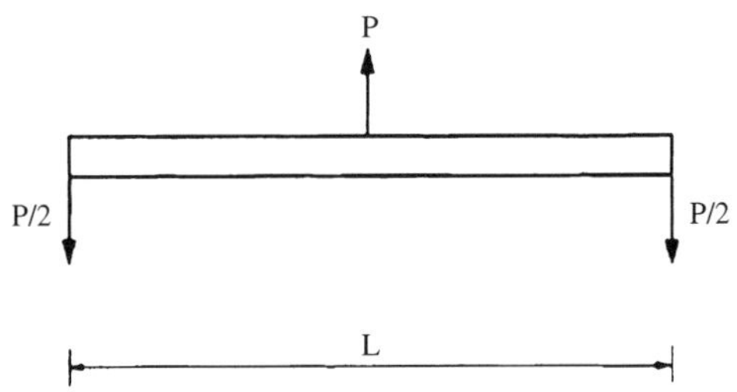

FIGURE 5.16 Spreader beam lifted at its midspan.

One other problem of significant interest is the buckling load of a simply supported member suspended from a cable attached to its top flange at its mid-span. Simplified solutions for this problem do not appear to be available. For this problem and other more general geometries, the solution must be obtained numerically.

5.2.11 Influence of Partial End Plates and Coped Ends on Lateral–Torsional Buckling

End connections such as partial end plates (Lindner, 1985) and coped ends (Yura and Cheng, 1985; Cheng et al., 1988; Cheng and Yura, 1988; Lindner, 1994) can influence the LTB resistance substantially. The influence of end copes should be considered when the depth of a cope is greater than or equal to $0.2d$, where d is the total depth of the cross section. Coped ends may be reinforced to offset this reduction in the LTB strength.

5.2.12 Overall System Lateral–Torsional Buckling

In certain cases, typically during construction, the overall or global buckling of a group of girders as a system may be more critical than the buckling of the individual girders between the diaphragms or cross-frames connecting the girders. Closely spaced I-girders interconnected by diaphragms or cross-frames are particularly susceptible to global LTB. Yura et al. (2008) have shown that the following equation provides an accurate estimate of the total global buckling moment for a twin simply supported prismatic I-girder system when the girder spacing S is greater than or equal to the girder depth h_o:

$$M_{crg} = C_b \frac{\pi^2 SE}{L_g^2} \sqrt{I_{ye} I_x} \tag{5.31}$$

where L_g is the overall span length of the girders, I_x is the major axis moment of inertia of the individual girders, and

$$I_{ye} = I_{yc} + \left(\frac{b}{c}\right) I_{yt} \tag{5.32}$$

where I_{yc} and I_{yt} are the lateral moments of inertia of the individual girder compression and tension flanges and b and c are the distances from the cross-sectional centroids to the tension and compression flanges, respectively. Yura et al. (2008) recommend a similar but more detailed equation when $S < h_o$.

In addition, the authors have shown:

- Top-flange loading effects have little impact on the twin girder global LTB resistance.
- Within practical ranges of design, the size and spacing of the cross-frames have little impact on the global buckling response.
- Three girder systems can be addressed in Eq. 5.31 by defining the distance between the exterior girders as S.
- For cases in which a girder system does not have adequate global buckling strength, a top-flange lateral bracing system can be provided between the girders over a short length at the ends of the overall span as an efficient way of increasing the strength (the authors provide recommendations for sizing these bracing elements).

It can be demonstrated that for equal-size doubly symmetric girders, global system LTB will tend to be more critical than LTB of the individual girders between the cross-frames or diaphragms when

$$\frac{L_b}{L_g} < 0.6\sqrt{\frac{b_f}{h_o}\frac{h_o}{S}} \tag{5.33}$$

For example, if $h_o/b_f = 4$ and $h_o/S = 1$, member buckling will tend to be more critical unless $L_b < 0.3L_g$ or, in other words, unless one provides at least three equally spaced intermediate cross-frames. The potential for system buckling is increased by lean-on framing, in which a number of girders are stabilized by other girder systems by tying the members together by top- and bottom-flange struts. Stability evaluation of lean-on bracing systems is discussed by Zhou (2006).

5.3 FUNDAMENTAL COMPARISON OF DESIGN STANDARDS, PRISMATIC I-SECTION MEMBERS

To provide an indication of the variation in nominal beam strengths used or recommended for design practice, Fig. 5.17 shows a comparison of representative LTB nominal resistance predictions pertaining to four current standards. The applicable formulas are listed below. The curves and equations correspond to the uniform bending case ($C_b = 1$). The different standards are calibrated in conjunction with their corresponding building codes and the load factors within these codes. In addition, various considerations enter into the development of each standard's strength curves, such as the targeted level of reliability and whether the level of reliability is varied according to the beam slenderness. In short, while the nominal strengths presented here illustrate the approaches for calculating the LTB resistance, they do not convey the entire picture. The ordinate of the curves in Fig. 5.17 is the normalized nominal moment capacity M_n/M_p. The cross-section elements of the selected W27 × 84 section are sufficiently stocky such that the maximum flexural resistance is equal to the plastic moment capacity in all of these standards. The abscissa of the curves in Fig. 5.17 is the normalized unbraced length L/L_p, where L_p is taken as $1.1r_t\sqrt{E/F_y}$. This is the unbraced length at which the mean nominal resistance of general I-section members reachers M_p in uniform major axis bending, based on a comprehensive assessment of experimental data (White and Jung 2008).

In general, Fig. 5.17 shows a considerable variation in the nominal LTB resistances used by the different standards. All of the curves are based upon the same elastic critical moment M_{ocr}. One difference is in the type of mapping from the elastic buckling resistance to the nominal strength. In addition, the two North

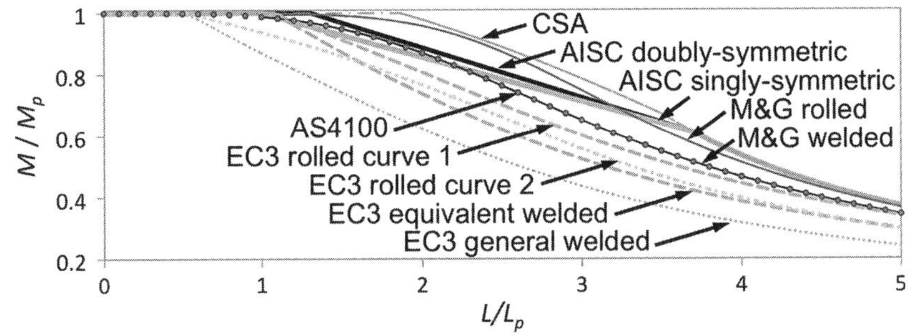

FIGURE 5.17 Comparison of nominal LTB resistances for different length W27x84 [F_y = 345 Mpa (50 ksi)] beams and equivalent section welded beams subjected to uniform bending moment.

American standards AISC and CSA implicitly assume that the beam has no initial out-of-straightness for long members that fail by elastic LTB. Conversely, the Australian and the European standards, AS4100 and EC3, provide a substantial penalty for geometric imperfections.

The Eurocode 3 (EC3) resistance equations corresponding to Fig. 5.17 are listed in Table 5.1. Similar to its handling of column buckling, EC3 (CEN, 2005) uses the Perry–Robertson formula (Robertson, 1925) for its characterization of beam LTB. Eurocode 3 provides two sets of coefficients for use with the equations shown in the table, one for general members and another for rolled I-section beams and equivalent welded beams. The coefficients $\overline{\lambda}_{LT,0}$, below which the resistance is constant at a maximum plateau level, and β, which affects the shape of the strength curve, are 0.2 and 1.0 for the general equation. For the special case of rolled I-section beams and equivalent welded beams, $\overline{\lambda}_{LT,0}$ may be increased to 0.4 and β can be reduced to 0.75. Eurocode 3 gives separate curves for relatively wide and narrow sections as well as for welded and rolled sections. These differences are expressed by different α_{LT} values as summarized in Table 5.2. Figure 5.17 shows the two sets of Eurocode 3 strength curves for a W27x84 rolled section as well as for a welded section with the same cross-sectional profile. The development of the Eurocode 3 equations is documented thoroughly in ECCS (2006) and is discussed further in Rebelo et al. (2009) and in Simoes da Silva et al. (2009). A key distinction between the Eurocode 3 and the AASHTO and AISC developments is in Eurocode 3 efforts to make extensive use of refined distributed plasticity

TABLE 5.1 Eurocode 3 Base Equations for Rolled or Welded Class 1 or Class 2 I-Section

$$M_n = \frac{M_p}{\Phi_{LT} + \sqrt{\Phi_{LT}^2 - \beta\overline{\lambda}_{LT}^2}} \le M_p$$

where: $\Phi_{LT} = 0.5\left[1 + \alpha_{LT}\left(\overline{\lambda}_{LT} - \overline{\lambda}_{LT,0}\right) + \beta\overline{\lambda}_{LT}^2\right]$

$$\overline{\lambda}_{LT} = \sqrt{\frac{M_p}{M_{ecr}}}$$

TABLE 5.2 Eurocode 3 LTB Curve Selection

		General Case		Rolled or Equivalent Welded Case	
Cross Section	Limits	Buckling Curve	α_{LT}	Buckling Curve	α_{LT}
Rolled I-sections	$d/b \le 2$	*a*	0.21	*b*	0.34
	$d/b > 2$	*b*	0.34	*c*	0.49
Welded I-sections	$d/b \le 2$	*c*	0.49	*c*	0.49
	$d/b > 2$	*d*	0.76	*d*	0.76
Other Cross-sections		*d*	0.76	*d*	0.76

analysis results based on representative idealized residual stresses and geometric imperfections, in addition to experimental test results.

The Australian Standard AS4100 (SAA, 1998) uses the following single set of equations to characterize the nominal LTB resistance:

$$M_n = \alpha_m \alpha_s M_s \leq M_s \tag{5.34a}$$

where α_m is an equivalent moment factor, which is in concept similar to the equations for C_b presented earlier in Section 5.2.3, M_s is the cross-sectional moment capacity, that is, the moment capacity of a fully braced member, and

$$\alpha_s = 0.6 \left(\sqrt{\left(\frac{M_s}{M_{ocr}} \right)^2 + 3} - \frac{M_s}{M_{ocr}} \right) \tag{5.34b}$$

As noted by Trahair (1993), the AS4100 equation provides a close fit to the lower bounds from 159 LTB tests of commercial hot-rolled steel I-beams. This equation is used with all values of the cross-section moment capacity M_s. For compact sections, M_s is the plastic moment capacity M_p. The parameter α_m is taken equal to 1.0 in Fig. 5.17.

The Canadian CSA S16.1 strength curve (CSA, 2001) for doubly symmetric Class 1 and 2 sections is defined by

$$M_n = 1.15 M_p \left(1 - 0.28 \frac{M_p}{M_{cr}} \right) \leq M_p \quad \text{if } M_{cr} \geq 0.67 M_p \tag{5.35a}$$

$$M_n = M_{cr} \tag{5.35b}$$

This curve is based on a statistical analysis conducted by Baker and Kennedy (1984) of Dibley's (1969) rolled I-section member tests. The parameter M_{cr} is the critical elastic buckling moment. A parameter equivalent to C_b is included in the calculation of this buckling value but is taken equal to 1.0 in Fig. 5.17. For doubly symmetric Class 3 sections, CSA S16.1 replaces M_p by M_y in Eqs. 5.35.

In recent developments, MacPhedran and Grondin (2009) have conducted statistical analysis of the experimental data collected by Greiner and Kaim (2001) on rolled and welded I-section beams and have proposed the following single nominal resistance equation for the CSA 16.1 standard:

$$M_n = M_s \left(1 + \left(\frac{M_s}{M_{cr}} \right)^n \right)^{-1/n} \tag{5.36}$$

where M_s is the moment capacity of a fully braced beam, taken equal to M_p for Class 1 or Class 2 cross sections and to the yield moment M_y for Class 3 sections, and where n is a shape parameter equal to 3.1 for rolled shapes and 1.9 for welded shapes. It is interesting to note that the curve recommended by MacPhedran and Grodin for welded I-section members is essentially identical to the AS4100 strength curve in Fig. 5.17.

TABLE 5.3 AISC/ANSI 360 Base Equations for Rolled or Welded Compact I-Section Members

$$M_n = M_p \qquad \text{for } L_b \le L_p$$

$$M_n = C_b\left[M_p - (M_p - F_L S_{xc})\left(\frac{L_b - L_p}{L_r - L_p}\right)\right] \le M_p \qquad \text{for } L_p < L_b \le L_r$$

$$M_n = C_b M_{ocr} \qquad \text{for } L_b > L_r$$

where: $L_p = 1.76 r_y \sqrt{\dfrac{E}{F_y}}$ for doubly-symmetric sections

$L_p = 1.1 r_t \sqrt{\dfrac{E}{F_y}}$ for singly-symmetric sections

$$L_r = 1.95 r_t \frac{E}{F_L}\sqrt{\frac{1}{X^2} + \sqrt{\frac{1}{X^4} + 6.76\left(\frac{F_L}{E}\right)^2}}$$

$F_L = 0.7F_y$ for $\dfrac{S_{xt}}{S_{xc}} \ge 0.7$

$F_L = F_y \dfrac{S_{xt}}{S_{xc}} \ge 0.5F_y$ for $\dfrac{S_{xt}}{S_{xc}} < 0.7$

X^2 is given by Eq. (5.9)

r_t is given by Eq. (5.7) or (5.8) as applicable

The base equations from the 2005 AISC Specification are summarized in Table 5.3. The 2005 AISC Specification curves use a flat plateau resistance for $L_b \le L_p$ and a linear function of the unbraced length L_b to represent the nominal inelastic LTB response. For compact I-section members, the plateau resistance is equal to M_p. The plateau resistance is reduced as a function of web slenderness for cases with noncompact and slender webs. For doubly symmetric I-section members, the AISC elastic-to-inelastic buckling transition point is taken at 70% of the yield moment, which implies a maximum residual stress effect of $0.3F_y$. As noted above, the AISC elastic LTB resistance is taken as the theoretical elastic LTB moment, which can be determined in general from any of the formulas in Section 5.2. In the case of singly-symmetric compact I-section members, AISC uses the second equation for L_p shown in the table. As mentioned above, this is the unbraced length at which the mean nominal resistance of general I-section members reaches M_p in uniform major axis bending, based on a comprehensive assessment of experimental data (White and Jung 2008).

A recent study (White, 2008; White and Jung, 2008; White and Kim, 2008) was conducted that has assisted the AASHTO and AISC specification committees in bringing their complete beam and girder design processes into a consistent formulation. These papers provide an overview of the conceptual developments for a full range of I-section members, including double and single symmetry of the cross section, hybrid girders in which the material strength of one or both flanges is different than the material strength of the web, composite girders in negative bending, and slender, noncompact and compact flanges and webs. The developments include extensive collection of, evaluation of, and calibration to available experimental test data, including prior data collected by Fukumoto and Kubo (1977), Fukumoto et al. (1980), and Fukumoto and Itoh (1981).

The AISC equations do not make any distinction between rolled I-section members and welded I-section members of the same geometry. In addition, there is no distinction between members with large versus small d/b_f other than the influence on the corresponding elastic LTB responses.

5.4 STEPPED, VARIABLE WEB DEPTH AND OTHER NONPRISMATIC I-SECTION MEMBERS

Section 3.8.3 outlines a generalized procedure, which is detailed in a recent MBMA/AISC Design Guide (Kaehler et al., 2009), for determining member axial resistances of nonprismatic members loaded in nonuniform axial compression. Kaehler et al. (2009) also provide similar generalized procedures for determining the flexural resistance of nonprismatic members. This section contains a summary of the basic concepts and procedures from this guide pertaining to LTB.

Although limit state checks such as flange local buckling or tension flange yielding can be handled on a cross section by cross section basis, the LTB resistance cannot be assessed solely on such a basis. This is because the LTB resistance depends on the cross-sectional properties along the entire unbraced length as well as the loading configuration (e.g., moment gradient) and the end conditions (e.g., continuity with adjacent unbraced lengths). These factors are similar to the factors that influence the member out-of-plane resistance in axial compression. As such, similar to the approach outlined in Section 3.8.3, the LTB resistance of a general nonprismatic I-section member may be determined by focusing on:

1. The ratio of the moments (or compression-flange stresses) at elastic LTB to the corresponding factored moments (or stresses):

$$\gamma_{e.\mathrm{LTB}} = \frac{M_{cr}}{M_u} = \frac{F_{cr}}{f_{bu}} \tag{5.37}$$

2. The largest ratio of the factored moment (or compression-flange stress) to the section yield strength:

$$\rho_{o.\max} = \left(\frac{M_u}{M_{yc}}\right)_{\max} = \left(\frac{f_{bu}}{F_{yc}}\right)_{\max} \tag{5.38}$$

for all the cross sections along the unbraced length.

The product of the above two parameters is equal to the ratio of the elastic buckling moment to the yield moment (M_{cr}/M_y) or the elastic buckling stress to the yield stress (F_{cr}/F_y), at the location corresponding to the largest flexural stress in the member. This value can be substituted into prismatic member resistance equations, written in terms of this ratio. This provides the mapping of the elastic buckling strength to the nominal design resistance.

Similar to the column-buckling calculations discussed in Section 3.8.3, there is only one $\gamma_{e.\text{LTB}}$ for a given unbraced length, although the compression-flange stresses and the corresponding moments vary in general from cross section to cross section along the member length. The above equations check the LTB resistance at the most highly stressed cross section. However, for elastic LTB, if $F_{cr} = \gamma_{e.\text{LTB}} f_{bu}$ and $M_{cr} = \gamma_{e.\text{LTB}} M_u$ were calculated based on any of the other cross sections, one can observe that the ratio $f_{bu}/F_{cr} = M_u/M_{cr}$ is $1/\gamma_{e.\text{LTB}}$.

For inelastic LTB, similar to the calculation of the inelastic column resistance in Section 3.8.3, the resistance is determined by mapping the nonprismatic member to an equivalent prismatic member that has the same $\gamma_{e.\text{LTB}}$ and a ρ_o equal to the above $\rho_{o.\max}$. This is illustrated by Fig. 5.18. This conceptual extension of the specification LTB resistance equations is the same as that invoked originally by Lee et al. (1981) in the context of AISC allowable stress design. Based on the above concepts, the LTB flexural capacity for a given unbraced length may be determined as follows:

1. Calculate $\rho_o = f_{bu}/F_{yc}$ or M_u/M_{yc} at the various cross sections along the unbraced length.
2. Determine the maximum value of ρ_o, that is, $\rho_{o.\max}$.
3. Calculate the ratio of the elastic LTB load level to the factored load level $\gamma_{e.\text{LTB}}$.
4. Determine the flexural capacity corresponding to the most highly stressed cross section by substituting the product of γ_e and $\rho_{o.\max}$ into the appropriate prismatic member resistance equations.

Kaehler et al. (2009) discuss various methods for calculating the ratio $\gamma_{e.\text{LTB}}$ for different member geometries. Potentially, $\gamma_{e.\text{LTB}}$ can be calculated most reliably

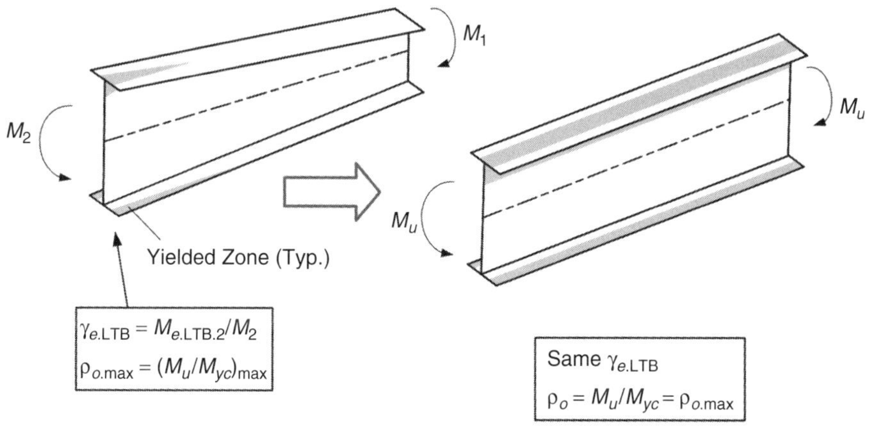

FIGURE 5.18 Conceptual mapping of a tapered-web I-section member subjected to bending moment to an equivalent prismatic member.

using computational software tools. Unfortunately, few analysis programs currently give accurate elastic LTB solutions for singly symmetric and/or nonprismatic members. Furthermore, there are a number of complexities associated with the proper definition of elastic LTB models, and most programs that have reliable capabilities for elastic LTB analysis are somewhat difficult to use in setting up these models. Even if this state of practice is rectified, there will always be a need for reliable simplified approximate solutions.

Yura and Helwig (2009) detail one such approximate solution procedure for practical unbraced lengths with linearly tapered web depths:

1. Calculate C_b using one of the equations in Section 5.2.3 but expressed in terms of the stresses in the flange under consideration rather than the cross-section moments.
2. Calculate the elastic LTB stress at the cross section having the largest $f_{bu}/F_{yc} = M_u/M_{yc}$ as

$$F_{cr} = C_b F_{cr.\text{mid}} \tag{5.39}$$

   where $F_{cr.\text{mid}}$ is the elastic LTB stress calculated from the elastic LTB equations for a prismatic member using the cross-section properties at the middle of the unbraced length.
3. Calculate F_{cr}/F_{yc} as the ratio of the above F_{cr} to the compression-flange yield strength at the cross section with the largest f_{bu}/F_{yc}, or alternatively, determine $\gamma_{e.\text{LTB}}$ by substituting F_{cr} and f_{bu} at the most highly stressed cross section into Eq. 5.37, determine $\rho_{o.\max}$ by substituting the elastic-flange stress and the compression-flange yield strength at this cross section into Eq. 5.38, and determine F_{cr}/E_{yc} as the product of $\gamma_{e.\text{LTB}}$ and $\rho_{o.\max}$.
4. Use the above F_{cr}/F_{yc} ratio to determine the nominal flexural resistance based on the applicable prismatic member equations, written in terms of this parameter.

White and Grubb (2003) give another approximate solution for unbraced lengths composed of prismatic segments with a single cross-sectional transition within the unbraced length. This procedure is adapted from Carskaddan and Schilling (1974) and Dalal (1969):

1. Calculate C_b using one of Eqs. 5.15 and assuming that the unbraced length is prismatic.
2. Calculate the elastic LTB stress corresponding to the section with the largest end moment M_2 as

$$F_{cr.2} = \chi \frac{C_b \pi^2 E}{(L_b/r_{t.2})^2} \tag{5.40}$$

   where $r_{t.2}$ is the radius of gyration of the compression flange plus one-third of the depth of the web in compression for the cross section corresponding to M_2 and χ is determined from the chart in Fig. 5.19. The procedure assumes

M_2 is the largest moment within the unbraced length under consideration and that this moment occurs in the larger of the two cross sections.

3. Calculate $\gamma_{e.\text{LTB}}$ as $F_{cr.2}/f_{bu.2}$, where $f_{bu.2}$ is the compression-flange factored stress at the cross section corresponding to M_2. Calculate $\rho_{o.\max}$ as the larger of $f_{bu.s}/F_{yc.s}$ and $f_{bu.2}/F_{yc.2}$, where $f_{bu.s}$ is the compression-flange factored stress at the smaller section at the cross-section transition, $F_{yc.s}$ is the corresponding compression-flange yield stress, and $F_{yc.2}$ is the compression-flange yield stress at the cross section corresponding to M_2.
4. Calculate $F_{e.\text{LTB}}/F_{yc}$ from the product of γ_e and $F_{cr.2}$, and use this ratio for determining the nominal flexural resistance from the applicable prismatic member resistance equations.

In the above, the parameter χ is the ratio $P_e/(\pi^2 EI_2/L_b^2)$ for the stepped column shown in Fig. 5.19. This ratio gives a slightly conservative estimate of the elastic LTB resistance for a stepped I-section member subjected to a moment gradient with the larger moment applied to the larger end cross section. Based on the behavior illustrated in Fig. 5.19, AASHTO (2007) articles C6.10.8.2.3 and CA6.3.3 allow transitions to a smaller cross section to be neglected in determining F_{cr} for unbraced lengths involving prismatic segments with

- $L_1/L_b \leq 0.2$,
- $I_1/I_b \geq 0.5$, and
- f_{bu}/f_{yc} maximum within the larger cross section.

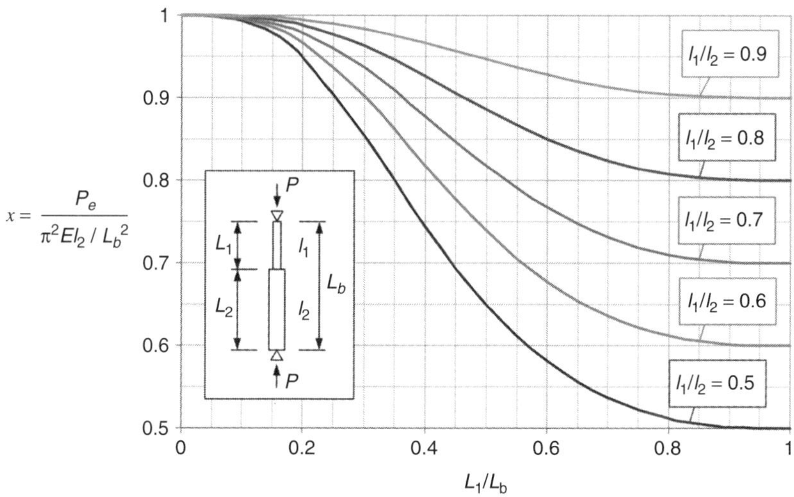

FIGURE 5.19 Ratio of elastic LTB stress at the section with the largest moment M_2 to the LTB stress determined assuming that the member is prismatic with the larger cross section throughout the unbraced length (Carskaddan and Schilling, 1974; Dalal, 1969).

where L_1 is the length of the segment with the smaller cross section and I_1 and I_2 are the individual flange moments of inertia about the plane of the web for the smaller and larger flanges, respectively (to be checked for both flanges).

Kaehler et al. (2009) provide several other elastic LTB solutions. This reference also provides specific recommended elastic LTB calculation procedures and an extensive annotated bibliography outlining additional calculation methods.

5.5 CONTINUOUS-SPAN COMPOSITE I-SECTION MEMBERS

Continuous composite beams are used widely in multistory buildings and in bridges. For the purpose of flexural strength calculations, these types of members may be subdivided into two regions, namely the regions of positive bending and negative bending. The flexural resistance of composite beams in positive bending is governed typically by the plastic strength of the composite steel beam and relatively simple calculations are capable of estimating the flexural capacity. The behavior, however, is more complex in the negative bending regions, because:

1. The contribution of the composite slab to the major axis bending moment can substantially increase the cross-sectional strength of the member in cases where flange local buckling or lateral buckling limit states do not govern.
2. The composite slab causes a shift in the neutral axis such that a relatively large depth of the web is in compression. This produces additional destabilizing effects on the web and the bottom flange.
3. The slab tends to restrain the twisting and lateral bending of the section at the top flange.
4. The distortional flexibility of the web can still lead to a failure mode involving lateral buckling of the bottom compression flange.

This problem has been addressed by numerous authors, including Svensson (1985), Williams and Jemah (1987), Bradford (1989, 1998) Johnson and Bradford (1983), Bradford and Johnson (1987), Weston et al. (1991), Kemp and Dekker (1991), Bradford and Gao (1992), Dekker et al. (1995), Kemp et al. (1995), and Vrcelj and Bradford (2009).

Lateral bracing from cross-frames is common in bridges but is less convenient in buildings, where the spacing between adjacent beams is typically wider relative to their depth. In additon, such bracing in buildings can interfere with required utilities and is often best avoided. For bridge construction, AASHTO (2007) uses a simplified approach in which a form of Eqs. 5.6 and 5.14 is employed with:

- r_t calculated based on depth of the web in compression,
- J calculated from the steel section alone (for compact and noncompact web sections), or taken equal to zero (for slender-web sections), and
- An effective S_{xc} determined based on the combined noncomposite and cracked composite cross-section properties and the extent of the noncomposite dead

load for slender-web members, and based solely on the cracked composite cross-section properties for noncompact and compact-web sections.

Given a typical moment gradient within the negative-moment region and common cross-frame spacings, the C_b values tend to place the response on the plateau of the LTB resistance curve. White and Kim (2004) report experimental strengths from nine composite beam tests ranging from 0.94 to 1.29 of the predictions using this method. Other tests conducted using cover plates on one side of the I-section to simulate the influence of a slab are predicted with good accuracy (White and Kim 2008).

For buildings, a more refined calculation is in order to handle cases where there is no bracing to the bottom flange within the span of the beam. For these types of members, Weston et al. (1991) recommend an expression for girders constructed of grade 43 or grade 50 structural steel that can be written as the following equivalent LTB unbraced length:

$$L_b = 23.7\left\{1.28\left[\left(\frac{L}{r_{yb}}\right)^{1/2}\left(\frac{h}{t_w}\right)^{1/3}\right] - 29\right\} r_{yb}\sqrt{\frac{F_{yb}}{E}} \tag{5.41}$$

where L is the total span length, r_{yb} the radius of gyration of the girder bottom flange (assumed prismatic in the authors' studies and equal to $b_f/\sqrt{12}$ assuming a rectangular bottom-flange plate), h the clear web depth between the flanges, t_w the web thickness, and F_{yb} the yield strength of the bottom flange.

It is important to note that the maximum hogging moment capacities are based on the cracked composite section in the study by Weston et al. (1991). Furthermore, these investigators observe some sensitivity to the ratio L/r_{yb} in girders they studied with larger L/r_{yb} values. This appears to indicate that other methods based on conclusions that the lateral distorsional buckling capacity is insensitive to this parameter may lead to overestimates of the strength. The authors' studies are based on an assumed gravity-load moment diagram with the hogging moment region extending approximately $0.24L$ from the support.

Bradford (1989) concludes that the lateral distortional buckling capacity is essentially independent of the moment gradient based on inelastic finite element studies. Based on the studies by Bradford and Weston et al., Oehlers and Bradford (1999) propose a modified form of the expression from Weston et al. (1991):

$$\frac{M_s}{M_{crd}} = \left[0.018\left(\frac{L}{r_{yb}}\right)^{1/2}\left(\frac{h}{t_w}\right)^{1/3} - 0.40\right]^2 \tag{5.42}$$

where M_s is the section strength of the girder and M_{crd} is the elastic distortional buckling moment. Oehlers and Bradford then suggest the following equation for the mapping of the above buckling strength ratio to the nominal resistance of the girder in the hogging moment region:

$$M_n = 0.8\left\{\sqrt{\left(\frac{M_s}{M_{crd}}\right)^2 + 3} - \frac{M_s}{M_{crd}}\right\} M_s \leq M_s \tag{5.43}$$

5.6 BEAMS WITH OTHER CROSS-SECTIONAL TYPES

5.6.1 Channel Section Beams

Channel sections loaded in bending possess structural properties that are different from wide-flange beams. The shear center and centroid do not coincide and both are not on a material point of the cross section (see Fig. 5.20). It is important to note that only when a section is loaded through its shear center will the section deflect without twisting. At a critical load, the sections will buckle suddenly in a lateral–torsional mode. It was found by Melcher (1994, 1999) and la Poutré et al. (2002) that these sections can be designed for inelastic lateral–torsional buckling by the procedures of Eurocode 3.

All channels loaded at the web are subject to an initial torsional load, in addition to bending, which will make the cross section twist. Consequently, the section reacts differently from channels loaded at the shear center. Three design approaches are proposed in the literature. Lindner and Glitsch (2004) propose an equation with three terms, the first for major axis bending, the second for minor axis bending, and the third for torsion. Kindmann and Frickel (2002) propose design methods that are similar to those used for doubly and singly symmetric sections. However, they introduce a correction term based on the nondimensional slenderness to reduce the buckling curves and account for torsion. Finally, la Poutré (2008) compares the two previous design procedures to the overall method of Eurocode 3 for the stability of structures under general loading. For the overall method, the plastic capacity in bending and torsion is needed, which currently can only be determined accurately with a finite element method (FEM). The overall method produces very accurate results but overestimates the inelastic buckling loads for members with low slendernesses. The methods of Lindner and Glitsch (2004) and Kindmann and Frickel (2002) work well for more slender channels but tend to underestimate the inelastic buckling loads for members with low slenderness.

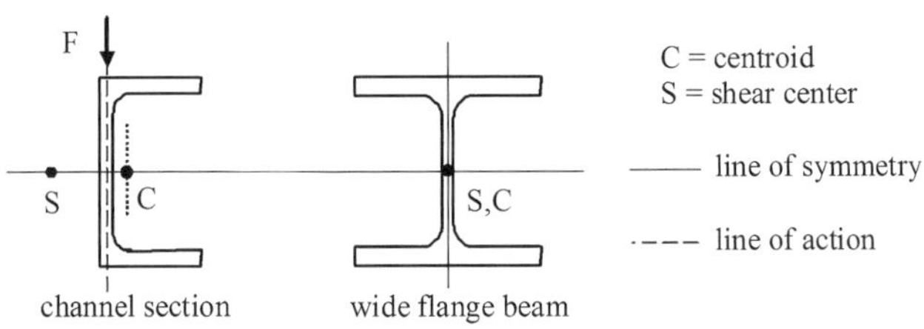

FIGURE 5.20 Channels versus wide-flange beams.

5.6.2 Tee Section Beams

For T-section or double-angle beams, $C_w = 0$ and β_x is determined from the formula in Fig. 5.2, with either $b_{ft} = t_w$ or $b_{fc} = t_w$, depending on whether the flange or the stem is in compression. Alternatively, β_x may be approximated by Eq. 5.12, where I_{yc} is taken equal to I_y if the flange is in compression and $I_{yc} = 0$ if the stem is in compression. Taking $h_o = d$ and substituting into Eq. 5.10 gives

$$M_{cr} = \frac{\pi\sqrt{EI_y GJ}}{L_b}\left(B + \sqrt{1 + B^2}\right) \tag{5.44a}$$

where

$$B = \pm 2.3\frac{d}{L_b}\sqrt{\frac{I_y}{J}} \tag{5.44b}$$

Generally, the moment gradient factor C_b should not be taken larger than 1.0 for tee-shaped beams and double-angle beams. Minor axis rotational end restraint may be considered by replacing L_b with $K_y L_b$ in the above expression.

5.6.3 Rectangular Solid and Box Section Beams

For rectangular solid and box section beams, the combination of Eqs. 5.2 and 5.14 may be used with W taken equal to zero. The moment gradient modifier C_b may be calculated using any of the expressions discussed in Section 5.2.3. Load height effects may be accounted for by using Eq. 5.16a with $R_m = 1.0$. Minor axis rotational end-restraint effects may be considered by replacing L_b by $K_y L_b$ in Eq. 5.2. Because of the high torsional resistance of typical box section beams, the LTB of these types of members usually does not control relative to other limit states.

5.6.4 Trusses

The lateral stability of trusses was investigated by Horne (1960), and approximate solutions and a lower-bound design equation for light factory-made trusses (steel joists) are given by Hribar and Laughlin (1968), Galambos (1970), and Minkoff (1975). Lateral buckling of the unsupported compression chord of trusses is discussed in Chapter 15.

5.7 DESIGN FOR INELASTIC DEFORMATION CAPACITY

Deformation capacity is a measure of the ductility of a member, which is a necessary consideration in plastic and seismic design (see Chapter 19). In order for the assumptions of these design philosophies to be satisfied, steel beams must display a ductility greater than that which is required for the intended level of moment redistribution (in the case of plastic design) or energy dissipation (in the case of seismic design).

Ductility is generally quantified as a function of the ratio between inelastic and elastic response at a given level of load (typically M_p). This response may also be quantified in terms of displacement, rotation, or curvature. For example, one commonly used measure of ductility is *rotation capacity*, which can be defined (Galambos, 1968b) as

$$R = \frac{\theta_{\max}}{\theta_p} - 1 \tag{5.45}$$

where R is the rotation capacity, $\theta_{\max}$ is the total end rotation at M_p during the unloading response of the beam, and θ_p is the elastic end rotation at M_p. These terms are defined in Fig. 5.21, which shows a typical moment–rotation behavior for a compact I-section beam. Here the solid line represents the actual response of the member and the dashed line idealizes this behavior by assuming elastic–perfectly plastic response.

While M_p provides a convenient reference point for evaluating ductility, ductility can be assessed at any moment level. The magnitude of moment considered is significant because beams exhibit larger rotation capacity based on the definition shown in Fig. 5.21 when R is determined at a smaller moment level. A consequence

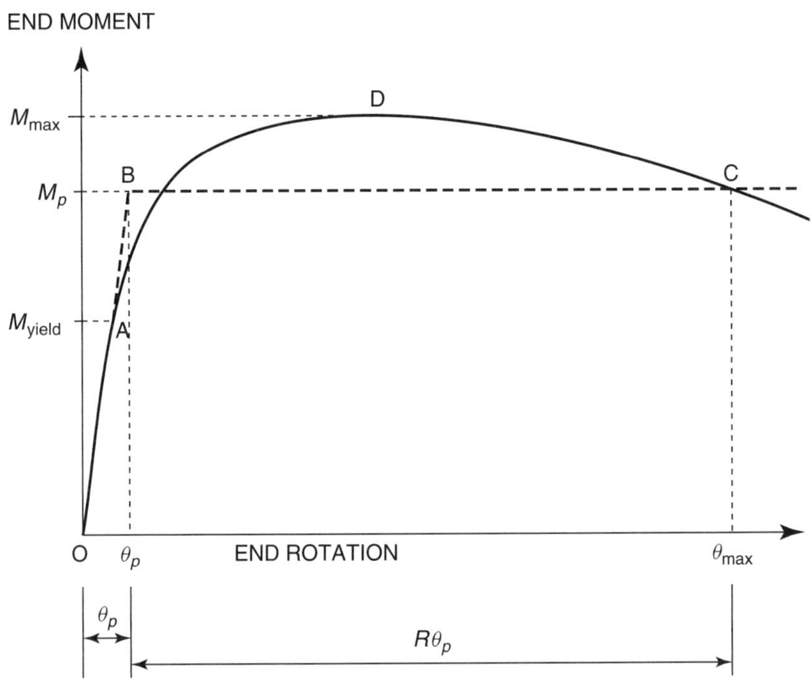

FIGURE 5.21 Definition of rotation capacity.

of defining ductility based on M_p is that beams with moment capacities less than M_p are considered to have zero ductility. Therefore this approach significantly restricts the class of sections that can be used in plastic/seismic design. To overcome this negative consequence of using M_p as a reference moment, Haaijer et al. (1980) suggested the concept of an effective plastic moment, M_{pe}. This suggestion was based on the observation that members with moment capacities less than M_p may also exhibit significant ductility, although at a reduced level of moment.

Regardless of the reference moment, there are several factors that contribute to an increased deformation capacity of a given beam. These factors have been identified via experimental testing and analytical study by numerous investigators over the past 50 years (Prasad and Galambos, 1963; Adams et al., 1964; Lukey and Adams, 1969; Lukey et al., 1969; Bansal, 1971; Climenhaga and Johnson, 1972; Holtz and Kulak, 1973, 1975; Grubb and Carskaddan, 1979, 1981; Carskaddan, 1980; Schilling, 1985; Schilling and Morcos, 1988; Tansil, 1991; Barth, 1996; Hartnagel, 1997; Yakel et al., 1999; Green et al., 2001, 2002; Kemp, 1986; Kato, 1989, 1990; Kemp and Dekker, 1991; Gioncu and Petcu, 1997a,b; Dinno and Birkemoe, 2001; Thomas and Earls, 2003; Lääne and Lebet, 2005; Righman 2005). As a result of these studies, the most influential parameters affecting rotation capacity have been shown to be flange slenderness, lateral brace spacing, and web slenderness (especially when quantified in terms of web depth in compression at M_p), where there is a decreasing amount of ductility with increasing slenderness values and lateral bracing spacing. Barth and White (1998) have synthesized the results of these experimental studies and complementary analytical studies with an empirical moment–rotation model for allows for estimating the rotation capacity of steel bridge I-girders at the nominal strength level M_n. In addition to flange slenderness and web slenderness, Barth and White also include the effect of the girder aspect ratio (which is defined as the web depth divided by the compression-flange width) and the parameter $\sqrt{E/F_y}$ (which accounts for the variation in compactness requirements with alternative grades of steel).

What is less well understood are the ductility requirements to which the actual ductility of a member should be compared to evaluate the suitability of a given design. In the United States, $R = 3$ is commonly accepted as sufficient for plastic and "low-seismic" (AISC, 2005) designs. This recommended rotation capacity was suggested by Yura et al. (1978) as sufficient for most civil engineering applications. This suggestion is largely based on research by Galambos (1968b) and Neal (1963) suggesting that a rotation capacity of $R = 2$ was generally sufficient for the building applications investigated but that higher rotation requirements may be necessary in some cases. Because plastic and seismic design have historically been limited to the use of compact-section beams, which generally exhibit greater ductility than required, more specific rotation requirements have not been widely investigated. Additional information on rotation capacity is provided in Chapter 19.

Furthermore, the determination of rotation requirements is sensitive to the application considered. The work of Li et al. (1995), Nethercot et al. (1995), and Kemp (1990) provides information pertinent to addressing rotation requirements for continuous-span beams. These studies, however, compute rotation requirements assuming a hinge exists at each support before load is placed on the girder, which potentially leads to overly conservative results. Righman (2005) has suggested empirical expressions for rotation requirements applicable to the inelastic design of continuous-span bridge girders.

5.8 CONCLUDING REMARKS

This chapter presents formulas for the solution of various cases of loading, beam geometry, and end restraint. These formulas are sufficient for developing design criteria in structural specifications and for solving most of the usual and several unusual cases encountered in practice. Because most of the cases of lateral–torsional buckling are not amenable to closed-form solution, the original solutions are numerical, many references are cited for acquiring these solutions, and the results have been are presented in the form of charts, tables, and special-purpose approximate expressions. Recourse can also be had to computational solutions. The available solution methods are in general based on (i) various energy methods (e.g., Bleich, 1952; Salvadori, 1955; Timoshenko and Gere, 1961; Vlasov, 1961; Andrade and Camotim 2003, 2005), (ii) finite difference methods (Galambos, 1968a; Vinnakota, 1977b), (iii) finite-integral methods (Brown and Trahair, 1968), and (iv) finite element methods (Yang and Yau, 1987; McGuire et al., 2000; Ronagh et al., 2000a,b; Boissonade and Maquoi, 2005; Alemdar and White, 2008). An important note regarding the analysis of tapered I-section members is that incorrect converged solutions generally are obtained using a piecewise prismatic discretization of a tapered I-section member. As rigorous computational tools become readily available to practicing structural engineers, it is certain that these tools will be used more and more in lieu of approximate equations. Thus, it is vital that structural engineers possess a thorough understanding of the basic aspects of LTB theory as it is explained in textbooks on stability theory.

Research on inelastic lateral–torsional buckling of beams has been performed vigorously since the 1940s, especially on problems related to continuous systems. The first paper on the subject was authored by Neal (1950), and the residual stress effect was first considered by Galambos (1963). The combined effects of residual stresses and geometrical imperfections on inelastic lateral–torsional buckling of beams were considered first by Lindner (1974), Vinnakota (1977a), and Celigoj (1979). A large number of analytical and experimental studies have been performed on the inelastic buckling of beams, with reviews given by Trahair (1977, 1983a, 1993), including a detailed recapitulation of the theory, the assumptions, the methods of analysis, experimental verifications, and tabulated and graphical results. In recent years, three-dimensional inelastic stability analyses of various types of open-section thin-walled beams accounting for a wide range of effects including spread of yielding through the member cross sections and along the length, residual stresses, initial camber and/or sweep, initial geometric imperfections, and general large displacement and large rotation effects have been conducted by numerous researchers (e.g., Izzuddin and Smith, 1996a,b; Teh and Clarke, 1998; Pi et al.,

2000; Bradford and Pi, 2001; Nukala and White, 2004; Boissonnade and Maquoi, 2005; Pi et al. 2007; Ziemian et al., 2008). These types of finite element based tools can serve to provide refined estimates of the maximum strength of physical imperfect beam members, assuming that the member cross sectional does not distort. Chapter 13 addresses the analysis of members where cross-sectional distortion has a significant effect on the response.

Finally, it should be noted that a challenge to part of the theory of monosymmetric beams, called the *Wagner hypothesis*, was presented by Ojalvo (1981). At the present time (2009), this challenge still has not been fully resolved.

REFERENCES

Adams, P. F., Lay, M. G., and Galambos, T. V. (1964), "Experiments on High Strength Steel Members,'' Fritz Engineering Laboratory Report No. 297.8, Fritz Engineering Laboratory, Lehigh University, Bethlehem, PA.

Allen, H. G., and Bulson, P. S. (1980), *Background to Buckling*, McGraw-Hill, Maidenhead, Berkshire, England.

American Association of State and Highway Transportation Officials. (AASHTO) (2007), *AASHTO LRFD Bridge Design Specifications*, 4th ed. with 2007 Interim Provisions, AASHTO, Washington, DC.

American Institute of Steel Construction. (AISC) (2005), *Specification for Structural Steel Buildings*, ANSI/AISC 360-05, AISC, Chicago, IL.

Alemdar, B.N. and White, D.W. (2008). "A New 3D Co-Rotational Beam-Column Element,'' *Proc. SSRC Annual Stability Conference*, Nashville, TN, pp. 175–194.

Anderson, J. M., and Trahair, N. S. (1972), "Stability of Monosymmetric Beams and Cantilevers,'' *ASCE J. Struct. Div.*, Vol. 98, No. ST1, pp. 269–286.

Andrade, A., and Camotim, D. (2003), "Lateral-Torsional Buckling of Singly Symmetric Web-Tapered I-Section Beams and Cantilevers,'' *Proceedings*, Annual Technical Session, Structural Stability Research Council, University of Missouri, Rolla, MO, pp. 233–254.

Andrade, A., and Camotim, D. (2005), "Lateral-Torsional Buckling of Singly Symmetric Web-Tapered Beams: Theory and Application,'' *J. Eng. Mech. ASCE*, Vol. 131, No. 6, pp. 586–597.

Andrade, A., Camotim, D., and Providencia e Costa, P. (2007), "On the Evaluation of Elastic Critical Moments in Doubly and Singly Symmetric I-Section Cantilevers,'' *J. Constr. Steel Res.*, Vol. 63, pp. 894–908.

Assadi, M., and Roeder, C. W. (1985), "Stability of Continuously Restrained Cantilevers,'' *J. Eng. Mech. ASCE*, Vol. 111, No. 12, pp. 1440–1456.

Austin, W.J. (1961), "Strength and Design of Metal Beam-Columns,'' *J. Struct. Div. ASCE*, Vol. 87, No. ST4, pp. 1–32.

Austin, W. J., Yegian, S., and Tung, T. P. (1957), "'Lateral Buckling of Elastically End-Restrained I-Beams,'' *Trans. ASCE*, Vol. 122, pp. 374–388.

Baker, K. A., and Kennedy, D. J. L. (1984), "Resistance Factors for Laterally Unsupported Steel Beams and Biaxially Loaded Steel Beam Columns,'' *Can. J. Civil Eng.*, Vol. 11, No. 4, pp. 1008–1019.

Bansal, J. P. (1971), "The Lateral Instability of Continuous Steel Beams,'' Ph.D. dissertation, University of Texas, Austin, TX.

Barth, K. E. (1996), "Moment-Rotation Characteristics for Inelastic Design of Steel Bridge Beams and Girders,'' dissertation, Purdue University, West Lafayette, IN.

Barth, K. E., and White, D. W. (1998), "Finite Element Evaluation of Moment-Rotation Characteristics,'' *Eng. Struct.*, Vol. 20, No. 8, pp. 761–778.

Bažant, Z. P., and Cedolin, L. (1991), *Stability of Structures—Elastic, Inelastic, Fracture and Damage Theories*, Oxford University Press, New York.

Bleich, F. (1952), *Buckling Strength of Metal Structures*, McGraw-Hill, New York.

Boissonnade, N., and Maquoi, R. (2005), "A Geometrically and Materially Non-Linear 3-D Beam Finite Element for the Analysis of Tapered Steel Members,'' *Steel Struct.*, Vol. 5, pp. 413–419.

Bradford, M. A. (1989), "Buckling Strength of Partially Restained I-Beams,'' *J. Struct. Eng. ASCE*, Vol. 115, No. 5, pp. 1272–1276.

Bradford, M. A. (1998), "Inelastic Buckling of I-Beams with Continuous Elastic Tension Flange Restraint,'' *J. Constr. Steel Res.*, Vol. 48, pp. 63–67.

Bradford, M. A., and Gao, Z. (1992), "Distortional Buckling Solutions for Continuous Composite Beams,'' *J. Struct. Eng. ASCE*, Vol. 118, No. 1, pp. 73–89.

Bradford, M. A., and Johnson, R. P. (1987), "Inelastic Buckling of Composite Bridge Girders Near Internal Supports,'' *Proc. Inst. Civ. Eng.*, Part 2, Vol. 83, pp. 143–159.

Bradford, M. A., and Pi, Y.-L. (2001), "Behavior of Unpropped Composite Girders Curved in Plan Under Construction Loading,'' *Eng. Struct.*, Vol. 23, pp. 779–789.

Brown, P. T., and Trahair, N. S. (1968), "Finite Integral Solutions of Differential Equations,'' *Civ. Eng. Trans. Inst. Eng. Aust.*, Vol. CE10, No. 2, pp. 193–196.

Brush, D. O., and Almroth, B. O. (1975), *Buckling of Bars, Plates and Shells*, McGraw-Hill, New York.

Canadian Standards Association (CSA) (2001), *Limit States Design of Steel Structures*, CAN/CSA-S16-01, CSA, Toronto, Ontario, Canada.

Carskaddan, P. S. (1980), "Autostress Design of Highway Bridges, Phase 3: Initial Moment Rotation Tests,'' Research Laboratory Report, United States Steel Corporation, Monroeville, PA.

Carskaddan, P. S., and Schilling, C. G. (1974), "Lateral Buckling of Highway Bridge Girders,'' Research Laboratory Report 22-G-001 (109-3), United States Steel Corporation, Monroeville, PA.

Celigoj, C. (1979), "Influence of Initial Deformations on the Carrying-Capacity of Beams,'' *Stahlbau,* Vol. 48, pp. 69–75, 117–121 (in German).

CEN (2002), Eurocode 3: *Design of Steel Structures,* Part 1-1: General Rules and Rules for Buildlings (prEN 1993-1-1), Stage 34 draft, European Committee for Standardization, Brussels, Belgium.

CEN (2005), Eurocode 3*: Design of Steel Structures, Part1.1—General Rules and Rules for Buildings*, ENV 1993-1-1, Comité Européen de Normalisation Brussels, Belgium.

Chajes, A. (1974), *Principles of Structural Stability Theory*, Prentice Hall, Upper Saddle River, NJ.

Chen, W. F., and Lui, E. M. (1987), *Structural Stability, Theory and Application*, Elsevier, New York.

Cheng, F. F., and Yura, J. A. (1988), "Lateral Buckling Tests on Coped Steel Beams," *ASCE J. Struct. Eng.*, Vol. 114, No. ST1, pp. 16–30.

Cheng, J. J., Yura, J. A., and Johnson, J. P. (1988), "Lateral Buckling of Steel Beams," *ASCE J. Struct. Eng.*, Vol. 114, No. ST1, pp. 1–15.

Clark, J. W., and Hill, H. N. (1960), "Lateral Buckling of Beams," *ASCE J. Struct. Div.*, Vol. 86, No. ST7, pp. 175–196.

Climenhaga, J. J., and Johnson, R. P. (1972), "Local Buckling in Continuous Composite Beams," *Struct. Eng.,* Vol. 50, No. 9, pp. 367–374.

Dalal, S. T. (1969), "Some Non-Conventional Cases of Column Design." *Eng. J. AISC*, Vol. 6, No. 1, pp. 28–39.

Dekker, N. W., Kemp, A. R, and Trinchero, P. (1995), "Factors Influencing the Strength of Continuous Composite Beams in Negative Bending," *J. Constr. Steel Res.,* Vol. 34, pp. 161–185.

Dibley, J. E. (1969), "Lateral Torsional Buckling of I-Sections in Grade 55 Steel," *Proc. Inst. Civil Eng*., Vol. 43, pp. 599–627.

Dinno, D. K., and Birkemoe, P. C. (2001), "The Rotational Capacity of Beams Exceeding Compact Section Slenderness Requirements," *Annual Technical Session, Structural Stability Research Council*, *Proceedings,* Ft. Lauderdale, FL May, pp. 291–312.

Dowswell, B. (2004), "Lateral-Torsional Buckling of Wide Flange Cantilever Beams," *Eng. J., AISC*, second quarter, Vol. 41, No. 2, pp. 85–91

Dux, P. F., and Kitipornchai, S. (1989), "Stability of I-Beams Under Self-Weight Lifting," *Steel Constr.*, Vol. 23, No. 2, pp. 2–11.

ECCS (2006), "Rules for Member Stability in EN 1993-1-1, Background Documentation and Design Guidelines," No. 119, ECCS Technical Committee 8—Stability, Brussels, Belgium, 259 pp.

Essa, H. S., and Kennedy, D. J. L. (1993), "Distortional Buckling of Steel Beams," Structural Engineering Report No. 185, Department of Civil Engineering, Univ. of Alberta, Edmonton, Alberta, Canada.

Essa, H. S., and Kennedy, D. J. L. (1994), "Design of Cantilever Steel Beams: Refined Approach," *J. Struct. Eng. ASCE*, Vol. 120, No. 9, pp. 2623–2636.

Essa, H. S., and Kennedy, D. J. L. (1995), "Design of Steel Beams in Cantilever-Suspended-Span Construction," *J. Struct. Eng. ASCE*, Vol. 121, No. 11, pp. 1667–1673.

Fukumoto, Y., and Itoh, Y. (1981), "Statistical Study of Experiments on Welded Beams," *ASCE J. Struct. Div.*, Vol. 107, No. ST1, pp. 89–104.

Fukumoto, Y., and Kubo, M. (1977), "An Experimental Review of Lateral Buckling of Beams and Girders," *Proc. Int. Colloq. Stab. Struct. Under Static Dyn. Loads*, Structural Stability Research Council, Bethlehem, PA, Mar.

Fukumoto, Y., Itoh, Y., and Kubo, M. (1980), "Strength Variation of Laterally Unsupported Beams," *ASCE J. Struct. Div.*, Vol. 106, No. ST1, pp. 165–182.

Galambos, T. V. (1963), "Inelastic Lateral Buckling of Beams," *ASCE J. Struct. Div.*, Vol. 89, No. ST5, pp. 217–244.

Galambos, T. V. (1968a), *Structural Members and Frames*, Prentice-Hall, Upper Saddle River, N.J.

Galambos, T. V. (1968b), "Deformation and Energy Absorption Capacity of Steel Structures in the Inelastic Range," Steel Research and Construction, AISI, Bulletin No. 8, Washington, DC.

Galambos, T. V. (1970), "Spacing of Bridging for Open-Web Steel Joists,'' Tech. Dig. No. 2, Steel Joist Institute, Myrtle Beach, SC.

Galambos, T. V. (Ed.) (1998), *Guide to Stability Design Criteria for Metal Structures*, 5th ed., Structural Stability Research Council, Wiley, New York.

Galambos, T.V., and Surovek, A. E. (2008), *Structural Stability of Steel, Concepts and Applications for Structural Engineers*, Wiley, New York.

Gioncu, V., and Petcu, D. (1997a), "Available Rotation Capacity of Wide-Flange Beams and Beam-Columns, Part 1: Theoretical Approaches'', *J. Constr. Steel Res.*, Vol. 45, pp. 161–217.

Gioncu, V., and Petcu, D. (1997b), "Available Rotation Capacity of Wide-Flange Beams and Beam-Columns, Part 2: Experimental and Numerical Tests,'' *J. Constr. Steel Res.*, Vol. 45, pp. 219–244.

Green, P. S., Ricles, J. M., and Sause, R. (2001), "The Inelastic Behavior of High Performance Steel Flexural Members,'' *Structural Stability Research Council Proceedings,* Ft. Lauderdale, FL.

Green, P. S., Sause, R., and Ricles, J. M. (2002), "Strength and Ductility of HPS Flexural Members,'' *J. Constr. Steel Res., Vol.* 58, pp. 907–941.

Greiner, R., and Kaim, P. (2001), "Comparison of LT-Buckling Design Curves with Test Results,'' ECCS TC 8 Report 23, European Convention for Constructional Steelwork, Brussels, Belgium.

Grubb, M. A., and Carskaddan, P. S. (1979), "Autostress Design of Highway Bridges, Phase 3: Initial Moment Rotation Tests,'' Research Laboratory Report, United States Steel Corporation, Monroeville, PA.

Grubb, M. A., and Carskaddan, P. S. (1981), "Autostress Design of Highway Bridges, Phase 3: Initial Moment Rotation Tests,'' Research Laboratory Report, United States Steel Corporation, Monroeville, PA.

Haaijer, G., Carskaddan, P. S., and Grubb, M. A. (1980), "Plastic Design with Noncompact Sections Including Composite Bridge Members,'' paper presented at the Structural Stability Research Council Annual Meeting. New York NY.

Hartmann, A. J. (1967), "Elastic Lateral Buckling of Continuous Beams,'' *ASCE J. Struct. Div.*, Vol. 93, No. ST4, pp. 11–28.

Hartnagel, B. A. (1997), "Inelastic Design and Experimental Testing of Compact and Noncompact Steel Girder Bridges,'' Ph.D. dissertation, University of Missouri-Columbia, Columbia, MO.

Heins, C. P. (1975), *Bending and Torsional Design in Structural Members*, Lexington Books, Lexington, MA.

Heins, C. P., and Potocko, R. A. (1979), "Torsional Stiffening of I-Girder Webs,'' *ASCE J. Struct. Div.*, Vol. 105, No. ST8, pp. 1689–1700.

Helwig, T. A., Frank, K. H., and Yura, J. A. (1997), "Lateral-Torsional Buckling of Singly Symmetric I-Beams,'' *J. Struct. Eng.*, Vol. 123, No. 9, pp. 1172–1179.

Holtz, N. M., and Kulak, G. L. (1973), "Web Slenderness Limits for Compact Beams,'' Struct. Eng. Rep. No. 43, Dept. Civ. Eng., University of Alberta, Edmonton, Alberta, Canada.

Holtz, N. M., and Kulak, G. L. (1975), "Web Slenderness Limits for Non-Compact Beams,'' Struct. Eng. Rep. No. 51, Dept. Civ. Eng., University of Alberta, Edmonton, Alberta, Canada.

Horne, M. R. (1954), "The Flexural–Torsional Buckling of Members of Symmetric I-Section Under Combined Thrust and Unequal Terminal Moments," *Q. J. Mech. Appl. Math.*, Vol. 7, No. 4, pp. 410–426.

Horne, M. R. (1960), "The Elastic Lateral Stability of Trusses," *Struct. Eng.*, Vol. 38, pp. 147–155.

Hribar, J. A., and Laughlin, W. P. (1968), "Lateral Stability of Welded Light Trusses," *ASCE J. Struct. Div.*, Vol. 94, No. ST3, pp. 809–832.

Izzuddin, B. A., and Smith, D. L. (1996a), "Large-Displacement Analysis of Elastoplastic Thin-Walled Frames. I: Formulation and Implementation," *J. Struct. Eng.*, Vol. 122, No. 8, pp. 905–914.

Izzuddin, B. A., and Smith, D. L. (1996b), "Large-Displacement Analysis of Elastoplastic Thin-Walled Frames. II: Verification and Application," *J. Struct. Eng.*, Vol. 122, No. 8, pp. 915–925.

Johnson, R. P., and Bradford, M. A. (1983), "Distortional Lateral Buckling of Stiffened Composite Bridge Girders," in *Instability and Plastic Collapse of Steel Structures*, Proceedings of the Michael R. Horne Conference (Ed.) L. J. Morris, Granada, London, pp. 569–580.

Johnston, B. G. (Ed.) (1976), *Guide to Stability Design Criteria for Metal Structures*, Wiley, New York

Kaehler, R. C., White, D. W., and Kim, Y. D. (2009), *Frame Design Using Web-Tapered Members*, Design Guide 25, Metal Building Manufacturers Association and American Institute of Steel Construction, Chicago, IL.

Kato, B. (1989), "Rotation Capacity of H-Section Members as Determined by Local Buckling," *J. Const. Steel Res.*, Vol. 13, pp. 95–109.

Kato, B. (1990), "Deformation Capacity of Steel Structures" *J. Constr. Steel Res.,* Vol. 17, pp. 33–94.

Kemp, A. R. (1986), "Factors Affecting the Rotation Capacity of Plastically Designed Members," *Struct. Eng.*, Vol. 64B, No. 2, pp. 28–35.

Kemp, A. R. (1990), "Quantifying Limit-States of Rotation Capacity in Flexural Members," Proceedings—Institution of Civil Engineers, Part 2: Research and Theory, Sept. 1990, pp. 387–406.

Kemp, A. R., and Dekker, N. W. (1991), "Available Rotation Capacity in Steel and Composite Beams," *Struct. Eng.*, Vol. 69, No. 5, pp. 88–97.

Kemp, A. R., Dekker, N. W., and Trinchero, P. (1995), "Differences in Inelastic Properties of Steel and Composite Beams," *J. Constr. Steel Res.*, Vol. 34, pp. 187–206.

Kim, Y. D., and White, D. W. (2007), "Practical Buckling Solutions for Web-Tapered Members," *Annual Proceedings*, Structural Stability Research Council, New Orleans, LA, pp. 259–278.

Kindmann, R., and Frickel, J. (2002), "Tragfähigkeit von U-Profilen bei Biegung und Torsion," *Rubstahl,* Vol. 1, pp. 1–2.

Kirby, P. A., and Nethercot, D. A. (1979), *Design for Structural Stability*, Wiley, New York.

Kitipornchai, S., Dux, P. F., and Richter, N. J. (1984), "Buckling and Bracing of Cantilevers," *J. Struct. Eng.*, Vol. 110, No. 9, pp. 2250–2262.

Kitipornchai, S., and Trahair, N. S. (1980), "Buckling Properties of Monosymmetric I-Beams," *ASCE J. Struct. Div.*, Vol. 106, No. ST5, pp. 941–958.

Lääne, A., and Lebet, J. (2005) "Available Rotation Capacity of Composite Bridge Plate Girders Under Negative Moment and Shear,'' *J. Constr. Steel Res.*, Vol. 61, pp. 305–327.

la Poutré, D. B., Snijder, H. H., and Hoenderkamp, J. C. D. (2002), Design Method for Lateral Torsional Buckling of Channel Sections,'' in *Annual Stability Conf. (SSRC)*, (Ed. N. Iwankiw), Seattle, Gainesville, FL, Structural Stability Research Council, pp. 311–326

la Poutré, D. B. (2008), "Lateral Torsional Buckling of Channel Sections Loaded in Bending,'' Structural Stability Research Council. *Annual Stability Conference*, Proceedings, Nashville, April 2–5, pp. 377–389.

Lee, G. C. (1960), "A Survey of the Literature on the Lateral Instability of Beams,'' *Weld. Res. Caunc. Bull. No. 63*, Aug.

Lee, G. C., Ketter, R. L., and Hsu, T. L. (1981), *The Design of Single Story Rigid Frames*, Metal Building Manufacturers Association, Cleveland, OH.

Li, T. Q., Choo, B. S., and Nethercot, D. A. (1995), "Determination of Ratation Capacity Requirements for Steel and Composite Beams,'' *J. Constr. Steel Res.*, Vol. 32, pp. 303–332.

Lindner, J. (1974), "Influence of Residual Stresses on the Load-Carrying Capacity of I-Beams,'' *Stahlbau*, Vol. 43, pp. 39–45, 86–91 (in German).

Lindner, J. (1985), "Influence of Parameters of Lateral Bracing on the Load-Carrying Capacity of Beams,'' *Proc. SSRC Annu. Tech. Session*, Cleveland, OH, Apr.

Lindner, J. (1994), "Influence of Construction Details on the Load-Carrying Capacity of Beams,'' *Proc. SSRC Annu. Tech. Session*, Bethlehem, PA, June, pp. 65–77.

Lindner, J. (1995), "Influence of Constructional Detailing on the Lateral–Torsional Buckling Capacity of Beam-Columns,'' in *Structural Stability and Design*, A. A. Balkema, Rottcrdam /Brookfield, pp. 61 66.

Lindner, J., and Gietzelt, R. (1983), "Influence of End Plates on the Ultimate Load of Laterally Unsupported Beams,'' in *Instability and Plastic Collapse of Steel Structures (M. R. Horne Conf.)*, Manchester, Granada, London, pp. 538–546.

Lindner, J., and Glitsch, T. (2004), "Vereinfachter Nachweis für I- und U-Träger beansprucht durch doppelte Biegung und Torsion,'' *Stahlbau,* Vol. 73, No. 9, pp. 704–715.

Lukey, A. F., and Adams, P. F. (1969), "Rotation Capacity of Beams Under Moment-Gradient'', *J. Struct. Div. ASCE*, Vol. 95, No. ST6, pp. 1173–1188.

Lukey, A. F., Smith, R. J., Hosain, M. U., and Adams, P. F. (1969), "Experiments on Wide-Flange Beams Under Moment Gradient,'' WRC Bulletin No. 142.

MacPhedran, I., and Grondin, G.Y. (2009), "A Proposed Simplified Canadian Beam Design Approach,'' Proceedings of the SSRC Annual Stability Conference, Phoenix, AZ.

Massonnet, C. (1947), "Buckling of Thin-Walled Bars with Open Cross Section,'' *Hommage Fac. Sci. Appl. Univ. Liege* (Ed. G. Thone) (in French).

McGuire, W., Gallagher, R., and Ziemian, R. D. (2000), *MatrixStructural Analysis,* 2nd ed, Wiley, New York.

Melcher, J. J. (1994), "Note to the Buckling of Beams with Monosymmetric Section Loaded Transversely to Its Plane of Symmetry,'' paper presented at the Annual Stability Conf (SSRC), Lehigh University, Bethlehem, PA, Structural Stability Research Council (SSRC), pp. 61–75.

Melcher, J. J. (1999), "Kippen von Trägern als Stabilitätsproblem zweier Gruppen von Querschnittypen,'' *Stahlbau,* Vol. 68, No. 1, pp. 24–29.

Minkoff, R. M. (1975), "Stability of Steel Joists During Erection,'' Research Report No. 39, Dept. of Civil Engineering, Washington University, St. Louis, MO.

Neal, B. G. (1950), "The Lateral Instability of Yielded Mild Steel Beams of Rectangular Cross Section,'' *Philos. Trans. R. Soc. London*, Vol. A242, pp. 197–242.

Neal, B. G. (1963), *The Plastic Methods of Structural Analysis* 2nd ed., Wiley, New York.

Nethercot, D. A. (1973a), "Buckling of Laterally and Torsionally Restrained Beams,'' *ASCE J. Eng. Mech. Div.*, Vol. 99, No. EM4, pp. 773–792.

Nethercot, D. A. (1973b), "The Effective Length of Cantilevers as Governed by Lateral Buckling,'' *Struct. Eng.*, Vol. 51, No. 5, pp. 161–168.

Nethercot, D. A. (1983), "Elastic Lateral Buckling of Beams,'' in *Beams and Beam Columns: Stability in Strength* (Ed. R. Narayanan), Applied Science Publishers, Barking, Essex, England.

Nethercot, D. A., and Al-Shankyty, M. A. F. (1979), "Bracing of Slender Cantilever Beams,'' in *Stability Problems in Engineering Structures and Components* (Ed. T. H, Richards), Applied Science Publishers, London, pp. 89–99.

Nethercot, D. A., Li, T. Q., and Choo, B. S. (1995), "Required Rotations and Moment Redistribution for Composite Frames and Continuous Beams,'' *J. Constr. Steel Res.*, Vol. 35, pp.121–163.

Nethercot, D. A., and Rockey, K. C. (1972), "A Unified Approach to the Elastic Lateral Buckling of Beams,'' *AISE Eng. J.*, Vol. 9, No. 3, pp. 96–107.

Nethercot, D. A, and Trahair, N. S. (1976), "Lateral Buckling Approximations for Elastic Beams,'' *Struct. Eng.*, Vol. 54, No. 6, pp. 197–204.

Nukala, P. K. V. V., and White, D. W. (2004), "A Mixed Finite Element for Three-Dimensional Nonlinear Analysis of Steel Frames,'' *Comput. Methods Appl. Mech. Eng.,* Vol. 193, pp. 2507–2545.

Oehlers, D. J., and Bradford, M. A. (1999), *Elementary Behavior of Composite Steel and Concrete Structural Members*, Butterworth Heinemann, Oxford.

Ojalvo, M. (1981), "Wagner Hypothesis in Beam and Column Theory,'' *ASCE J. Eng. Mech. Div.*, Vol. 107, No. EM4, pp. 649–658.

Ojalvo, M., and Chambers, R. S. (1977), "Effects of Warping Restraints on I-Beam Buckling,'' *ASCE J. Struct. Div.*, Vol. 103, No. ST12, pp. 2351–2360.

Ozdemir, K. M., and Topkaya, C. (2006), "Lateral Buckling of Overhanging Crane Trolley Monorails,'' *Eng. Struct.*, Vol. 28, pp. 1162–1172.

Pi, Y.-L., Bradford, M. A., and Trahair, N. S. (2000), "Inelastic Analysis and Behavior of Steel I-Beams Curved in Plan,'' *J. Struct. Eng.*, Vol. 126, No. 7, pp. 772–779.

Pi, Y.-L., Bradford, M. A. and Uy, B. (2007), "A Rational Elasto-Plastic Spatially Curved Thin-Walled Beam Element,'' Int. *J. Numer. Meth. Eng.*, Vol. 70, pp. 253–290.

Poley, S. (1956), "Lateral Buckling of Cantilevered I-Beams Under Uniform Load,'' *Trans. ASCE*, Vol. 121, pp. 786–790.

Powell, G., and Klingner, R. (1970), "Elastic Lateral Buckling of Steel Beams,'' *ASCE J. Struct. Div.*, Vol. 96, No. ST9, pp. 1919–1932.

Prasad, J., and Galambos, T. V. (1963), "The Influence of the Adjacent Spans on the Rotation Capacity of Beams,'' Fritz Engineering Laboratory Report No. 205H.12, Lehigh University, Bethlehem, PA.

Rebelo, C., Lopes, N., Simoes da Silva, L., Nethercot, D., and Vila Real, P. (2009), "Statistical Evaluation of the Lateral-Torsional Buckling Resistance of Steel I-Beams—Part 1: Variability of the Eurocode 3 Resistance Model, *J. Constr. Steel Res.*, Vol. 65, No. 4, pp. 818–831.

Righman, J. E. (2005), "Rotation Compatibility Approach to Moment Redistribution for Design and Rating of Steel I-Girders,'' Ph.D. dissertation, West Virginia University, Morgantown, WV.

Roberts, T. M., and Burt, C. A. (1985), "Instability of Monosymmetric I-Beams and Cantilevers,'' *Int. J. Mech. Sci.*, Vol. 27, No. 5, pp. 313–324.

Robertson, A. (1925), "The Strength of Struts,'' Selected Engineering Papers No. 28, Institution of Civil Engineers, London, England.

Ronagh, H. R., Bradford, M. A., and Attard, M. M. (2000a), "Nonlinear Analysis of Thin-Walled Members of Variable Cross-Section. Part I: Theory,'' *Comp. Struct.*, Vol. 77, pp. 285–299.

Ronagh, H. R., Bradford, M. A., and Attard, M. M. (2000b), "Nonlinear Analysis of Thin-Walled Members of Variable Cross-Section. Part I: Theory,'' *Comp. Struct.*, Vol. 77, pp. 301–313.

Salvadori, M. G. (1955), "Lateral Buckling of I-Beams,'' *Trans. ASCE*, Vol. 120, p. 1165 .

Salvadori, M. G. (1956), "Lateral Buckling of Eccentrically Loaded I-Columns,'' *Trans. ASCE*, Vol. 121, p. 1163 .

Schilling, C. G. (1985), "Moment-Rotation Tests of Steel Bridge Girders,'' Report on Project 188, American Iron and Steel Institute, Washington, DC.

Schilling, C. G., and Morcos, S. S. (1988), "Moment-Rotation Tests of Steel Girders with Ultracompact Flanges,'' Report on Project 188, American Iron and Steel Institute, Washington, DC.

Serna, M. A., Lopez, A., Puente, I., and Yong, D. J. (2006), "Equivalent Uniform Moment Factors for Lateral-Torsional Buckling of Steel Members,'' *J. Constr. Steel Res.*, Vol. 62. pp. 566–580.

Simitses, G. J., and Hodges, D. H. (2006), *Fundamentals of Structural Stability*, Elsevier, Amsterdam.

Simoes da Silva, L., Rebelo, C., Nethercot, D., Marques, L., Simoes, R., and Vila Real, P. M. M. (2009), "Statistical Evaluation of the Lateral-Torsional Buckling Resistance of Steel I-Beams, Part 2: Variability of Steel Properties,'' *J. Constr. Steel Res.*, Vol. 65, pp. 832–849

Standards Association of Australia (SAA) (1998), *Steel Structures*, AS4100-1998, SAA, Australian Institute of Steel Construction, Sydney, Australia.

Svensson, S. E. (1985), "Lateral Buckling of Beams Analysed as Elastically Supported Columns with Varying Axial Force,'' *J. Constr. Steel Res.*, Vol. 5, pp. 179–193

Szewczak, R. M., Smith, E. A., and DeWolf, J. T. (1983), "Beams with Torsional Stiffeners,'' *ASCE J. Struct. Div.*, Vol. 109, No. ST7, pp. 1635–1647.

Tanner, N. S. (1985), "Allowable Bending Stresses for Overhanging Monorails,'' *Eng. J. AISC*, 3rd quarter, pp. 133–138.

Tansil, T. C. (1991), "Behavior of a Composite Plate Girder in Negative Bending,'' M. S. thesis, University of Texas, Austin, TX.

Teh, L., and Clarke, M. (1998), "Plastic-Zone Analysis of 3D Steel Frames Using Beam Elements,'' *J. Struct. Eng.*, Vol. 125, No. 11, pp. 1328–1337

Thomas, S. J., and Earls, C. J. (2003), "Cross-Sectional Compactness and Bracing Requirements for HPS483W Girders,'' *ASCE J. Struct. Eng.*, Vol. 129, No. 12, pp. 1569–1583.

Timoshenko, S. P., and Gere, J. M. (1961), *Theory of Elastic Stability*, McGraw-Hill, New York.

Trahair, N. S. (1969), "Elastic Stability of Continuous Beams,'' *ASCE J. Struct. Div.*, Vol. 95, No. ST6, pp. 1295–1312.

Trahair, N. S. (1977), "Lateral Buckling of Beams and Beam-Columns,'' in *Theory of Beam-Columns*, Vol. 2 (Eds. W. F. Chen and T. Atsuta), McGraw-Hill, New York, Chap. 3.

Trahair, N. S. (1983a), "Inelastic Lateral Buckling of Beams,'' in *Beams and Beam-Columns: Stability and Strength* (Ed. R. Narayanan), Applied Science Publishers, Barking, Essex, England.

Trahair, N. S. (1983b), "Lateral Buckling of Overhanging Beams,'' in *Instability and Plastic Collapse of Steel Structures* (Ed. L. J. Morris), Granada Publishing, London, England, pp. 503–515.

Trahair, N. S. (1993), *Flexural-Torsional Buckling of Structures*, CRC Press, Boca Raton, FL.

Trahair, N. S. (2008), "Lateral Buckling of Monorail Beams,'' *Eng. Struct.*, Vol. 30, pp. 3213–3218

Trahair, N. S., and Bradford, M. A. (1988), *The Behaviour and Design of Steel Structures*, 2nd ed., Chapman & Hall, London.

Vacharajittiphan, P., and Trahair, N. S. (1974), "Warping and Distortion of I-Section Joints,'' *ASCE J. Struct. Div.*, Vol. 100, No. ST3, pp. 547–564.

Vacharajittiphan, P., and Trahair, N. S. (1975), "Analysis of Lateral Buckling in Plane Frames,'' *ASCE J. Struct. Div.*, Vol. 101, No. ST7, pp. 1497–1516.

Vinnakota, S. (1977a), "Inelastic Stability of Laterally Unsupported I-Beams,'' *Comput. Struct.*, Vol. 7, No. 3, pp. 377–389.

Vinnakota, S. (1977b), "Finite Difference Methods for Plastic Beam-Columns,'' in *Theory of Beam-Columns*, Vol. 2 (Eds. W. F. Chen and T. Atsuta), McGraw-Hill, New York, Chap. 10.

Vlasov, V. Z. (1961), *Thin-Walled Elastic Beams*, Israel Program for Scientific Translations, Jerusalem.

Vrcelj, Z., and Bradford, M. A. (2009), "Inelastic Restrained Distortional Buckling of Continuous Composite T-Beams,'' *J. Constr. Steel Res.*, Vol. 65, pp. 850–859.

Wang, C. M., and Kitipornchai, S. (1986), "On Stability of Monosymmetric Cantilevers,'' *Eng. Struct.*, Vol. 8, No. 3, pp. 169–180.

Wang, C. M., Kitipornchai, S., and Thevendran, V. (1987), "Buckling of Braced Monosymmetric Cantilevers,'' *Int. J. Mech. Sci.*, Vol. 29, No. 5, pp. 321–337.

Weston, G., Nethercot, D. A., and Crisfield, M. A. (1991), "Lateral Buckling in Continuous Composite Bridge Girders,'' *Struct. Eng.*, Vol. 69, No. 5, pp. 79–87.

White, D. W. (2006), "Structural Behavior of Steel,'' Structural Engineering Mechanics and Materials Report No. 2006-43, School of Civil and Environmental Engineering, Georgia Institute of Technology, Atlanta, GA.

White, D. W. (2008), "Unified Flexural Resistance Equations for Stability Design of Steel I-Section Members: Overview,'' *J. Struct. Eng. ASCE*, Vol. 134, No. 9, pp. 1405–1424.

White, D. W., and Grubb, M. A. (2003), "Lateral Torsional Buckling Resistance of Stepped Flanges," in *Three-Span Continuous Straight Composite I Girder, Load and Resistance Factor Design, Third Edition (Customary Units)*, Appendix C, National Steel Bridge Alliance, AISC, Chicago, IL, pp. C–1 to C–4.

White, D. W., and Jung, S.-K. (2004), "Unified Flexural Resistance Equations for Stability Design of Steel I-Section Members—Uniform Bending Tests," Structural Engineering, Mechanics and Materials Report No. 28, School of Civil and Environmental Engineering, Georgia Institute of Technology, Atlanta, GA.

White, D. W., and Jung, S.-K. (2007), "Effect of Web Distortion on the Buckling Strength of Noncomposite Discretely-Braced I-beams," *Eng. Struct.*, Vol. 29, No. 8, pp. 1872–1888.

White, D. W., and Jung, S.-K. (2008), "Unified Flexural Resistance Equations for Stability Design of Steel I-Section Members—Uniform Bending Tests," *J. Struct. Eng. ASCE*, Vol. 134, No. 9, pp. 1450–1470.

White, D. W., and Kim, Y. D. (2004), "Unified Resistance Equations for Stability Design of Steel I-Section Members—Moment Gradient Tests," Structural Engineering, Mechanics and Materials Report No. 26, School of Civil and Environmental Engineering, Georgia Institute of Technology, Atlanta, GA.

White, D. W., and Kim, Y. D. (2008), "Unified Flexural Resistance Equations for Stability Design of Steel I-Section Members—Moment Gradient Tests," *J. Struct. Eng ASCE*, Vol. 134, No. 9, pp. 1471–1486.

Williams, F. W., and Jemah, A. K. (1987), "Buckling Curves for Elastically Supported Columns with Varying Axial Force to Predict Lateral Buckling of Beams," *J. Constr. Steel Res.*, Vol. 7, pp. 133–148.

Wong, E., and Driver, R. G. (2010), "Critical Evaluation of Equivalent Moment Factor Procedures for Laterally Unsupported Beams," *Eng. J. AISC*, Vol. 47, No. 1.

Yakel, A. J., Mans, P., and Azizinamini, A. (1999), "Flexural Capacity of HPS-70W Bridge Girders," NaBRO Rep. University of Nebraska-Lincoln, Lincoln, NE.

Yang, Y. B. and Yau, J.-D. (1987), "Stability of Beams with Tapered I-Sections," *J. Eng. Mech. ASCE*, Vol. 113, No. 9, pp. 1337–1357.

Yura, J. A. (2001), "Fundamentals of Beam Bracing," *Eng. J. AISC*, Vol. 38, No. 1, pp. 11–26.

Yura, J. A., and Cheng, J. J. (1985), "Lateral and Local Instability of Coped Beams," *Proc. SSRC Annu. Tech. Session*, Cleveland, OH, Apr.

Yura, J. A., Galambos, T. V., and Ravindra, M. K. (1978), "The Bending Resistance of Steel Beams," *ASCE J. Struct. Div.*, Vol. 104, No. ST9, pp. 1355–1370.

Yura, J. A., and Helwig, T. A. (2009), "Bracing for Stability," Short course notes, Structural Stability Research Council, North American Steel Construction Conference, Phoenix, AZ, Apr.

Yura, J. A., Helwig, T. A., Herman, R. and Zhou, C. (2008), "Global Lateral Buckling of I-Shaped Girder Systems," *J. Struct. Eng. ASCE*, Vol. 134, No. 9, pp. 1487–1494.

Zhou, C. (2006), "Utilizing Lean-On Cross-Frame Bracing for Steel Bridges," Ph.D. dissertation, Department of Civil and Environmental Engineering, University of Houston, Houston, TX.

Ziemian, R. D., McGuire, W., and Seo, D. W. (2008), "On the Inelastic Strength of Beam-Columns under Biaxial Bending," *Proc. SSRC Annual Stability Conference*, Nashville, TN.

CHAPTER 6

PLATE GIRDERS

6.1 INTRODUCTION

This chapter deals with the buckling and strength of straight I-shaped plate girders. Curved and box girders are covered in Chapters 7 and 9, respectively. Although the term plate girder is often used for any section built up from plates, the conventional use of the term applies to members with slender webs. Considerations of buckling involve not only lateral–torsional buckling and local buckling of the flange, as in the case of beams, but buckling of the web as well. The web is often reinforced with transverse stiffeners, and occasionally with longitudinal stiffeners, to increase its resistance to buckling, and design involves finding a combination of plate thickness and stiffener spacing that will be economical in material and fabrication.

While column and beam members do not exhibit significant postbuckling strength, stiffened plates can often support stresses much higher than the buckling stress. Nevertheless, buckling was accepted as a basis for the design of plate girder webs almost exclusively until the early 1960s. This was due primarily to the fact that formulas for predicting buckling are relatively simple and have been known for many years, while suitable analyses of postbuckling strength were not available. The postbuckling strength, however, was acknowledged in most specifications with smaller factors of safety for web buckling than for yielding or failure of other elements.

The source of the postbuckling strength of stiffened plate girder webs in shear was explained by Wilson (1886). He had observed in his early experience (25 years before) that railroad plate girder bridges with webs $\frac{3}{16}$ in. thick and stiffeners at intervals of 5 ft were "bearing up well under use." He discovered: "By means of a paper model with a very thin flexible web, that when stiffeners were properly introduced, the web no longer resisted by compression, but by tension, the stiffeners taking up the duty of compressive resistance, like the posts of a Pratt truss, and dividing the girder into panels equivalent to those of an open truss, the web in each panel acting as an inclined tie." Wilson stated that using this theory he obtained "results that quite agreed with practical examples" but did not explain his analysis.

Wagner (1931) developed a diagonal-tension theory of web shear. Wagner's work was extended by Kuhn (1956) for applications in aircraft design. Extensive studies, both analytical and experimental, were made in the late 1950s by Basler and Thürlimann on the postbuckling behavior of web panels in bending as well as in shear (Basler and Thürlimann, 1960a,b, 1963; Basler et al., 1960; Yen and Basler, 1962; Basler, 1963a,b). Practical procedures were then developed and have been adopted in many specifications. Widespread interest in the subject resulted in a number of modifications to the Basler–Thürlimann approach to achieve better correlation between theory and tests.

Two approaches to the design of plate girder webs are used: (1) design based on buckling as a limiting condition, with a relatively low factor of safety to allow for postbuckling strength, and (2) design based on yielding or ultimate strength as a limiting condition, with the same factor of safety as for yielding or ultimate strength of other structural members.

The behavior of steel plate shear walls, an increasingly prevalent lateral-load-resisting system for buildings, has often been described using a plate girder analogy, wherein the building columns correspond to the girder flanges, the floor beams to the transverse stiffeners, and the infill plates to the girder web. Because slender infill plates are typically used in steel plate shear walls, the response of the system under lateral loading is dominated by postbuckling behavior of the plates. Moreover, in the case of seismic loading the infill plate is designed to deform into the inelastic range, and the overall response of the system tends to be highly ductile. As with plate girders, where the stiffeners and (to some degree) the flanges serve to anchor the postbuckling diagonal tension field in the web, the boundary frame members of a steel plate shear wall anchor the tension field in the infill plates. Therefore, the stiffness of the boundary members becomes an important design consideration to ensure the development of a relatively uniform postbuckling tension field and good overall performance of the system under lateral loading.

Girders with corrugated webs have been used in buildings in Sweden and Germany and in bridges in France (Combault, 1988). A summary of the research and development in beams and girders with corrugated webs was reported by Elgaaly and Dagher (1990). More recent analytical work on the subject has been completed by Machimdamrong et al. (2004). The corrugations increase the out-of-plane stiffness of the web and eliminate the need to use vertical stiffeners. Girders with corrugated webs were tested to failure and analyzed under shear (Elgaaly et al., 1992, 1993, 1995); the failure was due to buckling of the web. It was noted from the experimental and the analytical results that buckling of the web is generally local for coarse corrugations and global for dense corrugations. Simple buckling formulas to calculate the buckling load of the web were suggested which are based on local buckling of the corrugation subpanels as isotropic flat plates or global buckling of the entire web panel as an orthotropic plate.

6.2 PRELIMINARY SIZING

The economy of plate girders is realized by efficiently sizing the web and the flanges to meet the design requirements for the specific application. Because the flanges provide more significant contributions to the bending and torsional stiffness than the web, the web is generally sized primarily for shear requirements. In selecting the basic section geometry, however, the efficiency of the girder, in terms of the amount of material used, is dependent on several factors.

As a starting point for preliminary sizing, Yura (1987) developed a simple expression for the most efficient depth h_{eff} for the plate girder. He assumed that the web depth and overall depth were approximately equal and employed an approximation for the major axis moment of inertia of the section using

$$I_x = 2A_f \left(\frac{h}{2}\right)^2 + \frac{t_w h^3}{12} \tag{6.1}$$

where A_f is the area of the flange, h the girder depth, and t_w the web thickness. Since the elastic section modulus S_x is equal to $I_x/(h/2)$, the preliminary area of flange can be solved for as

$$A_f = \frac{S_x}{h} - \frac{ht_w}{6} \tag{6.2}$$

and the preliminary total area A_T as

$$A_T = 2A_f + ht_w = \frac{2S_x}{h} + \frac{2ht_w}{3} \tag{6.3}$$

To obtain the optimum depth that will minimize the amount of steel used in the cross section, Eq. 6.3 can be differentiated with respect to h and set equal to zero. To further optimize the section, however, the web should be proportioned so that its shear capacity requirements are just satisfied. The elastic shear buckling stress of the unstiffened web is given by

$$\tau_{cr} = k_s \frac{\pi^2 E}{12\left(1-\nu^2\right)\left(h/t_w\right)^2} = \frac{Ct_w^2}{h^2} \tag{6.4}$$

where all terms, except h and t_w, have been defined within the constant $C = k_s\pi^2 E/\left[12\left(1-\nu^2\right)\right]$. Because the shear strength V is often taken as $\tau_{cr} ht_w$, Eq. 6.4 can be rewritten in terms as

$$V = \frac{Ct_w^3}{h} \tag{6.5}$$

Solving for the web thickness results in

$$t_w = \left(\frac{Vh}{C}\right)^{1/3} \tag{6.6}$$

Substituting Eq. 6.6 into Eq. 6.3 yields the following expression for the total area:

$$A_T = \frac{2S_x}{h} + \frac{2}{3}h^{4/3}\left(\frac{V}{C}\right)^{1/3} \tag{6.7}$$

To determine the value of h that minimizes the total area A_T, the derivative of Eq. 6.7 with respect to h is set equal to zero and then solved for h, resulting in

$$h = \left(\frac{9}{4}\right)^{3/7} C^{1/7}\left(\frac{S_x^3}{V}\right)^{1/7} \tag{6.8}$$

For the case of an unstiffened steel web, the plate-buckling coefficient is $k_s = 5.34$, $E = 29{,}000$ ksi, and $\nu = 0.3$, the value for the constant $C = 140{,}000$ ksi. Employing a LRFD methodology, a resistance factor of $\phi = 0.9$ would be applied to the shear capacity, resulting in $C = 126{,}000$ ksi. For this situation, Eq. 6.8 simplifies to

$$h = 7.6\left(\frac{S_x^3}{V}\right)^{1/7} \tag{6.9}$$

To illustrate the use of Eq. 6.9, consider a steel plate girder with a design maximum moment of $M = 60{,}000$ kip-in., a maximum shear of $V = 300$ kips, and a material yield strength of $F_y = 50$ ksi. Assuming LRFD with $\phi = 0.9$, the required section modulus would be determined by assuming the required strength is equal to the factored yield moment, $S_x = M/(\phi F_y) = (60{,}000 \text{ kip-in.})/(0.9 \times 50 \text{ ksi}) = 1{,}333 \text{ in.}^3$ Substituting this value for S_x and the maximum shear of $V = 300$ kips into Eq. 6.9 suggests an initial girder depth of approximately $h = 73$ in. The corresponding web thickness can be solved for using Eq. 6.6 (recalling that $C = 126{,}000$), which results in $t_w = 0.55$ in. Because the exact plate thickness will not generally be readily available, the elastic shear buckling equation can be used to solve for required depth for the closest plate thickness. In this example, plate thicknesses of 0.5 and 0.5625 in. would be considered and the corresponding depths would be 52.5 and 75 in., respectively. Because a beam depth of 75 in. is close to the value predicted using Eq. 6.9, it should be more efficient than a 52.5-in.-deep girder. Although Eq. 6.9 provides a starting point for an efficient design, it has been shown that for other web depths the girder efficiency will not vary too much if the web is sized for the design shear force and the flanges are then sized for the design moment.

6.3 WEB BUCKLING AS A BASIS FOR DESIGN

When buckling is taken as the basis for the design of plate girder webs, the maximum stress in the web computed by conventional beam theory should not exceed the buckling stress divided by a factor of safety. A number of formulas and extensive charts and tables have been developed for the buckling analysis of stiffened and unstiffened plates. Buckling of rectangular plate panels is discussed in Chapter 4.

The geometric parameters that determine buckling of plate girder webs are the web thickness t, the web depth h between flanges, and the spacing a of transverse stiffeners. Four limiting values of web slenderness must be established:

1. limiting value of h/t to control flexural buckling of a web with no longitudinal stiffener
2. A limiting value of h/t to control flexural buckling of a web with a longitudinal stiffener
3. A limiting value of h/t to control shear buckling of a web with no transverse stiffeners
4. A limiting value of a/t to control shear buckling of a web with transverse stiffeners

The AISC specification (2005a) does not specifically limit web bend buckling (web buckling due to member bending) but instead applies a reduction factor R_{pg} to the bending strength when slender webs are used. This factor accounts for increases in the flange stresses due to web bend buckling. The expression is identical to the R_b factor that serves the same purpose in the *AASHTO LRFD Bridge Design Specifications* (2007). Based on the simplified version of the general plate-buckling expressions given previously in Eq. 4.1 or 6.4, the AASHTO specification also includes provisions to calculate the web bend buckling stress in sections without longitudinal stiffeners. Using a Poisson's ratio of $\nu = 0.3$, the buckling stress reduces to the following expression:

$$F_{cr} = 0.9k \frac{E}{(h/t_w)^2} \tag{6.10}$$

The plate-buckling coefficient is defined as follows:

$$k = \frac{9}{(h_c/h)^2} \tag{6.11}$$

where h_c is the depth of the web in compression. For a doubly symmetric section, the plate-buckling coefficient from Eq. 6.11 is $k = 36$, which is between the values of 23.9 and 39.6 corresponding to simply supported and fixed-end conditions (Fig. 4.4), respectively.

For webs with longitudinal stiffeners (case 2), the AASHTO specification uses Eq. 6.10 with the following plate-buckling coefficient, which is a function of the location of the stiffener relative to the compression flange, h_s (Frank and Helwig, 1995):

$$k = \begin{cases} \dfrac{5.17}{(h_s/h)^2} \leq \dfrac{9}{(h_c/h)^2} & \text{for } \dfrac{h_s}{h_c} \geq 0.4 \quad (6.12a) \\[2ex] \dfrac{11.64}{\left[(h_c - h_s)/h\right]^2} & \text{for } \dfrac{h_s}{h_c} < 0.4 \quad (6.12b) \end{cases}$$

Expressions for limiting web slenderness for shear buckling without transverse stiffeners (case 3) are relatively straightforward and can be obtained using the shear buckling provisions outlined in Chapter 4. The derivation of formulas for spacing of transverse stiffeners (case 4), however, is not as simple because of the dependence of k_s on the aspect ratio $\alpha = a/h$ (Gaylord and Gaylord, 1972). As an alternative, AREMA (2002) neglects this dependence and uses

$$\frac{a}{t} = \frac{10{,}500}{\sqrt{f_{v,\text{psi}}}} \tag{6.13a}$$

where f_v is the average shear stress in psi calculated service loads. By assuming that $k_s = 5(1 + 1/\alpha^2)$ for all values of α, AASHTO limits the allowable shear stress from web buckling in a web panel to

$$f_v = \frac{7 \times 10^7(1 + 1/\alpha^2)}{(h/t)^2} \leq \frac{1}{3}F_y \tag{6.13b}$$

Equation 6.13b can be rewritten in a form similar to Eq. 6.13a, resulting in

$$\frac{a}{t} = \frac{8370}{\sqrt{f_v - [8370/(h/t)]^2}} \tag{6.13c}$$

Because Eqs. 6.13b and 6.13c contain the web depth-to-thickness ratio h/t, these equations should provide more accurate results than Eq. 6.13a.

As a result of neglecting the postbuckling strength of the web, the equations given above for controlling web slenderness are generally conservative and, in some cases, excessively conservative.

6.4 SHEAR STRENGTH OF PLATE GIRDERS

In evaluating the behavior of a plate girder subjected to shear, it is usually assumed that the web is plane and the material is elastic–plastic. Based on these assumptions, the load at which web buckling occurs can be predicted theoretically and referred to

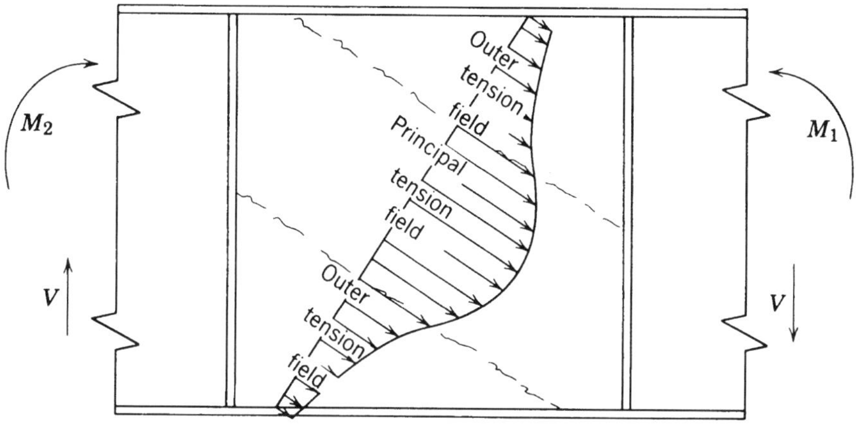

FIGURE 6.1 Tension-field action.

as the beam-action strength of the girder. Subsequent to buckling, the stress distribution in the web changes and considerable postbuckling strength may be realized because of the diagonal tension stresses that develop, often termed *tension-field action*. With or without transverse stiffeners, a plate girder can develop a shear stress at the ultimate load that is several times the shear buckling stress. Figure 6.1 shows the general distribution of the tension field that develops in a plate girder with transverse stiffeners. This stress distribution has been verified experimentally (Basler et al., 1960; Clark and Sharp, 1971; Steinhardt and Schröter, 1971).

The tension field in a girder with stiffeners is anchored by both the flanges and the stiffeners. The resulting lateral load on the flanges is illustrated in Fig. 6.2, and it should be clear that these tensile stresses cause the flanges to bend inward. Therefore, the nature of the tension field is influenced by the bending stiffness of the flanges. For example, if the stiffness of the flanges is large relative to the web, the tension field may be uniform over the entire panel. With continued increase in load, the tensile membrane stress combines with the shear buckling stress to cause yielding of the web. Failure of the panel is defined by the formation of a mechanism involving a yielded zone in the web and plastic hinges in the flanges. Figure 6.3 shows three possible failure modes, including (1) a beam mechanism in

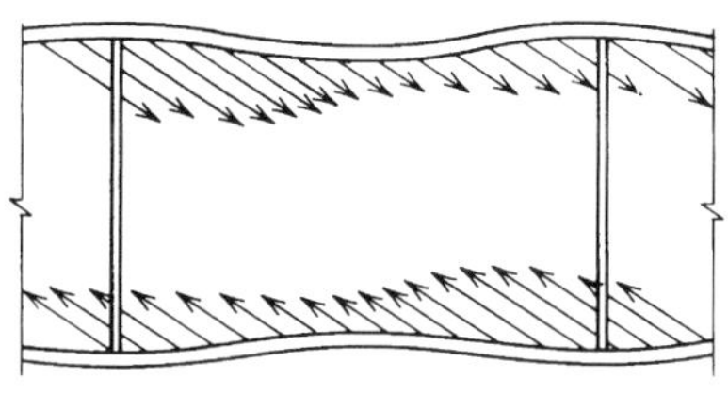

FIGURE 6.2 Flange resistance.

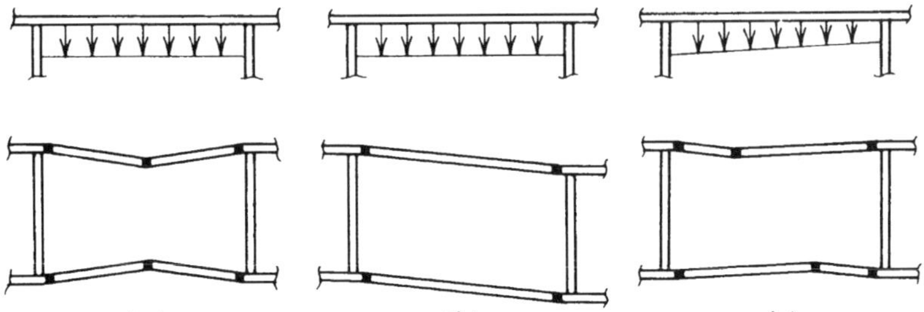

FIGURE 6.3 Frame action.

each flange, (2) a panel mechanism, and (3) a combined mechanism. The additional shear associated with the formation of a failure mechanism involving plastic hinges in the flanges is called *frame action* (Cescotto et al., 1981).

Wagner (1931) used a complete, uniform tension field to determine the strength of a panel in pure shear. The flanges are assumed to be rigid and the web very thin. This model corresponds to that shown in Fig. 6.3*b* except that there are no plastic hinges, since the flanges are infinitely stiff. The Wagner analysis has been found to be quite satisfactory for aircraft structures.

Basler and Thürlimann (1963) were the first to formulate a successful model for plate girders of the type used in civil engineering structures. They assumed that the flanges are too flexible to support a lateral loading from the tension field, and the resulting yield band shown in Table 6.1 controls the shear strength. The inclination and width of the yield band are defined by the angle θ, which is chosen so as to maximize the shear strength. The shear stress τ_u for the optimal value of θ was found to be[1]

$$\tau_u = \tau_{cr} + \tfrac{1}{2}\sigma_t \sin\theta_d \tag{6.14}$$

where τ_{cr} is the shear buckling stress, σ_t the tension-field stress, and θ_d the angle of panel diagonal with respect to the flange. Combining the beam shear τ_{cr} and the postbuckling tension σ_t and substituting the results in the von Mises yield condition give

$$\sigma_t = -\tfrac{3}{2}\tau_{cr}\sin 2\theta + \sqrt{\sigma_{yw}^2 + (\tfrac{9}{4}\sin^2 2\theta - 3)\tau_{cr}^2} \tag{6.15}$$

The maximum value of τ_u is then found by substituting σ_t from Eq. 6.15 into Eq. 6.14.

[1]Basler gives the result in terms of the aspect ratio $\alpha = a/h$ instead of θ_d, which is an angle that references the panel diagonal and obtained from $\tan(\theta_d) = 1/\alpha = h/a$.

To obtain a more simple solution, Basler assumed that σ_t and τ_{cr} were additive, as they would be if σ_t acted at 45°, and used the resulting combination of principal stresses in a linear approximation of the von Mises yield condition. This gives

$$\sigma_t = \sigma_{yw}\left(1 - \frac{\tau_{cr}}{\tau_{yw}}\right) \tag{6.16}$$

TABLE 6.1 Various Tension Field Theories for Plate Girders

Investigator	Mechanism	Web-Buckling Edge Support	Unequal Flanges	Longitudinal Stiffener	Shear and Moment
Basler (1963a)	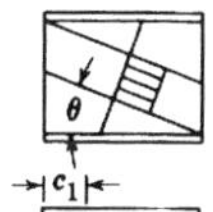	S S S S	Immaterial	Yes, Cooper (1965)	Yes
Takeuchi (1964)	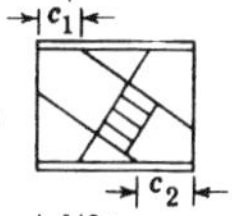	S S S S	Yes	No	No
Fujii (1968b, 1971)	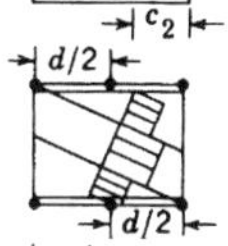	F S S F	Yes	Yes	Yes
Komatsu (1971)	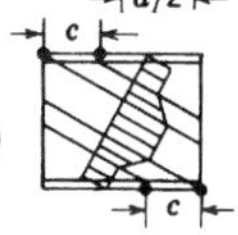	F S S F	No	Yes, at mid–depth	No
Chem and Ostapenko (1969)	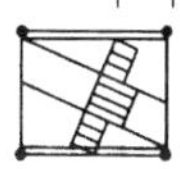	F S S F	Yes	Yes	Yes
Porter et al. (1975)	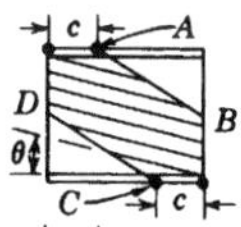	S S S S	Yes	Yes	Yes
Hoglund (1971a, b)	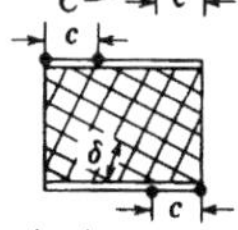	S S S S	No	No	Yes
Herzog (1974a,b)	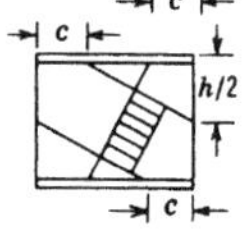	Web-buckling component neglected	Yes, in evaluating *c*	Yes	Yes
Sharp-and Clark (1971)	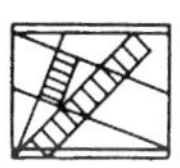	F/2 S S F/2	No	No	No
Steinhardt and Schröter (1971)	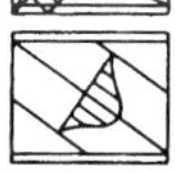	S S S S	Yes	Yes	Yes

where σ_{yw} and τ_{yw} are the web yield stresses in tension and shear, respectively. Values of τ_u by this approximation are less than those obtained by Eq. 6.15, but an investigation of girders with a wide range of proportions showed that the maximum difference was less than 10%. Substituting σ_t from Eq. 6.16 into Eq. 6.14 gives

$$\tau_u = \tau_{cr} + \tfrac{1}{2}\sigma_{yw}\left(1 - \frac{\tau_{cr}}{\tau_{yw}}\right)\sin\theta_d \tag{6.17}$$

Basler assumed that inelastic buckling will occur if τ_{cr} exceeds $0.8\tau_{yw}$ and recommended that the inelastic buckling stress $\tau_{cr,i,}$ be

$$\tau_{cr,i} = \sqrt{0.8\tau_{cr}\tau_{yw}} \qquad 0.8\tau_{yw} \leq \tau_{cr} \leq 1.25\tau_{yw} \tag{6.18}$$

This value is intended to be substituted for τ_{cr} in Eq. 6.17.

It was shown first by Gaylord (1963) and later by Fujii (1968a) and Selberg (1973) that Basler's formula gives the shear strength for a complete tension field instead of the limited band of Table 6.1. The correct formula for the limited band is

$$\tau_u = \tau_{cr} + \sigma_{yw}\left(1 - \frac{\tau_{cr}}{\tau_{yw}}\right)\frac{\sin\theta_d}{2 + \cos\theta_d} \tag{6.19}$$

Therefore, Basler's formula overestimates the shear strength of a girder that has flanges which are incapable of supporting lateral load from a tension field.

Many variations of the postbuckling tension field have been developed since the Basler–Thürlimann solution was published. The principal characteristics of most of these are shown in Table 6.1 and are discussed below. The table shows the tension field, the positions of the plastic hinges if they are involved in the solution, the edge conditions assumed in computing the shear buckling stress, and other features of the solutions. In all cases, except the Fujii and Herzog models, the shear buckling strength is added to the vertical component of the tension field to give the contribution of the web to the shear strength of the girder panel.

Takeuchi (1964) appears to have been the first to make an allowance for the effect of flange stiffness of the yield zone in the web. He located the boundaries of the tension field at the distances c_1 and c_2 from diagonally opposite corners of the panel (Table 6.1). These distances were assumed proportional to the respective flange stiffnesses I_{f1} and I_{f2} and were chosen to maximize the shear strength. Unfortunately, shear strengths determined in this way were not in good agreement with test results (Konishi et al., 1965). Lew and Toprac (1968) used this model in their investigation of hybrid plate girders and determined c_1 by a formula established to give agreement with available test results.

Fujii (1968a,b) assumed a tension field encompassing the whole panel, together with beam mechanisms in each flange with the interior hinge at midpanel (Table 6.1). The web compression in the direction perpendicular to the principal tension was assumed equal to the compression in that direction at the initiation of buckling. Tresca's yield condition was then used to determine the magnitude of the tension. If the flanges can resist the tributary web stress with the web in the yield condition, the web yields uniformly over the panel, but if they cannot, there is a central band of yielding with a smaller tension equal to that which the flange can support in the outer triangular portions. Inelastic shear buckling of the web was assumed to begin at $\tau_{cr} = 0.5\tau_{yw}$ and to vary parabolically from that point to τ_{yw} at $h/t = 0$. The theory was extended to include unsymmetrical girders (Fujii, 1971).

Komatsu (1971) gives formulas for four modes of failure. Failure in the first mode occurs in the manner shown in Table 6.1, where the inner band yields under the combined action of the buckling stress and the postbuckling tension field, while the smaller tension in the outer bands is the value that can be supported by the girder flange as a beam mechanism with the interior hinge at the distance c determined by an empirical formula based on tests. The inclination of the yield band is determined so as to maximize the shear, as in Basler's solution, but the optimum inclination must be determined by iteration. In the second mode, which is a limiting case of the first mode, the interior hinge develops at midpanel, and the web yields uniformly throughout the panel. In the third mode of failure, the flanges are assumed to remain elastic while allowing complete yielding of the web. An optimum value of the tension-field inclination must also be found by trial and error for this case. The fourth case is a limiting case in which a Wagner field develops along with a panel mechanism of the flanges.

Chern and Ostapenko (1969) proposed the tension field shown in Table 6.1, where the principal band is determined by yielding, taking into account the stress that exists at buckling. A panel mechanism is assumed to develop in the flanges. The resulting ultimate shear strength V_u is given as

$$V_u = \tau_{cr}A_w + \tfrac{1}{2}\sigma_t A_w[\sin 2\theta - (1-\rho)\alpha + (1-\rho)\alpha\cos 2\theta] + \frac{2}{a}(m_{pb} + m_{pt}) \tag{6.20}$$

where ρ is the ratio of outer-band tension to inner-band tension, m_{pb} and m_{pt} are the resisting moments of the bottom and top flanges, and α is the panel aspect ratio. Equation 6.15 is used to determine σ_t. In computing m_p the flange is assumed to act with an effective width of web b_e given by

$$b_e = 12.5t\left(0.8 - \frac{\tau_{cr}}{\tau_{yw}}\right) \quad \text{for } \frac{\tau_{cr}}{\tau_y} \leq 0.8 \tag{6.21}$$

Two categories of shear buckling are used to determine τ_{cr} in Eq. 6.15, including (1) elastic buckling, assuming the web panel fixed at the flanges and simply supported at the stiffeners and (2) inelastic buckling, which is assumed to occur when τ_{cr} for elastic buckling exceeds $0.5\tau_{yw}$. Chern and Ostapenko differentiate V_u with respect to θ to develop a formula for the optimum value of θ, which must be solved by trial.

The tension field of Porter et al. (1975) shown in Table 6.1 consists of a single band and is a development of one suggested earlier by Rockey and Škaloud (1972) in which the tension band was taken in the direction of the panel diagonal. The tensile membrane stress together with the buckling stress causes yielding, and failure occurs when hinges form in the flanges to produce a combined mechanism that includes the yield zone *ABCD*. The vertical component of the tension field is added to the shear at buckling and combined with the frame action shear. The resulting ultimate shear strength is

$$V_u = \tau_{cr}A_w + \sigma_t A_w \left(\frac{2c}{h} + \cot\theta - \cot\theta_d\right)\sin^2\theta + \frac{4m_p}{c} \tag{6.22}$$

The distance c of the plastic hinge is

$$c = \frac{2}{\sin\theta}\sqrt{\frac{m_p}{\sigma_t t}} \quad \text{for } 0 \le c \le a \tag{6.23}$$

where m_p is the plastic moment of the flange. The flange is assumed to act with an effective width of web b_e given by

$$b_e = 30t\left(1 - 2\frac{\tau_{cr}}{\tau_{yw}}\right) \quad \text{for } \frac{\tau_{cr}}{\tau_{yw}} \le 0.5 \tag{6.24}$$

Formulas are given to reduce m_p for the presence of axial force in the flange. Porter et al. suggest Eq. 6.15 for determining σ_t. The elastic shear buckling stress is calculated with the four edges of the panel simply supported, but if this value exceeds $0.8\tau_{yw}$, inelastic buckling is assumed to occur, with $\tau_{cr,i}$ given by

$$\frac{\tau_{cr,i}}{\tau_{yw}} = 1 - 0.16\frac{\tau_{yw}}{\tau_{cr}} \tag{6.25}$$

The maximum value of V_u must be found by iteration. For a given panel, however, θ is the only independent variable in Eq. 6.22, and the maximum V_u is not difficult to determine, because the optimal value of θ usually lies between $\theta_d/2$ and 45°, and τ_u is not sensitive to small changes from the optimum of θ. An assumption of $\theta = 0.667\theta_d$ will give either a very close approximation or an underestimation of the failure load.

If the flanges cannot develop moment, then $m_p = 0$ and Eq. 6.23 yields $c = 0$. Substituting this value in Eq. 6.22 and maximizing τ_u by differentiating with respect to θ results in the true Basler solution (Eq. 6.19). It is shown by Porter et al. (1975) that Eq. 6.22 includes several other existing solutions as special cases. A procedure for evaluating the effect on V_u of the reduction in m_p due to flange axial forces is also given.

Höglund (1971a,b, 1973) developed a theory for plate girders without intermediate transverse stiffeners that was later extended to girders with intermediate stiffeners. He used the system of bars shown in the figure in Table 6.1 as a model of the web. The compression bars in this system are perpendicular to the tension bars. When the angle δ between the tension bars and the flanges is decreased, the shear buckling load for the system is increased. If the load is uniformly distributed,

δ is varied along the girder because the shear varies. Calculated stresses in the bar system are in good agreement with stresses measured in test girders. The shear strength V_u is given by

$$V_u = V_w + \frac{4m_p}{c}\left(1 - \frac{M}{M_f}\right)^2 \quad \text{for } M \leq M_f \tag{6.26}$$

where V_w = shear strength of web (Table 6.2)
$c = a(0.25 + m_p/m_{pw})$
m_{pw} = plastic moment of web $(= \sigma_{yw} th^2/4)$
M = moment in panel
M_f = flange moment $= \sigma_{yf} A_f h$

The term that includes M/M_f accounts for the reduction in the flange moment m_p and depends on a mean value of M, but Höglund suggested that the largest value be used to simplify the problem.

It will be noted that V_w depends on the nature of the end stiffeners. End stiffeners can be considered rigid if the following conditions are satisfied:

$$e \geq 0.18h \tag{6.27a}$$

$$\frac{3000t}{\sqrt{\tau}} \geq g > 0.18h \tag{6.27b}$$

$$A_e \geq 0.1ht \tag{6.27c}$$

$$A_u \geq \frac{V_u}{\sigma_y} - 12t^2 \tag{6.27d}$$

$$A_{st} \geq \frac{16m_p}{a\sigma_{y(st)}} \tag{6.27e}$$

where e, g, A_e, A_u, and A_{st} are defined in Fig. 6.4. In the second of these equations τ is in kilograms per square centimeter and t in centimeters. The term $3000t/\sqrt{\tau}$ is identical with the right side of Eq. 6.13a; A_{st} is the intermediate-stiffener area required to develop the tension field in the adjacent panel.

Herzog (1974a,b) takes the boundary of the tension field from midheight of the panel at the stiffeners to the plastic hinges in the flanges (Table 6.1). The distance c is based on an average, relative flange stiffness to allow for unequal flanges, and a chart is given to determine this distance, which has been developed from a study of various test results, particularly those reported by Rockey and Škaloud (1968). He gives three relatively simple formulas for the ultimate shear, one for each of three ranges of panel aspect ratio, together with a coefficient by which the results are to be multiplied to account for a longitudinal stiffener.

Clark and Sharp (1971) and Sharp and Clark (1971) proposed a tension field for thin-web aluminum girders that consists of a Basler field on which a complete tension field inclined at 45° (Table 6.1) is superimposed. The flanges are assumed

TABLE 6.2 Values of V_W in Eq. 6.26[a]

$\sqrt{\tau_y/\tau_{cr}}$		
End Stiffeners Rigid[b]	End Stiffeners Nonrigid	V_w
Less than 0.8	Less than 0.8	$\tau_y ht$
0.8–2.75	0.8–1.25	$\tau_y ht \dfrac{1.8}{1+\sqrt{\tau_y/\tau_{cr}}}$
Over 2.75	—	$\tau_y ht \dfrac{1.32}{\sqrt{\tau_y/\tau_{cr}}}$
Over 2.75	Over 1.25	$\tau_y ht/\sqrt{\tau_y/\tau_{cr}}$

[a] τ_{cr} to be computed from Eq. 4.1 with k from Eq. 4.3.
[b] See Eq. 6.27 and Fig. 6.4.

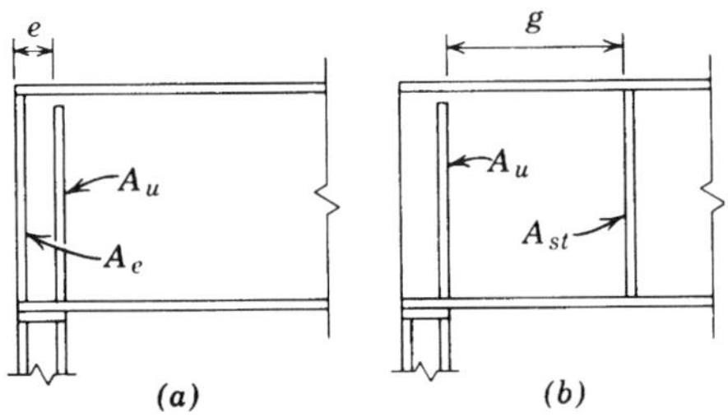

FIGURE 6.4 End stiffeners.

to be elastic beams continuous over the stiffeners and subjected to a uniform load from the 45° field. The shear strength is the sum of the vertical components of the two tension fields and the shear at buckling. Basler's approximation of the von Mises yield condition is used to determine the combination of web stresses that cause yielding. Because their procedure is based on general yielding of the web, it tends to give conservative results when compared with tests to ultimate load. It has been noted, however, that aluminum girders generally do not develop plastic hinges in the flanges, and failure usually occurs by flange or stiffener buckling or, in riveted and bolted girders, by cracks developing at web holes because of the diagonal tension.

Steinhardt and Schröter (1971) have also suggested a tension field for aluminum girders (Table 6.1). The tension-field band is in the direction of the panel diagonal and its boundaries intersect the midpanel points of the flanges. The tension-field loading on the flange is assumed to vary sinusoidally with a maximum value at the stiffeners. Assuming flange bending to be elastic the corresponding tension-field distribution shown in Table 6.1 is derived. The resultant shear is found by adding the vertical component of the tension field to the shear at buckling. The theory was extended to girders of unsymmetrical sections.

Tables 6.3A and 6.3B provide comparisons of predicted shear strengths according to Basler, Fujii, Ostapenko, Höglund, Rockey, and Komatsu with results of tests from several different sources. In computing the values, an axial force reduction in the flange plastic moment is taken into account by Höglund and Rockey but not by Fujii, Ostapenko, and Komatsu. Furthermore, in the five cases in Table 6.3A when M/M_f exceeds unity, Höglund has reduced his pure-shear values by Basler's shear–moment interaction formula. Two points should be kept in mind when evaluating the results presented in these tables: (1) definitions of ultimate strength differ among investigators and (2) girders that are identical in design and fabrication can sometimes differ considerably in ultimate strength. For example, the ultimate strengths of TG1 and TG1′ in Table 6.3B were 151 and 116.2 kN, respectively, yet these were tests on presumably identical halves of a single girder.

The results in Tables 6.3A and 6.3B are summarized in Table 6.4. Although Basler's formula gives the largest range of values, the others do not differ significantly from Basler's. Chern and Ostapenko's formulas tend to overestimate shear strength more than the others; this is due in part to the neglect of shear–moment interaction. All but one of the first 10 girders in Table 6.3A were subjected to fairly large moment, and when they were compared by Chern and Ostapenko (1970b) with the Chern–Ostapenko shear–moment interaction formulas, the following values of V_{ex}/V_u were obtained: 0.90, 0.92, 0.92, 0.97, 0.97, 0.95, 1.00, 0.96, and 0.91. The average of these 10 values is 0.95; the 10 values in Table 6.3A average 0.91.

Höglund's formulas give the most conservative results, particularly for slender webs (Table 6.3B). This conservatism is explained in part by the fact that he bases the axial force reduction in flange plastic moment (Eq. 6.26) on the largest moment in the panel. Rockey uses an average flange stress in computing this effect, and Basler found better correlation in predictions of shear–moment interaction when he used a reduced moment (Basler, 1963b).

Herzog presents in graphical form a comparison of shear strengths by his formulas with results of 96 tests by others, including all of those in Tables 6.3A and 6.3B. The mean value of the ratio V_{ex}/V_u is 1.036. The standard deviation, however, is 0.16, and in one test (not in Table 6.3A or 6.3B) V_{ex}/V_u is 1.41, while Höglund's formula gives 1.04.

The simplest formulas are Basler's, Höglund's, and Herzog's, and of these Höglund's gives the most consistent results except for slender webs ($h/t > 300$). Fujii's formulas and Komatsu's also give good results, but they are more complicated. The formulas by Rockey, Evans, and Porter give good results but require the optimum inclination of the tension field to be determined by trial. Chern and Ostapenko's formulas are complicated and also require a trial determination of the optimum inclination of the tension field.

Yonezawa et al. (1978) proposed an ultimate shear strength theory for plate girders with webs diagonally stiffened between vertical stiffeners. It is assumed that the ultimate shear strength consists of the sum of contributions from three sources, including (1) the beam shear force V_{cr} taken by the diagonally stiffened web, (2) the shear force V_t taken by the tension field in the web, and (3) the shear force V_s taken by the diagonal stiffener. Buckling coefficients, obtained from a finite

TABLE 6.3A Shear Strength of Plate Girders

Source	Test Number	$\frac{a}{h}$	$\frac{h}{t}$	V_{ex}/V_u				$\frac{M}{M_f}$
				Basler[a]	Fujii[b]	Ostapenko[a]	Höglund[c]	
Okumura et al.	G1	2.61	55	0.85	1.06	0.87	0.89	0.82
	G2	2.61	55	0.87	1.08	0.88	0.93	1.01
	G3	2.63	70	0.97	1.01	0.88	0.99	1.25
	G5	2.68	70	1.05	1.09	0.94	1.02	0.87
	G6	1.25	70	1.06	1.22	0.96	0.86	0.45
	G7	2.68	70	1.07	1.09	0.94	1.02	0.87
	G9	2.78	90	1.20	0.95	0.90	1.06	1.00
Nishino and Okumura	G1	2.67	60	1.04	0.94	0.98	1.03	0.97
	G2	2.67	60	0.98	1.11	0.93	0.98	1.25
	G3	2.67	77	1.00	0.98	0.86	0.96	1.09
Basler et al.	G6-T1	1.5	259	1.04	1.08	0.96	1.06	0.48
	G6-T2	0.75	259	0.95	0.97	0.94	1.06	0.62
	G6-T3	0.5	259	0.98	1.00	0.93	1.04	0.74
	G7-T1	1.0	255	0.98	1.05	0.96	1.12	0.59
	G7-T2	1.0	255	1.02	1.09	1.00	1.16	0.61
Cooper et al.	H1-T1	3.0	127	1.33	0.96	1.00	1.01	1.00
	H1-T2	1.5	127	1.08	0.92	0.97	1.06	0.60
Konishi et al.	B	1.0	267	0.81	1.02	0.79	1.18	0.53
Okumura and Nishino	G1-1	3.0	182	1.21	1.07	0.99	0.97	1.00
	G1-2	1.5	182	1.03	1.04	0.91	1.08	0.87
	G2-1	3.0	144	1.34	0.96	1.02	1.01	1.16
	G2-2	1.5	144	1.17	1.00	0.96	1.14	0.93
Mean value				1.05	1.03	0.94	1.03	
Standard deviation				0.14	0.07	0.05	0.08	

[a] From Chern and Ostapenko (1969).
[b] From Fujii (1971).
[c] From Höglund (1973).

TABLE 6.3B Shear Strength of Plate Girders

Source	Test Number	$\frac{a}{h}$	$\frac{h}{t}$	V_{ex}/V_u Rockey[a]	V_{ex}/V_u Höglund[b]	V_{ex}/V_u Komatsu[c]	$\frac{M}{M_f}$
Rockey and Škalound	TG 14	1.0	316	1.02	1.29	—	0.70
	TG 15	1.0	316	1.00	1.29	—	0.53
	TG 16	1.0	316	1.07	1.11	—	0.38
	TG 17	1.0	316	0.99	1.12	—	0.34
	TG 18	1.0	316	0.92	1.15	—	0.33
	TG 19	1.0	316	0.92	1.16	—	0.32
Basler et al.	G6-T1	1.5	259	1.02	1.06	1.03	0.48
	G6-T2	0.75	259	1.14	1.06	0.97	0.62
	G6-T3	0.5	259	1.12	1.04	0.99	0.74
	G7-T1	1.0	255	1.08	1.12	0.96	0.59
	G7-T2	1.0	255	1.04	1.16	1.00	0.61
Škalound	TG1[d]	1.0	400	—	1.27	1.10	0.65
	TG1′	1.0	400	—	0.97	0.84	0.50
	TG2	1.0	400	—	1.17	0.96	0.28
	TG2′	1.0	400	—	1.01	0.83	0.24
	TG3	1.0	400	—	1.13	1.00	0.20
	TG3′	1.0	400	—	1.12	1.00	0.20
	TG4	1.0	400	—	1.14	1.02	0.19
	TG4′	1.0	400	—	1.08	0.97	0.18
	TG5	1.0	400	—	1.11	0.99	0.15
	TG5′	1.0	400	—	1.08	0.96	0.14
Mean value				1.03	1.13	0.97	
Standard deviation				0.07	0.08	0.07	

[a] From Porter et al. (1975).
[b] From Höglund (1973).
[c] From Komatsu (1971).
[d] TG1, TG1′, and so on, are twin girders. End stiffeners nonrigid according to Höglund (Eq. 6.27).

TABLE 6.4 Comparison of Test Results in Tables 6.3A and 6.3B

Investigator	Mean	Standard Deviation	Range of V_{ex}/V_u	Ratio Highest/Lowest
Basler	1.05	0.14	1.33–0.84	1.64
Fujii	1.03	0.07	1.22–0.92	1.33
Chern and Ostapenko	0.94	0.05	1.02–0.79	1.29
Höglund (Table 6.3A)	1.03	0.08	1.18–0.86	1.37
Höglund (Table 6.3B)	1.13	0.08	1.29–1.01	1.28
Rockey	1.03	0.07	1.14–0.92	1.24
Komatsu	0.97	0.07	1.10–0.83	1.32

difference solution of the differential equation for a buckled web with a diagonal stiffener under shear, are presented for the computation of the beam shear force. Coefficients for panels with both compression- and tension-type stiffeners with either fixed- or pinned-plate boundaries are given. Expressions for the tension-field contribution are based on the work of Rockey et al. (1973, 1974). The shear force taken by the diagonal stiffener is determined by taking the vertical component of the force acting on the stiffener. Tests on two plate girders showed good agreement between theoretical and experimental ultimate loads. Studies on diagonally stiffened webs were also performed at Liege (Jetteur, 1984).

In the case of tapered plate girders, the axial forces in the inclined flanges have vertical components. These components of force may either increase or decrease the shear force carried by the web depending on the direction of taper and the direction of the applied shear. Shear buckling of simply supported plates of variable depth have been investigated by finite element analysis and charts for the buckling coefficient k have been prepared by Elgaaly (1973).

Two investigations have been reported on the ultimate shear strength of tapered web girders. Falby and Lee (1976) have proposed a method for estimating ultimate shear strength based on the Basler method. Their method is limited to small tapers and does not account for the effect on shear of the axial load in the inclined flange. Davies and Mandal (1979) extended Rockey's tension-field model to the case of tapered web girders. Their theory is not limited to small tapers and takes into account the influence of the axial force in the sloping flange. Good agreement is obtained between theoretical predictions and experimental collapse loads. The theory is applicable to loading conditions where the girder is loaded within the tip (the intersection point of the flanges), but it appears that modifications must be made to the theory in order to deal with the more common loading case associated with continuous plate girders with tapered webs at the supports.

6.5 GIRDERS WITH NO INTERMEDIATE STIFFENERS

Plate girders with bearing stiffeners at the supports and no intermediate stiffeners except for bearing stiffeners at heavy concentrated loads are of practical interest. The postbuckling strength of such girders can be significant. The shear stress at ultimate load in three tests with web slendernesses h/t of 210, 210, and 300 (B1, K1, and B4 of Table 6.5) were 3.69, 4.00, and 4.68 times the theoretical shear buckling stress assuming panel boundaries as simply supported. Höglund's formula (Eq. 6.26), which was developed originally for such girders, is in good agreement with test results (Table 6.5).

The formulas by Ostapenko and Chern are in good agreement with tests on girders with panel aspect ratios of 5.5 if the tension-field contribution is assumed to be zero (Table 6.5); in other words, the ultimate shear is the sum of the critical shear, assuming the web fixed at the flanges, and the flange plastic-hinge contribution.

TABLE 6.5 Shear Strength of Girders with Long Panels

Source	Test No.	$\frac{a}{h}$	$\frac{h}{t}$	V_{ex}/V_{th} Höglund[a]	V_{ex}/V_{th} Chern and Ostapenko[b]
Carskaddan (1968)	C-AC2	5.5	143	0.99	1.02
	C-AC3	5.5	71	0.97	0.94
	C-AC4	5.5	102	1.04	1.02
	C-AC5	5.5	103	0.96	0.96
	C-AHI	5.5	69	1.03	1.00
Höglund (1971b)	B1[c]	[d]	210	1.14	—
	K1	[d]	210	0.94	—
	B4	[d]	300	1.05	—

[a] From Höglund (1973).
[b] From Chern and Ostapenko (1969).
[c] End stiffeners nonrigid according to Höglund (Eq. 6.27).
[d] No intermediate transverse stiffeners.

6.6 STEEL PLATE SHEAR WALLS

Although the postbuckling behavior of thin plates under static loads has been studied for more than a century, investigations into this behavior in the context of what has become known as the unstiffened steel plate shear wall system are relatively recent. Early tests were conducted at the University of Alberta (Timler and Kulak, 1983; Tromposch and Kulak, 1987), the University of Maine (Elgaaly et al., 1991a; 1993a), and the University of Wales (Roberts and Sabouri, 1991) to study the monotonic and hysteretic characteristics of thin steel infill plates under shear loading. These test results indicated that the behavior of thin steel plates that are adequately supported along their boundaries and are subjected to cyclic shear loading is stable in the postbuckling domain. More recently, many large-scale tests have been conducted on steel plate shear walls fabricated in a variety of configurations (Driver et al., 1997, 1998a; Rezai, 1999; Lubell et al., 2000; Astaneh-Asl and Zhao, 2001; Berman and Bruneau, 2003; Behbahanifard et al., 2003; Park et al., 2007; Deng et al., 2008; Qu et al., 2008), all with an overarching objective of verifying the good performance of this system under extreme cyclic loading. Recent innovations arising out of steel plate shear wall research with the intent of optimizing the performance of the system in seismic applications include the use of reduced beam sections (Dastfan et al., 2008; Qu et al., 2008), composite columns (Astaneh-Asl and Zhao, 2001; Deng and Driver, 2007; Driver, 2008), and perforated and low-yield steel infill plates (Vian and Bruneau, 2004).

Because the stiffness of the boundary members for anchoring the postbuckling tension field is important to the performance of steel plate shear walls, North American design standards (AISC, 2005b; CSA, 2009) provide minimum member stiffness requirements. Based on the work of Wagner (1931) and Kuhn et al. (1952), which has been cited widely in conjunction with plate girder research, it has been determined that the minimum column moment of inertia, I_c, that should be provided to ensure a sufficiently uniform tension field is

$$I_c = 0.0031\frac{wh^4}{L} \tag{6.28}$$

where w is the infill plate thickness, h the storey height, and L the width of the wall, which is typically taken to the column centerlines. [The derivation of Eq. 6.28 from the work of Wagner (1931) and Kuhn et al. 1952 is demonstrated by Dastfan and Driver (2008).] Because the uniformity of the tension fields in the bottom and top infill panels of a steel plate shear wall depends on the stiffnesses of both the adjacent columns and the base or roof beam, Eq. 6.28 cannot be used for these panels. Dastfan and Driver (2008, 2009) have proposed that to achieve good behavior of the extreme panels, the boundary member flexibility parameter, ω_L, should be no greater than 2.0 for the panel at the base of the wall and 2.5 at the top of the wall, where

$$\omega_L = 0.7\left[\left(\frac{h^4}{I_c} + \frac{L^4}{I_b}\right)\frac{w}{4L}\right]^{0.25} \tag{6.29}$$

and I_b is the moment of inertia of the beam at the base or roof, as appropriate.

Thorburn et al. (1983) modeled the thin steel infill plate as a series of inclined tension strips, with the angle of inclination of the strips being determined as a function of the panel width and height, plate thickness, cross-sectional area of the surrounding beams and columns, and moment of inertia of the columns using the principle of least work. The use of this model to conduct a full nonlinear pushover analysis of a multistory steel plate shear wall using only an elastic plane frame structural analysis program was demonstrated by Driver et al. (1997, 1998b). Modifications to the original strip model have been proposed by several researchers. For example, Elgaaly et al. (1993b) proposed a trilinear stress–strain relationship up to the yield point to be used for the tension strips. Rezai (1999) suggested that for cases with flexible boundary members the strips should be splayed so that they converge at the beam-to-column joints. Lubell et al. (2000) proposed a strip model wherein the column and tension strips had trilinear stiffness parameters to account for both yield and postyield strain hardening. Shishkin et al. (2009) provide a means for improving modeling efficiency when using the strip model approach, while at the same time maintaining accurate analytical solutions. The latter study demonstrated that within the typical range of values the angle of inclination of the tension strips has little influence on the predicted inelastic behavior of a shear wall, and the authors recommend the use of 40° from the vertical to give accurate and generally conservative results in pushover analyses. The use of a constant value obviates the need to revise the angle in the model repeatedly as the boundary member sizes change through the normal design development process. In recent years, many researchers and designers have opted to use the finite element method to analyze steel plate shear walls, and with the sophisticated software tools now available very good results can be achieved. Nonlinear time history dynamic analyses that include both strain rate and $P-\Delta$ effects (Bhowmick et al., 2009a,b) have provided opportunities for comparing expected seismic forces with code-specified seismic design forces, which allow the ability to better assess the adequacy of current design methods.

Research needs and anticipated future directions for steel plate shear walls are discussed by Driver et al. (2001) and Berman et al. (2008).

6.7 BENDING STRENGTH OF PLATE GIRDERS

A plate girder subjected primarily to bending moment usually fails by lateral–torsional buckling, local buckling of the compression flange, or yielding of one or both flanges. Buckling of the compression flange into the web (vertical buckling) has been observed in many tests, and Basler and Thürlimann (1963) developed the following limiting value of the web slenderness *h/t* to preclude this mode of failure:

$$\frac{h}{t} \leq \frac{0.68E}{\sqrt{\sigma_y(\sigma_y + \sigma_r)}}\sqrt{\frac{A_w}{A_f}} \tag{6.30}$$

where A_w is the area of the web, A_f the area of one flange, and σ_r the residual tension that must be overcome to achieve uniform yield in compression. In tests in which vertical buckling was observed, it occurred only after general yielding of the compression flange in the panel. Therefore, Eq. 6.30 may be too conservative, or even unnecessary, for girders of practical proportions. Web slenderness, however, must be limited to facilitate fabrication and to avoid fatigue cracking under repeated loads due to out-of-plane web flexing.

As in the case of shear, buckling of the web due to bending does not exhaust the panel capacity. The distribution of the bending stress, however, does change in the postbuckling range and the web becomes less efficient. The solution to this problem presented by most investigators is based on the assumption that a portion of the web becomes ineffective. Basler and Thürlimann assumed a linear distribution of stress on the effective cross section (Fig. 6.5), with the ultimate moment being reached when the extreme fiber compression reaches the yield stress, or a critical stress if some form of buckling controls. The effective width b_e is assumed to be $30t$ for a web with $h/t = 360$. This is the limiting slenderness according to Eq. 6.30 with $A_w/A_f = 0.5$, $\sigma_y = 33$ ksi, and $\sigma_r = 16.5$ ksi. Bending strength is then assumed to increase linearly from the value for a girder that has a web that can reach yield stress in bending without buckling. This assumption gives

$$\frac{M_u}{M_y} = 1 - C\left[\frac{h}{t} - \left(\frac{h}{t}\right)_y\right] \tag{6.31}$$

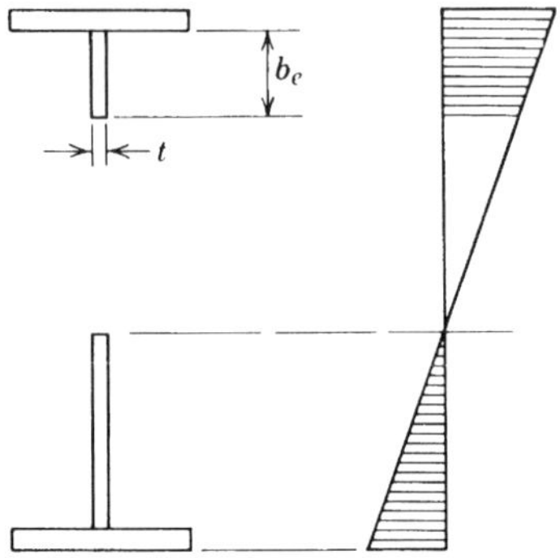

FIGURE 6.5 Bending stresses.

where C is a constant and $(h/t)_y$ is the web slenderness that permits yielding in bending without buckling. In particular, the following formula was suggested:

$$\frac{M_u}{M_y} = 1 - 0.0005\,\frac{A_w}{A_f}\left(\frac{h}{t} - 5.7\sqrt{\frac{E}{\sigma_y}}\right) \tag{6.32}$$

The value of $(h/t)_y$ in this formula is somewhat larger than the theoretical value for hinge-edged panels and was chosen to give the limiting slenderness 170 for steel with $\sigma_y = 33$ ksi (228 MPa), which was prescribed at that time by the AISC specification. Equation 6.32 has been found to be in good agreement with test results (Basler and Thürlimann, 1963; Maeda, 1971; Cooper, 1971) with girders with *h/t* values of 388, 444, and 751.

Höglund (1973) assumes the effective width b_e in Fig. 6.5 to be $0.76t\sqrt{E/\sigma_{yf}}$ and considers an additional strip of web of $1.64t\sqrt{E/\sigma_{yf}}$ immediately above the neutral axis to be effective. The following formula for the effective section modulus was derived:

$$S_{\text{eff}} = S\left[1 - 0.15\frac{A_w}{A_f}\left(1 - 4.8\frac{t}{h}\sqrt{\frac{E}{\sigma_{yf}}}\right)\right] \quad \text{for } \frac{h}{t} \leq 4.8\sqrt{\frac{E}{\sigma_{yf}}} \tag{6.33}$$

where S_{eff} is the effective section modulus and S is the section modulus of unreduced cross section. Equation 6.33 was found to give a slightly larger reduction than Eq. 6.32 and was in good agreement with 11 tests ($0.96 < M_{\text{exp}}/M_{th} < 1.04$). For hybrid girders, S_{eff} should be decreased by the amount ΔS given by

$$\Delta S = \frac{h^2 t}{12}\left(2 + \frac{\sigma_{yw}}{\sigma_{yf}}\right)\left(1 - \frac{\sigma_{yw}}{\sigma_{yf}}\right)^2 \quad \text{for } \frac{\sigma_{yw}}{\sigma_{yf}} \leq 1 \tag{6.34}$$

Fujii's formulas (1968b, 1971) for the ultimate moment are more complicated and are restricted to laterally supported girders. They involve the parameter $(t/h)\sqrt{E/\sigma_{yf}}$ of the Höglund formula and the web bend-buckling stress. A modification coefficient for hybrid girders is given. A comparison of predicted values with results of tests on 10 nonhybrid girders gave $0.94 < M_{\exp}/M_{th} < 1.11$.

Chern and Ostapenko (1970a) have developed formulas for hybrid girders with unequal flanges. The ultimate moment is the sum of the web-buckling moment based on an effective width similar to Basler's. The postbuckling moment is determined by yielding of the tension flange or by lateral or local buckling of the compression flange. A comparison of predicted values with results of tests gave $0.95 < M_{\exp}/M_{th} < 1.15$ for 14 nonhybrid girders and $0.86 < M_{\exp}/M_{th} < 1.13$ for 10 hybrid girders. In a later report (Ostapenko et al., 1971) the following modification of Basler's formula was proposed:

$$M_u = \frac{I}{y_c}\sigma_c \left\{1 - \frac{I_w}{I} + \frac{\sigma_{yw}}{\sigma_c}\left[\frac{I_w}{I} - 0.002\frac{y_c t}{A_c}\left(\frac{y_c}{t} - 2.85\sqrt{\frac{E}{\sigma_{yw}}}\right)\right]\right\} \tag{6.35}$$

where I = moment of inertia of cross section
I_w = moment of inertia of web about centroidal axis of cross section
y_c = distance from neutral axis to compression edge of web
A_c = area of compression flange
σ_c = compression-flange buckling stress

$$\frac{y_c}{t} \geq 2.85\sqrt{E/\sigma_{yw}}$$

$$\sigma_{yf} \geq \sigma_{yw}$$

This equation has been found to give good correlation with test results. The ultimate moment in terms of the tension flange is

$$M_u = \frac{I\sigma_{yt}}{y_t}\left[1 - \frac{I_w}{I}\left(1 - \frac{\sigma_{yw}}{\sigma_{yt}}\right)\right] \tag{6.36}$$

where y_t is the distance from the neutral axis to the tension edge of the web.

Herzog (1973, 1974b) gives a formula for the general case of an unsymmetrical hybrid girder with one or more longitudinal stiffeners which includes simple reduction coefficients for vertical, local, and lateral buckling of the compression flange. The ultimate moment is given by

$$M_u = \sigma_{yc}A_c\left(y_c + \frac{t_c}{2}\right) + \sigma_{yt}A_t\left(y_t + \frac{t_t}{2}\right) + \sigma_{ys}A_s y_s + \frac{\sigma_{yw}t_w}{6}[(1+\phi)y_c^2 + 2y_t^2] \tag{6.37}$$

where y and t are defined in Fig. 6.6 with c, t, s, and w signifying compression flange, tension flange, longitudinal stiffener, and web, respectively, and

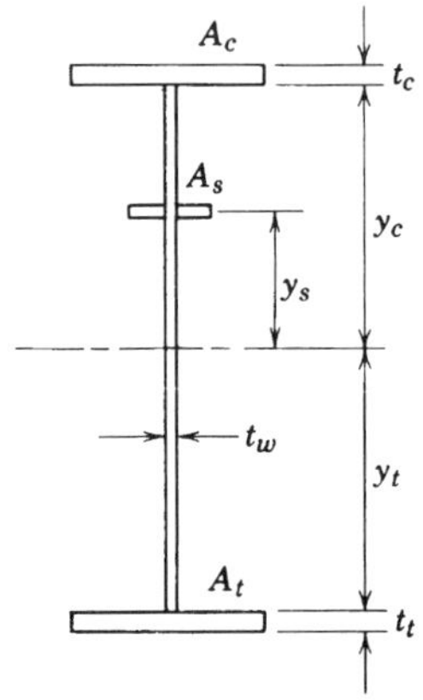

FIGURE 6.6 Plate girders with longitudinal stiffeners.

$\phi = (\sigma_{ys} \sum A_s)/(\sigma_{yw} A_w)$. The distance y_c to the compression edge of the web is

$$y_c = \frac{2}{3+\phi}\left[h - 2\left(\frac{\sigma_{yc}A_c + \sigma_{ys}A_s - \sigma_{yt}A_t}{\sigma_{yw}t_w}\right)\right] \tag{6.38}$$

A comparison of predicted values by Eq. 6.37 with results of tests on 23 girders without longitudinal stiffeners gave $0.91 < M_{\text{exp}}/M_{th} < 1.19$, with a mean value 1.005 and standard deviation 0.095. Comparisons with tests on 26 longitudinally stiffened girders gave $0.90 < M_{\text{exp}}/M_{th} < 1.16$, with a mean value 1.036 and standard deviation 0.105.

6.8 COMBINED BENDING AND SHEAR

Assuming the shear in a girder is resisted only by the web, as in Basler's solution, shear capacity is a maximum when the web is yielded uniformly, or, if shear buckling occurs at a smaller stress, when it has a fully developed tension field. These values are independent of the bending moment in the panel as long as the moment is less than $M_f = \sigma_{yf} A_f h$, which is the moment that can be carried by the flanges alone (*AB* in Fig. 6.7*a*). Any larger amount must be resisted in part by the web, which reduces the shear, until the shear capacity finally becomes zero for a panel in pure bending (*BC* in Fig. 6.7*a*). If the flange contribution to shear is taken into account, as in the more recent theories of shear strength, *AB* in Fig. 6.7*a* is not correct. This is because the flange axial force from the moment in the panel reduces the flange plastic moment m_p on which the flange contribution is based. This condition is represented by a line $A'B$, which can be defined, for example, by Höglund's formula (Eq. 6.26). Although expressions accounting for shear–moment interaction were considered in past design specifications, the expressions are not included in the 13th edition of the AISC specification (2005a) or the 3rd and 4th editions of *AASHTO LRFD Bridge Design Specification* (2004, 2007). These provisions were removed based upon work by White (2008) that showed that the tension-field design expressions sufficiently capture the behavior with a reasonable amount of accuracy relative to experimental test results. For completeness, however, past expressions on shear–moment interaction are provided in the remainder of this section.

Basler's (1963b) interaction diagram is shown in Fig. 6.7*b*. The segment BC, for which a formula is given, corresponds to BC in Fig. 6.7*a* with $M_u = M_p$, but is assumed to be invalid for thin-webbed girders when M exceeds M_y. The segment BC can be taken to be a straight line. Correlation with test results was good with M taken at $h/2$ from the high-moment end of the panel or at midpanel if $a < h$.

Herzog (1974a,b) assumes a trilinear diagram similar to Basler's. Fujii's interaction diagram is shown in Fig. 6.7*c*, where the additional point (M_f', V_{uo}) is for the case of a flange with no bending resistance (Fujii, 1971).

Chern and Ostapenko (1970b) assume the ultimate capacity will be dictated by failure of the web, instability of the compression flange, or yielding of the

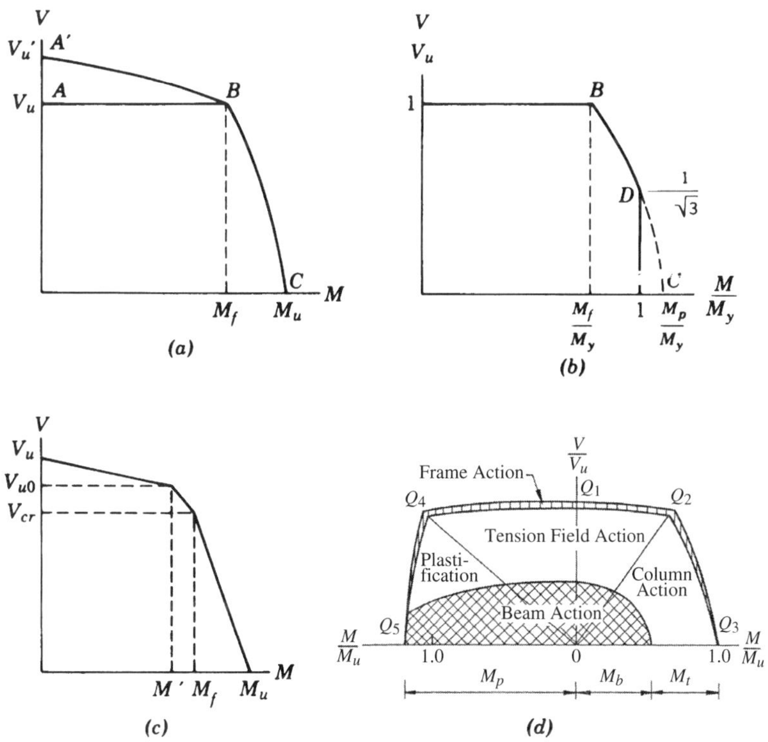

FIGURE 6.7 Shear–moment interaction diagrams.

tension flange. The complete interaction behavior is represented schematically by the interaction curve of Fig. 6.7*d*. Curve Q_2–Q_1–Q_4 represents failure of the web, curve Q_2–Q_3 represents buckling of the compression flange, and curve Q_4–Q_5 represents yielding of the tension flange.

When web failure controls (region Q_2–Q_1–Q_4 of Fig. 6.7*d*) the total shear resistance V_{wc} is

$$V_{wc} = V_{cr,c} + V_{ec} + V_{fc} \tag{6.39}$$

in which $V_{cr,c}$ is the beam-action shear, V_{ec} the tension-field shear, and V_{fc} the frame-action shear. The subscript c indicates that these shears are associated with combined stresses in the girder.

The beam-action shear is computed from

$$V_{cr,c} = A_w \tau_c \tag{6.40}$$

in which τ_c is obtained from the interaction equation

$$\left(\frac{\tau_c}{\tau_{cr}}\right)^2 + \frac{1+C}{2}\left(\frac{\sigma_{bc}}{\sigma_{cp}}\right) + \frac{1-C}{2}\left(\frac{\sigma_{bc}}{\sigma_{cp}}\right)^2 = 1 \tag{6.41}$$

in which C is the ratio of bending stresses in the tension and compression at the extreme fibers, σ_{bc} the maximum bending stress in the web, and σ_{cp} the critical buckling stress in the web under pure bending.

The tension-field shear V_{ec} in Eq. 6.39 is computed using the second term in Eq. 6.20 with $\rho = \frac{1}{2}$ and σ_t replaced by σ_{tc}. The tension-field stress σ_{tc} is found by using the combined stresses resulting from shear and bending in the yield criterion. As was done for the case of pure shear, an iteration procedure is used to find the maximum value of V_{wc}.

The frame-action shear is computed using the same frame mechanism as for the pure-shear case (the third term in Eq. 6.20). The flange plastic moments in the combined shear and bending case, however, will be affected by the axial forces in the flanges. The resulting frame-action shear is

$$V_{fc} = \frac{1}{a}(m_{cl} + m_{cr} + m_{tl} + m_{tr}) \tag{6.42}$$

in which m_{cl} and m_{cr} are the plastic moments in the compression flange at the left and right sides of the panel and m_{tl} and m_{tr} are the corresponding plastic moments in the tension flange. All of these moments are modified for the effect of axial force in the flange.

The other conditions for which failure occurs are depicted by curves Q_2–Q_3 and Q_4–Q_5 of Fig. 6.7*d*. Curve Q_2–Q_3 represents the case where buckling of the compression flange controls, in which case a completely developed tension field will not have formed. In the region represented by curve Q_4–Q_5, the tension flange starts yielding before the web plate reaches its ultimate shear strength. Yielding will penetrate into the cross section and the plastic strength of the girder panel will be the ultimate capacity.

The model postulated by Rockey et al. (Rockey, 1971a,b; Rockey and Škaloud, 1972; Rockey et al., 1973; Porter et al., 1975) for predicting the strength of girders without longitudinal stiffeners under bending and shear includes three additional factors not included in the pure-bending and pure-shear models. These factors are (1) the reduction in the shear buckling stress of the web due to the presence of bending stress, (2) the influence of the in-plane bending stress on the value of the diagonal tension-field stress at failure, and (3) the reduction of the magnitude of the plastic modulus of the flanges resulting from the axial compressive and tensile strength.

The buckling stress reduction is handled by using the interaction equation

$$\left(\frac{\sigma_{bc}}{\sigma_{cp}}\right)^2 + \left(\frac{\tau_c}{\tau_{cr}}\right)^2 = 1 \tag{6.43}$$

to determine the critical shear stress τ_c under combined bending and shear. Note that Eq. 6.43 is for the case of a symmetric cross section, which is a special case of Eq. 6.41 with $C = -1$.

After the panel has buckled, the tension-field shear is computed by using a postbuckling shear force modified to include the effect of combined bending and shear. This involves modifying the tension-field stress σ_1 to include both bending and shear stresses in the yield criterion, and it also involves using an interaction equation to find the flange plastic moments m_f when axial stress resulting from bending acts in the flanges. The equations previously given for pure shear are then used with an iterative approach to determine the strength of the girder panel.

6.9 PLATE GIRDERS WITH LONGITUDINAL STIFFENERS

Longitudinal stiffeners (Fig. 6.6) can greatly increase the bending strength of plate girders. This additional strength can be attributed to control of the lateral deflection of the web, which increases the flexural strength of the web and also improves the bending resistance of the flange due to greater web restraint. Rockey and Leggett (1962) have determined that the optimum location for a longitudinal stiffener used to increase the flexural buckling resistance of a panel is 0.22 times the web depth from the compression flange if the web is assumed to be fixed at the flanges and simply supported at all four edges. Accordingly, 0.20 of the depth has been adopted nearly universally by design specifications as the accepted location for a longitudinal stiffener. Where a longitudinal stiffener at 0.20 of the web depth from the compression flange is provided, the value of k in the plate-buckling formula is increased from 23.9 to 129, which means that the elastic critical bending stress is more than five times as large as for a girder with no longitudinal stiffener. Since an unstiffened web of mild steel with a slenderness of about 170 develops yield-stress moment without buckling, a stiffened web can do the same with a slenderness of $170\sqrt{123.9/23.9}$, or about 400. Tests show that an adequately proportioned longitudinal stiffener at $0.2h$ from the compression flange eliminates the bend buckling loss in girders with web slendernesses as large as 450, and hence the ultimate moment as determined by compression-flange buckling strength is attained (Cooper, 1967). Girders with larger slenderness are likely to require two or more longitudinal stiffeners to eliminate the web bend buckling loss. Of course, the increase in bending strength of a longitudinally stiffened thin-web girder is usually small because the web contribution to bending strength is small. Longitudinal stiffeners, however, can be important in a girder subjected to repeated loads because they reduce or eliminate the transverse bending of the web, which increases resistance to fatigue cracking at the web-to-flange juncture and allows more slender webs to be used (Yen and Mueller, 1966).

The optimum location of a longitudinal stiffener that is used to increase resistance to shear buckling is at middepth. In this case, the two subpanels buckle simultaneously and the increase in critical stress can be substantial. For example, the plate-buckling coefficients for a square panel and the corresponding subpanel are increased to $k = 9.34$ and $k = 6.34$, respectively, and the slenderness ratio *h/t* of the square panel is twice that of the subpanel. Therefore, the elastic shear buckling stress for the subpanel is 2.7 times as large as for the square panel. Of course, the larger subpanel buckles first in a web with a longitudinal stiffener not at middepth and at a smaller critical stress than that for the stiffener at middepth.

The postbuckling shear strength of longitudinally stiffened girders has been evaluated in two ways: (i) Cooper (1967) assumes that each subpanel develops its own tension field after buckling and (ii) Porter et al. 1975 assume that only one tension field is developed between the flanges and transverse stiffeners regardless of whether longitudinal stiffeners are used.

If each subpanel develops its own tension field as suggested by Cooper, the tension-field shears for a girder stiffened as shown in Fig. 6.8 are

$$V_{p1} = \frac{\sigma_{t1} b_1 t}{2\sqrt{1+\alpha_1^2}} \qquad \text{and} \qquad V_{p2} = \frac{\sigma_{t2} b_2 t}{2\sqrt{1+\alpha_2^2}} \tag{6.44}$$

in which b_1 and b_2 are the depths of the subpanels, $\sigma_t 1$ and $\sigma_t 2$ the diagonal tension stresses in the subpanels, and α_1 and α_2 the aspect ratios a/b_1 and a/b_2, respectively. Cooper goes on to develop the total shear strength after employing a modified form of the von Mises yield condition to evaluate the diagonal tension stresses. Chern and Ostapenko (1971) extended Cooper's model to include frame action of the flanges and of the longitudinal stiffener. The model used by Rockey et al. (1974) was suggested after it was observed from tests that an overall tension field develops in the web.

Figure 6.9 shows a test girder at failure. The inclined arrows painted on the web show the predicted angle of the tension field and are in good agreement with the axes of the buckles in both subpanels. The predicted positions of the flange plastic hinges, also painted on the girder, are in good agreement with the actual

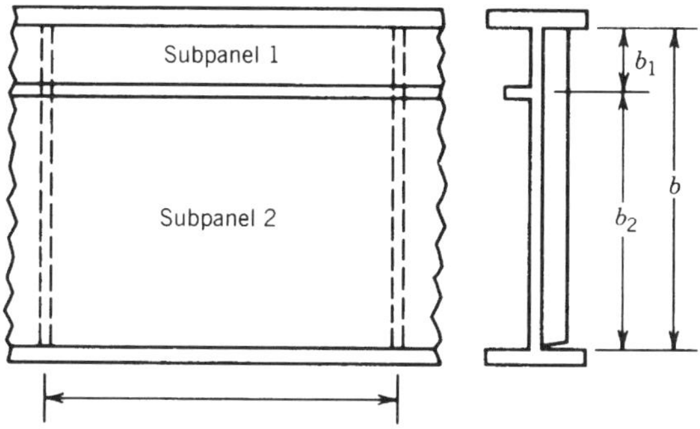

FIGURE 6.8 Typical longitudinally stiffened panel.

positions. In this specimen, the longitudinal stiffeners were located at $h/4$ from the compression flange in the right-hand panel. Although failure was in the right-hand panel, it is clear that failure in the other panel was imminent.

To obtain the tension-field shear resistance of a longitudinally stiffened girder, the same approach is used for an unstiffened girder, but in computing the tension-field stress associated with failure the critical beam-action shear corresponding to buckling of the largest subpanel is used in the yield criterion.

When longitudinal stiffeners are used in the panel, the shear strength under combined shear and bending is treated by Chern and Ostapenko (1971) and by

FIGURE 6.9 Longitudinally stiffened girder at ultimate load in shear (Rockey et al., 1974).

Rockey et al. (1974) in much the same way as if the panel were under pure shear. However, modifications to the buckling stresses, tension-field stresses, and plastic moments in the flanges to account for the combined stress effects are included. As noted previously, Chern and Ostapenko include both the contribution from the longitudinal stiffener in computing the frame-action strength and the modification of the stiffener's plastic moment capacity to account for the combined stress effects in this case.

Interaction equations to compute the critical buckling stress for each subpanel are presented by Chern and Ostapenko (1971). These equations are similar to Eq. 6.41. The subpanel tension-field stresses are then computed using the state of stress in each subpanel with the yield criterion. Finally, an iterative approach is used to find the shear strength.

Rockey et al. (1974) compute the buckling stress in the subpanels by an interaction equation approach applied to each subpanel stress condition as shown in Fig. 6.10. The state of stress in a subpanel is therefore a shear stress τ_c, a pure bending stress σ_{bc}, and an axial stress σ_{cc}. The buckling condition is defined by

$$\frac{\sigma_{cc}}{\sigma_{crc}} + \left(\frac{\sigma_{bc}}{\sigma_{cp}}\right)^2 + \left(\frac{\tau_c}{\tau_{cr}}\right)^2 = 1 \tag{6.45}$$

The postbuckling strength of the panel is determined assuming no longitudinal stiffeners are present.

Combined shear and bending of longitudinally stiffened girders with unequal flanges have been investigated by Chern and Ostapenko (1971). Their analysis for shear is based on their tension-field distribution for girders without longitudinal stiffeners with each subpanel treated independently as in Cooper's analysis and taking into account the plastic moments of the flanges and stiffener and the axial force reduction in the flange plastic moment. The girder strength is determined by shear until the shear strength curve intersects the curve for bending strength (Fig. 6.7*d*). The analysis involves some lengthy iterations, making it not suitable for hand computation.

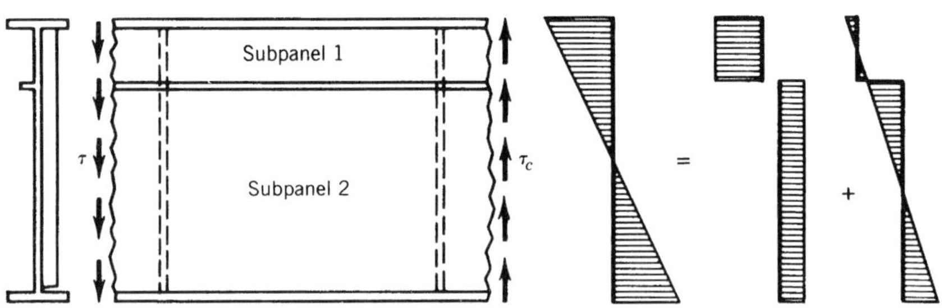

FIGURE 6.10 Stress distribution in panels of plate girder under shear and bending.

Of the three analyses discussed in this section, Cooper's is the most conservative and the easiest to use. The method by Rockey et al. requires trial and error to determine the optimum inclination of the tension field and, for each trial, an iteration to determine the axial force reduction in the flange plastic moment. Finally, Chern and Ostapenko's solution requires a computer to solve the significant number of iterative steps.

A large amount of experimental data is available from tests on transversely stiffened girders with and without longitudinal stiffeners. Many of these data can be found in the previously cited references.

The various theories for predicting the strength of plate girders in bending agree well with the experimental evidence. For example, Table 6.6 compares the theories of Cooper (1965) and Chern and Ostapenko (1970b) with experimental data for several girders with longitudinal stiffeners. As can be seen, there is a good correlation between theory and experimental results. For the nine bending test results compared with Cooper's theory, the average ratio of test ultimate load to predicted ultimate load is 0.99. For four bending tests, the Chern and Ostapenko ratio is 1.01.

Tables 6.7 through 6.9 give results of tests on longitudinally stiffened girders subjected to high shear. These results are compared with Cooper's (1967) theory in Table 6.7, Ostapenko and Chern's (1970) theory in Table 6.8, and the theory of Rockey et al. (1974) in Table 6.9. In Table 6.10, the aforementioned theories are compared for the series of tests reported by Cooper.

As can be seen, Cooper's theory conservatively estimates the shear strength consistent with his assumptions. The Chern–Ostapenko and Rockey et al. theories bracket the experimental data and are in good agreement with these tests. It is interesting to note that for the 15 tests reported, the average ratio of experimental to theoretical shears of Chern–Ostapenko equals 1.0. The standard deviation for these tests is 0.065. A similar comparison obtained using the theory of Rockey

TABLE 6.6 Bending Tests on Longitudinally Stiffened Girders[a]

				Experimental ÷ Theoretical	
Source of Data	Test Number	Aspect Ratio	Web Depth-to-Thickness Ratio	Cooper (1965)	Ostapenko and Chern (1970)
Massonnet (1962)	LB2	1.0	447	0.99	0.99
	LB3	1.0	447	1.00	1.00
	LB4	1.5	447	1.02	1.02
	LB5	0.75	447	1.02	1.02
	LB6	1.0	407	0.96	—
Dubas (1971)	D	0.6	299	1.00	—
	E	0.4	401	0.94	—
	3	0.6	300	1.02	—
	4	0.45	400	0.99	—

[a] All longitudinal stiffeners were located at one-fifth the web depth from the compression flange.

TABLE 6.7 Shear Test Results Compared with Cooper (1967) Theory

Test Number	Aspect Ratio	Web Depth-to-Thickness Ratio	Location of Longitudinal Stiffener from Compression Flange[a]	Experimental ÷ Theoretical Load
LS1-T2	1.0	256	0.33*b*	1.10
LS2-T1	1.0	275	0.33*b*	1.06
LS3-T1	1.5	276	0.33*b*	1.10
LS3-T2	1.5	276	0.33*b*	1.17
LS3-T3	0.75	276	0.33*b*	1.07
LS4-T1	1.0	260	0.2*b*	1.00
LS4-T2	1.0	260	0.5*b*	1.18

[a] b = web depth (Fig. 6.8).

TABLE 6.8 Comparison of Ostapenko and Chern (1970) Theory with Various Test Results

Source of Data	Test Number	Aspect Ratio	Web Depth-to-Thickness Ratio	Location of Longitudinal Stiffener from Compression Flange[a]	Experimental ÷ Theoretical Load
Cooper (1967)	LS1-T2	1.0	256	0.33*b*	1.00
	LS2-T1	1.0	275	0.33*b*	0.94
	LS3-T1	1.5	276	0.33*b*	1.01
	LS3-T2	1.5	276	0.33*b*	1.00
	LS3-T3	0.75	276	0.33*b*	0.93
	LS4-T1	1.0	260	0.2*b*	1.08
Škalound (1971)	UG5.1	1.77	400	0.26*b*	1.06
	UG5.2	1.15	400	0.26*b*	1.06
	UG5.3	1.46	400	0.26*b*	0.97
	UG5.4	1.77	264	0.26*b*	1.00
	UG5.5	0.83	264	0.26*b*	0.96
	UG5.6	1.77	264	0.26*b*	0.88
Porter et al. (1975)	F11-T1	1.39	365	0.2*b*	1.03
	F11-T2	1.20	365	0.2*b*	1.14
	F11-T3	1.00	365	0.2*b*	0.94

[a] b = web depth (Fig. 6.8).

TABLE 6.9 Comparison of Test Results with Theory of Rockey et al. (1974)

Girder Number	Aspect Ratio	Web Depth-to-Thickness Ratio	Location of Longitudinal Stiffener from Compression Flange[a]	Experimental ÷ Theoretical Load
SH1	1	387	0.50*b*	0.94
SH1	1	387	0.33*b*	0.99
SH2	1	387	0.20*b*	1.02
SH2	1	387	0.25*b*	1.02
SH2R	1	387	0.20*b*	0.93
SH2R	1	387	0.25*b*	0.93
SH4	1.33	400	0.25*b*	1.09
SH4	1.33	400	0.33*b*	1.07
SH5	1	370	0.25*b*	0.96
SH5	1	370	0.33*b*	0.98
SH7	1.67	295	0.33*b*	1.03
SH7	1.67	295	0.25*b*	1.03
SH8	1.67	295	0.50*b*	0.92
SH8	1.67	295	0.33*b*	1.05
SH9	1	375	0.50*b*	0.99
SH9	1	375	0.33*b*	1.04

[a] b = web depth (Fig. 6.8).

TABLE 6.10 Shear Strength of Longitudinally Stiffened Girders (Chern and Ostapenko, 1971; D'Apice et al., 1966)

Test Number (Cooper, 1967)	Aspect Ratio	Web Depth-to-Thickness Ratio	Location of Longitudinal Stiffener from Compression Flange[a]	Experimental ÷ Theoretical		
				Cooper (1967)	Rockey et al. (1974)	Ostapenko et al. (1971)
LS1-T2	1	256	0.33*b*	1.10	1.05	1.00
LS2-T1	1	275	0.33*b*	1.06	0.95	0.94
LS3-T1	1.5	276	0.33*b*	1.10	0.88	1.01
LS3-T2	1.5	276	0.33*b*	1.17	1.05	1.00
LS3-T3	0.75	276	0.33*b*	1.07	0.93	0.93
LS4-T1	1	260	0.20*b*	1.00	0.99	1.08
LS4-T2	1	260	0.50*b*	1.18	0.95	—

[a] b = web depth (Fig. 6.8).

et al. gives an average of 1.0 with a standard deviation of 0.051 for 16 tests. Additional work on the design of longitudinally stiffened plate girders is reported by Maquoi et al. (1983). The buckling of the webs of longitudinally stiffened girders with unsymmetric flanges is treated by Frank and Helwig (1995).

6.10 END PANELS

The tension field in a plate girder panel is resisted by the flanges and by the adjacent panels and transverse stiffeners. Because the panels adjacent to an interior panel are tension-field designed, they can be counted on to furnish the necessary support. An end panel, however, does not have such support and must be designed as a beam-shear panel unless the end stiffeners are designed to resist the bending effect of a tributary tension field. Basler (1963a) assumed that an end panel designed for beam shear can support a tension field in the adjacent interior panel, and this assumption has been generally accepted. This means that the end-panel stiffener spacing can be based on the shear buckling stress as discussed by Škaloud (1962). If the end panel is designed for tension-field action, an end post must be provided. A possible end post consists of the bearing stiffener and an end plate (Fig. 6.4*a*). According to Basler such an end post can be designed as a flexural member consisting of the stiffener, the end plate, and the portion of the web between, supported at the top and bottom flanges and subjected to the horizontal component of the tension field distributed uniformly over the depth. The required area A_e of the end plate, based on ultimate load considerations, is given by

$$A_e = \frac{(\tau - \tau_{cr})hA_w}{8e\sigma_y} \tag{6.46}$$

where e is the distance between the bearing stiffener and end plate (Fig. 6.4). The bearing stiffener itself is designed to support the end reaction. According to tests reported by Schueller and Ostapenko (1970), web shear may control the design of the end post.

6.11 DESIGN OF STIFFENERS

6.11.1 Transverse Stiffeners

Transverse stiffeners must be designed to preserve the straight boundaries that are assumed in computing the shear buckling of plate girder webs. Stein and Fralich (1950) developed a solution for this problem for an infinitely long web with simply supported edges and equally spaced stiffeners. Results were in fair agreement with tests on 20 specimens. Using their numerical data, Bleich developed a formula for the moment of inertia I of the stiffener that can be stated as

$$I = 2.5ht^3\left(\frac{h}{a} - 0.7\frac{a}{h}\right) \qquad a \le h \tag{6.47}$$

Using these variables, the AASHTO (2007) and AISC (2005a) formulas for I may be written as

$$I = Jat^3 \tag{6.48a}$$

in which

$$J = 2.5\left(\frac{h}{a}\right)^2 - 2 \geq 0.5 \tag{6.48b}$$

This is the same as Eq. 6.47 except that the coefficient of *a/h* in the second term in parentheses is 0.8 instead of 0.7.

In girders with longitudinal stiffeners, the transverse stiffener must also support the longitudinal stiffener as it forces a horizontal node in the bend buckling configuration of the web. According to an analysis by Cooper (1967), the required section modulus S_T of the transverse stiffener is given conservatively by $S_T = S_L h/a$, where S_L is the section modulus of the longitudinal stiffener. This requirement is reduced considerably in the AASHTO specification, where it appears as $S_T = S_L h/3a$.

Transverse stiffeners in girders that rely on a tension field must also be designed for their role in the development of the diagonal tension. In this situation they are compression members and therefore must be checked for local buckling. Furthermore, the transverse stiffeners must have cross-sectional area adequate to resist the axial force F that develops. The value of F_s in Basler's solution is

$$F_s = \tfrac{1}{2}\sigma_t at(1 - \cos\theta_d) \tag{6.49a}$$

Substituting the value of σ_t from Eq. 6.16 gives

$$F_s = \tfrac{1}{2}\sigma_{yw} at\left(1 - \frac{\tau_{cr}}{\tau_{yw}}\right)(1 - \cos\theta_d) \tag{6.49b}$$

The AISC formula for the cross-sectional area A_s of stiffeners symmetrical about the plane of the web is derived from Eq. 6.49b with $A_s = F_s/\sigma_{ys}$. If stiffeners are one-sided, A_s is multiplied by a factor to correct for eccentricity. The AASHTO and AISC formula, given here in the notation of this chapter, is a modification of Eq. 6.49b and is given by

$$A_s \geq \left[0.15Bht(1 - C)\frac{V}{V_u} - 18t^2\right]\frac{\sigma_{yw}}{\sigma_{ys}} \tag{6.50}$$

In Eq. 6.50, C replaces τ_{cr}/τ_y in Eq. 6.49b, B is a factor to correct for one-sided stiffeners, and the term $18t^2$is the area of a portion of the web assumed to act with the stiffener. The reduction in required area affected by V/V_u, where V is the design ultimate-load shear and V_u the shear strength of the panel, is also permitted by the AISC specification.

If longitudinally stiffened girders are assumed to develop independent tension fields in their subpanels, as Cooper assumes, then Eq. 6.50 can be used to determine the transverse stiffener area if h is taken to be the depth of the deeper subpanel. (The stiffener area requirement by this equation will always be larger for the deeper subpanel.) On the other hand, if a panel of a longitudinally stiffened girder develops the same tension field as it would without the longitudinal stiffener, as the work of Rockey et al. appears to show, then transverse-stiffener area requirements are the same for both.

In addition to the effects of axial force from the tension field, British Standard BS 5400: Part 3 (BSI, 1982) accounts for the destabilizing forces arising from the action of stresses in the plane of the girder web on the initial displacements of the transverse stiffeners. This destabilizing effect is assessed by assuming that adjacent transverse stiffeners have initial displacements that alternate on each side of the girder web in a "sawtooth" progression. Horne (1980) has shown that the magnification of displacements and the consequential bending stresses in the stiffeners by the longitudinal and shear stresses in the plate panels are equivalent to the effect produced by an imaginary axial force in the stiffeners. An approximate formula for a strut with an initial imperfection of *L*/750, where *L* is the strut length, is utilized for the design of the transverse stiffeners.

6.11.2 Longitudinal Stiffeners

A longitudinal stiffener must be stiff enough to maintain a node in a buckled web and must resist axial compression because of its location in the compression zone of the web. Therefore, both moment of inertia and cross-sectional area must be considered when determining its size. There have been a number of analytical investigations of this problem. Results are usually expressed in terms of three parameters, including (i) $\gamma^* = EI/hD$, where I is the moment of inertia of the stiffener and $D = Et^3/12(1 - \nu^2)$; (ii) $\delta = A_s/ht$, in which A_s is the area of the stiffener; and (iii) the panel aspect ratio a/h. Empirical formulas involving these parameters for various positions of the stiffener have been developed from numerical data, and charts (Klöppel and Scheer, 1960) are available.

The AASHTO (2004, 2007) formula for the moment of inertia of a longitudinal stiffener at $h/5$ from the compression flange is

$$I_s = ht^3(2.4\alpha^2 - 0.13) \tag{6.51}$$

Although Eq. 6.51 does not contain δ, it is in fairly good agreement with the values of γ^* given by Dubas (1948) for $\delta = 0.10$ in the range $0.5 \leq \alpha \leq 1.6$, and because γ^* decreases with a decrease in δ, it is a reasonable upper bound for girders of practical proportions. It is important to note that these values of γ^* are derived from linear buckling analysis and hence give a stiffness that guarantees only the critical load.

In calculating γ^*, a portion of the web should be considered as part of the cross section. Strain measurements reported by Massonnet (1962) showed a mean effective width of $20t$. Tests show that the theoretical values must be increased considerably to develop the ultimate strength. Massonnet found that they should be multiplied by a factor ranging from 3 for a longitudinal stiffener at middepth to 7 for one at $h/5$ from the compression flange. This conclusion has been verified by a number of other investigators, including Rockey (1971a,b) and Dubas (1971).

Further extensions were proposed by Maquoi et al. (1983). It is apparent, however, that there is a considerable range in the empirical multiplication factors. Difficulties may arise in practice in deciding on an appropriate factor to be used for a particular combination of applied stress and stiffener geometry.

An alternative approach to the design of a longitudinal web stiffener is to treat it as a column. This approach has been adopted in BS 5400: Part 3 (BSI, 1982). Axial loading of the strut includes the longitudinal stress due to girder bending moment plus an axial load equivalent to the destabilizing effect of shear stress and an in-plane stress transverse to the stiffener. For the design of longitudinal stiffeners, an effective width of web plate of $16t$ on each side of the stiffener is considered.

6.12 PANELS UNDER EDGE LOADING

6.12.1 Concentric Edge Loads with Respect to the Web Plane

Girders often support loads on the top flange which produce compression on the edge of the web. Such loads may be distributed over large distances, in some cases the length of the girder, or over relatively small distances, in which case they can be taken as concentrated loads. Bearing stiffeners may be needed for concentrated loads, but in some cases the web can carry them unaided.

A number of studies of buckling of plate elements under edge loading have been made (Girkmann, 1936; Zetlin, 1955; Wilkesmann, 1960; Klöppel and Wagemann, 1964; Warkenthin, 1965; Kawana and Yamakoshi, 1965; Bossert and Ostapenko, 1967; Ostapenko et al., 1968; Rockey and Bagchi, 1970; Khan and Walker, 1972; Khan et al., 1977). The ultimate strength of a web under edge loading may exceed the buckling load by a considerable margin, and several experimental investigations to determine ultimate strengths have been made.

Bossert and Ostapenko (1967) and Ostapenko et al. (1968) used a finite difference analysis to determine the web-buckling stress for a panel subjected to bending and an edge load distributed uniformly over the panel. The web was assumed to be fixed at the flanges and simply supported at the stiffeners. Ten tests were made on three plate girder specimens with aspect ratios varying from 0.8 to 1.6 and ratios of extreme-fiber bending stress to edge stress from 0 to 5. The resulting ultimate strengths are shown in Fig. 6.11. The horizontal line through the ordinate 3 gives a lower bound on the experimental strengths. The following formula for the allowable edge stress $\sigma_{e(all)}$ gives a factor of safety of 2 on the lower-bound ultimate edge stress:

$$\sigma_{e(\text{all})} = \frac{1}{\sqrt{a/h}} \frac{24,000}{(h/t)^2} \frac{K}{\sqrt{\sigma_y}} \tag{6.52}$$

where K is a coefficient given in Fig. 6.12.

Rockey et al. (1972), Elgaaly and Rockey (1973), Elgaaly (1975, 1977), and Bagchi and Rockey (1975) used the finite element method to determine the buckling

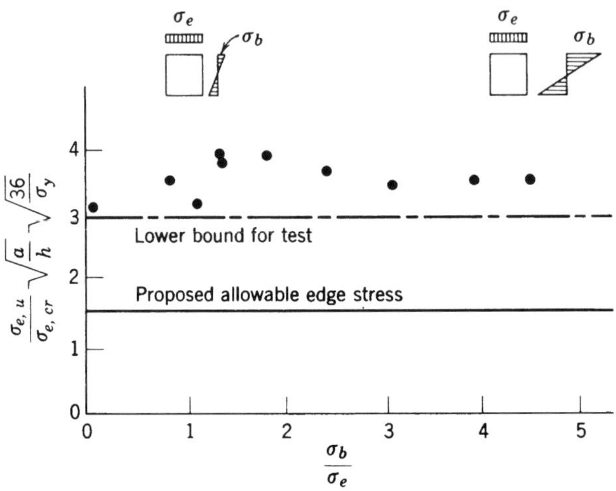

FIGURE 6.11 Web plates under edge loading.

load of a plate simply supported on all four edges and subjected to a partial edge loading (patch loading). Results were also obtained for the cases of a panel with patch load and shear and patch load and moment. Values of the coefficient k in the plate-buckling formula (Eq. 4.1) were determined.

Results of patch-load tests on 20 trough-section beams of 15 ft span are reported by Elgaaly and Rockey (1973). The objective of their study was to determine the interaction between edge loading and pure in-plane bending. The ratio c/a was 0.2 for all tests. It was found that the moment did not significantly reduce patch-loading strength until it exceeded 50% of the bending strength of the panel.

Elgaaly (1975) conducted tests on 18 trough-section beams of about 2 ft span, to determine the interaction between the ultimate strength under edge loading and shear. The ratio c/a was 0.2 for all tests and the panel slenderness ratio was 325 for 12 specimens and 200 for the remaining 6. The results from the tests show that the presence of shear will reduce the ultimate load-carrying capacity of the web under edge loading, and an approximate relationship for this reduction was established. It was also shown that the postbuckling strength increases with an increase in the panel slenderness ratio. Furthermore, panels subjected to the combination of edge loading and shear exhibit a higher ultimate strength-to-buckling strength ratio than panels subjected to edge loading only.

The effect of flange thickness on web capacity under direct in-plane loading was also examined (Elgaaly, 1977) and results of patch-load tests on five welded girders are reported. The web dimensions were kept identical in all five girders (aspect ratio 1 and slenderness ratio 250). The c/a ratio was 0.2 for all tests. The results from these tests are given in Fig. 6.13.

A theoretical study (Roberts and Rockey, 1979) was carried out to determine the influence of a longitudinal stiffener on the ultimate capacity of a concentrated edge load. The study was restricted to the consideration of the buckling of a square-plate

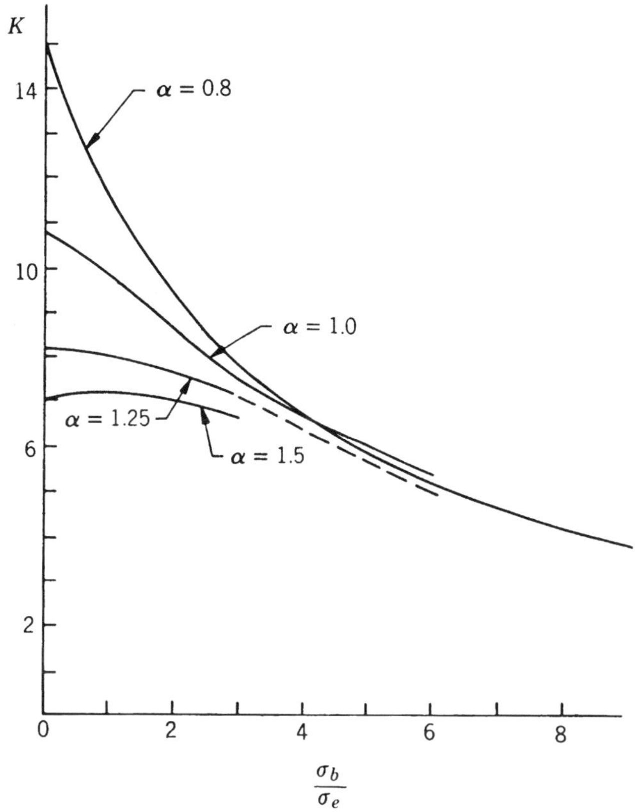

FIGURE 6.12 Values of K in Eq. 6.52.

panel reinforced by a single longitudinal stiffener at the one-fifth-depth position. It was shown that the presence of the stiffener significantly increased the ultimate edge load. Values of buckling coefficients relating buckling resistance of the plate and the flexural rigidity of the longitudinal stiffener are provided.

The research described above is applicable to either flangeless panels or beams with very thin flanges. For cases of plate girders and I-beams with reasonably thick flanges, it can be expected that the flanges will make a significant contribution to resisting local loading. On the basis of a preliminary series of tests, Granholm (1960) and Bergfelt and Hövik (1968) concluded that web thickness was the most important parameter affecting the ultimate concentrated edge load and proposed an empirical formula $P_u = 8.5t_w^2$ for mild steel (P_u is obtained in kilonewtons if t_w is given in millimeters). These tests had thin flanges and the failure mode was local yielding of the webs under the concentrated load. Further tests by Bergfelt (1976) on specimens with higher ratios of flange-to-web thickness showed that the failure load is also influenced by the bending stiffness of the flange and the corresponding failure mode included a combination of yielding and buckling of the web. The ultimate load for the "weak"-flange case in which web yielding controls may be predicted by

$$P_u = 13\eta t_{fi} t_w \sigma_y \tag{6.53}$$

where t_{fi} is an equivalent flange thickness given by $t_{fi} = t_f(b/25t_f)^{1/4}$ to account for b/t values other than 25 and η is a parameter dependent on t_{fi}/t_w with:

t_{fi}/t_w	0.5	1.0	1.5	2.0
η	0.55	0.65	0.85	2.00

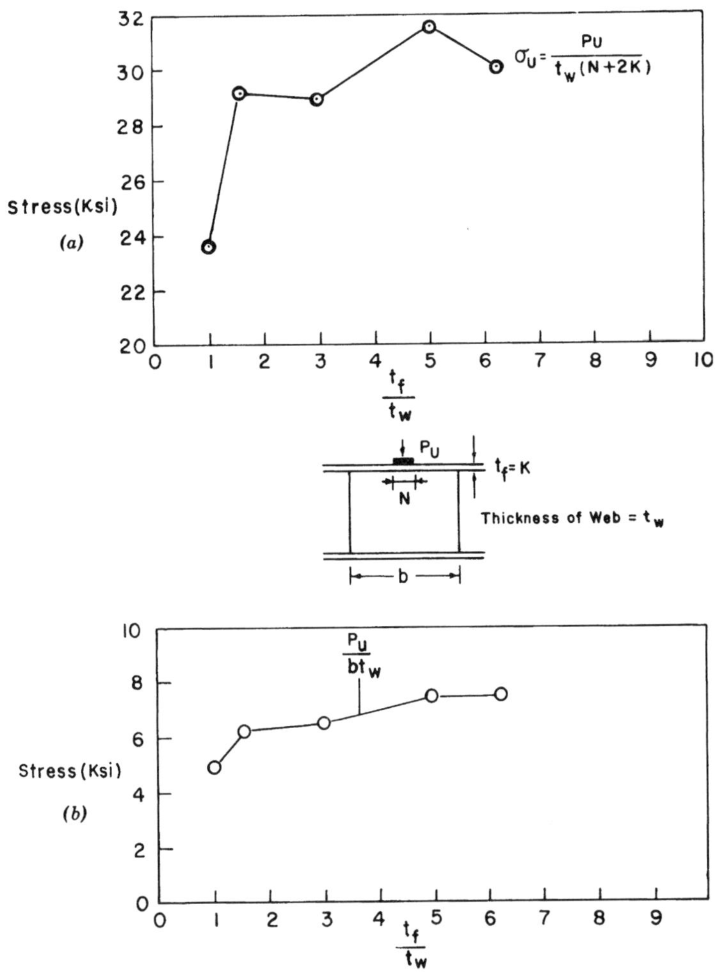

FIGURE 6.13 Test results (Elgaaly, 1977).

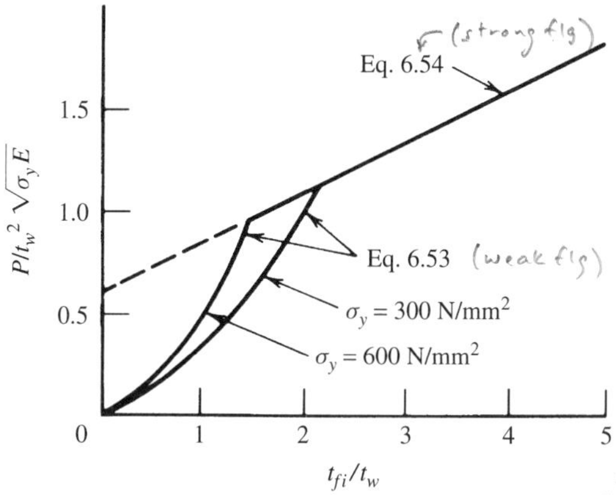

FIGURE 6.14 Ultimate concentrated loads from Eqs. 6.41 and 6.42.

The theoretical basis for Eq. 6.53 is the treatment of the flange as a beam on an elastic foundation with plasticity accounted for by adjustments to the spring constants. Use of the η values, as tabulated above, limits the ultimate load to a value associated with the start of rapidly increasing vertical deformation of the flange. For the "strong"-flange case the following empirical formula shows good correlation with test results:

$$P_u = 0.6t_w^2(E\sigma_y)^{1/2}\left(1 + \frac{0.4t_{fi}}{t_w}\right) \tag{6.54}$$

The ultimate concentrated loads from Eqs. 6.53 and 6.54 are shown in Fig. 6.14. When $t_{fi}/t_w = 2$ for mild steel, both equations predict a failure load in close agreement with Granholm's original formula. Bergfelt's formulas are applicable when the load is concentrated as a sharp edge load. When the load is distributed along a web length c, the ultimate load given by Eqs. 6.53 and 6.54 should be multiplied by a factor $f(c)$ given by

$$f(c) = \frac{\gamma}{1 - e^{-\gamma}\cos\gamma} \leq 1.3 \tag{6.55}$$

where $\gamma = c/2L$ and $L = 6.7t_f$.

The limitation of 1.3 on the factor $f(c)$ is imposed to reflect the observation that when the load is distributed along the web, buckling will occur before yielding and thus the distribution length has a small influence on the ultimate load. For combined bending moment and concentrated load it was observed that a reasonable lower bound to test results is obtained when the value of P_u from Eqs. 6.53 and 6.54

is multiplied by the following correction factor proposed by Djubek and Škaloud (1976):

$$\lambda = \left[1 - \left(\frac{\sigma_b}{\sigma_y}\right)^2\right]^{1/2} \tag{6.56}$$

Roberts and Rockey (1979) developed a method, which is based on the upper bound theorem of plastic collapse, for predicting the ultimate value of a concentrated edge load. The assumed failure mechanism, involving plastic hinges in the flange and web, is shown in Fig. 6.15*a*. Equating external and internal work gives an expression from which the ultimate load P_u may be determined if the web hinge location α is known. The latter is chosen empirically to agree with test results. The solution presented by Roberts and Rockey (1979) contains an error in the expression which was pointed out and corrected by Chatterjee (1980). The mechanism solution overestimates the experimental collapse load when the loaded length c becomes too large. Consequently, upper limits are placed on c so that the correlation with test data is satisfactory. For stocky webs failure may be initiated by direct yielding of the web resulting in the vertical descent of the applied load and a failure mechanism of the type shown in Fig. 6.15*b*. For this type of failure, the restrictions concerning the loaded length c do not apply. The actual value of the collapse load should be taken as the smaller of either the bending-type failure or the direct-compression failure of the web. For the tested girders, rotation of the flange was prevented by the loading arrangement, which is common to most tests using concentrated loads. Hence, the theoretical work in general is applicable to the case of flanges restrained against rotation. The mechanism solution was extended by Roberts and Chong (1981) to a loading condition in which the edge load is uniformly distributed between vertical stiffeners, as indicated in Fig. 6.15*c*.

To account for the interaction between coexisting global bending and local bending stresses, a correction factor of the type given in Eq. 6.56 could be applied to the values of P_u predicted by the mechanism solution. The provisions for patch loading contained in BS 5400; Part 3 (BSI, 1982) are based on a modified version of the mechanism solution of Roberts and Rockey.

Herzog (1974b,c) analyzed the results of the 72 tests reported by Bergfelt and Hövik (1968), Bergfelt (1971), and Škaloud and Novak (1972) and developed the empirical formula

$$P_u = 1430t^2\left[1.2 + 1.25\frac{I_f h}{I_w t}\left(1 + \frac{c}{h}\right)^2\left(0.85 + 0.01\frac{a}{h}\right)\right]\left[1 - \left(\frac{\sigma_b}{\sigma_y}\right)^2\right]^{1/8} \tag{6.57}$$

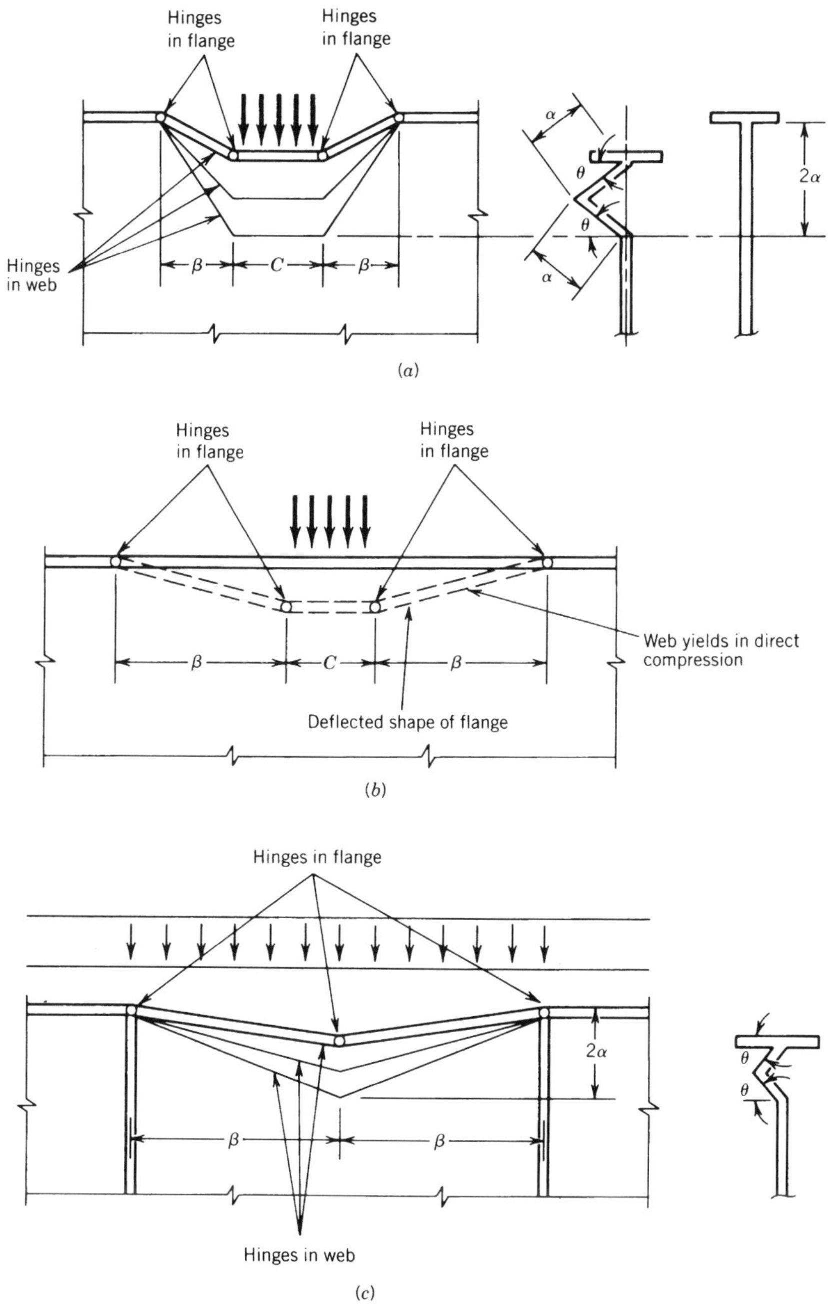

FIGURE 6.15 Failure mechanism models.

The mean value of the ratios of test load to predicted load by this formula was 1.001 with a standard deviation of 0.141.

Later, Aribert et al. (1981) investigated the effect of nearly concentrated edge loads on webs with an emphasis on concentrated direct loads on the web of a wide-flange column resulting from the end moment of a connected beam. The studies on crippling of plate girder webs subjected to patch or distributed edge loading have been summarized by Elgaaly (1983), and design recommendations to prevent web crippling were suggested.

Further tests and finite element analyses were performed by Elgaaly et al. (1991b, 1992) to supplement the data available. Results of recent investigations in Europe and Japan (Shimizu et al., 1989a,b; Drdacky, 1991; Höglund, 1991; Raoul et al., 1991; Spinassas et al., 1991) also add to the state of knowledge in this field. The formula to predict the ultimate capacity of webs when subjected to in-plane patch loading, developed by Roberts (1981), properly includes the effects of all the parameters that have influence on the capacity. This formula was adopted by AISC (1989, 1993) to determine the nominal strength of the web against crippling:

$$P_u = Kt_w^2\left[1 + 3\left(\frac{N}{d}\right)\left(\frac{t_w}{t_f}\right)^{1.5}\right]\left(\frac{F_{yw}t_f}{t_w}\right)^{0.5} \tag{6.58}$$

in which P_u is the ultimate capacity, K a given constant, t_w the web thickness, N the width of patch, d the beam or web depth, t_f the loaded flange thickness, and F_{yw} the yield stress of web material.

The constant K was taken to be 135 for loads applied at a distance not less than $d/2$ from the end of the member and 68 if the load is applied at a distance less than $d/2$ from the end. The choice of the distance $d/2$ appears to be arbitrary and creates a discontinuity. Furthermore, it was noted that for the case when the load is at a distance $0.5N$ from the end of the beam and N is less than d, the use of K equal to 68 is conservative. Some of this conservatism was eliminated in the AISC LRFD specification (AISC, 1993), based on Elgaaly et al.'s 1991b research results.

When a longitudinal stiffener is provided to increase the capacity of the web under in-plane bending, the stiffener is usually placed at a distance equal to 0.2 to 0.25 times the web depth from the compression flange. The effect of the presence of such a stiffener on the behavior of the web under in-plane compressive edge loads was investigated by Elgaaly and Salkar (1992), Shimizu et al. 1991, Bergfelt (1983), and Kutmanova et al. 1991. A general conclusion made from these studies is that the effect of such a stiffener on the ultimate capacity of the web, under in-plane compressive edge loading, is generally negligible. This is different from the conclusion reached by Rockey based on his theoretical studies on the effect of the longitudinal stiffener on the elastic web buckling under compressive edge loading, which was discussed earlier in this chapter.

Research work by Herzog, Skaloud, and Elgaaly reported by Elgaaly (1983) indicated that the following formula can be used to determine a reduction factor R

which should be multiplied by the ultimate capacity of the web under edge loading to include the effect of the presence of global in-plane bending,

$$R = [1 - (f_b/F_y)^3]^{1/3} \tag{6.59}$$

where f_b is the actual global bending stress. Research work by Elgaaly (1975) indicated that the presence of global shear will reduce the ultimate capacity of the web under discrete edge compressive loads. This was based on the results from 18 tests conducted on slender webs (d/t_w = 325 and 200) with N/d equal to 0.2. General recommendations to estimate the reduction in the crippling strength due to the presence of global shear cannot be made due to the limited results that are available. More research on the effect of the presence of global bending and/or shear on the ultimate capacity of the web under edge compressive loads is needed. The AISC specification does not address the effects of the presence of global bending and/or shear on the crippling load capacity of the web.

6.12.2 Eccentric Edge Loads with Respect to the Web Plane

Eccentricities with respect to the plane of the web are unavoidable in practice, and hence their impact on web strength must be examined. Research work was conducted by Elgaaly and Nunan (1989), Elgaaly and Salkar (1991), and Elgaaly et al. (1989, 1991b 1992) to study these effects. The maximum eccentricity that was considered equaled one-sixth of the flange width. The study includes both experimental and analytical work. From the test results, it was found that there is no reduction in the ultimate capacity due to the eccentricity when the load was applied through a thick patch plate placed eccentrically with respect to the plane of the web; in contrast, reductions did occur when the load was applied through a cylindrical roller, as shown in Fig. 6.16. To quantify the reduction in the strength due to the eccentricities, several rolled and built-up sections were tested, and finite element models of several beams were analyzed. As shown in Fig. 6.17, the reduction was found to be a function of t_f/t_w and varies linearly with e/b_f, where e and b_f are the eccentricity and the flange width, respectively.

A reduction factor R can be calculated using the formula

$$R = m\frac{e}{b_f} + c \tag{6.60}$$

where

$$R \leq 1.0$$

$$m = -0.45\left(\frac{t_f}{t_w}\right)^2 + 4.55\left(\frac{t_f}{t_w}\right) - 12.75$$

$$c = 1.15 - 0.025\left(\frac{t_f}{t_w}\right)$$

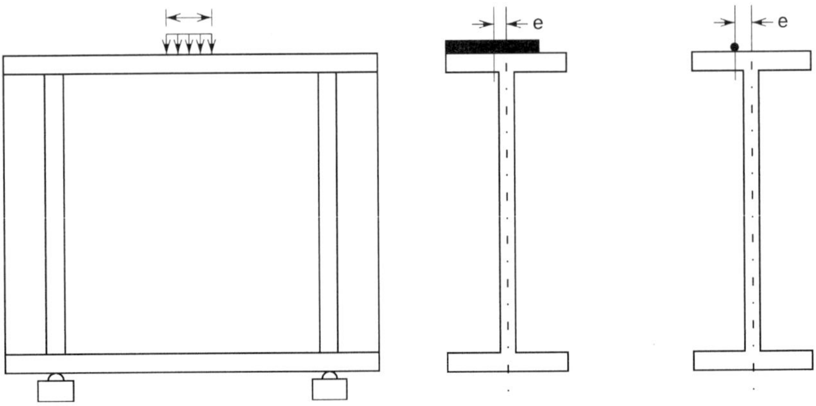

FIGURE 6.16 Eccentric loads with respect to the plane of the web.

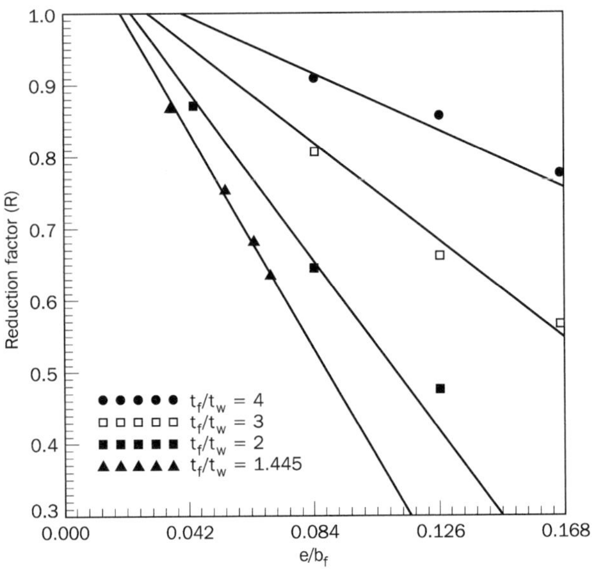

FIGURE 6.17 Reduction in ultimate capacity due to load eccentricity.

The linear equations for the reduction factor R and the corresponding test results are shown in Fig. 6.17. Because the reduction factor equation is empirical, it should be used only within the range of the test data (experimental and analytical models); namely, f_f/t_w equal to 1 to 4 and e/b_f less than $\frac{1}{6}$.

6.12.3 Bearing Stiffeners for Edge-Loaded Panels

It appears that no research work was published on the behavior of stiffened webs under concentrated compressive loads. The AISC specification requires that stiffeners to prevent web crippling should be double sided and extend at least one-half of the web depth. According to the specification, stiffeners under concentrated compressive loads should be designed as axially loaded columns with an effective length equal to 0.75 of the web depth. Furthermore, the cross section of such a column should include the pair of stiffeners and a strip of the web having a width equal to 25 times the web thickness.

Tests and finite element analysis were conducted to study the behavior of the stiffened webs under concentrated edge loads (Elgaaly et al., 1992; Elgaaly and Eash, 1993). The load was applied on the top flange using a roller, a patch plate, or a W-shape, as shown in Fig. 6.18. For the roller and W-shape tests, it was found that the behavior of the web and the failure loads were almost identical. In some tests, the load was applied with a small eccentricity e_1 with respect to the vertical axis of the stiffener. When the load was applied through a patch plate, the eccentricity had negligible effect on the results. The eccentricity, however, had an effect on the failure load and the web behavior when the load was applied through a roller.

In the case of deep stiffeners ($d_s/d > 0.75$), failure was due to the buckling of the stiffener (mostly global). It appears that the effective width of the web, which acts together with the stiffener, varies along the depth of the stiffener, and it is larger near the load. The reduction in the failure load due to an eccentricity of 0.5 in. was found to be about 15%. In the case of the shallow stiffeners ($d_s/d < 0.5$),

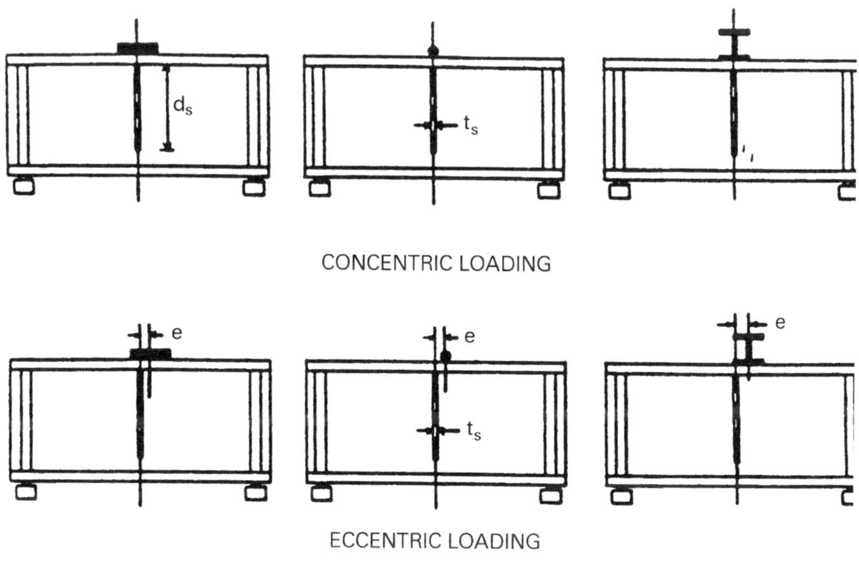

FIGURE 6.18 Method of load application.

failure was due to local buckling of the stiffener, buckling of the web below the stiffener, or a combination thereof. The reduction in the failure load due to a 0.5-in. eccentricity was found to be about 6 to 14% for thin to thick stiffeners, respectively.

The requirements in the AISC specification regarding the design of end stiffeners subjected to concentrated loads are similar to the requirements for the design of an intermediate stiffener. The only difference is in the strip width of the web, which is to be considered effective with the stiffener. The width of this strip in the case of end stiffeners is only 12 times the web thickness, instead of 25 times the web thickness in the case of intermediate stiffeners.

A total of 31 tests were conducted to examine the behavior of the stiffened web over the support (Elgaaly et al., 1992; Elgaaly and Eash, 1993). In a few tests, the support was eccentric with respect to the centerline of the stiffener. In Fig. 6.19, the failure loads are plotted versus the depth of the stiffener relative to the beam depth d_s/d for stiffener thickness t_s equal to $\frac{3}{16}$, $\frac{1}{4}$, and $\frac{5}{16}$ in. As can be observed from the figure, the ultimate capacity of the stiffened web reached a maximum value at a stiffener depth d_s equal to $0.75d$, and this maximum value was independent of the stiffener thickness. It should be noted that this ultimate capacity corresponds to the web panel shear capacity.

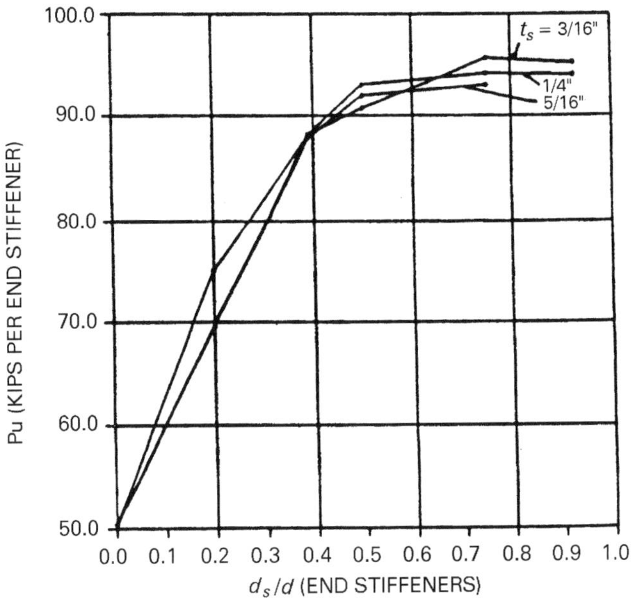

FIGURE 6.19 Web with end-stiffener ultimate capacity as a function of $d_s = d$ *and* t_s.

6.13 FATIGUE

Investigations have shown that the ultimate strength of thin-web plate girders under static load is not affected by initial out-of-flatness of the web (Shelestanko et al., 1970). On the other hand, fatigue cracks may develop at the web-to-flange juncture due to a lateral bending of the web under repeated loads (Maeda, 1971; Patterson et al., 1970). The magnitude of the initial deflection of the web and the extent to which the repeated stress exceeds the buckling stress appear to be the principal factors influencing the development of these cracks (Patterson et al., 1970; Parsanejad and Ostapenko, 1970). The factors are functions of the web slenderness and the panel aspect ratio.

According to the AASHTO specifications for load factor design, webs without longitudinal stiffeners must have slenderness that satisfy the formula

$$\frac{h}{t} \leq \frac{36,500}{\sqrt{F_{y,\mathrm{psi}}}} \tag{6.61}$$

If the girder has unequal flanges, h in this equation is replaced by $2y_0$, where y_0 is the portion of the web in compression. This limit may be somewhat conservative (Chern and Ostapenko, 1970a).

6.14 DESIGN PRINCIPLES AND PHILOSOPHIES

Despite the general recognition that classical buckling theory is an inadequate guide to the prediction of the strength of plate panels in shear or compression, it must also be realized that comprehensive ultimate-strength models are not yet fully developed. A situation exists, therefore, where some countries have opted to make use of ultimate-strength theory when justified, while others prefer to rely in varying degrees on the classical linear theory of plate buckling. Consequently, current design criteria for plate girders vary widely depending on the philosophy of design. It is interesting to compare the design provisions for plate girders adopted in various countries.

The current AISC specification and the allowable stress and load factor design methods of the AASHTO specification recognize the contribution of tension-field action. Design for shear is then based on simplified expressions of Basler's formulas. AASHTO allows the use of a single longitudinal stiffener located at one-fifth of the depth from the compression flange, its principal function being the control of lateral web deflections. This reduction of out-of-plane deflections both improves the fatigue life of the web and permits the use of thinner web plates. The existing AASHTO provisions for transverse and longitudinal stiffeners prescribe minimum rigidities based on elastic buckling theory. These provisions are intended for moderate-span plate girder bridges and are not adequate for long-span plate girders requiring very deep webs. For such cases, the problems requiring consideration include (1) the design of deep webs with multiple longitudinal stiffeners,

(2) the development of new design methods for stiffener proportioning, (3) the establishment of a procedure to account for axial load shedding from web to compression flange, and (4) the development of a general ultimate-strength method for transversely and longitudinally stiffened webs. Some of these problems have been investigated in a study carried out under the sponsorship of the Federal Highway Administration (FHWA, 1980).

In British Standard BS 5400: Part 3 (BSI, 1982) girders are classified into two categories: those without longitudinal web stiffeners and those with such stiffeners. For the former, experimental data were thought to be sufficiently complete to justify the use of tension-field theory and an adaptation of the Rockey model is used for the ultimate shear strength. For the latter, it was thought that ultimate-strength criteria had not been sufficiently established for longitudinal web (or box-girder flange) stiffeners to cope with the associated high shear deformations. Therefore, the design of webs in girders containing longitudinal stiffeners is based on web panel behavior that stops short of the tension-field mechanism. Individual plate panels are subject to checks for yielding and buckling. The buckling check uses coefficients based on a large-deflection, elastic–plastic numerical study of initially imperfect plates under various stress patterns (Harding and Hobbs, 1979). Stiffener design is based on a strength criterion that attempts to assess the function of the stiffeners in controlling the stability of the plate panels.

6.15 GIRDERS WITH CORRUGATED WEBS

Elgaaly and Hamilton (1993) performed 42 tests on 21 beams. Four different corrugation configurations (Table 6.11) and two thicknesses were used. All of the beams failed due to buckling of the web. The test results are given in Tables 6.12 and 6.13, where τ_e is the experimental shear stress at failure. Failure in all the specimens except those made of dense corrugation was initiated by local buckling of one of the corrugation fields, which subsequently propagated to other fields. The failure was sudden and the load-carrying capacity drops at the failure load and the specimen exhibits a residual strength of about 75% of its ultimate capacity. In the case of dense corrugation, the failure was due to global buckling of the panel as an orthotropic plate.

The ultimate shear stress τ_f from finite element analysis (Elgaaly et al., 1995) and the test results τ_e are given in Tables 6.12 and 6.13, where τ_y is the shear yield stress of the material. The tables indicate satisfactory agreement between the analytical and experimental results, with an average τ_f/τ_e ratio of 1.15. The primary reason that the analytical results are higher than the experimental results could be the presence of unavoidable out-of-plane initial imperfections in the test specimens.

TABLE 6.11 Corrugation Configurations for Test Specimens

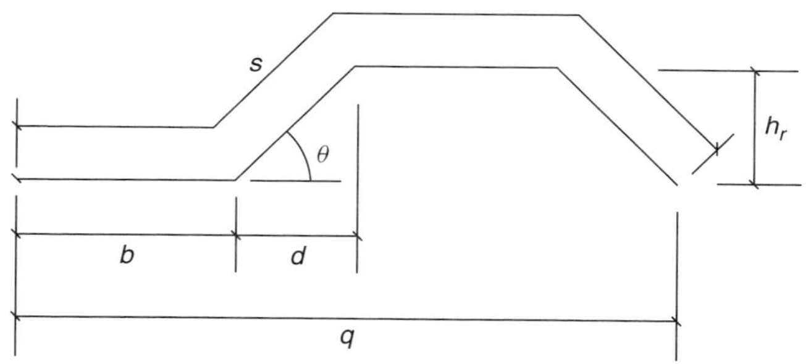

Panel Type	b [in. (mm)]	d [in. (mm)]	h_r [in. (mm)]	θ (deg)	s [in. (mm)]	q [in. (mm)]
UFS	0.78	0.47	0.56	50	3.0	2.5
	(19.8)	(11.9)	(14.2)		(76.2)	(63.5)
UF1X	1.50	1.00	1.00	45	5.83	5.0
	(38.1)	(25.4)	(25.4)		(148.1)	(127.0)
UFX-36	1.65	0.92	1.31	55	6.5	5.14
	(41.9)	(23.4)	(33.3)		(165.1)	(130.6)
UF2X	1.96	1.04	2.00	62.5	8.43	6.00
	(49.8)	(26.4)	(50.8)		(214.1)	(152.4)

TABLE 6.12 Stresses Based on Local Buckling Versus Stresses from Finite Element Analysis[a]

Panel	τ_f (ksi)	τ_e (ksi)	τ_e/τ_f	τ_y (ksi)	τ_{ssf} (ksi)	τ_f/τ_{ssf}	τ_{fx} (ksi)	τ_f/τ_{fx}	τ_f/τ_a
V121216A	44.92	37.35	0.83	56.58	41.02	1.10	54.87[b]	0.82	0.94
V121216B	52.71	54.54	1.04	55.71	51.57[b]	1.03	65.29[b]	0.95[c]	0.99
V121221A	39.95	34.95	0.88	55.71	33.23	1.20	48.96[b]	0.82	0.97
V121221B	52.72	43.96	0.83	55.71	47.95[b]	1.10	61.00[b]	0.95[c]	1.02
V121232A	31.68	30.59	0.97	55.71	18.76	1.69	30.16	1.05	1.30
V121232B	43.97	37.32	0.85	53.69	27.84	1.58	43.85[b]	1.00	1.23
V121832A	24.11	25.63	1.06	58.89	18.76	1.29	30.16	0.80	0.99
V121832B	43.26	27.62	0.64	47.05	38.17[b]	1.13	48.40[b]	0.92[c]	1.02
V122421A	35.18	30.48	0.87	51.96	38.23	0.92	50.71[b]	0.69	0.79
V122421B	50.49	37.23	0.74	53.41	46.80[b]	1.08	59.53[b]	0.95[c]	1.01
V122432A	27.31	23.15	0.85	59.76	18.76	1.46	30.16	0.91	1.12
V122432B	33.01	29.96	0.91	53.12	27.66	1.19	43.47[b]	0.76	0.93
V181216B	40.12	48.61	1.21	51.78	36.98	1.09	50.06[b]	0.80	0.92
V181216C	55.46	49.91	0.90	56.82	51.07[b]	1.09	65.33[b]	0.98[c]	1.03
V181221A	34.79	32.18	0.93	48.39	30.65	1.14	44.01[b]	0.79	0.93
V181221B	47.57	40.74	0.86	50.74	44.09[b]	1.08	56.33[b]	0.94[c]	1.00
V181232A	23.57	27.42	1.16	46.19	15.98	1.48	25.92	0.91	1.13
V181232B	33.33	33.90	1.02	50.44	25.18	1.32	40.60[b]	0.82	1.01
V181816A	41.69	37.33	0.90	49.53	39.88[b]	1.05	51.00[b]	0.84[c]	0.93
V181816B	51.44	41.42	0.81	51.40	47.12[b]	1.09	60.26[b]	1.00[c]	1.04
V181821A	35.13	28.22	0.80	46.19	33.26	1.06	44.79[b]	0.78	0.90
V181821B	45.08	40.23	0.89	49.91	47.27[b]	1.07	54.01[b]	0.90[c]	0.98
V181832A	27.69	27.55	1.00	57.74	16.67	1.66	27.03	1.02	1.27
V181832B	31.68	33.30	1.05	48.57	25.18	1.26	39.84[b]	0.80	0.98
V241216A	40.83	28.32	0.69	49.53	39.73[b]	1.03	50.95[b]	0.82[c]	0.92
V241216B	49.36	40.30	0.82	49.23	49.11[b]	1.01	62.99[b]	1.00[c]	1.00
V241221A	36.09	30.16	0.84	51.05	30.41	1.19	45.15[b]	0.80	0.96
V241221B	46.69	39.57	0.85	53.48	45.08[b]	1.04	57.77[b]	0.87[c]	0.95
V241232A	26.87	26.41	0.98	56.38	17.19	1.56	28.05	0.96	1.19
V241232B	33.53	31.68	0.95	48.93	25.78	1.30	40.58[b]	0.83	1.01

[a] $\tau_a = 0.5(\tau_{ssf} + \tau_{fx})$, 1 ksi = 6.895 MPa.
[b] Inelastic buckling.

TABLE 6.13 Stresses Based on Global Buckling Using Orthotropic Plate Theory Versus Stresses from Finite Element Analysis[a]

Panel	τ_f (ksi)	τ_e (ksi)	τ_e/τ_f	τ_y (ksi)	$\tau_{cr,e}$ (ksi)	$\tau_{cr,i}$ (ksi)	τ_{cr} (ksi)	τ_f/τ_{cr}
V121809A	48.59	42.57	0.88	47.92	129.71	70.52	47.92	1.01
V121809C	54.72	41.50	0.76	56.00	122.76	74.16	56.00	0.98
V122409A	44.34	38.55	0.87	49.07	130.42	71.55	49.07	0.90
V122409C	56.87	41.51	0.73	51.96	125.69	72.28	51.96	1.10
V181209A	57.93	45.96	0.79	57.74	51.29	48.67	48.67	1.19
V181209C	56.78	46.20	0.81	49.58	53.58	46.10	46.10	1.23
V181809A	51.46	42.82	0.83	51.77	53.58	47.11	47.11	1.09
V181809C	45.87	39.57	0.86	46.79	54.12	45.01	45.01	1.02
V241209A	32.48	27.06	0.83	50.74	30.46	—	30.46	1.07
V241209C	32.28	29.72	0.89	51.96	30.74	—	30.74	1.08

[a] τ_{cr} = minimum (τ_y, $\tau_{cr,e}$, $\tau_{cr,i}$), 1 ksi = 6.895 MPa.

In the local buckling mode, the corrugated web acts as a series of flat-plate subpanels that mutually support each other along their vertical (longer) edges and are supported by the flanges at their horizontal (shorter) edges. These flat-plate subpanels are subjected to shear, with the elastic buckling stress given by

$$\tau_{cr,e} = k_s \frac{\pi^2 E}{12(1-\nu^2)(w/t)^2} \tag{6.62}$$

where k_s is the buckling coefficient, which is a function of the panel aspect ratio (h/w) and the boundary support conditions, h the web depth, t the web thickness, w the flat-plate subpanel width (the horizontal or the inclined, whichever is larger), E the modulus of elasticity, and ν Poisson's ratio. The buckling coefficient k_s depends on the assumed edge support conditions and is given by:

Case 1: Longer edges simply supported and shorter edges fixed (clamped)

$$k = 5.34 + 2.31\left(\frac{w}{h}\right) - 3.44\left(\frac{w}{h}\right)^2 + 8.39\left(\frac{w}{h}\right)^3 \tag{6.63}$$

Case 2: All edges fixed (clamped)

$$k = 8.98 + 5.6\left(\frac{w}{h}\right)^2 \tag{6.64}$$

In situations when $\tau_{cr,e} > 0.8\tau_y$, inelastic buckling will occur and the inelastic buckling stress $\tau_{cr,i}$ can be calculated by $\tau_{cr,i} = (0.8\tau_{cr,e}\tau_y)^{0.5} \leq \tau_y$.

The local buckling stresses were calculated for the test specimens with the UF1X, UFX-36, and UF2X corrugation configurations, assuming the above two boundary conditions: for Case 1 the critical stress is denoted as τ_{ssf} and for case 2 as τ_{fx}. The calculated stresses are given in Table 6.12 and are compared with the corresponding stresses from the finite element analysis τ_f and the test results τ_e. An average local buckling stress $t_a = 0.5(\tau_{ssf} + \tau_{fx})$ was also calculated for all specimens and is included in the table. Based on these data, the average local buckling stresses τ_a agree reasonably well with the corresponding stresses from the finite element analysis τ_f ; the average ratio τ_f/τ_a was calculated to be 1.015 for all the values given in Table 6.12.

When global buckling controls, the buckling stress can be calculated for the entire corrugated web panel using orthotropic plate buckling theory. The global elastic buckling stress $\tau_{cr,e}$ can be calculated from

$$\tau_{cr,e} = \frac{k_s[(D_x)^{0.25}(D_y)^{0.75}]}{th^2} \tag{6.65}$$

where $D_x = (q/s)Et^3/12$
$D_y = EI_y/9$
$I_y = 2bt(h_r/2)^2 + [t(h_r)^3/6 \sin\theta]$
$k_s =$ buckling coefficient, equals 31.6 for simply supported boundaries and 59.2 for clamped boundaries
$t =$ corrugated plate thickness

and b, h_r, q, s, and θ are as shown in the figure portion of Table 6.11. When $\tau_{cr,e}$ calculated by Eq. 6.65 exceeds $0.8\tau_y$, inelastic buckling will control with inelastic buckling stress $\tau_{cr,i}$ given by $\tau_{cr,i} = (0.8\tau_{cr,e}\tau_y)^{0.5} \leq \tau_y$.

The global buckling stresses for the test specimens made of the UFS corrugation were calculated using the buckling formula for the orthotropic plate and a buckling coefficient k_s equal to 59. These results, the experimental results, and the results from the finite element analysis are given in Table 6.13. The average value of the ratio τ_f/τ_{cr} between the finite element analysis results and those obtained from the orthotropic plate theory is 1.067.

Beams and girders with corrugated webs are economical to use and can improve the aesthetics of the structure. Typical beams manufactured and used in Germany for buildings have webs with height-to-thickness ratios ranging from 150 to 260. The corrugated webs of two bridges built in France were 8 mm thick and the web height-to-thickness ratio was in the range 220 to 375. The failure modes for such slender webs are local and/or global buckling.

For corrugated webs subjected to shear, the strength can be calculated, with very good accuracy, using the buckling stress formulas for flat isotropic or orthotropic plates. When the corrugation is coarse, the capacity of the panel will be controlled by local buckling of the flat segments of the corrugation and, as the corrugation becomes dense, global buckling of the whole panel as an orthotropic plate controls. For practical applications it is recommended that local and global buckling values be calculated and the smaller value controls.

6.16 RESEARCH NEEDS

The following is a brief summary of the research needs related to plate girders.

Combined shear, moment, and axial force Plate girders in cable-stayed girder bridges, self-anchored suspension bridges, arches, and rigid frames are subjected to axial compression or tension in addition to shear and moment. Some of the analyses discussed in this chapter can be extended to cover this situation, but there is little experimental research on the ultimate strength of plate girder panels under combined shear, moment, and axial force. Consideration of multiple longitudinal stiffeners would also be an important aspect of such research.

Longitudinal stiffeners Design procedures for longitudinally stiffened girders should be studied with a view to simplifying the ultimate-strength analyses that show good correlation with test results. These investigations should cover multiple longitudinal stiffeners.

Stiffeners Current methods for analysis and design of stiffeners are probably adequate, but further investigations would be worthwhile.

Panels with variable depth More work is required to develop general design procedures for the ultimate strength of panels with variable depth.

Webs with holes The buckling strength of a thin square plate with a central circular hole clamped at its edges and subjected to pure shear was determined analytically and experimentally by Grosskurth et al. (1976). Stability of webs containing circular or rectangular middepth holes was investigated by Redwood and Venoya (1979). The postbuckling strength of webs with holes needs to be investigated. Narayanan and Der-Avanessian (1985) cite seven references on this topic and refer to over 70 tests performed at the University of Cardiff.

Fatigue More knowledge of the fatigue behavior of slender web plates is needed to take full advantage of ultimate-strength design of plate girder bridges and crane girders, particularly those with corrugated webs.

Corrugated webs Vertically corrugated webs may be an economical alternative to thin webs. There is little to no theoretical or experimental information on the ultimate strength of such panels under patch loading.

Composite girders The assumptions on which the analyses for shear and bending of plate girder panels are based can be expected to apply to composite girders. The large stiffness and low tensile strength of a concrete slab, compared to a steel flange, may result in some differences in the panel behavior. Additional tests on thin-webbed composite girders are needed.

REFERENCES

AASHTO (2004, 2007), *LRFD Bridge Design Specification*, American Association of State Highway and Transportation Officials, Washington DC.

AISC (1989), *Specification for Structural Steel Buildings — Allowable Stress Design and Plastic Design*, American Institute of Steel Construction, Chicago, IL.

AISC (1993), *Load and Resistance Factor Design Specification for Structural Steel Buildings*, American Institute of Steel Construction, Chicago, IL.

AISC (2005a), *Specification for Structural Steel Buildings*, American Institute of Steel Construction, Chicago, IL.

AISC (2005b), *Seismic Provisions for Structural Steel Buildings*, ANSI/AISC 341-05, American Institute of Steel Construction, Chicago, IL.

AREMA (2002), *Manual for Railway Engineering*, American Railway Engineering and Maintenance-of-Way Association, Washington DC.

Aribert, J. M., Lachal, A., and Elnawawy, O. (1981). "Modelisation Elasto-plastique de la Résistance d'un Profile en Compression Locale," *Constr. Met.*, Vol. 18 No. 2, pp. 3–26.

Astaneh-Asl, A. and Zhao, Q. (2001), "Cyclic Tests of Steel Shear Walls," Rep. No. UCB/CE-Steel-01/01, Department of Civil and Environmental Engineering, University of California, Berkeley.

Bagchi, D. K., and Rockey, K. C. (1975), "Postbuckling Behaviour of Web Plate Under Partial Edge Loading," *Proc. 3rd Int. Spec. Cold-Formed Steel Struct.*, University of Missouri– Rolla, Rolla, MO, Nov. 24–25, Vol. 1, pp. 351–384.

Basler, K. (1963a), "Strength of Plate Girders in Shear," *Trans. ASCE*, Vol. 128, Part II, p. 683 .

Basler, K. (1963b), "Strength of Plate Girders Under Combined Bending and Shear," *Trans. ASCE*, Vol. 128, Part II, p. 720 .

Basler, K., and Thürlimann, B. (1960a), "Carrying Capacity of Plate Girders," *IABSE 6th Congr., Prelim, Publ.*, Vol. 16.

Basler, K., and Thürlimann, B. (1960b), "Buckling Tests on Plate Girders," *IABSE 6th Congr., Prelim. Publ.*, Vol. 17.

Basler, K., and Thürlimann, B. (1963), "Strength of Plate Girders in Bending," *Trans. ASCE*, Vol. 128, Part II, p. 655 .

Basler, K., Yen, B. T., Mueller, J. A., and Thürlimann, B. (1960), "Web Buckling Tests on Welded Plate Girders," *Weld. Res. Counc. Bull.*, No. 64, Sept.

Behbahanifard, M. R., Grondin. G. Y., and Elwi, A. E. (2003), "Experimental and Numerical Investigations of Steel Plate Shear Walls," Struct. Eng. Rep. No. 254, Department of Civil and Environmental Engineering, University of Alberta, Edmonton, Alberta, Canada.

Bergfelt, A. (1971), "Studies and Tests on Slender Plate Girders without Stiffeners," paper presented at the IABSE Colloq. Des. Plate Box Girders Ultimate Strength, London.

Bergfelt, A. (1976), "The Behaviour and Design of Slender Webs Under Partial Edge Loading," paper presented at the Int. Conf. Steel Plated Struct., Imperial College, London.

Bergfelt, A. (1983), "Girder Web Stiffening for Patch Loading," Publ. S83:1, Chalmers University of Technology, Goteborg, Sweden.

Bergfelt, A., and Hövik, J. (1968), "Thin-Walled Deep Plate Girders Under Static Load," paper presented at the IABSE 8th Congr., Final Rep., New York, 1968.

Berman, J. W., and Bruneau, M. (2003), "Experimental Investigation of Light-Gauge Steel Plate Shear Walls for the Seismic Retrofit of Buildings," Tech. Rep. MCEER-03-0001, Multidisciplinary Center for Earthquake Engineering Research, Buffalo, NY.

Berman, J. W., Lowes, L. N., Okazaki, T., Bruneau, M., Tsai, K.-C., Driver, R. G., and Sabelli, R. (2008), "Research Needs and Future Directions for Steel Plate Shear Walls," *Proc., ASCE Structures Congress*, Apr., Vancouver, Canada.

Bhowmick, A. K., Driver, R. G., and Grondin, G. Y. (2009a), "Inelastic Seismic Response of Steel Plate Shear Walls Including P-delta Effects," *Proc., Structural Stability Research Council Annual Stability Conference*, Apr., Phoenix, AZ.

Bhowmick, A. K., Driver, R. G., and Grondin, G. Y. (2009b), "Seismic Analysis of Steel Plate Shear Walls Considering Strain Rate and P-delta Effects," *J. Constr. Steel Res.*, Vol. 65, pp. 1149–1159.

Bossert, T. W., and Ostapenko, A. (1967), "Buckling and Ultimate Loads for Plate Girder Web Plates Under Edge Loading," Fritz Eng. Lab. Rep. No. 319.1, Lehigh University, Bethlehem, PA, June.

BSI (1982), *Steel, Concrete and Composite Bridges*, British Standard BS 5400: Part 3, *Code of Practice for Design of Steel Bridges*, British Standards Institution, London.

Carskaddan, P. S. (1968), "Shear Buckling of Unstiffened Hybrid Beams," *ASCE J. Struct. Div.*, Vol. 94, No. ST8, pp. 1965–1992.

Cescotto, S., Maquoi, R., and Massonnet, C. (1981), "Sur Ordonnateur du Comportement a la Ruine des Putres a Âme Pleine Cisailles on Flèches," *Constr. Met.*, No. 2, pp. 27–40.

Chatterjee, S. (1980), "Design of Webs and Stiffeners in Plate and Box Girders," paper presented at the Int. Conf. Des. Steel Bridges, Cardiff, Wales.

Chern, C., and Ostapenko, A. (1969), "Ultimate Strength of Plate Girder Under Shear," Fritz Eng. Lab. Rep. No. 328.7, Lehigh University, Bethlehem, PA, Aug.

Chern, C., and Ostapenko, A. (1970a), "Bending Strength of Unsymmetrical Plate Girders," Fritz Eng. Lab. Rep. No. 328.8, Lehigh University, Bethlehem, PA, Sept.

Chern, C., and Ostapenko, A. (1970b), "Unsymmetrical Plate Girders Under Shear and Moment," Fritz Eng. Lab. Rep. No. 328.9, Lehigh University, Bethlehem, PA, Oct.

Chern, C., and Ostapenko, A. (1971), "Strength of Longitudinally Stiffened Plate Girders Under Combined Loads," paper presented at the IABSE Colloq. Des. Plate Box Girders Ultimate Strength, London.

Clark, J. W., and Sharp, M. L. (1971), "Limit Design of Aluminum Shear Web," paper presented at the IABSE Colloq. Des. Plate Box Girders Ultimate Strength, London.

Combault, J. (1988), "The Maupre Viaduct near Charolles, France," *Proc. AISC Eng. Conf.*, pp. 12.1–12.22.

Cooper, P. B. (1965), "Bending and Shear Strength of Longitudinally Stiffened Plate Girders," Fritz Eng. Lab. Rep. No. 304.6, Lehigh University, Bethlehem, PA, Sept.

Cooper, P. B. (1967), "Strength of Longitudinally Stiffened Plate Girders," *ASCE J. Struct. Div.*, Vol. 93, No. ST2, pp. 419–452.

Cooper, P. B. (1971), "The Ultimate Bending Moment for Plate Girders," paper presented at the IABSE Colloq. Des. Plate Box Girders Ultimate Strength, London.

CSA. (2009) *Limit States Design of Steel Structures*, CAN/CSA-S16-09, Canadian Standards Association, Mississauga, ON.

D'Apice, M. A., Fielding, D. J., and Cooper, P. B. (1966), "Static Tests on Longitudinally Stiffened Plate Girders," *Weld. Res. Counc. Bull.*, No. 117, Oct.

Dastfan, M., Deng, X., and Driver, R. G. (2008), "Large-Scale Tests of Steel Plate Shear Walls with PEC Columns," *Proc., North American Steel Construction Conference*, Apr., Nashville, TN.

Dastfan, M., and Driver, R. G. (2008), "Flexural Stiffness Limits for Frame Members of Steel Plate Shear Walls Systems," *Proc., Structural Stability Research Council Annual Stability Conference*, Apr., Nashville, TN.

Dastfan, M., and Driver, R. G. (2009), "Investigations on the Effect of Frame Member Connection Rigidity on the Behavior of Steel Plate Shear Walls Systems," *Proc., Behavior of Steel Structures in Seismic Areas (STESSA)*, Aug., Philadelphia, PA.

Davies, G., and Mandal, S. N. (1979), "The Collapse Behaviour of Tapered Plate Girders Loaded Within the Tip," *Proc. Inst. Civ. Eng.*, Vol. 67, Part 2.

Deng, X., Dastfan, M., and Driver, R. G. (2008), "Behaviour of Steel Plate Shear Walls with Composite Columns," *Proc., ASCE Structures Congress*, Apr., Vancouver, Canada.

Deng, X., and Driver, R. G. (2007), "Steel Plate Shear Walls Fabricated with Partially Encased Composite Columns", *Proc., Structural Stability Research Council Annual Stability Conference*, Apr., New Orleans, LA.

Djubek, J., and Škaloud, M. (1976), "Postbuckled Behaviour of Web Plates in the New Edition of Czechoslovak Design Specifications," paper presented at the Int. Conf. Steel Plated Struct., Imperial College, London.

Drdacky, M. (1991), "Non-Stiffened Steel Webs with Flanges Under Patch Loading," in *Contact Loading and Local Effects in Thin-Walled Plated and Shell Structures: IUTAM 1990 Symposium in Prague*, Springer-Verlag, Berlin.

Driver, R. G. (2008), "Development of Partially Encased Composite Columns for Use in Steel Shear Walls for Seismic Applications," *Proc., Seismic Eng. Intl. Conf. Commemorating the 1908 Messina and Reggio Calabria Earthquake (MERCEA)*, July, Reggio Calabria, Italy.

Driver, R. G., Grondin, G. Y., Behbahanifard, M., and Hussain, M. A. (2001), "Recent Developments and Future Directions in Steel Plate Shear Wall Research," *Proc., North American Steel Construction Conference*, May, Ft. Lauderdale, FL.

Driver, R. G., Kulak, G. L., Kennedy, D. J. L., and Elwi, A. E. (1997), "Seismic Behaviour of Steel Plate Shear Walls," Struct. Eng. Rep. No. 215, Department of Civil and Environmental Engineering, University of Alberta, Edmonton, Alberta, Canada.

Driver, R. G., Kulak, G. L., Kennedy, D. J. L., and Elwi, A. E. (1998a), "Cyclic Test of a Four-Story Steel Plate Shear Wall," ASCE *J. Struct. Eng.*, Vol. 124, No. 2, pp. 112–120.

Driver, R. G., Kulak, G. L., Elwi, A. E., and Kennedy, D. J. L. (1998b), "FE and Simplified Models of a Steel Plate Shear Wall," ASCE *J. Struct. Eng.*, Vol. 124, No. 2, pp. 121–130.

Dubas, C. (1948), "A Contribution to the Buckling of Stiffened Plates," paper presented at the IABSE 3rd Congr., Prelim. Publ., Liege, Belgium.

Dubas, P. (1971), "Tests on Postcritical Behaviour of Stiffened Box Girders," paper presented at the IABSE Colloq. Des. Plate Box Girders Ultimate Strength, London.

European Convention for Constructional Steelwork.

Elgaaly, M. A. (1973), "Buckling of Tapered Plates Under Pure Shear," *Proc. Conf. Steel Struct.*, Timisaora, Romania, Oct.

Elgaaly, M. (1975), "Failure of Thin-Walled Members Under Patch Loading and Shear," *Proc. 3rd Int. Spec. Conf. Cold-Formed Steel Struct.*, University of Missouri–Rolla, Rolla, MO, Nov. 24–25, Vol. 1, pp. 357–381.

Elgaaly, M. (1977), "Effect of Flange Thickness on Web Capacity Under Direct in-Plane Loading," *Proc. SSRC Annu. Tech. Session*, May.

Elgaaly, M. (1983), "Web Design Under Compressive Edge Loads," *AISC Eng. J.*, Vol. 20, No. 4, pp. 153–171.

Elgaaly, M., Caccese, V., and Chen, R. (1991a), "Cyclic Behavior of Unstiffened Thin Steel Plate Shear Walls," report submitted to the National Science Foundation, University of Maine, Orono, ME.

Elgaaly, M., Caccese, V., and Du, C. (1993b), "Postbuckling Behavior of Steel Plate Shear Walls Under Cyclic Loads," *ASCE J. Struct. Eng.*, Vol. 119, No. 2, pp. 588–605.

Elgaaly, M., Caccese, V., and Martin, D. (1993a), "Experimental Investigation of the Behavior of Bolted Thin Steel Plate Shear Walls," report submitted to the National Science Foundation, University of Maine, Orono, ME.

Elgaaly, M., and Dagher, H. (1990), "Beams and Girders with Corrugated Webs," *Proc. SSRC Annu. Tech. Session*, Lehigh University, Bethlehem, PA, pp. 37–53.

Elgaaly, M., and Eash, M. (1993) "Crippling of Beams over End Supports," report submitted to the National Science Foundation.

Elgaaly, M., and Hamilton, R. (1993), "Behavior of Welded Girders with Corrugated Webs," report submitted to the National Science Foundation, Civil Engineering Department, University of Maine, Orono, ME.

Elgaaly, M., and Nunan, W. (1989), "Behavior of Rolled Section Web Under Eccentric Edge Compressive Loads," *ASCE J. Struct. Eng.*, Vol. 115, No. 7, pp. 1561–1578.

Elgaaly, M., and Salkar, R. (1991a), "Behavior of Webs Under Eccentric Compressive Edge Loads," in *Contact Loading and Local Effects in Thin-Walled Plated and Shell Structures: IUTAM 1990 Symposium in Prague*, Springer-Verlag, Berlin.

Elgaaly, M., and Salkar, R. (1991b), "Web Crippling Under Edge Loading," *Proc. 1991 AISC Nat. Steel Constr. Conf.*

Elgaaly, M., and Salkar, R. (1992), "Web Behavior and Ultimate Capacity Under Inplane Compressive Loads" report submitted to the National Science Foundation.

Elgaaly, M., Salkar, R., and Du, C. (1991b), "Analytical Study of Web Crippling Under Edge Compressive Loads," *Proc. 5th Int. Conf. Comput. Exp. Meas.*, Montreal, Quebec, Canada, Elsevier, New York.

Elgaaly, M., Seshadri, A., and Hamilton, R. W. (1995), "Beams with Corrugated Webs, Research to Practice," *Proc. Conf. Research Transformed into Practice Implementation of NSF Research*, ASCE Press, New York.

Elgaaly, M., Smith, D., and Hamilton, R. (1992), "Beams and Girders with Corrugated Webs," *Proc. 1992 NSF Struct. Geomech. Build. Syst. Grantees' Conf.*, San Juan, Puerto Rico, pp. 126–128.

Elgaaly, M., Sturgis, J., and Nunan, W. (1989), "Stability of Plates Under Eccentric Edge Loads," *Proc. ASCE Struct. Congr.*

Elgaaly, M. A., and Rockey, K. C. (1973), "Ultimate Strength of Thin-Walled Members Under Patch Loading and Bending," *Proc. 2nd Spec. Conf. Cold-Formed Steel Struct.*, University of Missouri–Rolla, Rolla, MO, Oct.

Falby, W. E., and Lee, G. C. (1976), "Tension-Field Design of Tapered Webs," *Am. Inst. Steel Constr.*, Vol. 13, No. 1., p. 11 .

Federal Highway Administration (FHWA) (1980), "Proposed Design Specifications for Steel Box Girder Bridges," Rep. No. FHWA-TS-80-205, U.S. Department of Transportation, FHWA, Washington, DC.

Frank, K. H., and Helwig, T. A. (1995), "Buckling of Webs in Unsymmetric Plate Girders," *AISC Eng. J.*, Vol. 32, No. 2, pp. 43–53.

Fujii, T. (1968a), "On an Improved Theory for Dr. Basler's Theory," paper presented at the IABSE 8th Congr., Final Rep., New York.

Fujii, T. (1968b), "On Ultimate Strength of Plate Girders," *Jpn. Shipbuild. Mar. Eng.*, May, Vol. 3, No. 3, pp. 7–16.

Fujii, T. (1971), "A Comparison between Theoretical Values and Experimental Results for the Ultimate Shear Strength of Plate Girders" paper presented at the IABSE Colloq. Des. Plate Box Girders Ultimate Strength, London.

Gaylord, E. H. (1963), "Discussion of K. Basler 'Strength of Plate Girders in Shear,'" *Trans. ASCE*, Vol. 128, Part II, p. 712 .

Gaylord, E. H., and Gaylord, C. N. (1972), *Design of Steel Structures*, 2nd ed., McGraw-Hill, New York.

Girkmann, K. (1936), "Die Stabilität der Stegbleche vollwandiger Träger bei Berücksichtigung örtlicher Lastangriffe," paper presented at the IABSE 3rd Congr., Final Rep., Berlin.

Granholm, C. A. (1960), "Tests on Girders with Thin Web Plates," *Rapp. 202*, Institutionen for Byggnadsteknik, Chalmers Tekniska Hogskola, Goteborg, Sweden (in Swedish).

Grosskurth, J. F., White, R. N., and Gallagher, R. H. (1976), "Shear Buckling of Square Perforated Plates," *ASCE J. Eng. Mech. Div.*, Vol. 102, No. EM6, pp. 1025–1040.

Harding, J. E., and Hobbs, R. E. (1979), "The Ultimate Load Behaviour of Box Girder Web Panels," *Struct. Eng.*, Vol. 57B, No. 3, pp. 49–54.

Herzog, M. (1973), "Die Traglast versteifter, dünnwändiger Blechträger unter reiner Biegung nach Versuchen," *Bauingenieur*, Sept., Vol. 48, No. 9, pp. 317–322.

Herzog, M. (1974a) "Die Traglast unversteifter und versteifter, dünnwändiger Blechträger unter reinem Schub und Schub mit Biegung nach Versuchen," *Bauingenieur*, Oct., Vol. 49, No. 10, pp. 387–389.

Herzog, M. (1974b), "Ultimate Strength of Plate Girders from Tests," *ASCE J. Struct. Div.*, Vol. 100, No. ST5, pp. 849–864.

Herzog, M. (1974c), "Die Krüppellast sehr dünner Vollwandträgerstege nach Versuchen," *Stahlbau*, Vol. 43, pp. 26–28.

Höglund, T. (1971a), "Behavior and Load-Carrying Capacity of Thin-Plate I Girders," *R. Inst. Technol. Bull.*, No. 93, Stockholm (in Swedish).

Höglund, T. (1971b), "Simply Supported Thin Plate I-Girders without Web Stiffeners Subjected to Distributed Transverse Load," paper presented at the IABSE Colloq. Des. Plate Box Girders Ultimate Strength, London.

Höglund, T. (1973), "Design of Thin-Plate I Girders in Shear and Bending," *R. Inst. Technol. Bull.*, No. 94, Stockholm.

Höglund, T. (1991), "Local Buckling of Steel Bridge Girder Webs During Launching," in *Contact Loading and Local Effects in Thin-Walled Plated and Shell Structures: IUTAM 1990 Symposium in Prague*, Springer-Verlag, Berlin.

Horne, M. R. (1980), "Basic Concepts in the Design of Webs," paper presented at the Int. Conf. Des. Steel Bridges, Cardiff, Wales.

Jetteur, P. (1984), "Contribution a la Solution de Problème Particuliers d'Instabilité dans la Grand Poutres Metallique," Colloq. Publ. Fac. Sci. Appl. Liege No. 94.

Kawana, K., and Yamakoshi, M. (1965), "On the Buckling of Simply Supported Rectangular Plates Under Uniform Compression and Bending," *J. Soc. Nav. Archit. West Jpn.*, Vol. 29.

Khan, M. Z., Johns, K. C., and Hayman, B. (1977), "Buckling of Plates with Partially Loaded Edges," *ASCE J. Struct. Div.*, Vol. 103, No. ST3, pp. 547–558.

Khan, M. Z., and Walker, A. C. (1972), "Buckling of Plates Subjected to Localized Edge Loading," *Struct. Eng.*, Vol. 50, No. 6, pp. 225–232.

Klöppel, K., and Scheer, J. S. (1960), *Beulwerte ausgesteifter Rechteckplatten*, Wilhelm Ernst, Berlin.

Klöppel, K., and Wagemann, C. H. (1964), "Beulen eines Bleches unter einseitiger Gleichstreckenlast," *Stahlbau*, Vol. 33, pp. 216–220.

Komatsu, S. (1971), "Ultimate Strength of Stiffened Plate Girders Subjected to Shear," paper presented at the IABSE Colloq. Des. Plate Box Girders Ultimate Strength, London.

Konishi, I., et al. (1965), "Theories and Experiments on the Load Carrying Capacity of Plate Girders," *Rep. Res. Comm. Bridges Steel Frames Weld. Kansai Dist. Jpn.*, July (in Japanese).

Kuhn, P. (1956), *Stresses in Aircraft and Shell Structures*, McGraw-Hill, New York.

Kuhn, P., Peterson, J. P., and Levin, L. R. (1952), "A Summary of Diagonal Tension, Part 1: Methods of Analysis," Tech. Note 2661, National Advisory Committee on Aeronautics, Washington, DC.

Kutmanova, I., Škaloud, M., Janus, K., and Lowitova, O. (1991), "Ultimate Load Behavior of Longitudinally Stiffened Steel Webs Subject to Partial Edge Loading," in *Contact Loading and Local Effects in Thin-Walled Plated and Shell Structures: IUTAM 1990 Symposium in Prague*, Springer-Verlag, Berlin.

Lew, H. S., and Toprac, A. A. (1968), "Static Strength of Hybrid Plate Girders," SFRL Tech. Rep., University of Texas, p. 550–11, Jan.

Lubell, A. S., Prion, H. G. L., Ventura, C. E., and Rezai, M. (2000), "Unstiffened Steel Plate Shear Wall Performance Under Cyclic Loading," *ASCE J. Struct. Eng.*, Vol. 126, No. 4, pp. 453–460.

Machimdamrong, C., Watanabe, E., and Utsunomiya, T. (2004), "Shear Buckling of Corrugated Plates with Edges Elastically Restrained Against Rotation", *Int. J. Struct. Stability Dynam.*, Vol. 4, No. 1, pp 89–104.

Maeda, Y. (1971), "Ultimate Static Strength and Fatigue Behavior of Longitudinally Stiffened Plate Girders in Bending," paper presented at the IABSE Colloq. Des. Plate Box Girders Ultimate Strength, London.

Maquoi, R., Jetteur, P., Massonnet, C., and Škaloud, M. (1983), "Calcule des Âmes et Semelles Radier des Ponts en Acier," *Constr. Met.*, No. 4, pp. 15–28.

Massonnet, C. (1962), "Stability Considerations in the Design of Steel Plate Girders," *Trans. Am. Soc. Civ. Eng.*, Vol. 127, Part II, pp. 420–447.

Narayanan, R., and Der-Avanessian, N. G.-V. (1985), "Design of Slender Webs Having Rectangular Holes," *ASCE J. Struct. Eng.*, Vol. III, No. 4, pp. 777–787.

Ostapenko, A., Chern, C., and Parsaneiad, C. (1971), "Ultimate Strength Design of Plate Girders," *Fritz Eng. Lab. Rep.*, No. 328.12, Lehigh University, Bethlehem, PA, Jan.

Ostapenko, A., and Chern, C. (1970), "Strength of Longitudinally Stiffened Plate Girders," Fritz Eng. Lab. Rep. No. 328.10, Lehigh University, Bethlehem, PA, Dec.

Ostapenko, A., Yen, B. T., and Beedle, L. S. (1968), "Research on Plate Girders at Lehigh University," paper presented at the IABSE 8th Congr., Final Rep.

Park, H.-G., Kwack, J.-H., Jeon, S.-W., Kim, W.-K., and Choi, I.-R. (2007), "Framed Steel Plate Wall Behavior Under Cyclic Lateral Loading," *ASCE J. Struct. Eng.*, Vol. 133, No. 3, pp. 378–388.

Parsanejad, S., and Ostapenko, A. (1970), "On the Fatigue Strength of Unsymmetrical Steel Plate Girders," *Weld Res. Counc. Bull.*, No. 156, Nov.

Patterson, P. J., Corrado, J. A., Huang, J. S., and Yen, B. T. (1970), "Fatigue and Static Tests of Two Welded Plate Girders," *Weld. Res. Counc. Bull.*, No. 155, Oct.

Porter, D. M., Rockey, K. C., and Evans, H. R. (1975), "The Collapse Behavior of Plate Girders Loaded in Shear," *Struct. Eng.*, Vol. 53, No. 8, pp. 313–325.

Qu, B., Bruneau, M., Lin, C.-H., and Tsai, K.-C. (2008), "Testing of Full-Scale Two-Story Steel Plate Shear Wall with Reduced Beam Section Connections and Composite Floors," *ASCE J. Struct. Eng.*, Vol. 134, No. 3, pp. 364–373.

Raoul, J., Schaller, I., and Theillout, J. (1991), "Tests of Buckling of Panels Subjected to In-Plane Patch Loading," in *Contact Loading and Local Effects in Thin-Walled Plated and Shell Structures: IUTAM 1990 Symposium in Prague*, Springer-Verlag, Berlin.

Redwood, R. G., and Uenoya, M. (1979), "Critical Loads for Webs with Holes," *ASCE J. Struct. Div.*, Vol. 105, No. ST10, pp. 2053–2068.

Rezai, M. (1999), "Seismic Behaviour of Steel Plate Shear Walls by Shake Table Testing," Ph.D. dissertation, Dept. of Civil Engineering, University of British Columbia, Vancouver, Canada.

Roberts, T. M. (1981), "Slender Plate Girders Subjected to Edge Loading," *Proc. Inst. Civ. Eng.*, Part 2, p. 71 .

Roberts, T. M., and Chong, C. K. (1981), "Collapse of Plate Girders Under Edge Loading," *ASCE J. Struct. Div.*, Vol. 107, No. ST8, pp. 1503–1510.

Roberts, T. M., and Rockey, K. C. (1979), "A Mechanism Solution for Predicting the Collapse Loads of Slender Plate Girders When Subjected to In-Plane Patch Loading," *Proc. Inst. Civ. Eng.*, Vol. 67, Part 2.

Roberts, T. M., and Sabouri, S. (1991), "Hysteretic Characteristics of Unstiffened Plate Shear Panels," *Thin-Walled Struct.*, Vol. 12, No. 2, pp. 145–162.

Rockey, K. C. (1971a), "An Ultimate Load Method for the Design of Plate Girders," paper presented at the IABSE Colloq. Des. Plate Box Girders Ultimate Strength, London.

Rockey, K. C. (1971b), "An Ultimate Load Method of Design for Plate Girders," *Proc. Conf. Dev. Bridge Des. Constr.*, Crosby Lockwood, London.

Rockey, K. C., and Bagchi, D. K. (1970), "Buckling of Plate Girder Webs Under Partial Edge Loadings," *Int. J. Mech. Sci.*, Vol. 12, pp. 61–76.

Rockey, K. C., Elgaaly, M., and Bagchi, D. K. (1972), "Failure of Thin-Walled Members Under Patch Loading," *ASCE J. Struct. Div.*, Vol. 98, No. ST12, pp. 2739–2752.

Rockey, K. C., Evans, H. R., and Porter, D. M. (1973), "Ultimate Load Capacity of Stiffened Webs Subjected to Shear and Bending," *Proc. Conf. Steel Box Girders*, Institution of Civil Engineers, London.

Rockey, K. C., Evans, H. R., and Porter, D. M. (1974), "The Ultimate Strength Behavior of Longitudinally Stiffened Reinforced Plate Girders," paper presented at the Symp. Nonlinear Tech. Behav. Struct. Anal., Transport and Road Research Laboratory, Crowthorne, Dec.

Rockey, K. C., and Leggett, D. M. A. (1962), "The Buckling of a Plate Girder Web Under Pure Bending When Reinforced by a Single Longitudinal Stiffener," *Proc. Inst. Civ. Eng.*, Vol. 21.

Rockey, K. C., and Škaloud, M. (1968), "Influence of Flange Stiffness upon the Load Carrying Capacity of Webs in Shear," paper presented at the IABSE 8th Congr., Final Rep., New York.

Rockey, K. C., and Škaloud, M. (1972), "The Ultimate Load Behavior of Plate Girders Loaded in Shear," *Struct. Eng.*, Vol. 50, No. 1, pp. 29–48.

Schueller, W., and Ostapenko, A. (1970), "Tests on a Transversely Stiffened and on a Longitudinally Stiffened Unsymmetrical Plate Girder," *Weld Res. Counc. Bull.*, No. 156, Nov.

Selberg, A. (1973), "On the Shear Capacity of Girder Webs," Univ. Trondheim Rep.

Sharp, M. L., and Clark, J. W. (1971), "Thin Aluminum Shear Webs," *ASCE J. Struct. Div.*, Vol. 97, ST4, pp. 1021–1038.

Shelestanko, L. P., Dushnitsky, V. M., and Borovikov, V. (1970), "Investigation of the Influence of Limited Web Deformation on the Ultimate Strength of Welded Plate Girders," in *Res. Steel Compos. Superstruct. Bridges*, No. 76, Transport, Moscow (in Russian).

Shimizu, S., Yabana, H., and Yoshida, S. (1989a), "A Collapse Model for Patch Loaded Web Plates," *J. Constr. Steel Res.*, Vol. 13, pp. 61–73.

Shimizu, S., Horii, S., and Yoshida, S. (1989b), "The Collapse Mechanism of Patch Loaded Web Plates," *J. Constr. Steel Res.*, Vol. 14, pp. 321–337.

Shimizu, S., Horii, S., and Yoshida, S. (1991), "Behavior of Stiffened Web Plates Subjected to Patch Loads," in *Contact Loading and Local Effects in Thin-Walled Plated and Shell Structures: IUTAM 1990 Symposium in Prague*, Springer-Verlag, Berlin.

Shishkin, J. J., Driver, R. G., and Grondin, G. Y. (2009), "Analysis of Steel Plate Shear Walls Using the Modified Strip Model," *ASCE J. Struct. Eng.*, Vol. 135, No. 11, pp. 1357–1366.

Škaloud, M. (1962), "Design of Web Plates of Steel Girders with Regard to the Postbuckling Behavior (Analytical Solution)," *Struct. Eng.*, Vol. 40, No. 12, p. 409 .

Škaloud, M. (1971), "Ultimate Load and Failure Mechanism of Thin Webs in Shear," paper presented at the IABSE Colloq Des. Plate Box Girders Ultimate Strength, London.

Škaloud, M., and Novak, P. (1972), "Postbuckled Behavior and Incremental Collapse of Webs Subjected to Concentrated Load," paper presented at the IABSE 9th Congr., Prelim, Publ., Amsterdam.

Spinassas, I., Raoul, J., and Virlogeux, M. (1991), "Parametric Study on Plate Girders Subjected to Patch Loading," in *Contact Loading and Local Effects in Thin-Walled Plated and Shell Structures: IUTAM 1990 Symposium in Prague*, Springer-Verlag, Berlin.

Stein, M., and Fralich, R. W. (1950), "Critical Shear Stress of Infinitely Long Simply Supported Plate with Transverse Stiffeners," *J. Aeronaut. Sci.*, Vol. 17., p. 38 .

Steinhardt, O., and Schröter, W. (1971), "Postcritical Behavior of Aluminum Plate Girders with Transverse Stiffeners," paper presented at the IABSE Colloq. Des. Plate Box Girders Ultimate Strength, London.

Takeuchi, T. (1964), "Investigation of the Load Carrying Capacity of Plate Girders," M.S. thesis, University of Kyoto (in Japanese).

Thorburn, L. J., Kulak, G. L., and Montgomery, C. J. (1983), "Analysis of Steel Plate Shear Walls," Struct. Eng. Rep. No. 107, Department of Civil Engineering, University of Alberta, Edmonton, Alberta, Canada.

Timler, P. A. and Kulak, G. L. (1983), "Experimental Study of Steel Plate Shear Walls," Struct. Eng. Rep. No. 114, Department of Civil Engineering, University of Alberta, Edmonton, Alberta, Canada.

Tromposch, E. W., and Kulak, G. L. (1987), "Cyclic and Static Behavior of Thin Panel Steel Plate Shear Walls," Struct. Eng. Rep. No. 145, Department of Civil Engineering, University of Alberta, Edmonton, Alberta, Canada, Apr.

Vian, D., and Bruneau, M. (2004), "Testing of Special LYS Steel Plate Shear Walls," *Proc., 13th World Conference on Earthquake Engineering*, Vancouver Canada, Paper No. 978.

Wagner, H. (1931), "Flat Sheet Metal Girder with Very Thin Metal Web," Tech. Notes. 604, 605, 606, National Advisory Committee on Aeronautics, Washington, DC.

Warkenthin, W. (1965), "Zur Beulsicherheit querbelasteter Stegblechfelder," *Stahlbau*, Vol. 34, p. 28 .

White, D. W. (2008), "Unified Flexural Resistance Equations for Stability Design of Steel I-Section Members— Overview," *J. Struct. Eng. ASCE*, Vol. 134, No. 9, pp. 1405–1424.

Wilkesmann, F. W. (1960), "Stegblechbeulung bei Längsrandbelastung," *Stahlbau*, Vol. 29, No. 10, pp. 314–322.

Wilson, J. M. (1886), "On Specifications for Strength of Iron Bridges," *Trans. ASCE*, Vol. 15, Part I, pp. 401–403, 489–490.

Yen, B. T., and Basler, K. (1962), "Static Carrying Capacity of Steel Plate Girders," *Highw. Res. Board Proc.*, Vol. 41, pp. 173–182.

Yen, B. T., and Mueller, J. A. (1966), "Fatigue Tests of Large-Size Welded Plate Girders," *Weld. Res. Counc. Bull.*, No. 118, Nov.

Yonezawa, H., Miakami, I., Dogaki, M., and Uno, H. (1978), "Shear Strength of Plate Girders with Diagonally Stiffened Webs", *Trans. Jpn. Soc. Civ. Eng.*, Vol. 10, pp. 17–28.

Yura, J. A. (1987), "Course Notes for CE362N— Advanced Structural Steel Design," The University of Texas at Austin.

Zetlin, L. (1955), "Elastic Instability of Flat Plates Subjected to Partial Edge Loads," *Proc. ASCE*, Vol. 81, pp. 1–24.

CHAPTER 7

BOX GIRDERS

7.1 INTRODUCTION

Box girders are used for bridges (Fig. 7.1), heavy industrial buildings, offshore platforms, and other structures where large loads are frequently encountered. They are employed to best advantage when use is made of their considerable lateral and torsional stiffnesses, both during erection and in service. Box girders may have a variety of cross-sectional shapes (Fig. 7.2) ranging from deep narrow boxes to a wide shallow box, perhaps with many webs, some sloping or even curved. Those occurring most frequently have flanges that are wider and with thinner plates than in plate girder construction. The webs, on the other hand, may be of slenderness comparable with those of plate girders. It is the extensive use of slender plate construction for flanges, however, as well as webs that makes local stability considerations so important in the case of box girders. The need to consider torsional, distortional, and warping stresses is also peculiar to box girder design. Designers are encouraged to learn about and understand the specific behavior of box girders and to ensure that their analysis addresses all of the stresses and deformations unique to such structural systems. Furthermore, and although not specifically discussed in this chapter, box girders have many practical layout and detailing issues that make them unique and different from I-shaped steel girders, especially in specific regard to their local or regional design and construction practices and preferences (Coletti et al., 2005).

In this chapter aspects of the stability of box girders and their components, which are additional to those described in Chapter 6, are discussed. Most of these additional problems are a consequence of the use of wide flanges, for example, the influences of shear lag and the effects of transverse loading and different edge supports on the buckling of stiffened compression plates. Other problems relate to the need to provide diaphragms (plated, cross frames, or cross bracings) to transfer

FIGURE 7.1 Box girder bridges.

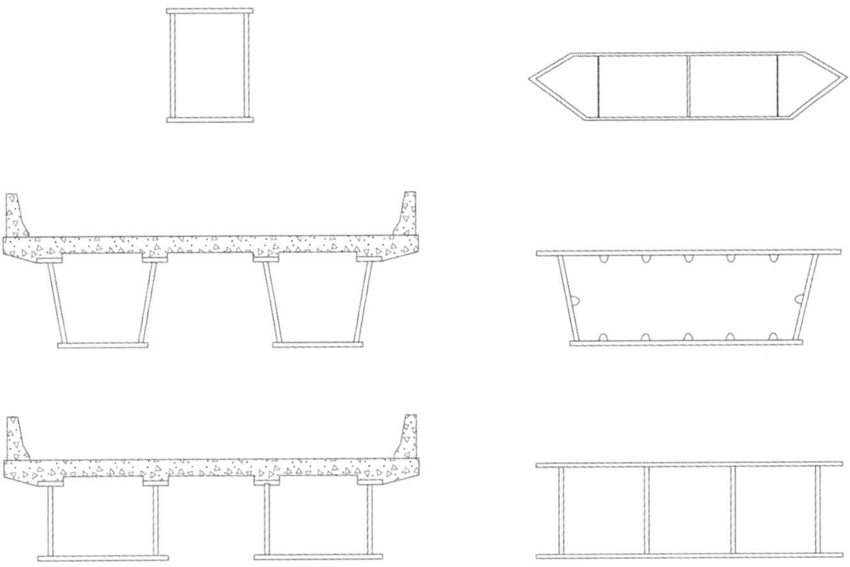

FIGURE 7.2 Bridge cross sections.

deck loads to the webs, to control distortion of a box, or to transfer forces in the webs to the support bearings. In the case of box girders, as opposed to I-shaped girders, the bearings may not be positioned directly beneath and in line with the webs. Much of the information on web stability provided already in the context of plate girders is applicable to box girders, but there are important differences that are discussed in the following sections. Some of the unresolved problems relating to box girders are also outlined.

It should be noted that this chapter is not meant to be a comprehensive specification or commentary for steel box girder design, although references to such design guidelines are provided. Instead, it is meant more to give the reader perspective of how some of the more important design issues are treated, either similarly or differently, by various specifications and guides from around the world. Unfortunately, one consequence of providing such a review is that notation, procedures, and formulas from different specifications are mixed and matched.

7.2 BASES OF DESIGN

Up until the end of the 1960s the design of box girders was not codified in detail in any country. Although a wide variety of approaches was used for the design of box girder bridges, the basis generally adopted was to use a factor of safety on the buckling stress, more often than not the *critical elastic buckling stress*, of the component being designed. These factors of safety varied from country to country and were intended to cover not only the as-built condition but also conditions met during construction. In some cases they also reflected an appreciation of the postbuckling reserve of plates, with lower factors of safety being used than would be the case where no such reserve was anticipated.

Three major collapses that occurred during the erection of box girder bridges, at Milford Haven (1969), West Gate (1970), and Koblenz (1971) as shown in Fig. 7.3, caused the entire basis of the design of box girders to be reviewed ("Inquiry," 1973) not only in Europe but also in the United States and worldwide. The causes of these failures were not similar, the first due to failure of a plated support diaphragm, the second due to attempts to overcome construction lack-of fit between two box girders, and the third due to local poor detail design. The failures did highlight, however, the need for more consistent and comprehensive design guidance. The drafting of new rules in the wake of these tragedies coincided with the trend toward limit-state format codes involving the consideration of both the serviceability and the ultimate limit states. Not surprisingly, therefore, emphasis was placed on the development of new design approaches based on the *ultimate*, or the *inelastic buckling*, strength of box girders and particularly of their steel-plated components (Horne, 1977).

Influence of Initial Geometric Imperfections and Residual Stresses

Much attention was paid to the role of initial imperfections, both geometric and residual stresses, during the investigations of the strength of box girders and their

FIGURE 7.3 Koblenz bridge collapse.

components (Dowling et al., 1977b). Extensive measurements were made of residual stresses produced during box girder fabrication by using both destructive, but mainly nondestructive, methods. These were carried out in model boxes fabricated using techniques representative of full-scale techniques and also in actual box girder bridges under construction. Numerous surveys of the magnitude and distribution of initial distortions in model and full-scale boxes were also made. Concurrent with these measurements, fundamental studies on the prediction of residual stresses, and on their effects, both on overall stresses in flanges and on local stresses in components together with the effects of geometric imperfections, were carried out in many research laboratories.

The weakening effect of both types of initial imperfection, separately and together, were explicitly treated in the Interim Design and Workmanship Rules ("Inquiry," 1973) and have subsequently been intrinsically allowed for in the various design methods produced to predict the buckling strength of plated structures. The weakening effect of component imperfections is most pronounced in the range of intermediate slenderness, that is, those slenderness ratios at which the critical buckling stress and the yield stress are roughly equal (Frieze et al., 1975). The decrease in strength is most marked for plates of intermediate slenderness subjected to compressive stressing and is least pronounced for shear-loaded cases, being fairly small for rectangular plates in pure shear.

Strength curves for plate panels used for design have normally been selected with allowances made for practical combinations of residual stresses and initial imperfections. Ultimate load design methods for stiffeners or stiffened plates also normally incorporate the effects of imperfections in their expressions. It should be noted that the construction tolerances for straightness and flatness specified in the workmanship rules to be used in conjunction with design rules differ for different countries and this may in part account for discrepancies between design strengths. Typically, a level of compressive residual stress in a component of the order of 10% of the yield stress, with an initial geometric imperfection that is related to fabrication tolerances, has been adopted by code writers. In other cases a geometric imperfection is chosen which is considered on the basis of theoretical or experimental evidence to incorporate the combined weakening effect of residual stress and practical initial distortions. It has been shown that provided that the magnitude of the initial distortion chosen is big enough (and this is often of the same order as specified fabrication tolerances), the additional weakening effect of even the largest residual compressive stresses is small for practical plates.

The option of specifying levels of residual stresses to be calculated by designers and measured by fabricators is not a realistic one in terms of the economics of normal fabricated box girders and is avoided in recent specifications. As a result of the efforts made to reconsider the basis of design for box girders, new rules were produced in the United Kingdom, the United States, and Germany, among other countries. The first to be fully codified were the British rules; as of 2007 most of the U.S. provisions were still in a proposal form (Wolchuk and Mayrbourl, 1980) although the current *AASHTO LRFD Bridge Design Specifications* (AASHTO, 2007) includes some of these provisions. For this reason the basis of the design approach contained in the current BS 5400: Part 3, Design of Steel Bridges (BSI, 2000), which has been minimally revised with respect to box girders since 1983, is summarized here in perhaps a bit more detail than is devoted to the other rules.

After 30 years an effort to harmonize design standards in European countries is coming to an end by the publication of the Eurocodes. They will replace National Standards in most European countries by 2010. Steel structures are covered by Eurocode 3, which has 20 parts giving general rules and specific rules for different types of structures. Bridges are considered in Part 2 (CEN, 2006a) and it refers back to general rules in Part 1-1 (CEN, 2005a) and special rules for plate buckling in Part 1-5 (CEN, 2006b). Composite bridges are treated in Eurocode 4, Part 2 (CEN, 2005b).

Eurocode 3, Parts 1-1 and 1-5 give generic rules for steel structures and the latter gives rules for buckling of unstiffened or stiffened plates, which are relevant for box girders other than for decks subject to transverse loading. Eurocode 3, Part

2 gives detailed rules for orthotropic plates concerning detailing and fabrication in order to avoid premature fatigue and empirical rules for the thickness of the deck plate.

Meanwhile, the U.S. codes have taken a more divergent path. Wolchuck and Mayrbourl (1980) prepared a complete set of box girder provisions that were intended to cover all box girder bridges, similar to the British provisions. For the purposes of this chapter, this document will be referred to as "the 1980 proposed U.S. code." This code was never officially adopted in the United States but was never officially voided either and is still referenced in some designs, particularly for large single box girder bridges. Meanwhile, the current official U.S. bridge design specification, the *AASHTO LRFD Bridge Design Specifications* (2007), includes many provisions applicable to box girder bridges, but they are primarily intended for use on multibox bridges (bridges that utilize multiple, smaller steel box girders with a composite concrete deck, rather than a single, very large/wide single- or multiple-cell steel box girder). For the purposes of this chapter, this document will be referred to as "the AASHTO LRFD specifications." These specifications have been continually updated and revised, particularly in recent years, to reflect significant new research into composite steel bridge behavior and design, and a number of the AASHTO LRFD specification provisions essentially supersede the 1980 proposed U.S. code. Because the AASHTO LRFD specifications do not comprehensively address all box girders, it is appropriate to include reference to both the 1980 proposed U.S. code and the AASHTO LRFD specifications in this chapter.

7.3 BUCKLING OF WIDE FLANGES

Box girder flanges are usually stiffened in both the longitudinal and transverse directions (Fig. 7.4). The unstiffened flanges of narrow box girders can be treated as plates, as described in Chapter 4, by using reduced effective widths to account for the effects of buckling and, according to some codes, shear lag. Possible modes of buckling occurring in an orthogonally stiffened plate are illustrated in Fig. 7.5. These include overall buckling of the stiffened flange, buckling of the longitudinally stiffened panels between transverse elements, various forms of local plate and stiffener buckling, and various combinations of these modes.

The linear elastic buckling of stiffened flanges with one or more stiffeners is covered in Chapter 4 and formed the basis of earlier design methods. Indeed, many codes still in use today are based on these methods together with the use of safety factors determined from test results (Dowling, 1981). In this chapter emphasis is placed on methods designed to predict the ultimate strength of such flanges, as these are being used to replace the elastic methods. It should be recognized, however, that even now the new rules for stiffened compression flanges are not based entirely on ultimate-strength considerations, as pointed out by Dowling (1981).

In considering the inelastic buckling of wide stiffened flanges, three basic approaches can be adopted, and each will be described in the following sections.

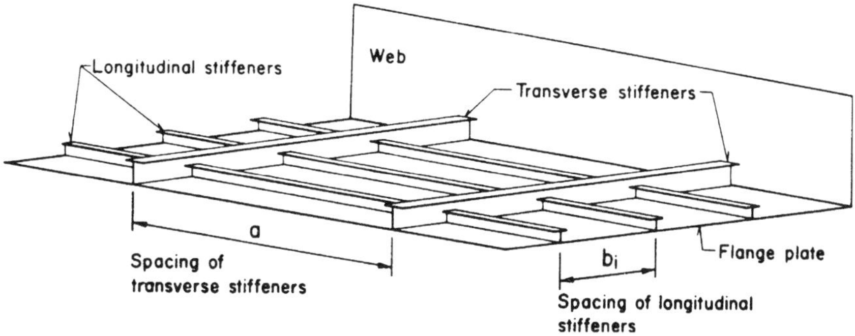

FIGURE 7.4 Sample stiffened panel (compression flange).

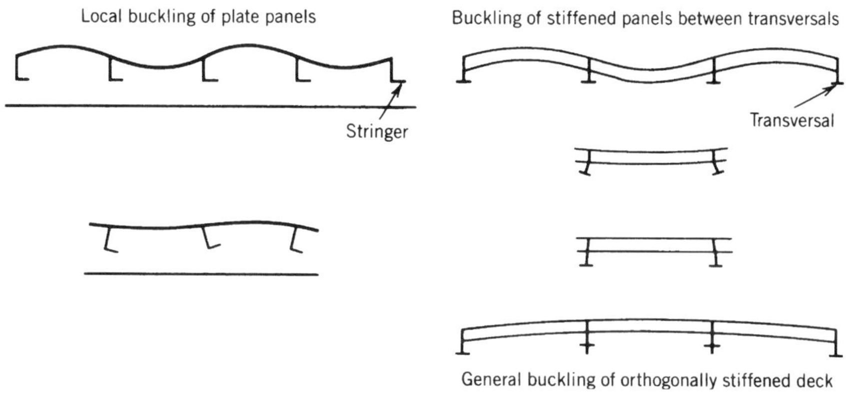

FIGURE 7.5 Various forms of local buckling of stringers and transversals.

In increasing order of complexity, the flanges can be treated as:

1. series of disconnected struts
2. An orthotropic plate
3. A discretely stiffened plate

For most stiffened flanges the strut approach is sufficiently accurate and is suitable for design purposes. When the flange stiffening is lighter than normal, advantage can be taken of the postbuckling reserve of the stiffened plate and many methods have been produced based on this type of modeling, notably those of Massonnet and his co-workers (Maquoi and Massonnet, 1971; Jetteur, 1983). The flange geometry, however, rarely justifies such an approach in practice, and the strut approach will, in any case, produce safe designs. The discretely stiffened-plate approach is of interest mainly for plates with one or two stiffeners. For such flanges Eurocode 3, Part 1-5 offers a simple design method that considers the restraint from the plate to the buckling of stiffeners. Each of these approaches is described in turn below.

7.3.1 Strut Approach

The basis of this approach is to treat a plate stiffened by several equally spaced longitudinal stiffeners as a series of unconnected compression members or struts, each of which consists of a stiffener acting together with an associated width of plate that represents the plate between stiffeners (Fig. 7.6). Where transverse stiffeners are present, they are designed to be sufficiently stiff to ensure that they provide nodal lines acting as simple rotationally free supports for the longitudinal struts. Thus the equivalent buckling length of the longitudinal stiffeners is effectively the distance between transverse elements. Such an approach was suggested by Ostapenko and Vojta (1967) and Dowling and his co-workers Moolani and Chaterjee (Dwight and Little, 1976; Dowling and Chatterjee, 1977; Dowling et al., 1977a; Chatterjee, 1978). Allowance is made for reduction in effectiveness due to buckling of the plate between stiffeners in calculating the overall buckling strength of the longitudinal stiffeners between transverse stiffeners.

Where the longitudinal stiffeners are designed as open sections such as flat bars, tees, or angles, limitations are placed on their cross-sectional geometry to ensure that local buckling of the stiffeners does not precede the attainment of the ultimate strength of the flange. A detailed discussion of stiffener cross-sectional limitations is provided in Section 7.3.6.

In accounting for the buckling of the repeating combined stiffener/plate "strut," allowance can be made for the effect of initial distortions and residual stresses caused by welding the stiffeners to the plate. The extensive amount of data now available on plate strength can be used to account for interstiffener plate buckling using either an effective-width or an effective-stress approach. Normally, a simple effective-width approach is used to account for both the reduced strength and stiffness of the compressed plate. Although there are approximations involved in selecting just one width to account for stiffness as well as strength, the gain in accuracy that could be obtained using a more complex approach is considered to be unnecessary for design, although the U.K. bridge assessment rules provide for the use of different widths for stiffness and strength calculations. The effective width used in BS 5400 is derived from the results of a parametric study and uses a strength-based effective width that accounts for practical levels of initial imperfections and compressive residual stresses. Figure 7.7 gives curves from BS 5400 that may be used to calculate the width of plate $K_c b$ considered to act effectively with a stiffener. Note that the value of K_c to be used is the higher of the values

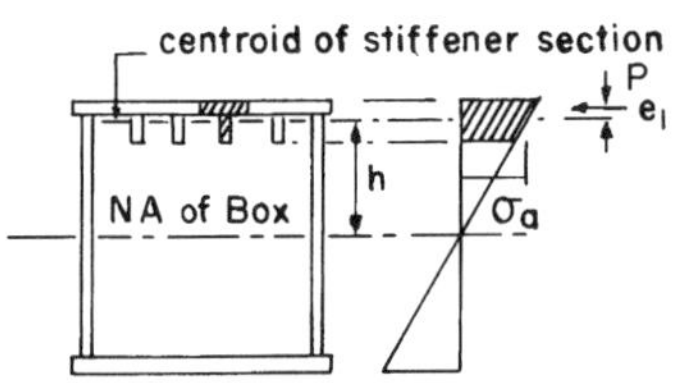

FIGURE 7.6 Curvature of box girder.

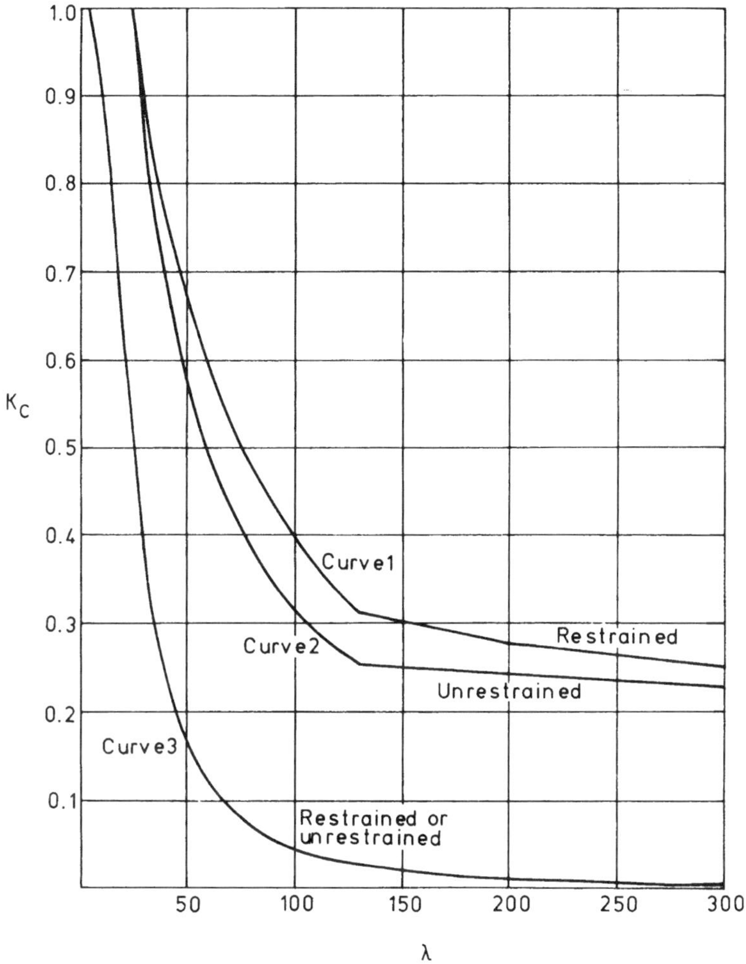

FIGURE 7.7 Coefficient K_c for plate panels under direct compression.

obtained using either (a) curve 1 or 2 as relevant with

$$\lambda = \frac{b}{t}\sqrt{\frac{\sigma_y}{355}} \tag{7.1}$$

or (b) curve 3 with

$$\lambda = \frac{a}{t}\sqrt{\frac{\sigma_y}{355}} \tag{7.2}$$

in which a is the panel dimension in the direction of stress considered and b is the panel dimension normal to the direction of stress. It is further noted that item (a) will always give the higher value for K_c where $a/b \geq 0.5$, and for $a/b < 0.5$, items (a) or (b) may give a higher value.

Knowing the properties of the strut, its buckling strength can be calculated from the Perry–Robertson formula that relates to a pin-ended, initially crooked, axially loaded column,

$$\frac{\sigma_{su}}{\sigma_y} = \frac{1}{2}\left\{\left[1 + (\eta + 1)\frac{\sigma_E}{\sigma_y}\right] - \sqrt{\left[1 + (\eta + 1)\frac{\sigma_E}{\sigma_y}\right]^2 - \frac{4\sigma_E}{\sigma_y}}\right\} \tag{7.3}$$

where σ_{su} = limiting applied axial stress on effective strut section
σ_E = Euler stress of effective strut
σ_y = available yield stress of compressive extreme fiber, $= \sigma_{ys}$ when checking the stiffener; $= \sigma_{ye}$ an effective yield stress allowing for the presence of shear when checking the flange
$\eta = \Delta y / r_{se}^2$
Δ = maximum equivalent initial imperfection (In BS5400 Δ is taken as related to the slenderness of the strut and to the residual stresses due to rolling and fabrication. The values have been derived by calibration against test results and were considered to be applicable to components within the construction tolerances in Part 6 of the U.K. bridge assessment standard. In the U.K. bridge assessment rules guidance is given on the derivation of effective stiffener imperfections for stiffened compression flanges.)
y = distance from centroid to compressive extreme fiber
r_{se} = radius of gyration of effective section about axis parallel to plate

Because the cross section is asymmetric, the possibility of failure in the two directions (one causing compressive failure of the plate, the other causing compressive yielding of the stiffener tip) must be checked. The magnitude of the initial crookedness in these two directions may well be different, although in BS 5400 the bow in the direction causing compression in the stiffener tip is taken to be as large as that in the opposite direction. This allows for the fact that compressive residual stresses caused by welding may be present at the stiffener tip, which are otherwise unaccounted for in the design procedure. The effect of end eccentricity of loading over the stiffener/plate strut is represented by a term in Δ given by $r_{se}^2/2y_{Bs}$. Here y_{Bs} is the distance from the centroid of the effective stiffener section to the neutral axis of the cross section of the complete beam (taken as positive when the neutral axis of the beam lies in the same direction from the centroid of the stiffener as that to the centroid of the beam cross section).

The design of the stiffened flange is checked to ensure that

$$\frac{\sigma_a + 2.5\tau_1 k_{s1}}{k_{l1}\sigma_{ye}} \leq 1 \tag{7.4}$$

$$\frac{\sigma_a + 2.5\tau_1 k_{s2}}{k_{l2}\sigma_{ye}} \leq 1 \tag{7.5}$$

using the curves plotted in Fig. 7.8. The first of these expressions checks for failure by yielding of the stiffener tip, and the second checks for buckling or yielding of the plate panel. If the longitudinal stress varies along the length of the stiffener, σ_a in the expressions above is taken at a point $0.4l$ from the end, where the stress is greater. In these formulas

σ_a = longitudinal stress at centroid of effective stiffener section
τ_1 = in-plane shear stress in flange plate due to torsion
σ_{ys} = yield stress of stiffener material
σ_{ye} = effective yield stress of flange plate material allowing for presence of shear
l = spacing of cross beams

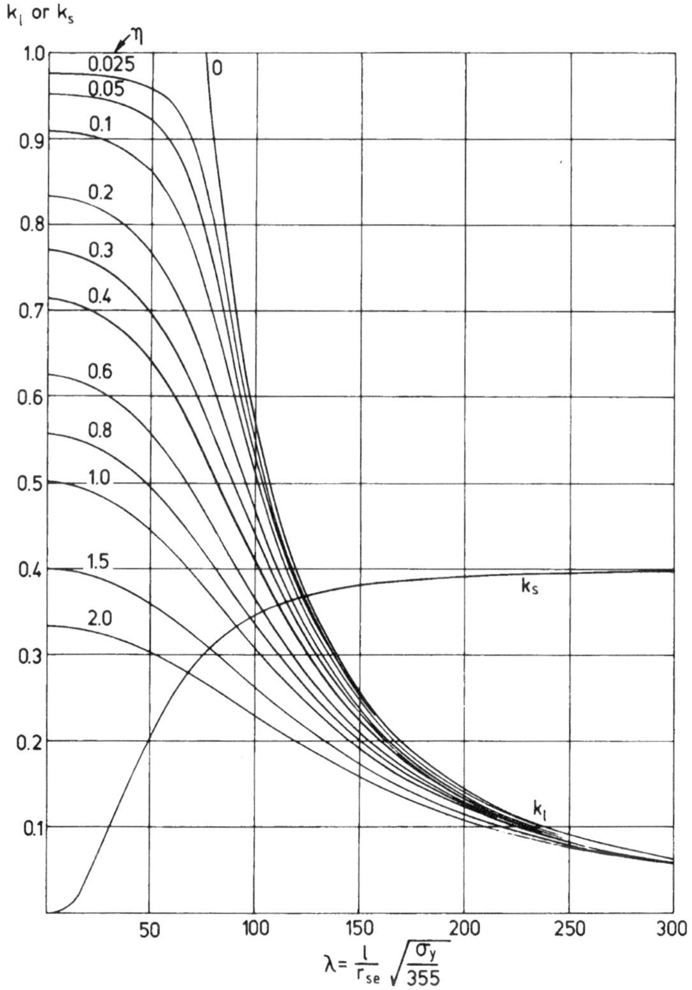

FIGURE 7.8 Parameters for the design of longitudinal flange stiffeners.

A method proposed by Wolchuk and Mayrbourl (1980) in the proposed U.S. code for box girders uses a format based on a German procedure (DA Stahlbau, 1978) for calculating stiffened-flange strength. Using the chart shown in Fig. 7.9, the method provides the opportunity to check chosen sections rapidly and determine the phenomenon controlling the strength (i.e., plate or stiffener buckling). The ultimate flange strength is computed as $P_u = F_u A_f$ or $F_u' A_f$, whichever is less, where A_f is the cross-sectional area of the stiffened flange; F_u is the ultimate strength of the stiffener strut (stiffener with full width of associated plating); F_u' the modified ultimate strength of stiffener strut under combined compression and shear, is given as

$$F_u' = \begin{cases} F_u & \text{for } f_v \leq 0.175\, F_y \quad (7.6) \\ 1.05 F_u \sqrt{1 - 3f_v^2/F_y^2} & \text{for } f_v \geq 0.175\, F_y \quad (7.7) \end{cases}$$

and f_v is the governing shear stress in the flange.

Design methods based primarily on the results of tests on stiffened panels have been suggested by Japanese researchers and were reviewed in the U.S.–Japan Seminar: Inelastic Instability of Steel Structures and Structural Elements held in Tokyo (Fujita and Galambos, 1981). One such proposed strength curve is given by

$$\frac{\sigma_u}{\sigma_y} = 1.24 - 0.54\bar{\lambda} \tag{7.8}$$

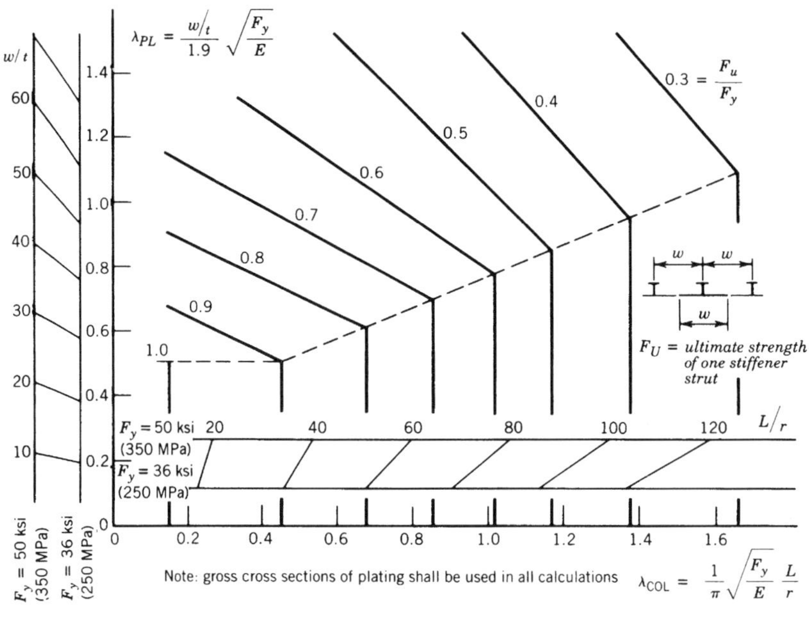

FIGURE 7.9 Strength of stiffened flange in compression (Wolchuk and Mayrbourl, 1980).

where

$$\bar{\lambda} = \frac{b}{t}\sqrt{\frac{\sigma_y}{E}\frac{12(1-\nu^2)}{\pi^2 k}} \tag{7.9}$$

and $k = 4n^2$ (n being the number of plate panels). It should be noted that this strength curve is chosen to agree with the *mean* of the test results obtained in Japan, as shown in Fig. 7.10.

In the case of stiffened flanges not subjected to transverse loading, transverse stiffeners are designed to provide sufficient stiffness to ensure that they act as supports which are effectively rigid against movements normal to the plane of the flange plate but offer little resistance to rotation of the longitudinal stiffeners at their junctions. Thus they ensure that the effective buckling length of the latter does not exceed their span between adjacent transverse members. When flanges form the deck of a box girder bridge that has to carry transverse loading, the cross girders are primarily proportioned to resist the transverse loading and typically have more than sufficient rigidity to constrain the stiffened panels in the manner envisioned.

The stiffness requirement for transverse members stipulated for design within BS 5400: Part 3 is based on elastic buckling theory combined with limited experimental observations and provides a factor of safety of 3 against overall linear elastic buckling. The approximate expression produced for the minimum second moment of area required of a transverse stiffener and its effective width of attached flange is

$$I_{be} = \frac{9\sigma_f^2 B^4 a A_f^2}{16KE^2 I_f} \tag{7.10}$$

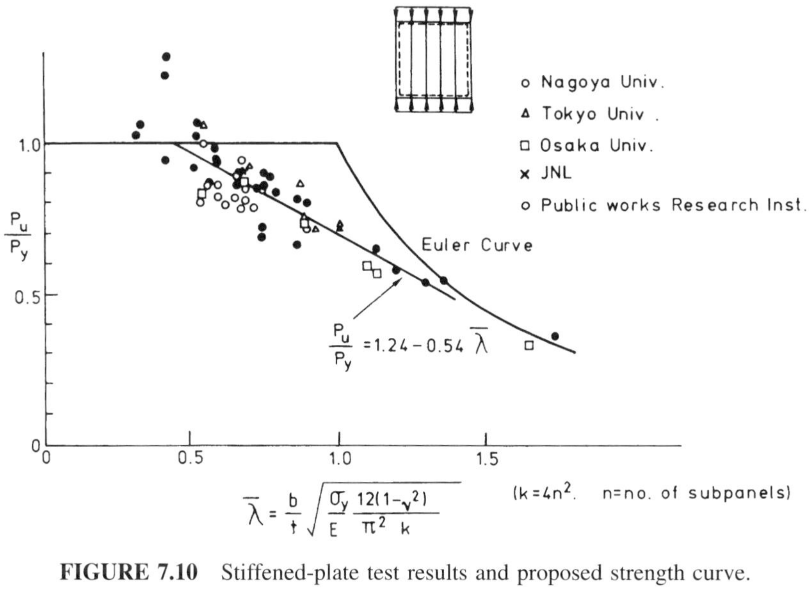

FIGURE 7.10 Stiffened-plate test results and proposed strength curve.

where σ_f = longitudinal compressive stress in flange
B = spacing of webs of main beam
a = cross-beam spacing
A_f = cross-sectional area per unit width of the flange
I_f = second moment of area per unit width of the flange
K = 24 for interior segments between webs of main beams and for other cases is determined from Fig. 7.11

It should be noted that the U.K. Bridge Assessment Standard (Highways Agency, 1996) provides treatment of cases in which the above stiffness criterion is not met.

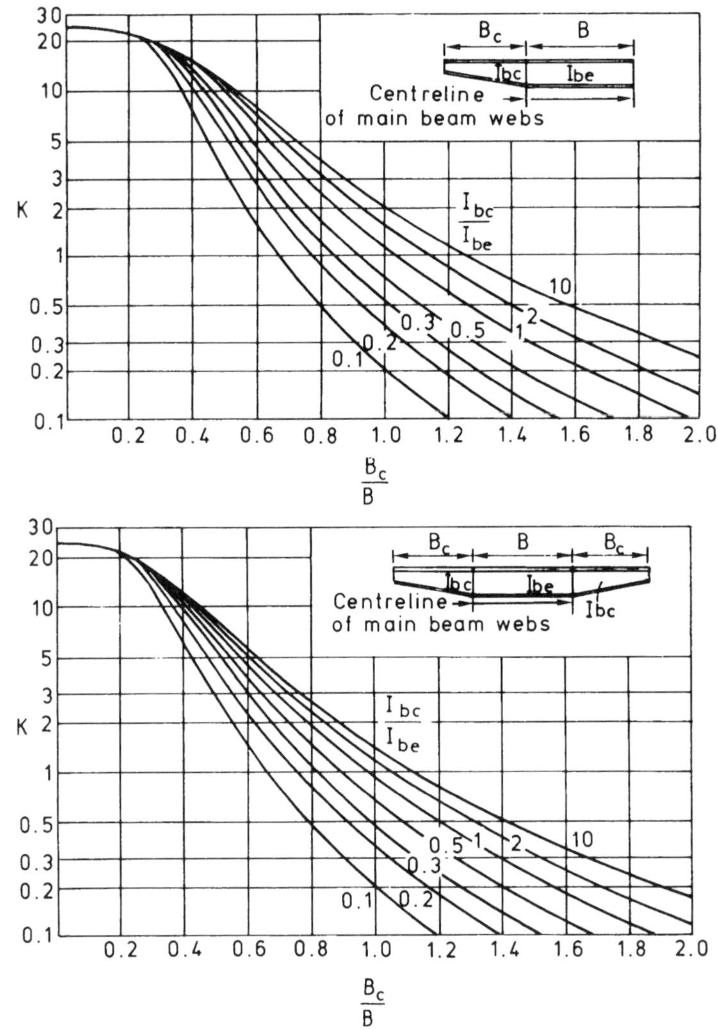

FIGURE 7.11 Buckling coefficient K for transverse members.

In addition, the transverse member must be designed to carry any locally applied transverse loading as well as the normal component of in-plane loading in the flange caused by lack of alignment of the transverse members. Two simplified loading cases, selected to cover the latter effect, are stipulated as follows:

1. uniformly distributed load per unit width of $\sigma_f A_f/200$
2. A concentrated load at a cantilever tip of $\sigma_f A_{sc}/160$, where A_{sc} is the cross-sectional area of the longitudinal stiffening member at the cantilever tip (σ_f and A_{sc} are in SI units)

In the 1980 proposed U.S. code, the destabilizing effect of flange geometric imperfections is taken into account by assuming the transverse stiffeners or cross-frame members to be loaded by a uniformly distributed load of 1% of the average factored compressive force in the flange. The rigidity requirement is again covered by an expression for the minimum moment of inertia I_t derived from a linear elastic stability analysis of spring-supported compression bars, but substituting a factored applied axial stress f for the critical stress gives

$$I_t = \frac{0.04 b^3 A_f f_s}{Ea} \tag{7.11a}$$

where I_t = moment of inertia of transverse member about an axis through its centroid
f_s = largest longitudinal flange stress in compression flange
E = Young's modulus
A_f = cross-sectional area of compression flange
a = longitudinal spacing of transverse flange stiffeners
b = spacing between webs

A similar requirement appears in the AASHTO LRFD specification:

$$I_t = 0.1\,(n+1)^3\,w^3\,\frac{f_s}{E}\frac{A_f}{a} \tag{7.11b}$$

in which the additional terms are
n = number of equally spaced longitudinal flange stiffeners
w = larger of the width of flange between longitudinal flange stiffeners or distance from a web to the nearest longitudinal flange stiffener

7.3.2 Orthotropic Plate Approach

When there are several stiffeners (more than three) in a flange, advantage can be taken of the orthotropic plate idealization in which the actual discretely stiffened plate is replaced by an orthotropic (*ortho*gonal aniso*tropic*) plate of constant thickness in which the stiffness of the stiffened plate is spread uniformly over its width.

The potential advantage of this method is that the inherent plate action, ignored by the strut approach, can be realized. This, of course, has a particular advantage in the postbuckling range when transverse tensile membrane stresses in the plate restrain the rate of growth of out-of-plane deflections in the stiffeners. As the equations describing postbuckling behavior are nonlinear, the solutions generally involve an iterative procedure to produce the ultimate strength. The latter is assumed to be reached when a particular collapse criterion is satisfied. Collapse criteria recognized include the onset of yield at the flange/web junction, the mean stress along that edge reaching yield, yield at a stiffener tip at the center of the flange, or in-plane yield in the plate in the same region. These methods have been described in some detail by Maquoi and Massonnet (1971) and Jetteur (1983) and are reviewed in the book *Behaviour and Design of Steel Plated Structures*, issued by the ECCS (Dubas and Gehri, 1986). They are adopted in the U.K. bridge assessment rules. The method has been further developed in Eurocode 3, Part 1-5, which makes a distinction between columnlike behavior and platelike behavior. The former is applicable to short and wide plates and the latter to long and narrow plates. The critical load for column buckling is denoted N_E and it is calculated for the plate assuming that the longitudinal edges are unsupported. The corresponding strength reduction factor for column buckling is denoted χ_c. The critical load for the orthotropic plate is denoted N_{cr} and the corresponding reduction factor for the gross area to effective area is denoted ρ. An interpolation between columnlike behavior and platelike behavior is done with the empirical formula

$$\rho_c = (\rho - \chi_c)\ \xi(2 - \xi) + \chi_c \tag{7.12}$$

where

$$\xi = (N_{cr}/N_E) - 1 \quad 0 \leq \xi \leq 1 \tag{7.13}$$

The factor ρ_c is used for reducing the effective area of the plate (already reduced for buckling between stiffeners) with respect to global buckling. If $N_{cr} = N_E$ the strength is governed by column buckling and if $N_{cr} > 2N_E$ it is a case of platelike behavior.

7.3.3 Discretely Stiffened Plate Approach

Eurocode 3, Part 1-5 describes a method for the design of plates with one or two stiffeners that is based on the theory of buckling of a strut on an elastic foundation. The plate is considered as parallel strips spanning in the transverse direction that constitute the elastic foundation. The fictitious strut consists of the stiffener and contributing plating similar to the system shown in Fig. 7.12. Using the notation defined Fig. 7.12, the spring stiffness can be written as

$$c = \frac{Et^3 b}{4(1 - \nu^2) b_1^2 b_2^2} \tag{7.14}$$

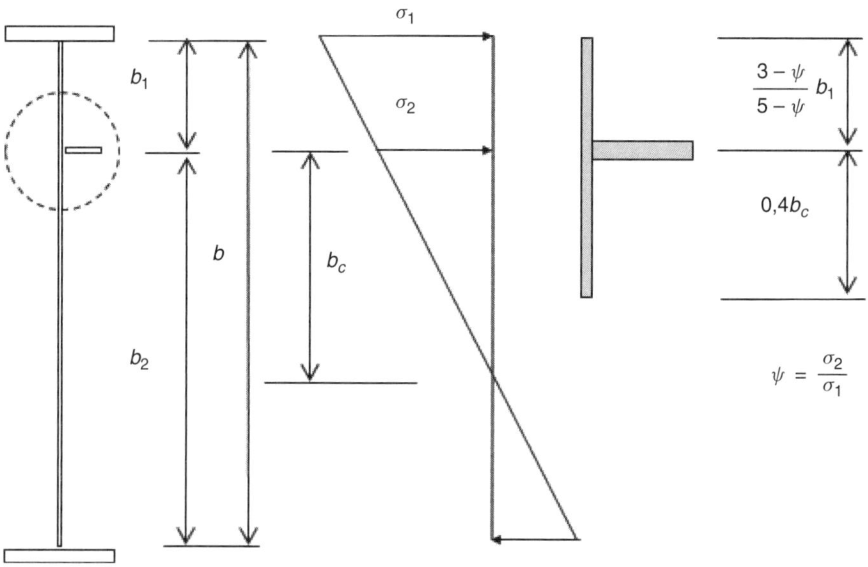

FIGURE 7.12 Plate with single stiffener and definition of fictitious strut.

A close approximation of the critical load of the fictitious strut with length a is

$$N_{cr} = \begin{cases} 1.05E\dfrac{\sqrt{I_{sl}t^3b}}{b_1b_2} & \text{if } a > a_c \qquad (7.15) \\[2ex] \dfrac{\pi^2EI_{sl}}{a^2} + \dfrac{Et^3ba^2}{35.7b_1^2b_2^2} & \text{if } a < a_c \qquad (7.16) \end{cases}$$

where

$$a_c = \pi\sqrt[4]{\frac{EI_{sl}}{c}} = 4.33\sqrt[4]{\frac{I_{sl}b_1^2b_2^2}{t^3b}} \tag{7.17}$$

I_{sl} = second moment of area of fictious strut

The critical load N_{cr} is then used in the procedure described at the end of Section 7.3.2.

7.3.4 Influence of Shear Lag on Flange Buckling

Shear lag arises because the in-plane shear straining of a flange causes those parts most remote from a web to develop smaller longitudinal stresses than those in the vicinity of the web. The resulting nonuniform distribution of longitudinal stresses across the flange width, with larger stresses occurring near flange/web junctions and smaller ones in the regions farthest removed from such junctions, is shown in Fig. 7.13. Shear lag effects are most marked in beams with flanges that are relatively wide compared to their length and therefore are of greater importance in box girder than in plate girder construction. The addition of stiffeners across the flange width results in the nonuniformity becoming even more pronounced, although the maximum stresses may be reduced. A systematic study of the factors influencing shear lag in box girder flanges has been reported by Moffatt and Dowling (1975, 1976) and forms the basis of the relevant rules in the proposed 1980 U.S. design specification (Fig. 7.14) and in BS5400: Part 3. Similar studies have been reported by Abdel-Sayed (1969) and others.

It is the nonuniform distribution of in-plane stresses in a box girder flange that is of interest from the point of view of structural stability. This form of distribution can increase or decrease the average stress, causing earlier buckling of the flange compared with the uniformly compressed case, depending on the degree of stiffening in the plate. On the other hand, it is clear that concentration of the applied stresses near the longitudinally compressed edges of the plate will encourage an earlier onset of yield at the edges than would uniform stressing across the flange width. The interaction of these two effects is complex and, indeed, is complicated further by the capacity of the stiffeners and plate panels to redistribute load. Tests (Dowling et al., 1977a; Frieze and Dowling, 1979) have shown, however, that for most practical cases shear lag can be ignored in calculating the ultimate compressive strength of stiffened or unstiffened flanges. This conclusion has been supported by the numerical studies of Lamas and Dowling (1980), Jetteur et al. (1984), and Burgan and Dowling (1985). Thus a flange may normally be considered to be loaded uniformly across its width in the ultimate limit state. Only in the case of flanges with particularly large aspect ratios, or particularly slender edge panels or stiffeners, is it necessary to consider the flange stability in greater detail.

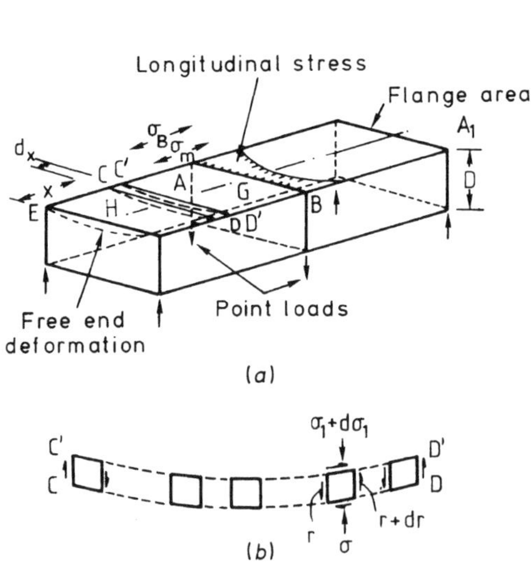

FIGURE 7.13 Definition of shear lag.

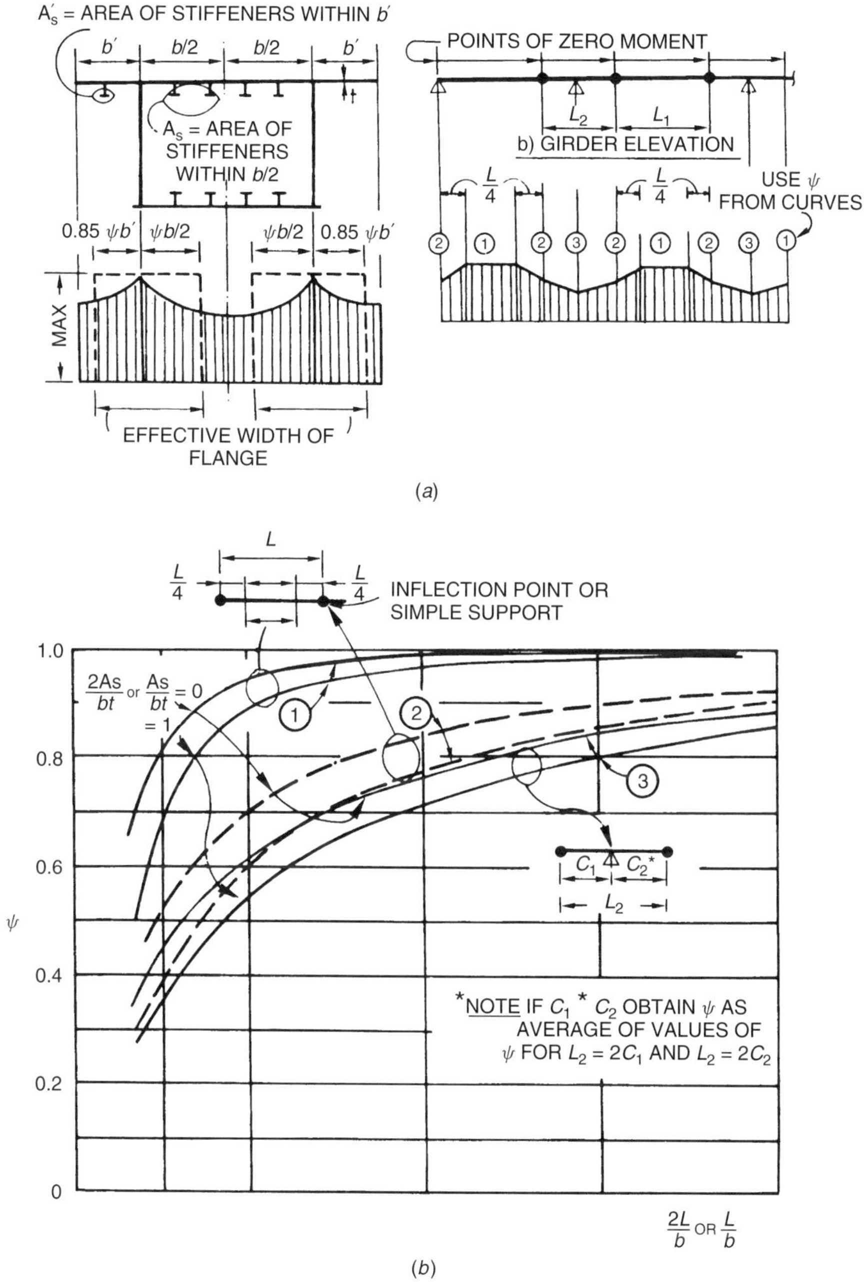

FIGURE 7.14 Effective width of flange: (*a*) longitudinal stress distribution across girder flange; (*b*) values of ψ.

The upper limit given in BS 5400: Part 3 for neglecting shear lag in the calculation of the strength of unstiffened flanges is for the case when the maximum longitudinal stress in the flanges is more than 1.67 times the mean longitudinal stress. For cases in which this limit is exceeded, the serviceability limit state has to be checked by including the effects of shear lag in the calculation of applied stresses.

Wolchuk, in the 1980 proposed U.S. code, suggests that shear lag may be ignored if the peak stress does not exceed the average stress by more than 20% and a uniform stress distribution is assumed for the calculation of the ultimate flange strength. Where these limits are exceeded it is suggested that the flange capacity at the web be increased to accommodate an additional force, which is computed as the load-factored stress in excess of 120% of the average stress multiplied by the flange area affected by the excess.

Eurocode 3, Part 1-5 gives rules for shear lag effects that distinguish between elastic shear lag and shear lag at the ultimate limit state. The rules for elastic shear lag give quite high stress peaks and they should be used for checking yielding in serviceability limit states and for fatigue design. For the ultimate limit state the shear lag effect is considered small and almost negligible. No specific criterion is given, noting that European countries often decide individually on how to treat shear lag in the ultimate limit state.

7.3.5 Influence of Transverse Loads

Transverse loads normal to a plated element can be of interest in the context of bridges when the deck constitutes the top flange of the box girder. In marine structures, such as a ship or an offshore platform, the transverse loading may be hydrostatic. In such cases, the effect of the transverse loading on the strength of the stiffened plate acting as a compression flange to a box girder should be considered.

Plates such as those used in bridge decks to resist traffic loading are normally stocky and designed primarily to limit deflections under the lateral loading to very small acceptable values (deflections $<<$ thickness of the plate) or to provide adequate fatigue life. The presence of in-plane stresses due to participation of the plate as part of the box girder flange magnifies the deflections and bending stresses by a factor of approximately $1/(1 - \sigma_a/\sigma_{cr})$. When appropriate, allowance can be made for the magnification of stresses in a compressed stiffened deck carrying transverse loading by also increasing the stresses due to the moments by this same amplification factor.

In BS 5400: Part 3 stiffened compression flanges used to carry local wheel loading, such as steel bridge decks, may be designed for the ultimate limit state ignoring the presence of the local loads. This is based on experimental evidence of a relatively limited nature. This is one of the few cases, however, for which it is necessary to carry out a serviceability check using elastic analysis. This check is done to ensure that yielding under working loads is prevented. The stresses in both the deck plate and the stiffeners are checked, including all elastic effects, such as shear lag, torsion, and in-plane stressing of the plate due to local bending of the stiffened plate under wheel loading. In calculating the stiffener stresses a distinction is made between the zones of hogging moment in the longitudinal stiffener over a transverse member and the sagging moment occurring between transverse members. Whereas in the latter case the full amplified stresses due to in-plane loading and local bending are combined, only the in-plane stresses in the former are amplified and added to the local bending stresses occurring at the longitudinal to transverse stiffener intersection.

The approximate expressions used to check the stiffener design are as follows: over transverse members,

$$\frac{\sigma_a + 2.5\tau_1 k_{s1}}{k_{l1}\sigma_{ys}} + \frac{\sigma_{f0}}{\sigma_{ys}} \leq 1 \tag{7.18}$$

and between transverse members,

$$\frac{\sigma_a + 2.5\tau_1 k_{s2}}{k_{l2}\sigma_{ye}} + \frac{\sigma_{fz}}{\sigma_{ye}} \leq 1 \tag{7.19}$$

where σ_{f0} is the stress due to local bending at a point on the stiffener farthest from the flange plate, σ_{fz} is the stress at the midplane of the flange plate due to local bending, and the remaining symbols are as defined previously.

The 1980 proposed U.S. code calls attention to the need to consider the combined effects of axial compression and bending under transverse loading when carrying out the design of orthotropic decks on the elastic basis, but the code offers no guidance on how this may be done.

Eurocode 3, Part 2, which is valid for steel bridges, states that longitudinal stiffeners with transverse load should be designed as beam-columns. There is also a recommendation on the combination of global and local effects that takes the full value of one effect and a reduced value of the other. The recommended combination factor ranges from 0.7 to 1 depending on the span. This is, however, a nationally determined parameter and some countries may prefer to set it to zero.

7.3.6 Cross-Sectional Limitations for Stiffeners

When the longitudinal stiffeners (or actually any stiffeners, including transverse flange and web stiffeners) are designed as open sections such as flat bars, tees, or angles, limitations are placed on their cross-sectional geometry to ensure that local buckling of the stiffeners does not precede the attainment of the ultimate strength of the flange. In the case of closed sections such as troughs or vees, used in orthotropic steel deck construction, the limitations on cross-sectional geometries are adjusted to allow for bending in the plane of the stiffener walls and to ensure that economical use of relatively thin walled sections is not precluded. Limitations of stiffener cross-sectional geometry for design are based on controlling the applied stresses under ultimate load to values that are fractions of the elastic critical buckling stresses. In BS 5400 the safety factor used is 2.25, which was derived as a suitable value for flat stiffeners but has been applied to all other types of stiffening. (The UK assessment rules provide a basis for calculating limiting stresses for stiffeners outside those limitations and with measured imperfections.) The limitations imposed in the proposed 1980 U.S. code (Wolchuk and Mayrbourl, 1980) are somewhat different and correspond to the requirement that the torsional buckling stress should be greater than the yield stress, with relaxations to allow for "thinning out" the stiffeners in low-stress zones. The AASHTO LRFD specifications provide limitations on the section properties associated with the longitudinal bottom-flange stiffener.

A summary of these criteria provide the following restrictions related to the geometry of a stiffener: In the 1980 proposed U.S. code

$$C_s \le \begin{cases} \dfrac{0.40}{\sqrt{F_y/E}} & \text{for } f_{\max} \ge 0.5F_y \qquad (7.20) \\ \dfrac{0.65}{\sqrt{F_y/E}} & \text{for } f_{\max} \le 0.5F_y \qquad (7.21) \end{cases}$$

where $f_{\max}$ is the maximum factored compression stress and the effective slenderness coefficient C_2 is given as

$$C_2 = \begin{cases} \dfrac{d}{1.5t_0} + \dfrac{w}{12t} & \text{for flats} \qquad (7.22a) \\ \dfrac{d}{1.35t_0 + 0.56r_y} + \dfrac{w}{12t} & \text{for tees or angles} \qquad (7.22b) \end{cases}$$

For any outstanding part of a stiffener,

$$\frac{b'}{t'} \le \frac{0.48}{\sqrt{F_y/E}} \tag{7.23}$$

In the AASHTO LRFD specifications, in addition to Eq. 7.23, the moment of inertia I_ℓ of each stiffener about an axis parallel to the flange and taken at the base of the stiffener shall satisfy

$$I_\ell \ge \psi w t_{fc}^3 \tag{7.24}$$

where $\psi = 0.125k^3$ (for $n = 1$); or $0.120k^3$ (for $n = 2$)

k = plate-buckling coefficient for uniform stress, $2.0 \le k \le 4.0$

n = number of equally spaced longitudinal flange stiffeners

w = larger width of the flange between longitudinal flange stiffeners or distance from a web to the nearest longitudinal flange stiffener

t_{fc} = thickness of compression flange

In BS 5400, for flats,

$$\frac{h_s}{t_s}\sqrt{\frac{\sigma_{ys}}{335}} \le 10 \tag{7.25}$$

or less than or equal to a higher value obtained from Fig. 7.15*a* when $b/t\sqrt{\sigma_y/355} \le 30$. For angles,

$$b_s \le h_s$$

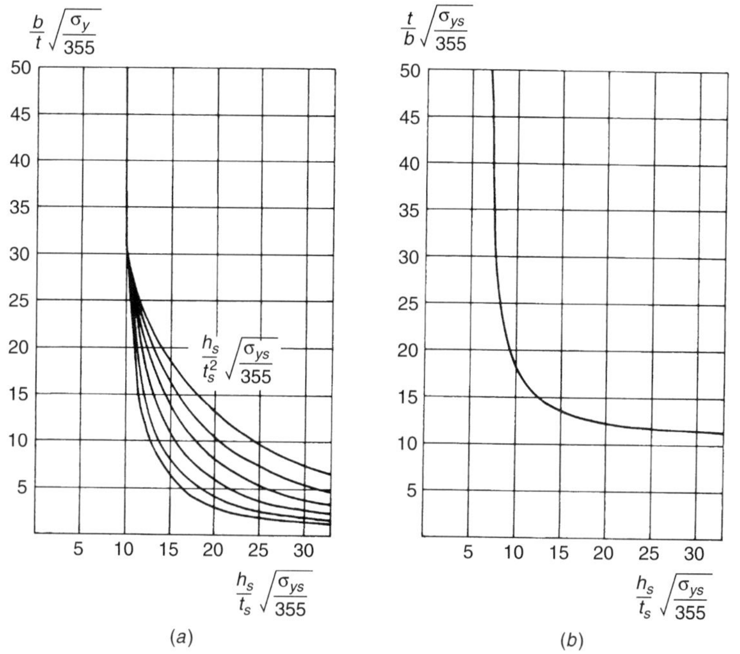

FIGURE 7.15 Limiting slenderness for: (*a*) flat stiffeners; (*b*) angle stiffeners.

$$\frac{b_s}{t_s}\sqrt{\frac{\sigma_{ys}}{355}} \leq 11 \tag{7.26}$$

$$\frac{h_s}{t_s}\sqrt{\frac{\sigma_{ys}}{355}} \leq 7$$

or less than or equal to a higher value obtained from Fig. 7.15*b* when $(l_s/b_s)\sqrt{\sigma_{ys}/355} \leq 50$.

In these formulas and as shown in Fig. 7.16,

d, h_s = stiffener depth
b_s = width of angle
t_0, t', t_s = stiffener thickness
t = plate thickness
w, b = spacing of stiffeners
l_s = span of stiffener between supporting members
r_y = radius of gyration of stiffener (without plate) about axis normal to plate
σ_{ys} = yield stress of stiffener, N/mm^2
σ_y = yield stress of plate, N/mm^2

Thus the design model relates to an orthogonally stiffened flange in which the controlling buckling mode envisioned is buckling of the longitudinally stiffened plate between transverse stiffeners, which may or may not be accompanied by local

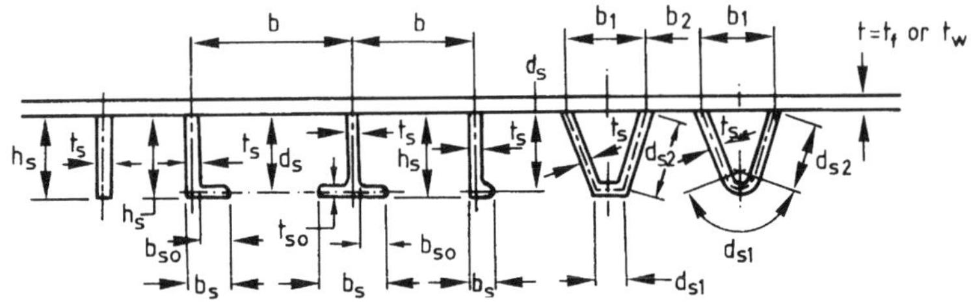

FIGURE 7.16 Flange and web stiffeners.

plate buckling between stiffeners. Local buckling of the stiffener is suppressed, however, together with any participation of the transverse members in the overall buckled mode.

7.4 BENDING STRENGTH OF BOX GIRDERS

The design of a box girder subjected primarily to bending moment is often controlled by local buckling of the compression flange. Unlike plate girders, lateral torsional buckling rarely governs for fully constructed box girders. Many composite box girder bridges, however, rely on the concrete deck as the "fourth side" of the box section in their final constructed state and utilize only a truss-type system of top-flange lateral bracing to form a "quasi-closed" section during construction. As a result, composite box girder bridges under erection can be sensitive to global stability concerns before the concrete slab is cast. Component girders have actually collapsed during casting due to lateral torsional buckling, with one such failure including the recent collapse of the Marcy Bridge in the United States (Yura and Widianto 2005). Although the methods described in Section 7.3 can be used to calculate the local strength of compression flanges, the design of composite box girders requires a more complete analysis of the entire girder system. This analysis should include careful consideration of the top-flange lateral bracing or, if limited or no such bracing is provided, the possibility of lateral torsional buckling of the girder. Section 7.9 provides a more detailed discussion of top-flange lateral bracing issues.

In the simplest case, it can be assumed that the bending strength of box girders is contributed by the flanges alone, and the moment resistance can be calculated on the basis of the design force multiplied by the distance separating the flange centroids. Such an approach is allowed within BS 5400: Part 3 and greatly simplifies design.

To allow for the contribution of the web in calculating the strength of the box girder, a linear distribution of stress over the depth of the cross section may be assumed. Effects of buckling in the web can be represented by using effective widths of web adjacent to the stiffeners as outlined in Chapter 4. The British code BS 5400 also allows the use of the effective-thickness approach for girders, which has the advantage that no recalculation of the position of the neutral axis is needed, a shortcoming of the effective-width approach. Cooper's expression for bending strength (1967) is used to give the effective thickness for unstiffened webs, and the full thickness is used for webs stiffened by effective longitudinal stiffeners. Failure is assumed to occur when the extreme fiber flange stress reaches the smaller of the calculated ultimate stress of the compression flange σ_{lc} and the yield stress of the tension flange σ_{yt}. Thus in BS 5400

$$M_D = Z_{xc}\sigma_{lc} \quad \text{or} \quad M_D = Z_{xi}\sigma_{yt} \tag{7.27}$$

where Z_{xc} and Z_{xt} are the elastic moduli of the effective section for the extreme compression and tension fibers, respectively.

7.5 NOMINAL SHEAR STRENGTH OF BOX GIRDERS

This section discusses the calculation of the nominal shear strength of box girders without consideration of bending–shear interaction. The effects of bending–shear interaction on the final shear strength of box girders are presented in Section 7.6.

The key differences between plate and box girders that may influence the shear strength of the webs is the use of relatively thin flanges in box girders at the boundaries of the webs and the treatment of web/diaphragm junctions. Caution is needed in applying available tension-field models, derived and verified in the context of plate girder webs, to the design of the webs of box girders. Of major concern is the relatively small amount of resistance against in-plane movement that may be contributed to the web by the thin flange of a box girder, in comparison to the restraint offered by the thicker and narrower flange of a corresponding plate girder. In the latter, the out-of-plane bending rigidity and in-plane extensional rigidity of the flange to resist movement perpendicular to and parallel to the flange/web junction, respectively, is more effectively mobilized than in the case of thin flanged box girders. In BS5400, this is accounted for by restricting the effective width of flange plate that may be assumed to provide restraint to the web. In the AASHTO LRFD specifications, this is addressed by flange proportion limits (essentially identical to those for I-shaped girders) that are intended to ensure that the flanges provide some restraint against web shear buckling and also that they will provide boundary conditions for the web that are consistent with assumptions inherent in the web bend-buckling and compression flange local buckling provisions of those specifications.

Eurocode 3, Part 1-5 also gives general rules for the shear buckling resistance of unstiffened and stiffened plates. The rules are applicable to I-girder webs as well as box girders. Strength provisions that assume no contribution from the flanges are based on a rotated stress theory developed by Höglund (1971) that is modified by curve fitting to test results. In contrast to tension-field theories, this approach takes as the starting point the resistance of a web with transverse stiffeners at the supports only. The theory has a basic assumption that there are no resulting transverse stresses in the web and hence it does not utilize any bending stiffness of the flanges. The theory only requires that the flanges have adequate axial stiffness to resist some compression to balance the longitudinal tensile stresses in the web. The strengths derived differ little from those from BS5400 for panel aspect ratios of 1.0

or less, but the method does provide an increased plateau of slenderness at which webs may be stressed to yield. The effect of transverse and longitudinal stiffeners is considered by their effect on the critical shear stress. For a web with intermediate transverse stiffeners, there is a small addition to the shear resistance from the flanges, although it has been suggested that this increase should be neglected for box girders. It should be noted that in following Eurocode 3 there is no requirement that transe stiffeners should be used and their effect is often so small that it is often more economical to avoid or minimize their use.

For the special case of box girders with sloping webs, the web must be designed for the component of vertical shear in the plane of the web.

7.5.1 Box Girders without Longitudinal Web Stiffeners

In BS 5400: Part 3 (BSI, 2000) the tension-field model of Rockey and his co-workers (Porter et al., 1975) has been modified for application to both plate and box girders without longitudinal stiffeners (see Chapter 6). Thus for box girders with unstiffened flanges but with transverse stiffeners in the webs (or flanges)—a form of construction associated with relatively small boxes— advantage can be taken of postbuckling strength using a tension-field model. Limited use is made of the plastic frame mechanism action in the tension field in order to keep shear deformations within the limits for the whole girder. To accomplish this, the maximum possible shear capacity is limited to the shear yield capacity of the web alone. In the case of a box with thin unstiffened flanges, the maximum width of a flange on either side of the web taken as effective is $10t_f\sqrt{355/\sigma_{yf}}$. When no flange projection occurs, the effect of frame action is neglected in the calculation of tension-field capacity (Harding and Dowling, 1981).

In addition to their use in increasing the shear capacity of web plates, transverse stiffeners in box girder bridges may require special treatment because they commonly form parts of cross frames that carry deck loads or are used to resist distortional effects. Otherwise, design methods for such stiffeners in plate girders are applicable provided that the flange bending contribution is limited by the requirements of BS5400.

Wolchuk and Mayrbourl (1980) suggest, in applying the tension-field model to the transversely stiffened webs of box girders, the solution of Basler (1961), which is based on the assumption of negligible flange bending rigidity. This corresponds roughly to neglecting the frame action in the Rockey solution and reflects the caution needed for box girders, where the flanges are generally more slender than in plate girders. Thus the 1980 proposed U.S. code suggests that

$$V_u = V_B + V_T \tag{7.28}$$

with

$$V_B = Dt_w F_{v,cr} \tag{7.29}$$

$$V_T = \frac{Dt_w F_T}{2(\sqrt{1 + \alpha^2} + \alpha)} \tag{7.30}$$

where D = depth of web between flanges, measured along web
d_o = transverse stiffener separation
$\alpha = d_o/D$
t_w = web thickness
$F_{v,cr}$ = critical buckling shear stress
F_T = tension-field stress

It should be noted that the provisions offered by Wolchuk and Mayrbourl (1980) in the 1980 proposed U.S. code require consideration of moment–shear interaction.

The AASHTO LRFD specifications (2007) reflect significant recent research, particularly into the postbuckling shear behavior of stiffened webs, and offer slightly different interpretations of Basler (1961). While these more recent provisions are still based largely on Basler's work, they offer modifications that eliminate the need to consider moment–shear interaction, thus simplifying the shear design of webs. The AASHTO LRFD specifications for box girder web shear design in large part defer to the web shear design provisions for I-shaped girders, which are presented in Chapter 6.

7.5.2 Box Girders with Longitudinal Web Stiffeners

In the case of box girders with longitudinal stiffeners there is limited experimental evidence available to support the application of tension-field theories to web design (Höglund, 1995) except that the rotated stress field theory can be adapted to consider the influence of longitudinal stiffeners. Rules for such a procedure are included in Eurocode 3, Part 1–5. To date, code writers have been cautious in addressing the interaction between thin longitudinally stiffened webs and thin box girder flanges. Conservative approaches to design have been suggested that still take advantage of postbuckling reserve, albeit to a lesser extent than might be possible with a full plastic tension-field treatment. The British codified method and the nearly identical method of the 1980 proposed U.S. code, however, take a different approach than the AASHTO LRFD specifications. Both approaches are described herein.

In the British codified method (BSI, 2000) the stiffened web is checked on a panel-by-panel basis. The design procedure consists of calculating the longitudinal stresses using simple bending theory and gross areas. Shear forces are assumed to be distributed uniformly down a cross section. Each panel is then checked for yielding under combined compression, bending, and shear using the interaction formula

$$\left(\frac{\sigma_1 + 0.77\sigma_b}{\sigma_{yw}}\right)^2 + 3\left(\frac{\tau}{\sigma_{yw}}\right) \leq 1 \tag{7.31}$$

and for buckling using

$$\frac{\sigma_1}{\sigma_{yw} K_1} + \left(\frac{\sigma_b}{\sigma_{yw} K_b}\right)^2 + 3\left(\frac{\tau}{\sigma_{yw} K_q}\right)^2 \leq 1 \tag{7.32}$$

where
σ_1 = mean longitudinal stress
σ_b = maximum longitudinal stress due to in-plane bending
τ = average shear stress
σ_{yw} = yield stress of web material
K_1, K_b, K_q = coefficients for ultimate plate strength

The coefficients were derived from large-deflection elastoplastic computer analysis of isolated plates, with results adjusted according to test data, and the interaction formula was shown to give lower bound solutions by similar analyses (Chatterjee,1981).

In checking yielding, any proportion of the longitudinal stresses σ_1 and σ_b, up to 60% maximum in a panel, can be assumed to be shed to the flanges while maintaining overall equilibrium. In checking stability, up to 60% of these stresses can be shed from the restrained inner panels, but none can be shed from outer panels, which are considered to be unrestrained.

The 1980 proposed U.S. code uses the same approach as for webs stiffened transversely only, except that $F_{v,cr}$ is now calculated for each subpanel bounded by longitudinal and transverse stiffening and assumes the lowest value.

The AASHTO LRFD specifications acknowledge in their commentary that the shear resistance of the entire panel can be taken as the sum of the shear resistance of the subpanels, but also note that longitudinal web stiffeners are often provided as a measure to improve flexural capacity, and when a longitudinal stiffener is located in the optimal position for flexure, its contribution to shear resistance is relatively small. As a result, it is specified that the influence of longitudinal stiffeners can be neglected in the computation of the nominal shear resistance of webs, and hence, the shear provisions are essentially identical for webs with and without longitudinal web stiffeners. Note that by taking this approach, the AASHTO LRFD specifications are able to implicitly allow consideration of postbuckling tension-field action in the computation of shear resistance in webs with longitudinal web stiffeners, but they conservatively do so without any beneficial contributions from those longitudinal web stiffeners.

7.6 STRENGTH OF BOX GIRDERS UNDER COMBINED BENDING, COMPRESSION, AND SHEAR

The interaction of bending and shear effects needs to be considered when designing the webs of box girders. In addition, the effect of mean longitudinal stresses in the webs should be accounted for in cases of girders with unequal flanges (particularly for girders with composite concrete flanges) and in cable-stayed bridges.

This section will expand nominal shear strength formulations given in Section 7.5 to include the interaction of bending and shear. The British and Eurocode provisions for box girders and the 1980 proposed U.S. code explicitly require consideration of bending–shear interaction in calculating shear strength. The AASHTO LRFD specifications, on the other hand, have recently been revised based on current research in such a way as to eliminate the need to consider bending–shear interaction. Designers are cautioned to fully comprehend the shear provisions of the specific design code being followed and to consider bending–shear interaction as appropriate in a manner consistent with the shear design provisions of the specific design code being followed.

7.6.1 Box Girders without Longitudinal Web Stiffeners

If, in the absence of mean compressive stresses, the bending and shear strengths have been calculated without any contributions from the web and flanges, respectively, there is evidence to suggest that girders can safely resist these magnitudes of moment M_R and shear V_R acting simultaneously, so no interaction needs to be considered. For box girders with flanges and webs unstiffened longitudinally, this approach provides a simple conservative estimate of combined bending and shear strength.

Webs, however, are often assumed to make some contribution to the bending strength M_D. Similarly, they are also typically designed to take advantage of post-buckling strength and, hence, the flanges may have been considered to contribute to the shear strength V_D through their framing action. In this regard, BS 5400 proposes the use of an interaction diagram (Fig. 7.17) that can be used for box girders with no longitudinal stiffeners. Unfortunately, no guidance is given on how to account for the effect of mean longitudinal compressive stresses.

For transversely stiffened webs of box girders, the 1980 proposed U.S. code (Wolchuk and Mayrbourl, 1980) uses an interaction equation to calculate the critical buckling stress for combined shear, bending, and compression, in terms of ratios

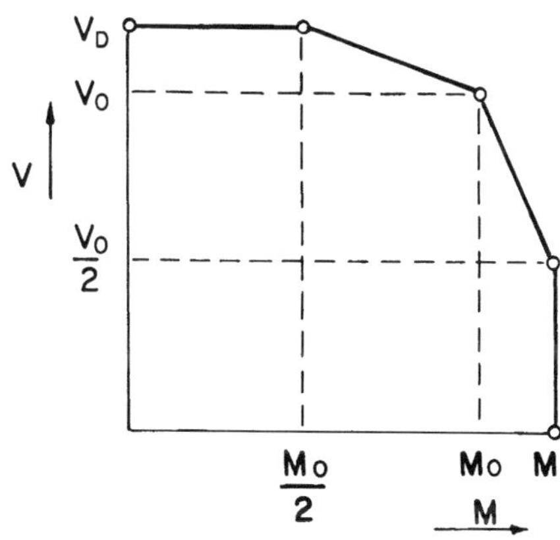

FIGURE 7.17 Interaction between shear force and bending moment.

of the individual stress components to their critical buckling values,

$$\left(\frac{F_{v,cr}}{F^0_{v,cr}}\right)^2 + \left(\frac{F_{b,cr}}{F^0_{b,cr}}\right)^2 + \frac{F_{c,cr}}{F^0_{c,cr}} = 1 \tag{7.33}$$

The stress components, however, are interdependent and may be related as follows:

$$F_{b,cr} = \frac{1-R}{2}\mu F_{v,cr} \tag{7.34}$$

$$F_{c,cr} = \frac{1+R}{2}\mu F_{v,cr} \tag{7.35}$$

where

$R = \dfrac{f_{2w}}{f_{1w}}$

$\mu = \dfrac{f_{1w}}{f_v}$

f_{1w} = governing axial compressive stress at longitudinal edge of web panel

f_{2w} = axial stress at opposite edge of panel

f_v = governing shear stress, $=V/D_{tw}$

These stresses are illustrated in Fig. 7.18. The buckling stresses $F^0_{v,cr}$, $F^0_{b,cr}$, and $F^0_{c,cr}$ are computed by assuming that only shear, bending, or compressive stresses, respectively, are present.

An additional force ΔF is added to the compression- and tension-flange forces computed in accordance with elastic analysis. It includes that portion that must be transferred to the flanges from the webs after buckling as well as additional force

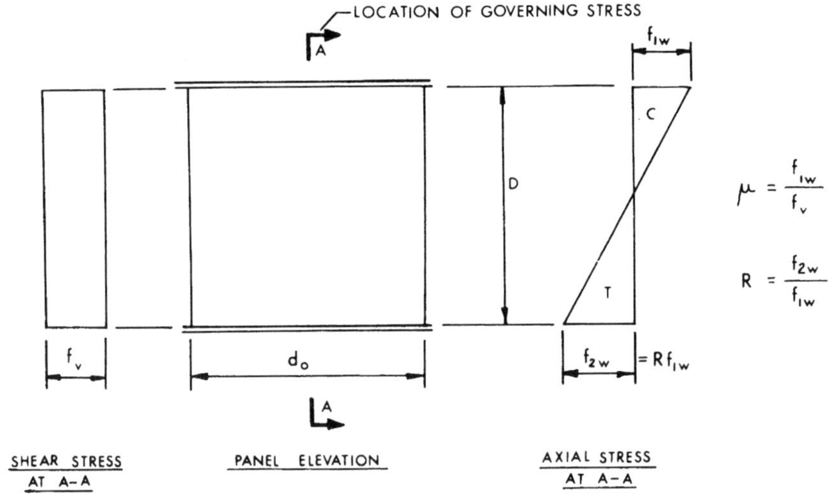

FIGURE 7.18 Definition of μ and R for unstiffened and transversely stiffened webs.

related to the use of tension-field action in designing the webs. The additional force for compression flanges is

$$\Delta F_1 = \left(1 - \frac{\sum V_B}{V_M}\right)\left[(f_{1R} - f_1)A_{fc} + \tfrac{1}{2}V_M \cot\left(\tfrac{1}{2}\theta_d\right)\right] \tag{7.36}$$

and for tension flanges

$$\Delta F_2 = \left(1 - \frac{\sum V_B}{V_M}\right)\left[(f_{2R} - f_2)A_{ft} - \tfrac{1}{2}V_M \cot\left(\tfrac{1}{2}\theta_d\right)\right] \tag{7.37}$$

where V_M = factored shear force acting coincident with maximum moment
$\sum V_B = \sum Dt_w F_{v,cr}$ = sum of buckling shear capacities of all webs at cross section considered
f_1, f_2 = stress in compression and tension flange, respectively, assuming fully participating webs
f_{1R}, f_{2R} = stresses assuming reduced moment of inertia of cross section
A_{fc}, A_{ft} = compression and tension-flange area, respectively
θ_d = angle of inclination of web panel diagonal to the horizontal

As mentioned previously the AASHTO LRFD specifications have recently been revised in such a way as to eliminate the need to consider bending–shear interaction.

7.6.2 Box Girders with Longitudinal Stiffeners

For box girders with longitudinally stiffened flanges and webs, the situation is complicated by the scarcity of research information, and in particular the lack of experimental data. Some redistribution of the longitudinal stress caused by bending or compression in a web is allowed within BS 5400: Part 3 and was noted in Section 7.5 on the design of longitudinally stiffened webs for shear. The resultant stress distribution after such shedding must be such that the whole of the applied bending moment and axial force (if any) is transmitted and equilibrium is maintained. The percentage reduction in stress in the web panels participating in the shedding can vary from panel to panel but is assumed to be uniform within any one panel. No shedding is permitted from panels containing holes larger than a specified size. Similarly, stresses that cause yielding of the tension flange, but not buckling or yielding of the compression flange, may be redistributed within certain restrictions as outlined in BS 5400: Part 3.

In summary, the interaction between moment, compression, and shear in box girders with longitudinal stiffeners is dealt with by relieving the web of some of the longitudinally destabilizing compressive stresses caused by bending and distributing the load to the compression flange while maintaining overall equilibrium of the cross section. In the 1980 proposed U.S. code, combined bending, compression, and shear are treated in the same way for webs with and without longitudinal stiffeners. In BS5400 separate rules are given for longitudinally stiffened webs with interaction relationships for the stress combinations. With longitudinal stiffening, however, the formulas in both these codes are applied to each subpanel in turn, rather than considering the overall web depth. Finally, and as previously noted in Section 7.5, the AASHTO LRFD specifications are developed so as to eliminate the need for considering bending shear interaction. The AASHTO LRFD specifications also indicate that the effects of longitudinal stiffeners should be neglected in the calculation of shear strength. Instead, they allow consideration of postbuckling tension-field action, but without consideration of any beneficial effects from longitudinal web stiffeners.

7.6.3 Rules in Eurocode 3

In considering bending–shear interaction, Eurocode 3 employs an approach that is fundamentally different from the British code, the 1980 proposed U.S. code, and the AASHTO LRFD specifications. The rules for interaction between bending and shear in Eurocode 3, Part 1-1 comprise a check of global action effects according to

$$\frac{M_{Ed}}{M_{pld}} + \left(1 - \frac{M_{fd}}{M_{pld}}\right)\left(\frac{2V_{Ed}}{V_{Rd}} - 1\right)^2 \leq 1 \tag{7.38}$$

where M_{Ed}, V_{Ed} = bending moment and shear of design loads, respectively
M_{pld} = plastic design resistance irrespective of cross section class
M_{fd} = bending moment that can be carried by the flanges alone
V_{Rd} = design resistance to shear buckling

If the girder carries an axial force simultaneously M_{pld} and M_{fd} should be reduced for the presence of the axial force.

Equation 7.38 is of a different form and less conservative than the BS interaction for webs without longitudinal stiffeners, and M_{pld} is used instead of the bending resistance of the beam, which in the BS code may be based on elastic design for noncompact sections. With regard to conservatism, it is worth noting that the BS rules have been shown to be very conservative when compared with finite element analysis results.

In addition to the interaction check, the bending resistance for the effective cross section should be checked. Although the criterion has been verified for unstiffened and longitudinally stiffened webs by tests and computer simulations of I-girders, it is believed to be sufficient for box girders. The buckling check is done at a distance of half the web depth from the cross section with maximum bending and unstiffened webs. Such a check rarely governs the design. For the cross section with the largest bending moment a check for yielding should also be done.

7.7 INFLUENCE OF TORSION ON STRENGTH OF BOX GIRDERS

The level of torsional stress developed in the webs and flanges of practical box girders does not of itself normally constitute an instability problem. Tests on small thin-walled boxes subjected to torsion, however, have shown that local instability of box corners can limit strength. Allowance for shear effects induced by torsion can be made in the design of flanges with stocky plates by further reducing the effective yield stress in the plate due to the presence of combined stresses. In the design of the webs and diaphragms, additional shear stresses caused by torsion may be added to those associated with bending when calculating the total stresses developed in a web or diaphragms.

It should be noted that in the specific stress analysis of box girders, it is important to assess the influence of torsional and distortional warping stresses. Warping effects do not, in and of themselves, constitute an instability problem per se, but they do contribute to longitudinal stresses in the cross section. Annex B in BS5400 and the U.K. bridge assessment rules contain informative material to assist designers in this respect. For the specific case of multibox bridges (bridges utilizing several small, separate box girders and a composite concrete deck), recently published research (Helwig and Fan, 2000) offers a helpful presentation on distortional warping stresses and how they can be controlled by use of internal intermediate diaphragms.

7.8 DIAPHRAGMS

Box girder bridges utilize a wide variety of diaphragms for many functions. For the purposes of discussion it is helpful to separate internal diaphragms from external diaphragms. Only internal diaphragms appear in bridges that use a single box section, whether it be single or multiple cell. Multibox bridges, which utilize two or more smaller box girders and a composite concrete deck, feature both internal diaphragms within each box girder and external diaphragms between adjacent girders.

Two types of internal diaphragms are typically encountered within box girders: intermediate diaphragms (Figs. 7.19 and 7.20) and load-bearing diaphragms (Fig. 7.21). The former carry distortional loads and limit cross-sectional deformation, while the latter are provided at points of support to give paths for vertical loads (web shears) and horizontal loads (flange shears) to the support bearings and to prevent buckling of the webs in the vicinity of these large concentrated loads.

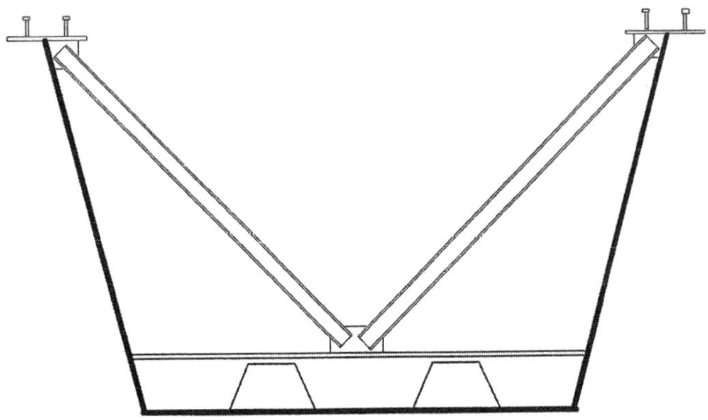

FIGURE 7.19 Intermediate diaphragm without transverse stiffeners in composite box girder bridge (permissible per Eurocode 3).

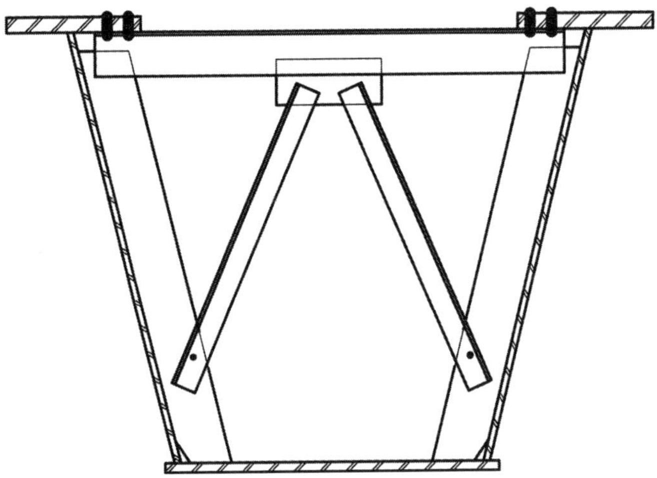

FIGURE 7.20 Intermediate diaphragm in composite box girder bridge (U.S.).

Internal load-bearing diaphragms are typically full depth and can be configured either as unstiffened (or lightly stiffened) plate diaphragms or as fully stiffened plate diaphragms.

Similarly, two types of external diaphragms are typically encountered between adjacent girders of multibox bridges: external intermediate diaphragms (Fig. 7.22) and external diaphragms at supports (Fig. 7.23). External intermediate diaphragms are provided primarily to address construction concerns of the girders prior to

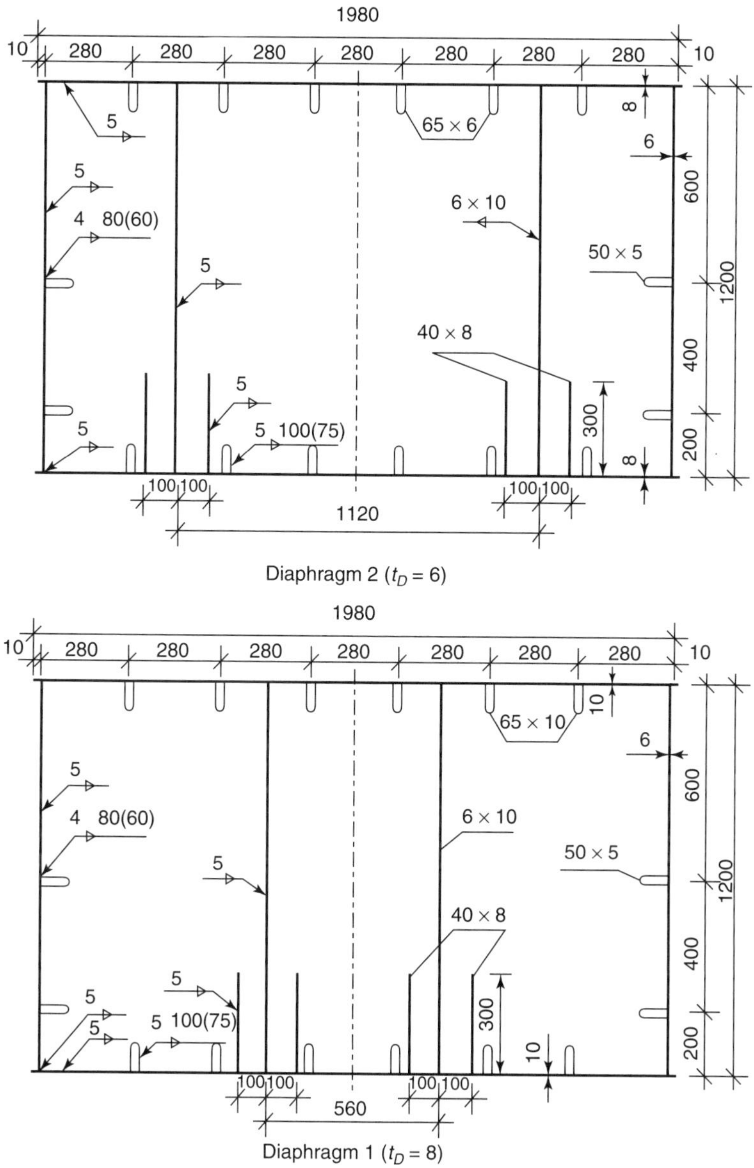

FIGURE 7.21 Details of load-bearing diaphragms.

hardening of the composite concrete deck. On the other hand, external diaphragms at supports are typically utilized to connect adjacent girders in order to develop a force couple to counter an overall system torsion, which often occurs in curved or skewed mutlibox bridges.

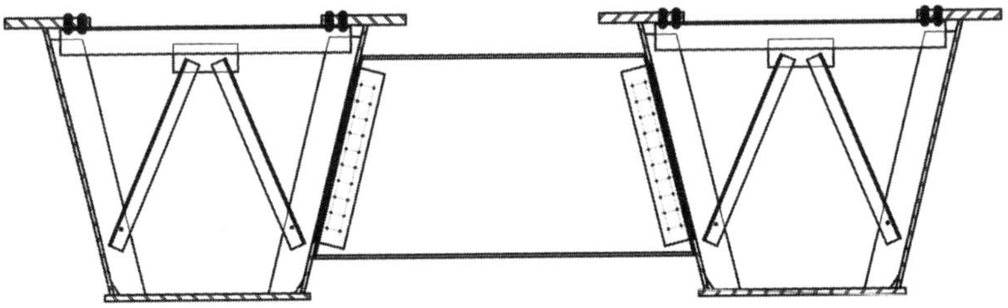

FIGURE 7.22 External intermediate diaphragm between girders in a multibox bridge.

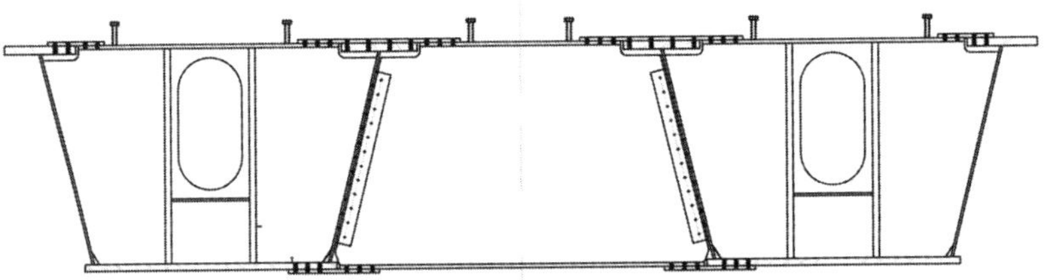

FIGURE 7.23 External diaphragm between adjacent box girders at supports of a multibox bridge.

7.8.1 Internal Intermediate Diaphragms

Methods of design of intermediate diaphragms are given in BS5400: Part3 (those for stiffened or unstiffened plated intermediate diaphragms being contained in the U.K. bridge assessment rules). The AASHTO LRFD specifications do not give specific design methods for intermediate diaphragms. Instead they require that internal intermediate diaphragm spacing be set to limit longitudinal warping stresses due to factored loads to approximately 10% of major axis bending stresses. They further limit the design of the specific internal intermediate diaphragm members to be appropriate for the intended use, typically a truss-type diaphragm that is more common in multibox bridges or a plate-type diaphragm more common in larger single box bridges. Although guidance on internal intermediate diaphragms is provided by U.S. Steel (1978), more recent guidance is provided by Helwig and Fan (2000) and Fan and Helwig (2002).

Plated diaphragms may involve substantial amounts of welding during fabrication and, hence, ring frames or truss-type diaphragms (also know as cross frames or cross bracings) provide an alternative solution that is more economical to fabricate. In the case of internal intermediate diaphragms, their function is to restrict distortion of the cross section and to control the effects of distortional warping. BS5400 gives general rules for the design of intermediate diaphragms and limitations on their layout and stiffnesses. A typical braced intermediate diaphragm for a composite box girder bridge is shown in Fig. 7.19. The vertical web stiffeners have been omitted and the resistance of the web to transverse compression has to be checked for opposite patch loading using rules that are provided in Eurocode 3, Part 1-5.

7.8.2 Load-Bearing Diaphragms

Load-bearing diaphragms need careful consideration in regard to buckling, just as support stiffeners do in the case of plate girders.

Most plated diaphragms are stiffened at the bearing locations with bearing stiffeners. These are often accompanied by short-length stub stiffeners immediately above the bearings, the role of which is to stiffen the diaphragm plate in the vicinity of the bearings so that localized redistribution of stress concentrations can occur through yielding of the plate. In longer diaphragms (Fig. 7.24), secondary stiffening can be provided to stabilize the diaphragm plate as an alternative to the use of a very thick plate.

Transverse secondary stiffening may also be necessary to resist any transverse compression caused by the diaphragm's behavior as a deep beam, an action that is accentuated by the use of sloping webs. As the aim of any sensible design must be to eliminate the possibility of diaphragm buckling limiting the strength of the box girder, it makes little sense to attempt to economize on diaphragm design and, hence, a conservative design approach is usually adopted.

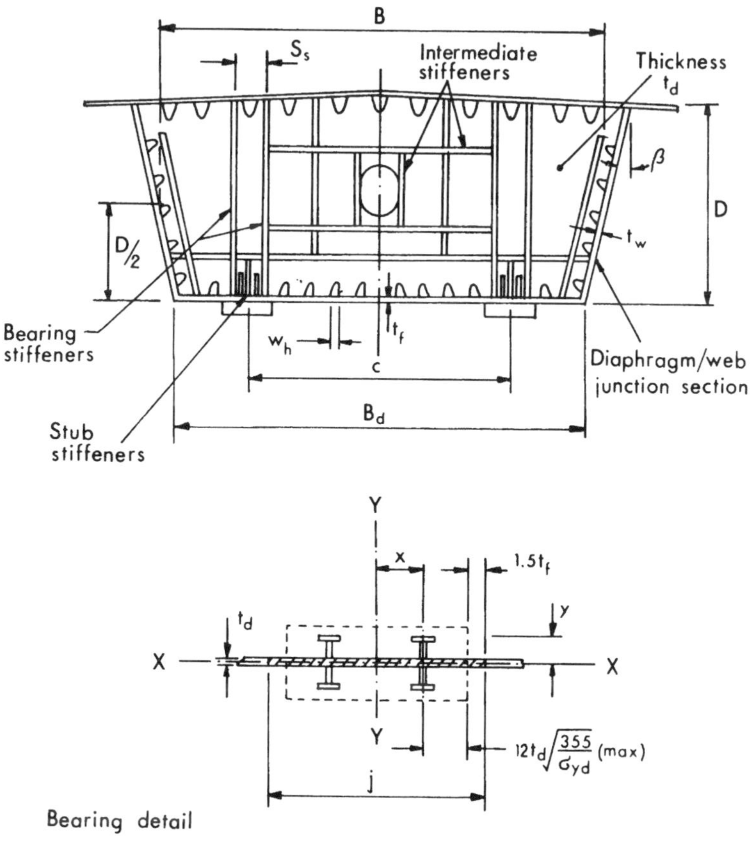

FIGURE 7.24 Geometric notation for diaphragms.

Support-diaphragm design is treated in detail in BS 5400: Part 3. Only single-cell boxes, without steeply sloping webs and with diaphragms normal to the girder axis, are considered. Two approaches to design are provided, one for simple diaphragms and the other for diaphragms of more complicated geometries. In each case the design procedure involves three main features:

1. Limitations of diaphragm geometry
2. Analysis of diaphragm stresses
3. Design checks on diaphragm yielding and buckling

The designer can choose to use diaphragms of simple layout, referred to as *unstiffened diaphragms*, even though they normally have full-height stiffeners at the bearing locations as well as a diaphragm plate (Fig. 7.21). The stiffeners, however, can be omitted in smaller boxes. For larger boxes the designer may opt for *stiffened diaphragms* with arrangements of bearing, stub, and secondary stiffening of the type referred to above.

The design method for the unstiffened diaphragms places greater restrictions on geometry and provides simple formulas for stress analysis as well as simple yielding and buckling expressions for use in the design checks. Stiffened diaphragms, while still subject to limitations of geometry, albeit less restrictive ones, normally require detailed computer analysis of stresses and detailed stiffener and plate panel failure checks.

7.8.3 Unstiffened or Lightly Stiffened Load-Bearing Diaphragms

Plated diaphragms without stiffeners, other than vertical stiffeners above the bearings, may be designed using rules contained in BS5400: Part 3. Vertical stiffeners, which must be the full height of the diaphragm and positioned symmetrically on both sides of the diaphragm, should be placed directly above bearings. These are provided solely to resist out-of-plane bending moments caused by eccentricity of the reactions with respect to the diaphragm midthickness. Such eccentricities may be due to fabrication tolerances or to longitudinal movements (temperature or otherwise) of the box girder. The stiffeners are loaded most heavily directly over the bearings with the loading tapering to zero at the intersection of the stiffener with the top flange, at which location they are normally attached to the flange stiffeners. The diaphragm plate resists all in-plane forces caused by bearing reactions, web and flange shear forces, and friction between diaphragm and bearings. The stiffeners are placed symmetrically to avoid addition to the stresses on the plate, which are calculated by ignoring the presence of the stiffeners.

Limitations on Diaphragm Geometry To prevent torsional instability of the stiffeners, the limitations that relate to stiffeners in general are applied, examples of which are given in Section 7.3.6. To control the effects of openings, needed for access and services, limitations are placed on their positioning and sizing. To avoid complicated calculations, rules are given that allow holes of certain dimensions to be used provided that they are sufficiently small not to affect plate stability. Holes are prohibited within the lower third of the diaphragm depth above bearings to allow some capacity for redistribution in a highly stressed area, a zone where the stresses may be further increased by the misalignment of bearings. Only one circular opening is allowed in each side of the vertical centerline of the diaphragm within the upper third of the depth, and its diameter is limited. Maximum sizes of cutouts for longitudinal stiffeners on the box walls are also given, and the stiffeners should be connected to the diaphragm plate.

Analysis of Diaphragm Stresses A simplified model of the diaphragm interaction with other box girder components is used in the calculation of diaphragm stresses. Portions of the flange are considered to act with the isolated diaphragm, which essentially behaves as a deep beam loaded by edge forces and supported by the bearings. The effective widths to be used in calculations are based on plane stress considerations (shear lag) alone and hence may only be used when the transverse stiffness of the compression flange is not significantly reduced by transverse stresses, that is, provided that the calculated stresses do not exceed:

1. $\frac{1}{4}$ of the flange longitudinal compressive strength
2. $\frac{1}{2}(t_f/b)^2E$, where t_f is the flange plate thickness and b is the spacing of longitudinal stiffeners

Otherwise, a reduced width of flange on either side must be used. It will be noted that no portion of the webs is included. This leads to a conservative estimate of the stresses and greatly simplifies what is in reality a complicated problem due to the high level of both shear and bending stresses, which may coexist in the web at that location if the diaphragm is at an internal support.

For simple diaphragms the shear flows are simplified as shown in Fig. 7.25. Advantage has been taken of the ability of both plate panels and welds connecting the diaphragm and webs to redistribute shears applied in the nonuniform manner protected by elastic theory. This has enabled the shear flows to be taken as uniform. By also including the effects of inclined webs, the resulting reference stress is calculated using

$$\sigma_{R2} = \left[\left(\frac{K_d \sum R_v}{2} + \frac{T}{B}\right) x_R + Q_{fv}\frac{l_f}{2}\right]\frac{1}{Z_e} + \frac{\sum R_v \tan\beta}{2A_e} \tag{7.39}$$

where, in addition to the symbols shown in Fig. 7.26,

K_d = 2 usually and allows for boundary shears
B = average width of diaphragm
$\sum R_v$ = total vertical force transmitted to bearings
Q_{fv} = vertical force transmitted to diaphragm by a change of flange shape
l_f = horizontal distance from reference point to nearest edge of bottom flange
T = torque transmitted to diaphragm

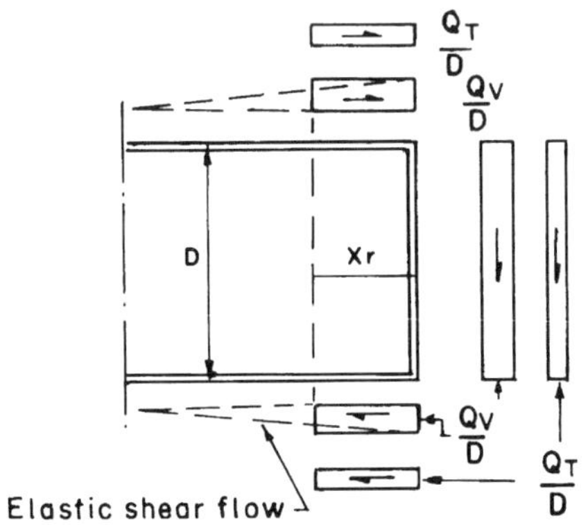

FIGURE 7.25 Shear flows assumed to derive K_D.

Z_e, A_e = effective section modulus and area, respectively, of diaphragm and flanges at the vertical section through the reference point

The vertical stress is calculated using

$$\sigma_{R1} = \frac{R_v(1 + 4e/t_d)}{\left(j - \sum w_h\right) t_d} \tag{7.40}$$

for twin symmetrical bearings, where t_d is the diaphragm thickness and j is the effective width of contact of the bearing pad allowing for load dispersion through the flange (Fig. 7.26). With a single central bearing the vertical stress has an additional component that is given by $0.77(T_b\, j/2I_{yd})$. Because maximum stressing in the vicinity of the bearings is very localized, a factor of 0.77 is used in the yield stress check to provide some allowance for plastic redistribution. In these formulas

R_v = total vertical load transmitted to one bearing

T_b = torsional reaction at single central bearing

$\sum w_h$ = sum of widths of cutouts for stiffeners immediately above the flange and within width j

I_{yd} = second moment of area of width j of diaphragm plate, excluding cutouts

e = eccentricity of bearing reaction along span

The shear stress is given by

$$\tau_R = \left(\frac{\sum R_v}{2} + Q_{fv} + \frac{T}{2B}\right)\frac{1}{A_{vea}} + \frac{Q_h}{A_{he}} \tag{7.41}$$

where Q_h is the shear force due to transverse horizontal loads transmitted from the top flange, A_{vea} the minimum net area of vertical cross section of diaphragm plating, and A_{he} the net area of horizontal cross section through the reference point. Thus, it is possible to calculate the combined stress at the reference point (Fig. 7.26).

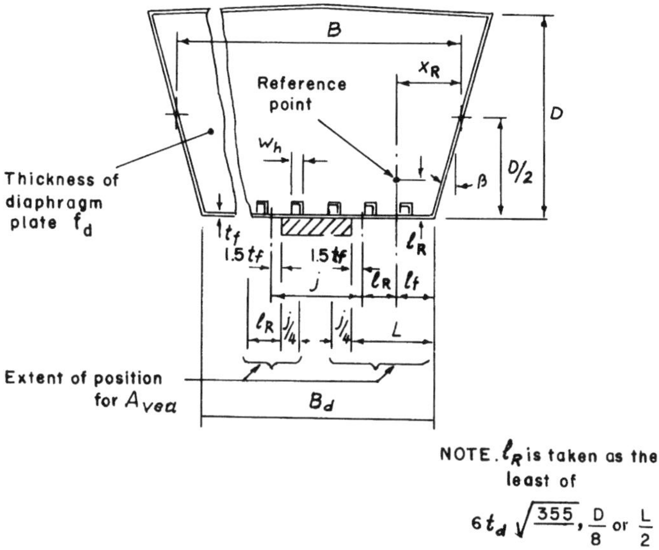

FIGURE 7.26 Reference point and notation for unstiffened diaphragms.

Design Checks on Diaphragm Yielding and Buckling. Because the plate is designed to take all in-plane forces from the webs and flanges, the bearing stiffener is checked for yielding but not buckling. In the check for yielding, however, the axial stresses above the bearings caused by the reactions at those locations are added to the stresses caused by the out-of-plane moments. The effective area of contact of plate and stiffeners to be used in calculating these stresses is shown in Fig. 7.24.

The diaphragm plate is checked for both yielding and buckling. Use is made of approximate expressions for the elastic critical buckling of rectangular and trapezoidal plates loaded along their edges in checking the buckling strength of the diaphragm plate. Rotational restraint provided by flanges and webs to the diaphragm at their intersections is conservatively neglected. The empirical expressions allow for slope of webs, spacing and width of bearings, panel aspect ratio, and influence of top-flange loading in coefficients K_1 to K_4. The complete buckling check is

$$\sum R_v + \frac{T}{l_b} \leq \frac{0.7KEt_d^3}{D} \tag{7.42}$$

where, in addition to the symbols for the dimensions defined in Figs. 7.24 and 7.25,

$K = K_1\ K_2\ K_3\ K_4$

$K_1 = 3.4 + 2.2\ D/Bd$

$K_2 = 0.4 + j/2B_d$ for single central bearing

$\quad = 0.4 + (c - j/3)/B_d$ for twin bearings

$K_3 = 1 - \beta/100$

$$K_4 = 1 - \frac{fP_d}{\sum R_v + T/l_b}\left(\frac{2B}{B_d} - 1\right)$$

c = distance between centers of bearings

$I_b = jK$ for single central bearings

$= c$ for twin bearings

$p_d = W_d \sum P/K_5$

W_d = uniformly distributed load applied to top of diaphragm

P = any local load applied to top of diaphragm

$K_5 = 0.4 + w/2B - B_d$

w = width of load P with allowance for dispersal through flange

$f = 0.55$ for $D/B \leq 0.7$

$= 0.86$ for $D/B \geq 1.5$

with linear interpolation for intermediate values

The rationale for the above buckling check is to ensure that the reactions applied to the diaphragm do not exceed 70% of the reactions causing critical buckling. This is based on the results of numerical parametric studies, which suggest that provided the average stress above the bearing is contained, no further yielding due to buckling will occur in other parts until the diaphragm is loaded beyond 70% of the critical load. The stress in the vicinity of the bearings may attain an average yield stress when the plate is loaded well below critical but is reduced as the stress approaches the cutoff level of 70% of the critical load (see Fig. 7.27).

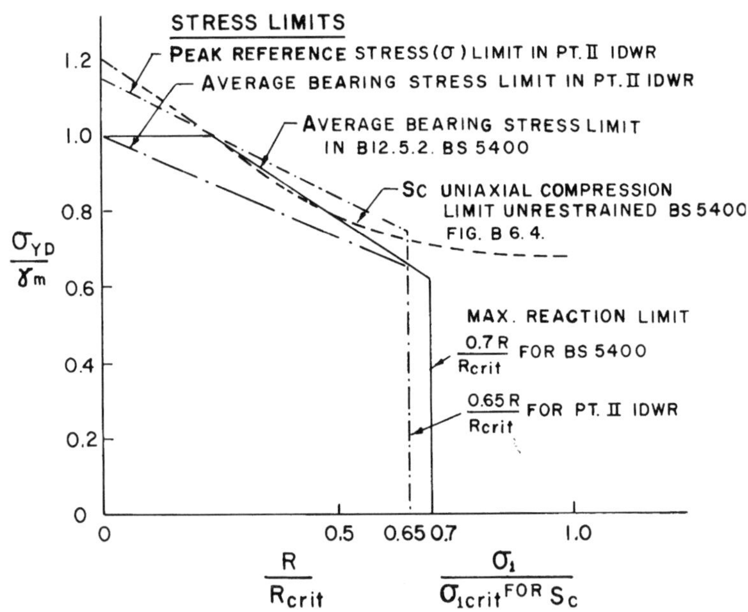

FIGURE 7.27 Stress limits in simple plate diaphragms.

7.8.4 Stiffened Load-Bearing Diaphragms

The basis for the design of stiffened diaphragms differs from that of the "simple" unstiffened diaphragms in that the bearing stiffeners are designed not only to resist any out-of-plane moments but also to act as load-bearing stiffeners to transmit vertical loads to the bearings. The plate panels are then designed for shear and transverse stress only. Secondary stiffeners are used to stabilize the plate by subdividing it into appropriately sized subpanels. Boundary stiffeners may also be provided at the web/diaphragm junctions to help resist tension-field forces in the webs or to transmit reactions from cross girders or floor beams, with or without cantilevers, into the diaphragm.

Limitations on Diaphragm Geometry The limitations on stiffeners to control stiffener buckling are the same as those for unstiffened diaphragms. Load-bearing stiffeners are not required to be symmetrical and can be welded to a single side, although they are still required to be full height. Unstiffened openings that might be used in unstiffened diaphragms may be replaced by larger holes framed by stiffeners designed to resist destabilizing forces. Access holes, required for future inspections, should be provided in internal diaphragms at supports. The placement of these access holes must consider the stress flow in the diaphragm due to bending and torsional moments. Several methods have been employed to assess the effects of access holes in internal solid web diaphragms, ranging from simply discounting the cross-sectional properties removed by the access hole, to treating the remaining sections of webs as local beams below the access hole, to calculation of redistribution stresses in the adjacent panels around the access hole. Darwin (1990) offers a comprehensive treatment of several of these methods. Bruhn (1973) offers a treatment specifically of the method involving stress redistribution in adjacent panels around the access hole.

Analysis of Diaphragm Stresses A rational analysis such as finite-element analysis is often used to determine stresses in a stiffened diaphragm. Such an analysis would model the presence of stiffeners and openings with sufficient accuracy for design purposes. Analytical modeling procedures and assumptions require careful choice and the results must be properly interpreted in their application to strength calculation. Some guidance on this is given in the British National Annex to EN1993-1-5, "Finite Element Methods of Analysis" (BSI, 2006a). Alternatively, a simplified analysis can be employed to determine the in-plane transverse stresses in the diaphragm plate. Secondary stiffeners are ignored in calculating diaphragm properties. Stresses must be calculated at the corners of each plate panel and any in-plane bending stresses due to Vierendeel action around large openings must be calculated separately and added to the other in-plane stresses. Load-bearing stiffeners, including a width of plating no more than 12 times the diaphragm thickness, are designed to carry linearly varying axial stresses compatible with the assumption of uniform shear flows and out-of-plane moments that decrease linearly with height.

Design Checks on Diaphragm Yielding and Buckling Plate panels are checked for yield at all critical parts, ignoring vertical stresses. The panels are checked for buckling using ultimate-strength interaction formulas originally derived for use in the design of longitudinally stiffened webs. Panels in the vicinity of the bearings are proportioned so that overall yield precedes buckling. Other panels are designed for buckling.

Load-bearing stiffeners are designed as struts, as in the case of longitudinally stiffened panels. A yield check is made for all sections, with some overstress, $1.33\sigma_{ys}$, permitted at the points of contact with bearings. Buckling checks are confined to the middle third of the height, where destabilizing effects are greatest in stiffeners of constant section. The destabilizing effects of transverse plate stresses are accounted for by means of an additional fictitious load in the stiffener. This load is taken as

$$P_{se} = \frac{\sigma_q l_s^2 t_d k_s}{a_{\max}} \tag{7.43}$$

where σ_q is the horizontal stress in the middle third of length l_s and l_s is the length of stiffener between points of effective restraint; $a_{\max}$ is the spacing of vertical stiffeners; k_s is found from Fig. 7.7 using $l = l_s$.

Intermediate stiffeners are assumed to be free of axial stresses and applied moments. They are proportioned to resist the destabilizing effect of transverse and shear stresses in the case of vertical stiffeners and shear stress alone in the case of horizontal stiffeners. A similar approach is used for stiffeners framing large holes, in which case the destabilizing terms are calculated assuming the hole to be absent.

7.8.5 External Intermediate Diaphragms

As indicated at the beginning of this section, external intermediate diaphragms and external diaphragms at supports are the two types of external diaphragms that are typically encountered between adjacent girders of multibox bridges. External intermediate diaphragms are provided primarily to address construction concerns of the girders prior to the hardening of the composite concrete deck. These external intermediate diaphragms aid in controlling torsional rotations of the girders (geometry control) and they also improve the overall stability of the noncomposite girders by allowing them to act as a more stable system rather than as less stable individual girders. Depending on the specific parameters of each bridge, these external intermediate diaphragms can sometimes be omitted. In cases when they are provided, designers and owners sometimes specify that external intermediate diaphragms be removed after the composite concrete deck is in place and hardened.

External intermediate diaphragms can take a number of forms, including full-depth plate diaphragms, partial-depth plate diaphragms, and truss-type diaphragms. Their internal forces are generally calculated based on an overall analysis of the bridge under all applicable loadings. Their specific designs are based on the type of diaphragm being used. For plate-type diaphragms, the rules for I-shaped girders are typically followed. Their design is typically straightforward due to the simple nature of their loading. Built-up plate girder sections are typically used to allow for tailoring of the specific diaphragm size, but rolled steel beams may be used as well. For truss-type diaphragms, typical truss design procedures are used. Designers should keep in mind, particularly for curved bridges, that the stiffness of these diaphragms affects the load distribution between adjacent girders and should be carefully modeled in the analysis.

7.8.6 External Diaphragms at Supports

External diaphragms at supports are typically utilized to connect adjacent girders in order to allow for the development of a force couple that can react to overall system torsion, behavior common in curved or skewed multibox bridges. These multibox bridges often utilize only a single bearing per box girder due to the small size of each of the box girders and thus would be torsionally unstable if not connected to adjacent girders at the supports.

External diaphragms at supports are typically full-depth plate-type diaphragms. Because the main purpose of these diaphragms is to develop a force couple between adjacent girders, full moment connections should be provided when connecting the diaphragms to the girders (and to the internal diaphragms at supports). Their design is typically straightforward due to the simple nature of their loading. Built-up plate girder sections are typically used to allow for tailoring of the specific diaphragm size, but rolled steel beams may be used as well.

7.9 TOP-FLANGE LATERAL BRACING OF QUASI-CLOSED SECTIONS

In many box girder bridges, a composite concrete deck is provided that forms the "fourth side" of the box section. This is especially common in multibox bridges (Fig. 7.28). For these types of bridges, the box section is not a closed section until the cast concrete deck hardens. For some cases of straight girder bridges, this may be acceptable given that lateral–torsional buckling stability is carefully investigated during the construction phase (prior to hardening of the concrete). On the other hand, it is necessary in many cases, including curved or skewed bridges and some straight girder bridges, to provide top-flange lateral bracing to form a "quasi-closed" section for each box girder. Providing this type of top-flange lateral bracing dramatically increases the torsional stiffness of the girder for the critical construction phase prior to the hardening of the concrete deck.

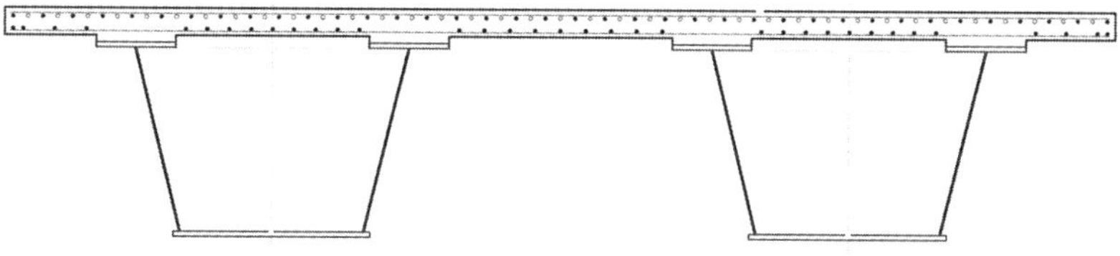

FIGURE 7.28 Multibox bridge with multiple adjacent box girders.

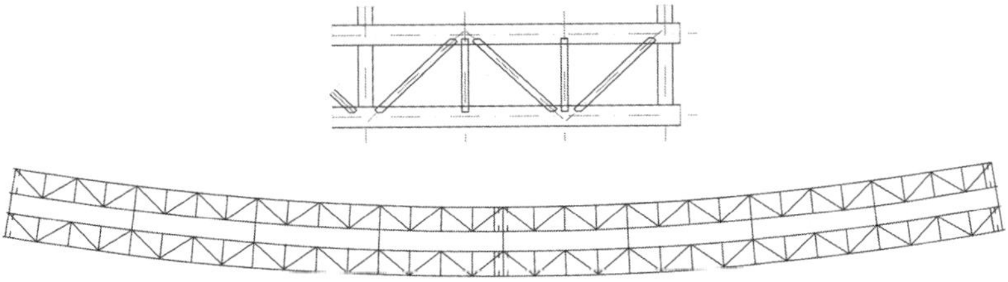

FIGURE 7.29 Top-flange lateral bracing system in a dual curved girder multibox bridge, with detail view above.

This top-flange lateral bracing most often takes the form of a horizontal truss, with the girder top flanges forming the chords of the truss and the lateral bracing forming the diagonals and verticals of the truss (Fig. 7.29). Careful consideration should be given to the layout of top-flange lateral bracing. The bay spacing of the bracing should be chosen to coincide with the internal intermediate diaphragm spacing or a fraction thereof, because the internal intermediate diaphragm top chord will also serve as a vertical in the top-flange lateral bracing's horizontal truss system. In addition, the bay spacing of the bracing should be selected to keep the angle of the diagonal members as close as possible to the ideal value of 45°. The width and lateral stiffness of the girder top flanges are also important parameters in the overall design of the top-flange lateral bracing system.

The analysis, design, and detailing of top-flange lateral bracing can be fairly complex. For the calculation of internal loads in this bracing system a number of approaches have been used. If direct-analysis results are desired for the bracing forces, a three-dimensional finite-element analysis is typically used, in which the flanges, webs, diaphragms, and top-flange lateral bracing are all modeled in detail. As an alternative, a two-dimensional finite-element analysis (sometimes called a "grid" or "grillage" analysis) can be performed, but then the top-flange lateral bracing forces must be calculated outside of the analysis model by hand calculations. Fan and Helwig (1999) provide guidance and design formulas for such calculations.

Once the internal forces in the top-flange lateral bracing system have been determined, its members and connections can be proportioned using typical truss design methods and assumptions. Care should be taken in addressing connection and member eccentricities, effective lengths, and so on, noting throughout the design process that the top-flange lateral bracing provides a critical and primary part of the overall stability of the box girders during the construction phase of the bridge.

Instead of a horizontal truss, trapezoidally corrugated sheeting that behaves as a diaphragm can be used. This sheeting can also be utilized as a "stay in place" or "lost" form for casting the concrete slab. In employing this design concept, it is imperative that both stiffness and strength be considered. Recent studies in the United States for using this type of system for plate girders have suggested that there is good potential for successful use in box girders, further noting that careful design of the connection details is critical (Helwig et al., 2005). One important issue is that distortion of the sheeting tends to reduce its stiffness substantially if it is fastened through the bottom of the troughs.

7.10 RESEARCH NEEDS

Despite the large amount of research carried out on box girders in the 1970s there are still some aspects of the stability of box girders that need further attention, some of which are summarized below.

Flange Buckling

- The limitations on cross-sectional shape to prevent local buckling of open and closed stiffeners could be defined more accurately.
- The limitations on redistribution of stresses caused by shear lag in the flange need to be defined more clearly for both simply and continuously supported box girders.
- A more rational ultimate load method for the design of transversely loaded flanges needs to be evolved.

Web Buckling

- Further research is needed into the buckling strength of transversely stiffened webs and their stiffeners.
- The application of corner stiffening to boxes as an aid to stabilizing the web and corner could be investigated.
- Design rules for curved webs are required.

Diaphragms

- Simplification of the design approach to unstiffened diaphragms should be sought based on further research.
- Methods for the design of diaphragms should be produced to cover the cases of skew diaphrams and twin-walled diaphragms.
- The effects of interactions among diaphragm, flange, and web need more consideration.

Behavior of Boxes

- The strength of longitudinally stiffened boxes under combined bending, compression, and shear needs closer attention in the future.
- The strength of stiffened boxes under combined bending, compression, shear, and torsion also needs attention.
- The behavior of box girder bridges, especially curved, during steel erection should be further investigated.

REFERENCES

Abdel-Sayed, G. (1969), "Effective Width of Steel Deck-Plate in Bridges," *ASCE J. Struct. Div.*, Vol. 95, p. 1459.

American Association of State Highway Transportation Officials, (AASHTO)(2007), *LRFD Bridge Design Specifications*, Washington, DC.

Basler, K. (1961), "Strength of Plate Girders in Shear," *ASCE J. Struct. Div.*, Vol. 87, p. 151.

Bruhn, E.F. (1973), *Analysis and Design of Flight Vehicle Structures*, Jacobs Publishing, Carmel, IN.

BSI (2000), *Steel, Concrete and Composite Bridges*, British Standard BS 5400: Part 3, *Code of Practice for Design of Steel Bridges*, British Standards Institution, London.

BSI (2006a), *Design of Steel Structures*, British Standard BS EN 1993-1-5:2006, Parts 1-5: *Plated Structural Elements*, British Standards Institute, London.

BSI (2006b), *Design of Steel Structures*, British Standard BS EN 1993-2:2006, Part 2: *Steel Bridges*, British Standards Institute, London.

Burgan, B. A., and Dowling, P. J. (1985), "The Collapse Behavior of Box Girder Compression Flanges—Numerical Modelling of Experimental Results," CESLIC Rep. BG 83, Imperial College, University of London.

CEN (2005a), Eurocode 3: *Design of Steel Structures*, Part 1-1: General Rules and Rules for Buildings, EN 1993-1-1:2005, Comité Européen de Normalisation (CEN), European Committee for Standardization, Brussels, Belgium.

CEN (2005b), Eurocode 4: *Design of Composite Steel and Concrete Structures*, Part 2: General Rules and Rules for Bridges , EN 1994-2:2005, Comité Européen de Normalisation (CEN), European Committee for Standardization, Brussels, Belgium.

CEN (2006a), Eurocode 3: *Design of Steel Structures*, Part 2: Steel Bridges, EN 1993-2:2006, Comité Européen de Normalisation (CEN), European Committee for Standardization, Brussels, Belgium.

CEN (2006b), Eurocode 3: *Design of Steel Structures*, Part 1-5: Plated Structural Elements, EN 1993-1-5:2006, Comité Européen de Normalisation (CEN), European Committee for Standardization, Brussels, Belgium.

Chatterjee, S. (1978), "Ultimate Load Analysis and Design of Stiffened Plates in Compression," Ph.D. thesis, Imperial College, University of London.

Chaterjee, S. (1981), "Design of Webs and Stiffeners in Plate and Box Girders," *The Design of Steel Bridges*, Granada, London.

Coletti, D., Fan, Z., Gatti, W., Holt, J., and Vogel, J., *Practical Steel Tub Girder Design*, National Steel Bridge Alliance, Chicago, IL, Apr. 2005.

Cooper, P. B. (1967), "Strength of Longitudinally Stiffened Plate Girders," *ASCE J. Struct. Div.*, Vol. 93, p. 419.

Darwin, D. (1990), *Steel and Composite Beams with Web Openings—Steel Design Guide Series, No. 2*, American Institute of Steel Construction (AISC), Chicago.

DA Stahlbau (1978), "Beulsicherheitsnachweise für Platten," *DASt Richtlinie 12*, Deutscher Ausschuss für Stahlbau, Cologne, Germany.

Dowling, P. J. (1981), "Codified Design Methods for Wide Steel Compression Flanges," *The Design of Steel Bridges*, Granada, London.

Dowling, P. J., and Chatterjee, S. (1977), "Design of Box Girder Compression Flanges," *2nd Int. Colloq. Stabil.*, European Convention for Constructional Steelwork, Brussels, Belgium, p. 153.

Dowling, P. J., Harding, J. E., and Frieze, P. A., (eds.) (1977a) *"Steel Plated Structures," Proc. Intl. Conf. Imperial College*, 1976, Crosby Lockwood, London.

Dowling, P. J., Frieze, P. A., and Harding, J. E. (1977b), "Imperfection Sensitivity of Steel Plates Under Complex Edge Loading" paper presented at the 2nd Int. Colloq. Stabil., Steel Structures, Liege, Belgium, p. 305.

Dubas, P., and Gehri, E. (Eds.) (1986), *Behaviour and Design of Steel Plated Structures*, ECCS Technical Committee 8.3, Publ. No. 44, European Convention for Constructional Steelwork, Brussels, Belgium.

Dwight, J. B., and Little, G. H. (1976), "Stiffened Steel Compression Flanges: A Simpler Approach," *Struct. Eng.*, Vol. 54, p. 501.

Fan, Z., and Helwig, T. A. (1999), "Behavior of Steel Box Girders with Top Flange Lateral Bracing," *ASCE J. Struct. Eng.*, Vol. 125, No. 8, pp. 829–837.

Fan, Z., and Helwig, T. A. (2002), "Distortional Loads and Brace Forces in Steel Box Girders," *ASCE J. Struct. Eng.*, Vol. 128, No. 6, June 2002, pp. 710–718.

Flint, A. R., and Wood, J. G. M. (1981), "Analysis for Box Girder Diaphragms Simplified for the New Code," *The Design of Steel Bridges*, Granada, London.

Frieze, P. A., and Dowling, P. J. (1979), "Testing of a Wide Girder with Slender Compression Flange Stiffeners Under Pronounced Shear Lag Conditions," CESLIC Rep. BG 49, Imperial College, University of London.

Frieze, P. A., Dowling, P. J., and Hobbs, R. E. (1975), "Parametric Study of Plates in Compression," CESLIC Rep. BG 39, Imperial College, University of London.

Fujita, Y., and Galambos, T. V. (Eds.) (1981), "Inelastic Instability of Steel Structures and Structural Elements," *U.S.–Jpn. Joint Sem.*, Tokyo.

Harding, J. E., and Dowling, P. J. (1981), "The Basis of the Proposed New Design Rules for the Strength of Web Plates and Other Panels Subject to Complex Edge Loading," in *Stability Problems in Engineering Structures and Components* (Eds. T. H. Richards and P. Stanley), Applied Science Publishers, Barking, Essex, England, p. 355.

Helwig, T., et al, (2005), "Lateral Bracing of Steel Girders by Permanent Metal Deck Forms," *Proceedings of the 2005 World Steel Bridge Symposium*, Orlando, FL.

Helwig, T. A., and Fan, Z. (2000), *"Field and Computational Studies of Steel Trapezoidal Box Girder Bridges," Texas Department of Transportation Research Project 0-1395*, University of Houston.

Highways Agency (1996), *Design Manual for Roads and Bridges*, Vol. 3: *Highways Structures, Assessment and Maintenance. Section 4. Assessment of Structures*, Part 12.BD56/96: The Assessment of Steel Highway Bridges, BA56/96, Advice Note and Commentary, London.

Höglund, T. (1971), "Design of Thin Plate I-Girders in Shear and Bending with Special Reference to Web Buckling," Bulletine No. 94 from Division of Building Statics and Structural Engineering, Royal Institute of Technology, Stockholm, 1973ICE (1971), "Steel Box Girder Bridges," *Proc. Int. Conf.*, Institution of Civil Engineers, London.

Höglund, T., (1995), "Strength of Steel and Aluminum Plate Girders—Shear Buckling and Overall Web Buckling of Plane and Trapezoidal Webs, Comparison with Tests," Royal Institute of Technology, Department of Structural Engineering, Technical Report 1995-4, Steel Structures.

Horne, M. R. (1977), "Structural Action in Steel Box Girders," *CIRIA Guide 3*, Construction Industry Research and Information Association, London.

"Inquiry into the Basis of Design and Method of Erection of Steel Box Girder Bridges" (1973), Report of the Committee (1973) and Appendices (1974), Her Majesty's Stationery Office, London.

Jetteur, P. (1983), "A New Design Method for Stiffened Compression Flanges of Box Girders," in *Thin-Walled Structures* (Eds. J. Rhodes and A. C. Walker), Granada, London, p. 189.

Jetteur, P., et al. (1984), "Interaction of Shear Lag with Plate Buckling in Longitudinally Stiffened Compression Flanges," *Acta Tech. CSAV*, No. 3, p. 376.

Lamas, A. R. G., and Dowling, P. J. (1980), "Effect of Shear Lag on the Inelastic Buckling Behaviour of Thin-Walled Structures," in *Thin-Walled Structures* (Eds. J. Rhodes and A. C. Walker), Granada, London, p. 100.

Maquoi, R., and Massonnet, C. (1971), "Théorie Non-lineaire de la Resistance Post-critique des Grandes Poutres en Caisson Raidies," *Mem. AIPC*, Vol. 31-II, p. 91.

Moffatt, K. R., and Dowling, P. J. (1975), "Shear Lag in Steel Box Girder Bridges," *Struct. Eng.*, Vol. 53, p. 439.

Moffatt, K. R., and Dowling, P. J. (1976), "Discussion," *Struct. Eng.*, Vol. 54, p. 285.

Ostapenko, A., and Vojta, J. F. (1967), "Ultimate Strength Design of Longitudinally Stiffened Plate Panels with Large *b/t*," Fritz Eng. Lab. Rep. No. 248.18, Lehigh University, Bethlehem, PA, Aug.

Porter, D. M., Rockey, K. C., and Evans, H. R. (1975), "The Collapse Behaviour of Plate Girders in Shear," *Struct. Eng.*, Vol. 53, p. 313.

Škaloud, M., and Novontny, R. (1965), "Uberkritisches Verhalten einer anfänglich gek-rümmten gleichformig gedrückten, in der Mitte mit einer lagsrippe versteiften Platte," *Acta Tech.* CSAV, p. 210.

U.S. Steel (1978), *Steel/Concrete Composite Box-Girder Bridges: A Construction Manual*, Pittsburgh, PA.

Wolchuk, R., and Mayrbourl, R. M. (1980), "Proposed Design Specification for Steel Box Girder Bridges," Rep. No. FHWA-TS 80205, U.S. Department of Transportation, Federal Highway Administration, Washington, DC.

Yura, J. A, and Widianto (2005), "Lateral Buckling and Bracing of Beams—a Re-Evaluation after the Marcy Bridge Collapse," *Structural Stability Research Council—Annual Stability Conference Proceeding*, Montreal, Quebec, Canada, April 6–9, pp 277–294.

CHAPTER 8

BEAM-COLUMNS

8.1 INTRODUCTION

Beam-columns are defined as members subjected to a combination of axial force and bending moment. They therefore provide a link between the column under pure axial load discussed in Chapter 3 and the beam loaded only by moments, which was the subject of Chapter 5. Indeed, an argument can be made for considering all members in frame structures as beam-columns, with columns and beams being the special cases that result when one load component becomes negligibly small. The bending moments in beam-columns may stem from transverse loading acting between the member ends, from loading on adjacent members in rigidly framed structures, or from the eccentricity of reactions and nominal axial forces in simply framed structures. When addressing the behavior of beam-columns in rigidly framed structures, it is usually necessary to also consider the influence of the surrounding members, which leads to the study of subassemblages and complete frames, the subject of Chapter 16. In the present chapter, the topic is treated mainly in terms of the response of an isolated member to a known system of end forces and moments.

Depending on the exact manner in which a beam-column is loaded and supported, its response may be categorized in a number of different ways. Perhaps the most fundamental feature is the presence (or absence) of a bracing system capable of preventing translation of one member end relative to the other. Problems involving sway are more appropriately considered in the context of the overall frame behavior. For nonsway beam-columns with doubly symmetric cross sections, three classes of problems may be identified (Fig. 8.1):

1. Members subjected to major axis bending and braced against minor axis flexure (or subjected to minor axis bending), which will collapse by excessive

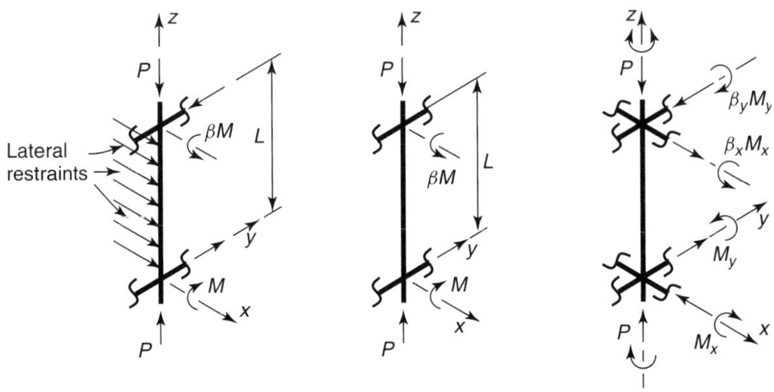

FIGURE 8.1 Classes of beam-column behavior.

in-plane bending deflections. This case corresponds to the interaction between column flexural buckling and simple uniaxial beam bending.

2. Unbraced members subjected to major axis bending, which will collapse by an interaction between column flexural buckling, beam lateral–torsional buckling, and uniaxial beam bending.
3. Unbraced members subjected to biaxial bending, which will collapse by an interaction between column flexural buckling, beam lateral–torsional buckling, and biaxial beam bending. Clearly, this constitutes the most general case—the previous ones corresponding to special cases of this one.

The analysis of various aspects of these problems has formed the subject of a very large number of research investigations. Initially these were confined to the elastic range, but the increasing availability of high-speed computers and sophisticated finite element software has meant that, by including inelastic material behavior in conjunction with such features as residual stresses and initial deformations, realistic ultimate-strength analyses may now be routinely conducted. The development of the theory of beam-columns has been summarized and presented in detail in the classic works by Massonnet (1976) and Chen and Atsuta (1977). A more general review of the most significant achievements prior to 1976 can be found in the third edition of this guide (Johnston, 1976), while reviews devoted specifically to the biaxial problem have been provided by Chen and Santathadaporn (1968) and Chen (1977a, 1981). In view of the availability of these comprehensive synopses of the general subject area of beam-columns, no in-depth historical review will be presented herein. Rather, the chapter will concentrate on an evaluation of the various design and analysis approaches that are available for several types of beam-column problems. In this way, it is hoped that the material will be of direct use to designers needing to work beyond the limitations of codes of practice, to specification-writing bodies engaged in the task of updating design codes, and to researchers concerned with advancing the understanding of particular aspects of the topic. Readers seeking a historical perspective of the subject are advised to consult the works referenced above, in which more than 300 relevant publications are listed.

A major international effort was undertaken by the SSRC to collect the stability design rules from all regions of the world (Beedle, 1991). Most code-writing authorities have changed to a limit-states design format in the last few decades, using a set of load factors and resistance factors that are based on probabilistic principles. This new direction provided an opportunity to also revise the specific rules for the design of beam-columns. Although the specification writers had access to the same basic research results, the detailed criteria may look very different in different parts of the world, reflecting the preferences of the engineering culture for which the standards were promulgated. The reasons for the differences have been discussed by Beedle (1991). Further discussion on the evolution of the beam-column design criteria can be found in Sputo (1993) and Kennedy et al. (1993). A worldwide specification has been proposed under the auspices of the International Organization for Standardization (ISO, 1994).

8.2 STRENGTH OF BEAM-COLUMNS

The load-carrying capacity of a beam-column depends on several factors, which may be conveniently categorized as *load related, member related*, and *imperfection related.* The first is readily appreciated if the strength of a beam-column subjected to any combination of axial load P, major axis moment M_x, and minor axis moment M_y is displayed on a three-dimensional interaction diagram, as shown in Fig. 8.2. Clearly, any point located on an axis corresponds only to a single type of loading, while the line or surface joining these endpoints defines the strength under two or three load components, respectively.

In constructing diagrams of the form of Fig. 8.2, the endpoints correspond to the member strengths as a column (P_u) and as a beam (M_{ux} and M_{uy}). Procedures for determining these strengths, which have been previously discussed in Chapters 3 (columns) and 5 (beams), make use of several properties of the member (e.g., geometrical proportions, material strength, unbraced length, or end-support conditions). These same properties will also have some effect on the exact shape of the interaction curves or surfaces (i.e., their degree of concavity or convexity).

The various "imperfections" (e.g., lack of straightness in either plane, residual stresses, or variation of material strength around the cross section) will also influence both the individual (component) strengths and the shape of the interaction curve or surface. Therefore, the estimation of the strength of a given beam-column requires knowing the shape of the interaction curve/surface (Fig. 8.2) associated with the particular set of member parameters under consideration. Subsequent sections of this chapter will address this problem in terms of both the design-oriented procedures that have evolved to meet this need and the evidence, theoretical and/or experimental, on which these have been based.

Traditionally, design methods for beam-columns fall into one of two categories: (1) those using charts or tables to provide safe combinations of the internal forces

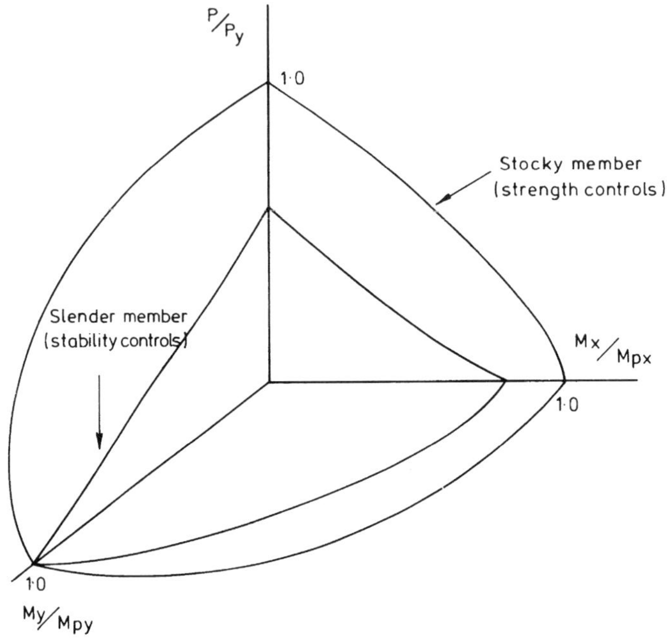

FIGURE 8.2 Interaction surfaces for beam-columns.

and moments generated by the applied loads, and (2) interaction formulas of the type

$$f\left(\frac{P}{P_u}, \frac{M_x}{M_{ux}}, \frac{M_y}{M_{uy}}\right) \leq 1.0 \tag{8.1}$$

which provide a smooth transition between the endpoints corresponding to strengths under only one internal force or moment (P_u, M_{ux}, or M_{uy})—of course, this design approach only leads to accurate beam-column strengths if the endpoints are reliably accurate. Although there are methods available to obtain accurate solutions for the problem types identified in Fig. 8.1, such methods always require the use of numerical procedures to follow the inelastic load–deflection behavior leading to the attainment of the "true" maximum strength. Thus, they cannot be directly used to develop design equations. Explicit forms of Eq. 8.1 for individual cases must therefore be developed either (1) as modifications to formulas derived from elastic analysis or (2) on a wholly empirical basis. The suitability of such formulas must then be verified against available theoretical and experimental data. Nevertheless, in the last couple of decades, the ever-growing availability of inexpensive personal computers has made it possible to perform rigorous analysis in special problems (e.g., beam-columns with unusual cross-section shapes, for which the application of methods developed for commonly used profiles might be considered questionable). Finally, it should be noted that several modern codes allow the use of "advanced analysis methods," that is, finite element-based numerical methods able to determine accurate beam-column maximum strengths.

8.3 UNIAXIAL BENDING: IN-PLANE STRENGTH

For members subjected to either (1) major axis bending and braced against minor axis flexure or (2) minor axis bending, a convenient form of Eq. 8.1, used as the starting point for several design formulas, is

$$\frac{P}{P_u} + \frac{M}{M_u} \leq 1.0 \tag{8.2}$$

where P is the applied axial force, P_u the ultimate strength of the centrally loaded column associated with buckling in the plane of the applied moment (taking into account column imperfections such as initial out-of-straightness and residual stresses), M the maximum applied bending moment, and M_u the ultimate beam strength (in the absence of the axial force).

If the design basis is to attain first yield in an initially stress-free member, then P_u and M_u must be chosen accordingly (Johnston, 1976). For ultimate-strength design, P_u and M_u represent the column strength, as discussed in Chapter 3, and the in-plane bending capacity defined in Chapter 5 (e.g., the fully plastic moment capacity M_p).

While the correct value for P is the applied axial load, the determination of the appropriate value for M in any particular case is more difficult because it will be affected by the bending deformation of the member. This deformation in turn will be a function principally of the form of the applied loading producing the moments, the level of axial force, the member slenderness, and the member support conditions. With reference to Fig. 8.3 the difference in behavior between slender and stocky simply supported beam-columns may be described as follows:

For the slender beam-column, the combined action of the axial load and the deflections produced by the primary (first-order) moments leads to further deflections and the so-called secondary moments (not to be confused with second-order moments, which equal the sum of the primary and secondary moments). The relative size of the primary and secondary moments will depend on the variables mentioned previously. In fact, in the particular case of simply supported beam-columns subjected to end moments causing double-curvature bending, the maximum second-order moment may occur at the end(s) of the member (even for high axial force values), thus being equal to the maximum first-order moment.

For the stocky beam-column, the maximum (second-order) moment may be taken as equal to the maximum primary (first-order) moment, which is calculated neglecting the axial load effects. Of course, it is often difficult to provide precise definitions for slender and stocky.

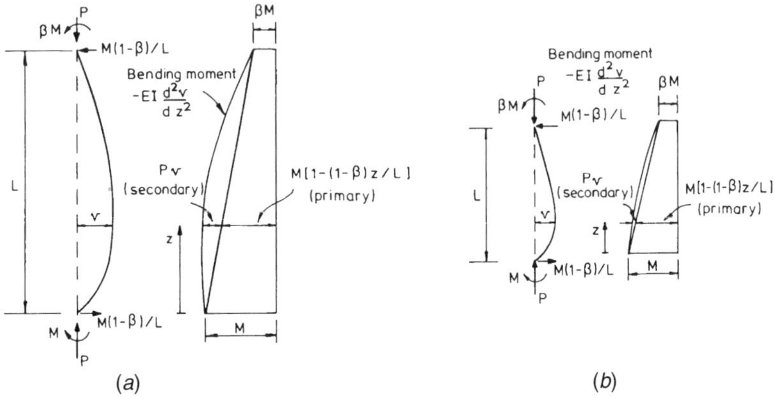

FIGURE 8.3 (a) Slender and (b) stocky simply supported beam-columns subjected to end moments.

The maximum moment at midlength in a simply supported beam-column subjected to compression P and equal and opposite end moment M_0 is given approximately[1] by

$$M_{\max} = M_0 \left(\frac{1}{1 - P/P_e} \right) \tag{8.3}$$

where P_e is the elastic critical load for buckling in the plane of the applied moments. The term in parentheses in Eq. 8.3 may be regarded as an amplification factor by which the first-order moment M_0 must be multiplied to obtain the second-order moment $M_{\max}$. It causes $M_{\max}$ to increase nonlinearly as shown in Fig. 8.4.

Although Eq. 8.3 was derived under the assumption of elastic behavior, its application in an ultimate-load context is well established. Substituting it into Eq. 8.2 gives the design formula

$$\frac{P}{P_u} + \frac{M_0}{M_u\,(1 - P/P_e)} \le 1.0 \tag{8.4}$$

which was first recommended by the SSRC in the third edition of this guide (Johnston 1976) and has since been included in several national codes. A direct comparison between Eq. 8.4 and the early numerical solution of Galambos and Ketter (1961), for a member comprised of a W8×31 section bent about its major axis, is provided in Fig. 8.5 (Massonnet and Save, 1965).

Additional solutions (see Table 8.1A) show that Eq. 8.4 provides an acceptable fit for all W-shapes, including those fabricated by welding (Galambos, 1964). Equation 8.4 may also be used for W-shapes bent about their minor axes, provided that P_u, M_u, and P_e are appropriately specified, and Fig. 8.6 compares its predictions

[1]Equation 8.3 is developed in the third edition of this guide (Johnston, 1976).

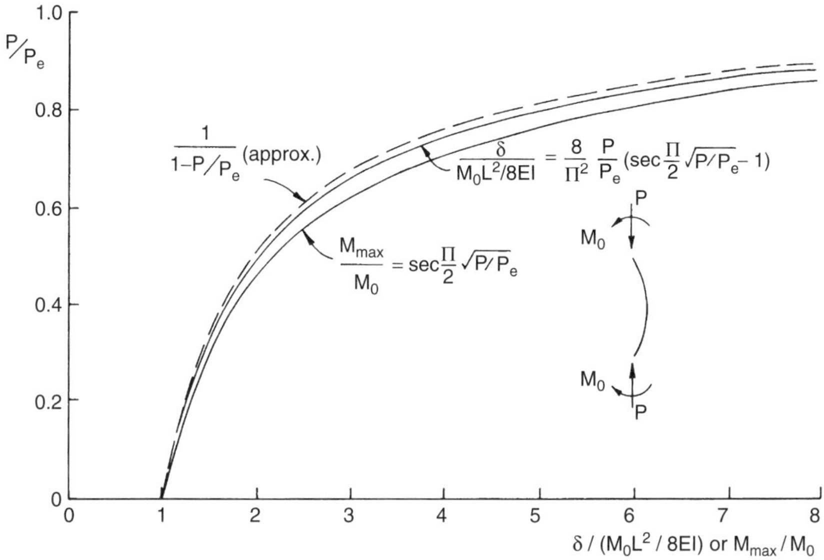

FIGURE 8.4 Maximum deflection and moment in elastic simply supported beam-columns subjected to equal end moments.

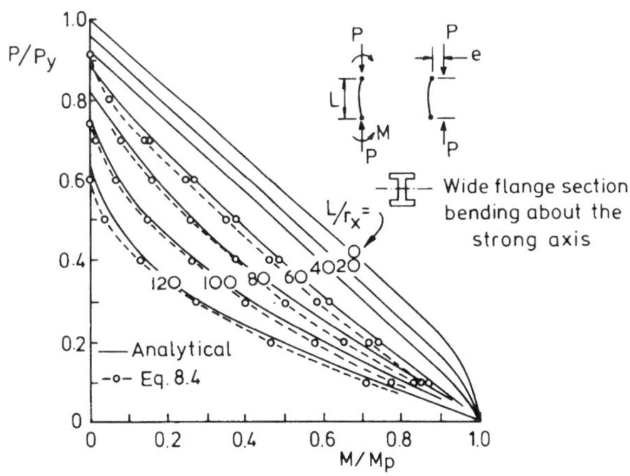

FIGURE 8.5 Comparison of Eq. 8.4 with numerical results of Galambos and Ketter for major axis bending of a W-shape (after Massonnet and Save, 1965).

with numerical data reported by Kanchanalai and Lu (1979). Test data for both cases, as well as for other cross-section shapes, are listed in Table 8.1B. Those listed for steel, however, are largely confined to rolled W-shapes, with only limited test data provided on other manufactured shapes (e.g., tubes) or welded sections.

TABLE 8.1A Theoretical Solutions for the In-Plane Strength of Beam-Columns

References	Results for Direct Use	Initial Deflection	Cross Section	Residual Stress	Load Cases	Comparison with Eq. 8.4
Galambos and Ketter (1961)	Yes	No	W, x–x axis	Pattern a	1 ($\beta = 1{,}0\ 0$)	Partial
Ketter (1962)	Yes	No	W, x–x axis	Pattern a	1 ($\beta = 1, 0.5, 0, -0.5, -1$)	Partial
	Yes	No	W, x–x axis	Pattern a	4	Partial
Lu and Kamalvand (1968)	Yes	No	W, x–x axis	Pattern a	4, 2. 5, 6	Yes
Chen (1971b)	Yes	No	W, x–x axis	None, pattern a	4,1 ($\beta = 1$)	No
	Yes	No	W, y–y axis	None, pattern a	4,1 ($\beta = 1$)	No
	Yes	No	Square tube	None	4,1 ($\beta = 1$)	No
Chen (1971a)	Yes	No	W, x–x axis	None	4,1 ($\beta = 1$)	No
	Yes	No	Rectangular tube	None	4,1 ($\beta = 1$)	No
Chen (1970)	Yes	No	W, x–x axis	Pattern a	3,1 ($\beta = 1$), 2	No
			Rectangular tube	None	3,1 ($\beta = 1$), 2	No
Lu et al. (1983)	Yes	Height/1000	W, x–x axis	Pattern a	1 ($\beta = 1, 0.\ 5, 0. - 0.5, -1$)	No
	Yes	Height/1000	W, y–y axis	Pattern a	1 ($\beta = 1, 0.\ 5, 0. - 0.5, -1$)	No
Chen and Atsuta (1972b)	Yes	No	W, x–x axis	Pattern a	1 ($\beta = 1$)	No
Cheong-Siat-Moy (1974a)	Yes	No	W, x–x axis	Pattern a	2, 4, 6	No
Cheong-Siat-Moy (1974b)	Yes	No	W, x–x axis	Pattern a	2, 7, 13	No
Ballio and Campanini (1981)	Yes	Height/1000	W, x–x axis	Pattern b	1 ($\beta = 1, 0, -1$)	Yes
	Yes	Height/1000	W, y–y axis	Pattern b	9, 2, 4, 8	Yes
	Yes	Height/1000	Square tube	Pattern b	3, 10, 7	Yes
	Yes	Height/1000	Circular tube	None	12, 11	Yes
Ballio et al. (1973)	Yes	Height/1000	W, x–x axis	Pattern b	1 ($\beta = 1$)	Yes
	Yes	Height/1000	W, y–y axis	Pattern b	1 ($\beta = 1$)	Yes
	Yes	Height/1000	Square tube	Pattern a	1 ($\beta = 1$)	Yes
	Yes	Height/1000	Square tube	Pattern b	1 ($\beta = 1$)	Yes
Yu and Tall (1971)	Yes	No	W, x–x axis	Pattern a	1 ($\beta = 1$)	No
			W, x–x axis	Pattern d	1 ($\beta = 1$)	No
Young (1973)	Yes	Height/1000	W, x–x axis	Pattern b and c	1 ($\beta = 1, 0, -1$)	No
			W, y–y axis	Pattern b and c	1 ($\beta = 1, 0, -1$)	No
			Square tube	Pattern b	1 ($\beta = 1, 0, -1$)	No

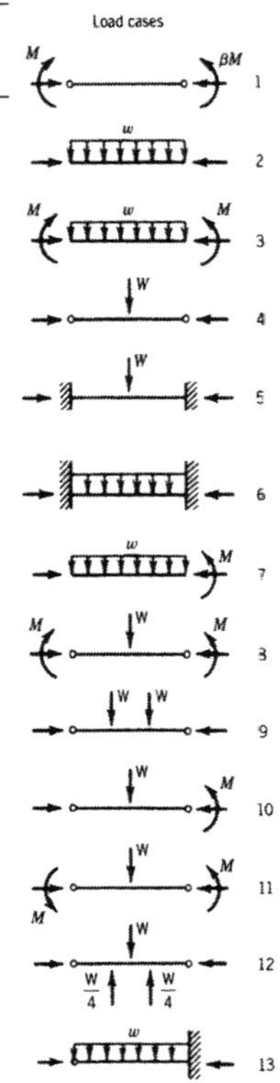

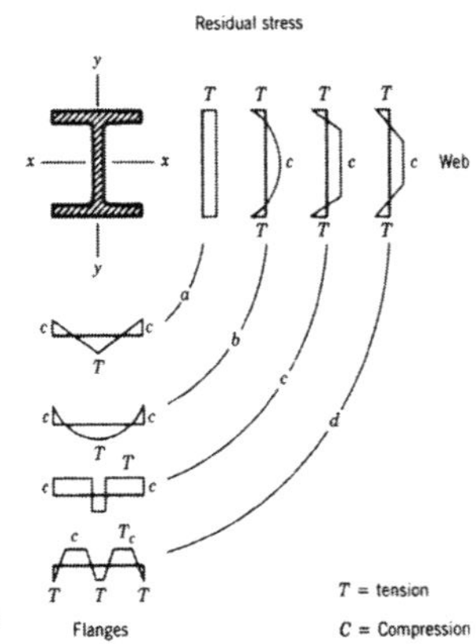

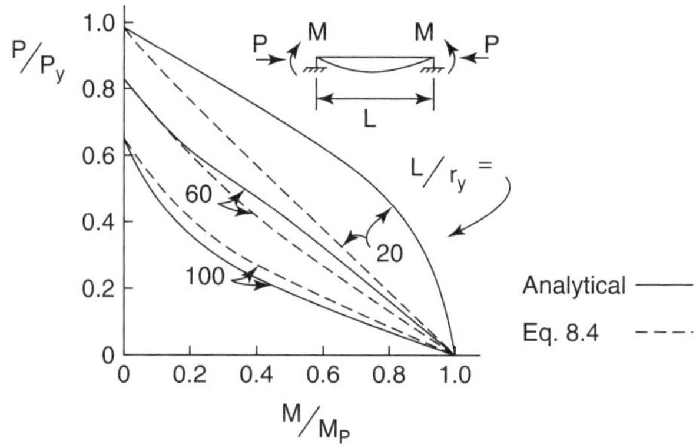

FIGURE 8.6 Comparison of Eq. 8.4 with numerical results of Kanchanalai and Lu (1979) for minor axis bending of a W-shape.

For aluminum sections the coverage was wider, including several results for monosymmetric sections (Klöppel and Barsch, 1973; Gilson and Cescotto, 1982). It is worth noting that (1) only a small amount of the theoretical studies reported in Table 8.1A include an allowance for initial curvature and (2) the available data appeared to be confined to either W-shapes or square and circular tubes.

For beam-columns subjected to unequal end moments and/or transverse loads, Eq. 8.4 may still be used provided that M_0 is replaced by an equivalent moment $M_{eq} = c_m M_0$, where C_m is an adjustment factor and M_0 is taken as the maximum first-order moment. Because the C_m factor is also applicable to the other types of beam-column problems illustrated in Fig. 8.1, a full discussion of its basis will be presented later in Section 8.5. The more general form of the interaction equation is, therefore,

$$\frac{P}{P_u} + \frac{C_m M_0}{M_u(1 - P/P_e)} \leq 1.0 \tag{8.5}$$

Equation 8.5 represents the conditions of failure stemming from instability due to excessive bending occurring within the member length. Most C_m-factor formulas, however, require that Eq. 8.5 must be supplemented with cross-section checks at the member ends. For example, in compact sections the end moments must be limited by M_{pc}, the plastic-hinge moment modified to account for the effect of axial compression. The determination of M_{pc}, by satisfying the equilibrium requirements, is simple but perhaps tedious for rectilinear shapes. Expressions for M_{pc} applicable to wide-flange shapes are available in the literature (ASCE, 1971). These may be extended to cover other orthogonal shapes, by applying the superposition procedure developed by Chen and Atsuta for biaxially bent beam-columns (Chen and Atsuta, 1972a, 1974, 1977). Expressions for M_{pc}, applicable to most wide-flange sections, are plotted nondimensionally in Figs. 8.7 and 8.8—the accuracy of the corresponding curves was verified through the comparison with experimental results. Simple approximate expressions for computing M_{pc} are also provided as dashed

TABLE 8.1B Experimental Data on the In-Plane Strength of Beam-Columns

References	Number of Tests	L/r Range	Cross Section	Residual Stress	Load Cases	Comparison with Eq. 8.4	Comments
Johnston and Cheney (1942)	30	23–122	W, x–x axis	Pattern a	$1(\beta = 1)$	Galambos and Ketter (1961)	Major axis failure
	30	26–126	W, y–y axis	Pattern a	$1(\beta = 1)$	No	
Galambos and Ketter (1961)	5	11–90	W, x–x axis	Pattern a	$1(\beta = 1)$	Yes	
Ketter et al. (1952, 1955)	14	28–120	W, x–x axis	Pattern a	$1(\beta = 1{,}0)$	Yes	Some out-of-plane failures
Galambos and Ketter (1961)							
Yu and Tall (1971)	2	40	W, x–x axis	Patterns a and d	$1(\beta = 1)$	No	Braced about y–y axis
Mason et al. (1958)	24	36–117	Hat shape	Unknown pattern	$1(\beta = 1)$	Yes	
van Kuren and Galambos (1964)	36	22–116	W, x–x axis	Pattern a	$1(\beta = 1, 0.5, 0, -1)$	Yes	10 tests braced about y–y axis, several out-of-plane failures for the rest
van Kuren and Galambos (1964)	1	72	W, y–y axis	Pattern a	$1(\beta = 1)$	Yes	
Dwyer and Galambos (1965)	3	38–80	Square tube	Unknown pattern	$1(\beta = 1)$	Yes	
Bijlaard et al. (1955)	18	40–50	Square tube	Unknown pattern	$1(\beta = 1)$	No	Elastic end restraints
			W, x–x axis	Pattern a	$1(\beta = 1)$	No	
Lay and Galambos (1965)	7	30–60	W, x–x axis	Pattern a	$1(\beta = 1)$	No	End restraint from beams, braced about y–y axis
Gilson and Cescotto (1982)	18	35–68	T-shape	Aluminum	$1(\beta = 1)$	No	Comparisons with ECCS
Hill et al. (1956)	9		Circular tube	Aluminum	$1(\beta = 1)$	Mazzolani and Frey (1983)	
Clark (1955)	28		Square tube Solid square	Aluminum	$1(\beta = 1)$		
Klöppel and Barsch (1973)	48	30–70	W, x–x axis	Aluminum	$1(\beta = 1)$	Mazzolani and Frey (1983)	
	48	30–82	Circular tube	Aluminum	$1(\beta = 1)$	Mazzolani and Frey (1983)	
	27	30–80	T-shape	Aluminum	$1(\beta = 1)$	Mazzolani and Frey (1983)	

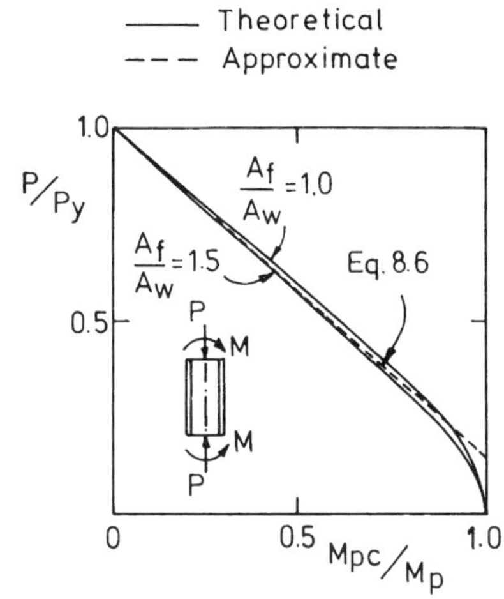

FIGURE 8.7 Approximate interaction equation for a W-shape (major axis bending, short column).

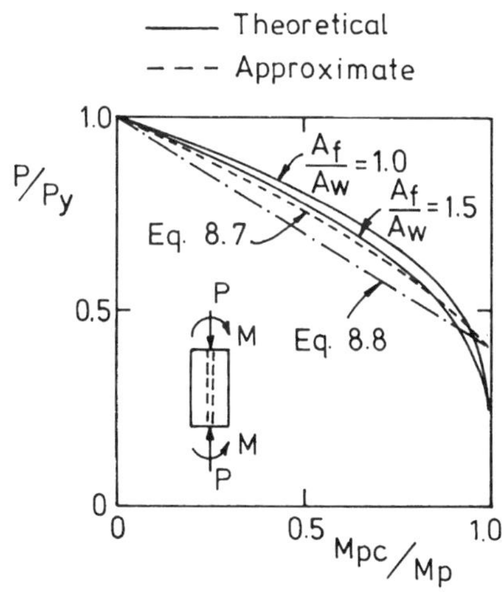

FIGURE 8.8 Approximate interaction equation for a W-shape (minor axis bending, short column).

lines in these figures. The cross-section strength condition $M_0 \leq M_{pc}$ leads to two additional interaction formulas as follows. For major axis bending,

$$\frac{P}{P_y} + 0.85\left(\frac{M_0}{M_p}\right) \leq 1.0 \quad \text{with} \quad M_0 \leq M_p \tag{8.6}$$

and for minor axis bending,

$$\left(\frac{P}{P_y}\right)^2 + 0.84\left(\frac{M_0}{M_p}\right) \leq 1.0 \quad \text{with} \quad M_0 \leq M_p \tag{8.7}$$

Pillai (1974) suggested

$$\frac{P}{P_y} + 0.6\left(\frac{M_0}{M_p}\right) \leq 1.0 \quad \text{with} \quad M_0 \leq M_p \tag{8.8}$$

as an alternative to Eq. 8.7. Due to the linear format, it is easier to incorporate into the biaxial interaction formulas that will be discussed in Section 8.6—its estimates are compared with the theoretically "exact" results in Fig. 8.8.

Several of the references listed in Tables 8.1A and 8.1B contain material that could be of direct use to the designer confronted with a problem falling outside the scope of standard design approaches. With this in mind, the most useful of them, together with other design-oriented references, are summarized in Table 8.1C.

Although Eqs. 8.6 through 8.8 enable the determination of the in-plane strength of beam-columns containing plastic hinges, such members must also possess sufficient ductility to perform satisfactorily in structures designed by inelastic analysis. Quantitative assessments of this ductility are usually expressed in terms of the rotation capacity R, which represents the ratio between the plastic rotation and the hypothetical rotation of an elastic member at the moment capacity given by Eqs. 8.6 through 8.8, as appropriate. For I-sections under major axis bending, and further assuming that $R = 3$ is a minimum acceptable value, Kemp (1984) proposed that the member slenderness should be limited to

$$\frac{L}{r_x} \leq \sqrt{\frac{\pi^2 E}{\sigma_y}(0.6 - 0.4\beta)\frac{1 - P/P_y}{1.5P/P_y}} \tag{8.9}$$

This expression is a development of the one suggested by Lay (1974), which was incorporated in an earlier version of the Australian code (SAA, 1998). A more general study on the interaction between axial force, end moment, slenderness, and rotation capacity has been published by van Manen (1982). The ductility of beam-columns was also investigated by Nakashima (1994) from the point of view of seismic behavior.

At this time it appears that the in-plane inelastic ultimate-strength behavior of doubly and singly symmetric beam-columns bent in the plane of symmetry is well understood. Efficient numerical methods are available to obtain solutions for beam-columns with any cross-section shape and subjected to any type of in-plane loading (Chen and Atsuta, Vol. 1, 1977). Extensive comparisons have been made between the various interaction design curves and equations published in the literature (Duan and Chen 1989, 1990; Duan et al. 1989; Cai and Chen, 1991; Sohal and Syed, 1992; Jaspart et al., 1993; Ofner, 1997; Kaim, 2004; Boissonnade et al., 2006). Equation 8.5, with some variations, still provides the basis for the design of beam-columns in most modern design specifications.

TABLE 8.1C Design-Oriented References on the In-Plane Strength of Beam-Columns

References	Direct Extension or Refinement of Eq. 8.4	Alternative to Eq. 8.4	Comments
Galambos and Ketter (1961)	Yes	—	Original basis for $\beta = 1, 0, -1$ cases
Ketter (1962)	Yes	—	Demonstrates dependence of C_m on axial load level
Lu and Kamalvand (1968)	Yes	—	Interaction curves and C_m values for lateral load cases
Chen and Atsuta (1977)	Yes	Yes	Numerous interaction curves, elastic C_m factors
Ballio and Campanini (1981)	Yes	—	C_m factors for a variety of cases, includes dependence of C_m on axial load level
Kanchananlai and Lu (1979)	Yes	Yes	Improvements for minor axis bending of I-sections, extension to members in unbraced frames
Massonnet and Save (1965)	—	—	Verification of Eq. 8.4 against theory and tests
Galambos (1964)	Yes	—	General discussion of Eq. 8.4, including direct comparison with tests, simple presentation of numerical approaches suitable for direct programming
Young (1973)	—	Yes	Alternative presentation of results as basis for a different type of design approach
Galambos (1981)	Yes	Yes	Discussion of Eq. 8.4 against more recent alternatives
Adams (1970)	Yes	—	Direct extension of Eq. 8.4 to 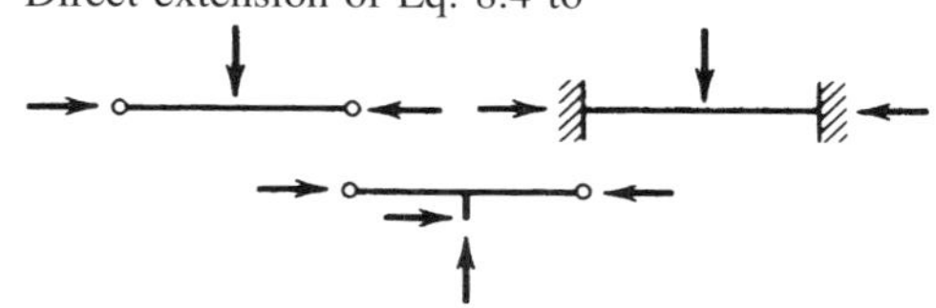
McLellan and Adams (1970), Adams (1974)	Yes	—	General discussion of Eq. 8.4 and its application to members in frames

TABLE 8.1C ***(Continued)***

References	Direct Extension or Refinement of Eq. 8.4	Alternative to Eq. 8.4	Comments
Mazzolani and Frey (1983)	Yes	Yes	Comparison with test data for aluminum beam-columns
Ojalvo and Fukumoto (1962)	—	Yes	Graphical presentation of theoretical results for $\beta = +1.0$, including full moment–rotation behavior
Galambos and Prasad (1962)	Yes	Yes	Tabular presentation of theoretical results for full range of P, β, L/r
Kemp (1984)	No	Yes	Considers beam-columns in plastically designed frames, provides limits on P, L/r, and β which will ensure satisfactory performance (rotation capacity)
Chen and Cheong-Siat-Moy (1980)	Yes	Yes	Covers application of interaction formula to members in unbraced frames
Cheong-Siat-Moy and Downs (1980)	Yes	Yes	Provides improved formulas for major and minor axis bending of members in unbraced frames
Djalaly (1975)	Yes	Yes	Suggests two modified amplification factors for major and minor axis bending of W-sections, provides C factors for several cases
Roik and Wagenknecht (1976), Roik and Bergmann (1977)	—	Yes	Interaction curves for 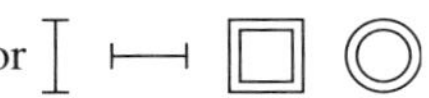 under various types of loading
Roik and Kindmann (1983)	—	Yes	Covers background to German beam-column interaction equations
Gonçalves and Camotim (2004)	Yes	—	Covers the application of Eurocode 3 beam-column interaction equations to non-simply supported members
Boisonnade et al. (2006)	Yes	Yes	Covers background to Eurocode 3 beam-column interaction equations

The in-plane member stability interaction equations presented below are taken from several representative design codes and concern doubly symmetric compact I-shaped nonsway members subjected to axial compression and uniaxial bending:

ANSI/AISC 360-05 (AISC, 2005—LRFD)

$$\frac{P_r}{P_c} + \frac{8}{9}\frac{M_r}{M_c} \leq 1.0 \text{ for } \quad \frac{P_r}{P_c} \geq 0.2 \tag{8.10a}$$

$$\frac{P_r}{2P_c} + \frac{M_r}{M_c} \leq 1.0 \text{ for } \quad \frac{P_r}{P_c} < 0.2 \tag{8.10b}$$

CAN/CSA-S16-01 (CSA, 2001)

$$\text{Major axis bending:} \qquad \frac{C_f}{C_{rx}} + \frac{0.85U_{1x}M_{fx}}{M_{rx}} \leq 1.0 \tag{8.11a}$$

$$\text{Minor axis bending:} \qquad \frac{C_f}{C_{ry}} + \frac{\beta U_{1y}M_{fy}}{M_{ry}} \leq 1.0 \tag{8.11b}$$

AS 4100-1998 (SAA, 1998)

$$\frac{N^*}{\phi N_c} + \frac{M^*}{\phi M_s} \leq 1.0 \tag{8.12}$$

Eurocode 3 (CEN, 2005)

$$\frac{N_{Ed}}{N_{b,Rd}} + k\frac{M_{Ed}}{M_{c,Rd}} \leq 1.0 \tag{8.13}$$

where P_r, C_f, N^*, N_{ED} = maximum axial load design value acting on the member
M_r, M_f, M^*, M_{Ed} = maximum bending moment design value acting on the member
$P_c, C_r, \phi N_c, N_{b,Rd}$ = factored buckling strength for axial compression
$M_c, M_r, \phi M_s, M_{c,RD}$ = factored cross-section plastic flexural strength

It should be noted that the American and Australian standard equations are based on second-order bending moments (i.e., M_r and M^* are obtained from second-order analysis). The Canadian and European equations are based on first-order moments and, therefore, must include amplification factors.

Because the nominal axial and flexural capacities, as well as the applied bending moment amplifications, are not the same in each specification, no general conclusions can be drawn concerning the relative merits of the various interaction equations—indeed, comparisons can only be made for specific members under given sets of loads. Nevertheless, the differences are always reasonable, as shown in the fifth edition of this guide and by White and Clarke (1997).

8.4 UNIAXIAL BENDING: LATERAL–TORSIONAL BUCKLING

When an I-section beam is bent about its major axis (i.e., in the plane of the web), there exists a tendency for it to fail by deflecting sideways and twisting, as explained in Chapter 5. When such a member behaves as a beam-column, the presence of a compressive axial force will accentuate this tendency, because the preferred mode of failure under pure compression would normally be by buckling about the minor axis. Beam-columns bent about the major axis therefore exhibit interaction between column and beam buckling.

The elastic critical load of a simply supported member subjected to compression P and equal and opposite end moments M_0, as shown in Fig. 8.1*b*, has been given by Hill and Clark (1951a,b), Horne (1956), Salvadori (1956), Campus and Massonnet (1956), and Chen and Atsuta (1977) and is defined by

$$\frac{M_0}{M_E} = \left[\left(1 - \frac{P}{P_{ey}}\right)\left(1 - \frac{P}{P_{\phi}}\right)\right]^{1/2} \tag{8.14}$$

where M_E is the beam elastic critical moment for lateral–torsional buckling (see Chapter 5), P_{ey} is the elastic critical load for minor axis flexural buckling, and P_{ϕ} is the elastic critical load for pure torsional buckling. Critical combinations of P and M_0 are plotted in Fig. 8.9.

If an approximate allowance is made for the effects of in-plane deflection, Eq. 8.14 becomes (Chen and Atsuta, 1977)

$$\frac{M_0}{M_E} = \left[\left(1 - \frac{P}{P_{ex}}\right)\left(1 - \frac{P}{P_{ey}}\right)\left(1 - \frac{P}{P_{\phi}}\right)\right]^{1/2} \tag{8.15}$$

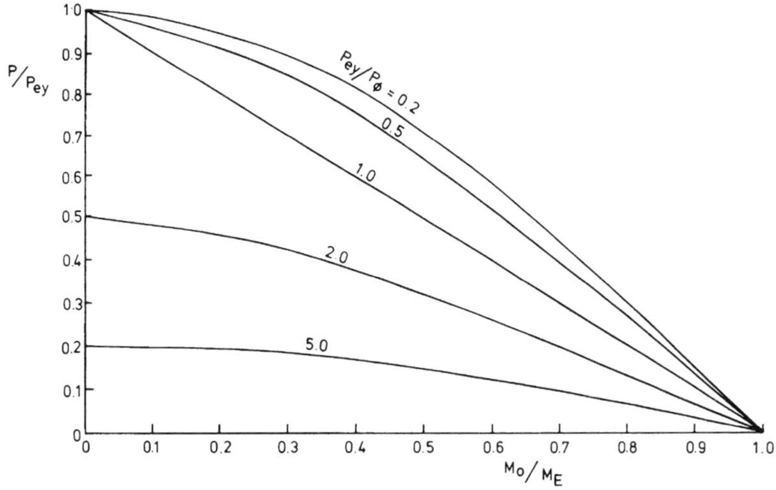

FIGURE 8.9 Elastic critical load combinations for beam-coluns with equal end moments.

where P_{ex} is the elastic critical load for major axis flexural buckling. A linearized and conservative alternative to Eq. 8.15 can be expressed as

$$\frac{P}{P_{ey}} + \frac{M_0}{M_E(1 - P/P_{ex})} \leq 1 \tag{8.16}$$

which has the same form as Eq. 8.4 for in-plane failure, except for the fact that the quantities appearing in the denominator, P_{ey} and M_E, are now related to out-of-plane failure.

By analogy with Eq. 8.5, the design version of Eq. 8.16 may be written as

$$\frac{P}{P_{uy}} + \frac{C_m M_0}{M_u(1 - P/P_{ex})} \leq 1 \tag{8.17}$$

where P is the applied axial load; P_{uy} is the axial load producing failure in the absence of bending moment, associated with minor axis buckling; M_0 is the maximum first-order applied moment; M_u is the moment producing failure in the absence of axial load, accounting for lateral–torsional buckling; and C_m is the factor addressed in Sections 8.3 and 8.5. Similarly to Eq. 8.5, it is also necessary to check the resistance of the beam-column cross-section strength at the member ends (e.g., using Eq. 8.6).

The problem of elastic and inelastic lateral–torsional buckling has received a great deal of attention from researchers—with several references listed in Table 8.2A. Although this list is by no means complete, it gives access to the formulas and charts required for most practical applications. Furthermore, the research on the lateral–torsional buckling of beam-columns has been summarized by Trahair (1993) in his book *Flexural–Torsional Buckling of Structures*. It should be noted that most of the studies on biaxially loaded beam-columns that will be discussed in Section 8.6 are able to handle the laterally unbraced member subjected to uniaxial bending as a special case. Several series of tests have been performed on laterally unbraced beam-columns, with a overview provided in Table 8.2B. In addition to these investigations, which were designed specifically to study the lateral–torsional buckling behavior, it is worth mentioning that some of the test series reported in Table 8.1B contain cases of lateral–torsional buckling failure, which usually stem from inadequate bracing. References expanding the use of Eq. 8.17 or presenting alternative design procedures are listed in Table 8.2C.

It should be noted that lateral–torsional buckling normally occurs in a torsionally weak section bent about its stiffer principal plane. Therefore, certain cross-section classes (e.g., tubes) are not susceptible to this mode of failure. In such cases bending about both principal axes is treated using the methods addressed in Section 8.3.

In order to take into account lateral–torsional buckling, the interaction equations presented earlier (Eqs. 8.10 to 8.13) must be modified or supplemented.

TABLE 8.2A Theoretical Solutions for the Lateral-Torsional Buckling Strength of Wide-Flange Beam-Columns

References	Approach	Residual Stress	Load Cases	Comparison with Eq. 8.17	Comments	Results for Direct Use
Vinnakota (1977)	Maximum strength	Pattern a	1 ($\beta = 1$)	No		Yes
Fukumoto and Galambos (1966)	Tangent modulus	Pattern a	1 ($\beta = 1.0$)	No		Yes
Galambos et al. (1965)	Tangent modulus	Pattern a	1 ($\beta = 1.0$)	No		Yes
Miranda and Ojalvo (1965)	Tangent modulus	None	1 ($\beta = 1$)	No	Allows for prior in-plane deflection	Yes
Lim and Lu (1970)	Tangent modulus	None	1	No	Approx. extrapolation to maximum strength; includes minor axis restraint, prior inplane deflection, and continuous members	
Lindner and Wiechart (1978)	Maximum strength	Pattern b	1 ($\beta = 1, 0, -0.7$) 4, 2	No		Yes
Abdel-Sayed and Aglan (1973)	Tangent modulus	Pattern a	1 ($\beta = 1$)	No	Allows for prior in-plane deflection, sample results for aluminum sections	
Bradford and Trahair (1985)	Maximum strength	Pattern b	Any in-plane loading	No	Finite element analysis (FEM)	Yes
Bradford and Trahair (1986)	Maximum strength	Measured in test	Continuous beam-column	No	FEM analysis of Cuk et al. (1986) tests	
Bradford et al. (1987)	Maximum strength	Pattern a and b	1,4	No	FEM analysis	Yes

Pi and Trahair (1992a, b)	Maximum strength	Any pattern	Any in-plane loading	No	Prebuckling deformations	Yes
Pi and Trahair (1994a, b)	Maximum strength	Any pattern	Any in-plane loading	No	All nonlinearity effects	Yes
Wang and Kitipornchai (1989)	Maximum strength	Any pattern	Any in-plane loading	No	Monosymmetric members	Yes
Kitiporchai and Wang (1988)	Elastic Bucking		In-plane loading	No	Elastic lateral–torsional buckling	Yes
Jingping et al. (1988)	Maximum strength	Pattern a	2–4	No	Transversely loaded beam-columns	Yes
Trahair (1993)	Buckling and maximum strength	Any pattern	Any plane of loading	No	Summary of all available work on subject	Yes

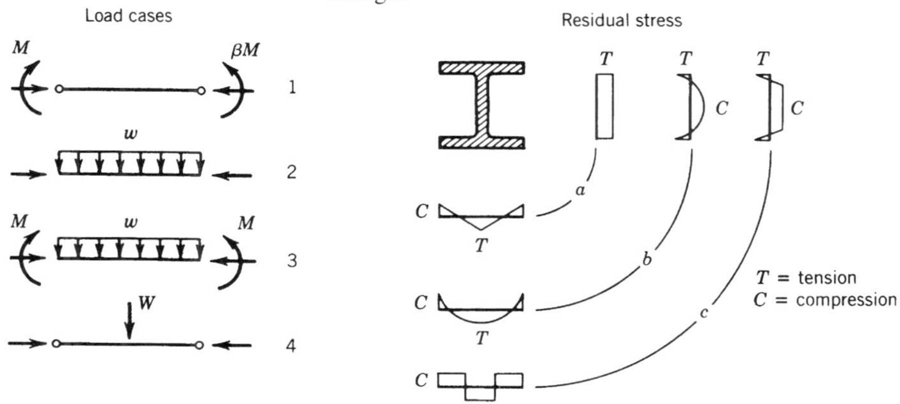

TABLE 8.2B Experimental Data for the Lateral-Torsional Buckling Strength of Wide-Flange Beam Columns

References	Number of Tests	L/r Range	Load Cases	Comparison with Eq. 8.17	Comments
Van Kuren and Galambos (1964)	26	22–116	 $\beta = 1, 0.5, 0, -1$	Yes	Some in-plane failures
Lindner and Kurth (1982)	70	58–169	$\beta = 1, 0.5, 0, -1$	No	High-strength steel
Campus and Massonnet (1956)	91	40–177	$\beta = 1, 0, -1$	Yes	See Nethercot (1983) for comparison
Djalaly (1971)	8	62–82	$\beta = 1, 0.5$	No	
Chagneau (1973)	8	60–80	$\beta = 1, 0.5$	No	
Hill and Clark (1951a, b)	59	28–183	$\beta = 1$	No	Aluminum
Gent and Sen (1977)	8	43–74	$\beta = 0.15$	No	Tests designed to investigate rotation capacity; some in-plane failures
Cuk et al. (1986)	14	$L/r_y = 98$	Framed member	No	Study effect of end restraint
Nakashima et al. (1990)	42	$L/r_y = 42$ to 91	Sway members	No	Study effect of lateral sway

TABLE 8.2C Design-Oriented References for the Lateral–Torsional Buckling Strength of Wide-Flange Beam-Columns

References	Direct Extension of Refinement of Eq. 8.17	Alternative to Eq. 8.17	Comments
Galambos (1981)	Yes	Yes	Discussion of Eq. 8.17 against more recent alternatives
Campus and Massonnet (1956)	Yes	—	Original basis for Eq. 8.17, derivation of comparison with test data
Mazzolani and Frey (1983)	Yes	Yes	Comparison with test data for aluminum beam-columns
Galambos et al. (1965)	—	Yes	Considers importance of in-plane deflections, alternative design approach
Djalaly (1973)	Yes	Yes	Eq. 8.17 and several alternatives compared with test data
Cuk and Trahair (1981)	Yes	—	Suggests that C_m depends on axial load level; basis is elastic critical loads
Bradford and Trahair (1985)	No	Yes	Formulas for Australian code (SAA, 1998)
Trahair and Bradford (1988)	No	Yes	General steel design text
Trahair (1993)	No	Yes	Summary of research on subject
Boisonnade et al. (2006)	—	Yes	Covers background to Eurocode 3 beam-column interaction equations

For beam-columns subjected to axial compression and uniaxial major axis bending these equations are:

ANSI/AISC 360-05 (AISC, 2005—LFRD)

As an alternative to using Eq. 8.10 with P_c defined as the out-of-plane buckling strength, two independent limit states may be considered:

In-plane buckling: Eq. 8.10, with P_c and M_c defined by in-plane strength

Out-of-plane buckling: $$\frac{P_r}{P_{co}} + \left(\frac{M_r}{M_{cx}}\right)^2 \leq 1.0 \tag{8.18}$$

CAN/CSA-S16-01 (CSA, 2001)

In-plane buckling: Eq. 8.11a

$$\text{Out-of-plane buckling:} \quad \frac{C_f}{C_{ry,T}} + \frac{0.85U_{1x}M_{fx}}{M_{rx}} \leq 1.0 \tag{8.19}$$

AS 4100-1998 (SAA, 1998)

In-plane buckling: Eq. 8.12

$$\text{Out-of-plane buckling:} \quad \frac{N^*}{\phi N_{cy}} + \frac{M^*}{\phi M_{bx}} \leq 1.0 \tag{8.20}$$

Eurocode 3 (CEN, 2005)

$$\text{In-plane buckling:} \quad \frac{N_{Ed}}{N_{b,y,Rd}} + k_{yy}\frac{M_{y,Ed}}{M_{b,Rd}} \leq 1.0 \tag{8.21}$$

$$\text{Out-of-plane buckling:} \quad \frac{N_{Ed}}{N_{b,z,Rd}} + k_{zy}\frac{M_{y,Ed}}{M_{b,Rd}} \leq 1.0 \tag{8.22}$$

where, for the above out-of-plane interaction equations:

$P_{co}, \phi N_{cy}, N_{b,z,Rd}$ = factored out-of-plane buckling strength for axial compression

$C_{ry,T}$ = factored out-of-plane flexural or torsional buckling strength for axial compression

$M_{cx}, M_{rx}, \phi M_{bx}, M_{b,Rd}$ = factored lateral–torsional buckling strength

The nominal axial and flexural capacities, as well as the amplified first-order applied moments, are determined differently in each of the above design standards. Therefore, meaningful comparisons can only be made for specific member conditions. As reviewed in the fifth edition of this guide (1998), White and Clarke (1997) have shown that the lateral–torsional capacity may differ considerably in the various specifications, particularly for low axial forces.

8.5 EQUIVALENT UNIFORM MOMENT FACTOR

When a simply supported beam-column is loaded by unequal end moments M_0 and βM_0, where $-1.0 \leq \beta \leq 1.0$, it is usually overly conservative to incorporate the larger end moment directly in the design formulas of Eq. 8.4 or 8.17. Because the design formulas are based on the case of uniform primary moment (i.e., $\beta = +1.0$), this is particularly true as β approaches -1.0 and the member tends to be bent in double curvature. Studies concerning both the in-plane case (Austin, 1961; Chen and Atsuta, 1977) and the lateral–torsional buckling problem (Horne, 1956; Salvadori, 1956; Campus and Massonnet, 1956) have shown that a simple and reasonably accurate correction may be obtained if M_0 is replaced by a reduced value $C_m M_0$, where

$$C_m = 0.6 + 0.4\beta \geq 0.4 \tag{8.23}$$

Equation 8.23 is a simplified version of the several different formulas that have been proposed by various researchers. For design purposes, it is particularly convenient to use the same expression for both problem types as well as to ignore the relatively small effect of P/P_e (member slenderness) on C_m.

For beam-columns subjected to other transverse loadings (e.g., transverse point loads acting between supports), extensive numerical results are available (e.g., Lu and Kamalvand, 1968; Chen and Atsuta, 1977; Ballio and Campanini, 1981) when failure occurs by excessive bending. In this case, a simple approximation consists of replacing M_0 in Eq. 8.4 by the maximum value taken from a straight-line envelope of the actual (primary) moment diagram. More accurate values, however, may be obtained if the C_m factors given in Table 8.3 are used. Because these factors allow for the use of moments below the maximum, it becomes necessary to (1) increase their values as the axial load decreases (i.e., the beam-column approaches a beam) and (2) perform the additional cross-section checks mentioned in Sections 8.3 and 8.4. Analogous conclusions were reached by Djalaly (1975), who provides C_m factors for nine different load cases.

For laterally unbraced beam-columns, similar refinements to Eq. 8.23 have been suggested by Cuk and Trahair (1981) on the basis of elastic buckling studies. Several other numerical studies have resulted in extensive tabular and chart solutions, which are summarized by Trahair (1993) and also appear in the references cited in Table 8.2A.

TABLE 8.3 C_m Factors for Use with Eq. 8.5 for In-Plane Strength of Beam-Columns

Load Case	Formula for C_m	Limits
	$0.6 - 0.4\beta$	$P/P_u \geq 0.4$
	$1 - \dfrac{1-(0.6-0.4\beta)}{0.4}\dfrac{P}{P_u}$	$P/P_u < 0.4$
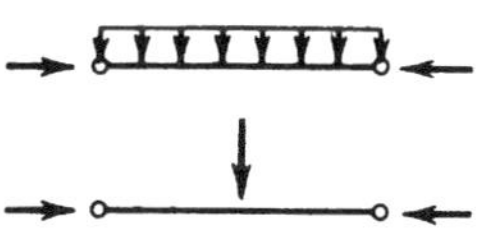	$0.5 + \dfrac{0.7M_m}{M_0}$	$P/P_e \geq 0.5$
	$1 - 2\left[1 - \left(0.5 + \dfrac{0.7M_m}{M_0}\right)\right]\dfrac{P}{P_e}$	$P/P_e < 0.5$
etc.	where M_0 = maximum moment in span, M_m = average moment	

8.6 BIAXIAL BENDING

Both of the problems discussed in Sections 8.3 and 8.4 are special cases of the more general beam-column problem illustrated in Fig. 8.1*c*. Contributions in this area prior to 1975 have been reviewed fully in the third edition of this guide, while subsequent general reviews are also available (Massonnet, 1976; Chen, 1977a, 1981; Chen and Atsuta, 1977). For design purposes it is convenient to address separately the short columns and intermediate/slender columns, which correspond to the outer and inner failure surfaces depicted in Fig. 8.2. Section 8.6.1 will focus on the strength of cross sections subjected to combined axial load and major and minor axis bending moments, a case relevant to stocky beam-columns, for which failure is likely to be governed by full yielding of the most heavily stressed cross section (assuming that the individual component plates are not susceptible to local buckling). Section 8.6.2 covers intermediate and slender beam-columns, which are likely to fail due to member buckling.

8.6.1 Strength of Short Beam-Columns

Provided that the cross section meets the compactness requirements necessary to prevent local buckling, the strength limit state of a short beam-column is defined by complete yielding of the cross section. For the uniaxial case, the location of the neutral axis may be determined from basic principles (e.g., Chen and Atsuta, 1977). Alternatively, Eqs. 8.6 and 8.7 for wide-flange sections, or the formulas of Table 8.4 for other common shapes, may be employed. No comparable simple expressions exist for the biaxial case. Moreover, the determination of the location of the neutral axis, which must satisfy the three conditions of equilibrium for the axial force P and the bending moments M_x and M_y, is very complex, thus making the use of such expressions prohibitive for anything other than research purposes.

Santathadaporn and Chen (1970) derived interaction curves relating M_x to M_y for a series of P values and wide-flange sections exhibiting varying width-to-depth and flange thickness-to-depth ratios. For design purposes the average set of curves given in Fig. 8.10 will normally prove adequate. If a single design equation is required, it can be provided only at the expense of accuracy, often leading to significantly conservative predictions of the beam-column ultimate strengths. A suitable linear expression is given by (Pillai, 1974)

$$\frac{P}{P_y} + 0.85\left(\frac{M_x}{M_{px}}\right) + 0.6\left(\frac{M_y}{M_{py}}\right) \leq 1 \tag{8.24}$$

in which the different coefficients for the two moment terms reflect the different shapes of the uniaxial interactions, as shown in Figs. 8.5 and 8.6. More accurate predictions may be obtained using the alternative equation (Chen and Atsuta, 1977; Tebedge and Chen, 1974)

$$\left(\frac{M_x}{M_{pcx}}\right)^{\alpha} + \left(\frac{M_y}{M_{pcy}}\right)^{\alpha} \leq 1 \tag{8.25}$$

TABLE 8.4 Full Plastic Interaction for Uniaxial Bending

Cross Section	M_P	Interaction of P and M_0
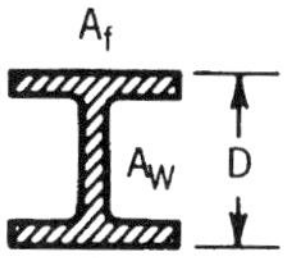	$\sigma_y A_f D\left(1-\frac{A_w}{4A_f}\right)$	$\left(\frac{P}{P_y}\right)^{1.2}+\frac{M_0}{M_p}=1$
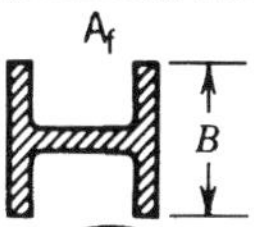	$\frac{\sigma_y A_f B}{2}$	$\left(\frac{P}{P_y}\right)^{2}+\left(\frac{M_0}{M_p}\right)^{1.2}=1$
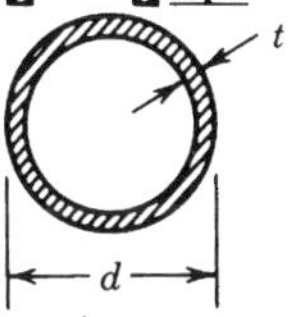	$\frac{1}{6}\sigma_y d^3\left[1-\left(1-\frac{2t}{d}\right)^3\right]$	$\left(\frac{P}{P_y}\right)^{1.7}+\frac{M_0}{M_p}=1$
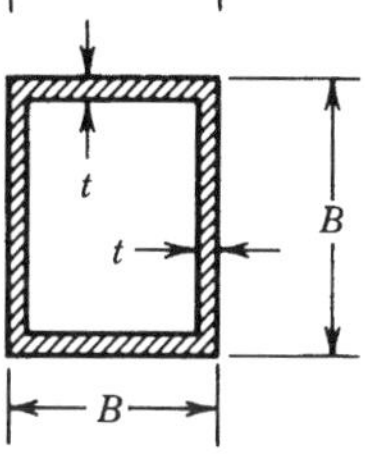	$\sigma_y t D^2\left(\frac{B}{D}+\frac{1}{2}\right)$	$\frac{D}{B}\geq 3$ $\left(\frac{P}{P_y}\right)^{2}+\frac{M_0}{M_p}=1$
		$1\leq\frac{D}{B}<3$ $\left(\frac{P}{P_y}\right)^{1.5}+\frac{M_0}{M_p}=1$
		$\frac{D}{B}<1$ $\left(\frac{P}{P_y}\right)^{1.2}+\frac{M_0}{M_p}=1$

in which M_{pcx} and M_{pcy} are the moment capacities (about the respective axes) reduced by the presence of the axial force P, which may, for example, be obtained from Eqs. 8.6 and 8.7 For I-shaped members, the value of the exponent α is given by

$$\alpha = \begin{cases} 1.0 & \text{for} \quad b_f/d < 0.5 \\ 1.6 - \dfrac{P/P_y}{2\ln(P/P_y)} & \text{for} \quad 0.5 \leq b_f/d < 1.0 \end{cases} \tag{8.26a}$$

and for boxed shape members

$$\alpha = 1.7 - \frac{P/P_y}{\ln(P/P_y)} \tag{8.26b}$$

Figure 8.11, which compares Eqs. 8.24 and 8.25 with the average "exact" results obtained by Santathadaporn and Chen (1970), shows how the interaction between the two moment terms becomes increasingly convex as the axial force increases. Therefore, in situations where beam-columns carry very large axial forces and also withstand small moments, Eq. 8.25 leads to significant material savings over Eq. 8.24.

Interaction curves for sections other than wide-flange shapes have been obtained by Chen and Atsuta (1974, 1977) using a superposition technique. Results for

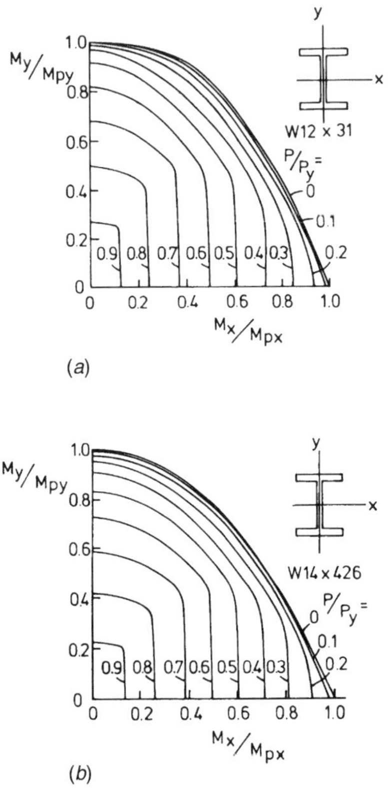

FIGURE 8.10 Interaction curves for strength of short beam-columns under biaxial loading: (*a*) light W-shape (*b*) heavy W-shape.

several different structural sections, including circular tubes and unsymmetrical shapes such as angles, are available in Chen and Atsuta (1974, 1977). These results show that the degree of symmetry exhibited by the interaction surface parallels the degree of symmetry of the section. The work of Chen and Atsuta (1977) also contains a procedure that can be used to develop these interaction surfaces, which is particularly suitable for programming. Finally, it should be noted that Bez (1983) provides a full set of interaction curves for standard angles.

All of the foregoing discussion has assumed that yielding occurs only as a result of the direct stresses produced by compression and bending. The presence of significant shear stresses (e.g., due to shear forces or St.-Venant torsion), however, can further reduce cross-section capacity. This effect may be accounted for in a particularly simple manner (Morris and Fenves, 1969; Chen and Atsuta, 1972a, 1977); by adopting the von Mises yield criterion, the normal yield stress can be reduced by the factor

$$\sqrt{1 - t^2} \tag{8.27}$$

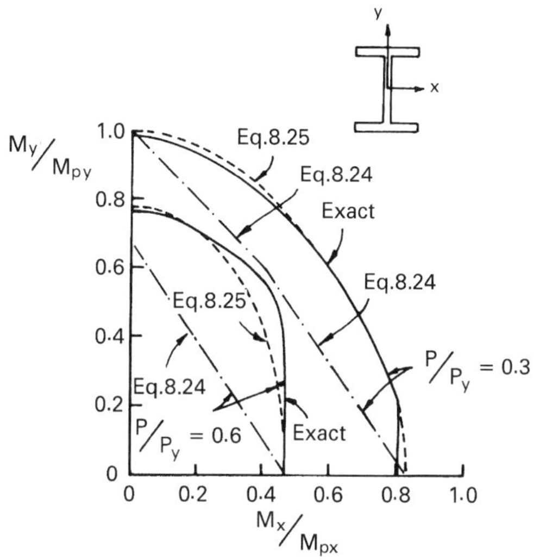

FIGURE 8.11 Comparison of Eqs. 8.24 and 8.25 with numerical data of Chen and Atsuta for strength of biaxially loaded beam-columns.

in which $t = \tau/\tau_y$ is the ratio of applied shear stress to the material shear yield stress.

8.6.2 Strength of Intermediate and Slender Beam-Columns

Slender beam-columns subjected to biaxial bending constitute a most formidable problem. Over the years, successive investigations have (1) produced increasingly refined analytical solutions, (2) conducted extensive and painstaking series of experiments, and (3) used this information as the basis for ever more reliable and accurate design procedures. Clearly, it is neither feasible nor appropriate to review all these contributions, a task that has been adequately performed by Chen and Santathadaporn (1968), Johnston (1976), Massonnet (1976), Chen and Atsuta (1977), Chen (1977a, 1981), Kaim (2004), and Boissonnade et al. (2006). Instead, selected design approaches are presented here, together with the discussion of their relationship to the available theoretical and experimental data. Tables 8.5A, 8.5B, and 8.5C also provide a guide to the sources of the very useful original data.

Because a design procedure for biaxial bending should reduce to that already recommended for uniaxial bending (in the absence of one of the moments), a logical starting point is Eqs. 8.5 and 8.17 As shown in Figs. 8.5 and 8.6, these expressions provide good descriptions of the P–M_x and P–M_y interactions for wide-flange sections. The second edition of this guide suggested an empirical combination of these two expressions to give

$$\frac{P}{P_u} + \frac{C_{mx}M_x}{M_{ux}(1 - P/P_{ex})} + \frac{C_{my}M_y}{M_{uy}(1 - P/P_{ey})} \leq 1 \tag{8.28}$$

TABLE 8.5A Theoretical Solutions for the Biaxial Bending Strength of Beam-Columns[a]

References	Cross Section	Residual Stress	Slender Column	Load Cases	Results for Direct Use	Comments
Djalaly (1975)	W-shape	Pattern b	Yes		Yes	Tabulated results for one section
Virdi (1981)	Square and circular tubes	None	Yes	Any loading	No	Applicable to cases where torsional effects can be neglected
Lindner (1972), Lindner and Gietzelt (1982)	W-shape	Pattern b	Yes	 $\beta = 1.0$	Yes	Tabulated results for one section: method explained in Lindner (1972): comparison with Eqs. 8.25, 8.28, 8.31
Hancock (1977)	W-shape	Pattern a	No		No	Cross-section analysis only
Razzaq and McVinnie (1982)	Square tube	None	Yes		No	Column deflection curves
Vinnakota (1977)	W-shape	Pattern a	Yes	Any loading	Yes	Detailed study of initial deflections
Rajasekaran (1977)	W-shape	None	Yes	Any loading	No	General finite element approach
Opperman (1983)	T, L, I, C shapes	None	Yes	Any loading	No	General finite element approach

Ramm and Osterrieder (1983)	Any open section	None	Yes	Any loading	No	General finite element approach
Duan and Chen (1989)	W-shape	Pattern a	Yes	Any loading	Yes	Examine interaction equations and compare to test results

[a]This topic is discussed fully by Chen and Atsuta (1977); papers incorporated in the text have not been included in the list above.

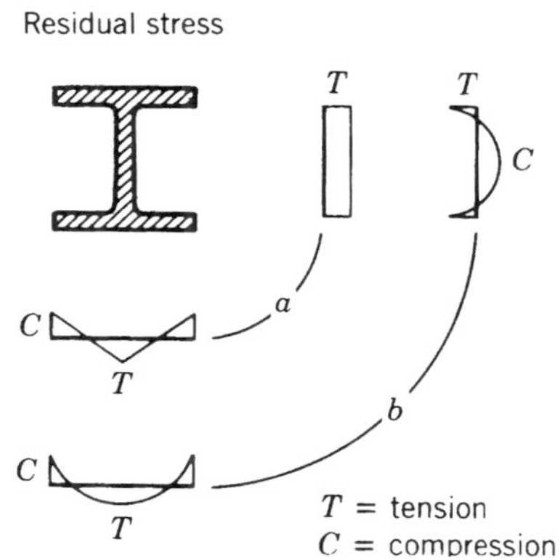

TABLE 8.5B Experimental Data for the Biaxial Strength of Beam-Columns

References	Number of Tests	Cross Section	L/r Range	Load Cases	Comments
Birnstiel (1968)	12	W-shapes	63–121	 $\beta_x = 1,\ \beta_y = 1$	Rolled, welded, and annealed sections
Chubkin (1959)	43	W-shapes	50–150	$\beta_x = 1,\ \beta_y = 1$	Residual stress data not available
Klöppel and Winkelmann (1962)	69	W-shapes	57–121	$\beta_x = 1,\ \beta_y = 1$	Variable residual stresses due to method of specimen manufacture
Anslijn (1983)	78	W-shapes	40–96	Various β_x, β_y	
Matthey (1982)	15	W-shapes		Various β_x, β_y	
Marshall and Ellis (1970)	29	Square tubes	27–135	Various β_x, β_y	Small-scale tests

TABLE 8.5C Design-Oriented References for the Biaxial Bending Strength of Beam-Columns

	Direct Extension or Refinement of Eqs.:				Alternative to Eqs.:				
References	8.24	8.25	8.28	8.31	8.24	8.25	8.28	8.31	Comments
Chen and Atsuta (1977)	Yes	Yes	Yes	Yes	—	—	—	—	Full discussion in Chapter 13, including review of all proposals up to 1976
Pillai (1980, 1981)	—	Yes	—	Yes	—	—	—	—	Assessment of Eqs. 8.25 and 8.31 against test data using various component data
Lindner and Gietzelt (1983)	—	Yes	Yes	Yes	Yes	Yes	Yes	Yes	Evaluation of various three-term formulas on basis of test data and theoretical results of Lindner and Gietzelt (1982)
Nethercot (1983)	—	Yes	—	Yes	Yes	Yes	Yes	Yes	Review of available interaction formulas with comparisons
Djalaly (1975)	—	Yes	—	Yes	Yes	Yes	Yes	Yes	
Chen and Liu (1983)	—	Yes	Yes	Yes	—	—	—	—	Proposed unification of beam-column and column design; discussion of C_m factor
Chen (1981)	—	Yes	Yes	Yes	Yes	Yes	Yes	Yes	Inclusion of external pressure effects
Maquoi and Rondal (1982)	Yes	Yes	Yes	Yes	Yes	Yes	Yes	Yes	Design-oriented reorganization of ECCS interaction formulas
Chen (1982)	—	Yes	—	Yes	Yes	Yes	Yes	Yes	Covers box sections and fabricated tubes, limited verification of C_m factors
Duan and Chen (1989)	—	—	Yes	Yes	—	—	Yes	Yes	Verify AISC (1993) interaction equations
Kaim (2004)	—	—	—	—	Yes	Yes	Yes	Yes	Covers W-shape, square tube, and monossymmetric cross sections as well as some special topics
Boissonnade et al. (2006)	—	—	—	—	Yes	Yes	Yes	Yes	Covers background to Eurocode 3 beam-column interaction equations

in which P_u is the ultimate strength of an axially loaded column and M_{ux} and M_{uy} are the ultimate bending moments of beams bent about the major (x) and minor (y) axes—in compact I-section beam-columns M_{uy} would be the minor axis full plastic moment capacity and M_{ux} would be reduced due to the possibility of lateral–torsional buckling. Pillai (1970) suggested an improvement for the particular case of hollow circular and square box sections as

$$\frac{P}{P_u} + C\left[\frac{C_{mx}M_x}{M_{ux}(1 - P/P_{ex})} + \frac{C_{my}M_y}{M_{uy}(1 - P/P_{ey})}\right] \leq 1 \tag{8.29}$$

in which

$$C = \frac{(e_x^2 + e_y^2)^{1/2}}{e_x + e_y}$$

and e_x and e_y are the eccentricities of the applied axial load P. For these two cross-sectional shapes, both M_{ux} and M_{uy} are, of course, the full plastic moment capacities.

Equation 8.28 can be alternatively expressed as (Chen, 1977b)

$$\frac{C_{mx}M_x}{M_{ucx}} + \frac{C_{my}M_y}{M_{ucy}} \leq 1 \tag{8.30}$$

in which M_{ucx} and M_{ucy} are the uniaxial moment capacities according to Eqs. 8.17 and 8.5, respectively (i.e., the ultimate values of M corresponding to a given P). It is worth noting that Eq. 8.30 implies a linear interaction between the moment terms, whereas both test data and theoretical solutions for beam-columns under biaxial bending indicate a convex interaction, as illustrated in Fig. 8.12.

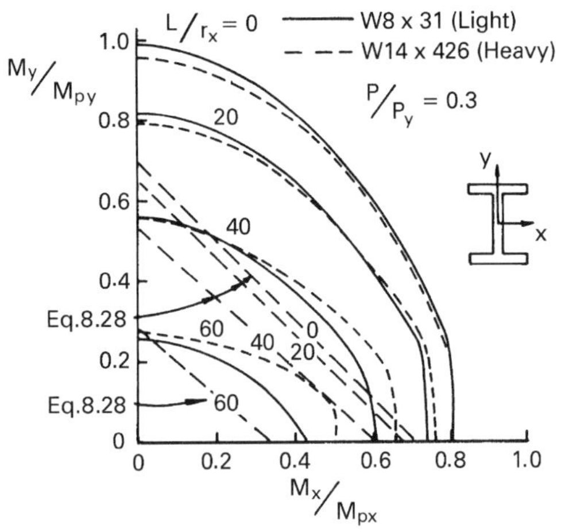

FIGURE 8.12 Comparison of Eqs 8.28 with numerical data of Chen and Atsuta for stability of biaxially loaded beam-columns.

As a development of their work on short columns, Tebedge and Chen (1974) proposed the use of a nonlinear expression similar to Eq. 8.25 for use with slender columns, which results in Eq. 8.30 being modified to

$$\left(\frac{C_{mx}M_x}{M_{ucx}}\right)^{\eta} + \left(\frac{C_{my}M_y}{M_{ucy}}\right)^{\eta} \leq 1.0 \tag{8.31}$$

Based largely on numerical studies covering I-shapes with $b_f/d = 0.3$ (Ross and Chen, 1976) as well as H-sections with $b_f/d = 1.0$ (Tebedge and Chen 1974), the exponent η is

$$\eta = \begin{cases} 0.4 + \dfrac{P}{P_y} + \dfrac{b_f}{d} & \text{for} \quad b_f/d \geq 0.3 \\ 1.0 & \text{for} \quad b_f/d < 0.3 \end{cases} \tag{8.32}$$

where b_f is the flange width and d the section depth. For square box sections, η should be taken as (Chen, 1977b)

$$\eta = 1.3 + \frac{1000}{(L/r)^2}\frac{P}{P_u} \geq 1.4 \tag{8.33}$$

The use of Eq. 8.31 with η values greater than unity amounts to recognizing the convexity of the interaction between the moment terms, as illustrated in Fig. 8.13.

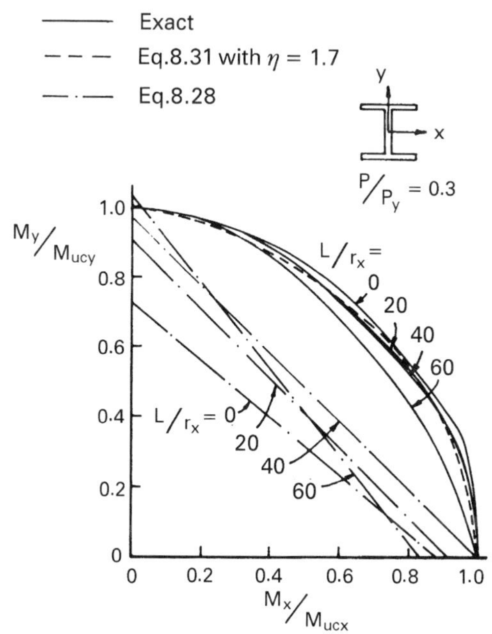

FIGURE 8.13 Comparison of Eqs. 8.28 and 8.31 with numerical data of Chen and Atsuta for stability of biaxially loaded beam-columns.

On the basis of their ability to predict the strengths of the specimens involved in four series of tests on biaxially loaded I-sections (Chubkin, 1959; Klöppel and Winkelmann, 1962; Birnstiel, 1968; Anslijn, 1983), Pillai (1980, 1981) reported a quantitative assessment of the accuracy of Eqs. 8.28 and 8.31. When conducting this comparative study, advantage was taken of the fact that the use of Eq. 8.31 is not tied to any particular method for determining M_{ucx} and M_{ucy}, as the concept of combining uniaxial moment capacities can be used with any reasonable moment expression. Pilllai found that using Eq. 8.25, together with the column and beam buckling predictions provided in the then current AISC specification, was able to predict quite accurately the results of 81 tests carried out by Anslijn (1983), as shown by the impressive mean value of $P_{\text{test}}/P_{\text{calc}} = 1.050$. The standard deviation obtained, however, equaled 0.101, which meant that Eq. 8.25 overpredicted 31% of the test results. A more recent evaluation of the comparisons between the AISC design equations, numerical predictions, and test results has been provided by Duan and Chen (1989).

Because theoretical procedures for assessing the beam-column strength under biaxial loading are rather complex, even within the elastic range, numerical solutions are required (Culver, 1966a,b). Thus, the results presented in any of the references listed in Table 8.5A will normally cover only a limited aspect of the problem. Several series of large-scale tests, however, have been conducted (see Table 8.5B) that have proved invaluable in the assessment of the various design approaches listed in Table 8.5C.

8.7 SPECIAL TOPICS

This chapter has focused primarily on beam-columns manufactured from hot-rolled steel wide-flange shapes. The behavior of beam-columns of cold-formed steel shapes is discussed in Chapter 14 of this guide. Many hot-rolled or welded beam-columns, however, may also be susceptible to the interaction between local and overall buckling. This very important topic is addressed, for example, in papers by Sohal and Chen (1988) and Zuyan and Zhang (1991), both concerning box-shaped beam-columns.

The design of beam-columns with tapered webs was the subject of an extensive set of research projects carried out at the University of Buffalo in New York, the outcome of which was reported in Chapter 9 of the fourth edition of this guide. Further relevant references on the stability of tapered steel beam-columns are the papers by Butler and Anderson (1963), Culver and Preg (1968), Hsu and Lee (1981), and Bradford (1988). More recently, the works of Andrade et al. (2005) and Boissonnade and Maquoi (2005) deserve to be mentioned. Finally, it is worth noting that the postbuckling behavior of beam-columns was investigated by Kounadis and Ioannidis (1994).

Determining the response of beam-columns to wave or earthquake motions is an important factor in assessing the strength of offshore structures or structures located in seismic areas. Therefore, considerable effort has also been directed towards developing some of the techniques for assessing the static strength listed in Tables 8.1C into efficient methods for the cyclic inelastic analysis of steel tubular beam-columns (Chen 1981; Han and Chen 1983a,b). Computer programs based on these techniques were verified against the results of large-scale tests and subsequently used to generate data on cyclic response (Han and Chen 1983a,b). A full account of this work can be found in the monograph by Chen and Han (1983).

REFERENCES

Abdel-Sayed, G., and Aglan, A. A. (1973), "Inelastic Lateral Torsional Buckling of Beam Columns," Publ. IABSE, No. 33–II, pp. 1–16.

Adams, P. F. (1970), "Segmental Design of Steel Beam-Columns," *Can. Struct. Eng. Conf.*, Toronto, Ontario, Canada.

Adams, P. F. (1974), *The Design of Steel Beam-Columns*, Canadian Steel Industries Construction Council, Toronto, Ontario, Canada.

American Institute of Steel Construction (AISC) (1993) "Load and Resistance Factor Design Specification for Structural Steel Buildings," American Institute of Steel Construction, Chicago IL.

American Institute of Steel Construction (AISC) (2005), "Specification for Structural Steel Buildings", ANSI/AISC 360-05, AISC, Chicago IL.

American Society of Civil Engineers, (ASCE) (1971), "Plastic Design in Steel: A Guide and Commentary," in *ASCE Manual of Engineering Practice*, No. 41, 2nd ed., ASCE, New York.

Andrade, A., Camotim D., Costa P. P. (2005), " Elastic Lateral-Torsional Buckling Behavior of Doubly Symmetric Tapered Beam-Columns", *Proceedings of Structural Stability Research Council (SSRC) Annual Stability Conference*, Montreal, April 6–9), pp. 445–468.

Anslijn, R. (1983), "Tests on Steel I Beam-Columns in Mild Steel Subjected to Thrust and Biaxial Bending," CRIF Rep. MT 157, Brussels, Belgium, Aug.

Austin, W. J. (1961), "Strength and Design of Metal Beam-Columns," *ASCE J. Struct. Div.*, Vol. 87, No. ST4, pp. 1–34.

Ballio, G., and Campanini, C. (1981), "Equivalent Bending Moments for Beam-Columns," *J. Constr. Steel Res.*, Vol. 1, No. 3.

Ballio, G., Petrini, V., and Urbano, C. (1973), "The Effect of the Loading Process and Imperfections on the Load Bearing Capacity of Beam Columns," *Meccanica*, Vol. 8, No. 1. pp. 56–67

Beedle, L. S., editor-in-chief (1991), *Stability of Metal Structures: A World View*, 2nd ed., Structural Stability Research Council, Bethlehem, PA.

Bez, R. (1983), "Diagrams d'Interaction de Cornières Metalliques," Publ. ICOM 111, Swiss Federal Institute of Technology, Lausanne, Switzerland, Mar.

Bijlaard, P. P., Fisher, G. P., and Winter, G. (1955), "Eccentrically Loaded, End-Restrained Columns," *Trans. Am. Soc. Civ. Eng.*, Vol. 120, p. 1070.

Birnstiel, C. (1968), "Experiments on H-Columns Under Biaxial Bending," *ASCE J. Struct. Div.*, Vol. 94, No. ST10, pp. 2429–2450.

Boissonnade, N., and Maquoi, R. (2005), "A Geometrically and Materially Non-Linear 3-D Beam Finite Element for the Analysis of Tapered Steel Members," *Int. J. Steel Struct.*, Vol. 5, No. 5, pp. 413–419.

Boissonnade, N., Greiner, R., Jaspart, J. P., and Lindner, J. (2006), "Rules for Member Stability in EN 1993-1-1 Background Documentation and Design Guidelines," Publication No. 119, Technical Committee 8 (Stability), European Convention for Constructional Steelwork, Brussels, Belgium.

Bradford, M. A. (1988), "Lateral Stability of Tapered Beam-Columns with Elastic Restraints," *Struct. Eng.*, Vol. 66, No. 22, pp. 376–384.

Bradford, M. A., Cuk, P. E., Gizejowski, M. A., and Trahair, N. S. (1987), "Inelastic Lateral Buckling of Beam-Columns," *ASCE J. Struct. Eng.*, Vol. 113, No. 11, pp. 2259–2277.

Bradford, M. A., and Trahair, N. S. (1985), "Inelastic Buckling of Beam-Columns with Unequal End-Moments," *J. Constr. Steel Res.*, Vol. 5, No. 3, pp. 195–212.

Bradford, M. A., and Trahair, N. S. (1986), "Inelastic Buckling Tests of Beam-Columns," *ASCE J. Struct. Eng.*, Vol. 112, No. 3, pp. 538–549.

Butler, D. J., and Anderson, G. C. (1963), "The Elastic Buckling of Tapered Beam-Columns," *Weld. J. Res. Suppl.*, Vol. 42, No. 1, pp. 29–36.

Cai, C. S., and Chen, W. F. (1991), "Further Verification of Beam-Column Strength Equation," *ASCE J. Struct. Eng.*, Vol. 117, No. 2, pp. 501–513.

Campus, F., and Massonnet, C. (1956), "Recherches sur le Flambement de Colonnes en Acier A37 a Profil en Double te Sollicitées Obliquement," *C. R. Rech. IRSIA*, Bulletin No. 17 Apr.

Canadian Standards Association (CSA) (2005), "Limit States Design of Steel Structures," CAN/CSA-S16-01, Mississauga, Ontario, Canada.

CEN (2005), Eurocode 3: *Design of Steel Structures*, Part 1.1: General Rules and Rules for Buildings, EN 1993-1-1, Brussels, Belgium.

Chagneau, A. (1973), "La barre dans la Structure," CTICM Report, Jan., Paris.

Chan, S. L., Kitipornchai, S., and Al-Bermani, F. G. (1991), "Elasto-Plastic Analysis of Box-Beam-Columns Including Local Buckling Effects," *ASCE J. Struct. Eng.*, Vol. 117, No. 7, pp. 1946–1962.

Chen, W. F. (1970), "General Solution of Inelastic Beam-Column Problem," *ASCE J. Eng. Mech. Div.*, Vol. 96, No. EM4, pp. 421–442.

Chen, W. F. (1971a), "Approximate Solution of Beam Columns," *ASCE J. Struct. Div.*, Vol. 97, No. ST2, pp. 743–751.

Chen, W. F. (1971b), "Further Studies of an Inelastic Beam-Column Problem," *ASCE J. Struct. Div.*, Vol. 97, No. ST2, pp. 529–544.

Chen, W. F. (1977a), "Theory of Beam-Columns: The State-of-the-Art Review," *Proc. Int. Colloq. Stabil. Struct. Under Static Dyn. Loads*, SSRC/ASCE, Washington, DC, Mar.

Chen, W. F. (1977b), "Design of Box Columns Under Biaxial Bending," *2nd Int. Colloq. Stabil. Steel Struct.. Prelim. Rep.*, Liege, Belgium.

Chen, W. F. (1981), "Recent Advances in Analysis and Design of Steel Beam-Columns in U.S.A.," *Proc. U.S.–Jpn. Sem. Inelastic Instabil. Steel Struct. Elements*, Tokyo, May.

Chen, W. F. (1982), "Box and Cylindrical Columns Under Biaxial Bending," in *Axially Compressed Structures: Stability and Strength* (Ed. R. Narayanan), Applied Science Publishers, Barking, Essex, England, Chap. 3.

Chen, W. F., and Atsuta, T. (1972a), "Interaction Equations for Biaxially Loaded Sections," *ASCE J. Struct. Div.*, Vol. 98, No. ST5, pp. 1035–1052.

Chen, W. F., and Atsuta, T. (1972b), "Column-Curvature Curve Method for Analysis of Beam-Columns," *Struct. Eng.*, Vol. 50, No. 6, pp. 233–240.

Chen, W. F., and Atsuta, T. (1974), "Interaction Curves for Steel Sections, Under Axial Load and Biaxial Bending," *EIC Eng. J.*, Vol. 17, No. A-3.

Chen, W. F., and Atsuta, T. (1977), *Theory of Beam-Columns*, Vols. 1 and 2, McGraw-Hill, New York.

Chen, W. F., and Cheong-Siat-Moy, F. (1980), *Limit States Design of Steel Beam-Columns*, Solid Mechanics Archives, Vol. 5, Noordhoff International Publishing, Leyden, The Netherlands.

Chen, W. F., and Han, D. J. (1983), *Tubular Members in Offshore Structures*, Pitman, London.

Chen, W. F., and Lui, E. M. (1983), "Design of Beam-Columns in North America," *Proc. 3rd Int. Colloq. Stab. Met. Struct.*, Toronto.

Chen, W. F., and Santathadaporn, S. (1968), "Review of Column Behaviour Under Biaxial Loading," *ASCE J. Struct. Div.*, Vol. 94, No. ST12, pp. 2999–3021.

Cheong-Siat-Moy, F. (1974a), "A Method of Analysis of Laterally Loaded Columns," *ASCE J. Struct. Div.*, Vol. 106, No. ST5, Proc. Pap. 10548, pp. 953–970.

Cheong-Siat-Moy, F. (1974b), "General Analysis of Laterally Loaded Beam-Columns," *ASCE J. Struct. Div.*, Vol. 100, No. ST6, Proc. Pap. 10625, pp. 1263–1278.

Cheong-Siat-Moy, F., and Downs, T. (1980), "New Interaction Equation for Steel Column Design," *ASCE J. Struct. Div.*, Vol. 106, No. ST5, pp. 1047–1062.

Chubkin, G. M. (1959), "Experimental Research on Stability of Thin Plate Steel Members with Biaxial Eccentricity," in *Analysis of Spatial Structures*, Vol. 5, Moscow, Paper 6, GILS.

Clark, J. W. (1955), "Eccentrically Loaded Aluminium Columns," *Trans. ASCE*, Vol. 120, p. 116.

Cuk, P. E., Rogers, D. E., and Trahair, N. S. (1986), "Inelastic Buckling of Continuous Steel Beam-Columns," *J. Constr. Steel. Res.*, Vol. 6, No. 1, pp. 21–52.

Cuk, P. E., and Trahair, N. S. (1981), "Elastic Buckling of Beam-Columns with Unequal End Moments," *Civ. Eng. Trans. Inst. Eng. Aust.*, Vol. CE23, No. 2, pp. 166–171.

Culver, C. G. (1966a), "Exact Solution of the Biaxial Bending Equations," *ASCE J. Struct. Div.*, Vol. 92, No. ST2, pp. 63–84.

Culver, C. G. (1966b), "Initial Imperfections in Biaxial Bending," *ASCE J. Struct. Div.*, Vol. 92, No. ST3, pp. 119–138.

Culver, G. C., and Preg, S. M., Jr. (1968), "Elastic Stability of Tapered Beam-Columns," *ASCE J. Struct. Div.*, Vol. 94, No. ST2, pp. 455–470.

Djalaly, H. (1971), "La Barre dans la Structure, CTICM Report, Mar., Paris.

Djalaly, H. (1973), "Calcul de la Résistance Ultime des Barres au Flambement par Flexion-torsion," *Constr. Met.*, No. 4. pp. 7–8

Djalaly, H. (1975), "Calcul de la Résistance Ultime des Barres Comprîmées et Flèchies," *Constr. Met.*, Vol. 12, No. 4, pp. 17–47.

Duan, L., and Chen, W. F. (1989), "Design Interaction Equation for Steel Beam-Columns," *ASCE J. Struct. Eng.*, Vol. 115, No. 5, pp. 1225–1243.

Duan, L., and Chen, W. F. (1990), "Design Interaction Equation for Cylindrical Tubular Beam-Columns," *ASCE J. Struct. Eng.*, Vol. 116, No. 7, pp. 1794–1812.

Duan, L., Sohal, I., and Chen, W. F. (1989), "On Beam-Column Moment Amplification Factor," *AISC Eng. J.*, Vol. 26, No. 4, pp. 130–135.

Dwyer, T. J., and Galambos, T. V. (1965), "Plastic Behaviour of Tubular Beam-Columns," *ASCE J. Struct. Div.*, Vol. 91, No. ST4, pp. 125–152.

Fukumoto, Y., and Galambos, T. V. (1966), "Inelastic Lateral-Torsional Buckling of Beam-Columns,' *ASCE J. Struct. Div.*, Vol. 92, No. ST2, pp. 41–62.

Galambos, T. V. (1964), "Combined Bending and Compression," in *Structural Steel Design* (Ed. L. Tall), Ronald Press, New York.

Galambos, T. V. (1981), "Beam-Columns," paper presented at the ASCE Conv., New York, May.

Galambos, T. V., Adams, P. F., and Fukumoto, Y. (1965), "Further Studies on the Lateral-Torsional Buckling of Steel Beam-Columns," Fritz Eng. Lab. Rep. No. 205A.36, Lehigh University, Bethlehem, PA.

Galambos, T. V., and Ketter, R. L. (1961), "Columns Under Combined Bending and Thrust," *Trans. ASCE*, Vol. 126, Part I, pp. 1–25.

Galambos, T. V., and Prasad, J. (1962), "Ultimate Strength Tables for Beam-Columns," *Weld. Res. Counc. Bull.*, No. 78, June.

Gent, A. R., and Sen, T. K. (1977), "The Plastic Deformation Capacity of H-Columns at High Axial Loads," Prelim. Rep. 2nd Int. Colloq. Stab. Steel Struct., Liege, Belgium.

Gilson, S., and Cescotto, S. (1982), "Experimental Research on the Buckling of Alualloy Columns with Unsymmetrical Cross-Section," Laboratorie de Mécanique des Materiaux et Théorie des Structures, Université de Liege, Liege, Belgium.

Gonçalves, R., and Camotim, D. (2004). "On the Application of Beam-Column Interaction Formulae to Steel Members with Arbitrary loading and Support Conditions, *J. Constr. Steel Res.*, Vol. 60, Nos. 3–5, pp. 433–450.

Han, D. J., and Chen, W. F. (1983a), "Behaviour of Portal and Strut Types of Beam-Columns," *Eng. Struct.*, Vol. 5, pp. 15–25.

Han, D. J., and Chen, W. F. (1983b), "Buckling and Cyclic Inelastic Analysis of Steel Tubular Beam-Columns," *Eng. Struct.*, Vol. 5, pp. 119–132.

Hancock, G. J. (1977), "Elastic-Plastic Analysis of Thin-Walled Cross-Sections," 6th Aust. Conf. Mech., Struct. Mater., Christchurch, New Zealand, Aug.

Hill, H. N., and Clark, J. W. (1951a), "Lateral Buckling of Eccentrically Loaded I- and H-Section Columns," *Proc. 1st Natl. Congr. Appl. Mech.*, ASME, New York.

Hill, H. N., and Clark, J. W. (1951b), "Lateral Buckling of Eccentrically Loaded I-Section Columns," *Trans. Am. Soc. Civ. Eng.*, Vol. 116, p. 1179.

Hill, H. N., Hartmann, E. C., and Clark, J. W. (1956), "Design of Aluminium Alloy Beam-Columns," *Trans. Am. Soc. Civ. Eng.*, Vol. 121, pp. 1–21.

Horne, M. R. (1956), "The Flexural-Torsional Buckling of Members of Symmetrical I-Section Under Combined Thrust and Unequal Terminal Moments," *Q. J. Mech. Appl. Math.*, Vol. 7. No. 4, pp. 410–426.

Hsu, T. L., and Lee, G. C. (1981), "Design of Beam-Columns with Lateral–Torsional End-Restraints," *Weld. Res. Counc. Bull.*, No. 272, Nov.

International Organization for Standardization (ISO) (1994), *Steel Structures: Materials and Design*, Draft ISO/TC 167/SC 1, ISO Paris.

Jaspart, J. P., Briquet, Ch., and Maquoi, R. (1993). "Étude Comparative des Diverses Formules d'Interaction des Barres Comprimées et Fléchies," *Revue Constr. Métall.*, No. 4, pp. 13–21 (French).

Jingping, L., Zaitian, G., and Shaofan, C. (1988), "Buckling of Transversely Loaded I-Beam-Columns," *ASCE J. Struct. Eng.* Vol. 114, No. 9, pp. 2109–2118.

Johnston, B. G. (1976), *Guide to Stability Design Criteria for Metal Structures*, 3rd ed., Wiley, New York.

Johnston, B. G., and Cheney, L. (1942), "Steel Columns of Rolled Wide Flange Sections," Prog. Rep. No. 2, American Institute of Steel Construction, Chicago, Nov.

Kaim, P. (2004), "Spatial Buckling Behaviour of Steel Members Under Bending and Compression," PhD thesis, Technical University of Graz, Austria.

Kanchanalai, T., and Lu, L. W. (1979), "Analysis and Design of Framed Columns Under Minor Axis Bending," *Eng. J. Am. Inst. Steel Constr.*, second quarter, pp. 29–41.

Kemp, A. R. (1984), "Slenderness Limits Normal to the Plane-of-Bending for Beam-Columns in Plastic Design," *J. Constr. Steel. Res.*, Vol. 4, pp. 135–150.

Kennedy, D. L. J., Picard, A., and Beaulieu, D. (1993), "Limit States Design of Beam-Columns: The Canadian Approach and Some Comparisons," *J. Constr. Steel Res.*, Vol. 25, No. 1–2, pp. 141–164.

Ketter, R. L. (1962), "Further Studies of the Strength of Beam-Columns," *Trans. Am. Soc. Civ. Eng.*, Vol. 127, Part II, pp. 224–466.

Ketter, R. L., Beedle, L. S., and Johnson, B. G. (1952), "Welded Continuous Frames and Their Components ," *Weld. J. Res. Suppl.*, Vol. 31, No. 12. pp. 607–622.

Ketter, R. L., Kaminsky, E. L., and Beedle, L. S. (1955), "Plastic Deformation of Wide-Flange Beam Columns," *Trans. Am. Soc. Civ. Eng.*, Vol. 120, p. 1028.

Kitipornchai, S., and Wang, C.-M. (1988), "Out-of-Plane Buckling Formulas for Beam-Columns/Tie Beams," *ASCE J. Struct. Eng.*, Vol. 114, No. 12, pp. 2773–2789.

Klöppel, K., and Barsch, W. (1973), "Versuche zum Kapitel Stabilittätsfälle der Neufassung von Din 4113," *Aluminium*, Vol. 49, No. 10, pp. 690–699.

Klöppel, K., and Winkelmann, E. (1962), "Experimentale und theoretische Untersuchungen über die Traglast von zweiachsig assussermittig gedrückten Stahlstäben," *Stahlbau*, Vol. 31, p. 33.

Kounadis, A. N., and Ioannidis, G. I. (1994), "Lateral Post-Buckling Analysis of Beam-Columns," *ASCE J. Eng. Mech.*, Vol. 120, No. 4, pp. 695–706.

Lay, M. G. (1974), *Source Book for the Australian Steel Structures Code AS1250*, Australian Institute of Steel Construction, Sydney, Australia.

Lay, M. G., and Galambos, T. V. (1965), "The Experimental Behaviour of Retrained Columns," *Weld. Res. Counc. Bull.*, No. 110, Nov.

Lim, L. C., and Lu, L. W. (1970), "The Strength and Behavior of Laterally Unsupported Columns," Fritz Eng. Lab. Rep. No. 329.5, Lehigh University, Bethlehem, PA, June.

Lindner, J. (1972), "Theoretical Investigations of Columns Under Biaxial Loading," *Proc. Int. Colloq. Column Strength*, IABSE-CRC-ECCS, Paris.

Lindner, J., and Gietzelt, R. (1982), "Design of Biaxially Loaded Steel Beam-Columns," Rep. 2061E, Technical University of Berlin, Nov.

Lindner, J., and Gietzelt, R. (1983), "Discussion of Interaction Equations for Members in Compression and Bending," *Proc. 3rd Int. Colloq. Stab. Met. Struct.*, Paris, Nov.

Lindner, J., and Kurth, W. (1982), "Zum Biegedrillknicken Von Stützen aus, Ste 690," *Stahlbau*, Vol. 51, No. 12, pp. 366–372.

Lindner, J., and Wiechart, G. (1978), "Zur Bemessung gegen Biegedrillknickung," in *Beitrag Festschr. Otto Jungbluth*, Technische Hochschule, Darmstadt, Germany.

Lu, L. W., and Kamalvand, H. (1968), "Ultimate Strength of Laterally Loaded Columns," *ASCE J. Struct. Div.*, Vol. 94, No. ST6, pp. 1505–1524.

Lu, L. W., Shen, Z. Y., and Hu, X. R. (1983), *Inelastic Instability Research at Lehigh University: Instability and Plastic Collapse of Steel Structures* (Ed. L. J. Morris), Granada, London.

Maquoi, R., and Rondal, J. (1982), "Sur la Force Portant des Poutres Colonnes," *Ann. Trav. Publics Belg.*, No. 5.

Marshall, P. J., and Ellis, J. S. (1970), "Ultimate Biaxial Capacity of Box Steel Columns," *ASCE J. Struct. Div.*, Vol. 96, No. ST9, pp. 1873–1888.

Mason, R. G., Fisher, G. P., and Winter, G. (1958), "Eccentrically Loaded, Hinged Steel Columns," *ASCE J. Eng. Mech. Div.*, Vol. 84, No. EM4, p. 1792.

Massonnet, C. (1976), "Forty Years of Research on Beam-Columns," *Solid Mech. Arch.*, Vol. 1, No. 1, pp. 27–157.

Massonnet, C., and Save, M. (1965), *Plastic Analysis and Design*, Vol. 1, *Beams and Frames*, Blaisdell, New York.

Matthey, P. A. (1982), "Flexion Gauche Non-symétrique de Colonnes en Double-te Recherche Expérimentale et Simulation," paper presented at the 3rd Int. Colloq. Stab. Met. Struct., Prelim. Rep., Paris.

Mazzolani, F. M., and Frey, F. (1983), "ECCS Stability Code for Aluminium-Alloy Members: Present State and Work in Progress," paper presented at the 3rd Int. Colloq. Stab. Met. Struct., Prelim. Rep., Paris.

McLellan, E. R., and Adams, P. F. (1970), "Design of Steel Crane Columns or Columns Subjected to Lateral Loads," *EIC Trans.*, Vol. 70, No. A-6.

Miranda, C., and Ojalvo, M. (1965), "Inelastic Lateral–Torsional Buckling of Beam-Columns," *ASCE J. Eng. Mech. Div.*, Vol. 91, No. EM6, pp. 21–38.

Morris, G. A., and Fenves, S. J. (1969), "Approximate Yield Surface Equations," *ASCE J. Eng. Mech. Div.*, Vol. 95, No. EM4, pp. 937–954.

Nakashima, M. (1994), "Variation of Ductility Capacity of Steel Beam-Columns," *ASCE J. Struct. Eng.*, Vol. 120, No. 7, pp. 1941–1960.

Nakashima, M., Takanashi, K., and Kato, H. (1990), "Tests of Steel Beam-Columns Subject to Sidesway," *ASCE J. Struct. Eng.*, Vol. 116, No. 9, pp. 2516–2531.

Nethercot, D. A. (1983), "Evaluation of Interaction Equations for Use in Design Specifications in Western Europe," *Proc. 3rd Int. Colloq. Stab. Met. Struct.*, Toronto.

Ofner R. (1997), "Traglasten Von Stäben aus Stahl bei Druck und Biegung, PhD thesis, Institut für Stahlbau, Holzbau und Fläschentragwerke, Graz Technical University, Austria.

Ojalvo, M., and Fukumoto, Y. (1962), "Nomographs for the Solution of Beam-Column Problems," *Weld. Res. Counc. Bull.*, No. 78, June.

Opperman, H. P. (1983), "Refined Beam-Column Analysis," *Proc. 3rd Int. Colloq. Stab. Met. Struct.*, Paris.

Pi, Y.-L., and Trahair, N. S. (1992a), "Prebuckling Deformations and Lateral Buckling: I. Theory," *ASCE J. Struct. Eng.*, Vol. 118, No. 11, pp. 2949–2966.

Pi, Y.-L., and Trahair, N. S. (1992b), "Prebuckling Deformations and Lateral Buckling: II. Applications," *ASCE J. Struct. Eng.*, Vol. 118, No. 11, pp. 2967–2985.

Pi, Y.-L., and Trahair, N. S. (1994a), "Non-Linear Inelastic Analysis of Steel Beam-Columns: I. Theory," *ASCE J. Struct. Eng.*, Vol. 120, No. 7, pp. 2041–2061.

Pi, Y.-L., and Trahair, N. S. (1994b), "Non-Linear Inelastic Analysis of Steel Beam-Columns: II. Application," *ASCE J. Struct. Eng.*, Vol. 120, No. 7, pp. 2062–2085.

Pillai, U. S. (1970), "Review of Recent Research on the Behaviour of Beam-Columns under Biaxial Bending," R. Mil. Coll. Can. Civ. Eng. Rep., Jan. Royal Military College Kingston, Ontario, Canada

Pillai, U. S. (1974), "Beam Columns of Hollow Sections," *J. Can. Soc. Civ. Eng.*, Vol. 1, No. 2, pp. 194–198.

Pillai, U. S. (1980), "Comparison of Test Results with Design Equations for Biaxially Loaded Steel Beam-Columns," R. Mil. Coll. Can. Dep. Civ. Eng. Res. Rep., No. 80–2, Aug. Royal Military College Kingston, Ontario, Canada

Pillai, U. S. (1981), "An Assessment of CSA Standard Equations for Beam-Column Design," *Can. J. Civ. Eng.*, Vol. 8, No. 22, pp. 130–136.

Rajasekaren, S. (1977), "Finite Element Method for Plastic Beam-Columns," in *Theory of Beam-Columns* (Eds. W. F. Chen and T. Atsuta), McGraw-Hill, New York, Chap. 12.

Ramm, E., and Osterrieder, P. (1983), "Ultimate Load Analysis of Three-Dimensional Beam Structures with Thin-Walled Cross Sections Using Finite Elements," *Proc. 3rd Int. Colloq. Stab. Met. Struct. Prelim. Rep.*, Paris.

Razzaq, Z., and McVinnie, W. W. (1982), "Rectangular Tubular Steel Columns Loaded Biaxially," *J. Struct. Mech.*, Vol. 10, No. 4, pp. 475–493.

Roik, K., and Bergmann, R. (1977), "Steel Column Design," Prelim. Rep. 2nd Int. Colloq. Stab. Steel Struct., Liege, Belgium.

Roik, K., and Kindmann, R. (1983), "Design of Simply Supported Members by Means of European Buckling Curves for Uniaxial Bending with Compression," Prelim. Rep. 3rd Int. Colloq. Stab. Met. Struct., Paris.

Roik, K., and Wagenknecht, R. (1976), "Traglastdiagramme zur Bemessung von Druckstaben mit doppellsymmetrischen Querschnitt aus Baustahl," *KIB/Berichte*, Vol. 27, Vulkan–Verlag, Essen, Germany.

Ross, D. A., and Chen, W. F. (1976), "Design Criteria for Steel I-Columns Under Axial Load and Biaxial Bending," *Can. J. Civ. Eng.*, Vol. 3, No. 3, pp. 466–473.

Salvadori, M. G. (1956), "Lateral Buckling of Eccentrically Loaded I-Columns," *ASCE Trans.*, Vol. 121, pp. 1163–1178.

Santathadaporn, S., and Chen, W. F. (1970), "Interaction Curves for Sections Under Combined Biaxial Bending and Axial Force," *Weld. Res. Counc. Bull.*, No. 148, Feb.

Sohal, I. S., and Chen, W. F. (1988), "Local and Post-Buckling Behavior of Tubular Beam-Columns," *ASCE J. Struct. Eng.*, Vol. 114, No. 5, pp. 1073–1090.

Sohal, I. S., and Syed, N. A. (1992), "Inelastic Amplification Factor for Design of Steel Beam-Columns," *ASCE J. Struct. Eng.*, Vol. 118, No. 7, pp. 1822–1839.

Sputo, T. (1993), "History of Steel Beam-Column Design," *ASCE J. Struct. Eng.*, Vol. 119, No. 2, pp. 547–557.

Standards Association of Australia (SAA) (1998), "Steel Structures", AS 4100-1998, SAA, Homebush, New South Wales, Australia.

Tebedge, N., and Chen, W. F. (1974), "Design Criteria for Steel H-Columns Under Biaxial Loading," *ASCE J. Struct. Div.*, Vol. 100, No. ST3, pp. 579–598.

Trahair, N. S. (1993), *Flexural-Torsional Buckling of Structures*, E&FN Spon, London, Chap. 11.

Trahair, N. S., and Bradford, M. A. (1988), *The Behaviour and Design of Steel Structures*, 2nd ed., Chapman & Hall, London, Chap. 7.

van Kuren, R. C., and Galambos, T. V. (1964), "Beam-Column Experiments," *ASCE J. Struct. Div.*, Vol. 90, No. ST2, pp. 223–256.

van Manen, S. E. (1982), "Plastic Design of Braced Frames Allowing Plastic Hinges in the Columns," *Heron*, Vol. 27, No. 2, pp. 35–45.

Vinnakota, S. (1977), "Finite Difference Method for Plastic Beam-Columns," in *Theory of Beam-Columns*, (Eds. W. F. Chen and T. Atsuta), McGraw-Hill, New York, Chap. 10.

Virdi, K. S. (1981), "Design of Circular and Rectangular Hollow Section Columns," *J. Constr. Steel. Res.*, Vol. 1, No. 4, pp. 35–45.

Wang, C. M., and Kitipornchai, S. (1989), "New Set of Buckling Parameters for Monosymmetric Beam-Columns/Tie Beams," *ASCE J. Struct. Eng.*, Vol. 115, No. 6, pp. 1497–1513.

White, D. W., and Clarke, M. (1997), "Design of Beam-Columns in Steel Frames I: Philosophies and Procedures," *ASCE J. Struct. Eng.*, Vol. 123, No. 12, Dec., pp. 1556–1564.

White, D. W., and Clarke, M. (1997), "Design of Beam-Columns in Steel Frames II: Comparison of Standards," *ASCE J. Struct. Eng.*,, Vol. 123, No. 12, Dec., pp. 1565–1574.

Young, B. W. (1973), "The In-Plane Failure of Steel Beam-Columns," *Struct. Eng.*, Vol. 51, No. 1, pp. 27–35.

Yu, C. K., and Tall, L. (1971), "A514 Steel Beam-Columns," *Publ. IABSE*, No. 31–II, pp. 185–213.

Zuyan, S., and Zhang, Q. (1991), "Interaction of Local and Overall Instability of Compressed Box-Columns," *ASCE J. Struct. Eng.*, Vol. 117, No. 11, pp. 3337–3355.

CHAPTER 9

HORIZONTALLY CURVED STEEL GIRDERS

9.1 INTRODUCTION

There has been a dramatic increase in the use of horizontally curved steel girders in highway bridges and interchanges throughout the world over the past several decades. The need for the smooth dissemination of congested traffic, right-of-way limitations, economic and environmental considerations, and increased emphasis on aesthetic considerations has encouraged this trend. When curved bridge superstructures were first introduced, they were generally composed of a series of straight girder chords. Over time these systems have steadily been replaced by structures containing curved sections. Although the cost of curved superstructures may be higher than that for their straight counterpart, the total cost of a curved girder bridge system has been found to be less than that of a similar straight system because a substantial portion of the substructure can be eliminated. Also, using continuous curved girders allows for shallower sections as well as a reduction in outside girder slab overhangs.

In the early years of modern bridge design, engineers were reluctant to use curved girders due to mathematical complexities associated with their design. Curved girders are subjected to not only major axis flexural stresses but also to significant torsional stresses, even under gravity loading. The overall effects of torsion on slender-web curved steel plate girders of dimensions commonly used in bridges can be approximated through torsional theory of beams; deflection, cross-sectional distortion, and deflection amplification (large-displacement) effects, however, may be much more pronounced in curved girder systems and may require more advanced methods to accurately predict behavior. The inherent rotation characteristics of horizontally curved girders require that diaphragms and bracing used in straight girder systems simply to prevent premature lateral buckling become very important (primary) load-carrying components. Despite these obstacles, the availability of digital computers over the past several decades to carry out complex analyses, along with advancements in fabrication and erection technology, have made horizontally curved girder superstructures a viable and cost-effective option for designers.

This chapter presents an overview of the behavioral issues associated with the stability and design of curved beams and girders. A brief historical review is initially provided. Because the primary thrust of recent research has been on developing resistance formulations for the design and construction of bridges comprised of curved steel superstructures, this chapter emphasizes the latest efforts conducted by U.S. researchers towards improving bridge design and construction methodology. Other important behavioral phenomena such as dynamics and fatigue are not addressed in this chapter.

9.2 HISTORICAL REVIEW

Research prior to the mid-1960s on the behavior of curved beams was generally limited to theoretical work on the linear elastic static behavior of isolated curved members, with the earliest work on curved beam theory attributed to St. Venant (1843) over 150 years ago. Since then, thousands of pages of technical papers, reports, and books have been published on the behavior and design of curved beams. These researchers include Gottfield (1932), Umanskii (1948), Dabrowski (1964, 1965, 1968), Vlasov (1961), Timoshenko (1905), Shimada and Kuranashi (1966), Watanabe (1967), and others. Comprehensive presentations of the theory of thin-walled beams including flexure torsion, distortion, and stress distribution are provided in several texts (Vlasov, 1961; Dabrowski, 1968; Kollbrunner and Basler, 1969; Heins, 1975; Nakai and Yoo, 1988).

Since the mid-1960s an emphasis on curved girder research in the United States and Japan has been placed on the practical use of curved beam theory in the design and construction of horizontally curved bridges. The first survey of published works pertaining to horizontally curved bridges was presented by McManus et al. (1969) and contained 202 references. Additional references were subsequently added by Ketcheck (1969), Tan et al. (1969), and Pandit et al. (1970). Several years later, the ASCE and AASHTO Task Committee on Curved Girders (AASHTO, 1978a) presented a state-of-the-art report that provides 106 references on horizontally curved box girders. The committee also presented results of a survey pertaining to the geometry, design, detail, construction, and performance of box-girder bridges constructed in the United States, Canada, Europe, and Japan (AASHTO, 1978b). Comprehensive reviews and syntheses of curved bridge girder research have been updated by others (Kulicki et al., 2006; Linzell et al., 2004b; White et al., 2001; Linzell, 1999; Zureick et al., 1994).

In 1963, U.S. Steel published a report that presented an approximate analysis technique for "open-framed" curved I-girder bridges (Richardson, Gordon, and Associates, 1963). The method presented in the 1963 report was viewed as too complicated and cumbersome for design, and hence in 1965 U.S. Steel published an approximate procedure that became known as "V-load analysis" for determining moments and shears in horizontally curved open-framed highway bridges (*Highway*

Structures Design Handbook, 1965). It is theoretically pure with regard to torsion due only to curvature and load distribution for static equilibrium. The method does not account for lateral bracing between girders in the plane of the flanges. Accuracy of the method with regard to live load depends upon the ability of the user to assign appropriate loads to the girders prior to the V-load analysis. It has been noted that the live-load distribution factors used in straight bridge design do not appropriately represent the distribution in curved bridges and researchers have subsequently proposed equations for curved bridge design (U.S. Steel Corporation, 1984; Heins and Jin, 1984; Brockenbrough, 1986).

In 1969, a comprehensive pooled funds research project, referred to as the Consortium of University Research Teams (CURT), sponsored by 25 participating U.S. state highway departments was initiated under the direction of the FHWA to study the behavior of curved bridges and to develop design requirements. The team consisted of researchers from Carnegie Mellon University, the University of Pennsylvania, the University of Rhode Island, and Syracuse University, whose efforts, along with those at the University of Maryland, resulted in the working stress design criteria and tentative design specifications. The ASCE-AASHTO Task Committee on Curved Girders (AASHTO, 1977c) compiled the results of CURT and others into a single set of recommendations for the design of curved I-girder bridges. The CURT project was followed by the development of the load factor design criteria (Stegmann and Galambos, 1976; Galambos, 1978), which, along with the working stress criteria, were adopted by AASHTO (1980) as the first *Guide Specifications for Horizontally Curved Highway Bridges,* subsequently referred to as the Guide Specifications (AASHTO, 1980, 1987). The Guide Specifications address both "I" and "box" shape girder bridge superstructures. Strength formulations for the web, flanges, and stiffeners are emphasized. Experience indicates that the strength formulations have been at least adequate in that there have been no reported failures of curved steel superstructure bridges in the United States due to overload. The original form of the Guide Specifications, however, is disjointed and difficult to use, and due to a lack of continuity with straight bridge standards, it was never adopted as an integral part of any of the AASHTO *Standard Specifications for Highway Bridges*. The only other bridge design document that specifically addresses curved bridge design is the Japanese *Guidelines for the Design of Horizontally Curved Girder Bridges*, published in draft form by the Hanshin Expressway Public Corporation (1988). Several researchers have demonstrated disparity in the strength formulations between the Japanese and American curved bridge design guides, which further emphasized the need for additional research. Changes to the original 1980 Guide Specifications were included in the AASHTO Interim Specifications for the years 1981, 1982, 1984, 1985, 1986, and 1990. A new edition was published in 1993 that included the interim and additional changes but did not reflect the extensive research on curved girder bridges that was conducted since 1980 or the many important changes in related provisions of the straight girder specifications.

A comprehensive research project administered in 1992 by the FHWA, Curved Steel Bridge Research Project (CSBRP), was initiated to conduct fundamental research into the behavior of curved steel flexural members and bridges that would lead to the development of LRFD Specifications for curved bridge design. The CSBRP included tests on a full-scale I-girder bridge (Figs. 9.1 and 9.2) and the research resulted in an expanded knowledge of the moment and shear capacities of horizontally curved I-girder bridges. The work of the CSBRP was extended in 1999 through a project, which was jointly sponsored by FHWA and the AISI, that used refined nonlinear finite element analyses to augment the knowledge gained through the experimental tests conducted under the CURT and CSBRP projects. This analytical work resulted in the unification of the design equations for straight and curved steel girders into the 2004 LRFD specifications (AASHTO, 2004; White and Grubb, 2005). Comprehensive summaries of the CSBRP are provided by Linzell (1999), White et al. (2001), Yoo and Davidson (2002), and Hartmann (2005). Other journal publications and conference presentations have resulted since the CSBRP began, many of which are highlighted below.

In 1993, NCHRP Project 12-38, "Improved Design Specifications for Horizontally Curved Steel Girder Highway Bridges," was initiated to reorganize the Guide Specifications using the current state-of-the-art information while retaining the Working Stress and Load Factor Design formats. Project 12-38 resulted in NCHRP Report 424, "Improved Design Specifications for Horizontally Curved Steel Girder Highway Bridges," and a 2003 edition update to the Guide Specifications (Hall et al., 1999). The 2003 Guide Specifications demonstrate the applicability of the guidelines through design examples for I- and box girders. In 1999, NCHRP Project 12-52, "LRFD Specifications for Horizontally Curved Steel Girder Highway Bridges," was initiated to prepare specifications for the design and construction of horizontally curved steel girder bridges (for both I- and box girders) in a calibrated LRFD format. Merging of the curved girder provisions into the straight girder design provisions was accomplished by statistically calibrating the curved bridge design provisions. NCHRP Project 12-52 culminated in NCHRP Report 563, "Development of LRFD Specifications for Horizontally Curved Steel Girder Bridges" (Kulicki et al., 2006), which documents the effort leading to the specifications, contains a compilation of abstracts of horizontally curved girder bridge research reports, and was used as the basis for updating the 2004 LRFD Bridge Design Specifications to include curvature effects.

9.3 FABRICATION AND CONSTRUCTION

Curvature induces many challenges from the very beginning of the bridge fabrication and construction process. Residual stresses formed during curved plate-girder fabrication can be much greater than those in straight plate girders and significantly affect the strength of the girder. Ensuring proper camber is difficult. Transporting and shipping girders with significant curvature becomes problematic because the girders must be properly restrained to prevent instability during shipping and overhangs must be carefully checked. Once at the job site, placing the girders becomes cumbersome.

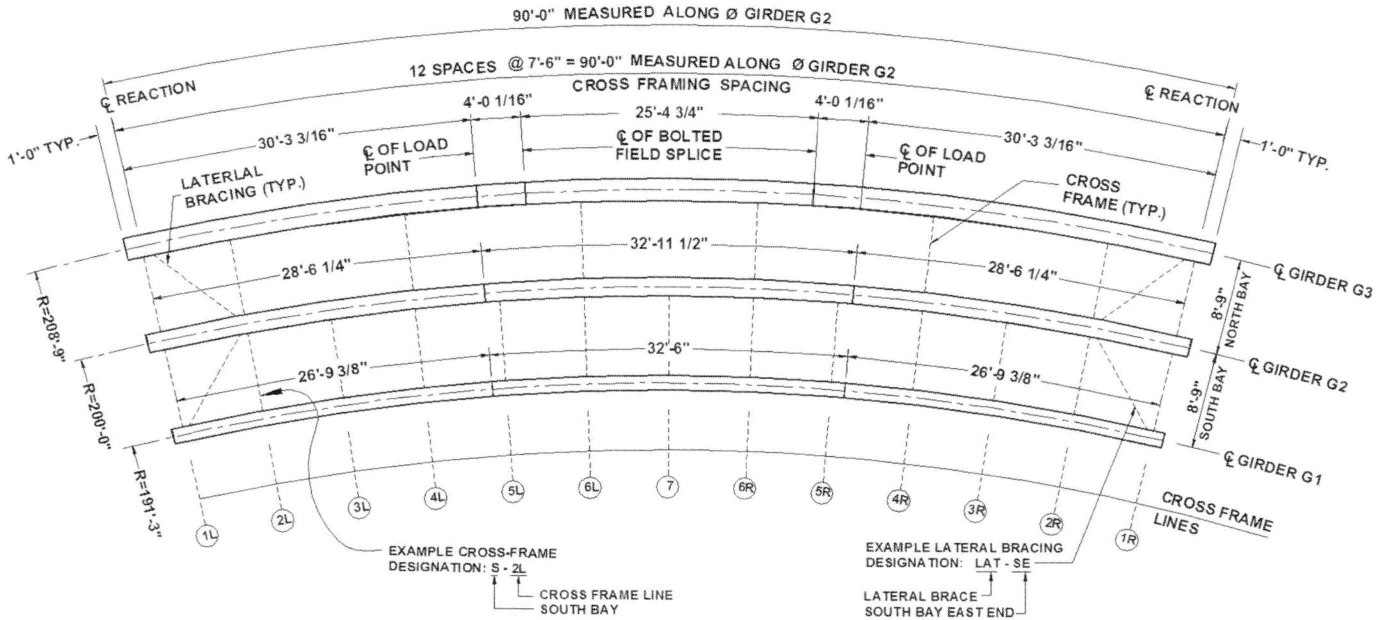

FIGURE 9.1 Plan view of the FHWA-CSBRP test frame (Hartmann, 2005).

FIGURE 9.2 FHWA-CSBRP test frame at the FWHA Laboratory (Hartmann, 2005).

9.3.1 Fabrication

Generally there are three methods of fabricating curved steel I-girders: (1) cut curving, (2) heat curving, and (3) cold curving. Cut curving involves flame cutting the flanges to the desired curvature from a standard steel plate. The advantage of this method is that there is no limit on the radius of curvature that can be obtained. This method of fabrication involves careful planning for economical cutting of the flange plates to minimize the amount of scrap generated. In addition, adhering to consistency in plate thickness and steel grades allows the fabricator to economize by combining and nesting plates (Grubb et al., 1996). Practically, all curved box girders are fabricated by cut curving. Extreme care must be exercised in the fabrication of box girders. The flexibility that engineers take advantage of during the field erection of I-girders is not available with box girders due to their very high torsional rigidity.

Heat curving is accomplished by simultaneously heating one side of the top and bottom flanges of a fabricated straight I-girder to introduce residual curvature after cooling. The application of heat can be continuous, strip, or V-type. In continuous heating, the flange edges are heated along their length. In strip heating, the flanges are heated in rectangular strips at regular intervals until the required curvature is attained. In V-type heating, the top and bottom flanges are heated in truncated triangular or wedge-shaped areas having their bases along the flange edge and spaced at regular intervals along each flange.

In the cold-curving process, a straight I-girder is bent plastically to obtain an over-deformed curvature and released. The relaxed configuration results in the required curvature. Cold curving is fast, efficient, and precise and could be one of the most economical methods of fabricating a curved I-girder. The possibility of fracture due to localized load effects, however, has raised concerns.

Regardless of the curving method used, plate-girder fabrication introduces residual stresses and camber loss. This is particularly true for heat-curved girders. Residual stresses are created whenever a member is permanently deformed or distorted in a nonuniform manner and persist in a material or a component under uniform temperature in the absence of externally applied loads. Very little information is available on the residual stresses developed in flame-cut and welded I-section curved girders (Bradford et al., 2001). Based on research by Kishima et al. (1969), Culver and Nasir (1969) suggested a residual stress pattern for welded I-sections. Brockenbrough (1970a,b,c) reported that the magnitude and distribution of residual stress in heat-curved girders is a function of dimensions and material properties of the straight girder and the curving procedure.

9.3.2 Transporting

Instability challenges that are exacerbated by curvature can occur while transporting girders from the fabrication plant to the construction site. Steel bridge girders can be transported by highways, railways, waterways, or a combination of the three, depending on where the bridge is to be constructed. The weight, height, length, and width of the girder sections may be limited by the transporting method chosen. The transporting mode must ensure that the girders can be delivered to the site without deforming the cross section and inducing additional stresses. The location of splices and the overall geometry of the section can be designed so that the girder will be easier to maneuver. Proper restraint against vertical, longitudinal, and transverse movement must be considered. For straight girders during transportation, lateral restraint is typically provided only at vertical supports, while curved girders may need to overhang and require additional support to prevent instability.

9.3.3 Erecting

Overall stability of single long slender girders during lifting is a major concern during the construction of highway bridges. Lifting of girders in straight bridge construction presents little difficulty as the center of gravity coincides with the centroidal axis of the beam cross section. A horizontally curved girder, however, introduces rotation during lifting as the center of gravity does not coincide with the centroidal axis of the beam cross section. Depending on the length of the beam, lateral–torsional buckling or significant nonlinear deflection behavior may occur, thus shifting the center of gravity and causing rigid-body instability.

The calculation of optimum pick locations for two lifting points can be approximated by treating the curved girder as a circular arc in plan and assuming that the section is prismatic (Grubb et al., 1996). Bridge girders, however, are often nonprismatic, which creates additional problems for locating the balance points. In practice, erectors often "weigh" a piece; the girder may be lifted a few inches and put down repeatedly until the balance points are located. Unfortunately for curved girders, this may take several trials. Wire rope slings or girder clamps are usually used for attaching the free edges of the top flange at the lifting points. Due to the inherent tendency of a curved girder to twist, high stresses may occur at the attachment locations. These intense stresses may occur on the inside (concave) or outside (convex) edge of the girder flange, depending on the direction in which the girder tends to rotate.

The lifting and support mechanism used depends on the length of the beam. While longer girders necessitate the use of spreader beams, shorter girders can be lifted with single- or double-cable slings. In addition, a component of the cable force in the horizontal plane that causes minor axis bending must be taken into account when inclined cables are used. Two parallel girders can be bolted together by the diaphragms or cross frames and lifted as one piece. Lifting of girders in pairs helps resist wind loads and may save time. Horizontal stiffening trusses can be added to the compression flanges before lifting. Of course, the ability to lift two girders at once ultimately depends on the crane capacity available at the jobsite.

9.3.4 Sequencing

Proper erecting and sequencing of curved girders is essential during construction. The placement sequence of the girders and diaphragms or cross frames should be carefully planned so that fit-up problems are minimized. Diaphragms or cross frames are bolted between each girder to provide stability and to control deflection of the girders. The fabricator normally assembles the bridge components prior to delivery to the jobsite to ensure that fit-up problems will not occur. Once the girders have been loaded for transporting and unloaded for placement, resulting changes in camber may cause further fit-up problems. In addition, the configuration of the partially completed structure must be stable. Unlike straight girders, curved girders depend on adjacent girders for stability.

Several approaches can be adopted for erecting and stabilizing curved girders. Grubb et al. (1996) describe three methods for proper erection and stabilization. In the first method, which assumes adequate crane capacity is available, paired erection is desirable. After erecting the first pair of girders, individual girders can be erected successively and connected to adjacent girders by cross frames. This increases the torsional stiffness, thereby adding stability to the system. A second method of erecting each girder is to use one crane to pick up the girder and place it while another crane supports the girder to which it is connected. While both cranes hold their girders, the diaphragms or cross frames are bolted into place. The addition of the second girder and cross bracing between them changes the governing instability mode from flexural–torsional buckling of a single girder to flexural buckling of two girders acting together as a unit. This method requires adequate area for mobilization of two cranes. A third method uses temporary false-work towers or bents to shore the girders, which requires a site space that is often not available.

The 2003 update to the Guide Specifications included sections that address curvature-specific constructability and construction loads. The design examples presented in the 2003 Guide also include construction sequence calculations. The

"Steel Bridge Erection Guide Specification," developed by Task Group 10 (Erection) of the AASHTO/NSBA Steel Bridge Collaboration and later adopted and published by AASHTO (2007), briefly addresses the fabrication and erection of curved girders. Given the frequency of problems encountered during construction, several researchers have studied the large-scale erection behavior of curved I-girders. Important recent studies include Linzell (1999, 2004a), Galambos et al. (1996, 2000), Simpson (2000), Chavel and Earls (2001, 2006a,b), Bell and Linzell (2007), and Chang and White (2008a).

9.4 ANALYSIS METHODS

Analysis methods for horizontally curved steel members can be classified into two major categories: approximate and refined methods. The approximate methods require minimal modeling effort and are adequate for preliminary analysis and design purposes. These include the plane grid, space frame, V-load, and M/R methods. The refined methods are more computationally intensive and time consuming in terms of modeling. Therefore, the methods that fall in this class are often used only for final analyses. Examples of such methods include the finite element method, finite strip method, finite difference method, slope deflection method, and analytical solutions to the governing differential equations. Software developed specifically for curved bridge design includes BSDI (2000), DESCUS (2002), and MDX (2000). A brief description of each of these analytical methods is presented below. The 2003 Guide Specifications also include a section and commentary on "approximate" and "refined" analysis methods. Additional details and demonstration of accuracy are provided by Modjeski and Masters (1989), Zureick et al. (1994), Zureick and Naquib (1999), McElwain and Laman (2000), Nevling et al. (2006), Bell and Linzell (2007), and Chang and White (2008a,b).

9.4.1 Approximate Methods

The following approximate methods of analysis have been found to be reasonably accurate for noncomposite dead loads and are capable of accounting for limited skew. There are many bridge geometries, however, where approximate methods do not provide adequate results.

Plane Grid Method The plane grid method idealizes the structure as an assemblage of two-dimensional grid members with one translational and two rotational degrees of freedom. It is frequently used for straight bridge design and analysis but does not account for nonuniform torsion (warping) and therefore can be used only for initial member sizing.

Space Frame Method This method was introduced in 1973 by Brennan and Mandel for the analysis of open and closed curved members. The curved members are idealized as three-dimensional straight members, while the diaphragms and lateral bracing are assumed as trusslike members that can carry only axial loads. The effects of warping are typically not included in this analysis, which again makes this method only applicable to initial design.

V-Load Method The V-load method idealizes the system as straight girders with span lengths equal to the arc lengths of the curved girders by accounting for curvature through self-equilibrating vertical shear forces (acting on diaphragm locations). It underestimates inner girder stresses and does not consider the bracing effect in the plane of the bottom flange. The magnitude of these loads depends on the radius of curvature, bridge width, and diaphragm spacing [U.S. Steel Corporation, 1984; Poellot 1987; Fiechtl et al., 1987; National Steel Bridge Alliance (NSBA), 1996].

M/R Method This method was presented in 1970 by Tung and Fountain for the approximate analysis of curved box-girder bridges. Torsion due to curvature is calculated for each girder. It is assumed that the girder is free to rotate without restraint of the deck or cross frames attached between boxes, that the box retains its shape, and that no other torsion is applied. These are rather strict limitations that may not always be applicable.

9.4.2 Refined Methods

Finite Element Method Described in more detail in Chapter 20, this approach discretizes the structure into small divisions (elements) where each element is defined by a specified number of nodes. The displacement field, stresses, and strains within the domain of each element are derived using an approximate function of its nodal displacements, which serve as the primary unknowns. This is one of the most general and accurate methods because it does not limit geometry, loads, or boundary conditions and therefore can be applied to open and closed girders. Static and dynamic analyses can be conducted. Geometric and material nonlinear effects, including residual stresses, may be included. The modeling and analysis efforts required for this method, however, often make it impractical for preliminary analyses.

Finite Strip Method The curved bridge is divided into narrow strips in the circumferential direction that are restrained in the radial direction. The analysis includes bending and membrane actions as well as warping and distortional effects (Hsu, 1989). Although this method provides simplicity over the finite element method because of the smaller number of unknowns that need to be solved for, it does not offer the flexibility and versatility of the finite element method.

Finite Difference Method A grid is superimposed on the structure and the governing differential equations (GDEs) are replaced by algebraic difference equations that are solved for each grid point.

Slope Deflection Method The partial differential equations are established in terms of slope-deflection equations and the solution is assumed to be a Fourier series. The analysis includes the effects of curvature, nonuniform torsion, and diaphragms.

Solution to the Governing Differential Equations An analytical solution to the GDE is obtained. The solution is usually a closed-form or a series solution, such as Fourier series.

Although these refined analysis methods are capable of predicting the behavior of each line girder, the finite element method is the only approach that can accurately describe the interaction between the adjacent girders, an effect that is rather dominant in horizontally curved girder bridges. Hence, the other refined methods may not provide adequate results.

9.5 STABILITY OF CURVED I-GIRDERS

9.5.1 Curvature Effects on Flange Plate Stability and Flange Slenderness Requirements

The combination of major axis (vertical) bending with warping torsion of curved I-shaped girders results in a nonuniform distribution of stress across the flange width. To establish the effect of curvature on flexural resistance contributed by the flanges, the normal stress is approximated as the linear addition of the bending normal stress, which is assumed to be uniform across the width of the flange, and the normal stress resulting from "lateral bending" of the flange, which can result from nonuniform torsion and/or lateral forces on the girder (Fig. 9.3). This results in a linear variation of normal stress that has a maximum value at one edge of the flange, an average value at the widthwise center (web juncture), and a minimum at the other edge. Therefore, the two sides of a slender flange will not buckle independently, nor is it realistic to assume that the full width of a compact flange can become plastic under combined bending and torsion.

Frampton (1968) and Culver and Frampton (1970) examined the elastic buckling case, in which each half of the flange is treated separately as an isotropic sector plate free on one edge and rotationally restrained along the other edge by the web and the other half of the flange. This investigation was later extended by Culver and Nasir (1969, 1971) to also cover the inelastic flange local buckling behavior. Culver (1971) presented a summary of his research related to proportioning the compression flange in a horizontally curved I-girder and pointed out that the total stress (warping plus bending) at the flange tip must be limited to $0.55F_y$ if the AASHTO b/t limit is used for curved girders. This approach was adopted by AASHTO in the Guide Specifications, where the flange local buckling criteria are based only on the b/t ratio.

Kang and Yoo (1990) conducted an analytical study that examined the allowable flexural stresses permitted by the Guide Specifications using three-dimensional

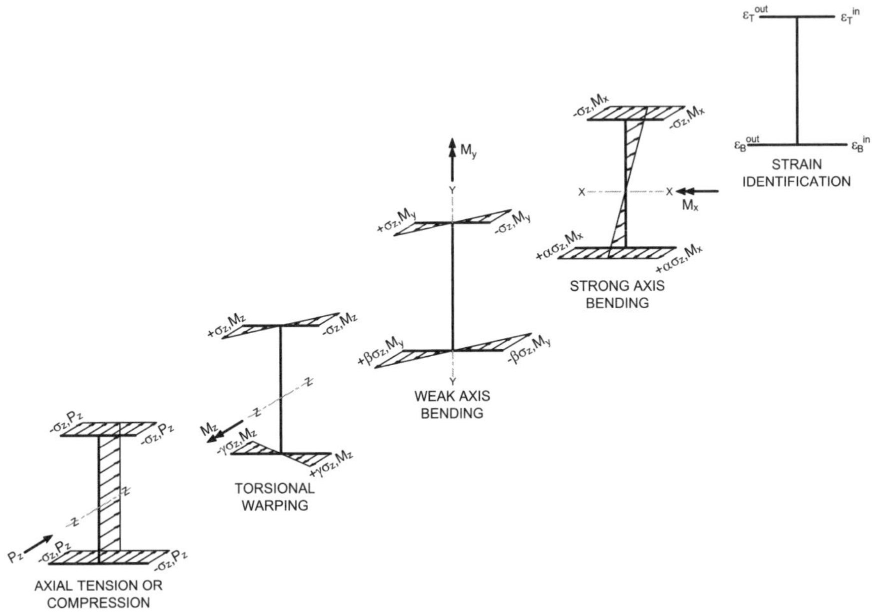

FIGURE 9.3 Components of longitudinal stress (Hartmann, 2005).

finite element models of the entire cross section. The results indicated that curvature may have a significant effect on local buckling. Using finite element analyses, Davidson and Yoo (1996) concluded that, contrary to the conclusions of Culver during the CURT program, the actual curvature of the flange plate was inconsequential for curvatures typical of bridge girders and the stress gradient and rotational restraint provided by the web are the dominating parameters affecting the local flange buckling of curved girder flanges. Finite element analyses were used to establish a slenderness equation that reflects the effect of curvature on local buckling.

Japanese researchers have also conducted analytical research on the local buckling behavior of curved compression flanges and have concluded that the influence of the stress gradient due to warping behavior cannot be omitted in evaluating the buckling strength of I-girders with substantial curvature (Nakai et al., 1981; Nakai and Yoo, 1988; Fujii and Ohmura, 1987). Initial deflections and residual compressive stresses were also considered (Komatsu et al., 1975; Komatsu and Kitada, 1981). This research suggests that the required curved compression flange thicknesses should be increased by approximately 30% in order to eliminate potential local buckling when warping effects are "predominant" (Komatsu and Kitada, 1981; Hanshin, 1988; Nakai and Yoo, 1988; Japan Road Association, 1990).

Madhavan and Davidson (2005, 2007) used a Galerkin series approach to define the effect of stress gradient on elastic buckling coefficients of centerline-stiffened flat plates subject to a compression stress gradient. Equations were developed for warping-to-bending stress ratios up to 10. The solution was related to slenderness requirements for the design of I-sections subjected to combined major axis bending and minor axis bending and/or torsion. Their results indicated that there is no practical need for b/t limits with curved compression flanges that are different from straight girder compression flanges and the reduction in elastic buckling strength for slender flanges is negligible. This work, however, did not consider residual stress effects.

The full-scale testing completed as part of the CSBRP included seven "bending specimens" of compression-flange slenderness varying from b/t of 13.6 to 33.6 (Hartmann, 2005). This testing was complimented by extensive finite element modeling that demonstrated excellent agreement with test results (White et al., 2001; Yoo and Davidson, 2002). The CSBRP bending tests and finite element work revealed that the compression-flange slenderness did affect the vertical bending capacity, with normalized performance decreasing with an increase in compression-flange slenderness. It also pointed out that the component of resistance provided by very slender flanges is sensitive to geometric imperfections. Overall, the CSBRP bending tests and finite element analyses supported the use of a "unified" design method. Based upon the analytical works described above and verified by the CSBRP testing, the consensus is that there is no need for more stringent flange slenderness requirements for curved bridge girders than those of straight girders. Therefore, the AASHTO (2004) LRFD specifications do not include curvature-specific flange slenderness requirements. A summary of b/t requirements for the various codes and guides is presented in Table 9.1.

9.5.2 Effect of Curvature on Web Plate Strength and Stability

The primary role of plate-girder webs in the region of high bending moment is to maintain the relative distance between flange plates. In regions near supports, webs resist most of the shear. Efficient design of plate girders therefore requires that the flange plates carry most of the primary bending moment and that the web be designed as "slender" as possible. Because of this, existing web depth thickness limitations, transverse stiffener spacing and rigidity, and longitudinal stiffener location and rigidity are largely based on buckling considerations.

TABLE 9.1 Compression Flange Slenderness Definitions b_f/t_f, Yield Strength 50 ksi (Kulicki et al., 2006)

Specifications	Compact	Slender
1993 Guide	14.3	19.7
2003 Guide	18.0	23.0
2004 LRFD	18.3	24.0

For straight girders, the bifurcation buckling behavior of the web plate is easily and accurately idealized using simplified boundary and loading conditions. For curved plate girders, however, curvature induces both warping of the cross section and, more importantly for web considerations, distortion of the web. Such distortion can cause the distribution of longitudinal membrane stresses in the web due to vertical bending to become nonlinear. Furthermore, the nonlinearity of this membrane stress distribution increases with an increase in curvature, which in turn results in an increase in flange normal stresses.

Pure Bending The curved I-girder web slenderness requirements presented in the Guide Specifications up to the 1993 edition were based largely on the analytical studies conducted by Culver et al. (1972a,b,c; Brogan, 1972). In these studies, the web panel was modeled as a series of isolated elastically supported cylindrical strips subject to fictitious radial loading, and a "spring foundation" boundary condition was employed that simulates the continuous curved plate under bending moment (Wachowiak, 1967). Numerical results were generated for the reduction in effective moment required to produce initial yield in the flanges based on curvature and web slenderness for an aspect ratio of 1.0 and a web-to-flange area ratio of 2.0. From the results, a maximum reduction of approximately 13% was noted for $a/R = 0.167$ and approximately 8% for $a/R = 0.10$ ($h/t_w = 150$), both of which correspond to extreme curvature (where a is the distance between transverse stiffeners, t_w web thickness, and R radius of curvature). To apply the parametric results to develop design criteria for practical curved girders, the deflections and web bending stresses that would occur for the girder with a curvature that corresponds to an initial imperfection out-of-flatness limit of $D/120$ were used, where D is the web depth. It is noted that this corresponds to a curvature of $a/R = 0.067$ for a panel with an aspect ratio of 1.0. The values of moment reduction using this approach were compared with those presented by Basler and Thurlimann (1961; Vincent, 1969).

An extension of this work was published a year later when Culver et al. (1973) checked the accuracy of the isolated elastically supported cylindrical strips by treating the panel as a two-way shell rather than as individual strips. The flange/web intersection was modeled as fixed and the boundaries at the transverse stiffeners were modeled as fixed and simple. Longitudinal stiffeners were modeled with moments of inertias as multiples of the AASHTO (1969) values for straight girders at the time. Using analytical results obtained for the slenderness required to limit the plate bending stresses in the curved panel to those of a flat panel with the maximum allowed out-of-flatness ($a/R = 0.067$) and with $D/t_w = 330$, an equation was developed for curved plate-girder web slenderness with one longitudinal stiffener. It was concluded that if longitudinal stiffeners are located in both the tension and compression regions, the reduction in D/t_w is not required. This work was continued in Mariani et al. (1973) in which the optimum transverse stiffener rigidity is determined analytically. During the same time, Abdel-Sayed (1973) studied the prebuckling and the elastic buckling behavior of curved web panels and proposed equations for estimating the critical load under pure normal loading (stress), pure shear, and combined normal and shear loading. It was demonstrated that the critical load was higher for curved panels over the comparable flat panel and increases with an increase in curvature.

Daniels et al. (1979a,b, 1980; Daniels and Batcheler, 1979; Daniels and Herbein, 1980; Zettlemoyer et al., 1980) summarized a Lehigh University five-year experimental research program on the fatigue behavior of horizontally curved bridges and concluded that the slenderness limits suggested by Culver are too severe. Equations for load factor design and allowable stress design were developed and implemented in early versions of the Guide Specifications.

Numerous analytical and experimental works on the subject have been published by Japanese researchers since the end of the CURT project. Mikami et al. presented work in Japanese journals (Mikami et al., 1980; Mikami and Furunishi, 1981) and later in the *ASCE Journal of Engineering Mechanics* (Mikami and Furunishi, 1984) on the nonlinear behavior of cylindrical web panels under bending and combined bending and shear. Significant conclusions included: (1) The compressive membrane stress in the circumferential direction decreases with increase in curvature; (2) a panel under combined bending and shear exhibits a lower level of the circumferential membrane stress when compared with a panel under pure bending and, as a result, the bending moment carried by the web panel is reduced; and (3) the plate bending stress under combined bending and shear is larger than that under pure bending. No formulations or recommendations for design use were made. Kuranishi and Hiwatashi (1981, 1983; Hiwatashi and Kuranishi, 1984) used the finite element method to demonstrate the elastic finite-displacement behavior of curved I-girder webs under bending using models with and without flange rigidities. The reduction in bending moment resistance was demonstrated; however, for slenderness in the design range, only a small reduction was noted. No formulations or recommendations for direct design use were made. Fujii and Ohmura (1985) presented research on the nonlinear behavior of curved webs using the finite element method. It was emphasized that the web panel model with no flange rigidity is inadequate in estimating the behavior of the curved panel under significant loading. No specific recommendations or formulations regarding the design of curved I-girder webs were made. Suetake et al. (1986) examined the influence of flanges on the strength of curved I-girders under bending using the mixed finite element approach. Conclusions indicated that the aspect ratio of the panel is of minor importance and the influence of the flange rigidity cannot be ignored. Also, observations were made on the torsional buckling behavior of the flanges. No quantitative formulations for design use were recommended.

Nakai et al. (1986) conducted analytical research on the elastic large-displacement behavior of curved web plates subjected to bending using the finite element method. The web plate panels were modeled with and without flange rigidity. Flanges were modeled with fixed and simple supports. The boundary conditions at the panel ends (transverse stiffener locations) were modeled as simple support. One and two levels of longitudinal stiffeners were also modeled. It was determined that including the flange stiffness is essential to extract reliable results for the behavior of the curved web panels and therefore all parametric results used flange rigidity models. It was further shown that increasing curvature has little effect on the resisting moment (less than 10% within the range of actual bridge parameters). This was attributed to the fact that the web contributes only a small portion to the flexural resistance compared to the flanges. It was also demonstrated that the maximum web deflection occurs in the vicinity of $0.25h$ from the compression flange, but that this transverse deflection is effectively eliminated when one or two longitudinal stiffeners are present. Web slenderness requirements based on the effects of curvature on displacement and stress were formulated by Nakai and proposed for adoption by the Hanshin (1988) *Guidelines for the Design of Horizontally Curved Girder Bridges* (Kitada et al., 1986). It was suggested that limiting values should be established so that the curved web plate transverse deflection and plate bending stress be limited to the maximum transverse deflection and bending stress that would occur in the straight girder with the same dimensions but, instead of curvature, with a maximum initial deflection of $D/250$, which was the maximum allowable initial deflection stipulated in the Japanese design code (Japan Road Association, 1990). Nakai et al. conducted other research pertaining to the behavior of curved I-girder webs, including a series of experimental studies on the behavior of curved I-girder webs under bending, shear, and combined bending and shear (Nakai et al., 1983, 1984a,b,c, 1985a,b,c).

Davidson et al. (1999a,b, 2000a,b) investigated the buckling, finite-displacement, and ultimate-strength behavior of curved I-girder web panels under pure shear, pure bending, and combined bending and shear using the finite element method. The finite element models confirmed that the elastic buckling stress for the curved panel is higher than that of the comparable flat panel under pure shear. Under pure bending, Davidson et al. (1999a,b) showed that the nonlinear transverse displacement effectively reduces the moment-carrying capacity of the curved section over that of a similar straight section. Based on a "lateral pressure analogy," curvature reduction equations on the design slenderness were developed. This work was extended to study the effects of longitudinal stiffeners on the strength and stability of curved web panels (Davidson et al., 2000b).

As a result of NCHRP Project 12-38, the web slenderness requirements of the 2003 guide specifications were adjusted to be more in line with the straight girder specifications. The curvature-dependent web slenderness equations resulting from the work of Culver and Daniels used in editions prior to the 2003 Guide Specifications was eliminated. The slenderness of unstiffened webs increases linearly from a D/t_w of 100 at 700 ft radius to 150 for radii of 2000 ft and greater. For transversely stiffened webs, D/t_w cannot exceed 150 and the spacing between stiffeners increases linearly from a spacing of D at 700 ft radius to $3D$ for radii of 2000 ft and beyond. For transversely and longitudinally stiffened webs, slenderness cannot exceed 300. The critical bending stresses for all cases are not dependent upon the curvature of the girder.

The bending tests conducted during the CSBRP include three different web slendernesses and one specimen that was not transversely stiffened. It was concluded from these tests that, although the most slender web section ($2D_c/t_w = 189$, singly symmetric) exhibited significant buckling, web slenderness does not affect bending capacity at first yield or ultimate moment. Therefore, the AASHTO (2004) LRFD Specifications do not include requirements reflecting curvature effects on web behavior in bending-dominant regions of the girder.

Pure Shear The web of a plate girder that is stiffened by flanges and transverse stiffeners has considerable postbuckling strength, and allowable stress design (ASD), load factor design (LFD), and LRFD codes have taken advantage of this strength for straight girders. According to Basler (1961), the ability of a plate girder to behave in a manner similar to a truss was recognized as early as 1898. The work of Basler led to a theory that agreed with tests and provides criteria to ensure that truss action can be developed within the web. By considering truss action, the shear strength is raised from that based on buckling to a condition corresponding to shear yield in classical beam theory.

Early analytical work on the elastic stability of stiffened cylindrical shells subjected to pure shear was conducted by Batdorf (1947) and Batdorf et al. (1947) and then by Stein and Fralich (1949) and Stein and Yeager (1949). Mariani et al. (1973) later extended the work of Stein and Yeager to include the case of the curved plate with multiple stiffeners under pure shear and developed an optimal stiffener spacing criterion to establish stiffener requirements for curved girder webs. Experimental research on the ultimate and postbuckling reserve strength of curved girders has been conducted by Ilyasevitch and Klujev (1971) and by Mozer et al. (1971) as part of the CURT project. It was observed that there is a decrease in the postbuckling strength with an increase in curvature, although the measured shear strengths are within 10% of the ultimate shear strength from straight girder theory, which could be considered to be within the range of acceptable experimental error. Also, the experimental investigation by Mozer et al. indicated that, in areas of negative bimoment (tending to bend the compression flange inward), the web behaves more like that of a straight girder and can carry the ultimate shear strength predicted for a straight girder with similar proportions.

The Japanese researchers conducted a series of experimental tests on the ultimate strength of curved web panels under pure shear, pure bending, and combined bending and shear (Nakai et al., 1984a,b,c, 1985a,b,c). Their results also indicate that curvature has little effect on the elastic critical shear load but that there is some decrease in ultimate strength. Like the Guide Specifications, the Japanese design specifications do not recognize postbuckling reserve strength for curved plate girders due to lack of research in this area (Hanshin, 1988; Japan Road Association, 1990).

Lee and Yoo presented experimental and analytical investigations on the behavior of flat thin-web plate girders subjected to shear (Lee and Yoo, 1998, 1999a,b, Lee et al., 2002, 2003). Yoo and Lee (2006) asserted that, although the currently used approaches for defining shear strength results in acceptable accuracy for design applications, underlying assumptions are flawed. They demonstrated that a self-equilibrating force system is developed in the web panel that does not depend on the flanges and stiffeners. Lee and Yoo (1998) studied the bifurcation buckling and ultimate-strength analysis of curved web panels subjected to pure shear. They used the finite element method with combined geometric and material nonlinear solution sequences to analyze typical plate-girder web panels of various curvatures. The aspect ratios of the panels were varied to compare the effects of transverse stiffener spacing for curved panels to that of straight. The results from the buckling analyses agreed with previous research that showed the elastic critical load of the curved panel is greater than that of the comparable flat panel. The analysis revealed that curved web panels are capable of developing considerable postbuckling strength after the bifurcation point and that most research findings on straight plate girders can be applied to curved I-girders. Results also suggest that the straight girder equations developed by Lee et al. (1994; Lee and Yoo, 1999b) can be used to predict the shear strength of curved web panels subjected to pure shear. Because transverse stiffeners improve the handling and shipping of long slender curved girders, it was recommended that the same provisions for straight girders be applied to curved girders.

Zureick et al. (2002) reported the results of four full-scale curved steel I-girder component tests conducted to examine their shear behavior and to determine their ultimate strengths. The girders were made of AASHTO M270 Grade 345 steel and had a nominal web depth and web thickness of 1219 mm (48 in.) and 8 mm ($\frac{5}{16}$ in.), respectively. The resulting nominal web slenderness ratio D/t_w is 154. Two of the girders had a nominal radius $R = 63{,}630$ mm (208.75 ft) and a transverse stiffener spacing such that the ratio d_o/D is 3 and 1.5 (producing d_o/R values of 0.0575 and 0.0287). The other two components were identical except that their radii were 36,580 mm (120 ft), resulting in $d_o/R = 0.10$, 0.050. All of the girders werc braced against radial deflections at intervals of 3658 mm (12 ft) along the girder arc. Therefore, the ratio L_b/R is equal to 0.0575 for two specimens and 0.10 for the other two, where L_b is the distance between the bracing systems along the girder arc. Jung and White (2006) reported results of finite element studies of the four curved steel I-girder shear components tested as well as parametric extensions of these tests. The models incorporated the measured material stress–strain relationships and section dimensions from the physical tests, detailed arrangments of the test boundary conditions, residual stresses due to flame cutting and welding, and initial geometric imperfections in the form of buckling mode shapes. The load transfer mechanisms of the test girders were investigated via elastic buckling and full nonlinear analyses. Parametric studies demonstrated the effects of residual stresses, geometric imperfections, and the influence of reduced flange size on the peak shear capacity and moment–shear interaction. These studies concluded that transversely stiffened curved I-girders can be designed for maximum-strength loading conditions based on the AASHTO (2004) LRFD I-girder shear capacity equations, including the consideration of postbuckling strength, at least up to the following limits of $d_o/D \leq 3$, $D/t_w \leq 160$, and $d_o/R \leq 0.10$. The results also supported the conclusion that moment–shear interaction need not be considered in straight and/or curved I-girders designed using the AASHTO (2004) LRFD specifications.

Combined Bending and Shear Although most regions of the girder will be subjected to both transverse shear and vertical bending moment, it is generally accepted for straight girders that the strength in bending is not influenced by shear, and vice versa. In slender webs where "bend buckling" may occur, the bending stress is redistributed so that the flanges carry an increased share. The ultimate shear strength, however, is not reduced as a result of bend buckling because most of the shear resistance results from tension-field action. In stockier webs bend buckling does not occur, but high web shear in combination with bending may cause yielding of the web adjacent to the flange.

Japanese researchers Mikami and Furunishi (1981) presented work including shear along with bending in Japanese journals and later in the *ASCE Journal of Engineering Mechanics* (1984). From these investigations it was concluded that the presence of shear along with bending adversely affects the moment-carrying capacity of the beam but no formulations for design use were presented. Abdel-Sayed (1973) studied the prebuckling and elastic buckling behavior of curved web panels under pure bending, pure shear, and combined bending and shear and showed that in all cases the elastic critical load of the curved panel was greater than that of the comparable flat panel. Nakai and co-researchers (Nakai et al., 1984a,b, 1985a,c) conducted experimental studies on the buckling and ultimate-strength behavior of the curved I-girder web panels under combined bending and shear. An interaction curve was fitted to the buckling values from the tests. Interaction curves were also derived for the ultimate strength of the curved girders involving the theoretical nominal strengths for pure shear and pure bending. Davidson et al. (2000a) used finite element simulations to demonstrate that the elastic buckling load under any combination of shear with vertical bending stresses results in higher critical loads for the curved panel over that of the straight. It was concluded that the use of design equations presented for pure bending will result in conservative designs up to $V/V_n = 0.6$, where V is the calculated shear force over the web and V_n is the nominal shear resistance defined for pure shear.

Lee et al. (2003a) conducted research on the ultimate shear strength of transversely stiffened curved web panels using nonlinear finite element analysis. They showed that the presence of high-intensity shear reduces the ultimate bending strength significantly in thick web panels when the shear strength is governed by the shear yield strength and the bending strength is governed by the yield moment. In the case of thin curved web panels, however, in which elastic shear buckling and bend buckling will occur prior to yielding, the presence of shear does not have any appreciable impact on bending if the bending moment is limited to the critical bending moment. They also found that the aspect ratio of the curved girder web panel can be extended to 3.0 or greater, whereas the 2003 guide specifications limit this ratio to 1.0.

Transverse Stiffener Rigidity Requirements Culver et al. (1973) and Mariani et al. (1973) studied curved web panels under pure shear using the Donnell shell equation and the Galerkin method. They concluded that the required stiffener rigidity for a curved web is less than that for a straight web if the panel aspect ratio a/d is less than 0.78, where a is the distance between transverse stiffeners and d is the depth of the web. For $0.78 \leq a/d \leq 1$, the required stiffener rigidity increases with curvature by the amount $1 + (a/d - 0.78)Z^4/1775$, where Z is a curvature parameter defined as $Z = (a^2/Rt)\sqrt{1 - \upsilon^2}$, with $a/d \leq 1$. In these equations, R is the radius to the centerline of the web, t the thickness of the web, and υ Poisson's ratio. The study was limited to curved girders in which $0 \leq Z \leq 10$.

Nakai et al. (1984b, 1985b) presented a beam-column model to estimate the strength of transverse stiffeners in curved girders. The results were compared with experiments conducted by Nakai et al. (1984a), which led to a recommendation that the relative rigidity parameter β, defined as the ratio between required rigidity of a transverse stiffener in horizontally curved girders to that in straight girders, must be:

For stiffeners attached to one side of the web plate

$$\beta = \begin{cases} 1.0 + (\alpha - 0.69)Z[9.38\alpha - 7.67 \\ \quad -(1.49\alpha - 1.78)Z] & \text{for } 0.69 \leq \alpha \leq 1.0 \\ 1.0 & \text{for } < 0.69 \end{cases} \tag{9.1}$$

and for stiffeners attached to both sides of the web plate

$$\beta = \begin{cases} 1.0 + (\alpha - 0.65)Z[12.67\alpha - 10.42 \\ \quad -(1.99\alpha - 2.49)Z] & \text{for } 0.65 \leq \alpha \leq 1.0 \\ 1.0 & \text{for } < 0.65 \end{cases} \tag{9.2}$$

where $\alpha = a/d$.

Using nonlinear finite element analyses, Lee et al. (2003a,b) and Yoo and Lee (2006) showed that the rigidity of the transverse stiffener should be increased six times the value obtained from AASHTO (2003) to develop the maximum potential postbuckling strength. In the case of thick web panels, however, when the shear design is governed by shear yielding, the stiffener rigidity does not have to be increased.

The behavior and design of one- and two-sided transverse stiffeners in straight and horizontally curved steel I-girders was also investigated by Kim et al. (2007) using nonlinear finite element analysis. Unified recommendations for design of transverse stiffeners were developed by combining the results from FEA studies with the results from a number of prior research studies. Furthermore, it was demonstrated that stiffener design based on providing adequate bending transverse stiffener stiffness and strength is a more important consideration in developing shear postbuckling resistance than the satisfaction of an area or axial force requirement.

9.5.3 Global Buckling and Lateral Bracing Requirements

Lateral–Torsional Buckling There have been many theoretical developments pertaining to the elastic lateral–torsional buckling behavior of curved members. A comprehensive review and comparison of these theories was presented by Kang (1992) and Kang and Yoo (1994a,b). Yoo (1982) and Rajasekaran and Ramm (1984) used the minimum potential energy principle to obtain solutions for the elastic flexural torsional buckling loads for in-plane and out-of-plane buckling modes of thin-walled curved beams that do not undergo local buckling. Yoo and Pfeiffer (1983; Pfeiffer, 1981) investigated the elastic buckling behavior of a thin-walled curved member through a variational based finite element formulation. Solutions to different cases pertaining to the stability of curved beams were obtained and compared to solutions presented by Timoshenko and Gere (1961), Vlasov (1961), and Culver and McManus (1971) and were shown to be significantly different. The discrepancies were attributed to incorrect formulations in the Timoshenko and Vlasov cases and to the fact that the governing differential equation was viewed as a deflection amplification problem rather than a classical eigenvalue problem. Later, Yoo and Pfeiffer (1984) presented a solution to the stability of curved beams with in-plane deformation and continued to assert their earlier conclusion related to the discrepancies with existing solutions, including the work of Vacharajittiphan and Trahair (1975), which is based on Vlasov's formulation. In 1985, Yoo and Carbine conducted a series of laboratory tests on 12 simply supported curved beam specimens subjected to concentrated loads. Papangelis and Trahair (1986, 1987) examined the work of Yoo (1982) and Yoo and Pfeiffer (1983, 1984) by conducting tests on circular aluminum arches. They concluded that the theoretical loads obtained from the work of Yoo differed substantially from analytical and experimental results of various researchers.

The conflict among curved beam theories was also discussed in a series of publications by Yang et al. (1991; Yang and Kuo, 1986, 1987; Rajasekaran et al., 1988), who derive the nonlinear differential equation of equilibrium for horizontally curved I-beams by making use of the principle of virtual displacements to establish the equilibrium of a bar in its buckled configuration. Numerical results were obtained and compared with those resulting from Yoo's (Yoo, 1982 and Yoo and Pfeiffer 1983) as well as Vlasov's (1961) theories. The discrepancy between Yang's results and Yoo's was attributed to the fact that Yoo not only neglected both the radial stress effect and the contribution of shear stresses to the potential energy but also substituted the curvature terms of the curved beam in the potential energy equation of a straight beam. Kuo and Yang (1991) further criticized the work of Vlasov and Yoo by solving numerically the buckling problem of a curved beam with a solid cross section under uniform bending and uniform compression.

Based on an elastic buckling finite element analysis study, Kang and Yoo (1990) showed that initial curvature and warping do not significantly affect the lateral–torsional buckling strength of curved girders with a subtended angle between two adjacent cross frames up to 0.1 radian, the maximum value allowed in the 1993 Guide Specifications. In 1994, Kang and Yoo (1994a,b) presented companion papers on the buckling and large-displacement behavior of thin-walled circular beams based on theory derived using the principle of minimum potential energy. Closed-form solutions were obtained for limited loading and boundary conditions and used for a comparison to the theory of other researchers.

As part of the CURT project, Culver and McManus (1971; McManus, 1971) presented a second-order analysis in which the equilibrium equations are formulated on the deformed structure. Results were compared to those of lateral buckling tests conducted by Mozer et al. (1971, 1975a,b; Mozer and Culver, 1975). The study recommended a set of formulas that were later adopted into the Guide Specifications.

Nishida et al. (1978) presented work that used the large-deflection theory of curved members to derive an expression for predicting the critical elastic moment for a horizontally curved beam subjected to equal end moments (Eq. 9.3). They showed that the critical moment approaches that of the straight girder as the radius of curvature approaches infinity. Their development, however, was derived for simple support end conditions with equal end moments and no intermediate lateral bracing, which limits its applicability. The equation appears as

$$M_{cr} = \sqrt{\left(1 - \frac{L^2}{\pi^2 R^2}\right)\left(\frac{\pi^2 E I_y}{L^2}\right)\left(GJ + \frac{\pi^2 E C_w}{L^2}\right)} \tag{9.3}$$

where L = unbraced length, R = radius of curvature, I_y = minor axis moment of inertia, E = Young's modulus, G = shear modulus, J = torsional rigidity of the section, C_w = warping constant, b = flange width, t_w = web thickness, and h = depth between flange centroids.

Kang (1992) demonstrated that a large variation in torsional rigidity ratio has little effect on the critical load ratio of the lateral buckling of horizontally curved girders loaded normal to the plane of curvature and that the subtended angle is the dominating parameter. The results from finite element analyses were used to form the following regression equation for the reduction in lateral–torsional buckling capacity of the curved girder over that of the straight girder (Yoo et al., 1996):

$$y = \left(1 - \gamma x^{\beta}\right)^{\alpha} \tag{9.4}$$

where y is the critical moment ratio (curved/straight), x the subtended angle in radians, $\alpha = 2.152$, $\beta = 2.129$, and $\gamma = 0.1058$.

For horizontally curved beams loaded normal to the plane of curvature, a large-displacement analysis gives more meaningful design information than the bifurcation buckling analysis. Significant deflections occur in such structures prior to reaching the predicted bifurcation load and, hence, the bifurcation buckling load represents an upper bound of elastic instability. Developments on the nonlinear theory of curved beams have been made by many of the researchers discussed above, with a comprehensive review and comparison of these theories presented by Kang (1992, 1994a). Subsequent to the curved beam element developments by Yoo and co-researchers, Pi and Trahair (1997) introduced a curved beam element formulation and used it to study the nonlinear behavior of elastic I-beams curved in a plan subjected to equal end moments.

Culver and McManus (1971) used second-order analyses and the finite difference technique to solve bifurcation buckling and large-deflection problems of curved beams. Their research became the basis of the ASD strength equations in the Guide Specifications (AASHTO, 1977a,b). As pointed out by Yoo and Pfeiffer (1983), the solution, however, cannot be accurate because the critical moment ratio (the ratio of the critical moment of the curved beam to that of the comparable straight beam) derived from their results approaches unity as the subtended angle approaches 90°.

Japanese researchers have also conducted analytical work on the large-deflection behavior. Nishida et al. (1978; Fukumoto et al., 1980; Fukumoto and Nishida, 1981) provided analytical results on the elastic lateral instability of horizontally curved beams based on large-deflection theory, and their work resulted in an equation that predicts critical moment. All of the lateral instability investigations involved the isolated curved member with simplified boundary conditions.

Brace Spacing In straight girder bridges, the primary function of the cross frames and diaphragms is to prevent premature lateral buckling of the girder; therefore, the cross-frame members are designed as secondary members. In curved bridge systems, however, the cross frames and diaphragms have the added responsibility of restraining the rotation of the girder, thereby reducing the warping stresses in the flanges and the vertical deflection of the system. Hence, the spacing interval between cross frames becomes a critical design parameter. Furthermore, there is growing sentiment in the bridge engineering community to minimize the number of cross frames due to increased fabrication and construction costs and fatigue concerns.

In 1986, Yoo and Littrell used a computer program to study the effects of cross-bracing in curved bridges. Bridge models were analyzed under dead and live loads for different curvatures. In 1989, Schelling et al. presented a study concerning the construction effects of bracing on curved I-girders. The investigation produced a set of equations that define the dead-load distributions throughout the superstructure system, which was analyzed by the two-dimensional grid method. Davidson et al. (1996a) investigated the cross-frame spacing requirements of horizontally curved I-girder bridges. The finite element method was used to determine dominant parameters and to develop an equation for the preliminary design of the cross-frame spacing needed to achieve a required warping-to-bending stress ratio (f_w/f_b). With a preliminary design target $f_w/f_b = 0.25$, the resulting equation is

$$S_{\max} = L\left[-\ln\left(\frac{Rb_f}{\alpha L^2}\right)\right]^{-1.52} \tag{9.5}$$

where S_{max} (m or ft) is the design spacing between cross frames, L (m or ft) the span length of the exterior girder, R (m or ft) the radius of curvature of the exterior girder, and b_f (mm or in.) the compression-flange width and $\alpha = 2000$ when metric units are used and $\alpha = 24$ with imperial units. Using existing bridge designs, results obtained from Eq. 9.5 were compared to that of a similar equation developed by Yoo and Littrell (1986) and to that obtained from using a simple distributed lateral load approximation. The comparisons revealed a good correlation is obtained between Eq. 9. 5 and the lateral load approximation but indicated that the Yoo–Littrell equation produces less accurate results.

9.5.4 Ultimate Bending Resistance and Strength Design

Analytical and experimental ultimate-strength investigations have been made by several researchers. Culver and McManus (1971) studied the inelastic behavior of horizontally curved girders and made design recommendations that were adopted into the LFD portion of the AASHTO guide specifications. Yoo and Heins (1972) studied the plastic collapse of horizontally curved girders and presented a yield criterion and design charts along with equations for practical applications. Yang et al. (1989; Yang and Fan, 1988) presented yield surface formulations for I-sections with nonuniform torsion and bimoments; similar work was presented by Imai and Ohto (1987) in Japan. A series of extensive analytical studies of horizontally curved beams that included the effects of large displacement and material nonlinear behavior was performed by Japanese investigators using the transfer matrix method and assuming ideal elastic–perfectly plastic material behavior (Yoshida and Imoto, 1973; Fukumoto and Nishida, 1981; Maegawa and Yoshida, 1981; Yoshida and Maegawa, 1983, 1984). Later, similar results were obtained using a flat six-degree-of-freedom triangular plate shell finite element developed by (Lee 1987; Lee and Yoo, 1988a).

The LFD portion of the Guide Specifications is based on research by Galambos (1978) as an extension of work performed in the CURT project at Carnegie Mellon University. The parts of the design equations that represent the reduction in strength due to curvature are complex and cumbersome and have been shown to be quite conservative. Hall et al. (1999) demonstrated that the lateral flange bending stress (f_w) due to curvature in the McManus–Culver predictor equation in the Guide Specifications is double counted. Therefore, the lateral flange bending stress due to curvature at the critical cross-frame location must be set equal to zero.

The Hanshin Expressway Public Corporation proposed an interaction formula for limiting the stresses in horizontally curved I-girders for adoption into its *Guidelines for the Design of Horizontally Curved Girder Bridges* (Hanshin, 1988; Nakai and Yoo, 1988). The equations represent an interaction for allowable stress in the compression flange and include the presence of warping and the reduced lateral buckling strength of the girder due to curvature. They are based on theoretical and experimental research in the elastic range by Nakai and co-researchers (Nakai and Kotoguchi, 1983; Nakai et al., 1983). In research by Fukomoto and Nishida

(1981) using the transfer matrix method for both the elastic and inelastic ranges, an approximate ultimate-strength formula was presented involving the plastic moment capacity of the section, the elastic buckling moment of the straight beam with the same length and cross-section dimensions, and the elastic buckling load of the entire section about the minor axis. Nakai et al. (1985a) presented an empirical equation for the ultimate moment based upon 19 tests in which the elements comprising the cross sections are classified as compact and the a/d ratio is less than 1 (where a is the distance between transverse stiffeners and d is girder depth).

In several investigations (Kang, 1992; Yoo and Pfeiffer, 1983; Yoo et al., 1996), it was demonstrated that (1) a large variation in the torsional rigidity ratio has little effect on the critical load ratio (curved/straight) for the lateral buckling of horizontally curved girders loaded normal to the plane of curvature and (2) the subtended angle is the dominating parameter. A curvature reduction formula was derived from regression of data resulting from an elastic finite element investigation using curved beam elements that include warping (Kang, 1992). Although this strength reduction equation was developed based upon an elastic analysis, it was proposed that the reduction in the critical moment of curved girders results from the presence of the rotational component of the girder behavior, and likewise there would be a similar reduction in ultimate moment capacity (Yoo et al., 1996). Ultimate strength tests by others on curved I-girders appear to verify this hypothesis (Yadlosky, 1993).

Yoo and Davidson (1997) presented yield interaction equations that are based on the static equilibrium of the I-shape girder under vertical moment and lateral flange moments. Seventeen interaction equations were developed encompassing noncomposite and composite sections that are doubly symmetric, singly symmetric, I-shaped compact, flange-compact, noncompact, cracked, and uncracked. It was assumed that only singly symmetric compact sections can be made hybrid. For compact-flange and noncompact sections, however, a homogeneous section was assumed because hybrid construction will not yield significantly higher moment capacities. The radius of curvature, cross-frame spacing, material properties, and cross-section geometry were included as variables of the interaction equations. A computer program was developed and the reduction due to curvature was demonstrated for a number of hypothetical cases.

Davidson and Yoo (2000) and Yoo and Davidson (2002) presented the results of finite element models of the curved three-girder FHWA-CSBRP test frame. Linear elastic static, buckling, and combined material and geometric nonlinear analyses were conducted using models that represent the test frame and component test specimens. The results were compared to various predictor equations developed from analytical work by other researchers, including Japanese research. Predictor equations were also compared for such parameters as warping stress, elastic flange buckling curvature reduction, and curved web maximum displacement and maximum stress.

Singapore researchers have conducted experimental and analytical investigations on the ultimate strength of I-beams curved in plan. Liew et al. (1995) used finite element models that were verified using available experimental results to study the ultimate strength of horizontally curved I-beams. Residual stress effects and intermediate bracing effects were considered, and an equation was derived for the ultimate resistance. Shanmugam et al. (1995) reported the results of experiments involving both rolled curved beams and built-up sections with varying curvatures. Observations and comparisons were presented regarding the impact of residual effects on the ultimate-strength behavior of rolled and built-up sections. Thevendran et al. (1999, 2000) reported experimental and analytical studies on the ultimate-load capacity of steel–concrete composite beams curved in plan. Shanmugam et al. (2003) conducted experimental and finite element studies on the ultimate strength of curved plate girders with varying curvatures and plate slenderness, with particular emphasis on the ability of the girders to develop tension-field action. Lian and Shanmugam (2003, 2004) conducted experimental and finite element studies on the ultimate strength of curved plate girders with varying-size circular openings.

Australian researchers have also conducted extensive experimental and analytical research on the ultimate strength of beams curved in plan. Pi and Trahair (1994) presented findings on the inelastic behavior of I-beams under combined bending and torsion based on the finite element method. They showed that the circular interaction equation provides a lower bound estimate of the ultimate strength of a straight I-beam under combined bending and nonuniform torsion only when lateral displacements of the beam are continuously prevented. When the lateral displacements of the beam are not fully prevented, the interaction equation may lead to an overestimate of the ultimate strength. Pi et al. (1999, 2000) presented a curved beam finite element model for the geometric and material nonlinear analysis of I-beams curved in plan. The von Mises yield criterion, the associated flow rule, and the isotropic strain-hardening rule were used in the development of an elastic–plastic incremental model. The model was validated using prior experimental results. It was demonstrated that, when the initial curvature of a curved beam is small, bending is the major action and the nonlinear inelastic behavior is similar to the inelastic flexural–torsional buckling of a straight beam. They also showed that when the curvature is more significant, both nonuniform torsion and bending are dominant with nonlinear inelastic behavior developing very early in the loading process. Pi and Bradford (2001) investigated the nonlinear elastic–plastic behavior of steel I-section beams curved in plan. The strength interactions between bending and torsion of continuously braced, centrally braced, and unbraced curved beams were studied. It was demonstrated that the conventional circular interaction equation overestimates the bending/torsion interaction strength of steel I-section curved beams because it does not consider the effects of secondary torsion and minor axis bending actions and the included angle of the curved beam. Formulas were proposed and validated against rational finite element results and existing experimental results.

As part of NCHRP Project 12-38, Hall et al. (1999) extended the interaction strength definitions presented by Nakai and Yoo (1988) and Schilling (1996). It was proposed that the allowable flange stress based on first-order elastic calculations be limited to the yield stress minus one-third of the lateral bending stress, which became known as the "one-third rule." This approach was adopted into the 2003 AASHTO Guide Specifications (AASHTO, 2003). White et al. (2001) extended the one-third rule concept and rigorously demonstrated its accuracy and applicability using finite element results from the CSBRP. Two limit-state definitions were proposed and subsequently adopted into the 2004 AASHTO Standard Specifications (AASHTO, 2004) as a "unified" approach for considering lateral flange bending:

$$f_{bu} + \tfrac{1}{3} f_\ell \leq \phi_f F_n \tag{9.6}$$

$$M_u + \tfrac{1}{3} f_\ell S_x \leq \phi_f M_n \tag{9.7}$$

where f_{bu} = flange major axis bending stress
f_l = flange lateral bending stress
$\phi_f F_n$ = factored flexural resistance in terms of the flange major axis bending stress
M_u = member major axis bending moment
S_x = elastic section modulus about the major axis of the section to the flange under consideration, taken generally as M_{yf}/F_{yf}
$\phi_f M_n$ = factored flexural resistance in terms of the member major axis bending moment
M_{yf} = yield moment corresponding to the flange under consideration, calculated as defined in Appendix D of AASHTO (2004) accounting for the influence of noncomposite, long-term composite, and short-term composite loadings
F_{yf} = specified minimum yield strength of the flange under consideration

Equation 9.6 is used to check slender-web noncomposite members, slender-web composite members in negative bending, and noncompact composite members in positive bending. In the limit that the flange lateral bending stress f_l is equal to zero, this equation reduces to the conventional format for checking the above types of members subjected to major axis bending only. The maximum potential value of $\phi_f F_n$ is $\phi_f F_{yf}$, but $\phi_f F_n$ can be less than $\phi_f F_{yf}$ due to slender-web bend buckling and/or hybrid-web yielding effects or due to compression-flange lateral–torsional or local buckling limit states.

Equation 9.7 is used to check the strength limit states of noncomposite members or composite members in negative bending that have compact or noncompact webs and for checking of compact composite members in positive bending. For these member types, $\phi_f M_n$ can be as large as $\phi_f M_p$, where M_p is the section plastic moment resistance.

9.6 STABILITY OF CURVED BOX GIRDERS

In recent years, many states in the United States have favored the use of curved box-girder bridges over open I-girders simply because of their superior torsional stiffness.[1] As a part of the CURT project, Culver and Mozer (1975a,b) and Mozer and Culver (1974) conducted a series of static tests that included stiffened and unstiffened compression flanges and trapezoidal and rectangular boxes. Most of the test results were reflected in the original guide specifications, with the 2003 edition retaining many of these test results, particularly those on local buckling.

9.6.1 Curvature Effects on Box-Girder Compression Flanges

The compression flange of a box girder in the high negative vertical bending moment zone is typically stiffened by longitudinal stiffeners. The gradient of the compressive normal stress (warping stress) in a horizontally curved box flange occurs due to warping caused by curvature and distortional stress. Because the longitudinal warping stress must not exceed 10% of the longitudinal stress due to vertical bending at the strength limit state, internal cross frames and intermediate bracing are necessary. Under this condition, Choi and Yoo (2005) found that the minimum stiffness requirements developed by Yoo et al. (2001) for straight box girders can also be used for curved box girders. Based on the AASHTO definition for the minimum moment of inertia of the longitudinal stiffeners, structural tees are the most efficient shape. The minimum moment of inertia of the longitudinal stiffener required is

$$I_s = 0.3\alpha^2\sqrt{n}wt_f^3 \tag{9.8}$$

where α = the aspect ratio of a subpanel bounded between adjacent longitudinal stiffeners or box web and adjacent transverse stiffeners

n = the number of longitudinal stiffeners

w = the width of the subpanel

t_f = the thickness of the compression flange

Choi et al. (2007) developed the optimum stiffness and spacing required for transverse stiffeners to be placed along with the longitudinal stiffeners mentioned above. It was found that the critical compressive stress of the compression flange is more effectively determined by the dated CRC (1960) column curve (inverse parabola), because the initial imperfection allowed by the American Welding Society (AWS, 2002) significantly reduces the critical compressive stress in the transition zone of the subpanel width-to-thickness ratio.

[1] On the other hand, this additional stiffness can present fit-up problems during erection, a situation which is not often encountered with the more flexible open I-girders. As such, any significant fabrication errors can only be corrected in the shop.

9.6.2 Effects of Curvature on Box-Girder Web Plate Strength and Stability

Pure Bending Material presented in Section 9.5.2 on pure bending is generally applicable to web plates of curved box girders. Longitudinal stiffeners must be continuous when they are placed on the curved box webs to prevent bend–buckling. Given that most curved box girders in the United States and Canada are trapezoidal, having a web inclination of 4 to 1, developing the proper longitudinal stiffener geometry often presents an additional fabrication challenge.

Pure Shear Again, material presented in Section 9.5.2 on pure shear is generally applicable. The St. Venant shear flow from torsion is additive to vertical bending shear in one of the two box-girder webs. Because the web in most box girders is inclined, the shear flow in the web resulting from vertical shear is increased slightly. For very thin web panels, it may be necessary to account for the true vertical bending shear flow characteristic of a closed box section.

9.6.3 Lateral Bracing and Intermediate Internal Bracing Requirements

Lateral Bracing Open-top tub girders have minimal torsional stiffness and strength. Field experience indicates that even in straight box girders significant torsional moments can develop due to eccentrically applied loads from the unhardened concrete deck and construction equipment. To provide adequate torsional rigidity during construction, the top opening must be laterally braced with double diagonals or Warren truss-type single diagonals, resulting in a pseudobox section (Fig. 7.29). Fan and Helwig (1999, 2002) presented one of the few comprehensive studies on the behavior of such a system and the forces that develop in the lateral bracing members. Kim and Yoo (2006b) extended the method adopted by Fan and Helwig (1999, 2002) to better predict the bracing forces in single diagonal-type lateral bracing members coupled with X-type internal cross frames.

Internal Cross Frames Kim and Yoo presented a complete interaction of top lateral and K-type internal bracing systems (2006d) and X-type internal bracing (2008). With the results of these two studies, it is now possible to design lateral and internal bracing systems in tub girders so that a complete interaction between these two systems is maintained. Contrary to prevailing ideas, there exists a strong interaction between these two systems, particularly in curved tub girders. The internal and lateral bracing systems of curved box girders cannot be designed separately without considering the complete interaction.

9.6.4 Ultimate Bending Strength

Nakai and Yoo (1988) summarized most of the early Japanese research on curved box girders. Nakai and Kitada (1992) and Nakai et al. (1990, 1992) continued their study on curved box girders including their experimental study on the ultimate strength of thin-walled box beams subjected to bending and torsion. Kim and Yoo (2006a, 2008) presented an improved ultimate-strength definition for box beams.

9.6.5 External Bracing between Box Girders

External bracing is provided between curved box girders to facilitate concrete deck placement by controlling the vertical deflection and rotation of the girder with respect to its longitudinal axis. External bracing, however, has shown to adversely affect the fatigue behavior of box girders. Therefore, many owners of curved bridges call for the removal of external bracing except those on abutments or piers, which can be costly. Kim and Yoo (2006c) demonstrated that only a single external bracing per span is effective in most horizontally curved box-girder bridges, thereby minimizing the need and cost of external bracing.

9.7 CONCLUDING REMARKS

Horizontally curved steel girders provide an efficient design approach for developing curved roadway alignments, and hence, the need for horizontally curved plate and box girders continues to grow as populations in major urban areas become denser. The simple addition of horizontal curvature, however, greatly affects the strength and stability of the system and therefore complicates behavior, design, and construction. This chapter outlines stability issues associated with curvature, provides a brief review of the most important literature, and describes the trends for improved design and construction of bridges using curved plate and box girders.

During the past 10 years, enormous progress has been achieved in defining the effect of curvature on the stability and resistance of curved girders, but much additional work remains. Future emphasis is expected to be on the use of hybrid girders for curved alignments, construction and erection issues, fatigue, system ultimate strength, and dynamic effects.

REFERENCES

AASHTO (1969), *Standard Specifications for Highway Bridges*, 10th ed., American Association of State Highway Officials, Washington, DC.

AASHTO (1980), *Guide Specifications for Horizontally Curved Highway Bridges*, American Association of State Highway Transportation Officials, Washington, DC.

AASHTO (1987), *Guide Specifications for Horizontally Curved Highway Bridges, 1980, as Revised by Interim Specifications for Bridges, 1981, 1982, 1984, 1985, 1986*, American Association of State Highway Transportation Officials, Washington, DC.

AASHTO (1993), *Guide Specifications for Horizontally Curved Highway Bridges*, as revised in 1981, 1982, 1984, 1985, 1986, 1990, and 1992. American Association of State Highway and Transportation Officials, Washington DC.

AASHTO (2003), *Guide Specifications for Horizontally Curved Highway Bridges*, American Association of State Highway and Transportation Officials, Washington DC.

AASHTO (2004), *LRFD Bridge Design Specifications, 3rd Ed. with 2005 Interim Provisions*, American Association of State Highway and Transportation Officials, Washington, DC.

AASHTO-ASCE Committee on Flexural Members (1973), "Survey of Curved Girder Bridges," *Civ. Eng.*, Vol. 43, No. 2, pp. 54–56.

ASCE-AASHTO Committee on Flexural Member of the Committee on Metals of Structural Division, Task Committee on Curved Girders (1977), "Curved I-Girder Bridge Design Recommendations," *Proc. ASCE*, Vol. 103, No. ST 5, pp. 1137–1167, May.

ASCE-AASHTO Committee on Flexural Member of the Committee on Metals of Structural Division, Task Committee on Curved Girders (1977), "Curved Steel Box Girder Bridges: A Survey," *Proc. ASCE*, Vol. 104, No. ST11, pp. 1697–1718.

AASHTO/NSBA Steel Bridge Collaboration (2007), "Steel Bridge Erection Guide Specification," AASHTO Document No: NSBASBEGS-1-OL, American Association of Highway and Transportation Officials, Washington, DC.

Abdel-Sayed, G. (1973), "Curved Webs under Combined Shear and Normal Stresses," *Journal of Structural Division*, ASCE, Vol. 99, No. ST3, March, pp. 511–525.

ASCE-AASHTO Task Committee on Curved Girders (1977), "Curved I-Girder Bridge Design Recommendations," *J. Struct. Div., ASCE*, Vol. 103, No. ST5, pp. 1137–1167.

ASCE-AASHTO Task Committee on Curved Girders (1978a), "Curved Steel Box-Girder Bridges: A Survey," *J. Struct. Div., ASCE*, Vol. 104, No. ST11, p. 1697 .

ASCE-AASHTO Task Committee on Curved Girders (1978b), "Curved Steel Box-Girder Bridges: State-of-the-Art," *J. Struct. Div.*, ASCE, Vol. 104, No. ST11, pp. 1719–1739.

AWS (2002) *Bridge Welding Code*, AASHTO/AWS D1.5M/D1.5, An American National Standard, American Association of Highway and Transportation Officials, Washington, DC.

Basler, K., (1961), "Strength of Plate Girders in Shear," *J. Struct. Div., ASCE*, Vol. 87, No. ST7, Proc. Paper 2967, Oct., pp. 151–180.

Basler, K., and Thurliman, B. (1961), "Strength of Plate Girders in Bending," *J. Struct. Div., ASCE*, Vol. 87, No. ST6, Proc. Paper 2913, August, pp. 153–181.

Batdorf, S. B. (1947a), "A Simplified Method of Elastic Stability Analysis for Thin Cylindrical Shells. II Modified Equilibrium Equation," NACA TN No. 1342, National Advisory Committee for Aeronautics, Washington, DC.

Batdorf, S. B., Stein, M., and Schildcrout, M. (1947b), "Critical Shear Stress of Curved Rectangular Plates," NACA TN No. 1342, National Advisory Committee for Aeronautics, Washington, DC.

Bell, B. J., and Linzell, D. G. (2007), "Erection Procedure Effects on Deformations and Stresses in a Large-Radius, Horizontally Curved, I-Girder Bridge," *J. Bridge Eng., ASCE*, Vol. 12, No. 4, July, pp. 467–476.

Bradford M.A., Uy B., and Pi Y.-L. (2001), "Behaviour of Unpropped Composite Girders Curved In-Plan under Construction Loading," *Eng. Struct.*, Vol. 23, No. 7, pp. 779–789.

Bradford M. A., and Pi, Y.-L. (2001), "Strength Design of Steel I-Section Beams Curved In-Plan," *J. Struct. Eng. ASCE*, Vol. 127, No. 6, pp. 639–646.

Brennan, P. J., and Mandel, J. A. (1973), "User's Manual: Program for Three-Dimensional Analysis of Horizontally Curved bridges," Rep. Res. Proj. HPR-2(111), Syracuse, University, Syracuse, NY. June.

Brockenbrough, R. L. (1970a), "Criteria for Heat Curving Steel Beams and Girders," *J. Struct. Div. ASCE*, Vol. 96, No. 10, pp. 2209–2226.

Brockenbrough, R. L. (1970b), "Theoretical Stresses and Strains from Heat-Curving," *J. Struct. Div. ASCE*, Vol. 96, No. 7, pp. 1421–1444.

Brockenbrough, R. L. (1970c), "Experimental Stresses and Strains from Heat-Curving," *J. Struct. Div. ASCE*, Vol. 96, No. 7, pp. 1305–1331.

Brockenbrough, R. L. (1986), "Distribution Factors for Curved I-Girder Bridges," *J. Struct. Eng. ASCE*, Vol. 112, No. 10, pp. 2200–2015.

Brogan, D. (1972), "Bending Behavior of Cylindrical Web Panels," Master's thesis, Carnegie Mellon University, Pittsburgh, PA.

Bridge Software Development International (BSDI) (2000), product scope, BSDI, Coopersburg, PA.

Chang, C.-J., and White D. W. (2008a), "Construction Simulation of Curved Steel I-Girder Bridge Systems," Final Report to Professional Services Industries, Inc., and Federal Highway Administration, George Institute of Technology, Atlanta, GA.

Chang, C.-J., and White D. W. (2008b), "An Assessment of Modeling Strategies for Composite Curved Steel I-Girder Bridges," *Eng. Struct.*, Vol. 30 Issue 11, pp. 2991–3002.

Chavel, B. W., and Earls, C. J. (2001), "Evaluation of Erection Procedures of the Curved Span of the Ford City Steel I-Girder Bridge," Rep. No. CE/ST 18, Dept. of Civil Engineering, Univ. of Pittsburgh, Pittsburgh, PA.

Chavel, B. W., and Earls, C. J. (2006a), "The Construction of a Horizontally Curved Steel I-Girder Bridge. Part I: Erection Sequence," *J. Bridge Eng.*, Vol. 11, No. 1, pp. 81–90.

Chavel, B. W., and Earls, C. J. (2006b), "The Construction of a Horizontally Curved Steel I-Girder Bridge. Part II: Inconsistent Detailing," *J. Bridge Eng.*, Vol. 11, No. 1, pp. 91–98.

Choi, B. H., Kang, Y. J., and Yoo, C. H. (2007), "Stiffness Requirements for Transverse Stiffeners of Compression Flanges," *Eng. Struct.*, Vol. 29, No. 9, pp. 2087–2096.

Choi, B. H., and Yoo, C. H. (2005), "Strength of Stiffened Flanges in Horizontally Curved Box Girders," *J. Eng. Mech. ASCE*, Vol. 131, No. 2, pp. 167–176.

Column Research Council (CRC) (1960), *Guide to Design Criteria for Metal Compression Members*, 2nd, (Ed. B. G. Johnston), Wiley, New York.

Culver, C. G. (1971), "Flange Proportions for Curved Plate Girders," *J. Highw. Res. Board*, Vol. 354, pp. 61–66.

Culver, C. G., Dym, C., and Brogan, D. (1972a), "Bending Behaviors of Cylindrical Web Panels," *J. Struct. Div. ASCE*, Vol. 98, No. ST10, pp. 2201–2308.

Culver, C. G., Dym, C., and Brogan, D. (1972b), "Instability of Horizontally Curved Members, Bending Behaviors of Cylindrical Web Panels," submitted to the Pennsylvania Department of Highways by the Department of Civil Engineering, Carnegie-Mellon University, Pittsburgh, PA.

Culver, C. G., Dym, C., and Brogan, D. (1972c), "Instability of Horizontally Curved Members, Shear Buckling of Cylindrical Web Panels," submitted to the Pennsylvania Department of Highways by the Department of Civil Engineering, Carnegie-Mellon University, Pittsburgh, PA.

Culver, C. G., Dym, C. L., and Uddin, T. (1973), "Web Slenderness Requirements for Curved Girders," *J. Struct. Div. ASCE*, Vol. 99, No. ST3, pp. 417–430.

Culver, C. G., and Frampton, R. E. (1970), "Local Instability of Horizontally Curved Members," *J. Struct. Div. ASCE*, Vol. 96, No. ST2, pp. 245–265.

Culver, C. G., and McManus, P. F. (1971), *Instability of Horizontally Curved Members: Lateral Buckling of Curved Plate Girders*, Department of Civil Engineering, Carnegie-Mellon University, Pittsburgh, PA.

Culver, C. G., and Mozer, J. D. (1975a), "Horizontally Curved Highway Bridges—Stability of Curved Box Girders—B1," FH-11-7389, Federal Highway Administration, Department of Transportation, Aug., Washington, DC.

Culver, C. G., and Mozer, J. D. (1975b), "Horizontally Curved Highway Bridges—Stability of Curved Box Girders—B2," FH-11-7389, Federal Highway Administration, Department of Transportation, Aug., Washington, DC.

Culver, C. G., and Nasir, G. (1969), "Instability of highway curved bridges: Flange buckling studies," Report 68-32, Carnegie-Mellon University, Pittsburgh, PA, and Pennsylvania Department of Transportation, Bureau of Public Roads, Harrisburg, PA.

Culver, C. G., and Nasir, G. (1971), "Inelastic Flange Buckling of Curved Plate Girders," *J. Struct. Div. ASCE*, Vol. 97, No. ST4, pp. 1239–1256.

Dabrowski, R. (1964), "The Analysis of Curved Thin Walled Girders of Open Sections," *Der Stahlbau*, Vol. 34, No. 12, pp. 364–372.

Dabrowski, R. (1965), "Warping Torsion of Curved Box Girders of Non-Deformable Cross Section," *Der Stahlbau*, Vol. 34, No. 5, pp. 135–141.

Dabrowski, R. (1968), *Curved Thin Walled Girders, Theory and Analysis*, translated from German by C. V. Amerongen, Cement and Concrete Association, Number 144, Springer Verlag, Berlin.

Daniels, J. H., and Batcheler R. P. (1979), "Fatigue of Curved Steel Bridge Elements-Effect of Heat Curving on the Fatigue Strength of Plate Girders," Report FHWA-RD-79-136, FHWA, U.S. Department of Transportation, Washington, DC. Aug.

Daniels, J. H., Fisher, J. W., Batcheler, R. P., and Maurer, J. K. (1979a), "Fatigue of Curved Steel Bridges Elements—Ultimate Strength Tests of Horizontally Curved Plate and Box Girders," Lehigh University DOT-FH-11-8198.7, Bethlehem, PA.

Daniels, J. H., Fisher, J. W., and Yen, B. T. (1980), "Fatigue of Curved Steel Bridge Elements, Design Recommendations for Fatigue of Curved Plate Girder and Box Girder Bridges," Federal Highway Administration, Offices of Research and Development Structures and Applied Mechanics Division, Report No. FHWA-RD-79-138, Federal Highway Administration, Washington, DC, Apr.

Daniels, J. H., and Herbein, W. C. (1980), "Fatigue of Curved Steel Bridge Elements—Fatigue Tests of Curved Plate Girder Assemblies," Lehigh University FHWA-RD-79-133, Bethlehem, PA.

Daniels, J. H., Zettlemoyer, N., Abraham, D., and Batcheler, R. P. (1979b), "Fatigue of Curved Steel Bridge Elements, Analysis and Design of Plate Girder and Box Girder Test Assemblies," Lehigh University DOT-FH-11-8198.1, Bethlehem, PA.

Davidson, J. S., Balance, S. R., and Yoo, C. H. (1999a), "Analytical Model of Curved I-Girder Web Panels Subjected to Bending," *J. Bridge Eng. ASCE*, Vol. 4, No. 3, pp. 204–212.

Davidson, J. S., Balance, S. R., and Yoo, C. H. (1999b), "Finite Displacement Behavior of Curved I-Girder Web Subjected to Bending," *J. Bridge Eng. ASCE*, Vol. 4, No. 3, pp. 213–220.

Davidson, J. S., Balance, S. R., and Yoo, C. H. (2000a), "Behavior of Curved I-Girder Webs Subjected to Combined Bending and Shear," *J. Bridge Eng. ASCE*, Vol. 5, No. 2, May, pp. 165–170.

Davidson, J. S., Balance, S. R., and Yoo, C. H. (2000b), "Effects of Longitudinal Stiffeners on Curved I-Girder Webs," *J. Bridge Eng. ASCE*, Vol. 5, No. 2, pp. 171–178.

Davidson, J. S., Keller, M. A., and Yoo, C. H. (1996), "Cross Frame Spacing and Parametric Effects in Horizontally Curved I-Girder Bridges," *J. Struct. Eng. ASCE,* Vol. 122, No. 9, pp. 1086–1096.

Davidson, J. S., and Yoo, C. H. (1996), "Local Buckling of Curved I-Girder Flanges," *J. Struct. Eng. ASCE*, Vol. 122, No. 8, pp. 936–947.

Davidson, J. S., and Yoo, C. H. (2000), "Evaluation of Strength Formulations for Horizontally Curved Flexural Members," *J. Bridge Eng. ASCE*, Vol. 5, No. 3, pp. 200–207.

DESCUS Software (2002), "Design and Analysis of Curved I-Girder Bridge System User Manual," Opti-mate, Inc., Macungie, PA.

Fan, Z., and Helwig, T. A. (1999), "Behavior of Steel Box Girders with Top Flange Bracing," *J. Struct. Eng. ASCE*, Vol. 125, No. 8, pp. 829–837.

Fan, Z., and Helwig, T. A. (2002), "Distortional Loads and Brace Forces in Steel Box Girders," *J. Struct. Eng. ASCE,* Vol. 128, No. 6, pp. 710–718.

Fiechtl, A. L., Fenves, G. L., and Frank, K. H. (1987), "Approximate Analysis of Horizontally Curved Girder Bridges," Final Rep. No. FHWA-TX-91-360-2F, University of Texas at Austin, Center for Transportation Research, Austin, TX, p. 96. .

Frampton, R. E. (1968), "Local Instability of Horizontally Curved Members," Master's thesis, Carnegie-Mellon University, Pittsburgh, PA.

Fujii, K., and Ohmura, H. (1985), "Nonlinear Behavior of Curved Girder Web Considering Flange Rigidities," *Proceedings of the Japanese Society of Civil Engineers, Struct. Eng./Earthquake Eng.*, Vol. 2, No. 1, pp. 45–55.

Fujii, K., and Ohmura, H. (1987), "Local Buckling and Width Thickness Ratio in Flanges of Curved I-Girders," *Proceedings of the Japanese Society of Civil Engineers, Struct. Eng./Earthquake Eng.*, No. 356/I-3, pp. 69–79.

Fukumoto, Y., Itoh, Y., and Kubo, M. (1980), "Strength Variation of Laterally Unsupported Beams," *J. Struct. Div. ASCE*, Vol. 106, No. ST1, pp. 165–181.

Fukumoto, Y., and Nishida, S. (1981), "Ultimate Load Behavior of Curved I-Beams," *J. Eng. Mech. Div. ASCE*, Vol. 107, No. EM-2, pp. 367–385.

Galambos, T. V. (1978), "Tentative Load Factor Design Criteria for Curved Steel Bridges," Res. Rep. No. 50, School of Engineering and Applied Science, Civil Engineering Department, Washington University, St. Louis, MO, May.

Galambos, T. V., Hajjar, J. F., Huang, W. H., Pulver, B. E., Leon, R. T., and Rudie, B. J. (2000), "Comparison of Measured and Computed Stresses in a Steel Curved Girder Bridge," *J. Bridge Eng. ASCE*, Vol. 5, No. 3, pp. 191–199.

Galambos, T. V., Hajjar, J. F., Huang, W. H., Pulver, B. E., and Rudie, B. J. (1996), "Stresses in a Steel Curved Girder Bridge," Report No. MN/RC-96/28, Minnesota Department of Transportation, St. Paul, MN.

Gottfeld, H. (1932), *The Analysis of Spatially Curved Steel Bridges*, Die Bautechnik (in German), Berlin.

Grubb, M. A., Yadlosky, J. M. and Duwadi, S. R. (1996), "Construction Issues in Steel Curved Bridges," Transportation Research Record 1544, TRB, National Research Council, Washington, DC, pp. 64–70.

Hall, D. H., Grubb, M. A., and Yoo, C. H. (1999), "Improved Design Specifications for Horizontally Curved Steel Girder Highway Bridges," NCHRP Report 424, National Cooperative Highway Research Program, Washington, DC.

Hanshin Expressway Public Corporation and Steel Structure Study Committee (1988), *Guidelines for the Design of Horizontally Curved Girder Bridges (Draft)*, The Hanshin Expressway Public Corporation, Oct. Osaka, Japan.

Hartmann, J. L. (2005), "An Experimental Investigation of the Flexural Resistance of Horizontally Curved Steel I-Girder Systems," dissertation submitted to the Faculty of the Graduate School of the University of Maryland, College Park, MD.

Heins, C. P. (1975), *Bending and Torsion Design in Structural Members*, Lexington Books, Lexington, MA.

Heins, C. P., and Jin, J. O. (1984), "Live Load Distribution on Braced Curved I-Girders," *J. Struct. Eng. ASCE*, Vol. 110, No. 3, pp. 523–530.

Highway Structures Design Handbook (1965), Vol. 1, U. S. Steel, ADUSS, 88-2222, Pittsburgh, PA.

Hiwatashi, S., and Kuranishi, S. (1984), "The Finite Displacement Behavior of Horizontally Curved Elastic I-Section Plate Girders Under Bending," *Proceedings of the Japanese Society of Civil Engineers, Struct. Eng./Earthquake Eng.*, Vol. 1, No. 2, pp. 59–69, October.

Hsu, Y. T. (1989), "The Development and Behavior of Vlasov Elements for the Modeling of Horizontally Curved Composite Box Girder Bridge Superstructures," Ph.D. dissertation, University of Maryland, College Park, MD.

Ilyasevitch, S., and Klujev, B. N. (1971), "Stabilitat der elastschen zylinder bauplatte bei querbiegung eines dunnwandigen stahltragers mit doppelter x formiger wand," report to the April 1971 Seminar of the International Association for Bridge and Structural Engineering, London, summarized in English in *IABSE Bull.*, Vol. 27, 1971, p. 17.

Imai, F., and Ohto, T. (1987), "Investigation on Yield Surface Equation of Curved I-Beams," *Trans. Jpn. Soc. Civil Eng.*, No. 380/I-7, Apr., pp. 349–354 (in Japanese).

Japan Road Association (1990), *Specifications for Highway Bridges, Part I: Common Specifications. Part II: Steel Bridge Specifications*, Maruzen, Tokyo, Feb.

Jung, S.-K., and White, D. W. (2006), "Shear Strength of Horizontally Curved Steel I-Girders—Finite Element Analysis Studies," *J. Constr. Steel Res.*, Vol. 62, pp. 329–342.

Kang, Y. J. (1992), "Nonlinear Theory of Thin Walled Curved Beams and Finite Element Formulation," Ph.D. dissertation, Auburn University, Auburn, AL.

Kang, Y. J., and Yoo, C. H. (1990), "Flexural Stress of Curved Bridge Girders," paper presented at the Structural Stability Research Council, 1990 Annual Technical Session, Stability of Bridges, St. Louis, MO, Apr. 10–11.

Kang, Y. J., and Yoo, C. H. (1994a), "Thin Walled Curved Beams. I: Formulation of Nonlinear Equations," *J. Eng. Mech. ASCE*, Vol. 120, No. 10, pp. 2072–2101.

Kang, Y. J., and Yoo, C. H. (1994b), "Thin Walled Curved Beams. II: Analytical Solutions for Buckling of Arches," *J. Eng. Mech. ASCE*, Vol. 120, No. 10, pp. 2072–2101.

Ketchek, K. F. (1969), "Discussion of 'Horizontally Curved Girders: State of the Art' by McManus, P. F. et al.," *J. Struct. Div. ASCE*, Vol. 95, No. ST12, pp. 2999–3001.

Kim, K., and Yoo, C. H. (2006a), "Ultimate Strength Interaction of Composite Box Girders in Positive Bending," *Adv. Struct. Eng.*, Vol. 9, No. 5, pp. 707–718.

Kim, K., and Yoo, C. H. (2006b), "Brace Forces in Steel Box Girders with Single Diagonal Bracing System," *J. Struct. Eng. ASCE*, Vol. 132, No. 8, pp. 1212–1222.

Kim, K., and Yoo, C. H. (2006c), "Effects of External Bracing on Horizontally Curved Box Girder Bridges During Construction," *Eng. Struct.*, Vol. 28, No. 12, pp. 1650–1657.

Kim, K., and Yoo, C. H. (2006d), "Interaction of Top Lateral and Internal Bracing Systems in Tub Girders," *J. Struct. Eng., ASCE,* Vol. 132, No. 10, pp. 1611–1620.

Kim, K., and Yoo, C. H. (2008), "Ultimate Strength of Steel Rectangular Box Beams Subjected to Combined Action of Bending and Torsion," *Eng. Struct,*, Vol. 30, No. 6, pp. 1677–1687.

Kim, Y. D., Jung, S.-K., and White D. W. (2007), "Transverse Stiffener Requirements in Straight and Horizontally Curved Steel I-Girders," *J. Bridge Eng., ASCE*, Vol. 12, No. 2, pp. 174–183.

Kishima, Y., Alpsten, G. A., and Tall, L. (1969), "Residual Stresses in Welded Shapes of Flame-Cut Plates in ASTM A572 (50) Steel," Fritz Engineering Laboratory Report No. 321.2, Lehigh University, Bethlehem, PA.

Kitada, T., Ohminami, R., and Nakai, H. (1986), "Criteria for Designing Web and Flange Plates of Horizontally Curved Plate Girders," *Proc. SSRC Annu. Tech. Session*, Stability of Plate Structures, Washington, DC, Apr. 15–16, pp. 119–130.

Kollbrunner, C., and Basler, K. (1969), *Torsion in Structures*, Springer-Verlag, Berlin.

Komatsu, S., and Kitada, T. (1981), "Ultimate Strength Characteristics of Outstanding Steel Plate with Initial Imperfection Under Compression," *Jpn. Soc. Civ. Eng.*, Vol. 314, pp. 15–27.

Komatsu, S., Kitada, T., and Miyazaki, S. (1975), "Elastic Plastic Analysis of Compressed Plate with Residual Stress and Initial Deflection," *Proceedings of the Japanese Society of Civil Engineers*, No. 244, pp. 1–14, Dec. (in Japanese).

Kulicki, J. M., Wassef, W. G., Kleinhans, D. D., Yoo, C. H., Nowak, A. S., and Grubb, M. (2006), "Development of LRFD Specifications for Horizontally Curved Steel Girder Bridges," National Cooperative Highway Research Program NCHRP Report 563, Transportation Research Board, Washington, DC.

Kuo, S. R., and Yang, Y. B. (1991), "New Theory on Buckling of Curved Beams," *J. Eng. Mech. Div. ASCE*, Vol. 117, No. 8, pp. 1698–1717.

Kuranishi, S., and Hiwatashi, S. (1981), "Elastic Behavior of Web Plates of Curved Plate Girders in Bending," *Proc. Jpn. Soc. Civ. Eng.*, Vol. 315, pp. 1–11.

Kuranishi, S., and Hiwatashi, S. (1983), "Nonlinear Behavior of Elastic Web Plates of Curved Plate Girders in Bending," *Proceedings of the 3rd International Colloquium on Stability of Metal Structures*, Paris France, Nov., pp. 305–312.

Lee, S. C. (1987), "Finite Element Analysis of Thin-Walled Assemblages," Ph.D. Dissertation, Auburn University, Auburn, AL.

Lee, S. C., and Yoo, C. H. (1988a), "A Novel Shell Element Including In-Plane Torque Effects," *Comp. Struct.*, Vol. 28, No. 4, 1988, pp. 505–522.

Lee, S. C., and Yoo, C. H. (1988b), "Interactive Buckling of Thin–Walled Members," *Proceedings of the 1988 Annual Technical Sessions, Structural Stability Research Council*, Minneapolis, MN., April 25–28, 1988.

Lee, S. C., Davidson, J. S., and Yoo, C. H. (1994), "Shear Buckling Coefficients of Plate Girder Web Panels," *Comput. Struct.*, Vol. 59, No. 5, pp. 789–795.

Lee, S. C., and Yoo, C. H. (1998), "Strength of Plate Girder Web Panels Under Pure Shear," *J. Struct. Eng. ASCE*, Vol. 124, No. 2, pp. 184–194.

Lee, S. C., and Yoo, C. H. (1999a), "Experimental Study on Ultimate Shear Strength of Web Panels," *J. Struct. Eng. ASCE*, Vol. 125, No. 8, pp. 838–846.

Lee, S. C., and Yoo, C. H. (1999b), "Strength of Curved I-Girder Web Panels Under Pure Shear," *J. Struct. Eng. ASCE*, Vol. 125, No. 8, pp. 847–853.

Lee, S. C., Yoo, C. H., Lee, D. S., and Yoon, D. Y. (2003a), "Ultimate Strength Behavior of Curved Girder Web Panels," *Proceedings of the Structural Stability Research Council*, 2003 Annual Technical Session and Meeting, Apr. 2–5, Baltimore, MD, pp. 255–278.

Lee, S. C., Yoo, C. H., and Yoon, D. Y. (2002), "Behavior of Intermediate Transverse Stiffeners Attached on Web Panel," *J. Struct. Eng. ASCE*, Vol. 128, No. 3, pp. 337–345.

Lee, S. C., Yoo, C. H., and Yoon, D. Y. (2003b), "New Design Rule for Intermediate Transverse Stiffeners Attached on Web Panels," *J. Struct. Eng. ASCE*, Vol. 129, No. 12, pp. 1607–1614.

Lian, V. T., and Shanmugam, N. E. (2003), "Openings in Horizontally Curved Plate Girders," *Thin-Walled Struct.*, Vol. 41, Nos. 2–3, pp. 245–269.

Lian, V. T., and Shanmugam, N. E. (2004), "Design of Horizontally Curved Plate Girder Webs Containing Circular Openings," *Thin Walled Struct.*, Vol. 42, No. 5, pp. 719–739.

Liew, J. Y. R., Thevendran, V., Shanmugam, N. E., and Tan, L. O. (1995), "Behaviour and Design of Horizontally Curved Steel Beams," *J. Constr. Steel Res.*, Vol. 32, No. 1, pp. 37–67.

Linzell, D. G. (1999), "Studies of Full-Scale Horizontally Curved Steel I-Girder Bridge Systems Under Self Weight," Ph.D. dissertation, School of Civil and Environmental Engineering, Georgia Institute of Technology, Atlanta, GA.

Linzell, D. G., Leon R. T., and Zureick, A. H. (2004a), "Experimental and Analytical Studies of a Horizontally Curved Steel I-Girder Bridge During Erection," *J. Bridge Eng. ASCE*, Vol. 9, No. 6, pp. 521–530.

Linzell, D. G., Hall, D. H., and White, D. W. (2004b), "A Historical Perspective on Horizontally Curved I-Girder Bridge Design in the United States," *J. Bridge Eng. ASCE*, Vol. 9, No. 3, pp. 218–229.

Madhavan, M., and Davidson, J. S. (2005), "Buckling of Centerline-Stiffened Plates Subjected to Uniaxial Eccentric Compression," *Thin-Walled Struct.*, Vol. 43, No. 8, pp. 1264–1276.

Madhavan, M., and Davidson, J. S. (2007), "Elastic Buckling of I-Beam Flanges Subjected to a Linearly Varying Stress Distribution," *J. Constr. Steel Res.*, Vol. 63, Issue 10, pp. 1373–1383.

Maegawa, K., and Yoshida, H. (1981), "Ultimate Strength Analysis of Curved I-Beams by Transfer Matrix Method," *Proceedings of the Japanese Society of Civil Engineers*, No. 312, pp. 27–42, Aug. (in Japanese).

Mariani, N., Mozer, J. D., Dym, C. L., and Culver, C. G. (1973), "Transverse Stiffener Requirements for Curved Webs," *J. Struct. Div. ASCE*, Vol. 99, No. ST4, pp. 757–771.

McElwain, B. A., and Laman, J. A. (2000), "Experimental Verification of Horizontally Curved I-Girder Bridge Behavior," *J. Bridge Eng. ASCE*, Vol. 5, No. 4, pp. 284–292.

McManus, P. F. (1971), "Lateral Buckling of Curved Plate Girders," PhD dissertation, Carnegie Mellon University, Pittsburgh, PA.

McManus, P. F., Nasir, G. A., and Culver, C. G. (1969), "Horizontally Curved Girders: State of the Art," *J. Struct. Div. ASCE*, Vol. 95, No. ST5, pp. 853–870.

MDX (2000), *Curved and Straight Steel Bridge Design and Rrating User Manual*, MDX Software, Columbia, MO.

Mikami, I., and Furunishi, K. (1981), "Nonlinear Behavior of Cylindrical Web Panels Under Bending and Shear," *Theor. Appl. Mech.*, Vol. 29, pp. 65–72.

Mikami, I., and Furunishi, K. (1984), "Nonlinear Behavior of Cylindrical Web Panels," *J. Eng. Mech. Div. ASCE*, Vol. 110, No. EM2, pp. 230–251.

Mikami, I., Furunishi, K., and Yonezawa, H. (1980), "Nonlinear Behavior of Cylindrical Web Panels Under Bending," *Proceedings of the Japanese Society of Civil Engineers*, No. 299, July, pp. 23–34 (in Japanese, English summary in *Trans. JSCE*, Vol. 12, pp. 65–66).

Modjeski and Masters, Inc. (1989), Review of computer programs for the analysis of girder bridges, Pennsylvania Department of Transportation, Harrisburg, PA.

Mozer, J. D., and Culver, C. G. (1975), "Horizontally Curved Highway Bridges, Stability of Curved Plate Girders-P1," Contract No. FH-11-7389, FHWA, Washington, DC.

Mozer, J. D., Cook, J., and Culver, C. G. (1975a), "Horizontally Curved Highway Bridges, Stability of Curved Plate Girders-P3," Contract No. FH-11-7389, FHWA, Washington, DC.

Mozer, J. D., Ohlson, R., and Culver, C. G. (1971), "Horizontally Curved Highway Bridges: Stability of Curved Plate Girders," Rep. No. P2, Res. Proj. HPR-2(111), Carnegie-Mellon University, Pittsburgh, PA.

Mozer, J. D., Ohlson, R., and Culver, C. G. (1975b), "Stability of Curved Plate Girders-P2," Contract No. FH-11-7389, FHWA, Washington, DC.

Mozer, J. D., and Culver, C. G. (1974), "Behavior of Curved Box Girders—Laboratory Observation," paper presented at the ASCE Specialty Conference on Metal Bridges, Nov. 12–13, St. Louis, MO.

Nakai, H., and Kitada, T. (1992), "On Ultimate Strength of Horizontally Curved Box Girder Bridges," *Proceedings of the first World Conference on Constructional Steel Design*, Acapulco, Mexico, pp. 108–114.

Nakai, H., Kitada, T., and Ohminami, R. (1983), "Experimental Study on Bending Strength of Horizontally Curved Girder Bridges," *Proceedings of the Japanese Society of Civil Engineers*, No. 340/I2, Dec., pp. 19–28 (in Japanese).

Nakai, H., Kitada, T., and Ohminami, R. (1984a), "Experimental Study on Ultimate Strength of Web Panels in Horizontally Curved Girder Bridges Subjected to Bending, Shear and Their Combinations," *Proc. SSRC Annu. Tech. Session, Stability Under Seismic Loading*, San Francisco, Apr. 10–11, pp. 91–102.

Nakai, H., Kitada, T., Ohminami, R., and Fukumoto, K. (1984b), "Experimental Study on Shear Strength of Horizontally Curved Plate Girders," *Jpn. Soc. Civ. Eng.*, Vol. 350, pp. 281–290.

Nakai, H., Kitada, T., Ohminami, R., and Fukumoto, K. (1984c), "A Proposition for Designing Transverse Stiffeners of Horizontally Curved Girders in Ultimate State," *Mem. Fac. Eng. Osaka City Univ.*, Vol. 25, pp. 111–131.

Nakai, H., Kitada, T., and Ohminami, R. (1985a), "Experimental Study on Bending Strength of Web Plates of Horizontally Curved Girder Bridges," *Jpn. Soc. Civ. Eng.*, Vol. 15, pp. 201–203.

Nakai, H., Kitada, T., and Ohminami, R. (1985b), "Proposition for Designing Intermediate Transverse Stiffeners in Web Plate of Horizontally Curved Girders," *Jpn. Soc. Civ. Eng.*, Vol. 361/I-4, pp. 249–257.

Nakai, H., Kitada, T., and Ohminami, R. (1985c), "Experimental Study on Buckling and Ultimate Strength of Curved Girders Subjected to Combined Loads of Bending and Shear," *Proceedings of the Japanese Society of Civil Engineers*, No. 356/I3, pp. 445–454, Apr. (in Japanese).

Nakai, H., Kitada, T., Ohminami, R., and Kawai, T. (1986), "A Study on Analysis and Design of Web Plates in Curved Bridges Subjected to Bending," *Proceedings of the Japanese Society of Civil Engineers*, No. 368/I5, pp. 235–44, Apr. (in Japanese).

Nakai, H., and Kotoguchi, H. (1983), "A Study on Lateral Buckling Strength and Design Aid for Horizontally Curved I-Girder Bridges," *Proceedings of the Japanese Society of Civil Engineers*, No. 339, pp. 195–204, Dec.

Nakai, H., Muramatsu, S., Yoshikawa, N., Kitada, T., and Ohminami, R. (1981), "A Survey for Web Plates of the Horizontally Curved Girder Bridges," *Bridge Found. Eng.* Vol. 15, pp. 38–45 (in Japanese).

Nakai, H., Murayama, Y., and Kitada, T. (1992), "An Experimental Study on Ultimate Strength of Thin-Walled Box Beams with Longitudinal Stiffeners Subjected to Bending and Torsion," *J. Struct. Eng. JSCE*, Vol. 38A, pp. 155–165 (in Japanese).

Nakai, H., Murayama, Y., Kitada, T., and Takada, Y. (1990), "An Experimental Study on Ultimate Strength of Thin-Walled Box Beams Subjected to Bending and Torsion," *J. Struct. Eng. JSCE*, Vol. 36A, pp. 63–70 (in Japanese).

Nakai, H., and Yoo, C. H. (1988), *Analysis and Design of Curved Steel Bridges*, McGraw-Hill, New York.

Nevling, D., Linzell, D., and Laman, J. (2006), "Examination of Level of Analysis Accuracy for Curved I-Girder Bridges through Comparisons to Field Data," *J. Bridge Eng.* Vol. 11, No. 2, pp. 160–168.

Nishida, S., Yoshida, H., and Fukumoto, Y. (1978), "Large Deflection Analysis of Curved Members with Thin Walled Open Cross Section," paper presented at the 24th Symposium of Structural Engineering, Feb., pp. 77–84.

National Steel Bridge Alliance (NSBA) (1996), "Chapter 12: V-Load Analysis—An Approximate Procedure, Simplified and Extended, for Determining Moments and Shear in Designing Horizontally Curved Openframe Highway Bridges," in *Highway Structures Design Handbook*, Vol. 1, NSBA, Chicago, IL Ch. 12.

Pandit, G. S., Ceradini, G., Grvarini, P., and Eremin, A. A. (1970), "Discussion of 'Horizontally Curved Girders: State of the Art' by McManus, P.F. et al.," *J. Struct. Div. ASCE*, Vol. 96, No. ST2, p. 433–436.

Papangelis, J. P., and Trahair, N. S. (1986), "Flexural-Torsional Buckling of Arches," *J. Struct. Div. ASCE*, Vol. 113, No. ST4, pp. 889–906.

Papangelis, J. P., and Trahair, N. S. (1987), "Flexural-Torsional Buckling Tests on Arches," *J. Struct. Div. ASCE*, Vol. 113, No. ST7, pp. 1433–1443.

Pfeiffer, P. A. (1981), "Elastic Stability of Curved Beams," Master's thesis, Marquette University, Milwaukee, WI.

Pi, Y.-L., and Bradford, M. A. (2001), "Strength Design of Steel I-Section Beams Curved in Plan," *J. Struct. Eng. ASCE*, Vol. 127, No. 6, pp. 639–646.

Pi, Y.-L., Bradford, M. A., and Trahair, N. S. (1999), "Inelastic Analysis and Behavior of Steel I-Beams Curved in Plan," UNICIV Rep., No. R-377, School of Civ. and Envir. Eng., University of New South Wales, Sydney, Australia.

Pi, Y.-L., Bradford, M. A., and Trahair N. S. (2000), "Inelastic Analysis and Behaviour of Steel I-Beams Curved in Plan," *J. Struct. Eng. ASCE*, Vol. 126, No. 7, pp. 772–779.

Pi, Y.-L., and Trahair, N. S. (1994), "Inelastic Bending and Torsion of Steel I-Beams," *J. Struct. Eng. ASCE*, Vol. 120, No. 12, pp. 3397–3417.

Pi, Y.-L., and Trahair, N. S. (1997), "Nonlinear Elastic Behavior of I-Beams Curved in Plan," *J. Struct. Eng. ASCE*, Vol. 123, No. 9, pp. 1201–1209.

Poellot, W. N. (1987), "Computer-Aided Design of Horizontally Curved Girders by the V-Load Method," *Eng. J. AISC*, Vol. 24, No. 1, pp. 42–50.

Rajasekaran, S., and Ramm, E. (1984), "Discussion of 'Flexural-Torsional Stability of Horizontally Curved Beams' by C. H. Yoo," *J. Eng. Mech. Div. ASCE*, Vol. 110, No. EM1, pp. 144–148.

Rajasekaran, S., Sundaraajan, T., and Rao, K. S. (1988), "Discussion of 'Static Stability of Curved Thin-Walled Beams' by Yang, Y. B. and Kuo, S. R.," *J. Struct. Div. ASCE*, Vol. 114, No. ST5, pp. 915–918.

Richardson, Gordon, and Associates (1963), "Analysis and Design of Horizontally Curved Steel Bridge Girder," Structural Report ADUSS 88-6003-01, U.S. Steel, Pittsburgh, PA.

Schelling, D., Namini, A. H., and Fu, C. C. (1989), "Construction Effects on Bracing on Curved I-Girders," *J. Struct. Div. ASCE*, Vol. 115, No. ST9, pp. 2145–2165.

Schilling, C. G. (1996), "Yield Interaction Relationships for Curved I-Girders," *J. Bridge Eng. ASCE*, Vol. 1, No. 1, pp. 26–33.

Shanmugam, N. E., Mahendrakumar, M., and Thevendran, V. (2003), "Ultimate Load Behaviour of Horizontally Curved Plate Girders," *J. Constr. Steel Res.*, Vol. 49, No. 4, pp. 509–529.

Shanmugam, N. E., Thevendran, V., Richard Liew, J. Y., and Tan, L. O. (1995), "Experimental Study on Steel Beams Curved in Plan," *J. Struct. Eng. ASCE*, Vol. 121, No. 2, pp. 249–259.

Shimada, S., and Kuranashi, S. (1966), *Formulas for Calculation of Curved Beam*, Gihodo, Tokyo (in Japanese).

Simpson, M. D. (2000), "Analytical Investigation of Curved Steel Girder Behavior," Ph.D. dissertation, Department of Civil Engineering, University of Toronto, Canada.

Stegmann, T. H., and Galambos, T. V. (1976), "Load Factor Design Criteria for Curved Steel Girders of Open Section," Res. Rep. 23, Civil Engineering Department, Washington University, St. Louis, MO, Apr.

Stein, M., and Fralich, R. W. (1949), "Critical Shear Stress of Infinitely Long, Simply Supported Plates with Transverse Stiffeners," NACA TN No. 1851, National Advisory Committee for Aeronautics, Washington, DC.

Stein, M., and Yeager, D. J. (1949), "Critical Shear Stress of a Curved Rectangular Panel with a Central Stiffener," NACA TN No. 1972, National Advisory Committee for Aeronautics, Washington, DC.

St. Venant, B. (1843), "Memoire sur le Calcul de la Resistance et la Flexion des Pieces Solides a Simple ou a Double Courbure, en Prenant Simultanement en Consideration les Divers Efforts Auxquels Eiles Peuvent Entre Soumises dans Tous les Sens," *Compts Rendus l'Academie des Sciences de Paris*, Vol. 17, pp. 1020–1031 (in French).

Suetake, Y., Hirashima, M., and Yoda, T. (1986), "Geometrical Nonlinear Analysis of Curved I-Girders Under Bending Using Mixed Finite Element Method," *Proceedings of the Japanese Society of Civil Engineers, Struct. Eng./ Earthquake Eng.*, Vol. 3, No. 2, pp. 207–214.

Tan, C. P., Shore, S., and Ketchek, K. (1969), "Discussion of 'Horizontally Curved Girders: State of the Art' by McManus, P. F. et al.," *J. Struct. Div. ASCE*, Vol. 95, No. ST12, pp. 2997–2999.

Tan, L. O., Thevendran, V., Richard Liew, J. Y., and Shanmugam, N. E. (1992a), "Analysis and Design of I-Beams Curved in Plan," *Steel Structures, J. Singapore Struct. Steel Soc.*, Vol. 3, No. 1, pp. 39–45.

Tan, L. O., Thevendran, V., Shanmugan, N. E. and Liew, J. Y. R. (1992b), "Analysis and Design of I-Beams Curved in Plan," *J. Singapore Struct. Steel Soc.*, Vol. 3, No. 1, pp. 39–45.

Thevendran, V., Chen, S., Shanmugam, N. E., and Liew, R. J. Y. (1999), "Nonlinear Analysis of Steel-Concrete Composite Beams Curved in Plan," *J. Finite Elements Anal. Design*, Vol. 32, pp. 125–139.

Thevendran, V., Shanmugam, N. E., Chen, S., and Liew, R. J. Y. (2000), "Experimental Study on Steel-Concrete Composite Beams Curved in Plan," *Eng. Struct.*, Vol. 22, pp. 877–879.

Timoshenko, S. P. (1905), "On the Stability of In Plane Bending of an I-Beam," Izvestiya St. Petersburg Politekhnicheskoqo Instituta, IV-V.

Timoshenko, S. P., and Gere, J. M. (1961), *Theory of Elastic Stability*, 2nd ed., McGraw Hill Book, New York.

Tung, D. H. H., and Fountain, R. S. (1970), "Approximate Torsional Analysis of Curved Box Girders by the M/R Method," *Eng. J. AISC*, July, pp. 65–74.

Umanskii, A. A. (1948), *Spatial Structures*, Moscow (in Russian).

U. S. Steel Corporation (1984), "V-Load Analysis, an Approximate Procedure, Simplified and Extended for Determining Moments and Shears in Designing Horizontally Curved Open Framed Highway Bridges," in *USS Highway Structures Design Handbook*, Vol. 1, U.S. Steel, Pittsburgh, PA, Chap. 12.

Vacharajittiphan, P., and Trahair, N. S. (1975), "Flexural-Torsional Buckling of Curved Members," *J. Struct. Div. ASCE*, Vol. 101, No. 6, pp. 1223–1238.

Vincent, G. S. (1969), "Tentative Criteria for Load Factor Design of Steel Highway Bridges," Bulletin No. 15, American Iron and Steel Institute, New York, Mar.

Vlasov, V. Z. (1961), *Thin Walled Elastic Beams*, 2nd ed., National Science Foundation, Washington, DC.

Wachowiak, T. (1967), "die Berechnung gekruemmter Stege duennwandigen traegern auf Grund der Shalentheorie," Dissertationarbeit, Politechnika Gdańska, Poland.

Watanabe, N. (1967), *Theory and Calculation of Curved Girder*, Gihodo, Tokyo (in Japanese).

White, D. W., and Grubb, M. A. (2005), "Unified Resistance Equations for Design of Curved and Tangent Steel Bridge I-Girders," *Proceedings of the 84th Annual Meeting*, Transportation Research Board, Washington, DC.

White, D. W., Zureick, A. H., Phoawanich, N., and Jung, S. (2001), "Development of Unified Equations for Design of Curved and Straight Steel Bridge I-Girders," prepared for American Iron and Steel Institute Transportation and Infrastructure Committee, Professional Services, Inc., and Federal Highway Administration, Georgia Institute of Technology, Atlanta, GA.

Yadlosky, J. M. (1993), "Curved Steel Bridge Research Project, Summary Work Plan (1.0)," HDR Interim Report DTFH61-92-C-00136, submitted to The Federal Highway Administration, Washington, DC.

Yang, Y. B., Chern, S. M., and Fan, H. T. (1989), "Yield Surfaces for I-Sections with Bimoments," *J. Struct. Eng. ASCE*, Vol. 115, No. 12, pp. 3044–3058.

Yang, Y. B, and Fan, H. T. (1988), "Yield Surface Modeling of I-Sections with Nonuniform Torsion," *J. Eng. Mech. ASCE*, Vol. 114, No. 6, pp. 953–972.

Yang, Y. B., and Kuo, S. R. (1986), "Static Stability of Curved Thin-Walled Beams," *J. Eng. Mech. Div. ASCE*, Vol. 112, No. ST8, pp. 821–841.

Yang, Y. B., and Kuo, S. R. (1987), "Effect of Curvature Stability of Curved Beams," *J. Struct. Div. ASCE*, Vol. 113, No. ST6, pp. 1185–1202.

Yang, Y. B., Kuo, S. R., and Yau, J. D. (1991), "Use of Straight-Beam Approach to Study Buckling of Curved Beams," *J. Struct. Div. ASCE*, Vol. 117, p. 1963.

Yoo, C. H. (1982), "Flexural-Torsional Stability of Curved Beams," *ASCE J. Eng. Mech. Div.*, Vol. 108, No. EM6, pp. 1351–1369.

Yoo, C. H., and Carbine, R. L. (1985), "Experimental Investigation of Horizontally Curved Steel Wide Flange Beams," *Proc. SSRC Ann. Tech. Session*, Stability Aspects of Industrial Buildings, Cleveland, OH, Apr. 16–17, pp. 183–191.

Yoo, C. H., Choi, B. H., and Ford, E. M. (2001), "Stiffness Requirements for Longitudinally Stiffened Box Girder Flanges," *J. Struct. Eng. ASCE*, Vol. 127, No. 6, pp. 705–711.

Yoo, C. H., and Davidson J. S. (1997), "Yield Interaction Equations for Nominal Bending Strength of Curved I-Girders," *J. Bridge Eng. ASCE*, Vol. 2, No. 2, pp. 37–44.

Yoo C. H., and Davidson J. S. (2002), "Curved Steel Bridge Research Project: Task D—Nominal Bending and Shear Strength," report submitted to the Federal Highway Administration, Turner-Fairbank Highway Research Center, as part of the Curved Steel Bridge Research Project, Contract No. DTFH61-92-C-00136, Auburn University, Highway Research Center, Auburn, AL.

Yoo, C. H., and Heins, C. P. (1972), "Plastic Collapse of Horizontally Curved Bridge Girders," *J. Struct. Eng. ASCE*, Vol. 98, No. ST 4, pp. 899–914.

Yoo, C. H., Kang, Y. J., and Davidson, J. S. (1996), "Buckling Analysis of Curved Beams by Finite Element Discretization," *J. Eng. Mech. ASCE*, Vol. 122, No. 8, pp. 762–770.

Yoo, C. H., and Lee, S. C. (2006), "Mechanics of Web Panel Postbuckling Behavior in Shear," *J. Struct. Eng. ASCE*, Vol. 132, No. 10, pp. 1580–1589.

Yoo, C. H., and Littrell, P. C. (1986), "Cross-Bracing Effects in Curved Stringer Bridges," *J. Struct. Div. ASCE*, Vol. 112, No. ST9, pp. 2127–2140.

Yoo, C. H., and Pfeiffer, P. A. (1983), "Elastic Stability of Curved Members," *J. Struct. Div. ASCE*, Vol. 109, No. ST12, pp. 2922–2940.

Yoo, C. H., and Pfeiffer, P. A. (1984), "Buckling of Curved Beams with In-Plane Deformation," *J. Struct. Div. ASCE*, Vol. 110, No. ST2, pp. 291–300.

Yoshida, H., and Imoto, Y. (1973), "Inelastic Lateral Buckling of Restrained Beams," *J. Eng. Mech. Div. ASCE*, Vol. 99, No. 2, pp. 343–366.

Yoshida, H., and Maegawa, K. (1983), "Ultimate Strength Analysis of Curved I-Beams," *J. Eng. Mech. Div. ASCE*, Vol. 109, No. 1, pp. 192–214.

Yoshida, H., and Maegawa, K. (1984), "Lateral Instability of I-Beams with Imperfections," *J. Struct. Eng. ASCE*, Vol. 110, No. 8, pp. 1875–1892.

Zettlemoyer, N., Fisher, J. W., and Daniels, J. H. (1980), "Fatigue of Curved Steel Bridge Elements—Stress Concentration, Stress Range Gradient, and Principal Stress Effects on Fatigue Life," FHWA-RD-79-132, Lehigh University, Bethlehem, PA.

Zureick, A., and Naqib, R. (1999), "Horizontally Curved Steel I-Girders, State-of-the-Art: Part I—Analysis Methods," *J. Bridge Eng. ASCE*, Vol. 4, No. 1, pp. 38–47.

Zureick, A., and Naqib, R., and Yadlosky, J. M. (1994), "Curved Steel Bridge Research Project, Interim Report I: Synthesis," FHWA-RD-93-129, Federal Highway Administration, U.S. Department of Transportation, Washington, DC.

Zureick, A., White, D. W., Phoawanich, N., and Park J. (2002), "Shear Strength of Horizontally Curved Steel I-Girders—Experimental Tests," final report to Professional Services Industries, Inc., and Federal Highway Administration, Washington, DC.

CHAPTER 10

COMPOSITE COLUMNS AND STRUCTURAL SYSTEMS

10.1 INTRODUCTION

Composite structural members are members in which steel and concrete act together through mechanical interlock, friction, and adhesion. Composite members are designed to maximize the efficiency of the two materials by using, whenever possible, the concrete in compression and the steel in tension. In addition to exploiting the stress–strain characteristics of the two materials to increase the ultimate capacity of the member, composite systems attempt to gain additional benefits from the synergy of their interaction. For example, the concrete can be used to limit global and local buckling problems in thinner steel elements, and the steel in tubular and round sections can be used to increase the confinement of the concrete and therefore help to maintain its strength in the postpeak region.

Buildings are seldom constructed only of composite members. Most often, composite columns are used in the lateral load-resisting systems in combination with either other composite members to form composite systems or other types of structural elements to form hybrid systems (Goel and Yamanuchi, 1993; Yamanuchi et al., 1993; Deierlein, 1995; Deierlein and Leon, 1996). Composite members are very popular in floor systems where composite beams, stub girders, and composite joists and trusses are the most economical alternatives in mixed-use structures (Viest et al., 1996). The stability benefits derived from the presence of a floor slab for floor members are not addressed in this chapter. It should be noted, however, that additional care needs to be taken in negative-moment regions of composite flexural members to ensure that the required rotational capacity is reached (Dekker et al., 1995). Thus, utilizing the beneficial aspects of composite action requires design checks that are different from those for typical reinforced-concrete or steel construction. Additional stability design provisions from composite beams with web openings and for composite joists and trusses are also available from an ASCE Task Group on Composite Design (ASCE Task Committee, 1994a, 1996).

In addition to floors systems, composite members are also being utilized as lateral load-resisting elements in braced and wall systems. In braced systems composite elements are desirable because they delay both global and local buckling, strengthening and stiffening the system significantly more than with conventional steel bracing (Liu and Goel, 1988). In wall systems, composite members are being used to facilitate the connection between reinforced-concrete walls and steel frames in hybrid systems and to delay shear cracking and improve hysteretic performance in reinforced-concrete systems (Harries et al., 1993; Shahrooz et al., 1993). Descriptions of composite and hybrid structural systems are given elsewhere (Griffis, 1986, 1992; FEMA 450, 2003), and this chapter is concerned primarily with stability effects related to composite columns and their connections. Composite columns are formed either by encasing a steel shape in concrete (called *SRC construction*) or by filling a structural pipe or tube with concrete (*CFT construction*). There are many possible variations of composite columns (Figs.10.1, 10.2, and 10.3), but they are generally used in the following situations:

- *Encased shapes in columns forming perimeter frames in high-rise structures*. Prefabricated steel column "trees," consisting of one- or two-story columns and short beam stubs on either side, are often used as erection columns and later encased in concrete. The encasement is used mainly to increase the stiffness of the columns and reduce the drift under wind or seismic loads. Examples of this type of application are the Gulf Tower and First City Tower in Houston (Griffis, 1992).
- *Very large encased shapes or round concrete-filled sections acting as corner columns for innovative structural systems in high-rise construction.* In these systems, the steel section is small and is used primarily for erection purposes in the case of encased shapes and for formwork in the case of concrete-filled tubes. Examples of this type of application are the Bank of China in Hong Kong, Two Union Square in Seattle, and the Norwest Center in Minneapolis (Griffis, 1992; Leon and Bawa, 1990).
- *Maximum reinforcement requirements.* Encased shapes are used in special situations where the amount of steel required would exceed the maximum permitted by current codes in a concrete section.
- *Transition columns between reinforced-concrete and steel columns.* This situation often arises in office buildings with a reinforced-concrete parking garage occupying the first few floors and a lighter steel frame making up most of the upper stories.
- *Reduce column slenderness.* Concrete-filled tubes are often employed in structures with high story heights where the additional stiffness provided by the concrete reduces the slenderness ratio of the column.
- *Impact and/or fire protection.* Composite columns are used in areas where impact and/or fire protection are crucial.

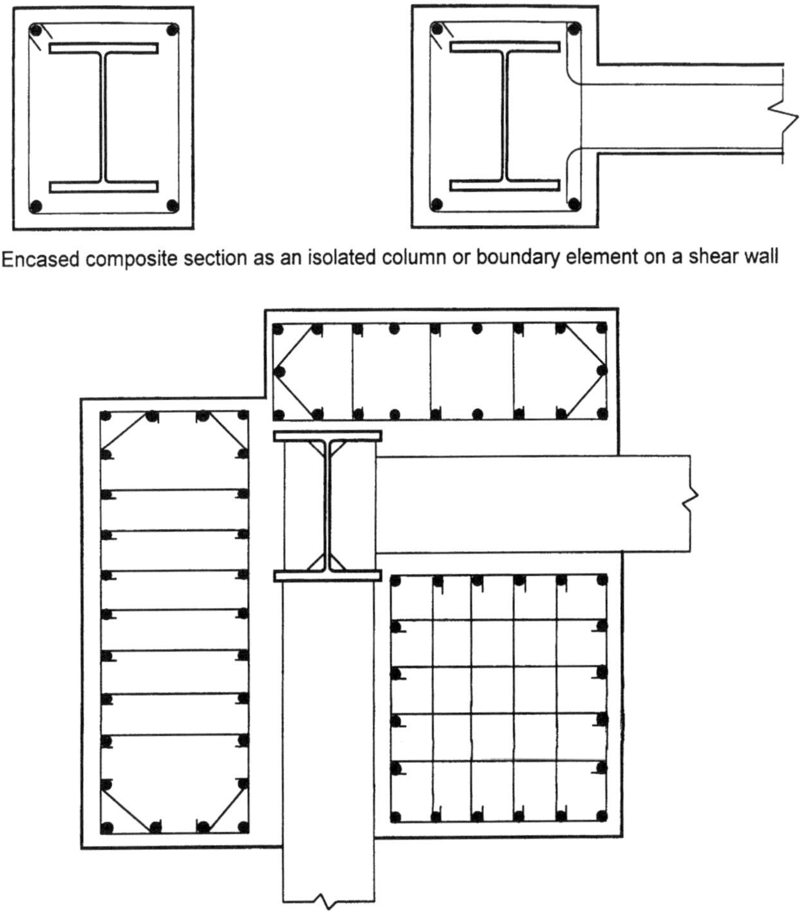

(*a*) Encased composite section as an isolated column or boundary element on a shear wall

(*b*) Large encased composite section where steel column is used primarily as an erection column. Section shows full moment connection to two steel beams and three large reinforcing cages (not all details shown)

FIGURE 10.1 Typical cross sections for encased composite columns.

- *Primary seismic load-resisting system*. Extensive use is made of both encased shapes and concrete-filled sections in other countries, particularly Japan, for seismic design.

Until recently most applications of composite columns have been in high-rise structures. Applications of composite columns in low-rise structures in the United States are scant because of the perceived poor cost-to-benefit ratio. For SRC construction this is due to the need to have several construction trades on site and to the expense of forming reinforcing cages in situ. For CFT construction most problems arise from connections and fire protection. Many of these objections can

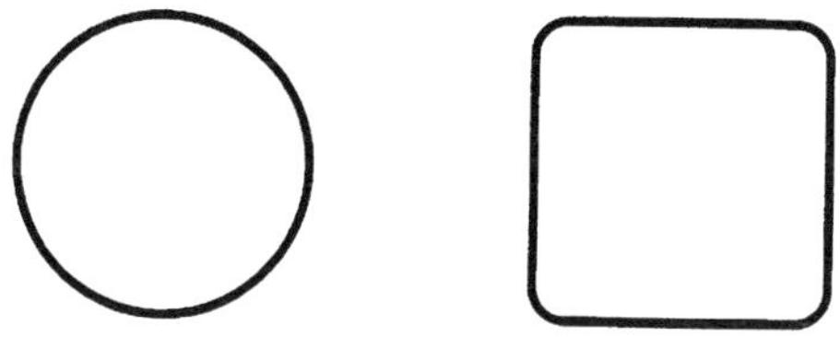

FIGURE 10.2 Typical cross sections for concrete-filled tubes.

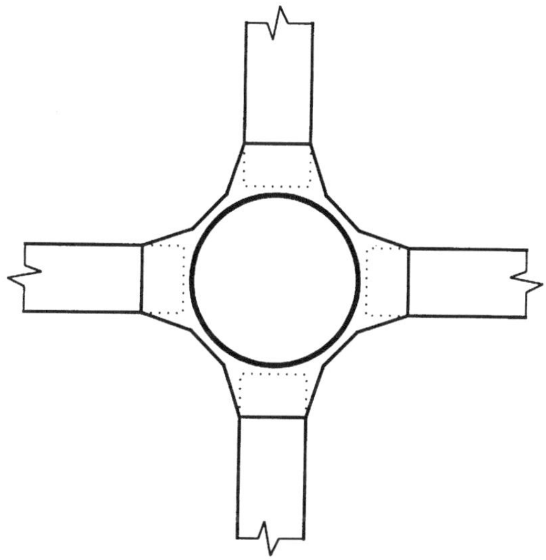

FIGURE 10.3 Bidirectional moment connection to a concrete-filled tube utilizing flared stiffeners.

be overcome with prefabrication and use of newer technologies (blind bolts for connections, for example). Moreover, as design provisions evolve and new design recommendations are adopted, it is likely that more extensive use of composite columns and other forms of composite construction will be made in the near future.

The advantages of composite construction were recognized early in the twentieth century (Talbot and Lord, 1912), and most multistory buildings were built with composite columns for fire protection until lightweight, sprayed fireproofing became available. Recently, composite members and systems are becoming popular again, primarily because of their stiffness characteristics (Griffis, 1992) and seismic resistance (Goel and Yamanuchi, 1993). Two excellent complementary references on composite construction for buildings have recently become available: one for fundamental mechanics issues (Oehlers and Bradford, 1995) and one for practical issues (Viest et al., 1996). Both provide more detailed discussions of many of the issues addressed in this chapter. In addition, the proceedings of several recent international conferences provide an up-to-date overview of recent experimental research and analytical advances (Roeder, 1985a; Buckner and Viest, 1988; Wakabayashi, 1991; Easterling and Roddis, 1993; Javor, 1994; Buckner and Shahrooz, 1997; Hajjar et al., 2002; Leon and Lange, 2006). No attempt is made in this chapter to summarize this extensive literature. For older research, on which many of the current design provisions are based, the third edition of this guide should be referenced. A complete history of the development of composite construction is provided by Viest et al. (1996, Chap. 1).

10.2 U.S.–JAPAN RESEARCH PROGRAM

More recently, the fifth phase of the U.S.–Japan cooperative earthquake research program focused on steel–concrete composite and hybrid structures and included several research projects on: (i) concrete-filled tube CFT systems, (ii) reinforced-concrete steel RCS systems, and (iii) hybrid wall systems. A special issue of the ASCE *Journal of Structural Engineering* (Volume 130, No. 2) was published which includes several articles summarizing the research findings from these projects. The editorial of this issue presents an overall summary of the articles. The issue also includes a forum discussing the goals and outcomes of the fifth phase of the U.S.–Japan program. This section summarizes some of the major findings from the U.S.–Japan program relevant to the stability perspective of this chapter on composite columns and systems.

10.2.1 Axially Loaded CFT Stub Columns

As part of the U.S.–Japan program, Sakino et al. (2004) conducted a comprehensive series of tests on 114 centrally loaded CFT *short* columns with a wide range of geometric and material parameters to: (i) establish a generally applicable design method for CFT columns, (ii) clarify the synergistic interactions between the steel tube and filled concrete, and (iii) characterize the load–deformation relationship of CFT columns. The parameters included in the tests were the tube shape (circular and square), steel yield strength (262 to 853 MPa), circular diameter-to-thickness ratio (17 to 152), square width-to-thickness ratio (18 to 74), and concrete strength (25 to 91 MPa).

The experimental results were used to develop the following equation for calculating the nominal axial strength P_o of circular CFT columns:

$$P_o = A_s \times 0.89\,\sigma_y + A_c \times \left(\gamma_u f_c' + 4.1 \times \frac{2t}{D - 2t} \times 0.19\,\sigma_y\right) \tag{10.1}$$

where A_s is the area of the steel tube, σ_y the steel yield strength, A_c the area of the concrete infill, f_c' the concrete compressive strength, D the tube diameter, and t the thickness. The strength reduction factor $\gamma_u = 1.67D^{-0.112}$ is used to account for the effects of size on the concrete compressive strength. The first term in Eq. 10.1 uses $0.89\sigma_y$ as the longitudinal stress capacity of the steel tube to account for the effects of tensile hoop stresses required to confine the concrete. The second term in the equation includes the effects of size and confinement on compression strength. Comparisons of experimental results with nominal strength predictions using Eq. 10.1 were quite favorable with a mean value of 0.989 and standard deviation of 0.052.

The experimental results were also used to develop an equation for calculating the nominal axial strength P_o of square CFT columns:

$$P_o = A_s \times \sigma_{cr} + A_c \times \gamma_u f_c' \tag{10.2}$$

where σ_{cr} is the longitudinal stress capacity of the square steel tube accounting for the effects of local buckling and is calculated from

$$\sigma_{cr} = \frac{\sigma_y}{0.698 + 0.128\left(\dfrac{b}{t}\right)^2 \dfrac{\sigma_y}{E} \times \dfrac{4.00}{6.97}} \leq \sigma_y \tag{10.3}$$

Equation 10.3 was developed using experimental results for hollow steel tube stub columns by modifying them to account for the change in the local buckling mode from hollow to CFT columns—hence, the term 4.00/6.97. The comparisons of experimental results with nominal strength predictions were quite favorable with a mean value of 1.032 and standard deviation of 0.058. These predictions were slightly conservative for columns with small width-to-thickness b/t ratios, which is probably due to the strain hardening of the thicker steel tubes.

10.2.2 Effective Stress–Strain Curves in Compression

Sakino et al. (2004) used the experimentally measured axial load–displacement responses to develop *effective* stress–strain curves for the steel tube and concrete infill of the CFT column in *compression*. These *effective* stress–strain curves implicitly accounted for the effects of steel tube local buckling and the transverse interaction between the steel tube and the concrete infill producing tensile hoop stresses in the steel tube and confinement of the concrete infill. They are quite comprehensive and applicable to CFT columns with the wide range of material and geometric parameters included in the experimental investigations.

The *effective* stress–strain curve for the concrete infill of circular and square CFT columns in compression is given by

$$Y = \frac{VX + (W - 1)X^2}{1 + (V - 2)X + WX^2} \tag{10.4}$$

in which $X = \varepsilon_c/\varepsilon_{cco}$, $Y = \sigma_c/\sigma_{cco}$, $V = E_c\varepsilon_c/\sigma_{cco}$, $W = 1.50 - 0.017\ f_c + 2.39\sqrt{\sigma_h}$, where $E_c = 6900 + 3320\sqrt{f_c}$ in megapascals, $\varepsilon_{co} = 0.94 \times (f_c)^{0.25} \times 10^{-3}$, and $f_c = f_c' \times 1.67D^{-0.112}$.

The parameters required to completely define the stress–strain behavior are included in Table 10.1, where f_c' is the concrete cylinder strength, f_c the concrete

TABLE 10.1 Parameters for CFT Concrete Stress–Strain Model in Compression

Parameters	Circular Columns	Square Columns
$\dfrac{\sigma_h}{\sigma_y}$	$\dfrac{4.1}{23} \times \left(\dfrac{2t \times 0.19}{D - 2t}\right)$	$\dfrac{2t^2(b-t)}{b^3}$
$\dfrac{\sigma_{cco}}{f_c}$	$1 + \dfrac{4.1}{f_c} \times \left(\dfrac{2t \times 0.19\sigma_y}{D - 2t}\right)$	1.0
$\dfrac{\varepsilon_{cco}}{\varepsilon_{co}}$	$1.0 + 4.7 \times \left(\dfrac{\sigma_{cco}}{f_c} - 1\right)$ for $\dfrac{\sigma_{cco}}{f_c} \leq 1.5$	1.0
	$3.4 + 20 \times \left(\dfrac{\sigma_{cco}}{f_c} - 1\right)$ for $\dfrac{\sigma_{cco}}{f_c} > 1.5$	

compressive strength accounting for scale effects, ε_{co} the strain at f_c, E_c the elastic modulus, σ_{cco} the confined concrete strength, ε_{cco} the strain at σ_{cco}, and σ_h the hoop stress in the steel tube.

Figure 10.4 shows examples of effective stress–strain curves developed using Eq. 10.4. Figure 10.4*a* shows effective stress–strain curves for circular CFTs with (i) f_c' equal to 55 MPa, (ii) D equal to 305 mm, (iii) D/t ratios equal to 50, 75, and 150, and (iv) σ_y equal to 317 and 552 MPa. Similarly, Fig. 10.4*b* shows effective stress–strain curves for square CFTs with (i) f_c' equal to 55 MPa, (ii) b equal to 305 mm, (iii) b/t ratios equal to 32, 48, and 64, and (iv) σ_y equal to 317 and 552 MPa. As shown, confinement improves the strain ductility of the concrete infill of square CFT columns, and it increases the strength and strain ductility of the concrete infill of circular CFT columns.

Sakino et al. (2004) proposed *elastic–plastic effective* stress–strain curves for the steel tubes of circular CFT columns in compression. The effective yield stress was equal to $0.89\sigma_y$. This reduction accounted for the effects of tensile hoop stresses σ_h required to provide confinement to the concrete infill. Figure 10.5 shows the effective stress–strain curves proposed for the steel tubes of square CFT columns in compression. The tubes are categorized into three types (also referred to as ranks) depending on the wall normalized slenderness ratio α_s, which can be calculated from

$$\alpha_s = \left(\frac{b}{t}\right)^2 \times \frac{\sigma_y}{E} \tag{10.5}$$

As shown in Fig. 10.5, type 1 or rank FA steel tubes undergo yielding and strain hardening before local buckling. Type 2 or rank FB steel tubes undergo local buckling at the yield stress, and type 3 or rank FC steel tubes undergo elastic local buckling before the yield stress is reached. After local buckling, all

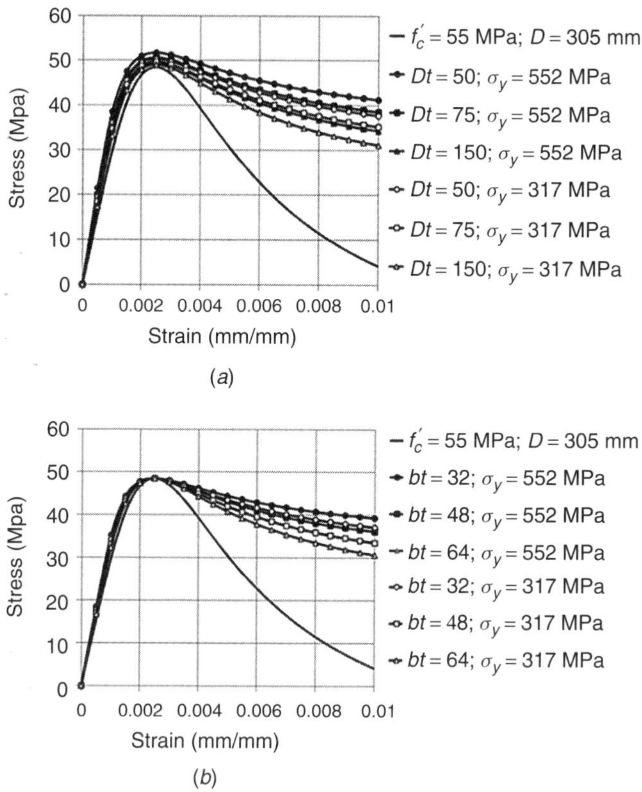

FIGURE 10.4 Concrete *effective* stress–strain curves for: (*a*) circular CFTs; (*b*) square CFTs.

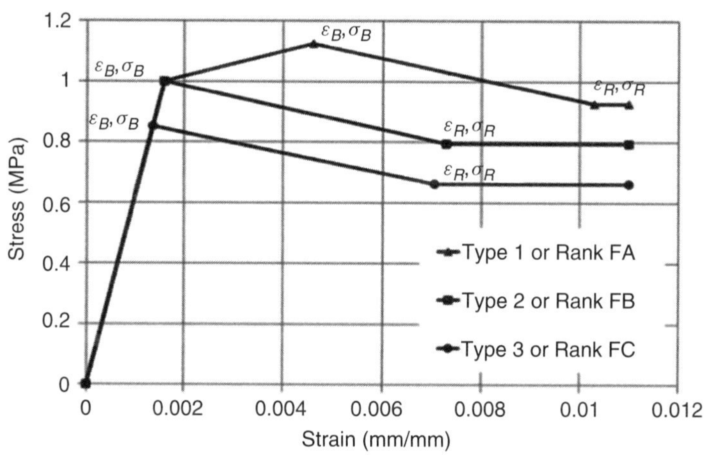

FIGURE 10.5 Effective stress–strain curves for steel tubes of square CFTs.

TABLE 10.2 Parameters for Steel Stress–Strain Model in Compression

Parameter	Type 1 or Rank FA $\alpha_s \le 2.37$	Type 2 or Rank FB $2.37 \le \alpha_s \le 4.12$	Type 3 or Rank FC $\alpha_s \le 4.12$
σ_B	$\dfrac{\sigma_y}{0.698 + 0.128\alpha_s}$	σ_y	$\dfrac{\sigma_y}{0.698 + 0.128\alpha_s \times \dfrac{4.00}{6.97}}$
ε_B	$\dfrac{\sigma_y}{E} \times \left(\dfrac{6.06}{\alpha_s^2} - \dfrac{0.801}{\alpha_s} + 1.10\right)$	$\dfrac{\sigma_y}{E}$	$\dfrac{\sigma_B}{E}$
σ_R	$1.19 - 0.207\sqrt{\alpha_s}$	$1.19 - 0.207\sqrt{\alpha_s}$	$1.19 - 0.207\sqrt{\alpha_s}$
ε_R	$3.59\dfrac{\sigma_y}{E} + \varepsilon_B$	$4.59\ \dfrac{\sigma_y}{E}$	$3.59\dfrac{\sigma_y}{E} + \varepsilon_B$

three types achieve some residual-stress capacity that remains constant with increasing strains. Table 10.2 summarizes the parameters required to define these multilinear effective stress–strain curves shown in Fig. 10.5 for all three types (or ranks) of steel tubes. In Table 10.2, σ_B and ε_B are the stress and strain corresponding to the onset of local buckling, and σ_R and ε_R are the residual stress and corresponding strain postlocal buckling.

Sakino et al. (2004) showed that the axial load–shortening responses predicted using these concrete and steel effective stress–strain curves in compression compared favorably with the experimental results for circular and square CFT columns for the wide range of material and geometric parameters included in the investigations.

10.2.3 Moment–Curvature Behavior of CFT Beam-Columns

As part of the U.S.–Japan program, Fujimoto et al. (2004) conducted extensive tests on eccentrically loaded circular and square CFT beam-columns to determine their fundamental moment–curvature (M–ϕ) behavior and to evaluate the effects of various material, geometric, and loading parameters. A total of 65 CFT beam-column specimens, including 33 circular and 32 square CFT specimens, were tested. The parameters varied within the experimental investigations and included: (i) steel tube yield stress (283 to 835 MPa), (ii) compressive strength of the concrete (20 to 80 MPa), (iii) diameter-to-thickness ratio of circular tubes (17 to 152), (iv) width-to-thickness ratio of square tubes (19 to 74), and (v) the axial load level (15 to 60% of axial load capacity P_o). The experimental results included the specimen M–ϕ relationships and estimates of moment capacity and ductility.

The experimental results indicated that: (i) using high-strength concrete reduces the ductility of CFT beam-columns, but this can be countered by using high-strength steel tubes or tubes with smaller b/t or D/t ratios; (ii) moment capacity enhancement due to concrete confinement by the steel tube is likely for circular CFTs with D/t ratios less than 75 and it is negligible for circular CFTs with D/t greater than 75 and for square CFT columns; and (iii) the effects of steel tube local buckling should be included when estimating the moment capacity of square CFTs with large (type 3 or rank FC) b/t ratios.

Fujimoto et al. (2004) also developed fiber models to predict the M–ϕ relationships of the tested specimens. These fiber models used the effective stress–strain curves in compression developed by Sakino et al. (2004) for steel and concrete. As described earlier, these effective stress–strain curves implicitly accounted for the effects of tube local buckling, concrete confinement, and scale effects. In tension, the concrete fibers were assumed to have zero stress capacity, and the steel fibers were assumed to have bilinear stress–strain behavior. The yield stress in tension was assumed to be equal to 1.08 and 1.10 times the nominal steel yield stress for circular and square steel tubes, respectively. The fiber models were found to predict the M–ϕ behavior and the moment capacities of the tested specimens with reasonable accuracy.

As part of the U.S.–Japan program, Varma et al. (2000, 2002, 2004, 2005) conducted experimental and analytical investigations to determine the behavior of high-strength square CFT columns and to evaluate the effects of various material, geometric, and loading parameters on their stiffness, strength, and ductility. The parameters included in the experimental investigations were the steel tube yield stress (266 to 630 MPa), concrete strength (110 MPa), tube b/t ratio (32 to 48), and axial load level for beam-columns (10 to 40% of the section axial load capacity P_o). Four CFT stub columns were tested under pure axial compression, eight CFT beam-columns were tested under constant axial loading (20 or 40% of P_o) and monotonically increasing flexural loading, and eight CFT beam-columns were tested under constant axial loading (10 to 30% of P_o) and cyclically increasing flexural deformations. The experimental results included the axial load–displacement responses, and the monotonic and cyclic axial force–moment–curvature (P–M–ϕ) responses of the high-strength CFT beam-columns. These were used to determine the axial stiffness, axial load capacity, flexural stiffness, moment capacity, and monotonic and cyclic ductility of high-strength CFTs.

The experimental results indicated that the axial stiffness can be predicted using the transformed section properties of the CFT section. The axial load capacity can be predicted with reasonable accuracy by superimposing the yield strength of the steel tube (A_sF_y) with 85% of the compressive strength of the concrete infill ($A_cf'_c$). The initial flexural stiffness of CFT beam-columns can be predicted using uncracked transformed section properties. The secant flexural stiffness corresponding to 60% of the moment capacity can be predicted using cracked transformed section properties. The deformation ductility of high-strength CFT beam-columns decreases significantly with increases in the applied axial force level and the steel tube b/t ratio. The steel tube yield stress does not seem to have a significant influence on the ductility. The moment capacity of high-strength CFT beam-columns can be predicted conservatively using ACI (2005) 318-05 recommendations and more accurately using the modified Architectural Institute of Japan (AIJ) method developed by Sakino et al. (2004). The Eurocode (CEN, 2004) provisions overpredicted the moment capacity of high-strength CFT beam-columns.

10.2.4 Modeling the Behavior of CFT Members

In addition to the U.S.–Japan research, Hajjar et al. (1997, 1998; Hajjar and Gourley, 1996, 1997; Tort and Hajjar, 2007) have conducted significant research on the behavior of CFT columns and beam-columns. The focus of Hajjar's research has been to develop and calibrate *macro*–finite elements that can be used to predict and model the behavior of CFT members with a wide range of geometric and material parameters and for various loading conditions. Their approach has been to first compile comprehensive databases of experimental research on CFT members (columns and beam-columns) conducted around the world. Then these databases, including extensive details of the experimental approach, observed behavior, and measured results, are used to develop, calibrate, and further validate macro–finite-element models.

For example, Gourley et al. (1995) compiled a comprehensive database of experimental research conducted on CFT members through 1995. Hajjar et al. (1997; Hajjar and Gourley, 1996, 1997) used this database to develop, calibrate, and validate a concentrated-plasticity-based three-dimensional cyclic nonlinear macro–finite element model for square CFT columns. This model consisted of a 12-degree-of-freedom elastic beam finite element with concentrated plastic hinges at the ends. The cyclic behavior of the concentrated plastic hinges was modeled using a two-surface bounding surface model in threee-dimensional stress-resultant space (axial load plus major and minor axis bending moments). The bounding surfaces were developed by fitting polynomial equations to the cross-section strengths of CFT columns that were determined from extensive fiber analyses. The fiber analyses used elastic–plastic stress–strain curves for the steel fiber and the effective stress–strain curves recommended by Tomii and Sakino (1979a,b) for the concrete fibers. Strength and stiffness degradation due to cyclic loading were modeled by kinematic hardening and isotropic hardening or softening of the bounding surfaces in stress-resultant space. The element was calibrated using experimental results reported by Sakino and Tomii (1981), Bridge (1976), Tomii and Sakino (1979a), Cederwall et al. (1990), and Shakir-Khalil (1991b). The calibrated element was limited to CFTs made from conventional strength materials.

More recently, an updated version of the database (v4.0) developed by Gourley et al. (2008) was published by Hajjar's research group. This is one of the most comprehensive databases of experimental research conducted on CFT members, connections, and systems available and includes the U.S.–Japan research. The databases include relevant details of the experimental approach, testing matrix, observed and measured limit states, and experimental results for major experimental programs. Tort and Hajjar (2007) used the results and limit states to develop a mixed finite element formulation for CFT members. The formulation included comprehensive cyclic constitutive models that accounted for confinement, concrete cracking and crushing, steel yielding and local buckling, and slip between the concrete and steel. It was developed and calibrated for CFT members made using conventional or high-strength materials. They also implemented a comparable steel finite element, where the formulation is suitable for both braced and unbraced frames with steel girders and steel or CFT braces framing into steel or CFT columns. The formulation is implemented in OpenSEES, which is an open-source finite element analysis program available from the website http://opensees.berkeley.edu. One to three elements have been found adequate for modeling members with highly nonlinear curvature and high axial compression. Tort and Hajjar (2007) used the finite element program to conduct parametric studies of damage limit states and to develop a framework for reliability-based, performance-based design of CFT structures.

10.3 CROSS-SECTIONAL STRENGTH OF COMPOSITE SECTIONS

The flexural strength M at any given axial load P can be calculated by assuming a position of the neutral axis, drawing the stress distributions, and summing their moments about the plastic neutral axis. This locus of points is shown as solid lines in Fig. 10.6, which schematically illustrates the ultimate strength of SRCs and CFTs. The lines correspond to the theoretically "exact" solution when nonlinear constitutive models are used for both steel and concrete, and the effect of confinement is appropriately modeled. To compute the capacities, a process analogous to that for any reinforced-concrete beam-column, in which the reinforcing bars are transformed into an equivalent thin steel section, can be used. Although this method is both tedious and computationally intensive, computer programs based on subdividing the cross section into small elements (finite elements of fiber models) are available for both SRC and CFT sections (Mahin and Bertero, 1977; Gourley and Hajjar, 1994; El-Tawil et al., 1995; Gourley et al., 1995). Many commercial reinforced-concrete computer design packages include some variation of the Mahin and Bertero (1977) approach as an option. In general, all these models assume *strain continuity* between the steel and concrete portions, an assumption that is not supported by much of the data from experimental programs in which the interface behavior was monitored. This assumption, however, which simplifies the problem considerably, appears to have negligible influence on the ultimate strength and a relatively small influence on the stiffness of the cross section.

The parabolic shape of the interaction curves is similar for SRCs and CFTs (Fig. 10.6). For SRCs, the difference in flexural capacity between the balance point (point D) and the pure flexure case (point B) decreases as the amount of structural steel and longitudinal steel reinforcement increase. Increases in the yield strength of the structural steel, for example from an A36 to an A572 material, significantly increases the capacity of the section at both points B and D. As one would expect, increases in concrete strength do not seem to have a major effect on the flexural strength at point B because steel yielding governs the strength, but they tend to increase the distance between points B and D in the interaction diagram. For most CFTs, the shape in Fig. 10.6 is preserved because in the limit the CFT section can be

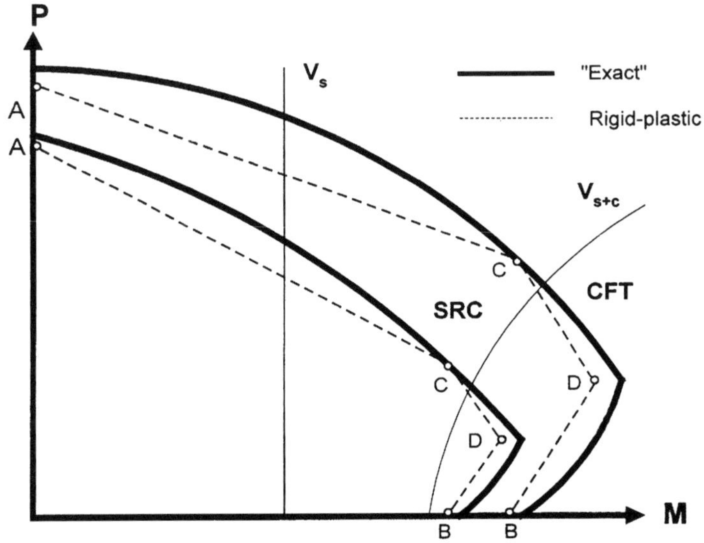

FIGURE 10.6 Interaction curve for a composite section.

modeled as a reinforced-concrete section with distributed steel. For circular CFTs, however, the balance point tends to lie higher in the axial load axis than that of a comparable SRC or rectangular CFT section. In addition, for CFTs the difference in moment capacity between the balance and no axial load points tends to be higher as the concrete strength increases and/or the tube slenderness increases because the addition of axial load increases the contribution of the concrete compression block. These differences are illustrated schematically by the theoretically "exact" CFT and exact SRC curves in Fig. 10.6. It should also be noted that for CFTs the effect of the wall slenderness (D/t ratio) has a significant impact on the postpeak strength as the axial load level increases (Bridge and Webb, 1993). While the presence of the infill concrete tends to increase the resistance to local buckling by a factor of up to 1.5 above that of hollow sections (Matsui and Tsuda, 1987), failures controlled by concrete crushing tend to be very brittle because the beneficial effect of the confinement provided by the steel cannot be maintained. For square hollow sections, Bridge and O'Shea (1996) found that the concrete infill provided restraint to inward local buckling that enhanced the tube strength over that for a hollow bare steel tube and that the enhancement could be taken into account using steel design specifications and codes that allowed for the increase in the elastic local buckling coefficient associated with the change in buckling mode. Contrary to some research, O'Shea and Bridge (1996) found that the concrete infill did not enhance the axial strength of circular tubes as the buckling mode was predominantly outward.

Most analytical studies show that the differences between the results of this type of exact approach and those given by a simplified theory using rigid-plastic stress blocks such as those shown in Fig. 10.7 are small and insignificant for design (Roik and Bergmann, 1992). This is shown schematically in Fig. 10.6. The axial and

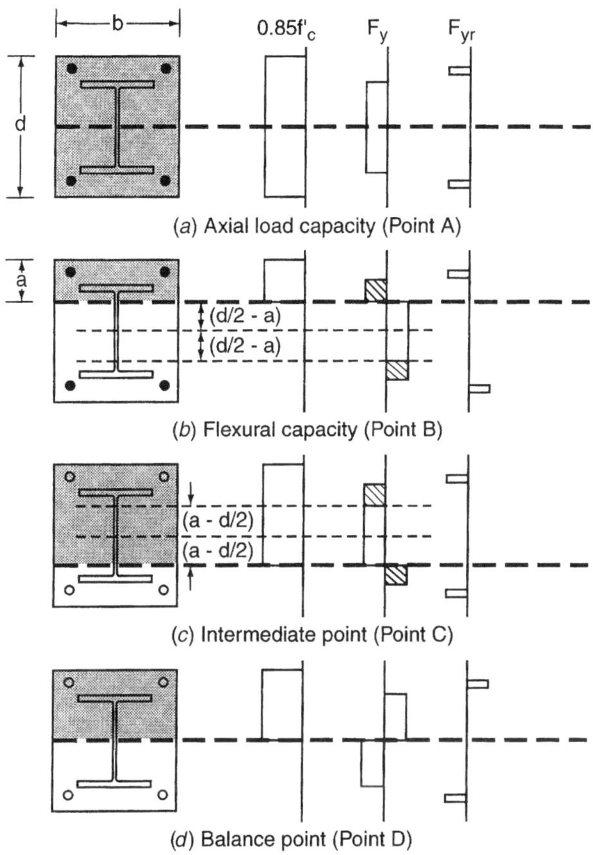

FIGURE 10.7 Stress distributions for key points on the interaction diagram.

flexural strengths are determined by assuming nominal yield strengths for the steel and 0.85 of the cylinder strength for the concrete. For encased shapes, Fig. 10.7 shows the plastic stress distribution for several of the points in the interaction surface shown in Fig. 10.6. The axial strength P_u, shown in Fig. 10.7a, is given by

$$P_u = 0.85\, f_c'(db - A_s - A_r) + A_s F_y + A_r F_{yr} \tag{10.6}$$

where f_c' is the concrete cylinder strength, d the overall depth, b the width, F_y the yield strength of the steel shape, F_{yr} the yield strength of longitudinal reinforcement, A_s the area of the steel shape, and A_r the area of longitudinal reinforcement.

A simple and elegant solution for other important points in the interaction surface for the rigid-plastic case can be found by following the procedure proposed by Roik and Bergmann (1989, 1992). Consider the typical case of an encased shape bent about its major axis and having only four bars as longitudinal reinforcement. Assuming that the neutral axis lies in the web of the steel beam, the plastic stress distributions for points B through D in Fig. 10.6 correspond to the stress distributions in Figs. 10.7b through d. Points B and C correspond, respectively, to the case of no axial load and to an apparently arbitrary point in the interaction diagram above the balance point. In fact, points B and C correspond to the same moment because the stress blocks lying within the distance h_n in both Figs. 10.7b and c have their centroid at the plastic neutral axis and thus do not contribute to the plastic bending capacity of the section. In addition, it needs to be recognized from both Figs. 10.7b and c that the axial forces from the reinforcement and the shaded portions of the forces from the steel shape cancel out.

Adding the stress distributions in Figs. 10.7b and c and considering axial loads only, the total axial load P_{conc} would still be that at point C. From superimposing the stress blocks, this axial load will be that given by the concrete section alone under a uniform stress of $0.85\,f_c'$ because the contributions from the steel shape within the distance h_n also cancel out, resulting in

$$P_{\text{conc}} = 0.85\,f_c'(bd - A_s - A_r) \tag{10.7}$$

The depth of the compression block a can then be calculated from Fig. 10.7b by assuming that the compressive force in the concrete is equal to the tensile force in the web of the steel shape within the distance $2h_n$, which leads to

$$a = \frac{t_w d F_y}{0.85\,f_c' b + 2t_w F_y} \tag{10.8}$$

A direct calculation for h_n, however, can be made by subtracting the stress distribution in Fig. 10.7b from that in Fig, 10.7c and again considering only axial forces. In this case, all stress blocks except those inside $2h_n$ disappear. Within this distance $2h_n$ the concrete will have a stress of $0.85\,f_c'$ while the steel will have a stress of $2F_y$. Because the total axial load is still P_{conc}, h_n is given by

$$h_n = \frac{P_{\text{conc}}}{2[0.85\,f_c' b + t_w(2F_y - 0.85\,f_c')]} \tag{10.9}$$

Knowing either h_n or a, the moment capacity of the section for both points B and C in Fig. 10.6 can easily be calculated.

The final point that needs to be defined is the balance point (point D). The maximum moment will be obtained when the neutral axis is at the centroid of the cross section (Fig. 10.7d), because in this case all the forces are additive with respect to moment. From Fig. 10.7d, it is clear that all contributions to the axial load from the steel shape and reinforcement cancel out and thus the axial load at this point corresponds to that of $0.85f_c'$ acting over half of the cross section, or $P_{\text{conc}}/2$. The balance moment M_{bal} is given by

$$M_{\text{bal}} = Z_x F_y + 0.5\,bd^2\,f_c' + \sum_{j=1}^{n} A_{ri} d_{ri} \tag{10.10}$$

where Z_x is the plastic section modulus, A_{ri} the area of any rebar, and d_{ri} its distance to the plastic neutral axis.

Equations 10.6 through 10.10 permit a very quick and accurate calculation for the key points in the interaction curve. Although Fig. 10.7 and Eqs. 10.6 through 10.10 provide a solution for only the simplest case, the above approach is general. It also applies for cases where the neutral axis is either in the flange of the steel shape or outside the steel shape and for any number and bar-type distribution. Equations for bending about both axes and for concrete-filled shapes are given by Roik and Bergmann (1992) in a much more general format. The two general approaches discussed above also apply for the case of biaxial bending. The exact approach only differs in that the assumed inclination of the neutral axis changes with respect to the principal axes of the cross section (El-Tawil et al., 1993). The validity of this approach has been verified experimentally (Morino et al., 1988, 1993) and analytically (Virdi and Dowling, 1973; Gourley and Hajjar, 1994). Although a rigid-plastic approach is also possible in this case, it is not possible to derive simple, general equations for this case. A simplified approach utilizing the rigid-plastic capacities computed for both principal axes has been proposed by Roik and Bergmann (1992) and adopted for the Eurocode (CEN, 2004).

10.4 OTHER CONSIDERATIONS FOR CROSS-SECTIONAL STRENGTH

The discussion presented in Section 10.3 is limited to composite cross sections made from conventional strength materials, that is, steel yield stress $F_y \leq 75$ ksi (525 MPa) and concrete compressive strength f_c' 10 ksi (69 MPa). The use of higher strength steels and concrete materials in CFT members has been studied extensively as part of the U.S.–Japan research program (Section 10.2). Using the results from these studies, Sakino et al. (2004) developed and validated a *modified AIJ* (MAIJ) method for calculating the cross-sectional strength of square CFT members. This method can be used for CFTs made from both conventional or high-strength materials because it accounts for the effects of steel tube local buckling (depending on wall slenderness) and the concrete compression blocks have been calibrated for both conventional and high-strength concrete.

The MAIJ method calculates the cross-sectional strength of CFT cross sections using the steel and concrete stress blocks shown in Fig. 10.8. As indicated in the figure, the method assumes strain compatibility and complete plastification of the composite cross section, which is similar to the rigid-plastic approach described in Section 10.3. The compressive stress capacity of steel is reduced to SF_y to account for the effects of local buckling. The reduction factor S, which is a function of the tube slenderness α_s (Eq. 10.5), is given by

$$S = \frac{1}{0.698 + 0.128 \times \dfrac{4.00}{6.97} \times \alpha_s} \leq 1.0 \qquad (10.11)$$

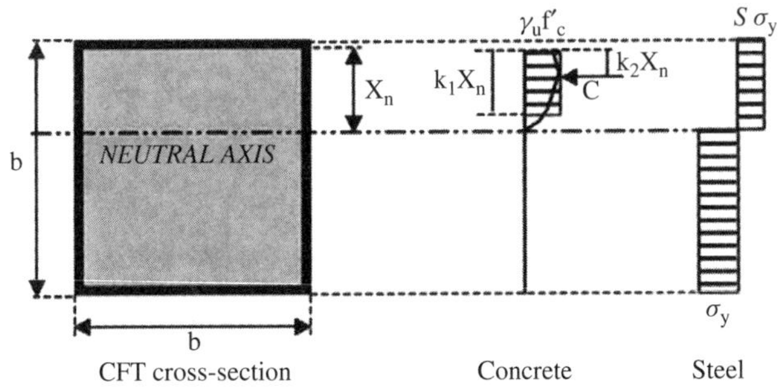

FIGURE 10.8 Modified AIJ method for calculating cross-sectional strength.

For values of α_s less than 4.12, there will be no reduction in the compressive stress capacity of the steel.

The concrete compression block parameters (k_1 and k_2) shown in Figure 10.8 were calibrated for both conventional and high-strength concrete (up to 120 MPa). The values of these parameters can be calculated from

$$k_1 = 0.831 - 0.076(\gamma_u f'_c/41.2) \geq 0.65 \tag{10.12a}$$

$$k_2 = 0.429 - 0.010(\gamma_u f'_c/41.2) \tag{10.12b}$$

in which the concrete compressive strength f'_c is in megapascals, and γ_u is the scale factor discussed earlier. The cross-sectional strength for any combination of axial load P and bending moment M can be obtained by establishing force equilibrium using the stress blocks shown in Fig. 10.8 and Eqs. 10.11 through 10.12.

The effect of transverse reinforcement on the concrete strength is typically ignored when calculating the cross-sectional strength, but designers should be conscious that transverse reinforcement will have a major effect on postpeak behavior. This is of particular importance under seismic loading, where strength and stiffness degradation will be influenced by the amount and distribution of transverse reinforcement. Designers should also understand that the confining effect decreases with increasing concrete strength because the amount of microcracking and resulting dilatatory behavior decreases with concrete strength (see Fig. 10.4*a*). In addition, the effect of transverse steel is dependent on its spacing and presence of crossties for the case of encased sections and on the wall thickness for the case of concrete-filled tubes (sees Fig. 10.4*b*). O'Shea and Bridge (1996) have found that for tube diameter-to-thickness ratios greater than 55 and concrete strengths in the range 110 to 120 MPa, the steel tube provides virtually no confinement to the concrete when both the steel and concrete are loaded together. Confinement effects could be obtained only if the concrete was loaded and the steel was not bonded to the concrete. Unbonded tube construction has been considered by Orito et al. (1988).

The effect of shear on the ultimate axial and flexural strength depends primarily on the shear span M/V ratio, the amount of axial load, and the detailing of the shear reinforcement. For a cross section with a large axial load and high shear span, it would appear that the shear resistance of the concrete and steel should be additive. Conversely, for a lightly loaded column with a low shear span, where a substantial portion of the concrete cross section may not be effective due to flexural or diagonal cracking, it would seem prudent to limit the shear strength to that of the steel section and the shear reinforcement. The differences in capacity computed from either set of assumptions are very large. This is illustrated schematically in Fig. 10.6, where the flexural capacity of any cross section can be limited by the capacity in shear given by either the steel alone V_s or the combined capacity V_{s+c}. In preparing this plot, it was assumed that the column was in double curvature and that the ultimate moment capacity could be reached at either end ($V_s = 2M_{\text{ult}}/L = A_w F_{yw}$). For V_{s+c}, it should be noted that the concrete capacity is dependent on the axial load, and thus the function is nonlinear.

The rigid-plastic stress distributions discussed in Section 10.3 are consistent with an assumption of uniform shear throughout the web. To calculate the ultimate strength properly, the web yield stress $F_{y,\text{web}}$ should be reduced to

$$F_{y,\text{web}} = \sqrt{F_y - 2\tau_w^2} \tag{10.13a}$$

$$\tau_w = \frac{V}{dt_w} \tag{10.13b}$$

where V is the shear at the cross section d in the depth of the steel member and t_w is the thickness of the web of the steel section. Note that shear will be the dominant failure mode in many columns used in seismic areas, in which case careful detailing for shear transfer is needed.

10.5 LENGTH EFFECTS

In principle, the design of a composite column for stability should be no different from that for a reinforced-oncrete or steel column. Two interrelated problems, however, one practical and one philosophical, arise when considering a composite section. The practical problem centers on how to compute an effective moment of inertia of the member for stability and drift calculations. This process is not straightforward because it is difficult to characterize the amount of cracking in the concrete. This cracking arises both from the type of loading and the long-term behavior of the concrete.

Insofar as loading is concerned, the amount of cracking that can occur is a function of the level of axial load, how the loads are introduced into the column (i.e., connection details and sequence of construction), and the tensile capacity of the concrete.

On the other hand, long-term effects such as creep and shrinkage are viscoelastic, time-dependent processes (Bradford and Gilbert, 1990; Oehlers and Bradford, 1995). Creep behavior is influenced primarily by the level of sustained axial load, the age at loading, and material properties (type of cement, water/cement ratio, and aggregate characteristics). Shrinkage is influenced by both material properties and curing conditions (temperature, humidity). In general, these phenomena are important at the service level but do not play a major role at ultimate loading (Bridge, 1979, 1988; Nakai et al., 1991; Leon and Bawa, 1990). Creep and shrinkage can be important at both the member and system levels. An example of the latter is the forces that arise as the result of differential creep and shrinkage in high-rise structures that incorporate a perimeter frame and internal core walls. In practice, many of the problems associated with creep and shrinkage can be mitigated by proper material selection and careful construction sequences. At the design stage, however, many of these details are not known, and thus highly simplified and conservative approaches need to be taken. Most codes resort to assuming empirically that only a portion of the transformed area is effective when the ultimate strength is reached.

The difficulty in assessing stability arises from the different approaches adopted by codes for reinforced concrete and steel (Deierlein and Leon, 1996). The stability provisions for reinforced-concrete structures (MacGregor, 1993) are based on calculating member forces using a second-order analysis in which the beams and columns are assigned reduced stiffnesses to model the frame behavior as the applied load approaches the structural stability limit point. The resulting forces therefore account for inelastic second-order effects directly, and the calculated member forces are compared to the beam-column cross-sectional strengths. On the other hand, the steel design procedures (ASCE Task Group, 1997) use a second-order elastic analysis combined with an axial force–moment interaction equation to account for geometric and material nonlinear behavior. The interaction equation accounts for member and frame stability either through the use of an effective buckling length (KL, with $K > 1$) or by a direct-analysis procedure that permits $K=1$ given that additional factors are accounted for including frame out-of-plumbness and partial yielding accentuated by the presence of residual stresses. Regardless of the details, the approaches differ fundamentally in terms of (i) the assumptions used conducting the second-order analysis and (ii) the method of checking the member strength based on the cross-sectional capacity versus the member-buckling capacity. In U.S. design practice, there are two additional complicating factors that arise from code differences. The first is that there is no set of consistent load factors for steel and concrete except in model codes for seismic design [Federal Emergency Management Agency (FEMA) 450, 2003]. The second is that there is no consistent methodology for handling the very common case of hybrid structural systems. Further research is needed to resolve these issues.

10.6 FORCE TRANSFER BETWEEN CONCRETE AND STEEL

Interaction between the steel and concrete portions of composite members results from a combination of chemical adhesion, friction, and mechanical interlock (mostly bearing). The most dependable composite action arises from the use of mechanical shear connectors (generally headed shear studs) that transfer the forces between the two materials by direct bearing and shear. From the strength standpoint, the studs need to be designed to transfer all the shear forces at the interface between the steel and concrete that are consistent with the development of the plastic capacity of the cross section. This is typically defined as a full-strength shear connection. When fewer studs than these are provided, the system is said to be partial strength.

Although the concepts of full and partial strength are useful to describe the behavior of members such as composite beams where the two materials are in contact at a small and well-defined boundary, their application to composite columns is not as straightforward. Studies have shown that the ultimate strength given by a rigid-plastic approach is usually achieved by short composite columns under monotonic loading irrespective of whether or not mechanical force transfer is provided. Thus full strength is not as meaningful a design parameter in composite columns as it is in beams, except for the case where fatigue or seismic loadings govern the design. In such cases, mechanical shear connectors should be provided because adhesion and friction are not reliable force transfer mechanisms under these types of loads.

Full and partial strength are different concepts from full and partial interaction. Full interaction implies continuity of strains and curvature across the steel concrete boundaries. Because most shear connectors provide a nonlinear shear strength–slip behavior (i.e., some slip is needed before the resistance builds up) full interaction cannot be achieved in practice even at service load levels. In most practical cases, the achievement of ultimate strength in composite systems requires substantial slip at the steel-to-concrete interface, resulting in severe discontinuities in the strain profile between the steel and concrete. Because most computer programs developed to calculate ultimate cross-sectional capacity rely on an assumption of strain compatibility between the materials, large discrepancies would be expected between their predictions and available experimental results. This, however, is not the case because at ultimate capacity, when strain discontinuities do exist, their effect on the stresses is small because both materials are on fairly flat portions of their respective stress–strain curves. A powerful argument for proposing the use of rigid-plastic stress distributions for calculating composite member strength is that they circumvent the need to account for these discontinuities in interface strains. As long as the strains are larger than both the yield strain for the steel (F_y/E) and the strain consistent with the attainment of the maximum uniaxial strength for the concrete (about 0.002 to 0.003), the stress distributions in Fig. 10.7 can be achieved. The use of rigid-plastic stress distributions also obviates the need to design for the perpendicular forces that develop at the interface if the steel and concrete are assumed to remain in contact. These forces are small except for areas adjacent to large concentrated loads (Robinson and Naraine, 1988) and can easily be handled by the horizontal projections (heads) of most mechanical connectors.

In both axially loaded specimens and flexural specimens, test results can be interpreted to imply that chemical adhesion provides a substantial contribution to the shear transfer in the service load range (Roeder, 1985b; Wium and Lebet, 1990a,b). Quantification of this effect is impossible given both the difficulties in measuring this property and the large scatter that can be expected from varying surface conditions, loading history, casting position, size effects, and concrete mix proportions, to name a few of the relevant variables.

The longitudinal bond between the steel tube and the concrete infill of a CFT column has been investigated by several researchers, including Virdi and Dowling (1980), Shakir-Khalil (1991a, 1993a,b), Morishita and Tomii (1982), and Roeder et al. (1999). These studies indicate that the bond strength increases with the roughness of the steel at the steel–concrete interface. The bond strength does not change when shear connectors are used, and the shear connectors contribute to the load-carrying mechanism only after slip has occurred. The bond strength for square CFTs is smaller than the bond strength for circular CFTs, and the concrete compressive strength does not seem to have a consistent effect on the bond strength.

Roeder et al. (1999) compiled numerous bond strength test results from previous researchers, including 104 circular and 49 square CFT bond tests, and combined them with their own experimental results. These studies indicate that the bond strength decreases dramatically as the width b or width-to-thickness ratio increases. The width b or diameter d of the CFT column has a significant influence on the shrinkage of the concrete infill, which makes it an important parameter for bond strength. A linear regression analysis was performed using all of the experimental data, and the results were used to develop the following expression for average maximum bond strength ($f_{2\sigma}$) in megapascals:

$$f_{2\sigma} = 2.109 - 0.026(d/t) \tag{10.14}$$

with 97.5% of the specimens having an average maximum bond strength greater that that predicted by Eq. 10.14.

Equation 10.14 predicts larger bond stresses for stocky tubes with smaller d/t ratio and smaller bond stresses for thinner tubes with larger d/t ratio. At the ultimate load, the bond stress can be assumed to be uniformly distributed around the periphery of the interface and along a length equal to the lesser of the column length or 3.5 times the diameter of the tube, as shown in Fig. 10.9. Due to bond deterioration considerations under cyclic loading after initial slip at the serviceability load, the bond stress can be assumed to be triangularly distributed over a length equal to one-half of the tube diameter due to bond deterioration considerations under cyclic loading after initial slip (see Fig. 10.9). The bond stress evaluation is needed for axial load transfer; bending moment develops binding action that enhances the local bond stress capacity. Because shear connectors do not work well with natural bond stress, axial load transfer should be accomplished entirely by bond stress or other mechanical connectors.

At ultimate capacity and in the absence of mechanical shear connectors and adhesion, all forces must be transferred by friction. The normal forces necessary for friction arise primarily from shear stresses and the differential expansion of the two materials under load. Shear stresses arise from moment gradients and the assumption of no vertical separation between the steel and concrete components (i.e., the assumption of equal curvatures in the steel and concrete portions). The

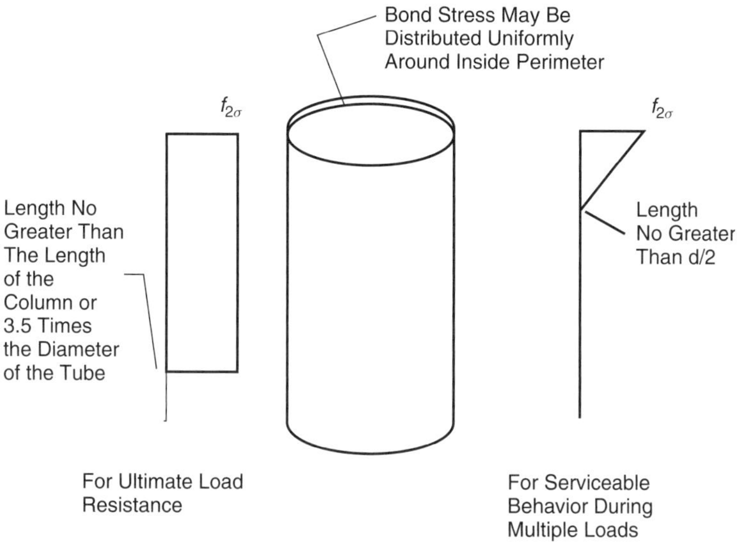

FIGURE 10.9 Bond Stress distribution models (Roeder et al., 1999).

differential expansion results from different values of Poisson's ratio (ν) and is dependent on the level of stress and the mix proportions. This effect is very different for SRCs and CFTs. At low levels of stress, the steel expands more than the concrete ($\nu \approx 0.3$ for steel and $\nu \approx 0.15$ to 0.20 for concrete in the elastic range). This does not result in any appreciable development of frictional forces unless the entire load is introduced directly to the steel shape. As the concrete stresses increase over $0.5f'_c$, the dilatational behavior of the concrete begins to take over as microcracking progresses and the apparent ν of the concrete increases over that of the steel. In SRCs the confinement effect provided by ties is insignificant because the amount of transverse reinforcement is volumetrically small and cross ties are typically not used. In CFTs, on the other hand, the expansion of the concrete is controlled by the steel section. For round tubes this results in large hoop stresses and the development of a very efficient confinement effect. This increases the nominal crushing strength of the concrete and helps maintain the strength in the postpeak region of the stress–strain curve. The beneficial effects of the encasement are present even in extreme cases when no shear connection is present and adhesion has been prevented with the use of lubricants (Orito et al., 1988). The hoop stresses, on the other hand, result in biaxial state of stress in the tube wall that can lead to early yielding or buckling. For rectangular and square CFTs, the effect of confinement will be smaller because of the ineffectiveness of this type of cross section in developing hoop stresses. The amount of friction that can be developed depends primarily on the surface conditions, degree of compaction of the concrete, and any longitudinal out-of-straightness. While experimental results for CFTs show this effect clearly (Virdi and Dowling, 1980), test results for SRCs are somewhat inconclusive in this regard (Bryson and Mathey, 1973; Dobruszkes and Piraprez, 1981).

It should be clear from the above discussion that the force transfer between the steel and concrete portions of a composite member is a complex phenomenon. At this time, there appear to be no mechanistic models that adequately address the numerous interactions described above, and the experimental data needed to calibrate such models are not available. It is clear from the vast majority of the experimental results, however, that a model that incorporates a sophisticated interface force transfer mechanism is not necessary to accurately predict the ultimate strength of a composite cross section. Rigid-plastic models such as the one described above or finite elements with fiber models that assume no slip between the concrete and steel predict the ultimate strength equally well provided that the strains achieved in the model are large.

10.7 DESIGN APPROACHES

Over the years, a large number of design approaches have been proposed for composite columns. Good summaries of these can be found in Furlong (1968, 1974, 1983, 1988), Basu and Sommerville (1969), Virdi and Dowling (1976), Roberts and Yam (1983), Roik and Bergmann (1985), Shakir-Khalil (1988), Bradford (1995), and Varma et al. (2002, 2004). In the United States the design of composite columns by the AISC (2005) specification is based on the work of Aho and Leon (1997). The Eurocode 4 (CEN, 2004) procedure is based primarily on the work of Roik and Bergmann (1992), based on calibrations by Roik and Bergmann (1989) and Johnson (1997). The Japanese provisions are based on the work of Sakino et al. (2004).

10.7.1 AISC Composite Column Design

The AISC provisions recommend two methods for determining the nominal strength of composite sections: (1) the *plastic stress distribution method* and (2) the *strain compatibility method*. The tensile strength of concrete is not included in the determination of nominal strength. For the plastic stress distribution method, the nominal strength is computed assuming that the steel components reach the yield stress of F_y in tension or compression, and the concrete components in compression reach a stress of $0.85f_c'$. For round CFTs, the concrete components can achieve up to $0.95f_c'$; with the increase due to confinement effects. For the strain compatibility method, a linear distribution of strains across the section is assumed. The maximum compressive strain is equal to 0.003 in./in., and the stress–strain relationships can be taken from tests or published results. The AISC provisions specify that for the determination of nominal strength, the compressive strength of normal-weight concrete is limited to values greater than 3 ksi (21 MPa) and less than 10 ksi (70 MPa). The compressive strength of lightweight concrete is limited to values greater than 3 ksi (21 MPa) and less than 6 ksi (42 MPa). The specificed minimum yield stress of structural steel and reinforcing bars is limited to 75 ksi (525 MPa).

For encased composite columns, the cross-sectional area of the steel core has to be greater than or equal to 1% of the total composite section area. The concrete encasement of the steel core has to be reinforced with continuous longitudinal bars and lateral ties or spirals. The minimum *transverse* reinforcement area is 0.009 in.^{2}/in.(6 mm^2/mm) of tie spacing. The minimum reinforcement ratio for continuous longitudinal reinforcing (area of reinforcement/gross section area) is 0.4%. For CFT columns, the cross-section area of the steel tube section has to be greater than 1% of the total composite area. The maximum b/t ratio for the steel tube of a square CFT is limited to $2.26\sqrt{E/F_y}$. The maximum D/t ratio for the steel tube of a round CFT is limited to $0.15E/F_y$.

The nominal compressive strength P_n for axially loaded composite columns is determined for the limit state of flexural buckling according to

$$P_n = P_o\left[0.658^{(P_o/P_e)}\right] \qquad \text{for } P_e \geq 0.44P_o \tag{10.15}$$

$$P_n = 0.877P_e \qquad \text{for } P_e < 0.44P_o \tag{10.16}$$

where $P_e = \pi^2\, EI_{\text{eff}}/KL^2$ and the parameters P_o and EI_{eff} are defined in Table 10.3 with

A_s = area of steel shape, in.^{2}

A_c = area of concrete, in.^{2}

A_{sr} = area of longitudinal reinforcing bars, in.^{2}

E = modulus of elasticity of steel, ksi

E_c = modulus of elasticity of concrete $= w_c^{1.5}\sqrt{f'_c}$, ksi

w_c = unit weight of concrete, lb/ft^3

f'_c = specified compressive strength of concrete, ksi

F_y = specified minimum yield stress of the steel section, ksi

F_{yr} = specified minimum yield stress of the longitudinal reinforcing bars, ksi

I_s = moment of inertia of steel shape, in.^{4}

I_c = moment of inertia of concrete section, in.^{4}

I_{sr} = moment of inertia of longitudinal reinforcing bars, in.^{4}

The P–M interaction curve for the axial force–moment capacity of the composite cross section can be developed using the rigid-plastic approach described in Section 10.3. The "exact" P–M interaction curve can be approximated using straight lines to connect points B, D, C, and A, as identified in Fig. 10.6. The axial forces and moment capacities corresponding to these points can be calculated using the stress block distributions and equations shown in Fig. 10.7. The cross-section

TABLE 10.3 Parameters for AISC LRFD Column Equations

For SRC Columns	For CFT Columns
$P_o = A_sF_y + A_{sr}F_{yr} + 0.85\,A_cf'_c$	$P_o = A_sF_y + A_{sr}F_{yr} + C_2\,A_cf'_c$ $C_2 = 0.85$ for rectangular and 0.95 for circular
$EI_{\text{eff}} = E_sI_s + 0.5\,E_sI_{sr} + C_IE_cI_c$	$EI_{\text{eff}} = E_sI_s + E_sI_{sr} + C_3E_cI_c$
$C_1 = 0.1 + 2\left(\dfrac{A_s}{A_c + A_s}\right) \leq 0.3$	$C_3 = 0.6 + 2\left(\dfrac{A_s}{A_c + A_s}\right) \leq 0.9$

P–M curve may be further simplified to a bilinear interaction curve defined by two lines connecting points B, C, and A.

This bilinear P–M interaction curve can be modified to account for length effects, as shown in Fig. 10.10. The section axial capacity corresponding to point A is reduced to A_λ to account for length effects using Eqs. 10.15 and 10.16. For design by LRFD, point $A_\lambda$would be multiplied by the resistance factor ϕ_c for composite columns to reduce it to A_D. Similarly, flexural capacity corresponding to point B is multiplied by the bending resistance factor ϕ_b for composite beams to reduce it to B_D. Point C is adjusted by the same length effect reduction factor applied to point A to obtain C_λ. For LRFD, point C_λ is then adjusted downward by multiplying it by ϕ_c and to the left by multiplying it by ϕ_b to obtain the point C_D. The LRFD composite member P–M interaction curve accounting for length effects can then be approximated using straight lines joining B_D, C_D, and A_D, as shown in Fig. 10.10.

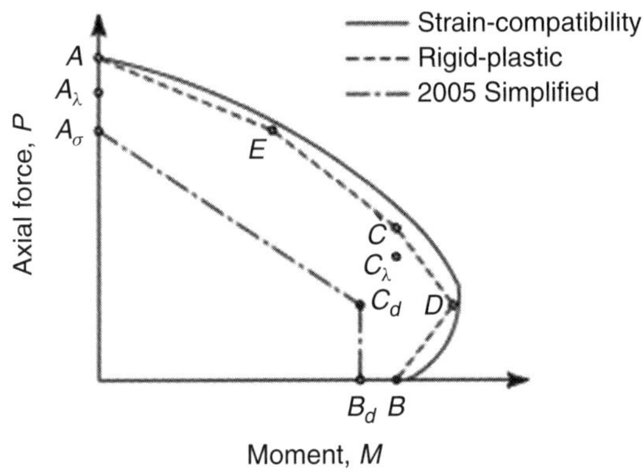

FIGURE 10.10 Axial force–moment interaction accounting for length effects.

10.7.2 Eurocode 4 (CEN, 2004)

The Eurocode approach to designing composite columns is based on the strength design method. Starting with a squash load calculated by combining the resistance of the steel, concrete, and reinforcement, this strength is then modified for column slenderness as necessary. Rather than using an overall resistance ϕ factor as AISC does, the Eurocode employs partial safety factors during the course of the design calculations. The squash load $N_{pl,rd}$ is calculated by

$$N_{pl,rd} = A_s \frac{F_y}{\gamma_s} + A_c \frac{0.85 f_c'}{\gamma_c} + A_r \frac{F_{yr}}{\gamma_r} \quad \text{for encased shapes} \tag{10.17}$$

$$N_{pl,rd} = A_s \frac{F_y}{\gamma_s} + A_c \frac{f_c'}{\gamma_c} + A_r \frac{F_{yr}}{\gamma_r} \quad \text{for concrete-filled tubes} \tag{10.18}$$

$$N_{pl,rd} = A_s \frac{F_y}{\gamma_s} \eta_2 + A_c \frac{f_c'}{\gamma_c} \left(1 + \eta_1 \frac{t}{d} \frac{F_y}{f_c'}\right) + A_r \frac{F_{yr}}{\gamma_r} \quad \text{for concrete-filled circular tubes} \tag{10.19}$$

where A_s = area of steel
A_c = area of concrete
A_r = area of longitudinal reinforcing bars
F_y = specified minimum yield stress of the steel shape, pipe, or tube
F_{yr} = specified minimum yield stress of the longitudinal reinforcing bars
f_c' = specified compressive strength of concrete
γ_s = partial safety factor for the structural steel = 1.1
γ_c = partial safety factor for the concrete = 1.5
γ_r = partial safety factor for the reinforcing steel = 1.15

Note that the 0.85 appearing in Eq. 10.17 can be increased to 1.0 for concrete-filled tubes with $\lambda < 0.5$, where λ is defined below.

The benefit of good confinement conditions can only be taken into account for concrete-filled circular tubes (Eq. 10.19) and if the column is relatively stocky ($\lambda < 0.5$). The coefficients η_1 and η_2, which account for concrete confinement, are functions of the slenderness and the ratio of the eccentricity e of the axial load to the tube diameter d (which for concentrically loaded columns $e/d = 0$) and are defined by

$$\eta_1 = \eta_{10} \left(1 - 10\frac{e}{d}\right) \geq 0.0 \tag{10.20}$$

$$\eta_2 = \eta_{20} + (1 - \eta_{20})\, 10\frac{e}{d} \leq 1.0 \tag{10.21}$$

where

$$\eta_{10} = 4.9 - 18.5\lambda + 17\lambda^2 \geq 0 \tag{10.22}$$

$$\eta_2 = 0.25\,(3 + 2\lambda) \leq 1.0 \tag{10.23}$$

Length effects are handled by the eurocode with a different approach than in U.S. codes. In the AISC specification, length effects for composite columns are accounted for by finding a reduced interaction diagram, similar to the approach employed for a noncomposite steel column (Fig. 10.10). In Eurocode, on the other hand, the slenderness parameter of the column is defined by

$$\lambda = \sqrt{\frac{A_s F_y + 0.85 A_c f_c' + A_r F_{yr}}{P_e}} \leq 2.0 \tag{10.24}$$

$$P_e = \frac{(EI)_{\text{eff}} \pi^2}{(KL)^2} = N_{cr} \tag{10.25}$$

where KL = effective length of the column
$(EI)_{\text{eff}} = E_s I_s + E_c I_c + E_r I_r$
E_s = modulus of elasticity of steel
E_c = modulus of elasticity of concrete $= 0.8E_{cm}/\gamma_c$, with $\gamma_c = 1.35$
E_{cm} = secant modulus of concrete
E_r = modulus of elasticity of reinforcing steel
I_s = moment of inertia of steel
I_c = moment of inertia of concrete (assumed uncracked)
I_r = moment of inertia of reinforcing steel

The Eurocode calculates the axial strength of a column by reducing its squash load $N_{pl,rd}$ by a factor κ that accounts for the slenderness of the column (Fig. 10.11); the governing design equation is

$$P_u \leq \kappa N_{pl,rd} \tag{10.26}$$

where P_u is the design axial load. The reduction factor κ is based on the European strut curves (Fig. 10.12) and is calculated by

$$\kappa = f_k - \sqrt{f_k^2 - \frac{1}{\lambda^2}} \leq 1.0 \tag{10.27}$$

with

$$f_k = \frac{1 - \alpha(\lambda - 0.2) + \lambda^2}{2\lambda^2} \tag{10.28}$$

and $\alpha = 0.21$ for concrete-filled circular and rectangular hollow sections (curve a); $\alpha = 0.34$ for completely or partly concrete-encased I-sections with bending about the major axis of the profile (curve b); and $\alpha = 0.49$ for completely or partly concrete-encased I-section with bending about the minor axis of the profile (curve c). Thus, the eurocode uses multiple-column curves specifically derived for various composite sections.

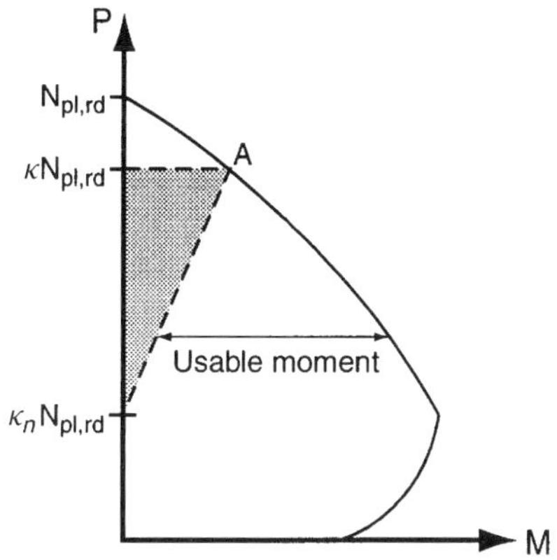

FIGURE 10.11 Eurocode approach for accounting for slenderness effects.

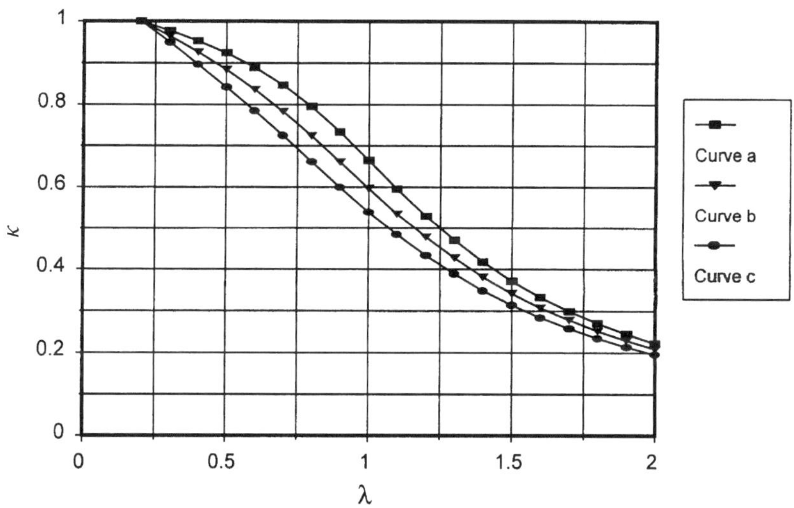

FIGURE 10.12 Strut curves used in Eurocodes.

The slenderness effects are then taken as the shaded region in Fig. 10.11, leaving the usable flexural strength as the distance between the line joining $\kappa_n N_{pl,rd}$ and point A in the failure envelope. Point $\kappa_n N_{pl,rd}$ accounts for the effects of variable end moments and is given as

$$\kappa_n = \kappa \frac{1 - r}{4} \quad \text{with} - 1 \leq r \leq 1 \tag{10.29}$$

where r is the ratio of the end moments $M_{\min}/M_{\max}$. For slender columns with small eccentricities the Eurocode also requires that the effect of creep be included in the calculations by modifying the modulus of elasticity of the concrete.

Second-order analysis should be used if the second-order moments are a significant (more than 10% of the first-order moments). The second-order moments can be obtained by multiplying the first-order moments by the amplification factor k, where k is defined as

$$k = \frac{\beta}{1 - N/N_{cr}} \geq 1.0 \tag{10.30}$$

and

$$\beta = \begin{cases} 1.0 \text{ for bending moments from lateral loads in isolated (pinned) columns} \\ 0.66 + 0.44r \geq 0.44 \end{cases}$$

The Eurocode also has checks for local buckling of the steel in compression members. The limits for local buckling are based on a depth-to-thickness ratio of the section. For rectangular hollow steel sections with h being the greater overall dimension of the section,

$$\frac{h}{t} \leq 52\varepsilon \tag{10.31}$$

for circular hollow steel sections,

$$\frac{d}{t} \leq 90\varepsilon^2 \tag{10.32}$$

and for partially encased I-sections,

$$\frac{b}{t_f} \leq 44\varepsilon \tag{10.33}$$

where ε is based on the yield strength of the steel and is defined as

$$\varepsilon = \sqrt{\frac{34.08}{F_y}} \tag{10.34}$$

with F_y in ksi. The 34.08 is the result of a conversion of units to ensure that ε is nondimensional. These calculations can be ignored for a section entirely encased in concrete, as the concrete is assumed to prevent local buckling from occurring.

10.8 STRUCTURAL SYSTEMS AND CONNECTIONS FOR COMPOSITE AND HYBRID STRUCTURES

The current International Code Council (ICC, 2005) code references FEMA (2003) 450, which recognizes the following different types of composite structural systems. Because many of them have a counterpart in steel and concrete, the prefix "C-" has been used to name the corresponding composite system.

Composite Partially Restrained Frames (C-PRF) C-PRFs consist of steel columns and composite beams joined by composite semirigid connections (Leon, 1990, 1994; Johnson and Hope-McGill, 1972; Kato and Tagami, 1985). In this case the connections utilize the slab and its reinforcement to provide negative-moment capacity and stiffness to the system.

Composite Ordinary Moment Frames (C-OMF) C-OMFs include a variety of configurations in which steel or composite beams are combined with steel, composite, or reinforced-concrete columns (Deierlein et al., 1989; Sheikh et al., 1989; ASCE Task Committee, 1994a). The term *ordinary* is used to indicate that little of the detailing required for critical structures is envisioned in this type of structure.

Composite Special Moment Frames (C-SMF) C-SMFs are similar to C-OMFs except that much more stringent detailing is required to provide behavior similar to that of a steel SMF (Minami, 1985). In this case, any composite columns are required to meet all AISC requirements for b/t and h/t ratios and to have all the transverse reinforcement required for columns by Chapter 21 of ACI (2005). As in most ductile frames, the columns and joints are required to develop the full strength of the beams and hence provide for the development of a stable strong column–weak beam mechanism.

Composite Concentrically Braced Frames (C-CBF) C-CBFs are similar to their steel counterparts except that some of the members (beams, columns, and braces) are composite. There is considerable debate on the applicability of braced frames in areas of high seismicity because the tendency of the braces to buckle results in poor energy dissipation characteristics, especially if the structure undergoes inelastic behavior. To alleviate the buckling problem, several researchers have proposed utilizing buckling restrained braces or composite braces (either encased shapes or concrete-filled tubes) where the stiffening effect of the concrete prevents local buckling (Liu and Goel, 1988).

Composite Eccentrically Braced Frames (C-EBF) When the eccentrically braced frame concept was originally developed, there was some concern as to whether the floor beams, which are in effect composite beams, could accommodate the large rotational ductilities demanded by the system without causing local failures. Extensive research has been carried out in this area, indicating that the floor elements are capable of withstanding the very large shear deformations required by short links (Ricles and Popov, 1989).

RC Walls Composite with Steel Elements At least three possible variations of this system exist, and they correspond to cases of hybrid structures. The first utilizes concrete panels as infills in steel or composite frames. The second is where large steel sections are used as boundary elements in concrete shear walls. The third one is where steel or encased composite beams are used to tie two reinforced-concrete shear walls (Shahrooz et al., 1993; Harries et al., 1993).

Steel Plate-Reinforced Shear Walls Since the early 1980s, the concept of utilizing steel plate shear walls has been popular (Thorburn et al., 1983). The concept is similar to the use of plate girders in bridges, except that the main element is vertical rather than horizontal. These systems have been used successfully as retrofits in critical steel structures (e.g., hospitals), where access to the structure was severely limited by the need to keep it operational during the retrofit construction. The system basically behaves as a CBF with the tension field action taking the lateral loads. Composite steel shear walls, in which the steel plate is covered with concrete and composite action activated by mechanical connectors, have been postulated as a system with better energy dissipation capacity. Another variation could be a sandwich configuration where the space between two thin steel plates containing studs is filled with concrete. In this case, the steel plates act as the formwork and could be welded directly to an existing steel frame. Great care is needed in connecting the plates to the boundary elements because the shear wall is such an efficient structural element that it can easily overstress the adjacent columns and beams.

The structural systems discussed above give a flavor of the many variations that can be developed by combining composite elements. The stability design of many of these systems has not been investigated extensively and thus the suggested design procedures are conservative (FEMA 450, 2003). The design of connections in composite construction has recently been reviewed by Leon and Zandonini (1992) and Gourley et al. (2008).

10.9 SUMMARY

Significant experimental and analytical research of composite columns has been conducted over the past few years by researchers from many countries. These experimental and analytical results have been used to develop comprehensive design guidelines for composite columns with a wide range of geometric and material properties including high-strength materials. Some of these design guidelines have been adopted in newer versions of design codes, while some of them are still being assessed. There are some philosophical and design questions regarding length effects, especially the behavior and design of long slender composite columns. In all regards, research into the stability of composite systems continues to provide for advances in the design of composite and hybrid structures.

REFERENCES

ACI (2005), *Building Code Requirements for Reinforced Concrete and Commentary*, ACI 318–05, American Concrete Institute, Detroit, MI.

Aho, M. F. (1996), "A Database for Encased and Concrete-Filled Columns," M.S. thesis, School of Civil and Environmental Engineering, Georgia Institute of Technology, Atlanta, GA.

Aho, M. F., and Leon, R. T. (1997), "A Database for Encased and Concrete-Filled Columns," Report No. 97-01, Georgia Institute of Technology, School of Civil and Environmental Engineering, Atlanta, GA.

AIJ (1987), *Structural Calculations of Steel Reinforced Concrete Structures*, Architectural Tokyo.

AISC (2005), *Load and Resistance Factor Design Specification for Structural Steel Buildings*, American Institute of Steel Construction, Chicago, IL.

ASCE Task Committee on Design Criteria for Composite Structures in Steel and Concrete (1994a), "Proposed Specification for Structural Steel Beams with Web Openings," *ASCE J. Struct. Eng.*, Vol. 118, No. 12, pp. 3315–3324.

ASCE Task Committee on Design Criteria for Composite Structures in Steel and Concrete (1994b), "Guidelines for Design of Joints between Steel Beams and Reinforced Concrete Columns," *ASCE J. Struct. Eng.*, Vol. 120, No. ST8, pp. 2330–2357.

ASCE Task Committee on Design Criteria for Composite Structures in Steel and Concrete (1996), "Proposed Specification and Commentary for Composite Joists and Trusses," *ASCE J. Struct. Div.*, Vol. 122, No. ST4, pp. 350–358.

ASCE Task Group on Effective Length (1997), *Effective Length and Notional Load Approaches for Assessing Frame Stability: Implications for American Steel Design*, Committee on Load and Resistance Factor Design, American Society of Civil Engineers, New York.

Basu, A. K., and Sommerville, W. (1969), "Derivation of Formulae for the Design of Rectangular Composite Columns," *Proc. Inst. Civ. Eng.*, Suppl. Vol., pp. 233–280.

Bradford, M. A. (1995), "Design Strength of Slender Concrete-Filled Rectangular Steel Tubes," *ACI Struct. J.*, Vol. 93, No. 2, pp. 229–235.

Bradford, M. A., and Gilbert, R. I. (1990), "Time-Dependent Analysis and Design of Composite Columns," *ASCE J. Struct. Eng.*, Vol. 116, No. ST2, pp. 3338–3357.

Bridge, R. Q. (1976), "Concrete Filled Steel Tubular Columns," Research Report No. R283, University of Sydney, School of Civil Engineering, Sydney, Australia.

Bridge, R. Q. (1979), "Composite Columns Under Sustained Loads," *ASCE J. Struct. Eng.*, Vol. 105, No. ST3, pp. 563–576.

Bridge, R. Q. (1988), "The Long-Term Behavior of Composite Columns," in *Composite Construction in Steel and Concrete* (Eds. D. Buckner and I. M. Viest), American Society of Civil Engineers, New York, pp. 460–471.

Bridge, R. Q., and O'Shea, M. D. (1996), "Local Buckling of Square Thin-Walled Steel Tubes Filled with Concrete," *Proc. 5th Int. Colloq.*, Structural Stability Research Council, Chicago, IL, pp. 63–72.

Bridge, R. Q., and Webb, J. (1993), "Thin Walled Circular Concrete Filled Steel Tubular Columns," in *Composite Construction in Steel and Concrete,* Vol. 2 (Eds. W. S. Easterling and W. M. K. Roddis), American Society of Civil Engineers, New York, pp. 634–649.

Bryson, J. O., and Mathey, R. G. (1973), "Surface Condition Effect on Bond Strength of Steel Beams Embedded in Concrete," *ACI J.*, Vol. 59, No. 3, pp. 39–46.

Buckner, C. D., and Shahrooz, B. (Eds.) (1997), *Composite Construction in Steel and Concrete*, Vol. 3, American Society of Civil Engineers, New York.

Buckner, C. D., and Viest, I. M. (Eds.) (1988), *Composite Construction in Steel and Concrete*, American Society of Civil Engineers, New York.

Cederwall, K., Engstrom, B., and Grauers, M. (1990), "High-Strength Concrete Used in Composite Columns," *Proc. 2nd Int. Symp. Utiliz. High-Strength Concrete* (Ed. W. T. Hester), Berkeley, CA, pp. 195–214.

CEN (2004), Eurocode 4: Design of Composite Steel and Concrete Structures, Part 1–2, General Rules and Rules for Buildings, Comité Europée de Normalisation (CEN), European Committee for Standardization, Brussels, Belgium.

Deierlein, G. G. (1995), "An Overview of the 1994 NEHRP Recommended Provisions for the Seismic Design of Composite Structures," *Proc. Struct. Congr. XIII* (Ed. M. Sanayaei), Vol. 2, pp. 1305–1308.

Deierlein, G. G., and Leon, R. T. (1996), "Design Criteria for Composite Steel–Concrete Structures: Current Status and Future Needs," in *ACI SP*: *Hybrid and Composite Structures* (Eds. B. Sharooz and G. Sabnis), American Concrete Institute, Farmington Hius, MI.

Deierlein, G. G., Sheikh, T. M., Yura, J. A., and Jirsa, J. O. (1989), "Beam-Column Moment Connections for Composite Frames, Part 2," *ASCE J. Struct. Eng.*, Vol. 115, No. ST11, pp. 2877–2896.

Dekker, N. W., Kemp, A. R., and Trinchero, P. (1995), "Factors Influencing the Strength of Continuous Composite Beams in Negative Bending," *J. Constr. Steel Res.*, Vol. 34, Nos. 2–3, pp. 161–186.

Dobruszkes, A., and Piraprez, E. (1981), "Détermination de l'Adhérence Naturelle Acierbéton dans le Cas de Poutres Mixtes Composées d'un Profilé Métallique Enrobe de Béton," Rep. MT 241, CRST, Brussels, Belgium.

Easterling, W. S., and Roddis, W. M. K. (Eds.) (1993), *Composite Construction in Steel and Concrete*, Vol. 2, American Society of Civil Engineers, New York.

El-Tawil, S., Sanz-Picon, C. F., and Deierlein, G. G. (1993), "Evaluation of ACI 318 and AISC (LRFD) Strength Provisions for Composite Beam-Columns," *J. Constr. Steel Res.*, Vol. 34, No. 1, pp. 103–123.

El-Tawil, S., Sanz-Picon, C. F., and Deierlein, G. G. (1995), "Evaluation of ACI 318 and AISC (LRFD) Strength Provisions for Composite Columns," *J. Constr. Steel Res.*, Vol. 34, pp. 103–123.

FEMA 450 (2003), *NEHRP Recommended Provisions and Commentary for Seismic Regulations for New Buildings and Other Structures.* Federal Engineering Management Agency, Washington, DC, 752 pp.

Fujimoto, T., Mukai, A., Makato, K., Tokiniya, H., and Fukumoto, T. (2004), "Behavior of Concrete-Filled Steel-Tube Beam Columns," *J. Struct. Eng. ASCE*, Vol. 130, No. 2, pp. 189–202.

Furlong, R. W. (1968), "Design of Steel Encased Concrete Beam Columns," *ASCE J. Struct. Div.*, Vol. 94, No. ST1, pp. 267–281.

Furlong, R. W. (1974), "Concrete Encased Steel Beam Columns: Design Tables," *ASCE J. Struct. Div.*, Vol. 100, No. ST9, pp. 1865–1883.

Furlong, R. W. (1983), "Column Rules of ACI, SSLC, and LRFD Compared," *ASCE J. Struct. Eng.*, Vol. 109, No. ST8, pp. 2375–2386.

Furlong, R. W. (1988), "Steel–Concrete Composite Columns: II," in *Steel–Concrete Composite Structures* (Ed. R. Narayanan), Elsevier Applied Sciences, New York, pp. 195–220.

Goel, S. C., and Yamanuchi, H. (Eds.) (1993), "Proceedings of the, 1992 U.S.–Japan Workshop on Composite and Hybrid Structures," Res. Rep. UMCEE 92-29, Department of Civil and Environmental Engineering, University of Michigan, Ann Arbor, MI.

Gourley, B. C. and Hajjar, J. F. (1994), "Cyclic Non-Linear Analysis of Concrete-Filled Beam Columns and Composite Frames," Rep. No. ST-94-3, Department of Civil Engineering, University of Minnesota, Minneapolis, MN.

Gourley, B. C., Hajjar, J. F., and Schiller, P. H. (1995), "A Synopsis of Studies of the Monotonic and Cyclic Behavior of Concrete-Filled Steel Tube Beam-Columns," Rep. No. ST-93–5, Department of Civil and Mineral Engineering, University of Minnesota, Minneapolis, MI.

Gourley, B. C., Tort, C., Denavit, M. D., Schiller, P. H., and Hajjar, J. F. (2008), "A Synopsis of Studies of the Monotonic and Cyclic Behavior of Concrete-Filled Steel Tube Beam-Columns," Report No. UILU-ENG-2008-1802, Newmark Structural Laboratory Report Series (ISSN, 1940-9826), Dept. of Civil and Env. Eng., Univ. of Illinois at Urbana-Champaign, Urbana, Il, Apr.

Griffis, L. G. (1986), "Some Design Considerations for Composite-Frame Structures," *AISC Eng. J.*, Vol. 23, No. 2, pp. 59–64.

Griffis, L. G. (1992), "Composite Frame Construction." in *Constructional Steel Design* (Eds. P. J. Dowling et al.), Elsevier Applied Science, London, pp. 523–553.

Hajjar, J. F., and Gourley, B. C. (1996), "Representation of Concrete-Filled Steel Tube Cross-Section Strength," *J. Struct. Eng. ASCE*, Vol. 122, No. 11, pp. 1327–1336.

Hajjar, J. F., and Gourley, B. C. (1997), "A Cyclic Nonlinear Model for Concrete-Filled Tubes. I. Formulation," *J. Struct. Eng. ASCE*, Vol. 123, No. 6, pp. 736–744.

Hajjar, J. F., Gourley, B. C., and Olson, M. C. (1997), "A Cyclic Nonlinear Model for Concrete-Filled Tubes. II. Verification," *J. Struct. Eng. ASCE*, Vol. 123, No. 6, pp. 745–754.

Hajjar, J. F., Hosain, M., Easterling, W. S., and Shahrooz, B. M. (2002). "Composite Construction in Steel and Concrete IV," *Proceedings from the Composite Construction in Steel and Concrete IV Conference,* May 30–June 2, 2000, Banff, Alberta, Canada, ASCE, Reston, VA.

Hajjar, J. F., Molodan, A., and Schiller, P. H. (1998), "A Distributed Plasticity Model for Cyclic Analysis of Concrete-Filled Steel Tube Beam-Columns and Composite Frames," *Eng. Struct.*, Vol. 20, Nos. 4–6, pp. 398–412.

Harries, K., Mitchell, D., Cook, W. D., and Redwood, R. G., (1993), "Seismic Response of Steel Beams Coupling Concrete Walls," *ASCE J. Struct. Eng.*, Vol. 119, No. 12, pp. 3611–3629.

ICC (2000), *International Building Code*, International Code Council, Inc., Falls Church, VA.

Javor, T. (Ed.) (1994), *Steel–Concrete Composite Structures, Proc. 3, EXPERTCENTRUM*, Bratislava, Slovakia.

Johnson, R. P. (1997), "Statistical Calibration of Safety Factors for Encased Concrete Columns," in *Composite Construction in Steel and Concrete*, Vol. 3 (Eds. D. Buckner and B. Shahrooz), American Society of Civil Engineers, New York.

Johnson, R. P., and Hope-McGill, M. (1972), "Semi-Rigid Joints in Composite Frames," Prelim. Rep., 9th Congr. IABSE, pp. 133–144.

Kato, B., and Tagami, J. (1985), "Beam-to-Column Connection of a Composite Structure," in *Composite and Mixed Construction* (Ed. C. Roeder), American Society of Civil Engineers, New York, pp. 205–214.

Leon, R. T. (1990), "Semirigid Composite Construction," *J. Constr. Steel Res.*, Vol. 15, No. 2, pp. 99–120.

Leon, R. T. (1994), "Composite Semi-Rigid Construction," *AISC Eng. J.*, Vol. 31, 2nd quarter, pp. 57–67.

Leon, R. T., and Bawa, S. (1990), "Performance of Large Composite Columns," in *Mixed Structures, Proc IABSE Symp.*, Brussels, Vol. 50, IABSE, Zurich, pp. 179–184.

Leon, R. T., and Lange, J. (2006). "Composite Construction in Steel and Concrete V," *Proceedings of the Fifth International Conference on Composite Construction in Steel and Concrete*, Kruger National Park, South Africa, July 18–23, 2004, ASCE, Reston, VA.

Leon, R. T., and Zandonini, R. (1992), "Composite Connections," in *Constructional Steel Design* (Eds. P. Dowling et al.), Elsevier Science Publishers, New York, pp. 501–520.

Liu, Z., and Goel, S. C. (1988), "Cyclic Load Behavior of Concrete-Filled Tubular Braces," *ASCE J. Struct. Eng.*, Vol. 114, No. 7, pp. 1488–1506.

MacGregor, J. G. (1993), "Design of Slender Concrete Columns—Revisited," *ACI Struct. J.*, Vol. 90, No. 3, pp. 302–309.

Mahin, S. A., and Bertero, V. V. (1977), *RCCOLA: A Computer Program for R. C. Column Analysis: Users' Manual and Documentation*, Department of Civil Engineering, University of California, Berkeley, CA.

Matsui, C., and Tsuda, K. (1987), "Strength and Behavior of Concrete-Filled Steel Square Tubular Columns with Large Width-Thickness Ratios," *Proc. Pac. Conf. Earthquake Eng.*, Vol. 2, Waimkei, New Zealand, pp. 1–9.

Minami, K. (1985), "Beam to Column Stress Transfer in Composite Structures," in *Composite and Mixed Construction* (Ed. C. Roeder), ASCE, New York, pp. 215–226.

Morino, S., Kawaguchi, J., Yazuzaki, C., and Kanazawa, S. (1993), "Behavior of Concrete-Filled Steel Tubular Three-Dimensional Subassemblages," in *Composite Construction in Steel and Concrete*, Vol. 2 (Eds. W. S. Easterling and W. M. K. Roddis), American Society of Civil Engineers, New York, pp. 726–741.

Morino, S., Uchida, Y., and Ozaki, M. (1988), "Experimental Study of the Behavior of SRC Beam-Columns Subjected to Biaxial Bending," in *Composite Construction in Steel and Concrete* (Eds. D. Buckner and I. M. Viest), American Society of Civil Engineers, New York, pp. 753–777.

Morishita, Y., and Tomii, M. (1982), "Experimental Studies on Bond Strength between Square Steel Tube and Encased Concrete Core Under Cyclic Shearing Force and Constant Loads," *Trans. Jpn. Concrete Inst.*, Vol. 4, pp. 115–122.

Nakai, H., Kurita, A., and Ichinose, L. H. (1991), "An Experimental Study on Creep of Concrete Filled Steel Pipes," *Proc. 3rd Int. Conf. Steel–Concrete Compos. Struct* (Ed. M. Wakabayashi), Fukuoka, Japan, pp. 55–60.

Oehlers, D. J., and Bradford, M. A. (1995), *Composite Steel and Concrete Structural Members: Fundamental Behavior*, Pergamon Press, Elmsford, NY.

Orito, Y., Sato, T., Tanaka, N., and Watanabe, Y. (1988), "Study of Unbonded Steel Tube Concrete Structure," in *Composite Construction in Steel and Concrete* (Eds. D. Buckner and I. M. Viest), American Society of Civil Engineers, New York, pp. 786–804.

O'Shea, M. D., and Bridge, R. Q. (1996), "Circular Thin-Walled Tubes with High Strength Concrete Infill," in *Composite Construction in Steel and Concrete*, Vol. 3 (Eds. D. Buckner and B. Shahrooz), American Society of Civil Engineers, New York.

Ricles, J. M., and Popov, E. P. (1989), "Composite Action in Eccentrically Braced Frames," *ASCE J. Struct. Eng.*, Vol. 115, No. 8, pp. 2046–2066.

Roberts, E. H., and Yam, L. C. P. (1983), "Some Recent Methods for the Design of Steel, Reinforced Concrete and Composite Steel–Concrete Columns in the UK," *ACI J.*, Vol. 80, No. 2, pp. 139–149.

Robinson, H., and Naraine, K. S. (1988), "Slip and Uplift Effects in Composite Beams," in *Composite Construction in Steel and Concrete* (Eds. D. Buckner and I. M. Viest), American Society of Civil Engineers, New York, pp. 487–497.

Roeder, C. W. (Ed.) (1985a), *Composite and Mixed Construction*, American Society of Civil Engineers, New York.

Roeder, C. W. (1985b), "Bond Stress of Embedded Steel Shapes in Concrete," in *Composite and Mixed Construction* (Ed. C. Roeder), ASCE, New York, pp. 227–240.

Roeder, C.W., Cameron, B., and Brown, C.B. (1999), "Composite Action in Concrete Filled Tubes," *J. Struct. Eng. ASCE*, Vol. 125, No. 5, pp. 477–484.

Roik, K., and Bergmann, R. (1985), "Composite Columns: Design and Examples for Construction," in *Composite and Mixed Construction* (Ed. C. Roeder), ASCE, New York, pp. 267–278.

Roik, K., and Bergmann, R. (1989), "Report on Eurocode 4: Clause 4.8 and 4.9—Composite Columns," EC4/6/89, Miniter für Raumordnung, Bauwesen, und Stadtgebau der Bundesrepublik Deutschland, RSII 1-6741028630, Bonn, Germany.

Roik, K., and Bergmann, R. (1992), "Composite Columns," in *Constructional Steel Design* (Eds. P. Dowling et al.), Elsevier Science Publishers, New York, pp. 443–470.

Sakino, K., Nakahara, H., Morino, S., and Nishiyama, I. (2004). "Behavior of Centrally Loaded Concrete-Filled Steel-Tube Short Columns," *J. Struct. Eng. ASCE*, Vol. 130, No. 2, pp. 180–188.

Sakino, K., and Tomii, M. (1981), "Hysteretic Behavior of Concrete Filled Square Steel Tubular Beam-Columns Failed in Flexure," *Trans. Jpn. Concrete Inst.*, Vol. 3, pp. 439–446.

Shahrooz, B. M., Remmetter, M. E., and Qin, F. (1993), "Seismic Design and Performance of Composite Shear Walls," *ASCE J. Struct. Eng.*, Vol. 119, No. ST11, pp. 3291–3309.

Shakir-Khalil, H. (1988), "Steel–Concrete Composite Columns: I," in *Steel–Concrete Composite Structures* (Ed. R. Narayanan), Elsevier Applied Sciences, New York, pp. 163–194.

Shakir-Khalil, H. (1991a), "Bond Strength in Concrete-Filled Steel Hollow Sections," in *International Conference on Steel and Aluminum Structures*, Elsevier Science, pp. 157–168.

Shakir-Khalil, H. (1991b), "Tests on Concrete-filled Hollow Section Columns," in *Proceedings of 3rd International Conference on Steel-Concrete Composite Structures* (Ed. M. Wakabayashi), Association for International Cooperation and Research in Steel-Concrete Composite Structures, Fukuoha, Japan, pp. 89–95.

Shakir-Khalil, H. (1993a), "Pushout Strength of Concrete-Filled Steel Hollow Sections," *Struct. Eng.*, Vol. 71, No. 3, pp. 230–233.

Shakir-Khalil, H. (1993b), "Resistance of Concrete-filled Steel Hollow Tubes to Pushout Forces," *Struct. Eng.*, Vol. 71, No. 3, pp. 234–243.

Sheikh, T. M., Deierlein, G. G., Yura, J. A., and Jirsa, J. O. (1989), "Beam-Column Moment Connections for Composite Frames, Part 1," *ASCE J. Struct. Div.*, Vol. 115, No. ST11, pp. 2858–2876.

Talbot, A. N., and Lord, A. R. (1912), "Tests of Columns: An Investigation of the Value of Concrete as Reinforcement for Structural Steel Columns," Bull. 56, University of Illinois Engineering Experiment Station, Champaign-Urbana, IL.

Thornburn, L. J., Kulak, G. L., and Montgomery, C. J. (1983), "Analysis of Steel Plate Shear Walls," Struct. Eng. Rep. No. 107, University of Alberta, Edmonton, Alberta, Canada.

Tomii, M., and Sakino, K. (1979a), "Experimental Studies on the Ultimate Moment of Concrete Filled Square Steel Tubular Beam-Columns," *Trans. Architect. Inst. Jpn.*, No. 275, pp. 55–63.

Tomii, M., and Sakino, K. (1979b), "Elasto-Plastic Behavior of Concrete Filled Square Steel Tubular Beam-Columns," *Trans. Architect. Inst. Jpn.*, No. 280, pp. 111–120.

Tort, C., and Hajjar, J. F. (2007). "Reliability-Based Performance-Based Design of Rectangular Concrete-Filled Steel Tube (RCFT) Members and Frames," Structural Engineering Report No. ST-07-1, Department of Civil Engineering, Institute of Technology, University of Minnesota, Minneapolis, MN.

Varma, A. H., Ricles, J. M., Sause, R., Hull, B., in and Lu, L. W. (2000), "An Experimental Evaluation of High Strength Square CFT Columns," in *ACI SP-196: Composite and Hybrid Systems*, (Eds. R. S. Aboutaha and J. M. Bracci), American Concrete Institute, Farmington Hills, MI, pp. 51–86.

Varma, A. H., Ricles, J. M, Sause, R., and Lu, L. W. (2002), "Experimental Behavior of High Strength Square Concrete Filled Steel Tube (CFT) Columns," *J. Struct. Eng. ASCE*, Vol. 128, No. 3, pp. 309–318.

Varma, A. H., Ricles, J. M., Sause, R., and Ream, A. (2004), "Seismic Behavior and Design of High Strength Square Concrete Filled Steel Tube Beam-Columns," *J. Struct Eng. ASCE*, Vol. 130, No. 2, pp. 169–179.

Varma, A. H., Sause, R., Ricles, J. M., and Qinggang, L. (2005), "Development and Validation of Fiber Models for High Strength Square CFT Beam-Columns," *ACI Struct. J.*, Vol. 102, No. 1, pp. 73–85.

Viest, I. M. Colaco, J. P., Furlong, R. W., Griffis, L. G., Leon, R. T., and Wyllie, L. A. (1996), *Composite Construction: Design for Buildings*, ASCE-McGraw Hill, New York.

Virdi, K. S., and Dowling, P. J. (1973), "The Ultimate Strength of Composite Columns in Biaxial Bending," *Proc. Inst. Civ. Eng.*, Part 2, Vol. 55, pp. 251–272.

Virdi, K. S., and Dowling, P. J. (1976), "A Unified Design Method for Composite Columns," IABSE Pub. 36, International Association for Bridge and Structural Engineering, IABSE, Zurich Switzerland, pp. 165–1843.

Virdi, K. S., and Dowling, P. J. (1980), "Bond Strength in Concrete-Filled Steel Tubes," IABSE Per. 3/1980. pp. 125–139

Wakabayashi, M. (Ed.) (1991), *Steel–Concrete Composite Structures*, AICRSCCS, Fukuoka, Japan.

Wium, J. A., and Lebet, J. P. (1990a), "Pushout Tests on Embedded Steel Plates," Publ. ICOM 240, Ecole Polytechnique Fédérale de Lausanne, Lausanne, Switzerland.

Wium, J. A., and Lebet, J. P. (1990b), "Test for the Application of Forces on Short Composite Columns," Publ. ICOM 241, Ecole Polytechnique Fédérale de Lausanne, Lausanne, Switzerland.

Yamanuchi, H., Mahin, S. A., Goel, S. C., and Nishiyama, I. (1993), "U.S.–Japan Cooperative Earthquake Research Program on Composite and Hybrid Structures," *Proc., 1992 U.S.-Japan Workshop Compos. Hybrid Struct.*, U.S.-Japan Cooperative Research Program.

CHAPTER 11

STABILITY OF ANGLE MEMBERS

11.1 INTRODUCTION

Single angles in compression are used in many applications, such as web members in steel joists and trusses, members of latticed transmission towers or communication structures, elements of built-up columns, and bracing members. Bolted or welded members made of single angles are common in industrial plants to support pipes, cable trays, and heating, ventilating, and air-conditioning systems.

Single angles are popular because they can easily be connected to other structural members. They are almost always connected eccentrically at their ends. In most practical applications, angles are loaded through one leg only, thus introducing eccentricity with respect to the centroid of the cross section. At the same time, restraining moments at the ends of the member are also present due to the flexural rigidity of the connected elements. In a rational design philosophy, therefore, the detrimental effect of the end eccentricity and the advantageous effect of the end restraint on the compression capacity of the angle member should be addressed. Furthermore, due to the asymmetry of the angle cross section, the determination of the compression capacity under eccentric loading along with end restraints is complex. The end eccentricities cause biaxial flexural deformation of the strut at any given load; thus the problem is that of load–deformation response.

In the design of open-web steel joists, eccentricity in the web members occurs when the centroid of the web cross section is not in the same plane as that of the chord members. One means of eliminating this eccentricity is "crimping" the end of the angle and rotating the member such that the web centroid lies within the centroidal plane of the chords. From a design context, it is advantageous to crimp the member ends and theoretically eliminate the bending stresses caused by eccentric loading. A crimped web member, however, does have a reduced minor axis bending stiffness at its ends, which may need to be included in the design process.

Double angles are also frequently used as members in trusses and as bracing members. The ease with which connections can be made contributes to the popularity of their use. Angles are often used as beams in walkways, racks, equipment support frames, pipe support systems, and other lightly loaded framing. Angles used in joists and trusses are often subjected to flexure from transverse loading.

In this chapter, single- and multiple-angle members subjected to compression and/or flexure are addressed. Experimental and analytical research and standard industry practice in the United States and abroad are also discussed.

11.2 REVIEW OF EXPERIMENTAL AND ANALYTICAL RESEARCH

The earliest tests on angles in compression in North America were performed by the National Bureau of Standards (1924). After a gap of 24 years, Foehl (1948) reported tests on seven single-angle steel columns. One leg at the end of each of these columns was welded to the stem of a structural tee section, thus simulating the arrangement in the web of a long-span joist. The main objectives of these tests were to determine the effective slenderness ratios of these columns and to ascertain which leg of an unequal-leg angle should be placed perpendicular to the plane of the truss. It was concluded that the long leg should be placed in this position. At Washington University theoretical and experimental investigations were carried out on single-angle columns under biaxially eccentric loading (Trahair, 1969; Usami, 1970; Usami and Galambos, 1971; Leigh and Galambos, 1972). The test specimens were representative of the web members used in standard long-span steel joists, with their ends welded to structural T-sections to simulate the chords of such joists. An analytical investigation was carried out by assuming that the column is made of an elastic–perfectly plastic material and representing the out-of-plane end restraint by an elastic–plastic rotational spring and the in-plane end restraint by an elastic rotational spring; good correlation between the theoretical and experimental results was obtained. Similar tests were conducted by Mengelkoch and Yura (2002). Usami and Fukumoto (1972) studied the behavior of angle members and found that the effect of residual stresses was not significant, a conclusion which agrees with the earlier finding of Ishida (1968). It was recommended that the maximum load of an eccentrically loaded angle bracing member can be taken as 58% of the maximum load of a corresponding concentrically loaded member.

The influence of end connections on the load-carrying capacity of web angle members was experimentally studied by Lorin and Cuille (1977). They found that increasing the yield strength of the gusset plate did not increase the load-carrying capacity of the member; doubling the thickness of the gusset plate from 0.4 to 0.8 in., however, increased the buckling load by approximately 40%.

El-Tayem and Goel (1986) reported on tests performed on cross-bracing systems made of single-angle members. Only one-half of the compression diagonal buckled about the minor principal axis. A theoretical model, including the end-restraint effect of the gusset connection and the restraint effect of the tension diagonal, was also presented to depict the results of the tests. The test results on the capacity of the compression diagonal was compared with the AISC specification (1989a) compression capacity using the minor principal axis slenderness ratio and one-half the length of the compression diagonal. The results indicate that the use of an effective-length factor K equal to 0.85 is acceptable without including the effect of the end eccentricity in the computation of the allowable loads.

Works by Culver (1966) and Dabrowski (1961) can be used to predict the elastic capacity of eccentrically connected single-angle struts. The effect of symmetrical end restraint (i.e., the same end restraint at both ends of the strut) was investigated by Trahair (1969) for the elastic case. Both equal- and unequal-leg angles were included in the study. The applicable equilibrium equations were solved using the finite integral method. The end condition considered is shown in Fig. 11.1. Line AB in this figure represents a knife-edge support; load P is assumed to be applied anywhere along line AB. The results of the analysis were compared with tests by Foehl (1948), and good agreement was found between the analysis and test results. It has to be noted that this investigation did not include the effects of initial deformations on the load-carrying capacity of the angle struts.

The results of tests conducted by Elgaaly et al. (1991, 1992) indicated that the design specifications for single-angle members (AISC, 1989a) can be conservative, while the ASCE (1988) design recommendations for transmission towers can be unconservative. Based on available test results, Adluri and Madugula (1992) reached the same conclusions with respect to the requirements of the AISC specifications.

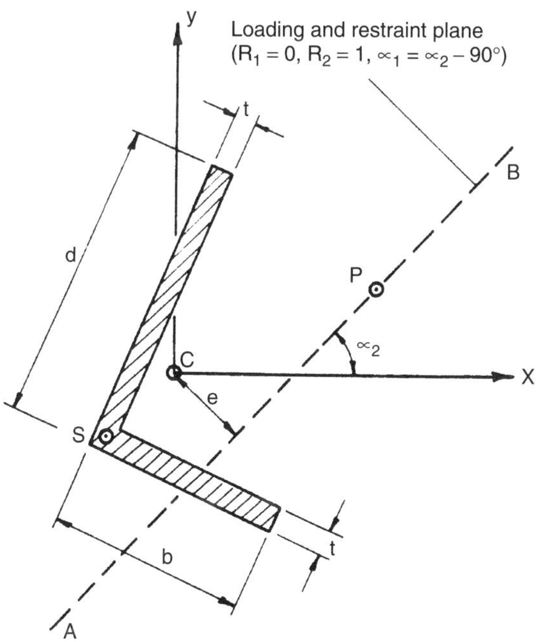

FIGURE 11.1 Eccentrically loaded single-angle strut (Trahair, 1969).

Finite element analysis provides a general tool for the determination of the load-carrying capacity of eccentrically connected angle struts. Two basic alternatives are available. The strut and its connections can be modeled using plate elements. This type of modeling has the advantage of including local plate buckling and cross-sectional distortion effects. The modeling and computational efforts involved in such an analysis, however, can be justified only for isolated strut problems. The second approach relies on the beam-column theory of the open cross section. Beam elements are used in this method, which permits the analysis of relatively large frame assemblies with minimum modeling and computation effort. Chunmei (1984) studied the inelastic buckling of single-angle struts using large deformation plate element analysis. The analysis results were in good agreement with test results.

A general finite beam element solution developed at Lehigh University (Hu et al., 1982; Lu et al., 1983) was used to predict the strength of unequal-leg angle struts loaded through gusset plates in the center of the outstanding leg. Figures 11.2 and 11.3 show the results from analysis of a $3 \times 2 \times \frac{1}{4}$ in. angle for two cases, with the short leg outstanding and with the long leg outstanding. A comparison of the results indicated that the long-leg-outstanding arrangement is not always the more favorable arrangement. In fact, higher ultimate capacity may be obtained with the short-leg-outstanding arrangement for relatively short members. The results also indicated that end restraint can significantly increase the strength of an eccentrically loaded single-angle strut. Kitipornchai and Chan (1987) presented a finite

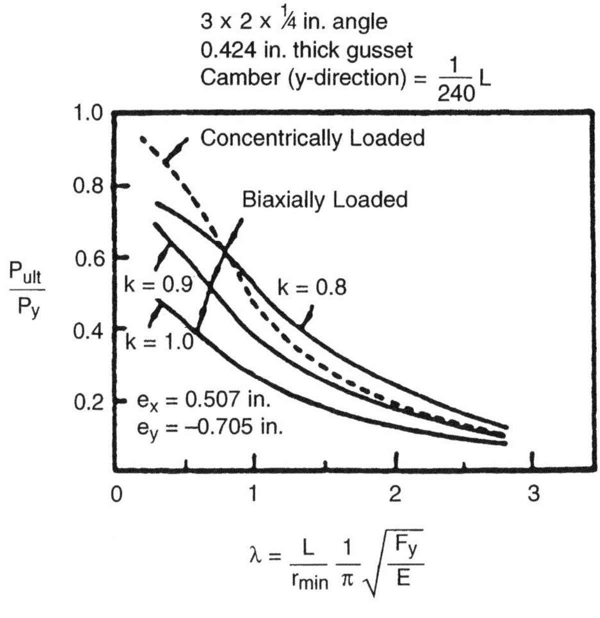

FIGURE 11.2 Ultimate strength of single-angle columns loaded through gusset plates, short leg outstanding (Lu et al., 1983).

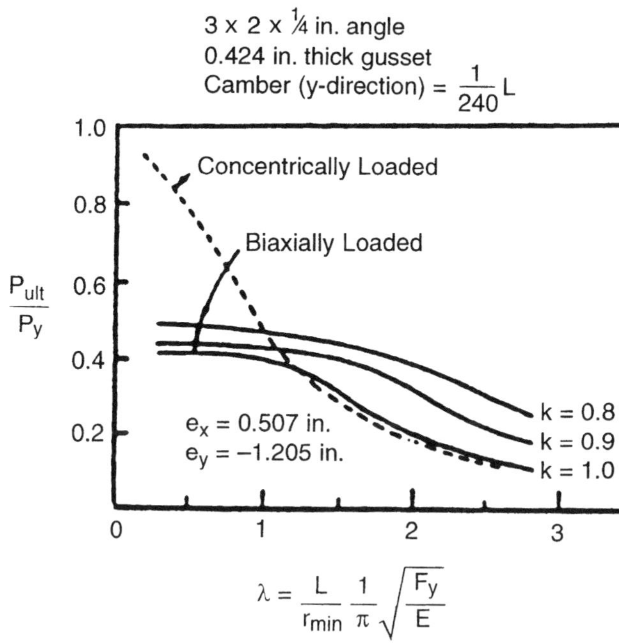

FIGURE 11.3 Ultimate strength of single-angle columns loaded through gusset plates, long leg outstanding (Lu et al., 1983).

beam element analysis for single-angle cross sections in the elastic range. End eccentricities and end restraints were included in the analysis. The formulation used the conventional beam stiffness matrix along with a geometry matrix specifically derived for the angle cross section. The equilibrium equations were solved in load steps accounting for the P–δ effect. The results were compared with those of Trahair (1969), which did not include the P–δ effect. Figure 11.4 shows that the P–δ effect can have a significant influence on the load-carrying capacity of the eccentrically connected single-angle struts.

Practical compactness and bracing provisions for the design of single-angle beams were investigated by Earls and Galambos (1998). Both experimental and analytical studies were carried out on single-angle members subjected to geometric axis flexure (Madugula et al., 1995, 1996; Earls and Galambos, 1998; Earls, 2001a–d). Experimentally verified nonlinear finite element modeling techniques were applied to the study of equal-leg, single-angle flexural compactness associated with bending about a geometric axis, such that the horizontal leg is in tension while the outstanding vertical leg is in compression. Interpretation of the finite element results indicated that any accurate prediction of compactness for this single-angle flexural case must address three parameters: (1) cross-sectional plate slenderness; (2) beam slenderness; and (3) steel grade. A design compactness equation combining the three previously mentioned influences was presented by Earls (2001a–e).

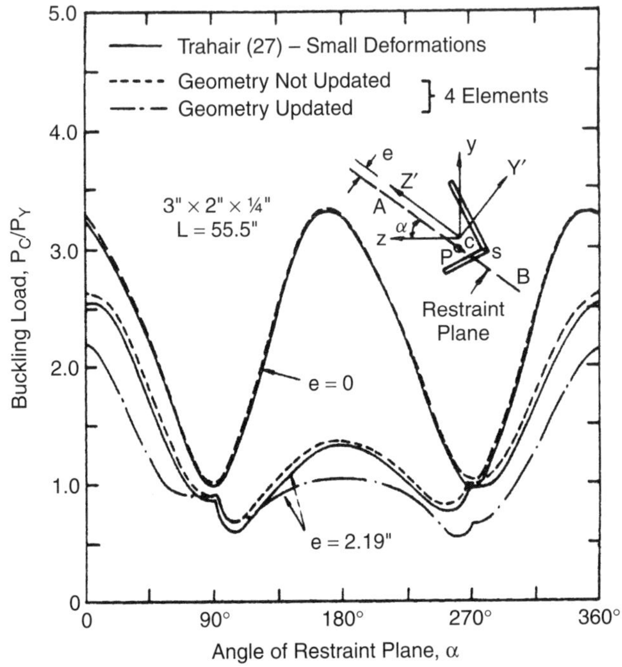

FIGURE 11.4 Buckling of eccentrically loaded and restrained angle beam-columns (Kitipornchai and Chan, 1987).

More recently, the behavior and design of single-angle section steel beams were studied extensively by Trahair (2002a,b, 2003, 2005a,b,c, 2007). Initially, a first-order (small-deformation) elastic analysis of the biaxial bending (without torsion) of angle section beams that include the effects of elastic restraints was investigated. Provisions were developed and suggested for section moment capacities that approximate the effects of full plasticity in compact sections, first yield in semicompact sections, and local buckling in slender sections (Trahair, 2002a). In a companion paper, Trahair (2002b) developed provisions for the bearing, shear, and uniform torsion section capacities. The results of Trahair's work may be used to design steel single-angle section beams, which are laterally restrained and consequently lateral buckling or second-order effects are negligible. The lateral buckling strengths of unrestrained steel single-angle section beams, which are loaded in the major principal plane producing no primary minor axis bending or torsion effects, were also considered (Trahair, 2003). An approximate method of predicting the elastic second-order twist rotations of an equal-angle beam in uniform bending was also developed. The accuracy of this method was investigated subsequently in a study of the elastic nonlinear large-rotation behavior of beams under nonuniform torsion (Trahair, 2005c). The approximate predicted rotations were used with the earlier formulations of the biaxial bending moment capacities to investigate the lateral buckling strengths of angle section beams with initial twist rotations, which were intended to approximate the effects of geometrical imperfections (initial out-of-straightness and twist) and residual stresses (Trahair, 2002a). The investigation included simple design procedures that approximated the predicted lateral buckling strengths.

Later, Trahair (2004) considered the biaxial bending of unrestrained steel angle section beams, which are loaded through the shear center and thus produce no primary torsion actions. An approximate analysis method for calculating the elastic second-order twist rotations of an equal-angle section beam in uniform bending was proposed, and simple approximations for the biaxial bending design strengths of equal- or unequal-angle section beams were developed using these predicted rotations. As an extension on the lateral buckling strengths of single-angle section beams, the beam was considered to be loaded in a plane parallel to the major principal plane but eccentric from the shear center, thus resulting in primary major axis bending and torsion actions but no primary minor axis bending actions (Trahair, 2005a,b). Similar to his previous work, an approximate analysis was developed for predicting the elastic second-order twist rotations of an equal-angle section beam under idealized loading, and these rotations were used to predict the effects of torsion on beam lateral buckling strengths and to develop simple proposals for the design of beams under major axis bending and torsion.

The lateral buckling and torsion of single equal-angle section beams was further investigated by Trahair (2007), and in this research the beam is assumed loaded in a plane inclined to the major principal plane and eccentric from the shear center, thus generating primary biaxial bending and torsion actions (Trahair, 2007). Again, approximate methods for predicting elastic second-order twist rotations are provided, and they are used to predict the effects of biaxial bending and torsion on beam design strengths.

The research reported by Earls (2001a–d) applied experimentally verified, nonlinear finite element modeling techniques to the study of geometric axis flexural ductility for the case when the horizontal leg of the angle cross section is in tension. It is shown that compactness criteria must be given in terms of both cross-sectional plate slenderness as well as beam slenderness. It is further observed that the grade of steel can significantly impact compactness requirements. Based on the results of this study, a practical design equation is proposed as a means for predicting equal-leg, single-angle compactness for this flexural orientation.

Open-web steel joists constructed from single-angle web members that have uncrimped and crimped ends have been studied experimentally and analytically by Yost et al. (2004). All K-series joists tested failed at a web compression member by either buckling at the midspan or buckling at the crimped transition zone (Fig. 11.5). For the joist configurations tested, it was determined that the effective length factor K used in determination of the maximum load capacity of a joist can be reduced from the typically assumed $K = 1.0$ (pin-ended condition) to a value of $K = 0.80$. Yost et al. (2006) further experimentally investigated the performance of K-series joists of three different depths with uncrimped and crimped web members. Failure modes were similar to those reported in Yost et al. (2004). It was determined that crimping the web angles increases the member's compression strength by 11 to

(*a*) (*b*)

FIGURE 11.5 Typical web buckling: (*a*) at midspan of member; (*b*) at crimp at member end (Yost et al., 2004).

36% relative to the companion uncrimped strength. It was also shown that the column analysis currently mandated by the Steel Joist Institute (SJI, 2005) with an effective length factor of 1.0 for crimped angle web members is overly conservative for both LRFD and ASD methodologies. In these studies, the critical web members were instrumented with four strain gages equally spaced on each 1-in. leg of the angle at midspan. Figure 11.6 shows the strain gage results for a crimped and uncrimped angle for the 30-in.-deep joists at approximate load intervals of 500 lb. A pure axial load condition would be represented by a horizontal line. Based on measured strains at midspan, internal bending is present in crimped web members, and, as expected, measured internal bending for uncrimped web members is greater than that for companion crimped web members.

Additional experimental test results of open-web steel joists with crimped end critical web members were reported by Buckley et al. (2007, 2008). Specimens included thirty 8-ft joists, thirty 22-ft joists, and thirty 28-ft joists manufactured by

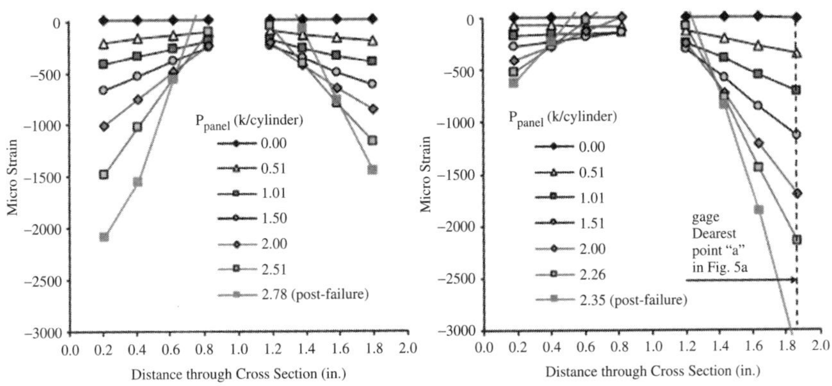

(*a*) (*b*)

FIGURE 11.6 Strain distribution at midspan of critical web member: (*a*) crimped member; (*b*) uncrimped member (Yost et al., 2006).

three different SJI member companies. In many cases, unexpected failures of joist members that were designed to be noncritical indicated that critical web members were able to resist significantly larger forces than anticipated. This study also showed that joists which failed locally, due either to web member buckling or weld fracture, could not resist loads large enough to produce a global mode of flexural buckling.

11.3 SINGLE-ANGLE COMPRESSION MEMBERS

11.3.1 Elastic Behavior

The elastic behavior is a special case of the stability of thin-walled members (Bleich, 1952, Chap. 3; Vlasov, 1961, Chap. 6; Timoshenko and Gere, 1961, Chap. 5). Because the shear center is located at the intersection of the two angle legs, there is practically no warping rigidity and the warping constant C_w can be assumed to be zero. Angle members loaded through the centroid by a compressive axial force will buckle in flexural buckling about the minor principal axis of the cross section or in a flexural–torsional mode. The critical load P (Galambos, 1968) is the lowest root of the equation

$$(P_u - P)(P_z - P)(P_t - P) - P^2[(P_u - P)z_o^2 + (P_z - P)u_o^2]\frac{1}{r_o^2} = 0 \qquad (11.1)$$

where

$$P_u = \frac{\pi^2 EI_u}{L^2} \qquad (11.2)$$

$$P_z = \frac{\pi^2 EI_z}{L^2} \qquad (11.3)$$

$$P_t = \frac{GJ}{r_o^2} \qquad (11.4)$$

The terms in these equations are defined in Fig. 11.7, and L is the length of the column. For equal-leg angles, $z_o = 0$ and Eq. 11.1 can be reduced to

$$P = \min(P_z, P_o) \qquad (11.5)$$

where

$$P_o = \frac{(P_t + P_u) \pm \left[(P_t + P_u)^2 - 4P_u P_t \left(1 - u_o^2/r_o^2\right)\right]^{1/2}}{2(1 - u_o^2/r_o^2)} \qquad (11.6)$$

11.3.2 End Restraint

Equations 11.2 and 11.3 apply only for pin-ended columns. Inevitably, some end restraint is introduced by the method of attachment of the angle to other members of the structure. Because this attachment is often to one leg only, the modeling of

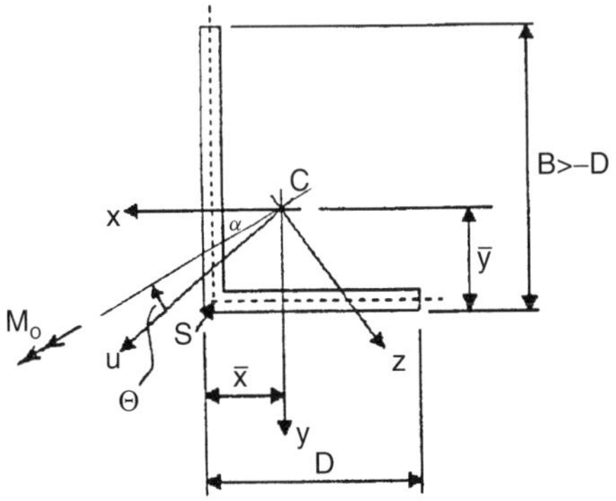

from the Handbook: α, $\bar{x}$, $\bar{y}$, I_x, I_y, r_z, A

C: centroid (0,0)

S: shear center (u_o, z_0)

x, y: non-principal axes

u, z: principal axes

t: thickness of the angle

$$x_o = \bar{x} - \frac{t}{2}$$

$$y_o = \bar{y} - \frac{t}{2}$$

$$u_o = y_o \cdot \sin(\alpha) + x_o \cdot \cos(\alpha)$$

$$z_o = y_o \cdot \cos(\alpha) - x_o \cdot \sin(\alpha)$$

$$d = D - \frac{t}{2}$$

$$b = B - \frac{t}{2}$$

$$I_z = A \cdot r_z^2$$

$$I_u = I_x + I_y - I_z$$

$$J = \frac{A \cdot t^2}{3}$$

$$r_o^2 = u_o^2 + z_o^2 + \frac{I_u + I_z}{A}$$

$$C_1 = \frac{x_o^2}{2} \cdot [\, y_o^2 - (\, y_o - b)^2] + \frac{y_o^4 - (y_o - b)^4}{4} + \frac{y_o}{3} \cdot [x_o^3 - (x_o - d)^3] + y_o^3 \cdot d$$

$$C_2 - \frac{x_o}{3} \cdot [\, y_o^3 - (\, y_o - b)^3\,] + x_o^3 \cdot b + \frac{x_o^4 - (x_o - d)^4}{4} + \frac{y_o^2}{2} \cdot [x_o^2 - (x_o - d)^2]$$

$$\beta_z = \frac{t \cdot (c_1 \cdot \sin(\alpha) + c_2 \cdot \cos(\alpha))}{I_z} - 2 \cdot u_o$$

$$\beta_u = \frac{t \,.\, (c_1 \,.\, \cos(\alpha) - c_2 \cdot \sin(\alpha))}{I_u} - 2 \cdot z_o$$

FIGURE 11.7 Definition of cross-section properties.

the end-restraint effect becomes complicated. Only a few solutions are available (Trahair, 1969; Kitipornchai and Lee, 1986). An acceptable design office solution is the use of effective-length factors so that Eqs. 11.2 and 11.3 are

$$P_u = \frac{\pi^2 E I_u}{(K_u L)^2} \tag{11.7}$$

$$P_z = \frac{\pi^2 E I_z}{(K_z L)^2} \tag{11.8}$$

where $K_u L$ and $K_z L$ are the effective length in the u and z directions, respectively.

11.3.3 Inelastic Behavior

Equations 11.2 and 11.3 (or Eqs. 11.7 and 11.8) are valid only in the elastic range. Inelastic behavior of angle columns has been investigated by Kitipornchai and Lee (1986) using the finite element method. A somewhat less elaborate solution is to replace the elastic modulus E in Eqs. 11.2 and 11.3 (or Eqs. 11.7 and 11.8) by the tangent modulus E_t. The shear modulus G in Eq. 11.4 remains unchanged. Approximate tangent-modulus representations for steel and aluminum are given in Chapter 3 of this guide. Use of the tangent modulus is complicated by the need for iteration. A simpler approximation for singly symmetric columns is recommended in Appendix E3 of the AISC LRFD Specifications (AISC,1999) and is described as follows:

1. Determine the elastic critical load P by Eq. 11.1 or 11.6, as appropriate.
2. Compute an equivalent slenderness parameter:

$$\lambda_{\text{eq}} = \frac{1}{\pi}\left(\frac{L}{r}\right)_{\text{eq}}\sqrt{\frac{F_y}{E}} = \sqrt{\frac{AF_y}{P}} \tag{11.9}$$

3. Determine the buckling load using the formulas in Section E2 of the 1999 AISC LRFD specification.

It should be noted, however, that the 2000 AISC specification on single angles (AISC, 2000), does not require flexural–torsional buckling to be considered for hot-rolled single angles.

Kitipornchai (1983) suggested the following approximations for the equivalent slenderness ratio from curve-fitting solutions to Eq. 11.1 or 11.6: For equal-leg angles

$$\left(\frac{L}{r}\right)_{\text{eq}} = 0.05\left(\frac{L}{r_z}\right) + 0.48\alpha_2 \le \frac{L}{r_z} \tag{11.10}$$

and for unequal-leg angles

$$\left(\frac{L}{r_z}\right)_{\text{eq}} = \left[\left(\frac{L}{r_z}\right)^3 - 8\,(\alpha_1 - 0.5)\,\alpha_2^2\frac{L}{r} + 0.76\alpha_2^3\right]^{1/3} \tag{11.11}$$

where $\alpha_1 = D/B$ and $\alpha_2 = B/t$.

Tests performed by Kennedy and Murty (1972) on hot-rolled angles showed that this method provided a satisfactory prediction of the strength. Similar confirmation has been provided by Marsh (1969), who found that his tests on slender equal-leg aluminum angles with single- and double-bolt connections were adequately predicted by

$$P_{cr} = \frac{0.9\pi^2 E}{(L/r)_{\text{eq}}^2} \tag{11.12}$$

where

$$\left(\frac{L}{r}\right)_{\text{eq}} = \sqrt{\left(\frac{5b}{t}\right)^2 + \left(\frac{KL}{r_y}\right)^2} \tag{11.13}$$

Upper limits of $0.50P_y$ (single bolts) and $0.67P_y$ (double bolts) were placed on P_{cr} ($P_y = AF_y$).

11.3.4 Local Buckling

If the slenderness ratio b/t of the angle legs is large, the compressive capacity can be reduced by local buckling. Table 11.1 provides plate-buckling coefficients and several stress variations encountered with equal-leg angles. For the case of uniform compression on a plate element with one edge free and the other hinged, the plate-buckling coefficient is 0.43.

It should be noted that for angle legs, the AISC LRFD specification (AISC, 1993) sets the critical elastic stress at $0.534E/(b/t)^2$. This limit is used for axial load and for flexural loading when the tip of the angle has the maximum stress. Although this stress value appears to be unconservative for axial load when compared to the $0.389E/(b/t)^2$value computed from Table 11.1, the use of overall leg length for b and the additional torsional stiffness due to the inside radius of hot-rolled angles increase the actual capacity above the theoretical. In addition, for unequal-leg angles the shorter leg provides restraint to the critical longer leg, whereas the theoretical values consider a hinge at the supported edge.

TABLE 11.1 Plate-Buckling Coefficients for Equal-Leg Angles

Case	Stress Variation	K Coefficient	Steel $f_{cr}(b/t)^2$, ksi	Angle Orientation
1	$f_2 = f_1$	0.43	11,270	
2	$f_2 = 0$	0.57	14,940	
3	$f_2 = -f_1$	0.85	22,280	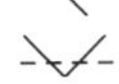
4	$f_1 = 0$	1.70	44,560	

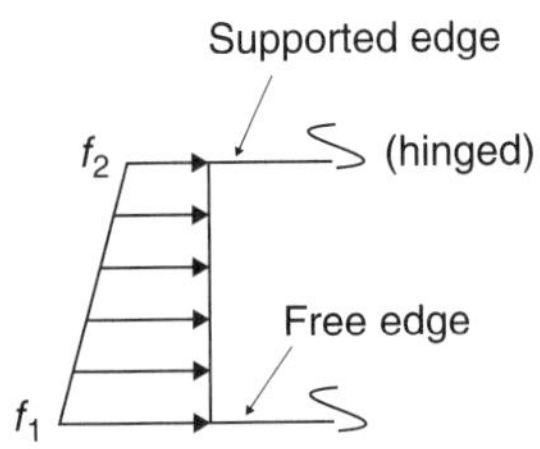

$$f_{cr} = \frac{K\pi^2 E}{12(1-\nu^2)(b/t)^2}$$

11.3.5 Eccentric Loading

The exact elastic solution of an angle with restrained ends and eccentric compressive load is complicated, and hence numerical methods must be employed (Trahair, 1969; Kitipornchai and Chan, 1987). Exact solutions for biaxially loaded columns with equal end eccentricities can be obtained by the method given by Culver (1966). If such columns have pinned ends, a simpler solution can be obtained by assuming the deflected shape to be a half sine wave (Dabrowski, 1961). This method is illustrated in Figs. 11.8 and 11.9, and the computations will result in midheight deformations, from which deflection limits can be checked and stresses at the extreme fibers can be calculated.

When the axial compressive load is applied with eccentricities e_u and e_z (Fig. 11.8), the problem is no longer a bifurcation problem because deformation (bending) will occur with any axial load. It then becomes necessary to determine the resulting stresses, which may become amplified or perhaps even reduced, as was found for single-angle beams by Thomas et al. (1973) due to the deflection and twist of the section. The problem of eccentrically loaded and end-restrained single-angle struts for the special case of end restraint provided by tee stubs, which approximate the double-angle chord of a truss, was examined for the elastic case by Trahair (1969) for the limit state $\sigma_{\max} = \sigma_y$ and for the inelastic case by Usami and Galambos (1971). Trahair compared his results with tests performed by Foehl (1948), and Usami and Galambos (1971) reported comparisons with a series of experimental tests they had performed and with Foehl's tests. In both cases, good correlation was achieved between test and prediction. In a series of

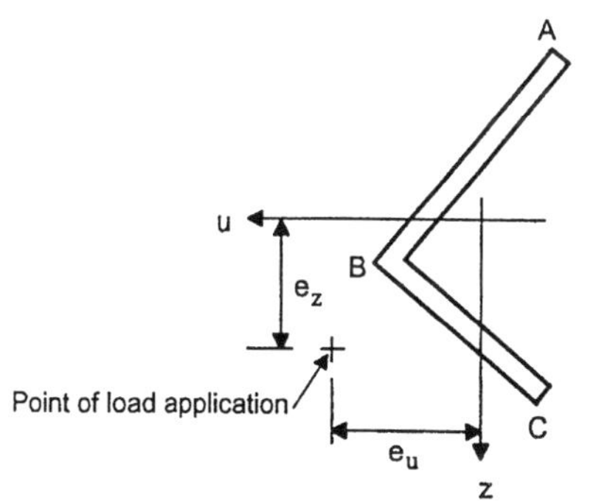

POINT	x	y
A	$\bar{x}$	$\bar{y}$-B
B	$\bar{x}$	$\bar{y}$
C	$\bar{x}$-D	$\bar{y}$

$$u = y * \sin(\alpha) + x * \cos(\alpha)$$

$$z = y * \cos(\alpha) - x * \sin(\alpha)$$

FIGURE 11.8 Coordinates of the cross section in biaxial loading.

Given: L, P, e_u, e_z, cross-sectional dimensions

Step 1: Solve for A_o, B_o, C_o from the following three simultaneous equations

$$\begin{bmatrix} (P_z - P) & 0 & [-P\cdot(e_z - z_o)] \\ 0 & (P_u \quad P) & [P\cdot(e_u - u_o)] \\ [-P\cdot(e_z - z_o)] & [-P\cdot(e_u - u_o)] & [(P_t - P)\cdot r_o^2 - P\cdot(\beta_u\cdot e_z + \beta_z\cdot e_u)] \end{bmatrix} \cdot \begin{pmatrix} A_o \\ B_o \\ C_o \end{pmatrix} = \begin{bmatrix} \dfrac{-4\cdot P\cdot e_u}{\pi} \\ \dfrac{4\cdot P\cdot e_z}{\pi} \\ 0 \end{bmatrix}$$

A_o, B_o, C_o are deflections in the u and z directions and the angle of twist, respectively, at the column center.

$$P_z = \frac{\pi^2\cdot E\cdot I_z}{L^2} \qquad P_u = \frac{\pi^2\cdot E\cdot I_u}{L^2} \qquad P_z = \frac{G\cdot J}{r_o^2}$$

Step 2: compute the stresses at the extreme fibers (@ points A,B,C in Fig. 11.8)

$$\sigma = -\frac{P}{A} + \frac{P\cdot u}{I_z}\cdot[-e_u + (e_z - z_o)\cdot C_o + A_o] - \frac{P\cdot z}{I_u}\cdot[e_z + (e_u - u_o)\cdot C_o + B_o]$$

FIGURE 11.9 Method of determining the stresses at any point in the section.

unpublished papers by Usami (1970) and Leigh and Galambos (1972), tests on the strength of single-angle eccentric webs in trusses were reported, and the results from the interaction equation

$$\frac{P}{P_u} + \frac{Pe_u}{M_{uY}\,(1 - P/P_u)} + \frac{Pe_z}{M_{zY}\,(1 - P/P_u)} = 1 \tag{11.14}$$

were compared to test results and theoretical predictions. It was found that Eq. 11.14 gives a satisfactory if somewhat conservative prediction of the actual capacity, provided that the end eccentricities were reduced to account for end restraint from the other web members entering the panel point. The axial force P_u in Eq. 11.14 are defined by Eqs. 11.2 and 11.3, e_u and e_z are the eccentricities shown in Fig. 11.8, and M_{uY} and M_{zY} are the moments required to produce compressive yield in the extreme fiber when $P = 0$. Equation 11.14 has also been recommended for the design of single-angle web members of trusses with ends connected to the chords by welding or by using multiple bolted connections (Woolcock and Kitipornchai, 1980, 1986). Care is needed in deciding on appropriate end-moment values if these are to reflect the effects of the load eccentricity accurately. A modification of Eq. 11.14 containing an additional torsional term has been suggested by Marsh (1972), who found that it gave good predictions of his tests on equal-leg angles under biaxially eccentric loading. While further clarification of the application of Eq. 11.14 would be useful, the evidence available suggests that it constitutes a reasonable basis for the design of angles required to function as beam-columns.

Designers of transmission towers use an alternative approach by assuming single-angle struts are connected at their ends on one leg by bolts or by welding. The axial load is thus applied eccentrically, and the ends are restrained. For the design solution, the slenderness ratio is modified for use in an appropriate column formula so as to account empirically for both the end eccentricity and the end restraints. Examples of this approach are given by the ASCE (1988, 2000), ECCS (1976), and AISC (2005) recommendations. These methods have been reviewed by Kennedy and Madugula (1982). Support for the possible extension of this approach to a wider range of structures has been provided by Haaijer et al. (1981) on the basis of detailed finite element calculation for angles loaded (and restrained) through one leg.

In summary, present information on eccentrically loaded single-angle struts applicable for design office use indicates the utilization of either an interaction equation (modified by the Q-factor reduction scheme of the AISC specification for local buckling, as applicable) or, for triangulated towers in particular, the use of an empirical effective-length factor approach seems appropriate. Theory which has been validated by tests is available and could be used for the development of more accurate design criteria.

11.4 CURRENT INDUSTRY PRACTICE FOR HOT-ROLLED SINGLE-ANGLE MEMBERS IN THE UNITED STATES

11.4.1 ASCE Manual 10-97: Design of Latticed Steel Transmission Structures

For single-angle members used in transmission towers, the ASCE (1988, 2000) recognizes the importance of the eccentricity of load and the rotational restraints at the connections on the axial load-carrying capacity. An effective-length factor K is included to accommodate the various end conditions of a single-angle member. Based on a review of many years of tower industry experience and the results of laboratory and full-scale tower tests, the ASCE design standard provides the following recommendations regarding effects of end eccentricity and rotational restraint on angle capacity:

(a) For members with slenderness ratios less than 120, the end eccentricity plays the most predominant role. Conversely, compressive capacity of single-angle members with slenderness ratios greater than 120 is influenced primarily by the end rotational restraint. For members partially restrained against rotation at only one end,

$$\frac{Kl}{r} = 28.6 + 0.762\left(\frac{l}{r}\right); \qquad 120 < \frac{l}{r} < 225 \qquad (11.15)$$

and for members partially restrained against rotation at both ends,

$$\frac{Kl}{r} = 46.2 + 0.615\left(\frac{l}{r}\right); \qquad 120 < \frac{l}{r} < 250 \tag{11.16}$$

The same provisions are recommended for single-angle redundant members except that higher slenderness ratio limits are permitted. For a member to qualify for the use of the effective slenderness ratio given above, the ASCE standard states that (1) the restrained member must be connected to the restraining member with at least two bolts and (2) the restraining member must have a stiffness factor (*I/L*) in the stress plane that equals or exceeds those of the restrained member. In other words, a single-bolt connection at either end of the single-angle member or at a point of intermediate support will not be considered as furnishing adequate rotational restraint. Only a multiple-bolt connection detailed to minimize eccentricity can be considered to offer partial rotational restraint if the restraining members are capable of resisting the rotation of the joint. Even with partial restraint, its impact on the allowable load is very significant. By using the provision above, a reduction of effective length up to 20% (i.e., $K = 0.8$) is permitted for members with qualified partial restraint at both ends.

(b) For single-angle members that are connected by only one leg and have a slenderness ratio of less than 120, the ASCE design standard recommends that the effective slenderness ratio for members with a concentric load at one end and "normal" framing eccentricity at the other end be taken as

$$\frac{Kl}{r} = 30 + 0.75\left(\frac{l}{r}\right) \tag{11.17}$$

and for members with "normal" framing eccentricities at both ends

$$\frac{Kl}{r} = 60 + 0.5\left(\frac{l}{r}\right) \tag{11.18}$$

Normal framing eccentricity at load transfer points implies that the centroid of the bolt pattern, except for some of the smaller angle sizes, is located between the heel of the angle and the centerline of the connected leg. These two design criteria imply that the normal-end eccentricity may cause a reduction of axial load capacity by up to 20% for stocky single-angle struts. An even larger reduction of load-carrying capacity is possible when joint eccentricities exceed the normal framing eccentricity. In such a case, due consideration should be given to the additional bending stresses introduced in the member.

(c) To guard against local buckling, the ratio of the flat width-to-thickness ratio *b/t* of steel angle legs is limited to $80/\sqrt{F_y}$. Where this limit is exceeded, the effect of reduced local-buckling strength on the flexural buckling strength must be considered.

In some truss-type frames, single-angle bracing members are used as cross braces and connected by a single bolt at the crossover point of the two brace members. Tests have shown that if the tension member of the cross bracing has at least 80% of the load of the compression member, the tension member provides partial support for the outstanding leg of the compression member. Under this condition, the controlling radius of gyration of equal-leg angles will be r_z. For unequal-leg angles, with the long leg connected, the controlling radius of gyration may be r_z or r_y. If the load in the tension member is less than 80% of the load in the compression member, partial support for the outstanding leg at the centerline connection point is questionable.

11.4.2 Steel Joist Institute (2005): Standard Specifications for Steel Joists and Joist Girders

The SJI *Standard Specification for Open Web Steel Joists, K-Series*, establishes maximum slenderness ratios for various components that comprise a joist: top-chord interior panels $L/r \leq 90$; top-chord end panels $L/r \leq 120$; compression members other than top chord $L/r \leq 200$; and tension members $L/r \leq 240$. The effective-length factor K is taken as 1.0 and no rotational restraint effects are considered. The maximum slenderness ratios found in the *Standard Specification for Joist Girders* are the same as those defined in the K-series standard specification. The *Standard Specification for Longspan Steel Joists, LH-Series, and Deep Longspan Steel Joists, DLH-Series*, also have the same maximum slenderness ratios for the top-chord interior panel, top-chord end panel, compression web members, and tension members (chords and webs) as the K-series standard specification but introduces effective-length factors that need to be considered in calculating the nominal stresses, F_{cr} and F'_e. The effective-length factor for a top-chord interior panel is taken as 1.0 for out-of-plane buckling and 0.75 for in-plane buckling. The effective-length factor for a compression web member is taken as 1.0 for out-of-plane buckling and 0.75 for in-plane buckling when a moment-resistant weld group is used at the ends of a crimped angle member. Otherwise, 1.2 should be used as the effective-length factor for in-plane buckling. An effective length less than the distance between intersection points could be justified because of the out-of-plane restraining influence of adjacent web members as the web member buckles. Because web members are not always rigidly connected to each other at the joints, however, it is prudent not to count on out-of-plane end restraint in determining their effective length.

11.4.3 AISC Manual of Steel Construction, 13th Edition

In 2005, a new AISC specification (AISC, 2005) was issued that provides design guidance for both ASD and LRFD of single-angle members. Using provisions from the 1993 and 2000 *Specification for Load and Resistance Factor Design of Single-Angle Members* (AISC, 1993, 2000), the 2005 specification considers the design of hot-rolled, single-angle members with equal and unequal legs in tension, shear, compression, flexure, and combined stresses.

The 1993 specification for single angles (AISC, 1993) addressed general flexural behavior as well as axial load capacity. Using the LRFD procedure for axially loaded angle members, only flexural buckling is considered. While the specification recommends the use of the AISC LRFD interaction formula to address the effects of biaxial bending, it indicates that the terms should reflect the stresses at a particular location on the cross section.

In 2000, the single-angle specification was updated (AISC, 2000). The permitted shape factor for flexure was made less conservative and specific local buckling limits for flexure were introduced. For axially loaded single-angle members, strength limit states were based only on flexural buckling because flexural–torsional buckling was deemed not critical for hot-rolled angles.

Five years later, the single-angle provisions were incorporated into the *Specification for Structural Steel Buildings* (AISC, 2005). The provisions remain unchanged for the general determination of the axial and flexural capacity of single angles, except for the modification of the interaction expression intended for single-angle members. An empirical effective-length procedure, however, was introduced for the design of single-angle (equal-leg and some unequal-leg) struts in which one leg at each end is attached to a truss chord. The procedure covers two conditions:

1. Box trusses in which the chords are well restrained torsionally:

$$\frac{KL}{r} = \begin{cases} 60 + 0.8\dfrac{L}{r_x} & \text{for } 0 \leq \dfrac{L}{r_x} \leq 75 \quad (11.19a) \\[2ex] 45 + \dfrac{L}{r_x} \leq 200 & \text{for } \dfrac{L}{r_x} > 75 \quad (11.19b) \end{cases}$$

(Note that these equations are equivalent to Eqs. 11.18 and 11.16, respectively, for equal-leg angles in transmission towers.)

2. Planar trusses in which the chords are considered to be less torsionally restrained:

$$\frac{KL}{r} = \begin{cases} 72 + 0.75\dfrac{L}{r_x} & \text{for } 0 \leq \dfrac{L}{r_x} \leq 80 \quad (11.20a) \\[2ex] 32 + 1.25\dfrac{L}{r_x} \leq 200 & \text{for } \dfrac{L}{r_x} > 80 \quad (11.20b) \end{cases}$$

For both conditions, the expressions are written in terms of r_x, where the x-axis is parallel to the truss chord, and consequently they can be employed for unequal-leg angles. Note that the additional slenderness modifications and the specific slenderness upper limits for unequal-leg angles are not given here.

In summary, the 2005 AISC *Steel Construction Manual* tabulates the axial capacity of hot-rolled single angles concentrically and eccentrically loaded, with the latter assuming no end restraint. In addition, the specification (AISC, 2005) provides the empirical effective-length method noted above so that eccentrically loaded single angles can be designed as axially loaded members with the end restraint considered.

11.5 DESIGN CRITERIA FOR HOT-ROLLED ANGLE COLUMNS IN EUROPE, AUSTRALIA, AND JAPAN

11.5.1 Eurocode 3 Design of Steel Structures

The European criteria for the design of single-angle compression members (angle columns) are similar to the rules in the AISC specification; if both ends of the column are welded or connected by at least two bolts to the connecting members, effective-length formulas are given for use in the applicable column formulas. For other end conditions and loadings, the angle columns are designed for the actual end eccentricities and end restraints using the appropriate interaction equations. European Standard prEN 1993-1-1, the Eurocode, specifies the following formulas for the effective-length factors:

$$\overline{\lambda}_{\text{eff},z} = 0.35 + 0.7\overline{\lambda}_z \quad \text{for buckling about the } z \text{ axis} \tag{11.21a}$$

$$\overline{\lambda}_{\text{eff},y} = 0.50 + 0.7\overline{\lambda}_y \quad \text{for buckling about the } y \text{ axis} \tag{11.21b}$$

$$\overline{\lambda}_{\text{eff},x} = 0.50 + 0.7\overline{\lambda}_x \quad \text{for buckling about the } x \text{ axis} \tag{11.21c}$$

where

$$\overline{\lambda} = \sqrt{\frac{AF_y}{P_e}} = \frac{L}{\pi r}\sqrt{\frac{E}{F_y}} \tag{11.21d}$$

The subscripts z, y, and x refer to the minor principal axis and to the geometric axes, respectively, as shown in Fig. 11.7. The Eurocode column formula is

$$P_{cr} = \chi A F_y \tag{11.22}$$

where $\chi = 1/\left(\Phi + \sqrt{\Phi^2 - \overline{\lambda}^2}\right) \le 1.0$, $\Phi = 0.5\left[1 + \alpha\left(\overline{\lambda} - 0.2\right) + \overline{\lambda}^2\right]$ and for angle columns the "imperfection factor" is taken as $\alpha = 0.34$ (European column curve b, see Chapter 3)

11.5.2 Australian Practice

A design approach was proposed by Woolcock and Kitipornchai (1986) to design single-angle web members. Basically, they recommended the combined stress interaction equation to be used for buckling and bending in the plane perpendicular to the truss. The out-of-plane bending moment at the end of each strut is calculated considering the eccentricity of the connection and the influence of adjacent web members. The current steel standard AS 4100-(SAA, 1998) also follows this approach.

11.5.3 Japanese Practice

The *Design Standard for Steel Structures* of the Architectural Institute of Japan recommends that single-angle web compression members of a truss be designed as pin-ended centrally compressed members subjected to buckling about the minor principal axis. Where the end connection of web members is accomplished by a single fastener, the allowable compressive stress in these members will be reduced by one-half to account for the effect of end eccentricity and loss of end restraint.

11.6 DESIGN OF AXIALLY LOADED COLD-FORMED SINGLE ANGLES

The provisions of the ANSI–ASCE Standard 10–97 (*Design of Latticed Steel Transmission Structures*) (ASCE, 2000), the *North American Specification for the Design of Cold-Formed Steel Structural Members*, and the ECCS (1976, 1978, 1985, 1992) recommendations are summarized in this section.

11.6.1 ASCE Standard 10-97 (2000)

Historically, most transmission structures have been fabricated from hot-rolled steel angles. In contrast to hot-rolled sections, cold-formed angles are more readily available in thinner and smaller sections and in turn provide a feasible alternative for more economical structures. An excellent summary on the design of cold-formed angles for transmission towers is given by Gaylord and Wilhoite (1985).

Unlike hot-rolled sections, cold-formed angles are available in more varieties of shapes. Three of the more commonly used configurations are 90° angle, 60° angle, and lipped angle. The recommended design criteria for compression members using cold-formed angles are covered by the updated version of ASCE Manual 52, identified as ANSI–ASCE 10-97.

Design of cold-formed plain angles is identical to that of hot-rolled angles, except that the flat width should be taken as the leg width minus the sum of the angle thickness and a maximum inside bend radius of two times this thickness. The ASCE standard (ASCE 10–97) requires the member to be checked for flexural and local buckling. The validity of these design criteria have been verified experimentally for cold-formed angles loaded concentrically or eccentrically (Gaylord and Wilhoite, 1985).

For symmetrical lipped angles, the member must be checked for flexural and flexural–torsional buckling. This is because the local buckling strength of the lipped-angle leg is not equivalent to the torsional buckling strength. The flexural–torsional strength can be determined by the AISI specification for the design of cold-formed steel structural members or using an equivalent radius of gyration r_{tf} which can be determined from

$$\frac{2}{r_{tf}^2} = \frac{1}{r_t^2} + \frac{1}{r_u^2} + 4\left(\frac{U_o}{r_t r_u r_{ps}}\right)^2 \sqrt{\left(\frac{1}{r_t^2} - \frac{1}{r_u^2}\right)^2} \tag{11.23}$$

where r_t = equivalent radius of gyration for torsional buckling

$r_t = \sqrt{C_w + 0.4J\,(KL)^2\,/I_{ps}}$

C_w = warping constant

J = St. Venant torsional constant

KL = effective length

I_{ps} = polar moment of inertia of cross section about shear center

r_u = radius of gyration of cross section about U axis

U_o = distance between centroid and shear center

$r_{ps} = \sqrt{I_{ps}/A}$ = polar radius of gyration about shear center

In addition, the edge stiffener should consist of a single lip bent at right angles to the stiffened element and meet a minimum lip depth requirement. Evaluation of flexural–torsional buckling requires determination of certain geometrical properties, which are not encountered in flexural buckling. Procedures for computing the torsional constant, warping constant, location of the shear center, polar moment of inertia about the shear center, and other properties of cold-formed sections are given in the AISI *Cold-Formed Steel Design Manual* (2001), and in Madugula and Ray (1984a).

11.6.2 North American Specification for the Design of Cold-Formed Steel Structural Members (AISI 2001)

This specification is a joint effort of AISI, the Canadian Standards Association (CSA S136-01), and CANACERO of Mexico. In this specification, cold-formed angles are to be designed for both axial load and flexure (see Chapter 13 for details of this procedure). An eccentricity of $L/1000$ about the minor axis of the angle is considered when evaluating the axial capacity of the single angle when the section is not fully effective. The Australian cold-formed steel standard is similar to the AISI method.

11.6.3 European Practice

According to ECCS (1978) recommendations, the axial strength of cold-formed angles is determined in the same way as for hot-rolled angles, except that the modulus of elasticity of cold-formed steel is taken as 210 GPa. Based on the work of Madugula et al. (1983) and Madugula and Ray (1984b), the computed failure loads using ECCS recommendations are either less than or very close to the experimental failure loads for eccentrically loaded cold-formed angles for all slenderness ratios. They are generally more conservative than those given in ASCE 10-97 (2000). Further, the ECCS recommendations give essentially the same results as the ASCE 10-90 for the concentrically loaded cold-formed angles.

11.7 CONCLUDING REMARKS ON THE COMPRESSIVE STRENGTH OF ECCENTRICALLY LOADED SINGLE-ANGLE MEMBERS

The compressive strength of eccentrically loaded single-angle members has been studied over the years both analytically and experimentally. For cases when the end restraints and eccentricities are clearly defined, the strength of the column can be predicted with good accuracy using either classical stability analysis or the finite element method. The latter method has the benefit of providing very good results in both the elastic and inelastic ranges.

Both the experimental and analytical studies show that the compressive strength of single-angle members is reduced by end eccentricity in the lower slenderness ratio range. It has also been concluded that the end restraint can significantly increase the ultimate strength of single-angle members of higher slenderness ratios. One of the most difficult tasks for designers is to judge or to determine the end restraint and eccentricity condition for their specific application. Failure to consider the restraint effect could lead to an uneconomical design. On the other hand, ignoring the end eccentricity may sometimes result in an unsafe design. To ensure a safe and economical design of a structure made of single-angle members, designers must pay attention to minimizing framing eccentricity in detailing angle connections, while the beneficial effects of rotational restraints, where applicable, should not be ignored.

The empirical approach to include the effects of end restraints and joint eccentricities by modifying the member's effective slenderness ratio, as adopted by ASCE Standard 10-97, the 2005 AISC specification, ECCS recommendations, and BS 5950, is very convenient to use. Application of these design criteria to structures with uncertain end restraints or more excessive eccentricities, however, should be exercised with great care. When end restraint is poor and joint eccentricities are large, eccentricities must be taken into consideration by considering biaxial bending. For those cases, the interaction equation in the AISC specification is recommended.

It appears that more research is needed to develop a general method for quantifying the effects of both end eccentricity and end restraint on member strength. Such results would be useful in helping designers evaluate single-angle members with excessive eccentricities or with end restraints that are not covered by the various design specifications.

11.8 MULTIPLE ANGLES IN COMPRESSION

11.8.1 Introduction

Several different configurations of double angles are possible. The most popular is to arrange the angles back to back, essentially forming a tee section (Fig. 11.10*a*). A second configuration, now used less frequently, is to connect the edges of the angles to form a boxed section (Fig. 11.10*b*). Finally, a third configuration, which seems to be gaining in popularity, is the starred-angle compression member (Fig. 11.10*c*).

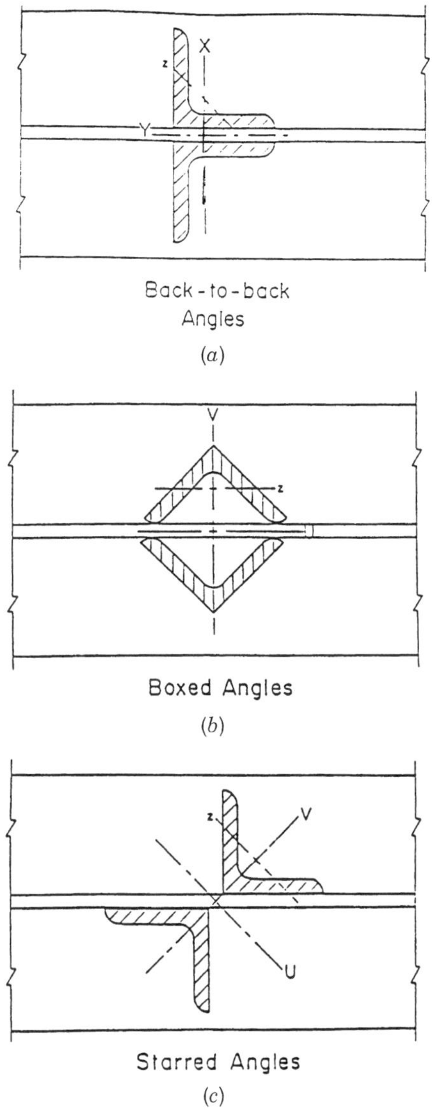

FIGURE 11.10 Double-angle configurations.

The advantages and disadvantages of each arrangement should be considered before a designer selects the appropriate double-angle configuration and may include:

1. The back-to-back and boxed double angles have interior surfaces that are virtually inaccessible. This may be undesirable when maintenance is required, such as in the chemical industry, where a corrosive atmosphere exists, or where building hygiene is important, such as in the food or pharmaceutical

industries. In contrast, with starred angles all the surfaces are accessible for maintenance.

2. In back-to-back angles, bolts in connections can be used in double shear. In starred angles, bolts can be used only in single shear. Boxed-angle connections have to be welded.
3. Boxed and starred angles tend to make more effective use of equal-leg angles than does the back-to-back configuration. The minimum moment of inertia of a boxed or starred angle made from equal-leg angles is about 60% greater than for the same angles in the back-to-back configuration.
4. Boxed angles seem to be preferred by some engineers and architects for aesthetic reasons; from a distance a boxed section has the appearance of a structural tube.
5. Starred angles are used as vertical web members in primary trusses which support secondary trusses in large industrial buildings. The nature of the cross section and the ease with which the connection can be made make these members desirable for this application. Such a connection at the exterior of a building is shown in Fig. 11.11*a*, and the cross section of the starred angle at the connection is shown in Fig. 11.11*b*.

11.8.2 Interconnection

Interconnectors In some double-angle web members no interconnectors are used, except, of course, for the connection of the angles to the chords of the truss. The load-carrying capacity of such double angles is calculated as twice that of a single angle using the radius of gyration about the minor principal axis, the z-axis. This tends to be very conservative for slender back-to-back angles (Temple and Tan, 1988). It was found that the average failure load was as high as 2.2 times the load calculated according to the North American standards and specifications for a slender member and about 1.3 times the load calculated for angles of intermediate length. Because the double-angle web member is attached to the chord of the truss by one leg, the individual angle does not buckle about the z axis, and failure occurs with the displacement in the y direction at midheight (see Fig. 11.10*a*) approximately five times that in the x direction.

In most cases double angles are designed to act as an integral unit. Thus interconnectors, which normally consist of small bars but could be short lengths of angles, tubes, and so on, are used to connect the angles intermittently along the length of the member. Figure 11.12 illustrates various methods of connecting double angles. It seems that fabricators prefer small pieces of plates to other connectors, such as short lengths of angles, simply from the point of view of economy. Welded connections are primarily used, with bolted connectors limited to bracing members.

Tests on starred angles (Temple et al., 1986) demonstrated that the type of connector did not affect the load-carrying capacity as long as the angles were firmly connected at the points of interconnection. It seems that fabricators will use a connector that is similar in thickness to the angles being connected and of

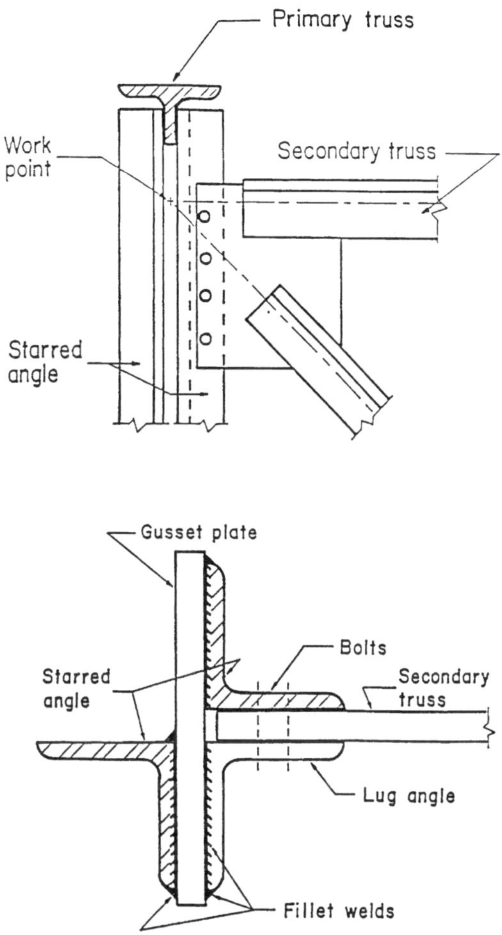

FIGURE 11.11 End connections of starred-angle columns.

dimensions that are of "manageable proportions." The connectors are then welded to the legs of the angles with the minimum-size weld specified by the applicable standard or specification for the thickness of material being used.

Interconnector versus Batten It has been proposed by Temple (1990) that when buckling of back-to-back angles occurs by bending about the x axis (Fig. 11.10a) the connector be termed an *interconnector*. The function of an interconnector is to make the two angles act together, that is, to deflect about the same amount in the y direction, to maintain the original back-to-back separation between the angles, and to help restrain any rotation of the individual angles that might occur if connectors were not present. Tests have shown (Temple and Tan, 1988) that the forces and moments in the interconnectors are very small.

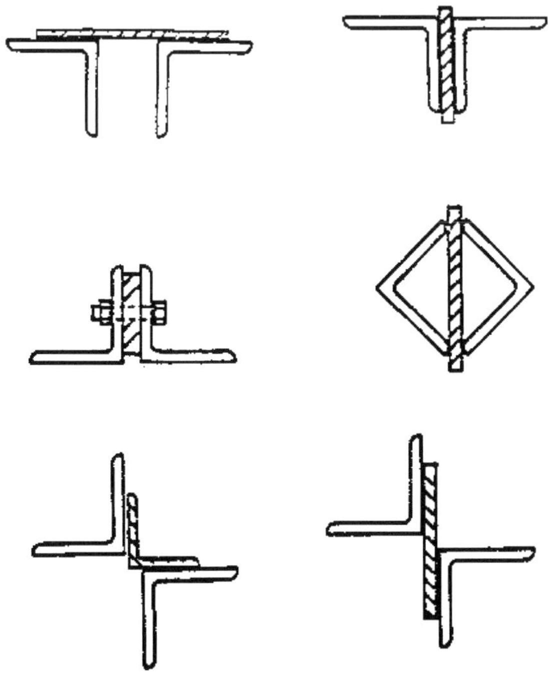

FIGURE 11.12 Intermediate connectors between double angles.

When flexural–torsional buckling of the back-to-back angles occurs about the y axis, the connector behaves like a batten and hence should be designed for the appropriate moments and shears shown in Fig. 11.13 (Temple, 1990). It is realized that these forces are developed for flexural buckling only and not for torsional–flexural buckling. It appears that the shear force, expressed as a percentage of the compressive resistance, normally used in such calculations would be sufficient to handle the shear and moments developed in the connector from the flexural as well as torsional effects. The Canadian Standard CAN-S16.94 (CSA, 1994) has specified a shear force of 2.5% of the axial compressive force in the member under factored loads. The AISC specification (AISC, 2005) does not specify any requirements for battens. During a research project to determine the interconnection requirements of starred angles, Temple et al. (1986) determined the force that must be resisted by the connectors in such double angles. It was concluded that the force carried by such connectors was not large, and thus it is recommended that these connectors can be treated as interconnectors. As part of a research project on the interconnection of boxed-angle compression members (Temple et al., 1987), connectors at the third points were instrumented with strain gauges. It was determined that the connectors were under small compressive forces and could once again be treated as interconnectors.

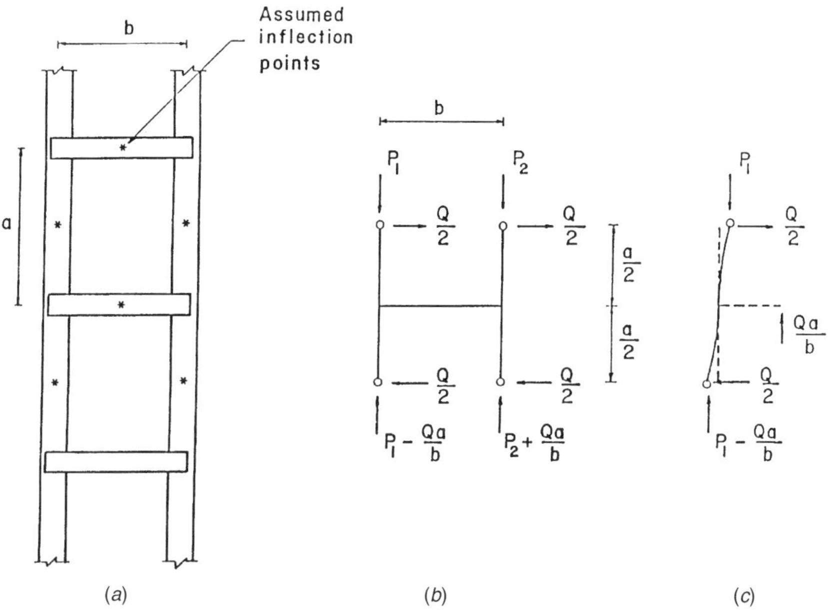

FIGURE 11.13 Calculation of forces in intermediate connectors between double angles.

Effect of Weld Pattern In most cases, connectors are welded to the legs of back-to-back angles, as shown in Fig. 11.14*a*. In some cases, however, it is not convenient during fabrication to apply this "normal" weld pattern (Temple and Tan, 1988), and the welds are placed between the angles only as shown in Fig. 11.14*b*, referred to as a "special" weld pattern. Tests indicate that the load-carrying capacity is not affected significantly by the weld pattern. It would seem to be conservative to avoid, if possible, the special weld pattern, as it does not provide as much rotational restraint to the individual angles as does the normal weld pattern.

11.8.3 Interconnection Requirements in the North American Standards

The Canadian Standard S16.1 (CSA, 1994) and the American specification (AISC, 2005) have similar requirements for the connection of double-angle compression members. They require that the slenderness ratio of the individual angle between points of connection cannot exceed all of (CSA) or three-fourths of (AISC) the slenderness ratio of the built-up member. The AISC specification provides an equation for an increased equivalent slenderness ratio to account for shear effects in separated members (AISC, 2005).

11.8.4 Connection Test Results

Test results for starred- and boxed-angle compression members (Temple et al. 1986, 1987) indicate that the current Canadian requirements for the interconnection of

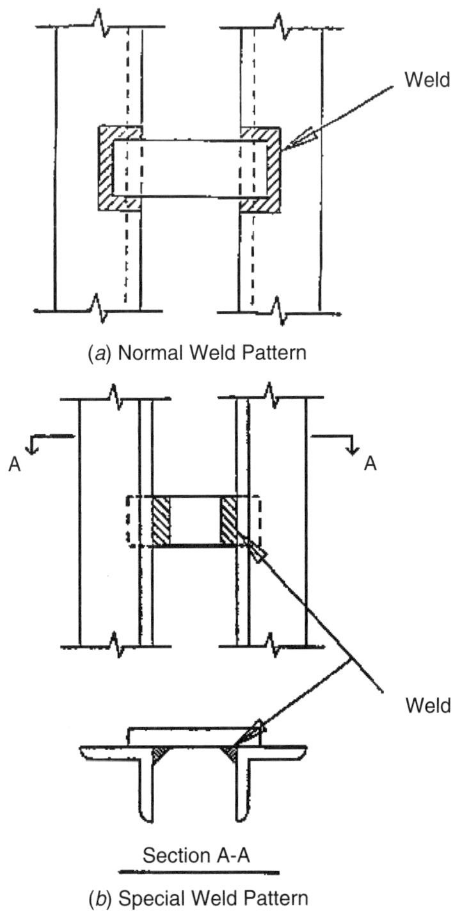

FIGURE 11.14 Welding of intermediate connectors between double angles.

these members are not adequate. For slender starred angles ($L/r = 122$) with zero interconnectors, which means that the angles were connected at the ends only, and with one interconnector at midheight, the failure loads were less than those specified by the applicable standards. The failure mode for these specimens consisted of buckling about the u axis (Fig. 11.10c), but obviously not as an integral unit. With two or more interconnectors there was an abrupt change in the magnitude of the failure loads and mode. The failure loads were greater than those specified and the failure mode was one of flexural buckling about the v axis, the minor principal axis of the composite cross section. The number of interconnectors in these tests was varied from zero to five, but the use of more than two connectors did not significantly affect the failure loads.

In these tests, the type of connector was varied and it was noted that the type of interconnector did not significantly affect the load-carrying capacity or failure mode as long as the angles were firmly attached at the point of connection. For starred angles of intermediate length ($L/r = 80$ or 40) there was not much difference in the load-carrying capacity as the number of connectors was changed from zero to one or two. There was, however, a difference in the failure mode, which makes the use of two connectors desirable. These intermediate-length starred angles with two connectors buckled about the minor principal axis of the cross section, the v axis, whereas with zero and one connector the starred angles failed by buckling about the u axis.

The research project on the interconnector of starred-angle compression members made from equal-leg angles concluded that two interconnectors, one at each of the third points, should be used. The research project on the connection requirements of boxed angles (Temple et al., 1987) concluded with results that are very similar to those in the starred-angle research project. This is not surprising because rotating the individual angles by 180° results in the starred-angle configuration. Again, it was concluded that two interconnectors, one at each of the third points, should be used. In a research project on back-to-back angles, Temple and Tan (1988) demonstrated that for angles that buckle about the x axis one interconnector at midlength was sufficient to give back-to-back angles made from equal-leg angles a load-carrying capacity equal to or greater than that specified by the CSA (1994) Standard. It should be pointed out, however, that in these tests, when one interconnector was used, the slenderness ratio of the individual angles between points of connection was only 78% of that for the double-angle member.

11.8.5 Load-Carrying Capacity

For singly symmetric sections such as back-to-back angles (Fig. 11.10*a*), instability can occur due to flexural buckling about the x axis or to flexural–torsional buckling. If the section is doubly symmetric, as with boxed and starred angles (Figs. 11.10*b* and *c*), buckling involves flexure about the x or y axis, or torsion about the shear center, which coincides with the centroid of the composite cross section. For doubly symmetric sections the three failure modes are independent and the minimum load determines the load-carrying capacity. Tests conducted by Kennedy and Murty (1972) on double-angle struts with hinged- and fixed-end conditions under concentric loading confirmed the theoretical solutions commonly used to calculate flexural and flexural–torsional buckling loads of these specimens. Temple and Tan (1988) demonstrated that the load-carrying capacity of back-to-back angles buckling about the x axis can be preferred by the then current North American standards and specifications. The flexural–torsional buckling load can be predicted by the procedure recommended in the AISC LRFD Specification (AISC, 2005). This procedure is demonstrated for back-to-back angles in a paper by Brandt (1988). He demonstrates that a simple comparison of the slenderness ratios about x and y for double angles can result in incorrect conclusions. Libove (1985) has studied sparsely connected built-up columns where the two identical elements are slightly separated and connected at a few points along their length. This includes back-to-back angles where buckling occurs at right angles to the axis of symmetry. Libove shows that, satisfying the common rule, the slenderness ratio between points of connection should not exceed the slenderness ratio of the built-up member may result in a load-carrying capacity considerably less than the buckling load predicted using the composite cross section. Libove also cautions that the postbuckling behavior is unstable; that is, the load required to maintain a buckling configuration is less than that required to initiate it. New design provisions in the AISC specification (AISC, 2005) were introduced to counteract this unconservatism.

Lorin and Cuille (1977) demonstrated that the substitution of a double angle for a larger angle, at relatively the same cost, makes for an improvement in the load-carrying capacity of 30 to 70%, with the higher values obtained when the thickness of the gusset plates was doubled. Astaneh-Asl and Goel (1984) studied double-angle bracing members subjected to in-plane buckling due to severe cyclic load reversals. Eight full-size test specimens made of back-to-back double-angle sections were tested under constant-amplitude cyclic loading. Some specimens that were designed by code procedures failed during early cycles of loading. New design procedures were proposed, and these were adopted in the AISC specification (AISC, 2005).

11.9 ANGLES IN FLEXURE

Angles are sometimes used as purely flexure members or as struts subjected to transverse loading. When angles are employed as beams, they often appear in walkways, racks, equipment support frames, pipe support systems, and other lightly loading framing. Pipe support systems composed of angles are used extensively in the nuclear power industry. The angles used are also often subjected to flexure from transverse loading. Support frames are often composed of members that become struts with small transverse load or beams with small axial load when subjected to seismic or wind loads.

11.9.1 Limit States

The ultimate strength of angle flexure members are determined by one of the following limit states: (1) yielding of the section, (2) local buckling of an angle leg, and (3) lateral–torsional buckling. Deflection or vibration of angle flexural members represents a serviceability limit state that must be considered and may govern over the strength limit state.

11.9.2 Yielding of the Section

Multiple angles in back-to-back, starred (or cruciform) configurations will yield at the tip of one of the angle legs, similar to a single-angle member, provided that one of the other limit states does not control the strength of the section. Each of these shapes has a shape factor of at least 1.5, which means that the section has the potential of achieving an ultimate or plastic moment at least 50% more than the moment at first yield. It should be noted that in the multiple-angle sections substantial interconnection of the components is required to achieve the full plastic moment capacity. For the boxed double-angle section (composed of two equal-leg angles), first yield may occur along two faces simultaneously, in which case the shape factor is approximately 1.15. On the other hand, if the section bends such that the neutral axis is along the diagonal, the shape factor is 1.5. Note that the plastic moment is about the same for both orientations, but the elastic section modulus is lower in the diagonally loaded orientation.

Based on the postyield behavior indicated above, the AISC specification (AISC, 2005) defines the nominal moment strength for single angles to be $1.5F_yS$.

11.9.3 Local Buckling

As indicated in Secion 11.3.4, local buckling may limit the section flexural capacity when one of the legs of a single- or multiple-angle section has its tip at the maximum compression stress. Table 11.1 provides several cases which are a function of the orientation of the bending neutral axis for calculating the elastic critical buckling stress. As the compression stress condition across the element becomes less severe, the buckling coefficient (and hence the elastic local buckling load) increases significantly. Given that rotational restraint at the supported edge will increase the buckling stress for each case, the AISC specification (AISC, 2005) sets the critical elastic stress at $0.71E/(b/t)^2$ for situations that pertain to cases 2 and 3 of Table 11.1. A situation where the maximum flexural stress is at the heel is not considered to be a local buckling concern for hot-rolled angles.

11.9.4 Excessive Deflection

As with all structural members in flexure, deflection may be a controlling design condition for a member. The deflection-limiting L/b ratio varies significantly with orientation of the bending axis and with end-restraint conditions. For simply supported spans with $L/175$ deflection limit and major axis bending of an equal-leg angle, L/b can reach a value of about 50 before deflection controls over stress (which also assumes A36 steel). If a total deflection of $L/360$ is used for an equal-leg angle bent about its minor axis, deflection will control when L/b exceeds about 12.

Often, angles are loaded parallel to one of their legs. If such an angle is laterally unbraced, it will deflect not only in the direction of the load but also perpendicular to that direction. The total deflection will be larger than that for a laterally braced angle. For an equal-leg angle that is laterally unbraced with no rotational restraint at the ends parallel to the direction of load, the deflection in the loaded direction is 1.56 times that of the same angle when it is laterally braced along its length. In addition, the unbraced equal-leg angle will deflect perpendicular to the loaded direction 60% of the loaded direction amount. The resultant represents an 82% larger deflection than the laterally braced angle.

Twisting of angles can also be of concern. The shear center is at the intersection of the angle legs. Obviously, increasing load eccentricity and using relatively thin angle legs will increase angle rotation. Angle rotation, in turn, can increase the normal stress as demonstrated by Leigh and Lay (1969) for angles bent about an axis parallel to a leg. For angles bent about a geometric axis, however, no increase in stress from twist will occur if L/t is kept below 800 for A36 steel.

11.9.5 Lateral–Torsional Buckling

The general case is that of an unequal-leg angle that can be bent about any axis. This is illustrated in Fig. 11.7 with moment applied at angle θ from the u axis. For the case of uniform bending moment M_0, and ignoring the warping stiffness such that the warping constant C_w can be assumed zero, the general equation for lateral–torsional buckling becomes

$$\phi''\left(GJ + \beta_u M_0 \cos\theta - \beta_z M_0 \sin\theta\right) + \phi \frac{M_0^2}{E}\left(\frac{\cos^2\theta}{I_z} + \frac{\sin^2\theta}{I_u}\right) + \frac{M_0^2 \sin\theta \cos\theta}{E}\left(\frac{1}{I_u} - \frac{1}{I_z}\right) = 0 \tag{11.24}$$

For the case of equal-leg angles, the value of β_u becomes zero. Typically I_z, I_u, and J are expressed in terms of b and t (see Fig. 11.7 for definitions). By approximating the section as two rectangular areas of length b and width t, one can obtain $I_u = tb^3/3$, $I_z = tb^3/12$, $J = 2bt^3/3$, and $\beta_z = \sqrt{2}b$. In this case, the solution of Eq. 11.24 can be expressed in the following form:

$$M_0 = \frac{\sqrt{2}\pi^2 E b^4 t}{6\left(1 + 3\cos^2\theta\right)L^2}\left[\sqrt{\sin^2\theta + \frac{4G\left(1 + 3\cos^2\theta\right)L^2 t^2}{\pi^2 E b^4}} - \sin\theta\right] \tag{11.25}$$

If M_0 is applied about the u, or major principal, axis ($\theta = 0$), this expression further simplifies to

$$M_0 = \frac{\sqrt{2}\pi}{6}\sqrt{GE}\frac{b^2 t^2}{L} \tag{11.26}$$

In terms of critical maximum bending stress F_{ob} for equal-leg angles, using I_u given above and $b/\sqrt{2}$ as the distance from the neutral axis,

$$F_{ob} = \frac{\pi\sqrt{GE}}{2\left(L/t\right)} \tag{11.27}$$

In the case of unequal-leg angles bent about the principal u axis ($\theta = 0$), the critical moment from Eq. 11.25 can be used to determine the critical moment M_0 as

$$M_0 = \frac{\pi^2 EI_z}{2L^2}\left[\sqrt{\beta_u^2 + \frac{4G}{3\pi^2 E}\left(\frac{Lt}{r_z}\right)^2} + \beta_u\right] \tag{11.28}$$

where β_u is the section property defined in Fig. 11.7. Its value should be taken as positive when the shorter angle leg (length D) is in compression and negative when the longer angle leg (length B) is in compression. Equations 11.27 and 11.28 both reflect the combined effects of lateral and torsional flexibility of the angle.

Often, angles are bent about one of the geometric axes (x or y). If this bending produces compression at the tip of the angle leg, lateral–torsional instability remains a potential problem. In Fig. 11.7, this situation occurs when θ equals α or $\alpha + 90^\circ$ (i.e., when M_0 is applied about the x or y axis to produce compression at the top of the angle shown).

The critical moment for equal-leg angles can readily be evaluated using Eq. 11.25 by setting $\theta = 45^\circ$ and can be expressed as

$$M_{0x} = \frac{\pi^2 Eb^4 t}{15L^2}\left(\sqrt{1 + \frac{20GL^2t^2}{\pi^2 Eb^4}} - 1\right) \tag{11.29}$$

The critical lateral–torsional moment can also be evaluated from Eq. 11.24 for an equal-leg angle bent about its z axis. If $\theta = 90^\circ$, the tips of the legs will be in compression, leading to

$$M_{0z} = \frac{\sqrt{2}\pi^2 Eb^4 t}{6L^2}\left(\sqrt{1 + \frac{16GL^2t^2}{\pi^2 Eb^4}} - 1\right) \tag{11.30}$$

The equations and procedures above are the bases for the provisions in the AISC specification (AISC, 2005) for determining the elastic lateral–torsional buckling moment of single-angle members. In the inelastic range of buckling, the specification (AISC, 2005) uses a transition relationship with

$$M_n = \begin{cases} \left(0.92 - 0.17\dfrac{M_0}{M_y}\right) M_0 & \text{for } M_0 \le M_y \quad (11.31) \\[2ex] \left(1.92 - 1.17\sqrt{\dfrac{M_y}{M_0}}\right) M_y \le 1.5 M_y & \text{for } M_0 > M_y \quad (11.32) \end{cases}$$

where M_n is the nominal moment capacity, M_y the yield moment, and M_0 the elastic lateral–torsional buckling moment determined from the equations above.

REFERENCES

Adluri, S. M. R., and Madugula, M. K. S. (1992), "Eccentrically Loaded Steel Angle Struts," *AISC Eng. J.*, Vol. 29, No. 2, pp. 59–66.

AISC (1989a), *Specification for Allowable Stress Design of Single-Angle Members*, American Institute of Steel Construction, Chicago, IL.

AISC (1989b), *Specification for Structural Steel Buildings, Allowable Stress Design and Plastic Design*, American Institute of Steel Construction, Chicago, IL.

AISC (1993), *Specification for Load and Resistance Factor Design of Single-Angle Members*, American Institute of Steel Construction, Chicago, IL.

AISC (1999), *Specification for Load and Resistance Factor Design of Single-Angle Members*, American Institute of Steel Construction, Chicago, IL.

AISC (2000), *Specification for Load and Resistance Factor Design of Single-Angle Members*, American Institute of Steel Construction, Chicago, IL.

AISC (2005), *Specification for Structural Steel Buildings*, American Institute of Steel Construction, Chicago, IL.

AISI (2002), *Cold-Formed Steel Design Manual*, American Iron and Steel Institute, Washington, DC.

AISI (2001), *North American Specification for the Design of Cold-Formed Steel Structural Members*, American Iron and Steel Institute, Washington, DC.

ASCE (1988), *Guide for Design of Steel Transmission Towers*, ASCE Manual Eng. Pract. No. 52, American Society of Civil Engineers, New York.

ASCE (2000), *Design of Latticed Steel Transmission Structures*, ASCE Standard 10-97, American Society of Civil Engineers, New York.

Astaneh-Asl, A., and Goel, S. C. (1984), "Cyclic In-plane Buckling of Double Angle Bracing," *ASCE J. Struct. Div.*, Vol. 110, No. ST9, pp. 2036–2055.

Bleich, F. (1952), *Buckling Strength of Metal Structures*, McGraw-Hill, New York.

Brandt, G. D. (1988), "Flexural–Torsional Buckling for Pairs of Angles Used As Columns," *AISC Eng. J.*, first quarter, pp. 39–40.

British Standards Institute (BSI) (1985), *Structural Use of Steelwork on Building*, BS5950, Part 1, BSI, London.

Buckley, E. T., Dinehart, D. W., Green, P. S., Gross, S. P., and Yost, J. R. (2008), "Buckling of Crimped Web Members in Open Web Steel Joists," *Proceedings of the 2008 Annual Stability Conference*, SSRC, Nashville, TN, Apr.

Buckley, E. T., Dinehart, D. W., Gross, S. P., and Yost, J. R. (2007), "Experimental Study of Open Web Steel Joists with Crimped Critical Web Members," Villanova University Research Report No. 1 to Steel Joist Institute, Villanova, PA, July.

Chunmei, G. (1984), "Elastoplastic Buckling of Single Angle Columns," *ASCE J. Struct. Eng.*, Vol. 110, No. 6, pp. 1391–1395.

CSA (1989), *Cold-Formed Steel Structural Members*, Canadian Standards Association, Toronto, Ontario, Canada.

CSA (1994), *Limit States Design of Steel Structures*, S16.1, Canadian Standards Association, Toronto, Ontario, Canada.

Culver, C. G. (1966), "Exact Solution of the Biaxial Bending Equations," *ASCE J. Struct. Eng.*, Vol. 92, No. ST2, pp. 63–83.

Dabrowski, R. (1961), "Thin-Walled Members Under Biaxial Eccentric Compression," *Stahlbau*, Vol. 30, Dec., pp. 360–365.

Earls, C. J. (1999), "On Single Angle Major Axis Flexure," J. Constr. Steel Res., Vol. 51, No. 2, pp. 81–97.

Earls, C. J. (2001a), "Geometric Axis Compactness Criteria for Equal Leg Angles: Horizontal Leg Compression," J. Constr. Steel Res., Vol. 57, No. 4, pp. 351–373.

Earls, C. J. (2001b), "Single Angle Geometric Axis Flexural Compactness Criteria: Horizontal Leg Tension," J. Struct. Eng. ASCE, Vol. 127, No. 6, pp. 616–624.

Earls, C. J. (2001c), "Single Angle Geometric Axis Flexure I: Background and Model Verification," J. Constr. Steel Res., Vol. 57, No. 2, pp. 602–622.

Earls, C. J. (2001d), "Single Angle Geometric Axis Flexure II: Design Recommendation," J. Constr. Steel Res., Vol. 57, No. 2, pp. 623–646.

Earls, C. J. (2001e), "Compactness and Bracing Recommendations for Equal Leg Single Angle Beams," Eng. J. AISC, Vol. 38, No. 1, pp. 204–217.

Earls, C. J. and Galambos, T. V. (1998) "Practical Compactness and Bracing Provisions for the Design of Single Angle Beams," Eng. J. AISC, Vol. 35, No. 1, pp. 19–25

Elgaaly, M., Dagher, H., and Davids, W. (1991), "Behavior of Single Angle Compression Members," *ASCE J. Struct. Eng.*, Vol. 117, No. 12, pp. 3720–3741.

Elgaaly, M., Davids, W., and Dagher, H., (1992), "Nonslender Single Angle Struts," *AISC Eng. J.*, Vol. 29, No. 2, pp. 49–58.

El-Tayem, A., and Goel, S. C. (1986), "Effective Length Factor for the Design of X-Bracing Systems," *AISC Eng. J.*, Vol. 23, No. 1, pp. 41–45.

ECCS (1992), *Design of Steel Structures*, European Committee on Standardization, Brussels, Belgium.

ECCS (1976), "Manual of the Stability of Steel Structures," *2nd Colloq. Stab. Steel Struct.*, European Convention for Constructional Steelwork, Liege, Belgium.

ECCS (1978), *European Recommendations for Steel Construction*, European Convention for Constructional Steelwork, Brussels, Belgium.

ECCS (1985), *Recommendations for Angles in Lattice Transmission Towers*, European Convention for Construction Steelwork, Brussels, Belgium.

Foehl, F. P. (1948), "Direct Method of Designing Single Angle Struts in Welded Trusses," in *Design Book of Welding*, Lincoln Electric Co., Cleveland, OH.

Galambos, T. V. (1968), *Structural Members and Frames*, Prentice Hall, Upper Saddle River, NJ.

Gaylord, E. H., and Wilhoite, G. M. (1985), "Transmission Towers: Design of Cold-Formed Angles," *ASCE J. Struct. Eng.*, Vol. 111, No. 8, pp. 1810–1825.

Haaijer, G., Carskaddan, P. S., and Grubb, M. A. (1981), "Eccentric Load Test of Angle Column Simulated with MSC/Nastran Finite Element Program," *Proc. SSRC Annu. Tech. Session*, Chicago, IL.

Hu, X. R., Shen, Z. Y., and Lu, L. W. (1982), "Inelastic Stability Analysis of Biaxially Loaded Beam Columns by the Finite Elements Methods," *Proc. Int. Conf. Finite Element Methods*, Vol. 2, Shanghai, China, pp. 52–57.

Ishida, A. (1968), "Experimental Study on Column Carrying Capacity of SHY Steel Angles," Yawata Tech. Rep. No. 265, Dec., Yawata Iron & Steel Co., Tokyo, pp. 8584–8582, 8761–8763.

Kennedy, J. B., and Madugula, M. K. S. (1982), "Buckling of Single and Compound Angles," in *Axially Compressed Structures: Stability and Strength* (Ed. R. Narayanan), Applied Science Publishers, Barking, Essex, England, Chap. 6.

Kennedy, J. B., and Murty, M. K. S. (1972), "Buckling of Steel Angle and Tee Struts," *ASCE J. Struct. Div.*, Vol. 98, No. ST11, pp. 2507–2522.

Kitipornchai, S. (1983), "Torsional–Flexural Buckling of Angles: A Parametric Study," *J. Constr. Steel Res.*, Vol. 3, No. 3, pp. 27–31.

Kitipornchai, S., and Chan, S. L. (1987), "Nonlinear Finite Element Analysis of Angle and Tee Beam-Columns," *ASCE J. Struct. Eng.*, Vol. 113, No. 4, pp. 721–739.

Kitipornchai, S., and Lee, H. W. (1986), "Inelastic Buckling of Single Angle Tee and Double Angle Struts," *J. Constr. Steel Res.*, Vol. 6, No. 1, pp. 3–20.

Leigh, J. M., and Galambos, T. V. (1972), "The Design of Compression Webs for Longspan Steel Joists," Res. Rep. No. 21, Department of Civil Engineering, Washington University, St. Louis, MO.

Leigh, J. M., and Lay, M. G., (1969), "The Design of Laterally Unsupported Angles," BHP Tech. Bull., Melbourne, No. 3, Vol. 13, Australia, Nov.

Libove, C. (1985), "Sparsely Connected Built-Up Columns," *ASCE J. Struct. Div.*, Vol. 111, No. ST3, pp. 609–627.

Lorin, M., and Cuille, J. P. (1977), "An Experimental Study of the Influence of the Connections of the Transmission Tower Web Members on Their Buckling Resistance," *Proc. 2nd Int. Colloq. Stab. Prelim. Rep.*, Liege, Belgium, pp. 447–456.

Lu, L. W., Shen, Z. Y., and Hu, X. R. (1983), "Inelastic Instability Research at Lehigh University," paper presented at the Michael R Horne Int. Conf. Instab. Plastic Collapse Steel Struct., University of Manchester, England.

Madugula, M. K. S., Kojima, T., Kajita, Y., and Ohama, M. (1995), "Minor Axis Strength of Angle Beams," in *Structural Stability and Design*, Kitipornchai, S., Hancock, G., and Bradford, M. (eds.) Balkema, Rotterdam, pp. 73–78.

Madugula, M. K. S., Kojima, T., Kajita, Y., and Ohama, M. (1996), "Geometric Axis Bending Strength of Double-Angle Beams," *J. Constr. Steel Res.*, Vol. 38, No. 1, pp. 23–40.

Madugula, M. K. S., Prabhu, T. S., and Temple, M. C. (1983), "Ultimate Strength of Concentrically Loaded Cold-Formed Angles," *Can. J. Civ. Eng.*, Vol. 10, No. 1, pp. 60–68.

Madugula, M. K. S., and Ray, S. K. (1984a), "Cross-Sectional Properties of Cold-Formed Angles," *Can. J. Civ. Eng.*, Vol. 11, No. 3, pp. 649–655.

Madugula, M. K. S., and Ray, S. K. (1984b), "Ultimate Strength of Eccentrically Loaded Cold-Formed Angles," *Can. J. Civ. Eng.*, Vol. 11, No. 2, pp. 225–233.

Marsh, C. (1969), "Single Angle Members in Tension and Compression," *ASCE J. Struct. Div.*, Vol. 95, No. ST5, p. 1043.

Marsh, C. (1972), "Lateral Stability of Single Angle Beam-Columns," CSSIC Proj. 712, Sir George Williams University, Montreal, Quebec, Canada, Mar.

Mengelkoch, N. S., and Yura, J. A. (2002), "Single-Angle Compression Members Loaded Through One Leg," *Proceedings, Structural Stabiltiy Research Council Annual Stabiltiy Conference.*

National Bureau of Standards (1924), "Results of Some Compression Tests of Structural Steel Angles," Tech. Pap. No. 281, U.S. Government Printing Office, Washington, DC.

SAA (1998), *Steel Structures*, AS 4100, Standards Association of Australia, North Sydney, New South Wales, Australia.

SJI (2005), 42nd Edition Catalog containing *Standard Specifications, Load Tables and Weight Tables for Steel Joists and Joist Girders: K-Series, LH-Series and DLH-Series, and Joist Girders*, Steel Joist Institute, Myrtle Beach, SC.

Temple, M. C. (1990), "Connectors for Double Angles: Interconnectors or Battens," *Can. J. Civ. Eng.*, Vol. 17, No. 1, pp. 8–11.

Temple, M. C., McCloskey, D. C., and Calabresse, J. M. (1987), "The Interconnection of Boxed Angle Compression Members," *Can. J. Civ. Eng.*, Vol. 14, No. 4, pp. 534–541.

Temple, M. C., Schepers, J. A., and Kennedy, D. J. L. (1986), "Interconnection of Starred Angle Compression Members," *Can. J. Civ. Eng.*, Vol. 13, No. 6, pp. 606–619.

Temple, M. C., and Tan, J. C. (1988), "Interconnection of Widely Spaced Angles," *Can. J. Civ. Eng.*, Vol. 15, No. 4, pp. 732–741.

Thomas, B. F., Leigh, J. M., and Lay, M. G. (1973), "The Behavior of Laterally Unsupported Angles," *Civ. Eng. Trans, Inst. Eng. Australia.*, pp. 103–110.

Timoshenko, S. P., and Gere, J. M. (1961), *Theory of Elastic Stability*, Engineering Societies Monograph, McGraw-Hill, New York.

Trahair, N. S. (1969), "Restrained Elastic Beam-Columns," *ASCE J. Struct. Div.*, Vol. 95, No. ST12, pp. 2641–2664.

Trahair, N. S. (2002a), "Moment Capacities of Steel Angle Sections," *J. Struct. Eng.*, Vol. 128, No. 11, pp. 1387–1393.

Trahair, N. S. (2002b), "Bearing, Shear, and Torsion Capacities of Steel Angle Sections," *J. Struct. Eng.*, Vol. 128, No. 11, pp. 1394–1398.

Trahair, N. S. (2003), "Lateral Buckling Strengths of Steel Angle Section Beams," *J. Struct. Eng.*, Vol. 129, No. 6, pp. 784–791.

Trahair, N. S. (2004), "Biaxial Bending of Steel Angle Section Beams." *J. Struct. Eng.*, Vol. 130, No. 4, pp. 554–561.

Trahair, N. S. (2005a), "Buckling and Torsion of Steel Equal Angle Section Beams." *J. Struct. Eng.*, Vol. 131, No. 3, pp. 467–473.

Trahair, N. S. (2005b). "Buckling and Torsion of Steel Unequal Angle Section Beams." *J. Struct. Eng.*, Vol. 131, No. 3, pp. 474–480.

Trahair, N. S. (2005c), "Nonlinear Elastic Nonuniform Torsion," *J. Struct. Eng.*, Vol. 131, No. 7, pp. 1135–1142.

Trahair, N. S. (2005d), "Biaxial Bending and Torsion of Steel Equal Angle Section Beams," Research Report No. R852, Dept. of Civil Engineering, Univ. of Sydney, Sydney, Australia.

Trahair, N. S. (2007), "Biaxial Bending and Torsion of Steel Equal Angle Section Beams," *J. Struct. Eng.*, Vol. 133, No. 1, pp. 78–84.

Usami, T. (1970), "Restrained Single-Angle Columns Under Biaxial Bending," Res. Rep. No. 14, Civil Engineering Department, Washington University, St. Louis, MO, June.

Usami, T., and Fukumoto, Y. (1972), "Compressive Strength and Design of Bracing Members with Angle or Tee Sections," *Proc. Jpn. Soc. Civ. Eng.*, Vol. 201, May, pp. 31–50 (in Japanese).

Usami, T., and Galambos, T. V. (1971), "Eccentrically Loaded Single Angle Columns," *International Association for Bridge and Structural Engineering*, Zurich, Switzerland, Publ. 31-/ii, pp. 153–174.

Vlasov, V. Z. (1961), *Thin-Walled Elastic Beams*, Israel Program for Scientific Translation for NSF, Jerusalem (original Russian edition, 1959).

Woolcock, S. T., and Kitipornchai, S. (1980), "The Design of Single Angle Struts," *Steel Constr.*, Vol. 14, No, 4, p. 2.

Woolcock, S. T., and Kitipornchai, S. (1986), "Design of Single Angle Web Struts in Trusses," *ASCE J. Struct. Eng.*, Vol. 112, No. 6, pp. 1327–1345.

Yost, J. R., Dinehart, D. W., Gross, S. P., Pote, J. J., and Gargan, B. (2004), "Strength and Design of Open Web Steel Joists with Crimped-End Web Members", *ASCE J. Struct. Eng.*, Vol. 130, No. 5, pp. 715–724.

Yost, J. R., Gross, S. P., Dinehart, D. W., Pote, J., and Deeney, J. (2006), "Buckling Strength of Single Angle Compression Members in K-Series Joists," *AISC Eng. J.*, Vol. 43, No. 2, Q2, pp. 141–152.

CHAPTER 12

BRACING

12.1 INTRODUCTION

Stability bracing requirements first appeared in the early 1900s related to the design of lacing in the built-up members of trusses (Waddell, 1916). Numerous railroad truss failures prompted the development of the 2% rule—the lacing shear force equals 2% of the force in the column. The lacing rule was most likely simply extended by structural engineers to all stability bracing situations, primarily as a result of steel design specifications in the United States not containing general bracing requirements until 1999. In the 1970s, the New York City building code contained the 2% rule for stability bracing but no stiffness requirements. Timoshenko's 1936 book, *Theory of Elastic Stability*, contained solutions for columns with flexible supports (brace points). He showed that if the flexible supports had a certain minimum stiffness, a straight column would behave as if the supports were rigid (no movement). Making the brace stiffness greater than the minimum did not affect the column strength. He also gave a simple technique for determining the *minimum* (later called *ideal*) stiffness for column bracing. Winter (1960) extended Timoshenko's solution to columns with initial crookedness (real columns) and to beams. Winter introduced the concept that stability bracing strength and stiffness requirements are interconnected. He showed that the design stiffness needed to be twice the ideal stiffness to keep brace forces small. The recommendations found in this chapter follow and expand on Winter's concept.

An adequate brace system requires both strength and stiffness (Winter, 1960). A simple brace design formulation, such as designing the brace for 2% of the member compressive force, addresses only the strength criterion. The magnitude of the initial out-of-straightness of the members to be braced has a direct effect on the bracing force. The brace stiffness also affects the brace force. Many published solutions provide stiffness recommendations only for perfectly straight structural systems. Such recommendations should not be used directly in design because very large brace forces may result, as will be shown subsequently.

A general design guide for stability bracing of columns, beams, and frames is presented. The focus is on simplicity, not exact formulations. The design recommendations cover four general types of bracing systems: relative, discrete, continuous, and lean-on, as illustrated in Fig. 12.1.

(a) A *relative brace* controls the relative movement of adjacent stories or of adjacent points along the length of the column or beam. If a cut everywhere along the braced member passes through the brace itself, then the brace system is relative, as illustrated by diagonal bracing, shear walls, or truss bracing.

(b) A *discrete brace* controls the movement only at that particular brace point. For example, in Fig. 12.1*b* the column is braced at points 1 by cross beams. A cut at the column midheight does not pass through any brace, so the brace system is not relative but discrete. Two adjacent beams with diaphragms or cross frames are discretely braced at the cross-frame locations.

(c) *Continuous bracing* is self evident; there is no unbraced length. The special case of diaphragm-braced columns and beams are discussed in Chapter 13.

(d) A beam or column that relies on adjacent structural members for support is braced in a *lean-on* system. Structural members that are tied or linked together, such that buckling of the member would require adjacent members to buckle with the same lateral displacement, characterize lean-on systems, as shown in Fig. 12.1*d*. In the sway mode member A leans on member B, that is, member B braces member A.

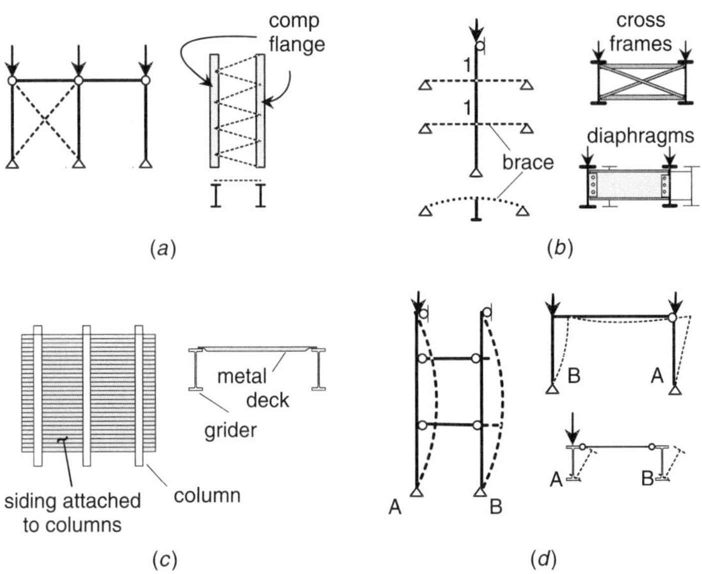

FIGURE 12.1 Types of bracing systems: (*a*) relative; (*b*) discrete; (*c*) continuous; (*d*) lean-on.

12.2 BACKGROUND

A general discussion of stability bracing for beams, columns, and frames has been provided by Trahair and Nethercot (1984), Chen and Tong (1994), and Yura (1995). Before presenting the various bracing recommendations, some background material on the importance of initial out-of-straightness, connection stiffness, and member inelasticity on bracing effects is discussed along with the limitations of the design criteria.

12.2.1 Member Out-of-Straightness

Winter (1960) derived the interrelationship between bracing strength and stiffness using simple models. He showed that the brace force is a function of the initial column out-of-straightness, Δ_o, and the brace stiffness, β. The concept is illustrated for the relative brace system shown in Fig. 12.2, where the brace, represented by the spring at the top of the column, controls the movement at the column top, Δ, relative to the column base. The unbraced length is defined as L. Summation of moments about point A gives $P\Delta_T = \beta L(\Delta_T - \Delta_o)$ where $\Delta_T = \Delta + \Delta_o$. If $\Delta_o = 0$ (an initially perfectly plumb member), then $P_{cr} = \beta L$, which indicates that the critical load increases with an increase in brace stiffness. The brace stiffness required in the sway mode to reach the load corresponding to Euler buckling between brace points, P_o, is called the *ideal stiffness*, β_i, where $\beta_i = P_o/L$ in this case.

For the out-of-plumb column, the relationship between P, β, and Δ_T is plotted in Fig. 12.3*a*. If $\beta = \beta_i, P_o$ can be reached only if the sway deflection gets very large. Unfortunately, such large displacements produce large brace forces, F_{br}, as shown in Fig. 12.3*b* because $F_{\text{br}} = \beta\Delta$. For practical design, Δ must be kept small at the maximum factored load level. This can be accomplished by specifying $\beta > \beta_i$. For example, if $\beta = 2\beta_i$, then $\Delta = \Delta_o$ at P_o as shown in Fig. 12.3*b*. The larger the brace stiffness, the smaller the brace force. For very stiff brace systems the brace force approaches $F_{\text{br}} = P_o\Delta_o/L$. The brace force is a linear function of the initial out-of-plumbness. The recommendations given later will assume a

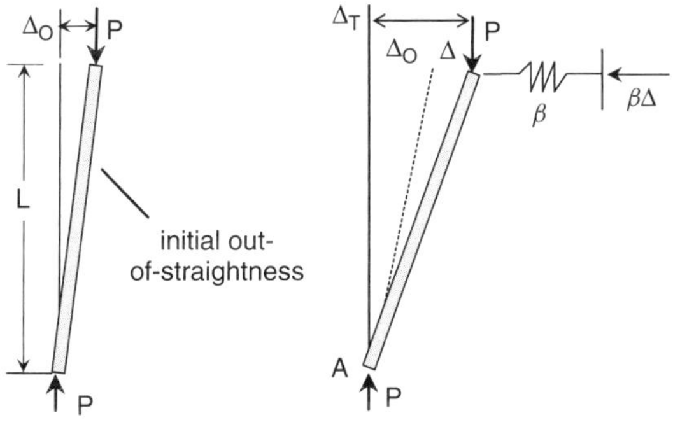

FIGURE 12.2 Relative brace.

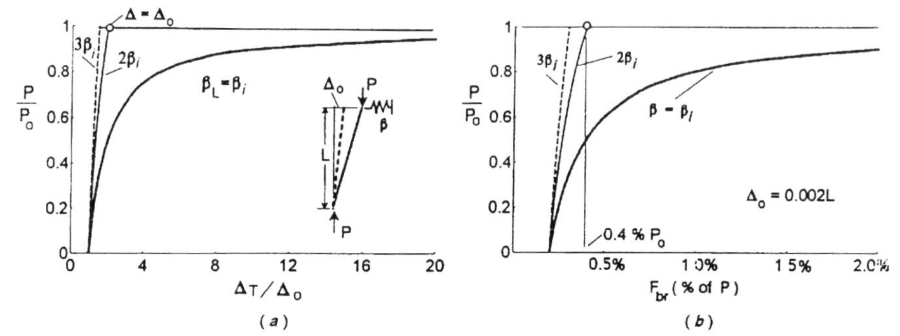

FIGURE 12.3 Effect of initial out-of-plumbness.

particular out-of-straightness and a brace stiffness at least twice the ideal stiffness. The effects of the magnitude and shape of the initial imperfection pattern on the brace forces in beams are discussed by Wang and Helwig (2005).

12.2.2 Member Inelasticity

Most research on bracing requirements for structures are based on elastic concepts (Trahair and Nethercot, 1984). The design requirements for relative braces, however, are merely a function of the load on the member and the distance between braces, as illustrated above, not column elasticity or inelasticity. For discrete bracing systems, Pincus (1964) used a simple theoretical model to demonstrate that the bracing stiffness requirements for inelastic columns are greater than those for elastic columns. Gil and Yura (1999), however, showed experimentally and analytically that an inelastic column with a midspan discrete brace showed no effect of column inelasticity on the bracing requirements. Also, Ales and Yura (1993) cast doubt on the Pincus solution, and their experiments on discrete bracing of inelastic beams verified Winter's approach. Nakamura (1988) presents a few beam experiments that also appear to follow the trends suggested by Winter's approach. Wang and Nethercot (1989) conducted a theoretical study of brace stiffness and strength requirements for beams with a concentrated load at midspan. Their study further verified the Winter approach, especially on the need to use at least twice the ideal full bracing stiffness in order to reduce the brace forces. The brace forces were less than 1% of the flange force when the recommended stiffness was provided. The results appear to verify Winter's approach for use with inelastic beams, but the loading condition considered involved only a small amount of inelasticity near midspan.

For beams in the inelastic range, most research has been concerned with the spacing of the braces, not the properties of the braces. *Commentary on Plastic Design in Steel* (ASCE, 1971) gives requirements for bracing at plastic hinge locations. In the ASCE recommendations, the lateral brace must have axial strength, axial stiffness, and flexural stiffness. Experiments on simply supported beams do not verify the need for flexural stiffness in the lateral braces. Yura and Li (2002) studied bracing requirements for beams in the plastic range for steels with $F_y \leq 50$ ksi. They found that brace stiffness requirements are not sensitive to rotation capacity, but brace forces do increase as rotation capacity increases and when local flange and web buckling occur. The bracing recommendations presented herein can be used in plastic design with $F_y \leq 50$ ksi, but not for applications requiring rotation capacities greater than 3. Thomas and Earls (2003) found that the bracing requirements herein were inadequate for the plastic design of A709 Gr. HPS483W high-performance steel girders with current compactness limits.

The few documented studies on discrete bracing requirements for inelastic beams and columns cited above indicate that inelasticity in the main members does not affect the bracing requirements unless large rotation capacities are required. Undocumented bracing failures of test setups in experiments when instability occurs in the inelastic range has contributed to the notion that inelastic structures require larger braces than elastic structures. When a lateral bracing failure occurs in a load test into the inelastic range, however, it usually happens *after* a local flange or web buckle occurs, which causes the W-shape beam to become unsymmetric. The loss of symmetry of the section causes shifts and inclinations of the principal bending axes that can cause very substantial lateral and torsional forces, much like those in channel sections not loaded through the shear center. Lateral bracing forces caused after local buckling occurs are very substantial (Yura and Li, 2002). Because most local buckling occurs in the plastic range, however, bracing failures are often associated with inelasticity rather than local buckling.

In continuous and lean-on brace systems, the brace requirements are based on the elastic and/or inelastic stiffness of the members to be braced, as will be given later. In these stability problems the effect of member inelasticity on the buckling solution can be reasonably approximated by representing the stiffness with the tangent modulus E_T (with $E_T = \tau E$, where τ is the inelastic stiffness reduction factor) instead of the elastic modulus, E. The elastic range is defined by the axial stress in the member, not the slenderness ratio. A member with low slenderness ratio (L/r) will respond elastically if the axial stress is low. In the AISC 2005 Specification, an axial stress less than $0.35F_y$ places the column in the elastic range. The AISC Manual (AISC, 2005) tabulates the stiffness reduction factor for various P/A stress levels. In LRFD, $\tau = -6.97(P/P_y)\log(1.111P/P_y)$, where P is the factored column load and P_y is the yield load, F_yA. The potential axial buckling capacity of a column is $\phi\tau(0.877)\pi^2 EI/(KL)^2$ for $P/P_y \geq 0.35$. For $P/P_y < 0.35$, $\tau = 1.0$. This τ factor will be used in some of the following example problems.

12.2.3 Limitations

The brace requirements presented below will enable a member to reach the Euler buckling load between the brace points (i.e., use $K = 1.0$). Because the ideal brace stiffness $\beta_i = 1.0P_e/L$ corresponds to $K = 1.0$, this is not the same as the no-sway buckling load as illustrated in Fig. 12.4 for the braced cantilever with rigid rotational base restraint. For a brace with twice the ideal stiffness, the buckling load is only 75% ($K = 0.81$) of the no-sway case. A brace with six times the ideal stiffness is

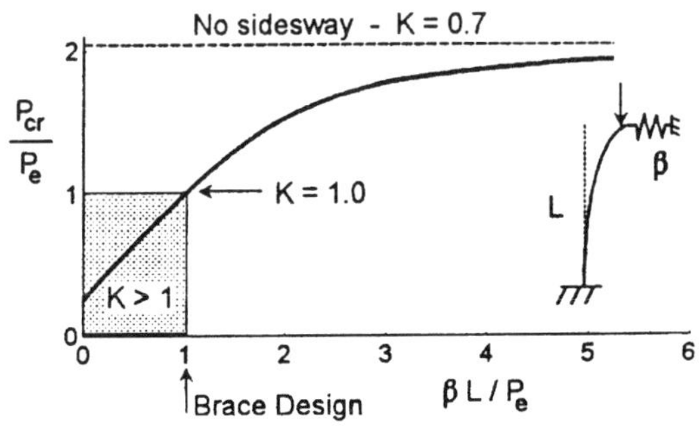

FIGURE 12.4 Braced cantileveler.

necessary to reach 95% of the $K = 0.7$ load limit. Theoretically, an infinitely stiff brace is required to reach the no-sway limit.

12.2.4 Brace System Stiffness

If they are flexible, brace connections should be considered in the evaluation of the bracing stiffness as follows:

$$\frac{1}{\beta_{\text{sys}}} = \frac{1}{\beta_{\text{conn}}} + \frac{1}{\beta_{\text{brace}}} \tag{12.1}$$

The brace system stiffness, β_{sys}, is less than the smaller of the connection stiffness, β_{conn}, or the stiffness of the brace, β_{brace}. When evaluating the bracing of rows of columns or beams, consideration must be given to the accumulation of the brace forces along the length of the brace, which results in a different displacement at each beam or column location. Medland and Segedin (1979) and Tong and Chen (1989) have studied interbraced structures. The solutions are fairly complex for use in design. In general, bracing forces can be minimized by increasing the number of braced bays and using stiff braces. Chen and Tong (1994) recommend bracing at least every eight bays.

12.3 SAFETY FACTORS, ϕ FACTORS, AND DEFINITIONS

The recommendations presented are based on ultimate strength. Column and beam loads are assumed to be factored loads. For brace stiffness formulations, a value of $\phi = 0.75$ is recommended in LRFD. If the load calculations are based on service loads as in ASD, a factor of safety of 2.0 can be applied to the factored load stiffness requirements. The strength requirements use the built-in safety factors or ϕ factors within each design specification. In LRFD, the design brace force will be based on factored loads and compared to the design strength of the brace and its connections. In ASD, the brace force will be a function of the applied service

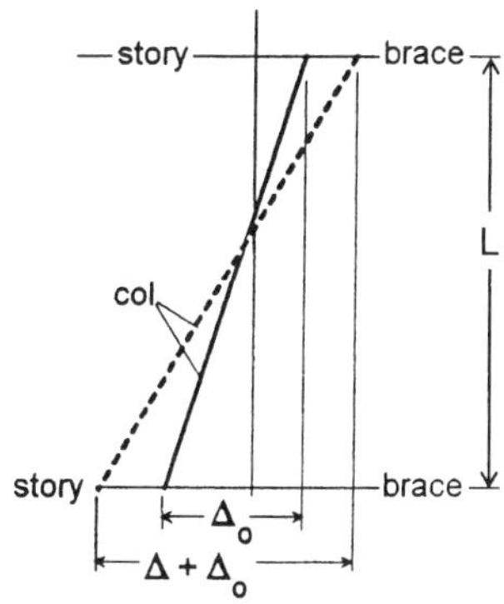

FIGURE 12.5 Definitions.

loads, and this force will be compared to the allowable brace loads and connection capacity.

The initial displacement Δ_o for relative and discrete braces is defined with respect to the distance between adjacent braces as shown in Fig. 12.5. In frames, P is the sum of the column loads in a story to be stabilized by the brace. In the case of a discrete brace for a member, P would be the average load in the compression member above and below the brace point. The initial displacement Δ_o is a small displacement from the straight position at the brace points caused by sources other than the gravity loads or compressive forces. For example, Δ_o would be a displacement caused by wind or other lateral forces, erection tolerance (initial out-of-plumb), and so on. In all cases, the brace force recommendations are based on an assumed $\Delta_o = 0.002L$, with direct proportion permitted for other Δ_o values. For torsional bracing of columns or beams, an initial twist β_o of $0.002L/h$ is used where h is the distance between the flange centroids. For cases when n columns, each with a random Δ_o, are to be stabilized by a brace system, Chen and Tong (1994) recommend an average $\Delta_o = 0.002L/\sqrt{n}$ value to account for the variation in initial out-of-straightness.

12.4 RELATIVE BRACES FOR COLUMNS OR FRAMES

Design Recommendation Based on an initial out-of-plumbness of $\Delta_o = 0.002L$ and a brace stiffness twice the ideal value, $\beta_i = P/L$, the design (LRFD) recommendation is

$$\phi = 0.75 \qquad \beta_{\text{req}} = \frac{2\sum P}{\phi L} \qquad F_{\text{br}} = 0.004\sum P$$

Example 1 illustrates the bracing design for a typical interior portion of a building with bracing every third frame. Each interior brace must stabilize 1500 kips. The floor is assumed to act as a rigid diaphragm and all Δ_o are equal. It is also assumed that only the tension diagonal brace, taken as a threaded rod, controls the lateral flexibility of the structure. The cosine functions are necessary to convert the diagonal brace to an equivalent brace perpendicular to the column(s). Stiffness controls the design in this case. If Δ_o is different from $0.002L$, F_{br} should be changed in direct proportion to the actual Δ_o. If the brace stiffness provided, β_{act}, is different from β_{req}, F_{br} can be modified as follows:

$$F_{\mathrm{br}} = 0.004 \sum P \frac{1}{2 - (\beta_{\mathrm{req}}/\beta_{\mathrm{act}})} \tag{12.2}$$

Example 12.1: Relative Brace–Tension System A typical brace with $F_y = 36\,\mathrm{ksi}$ must stabilize three bents. The factored load for each bent is

$$150 + 250 + 100 = 500\,\mathrm{kips}$$

Design recommendations assume that F_{br} and Δ are perpendicular to the column:

- Brace force

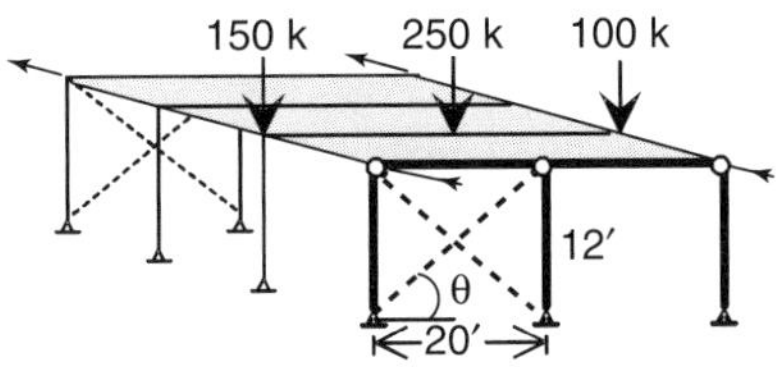

$$F_{\mathrm{br}} = \frac{0.004(3 \times 500)}{\cos\theta} = 6.99\,\mathrm{kips}$$

$\frac{5}{8}$-in. threaded rod OK

- Brace stiffness

$$\frac{A_b E}{L_b}\cos^2\theta = \beta_{\mathrm{req}} = \frac{2(3 \times 500\ \mathrm{kips})}{0.75(12)} \qquad \text{gives } A_{b_{\mathrm{gross}}} = 0.364\,\mathrm{in.}^2$$

Use $\frac{3}{4}$-in. rod, $A_g = 0.44\ \mathrm{in.}^2$

12.5 DISCRETE BRACING SYSTEMS FOR COLUMNS

Design Recommendation The design (LRFD) recommendation for discrete bracing is

$$\phi = 0.75 \qquad \beta_{\mathrm{req}} = N_i \frac{2P}{\phi L} \qquad F_{\mathrm{br}} = 0.01P$$

in which P is the factored load, L the required brace spacing, and n the number of braces, $N_i \approx 4 - (2/n)$.

Basis Discrete bracing systems can be represented by the model shown in Fig. 12.6 for three intermediate braces. The exact solution taken from Timoshenko and Gere (1961) shows the relationship between P_{cr} and the brace stiffness, β. With no bracing $P_{cr} = \pi^2 EI/(4L)^2$. At low brace stiffness the buckling load increases substantially with the buckled shape a single (first-mode) wave. As the brace stiffness is further increased, the buckled shape changes and additional brace stiffness becomes less effective. Full bracing occurs at $\beta L/P_e = 3.41 = N_i$. This ideal nondimensionalized stiffness factor N_i varies for equally spaced braces between 2.0 for one brace to 4.0 for a large number of braces. Thus 4.0 can be used conservatively for all cases. The above design recommendation is based on full bracing assuming the load is at P_e. If P varies along the length, the design of a brace can be based on the average load in the two adjacent unbraced segments.

The discrete brace force requirement (Yura, 1993) was developed initially from Winter's rigid member model assuming zero moment at the node points, which gives $F_{\text{br}} = 0.8\%$ of P from solutions similar to those shown in Fig. 12.3. Tong and Chen (1987) and Plaut (1993) showed that Winter's model was unconservative for the case of a single brace at midspan, and hence, it is recommended that $F_{\text{br}} = 1\%$ of P. This force assumes that a brace stiffness twice the ideal value is used. For other brace stiffnesses, the adjustment factor given in Eq. 12.2 can be used.

Typically, P may be less than P_e so it is conservative to use the actual column load P to derive the design stiffness represented by the dotted line in Fig. 12.6. Note that the required brace stiffness is inversely proportional to the brace spacing L. In many applications there are more potential brace points than necessary to support the required member forces. Closer spaced braces require more stiffness because the derivations assume that the unbraced length provided is just sufficient to support the column load. For example, suppose three girts are available to provide minor axis bracing to the columns and that the column load is such that only a single full brace at midspan would suffice. The required stiffness of the three-brace arrangement

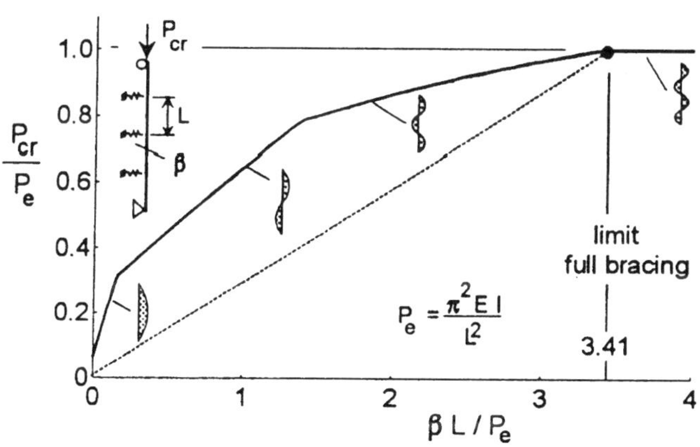

FIGURE 12.6 Three discrete braces.

could then be conservatively estimated by using the permissible unbraced length in the brace stiffness equation rather than the actual unbraced length. It should be noted that the continuous bracing formula given in the next section more accurately represents the true response of Fig. 12.6 for less than full bracing.

The design recommendation is based on twice the ideal stiffness to account for initial out-of-straightness. Example 12.2 illustrates the design procedure for a single discrete brace at the column midheight. The value of N_i is based on equal brace spacing and is unconservative for unequal spacing. For unequal spacing, N_i can be derived simply by using a rigid-bar model between braces (Yura, 1994). For a single discrete brace at any location along the column length, with the longest segment defined as L and the shorter segment as aL, N_i can be determined as follows:

$$N_i = 1 + \frac{1}{a} \tag{12.3}$$

Example 12.2: Discrete Brace at Midheight A cross member braces the minor axis of W16×26 at midheight. Factored loads are shown.

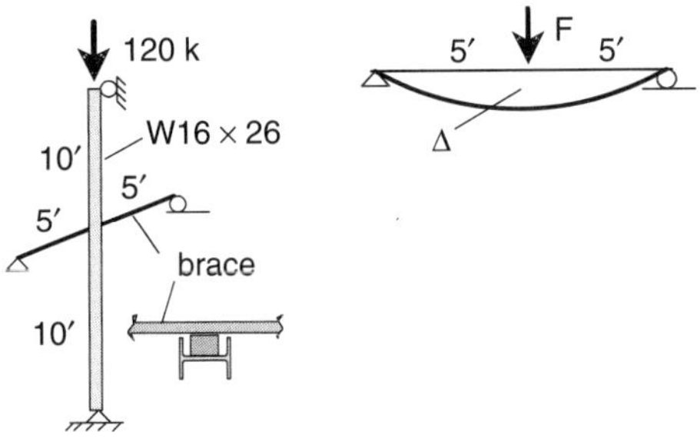

$$n = 1 \quad N_i = 2 \qquad \beta_{\text{req}} = 2\left[\frac{2(120)}{0.75(120)}\right] = 5.33\,\text{kips/in.}$$

$$\beta = \frac{F}{\Delta} = \frac{48EI}{(10 \times 12)^3}$$

$$I_{\text{req}} = \frac{5.33(120)^3}{48(29{,}000)} = 6.6\,\text{in.}^4$$

Try a C5×6.7:

$$I_x = 7.5\ \text{in.}^4 \qquad S_x = 3.5\ \text{in.}^3 \qquad F_{\text{br}} = 0.01(120) = 1.2\,\text{kips}$$

$$F_y = 36\,\text{ksi} \qquad f_b = \frac{1.2(120)}{4(3.5)} = 10.3\,\text{ksi} \qquad \text{OK}$$

12.6 CONTINUOUS COLUMN BRACING

For a column braced continuously, Timoshenko and Gere (1961) give

$$P_{cr} = P_e \left(n^2 + \frac{\overline{\beta} L^2}{n^2 \pi^2 P_e} \right) \tag{12.4}$$

where n is the number of half sine waves in the buckled shape as shown by the solid line in Fig. 12.7. As the brace stiffness per unit length $\overline{\beta}$ increases, the buckling load and n also increase. The switch in buckling modes for each n occurs when $\overline{\beta} L^2 / \pi^2 P_e = n^2 (n+1)^2$. Substituting this expression for n into Eq. 12.4 gives

$$P_{cr} = P_e + \frac{2L}{\pi} \sqrt{\overline{\beta} P_e} \tag{12.5}$$

Equation 12.5 is an approximate solution, shown dashed in Fig. 12.7, which gives the critical load for any value of $\overline{\beta}$ without the need to determine n. In the inelastic range use τP_e for P_e in Eq. 12.5.

Equation 12.5 can also be used for discrete braces by defining $\overline{\beta} \equiv \beta \times$ (number of braces)/L and limiting $P_{cr} \leq \pi^2 EI / l^2$, where l is the distance between braces. This approach is accurate for two or more braces. For example, if there are two discrete-braces, the ideal discrete-brace stiffness is $\beta = 3P_{cr}/l$, where $l = L/3$ and $P_{cr} = \pi^2 EI / l^2$. Using Eq. 12.5 with $\overline{\beta} = 2(3P_{cr}/l)/L$ gives $P_{cr} = 1.01(\pi^2 EI / l^2)$.

The bracing design recommendation given below is based on Eq. 12.5 with $\overline{\beta}$ adjusted by a factor of 2 to limit the brace forces, adding a $\phi_{br} = 0.75$, and using $P_o = \phi_c (0.877) \tau P_e$, which is the AISC LRFD column design strength. The

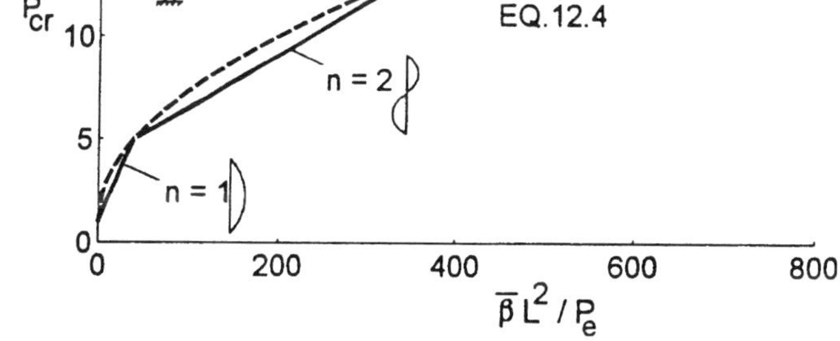

FIGURE 12.7 Continuous bracing.

brace strength requirement $\overline{F_{\text{br}}} = \pi^2 P \Delta_T / L_o^2$, where L_o is the maximum theoretical unbraced length that can support the column load, was developed by Zuk (1956). Taking $\Delta_T = 2\Delta_o$ and$\Delta_o = 0.002L_o$ gives $\overline{F_{\text{br}}} = 0.04P/L_o$

Design Recommendation The design (LRFD) recommendation for continuous bracing is

$$\phi_c P_{cr} = P_o + \frac{L}{\pi}\sqrt{2\phi_{\text{br}}\overline{\beta} P_o} \qquad F_{\text{br}} = 0.04P/L_o$$

in which $P_o = \phi_c(0.877)\tau P_e$, $\phi_c = 0.90$, and $\phi_{\text{br}} = 0.75$.

12.7 LEAN-ON SYSTEMS

When some members lean on adjacent members for stability support (bracing), the $\sum P$ concept (Yura, 1971) can be used to design the members. The approach will be explained using the problem shown in Fig. 12.8, in which column A has a load P with three connecting beams attached between columns A and B. There are two principal buckling modes for this structure, the no-sway and the sway modes.

If column B is sufficiently slender, the system will buckle in the sway mode, shown by the dot-dash line in Fig. 12.8*a*. In the sway mode the buckling strength involves the sum ($\sum P_{cr}$) of the buckling capacity of the two columns because each column has the same deformation pattern. The system is stable in the sway mode if the sum of the applied loads ($\sum P$) is less than the $\sum P_{cr}$. This, of course, assumes that all the columns have the same height. If column B is sufficiently stiff,

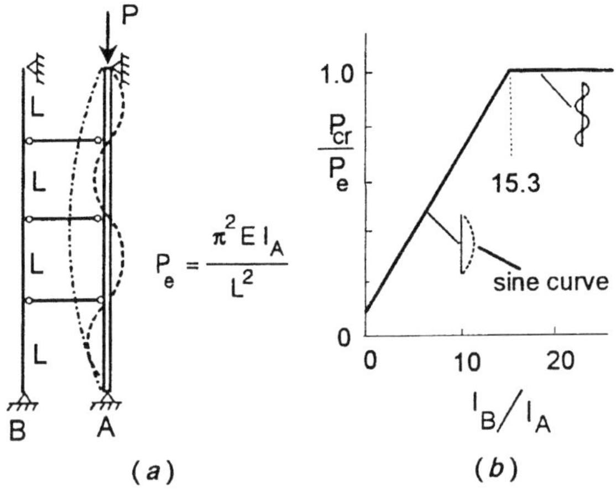

FIGURE 12.8 Lean-on bracing: (*a*) sway and no-sway buckling modes; (*b*) impact of relative column stiffnesses.

the buckling capacity may be controlled by the no-sway mode shown dashed. Both modes must be checked in design.

An exact elastic solution, developed with nonlinear finite element analysis software (ANSYS), shows that as I_B (the bending moment of inertia of column B) increases, P_{cr} increases linearly in the sway mode. For $I_B/I_A \geq 15.3$, column A buckles in the no-sway mode. The I_B required to develop full bracing can be approximated using the $\sum P$ concept. In the sway mode, the elastic capacities of columns A and B are $\pi^2 EI_A/(4L)^2$ and $\pi^2 EI_B/(4L)^2$, respectively. The desired P_{cr} corresponding to the no-sway mode is $\pi^2 EI_A/L^2$. Equating the sum of the sway capacities to the P_{cr} in the no-sway mode,

$$\frac{\pi^2 E(I_A + I_B)}{(4L)^2} = \frac{\pi^2 EI_A}{L^2} \tag{12.6}$$

gives $I_B = 15I_A$, which is close to the exact solution of $I_B = 15.3\ I_A$. In the inelastic range, τ_i is used where τ_i is based on the axial load in each column, P_i. There can be axial load on all the columns.

Example 12.3, which is similar to a problem solved by Lutz and Fisher (1985), shows a W12×40 with its minor axis in-plane supported by an adjacent column W12×26 with the major axis in-plane. Only in-plane buckling is considered. The tie beams have shear-only (pinned) end connections, so it is assumed that the tie beams do not contribute to the sway stiffness of the system. Sway is prevented at the top of the columns. The W12×40 has been sized based on buckling between the supports, spaced at $L = 8$ ft. The calculations show that the elastic W12×26 adjacent column can brace the minor axis column, which is in the inelastic range.

Example 12.3: Lean-On System Confirm that the W12×26 is capable of bracing the W12×40. Assume $F_y = 50$ ksi, factored loads are given, and the AISC LRFD specification governs.

From the AISC manual, $\phi P_n = 439$ kips for $L = 8$ ft.

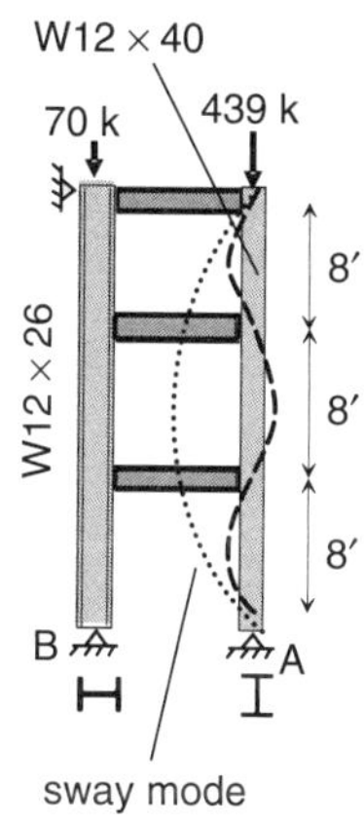

(continued)

$$\sum P \text{ concept:} \quad W\,12\times40, A = 11.7\,\text{in.}^2, I_y = 44.1\,\text{in.}^4$$
$$W\,12\times26, A = 7.65\,\text{in.}^2, I_x = 204\,\text{in.}$$

Column A :

$$\frac{P_A}{F_yA} = \frac{439}{(50 \times 11.7)}$$
$$= 0.750 > 0.35 \quad \therefore \text{inelastic}$$
$$\tau = -6.97(0.750)\log(1.111 \times 0.750) = 0.414$$
$$\phi P_A = \frac{0.90(0.414)(0.877)\pi^2(29{,}000)(44.1)}{(288)^2}$$
$$= 49.7\,\text{kips}$$

Column B :

$$\frac{P_B}{F_yA} = \frac{70}{50 \times 7.65} = 0.183 < 0.35 \quad \therefore \tau = 1.0$$
$$\phi P_B = \frac{0.90(0.877)\pi^2(29{,}000)(204)}{(288)^2} = 566\ \text{kips}$$
$$\sum P_{cr} = 50 + 556 = 606 > \sum P = 439 + 70 = 509\,\text{kips} \qquad \text{OK}$$

12.8 COLUMNS BRACED ON ONE FLANGE

Doubly symmetric columns will buckle in a flexural mode between brace points if the braces prevent both twist and displacement. If the brace detail does not prevent twist, such as rod bracing framing into the center of the web, then the column can buckle in a torsional mode. Another common bracing detail that can result in twist of the section is shown in Fig. 12.9. Girts frame into the column flange, which restrains minor axis lateral displacement near the flange. If the girts are discontinuous, they will not provide any torsional restraint and the column may buckle by twisting about the lateral brace point as shown in Fig. 12.9*b*.

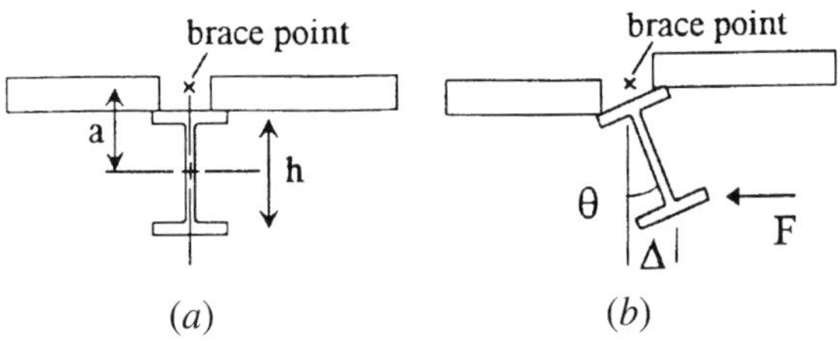

FIGURE 12.9 Buckling about a restrained axis: (*a*) lateral brace at flange; (*b*) buckled shape.

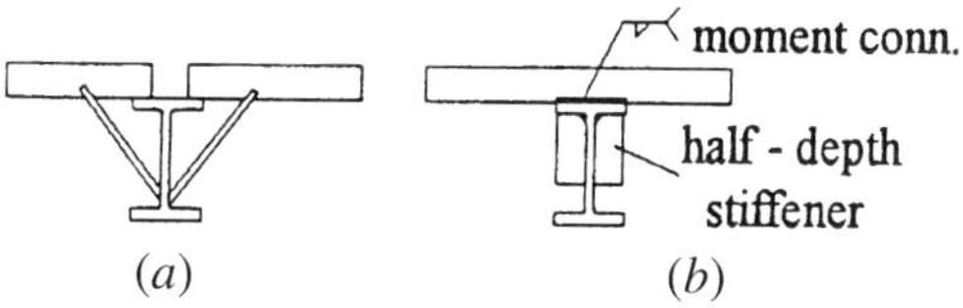

FIGURE 12.10 Typical torsional brace details: (*a*) using struts; (*b*) using moment connection with stiffener.

The torsional buckling load, P_T, for a column with a lateral restraint (Timoshenko and Gere, 1961) is

$$P_T = \frac{\tau P_{ey}\left(h^2/4 + a^2\right) + GJ}{a^2 + r_x^2 + r_y^2} \tag{12.7}$$

where a is the distance between the restrained axis and the centroid, r_x and r_y the principal radii of gyration, h the distance between the flange centroids, P_{ey} the Euler load based on the column length between points with zero twist, and G and J the material shear modulus and cross-sectional torsion constant, respectively. An infinitely stiff lateral brace at the brace point (zero displacement) was assumed in the derivation of Eq. 12.7. To compensate for finite stiffness, the maximum factored column load should not exceed 90% of P_T. Horne and Ajmani (1971, 1972) studied the more complex problem of beam-columns braced on one flange.

When the applied factored load is greater than P_T, torsional bracing must be provided. Two typical bracing schemes are shown in Fig. 12.10. When a moment connection is used, a partial-depth web stiffener is recommended to prevent web distortion. The design requirements for the torsional braces for columns are given by Helwig and Yura (1999).

12.9 BEAM BUCKLING AND BRACING

Before presenting the beam-bracing recommendations, the suitability of assuming the inflection point as a brace point to define L_b in restrained beams will be discussed. In many cases when this issue is raised, the top flange is laterally braced by the slab or joists along the full span length while the bottom flange is unbraced. An inflection point *cannot* be considered a brace point as illustrated by the example shown in Fig. 12.11. One beam has a moment at one end ($C_b = 1.67$) with $L_b = L$, and the other beam has an inflection point at midspan ($C_b = 2.3$) with $L_b = 2L$. The $2L$ span with the inflection point will buckle at a load that is 68% of the beam with span L. If the inflection point were a brace point, the critical moment of both beams would be the same. The buckled shape of the $2L$ beam shows that the top flange and bottom flange move laterally in opposite directions at midspan. It should be noted that an actual brace on one flange at the inflection point still does not provide effective bracing at midspan (Yura, 1993).

Lateral buckling solutions from finite element analysis for beams with top-flange lateral bracing were obtained and the approximate C_b formulas developed are given

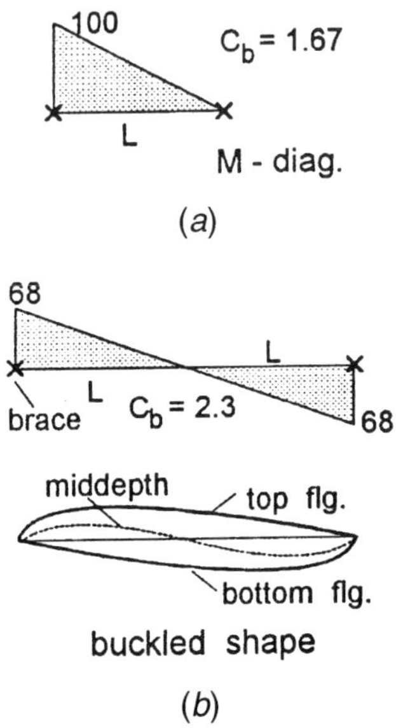

FIGURE 12.11 Comparison of buckling behavior: (*a*) beam braced at both ends, $L_b = L$; (*b*) beam with inflection point at midspan, $L_b = 2L$.

in Fig. 12.12. Two general cases are derived, top flange laterally braced with top-flange gravity loading and top flange braced with uplift loading. These C_b values can be used in design with L_b equaling the span length if twist is positively controlled only at the supports. Torsional restraint along the top flange was neglected. Essa and Kennedy (1995) have presented design charts for suspended construction which also consider the torsional restraint provided by joists attached to the top flange.

12.10 BEAM BRACING

There are two general types of beam bracing, lateral and torsional. Bracing systems for beams must prevent the *relative* displacement of the top and bottom flanges (i.e., twist of the section). Lateral bracing (joists attached to the compression flange of a simply supported beam) and torsional bracing (cross frames or diaphragms between adjacent girders) can effectively control twist. Some bracing systems restrain lateral movement and twist simultaneously (slab attached to the top flange with shear studs). Mutton and Trahair (1973) and Tong and Chen (1988) have shown that combined lateral and torsional bracing is more effective than either lateral or torsional bracing acting alone for beams under uniform moment. Deck systems that are attached directly to the top flange of a beam and act as shear diaphragms can also improve beam stability. Such systems provide mainly warping restraint to the top flange rather than lateral or torsional restraint. Design recommendations for diaphragm-braced beams given in Helwig and Yura (2008) indicate that the diaphragm strength requirement, which is limited by the fastener capacity, generally controls the design.

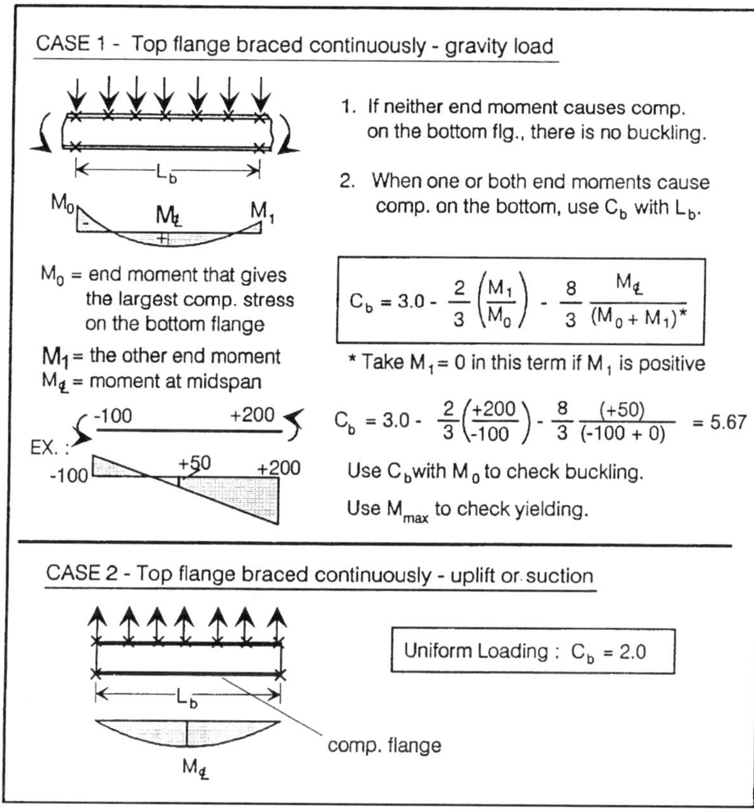

FIGURE 12.12 C_b for braced beams.

A general discussion of beam lateral and torsional bracing and the development of the design recommendations herein are presented elsewhere (Yura, 1993). The provisions are limited to doubly and singly symmetric members loaded in the plane of the web. Lateral bracing can be relative, discrete, continuous, or lean-on. Only relative and discrete lateral bracing requirements are presented here. Continuous lateral bracing is addressed by Trahair and Nethercot (1984) and Yura and Phillips (1992). Beams that are linked together lean on each other and the lateral buckling cannot occur at the links unless all the members buckle. In this case the beams in the structural system cannot buckle until the sum of the maximum moment in each beam exceeds the sum of the individual buckling capacities of each beam (Yura et al., 1992). Buckling of an individual beam can occur only between the cross members in a lean-on system. No additional bracing requirements are necessary in lean-on systems.

Torsional bracing can be either discrete or continuous. If two adjacent beams are interconnected by a properly designed cross frame or diaphragm at midspan, that point can be considered a braced point when evaluating the beam-buckling strength. Because the beams can move laterally at midspan, the effectiveness of such a bracing system is sometimes questioned. As long as the two flanges move laterally by the same amount, there will be no twist. If *twist* is prevented, the beam can be treated as braced. Tests and theory confirm this approach (Flint, 1951; Yura et al., 1992).

12.10.1 Lateral Bracing

The effectiveness and size of a lateral brace depend on its location on the cross section, the moment diagram, the number of discrete braces in the span, and the location of the load on the cross section (Yura, 1993). These factors have been included in the following recommendations. Lateral bracing is most effective when it is attached near the compression flange. The exception to this is for cantilevers where top (tension) flange bracing is most effective. Lateral bracing near the centroid of the cross section is rather ineffective.

The relative and discrete brace design provisions presented below, which are based on Winter's approach, are applicable only for bracing attached near the top flange. The provisions assume top-flange loading, which is a worse case scenario and can be used for any number of discrete braces. The compressive force is conservatively approximated as M_f/h. When the beam has an inflection point, lateral bracing must be attached to both flanges and the stiffness requirements are greater as given by the C_d factor in the brace requirements. For example, for a beam in reverse curvature as shown in Fig.12.11*b*, a brace on both the top and the bottom flanges at midspan will require twice as much stiffness for each brace as a similar length beam with compression on only one flange. The brace force provisions are twice those for columns (Sections 12.4 and 12.5) because top-flange loading is assumed. A brace stiffness of twice the ideal value has been used in the development of these recommendations. When L_b is smaller than the unbraced length, L_q, needed to support the factored loads, then L_q can be substituted for L_b in the stiffness requirement for discrete bracing, as discussed in Section 12.5 for discrete bracing of columns.

Design Recommendation The lateral brace design (LRFD) recommendations are

	Relative	Discrete
Stiffness:	$\beta_L = \dfrac{4M_f C_d}{\phi L_b h}$	$\beta_L = \dfrac{10M_f C_d}{\phi L_b h}$
Strength:	$F_{\text{br}} = \dfrac{0.008M_f C_d}{h}$	$F_{\text{br}} = \dfrac{0.02M_f C_d}{h}$

where M_f is the maximum moment, h the distance between flange centroids, L_b the unbraced length, $\phi = 0.75$, and $C_d = 1.0$ for single curvature and $C_d = 2.0$ for reverse curvature.

The lateral bracing provisions are illustrated in Example 12.4 where a top-flange relative brace truss system is used to stabilize the compression flange during construction of the composite plate girders. Each truss system is designed to stabilize two and one-half girders. The diagonal braces are assumed to support tension only.

Example 12.4: Relative Lateral Brace System Design the diagonals of the top-flange horizontal truss to stabilize the five 80-ft girders with the factored moments shown. Assume $F_y = 36$ ksi.

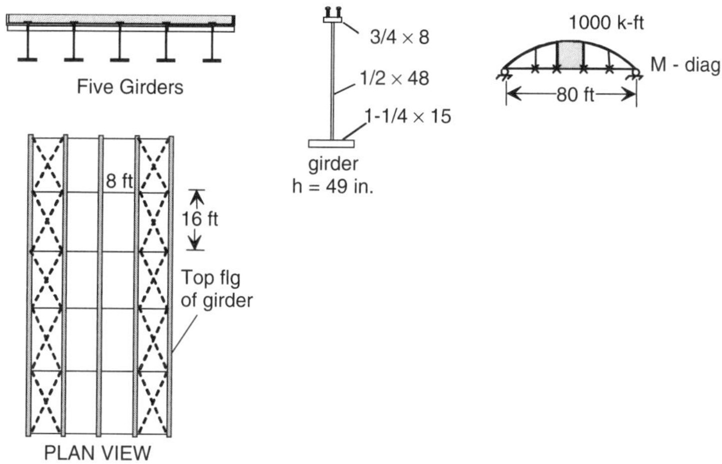

- Stiffness

$$\beta_L = \frac{4.0(1000 \times 12)}{0.75(16 \times 12)49}$$

$$= 6.80 \text{ kips/in. for each girder}$$

$$\times\ 2.5 \text{ girders} = 17.0 \text{ kips/in}$$

$$\left(\frac{AE}{L}\right)_b \cos^2\theta = \frac{A_b(29{,}000)}{8 \times 12 \times \sqrt{5}}\left(\frac{1}{\sqrt{5}}\right)^2 = 17.0$$

$$\text{A}_b = 0.629 \text{ in}^2 \leftarrow \text{controls}$$

- Strength

$$F_{\text{br}} = \frac{0.008(2.5)(1000 \times 12)}{49} = 4.90 \text{ kips}$$

$$A_b = \frac{4.90\sqrt{5}}{0.9 \times 36} = 0.34 \text{ in.}^2$$

Use L2×2×$\frac{3}{16}$; $\quad A = 0.715$ in.^{2}

12.10.2 Torsional Bracing

Cross frames or diaphragms at discrete locations or continuous bracing provided by the floor system in through-girders or pony trusses or by metal decks and slabs represent torsional bracing systems. In the development of the design recommendations (Yura, 1993), which are based on the work of Taylor and Ojalvo (1966), it was determined that factors that had a significant effect on lateral bracing had a substantially reduced effect on torsional bracing. The effects of the number of braces, top-flange loading, and brace location on the cross section are relatively unimportant when sizing a torsional brace. A torsional brace is equally effective if it is attached to the tension flange or the compression flange. A moment diagram with compression in both flanges (reverse curvature) does not alter the torsional brace requirements.

On the other hand, the effectiveness of a torsional brace is greatly impacted by cross-sectional distortion at the brace point, as illustrated in Fig. 12.13. The top flange is prevented from twisting by the torsional brace, but the web distortion permits a relative displacement between the two flanges. A stiffener at the brace location can be used to prevent the distortion. The design method presented below considers web distortion and any stiffeners required.

Discrete braces and continuous bracing use the same basic design formula with the continuous brace stiffness taken as $\overline{\beta}_T = \beta_T n/L$, where β_T is the discrete brace stiffness, n the number of braces, and L the span length. In this case, β_T and $\overline{\beta}_T$ are defined as the torsional stiffnesses of the discrete and continuous bracing systems, respectively. The system stiffness β_T is primarily related to the stiffness of the brace β_b and the stiffness of the web plus any stiffeners β_{sec} by

$$\frac{1}{\beta_T} = \frac{1}{\beta_b} + \frac{1}{\beta_{\text{sec}}} \tag{12.8}$$

The β_b for diaphragm systems is given in Fig. 12.14. The discrete web stiffener detail can vary over the web as shown in Fig. 12.15. The term $1/\beta_{sec} = \sum(1/\beta_i)$ with the stiffness of each portion i of the web given by

$$\beta_c, \beta_s, \beta_t = \frac{3.3E}{h_i}\left(\frac{h}{h_i}\right)^2\left(\frac{(1.5h_i)\,t_w^3}{12} + \frac{t_s b_s^3}{12}\right) \tag{12.9}$$

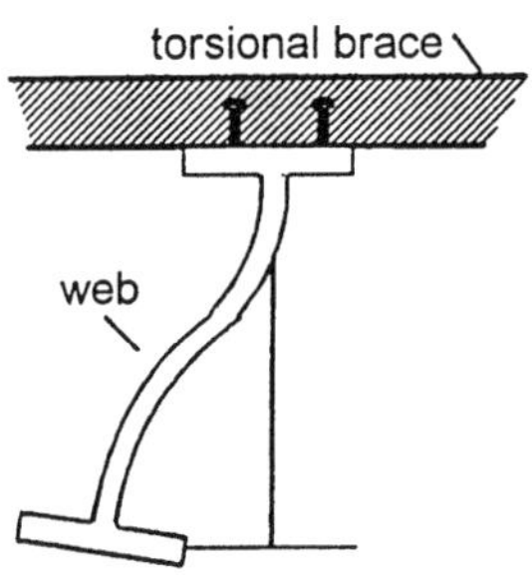

FIGURE 12.13 Web distortion.

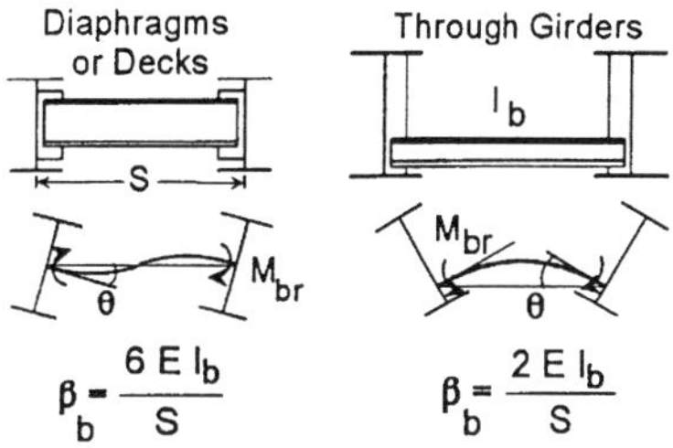

FIGURE 12.14 Diaphragm β_b.

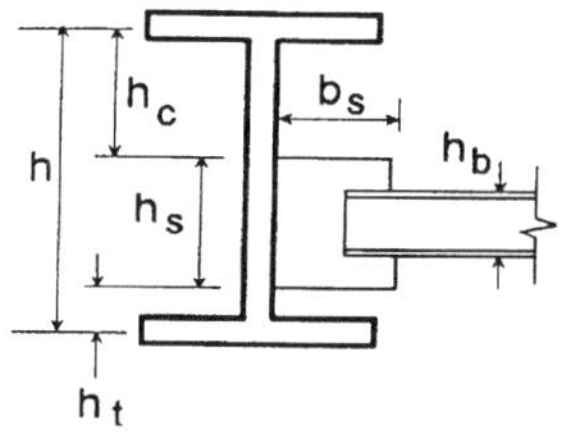

FIGURE 12.15 Partially stiffened webs.

where t_s is the thickness of the stiffener. For continuous bracing, it is recommended that $1.5h$ be replaced with 1 in. and the t_s term be neglected if there is no stiffener. The portion of the web within h_b can be considered infinitely stiff. Equations 12.8 and 12.9 were developed from Milner and Rao (1978) and finite element buckling analysis that considers cross-sectional distortion (Akay et al., 1977). For rolled sections ($h/t_w < 60$) cross-sectional distortion will not be significant if the diaphragm connection extends at least three-fourths of the web depth.

Design Recommendation The torsional brace design (LRFD) recommendations are

$$\text{Stiffness:}\quad \beta_T = \frac{\overline{\beta}_{TL}}{n} = \frac{2.4LM_f^2}{\phi nEI_{\text{eff}}C_{bb}^2} \tag{12.10}$$

$$\text{Strength:}\quad M_{\text{br}} = F_{\text{br}}h_b = \frac{0.005L_bLM_f^2}{nEI_{\text{eff}}C_{bb}^2h} \tag{12.11}$$

where M_f is the maximum moment, $I_{\text{eff}} = I_{yc} + (t/c)I_{yt}$, L the span length, L_b the unbraced length, n the number of span braces, h the distance between flange centroids, and C_{bb} the moment modification factor for the full-bracing condition. For a singly symmetric section, I_{yc} and I_{yt} are the out-of-plane moments of inertia of the compression and tension flanges (Fig. 12.16), respectively. If the cross section is doubly symmetric, I_{eff} becomes I_y. The 2.4 factor in the stiffness requirement comes from using twice the ideal stiffness and an additional 20% increase to account for top-flange loading. The brace strength provision, M_{br}, assumes an initial twist of $0.002L_b/h$ and is consistent with the imperfection used for lateral bracing (Helwig et

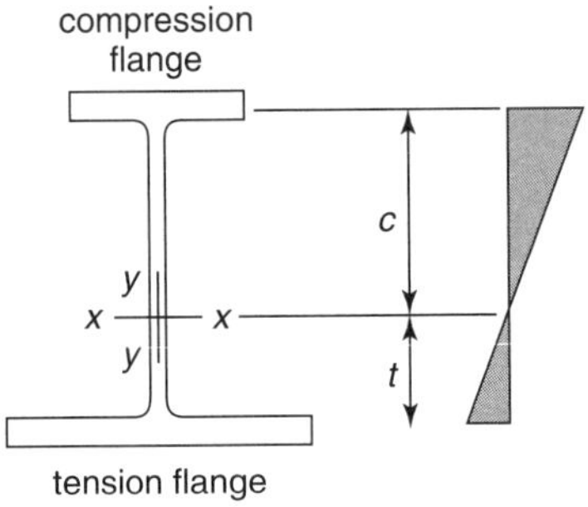

FIGURE 12.16 Singly symmetric section.

al., 1993). When the values of the variables in the two unbraced segments adjacent to a brace are different, the brace can be designed for the average of the strengths and stiffnesses determined for both segments.

In Example 12.5 a diaphragm torsional bracing system is used for the problem given in Example 12.4. The C9×13.4 diaphragm will not brace the girders if a stiffener is not used. Even a much larger diaphragm cannot work without web stiffeners because of the web distortion. Similar example problems using cross frames are given by Yura (1993).

Example 12.5: Torsional Beam Bracing

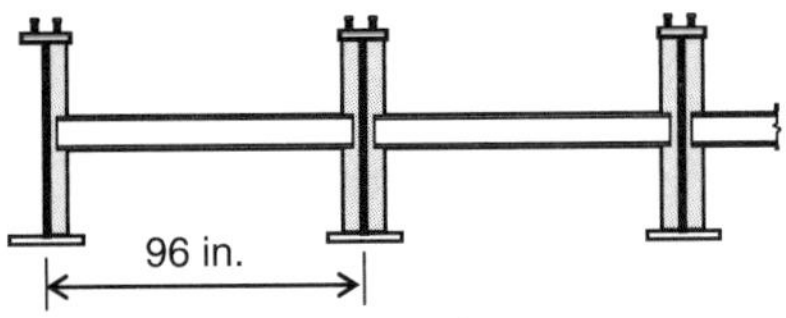

Same as Example 12.4 but use the diaphragm system shown. Assume $M_{\max} =$ 1000 kip-ft, $C_b = 1.0$; four braces, $F_y = 36$ ksi, $L = 80$ ft. The girder properties are as follows:

$$h = 49.0\text{ in.} \qquad c = 30.85\text{ in.} \qquad t = 18.15\text{ in.}$$

$$I_x = 17{,}500\text{ in.}^4 \qquad I_{yc} = 32.0\text{ in.}^4$$

$$I_{yt} = 352\text{ in.}^4 \qquad I_{\text{eff}} = 32 + \frac{18.15}{30.85}352 = 239\text{ in.}^4$$

The strength is given by

$$M_{\text{br}} = \frac{0.005(80 \times 12)(1000 \times 12)^2(16 \times 12)}{4(29{,}000)239(1.0)^2 49} = 97.7\text{ in.-kips}$$

$$S_{x\ \text{req}} = \frac{97.7}{(0.9) \times (36)} = 3.02\text{ in.}^3$$

The stiffness of the diaphragms on the exterior girders is $6EI_{br}/S$. Because there are diaphragms on both sides of each interior girder, the stiffness is $2 \times 6EI_{br}/S$. The average stiffness available to each girder is $(2 \times 6 + 3 \times 12)/5 = 9.6EI_{br}/S$.

$$\beta_{T\ \text{req}} = \frac{2.4(80 \times 12)(1000 \times 12)^2}{(0.75)4(29{,}000)239(1.0)^2} = 15{,}960\ \text{in.-kips/rad}$$

$$I_{\text{br}_{\min}} = \frac{15{,}960(96)}{(9.6)29{,}000} = 5.50\ \text{in.}^4$$

Try C9×13.4: $S_x = 12.5\ \text{in.}^3 > 3.02, \quad I_x = 47.9\ \text{in.}^4$

$$\beta_b = \frac{9.6(29{,}000)47.9}{96} = 138{,}900\ \text{in.-kips/rad}$$

$$\frac{1}{15{,}960} = \frac{1}{138{,}900} + \frac{1}{\beta_{\text{sec}}} \qquad \beta_{\text{sec}} = 17{,}900\ \text{in.-kips/rad}$$

$$\frac{1}{17{,}900} = \frac{2}{\beta_c} \qquad \beta_c = 2(17{,}900) = \frac{3.3(29{,}000)}{20}\left(\frac{49}{20}\right)^2 \times \left[\frac{1.5(20)(0.5)^3}{12} + \frac{0.375b_s^3}{12}\right]$$

$$b_s = 3.10\ \text{in.}$$

Use a $\frac{3}{8} \times 3\frac{1}{2}$-in.stiffener.

REFERENCES

AISC (2005), *Steel Construction Manual*, American Institute of Steel Construction, 13th ed., Chicago, IL.

Akay, H. U., Johnson, C. P., and Will, K. M. (1977), "Lateral and Local Buckling of Beams and Frames," *ASCE J. Struct. Div.*, Vol. 103, No. ST9, pp. 1821–1832.

Ales, J. M., and Yura, J. A. (1993), "Bracing Design for Inelastic Structures," *Proc., SSRC Conf. "Is Your Structure Suitably Braced?"* Milwaukee, WI., Apr., pp. 29–37.

ASCE (1971), *Commentary on Plastic Design in Steel*, ASCE Manual No. 41, 2nd ed., American Society of Civil Engineers, New York.

Chen, S., and Tong, G. (1994), "Design for Stability: Correct Use of Braces," *Steel Struct. J. Singapore Struct. Steel Soc.*, Vol. 5, No. 1, pp. 15–23.

Essa, H. S., and Kennedy, D. J. L. (1995), "Design of Steel Beams in Cantilever-Suspended-Span Construction," *J. Struct. Eng.*, Vol. 121, No. 11, pp. 1667–1673.

Flint, A. R. (1951), "The Stability of Beams Loaded through Secondary Members," *Civ. Eng. Public Works Rev.*, Vol. 46, No. 537-8, pp. 175–177, 259–260.

Gil, H., and Yura, J. A. (1999), "Bracing Requirements of Inelastic Columns," *J. Constr. Steel Res.*, Vol. 51, No. 1, pp. 1–19.

Helwig, T. A., and Yura, J. A. (1999), "Torsional Bracing of Columns," *J. Struct. Eng.*, Vol. 125, No. 5, pp. 547–555.

Helwig, T. A., and Yura, J. A. (2008), "Shear Diaphragm Bracing of Beams. II: Design Requirements," *J. Struct. Eng.*, Vol. 134, No. 3, pp. 357–363.

Helwig, T. A., Yura, J. A., and Frank, K. H., (1993), "Bracing Forces in Diaphragms and Cross Frames," *Proc., SSRC Conf., "Is Your Structure Suitably Braced?"* Milwaukee, WI., Apr., pp. 129–140.

Horne, M. R., and Ajmani, J. L. (1971), "Design of Columns Restrained by Side Rails," *Struct. Eng.*, Vol. 49, No. 8, pp. 329–345.

Horne, M. R., and Ajmani, J. L. (1972), "Failure of Columns Laterally Supported on One Flange," *Struct. Eng.*, Vol. 50, No. 9, pp. 355–366.

Lutz, L. A., and Fisher, J. M. (1985), "A Unified Approach for Stability Bracing Requirements," *AISC Eng. J.*, Vol. 22, No. 4, pp. 163–167.

Medland, I. C., and Segedin, C. M. (1979), "Brace Forces in Interbraced Column Structures," *ASCE J. Struct. Div.*, Vol. 105, No. ST7, pp. 1543–1556.

Milner, H. R., and Rao, S. N. (1978), "Strength and Stiffness of Moment Resisting Beam-Purlin Connections," *Civil Eng. Trans. Inst. Eng., Aust.*, Vol. CE 20, No. 1, pp. 37–42.

Mutton, B. R., and Trahair, N. S. (1973), "Stiffness Requirements for Lateral Bracing," *ASCE J. Struct. Div.*, Vol. 99, No. ST10, pp. 2167–2182.

Nakamura, T. (1988), "Strength and Deformability of H-Shaped Steel Beams and Lateral Bracing Requirements," *J. Const. Steel Res*, Vol. 9, pp. 217–228.

Pincus, G. (1964), "On the Lateral Support of Inelastic Columns," *AISC Eng. J.*, Vol. 1, No. 4, pp. 113–115.

Plaut, R. H. (1993), "Requirements for Lateral Bracing of Columns with Two Spans," *J. Struct. Eng.*, Vol. 119, No. 10, pp. 2913–2931.

Taylor, A. C., and Ojalvo, M. (1966), "Torsional Restraint of Lateral Buckling," *ASCE J. Struct. Div.*, Vol. 92, No. ST2, pp. 115–129.

Thomas, S. J., and Earls, C. J. (2003), "Cross-Sectional Compactness and Bracing Requirements for HPS483W Girders," *J. Struct. Eng.*, Vol. 129, No. 12, pp. 1569–1581.

Timoshenko, S. P. (1936), *Theory of Elastic Stability*, McGraw-Hill, New York.

Timoshenko, S. P., and Gere, J. M. (1961), *Theory of Elastic Stability*, McGraw-Hill, New York.

Tong, G., and Chen, S. (1987), "Design Forces of Horizontal Inter-Column Braces" *J. Constr. Steel Res.*, Vol. 7, pp. 363–370.

Tong, G., and Chen, S. (1988), "Buckling of Laterally and Torsionally Braced Beams," *J. Constr. Steel Res.*, Vol. 11, pp. 41–55.

Tong, G., and Chen, S. (1989), "The Elastic Buckling of Interbraced Girders," *J. Constr. Steel Res.*, Vol. 14, pp. 87–105.

Trahair, N. S., and Nethercot, D. A. (1984), "Bracing Requirements in Thin-Walled Structures," in *Developments in Thin-Walled Structures*, Vol. 2 (Eds. J. R. Rhodes and A. C. Walker), Elsevier, New York, pp. 93–130.

Waddell, J. A. L. (1916), *Bridge Engineering*, Vols. I and II, Wiley, New York.

Wang, L., and Helwig, T. A. (2005), "Critical Imperfections for Beam Bracing Systems," *J. Struct. Eng.*, Vol. 131, No. 6, pp. 933–940.

Wang, Y. C., and Nethercot, D. A. (1989), "Ultimate Strength Analysis of Three-Dimensional Braced I-Beams," *Proc. Inst. of Civ. Eng*, Part 2, Vol. 87, pp. 87–112.

Winter, G. (1960), "Lateral Bracing of Columns and Beams," *Trans. ASCE*, Vol. 125, Part 1, pp. 809–825.

Yura, J. A. (1971), "The Effective Length of Columns in Unbraced Frames," *AISC Eng. J.*, Vol. 8, No. 2, pp. 37–42.

Yura, J. A. (1993), "Fundamentals of Beam Bracing," *Proc. SSRC Conf., "Is Your Structure Suitably Braced?"* Milwaukee, WI., Apr. Reprinted and updated in *AISC Eng. J.*, Vol. 38, No. 1, 2001, pp. 11–26.

Yura, J. A. (1994), "Winters Bracing Model Revisited," *50th Anniv. Proc. SSRC,* Bethlehem, PA, pp. 375–382.

Yura, J. A. (1995), "Bracing for Stability-State-of-the-Art," *Proc. Struct. Congr. XIII*, ASCE, Boston, Apr., pp. 88–103.

Yura, J. A., and Li, G. (2002), "Bracing Requirements for Inelastic Beams," *Proc., SSRC Conf.*, Seattle, WA, pp 53–73.

Yura, J. A., and Phillips, B. A. (1992), "Bracing Requirements for Elastic Steel Beams," Research Report 1239-1, Center for Transportation Research, Univ. of Texas, Austin, TX, May.

Yura, J. A., Phillips, B., Raju, S., and Webb, S. (1992), "Bracing of Steel Beams in Bridges," Research Report 1239-4F, Center for Transportation Research, Univ. of Texas, Austin, TX, Oct.

Zuk, W. (1956), "Lateral Bracing Forces on Beams and Columns," *ASCE J. Eng. Mech. Div.*, Vol. 82, No. EM3, pp 1032-1–1032-11.

CHAPTER 13

THIN-WALLED METAL CONSTRUCTION

13.1 INTRODUCTION

Thin-walled metallic members are used in a significant number of structural applications, including buildings, bridges, storage tanks, cars, ships, and aircraft. Thin-walled members are members composed of elements with large width-to-thickness ratios and so may be thought of as slender-element members. The elements need not be especially thin but rather need to have a small thickness relative to their width. Within this chapter, thin-walled metal construction refers primarily to civil engineering applications, especially building structures. Thin-walled metallic members provide the primary framing system in low-rise and midrise construction and the secondary framing system in high-rise or long-span construction. Such members are also commonly used in specialty structures such as storage racks and greenhouses.

Cold-formed steel, stainless steel, and aluminum members, either by the use of slender elements or the inherent nature of the stress–strain response of the base material, all qualify as thin-walled metallic members. Example cross sections include tubular members (Fig. 4.1*b*), I-sections (Fig. 4.1*h*), channels, Z-sections, hat sections (Fig. 4.1*i*), T-sections (Fig. 4.1*m*), and panels (Figs. 4.1*o*, *p*, *q*). The depth of such members generally ranges from about 1 to 12 in. (25 to 300 mm), and the thickness of the material ranges from about 0.003 to 0.5 in. (0.08 to 12 mm). Members are often made of sheet metal, but may also be extruded, especially in aluminum.

In thin-walled metal construction understanding stability behavior and accounting for, or mitigating, this behavior in design play a dominant role in successful engineering. In a thin-walled member, local buckling and cross-section distortion are an essential part of member design. These complications also provide certain opportunities, as local buckling, in particular, has the capacity for beneficial post-buckling reserve that can provide greater strength than the buckling strength. As a result, the ultimate efficiency, for example, strength-to-weight ratio, can be quite high for thin-walled metallic members. The challenge for any design method is to incorporate as many of these complicated phenomena, which are often ignored in the conventional design of "compact" metal sections, into as simple and familiar a design method as possible. Further complicating the creation of simple design methods for thin-walled members is the lack of symmetry in many cross sections, the enhanced possibility of limit states related directly to the use of thin sheets such as web crippling, and other unique characteristics of their manufacture and application.

Thin-walled cold-formed steel enjoys a wide and growing base of application in civil engineering structures. For a number of years, cold-formed steel members have been a mainstay of metal building systems serving as purlins, girts, and the building skin. Also, in low- and high-rise construction cold-formed steel panels are widely used as floor decking. Today high-rise construction also uses a significant amount of cold-formed steel for partition walls. In addition, load-bearing cold-formed steel for low-rise and midrise buildings has seen significant growth in the last two decades, where cold-formed steel members frame the walls, floors, and roof (including trusses). Along with this increase in applications has come research to support its innovative use and overcome new challenges.

The emphasis in this chapter is on design and stability related to cold-formed steel construction. Stainless steel and aluminum members are covered in the last two sections of this chapter. The first requirement for any thin-walled metallic member is to determine the elastic stability modes of the member, including local, distortional, and global—this is covered in Section 13.3. Two design methods currently exist for cold-formed steel, the classical *effective width method* and the newly developed *direct strength method*. These important design approaches are the focus of Sections 13.3 and 13.4, respectively. Other design approaches such as *reduced stress*, *effective thickness*, the *form-factor* (or Q-factor) *approach*, and more recently, the *erosion of critical bifurcation load* (Ungureanu and Dubina, 2004) are not detailed here. As the focus of the presented design methods (Sections 13.3 and 13.4) is on columns and beams, Section 13.6 provides the additional stability and strength limit states that must also be considered for a successful cold-formed steel member. With most of the current research focusing on cold-formed steel systems, the stability and strength of such assemblies are discussed in Section 13.7.

13.2 MEMBER STABILITY MODES (ELASTIC)

A distinguishing feature of thin-walled members is that cross-section stability must be considered in their design, because it often contributes to, or dominates, the observed behavior under load. This section presents elastic stability modes of thin-walled members, namely local, distortional, and global buckling. Historically, closed-form expressions have been employed by engineers in design, and today this trend continues, though some relief using computational methods is typically allowed. Given the historical importance of the closed-form expression for the stability modes, Sections 13.2.2, 13.2.3, and 13.2.4 cover the analytical expressions in use for local, distortional, and global buckling, respectively. Finally, Section 13.3.5 addresses computational tools for thin-walled member stability, with particular emphasis on the finite strip method.

13.2.1 Local, Distortional, and Global Buckling

Thin-walled members typically have at least three stability modes of interest in design and include local, distortional, and global buckling. The AISI Specification (2007a) provides definitions for the three buckling modes of a flexural member as follows:

- *Local Buckling*: Buckling of elements only within a section, where the line junctions between elements remain straight and angles between elements do not change.
- *Distortional Buckling*: A mode of buckling involving change in cross-sectional shape, excluding local buckling.
- *Lateral–Torsional Buckling*: Buckling mode of a flexural member involving deflection out of the plane of bending occurring simultaneously with twist about the shear center of the cross section.

As an example consider the three stability modes for a cold-formed steel lipped channel in bending as provided in Fig. 13.1. In the example, the local buckling moment is 67% of the yield moment (M_y), and the buckling mode shape has a half-wavelength of 5 in., the distortional mode is 65% of M_y with a 25-in. half-wavelength, and the global mode is lateral–torsional buckling with only one half-wave along the length relevant. Given the magnitude of the buckling moments (significantly less than M_y), all three buckling modes, and potentially their interactions, may be involved in the design of this member. This is typical in thin-walled member design.

Examination of the buckled shapes provided in Fig. 13.1 provides support for the AISI definitions. The provided definitions, however, are limited in their application, particularly for distortional buckling. From a practical standpoint, the modes may often be identified by the characteristics of the buckled shape and their appearance at a given half-wavelength. Getting beyond heuristics, mechanics-based definitions of the three modes have been proposed and implemented in the context of generalized beam theory (Silvestre and Camotim, 2002a,b) and the finite strip method (Adany and Schafer, 2006a,b). Both these methods provide the potential for automatic identification of the three primary buckling modes.

The "signature curve" of a thin-walled cross section as given in Fig. 13.1 was pioneered and popularized by Hancock [e.g., see Hancock (1978) and the figures in Chapter 4, Section 4.6]. This signature curve of a cross section has become an organizing principle for understanding the behavior of thin-walled sections. This curve was generated using the finite strip method and is discussed further in Section

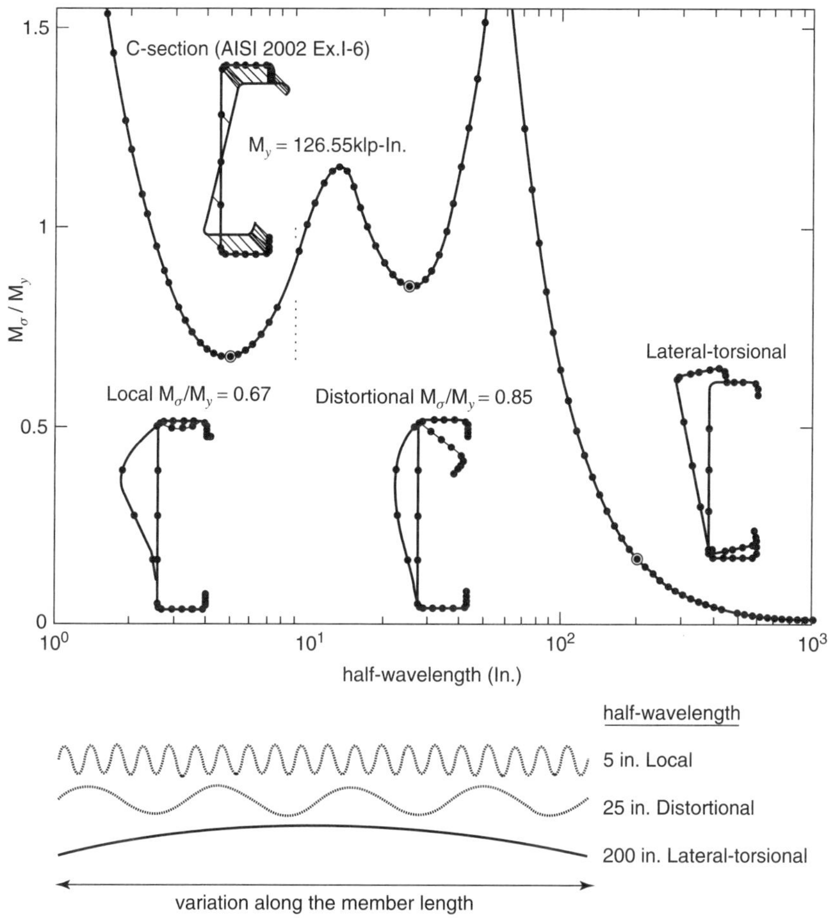

FIGURE 13.1 Normalized buckling moment vs. buckling mode half-wavelength for a lipped channel.

13.2.5. Further, the implications of and ease with which such information can now be obtained are the motivating factors for a new design method, the *direct strength method*, discussed in Section 13.4.

13.2.2 Local Buckling via Plate Stability

As discussed fully in Chapter 4, the classical method for determining the local stability of thin-walled cross sections is to break the section into a series of plates. For a lipped channel in compression, the procedure is illustrated in Fig. 13.2. The web (part A) is a stiffened element and may be idealized as a plate simply supported

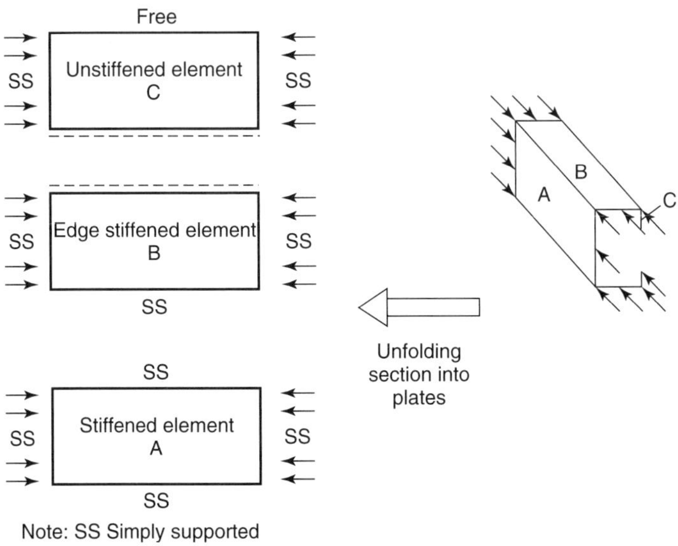

FIGURE 13.2 Breakdown of a thin-walled cross section into elements for plate-buckling evaluation.

on all four sides. The plate-buckling stress is

$$f_{cr,l} = k \frac{\pi^2 E}{12\left(1 - \nu^2\right)} \left(\frac{t}{w}\right)^2 \tag{13.1}$$

where $k = 4.0$, via case 1 of Fig. 4.2, and E is Young's modulus, ν the Poisson's ratio, t the thickness, and w the plate width. The lip (part C) is an unstiffened element and may be idealized as a simply supported plate with one longitudinal edge free, $k = 0.425$, via case 4 of Fig. 4.2.

The flange (part B) is an edge-stiffened element and represents a potentially more complicated condition. For pure local buckling, the buckling mode shape of the flange (see inset of Fig. 13.1) is similar to the web—and may be approximated as a plate simply supported on four sides with a k of 4.0. If the lip is short, however, then the edge-stiffened element may behave more like an unstiffened element; hence the reason the flange is referred to as an "edge-stiffened" element instead of simply a "stiffened" element. Treatment of this mode of buckling has varied significantly over the years but today is known as distortional buckling (again, see inset buckled shapes of Fig. 13.1) and will be discussed in detail in the next section.

If the lipped channel of Fig. 13.2 is under bending instead of compression, then the plate-buckling coefficients (k's) must be suitably modified. Stiffened elements under stress gradients and unstiffened elements under stress gradients are handled by the expressions reported in Section 4.2.2 (Bambach and Rasmussen, 2004; Peköz, 1987). In addition, the influence of moment gradient on the local plate buckling may also be considered as reported in Section 4.2.7 (Yu and Schafer, 2007a).

Interaction of Elements in Local Buckling As indicated in Section 4.6, the classical approach to local buckling using isolated plates violates equilibrium and compatibility of the cross section. Consider, for instance, the cross section of Fig. 13.1, but now under the simpler case of pure compression. The centerline dimensions (ignoring corner radii) are $h = 8.94$ in. (227.1 mm), $b = 2.44$ in. (62.00 mm), $d = 0.744$ in. (18.88 mm), and $t = 0.059$ in. (1.499 mm), and the critical buckling stresses f_{cr} of each element using the classical plate stability approach are

Lip: $k = 0.43, \quad f_{cr,l\text{-lip}} = 0.43[\pi^2 E/(12(1-\nu^2))](t/d)^2$
$= 72.1$ ksi (497 MPa)

Flange: $k = 4, \quad f_{cr,l\text{-flange}} = 4.0[\pi^2 E/(12(1-\nu^2))](t/b)^2$
$= 62.4$ ksi (430 MPa)

Web: $k = 4, \quad f_{cr,l\text{-web}} = 4.0[\pi^2 E/(12(1-\nu^2))](t/h)^2$
$= 4.6$ ksi (32.0 MPa)

Each element predicts a different buckling stress, even though the member is a connected (continuous) group. The relatively high flange- and lip-buckling stresses have little relevance given the low web-buckling stress. A finite strip analysis (see Section 13.2.5 for more on computational solutions), which includes the interaction among the elements, shows that the flange aids the web significantly in local buckling, increasing the web-buckling stress from 4.6 ksi (32.0 MPa) to 6.6 ksi (45.4 MPa), but the buckling stress in the flange and lip are significantly reduced due to this same interaction.

It is possible to approximate the local buckling stress as the minimum of the element-buckling stresses, but this method is typically overly conservative, as demonstrated in the preceding example and previously discussed in Section 4.6. The nomographs and references of Section 4.6 provide a means to determine the local buckling stress, including the interaction for many simple shapes. Analytical expressions for flange–web and flange–lip local buckling interaction do exist for beams (Schafer and Peköz, 1999) and columns (Schafer, 2002). The computational methods discussed in Section 13.2.5, however, are by far the simplest method for including the interaction of the connected elements (plates) in local buckling.

13.2.3 Distortional Buckling Expressions

Edge Stiffeners Analytical models for the elastic critical distortional buckling stress (see Fig. 13.1) have proven to be relatively complicated. The deformations involved include both membrane deformations, primarily in the flange and lip, and bending deformations, primarily in the web. Desmond et al. (1981a) provided expressions for the plate-buckling coefficient (k) of the flange that are a function of the flange width and lip stiffener moment of inertia, and these are provided in AISI (2007a). They have been shown, however, to be a poor predictor of the elastic distortional buckling stress (Schafer and Peköz, 1999) and are only intended to be used in conjunction with the specific effective width expressions provided [that is to say the k's provided by Desmond et al. (1981a) are not actually elastic buckling k's, but empirically modified].

An alternative approach has been to account for the distortional mode of buckling as a compressed strut on an elastic foundation, which in turn can be represented by a spring that depends upon the bending stiffness of adjacent parts of plane elements and the boundary conditions of the element. This procedure has been adopted in Eurocode 3, Part 1.3 (Eurocode, 2004). The method accounts for the elastic restraint of all elements in the section, including the web, by incorporation of their flexibility in the elastic spring restraint. A detailed discussion of this method applied to channel sections is given in Buhagiar et al. (1992). Another interesting alternative approach is the analytical application of *generalized beam theory* to determine closed-form expressions for distortional buckling (Silvestre and Camotim, 2004a,b,c).

The analytical model in widest use is Lau and Hancock's (1987) and is based primarily on the assumption that the flange acts as an isolated column undergoing flexural–torsional buckling, while the web provides elastic restraint to the flange, as shown in Fig. 13.3. This model was first considered by aluminum researchers (Sharp, 1966) and subsequently improved to include more consistent treatment of the web (Lau and Hancock, 1987). Schafer and Peköz (1999) further developed this model to allow for the impact of applied stresses on the web's rotational stiffness, thus allowing for the case when distortional buckling is triggered by instability of the web as opposed to the flange. Teng et al. (2003) examined the method's application to beam-columns. Lau and Hancock's treatment is used in the Australian cold-formed steel standard (AS/NZS, 2005) and Schafer and Peköz's in the AISI (2007a) Specification.

Treatment of the flange as a column requires that separate section properties (I, J, C_w, etc.) for the flange–lip components be calculated. These section properties

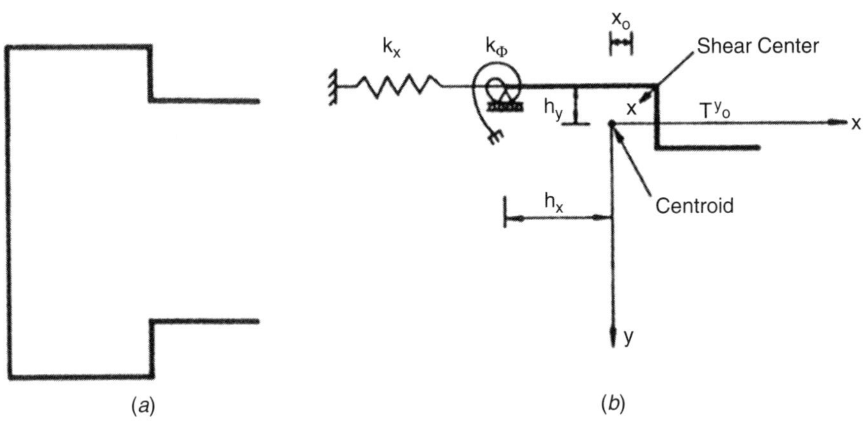

FIGURE 13.3 Lau and Hancock's model for distortional buckling. Flange elastically restrained along flange/web junction: (*a*) original column section; (*b*) isolated flange (Lau and Hancock, 1987).

are then used in the torsional–flexural buckling problem, which itself requires the solution to a quadratic equation (see Eq. 13.3). The involved nature of these calculations make computational solutions far more attractive, and indeed the AISI (2007a) Specification recognizes this by making it explicitly clear that rational elastic buckling analysis is permitted for determining the elastic distortional buckling stress.

Intermediate Stiffeners To a certain extent the elastic distortional buckling treatment of intermediate stiffeners has developed in a manner similar to edge stiffeners. Desmond et al. (1981b) provided a method for predicting k as a function of the plate width and stiffener moment of inertia. This method was employed from the 1986 to 1996 versions of the AISI Specification. Eurocode employs the model of a compressed strut on an elastic foundation (Eurocode, 2004). Based on the classical expressions for stiffened plates (see Section 4.4), AISI (2001) adopted new expressions for plates with intermediate stiffeners (Schafer and Peköz, 1998a) that are provided in Section 4.4.1. The primary difference in expressions for elastic distortional buckling of intermediate stiffeners, as opposed to edge stiffeners, is that the interaction of the elements (e.g., web–flange) is typically ignored and the focus remains only on the element with the intermediate stiffeners (the classic stiffened plate).

Edge and Intermediate Stiffeners Analytical expressions for distortional buckling of sections with both intermediate and edge stiffeners are essentially too involved to be practical. The only potential exception to this is the case of a single-mode generalized beam theory solution—if the generalized beam theory cross-section parameters are known, then the resulting analytical expressions are tractable (Silvestre and Camotim, 2004c). Even the calculation of the cross-section properties, however, is essentially a computational method. Computational solutions for sections with edge and intermediate stiffeners pose no particularly unique problem for computational solutions (as discussed further in Section 13.1.5) and are recommended for the design of such sections.

13.2.4 Global/Flexural–Torsional Buckling

Columns Concentrically loaded columns can buckle by flexure about one of the principal axes (flexural buckling), twisting about the shear center (torsional buckling), or a combination of both flexure and twisting (flexural–torsional buckling). Torsional buckling is a possible failure mode for point symmetric sections. Flexural–torsional buckling must be checked for open sections that are singly symmetric and for sections that have no symmetry. Because it is assumed that no bending occurs prior to buckling, open sections that are doubly symmetric or point symmetric are not subject to flexural–torsional buckling because their shear center and centroid coincide. Closed sections also are immune to flexural–torsional buckling. Flexural–torsional buckling is common in thin-walled construction and will be discussed below (see Chapter 3 for further discussion on flexural column buckling).

One can explain the nature of flexural–torsional buckling with the aid of Fig. 13.4. At buckling, the axial load can be visualized to have a lateral component $q\,dz$ as a consequence of the column deflection. The torsional moment of this lateral component about the shear center of the open section shown in Fig. 13.4 causes twisting of the column. The degree of interaction between the torsional and flexural deformations determines the amount of reduction of the buckling load in comparison to the flexural buckling load. Therefore, as the distance between the shear center and the point of application of the axial load increases, the twisting tendency increases and the flexural–torsional buckling load decreases. Flexural–torsional buckling can be a critical mode of failure for thin-walled open sections because of their low torsional rigidity. The theory of elastic flexural–torsional instability is well developed (Goodier, 1942; Vlasov, 1959; Timoshenko and Gere, 1961; Galambos, 1968). Flexural–torsional buckling of singly symmetric thin-walled open sections under concentric and eccentric loading also has been studied in detail, and design aids have been devised (Klöppel and Schardt, 1958; Pfluger, 1961; Chajes and Winter, 1965; Chilver, 1967; Peköz, 1969; Peköz and Winter, 1969). Since 1980, the AISI specification has contained flexural–torsional buckling provisions based on the work of Chajes et al. (1966), Peköz (1969), and Peköz and Winter (1969).

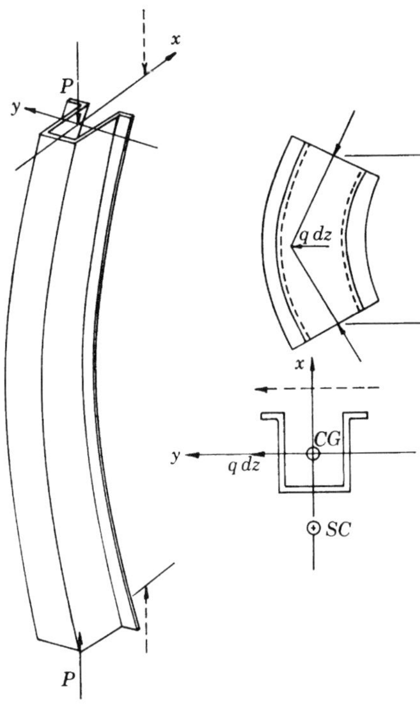

FIGURE 13.4 Forces causing flexural–torsional buckling.

Differential equations of equilibrium for the general case of biaxial eccentricities have been solved by Thurlimann (1953), Vlasov (1959), Dabrowski (1961), Prawel and Lee (1964), Culver (1966), and Peköz and Winter (1969) using different solution procedures. If the section is singly symmetric, such as the sections shown in Fig. 13.5, and is acted on by an axial load not in the plane of symmetry or if the section is not symmetric, the solution of the differential equations indicates that as the axial load increases the member continuously twists and deflects biaxially. The principal axes, twist angle ϕ, and deflections u and v are shown in Fig. 13.6. Analogous to small-deflection flexural beam-column theory, infinite deflections and rotation are predicted for a certain value of the axial load.

If the section is singly symmetric and the axial load is applied through the centroid, however, the behavior of the member is described by three homogeneous differential equations, two of which are coupled. If the member is assumed to be hinged at both ends, namely, $u'' = v'' = \phi'' = 0$, the solution of the one uncoupled equation gives the critical load for buckling in the direction of the symmetry axis

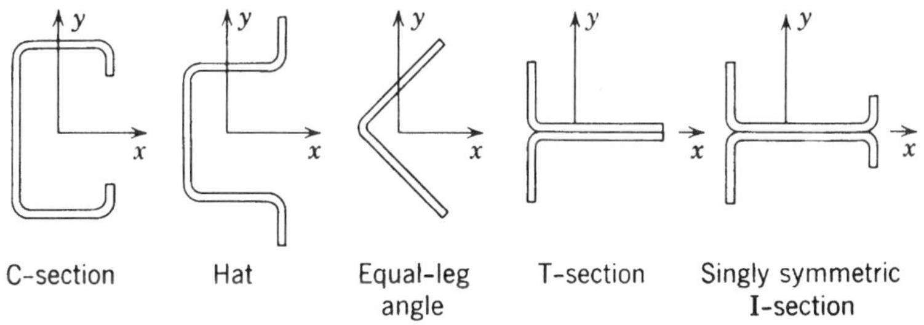

FIGURE 13.5 Several singly symmetric sections and coordinate axis orientation.

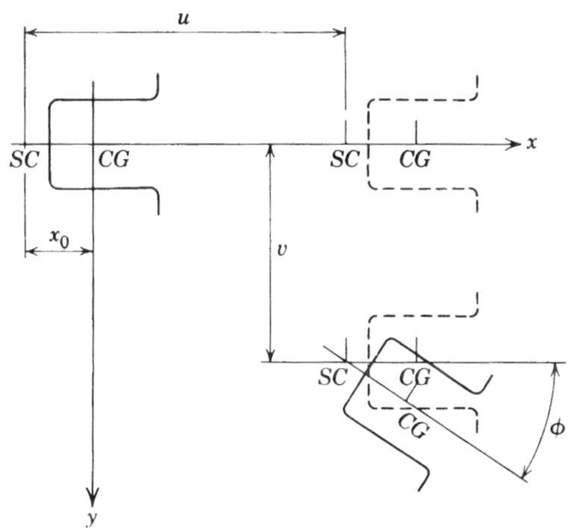

FIGURE 13.6 Principal axes and deflection components.

(taken here as the x axis) as

$$P_{ye} = K_{11} EI_y \frac{\pi^2}{L^2} \tag{13.2}$$

where I_y is the moment of inertia about the y axis and L is the length of the column. The K_{11} and other K values determined by the Galerkin method for various boundary conditions are given by Peköz (1969). The discussion here will be limited to hinged ends.

The two coupled equations describing deformations v and ϕ result in a single buckling load P_{TF} for the flexural–torsional mode. The same buckling mode also occurs in the more general case of the load acting eccentrically in the plane of symmetry. Then the member continuously deflects as a beam-column in the plane of symmetry (x direction) but is subject to flexural–torsional buckling out of this plane under load P_{TF} given in this case by Eq. 13.3. (The solution for a concentric load is obtained by setting $e_s = 0$ in this equation.)

$$P_{\mathrm{TF}} = \frac{P'_{\phi e} + \alpha P_{xe} \pm \sqrt{P'_{\phi e} + \alpha P_{xe}^2 - 4\gamma P_{xe} P'_{\phi c}}}{2\gamma} \tag{13.3}$$

where

$$\alpha = 1 + e_x \beta_2 \frac{A}{I_0} \tag{13.4}$$

$$\gamma = \frac{A}{I_0}(x_0 - e_x)^2 + \mathrm{e}_x \beta_2 \frac{A}{I_0} + 1 \tag{13.5}$$

$$P'_{\phi e} = P_{\phi e} \alpha \tag{13.6}$$

$$P_{\phi e} = \frac{A}{I_0}\left(EC_w \frac{\pi^2}{L^2} + GJ\right) \tag{13.7}$$

$$P_{xe} = EI_x \frac{\pi^2}{L^2} \tag{13.8}$$

$$\beta_2 = \frac{1}{I_y}\left(\int_A x^2 dA + \int_A xy^2 dA\right) - 2x_0 \tag{13.9}$$

in which e_x is the eccentricity with respect to the center of gravity; X_0 the x coordinate of the shear center; I_x the moment of inertia about the x axis; I_0 the polar moment of inertia about the shear center; A the area of the cross section; and C_w, J the warping and St.-Venant torsional constants for the cross section, respectively.

The parameter $P_{\phi e}$ has the physical meaning that it is the concentric torsional buckling load if the displacements u and v are prevented; $P'_{\phi e}$ is the corresponding value for eccentric loading, and P_{xe} designates the load for buckling in the direction of the y axis if displacements ϕ and u are prevented. A simplified expression for P_{TF} is employed in the AISI specification.

Beams For doubly and monosymmetric sections global lateral–torsional buckling is discussed in Chapter 5. For thin-walled metal construction, the critical stress for lateral buckling of an I-beam having unequal flanges can be determined by the formula (Winter, 1943a, 1970)

$$\sigma_e = \frac{\pi^2 Ed}{2S_{xc}L^2}\left(I_{yc} - I_{yt} + I_y\sqrt{1 + \frac{4GJL^2}{\pi^2 EI_y d^2}}\right) \tag{13.10}$$

in which S_{xc} is the compressive section modulus of the entire section about the major axis; I_{yc}, I_{yt} the moments of inertia of the compression and tension portion, respectively, of a section about its centroidal axis parallel to the web; E the modulus of elasticity; G the shear modulus; J the torsional constant of the section; d the depth of the section; and L the unbraced length.

For thin-walled steel sections, the fraction (second term) under the radical in Eq. 13.10 is usually much less than 1.0 (Winter, 1947a). If the second term is omitted and considering that $I_y = I_{yc} + I_{yt}$, the following equation can be obtained for determination of critical stress for lateral buckling in the elastic range,

$$\sigma_c = \pi^2 EC_b\left(\frac{dI_{yc}}{L^2 S_{xc}}\right) \tag{13.11}$$

where C_b is a bending coefficient to account for moment gradient (see Chapter 5). In the AISI Specification the lateral buckling strength for I-section beams in the elastic range is based on Eq. 13.11. A conservative approximation to the actual lateral–torsional buckling stress of Z-section beams is obtained by dividing Eq. 13.11 by 2.

For singly symmetric sections (Fig. 13.5) the torsional–flexural buckling solution presented in the above for columns can be extended. When there is no axial load, the Galerkin method solution (Peköz, 1969) of the general differential equations of equilibrium gives the expression for the critical moments M_{cr} as

$$M_{cr} = -\frac{P_{xe}\beta_2}{2}\left(1 \pm \sqrt{1 + \frac{4I_0 R}{\beta_2^2 A}}\right) \tag{13.12}$$

where $R = P_{\phi e}/P_{xe}$. For singly and doubly symmetric section bending about the symmetry axis perpendicular to the web, introducing the notation used in the AISI Specification (as opposed to Peköz, 1969) and including the moment gradient C_b factor, then the elastic critical lateral–torsional buckling moment may be expressed as

$$M_{cr,e} = C_b r_0 A\sqrt{\sigma_{ey}\sigma_t} \tag{13.13}$$

where

$$r_0 = \sqrt{r_x^2 + r_y^2 + x_0^2} \tag{13.14}$$

$$\sigma_{ey} = \frac{\pi^2 E}{(K_y L_y / r_y)^2} \tag{13.15}$$

$$\sigma_t = \frac{GJ + \pi^2 EC_w/(K_t L_t)^2}{Ar_0^2} \tag{13.16}$$

in which r_x and r_y are radii of gyration of the cross section about the centroidal principal axis; x_0 the distance from shear center to centroid along the principal x axis, taken as negative; A the full cross-sectional area; K_y, K_t the effective-length factors for bending about the y axis and for twisting; and L_y, L_t the unbraced lengths for bending about the y axis and for twisting.

For laterally unbraced hat sections bent about the x axis, no stress reduction is necessary if $I_y > I_x$, because there is no tendency to buckle. When $I_y < I_x$ a conservative estimate of the elastic critical stress may be determined by regarding the compression portion of the section as an independent strut, which gives

$$\sigma_e = \frac{\pi^2 E}{(L/r_y)^2} \tag{13.17}$$

where r_y is the radius of gyration about the vertical axis of that portion of the hat section which is in compression. A more accurate analysis for such hat-shaped sections and for any other singly symmetric section is to use the equations given in Chapter 5. This is the approach required in the AISI specification for the design of a singly symmetric section. The AISI Design Manual (2002) and Yu (2000) provide design aids for calculation of the necessary section properties for common thin-walled shapes. Due to the involved nature of the preceding expressions, however, it is not uncommon for computational tools to be employed to (a) determine the necessary section properties and (b) implement and provide solutions for P_{TF} and M_{cr}.

13.2.5 Computational Elastic Buckling Solutions

Although the underlying mechanics is relatively straightforward for obtaining complete analytical expressions for local, distortional, and global buckling, the resulting expressions are significantly involved and, more importantly, apply only to elastic bifurcation buckling (large deformations and inelasticity were not accounted for). Today, computational solutions offer a powerful alternative—particularly for elastic buckling where the solution sensitivity is small. A variety of numerical methods: finite element, finite difference, boundary element, generalized beam theory, finite strip analysis, and others, provide accurate elastic buckling solutions for thin-walled beams and columns.

Traditional finite element analysis using thin plate or shell elements may be used for elastic buckling prediction (see Chapter 20 for additional discussion). Due to the common practice of using polynomial shape functions, the number of elements required for reasonable accuracy can be significant. Finite element analysis books such as Cook et al. (1989) and Zienkiewicz and Taylor (1989, 1991) explain the basic theory, while a number of commercial implementations can provide accurate elastic buckling answers if implemented with a modicum of care. Finite difference solutions for plate stability are implemented by Harik et al. (1991) and others. Elzein (1991) shows that the boundary element method may also be used for elastic stability. Generalized beam theory, developed originally by Schardt (1989, 1994) with contributions from Davies et al. (1994, 1998) and Davies and Jiang (1996), and significant extensions in recent years by Silvestre and Camotim (2002a,b), who have also recently provided a free user-friendly software implementation (Bebiano et al., 2008), also provide an attractive alternative.

Finite strip analyses, with results presented in Figs. 13.1 and 13.7, are a specialized variant of finite element analysis. For elastic stability of thin-walled sections,

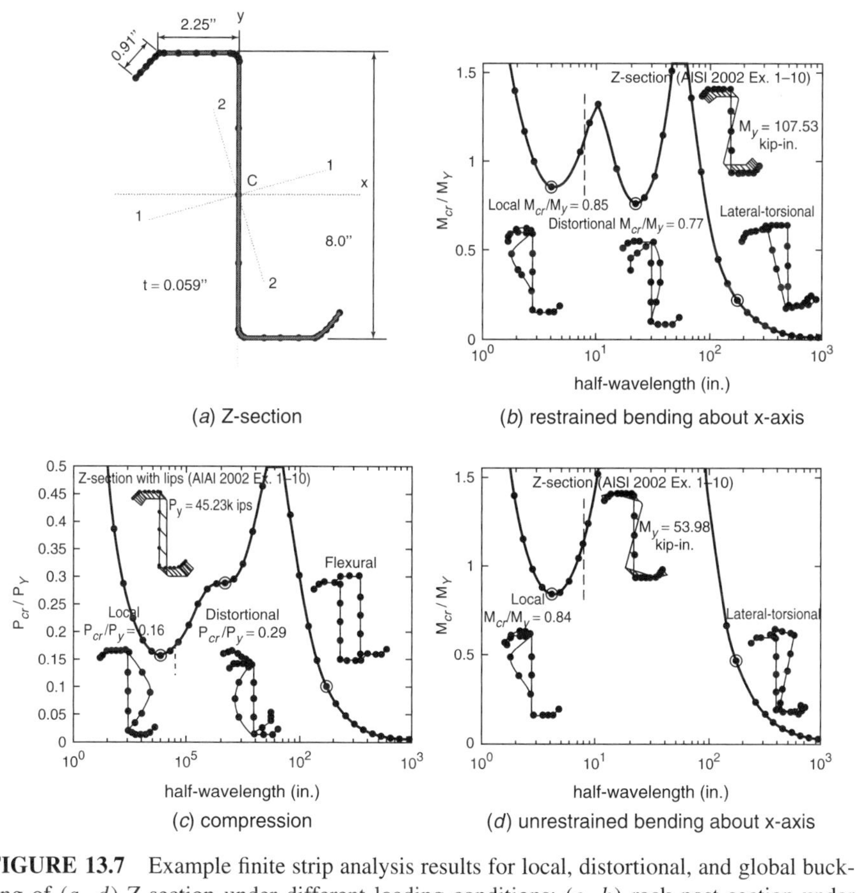

FIGURE 13.7 Example finite strip analysis results for local, distortional, and global buckling of $(a-d)$ Z-section under different loading conditions; $(e-h)$ rack post section under different loading conditions.

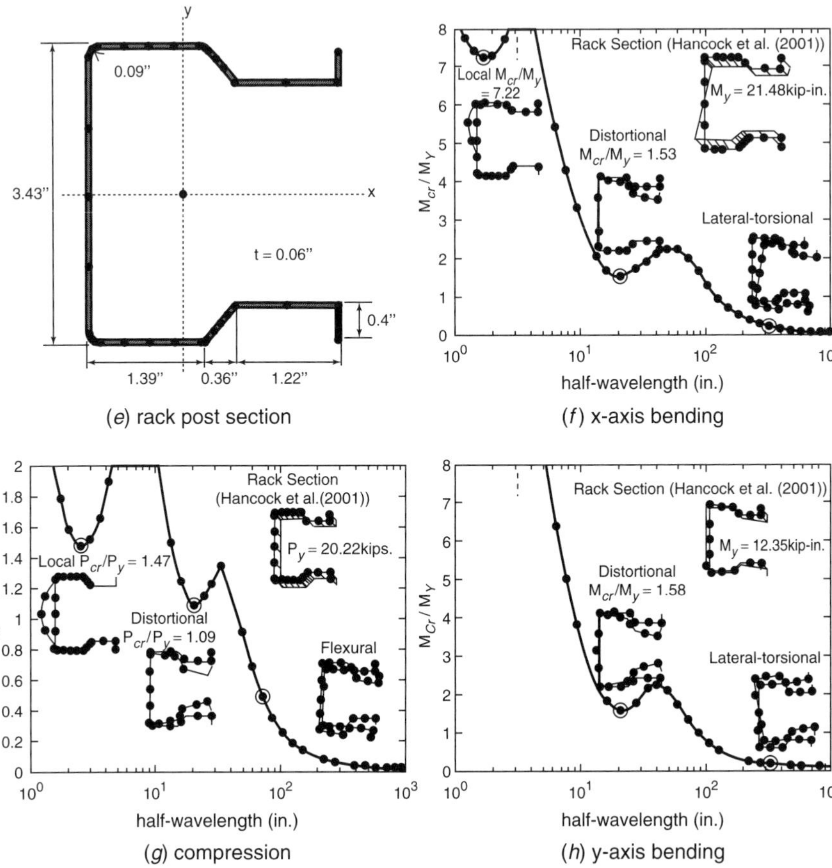

FIGURE 13.7 *(Continued)*

it is one of the most efficient and popular methods. The specific version of the finite strip method that can account for both plate flexural buckling and membrane buckling in thin-walled members was developed by Plank and Wittrick (1974). Cheung and Tham (1998) explain the basic theory. Hancock and his collaborators [see Hancock et al. (2001) for full references and descriptions] pioneered the use of finite strip analysis for stability investigations of cold-formed steel members. They also demonstrated convincingly the important potential of finite strip analysis in both cold-formed steel design and interpreting behavior. Consider, for example, the results of Fig. 13.7; local, distortional, and global buckling loads (or moments) along with the corresponding mode shapes are identified in the figure. All of the instabilities that need to be considered in basic thin-walled member design of these members are presented in one compact result.

Finite strip analysis is a general tool that provides accurate elastic buckling solutions with minimum effort and time. The method, as implemented in conventional software programs, does have limitations, with the two most important being

(1) the model assumes the ends of the member are simply supported and (2) the cross section may not vary along its length. These are limitations not of the method per se but rather of the implementations that are commonly available (Papangelis and Hancock, 1995; Schafer and Adany, 2006). Despite these limitations, the tool is useful and a major advance over plate-buckling solutions and plate-buckling coefficients (k's), which only partially account for the important stability behavior of thin-walled members.

13.3 EFFECTIVE WIDTH MEMBER DESIGN

Since the release of the first cold-formed steel specifications, the concept of effective width reductions to account for local buckling and postbuckling strength has been central to capacity determination of members. In addition to *effective width*, *reduced stress* and *form factor* (or *Q-factor*) approaches were also in use until the 1986 AISI Specification. Reduced-stress methods were then replaced with *effective width methods* for webs and unstiffened elements, and the *Q-factor* approach for local–global interaction was replaced with a novel extension to the effective width approach, under what is now known as the *unified effective width method* (Peköz, 1987).

Today, the AISI Specification (2007a) provides two alternative procedures for strength determination of cold-formed steel members: the unified *effective width method* appearing in the main body of the specification and the *direct strength method* in its Appendix 1. There is significant overlap between the two approaches. Further, not all material in the main body of the specification directly uses the concept of effective width. This section provides an examination of beams and columns by the *effective width method*, while the following section presents the *direct strength method*.

The focus of this section is to provide the *effective width method* solution to the strength of columns and beams that are subject to the member elastic stability modes addressed in Section 13.2. Namely, how does one determine the strengths associated with local, distortional, and global buckling—as well as the potential interactions among these modes? Local buckling, which is the focus of the effective width method, is treated first, followed by global buckling, local–global interaction, and finally distortional buckling. Comparison between the *effective width method* and the *direct strength method* is provided in Section 13.4.

13.3.1 Local Buckling Strength

Columns The role of local buckling in cold-formed steel columns has been studied since the 1940s with Winter (1949) summarizing U.S. contributions (e.g., Winter, 1940, 1943b) and Chilver (1951, 1953) and Harvey (1953) summarizing work in the U.K. After 60 years of progress, modern column research is still similar to Chilver's work with the use of elastic stability solutions for local plate buckling and "effective width" for the ultimate strength. The elastic plate-buckling solution of Chilver and Harvey was based on Lundquist and Stowell (1943), who extended the work of Timoshenko and Gere (1936) by providing practical methods for calculating the stability of connected plates. The effective width solution was based on von Kármán et al. (1932) and the experimental corrections of Winter (1947a). Notably, both Chilver and Harvey properly included the interaction of elements in determining the local buckling stress.

The basic premise of the *effective width method* for local buckling is to reduce each plate comprising a cross section to an effective plate (Fig. 13.8). The effective plate is an approximation of the longitudinal normal stress distribution in the real plate, in which the effective plate can carry the full applied stresses, but only in the effective portions. This method has long been used for flat plates in compression and is fully detailed in Section 4.3.3.

The actual expressions used to determine the effective width b of a given element with gross width w (for example, see Fig. 13.2 or 13.8 for typical elements of a cross section) are given, for instance, by AISI (2007a) as

$$b = \begin{cases} w & \text{when } \lambda \leq 0.673 \\ \rho w & \text{when } \lambda > 0.673 \end{cases} \tag{13.18}$$

where

$$\rho = (-0.22/\lambda)/\lambda \tag{13.19}$$

$$\lambda = \sqrt{f/f_{cr,l}} \tag{13.20}$$

$$f_{cr,l} = k\frac{\pi^2 E}{12(-\nu^2)}\left(\frac{t}{w}\right)^2 \tag{13.21}$$

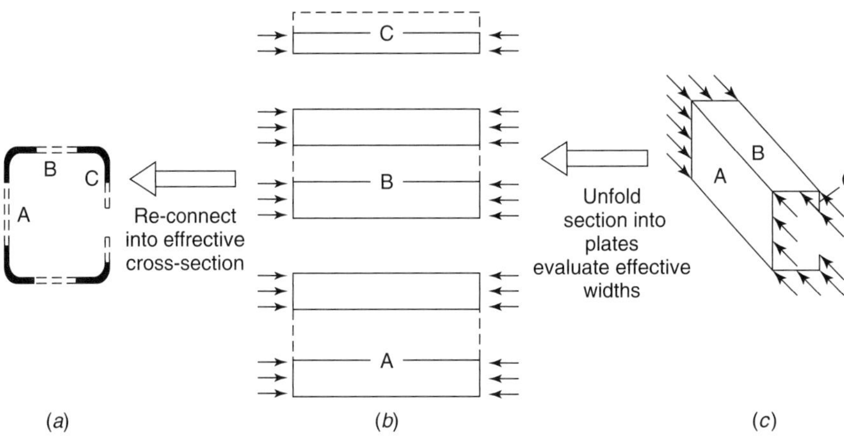

FIGURE 13.8 Lipped channel: (*a*) effective section; (*b*) effective width of component plates; (*c*) gross section.

in which f is the uniform stress in the effective portions ($f = f_y$ for the maximum plate or section capacity); k the plate-buckling coefficient; E the modulus of elasticity; v Poisson's ratio; and t the plate thickness. The expression for ρ is equivalent to Eq. 4.30. The key step in the implementation of this effective width is the determination of the plate-buckling coefficient k (which is provided in Section 4.2). Section 4.3.3 provides the complete history of the development of this expression, and the fifth edition of this guide provides further information on minor changes in formatting with regard to the presentation of the effective width expression over the years in the AISI Specification.

The summation of the effective plates also creates an effective cross section, as shown in Fig. 13.8. This effective cross section provides an initial means to understand why local–global interaction is so important in thin-walled members; as the moment of inertia of the section is effectively reduced, so must the global flexural buckling stress (additional details are provided in Section 13.3.3). If the applied stress on the elements f is set to the yield stress f_y, then the effective cross section is an estimate of the stiffness at collapse. If the applied stress on the elements f is set to the stresses on the section under service loads, then the effective cross section is an estimate of the stiffness for serviceability considerations.

Beams Use of effective width expressions for beams actually came significantly later than that for columns and was initiated by an extensive experimental program (LaBoube and Yu, 1982). The traditional approach, prior to the advent of the unified *effective width method* (Peköz, 1987), was to use effective width for compression flanges and average stress for webs. The unified *effective width method* provided expressions for the effective width of webs (i.e., elements under a stress gradient) based on Cohen and Peköz (1987) that were adopted by AISI. During development of the North American version of the cold-formed steel specification (AISI, 2001), it was determined that significant differences existed between the methods adopted from Cohen and Peköz (1987)—the United States (AISI, 1996) had adopted a more liberal set of effective width expressions than Canada (CSA, 1994) though both were from the same source document. A comprehensive study comparing strength predictions with test data demonstrated unconservative predictions for AISI (1996) with members with tall webs and narrow flanges (Schafer and Trestain, 2002) and as a result the AISI (1996) method was adopted for $h/b < 4$ and the CSA (1994) method for $h/b > 4$.

A complication of the *effective width method* applied to beams is the necessity for iteration. Once effective widths have been introduced into all the plate elements, the gross neutral axis is unlikely to be at the same location as the effective neutral axis. Because the effective width of the web is a function of the neutral axis location, iteration is required. Complete design examples that employ iterative procedures are provided in texts by Yu (2000) and Hancock et al. (2001) as well as the AISI *Cold-Formed Steel Design Manual* (AISI, 2002).

13.3.2 Global Buckling Strength

The global buckling strength of cold-formed steel columns and beams is generally treated in two steps. First, the strength without consideration of local or distortional buckling is determined. Then, the strength considering these interactions is examined. Because cold-formed steel sections often offer torsionally weak, open profiles, and singly, point, and unsymmetric sections are common, even determining the basic (compact cross-section) global buckling strength can be relatively involved. Noting that the basic global buckling strength is independent of the design approach employed, whether it be the *effective width method* or the *direct strength method*, this topic is presented in this section because it is found in the main body of the AISI Specification (as is the *effective width method*).

Columns The global buckling strength of thin-walled cold-formed steel columns (according to AISI) uses the same simplification of SSRC curve 2P (Chapter 3) for all columns as AISC (2005). That is, the global buckling stress F_n (inelastic or elastic) is expressed at stress as

$$F_n = 0.658^{\lambda_c^2} F_y \qquad \text{for } \lambda_c \leq 1.5 \tag{13.22}$$

$$F_n = \left(0.877/\lambda_c^2\right) F_y = 0.877 F_e \quad \text{for } \lambda_c > 1.5 \tag{13.23}$$

where

$$\lambda_c = \sqrt{F_y/F_e} \tag{13.24}$$

and F_y is the yield stress and F_e is the least of the elastic flexural, torsional, and flexural–torsional buckling stresses of the section. Equations 13.22 and 13.23 were shown to provide adequate strength predictions when originally adopted (Peköz, 1987) as well as in more recent studies (Schafer, 2002). This is somewhat surprising given the vastly different state of initial imperfections and residual stresses in cold-formed steel members (Moen et al., 2008; Schafer and Peköz, 1998b; Weng and Peköz, 1990) when compared with hot-rolled steel members, as well as the prevalence of flexural–torsional limit states in cold-formed steel members as opposed to flexural limit states in hot-rolled steel members. As Fig. 3.22 illustrates, however, a single column curve is by definition a crude instrument when compared against real data, and hence the choice of the same column curve for hot-rolled (AISC) and cold-formed (AISI) steel is as much a matter of convenience on the part of AISI as a demonstration of theoretical agreement.

Suggesting that global buckling strength begins and ends with the selection of an empirical column curve is to miss the real complexities that underlie the behavior of such sections, particularly for the common singly symmetric sections undergoing flexural–torsional buckling. For the inelastic domain, approximate approaches have been developed and adequately substantiated by tests (Chajes et al., 1966). Since 1980, the AISI Specification contains flexural–torsional buckling provisions based on the work of Chajes and Winter (1965), Peköz (1969), and Peköz and Winter (1969).

It is assumed that for the thin-walled sections being considered, the attainment of the yield stresses represents the limit of load-carrying capacity; that is, the plastic reserve capacity, if any, is negligible in torsional–flexural buckling. This point has been verified experimentally (Peköz, 1969; Peköz and Winter, 1969). Therefore, elastic flexural–torsional buckling is a possible mode of failure only if the axial load P_{yd} that causes incipient yielding (e.g., as predicted by the secant formula) is larger than P_{TF} (per Eq. 13.3). Extensive numerical studies were carried out on a variety of singly symmetric open sections and are reported by Peköz (1969). Figure 13.9 is a typical sample of the plots given in that reference that illustrates the complex behavior of such compression members.

For positive eccentricities, numerical studies indicate that both yielding and instability need to be considered. The following expression is shown to give more than acceptable results:

$$\frac{1}{P_{\mathrm{TF},O}} + \frac{e_x}{M_{cr+}} = \frac{1}{P_{\mathrm{TF}}} \tag{13.25}$$

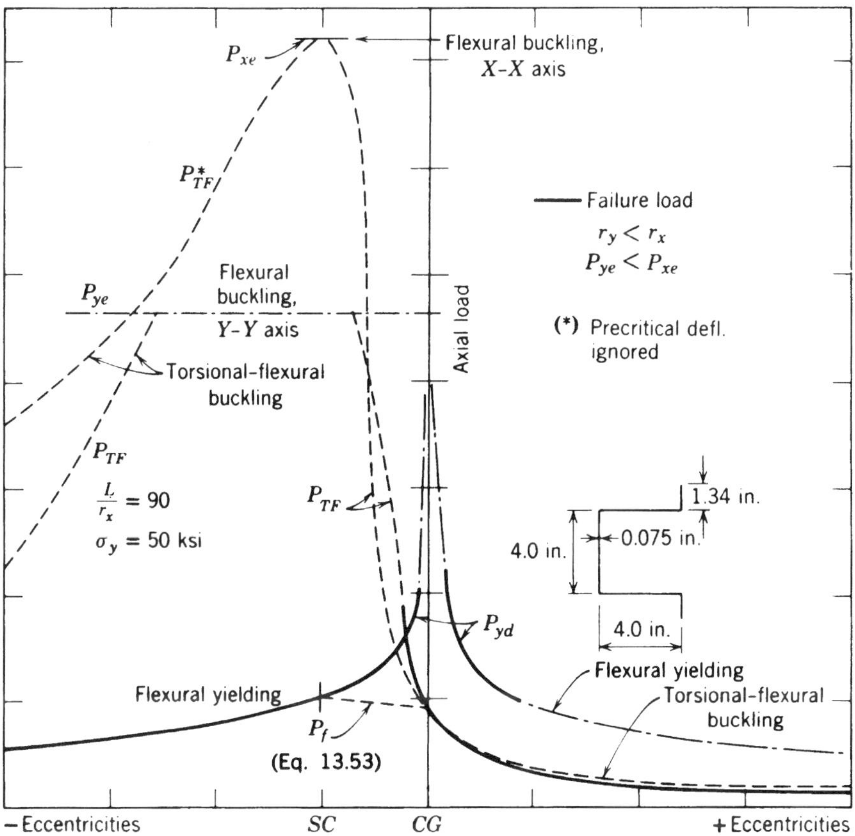

FIGURE 13.9 Failure modes and corresponding loads (Peköz, 1969).

in which P_{TF} is the flexural–torsional buckling load for an eccentricity e_x; $P_{\mathrm{TF},O}$ the flexural–torsional buckling load for concentric loading, regardless of whether it is the governing mode; and M_{cr+} the positive critical moment when there is no axial load, regardless of whether it is the governing mode (see Eq. 13.12). With the aid of charts given by Peköz (1969) for computing $P_{\mathrm{TF},O}$ and M_{cr+}, this equation is much more convenient to use than Eq. 13.3.

For negative eccentricities greater than x_0—that is, if the point of application of the axial load is on the side of the shear center opposite from the center of gravity—numerical studies on hat, channel, lipped channel, angle, and lipped angle sections of typical dimensions and yield stresses below 50 ksi (345 MPa) indicate that flexural–torsional buckling is not a governing mode of failure. For such eccentricities, these members fail by yielding after deflecting in the direction of the symmetry axis as a beam-column. For singly symmetric I-sections, however, both yielding and flexural–torsional buckling need to be investigated and the following interaction equation may be used:

$$\frac{1}{P_{xe}} + \frac{e_x - x_0}{M_{cr-}} = \frac{1}{P_{\mathrm{TF}}} \tag{13.26}$$

in which M_{cr-} is the negative critical moment when there is no axial load, regardless of whether it is the governing mode (see Eq. 13.12).

If when concentrically loaded a section can fail in flexural–torsional buckling, then flexural–torsional buckling is also a possible mode of failure for some range of eccentricities between the centroid and the shear center. As shown Fig. 13.9, the two branches of the failure curve in this region (yielding on the left and flexural–torsional buckling on the right) show a definite and sharp peak. This means that small changes or inaccuracies in eccentricity can produce large reductions in load capacities. For this reason it seems reasonable and conservative to disregard the uncertain high carrying capacity in the region of the peak and to base design values on the dashed straight cutoff shown in Fig. 13.9. In this range of eccentricities, the following linear interpolation formula between the axial load P_S applied at the shear center, which causes yielding or buckling, and the concentric flexural–torsional buckling load $P_{\mathrm{TF},O}$ gives a realistic and conservative flexural–torsional buckling load P_F,

$$P_F = P_{\mathrm{TF},O} + \frac{e_x}{x_0}\left(P_S - P_{\mathrm{TF},O}\right) \tag{13.27}$$

For singly symmetric I-sections, P_S is the smaller of the yield load P_{yd} or the buckling load P_{xe}, whereas for the other open sections, only yielding need be considered, as explained previously.

In addition to the points discussed in the preceding section, the following need to be considered in the design of thin-walled members to resist flexural–torsional buckling:

Inelastic Stability Behavior for Members of Relatively Low Slenderness Ratios. Chajes et al. (1966) studied this problem and reported that an expression similar to SSRC curve 2P is satisfactory for concentric flexural–torsional buckling. The AISI specifications reflect this approach for both concentric and eccentric loading.

Unequal Eccentricities at Opposite Ends of the Member. Peköz (1969) presents the results of an extensive study of this subject and makes the conclusion that application of a modification factor C_{TF} to the second term of Eq. 13.25 is quite accurate. The value C_{TF} is the same as C_m discussed in Chapter 8, except that it does not have 0.4 as its lower limit.

Influence of Precritical Beam-Column Deflection Flexural–Torsional Buckling. Again, on the basis of an analytical and experimental treatment of the subject, Peköz (1969) recommends the use of an amplification factor $1/(1 - P/P_{ye})$ for the moments.

Wandering Centroid Problem. Centrally loaded, singly symmetric columns become beam-columns upon local buckling and shifting of the neutral axis. In an extensive statistical study, Peköz (1987) established good correlation with test results if a concentrically loaded column is defined as a member loaded through its effective centroid. The effective centroid is calculated at the reduced column stress F_n by Eq. 13.22 or 13.23. This is the approach used in the AISI Specification.

Behavior of Biaxially Loaded Beam-Columns. Prior to 1986, the AISI Specification did not permit the analysis of singly symmetric beam-columns bending about the symmetry axis. The designer had to resort to experimental tests. Based primarily on the work of Loh (1985), Mulligan and Peköz (1983), and Peköz (1987), the AISI specification now determines the capacity of biaxially loaded open sections using an interaction equation with eccentricities measured from the effective centroid. [The method is similar to procedures adopted previously by the Rack Manufacturers Institute (RMI, 1979) standard and followed in the CSA (1989) Standard.] Lipped channel sections used in the endwalls of metal buildings and columns in industrial storage racks are examples of members subjected to such loads.

Beams The lateral–torsional buckling strength can be based on empirical expression similar in spirit to the column curve, consisting of a yield plateau and inelastic and elastic buckling regions. The strength is expressed as an extreme fiber stress F_n with

$$F_n = \begin{cases} F_y & \text{for } F_e \geq 2.78F_y \quad (13.28) \\ \dfrac{10}{9}F_y\left(1 - \dfrac{10F_y}{36F_e}\right) & \text{for } 0.56F_y < F_e < 2.78F_y \quad (13.29) \\ F_e & \text{for } F_e \leq 0.56F_y \quad (13.30) \end{cases}$$

where F_y is the yield stress and F_e is the elastic lateral–torsional buckling stress. This strength may also be expressed in terms of the critical moment $M_{cr} = S_g F_n$, with S_g gross elastic section modulus. The solid curve in Fig. 13.10 shows the variation of this moment with the unbraced length. The three regions of yielding (Eq. 13.28), inelastic buckling (Eq. 13.29), and elastic buckling (Eq. 13.30) are clearly delineated.

For a given $L/\sqrt{dI_{yc}}$ ratio, a Z-section (or any point-symmetric section) will buckle laterally at a lower stress than that of an I- (doubly symmetric) or a channel (singly symmetric) section. As indicated earlier, a conservative design approach has been used in the AISI Specification since 1986, in which the critical moments for Z-sections in the elastic range are one-half of those permitted for I-beams or channels at the same $L/\sqrt{dI_{yc}}$ ratio. The lateral buckling curve for Z-shaped beams is shown as the dashed line in Fig. 13.10. Functionally, this is enabled in the AISI Specification by using $\frac{1}{2}F_e$ for the lateral–torsional buckling stress of point-symmetric sections.

Figure 13.10 shows a yield plateau at the yield moment, M_y, as opposed to the fully plastic moment M_p (common in compact cross sections where local buckling will not control). This is not a completely accurate representation of the inelastic reserve capacity in thin-walled sections. Larger capacities are possible when the cross section can sustain the larger strains associated with capacities greater than M_y. Such inelastic reserve ($M > M_y$) can be treated as a strength reserve in local buckling, and provisions are provided in the AISI Specification on a component basis for stiffened elements (Yener and Peköz, 1985) and unstiffened elements (Bambach and Rasmussen, 2004).

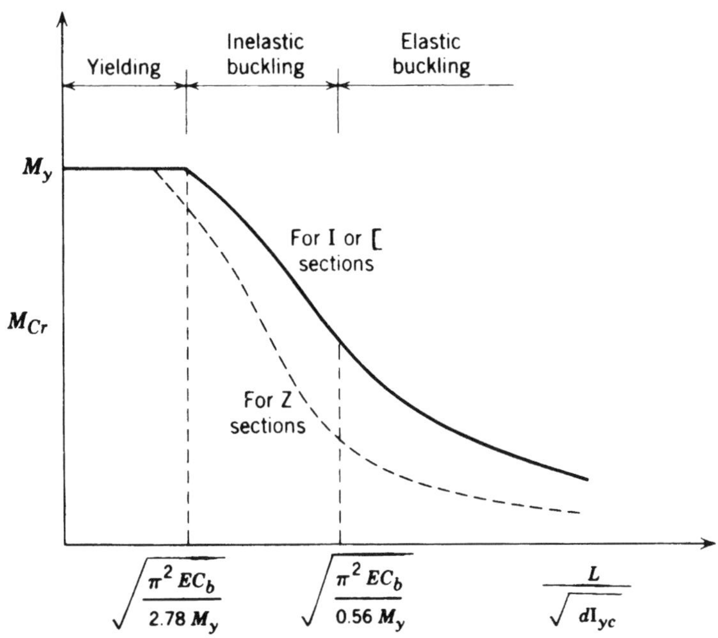

FIGURE 13.10 Allowable compressive stress for lateral buckling of beams (Winter, 1970).

13.3.3 Local–Global Interactive Strength

Columns Although local buckling may be followed by a significant postbuckling reserve, the deformations and redistribution of stresses that are associated with that reserve change the global buckling response of the member (whether it be a column or beam). Study of this local–global interaction has included experimental testing and nonlinear finite element analysis using shell element models. Recognizing and accounting for local–global interaction in the design of thin-walled members was one of the fundamental steps toward making thin-walled construction practical.

Results of early studies on interaction between local and global buckling were presented by Bijlaard and Fisher (1952a,b). Across the world, column research in the 1970s focused on the interaction between local and overall (i.e, global–flexural, torsional, flexural–torsional) buckling modes (DeWolf, 1974; Klöppel and Bilstein, 1976; Rhodes and Harvey, 1977; Peköz, 1977; Loughlan, 1979). In the 1980s, Hancock (1981) and Sridharan and Benito (1984) investigated this interaction using the finite strip method. Mulligan (1983) specifically examined local buckling interaction. In Europe, researchers continued to provide strong evidence for interaction of local and overall column buckling (Batista et al., 1987; Rhodes and Loughlan, 1980; Zaras and Rhodes, 1987). More recently, Rasmussen and Hancock (1991) showed the importance of different end fixity on the postbuckling behavior, and Young (1997) experimentally demonstrated that fixed-ended columns do not suffer the same interaction problems as pin-ended columns.

The first widely used design method, and the one still employed in the AISC (2005) Specification, is the Q-factor, or form factor, approach. The third edition of this guide provides a detailed summary of this method. The basic idea is to increase the column global slenderness used in the column strength curve (λ_c in Eq. 13.24) according to the local slenderness of the stiffened and unstiffened elements. The procedure is simple and as implemented is typically conservative, with the difficulties primarily arising with unstiffened elements and the nature of the reductions. Comparisons with cold-formed steel columns demonstrated that the method could be either excessively conservative (particularly for sections with slender unstiffened elements) or even unconservative, and hence, the approach was abandoned by the AISI Specification in favor of an effective width method (Peköz, 1987).

The original adaptation of the effective width method for local–global interaction envisioned using effective section properties, primarily the effective moment of inertia, in the calculation of the column-buckling stress; and then using that buckling stress in the typical column strength curve (e.g., DeWolf et al., 1974, 1976; Kalyanaraman et al., 1977). Such a procedure requires iteration because the effective cross-section properties are a function of the applied stresses. What is now known as the unified effective width method was originally suggested by Springfield and Trestain (Trestain, 1982), who developed a column design method for the 1984 CSA *Cold-Formed Steel Standard* (CAN3-S136-M84). For design office use, the iterative approach utilizing effective section properties was deemed unsuitable. The key features of the new method were the use of gross section properties to determine slenderness and buckling stress and the determination of effective area at this buckling stress rather than at F_y as used previously. Another unique feature was the use of effective width for both stiffened and unstiffened elements. The column curve developed by Lind for the 1974 CSA standard was retained; this curve is geometrically similar to the AISC column curve. The method predicted the test results of DeWolf et al. (1974) with remarkable accuracy (Trestain, 1982).

The 1984 CSA method was incorporated into the unified approach proposed by Peköz (1987) for the 1986 AISI Specification. Since then, the AISI Specification follows this method using an effective cross-section area to determine column strength P_n according to

$$P_n = A_e F_n \tag{13.31}$$

where A_e is the effective area (which is the summation of the effective width of the elements times the thickness) at stress F_n and F_n is the ultimate stress determined from either Eq. 13.22 or Eq. 13.23. Further evaluation by Peköz led to refinements, including the use of the effective centroid rather than the gross centroid as the origin for determining the eccentricity of load.

A singly symmetric column that is not fully effective presents a unique and difficult problem. Not only are the effective section properties reduced by local buckling, but the effective centroid shifts along the axis of symmetry. Thus an initially concentrically loaded column becomes a beam column. Testing such a column that is truly concentrically loaded throughout its loading history appears difficult, if not impossible. Furthermore, centrally loaded singly symmetric columns appear to exist in practice only if its section is fully effective and is loaded at its ends uniformly around the periphery. In practice, it may be difficult to be assured that such conditions will exist. Consequently, many columns that have no obvious moment applied to their ends may be, in fact, beam columns.

The method represented by Eq. 13.31 works well for local–global interaction but has been shown to be a poor predictor when distortional buckling is involved (see Sections 13.3.4 and 13.4.1).

Beams Based on the unified method (Peköz, 1987), the effective width approach to local–global interaction in beams is essentially the same as that for columns. The bending strength M_n is taken as

$$M_n = S_e F_n \tag{13.32}$$

in which the effective section modulus S_e is calculated at the design stress F_n and uses an appropriate summation of the effective width of all of the elements, and the design stress F_n is a function of the global beam slenderness and is determined from the beam curve given by Eqs. 13.28 to 13.30 using gross cross-section properties. The method generally works well for local–global interaction but has been shown to be inaccurate when distortional buckling is involved (as shown below and also discussed in Section 13.4.1).

13.3.4 Distortional Buckling Strength

Extension of the effective width method for distortional buckling has proven difficult in many situations. The same complications with predicting the elastic buckling stress are exacerbated when determining the strength, because the deformations involved include both membrane deformations, primarily in the flange and lip, and bending deformations, primarily in the web. Effective width expressions for edge stiffeners and intermediate stiffeners were developed in the 1980s and were in use in the AISI specification until 2001 (Desmond et al., 1981a,b). The expressions, however, were found to be in poor agreement with columns and beams failing in pure distortional buckling (Hancock et al., 1994; Kwon and Hancock, 1992; Rogers and Schuster, 1997; Schafer, 2001; Schafer and Peköz, 1998a, 1999; Yu and Schafer, 2006; 2007b). New distortional buckling provisions, adopted in the 2007 AISI Specification, follow a methodology that is consistent with the *direct strength method* and are described in Section 13.3.1. The Eurocode method uses a combination of effective width and reduced-thickness methods to determine the distortional buckling strength; essentially, they extend their beam on an elastic foundation model and assume the stiffeners follow a basic column curve for strength (thus, no postbuckling capacity is allowed).

Distortional–Global, Local–Distortional Interaction If distortional buckling is treated with effective widths, as was done for a number of years in the AISI Specification, one advantage is automatic inclusion of the potential for distortional–global interaction, at least in some approximate form. The distortional buckling effective widths are calculated at the long (global) column or beam stress F_n, and thus the potential for this interaction is allowed.

Local–distortional interaction is more difficult to include in the effective width method, if an effective width for local buckling and an effective width for distortional buckling are both to be determined for the same element. Eurocode combines a reduced-thickness approach with a traditional effective width local buckling reduction to account for this potential phenomenon.

13.4 DIRECT STRENGTH MEMBER DESIGN

In contrast to the *effective width method*, which focuses on the individual elements that comprise a cross section, the key to the *direct strength method* is an accurate elastic stability analysis that will provide local distortional, and global buckling results for the member (Figs. 13.1 and 13.7). The premise of the method is that member strength can be defined in terms of the elastic instabilities for the gross cross section and the force or moment that causes the section to yield. For example, column strength P_n can be determined from simple "direct strength" equations that are a function of (1) the elastic buckling loads, including local $P_{cr,l}$ distortional $P_{cr,d}$, and global $P_{cr,e}$ modes, and (2) the yield load P_y,

$$P_n = f\left(P_{cr,l}, P_{cr,d}, P_{cr,e}, P_y\right) \tag{13.33}$$

For the specific case of global (long) column buckling, the "Direct strength" prediction equations could be defined as a function of $P_{cr,e}$ and P_y by simply multiplying the stresses F_e, F_y, and F_n in Eqs. 13.22 and 13.23 by the gross area of the cross section A_g. Again, the key to this method is the use of an accurate elastic stability analysis to provide $P_{cr,e}$, which in this case would account for the possibility of flexural, torsional, and flexural–torsional buckling modes. Using a similar approach, the *direct strength method* may be extended to predict column (and beam) strengths that account for local and distortional buckling instabilities with appropriate consideration of postbuckling reserve and interaction in these modes. In 2004, the AISI Specification adopted the *direct strength method* as an alternative design method; see Appendix 1 of AISI (2004, 2007a).

It is important to recognize in any discussion regarding the *effective width method, the direct strength method*, or other semiempirical design methods for thin-walled construction that none of these design methods are theoretically correct. Rather, a complicated nonlinear problem is simplified in some manner so that engineers may have a working model to design from without resorting to detailed analyses or experimental testing of each individual member. Of course, it is essential that these simplified models apply some degree of reliability theory to incorporate uncertainty in their ability to predict strength. A full discussion of reliability, development of the *direct strength method*, and summary of current related research is proved in Schafer (2008).

Distortional buckling is covered first in this section because it is of special significance in the *direct strength method*. The design prediction equations, which were proposed by Hancock for strength in distortional buckling modes, may be viewed as the genesis for the *direct strength method*. Distortional buckling is followed by global buckling in Section 13.4.2 and local and local–global interaction in Section 13.4.3. Because the *direct strength method* is a relatively new procedure, it is compared directly with the *effective width method* in Section 13.4.3. Finally, this discussion on the *direct strength method* closes with a look at current research, including the consideration of interaction among the buckling modes.

13.4.1 Distortional Buckling Strength

In any explanation of the *direct strength method*, distortional buckling is of prominent importance because much of the *direct strength method* development is centered around finding adequate design solutions to the complicated problem of elastic buckling and strength determination in open singly and point-symmetric cold-formed steel cross sections undergoing distortional buckling.

Columns Currently, the distortional buckling strength P_{nd} may be predicted with the simple expression

$$P_{nd} = \begin{cases} P_y & \text{for } \lambda_d \le 0.561 \quad (13.34) \\ \left(1 - 0.25\left(\dfrac{P_{cr,d}}{P_y}\right)^{0.6}\right)\left(\dfrac{P_{cr,d}}{P_y}\right)^{0.6} P_y & \text{for } \lambda_d > 0.561 \quad (13.35) \end{cases}$$

where

$$\lambda_d = \sqrt{P_y/P_{cr,d}} \tag{13.36}$$

and

$P_y = A_g F_y$, the yield load for the column
$P_{cr,d}$ = critical elastic distortional column buckling load

Seemingly, the only complication of significance in Eqs. 13.34 to 13.36 is the determination of the critical elastic distortional column buckling load $P_{cr,d}$. The research and design decisions required to arrive at this point, however, are far more significant than the equations reveal. Using the preceding approach [which has been adopted in AISI (2007a) in both the main specification as an additional check on local buckling and in Appendix 1, the *direct strength method*] assumes the following: (a) distortional buckling may be treated on the whole section as opposed to its component elements, (b) postbuckling in distortional buckling is specific to the mode (and not the same as local or global buckling), (c) interaction of distortional buckling with other modes need not be considered. The research to arrive at these conclusions includes a number of significant contributions as summarized in the following, but point (c) is still actively under study and will be discussed further in Section 13.4.4.

Distortional buckling has been long recognized as a potential problem; Chilver (1951, 1953) stated that the reinforcing "lip" in a lipped channel should be sufficiently stiff to ensure local buckling controls (and thus avoid distortional buckling) but gave no criteria for achieving this. Desmond et al. (1981a,b) studied the distortional mode in detail, referring to the mode as stiffener buckling. Desmond et al. recognized that elastic buckling criteria, such as ensuring that stiffener buckling is a higher critical stress than local buckling, does not meet Chilver's criteria. Instead, Desmond et al. used experimental data to empirically formulate rules for an "adequate" stiffener and the plate-buckling coefficient k when the stiffener is only partially effective. As a result, distortional buckling was incorporated into the AISI Specification (1986–2004) as a local mode, implicitly assuming postbuckling reserve in distortional buckling was the same as in local buckling. Additionally, the experimental work by Desmond et al. investigated members with back-to-back specimens, which avoided web local buckling but artificially increased the distortional buckling stress.

Distortional buckling has long been observed in testing but is often removed before given further consideration. In Sweden, Thomasson (1978) performed experiments on lipped channels with slender webs. To increase the local buckling stress of the webs, small intermediate (groove) stiffeners were folded in. This eliminated local buckling but created what Thomasson called a "local-torsional" problem, which was actually distortional buckling. Thomasson considered this local-torsional mode undesirable and responded by placing closely spaced braces from lip to lip, ensuring that distortional buckling did not occur and therefore making the local mode again dominant. Mulligan (1983) encountered the same local-torsional mode in testing and observed that the adequate stiffener criteria of Desmond et al. did not appear to restrain distortional buckling in many cases. Subsequently, Mulligan followed Thomassons's experimental modification and provided braces that restricted distortional buckling.

Another attempt at understanding distortional buckling focused on isolating the flange from the rest of the section (as is traditionally done in effective width design). Isolated edge stiffeners were studied experimentally and analytically by physically replacing the web/flange juncture with a simple support, thus providing known boundary conditions (Klöppel and Unger, 1970; Lim, 1985; Lim and Rhodes, 1986). The "torsional" mode (distortional buckling) for these flanges may be accurately predicted due to the special boundary conditions.

Eurocode has taken another approach to distortional buckling calculations. Eurocode 3, Part 1.3 (1996, 2004) provides a method for predicting the distortional buckling of simple lipped sections, such as channels, by using a strut to account for the elastic restraint provided by the web and flange to the buckling of the lip. This method approximately accounts for the distortional deformations of the web and flange but used a global column curve for the failure of the lip, thus assuming that there was no postbuckling reserve in the distortional mode.

At the University of Sydney distortional buckling was directly studied, motivated greatly by the prevalence of distortional buckling in high-strength cold-formed steel storage racks (e.g., Lau, 1988). This work led to the refinement of the finite strip method in elastic distortional buckling (Lau and Hancock, 1990), the development of unique analytical methods (Lau and Hancock, 1987), and strength curves for distortional buckling (Hancock et al., 1994). Based on the Hancock et al. strength curves, the Australian Standard for Steel Storage Racking (Standards Australia, 1993) and the Australian/New Zealand Standard for Cold-Formed Steel Structures (AS/NZS, 1996, 2005) were the first to adopt explicit design rules for distortional buckling in compression members. Equation 13.35 adopted in the *direct strength method* (AISI, 2004, 2007a) is identical to an equation presented in Hancock et al. (1994, Eq. 4b).

In Kwon and Hancock's (1992) experiments on lipped channels, with and without groove stiffeners in the web, the distortional mode was unrestricted and the tests showed that the interaction of distortional buckling with other modes was weak. In later testing at Sydney, Young (1997) also experimentally observed a weak interaction of distortional buckling with other modes. Using generalized beam theory, Davies and Jiang (1996) argued that distortional buckling has weak interactions with other modes and endorsed the strength curves of Hancock et al. (1994) for ultimate-strength prediction, which are given by Eq. 13.35. Further examination against a wider body of test data also showed a weak interaction of distortional with other modes (Schafer, 2002).

In the University of Sydney tests, distortional buckling was experimentally observed to have postbuckling capacity (thus leading to the rejection of the Eurocode strength methodology), but to a lesser extent than local buckling (essentially a rejection of the prevailing AISI method of the time based on Desmond et al.'s work). The presence of postbuckling capacity indicates that the distortional mode should not be treated as a global mode for strength, while the limitations to the available postbuckling capacity indicate that inclusion in a local-buckling-based effective width approach must be done with care, because use of the standard expressions will result in overprediction of the postbuckling reserve. One explanation for the limited postbuckling behavior is that the membrane stress at the flange tips of edge-stiffening lips increases dramatically after distortional buckling (Sridharan, 1982). Using finite strip and nonlinear finite element analyses, reduced postbuckling capacity and increased imperfection sensitivity for distortional buckling failures, when compared to local buckling, was observed both for edge-stiffened and intermediately stiffened beams and columns (Schafer, 2002; Schafer and Peköz, 1999).

In parallel to the extensive work at the University of Sydney, studies at the University of Strathclyde were made on "torsional" buckling, that is, distortional buckling (Seah, 1989; Seah et al., 1993; Seah and Rhodes, 1993; Chou et al., 1996). Hats and channels with compound lips were investigated experimentally, hand methods for the prediction of distortional buckling were provided, and the ultimate strength in the distortional mode was treated via the effective width approach (rather than the column curve approach proposed by Sydney researchers). Tests were also completed in Finland (Salmi and Talja, 1993), in Japan on more complicated polygonal cross sections (Hikosaka et al., 1987, Takahashi, 1988), and in the United States on Z-section columns in distortional buckling (Polyzois and Sudharampal, 1990; Purnadi et al., 1990; Polyzois and Charvarnichborikarn, 1993).

In 2002 an attempt was made to investigate distortional buckling using the entirety of the available existing experimental data (Schafer, 2002). For the variety of cross sections tested at the University of Sydney, Hancock et al. (1994) had demonstrated that when a section was known to fail in distortional buckling, the measured compressive strength correlated well with the slenderness in the elastic distortional mode. Data in Schafer (2002) provided additional experimental agreement for Hancock's design expression (Fig. 13.11). In addition, the failure mode for specimens collected in Schafer (2002) was generally unknown, and thus a variety of methods were examined for strength calculation, considering local, distortional, and global buckling and their potential interactions. The result was a method that both identified the failure mode and predicted the strength. The extension to prediction of all failure modes by appropriate expressions correlated to elastic slenderness in the various modes (and combinations) was the essential step in the development of the *direct strength method* as a general approach.

The work by Hancock et al. provided the essential curve for distortional buckling by the *direct strength method*. As is often the case with attempts to determine an origin, one can go back even further as Hancock often attributes his methodology to Trahair's work on the strength prediction of columns undergoing flexural–torsional buckling. In this regard, it should become clear that the *direct strength method* is not a new idea, rather the extension of an old one to new instability limit states.

Beams The first mention of the *direct strength method* occurs in Schafer and Peköz (1998c) and was closely coupled to the development of a design method for

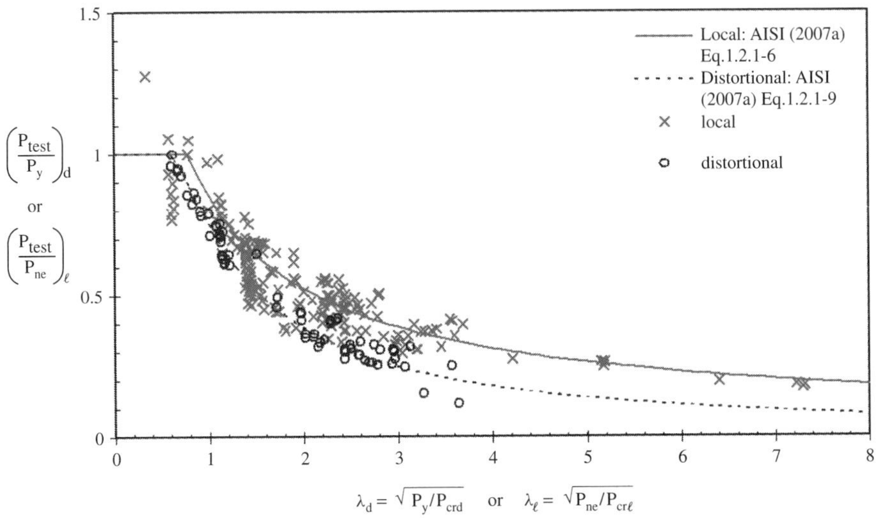

FIGURE 13.11 Comparison of the *direct strength method* with test data for columns (equation numbers refer to Appendix 1 of AISI, 2007a).

beams. This research focused on the application of a large database of sections that was collected to explore two problems, including distortional buckling in C- and Z-section beams and local and distortional buckling in deck sections with multiple longitudinal intermediate stiffeners in the compression flange. At the same time, Hancock and related researchers at the University of Sydney demonstrated that distortional buckling failures for a wide variety of failures were well correlated with the elastic distortional slenderness (Hancock et al., 1994, 1996).

The form of the presentation of the *direct strength method* for beams evolved somewhat from Schafer and Peköz (1998c, 1999). In particular, curve 2 in their work is identical to the distortional buckling expression (Eq. 4b) provided in Hancock et al. (1994). With this in mind, the nominal flexural strength for distortional buckling M_{nd}, which appears in the *direct strength method* provisions (AISI, 2004, 2007a), is

$$M_{nd} = \begin{cases} M_y & \text{for } \lambda_d \leq 0.673 \quad (13.37) \\ \left(1 - 0.22\left(\dfrac{M_{cr,d}}{M_y}\right)^{0.5}\right)\left(\dfrac{M_{cr,d}}{M_y}\right)^{0.5} M_y & \text{for } \lambda_d > 0.673 \quad (13.38) \end{cases}$$

where

$$\lambda_d = \sqrt{M_y/M_{cr,d}} \tag{13.39}$$

and

$M_y = S_g F_y$ and S_g is the gross section modulus referenced to the fiber at first yield

$M_{cr,d}$ = Critical elastic distortional buckling moment

Similar to the work on columns in distortional buckling, the test database was expanded beyond the original University of Sydney results reported in Hancock et al. (1994) and a method for local buckling, which will be presented in Section 13.4.3, was also developed by Schafer and Peköz (1999). Both of these expressions are compared with the expanded test database in Fig. 13.12. It should be emphasized that in providing an expression for determining the distortional buckling strength of an arbitrary cross section, a distinction can be made between limit states of local and distortional buckling.

In the development of the *direct strength method* for C- and Z-section beams, separation of local and distortional buckling failure modes was initially complicated by the bracing and boundary conditions used in the testing, which could only partially (but not fully) restrain distortional buckling. Nonetheless, expressions were developed and adopted in AISI (2004, 2007a). More recently, a combination of flexural tests and finite element analysis studies on a variety of C- and Z-sections failing in local buckling (Yu and Schafer, 2003, 2007b; Yu, 2005) and distortional buckling (Yu, 2005; Yu and Schafer, 2006, 2007b) were able to isolate the two

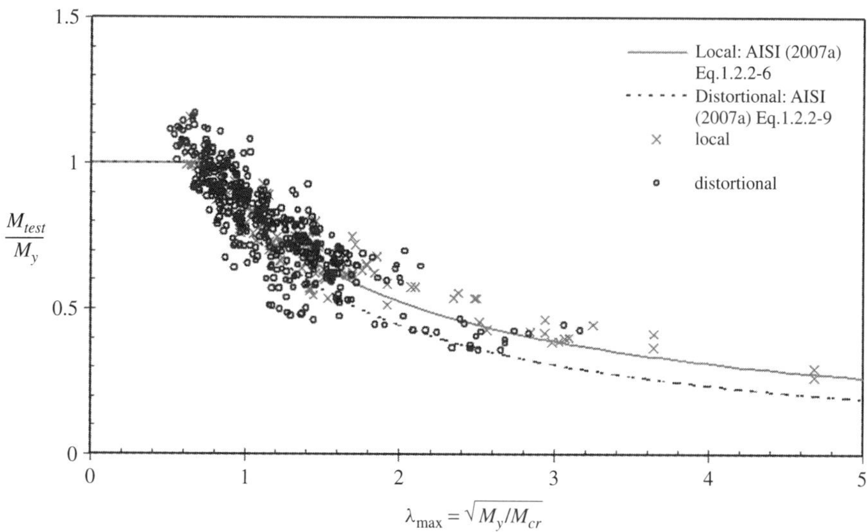

FIGURE 13.12 Comparison of the *direct strength method* with tests data for laterally braced beams (equation numbers refer to Appendix 1 of AISI 2004, 2007a).

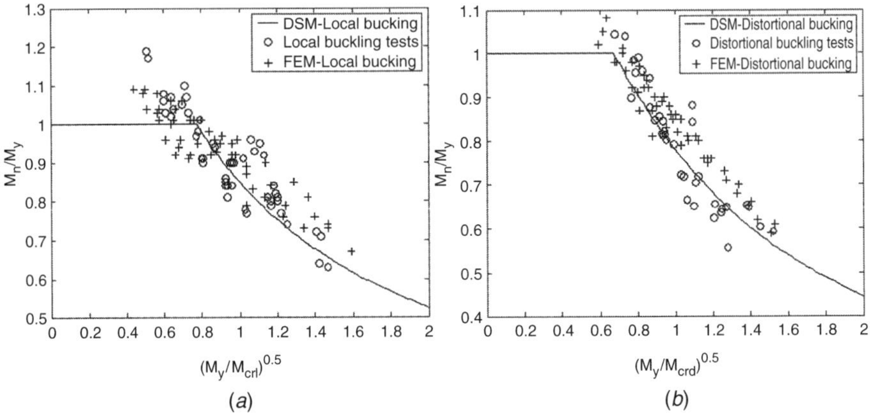

FIGURE 13.13 Comparison of *direct strength method* for beams to tests and nonlinear finite element results for C- and Z-sections in (*a*) local and (*b*) distortional buckling as reported in Yu and Schafer (2007b).

modes. This work demonstrated unequivocally the robustness of the *direct strength method* predictions for C- and Z-sections failing in either mode. A summary of the performance of these sections is provided in Fig. 13.13.

13.4.2 Global Buckling Strength

The global buckling strength of columns and beams in the *direct strength method* is the same as in the main body of the AISI Specification (i.e., Eqs. 13.22, 13.24, and 13.28 to 13.30), with two exceptions: (1) the provisions are provided in terms of load and moment, instead of stress, and (2) no method is prescribed for determining the elastic global buckling load or moment. With respect to the latter, the *direct strength method* requires a "rational" analysis for the accurate calculation of these quantities. Such analysis methods could range from numerically solving classical analytical expressions to employing finite strip or finite element methods. A variety of analysis methods are presented in the *Direct Strength Method Design Guide* (Schafer, 2006).

Columns The nominal axial strength P_{ne} for flexural, torsional, or torsional–flexural buckling is

$$P_{ne} = \begin{cases} 0.658^{\lambda_c^2} P_y & \text{for } \lambda_c^2 \le 1.5 \quad (13.40) \\ \dfrac{0.877}{\lambda_c^2} P_y = 0.877 P_{cr,e} & \text{for } \lambda_c^2 > 1.5 \quad (13.41) \end{cases}$$

where

$$\lambda_c = \sqrt{P_y / P_{cr,e}} \tag{13.42}$$

$$P_y = A_g F_y \text{ yield load for column} \tag{13.43}$$

and $P_{cr,e}$ is the minimum of the critical elastic flexural, torsional, or torsional–flexural buckling loads.

Beams The nominal flexural strength M_{ne} for lateral–torsional buckling is

$$M_{ne} = \begin{cases} M_{cr,e} & \text{for } M_{cr,e} < 0.56M_y \quad (13.44) \\ \dfrac{10}{9}M_y\left(1 - \dfrac{10M_y}{36M_{cr,e}}\right) & \text{for } 0.56M_y \leq M_{cr,e} \leq 2.78M_y \quad (13.45) \\ M_y & \text{for } M_{cr,e} > 2.78M_y \quad (13.46) \end{cases}$$

where

$$M_y = S_f F_y \tag{13.47}$$

S_f = gross elastic section modulus referenced to the extreme fiber at first yield
$M_{cr,e}$ = critical elastic distortional lateral–torsional buckling moment

13.4.3 Local and Local–Global Interactive Strength

Columns The basis of the *direct strength method* for local buckling can be understood through an examination of Winter's effective width expression for local buckling of plates. If the slenderness parameter λ of Eq. 13.20 is substituted into Eq. 13.19 and the resulting expression for ρ into Eq. 13.18, then the effective width b may be expressed as a function of the gross width w,

$$b = \left(1 - 0.22\sqrt{\frac{f_{cr}}{f_y}}\right)\sqrt{\frac{f_{cr}}{f_y}}w \quad \text{for}\sqrt{\frac{f_y}{f_{cr}}} > 0.673 \tag{13.48}$$

This shows that the effective width (the strength in local bucking) can be stated as a function of f_{cr} and f_y—the basic premise of the *direct strength method*. Consider now that the local plate critical buckling stress f_{cr} in Eq. 13.48 is replaced with the cross-section local buckling stress $f_{cr,l}$ (e.g., from Fig. 13.7) and the effective width and gross width are replaced by the effective area and gross area,

$$A_e = \left(1 - 0.22\sqrt{\frac{f_{cr,l}}{f_y}}\right)\sqrt{\frac{f_{cr,l}}{f_y}}A_g \quad \text{for}\sqrt{\frac{f_y}{f_{cr,l}}} > 0.673 \tag{13.49}$$

Multiplying all stress terms by the gross area A_g and then multiplying both sides of the resulting equation by f_y provide direct strength prediction for column local buckling strength P_{nl} as

$$P_{nl} = \left(1 - 0.22\sqrt{\frac{P_{cr,l}}{P_y}}\right)\sqrt{\frac{P_{cr,l}}{P_y}}P_y \quad \text{for}\sqrt{\frac{P_y}{P_{cr,l}}} > 0.673 \tag{13.50}$$

Comparison with experimental data (Schafer, 2002) indicates that this expression can be slightly conservative. Calibration of Eq. 13.50 to the data results in

$$P_{nl} = \left(1 - 0.15\left(\frac{P_{cr,l}}{P_y}\right)^{0.4}\right)\left(\frac{P_{cr,l}}{P_y}\right)^{0.4} P_y \quad \text{for} \quad \sqrt{\frac{P_y}{P_{cr,l}}} > 0.776 \qquad (13.51)$$

Local–global interaction is fundamental to thin-walled members. To account for this in the effective width approach (see Section 13.3.3), f_y in Eq. 13.49 is replaced with f_n (which would be calculated from the column curve of Eqs. 13.22 and 13.23). The result is

$$b = \left(1 - 0.22\sqrt{\frac{f_{cr}}{f_n}}\right)\sqrt{\frac{f_{cr}}{f_n}}\,w \quad \text{for} \quad \sqrt{\frac{f_n}{f_{cr}}} > 0.673 \qquad (13.52)$$

Using the same substitutions and calibration in moving from Eq. 13.48 to 13.51, *the direct strength method* format of Eq. 13.52 is

$$P_{nl} = \left(1 - 0.15\left(\frac{P_{cr,l}}{P_{ne}}\right)^{0.4}\right)\left(\frac{P_{cr,l}}{P_{ne}}\right)^{0.4} P_y \quad \text{for} \quad \sqrt{\frac{P_y}{P_{ne}}} > 0.776 \qquad (13.53)$$

where $P_{ne} = A_g f_n$ and f_n is from Eqs. 13.22 and 13.23, or equivalently P_{ne} is given by Eqs. 13.40 and 13.41.

Through this simple substitution of f_n for f_y, local–global interaction has been included. With the maximum strength in local buckling capped by the global buckling strength P_{ne}, further reductions are made depending on the local slenderness of the cross section. As appearing in the AISI Specification (2004, 2007a), the nominal axial strength P_{nl} for local buckling is

$$P_{nl} = \begin{cases} P_{ne} & \text{for } \lambda_l \le 0.776 \quad (13.54) \\ \left(1 - 0.15\left(\frac{P_{cr,l}}{P_{ne}}\right)^{0.4}\right)\left(\frac{P_{cr,l}}{P_{ne}}\right)^{0.4} P_{ne} & \text{for } \lambda_l > 0.776 \quad (13.55) \end{cases}$$

where

$$\lambda_l = \sqrt{P_{ne}/P_{cr,l}} \qquad (13.56)$$

and

P_{ne} = global buckling strength as defined by Eqs. 13.40 and 13.41
$P_{cr,l}$ = critical elastic local column buckling load

Agreement of Eqs. 13.54 and 13.55 with available test data is shown in Fig. 13.11 (Schafer, 2002). Overall trends are clear with some significant scatter prevalent (though not more than in effective width implementations). The choicc of the coefficients and exponents of Eq. 13.56 were influenced by the solution for the *direct strength method* for the local buckling strength of beams, which preceded the column research and is discussed below.

Beams Development of the initial direct strength prediction equations for local buckling of flexural members follows the same basic progression as described in the above for columns. Beginning with Winter's basic effective width expression in the form of Eq. 13.48, f_{cr} is replaced with the cross-section local buckling stress $f_{cr,l}$ (referenced to the compression flange). The effective width and gross width are replaced with the effective section modulus S_e and gross section modulus S_g, and both sides of the equation are multiplied by the yield stress:

$$S_e f_y = \left(1 - 0.22\sqrt{\frac{f_{cr,l}}{f_y}}\right)\sqrt{\frac{f_{cr,l}}{f_y}} S_g f_y \quad \text{for} \quad \sqrt{\frac{f_y}{f_{cr,l}}} > 0.673 \tag{13.57}$$

Stresses under the radical are multiplied by S_g and the products $S_e f_y$ and $S_g f_y$ are defined as the nominal local buckling capacity M_{nl} and yield moment M_y, respectively. Thus Winter's effective width expression for the bending strength is rewritten as

$$M_{nl} = \left(1 - 0.22\sqrt{\frac{M_{cr,l}}{M_y}}\right)\sqrt{\frac{M_{cr,l}}{M_y}} M_y \quad \text{for} \quad \sqrt{\frac{M_y}{M_{cr,l}}} > 0.673 \tag{13.58}$$

Schafer and Peköz (1998b) compared the results of Eq. 13.58 with the available experimental test data and determined that a basic trend for agreement exists, but the expression is overly conservative. As a result, curve 3 suggested in Schafer and Peköz (1998b) was adopted by the AISI (2004, 2007a), specification and the nominal flexural strength for local buckling M_{nl} for use in the *direct strength method* appears as

$$M_{nl} = \begin{cases} M_{ne} & \text{for } \lambda_l \leq 0.776 \quad (13.59) \\ \left(1 - 0.15\left(\dfrac{M_{cr,l}}{M_{ne}}\right)^{0.4}\right)\left(\dfrac{M_{cr,l}}{M_{ne}}\right)^{0.4} M_{ne} & \text{for } \lambda_l > 0.776 \quad (13.60) \end{cases}$$

where

$$\lambda_l = \sqrt{M_{ne}/M_{cr,l}} \tag{13.61}$$

and

M_{ne} = gobal buckling strength as defined by Eqs. 13.44 to 13.46
$M_{cr,l}$ = critical elastic local buckling moment

Agreement of Eqs. 13.59 and 13.60 with experimental data is demonstrated in Figs. 13.12 and 13.13*a*. It should be noted that for the beam data (in contrast to the column data), all of the M_{test} values are normalized against the moment at first yield M_y. This is due to the fact that all of the test data employed was for laterally braced members. It should be further noted that while local–global interaction was experimentally examined for columns, and the same methodology was applied for beams, local–global, distortional–global, and local–distortional interactions have not been experimentally examined in the context of the *direct strength method* for beams. Based on the findings for columns, local–global interaction has been included and local–distortional and distortional–global interactions ignored. Modal interaction is further discussed in Section 13.4.4. The performance of laterally unbraced beams deserves further study, not only in the context of the *direct strength method* and potential interactions, but also to better understand how warping torsion should be treated. For moderate rotations, the influence of the torsional stress on local and distortional buckling modes is real (Gotluru et al., 2000) and its potential inclusion in the *direct strength method* (as well as in the main body of the AISI Specification) is worthy of further study.

Finally, the testing to date on C- and Z-section beams has focused on major axis bending and associated buckling, with extension to minor axis bending only assumed. This assumption is justified in part by the inclusion of hats and decks in the experimental database; such sections are bent about their minor axis, and their behavior is similar to a C-section in minor axis bending. Further, the major axis bending modes are considered more critical because global lateral–torsional buckling modes are eliminated for minor axis bending.

Element Interaction A significant difference between the *effective width method* and the *direct strength method* is the replacement of element- or plate-buckling stress f_{cr} with cross-section local bucking stress $f_{cr,l}$. In the elastic region, the use of cross section $f_{cr,l}$ ensures that equilibrium and compatibility around the cross section are maintained. Although both design methods have reasonable levels of overall reliability (Schafer, 2008), it is obtained in different ways.

In Fig. 13.14 the strength predictions of the *effective width method* and the *direct strength method* are compared as a function of the web slenderness of a C-section column. As web slenderness increases, the *effective width method* solution becomes systematically unconservative. This behavior is exacerbated for typically available C-sections by the fact that as the web becomes deeper the flange width remains at approximately the same width; hence, high web slenderness is strongly correlated with high web-to-flange width ratios (i.e., narrow C-sections). This detrimental behavior is primarily the result of local web-to-flange interaction, and not distortional buckling. Because the *effective width method* uses an element approach, the slenderness of the web (no matter how large it becomes) has no effect on the solution for the flange. In contrast, the *direct strength method*, which includes element interaction in local buckling (i.e., interaction between the flange and the web), performs accurately over the full range of web slenderness. Obviously, proper inclusion of element interaction is necessary for accurate strength prediction of these columns.

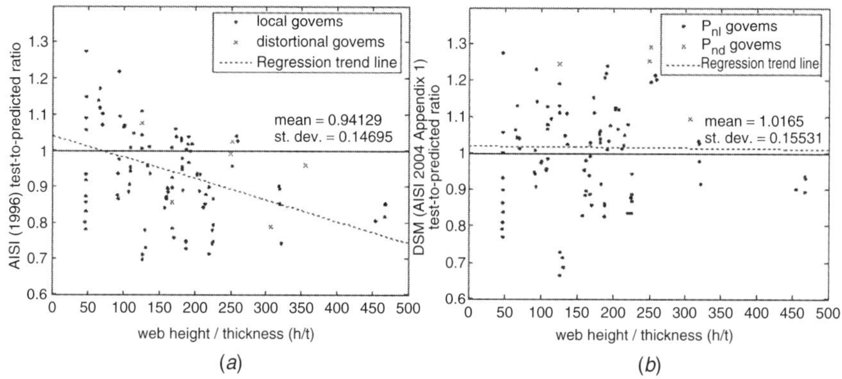

FIGURE 13.14 Test-to-predicted ratio for (*a*) the *effective width method* of AISI (1996) and (*b*) the *direct strength method* of AISI (2004, App. 1) for all lipped channel columns used in the development of *direct strength method* predictor equations plotted as a function of web slenderness (h/t).

Taken to the extreme, inclusion of elastic element interaction can also work against the *direct strength method*, making the method overly conservative. This fundamental limitation of the *direct strength method* was reported in the first paper to propose the approach (Schafer and Peköz, 1998). When one part (element) of the cross section becomes extraordinarily slender that element will drive the member elastic critical buckling stress to approach zero. The *direct strength method* will assume the member strength, like the member elastic critical buckling stress, will also approach zero. In contrast, the *effective width method* presumes only that the element itself (not the member) will have no strength in such a situation. Deck or hat sections in bending with low yield stress and very slender (wide) compression flanges without intermediate stiffeners tend to fall in this category and thus have unduly conservative predictions by the *direct strength method*. On the other hand, quite reasonable predictions are obtained wih the *effective width method*. Ignoring interelement interaction, however, as the *effective width method* traditionally does, is not universally warranted, as illustrated for the C-section columns of Fig. 13.14.

For optimized deck sections with multiple longitudinal intermediate stiffeners in the web and the flange (e.g., see Höglund, 1980), the *direct strength method* is highly desirable over the *effective width method,* with the primary benefit being one of convenience. If a computational solution is employed for determining the elastic buckling stresses (moments), an optimized deck section is no more complicated than a simple hat for strength determination. For the *effective width method*, on the other hand, the calculation of effective section properties and accurately handling the effective width of the numerous subelements leads to significant complication without increased accuracy; or worse in the case of many specifications (e.g., AISI, 1996, 2001) no design approach is even available for such a section using the *effective width method*. In general, as sections are optimized, the *direct strength method* provides a simpler design methodology with more universal applicability than the *effective width method*.

13.4.4 Modal Interactions

If thin-walled cross sections can be generally characterized as having local, distortional, and global buckling modes, then there exists a potential for any of these modes to interact. For example, local–global interaction is known to occur and be of significance in many cases and is thus accounted for specifically in design (e.g., Eq. 13.31 or 13.54 to 13.56). Interaction of the other modes, such as local–distortional, distortional–global, and local–distortional–global, has been and continues to be the subject of much research.

No definitive consensus exits on what it means for a mode to interact, nor when such interactions will occur. Definitions depend on whether one is considering a mathematical interaction, or even the coupling of multiple modes, or one is considering observed behavior where one buckling mode influences the deformations and strength of a second mode. When such interaction will occur is complicated by the varied degree of postbuckling in the modes, wavelength of the modes, and dependence of postbuckling on material properties (e.g., yield stress) as buckling modes trigger plastic mechanisms and ultimately failure.

The *effective width method* (as implemented in AISI, 2007a) explicitly includes local–global interaction. Local–distortional, distortional–global, and local–distortional–global interaction are assumed to not occur or be irrelevant for design in the *effective width method.*[1] The *direct strength method* (as implemented in AISI, 2007a) includes only local–global interaction, and ignores local–distortional, distortional–global, and local–distortional–global interaction. These conclusions were drawn from conflicting data, which nonetheless largely point out that if interactions are included for all modes the resulting capacities are not consistent with observations.

Local–Distortional/Distortional–Global Interaction As summarized in Section 13.4.1, initial analytical and experimental investigations of distortional buckling largely indicated that interaction of distortional buckling with other buckling modes (local, global) is generally weak. The tests by Kwon and Hancock (1992) on lipped channels, with and without groove stiffeners in the web, were designed to determine whether adverse interaction occurred if local and distortional buckling were simultaneous or nearly simultaneous. No adverse interaction was found between local and distortional buckling for the channel sections tested. Tests of trapezoidal decks by Bernard et al (1992a,b, 1993a,b), however, included sections that underwent local buckling before and after distortional buckling.

[1]This statement is subject to some interpretation, as the empirical expressions of Desmond et al., which are still used for the effective width of edge-stiffened elements in AISI (2007), includes some amount of local–distortional interaction in their development; it may be shown. however, that the distortional buckling strength using Eqs. 13.34 to 13.36 or 13.37 to 13.39 results in lower capacities than Desmond's expressions— thus the statement that local–distortional interaction is not meaningfully included in AISI (2007).

Sections that underwent local buckling well before distortional buckling needed to account for local–distortional interaction, while in sections where distortional buckling occurred first no similar reduction for local buckling was required. As indicated above, research by Young (1997), Davies and Jiang (1996), and Schafer (2002) show that the interaction of distortional buckling with other modes is weak.

Using a wide body of test data (Schafer, 2002), interaction of the buckling modes was systematically studied for local–global, distortional–global, and local–distortional buckling of the columns. Based on overall test-to-predicted ratios, and when available the failure modes noted by the researchers in their testing, it was determined that local–global interaction should be included, and distortional–global and local–distortional interaction could be neglected. For example, the local–distortional interactive strength was formulated by replacing P_{ne} of Eqs. 13.54 to 13.56 with P_{nd} of Eqs. 13.34 to 13.36. Such a strength check results in overly conservative predictions, with 169 of the 187 tests studied being identified to fail in local–distortional interaction, and the average test-to-predicted ratio would be increased to 1.35 (Schafer, 2000, 2002). Neither the failure mode nor the strength prediction is consistent with the observations from tests when local–distortional interaction is included for all columns. As a result, it was recommended to only include local–global interaction in the *direct strength method*.

Recent experimental and analytical work has left this conclusion somewhat in question. Hancock and his research team have continued to study the distortional buckling mode for high-strength steel sections, including lipped channels with intermediate stiffeners (Yang and Hancock, 2004; Yap and Hancock, 2008b) and unique cross-shaped open sections with multiple distortional buckling modes (Yap and Hancock, 2008a). They have found that the strength for these sections does not follow the expected distortional strength curve, that is, *direct strength method,* Eqs. 13.34 to 13.36 or Eq. 4(b) of Hancock et al. (1994), but is somewhat reduced. (Postbucking strength is still observed.) It is unclear if the reduction should be attributed to local–distortional or distortional–global interaction. The best numerical (strength prediction) agreement is found with ignoring local–distortional interaction in the calculation and including distortional–global interaction by replacing P_y in Eqs. 13.34 to 13.36 with P_{ne}. Local–distortional interaction, however, is visually observed in the testing.

In addition to studying the mechanics of postbucking in local and distortional buckling, Silvestre et al. (2006) have also been studying local–distortional interaction in lipped channels. Their approach has been to examine cross sections where $P_{cr,l}$ and $P_{cr,d}$ are at the same or nearly the same elastic critical buckling load (or moment) and then examine the ultimate strength through the use of a nonlinear finite element analysis. Comparisons with the *direct strength method* show a decrease from the expected postbuckling strength. Based on their analyses, an experimental study is now underway to confirm the exact nature of the reductions and when they should be applied in design.

13.5 ADDITIONAL DESIGN CONSIDERATIONS

The two previous sections cover the design of thin-walled axial and flexural members by either the *effective width method* or the *direct strength method*. Regardless of the method selected, a number of additional design considerations must be considered in construction using thin-walled members, including shear, inelastic reserve capacity, web crippling, and interactions between bending and web crippling, bending and shear, and bending and axial load.

13.5.1 Shear

The design expressions for cold-formed steel members in shear closely parallel those of the AISC specification and are based primarily on the experimental work of LaBoube and Yu (1978). The fifth edition of this guide provides complete details of the expressions used prior to 2001. In 2001, the shear design expressions were modified slightly so that a uniform resistance factor could be employed through the three regions of yield, inelastic buckling, and elastic buckling (Craig and Schuster, 2000). Development of a *direct strength method* for shear has sparked more recent work (Pham and Hancock, 2008; Schafer, 2008).

13.5.2 Bending: Inelastic Reserve

It is possible, and indeed relatively common, for thin-walled cold-formed steel beams to have bending capacities in excess of the moment at first yield. One method for determining the capacity in such a situation is to determine the strains that the compressed elements can sustain, and then based on that strain back-calculate the capacity in excess of M_y. The AISI Specification provides the strain capacity of stiffened elements (Yener and Peköz, 1985) and unstiffened elements (Bambach and Rasmussen, 2004) that may be utilized for this calculation. In addition, work has begun on a *direct strength method* approach to account for inelastic reserve (Shifferaw and Schafer, 2007). Additional inelastic reserve capacity due to the redistribution of moments in statically indeterminate beams and profiled decks was studied by Unger (1973), Yener and Peköz (1980), Yu (1981), and Bryan and Leach (1984). The post-local-buckling behavior of continuous beams is discussed by Wang and Yeh (1974).

13.5.3 Web Crippling

The necessity to check web crippling is an important distinction for thin-walled members. The thin webs of beams may cripple due to the high local stresses caused by concentrated loads or reactions. Figure 13.15 shows the types of deformation that occur due to crippling of unrestrained single webs and restrained double webs.

Early experimental work (Winter and Pian, 1946; Rockey et al., 1972; Hetrakul and Yu, 1978) indicated that the web-crippling strength of thin-walled beams depends on N/t, h/t, R/t, and F_y, where t is the web thickness, N the bearing length,

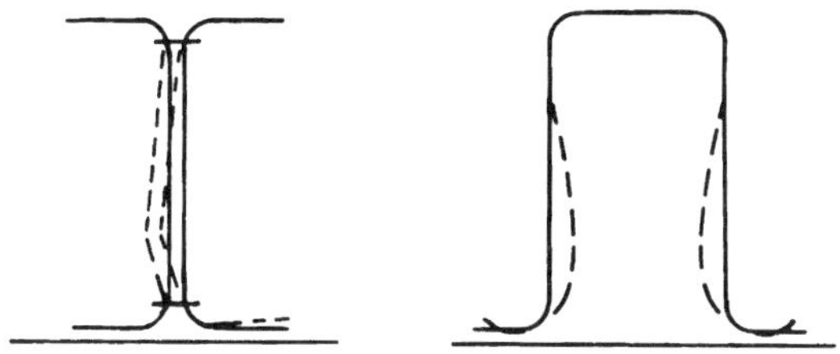

FIGURE 13.15 Web crippling of beams.

h the flat width of the web, R the inside bend radius, and F_y the yield stress of the steel. In 2001, all of the web-crippling design expressions were brought together into one format in the AISI Specification (Prabakaran and Schuster, 1998) and appear as

$$P_n = Ct^2 F_y \sin\theta \left(1 - C_R\sqrt{\frac{R}{t}}\right)\left(1 - C_N\sqrt{\frac{N}{t}}\right)\left(1 - C_h\sqrt{\frac{h}{t}}\right) \tag{13.62}$$

in which P_n is the web-crippling capacity, θ is the web angle, and C, C_R, C_N, and C_h are empirical coefficients that are cross section and loading dependent.

The empirical coefficients, C's, are based on the extensive experimental testing that has been completed on web crippling, including significant recent testing (Beshara and Schuster, 2000; Gerges and Schuster, 1998; Holesapple and LaBoube, 2003; Prabakaran and Schuster, 1998; Santaputra et al., 1989; Wallace and Schuster, 2005; Young and Hancock, 2001, 2004) as well as earlier testing as reported in the AISI specification. Equation 13.62 has been shown to provide reliable predictions across the broad data set (Beshara and Schuster, 2002).

Experimental work is now moving away from testing individual cross sections and to testing structural systems. This trend is particularly evident in lightweight steel framing systems with tests completed on web crippling in trusses (Ibrahim et al., 1998), headers (Stephens and LaBoube, 2000), stud-to-track connections (Fox and Schuster, 2002), joist-to-rim connections (Serrette, 2002), and framing members with holes (LaBoube et al., 1999, 1997; Langan et al., 1994).

Although Eq. 13.62 is quite useful and simple, it is not without its drawbacks. The empirical coefficients vary significantly over the 46 different categories provided in the specification and not always in a rational fashion. Alternatives to experimental testing include nonlinear finite element analysis (Fox and Brodland, 2004; Ren et al., 2006b; Sivakumaran, 1989) as well as yield-line theory (Bakker and Stark, 1994). To date these alternatives appear too involved for regular design use.

13.5.4 Bending and Web Crippling

Although web crippling is generally associated with shear loads, bending demands can also decrease the web-crippling capacity. The AISI Specification (2007a) provides a series of interaction equations for checking bending and web crippling, with the equations in current use summarized in Wallace et al. (2002). In addition to the experimental work also summarized by Wallace et al. (2002), recent works on isolated channels (Young and Hancock, 2002) and headers in light steel framing (Stephens and LaBoube, 2002, 2003) have investigated bending and web-crippling interaction. Work continues on analytical models to understand web crippling and its interaction with bending (Hofmeyer et al., 2001) as well as provide fully computational solutions (Ren et al., 2006a).

13.5.5 Bending and Shear

When high bending stresses and high shear stresses act simultaneously, as in cantilever beams and at supports of continuous beams, the webs of such members will buckle at a lower stress than if only one component of stress were present. For a combination of bending and shear, Eq. 4.8 can be used to predict buckling.

$$\left(\frac{\sigma_{cb}}{\sigma^*_{cb}}\right)^2 + \left(\frac{\tau_c}{\tau^*_c}\right)^2 \leq 1 \qquad (13.63)$$

where σ_{cb} is the actual compressive stress at the junction of flange and web, τ_c is the actual average shear stress, and σ^*_{cb} and τ^*_c are the critical stresses for bending and shear, respectively. Equation 13.63 has been modified for design use by (1) replacing the critical stresses for bending and shear with permissible design strength values and (2) presenting it in a load format by replacing the shear stress with shear force and bending stress with bending moment,

$$\left(\frac{M}{M_n}\right)^2 + \left(\frac{V}{V_n}\right)^2 \leq 1 \qquad (13.64)$$

where M and V are the applied moments and shears and M_n and V_n are the nominal strengths. Webs with transverse stiffeners can develop strength in excess of Eq. 13.64 (LaBoube and Yu, 1978) and a slightly more liberal interaction equation is provided for that case in AISI (2007a).

13.5.6 Bending and Axial Load

The design of members under bending and axial load, beam-columns, has received much study in recent years; for example, see Chapters 8 and 16 of this guide. Similar to hot-rolled steel construction, cold-formed steel now provides a traditional K-factor approach in the main body of the specification (AISI, 2007a) and a second-order analysis approach in the Appendix (AISI, 2007, App. 2).

The main specification approach uses a simple linear interaction equation with classical moment amplification for $P-\Delta$ and $P-\delta$ moments,

$$\frac{P}{\phi_c P_n} + \frac{M_x}{\phi_b M_{nx}} + \frac{M_y}{\phi_b M_{ny}} \leq 1.0 \qquad (13.65)$$

where P, M_x, M_y are the second-order demands and P_n, M_{nx}, M_{ny} the capacities and ϕ is the resistance factor. Demands M_x and M_y reflect P–Δ and P–δ contributions through amplification of the first-order moments M_o by the traditional approach

$$M = C_m \left(\frac{1}{1 - P/P_{cr}} \right) M_o \tag{13.66}$$

where C_m is a reduction factor to account for the typical case of a nonuniform moment distribution in the amplification of M_o and P_{cr} is the buckling load determined through knowledge of the effective length *KL*. Specific provisions are not provided for determining K, but reference is given to the AISI commentary and earlier editions of this guide.

New to the 2007 AISI Specification is Appendix 2, which provides guidelines for applying second-order analysis to determine M_x and M_y, as opposed to employing Eq. 13.66 (Sarawit and Peköz, 2006). Initial imperfections, or equivalent notional loads, are added to a frame model of the structure and the analysis is conducted to determine the structural demands according to equilibrium on the deformed geometry. Because the impact of member out-of-straightness is inherently assumed in the P_n capacity of Eqs. 13.22 and 13.23 or 13.40 and 13.41, such initial imperfections are not required to be explicitly modeled in the second-order analysis. In addition, and following the equivalent procedure in AISC (2005), the elastic stiffness E in the analysis model is reduced (that is, the bending stiffness *EI* and axial stiffness *EA* are reduced) in an attempt to account for member reliability and reduced stiffness due to partial yielding under large axial loads. Further details are provided in the AISI commentary, the AISC commentary, and Chapter 16 of this guide.

13.6 STRUCTURAL ASSEMBLIES

13.6.1 Built-Up Sections

Although built-up sections are common in light steel framing, the AISI Specification provisions for them remain somewhat rudimentary. To account for the reduced shear rigidity in built-up sections with discrete fastening, AISI (2007a) prescribes for built-up columns a modified slenderness $(KL/r)_m$ for use in flexural and flexural–torsional-buckling calculations:

$$\left(\frac{KL}{r}\right)_m = \sqrt{\left(\frac{KL}{r}\right)_o^2 + \left(\frac{a}{r_i}\right)^2} \tag{13.67}$$

where $(KL/r)_o$ is the overall slenderness ratio of the entire section about the built-up member axis, a the fastener spacing, and r_i the minimum radius of gyration of an individual shape in the built-up member. In addition, the fasteners must be designed for 2.5% of P_n, the faster spacing is limited, and end details are prescribed. The design of built-up cold-formed steel members is derived largely from the extension of similar research for built-up hot-rolled steel members and further work in this area is needed.

13.6.2 Bracing

As discussed in Chapter 12, structural bracing may be divided according to its function into two general types: (1) bracing provided to resist secondary loads on structures, such as wind bracing, and (2) bracing provided to increase the strength of individual structural members by preventing them from deforming in their weakest direction (Winter, 1958). There are two basic cases of the second type: (1) bracing applied to prevent buckling and thereby increase the unstable strength of the member and (2) bracing applied to counteract stable but detrimental types of deformation. As examples of the latter, C- and Z-shaped beams loaded in the plane of the web twist or deflect laterally, with consequent loss of strength unless they are properly braced (Murray and Elhouar, 1985). For C-sections, the eccentricity between the shear center and the plane of loading creates a torsion demand (a tendency to roll), while for Z-sections the difference between the principal (inclined) axes and the geometric axis creates a similar torsion demand. In either case bracing may be used to restrict the movement, and appropriate expressions for the developed forces are provided in AISI (2007a).

Bracing may be *continuous*, such as that provided by wall panels, roof decking, or floor systems, or it may be noncontinuous or *discrete*, such as cross-bracing. For discrete bracing, the spacing of the braced points also is important. Finally, bracing may also be distinguished according to its behavior—that which provides restraint through resistance to axial deformation, such as cross-bracing, and that which provides restraint through resistance to shear deformation, such as diaphragms (Fig. 13.16).

For bracing against buckling to be effective in an actual situation, it must possess not only the requisite strength but also a definite minimum stiffness. The required strength can only be computed uniquely on the basis of assumed imperfections of shape and/or loading of the member to be braced (Winter, 1958). Recognizing this fact has led to the development of the recent AISC bracing provisions that are discussed in Chapter 12. For cold-formed steel, explicit adoption of this design philosophy is still underway. In many cases the stiffness is assumed and the brace force is assumed to be 2% of the axial load. For axial loaded members undergoing flexural buckling, however, research by Green et al. (2006) has led to the recent adoption of strength and stiffness bracing provisions in AISI (2007a). These provisions require an axial brace capacity of 1% of the member axial capacity and a stiffness based on the discrete brace spacing developed directly from Winter's work. Because diaphragm bracing and continuous bracing can be quite application specific in thin-walled steel construction, specific cases, as opposed to a general treatment, are covered in AISI (2007a).

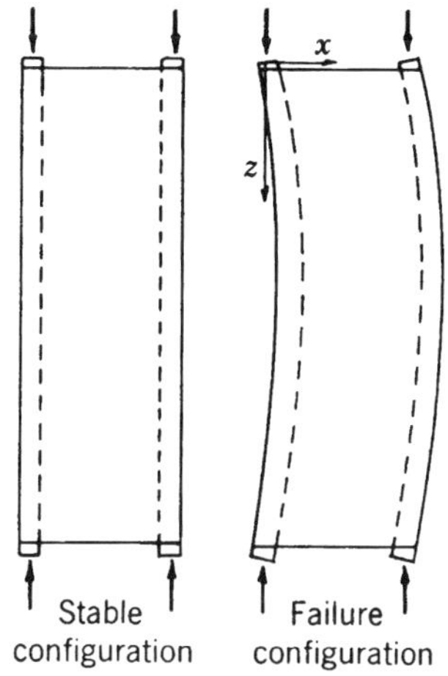

FIGURE 13.16 Members with diaphragm bracing.

13.6.3 Light-Frame Construction

Since the last edition of this guide, the practice of light-frame cold-formed steel construction has evolved significantly, particularly with respect to design standards. A significant amount of research has gone into the development of the North American standards for cold-formed steel framing, which include general provisions (AISI, 2007b), product data (AISI, 2007c), floor and roof system design (AISI, 2007d), wall stud design (AISI, 2007e), header design (AISI, 2007f), lateral design (AISI, 2007g), and truss design (AISI, 2007h). Whenever possible, these standards attempt to treat the system as opposed to the individual members.

The general provisions (AISI, 2007b) and product data standards (AISI, 2007c) are essentially self-explanatory. The floor and roof system design standard (AISI, 2007d) covers both discretely braced design and continuously braced design philosophies. For continuously braced design, prescriptive sheathing requirements are provided (ensuring a level of stiffness for the brace) along with the forces required to counteract rolling of the joists. The floor and roof system design standard also provides a simple means to design clip angle bearing stiffeners, which is based on recent research by Fox (2006).

The wall stud design standard (AISI, 2007e) contains provisions similar to the floor and roof standard and covers the all-steel design and sheathing braced design philosophies. Prescriptive load limits are provided for gypsum sheathed designs based on experimental testing (Miller and Peköz, 1994). Sheathing braced design does not imply diaphragm-based design methods, which were essentially abandoned for sheathed walls, based on the observation that local fastener deformations, and not sheathing in shear, dominates the response. Also covered in this standard are stud-to-track connection strength, including web crippling (Fox and Schuster, 2000), and deflection track strength.

Although it was indicated in Section 13.6.1 that the general built-up section provisions in AISI (2007a) are rudimentary, significant research has been conducted on built-up headers used in light-frame construction (Elhajj and LaBoube, 2000; Stephens and LaBoube, 2000, 2003). This research has led to provisions for box headers, double-L headers, and single-L headers covering web crippling, bending and web crippling, and simplified moment calculations in the header design standard (AISI, 2007f).

The lateral design standard (AISI, 2007g) has had a significant impact on practice, as a result of this standard providing a means to determine the lateral strength of cold-formed steel systems used in wind and seismic demands. The standard provides compiled test results for cold-formed steel shear walls and diaphragms with a variety of sheathing, fastener spacing, stud spacing, and so on. Specific seismic detailing provisions are provided, for example, for strap-braced shear walls (Al-Kharat and Rogers, 2007). In addition to strength, expressions are provided, for deflection calculations (Serrette and Chau, 2006).

The truss design standard (AISI, 2007h) provides specific guidance on beam-column design for chord and web members of cold-formed steel trusses. Due to the presence of concentrated loads (Ibrahim et al., 1998) in locations with compression and bending a unique interaction equation check for compression, bending, and web crippling is also provided. Finally, specific guidance is provided for gusset plate design (Lutz and LaBoube, 2005) and methods for testing trusses.

13.6.4 Diaphragm Construction (Metal Roof and Wall Systems)

Thin-walled metal panels are often used as wall cladding, roof decking, and floor decking, where their primary structural function is to carry loads acting normal to their surface. Properly designed and interconnected metal roof, wall, and floor systems are also capable of resisting shear forces in their own planes, which is typically referred to as diaphragm action. Thus, primary components of the lateral force resisting system for wind or seismic may be a properly detailed floor and/or roof diaphragm composed of fastened and/or welded thin-walled metal panels. Such a system can be highly advantageous; for example, procedures have been developed that recognize the ability of diaphragms assembled from such panels to transfer load from a heavily loaded frame to less heavily loaded adjacent frames in a single-story structure, thus reducing the required maximum frame size (Luttrell, 1967; Bryan and Davies, 1981). The shear strength and stiffness of thin-walled panels can be utilized in folded plate structures (Nilson, 1960), hyperbolic paraboloids (Gergely et al., 1971), and other shell roof structures (Bresler et al., 1968). In addition, theory and test results both have shown that the shear strength and stiffness of properly connected diaphragms can be effective as bracing for individual beams and columns.

Diaphragm-Braced Columns In the elastic range, the minor axis elastic buckling load of an ideal axially loaded I-section column or wall stud with directly attached symmetrical diaphragm bracing (Larson, 1960; Pincus and Fisher, 1966) is determined as

$$P = P_{yy} + Q \tag{13.68}$$

where P_{yy} is the minor axis buckling load of the unbraced column and Q is the shear rigidity of the diaphragm contributing to the support of the column. The shear rigidity can be expressed as

$$Q = A_d G_{\text{eff}} \tag{13.69}$$

where A_d is the cross-sectional area of the diaphragm normal to the column axis and contributing to the support of the member and G_{eff} is the effective shear modulus of the diaphragm. Similar expressions are also possible for columns undergoing flexural–torsional buckling, having sheeting on only one side or dissimilar sheeting (Simaan and Peköz, 1976). Of course, Eq. 13.68 only predicts the increased buckling load for ideal members; consideration of initial imperfections and a full treatment of the fastener and diaphragm stiffness are still required.

Due to the complexity of predicting the strength of diaphragm-braced columns the AISI Specification (2007a) provides empirical methods for compression members with one flange attached to a metal deck (Glaser et al., 1994) or standing seam roof systems (Stolarczyk et al., 2002). Strict prescriptive limits define the system and details for the application of these expressions. The more general treatment of Simaan and Peköz (1976) was adotped in the AISI Specification for wall studs (with different sheathing types) from 1981 to 2004 but has been replaced by rational engineering analysis in the wall stud standard (AISI, 2007e). A key requirement for diaphragm action to occur is that the fasteners must be stiff enough to engage the diaphragm; noting that in many traditional sheathing materials (as opposed to thin-walled metal panels) this does not appear readily possible with conventional detailing (Miller and Peköz, 1993, 1994), and thus it may be difficult to get full diaphragm action engaged.

Diaphragm-Braced Beams The same type of diaphragm action is also useful in counteracting lateral–torsional buckling of beams. For ideal I-section beams braced directly by diaphragms on the compression flange, the critical lateral buckling moment can be estimated as (Errera et al., 1967)

$$M_{cr} = M_0 + 2Qe \tag{13.70}$$

where M_0 is the lateral buckling moment of the unbraced beam, e the distance between the center of gravity of the beam cross section and the plane of the diaphragm, and Q the shear rigidity of the diaphragm contributing to the support of the member, as defined previously. Similar to columns, it is emphasized that Eq. 13.70 predicts the buckling load of an ideal member. For real members, initial imperfections and the strength of the bracing must be taken into consideration.

As discussed in Section 13.6.2, C- and Z-shaped beams have a tendency to twist under lateral load that is applied through the web. When both flanges of such beams are connected to deck or sheathing material in such a manner as to restrain lateral deflection effectively, no further bracing is required. When only one flange is connected, however, the problem can be complex. For purlins and girts with one flange connected to sheeting as commonly found in metal buildings, design for both conventional metal sheeting and standing seam roof systems may proceed either by experimental test methods or through empirical procedures (valid only within prescriptive limits) as provided by the AISI Specification (2007a).

Proper performance of a diaphragm requires adequate anchorage to the structure. Determination of the anchorage forces for sloped diaphragms common in metal building systems has been investigated significantly in recent years. The key to recent improvements is to consider the stiffness of the full system, including diaphragm, purlin, and anchorage when determining the forces. Considering anchorage stiffness (as opposed to ideally rigid) reduces the forces that the anchorage must be designed for and allows for adequate performance provided that a minimum stiffness is provided. Development of such a design method included experiments and shell finite element analysis (Seek and Murray, 2005) as well as the creation of a complete rational engineering analysis method that is capable of handling essentially all situations encountered in practice (Seek and Murray, 2006, 2007, 2008). Ultimately this general method was simplified somewhat (Sears and Murray, 2007) to that which is provided in the AISI Specification (2007a).

13.7 STAINLESS STEEL STRUCTURAL MEMBERS

Cold-formed stainless steel sections have been widely used for architectural purposes in buildings because of their superior corrosion resistance, ease of maintenance, and pleasing appearance. Typical applications include column covers, curtain-wall panels, mullions, door and window framing, roofing and siding, stairs, elevators and escalators, flagpoles, signs, and many others. Since 1968, their use for structural load-carrying purposes has been increased due to the availability of the AISI (1974) and ASCE Design Specifications (1990, 2002).

The main reason for having a different specification for stainless steel structural members is because the material properties of the steel differ from those of carbon steel in many ways, including:

1. Anisotropy
2. Nonlinear stress–strain relationship
3. Low proportional limit
4. Pronounced response to cold work

Figure 13.17 shows the stress–strain curves of annealed, half-hard, and full-hard stainless steels. Because of the differences in mechanical properties and structural

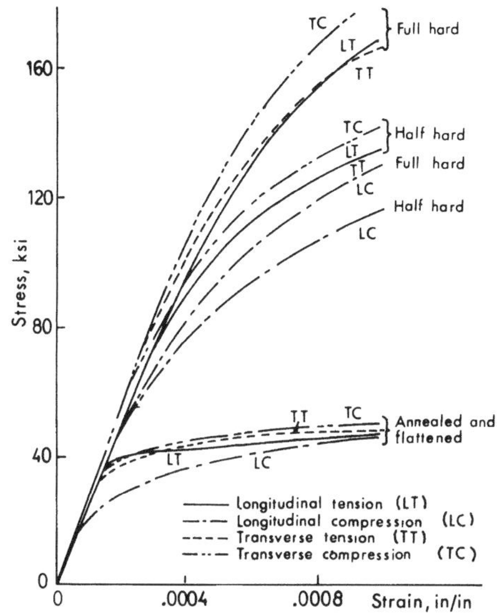

FIGURE 13.17 Stress–strain curves of annealed half-hard and full-hard stainless steels (Wang, 1969; Johnson and Kelsen, 1969).

uses between stainless steel and carbon steel, the ASCE Specification for stainless steel design contains modified design provisions for local buckling of flat elements, *w/t* slenderness limitations, deflection calculations, service stress limitations, lateral buckling of beams, column buckling, and connections. In general, the factors of safety used for the allowable stress design of stainless steels are somewhat larger than those used for carbon steel. Due consideration has been given to the development of the load and resistance factor design criteria (Lin et al., 1992).

The first edition of the *Specification for the Design of Cold-Formed Stainless Steel Structural Members* was issued by American Iron and Steel Institute in 1968 on the basis of the research conducted by Johnson and Winter (1966) at Cornell University. This specification was revised in 1974 to reflect the results of additional research (Wang and Errera, 1971) and the improved knowledge of material properties and structural applications. This edition of the AISI specification contained design information on annealed and cold-rolled grades of sheet and strip stainless steels, types 201, 202, 301, 304, and 316. In 1990, a new standard, *Specification for the Design of Cold-Formed Stainless Steel Structural Members* (ASCE, 1990; Lin et al., 1992; Yu and Lin, 1992), was published by the ASCE to supersede the AISI Specification. The new ASCE specification is based on both limit-state design and allowable stress design and is applicable to the use of four types of austenitic stainless steels (types 201, 301, 304, and 316) and three types of ferritic stainless steels (annealed types 409, 430, and 439). In 2002, the ASCE Specification was updated, largely in an effort to keep the standards in-line with the AISI cold-formed carbon steel standards as appropriate. Austenitic Unified Numbering System (UNS) Designation S20400 (annealed and 1/4 hard) stainless steels were added to the 2002 Specification. Fully developed stainless steel standards also exist in Europe, South Africa, and Australia/New Zealand, just to name a few.

Research on stainless steel members remains active. Fundamental experimental work on beam and column strength of stainless steel tubular and hollow sections including high-strength sections has provided a wealth of new reliable data for better understanding cross-sectional strength (Gardner and Nethercot, 2004a,b; Rasmussen, 2000; Rasmussen and Hancock, 1993a,b; van den Berg, 1998, 2000; Young, 2008; Young and Hartono, 2002; Young and Liu, 2003; Young and Lui, 2005, 2006; Zhou and Young, 2005). Analytical methods for material modeling, including treatment of the highly worked corner regions (Ashraf et al., 2005; Rasmussen, 2003) and complete summaries of imperfections and residual stresses (Cruise and Gardner, 2006), are significantly improving. Shell finite element models are being used successfully by a wide variety of researchers to extend parametric studies and more closely examine the impact of material differences between stainless steel and carbon steel on the behavior and strength of structural members (Ashraf et al., 2007; Ellobody and Young, 2005; Gardner and Nethercot, 2004c; Rasmussen et al., 2003; Young and Ellobody, 2006). Web crippling in stainless steel sections has seen recent study (Zhou and Young, 2006, 2007a,b) as has the performance of stainless steel sections at elevated temperatures, that is, fire conditions (Chen and Young, 2006; Gardner and Ng, 2006; Ng and Gardner, 2007; To and Young, 2008).

Development of design methods for stainless steel also remains active, with researchers looking for robust methods that can accommodate the unique properties of stainless steel while at the same time providing simple solutions for use in conventional design. For example, alternatives to ASCE's iterative column design methods are available with similar reliability (Rasmussen and Rondal, 1997). Improvements to the *effective width method* to better handle gradual yielding behavior of stainless steel sections have been proposed (Rasmussen et al., 2004). Other design method improvements have been proposed, such as basing the resistance on deformation capacity and the use of continuous instead of discrete cross-section classification methods, such as the procedure used in Eurocode (Ashraf et al., 2008; Gardner and Nethercot, 2004d; Gardner et al., 2006; Gardner and Theofanous, 2008). Finally, the initial development of the *direct strength method* for stainless steel has also been completed (Becque et al., 2008).

13.8 ALUMINUM STRUCTURAL MEMBERS

13.8.1 Effective Widths

Postbuckling strengths of thin aluminum elements are generally based on the von Kármán concept that the effective width b_e, for which the elastic buckling stress (Eq. 4.1)

$$\sigma_{cr} = k \frac{\pi^2 E}{12\left(1 - \nu^2\right)(b/t)^2} = \frac{\pi^2 E}{(mb/t)^2} \tag{13.71}$$

is equal to the yield stress, $\sigma_{cr} = \sigma_y$, gives a limiting capacity that remains constant for all other widths at

$$b_e = \frac{\pi t}{m}\sqrt{\frac{E}{\sigma_y}} \tag{13.72}$$

For a simply supported plate with the buckling coefficient $k = 4$ and an aluminum material with $\nu = 0.33$, $m = \sqrt{12\left(1-\nu^2\right)/k} = 1.63$ and $b_e = 1.93t\sqrt{E/\sigma_y}$.

In the inelastic range, the *Specification for Aluminum Structures* (AA, 2005) uses the equivalent slenderness ratio *mb/t* in the expression

$$\sigma_c = \left(B_p - \frac{D_p mb}{t}\right) \tag{13.73}$$

This expression represents an approximation of the true limiting stress (Jombock and Clark, 1968) and can be used to define the maximum load P resisted by a plate element in the inelastic range as

$$P = \sigma_c bt = \left(B_p - \frac{D_p mb}{t}\right) bt \tag{13.74}$$

The effective width, defined as that width which when multiplied by the yield stress and the thickness gives the failure load for the element, then becomes

$$b_e = \frac{b}{\sigma_y}\left(B_p - \frac{D_p mb}{t}\right) \tag{13.75}$$

As b increases, the maximum load that can be resisted occurs when the derivative of Eq. 13.74 equals zero. By setting $dP/db = 0$, the corresponding slenderness is

$$\frac{b}{t} = \frac{B_p}{2D_p m} \tag{13.76}$$

The maximum load P can then be determined by substituting this slenderness into Eq. 13.74,

$$P = \frac{B_p^2 t^2}{4D_p m} \tag{13.77}$$

This load capacity remains reasonably constant for all higher values of *b/t*. Jombock and Clark (1968) provide values for B_p and D_p which can be represented as

$$B_p = \sigma_y \alpha^2 \tag{13.78}$$

$$D_p = \frac{\sigma_y k \alpha}{g} \tag{13.79}$$

where

$\alpha = (1 + 2/g^{2/3})^{1/2}$ for fully heat-treated alloy
$\alpha = (1 + 3/g^{2/3})^{1/2}$ for other alloys
$g = (E/\sigma_y)^{1/2}$
$k = 0.1$ for fully heat-treated alloys
$k = 0.12$ for other alloys

The effective width is then

$$b_e = b\alpha(\alpha - kmb/t) \tag{13.80}$$

with a maximum value of

$$b_e = \alpha^3 gt/4mk \tag{13.81}$$

On the basis of limiting strain, Jombock and Clark (1968) provided a theoretical foundation for the model above, which has been adopted in the U.S. specification (AA, 2005). This model gives a continuous transition from compact to thin-walled elements without discontinuities or changes in the form of the design expression.

Postbuckling behavior of elements supported on both longitudinal edges (to which the model above applies) differs from that of outstanding flange-type elements. For the latter, initial elastic buckling precipitates the collapse of single unsymmetrical elements, while there may be some reserve capacity in symmetrical sections which can be represented by the treatment for edge-supported elements using an appropriate value of the coefficient m, usually 5.

The limiting stress on unstiffened elements of an unsymmetrical section such as a channel or Z is obtained using the expression

$$\sigma = \begin{cases} B - D\left(\dfrac{mb}{t}\right) & \text{for } \dfrac{mb}{t} \le S_2 \quad (13.82) \\[2ex] \pi^2 E \bigg/ \left(\dfrac{mb}{t}\right)^2 & \text{for } \dfrac{mb}{t} > S_2 \quad (13.83) \end{cases}$$

where the value of m lies between 3 and 5, depending on the degree of edge restraint, and S_2 is the slenderness parameter separating elastic and inelastic buckling.

In symmetrical sections, such as I- and double-channel shapes, the effective width will be given by Eqs. 13.80 and 13.81 using $m = 5$, unless local buckling can precipitate overall flexural or lateral–torsional buckling, which will occur when the critical stresses for the different buckling modes are close in value.

13.8.2 Effective Section at Service Loads

The ratios of service loads to the ultimate loads usually exceed 1.5; thus, the extent of any local buckling in service can be considered small and confined to zones of maximum moment. For this reason, the influence of local buckling on deflections under service loads has been neglected in some codes. If a calculation is to be made, the difficulty of computing deflections with a varying and initially unknown effective moment of inertia is usually resolved by assuming conservatively that the effective section at the point of highest moment applies throughout. The effective width b_e of the elements comprising the effective section are obtained using the actual width b and the ratio of the critical stress σ_c to the applied stress σ computed on the basis of a fully effective section (AA, 2005); thus

$$b_e = b\left(\frac{\sigma_c}{\sigma}\right)^{1/2} \qquad (13.84)$$

13.8.3 Interaction between Local Buckling and Global Buckling

If the local buckling strength is less than the global buckling strength, the reduced stiffness that accompanies local buckling may reduce the global buckling strength. Bijlaard and Fisher (1952a) and Sharp (1970) investigated this effect for aluminum members. If the elastic local buckling stress of an element of the shape F_{cr} is less than the global buckling stress, *Specification for Aluminum Structures* (AA, 2005) gives the reduced stress corresponding to the column's strength as ${F_{ec}}^{1/3}{F_{cr}}^{2/3}$, where F_{ec} is the elastic global buckling stress.

13.8.4 Alternative Method for Determining Local Buckling Strength

Specification for Aluminum Structures (AA, 2005) gives an alternative method for determining the local buckling strength. This is especially useful for aluminum members, because their sections vary widely because they can be extruded or roll formed from sheet.

This method requires determining the elastic local buckling stress F_{cr} and determining an equivalent slenderness ratio from it as

$$\lambda_{\text{eq}} = \pi\sqrt{\frac{E}{F_{cr}}} \qquad (13.85)$$

The buckling stress of flat elements in uniform compression is computed using λ_{eq} with $m = 1$, and the buckling stress of flat elements in flexure is computed using λ_{eq} with $m = 1$. The lesser of these may be used as the local buckling stress or a weighted average local buckling stress can be computed from these.

13.9 TORSIONAL BUCKLING

Angles As indicated in Chapter 11, a single angle may fail by flexure or flexural–torsional buckling; only by a special proportion of heavy root bulbs and very thin legs can local buckling control. To optimize shapes, bulbs and root fillets may be added to increase the torsional stiffness such that the equivalent slenderness ratio for torsional buckling is around 60. Because of the interaction of torsion with flexure about the stronger axis, is it not effective to design sections of equal inertia about the two principal axes, and the optimum is an equal-leg right-angle section for both plain and bulb shapes. This is also true of double angles, designed to balance torsional and flexural buckling, in which case equal-leg angles are again very close to the optimum.

Eccentrically Loaded Columns Unsymmetrical open sections loaded axially fail in combined torsion and flexure. Should they be loaded through the shear center, the modes are uncoupled and torsional buckling can be eliminated. Use has been made of this in T-sections for diagonals which, when bolted through the flanges, are loaded through the shear center. This allows much thinner sections to be used with a considerable increase in efficiency despite the moment due to the eccentricity. The optimum form is a lipped shape to control local buckling of the flanges.

Single angles loaded through one leg fail by lateral–torsional buckling in the manner of a beam-column (Marsh, 1969) and the design procedure adopted by CSA (1983) and ASCE Task Committee (1972) treat this interaction by using an effective slenderness ratio

$$\left(\frac{KL}{r}\right)_{\text{eff}} = \left[\left(\frac{5b}{t}\right)^2 + \left(\frac{KL}{r_v}\right)^2\right]^{1/2} \tag{13.86}$$

where $5b/t$ is the slenderness ratio for torsional buckling of angles and KL/r_v is the slenderness ratio for flexural buckling.

Postbuckling Strength In general, the critical stress for a column failing by torsional buckling represents the maximum capacity of the member. This is always true of pin-ended unsymmetrical sections, as the application of the load through the centroid requires a uniform stress in the section for equilibrium. Should the column be loaded by fixed platens, the axis of load application can shift as the member twists, causing an increase in stress toward the shear center (Smith, 1955). Symmetrical sections, such as a cruciform, even when pin ended, can accept a higher stress at the center as the member twists, giving a higher critical load than that obtained for a uniform stress.

REFERENCES

AA (2005), *Specification for Aluminum Structures*, Aluminum Association, Arlington, VA.

Adany, S., and Schafer, B. W. (2006a), "Buckling Mode Decomposition of Single-Branched Open Cross-Section Members via Finite Strip Method: Application and Examples," *Thin-Walled Struct.*, Vol. 44, No. 5, pp. 585–600.

Adany, S., and Schafer, B. W. (2006b), "Buckling Mode Decomposition of Single-Branched Open Cross-Section Members via Finite Strip Method: Derivation," *Thin-Walled Struct.*, Vol. 44, No. 5, pp. 563–584.

AISC (2005), *Specification for Structural Steel Buildings*, ANSI/AISC 360-05, American Institute of Steel Construction, Chicago, IL.

AISI (1974), *Stainless Steel Cold-Formed Structural Design Manual*, American Iron and Steel Institute, Washington, DC.

AISI (1980, 1986, 1989, 1996), *Specification for the Design of Cold-Formed Steel Structural Members*, American Iron and Steel Institute, Washington, DC.

AISI (1986, 1996, 2002), *Cold-Formed Steel Design Manual*, American Iron and Steel Institute, Washington, DC.

AISI (2001), *North American Specification for the Design of Cold-Formed Steel Structural Members*, American Iron and Steel Institute, Washington, DC.

AISI (2002), *Cold-Formed Steel Design Manual*, American Iron and Steel Institute, Washington, DC.

AISI (2004), "Appendix 1: Design of Cold-Formed Steel Structural Members Using the Direct Strength Method," *2004 Supplement to the North American Specification for the Design of Cold-Formed Steel Structures*, American Iron and Steel Institute, Washington, DC.

AISI (2007a), *North American Specification for the Design of Cold-Formed Steel Structural Members*, AISI-S100-07, American Iron and Steel Institute, Washington, DC.

AISI (2007b), *North American Standard for Cold-Formed Steel Framing: General Provisions*, ANSI/AISI-S200-07, American Iron and Steel Institute, Washington, DC.

AISI (2007c), *North American Standard for Cold-Formed Steel Framing: Product Data*, ANSI/AISI-S201-07, American Iron and Steel Institute, Washington, DC.

AISI (2007d), *North American Standard for Cold-Formed Steel Framing: Floor and Roof System Design*, ANSI/AISI-S210-07, American Iron and Steel Institute, Washington, DC.

AISI (2007e), *North American Standard for Cold-Formed Steel Framing: Wall Stud Design*, ANSI/AISI-S211-07, American Iron and Steel Institute, Washington, DC.

AISI (2007f), *North American Standard for Cold-Formed Steel Framing: Header Design*, ANSI/AISI-S212-07, American Iron and Steel Institute, Washington, DC.

AISI (2007g), *North American Standard for Cold-Formed Steel Framing: Lateral Design*, ANSI/AISI-S213-07, American Iron and Steel Institute, Washington, DC.

AISI (2007h), *North American Standard for Cold-Formed Steel Framing: Truss Design*, ANSI/AISI-S214-07, American Iron and Steel Institute, Washington, DC.

Al-Kharat, M., and Rogers, C. A. (2007), "Inelastic Performance of Cold-Formed Steel Strap Braced Walls," *J. Constr. Steel Res.*, Vol. 63, No. 4, pp. 460–474.

AS/NZS:4600 (1996, 2005), *Cold-Formed Steel Structures*, AS/NZS:4600, Standards Australia/Standards New Zealand, Sydney Australia.

ASCE (1990), *Specification for the Design of Cold-Formed Stainless Steel Structural Members*, ANSI/ASCE 8-90, American Society of Civil Engineers, NY.

ASCE (2002), *Specification for the Design of Cold-Formed Stainless Steel Structural Members*, ANSI/ASCE 8-02, American Society of Civil Engineers, NY.

ASCE Task Committee on Light-Weight Alloys (1972), "Guide for the Design of Aluminum Transmission Towers," *ASCE J. Struct. Div.*, Vol. 98, No. ST12, pp. 2785–2804.

Ashraf, M., Gardner, L., and Nethercot, D. A. (2005), "Strength Enhancement of the Corner Regions of Stainless Steel Cross-Sections," *J. Constr. Steel Res.*, Vol. 61, No. 1, pp. 37–52.

Ashraf, M., Gardner, L., and Nethercot, D. A. (2007), "Finite Element Modelling of Structural Stainless Steel Cross-sections," *Thin-Walled Struct.*, Vol. 44, No. 10, pp. 1048–1062.

Ashraf, M., Gardner, L., and Nethercot, D. A. (2008), "Structural Stainless Steel Design: Resistance Based on Deformation Capacity," *J. Struct. Eng.*, Vol. 134, No. 3, pp. 402–411.

Bakker, M. C. M., and Stark, J. W. B. (1994), "Theoretical and Experimental Research on Web Crippling of Cold-Formed Flexural Steel Members," *Thin-Walled Struct.*, Vol. 18, No. 4, 261–290.

Bambach, M. R., and Rasmussen, K. J. R. (2004), "Design Provisions for Sections Containing Unstiffened Elements with Stress Gradient," *J. Struct. Eng.*, Vol. 130, No. 10, pp. 1620–1628.

Batista, E., Rondal, J., and Maquoi, R. (1987), "Column Stability of Cold-Formed U and C Sections," *Proc. of the Int. Conf. on Steel and Aluminum Structures,* Cardiff, U.K.

Bebiano, R., Slivestre, N., and Camotim, D. (2008), "GBTUL— A Code for the Buckling Analysis of Cold-Formed Steel Members," *19th Int. Spec. Conf.: Recent Research and Developments in Cold-Formed Steel Design and Construction*, St. Louis, MO, Vol. 64, No. 11, pp. 61–80.

Becque, J., Lecce, M., and Rasmussen, K. J. R. (2008), "The Direct Strength Method for Stainless Steel Compression Members," *J. Constr. Steel Res.*, Vol. 64, No. 11, pp. 1231–1238.

Bernard, E. S., Bridge, R. Q., and Hancock, G. J. (1992a), "Tests of Profiled Steel Decks with V-Stiffeners," *11th Int. Spec. Conf. Cold-Formed Steel Struct.*, St. Louis, MO.

Bernard, E. S., Bridge, R. Q., and Hancock, G. J. (1992b), "Intermediate Stiffeners in Cold Formed Profiled Steel Decks: Part I. 'Flat Hat' Shaped Stiffeners," Res. Rep. No. R658, School of Civil and Mining Engineering, University of Sydney, Sydney, Australia.

Bernard, E. S., Bridge, R. Q., and Hancock, G. J. (1993a), "Intermediate Stiffeners in Cold Formed Profiled Steel Decks: Part I. 'V' Shaped Stiffeners," *ASCE, J. Struct. Eng.*, Vol. 119, No. 8, pp. 2277–2293.

Bernard, E. S., Bridge, R. Q., and Hancock, G. J. (1993b), "Design of Decking Panels with Intermediate Stiffeners," Res. Rep. No. R676, School of Civil and Mining Engineering, University of Sydney, Sydney, Australia, Apr.

Beshara, B., and Schuster, R. M. (2000), "Web Crippling of Cold Formed Steel C- and Z-Sections," *15th Int. Spec. Conf. Cold-Formed Steel Struct.: Recent Research and Developments in Cold-Formed Steel Design and Construction* St. Louis, MO, pp. 23–42.

Beshara, B., and Schuster, R. M. (2002), "Web Crippling of Cold-Formed Steel Members," *16th Int. Spec. Conf. Cold-Formed Steel Struct.: Recent Research and Developments in Cold-Formed Steel Design and Construction*, Orlando, FL, pp. 219–236.

Bijlaard, P. P., and Fisher, G. P. (1952a), "Interaction of Column and Local Buckling in Compression Members," NACA Tech. Note No. 2640, Mar.

Bijlaard, P. P., and Fisher, G. P. (1952b), "Column Strength of H-Sections and Square Tubes in Post-buckling Range of Component Plates," NACA Tech. Note No. 2994, Aug.

Bresler, B., Lin, T. Y., and Scalzi, J. B. (1968), *Design of Steel Structures,* Wiley, New York.

Bryan, E. R., and Davies, J. M. (1981), *Steel Diaphragm Roof Decks*, Wiley, New York.

Bryan, E. R., and Leach, P. (1984), "Design of Profiled Sheathing as Permanent Formwork," Tech. Note No. 116, Construction Industry Research and Information Association, London.

Buhagiar, D., Chapman, I. C., and Dowling, P. I. (1992), "Design of C-Sections against Deformational Lip Buckling," *Proc. 11th Int. Spec. Conf. Cold-Formed Steel Struct.*, St. Louis, MO, pp. 75–94.

Canadian Standards Association (CSA) (1983), *Strength Design in Aluminum*, CAN3-S157-M83, CSA, Rexdale, Ontario, Canada.

Canadian Standards Association (CSA) (1984 1989, 1994), *Cold-Formed Steel Structural Members*, CAN3S136-M84, CSA, Rexdale, Ontario, Canada.

Chajes, A., Fang, J., and Winter, G. (1966), "Torsional Flexural Buckling, Elastic and Inelastic, of Cold Formed Thin Walled Columns," *Cornell Eng. Res. Bull.*, Vol. 66, No. 1.

Chajes, A., and Winter, G. (1965), "Torsional-Flexural Buckling of Thin-Walled Members," *ASCE J. Struct. Div.*, Vol. 91, No. ST4, pp. 103–124.

Chen, J., and Young, B. (2006), "Stress-Strain Curves for Stainless Steel at Elevated Temperatures," *Eng. Struct.*, Vol. 28, No. 2, pp. 229–239.

Cheung, Y. K., and Tham, L. G. (1998), *The Finite Strip Method*, CRC Press, Boca Raton, FL.

Chilver, A. H. (1951), "The Behavior of Thin-Walled Structural Members in Compression," Vol. 172, No. 4466, *Engineering*, pp. 281–282.

Chilver, A. H. (1953), "The Stability and Strength of Thin-Walled Steel Struts," *Engineer*, Vol. 196, No. 5089, pp. 180–183.

Chilver, A. H. (1967), *Thin-Walled Structures*, Wiley, New York.

Chou, S. W., Seah, L. K., and Rhodes, J. (1996), "Accuracy of Some Codes of Practice in Predicting the Load Capacity of Cold-Formed Columns," *J. Constr. Steel Res.,* Vol. 37, No. 2, pp. 137–172.

Cohen, J. M., and Peköz, T. B. (1987), "Local Buckling Behavior of Plate Elements," Res. Rep., Department of Structural Engineering, Cornell University, Ithaca, NY.

Cook, R. D., Malkus, D. S., and Plesha, M. E. (1989), *Concepts and Applications of Finite Element Analysis,* 3rd ed., Wiley, New York.

Craig, B., and Schuster, R. M. (2000), "Calibration of Cold Formed Steel Shear Equations," *15th Int. Spec. Conf. Cold-Formed Steel Struct.: Recent Research and Developments in Cold-Formed Steel Design and Construction* St. Louis, MO, pp. 207–221.

Cruise, R. B., and Gardner, L. (2006), "Measurement and Prediction of Geometric Imperfections in Structural Stainless Steel Members," *Struct. Eng. Mech.*, Vol. 24, No. 1, pp. 63–89.

Culver, C. G. (1966), "Exact Solution of the Biaxial Bending Equations," *ASCE J. Struct. Div.*, Vol. 92, No. ST2, pp. 63–84.

Dabrowski, R. (1961), "Dünwandige Stabe unter zweiachsig aussermittigem Druck," *Stahlbau*, Vol. 30, p. 360 .

Davies, J. M., and Jiang, C. (1996), "Design of Thin-Walled Beams for Distortional Buckling," *13th Int. Spec. Conf. Cold-Formed Steel Struct.,* St. Louis, MO, pp. 141–154.

Davies, J. M., Jiang, C., and Ungureanu, V. (1998), "Buckling Mode Interaction in Cold-Formed Steel Columns and Beams," *14th Int. Spec. Conf. Cold-Formed Steel Struct*., St. Louis, MO, pp. 53–68.

Davies, J. M., Leach, P., and Heinz, D. (1994) "Second-Order Generalised Beam Theory," *J. Constr. Steel Res.*, Vol. 31, Nos. 2/3, pp. 221–242.

Desmond, T. P., Peköz, T., and Winter, G. (1981a), "Edge Stiffeners for Thin-Walled Members," *ASCE J. Struct. Div*, Vol. 107, No. 2, pp. 329–353.

Desmond, T. P., Peköz, T., and Winter, G. (1981b), "Intermediate Stiffeners for Thin-Walled Members," *ASCE J. Struct. Div.*, Vol. 107, No. 4, pp. 627–648.

DeWolf, J. T., Peköz, T., and Winter, G. (1974), "Local and Overall Buckling of Cold Formed Steel Members," *ASCE J. Struct. Div.*, Vol. 100, No. ST10, pp. 2017–2036.

DeWolf, J. T., Peköz, T., and Winter, G. (1976), "Closure to 'Local and Overall Buckling of Cold-Formed Steel Members,'" *ASCE J. Struct, Div.*, Vol. 102, No. ST2, pp. 451–454.

Elhajj, N. R., and LaBoube, R. A. (2000), "L-Header Testing, Evaluation and Design Methodology," *15th Int. Spec. Conf. Cold-Formed Steel Struct.: Recent Research and Developments in Cold-Formed Steel Design and Construction* St. Louis, MO, pp. 437–456.

Ellobody, E., and Young, B. (2005), "Structural Performance of Cold-Formed High Strength Stainless Steel Columns," *J. Constr. Steel Res.*, Vol. 61, No. 12, pp. 1631–1649.

Elzein, A. (1991), *Plate Stability by Boundary Element Method,* Springer-Verlag, New York.

Errera, S. J., Pincus, G., and Fisher, G. P. (1967), "Columns and Beams Braced by Diaphragms," *ASCE J. Struct. Div.*, Vol. 93, No. ST1, pp. 295–318.

Eurocode (1996), *Eurocode 3 Design of Steel Structures: Part 1.3 Supplementary Rules for Cold Formed Members and Sheeting,* EN 1993-1-3:1996, Brussels, Belgium.

Eurocode (2004), *Eurocode 3 Design of Steel Structures: Part 1.3 Supplementary Rules for Cold Formed Members and Sheeting,* EN 1993-1-3:2004, Brussels, Belgium.

Fox, S. R. (2006), "The Strength of CFS Floor Assemblies with Clip Angle Bearing Stiffeners," *18th Int. Spec. Conf. Cold-Formed Steel Struct.: Recent Research and Developments in Cold-Formed Steel Design and Construction* Orlando, FL, pp. 239–252.

Fox, S. R., and Brodland, G. W. (2004), "Design Expressions Based on a Finite Element Model of a Stiffened Cold-Formed Steel C-Section," *J Struct. Eng.*, Vol. 130, No. 5, pp. 708–714.

Fox, S. R., and Schuster, R. M. (2000), "Strength of Bearing Stiffeners in Cold Formed C-Sections," *15th Int. Spec. Conf. Cold-Formed Steel Struct.: Recent Research and Developments in Cold-Formed Steel Design and Construction* St. Louis, MO, pp. 135–147.

Fox, S. R., and Schuster, R. M. (2002), "Strength of Wind Load Bearing Wall Stud-to-Track Connections," *Can. J. Civil Eng.*, Vol. 29, No. 5, pp. 777–786.

Galambos, T. V. (1968), *Structural Members and Frames*, Prentice Hall, Upper Saddle River, NJ.

Gardner, L., and Nethercot, D. A. (2004a), "Experiments on Stainless Steel Hollow Sections—Part 1: Material and Cross-Sectional Behaviour," *J. Constr. Steel Res.*, Vol. 60, No. 9, pp. 1291–1318.

Gardner, L., and Nethercot, D. A. (2004b), "Experiments on Stainless Steel Hollow Sections—Part 2: Member Behaviour of Columns and Beams," *J. Const. Steel Res.*, Vol. 60, No. 9, pp. 1319–1332.

Gardner, L., and Nethercot, D. A. (2004c), "Numerical Modeling of Stainless Steel Structural Components—A Consistent Approach," *J. Struct. Eng.*, Vol. 130, No. 10, pp. 1586–1601.

Gardner, L., and Nethercot, D. A. (2004d), "Structural Stainless Steel Design: A New Approach," *Struct. Eng.*, Vol. 82, No. 21, pp. 21–28.

Gardner, L., and Ng, K. T. (2006), "Temperature Development in Structural Stainless Steel Sections Exposed to Fire," *Fire Safety J.*, Vol. 41, No. 3, pp. 185–203.

Gardner, L., Talja, A., and Baddoo, N. R. (2006), "Structural Design of High-Strength Austenitic Stainless Steel," *Thin-Walled Struct.*, Vol. 44, No. 5, pp. 517–528.

Gardner, L., and Theofanous, M. (2008), "Discrete and Continuous Treatment of Local Buckling in Stainless Steel Elements," *J. Constr. Steel Res.*, Vol. 64, No. 11, pp. 1207–1216.

Gergely, P., Banavalkar, P. V., and Parker, J. E. (1971), "The Analysis and Behavior of Thin-Steel Hyperbolic Paraboloid Shells," Cornell Univ. Dept. Struct. Eng. Rep. No. 338, Sept.

Gerges, R. R., and Schuster, R. M. (1998), "Web Crippling of Single Web Cold Formed Steel Members Subjected to End One-Flange Loading," *14th Int. Spec. Conf. Cold-Formed Steel Struct.: Recent Research and Developments in Cold-Formed Steel Design and Construction* St. Louis, MO, pp. 165–192.

Glaser, N., Kaehler, R., and Fisher, J. M. (1994), "Axial Load Capacity of Sheeted C and Z Members," *12th Int. Speci. Conf. Cold-Formed Steel Struct.: Recent Research and Developments in Cold-Formed Steel Design and Construction* St. Louis, MO, pp. 451–469.

Goodier, J. N. (1942), "Flexural-Torsional Buckling of Bars of Open Section," Cornell Univ. Eng. Exp. Sta. Bull. No. 28, Jan.

Gotluru, B. P., Schafer, B. W., and Peköz, T. (2000), "Torsion in Thin-Walled Steel Beams," *Thin-Walled Struct.*, Vol. 37, No. 2, pp. 127–145.

Green, P. S., Sputo, T., and Urala, V. (2006), "Strength and Stiffness of Conventional Bridging Systems for Cold-Formed CEE Studs," *J. Struct. Eng.*, Vol. 132, No. 4, pp. 550–557.

Hancock, G. J. (1978), "Local, Distortional, and Lateral Buckling of I-Beams," *ASCE J. Struct. Div.*, Vol. 104, No. 11, pp. 1787–1798.

Hancock, G. J. (1981), "Nonlinear Analysis of Thin Sections in Compression," *ASCE J. Struct. Div.*, Vol. 107, No. ST3, pp. 455–472.

Hancock, G. J., Kwon, Y. B., and Stefan Bernard, E. (1994), "Strength Design Curves for Thin-Walled Sections Undergoing Distortional Buckling," *J. Constr. Steel Res.*, Vol. 31, Nos. 2/3, pp. 169–186.

Hancock, G. J., Murray, T. M., and Ellifritt, D. S. (2001), *Cold-Formed Steel Structures to the AISI Specification,* Marcell-Dekker, New York.

Hancock, G. J., Rogers, C. A., and Schuster, R. M. (1996), "Comparison of the Distortional Buckling Method for Flexural Members with Tests," paper presented at the *13th Int. Spec. Conf. Cold-Formed Steel Struct.: Recent Research and Developments in Cold-Formed Steel Design and Construction* St. Louis, MO, pp. 125–139.

Harik, I. E., Liu, X., and Ekambaram, R. (1991), "Elastic Stability of Plates with Varying Rigidities," *Comput. Struct.*, Vol. 38, No. 2, pp. 161–168.

Harvey, J. M. (1953), "Structural Strength of Thin-Walled Channel Sections," *Engineering*, Vol. 175, No. 4545, pp. 291–293.

Hetrakul, N., and Yu, W. W. (1978), "Structural Behavior of Beam Webs Subjected to Web Crippling and a Combination of Web Crippling and Bending," Rep. No. 78-4, University of Missouri—Rolla, Rolla, MO, June.

Hikosaka, H., Takami, K., and Maruyama, Y. (1987), "Analysis of Elastic Distortional Instability of Thin-Walled Members with Open Polygonal Cross Section," *Proceedings of Japan Society of Civil Engineers, Struct. Eng./Earthquake Eng.*, Vol. 4, No. 1, pp. 31–40.

Hofmeyer, H., Kerstens, J. G. M., Snijder, H. H., and Bakker, M. C. M. (2001), "New Prediction Model for Failure of Steel Sheeting Subject to Concentrated Load (Web Crippling) and Bending," *Thin-Walled Struct.*, Vol. 39, No. 9, pp. 773–796.

Höglund, T. (1980), "Design of Trapezoidal Sheeting Provided with Stiffeners in the Flanges and Webs," D28:1980, Swedish Council for Building Research, Stockholm, Sweden.

Holesapple, M. W., and LaBoube, R. A. (2003), "Web Crippling of Cold-Formed Steel Beams at End Supports," *Eng. Struct.*, Vol. 25, No. 9, pp. 1211–1216.

Ibrahim, T. M., LaBoube, R. A., and Yu, W. W. (1998), "Behavior of Cold-Formed Steel Roof Trusses Under Concentrated Loads," *J. Const. Steel Res.*, Vol. 46, Nos. 1–3, pp. 176–177.

Johnson, A. L., and Kelsen, G. A. (1969), "Stainless Steel in Structural Applications," *Stainless Steel for Architecture*, ASTM STP, American Society for Testing and Materials, Philadelphia, p. 454 .

Johnson, A. L., and Winter, G. (1966), "Behavior of Stainless Steel Columns and Beams," *ASCE J. Struct. Div.*, Vol. 92, No. ST5, pp. 97–118.

Jombock, J. R., and Clark, J. W. (1968), "Bending Strength of Aluminum Formed Sheet Members," *ASCE J. Struct. Div.*, Vol. 94, No. ST2, pp. 511–528.

Kalyanaraman, V., Peköz, T., and Winter, G. (1977), "Unstiffened Compression Elements," *J. Struct. Div. ASCE*, Vol. 103, No. ST9, pp. 1833–1848.

Kim, Y., and Peköz, T. (2004), "Flexural Strength of Aluminum Extrusions of Arbitrary Cross-Section," *Proceedings of the Annual Stability Conference*, Structural Stability Research Council, pp. 455–474.

Klöppel, K., and Bilstein, W. (1976), "Untersuchungen zur linearn und nichtlinearen Beultheorie mit Beulwerttafeln fur dunnwandige U, C und Hut-Profile und Tafeln fur mitwirkende Breiten und Tragspannungen von dreiseitig und vierseitig gelenkig gelagerten Reckteckplatten nach der nichtlinearen Beultheorie," *Stahlbau*, Vol. 45, No. 2, pp. 33–38.

Klöppel, K., and Schardt, R. (1958), "Beitrag zur praktischen Ermittlung der Vergleichsschlankheit von mittig gedrückten Stäben mit einfachsymmetrischem offenem dünnwandigem Querschnitt," *Stahlbau*, Vol. 27, p. 35 .

Klöppel, K., and Unger, von B. (1970), "Das Ausbeulen einer am freien Rand verstuiften, dreiseitig momentenfrei gelagerten Platte unter Verwendung der nichtlinearen Beultheorie, Teill II: Experimentelle Untersuchungen, Bergleich der experimentellen mit den theoretischen Ergebnissen," *Der Stahlabu*, Vol. 39, pp. 115–123.

Kwon, Y. B., and Hancock, G. J. (1992), "Tests of Cold-Formed Channels with Local and Distortional Buckling," *J. Struct. Eng. N.Y.*, Vol. 118, No. 7, pp. 1786–1803.

LaBoube, R. A., and Yu, W.-W. (1978), "Cold-Formed Steel Web Elements Under Combined Bending and Shear," *Proceedings—Annual Public Water Supply Engineers' Conference*, Vol. 1, pp. 219–251.

LaBoube, R. A., and Yu, W. W. (1982), "Bending Strength of Webs of Cold-Formed Steel Beams," *ASCE J Struct. Div.*, Vol. 108, No. ST7, pp. 1589–1604.

LaBoube, R. A., Yu, W. W., Deshmukh, S. U., and Uphoff, C. A. (1999), "Crippling Capacity of Web Elements with Openings," *J. Struct. Eng.*, Vol. 125, No. 2, pp. 137–141.

LaBoube, R. A., Yu, W. W., Langan, J. E., and Shan, M. Y. (1997), "Cold-Formed Steel Webs with Openings: Summary Report," *Thin-Walled Struct.*, Vol. 27, No.1, pp. 79–84.

Langan, J. E., LaBoube, R. A., and Yu, W. W. (1994), "Perforated Webs Subjected to End-One-Flange Loading," paper presented at the *12th Int. Spec. Conf. Cold-Formed Steel Struct.: Recent Research and Developments in Cold-Formed Steel Design and Construction* St. Louis, MO, pp., 181–194.

Larson, M. A. (1960), "Discussion of 'Lateral Bracing of Columns and Beams' by George Winter," *Trans. ASCE*, Vol. 125, pp. 830–838.

Lau, S. C. W. (1988), "Distortional Buckling of Thin-Walled Columns," Ph.D. thesis, University of Sydney, Sydney, Australia.

Lau, S. C. W., and Hancock, G. J. (1987), "Distortional Buckling Formulas for Channel Columns," *J. Struct. Eng. N.Y.*, Vol. 113, No. 5, pp. 1063–1078.

Lau, S. C. W., and Hancock, G. J (1990), "Inelastic Buckling of Channel Columns in the Distortional Mode," *Thin-Walled Struct.,* Vol. 10, No. 1, pp. 59–84.

Lim, B. S. (1985), "Buckling Behaviour of Asymmetric Edge Stiffened Plates," Ph.D. thesis, University of Strathclyde, Glasgow.

Lim, B. S., and Rhodes, J. (1986), "Buckling Behaviour of Asymmetric Edge Stiffened Plates," in *Applied Solid Mechanics*, Elsevier, Vol. 1, New York, pp. 351–375.

Lin, S.-H., Yu, W. W., and Galambos, T. V. (1992), "ASCE LRFD Method for Stainless Steel Structures," *J. Struct. Eng. N.Y.*, Vol. 118, No. 4, pp. 1056–1070.

Loh, T. S. (1985), "Combined Axial Load and Bending in Cold-Formed Steel Members," Cornell Univ. Dept. Struct. Eng. Rep. No. 85-3, Feb.

Loughlan, J. (1979), "Mode Interaction in Lipped Channel Columns under Concentric or Eccentric Loading, " Ph.D. thesis. University of Strathclyde, Glasgow.

Lundquist, E. E., and Stowel, E. Z. and Schuette, E. Z. (1945), "Principles of Moment Distribution Applied to the Stability of Structures Composed of Bars or Plates," NACA Rep. No. 809.

Luttrell, L. C. (1967), "Strength and Behavior of Light Gage Steel Shear Diaphragm," Cornell Univ. Eng. Res. Bull. No. 67–1.

Lutz, D. G., and LaBoube, R. A. (2005), "Behavior of Thin Gusset Plates in Compression," *Thin-Walled Struct.*, Vol. 43, No. 5, pp. 861–875.

Marsh, C. (1969), "Single Angles in Tension and Compression," *ASCE J. Struct. Div. Tech. Note*, Vol. 95, No. ST5, pp. 1043–1049.

Miller, T. H., and Peköz, T. (1993), "Behavior of Cold-Formed Wall Stud-Assemblies," *J. Struct. Eng. N.Y.*, Vol. 119, No. 2, pp. 641–651.

Miller, T. H., and Peköz, T. (1994), "Behavior of Gypsum-Sheathed Cold-Formed Steel Wall Studs," *J. Struct. Eng. N.Y.*, Vol. 120, No. 5, pp. 1644–1650.

Moen, C. D., Igusa, T., and Schafer, B. W. (2008), "Prediction of Residual Stresses and Strains in Cold-Formed Steel Members," *Thin-Walled Struct.*, Vol. 46, No. 11, pp. 1274–1289.

Mulligan, G. P. (1983), "The Influence of Local Buckling on the Structural Behavior of Singly-Symmetric Cold-Formed Steel Columns," Ph.D. thesis, Cornell University, Ithaca, NY.

Mulligan, G. P., and Peköz, T. (1983), "Influence of Local Buckling on the Behavior of Singly Symmetric Cold-Formed Steel Columns," Cornell Univ. Dep. Struct. Eng. Rep. No. 83-2, Mar.

Murray, T. M., and Elhouar, S. (1985), "Stability Requirements of Z-Purlin Supported Conventional Metal Building Roof Systems," *Proc. Annu. SSRC Tech. Session*, Structural Stability Research Council, Bethlehem, PA.

Ng, K. T., and Gardner, L. (2007), "Buckling of Stainless Steel Columns and Beams in Fire," *Eng. Struct.*, Vol. 29, No. 5, pp. 717–730.

Nilson, A. H. (1960), "Shear Diaphragms of Light Gage Steel," *ASCE J. Struct. Div.*, Vol. 86, No. ST11, pp. 111–140.

Papangelis, J. P., and Hancock, G. J. (1995), "Computer Analysis of Thin-Walled Structural Members," *Comput. Struct.*, Vol. 56, No. 1, pp. 157–176.

Peköz, T. (1977), "Post-Buckling Interaction of Plate Elements: Progress Report to Swedish Building Research Council," Dept. of Structural Eng., Cornell University, Ithaca, NY.

Peköz, T. (1987), *Development of a Unified Approach to the Design of Cold-Formed Steel Members*, American Iron and Steel Institute, Washington, DC.

Peköz, T. B. (1969) (with a contribution by N. Celebi), "Torsional–Flexural Buckling of Thin-Walled Sections Under Eccentric Load," Cornell Eng. Res. Bull. No. 69.1, Sept.

Peköz, T. B., and Winter, G. (1969), "Torsional–Flexural Buckling of Thin-Walled Sections Under Eccentric Load," *ASCE J. Struct. Div.*, Vol. 95, No. ST5, pp. 941–964.

Pfluger, A. (1961), "Thin-Walled Compression Members," Mitt. Inst. Statik Tech. Hochsch. Hannover, Part 1 (1959); Part 2 (1959); Part 3 (1961),

Pham, C. H., and Hancock, G. J. (2009), "Shear Buckling of Thin-Walled Channel Sections," *J. Constr. Steel Res.*, in press. Vol. 65, No. 3, pp. 578–585.

Pincus, G., and Fisher, G. P. (1966), "Behavior of Diaphragm-Braced Columns and Beams," *ASCE J. Struct. Div.*, Vol. 92, No. ST2, pp. 323–350.

Plank, R. J., and Wittrick, W. H. (1974), "Buckling under Combined Loading of Thin, Flat-Walled Structures by a Complex Finite Strip Method," *Int. J. Num. Meth. Eng.,* Vol. 8, pp. 323–339.

Polyzois, D., and Charnvarnichborikarn, P. (1993), "Web-Flange Interaction in Cold-Formed Steel Z-Section Columns," *J. Struct. Eng. ASCE*, Vol. 119, No. 9, pp. 2607–2628.

Polyzois, D., and Sudharmapal, A. R. (1990), "Cold-Formed Steel Z-Sections with Sloping Edge Stiffness under Axial Load," *J. Struct. Eng. ASCE*, Vol. 116, No. 2, pp. 392–406

Prabakaran, K., and Schuster, R. M. (1998), "Web Crippling of Cold Formed Steel Members," *14th Int. Spec Conf. Cold-Formed Steel Struct.: Recent Research and Developments in Cold-Formed Steel Design and Construction* St. Louis, MO, pp. 151–164.

Prawel, S. P., and Lee, G. C. (1964), "Biaxial Flexure of Columns by Analog Computers," Proc. Pap. 3805, *ASCE J. Eng. Mech. Div.*, Vol. 90, No. EM1, pp. 83–112.

Purnadi, R. W., Tassoulas, J. L., and Polyzois, D. (1990), "Study of Cold-Formed Z-section Steel Member under Axial Loading," *10th Int. Spec. Conf. Cold-Formed Steel Struct.,* St. Louis, MO.

Rack Manufacturers Institute (RMI) (1979), *Specification for the Design, Testing and Utilization of Industrial Steel Storage Racks*, Charlotle, NC. RMI.

Rasmussen, K. J. R. (2000), "Recent Research on Stainless Steel Tubular Structures," *J. Constr. Steel Res.*, Vol. 54, No.1, pp. 75–88.

Rasmussen, K. J. R. (2003), "Full-Range Stress-Strain Curves for Stainless Steel Alloys," *J. Constr. Steel Res.*, Vol. 59, No. 1, pp. 47–61.

Rasmussen, K. J. R., Burns, T., and Bezkorovalny, P. (2004), "Design of Stiffened Elements in Cold-Formed Stainless Steel Sections," *J. Struct. Eng.* Vol. 130, No. 11, pp. 1764–1771.

Rasmussen, K. J. R., Burns, T., Bezkorovainy, P., and Bambach, M. R. (2003), "Numerical Modelling of Stainless Steel Plates in Compression," *J. Const. Steel Res.*, Vol. 59, No. 11, pp. 1345–1362.

Rasmussen, K. J. R., and Hancock, G. J. (1991), "The Flexural Behavior of Thin-Walled Singly Symmetric Columns," paper presented at the Int. Conf. on Steel and Aluminum Structures, Singapore.

Rasmussen, K. J. R., and Hancock, G. J. (1993a), "Design of Cold-Formed Stainless Steel Tubular Members. I. Columns," *J. Struct. Eng. N.Y.*, 119 (8), 2349–2367.

Rasmussen, K. J. R., and Hancock, G. J. (1993b), "Design of Cold-Formed Stainless Steel Tubular Members. II. Beams," *J. Struct. Eng. N.Y.*, Vol. 119, No. 8, pp. 2368–2386.

Rasmussen, K. J. R., and Rondal, J. (1997), "Explicit Approach to Design of Stainless Steel Columns," *J. Struct. Eng.*, Vol. 123, No. 7, pp. 857–863.

Ren, W. X., Fang, S. E., and Young, B. (2006a), "Analysis and Design of Cold-Formed Steel Channels Subjected to Combined Bending and Web Crippling," *Thin-Walled Struct.*, Vol. 44, No. 3, pp. 314–320.

Ren, W. X., Fang, S. E., and Young, B. (2006b), "Finite element Simulation and Design of Cold-Formed Steel Channels Subjected to Web Crippling," *J. Struct. Eng.*, Vol. 132, No. 12, pp. 1967–1975.

Rhodes, J., and Harvey, J.M. (1977), "Interaction Behaviour of Plain Channel Columns under Concentric or Eccentric Loading," *Proc. of the 2nd Int. Colloquium on the Stability of Steel Structures,* ECCS, Liege, pp. 439–444.

Rhodes, J., and Loughlan, J. (1980), "Simple Design Analysis of Lipped Channel Columns," paper presented at the *5th Int. Spec. Conf. Cold-Formed Steel Struct.,* St. Louis, MO.

Rockey, K. C., Elgaaly, M. A., and Bagchi, D. K. (1972), "Failure of Thin-Walled Members Under Patch Loading," *ASCE J. Struct. Div.*, Vol. 98, No. ST12, pp. 2739–2752.

Rogers, C. A., and Schuster, R. M. (1997), "Flange/Web Distortional Buckling of Cold-Formed Steel Sections in Bending," *Thin-Walled Struct.*, Vol. 27, No. 1, pp. 13–29.

Salmi, P., and Talja, A. (1993), "Design of Cold-Formed HSS Channels for Bending and Eccentric Compression," VTT Research Notes, 1505, Technical Research Centre of Finland, Espoo.

Santaputra, C., Parks, M. B., and Yu, W. W. (1989), "Web-Crippling Strength of Cold-Formed Steel Beams," *J. Struct. Eng. N.Y.*, Vol. 115, No. 10, pp. 2511–2527.

Sarawit, A. T., and Peköz, T. (2006), "Notional Load Method for Industrial Steel Storage Racks," *Thin-Walled Struct.*, Vol. 44, No. 12, pp. 1280–1286.

Schafer, B.W. (2000), "Distortional Buckling of Cold-Formed Steel Columns," Final Report to the American Iron and Steel Institute, Washington, DC.

Schafer, B. W. (2001), "Thin-Walled Column Design Considering Local, Distortional and Euler Buckling," *Proceeding—Annual Technical Session, Structural Stability Research Council*, pp. 419–438.

Schafer, B. W. (2002), "Local, Distortional, and Euler Buckling of Thin-Walled Columns," *J. Struct. Eng.*, Vol. 128, No. 3, pp. 289–299.

Schafer, B. W. (2006), *Direct Strength Method Design Guide.* CF06-1, American Iron and Steel Institute, Washington, DC.

Schafer, B. W. (2008), "Review: The Direct Strength Method of Cold-Formed Steel Member Design," *J. Constr. Steel Res.*, Vol. 64, Nos. 7/8, pp. 766–778.

Schafer, B. W., and Adany, S. (2006), "Buckling Analysis of Cold-Formed Steel Members Using CUFSM: Conventional and Constrained Finite Strip Methods," *18th Int. Spec. Conf. Cold-Formed Steel Struct. Recent Research and Developments in Cold-Formed Steel Design and Construction* Orlando, FL, pp. 39–54.

Schafer, B. W., and Peköz, T. (1998a), "Cold-Formed Steel Members with Multiple Longitudinal Intermediate Stiffeners," *J. Struct Eng*, Vol. 124, No. 10, pp. 1175–1181.

Schafer, B. W., and Peköz, T. (1998b), "Computational Modeling of Cold-Formed Steel: Characterizing Geometric Imperfections and Residual Stresses," *J. Constr. Steel Res.*, Vol. 47, No. 3, pp. 193–210.

Schafer, B. W., and Peköz, T. (1998c), "Direct Strength Prediction of Cold-Formed Steel Members using Numerical Elastic Buckling Solutions," *14th Int. Spec. Conf. Cold-Formed Steel Struct.,* St. Louis, MO, pp. 69–76.

Schafer, B. W., and Peköz, T. (1999), "Laterally Braced Cold-Formed Steel Flexural Members with Edge Stiffened Flanges," *J. Struct. Eng.* Vol. 125, No. 2, pp. 118–127.

Schafer, B. W., and Trestain, T. W. J. (2002), "Interim Design Rules for Flexure in Cold-Formed Steel Webs," *16th Int. Spec. Conf. Cold-Formed Steel Structures: Recent Research and Developments in Cold-Formed Steel Design and Construction* Orlando, FL, pp. 145–160.

Schardt, R. (1989), *Verallgemeinerte Technische Biegetheorie,* Springer-Verlag, Berlin.

Schardt, R. (1994), "Generalized Beam Theory—An Adequate Method for Coupled Stability Problems," *Thin-Walled Struct.,* Vol. 19, Nos. 2–4, pp. 161–180.

Seah, L. K. (1989), "Buckling Behaviour of Edge Stiffeners in Thin-Walled Sections," Ph.D. thesis, University of Strathclyde, Glasgow.

Seah, L. K., Rhodes, J. (1993), "Simplified Buckling Analysis of Plate with Compound Edge Stiffners," *J. Eng. Mech. ASCE*, Vol. 119, No. 1, pp. 19–38.

Seah, L. K., Rhodes, J., and Lim, B. S. (1993), "Influence of Lips on Thin-Walled Sections," *Thin-Walled Struct.,* Vol. 16, Nos. 1–4, pp. 145–178.

Sears, J. M., and Murray, T. M. (2007), "Proposed Method for the Prediction of Lateral Restraint Forces in Metal Building Roof Systems," *Structural Stability Research Council—Proceedings of the 2007 Annual Stability Conference*, pp. 521–540.

Seek, M. W., and Murray, T. M. (2005), "Computer Modeling of Sloped Z-Purlin Supported Roof Systems to Predict Lateral Restraint Force Requirements," *17th Int. Spec. Conf. Cold-Formed Steel Struct.: Recent Research and Developments in Cold-Formed Steel Design and Construction* Orlando, FL, pp. 667–678.

Seek, M. W., and Murray, T. M. (2006), "Component Stiffness Method to Predict Lateral Restraint Forces in End Restrained Single Span Z-Section Supported Roof Systems with One Flange Attached to Sheathing," *18th Int. Spec. Conf. Cold-Formed Steel Struct.: Recent Research and Developments in Cold-Formed Steel Design and Construction* Orlando, FL, pp. 491–505.

Seek, M. W., and Murray, T. M. (2007), "Lateral Brace Forces in Single Span Z-Section Roof Systems with Interior Restraints Using the Component Stiffness Method," *Structural Stability Research Counci—Proceedings of the 2007 Annual Stability Conference*, pp. 339–354.

Seek, M. W., and Murray, T. M. (2008), "Lateral Restraint Forces in Z-Section Roof Systems Using the Component Stiffness Method," *J. Constr. Steel Res.*, Vol. 64, No. 12, pp. 1366–1378.

Serrette, R. L. (2002), "Web Crippling Behavior of Floor Joist/Rim Members in Built Assemblies," *J. Struct. Eng.*, Vol. 128, No. 9, 1177–1185.

Serrette, R. L., and Chau, K. (2006), "Estimating Drift in Cold-Formed Steel Frame Structures," *Proceedings of the Structures Congress and Exposition*, p. 53 .

Sharp, M. L., (1966), "Longitudinal Stiffeners for Compression Members," *J. Struct. Div. ASCE*, Vol. 92, No. ST5, pp. 187–211.

Sharp, M. L. (1970), "Strength of Beams or Columns with Buckled Elements," *J. Struct. Div. ASCE*, Vol. 96, No. ST5, pp. 1011–1015

Shifferaw, Y., and Schafer, B. W. (2007), "Inelastic Bending Capacity in Cold-Formed Steel Members," *Structural Stability Research Council—Proceedings of the 2007 Annual Stability Conference*, pp. 279–300.

Silvestre, N., and Camotim, D. (2002a), "First-Order Generalised Beam Theory for Arbitrary Orthotropic Materials," *Thin-Walled Struct.*, Vol. 40, No. 9, pp. 755–789.

Silvestre, N., and Camotim, D. (2002b), "Second-Order Generalised Beam Theory for Arbitrary Orthotropic Materials," *Thin-Walled Struct.*, Vol. 40, No. 9, pp. 791–820.

Silvestre, N., and Camotim, D. (2004a), "Distortional Buckling Formulae for Cold-Formed Steel C- and Z-Section Members: Part II—Validation and Application," *Thin-Walled Struct.*, Vol. 42, No. 11, pp. 1599–1629.

Silvestre, N., and Camotim, D. (2004b), "Distortional Buckling Formulae for Cold-Formed Steel C and Z-Section Members: Part I—Derivation," *Thin-Walled Struct.*, Vol. 42, No. 11, pp. 1567–1597.

Silvestre, N., and Camotim, D. (2004c), "Distortional Buckling Formulae for Cold-Formed Steel Rack-Section Members," *Steel Composite Struct.*, Vol. 4, No. 1, pp. 49–75.

Silvestre, N., and Camotim, D. (2006), "Direct Strength Method for Lipped Channel Columns Beams Affected by Local-Plate/Distortional Interaction," paper presented at the *18th Int. Spec. Conf. Cold-Formed Steel Struct.,* Orlando, FL, pp. 17–37.

Silvestre, N., Dinis, P. B., and Camotim, D. (2006), "Direct Strength Method for Lipped Channel Columns and Beams Affected By Local-Plate/Distortional Interaction," paper presented at the *18th Int. Spec. Conf. Cold-Formed Steel Struct.: Recent Research and Developments in Cold-Formed Steel Design and Construction* Orlando, FL, pp. 17–37.

Simaan, A., and Peköz, T. (1976), "Diaphragm-Braced Members and Design of Wall Studs," *ASCE J. Struct. Div.*, Vol. 102, No. ST1, pp. 77–92.

Sivakumaran, K. S. (1989), "Analysis for Web Crippling Behaviour of Cold-Formed Steel Members," *Comput. Struct.*, Vol. 32, Nos. 3/4, pp. 707–719.

Smith, R. E. (1955), "Column Tests on Some Proposed Aluminium Standard Structural Sections," Rep. No. 25, Aluminum Development Association, London.

Sridharan, S. (1982), "A Semi-Analytical Method for the Post-Local-Torsional Buckling Analysis of Prismatic Plate Structures," *Int. J. Num. Meth. Eng.,* Vol. 18, pp. 1685–1697.

Sridharan, S., and Benito, R. (1984), "Columns: Static and Dynamic Interactive Buckling," *ASCE J. Eng. Mech. Div.*, Vol. 110, No. EM1, pp. 49–65.

Standards Australia (SA) (1993), "Steel Storage Racking," AS 4084, SA, Sydney.

Stephens, S. F., and LaBoube, R. A. (2000), "Structural Behavior of Cold-Formed Steel Header Beams for Residential Construction," paper presented at the *15th Int. Spec. Conf. Cold-Formed Steel Struct.: Recent Research and Developments in Cold-Formed Steel Design and Construction* St. Louis, MO, pp. 423–436.

Stephens, S. F., and LaBoube, R. A. (2002), "Web Crippling and Combined Bending and Web Crippling of Cold-Formed Steel Box-Beam Headers," paper presented at the *16th Int. Spec. Conf. Cold-Formed Steel Struct.: Recent Research and Developments in Cold-Formed Steel Design and Construction* Orlando FL, pp. 251–269.

Stephens, S. F., and LaBoube, R. A. (2003), "Web Crippling and Combined Bending and Web Crippling of Cold-Formed Steel Beam Headers," *Thin-Walled Struct.*, Vol. 41, No. 12, pp. 1073–1087.

Stolarczyk, J. A., Fisher, J. M., and Ghorbanpoor, A. (2002), "Axial Strength of Purlins Attached to Standing Seam Roof Panels," *16th Int. Speci. Conf. Cold-Formed Steel Struct.: Recent Research and Developments in Cold-Formed Steel Design and Construction* Orlando, FL, pp. 550–560.

Takahashi, K. (1988), "A New Buckling Mode of Thin-Walled Columns with Cross-Sectional Distortions," *Applied* in *Solid Mechanics,* Vol. 2 (Eds. A. S. Tooth and J. Spencer), Elsevier, pp. 553–573.

Teng, J. G., Yao, J., and Zhao, Y. (2003), "Distortional Buckling of Channel Beam-Columns," *Thin-Walled Struct.*, Vol. 41, No. 7, pp. 595–617.

Thomasson, P.-O. (1978), "Thin-Walled C-Shaped Panels in Axial Compression," D1:1978, Swedish Council for Building Research, Stockholm, Sweden.

Thürlimann, B. (1953), "Deformation of and Stresses in Initially Twisted and Eccentrically Loaded Columns of Thin-Walled Open Cross-Section," Rep. No. E696-3, Brown University, Providence, RI, June.

Timoshenko, S. P., and Gere, J. M. (1936), *Theory of Elastic Stability*. McGraw-Hill, New York.

Timoshenko, S. P., and Gere, J. M. (1961), *Theory of Elastic Stability*, 2nd ed., McGraw-Hill, New York.

To, E. C. Y., and Young, B. (2008), "Performance of Cold-Formed Stainless Steel Tubular Columns at Elevated Temperatures," *Eng. Struct.*, Vol. 30, No. 7, pp. 2012–2021.

Trestain, T. W. J. (1982), "A Review of Cold Formed Steel Column Design," Rep. No. 81109-1, CSCC Proj. DSS 817, Canadian Sheet Steel Building Institute, Toronto, Ontario, Canada, Dec.

Unger, von B. (1973), "Ein Beitrag zur Ermittlung der Traglast von querbelasterten Durchlaufträgern mit dünnwandigem Querschnitt, insbesondere von durchlaufenden Trapezblechen für Dach und Geschossdecken," *Stahlbau*, Vol. 42, No. 1, pp. 20–24.

Ungureanu, V., and Dubina, D. (2004), "Recent Research Advances on ECBL Approach. Part I: Plastic-Elastic Interactive Buckling of Cold-Formed Steel Sections," *Thin-Walled Struct.*, Vol. 42, No. 2, pp. 177–194.

van den Berg, G. J. (1998), "Behaviour of Cold-Formed Stainless Steel Structural Members," *J. Const. Steel Res.*, Vol. 46, Nos. 1–3, pp. 463 .

Van Den Berg, G. J. (2000), "The Effect of the Non-Linear Stress-Strain Behaviour of Stainless Steels on Member Capacity," *J. Constr. Steel Res.*, Vol. 54, No. 1, pp. 135–160.

Vlasov, V. Z. (1959), *Thin-Walled Elastic Beams*, 2nd ed. (translation from Russian), Office of Technical Services, U.S. Department of Commerce, Washington, DC.

von Kármán, T., Sechler, E. E., and Donnell, L. H. (1932), "The Strength of Thin Plates in Compression," *Trans. ASME*, Vol. 54, pp. 53–57.

Wallace, J. A., and Schuster, R. M. (2005), "Web Crippling of Cold Formed Steel Multi-Web Deck Sections Subjected to End One-Flange Loading," *17th Int. Spec. Conf. Cold-Formed Steel Struct.: Recent Research and Developments in Cold-Formed Steel Design and Construction* Orlando, FL, pp. 171–185.

Wallace, J. A., Schuster, R. M., and LaBoube, R. A. (2002), "Bending and Web Crippling Interaction of Cold-Formed Steel Members," *16th Int. Spec. Conf. Cold-Formed Steel Structures: Recent Research and Developments in Cold-Formed Steel Design and Construction* Orlando, FL, pp. 237–250.

Wang, S. T. (1969), "Cold-Rolled Austenitic Stainless Steel: Material Properties and Structural Performance," Cornell Univ. Dept. Struct. Eng. Rep. No. 334, July.

Wang, S. T., and Errera, S. J. (1971), "Behavior of Cold Rolled Stainless Steel Members," *First Spec. Conf. Cold-Formed Steel Struct.*, University of Missouri—Rolla, Rolla, MO.

Wang, S. T., and Yeh, S. S. (1974), "Post-Local-Buckling Behavior of Continuous Beams," *ASCE J. Struct. Div.*, Vol. 100, No. ST6, pp. 1169–1188.

Weng, C. C., and Peköz, T. (1990), "Residual Stresses in Cold-Formed Steel Members," *J. Struct. Eng. N.Y.*, Vol. 116, No. 6, pp. 1611–1625.

Winter, G. (1940), "Tests on Light Studs of Cold-Formed Steel," Third Progress Report, Cornell University (unpublished).

Winter, G. (1943a), "Lateral Stability of Unsymmetrical I-Beams and Trusses in Bending," *Trans. ASCE*, Vol. 108, pp. 247–260.

Winter, G. (1943b), "Tests on Light Studs of Cold-Formed Steel," Second Summary Report, Cornell University (unpublished).

Winter, G. (1947a) "Strength of Thin Steel Compression Flanges," *Trans. ASCE*, Paper No. 2305, Vol. 112, pp. 527–576.

Winter, G. (1947b), "Discussion of 'Strength of Beams as Determined by Lateral Buckling," by Karl deVries," *Trans. ASCE*, Vol. 112, pp. 1212–1276.

Winter, G. (1949), "Performance of Compression Plates as Parts of Structural Members," *Research*, Engineering Structures Supplement, (*Colston Papers*, Vol. II), publisher Butlerwont pp. 179–184, London.

Winter, G. (1958), "Lateral Bracing of Columns and Beams," *ASCE J. Struct. Div.*, Vol. 84, No. ST2, p. 1561 .

Winter, G. (1970), *Commentary on the 1968 Edition of Light Gage Cold-Formed Steel Design Manual*, American Iron and Steel Institute, Washington, DC.

Winter, G., and Pian, R. H. J. (1946), "Crushing Strength of Thin Steel Webs," Cornell Univ. Eng. Exp. Sta. Bull. No. 35, Part 1.

Yang, D., and Hancock, G. J. (2004), "Compression Tests of High Strength Steel Channel Columns with Interaction Between Local and Distortional Buckling," *J. Struct. Eng.*, Vol. 130, No. 12, pp. 1954–1963.

Yap, D. C. Y., and Hancock, G. J. (2008a), "Experimental Study of Complex High-Strength Cold-Formed Cross-Shaped Steel Section," *J. Struct. Eng.*, Vol. 134, No. 8, pp. 1322–1333.

Yap, D. C. Y., and Hancock, G. J. (2008b), "Experimental Study of High Strength Cold-Formed Stiffened Web Steel Sections," R889, School of Civil Engineering, University of Sydney, Sydney, Australia.

Yener, M., and Peköz, T. (1980), "Inelastic Load Carrying Capacity of Cold-Formed Steel Beams," *5th Int. Conf. Cold-Formed Steel Struct.*, St. Louis, MO, Nov.

Yener, M., and Peköz, T. (1985), "Partial Stress Redistribution in Cold-Formed Steel," *J. Struct. Eng.*, Vol. 116, No. 6, pp. 1169–1186.

Young, B. (1997), "The Behaviour and Design of Cold-Formed Channel Columns," Ph.D. thesis, University of Sydney, Sydney, Australia.

Young, B. (2008), "Experimental and Numerical Investigation of High Strength Stainless Steel Structures," *J Const. Steel Res.*, Vol. 64, No. 11, pp. 1225–1230.

Young, B., and Ellobody, E. (2006), "Column Design of Cold-Formed Stainless Steel Slender Circular Hollow Sections," *Steel Composite Struct.*, Vol. 6, No. 4, pp. 285–302.

Young, B., and Hancock, G. J. (2001), "Design of Cold-Formed Channels Subjected to Web Crippling," *J. Struct. Eng.*, Vol. 127, No. 10, pp. 1137–1144.

Young, B., and Hancock, G. J. (2002), "Tests of Channels Subjected to Combined Bending and Web Crippling," *J. Struct. Eng.*, Vol. 128, No. 3, pp. 300–308.

Young, B., and Hancock, G. J. (2004), "Web Crippling of Cold-Formed Unlipped Channels with Flanges Restrained," *Thin-Walled Struct.*, Vol. 42, No. 6, pp. 911–930.

Young, B., and Hartono, W. (2002), "Compression Tests of Stainless Steel Tubular Members," *J. Struct. Eng.*, Vol. 128, No. 6, pp. 754–761.

Young, B., and Liu, Y. (2003), "Experimental Investigation of Cold-Formed Stainless Steel Columns," *J. Struct. Eng.*, Vol. 129, No. 2, pp. 169–176.

Young, B., and Lui, W. M. (2005), "Behavior of Cold-Formed High Strength Stainless Steel Sections," *J. Struct. Eng.*, Vol. 131, No. 11, pp. 1738–1745.

Young, B., and Lui, W. M. (2006), "Tests of Cold-Formed High Strength Stainless Steel Compression Members," *Thin-Walled Struct.*, Vol. 44, No. 2, pp. 224–234.

Yu, C. (2005), "Distortional Buckling of Cold-Formed Steel Members in Bending," Ph.D. thesis, Johns Hopkins University, Baltimore, MD.

Yu, C., and Schafer, B. W. (2003), "Local Buckling Tests on Cold-Formed Steel Beams," *Struct. Eng.*, Vol. 129, No. 12, pp. 1596–1606.

Yu, C., and Schafer, B. W. (2006), "Distortional Buckling Tests on Cold-Formed Steel Beams," *J. Struct. Eng.*, Vol. 132, No. 4, pp. 515–528.

Yu, C., and Schafer, B. W. (2007a), "Effect of Longitudinal Stress Gradients on Elastic Buckling of Thin Plates," *J. Eng. Mech.*, Vol. 133, No. 4, pp. 452–463.

Yu, C., and Schafer, B. W. (2007b), "Simulation of Cold-Formed Steel Beams in Local and Distortional Buckling with Applications to the Direct Strength Method," *J. Constr. Steel Res.*, Vol. 63, No. 5, pp. 581–590.

Yu, W. W. (1981), "Web Crippling and Combined Web Crippling and Bending of Steel Decks," Civ. Eng. Study 81-2, University of Missouri—Rolla, Rolla, MO.

Yu, W. W. (1991), *Cold-Formed Steel Design*, 2nd ed., Wiley-Interscience, New York.

Yu, W.W. (2000), *Cold-Formed Steel Design*, Wiley, New York.

Yu, W. W., and Lin, S. H. (1992), "Development of Design Standards for Stainless Steel Structures," in *Applications of Stainless Steel '92*, Vol. 1, Stockholm, Sweden, pp. 555–564.

Zaras, J., and Rhodes, J. (1987), "Carefully Controlled Compression Tests on Thin-Walled Cold-Formed Sections," *Appl. Solid Mech.*, Vol. 2, pp. 519–551.

Zhou, F., and Young, B. (2005), "Tests of Cold-Formed Stainless Steel Tubular Flexural Members," *Thin-Walled Struct.*, Vol. 43, No. 9, pp. 1325–1337.

Zhou, F., and Young, B. (2006), "Cold-Formed Stainless Steel Sections Subjected to Web Crippling," *J. Struct. Eng.*, Vol. 132, No. 1, pp. 134–144.

Zhou, F., and Young, B. (2007a), "Cold-Formed High-Strength Stainless Steel Tubular Sections Subjected to Web Crippling," *J. Struct. Eng.*, Vol. 133, No. 3, pp. 368–377.

Zhou, F., and Young, B. (2007b), "Experimental Investigation of Cold-Formed High-Strength Stainless Steel Tubular Members Subjected to Combined Bending and Web Crippling," *J. Struct. Eng.*, Vol. 133, No. 7, pp. 1027–1034.

Zienkiewicz, O. C., and Taylor, R.L. (1989), *The Finite Element Method,* Vol. 1: *Basic Formulations and Linear Problems,* 4th ed., McGraw-Hill, New York.

Zienkiewicz, O. C., and Taylor, R. L. (1991), *The Finite Element Method,* Vol. 2: *Solid and Fluid Mechanics Dynamics and Non-Linearity*. 4th ed., McGraw-Hill, New York.

CHAPTER 14

CIRCULAR TUBES AND SHELLS

14.1 INTRODUCTION

A variety of structures include thin-walled cylinders that are susceptible to buckling. A round cylinder provides the most efficient shape available for centrally loaded columns with no lateral support between the ends. Such columns are typically used in three-dimensional loading applications such as transmission towers, reticulated shells, and offshore platforms. Stiffened and unstiffened cylindrical shells (cylinders with large diameter-to-thickness ratios) are used as grain storage bins, liquid storage tanks, pressure vessels, and caissons for underwater construction.

Tubes and shells may be subject to axial compression, bending, twisting, or external or internal pressures, any one of which can cause failure. Depending on the dimensions of the cylinder, either local or overall buckling failures can occur. If the diameter of the cylinder is relatively large, longitudinal and/or ring stiffeners are often used to provide additional strength. General notation for the geometric parameters defining both an unstiffened and stiffened cylinder is given in Fig. 14.1.

Cylinders with relatively small diameter-to-thickness D/t ratios are usually referred to as tubes or pipes, and cylinders with large D/t ratios most often are called shells. Typically, shells are stiffened. When a single descriptive term is required, a *cylinder* or *tubular member* is used.

14.1.1 Production Practice

The behavior of a tubular member is influenced by whether it is manufactured in a pipe (or tubing) mill or fabricated from a plate. The distinction is important primarily because of the differences in geometric imperfections and residual stress levels that result from the two production methods (Schilling, 1965). In general, fabricated cylinders have considerably larger imperfections (in diameter, ovality, and lack of straightness) than the mill products.

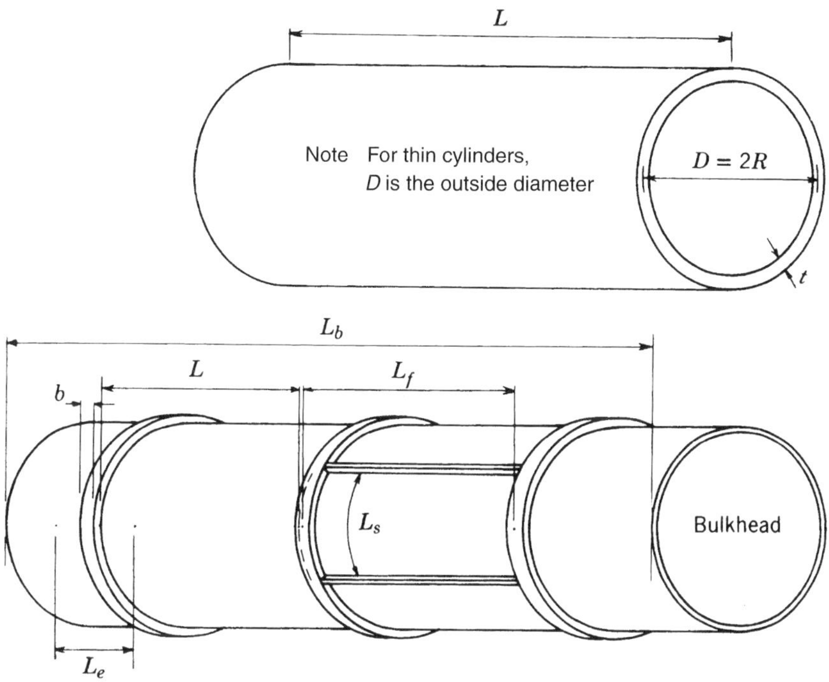

FIGURE 14.1 Dimensional notation.

Manufactured cylinders are made as seamless pipe or with continuous seam welds of various types or, in the case of aluminum, by extrusion. Each of these methods includes a hot- or cold-formed finishing treatment to obtain the proper size and circular shape (Graham, 1965; U.S. Steel, 1964; Sherman, 1992).

Fabricated tubes and shells are produced by welding or mechanically joining plates of cold- or hot-formed materials such as carbon steel, high-strength low-alloy steel, constructional alloy steel, or structural aluminum alloys. Fabricated structural members are frequently made by butt welding a series of short cans with the longitudinal welds on adjacent cans separated by rotating the cans.

14.1.2 Stress–Strain Curves and Residual Stresses

The basic stress–strain curve of a tubular section can be either (1) linear up to a yield-point stress with subsequent plastic straining at essentially constant stress or (2) linear up to a proportional limit less than the yield strength, with subsequent gradual nonlinear transition to a yielding plateau or nonlinear softening prior to failure (Schilling, 1965). These two general categories of stress–strain behavior are illustrated by the solid curves in Fig. 14.2. The presence of residual stresses will change the effective stress–strain relationships to the dashed curves in Fig. 14.2. Generally, hot-finished cylinders have sharp yielding stress–strain

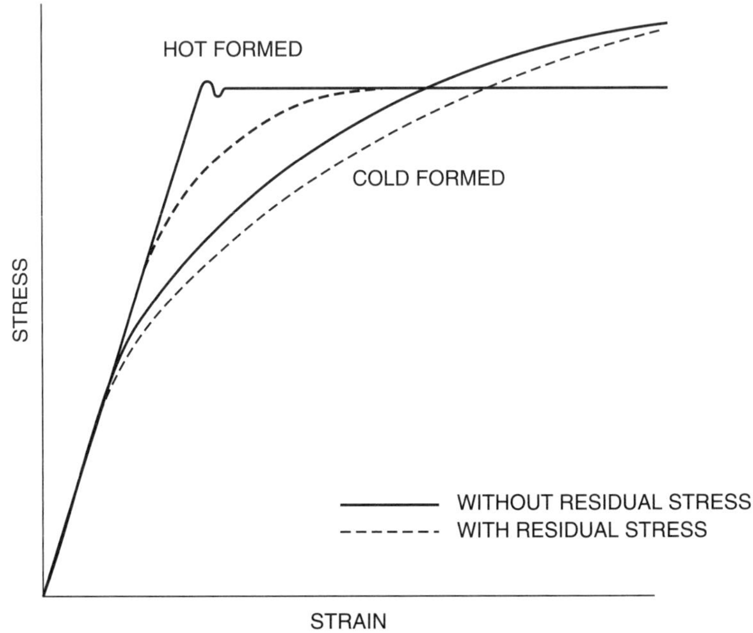

FIGURE 14.2 Two general types of stress–strain curves.

curves, whereas cold-finished cylinders exhibit gradual yielding behavior. Cold work in any cold-finished operation causes a change in stress–strain behavior from the basic material properties. Cylinders made of materials such as certain stainless steels and aluminum alloys have gradual yielding stress–strain curves, regardless of the production practice employed.

Residual stresses most commonly arise from the cooling effects after hot finishing, from the welding practices employed, or by the prevention of springback introduced during forming operations. Longitudinal residual stresses in manufactured cylinders may result from nonuniform plastic flow through the thickness of the cylinder wall. Because of the foregoing, the exact shape of the stress–strain curve, the proportional limit, and the yield strength of a tubular member are rather unpredictable.

Residual stresses can be measured (Denton and Alexander, 1963a,b; Sherman, 1969) and, in certain cases, the effects of cold work can be determined (Kuper and Macadam, 1969). Measurements on members fabricated for a column testing program (Chen and Ross, 1978) gave the longitudinal and through-thickness circumferential residual stress patterns shown in Fig. 14.3. The pattern and general magnitude have been confirmed (Prion and Birkemoe, 1988). Manufactured cold-formed pipes have similar patterns unless the finishing operation has a stress-relieving effect. Hot-formed, hot-finished, and stretch-straightened extruded aluminum cylinders generally have very low residual stresses.

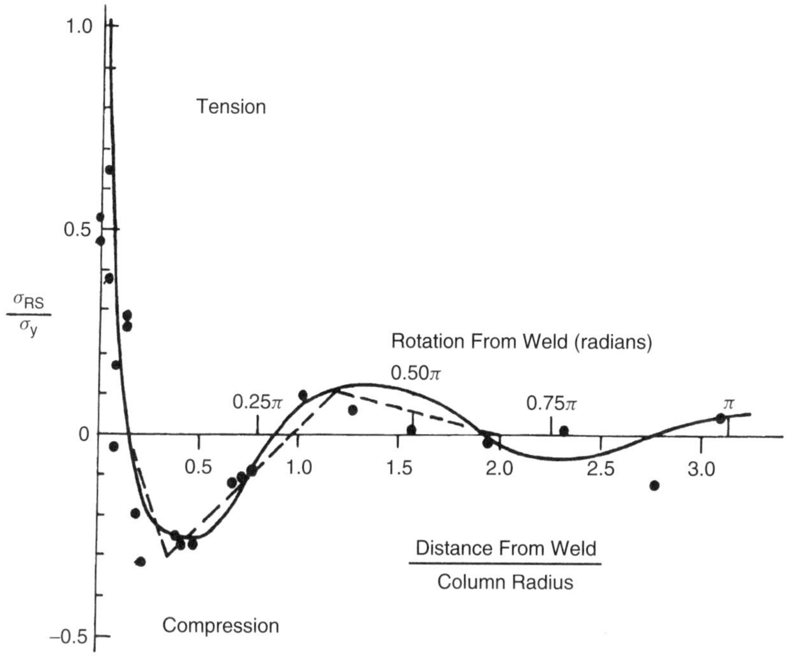

(a)

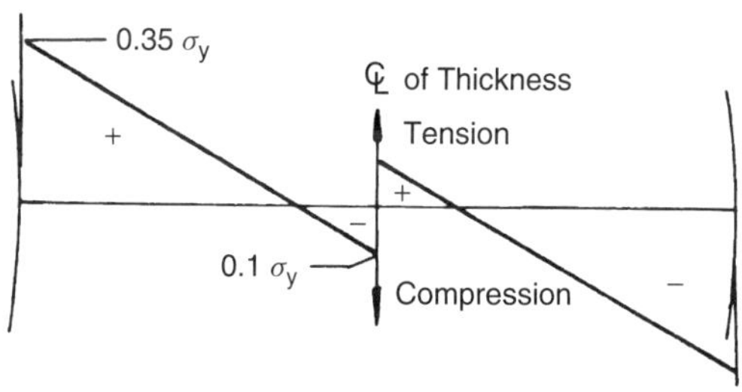

(b)

FIGURE 14.3 Measured residual stresses in fabricated pipe: (*a*) longitudinal residual stress distribution obtained from method of sectioning; (*b*) average circumferential residual stress pattern (Chen and Ross, 1978).

14.2 DESCRIPTION OF BUCKLING BEHAVIOR

14.2.1 Factors Affecting Buckling

Buckling load predictions for cylindrical members are more complicated than for wide-flange sections due to the larger number of variables. The variety of geometric proportions is quite large because the cylindrical shape is used as a structural member with large length-to-diameter (L/D) and small D/t ratios or as a shell structure where L/D is frequently less than 1 and D/t may be 100 times larger than for some unstiffened members. In many applications, circumferential and/or longitudinal stiffeners are used to increase the buckling strength. Their size, spacing, and position on the inside or outside of the cylinder affect the performance of the cylinder.

Cylinders are made by several different procedures, which result in different magnitudes and patterns of both longitudinal and circumferential residual stresses. As in any structural element, the residual stresses can play an important role in determining the buckling strength. The method of producing the cylinder and the acceptable tolerances also influence the degree of initial imperfection that is present. Tests and theory have shown that the elastic local buckling strength of thin-walled cylinders is very sensitive to geometric imperfections and in some cases to the boundary conditions. Tests of thin cylinders under axial loading resulted in failures at buckling strengths considerably less than the theoretical elastic capacity. These discrepancies have generally been attributed to specimen imperfections and poorly modeled boundary conditions; in many cases, however, the observed buckling capacities cannot be totally correlated with measured imperfection levels.

Quantitative predictions of buckling strengths must include all the loadings associated with metal members, including axial compression, flexure and shear, and torsion. In addition, the enclosed nature of the cylinder allows for pressure to be an important type of loading. Because circular beams are not subject to lateral–torsional buckling, only axial loadings cause an overall general buckling of a cylindrical member. Because tubular columns and beam-columns were presented in Chapters 3 and 8, this chapter will focus only on the local buckling modes, except as they may interact with general buckling. In unstiffened cylinders, the local buckling mode is a type of shell buckling. For cylinders with circumferential and/or longitudinal stiffeners, the local buckling mode may be shell buckling or panel buckling between heavy stiffeners or may include buckling of light stiffeners in a wave configuration.

14.2.2 Buckling Equations

The stability of cylinders has been studied analytically and experimentally for many years. As a result of these studies and the wide variation in types of cylinders investigated, a large number of empirical and semiempirical expressions have been proposed for predicting their buckling strength. Many of these were presented in the third edition of this guide (Johnston, 1976) as part of a historical description of progress. This edition presents only those that are best supported by recent research. These equations represent the classical approach to the stability of cylinders.

Miller (1983) and Miller et al. (1983) have proposed the use of a unified equation format. In this format, the linear elastic bifurcation buckling stress is reduced by a series of reduction ("knockdown") factors to obtain a realistic critical stress:

$$F_i = \eta K_i \alpha_{ij} \sigma_{iej} \tag{14.1}$$

where σ_{iej} is the theoretical elastic buckling stress for a particular loading, i, with j being either x or c representing axial and circumferential stresses, respectively.

The reduction factors include:

1. A *plasticity factor* η that reflects the residual stress levels and shape of the stress–strain curve. This factor is unity if buckling is purely elastic.
2. A *slenderness factor* K that accounts for the length and theoretical boundary conditions.
3. A *capacity reduction factor* α used to adjust for deviations between theory and tests. It accounts for the effects of imperfections in the boundary conditions and shell dimensions.

The reduction factors are derived empirically and are normally a lower bound on test data. This approach requires an extensive database to account for all the parameters that exist in cylinders. The effects of various parameters, however, are included in specific terms. In addition to clarifying the influence of the parameters, this approach also facilitates modifications as additional research becomes available. Miller (1984) presents a large number of design equations for the factors applicable to various loading conditions in unstiffened and stiffened cylinders. Sections 14.3 through 14.7 represent the classical approach.

14.3 UNSTIFFENED OR HEAVY-RING-STIFFENED CYLINDERS

14.3.1 Axial Compression

The instability modes for an axially compressed cylinder are overall column buckling and local wall buckling, either of which can be elastic or inelastic. The type of buckling to which a particular cylinder is susceptible depends on both the ratio of length to radius of gyration, L/r, and the ratio of cylinder diameter to wall thickness, D/t. Column buckling is controlled by the L/r ratio, while local buckling depends on the D/t ratio. For example, a cylinder with a large L/r and sufficiently small D/t will buckle as an elastic column, whereas a cylinder with a moderate to large D/t can buckle in either an inelastic or an elastic local buckling mode. In many cases it is difficult to predict which of the four types of buckling a particular cylinder will exhibit.

Cylinders with heavy ring stiffeners that do not distort with the local buckle exhibit behavior between the rings which is similar to that of unstiffened cylinders. Closely spaced rings can enhance the local buckling strength. They will influence column buckling only in slenderness ranges where an interaction between column and local buckling exists. The general elastic stability theory for cylinders includes the spectrum from pure column to pure local buckling. The specific topic of realistic column buckling that reflects the stress–strain characteristics, residual stresses, and out-of-straightness of cylindrical columns is covered in Chapter 3. Therefore, this section is devoted to the discussion of local buckling of unstiffened cylinders and of cylindrical shells between theoretically rigid ring stiffeners. Guidelines for sizing fully effective ring stiffeners are provided in Section 14.4, and Section 14.7 contains a discussion of the interaction between column and local buckling.

Elastic Shell Buckling The buckling of axially compressed cylindrical members was first approximately analyzed by Lorenz in 1908 and then in succeeding years more accurately by Timoshenko (1910), Southwell (1914), and Flügge (1932). Test results indicated, however, that actual cylinders buckled at loads well below those predicted by these early theoretical solutions. All of these solutions were based on small-deformation theory. In 1934, Donnell realized that a linear analysis was inadequate and suggested the need for a method of analysis that would account for large deformations. The first correct large-deflection solution was obtained by von Kármán and Tsien in 1941. Since then, numerous large-deflection investigations of axially compressed cylinders have been carried out. Of paramount importance among these was the analysis of Donnell and Wan (1950), who showed that initial imperfections are responsible for much of the discrepancy between linear theory and experimental results. The entire development of the theory of axially compressed cylindrical members was reviewed by Hoff (1966), who made several important contributions to the subject. A comprehensive study of all aspects of the buckling of cylindrical shells has been prepared by Gerard and Becker (1957). It includes theories, test results, and design recommendations.

The surface of a short cylinder buckles like an infinitely wide plate. The critical stress depends on L/D, D/t, and the boundary conditions of the edges. As L/D decreases, the critical stress approaches that for a plate strip of unit width discussed in Chapter 4. Longer cylinders buckle into a series of diamond-shaped bulges and the critical stress depends only on D/t. Still longer cylinders buckle as Euler columns where L/r is the parameter.

The theoretical elastic buckling stresses are summarized below with

$$Z = 2\left(\frac{L}{D}\right)^2\left(\frac{D}{t}\right)\sqrt{1-\nu^2} \tag{14.2}$$

used as a parameter to delineate the regions of behavior. For platelike buckling,

$$Z < 2.85, \sigma_{xc} = k_c \frac{\pi^2 E}{12\left(1-\nu^2\right)(L/t)^2} \tag{14.3a}$$

where

$$k_c = \begin{cases} \dfrac{12Z^2}{\pi^4} & \text{for simply supported edges} \quad (14.3b) \\ 4 + \dfrac{3Z^2}{\pi^4} & \text{for fully clamped edges} \quad (14.3c) \end{cases}$$

For diamond-shaped bulges,

$$2.85 \le Z < \frac{1.2\,(D/t)^2}{C}, \quad \sigma_{xc} = \frac{2CE}{D/t} \tag{14.4a}$$

where

$$C = \frac{1}{\sqrt{3\left(1-\nu^2\right)}} \tag{14.4b}$$

For column buckling,

$$Z \geq \frac{1.2(D/t)^2}{C} \qquad \sigma_{xc} = \frac{\pi^2 E}{(L/r)^2} \tag{14.5}$$

Boundary conditions at the ends have little influence on Eq. 14.4 except when the edges are simply supported but can move freely in the tangential direction (Batdorf et al., 1947a; Almroth, 1966), in which case the buckling stress is half as large as the classical one.

Imperfections The classical buckling stress given by Eq. 14.4 is a theoretical value assuming a geometrically perfect cylinder. The results of numerous compression tests (Lundquist, 1933; Donnell, 1956; Stein, 1968) show that actual cylinders may buckle elastically and fail at stresses as low as 30% of the critical stress given by Eq. 14.4. This discrepancy is due to the unstable postbuckling strength of such shells, which makes them extremely sensitive to small initial imperfections (such as deviations from the perfect geometrical shape) or due to residual stresses. The load that an axially loaded shell can support drops sharply subsequent to the onset of buckling, and the maximum load attained by the imperfect specimen is significantly below the critical load given by classical theory. Consequently, different values of C are recommended for applications where the normal degree of imperfections differs.

For manufactured or fabricated structural members, the value of C in Eq. 14.4a should be 0.165. This is approximately one-fourth of the theoretical value of Eq. 14.4b and was recommended by Plantema in 1946. Plantema's value was based on tests of manufactured members and more recent tests of fabricated members with D/t in the range of 350 to 450 (Stephens et al., 1982, 1983) correlate with this value. The test members that form the basis for the recommended values had an out-of-roundness $(D_{\max} - D_{\min})/D_{\text{nominal}}$ limit of 1%. Usually, structural members do not have proportions for which the critical local buckling stress would be increased by boundary conditions as in Eq. 14.3.

Considerable research has been conducted to determine realistic values of C to be used in place of Eq. 14.4b for cylinders with large D/t ratios and aerospace-quality tolerances. Donnell and Wan (1950) developed theoretical curves for imperfect cylinders that are shown in Fig. 14.4. The parameter U in their curves is a measure of the initial imperfection of the cylinder.

An empirical curve (also shown in Fig. 14.4) developed by Batdorf et al. (1947a) for cylinders with D/t greater than 1000 merges with the Donnell–Wan curve for $U = 0.00025$. The NASA publication (Weingarten et al., 1968) suggests that

$$C = 0.6[1.0 - 0.9(1.0 - e^{-\theta})] \tag{14.6a}$$

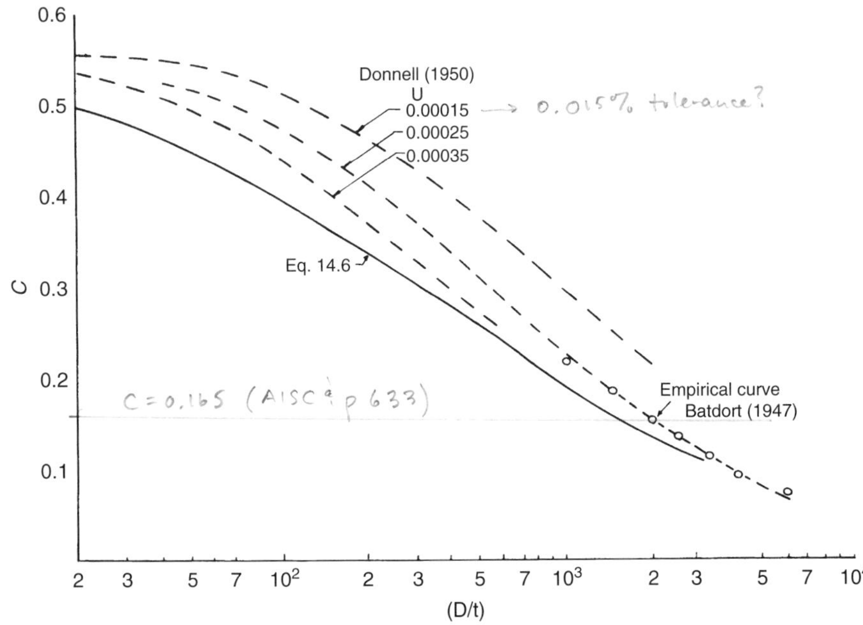

FIGURE 14.4 Elastic buckling coefficient, C, for axially compressed cylinders.

where

$$\theta = 0.0442\sqrt{\frac{D}{t}} \tag{14.6b}$$

for D/t less than 3000. This relation is also plotted in Fig. 14.4 and is somewhat more conservative than the empirical curve given by Batdorf et al. Figure 14.4 further shows that the value of 0.165 recommended for C in structural members is conservative for D/t values less than 1000.

The NASA publication (Weingarten et al., 1965) also suggests that the effect of typical imperfections in short cylinders can be included by using $1.67CZ$ in place of Z in the K coefficients of Eqs. 14.3b and 14.3c. An earlier NACA guide (Gerard, 1957) presented different expressions for the short-cylinder coefficients, but these differ by only a few percent from the values obtained using the modified Z.

Clark and Rolf (1964) proposed a design procedure for aluminum tubes based on the empirical relation

$$C = \frac{1}{\sqrt{3\left(1-\nu^2\right)}\left(1+0.02\sqrt{D/t}\right)^2} \tag{14.7}$$

Because $\nu = 0.33$ for aluminum

$$C = \frac{0.612}{\left(1+0.02\sqrt{D/t}\right)^2} \tag{14.8}$$

They suggested the following formula for moderately long cylinders:

$$\sigma_{xc} = \frac{\pi^2 E}{8\,(D/t)\left(1 + 0.02\sqrt{D/t}\right)^2} \tag{14.9}$$

For a cylinder whose length is less than its mean radius, they suggest using either Eq. 14.9 or the buckling equation for a flat-plate column, whichever gives the higher stress. Equation 14.9 is used in the Aluminum Association specification (AA, 2005).

Inelastic Shell Buckling The inelastic buckling stress of cylindrical shells and tubes is usually obtained in one of two ways. Either the elastic formula is used with an effective modulus in place of the elastic modulus or empirical relations are developed for specific classes of materials. The former approach is applicable only when the material stress–strain curve varies smoothly. This method has a long history of use and discussion (Gerard, 1956; Harris et al., 1957; Clark and Rolf, 1964; Weingarten et al., 1968). When the cylinder geometry and material properties are such that the computed buckling stress is in the plastic range, it is suggested that E in the elastic buckling equations be replaced by an effective modulus

$$E_{\text{eff}} = \sqrt{E_s E_t} \tag{14.10}$$

where E_s and E_t are the secant and tangent moduli, respectively.

As is usually the case when the effective-modulus approach is used, a direct solution for the critical stress is not possible, and a graphical or spreadsheet trial-and-error approach is helpful.

The effective-modulus approach is applicable to homogeneous materials such as aluminum alloys and stainless steels, while the inelastic buckling of cylinders made from structural steels is more conveniently handled with an empirical formula. Even for aluminum alloys, however, Clark and Rolf (1964) pointed out that a shortcoming of the effective-modulus approach is that each different alloy requires its own design curve. They therefore proposed that the following single equation be used for all aluminum alloys:

$$\sigma_{xc} = B_t - D_t\sqrt{\frac{D}{2t}} \tag{14.11a}$$

where

$$B_t = \sigma_2\left[1 + 4.6\left(\frac{1000\sigma_2}{E}\right)^{0.2}\left(\frac{\sigma_2}{\sigma_1} - 1\right)\right] \tag{14.11b}$$

$$D_t = \frac{B_t}{0.9}\left(\frac{B_t}{E}\right)^{1/3}\left(\sqrt{\frac{\sigma_2}{\sigma_1}} - 1\right) \tag{14.11c}$$

The quantities σ_1 and σ_2 are the values of the compressive yield strength at 0.1 and 0.2% offset (k/in.^{2}), respectively. To avoid curves that do not intersect the elastic curve, the ratio of σ_2/σ_1 is taken to be 1.06 in Eqs. 14.11b and 14.11c for those cases when the actual ratio exceeds this value. Equation 14.11a is applicable to any material with a stress–strain curve that can be represented by the Ramberg–Osgood–Hill three-parameter relation

$$\varepsilon = \frac{\sigma}{E} + 0.002\left(\frac{\sigma}{\sigma_2}\right)^n \tag{14.12a}$$

where

$$n = \frac{0.301}{\log_{10}(\sigma_2/\sigma_1)} \tag{14.12b}$$

Equation 14.11 is shown by Clark and Rolf (1964) to agree well with both test results and the inelastic design criteria developed by Gerard (1956).

When dealing with cylinders made from carbon or low-alloy steel, the empirical approach is used exclusively. Several major experimental programs have been conducted since 1933. Figure 14.5 shows the experimental test data from Wilson and Newmark (1933), Wilson (1937), Plantema (1946), Ostapenko and Gunzelman (1976, 1978), Chen and Ross (1978), Ostapenko and Grimm (1980), Marzullo and Ostapenko (1978), Birkemoe et al. (1983) and Eder et al. (1984) and, for thin-walled cylinders, the work of Stephens et al. (1982, 1983) Prion and Birkemoe (1988) and O'Shea and Bridge (1997). All data from Wilson (1937) and Wilson and Newmark (1933) are shown for cylinders having thicknesses greater than or equal to $\frac{1}{4}$ in. while only the data for thicknesses less than $\frac{1}{4}$ in. that failed at a stress below the proportional limit are shown.

The considerable scatter in the data is probably due to differences in methods of fabricating the cylinder that result in variation in imperfections, residual stresses, and material characteristics. There may also have been differences in test methods, data interpretation, and particularly how the yield strength was determined.

As a result of the scatter in the test data, a number of different equations applicable to intermediate-length cylinders have been proposed for predicting the critical stress as a function of the wall slenderness. The equations vary in complexity and values of the critical stress depending on the philosophy of the proposer and the database used. The slenderness parameters are usually either D/t or a nondimensional local buckling parameter, α, which for a circular cylinder can be expressed as

$$\alpha = \frac{E/\sigma_y}{D/t} \tag{14.13}$$

Table 14.1 contains a listing of the various proposed equations. They are compared to one another and to the test data in Fig. 14.5. These equations have been converted from the original sources to express the ultimate strength in terms of α.

TABLE 14.1 Various Equations Proposed for the Critical Stress in Axially Loaded Steel Cylinders

References	Equation for σ_{xc}/σ_y	Limits	Equation Number
Plantema (1946)	1.0	$\alpha \geq 8$	14.14a
	$0.75 + 0.031\alpha$	$2.5 \leq \alpha < 8$	14.14b
	0.33α	$\alpha < 2.5$	14.14c
AISI (2005) specification	1.0	$\alpha \geq 9.1$	14.15a
	$0.665 + 0.0368\alpha$	$2.27 \leq \alpha \leq 9.1$	14.15b
SSRC Guide, 3rd ed. (1976)	1.0	$\alpha \geq 9.1$	14.16a
	$0.61 + 0.043\alpha$	$2.57 \leq \alpha < 9.1$	14.16b
	0.28α	$\alpha < 2.57$	14.16c
AWWA (1967)		$\sigma_y = 30$ ksi, $t \geq \frac{1}{4}$in.	
Timoshenko (1910)	1.0	$\alpha \geq 7.25$	14.17a
Tokugawa (1929)	$0.27\alpha - 0.019\alpha^2$	$\alpha > 7.25$	14.17b
Marshall (1971)	1.0	$\alpha \geq 9.1$	14.18a
	$0.22\alpha - 0.0121\alpha^2$	$\alpha < 9.1$	14.18b
API (1989)	1.0	$\alpha \geq \dfrac{E/\sigma_y}{60}$	14.19a
	$1.64 - 0.23/(\sigma_y\alpha/E)^{0.25}$ not to exceed 0.3α	$\alpha > \dfrac{E/\sigma_y}{60}$	14.19b
Ostapenko and Grimm (1980)	1.0	$\alpha \geq 0.07/\gamma$	14.20a
	$38(\gamma\alpha) - 400(\gamma\alpha)^2 + 2020(\gamma\alpha)^3$ where $\gamma = (\sigma_y/E)^{2/3}$	$\alpha < 0.07/\gamma$	14.20b

Equations 14.14 through 14.16 (appearing Table 14.1) have a similar form with a linear dependence of α. In 1946, Plantema proposed Eq. 14.14 based on tests of manufactured tubes, and the AISI adopted the more conservative Eq. 14.15 after considering a larger database that included manufactured cylinders. Equation 14.16 is an even more conservative relation fitted to the pre-1976 data. The American Water Works Association (AWWA, 1967) adopted Eq. 14.17 with a safety factor of 2 on the basis of the fabricated cylinder tests conducted in the 1930s. Equation 14.18 also uses the quadratic form but is closer to a lower bound on the test data. The API equation (14.19) and the more recent equation (14.20), which was derived as a best fit to a series of tests on fabricated cylinders with several strength levels, contain the yield strength as part of the slenderness parameter in addition to α.

The tests on fabricated cylinders conducted since 1976 fall considerably higher in Fig. 14.5 than test results from the 1930s. This may reflect improved fabricating technology and indicates that lower bound equations for the total data base may be overly conservative. At the same time, the scatter in the data probably does not warrant a complex best-fit expression for predicting strength. Therefore, Eq. 14.15 is recommended as a reasonable estimation of a lower bound for the inelastic

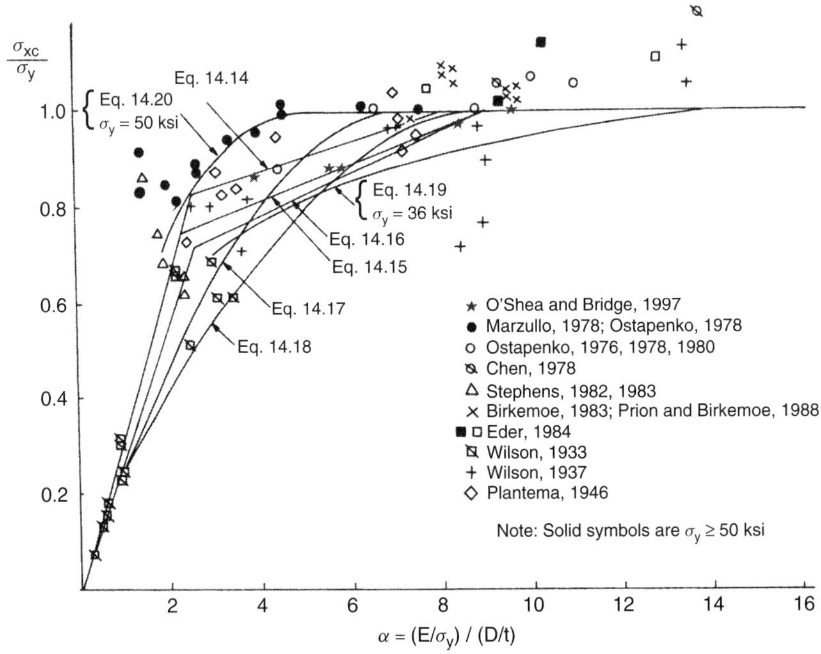

FIGURE 14.5 Inelastic buckling equations and data for axially loaded cylinders.

axial buckling stress of currently produced fabricated or manufactured steel tubes and pipe.

As an alternative to specifying σ_{xc}, the buckling strength of a member in axial compression can be expressed in terms of an effective area,

$$P_{cr} = \sigma_c A_e$$

If σ_c is taken as the yield stress, P_{cr} is the local buckling load of the section (Bradford et al., 2001). If σ_c is the axial stress corresponding to the flexural buckling load of the column, the interaction between local and column buckling is obtained (AISI, 2007). Bradford et al. propose the following simple expression for A_e:

$$A_e = \frac{100}{(D/t)\left(800\sigma_y/E\right)} A \quad \leq A$$

which has been shown to be conservative when compared with recent local buckling tests and closely approximates the D/t local buckling limits in current specifications.

14.3.2 Cylindrical Shells Subjected to Bending

The buckling behavior of cylinders in flexure differs from that of axially compressed cylinders in that bent cylinders have a stress gradient which is not present in axially compressed cylinders and tends to ovalize the cross section (Brazier, 1927; Ades, 1957; Gellin, 1980). Donnell (1934) found that the elastic buckling stress in bending is somewhat higher than the critical stress for axial compression. Flügge (1932) and Timoshenko and Gere (1961) reached the same conclusion. Other investigators (Wilson and Olson, 1941; Weingarten and Seide, 1961; Yao, 1962) have indicated that there is not much difference between the critical stress in bending and in axial compression. Until this disagreement is resolved, it is recommended that Eq. 14.4 with $C = 0.165$ as in axial compression be used for determining the critical bending stress of cylinders that buckle elastically.

Inelastic buckling in flexure includes not only the nonlinear behavior below the material yield stress as in axial compression but also the region between the yield moment and where the maximum stress is at yield while the strain level increases. This is a significant region because of the relatively high shape factors in cylinders, given by

$$\frac{Z}{S} = \frac{4}{\pi}\left(1 + \frac{t}{D}\right) \tag{14.21}$$

where Z is the plastic-section modulus and S is the elastic-section modulus. Typical values for tubes listed in the AISC manual range from 1.30 upward. About 96% of the plastic moment can be developed at only twice the yield strain.

Data for the inelastic flexural capacity of steel tubes with a uniform bending moment are plotted in Fig. 14.6. Because local buckling after the yield moment has been reached is a function of strain rather than stress, the plot is in terms of normalized moment capacity rather than critical stress. The earliest tests on hot-formed pipe reported by Schilling (1965) indicated that the plastic moment could be achieved for $\alpha > 8.33$, which is approximately the same limit as that used in Eq. 14.14 for achieving the full yield strength of axially loaded cylinders. Although this was used to define a plastic moment for a number of years, the length that was machined to obtain the reduced thickness was too short to allow free ovalization and, as shown by Sherman (1986) in tests for moment capacity at fixed ends, this resulted in high moment capacities. Later tests that also included cold-formed (Jirsa et al., 1972; Sherman, 1976; Korol, 1978) and fabricated pipe (Sherman, 1992) did not always reach the plastic moment when α exceeded 8.33. This is logical because the inelastic strain required for a plastic moment is much greater than that required for the yield capacity under axial load. In addition, pipes ovalize somewhat in bending, thereby producing additional geometric imperfection and biaxial stress conditions.

The moment capacities in Fig. 14.6 can be reasonably represented by a "best-fit" linear expression in terms of α similar to the form of Eq. 14.15 for critical axial stresses:

$$\frac{M_u}{M_p} = \begin{cases} 1.0 & \text{for } \alpha \geq 14 \quad (14.22a) \\ 0.775 + 0.016\alpha & \text{for } \alpha < 14 \quad (14.22b) \end{cases}$$

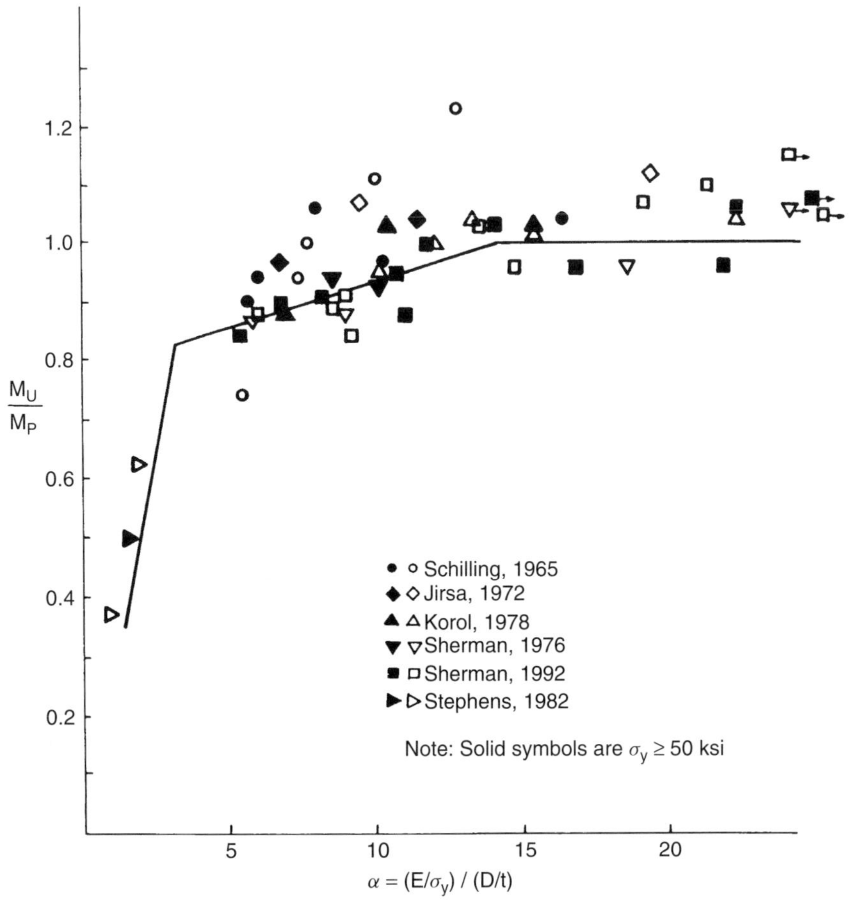

FIGURE 14.6 Moment capacity data.

Equation 14.22b is valid until the moment capacity is determined by elastic buckling,

$$M_u = \sigma_{xc} S \tag{14.22c}$$

where σ_{xc} is from Eq. 14.4. A lower bound to the data is obtained if the values in Eqs. 14.22a and 14.22b are multiplied by 0.9. Kulak (1994) has proposed an alternative equation that is based on the slenderness parameter $(E/F_y)^{0.5}(t/R)^{1.5}$ and results in a continuous curve.

Clark and Rolf (1964) developed the following criterion for aluminum and materials with a similar stress–strain curve,

$$\sigma_{tb} = B_{tb} - 0.707 D_{tb} \sqrt{\frac{D}{t}} \tag{14.23}$$

where σ_{tb} is the bending stress in the tube (maximum value of bending moment divided by elastic section modulus) and B_{tb} and D_{tb} are coefficients given by

$$B_{tb} = 1.5\left[1 + 4.6\left(\frac{1000\sigma_y}{E}\right)^{0.2}\left(\frac{\sigma_2}{\sigma_1} - 1\right)\right]\sigma_y \tag{14.24a}$$

$$D_{tb} = \frac{B_{tb}}{2.7}\left(\frac{B_{tb}}{E}\right)^{1/3} \tag{14.24b}$$

In the formulas above, σ_y (ksi) is either the tensile or compressive yield strength in the axial direction, whichever is lower, and σ_1 and σ_2 (ksi) are the values of the compressive yield strength at 0.1 and 0.2% offset, respectively. The ratio σ_2/σ_1 should be taken as 1.06 for those cases where the actual ratio exceeds this value. When $D/2t > (B_{tb} - B_t)^2/(D_{tb} - D_t)^2$, the formulas for axial compression should be used instead of Eq. 14.23. Note that the coefficients B_t and D_t are defined in Eqs. 14.11b and 14.11c.

14.3.3 Cylindrical Shells Subjected to Torsion

The elastic shear buckling stress, τ_c, due to torsion in cylinders of any length can be expressed as (Batdorf et al., 1947b)

$$\tau_c = K_s \frac{\pi^2 E}{12\left(1 - \nu^2\right)(L/t)^2} \tag{14.25a}$$

As in the case of axial loading, the value of K_s depends on the cylinder proportions and, for shorter cylinders, on the boundary conditions. The value of K_s can be expressed in terms of the parameter Z defined in Eq. 14.2.

For short cylinders ($Z < 50$), end conditions are of major importance. For a simply supported short cylinder,

$$K_s = 5.35 + 0.213Z \tag{14.25b}$$

and for a short tube having full end fixity,

$$K_s = 8.98 + 0.10Z \tag{14.25c}$$

where Z is defined by Eq. 14.2.

For intermediate-length cylinders, with $100 \leq Z \leq 19.2(1 - \nu^2)(D/t)^2$,

$$K_s = 0.85Z^{0.75} \tag{14.25d}$$

for all end conditions.

For very long cylinders, Timoshenko and Gere (1961) recommend

$$K_s = \frac{0.406Z}{\left(1 - \nu^2\right)^{0.25} (D/t)^{0.5}} \tag{14.25e}$$

Schilling (1965) presents test data for some alloy steels whose stress–train curve approaches that of a sharp-yielding material. From these data he concludes that the critical shear stress for these steels can be approximated by the shear yield strength τ_y when the parameter $(\tau_y/E)(D/t)^{1.25}(L/D)^{0.5}$ is less than 1.076 and by the elastic shear buckling stress when the same parameter exceeds 1.076.

Studies of postbuckling behavior of intermediate-length cylinders loaded in torsion (Nash, 1957) show that the maximum load that an initially imperfect cylinder can resist is less than the classical shear buckling load. The drop in load subsequent to buckling, however, is very small compared to that which takes place with axial compression. Hence, the failure stress of an actual imperfect cylinder is only slightly lower than the critical stress predicted by linear theory. Schilling (1965) recommends that the theoretically obtained torsional buckling stress of intermediate-length cylinders loaded in torsion be reduced by 15% to account for initial imperfections.

For materials such as aluminum alloys and stainless steels, which have gradually yielding stress–strain curves, the buckling stress in the inelastic range can be obtained by replacing E in the elastic formulas by E_s, the secant modulus at $\sigma = 2\tau_c$ (Gerard, 1957).

In the AA (2005) specification, an equivalent slenderness ratio is used to obtain the shear buckling stress for tubes subjected to torsion. In the elastic range the buckling stress is given by

$$\tau_c = \frac{\pi^2 E}{\lambda^2} \tag{14.26a}$$

in which λ, the equivalent slenderness ratio, is approximated by

$$\lambda = 3.73 \left(\frac{D}{t}\right)^{0.75} W^{0.5} \tag{14.26b}$$

The coefficient W is equal to unity for long, unstiffened tubes, and

$$W = 0.561 \frac{\sqrt{L/D}}{(D/t)^{0.25}} \tag{14.26c}$$

for tubes where the clear length L between circumferential stiffeners is such that the value of W from Eq. 14.26c is less than unity.

For the inelastic buckling of aluminum tubes, Clark and Rolf (1964) propose the following relation:

$$\tau_c = B_s - D_s \lambda \tag{14.27a}$$

in which λ is given by Eq. 14.26b and B_s and D_s are coefficients defined as follows:

$$B_s = \tau_2 \left[1 + 5.8 \left(\frac{1000\tau_2}{E} \right)^{1/3} \left(\frac{\tau_2}{\tau_1} - 1 \right) \right] \tag{14.27b}$$

$$D_s = \frac{B_s}{2} \sqrt{\frac{B_s}{E} \left(\frac{\tau_2}{\tau_1} - 1 \right)} \tag{14.27c}$$

in which τ_1 and τ_2 (ksi) are the shear yield strengths at 0.1 and 0.2% offsets, respectively. To avoid inelastic buckling curves that do not intersect the elastic curve, the ratio τ_2/τ_1 is taken to be 1.06 for those cases where the actual ratio exceeds this value. It is permissible to substitute σ_2/σ_1 for τ_2/τ_1 where σ_1 and σ_2 are determined from the compressive stress–strain curve. Good agreement is shown to exist between Eq. 14.27 and test results for five different aluminum alloys.

14.3.4 Cylindrical Shells Subjected to Transverse Shear

Little information is available on the subject of local instability of shells or tubes subjected to transverse shear. It seems logical, however, that because of the presence of a stress gradient, tubes loaded in transverse shear will buckle at a higher stress than similar tubes loaded in torsion. Schilling (1965) suggests that for manufactured tubes the critical shear stress in transverse shear be taken as 1.25 times the critical stress in torsion when buckling is elastic. In the inelastic range, he advises using the same critical stress for transverse shear as is used for torsion.

14.3.5 Cylindrical Shells Subjected to Uniform External Pressure

The strength of a shell under external pressure depends on its L/D and D/t ratios and upon the physical properties of the material. It also depends on the amount of deviation of the shell from a true circular form. Failure of a shell can occur by yielding or by buckling at stresses that may be considerably below the yield point. The effective length of the shell can be reduced by the addition of circumferential stiffeners.

A thinness factor K has been presented by Windenburg and Trilling (1934) that is indicative of the mode of failure to be expected:

$$K = \left(\frac{D}{t} \right)^{0.75} \sqrt{\left(\frac{L}{D} \right) \left(\frac{\sigma_y}{E} \right)} \tag{14.28}$$

where L is the unsupported length of shell between stiffeners or between the ends of the cylinder (Fig. 14.1) and D is the diameter-to-midthickness ratio of the cylindrical shell. For approximation purposes, elastic instability is likely to occur in the range $K > 1.2$, inelastic-shell instability in the range 0.8 to 1.2, and shell yielding if $K < 0.8$.

If the shell stiffeners are placed a large enough distance apart, the shell region between stiffeners will behave under pressure as though no stiffeners were present. The shortest length of the cylinder for which the strengthening effect of the stiffeners can be ignored is defined as the *critical length*.

A distinction can also be made in regard to the support conditions at the ends of the cylinder. If the pressure produces longitudinal stresses in addition to circumferential stresses, the cylinder is hydrostatically loaded. If end conditions, however, do not produce longitudinal pressure stresses, the cylinder is defined as being loaded by lateral pressure only.

Elastic Buckling Solutions for the critical elastic pressure of cylinders with finite length were first developed in the early 1900s (Southwell, 1915; von Mises, 1931, 1933). In the intervening years, modifications have been made to account for realistic boundary conditions (Von Sanden and Tolke, 1949) and to provide simpler equations that closely approximate the exact solution in certain ranges of cylinder proportions. The most exact theory is that of Reynolds (1962), which includes the influence of elastic stiffening rings at the boundaries on both the prebuckling and buckling deformations. This theory agrees well with the test results presented by Hom and Couch (1961) and Reynolds (1962) and should be used for comparison of theory and experiment. For general use, however, the simpler but more conservative equations of von Mises are recommended.

The von Mises equation for lateral pressure (Windenburg and Trilling, 1934, Eq. 2) is given by

$$p_c = 2E \times \frac{t}{D}\left\{\frac{(t/D)^2}{3(1-\nu^2)}\left[\left(n^2-1\right)+\frac{\lambda^2\left(2n^2-1-\nu\right)}{n^2-\lambda^2}\right]+\frac{\lambda^4}{\left(n^2-1\right)\left(n^2+\lambda^2\right)^2}\right\} \tag{14.29a}$$

where n is the number of circumferential lobes formed at collapse and

$$\lambda = \frac{\pi D}{2L} \tag{14.29b}$$

The von Mises equation for the more common hydrostatic pressure case can be approximated by

$$p_c = \frac{2E\,(t/D)}{n^2+(\lambda^2/2)-1}\left\{\frac{(t/D)^2}{3(1-\nu^2)}\left[(n^2+\lambda^2)^2-2n^2+1\right]+\frac{\lambda^4}{\left(n^2+\lambda^2\right)^2}\right\} \tag{14.30}$$

Equation 14.30 is a simplified version of Eq. 6 given by Windenburg and Trilling (1934) and shows good agreement for most shell geometries. The correct value of n in Eq. 14.29 or 14.30 is that which makes p_c a minimum.

In certain ranges of L/D, the number of lobes in the buckling mode may be known or one of the terms in the expression for critical pressure becomes negligible.

TABLE 14.2 Summary of Equations for the Elastic Buckling of a Perfect Cylinder with Poisson's Ratio $\nu = 0.3^a$

Pressure				Reference		Critical Pressure, P_c	
Lateral External				Windenburg and Trilling (1934)		$\frac{2E}{D/t}\left\{\frac{0.366}{(D/t)^2}\left[(n^2-1)+\frac{\lambda^2(2n^2-1.3)}{n^2-\lambda^2}\right]+\frac{\lambda^4}{(n^2-1)(n^2+\lambda^2)^2}\right\}$	Eq. 14.29a
Hydrostatic External							
θ		L/D					
From	To	From	To			Equation	Equation Number
0	0.8	0	$\frac{0.44}{\sqrt{D/t}}$		Very short	$\frac{3.615E}{(D/t)^3(L/D)^2}$	14.31
0.8	1.4	$\frac{0.44}{\sqrt{D/t}}$	$\frac{0.77}{\sqrt{D/t}}$	SSRC Guide, 3rd ed. (1976)		$\frac{3.615E}{(D/t)^2}\left[0.448(L/D)^2(D/t)+\frac{1}{(L/D)^2(D/t)}\right]$	14.32
1.4	2.0	$\frac{0.77}{\sqrt{D/t}}$	$\frac{1.1}{\sqrt{D/t}}$	Windenburg and Trilling (1934)	Short	$\frac{2.0E}{(D/t)(n^2+0.5\lambda^2-1)}\left\{\frac{0.367}{(D/t)^2}[(n^2+\lambda^2)^2-2n+1]+\frac{\lambda^4}{(n^2+\lambda^2)^2}\right\}$	14.30
2.0	10.0	$\frac{1.1}{\sqrt{D/t}}$	$\frac{5.5}{\sqrt{D/t}}$	Windenburg and Trilling (1934)		$\frac{2.6E}{(D/t)^{2.5}[(L/D)-0.45(t/D)^{0.5}]}$	14.33
10	D/t	$\frac{5.5}{\sqrt{D/t}}$	$0.55\sqrt{D/t}$		Intermediate	$\frac{2.6E}{(L/D)(D/t)^{2.5}}$	14.34
D/t	$4D/t$	$0.55\sqrt{D/t}$	$2.1\sqrt{D/t}$	SSRC Guide, 3rd ed. (1976)		$\frac{2.0E}{(D/t)(3+0.5\lambda^2)}\left\{\frac{0.367}{(D/t)^2}[(4+\lambda^2)^2-7]+\frac{\lambda^4}{(4+\lambda^2)^2}\right\}$	14.35
$4D/t$	–	$2.1\sqrt{D/t}$	–	Bryant (1954)	Long	$\frac{2.2E}{(D/t)^3}$	14.36

[a] L is the distance between a major bulkhead and a ring stiffener or successive ring stiffeners, n is the integer at which p_c is a minimum, $\lambda = 0.5\pi/(L/D)$, $\theta = 1.818(L/D)(\sqrt{D/t})$.

This leads to reasonable simplifications of the more complex equations so that iterative solutions for n can be avoided. Table 14.2 contains a summary of these equations when Poisson's ratio is equal to 0.3. Simplified equations are provided only for the case of hydrostatic external pressure. The range of validity for each of the equations is defined in terms of L/D, or more conveniently by the parameter

$$\theta = [12(1-\nu^2)]^{0.25}\frac{L}{D}\sqrt{\frac{D}{t}} \tag{14.37a}$$

or

$$\theta = 1.818\frac{L}{D}\sqrt{\frac{D}{t}} \quad \text{for } \upsilon = 0.3 \tag{14.37b}$$

In a few cases, the equations are still relatively complex, but approximate values of the critical pressure can be obtained by using a simpler expression from an adjacent range, as is evident in Fig. 14.7.

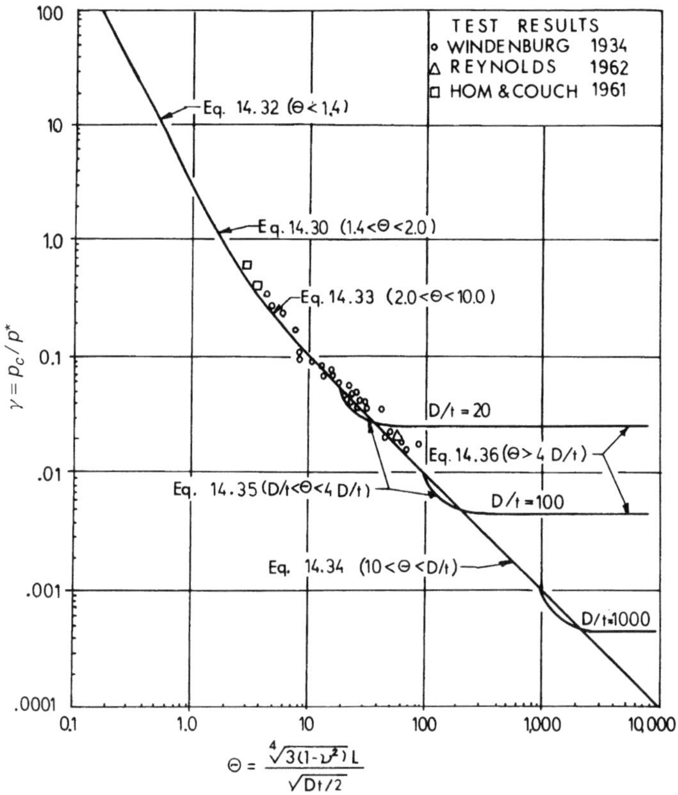

FIGURE 14.7 Elastic buckling coefficients for circular cylinders under hydrostatic pressure ($\nu = 0.3$).

Figure 14.7 is a plot of the buckling pressure coefficient γ as determined from Eqs. 14.30 through 14.36 for values of the shell geometry parameter θ where

$$\gamma = \frac{p_{\phi c}}{p^*} \tag{14.38a}$$

and

$$p^* = \frac{8E\,(t/D)^2}{\sqrt{3\left(1-\nu^2\right)}} \tag{14.38b}$$

The value p^* is the theoretical critical end pressure for a cylinder compressed only at its ends. The buckling coefficient γ is single valued for all values of θ, except when approaching D/t values where $n = 2$. Test results from Windenburg and Trilling (1934), Hom and Couch (1961), and Reynolds (1962) are also shown in Fig. 14.7.

A hemispherical head, rather than a flat plate, is often used to close the ends of a cylinder. If the heads remains stable while the cylindrical section buckles, Harari and Baron (1970, 1971) have shown that the cylinder can be treated as a longer cylinder with a length of $D/2n$ added for each head, where n is the buckling mode of the equivalent cylinder.

Inelastic Buckling During the 1960s many studies were conducted on the inelastic collapse of cylinders under external pressure (Holmquist and Nadai, 1939; Gerard and Becker, 1957; DeHeart and Basdekas, 1960; Reynolds, 1960; Hom and Couch, 1961; Lunchick, 1961a,b, 1963; Krenzke and Kiernan, 1963; Pulos, 1963; Boichot and Reynolds, 1964; Heise and Esztergar, 1970). These have included theoretical and experimental studies to develop predictive equations or semigraphical procedures. The most general approach consistent with plastic buckling theory, however, is to substitute a reduced modulus, $\sqrt{E_s E_t}$, or a tangent modulus, E_t, for E in the elastic equations of Table 14.2. Determination of the tangent or secant moduli is complicated by the biaxial stress condition that exists under hydrostatic pressure. One approach is to assume that the distortion energy criterion applies to the plastic range, so that the stress intensity σ_i can be defined as

$$\sigma_i = \left(\sigma_\phi^2 + \sigma_x^2 - \sigma_\phi \sigma_x\right)^{0.5} \tag{14.39a}$$

where

$$\sigma_\phi = -\frac{pD}{2t} \quad \text{and} \quad \sigma_\chi = \frac{-pD}{4t} \tag{14.39b}$$

From a representative stress–strain curve of the material used in the cylinder, the values of the desired moduli that correspond to σ_i are determined.

As in the case of columns, the use of the tangent modulus provides the lower limit load when it is used in the elastic buckling expression (Heise and Esztergar, 1970). Another limit is the plastic collapse pressure when the circumferential stress is at yield,

$$p_p = \frac{2\sigma_y}{D/t} \tag{14.40}$$

One practical way of determining the critical pressure is to use the charts in the ASME (1980) code. The charts include elastic buckling, elastic–plastic buckling, and plastic collapse for several different materials. They have been developed from the theoretical elastic equations, with the tangent modulus for the particular material and a uniform factor of safety of 3. Figure 14.8*a* is a chart that includes the length and D/t terms in the critical-stress equations. Knowing L/D and D/t for the cylinder, the factor A is determined, and this corresponds to σ_θ/E. The material curve in Fig. 14.8*b* is for carbon steel or low-alloy steels with yield strength greater than 30 ksi. Entering the chart with factor A, factor B is determined. In the ASME code, the allowable pressure is computed from B. Multiplying by 3, however, to remove the factor of safety, the critical pressure is

$$p_c = \frac{4B}{D/t} \tag{14.41}$$

or the critical circumferential stress is

$$\sigma_{\phi c} = 2B \tag{14.42}$$

The reduced modulus is given by $E_R = 2B/A$.

The horizontal portion of the material line in Fig. 14.8*b* represents the plastic collapse at a circumferential yield stress of 35 ksi. The other extreme of elastic buckling of long cylinders is also represented. The vertical portions of the D/t lines in Fig. 14.8*a* correspond to long cylinders, and the value of A for this case is given by

$$A = \frac{1.1}{D/t^2} \tag{14.43}$$

If this is substituted in the equation for the elastic portion of the material line in Fig. 14.8*b*, Eq. 14.36 in Table 14.2 is obtained.

The difficulty with the ASME charts is that a class of materials is included in one chart regardless of the yield strength. For example, Fig. 14.8*b* is for carbon and low-alloy steels with specified yield strengths of 30 ksi and over. To obtain a better estimate for the inelastic buckling pressure of cylinders with different yield strengths, new curves would have to be estimated between the elastic limit and yield.

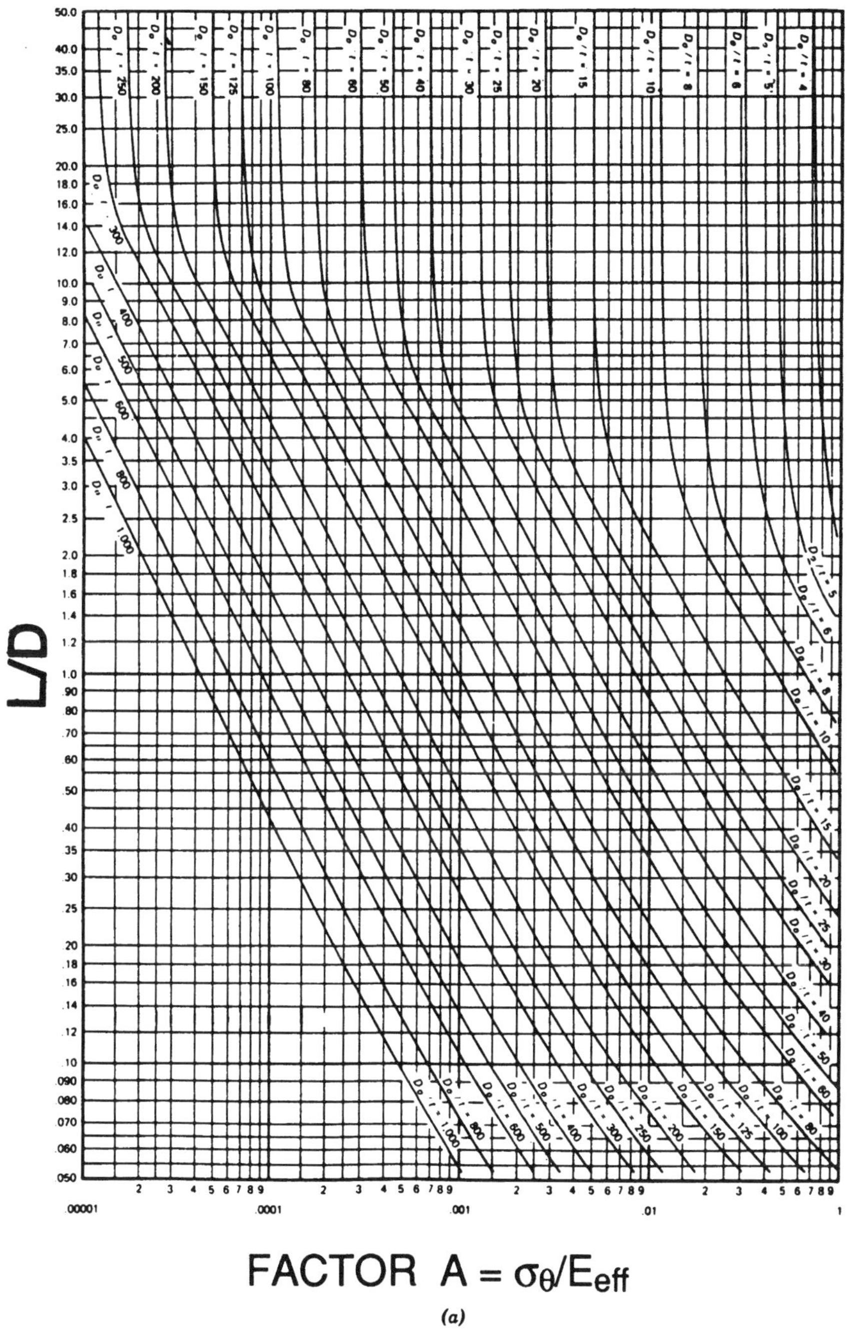

(a)

FIGURE 14.8 ASME charts for buckling under hydrostatic pressure: *(a)* geometry terms from elastic equations: *(b)* material curve for carbon and low-alloy steels with $\sigma_y \geq 30$ ksi.

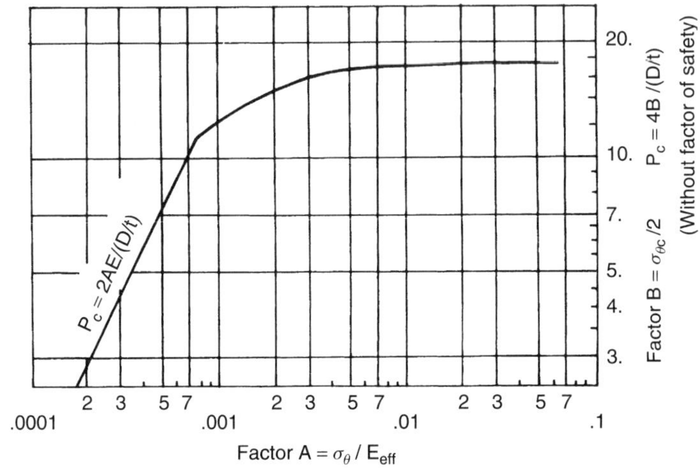

(*b*)

FIGURE 14.8 (*Continued*)

Effects of Imperfections The effect of initial imperfections has been studied by several investigators. Most analyses of the effects of initial imperfections are based on the assumption that the initial out-of-roundness is similar in form to one of the assumed buckling modes and the critical pressure is assumed to be the pressure at which the extreme fibers of the shell begin to yield.

Timoshenko and Gere (1961, Eq. e, p. 296) have developed a formula for determining the elastic critical pressure for cylinders of infinite length having a definable eccentricity. The hydrostatic pressure p_y at which yielding begins can be determined from

$$p_y^2 - \left[\frac{2\sigma_y t}{D} + \left(1 + \frac{1.5De_0}{t}\right)p_c\right]p_y + \frac{2\sigma_y t}{D}p_c = 0 \tag{14.44}$$

where e_0 is the out-of-roundness equal to $(D_{\max} - D_{\min})/D = 4e/D$, in which e is the radial eccentricity and p_c is the critical pressure determined by Eqs. 14.30 through 14.36 (with E or E_t as appropriate). This equation is applicable only when the buckling mode is such that $n = 2$. For situations where the critical n is greater than 2, Kendrick's (1953b) equations should be used in place of Eq. 14.44.

The use of the equations with out-of-roundness produces more conservative approximations of test data than the theoretical elastic or inelastic equations. They do not, however, eliminate the scatter, as would be the case if imperfections were its only source (Heise and Esztergar, 1970). It has been shown (Heise and Esztergar, 1970) that imperfections have less influence on cylinders with low D/t that buckle inelastically than on those that buckle elastically.

A value of $e_0 = 0.01$ has been adopted for fabricated tubes by the API (1977) and the ASME (1980). This value of e_0 does not appear explicitly in the design equations but it has been considered in establishing the uniform factor of safety in both specifications.

Expressions and procedures for considering the out-of-roundness effect in shells with finite length have been developed by several investigators (Sturm, 1941; Holt, 1952; Bodner and Berks, 1952; Donnell, 1956; Galletly and Bart, 1957). The problem of an initial deflection of the shell in the longitudinal direction (out-of-straightness) has also been analyzed (Wu et al., 1953; Lunchick and Short, 1957).

An important aspect is the method of determining or defining the initial imperfection. Several methods have been proposed (Galletly and Bart, 1957), but the most satisfactory from a theoretical point of view is to use the maximum radial deviation from a perfect circle, e, measured over an arc length, A, corresponding to one-half lobe length:

$$A = \frac{\pi D}{2n} \tag{14.45}$$

where n is the lobar number in Eq. 14.30. Based on this imperfection measurement, the corrective method of Galletly and Bart (1957) is the most effective, although it does not account for all test data scatter (Heise and Esztergar, 1970).

For design purposes, imperfections are usually considered by specifying permissible out-of-roundness. An empirical expression (Windenburg, 1960) for the maximum e/t value so that the collapse pressure will not be less than 80% of that corresponding to a perfect shell is

$$\frac{e}{t} = \frac{0.018}{n}\frac{D}{t} + 0.015n \tag{14.46}$$

This procedure has been incorporated into the ASME (1980) code, where the maximum permissible deviation and the length over which it is to be measured are presented in graphical form.

14.4 GENERAL INSTABILITY OF RING-STIFFENED CYLINDERS

Under static conditions, a ring-stiffened shell may fail in one or more of the three possible instability modes shown in Fig. 14.9, as follows:

1. Axisymmetric collapse of the shell between adjacent ring stiffeners. This mode is a combination of yielding and axisymmetric buckling and is characterized by accordion-shaped pleats around the circumference of the shell.
2. Asymmetric or lobar buckling of the shell between adjacent ring stiffeners. This mode is characterized by the forming of two or more dimples of lobes around the circumference.

(b) (c)

FIGURE 14.9 Buckling modes for ring-stiffened cylinders: (*a*) local buckling (axisymmetric); (*b*) local buckling (asymmetric); (*c*) general instability (overall collapse).

3. General instability or overall collapse of the entire shell and stiffeners (not column buckling). This mode is characterized by large dished-in portions of the stiffened shell, wherein the shell and stiffeners deflect together. The occurrence of this mode is strongly influenced by the shape, moment of inertia, and circularity of the stiffening rings and the ratio of the overall length to the radius of the cylinder.

The first two instability modes were the topic of the preceding section of this chapter and the third is the topic of this section.

Ring stiffeners have little influence on column-buckling strengths of cylinders and the column curves for fabricated cylinders in Chapter 3 are applicable. Therefore, the purpose of ring stiffeners is primarily to enhance the local buckling strength of a cylinder. To accomplish this, their spacing L must be small enough to influence the strength predictions in the equations of Section 14.3 (i.e., L must be a term in the equation).

If the stiffeners have sufficient stiffness and strength, the failure will be in one of the first two instability modes and the collapse load a function of L. Lighter stiffeners, on the other hand, will lead to the third instability mode and its collapse load will be between that of an unstiffened long cylinder and one whose stiffeners are spaced L.

14.4.1 Nonpressure Loadings

Very little information is available on the general instability of ring-stiffened cylinders under nonpressure loads. This is probably due to the infrequent occurrence of conditions that would produce this mode. Under axial load, the critical length implied by the limit for Eq. 14.30 is $L/D < 1.22/\sqrt{D/t}$. This is an extremely small spacing relative to the critical length for pressure loading and would seldom be encountered in practice. Ring stiffeners are a very inefficient method of reinforcing a cylinder for axial loads. Ring stiffeners can be more effective in flexure because they reduce ovalization prior to local buckling. The spacing, however, would have to be very small, as in the case of axial load, and would be considered only for regions of steep moment gradients. In the event that closely spaced ring stiffeners are present, their stiffness should be considered as uniformly distributed and the general buckling strength under axial load or flexure estimated using orthotropic shell theory.

The maximum (critical) spacing beyond which properly sized rings are no longer effective indicated in Eq. 14.25 for torsional loading is $L/D = 3.03\sqrt{D/t}$, which is considerably greater than that for the axial load requirements of Eq. 14.5. Therefore, it would appear that stiffeners could be effective in increasing the local buckling strength of torsion and similarly, cylinders subject to transverse shear. Little published information, however, is readily available to determine the minimum size for a stiffener to be fully effective or to determine the general buckling strength if lighter stiffeners are used for either case. As a practical matter, it would seem that ring stiffeners are seldom used, or needed, to increase the buckling capacity of cylinders in torsion or shear.

14.4.2 Uniform External Pressure

Ring stiffeners are most frequently used to increase the local buckling strength of cylinders subject to external pressure. According to Table 14.2, the critical spacing is $L/D = 2.1\sqrt{D/t}$ for the hydrostatic loading case. A review of the state of the art in the design of ring-stiffened cylindrical shells under hydrostatic pressure was presented by Pulos (1963). This report includes those formulations that have found extensive use. A later survey was made by Basdekas (1966) of the analytical and empirical procedures for the determination of dynamic as well as static strengths of cylindrical, spherical, and spheroidal shells. Meck (1965) also presents, in a concise form, the elastic and inelastic solutions for shell buckling and general instability.

Elastic General Instability The elastic instability of ring-stiffened cylinders has been considered in the orthotropic shell theory (Bodner, 1957; Becker, 1958; Timoshenko and Gere, 1961, p. 499; Ball, 1962) and by a *split-rigidity method* (Tokugawa, 1929; Bryant, 1954) in which the formula for critical pressure consists of a shell term and a stiffener term. The most acceptable solution is that of Kendrick (1953a), whose theoretical predictions have been confirmed experimentally (Reynolds, 1957). These equations, however, cannot be readily solved without the aid of a computer or graphical procedures (Reynolds, 1957; Reynolds and Blumenberg, 1959; Ball, 1962).

Although somewhat less exact than Kendrick's solution, the most widely used design equation is that of Bryant (1954) modified for the position of the stiffener on the inside or outside of the shell (Krenzke and Kiernan, 1963):

$$p_c = \frac{2E}{D/t}\,\frac{\lambda^4}{\left(n^2 + (\lambda^2/2) - 1\right)\left(n^2 + \lambda^2\right)^2} + \frac{EI_e\left(n^2 - 1\right)}{L_f R_0 R_c^2} \tag{14.47}$$

where $\lambda = \pi D/2L_b$

R_0 = outside radius of shell

R_c = radius to centroidal axis of the combined stiffeners and shell of effective width L_e (see Fig. 14.1)

n = number of circumferential lobes existing at collapse

L_f = center-to-center spacing of stiffening rings

L_b = length of shell between bulkheads (or stiffening elements with sufficient stiffness to act as bulkheads) (see Fig. 14.1)

I_e = effective moment of inertia about the centroid of a section comprising one stiffener plus an effective width of shell, L_e

The effective width of a shell acting as part of the stiffener may be determined by

$$L_e = F_1 L + b \tag{14.48a}$$

where L is the unsupported length of the shell between stiffeners and

$$F_1 = \frac{2}{\theta} \frac{\cosh\theta - \cos\theta}{\sinh\theta + \sin\theta} \tag{14.48b}$$

where $\theta = 1.818(L/D)\sqrt{D/t}$ when $\nu = 0.3$ (Eq. 14.37) and b is the contact width between stiffener and shell. The value L_e can be approximated by $1.1\sqrt{Dt} + b$ when $\theta > 2$ and $L + b$ when $\theta < 2$.

The first term in Eq. 14.47 is the shell term and is identical to the last term of Eq. 14.30. It is important only for shells with large D/t and low L/D ratios. The second term of Eq. 14.47 is the Levy formula for determining the critical uniform radial load on a circular ring [see discussion of Timoshenko and Gere (1961, pp. 287–392)] and in many cases is sufficiently accurate for design purposes.

The correct value of n is that which gives a minimum value of p in Eq. 14.47. The number of waves in a buckle pattern is determined by the restraint of adjoining stiffeners, heads, or diaphragms and the distance between them (spacing). When there are no effective restraints, $n = 2$. Where heads, diaphragms, and/or large stiffeners are used to restrain the ends or are spaced along the length of the cylinder between which intermediate smaller rings are attached, n becomes greater but will be less than 10 for most shells of interest. The ASME (1980) code does not recognize the restraint at the ends of a vessel and assumes that $n = 2$. Use of this approximation in design can lead to very conservative ring sizes.

Except for large values of θ where n equals 2 and the second term of Eq. 14.47 is dominant, the desired stiffener size for a particular pressure must be determined by iterative procedures. The largest effective stiffener size, however, is achieved when the critical general buckling pressure of Eq. 14.47 is equal to that from the appropriate equation in Table 14.2 for buckling of the shell between stiffeners.

The formula for general instability developed by Bryant (1954) is dependent on the distance between large rings (any diaphragm or bulkhead is equivalent to a large ring in preventing buckling). In practice, many designs involve small, uniform rings evenly spaced, and at greater intervals of spacing intermediate heavy rings may be incorporated, again of uniform but different cross section. Although theoretical solutions for the critical pressure with this stiffening arrangement have been formulated (Kendrick, 1953a,b; Reynolds, 1957), an empirical equation (Blumenberg, 1965) that agrees quite well with test results is recommended:

$$p_c = \frac{(I_E - I_c)\,(p_F - p_B)}{I_{\rm FE} - Ie} + p_B \tag{14.49}$$

where

$$I_{\rm FE} = \frac{p_F L_f R_0 R_d^2}{E\left(n^2 - 1\right)}$$

and

n = number of buckling lobes as determined by using p_B

I_E = moment of inertia of large stiffener plus effective width of shell L_E

I_{FE} = value of I_E that makes the large stiffener fully effective, that is, equivalent to a bulkhead

L_e = L_e of Eq. 14.48

L_E = $F_1 L(A_s/A_S) + b$

L_f = center-to-center spacing of large stiffeners

p_F = critical pressure determined by Eq. 14.47, where $L_b = L_f$

p_B = critical pressure determined by Eq. 14.47 assuming the large stiffeners are the same size as the small stiffeners

R_d = radius to centroidal axis of the large stiffener plus effective width of shell L

A_s = area of small stiffener plus total area of shell between small stiffeners

A_S = area of large stiffener plus total area of shell between small stiffeners

R_0 = b, and I_c as in Eq. 14.47 and F_1 from Eq. 14.48b.

Inelastic General Instability Reynolds (1971) states that all the inelastic results can be summed up by one simple formula that came from the work of Bijlaard (1949). If the elastic buckling pressure of a structure can be expressed in the form

$$p_c = E(C_a + C_b) \tag{14.50}$$

in which C_a is a term that represents the membrane stiffness and C_b the bending stiffness of a stiffened cylinder, the inelastic buckling pressure can be written as

$$p_c = \sqrt{E_s E_t}C_a + E_t C_b \tag{14.51}$$

where E_s and E_t are the secant and tangent moduli, respectively, in the inelastic region. Therefore, $\sqrt{E_s E_t}$ would be used in place of E in the first term of Eq. 14.47, whereas E_t would replace E in the second term of Eq. 14.47.

Determination of the moduli to be used in Eq. 14.51 requires knowledge of the stress field at the stiffener and in the shell midway between stiffeners. Because first yielding at the stiffeners is localized, it does not appreciably alter the elastic distribution of stresses between the stiffener region and the shell between stiffeners (Lunchick, 1959). Therefore, using $v = 0.3$ and assuming that the stiffener width is negligible compared to the spacing between stiffeners, the circumferential stresses are given by (Pulos and Salerno, 1961).

$$\sigma_\phi = \begin{cases} -\dfrac{0.85pR_0L_e}{A_f + L_e t} & \text{at the stiffener} \quad (14.52a) \\[2ex] -\dfrac{pD}{2t}\left[1 - \dfrac{0.85A_f}{A_f + L_e t}F(\theta)\right] & \text{in the shell} \quad (14.52b) \end{cases}$$

where

$$F(\theta) = \begin{cases} 1 & \text{for } \theta \leq 1 \\ 1.33 - 0.33\theta & \text{for } 1 < \theta < 4 \\ 0 & \text{for } \theta \leq 4 \end{cases}$$

and θ is defined in Eq. 14.37.

The longitudinal stresses at both locations are given by

$$\sigma_x = -\frac{pD}{4t} \tag{14.52c}$$

As in the case of inelastic buckling of unstiffened cylinders, the stress intensity of Eq. 14.39 based on the distortion energy yield criterion is calculated at the two locations and the moduli are determined from a representative stress–strain curve of the material. The ASME charts similar to Fig. 14.8*b* may also be used to determine a reduced modulus $E_r = 2B/A$.

More precise equations and charts for determining the circumferential stress when the stiffener width is not negligible are available from Krenzke and Short (1959) and Pulos and Salerno (1961) and are to be used in place of Eqs. 14.52a and 14.52b. Also, a more exact formulation of a collapse pressure based on a three-hinge mechanism of failure of the shell between stiffeners is presented by Lunchick (1959).

Effect of Imperfection In addition to the effects of shell imperfections discussed under unstiffened cylinders, the initial out-of-roundness of the stiffener rings is equally important in determining the impact on the general instability strength of a stiffened shell under external pressure. Analytical studies of the effects of stiffener out-of-roundness were first considered by Kendrick (1953a,b). Assuming a more realistic out-of-roundness function, Hom (1962) developed a method for determining the maximum bending stress introduced in the stiffener flange by stiffener eccentricity. Hom also developed an approximate formula for this bending stress, which is applicable only in the elastic range,

$$\sigma_b = \pm\frac{16}{\pi}\left[\frac{Ee_f e}{D^2}\left(n^2 - 1\right)\frac{p}{p_c - p}\right] \tag{14.53}$$

where n = number of circumferential buckling lobes (corresponding to n from Eqs. 14.47 and 14.49)

e_f = distance from midthickness of the shell to the extreme fiber of stiffener, positive for internal stiffener and negative for external stiffener

p = applied pressure

p_c = critical pressure for a perfect ring-stiffened cylinder (Eq. 14.47 or 14.49)

e = radial eccentricity from a true circle

Another type of stress is introduced if the stiffener is initially tilted or twisted. In this case, the radial forces do not pass through the centroid of the cross section of the stiffener and a twisting moment results. When the radial stress in the web and the circumferential stress in the flange become excessive, yielding can result and cause crippling of the stiffeners, thereby precipitating collapse in the general instability mode. For outside stiffeners, these radial forces induce a twist that tends to reduce the initial tilt, whereas for internal stiffeners this radial force tends to increase the tilt.

The total stress in the flange is the sum of the hoop stress σ_ϕ, the bending stress σ_b in the plane of the ring, and any tilt-induced stress. A method for calculating the latter stress is given by Wenk and Kennard (1956).

Torsional buckling of the stiffeners has also been considered by Farmer (1966) and Wah (1967). They make the overly conservative assumptions that the shell offers no resistance to twisting and the torsional failure of the stiffener is independent of the in-plane stiffener buckling mode. A comparison of their predicted buckling pressures with test results (Blumenberg, 1965) indicates that their formula is ultraconservative. Experience indicates that if the stiffeners are symmetrical sections and are proportioned to satisfy the compact section requirements of the AISC specification, torsional buckling will not occur.

Inelastic Behavior Because of the complexity of the effects that inelastic behavior has on the correct prediction of critical buckling pressures of imperfect cylinders, there appears to be no definitive guidance available to the designer.

14.5 STRINGER- OR RING-AND-STRINGER-STIFFENED CYLINDERS

Stringer (longitudinal) stiffeners are primarily used to increase the axial or bending load capacities of cylinders. They can be used either alone or in combination with ring stiffeners. Due to the greater number of parameters involved and the multiple potential mechanisms of failure, it is virtually impossible to achieve a universal set of design formulas substantiated by tests. Therefore, this section of the chapter is limited to a general discussion of methods that can be used to evaluate the critical axis stress and critical pressure.

14.5.1 Nonpressure Loadings

Stiffeners may be placed on the outside of the shell (positive eccentricity) or on the inside of the shell (negative eccentricity) or a combination of both may be employed, such as stringers on the outside and rings on the inside. Hutchinson and Amazigo (1967), however, point out that based on large-deflection theory, shells with external stiffening are more sensitive to imperfections. Quite to the contrary, Singer (1967) claims that stiffened shells are relatively insensitive to imperfections. The conflicting results are typical and illustrate why only a general discussion of the behavior of stiffened cylinders subject to axial loading is included.

To check the design of a stiffened cylindrical shell subjected to axial compression, the following forms of failure should be considered:

1. Overall column buckling or yielding
2. Local buckling encompassing several stiffeners
3. Local buckling between stiffeners (panel buckling)
4. Buckling of individual stiffeners
5. Local yielding of the shell or the stiffeners

The possibility of overall column buckling can be studied by including the stringers in the calculation of the radius of gyration of the column cross section. As indicated earlier, circumferential stiffeners have no direct effect on the overall column-buckling mode.

In many cases, close stringer spacing allows them to be treated as if they were uniformly distributed on the shell's circumference. Under this assumption, the stiffened shell can be modeled as an equivalent orthotropic shell. This approach is often used to investigate the possiblity of local buckling encompassing several stiffeners. If ring stiffeners are closely spaced, they can similarly be considered as being uniformly distributed along the axial direction; their spacing, however, is normally such that they should be considered discretely spaced.

Spacing of longitudinal stiffeners (stringers) must be close enough to prevent panel buckling between the stiffeners. If both the stringers and the rings are sufficiently rigid, the cylinder can be treated as a series of curved panels each of which is supported along four edges. The buckling behavior of these curved panels is very similar to that of an entire cylinder. If the panel is short and its curvature is small, the panel buckles essentially as if it were a flat plate (as treated in Chapter 4) and the critical stress is given by

$$\sigma_{sc} = k \frac{\pi^2 E}{12\left(1 - \nu^2\right)(L_s/t)^2} \tag{14.54a}$$

where

$$k = \begin{cases} \left(\dfrac{L_s}{L} + \dfrac{L}{L_s}\right)^2 & \text{for } L/L_s < 1 \quad (14.54\text{b}) \\ 4.0 & \text{for } L/L_s \geq 1 \quad (14.54\text{c}) \end{cases}$$

where L is the axial distance between circumferential stiffeners and L_s is the circumferential distance between longitudinal stringers (Fig. 14.1).

Long panels of sufficient curvature behave in the same manner as do moderately long cylinders. To obtain a rough estimate of the elastic buckling strength of moderately curved panels, Timoshenko and Gere (1961) suggest that Eq. 14.4 be used. When the axial and circumferential dimensions of the panel are about equal and the central angle subtended by the panel is less than $\frac{1}{2}$ rad, C may be taken equal to 0.6.

Buckling of individual stiffeners can be investigated by treating the stiffener and an effective width of the shell as a column. For example, the critical load of a typical stringer and the effective shell skin is obtained by assuming the stringer to behave like a column on an elastic foundation. The circumferential rings act as the foundation, and depending on their spacing and area, the foundation is considered to be continuous or made up of elastic or rigid point supports.

Local yielding of the shell or stiffeners is perhaps more of a stress analysis problem than a stability problem, but it should be considered. The designer must keep one point in mind when adding stiffeners to overcome a shell stability problem. It is possible that the added stiffener is so rigid compared to the shell (which it is supposed to be reinforcing) that the stiffener becomes the main load-carrying component of the assemblage. For a further discussion of stiffened-plate behavior, refer to Chapter 4.

14.5.2 Uniform External Pressure

Stringer stiffeners are not generally used to increase the critical external pressure of a cylinder. Ring stiffeners are preferred for pressure loading, but in cases when the cylinder must withstand several loading conditions, stringer stiffeners may also be present. If stringers are spaced more closely than the buckling wavelength of the shell, they will increase the critical pressure. Test programs are currently underway to study the influence of size and spacing of stringer stiffeners on the critical pressure, but little definitive information is presently available.

14.6 EFFECTS ON COLUMN BUCKLING

The basic column curves for tubular members subjected to axial compression are discussed in Chapter 3. The designer should, however, be aware of situations when these require some modification or interpretation to ensure that the proper critical axial stress is predicted.

14.6.1 Interaction between Column and Local Buckling

For many practical applications, when the D/t ratio of a member is such that local buckling is probable, the KL/r ratio of the member may be such that column buckling is also an important consideration. One way to establish the allowable strength for such a member is to use an allowable load based on the smaller of the loads derived from local buckling and overall column behavior.

Marshall (1971) presented a different approach for handling flexural–local buckling interaction when designing light-gage cold-formed steel structural members. This approach is best explained with an example. Referring to Fig. 14.10, column curve A defines the flexural buckling strength of the compression member, with a maximum strength limited to the yield strength $F_y = 50$ ksi. If it is further determined that the theoretical, local buckling strength of the column equals a value

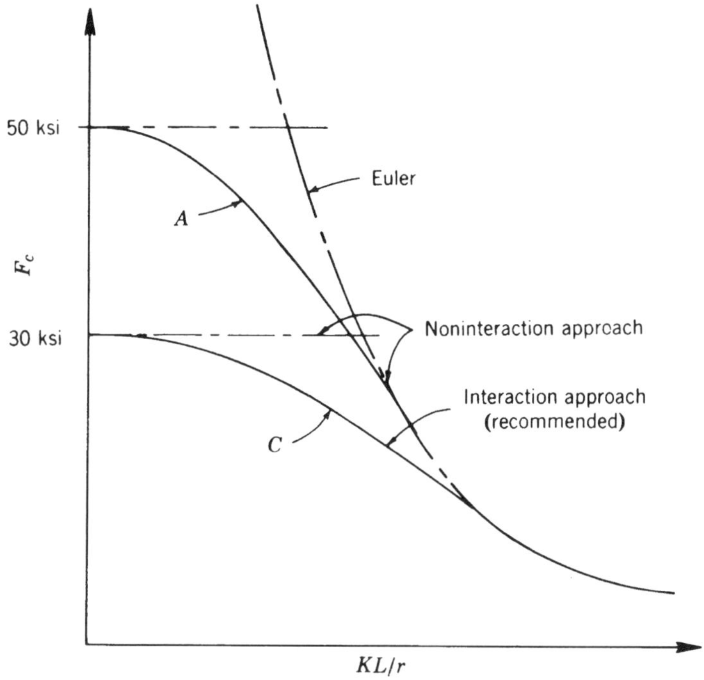

FIGURE 14.10 Interaction of local buckling and column buckling.

less than the yield strength, for example, $F_{cr} = 30$ ksi, then the local buckling strength F_{cr} is substituted for F_y in the column curve formula resulting in curve C. A similar approach is taken in Section E7 of the specification AISC (2005). For a more complete discussion of these and other methods of analysis that account for such interaction, refer to Chapters 4 and 13.

14.6.2 Effect of Fluid Pressure on Overall Column Buckling

The influence of fluid pressure on the overall stability of tubular members is normally not a structural consideration. In the oil industry, however, tubular members transmitting fluids from deep oil and gas wells have experienced failures as a result of this type of loading (Lubinski, 1951; Lubinski et al., 1952). Under these conditions, the cylinder is not closed ended and the pressure does not create longitudinal stresses.

The fact that pressures can influence column stability was recognized by Prescott (1946) and was discussed more recently in publications with theoretical derivations (Seide, 1960; Flügge, 1973), which have been experimentally confirmed (Palmer and Baldry, 1974).

A free-body diagram of an element of an open-ended cylindrical member subjected to internal pressure is shown in Fig. 14.11. It can be seen that an effective axial compressive force on the element results from the internal pressure. This

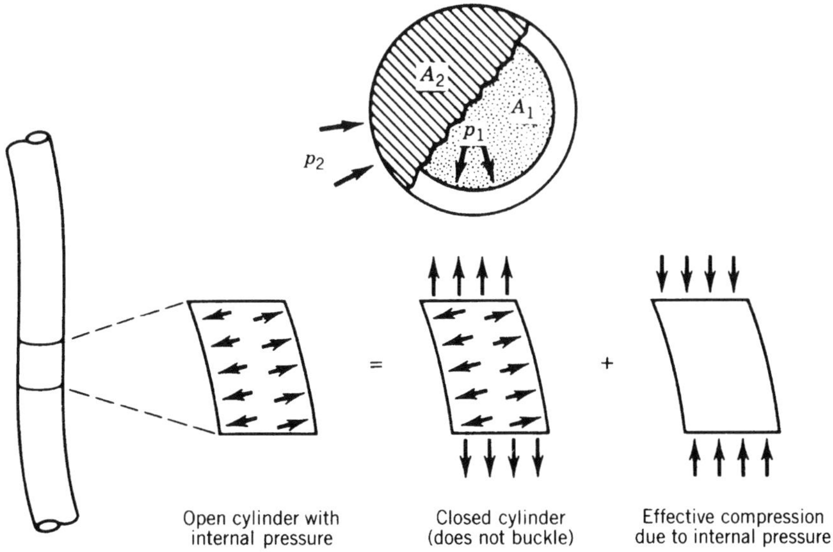

FIGURE 14.11 Equilibrium of column subjected to internal pressure.

force can influence the stability of the column by adding to or subtracting from any preexisting axial force.

Figure 14.12 shows an example where internal pressurization alone can cause a cylindrical tube to buckle. In this illustration, two frictionless pistons at the tube ends contain the fluid within the tube. The pressure force on the pistons is restrained by the cable connecting the two pistons. With no external pressure or applied load, the effective buckling force is $p_1 A_1$. In this illustration, the end conditions are essentially pinned and the elastic buckling pressure will be given by the corresponding Euler formula

$$p_1 = \frac{\pi^2 EI}{A_1 L^2} \tag{14.55}$$

The buckled shape will be sinusoidal until, with continued fluid injection, the inner wall contacts the cable. At that point the buckling will stop in the plane shown and additional pressurization will cause a deformation out of plane that leads to a helical shape for the buckled tube.

In contrast, a close-ended cylinder subjected to internal pressure will not suffer general instability as the compressive force will be balanced by the tensile force from the end closures. The resulting effective force is zero.

In a membrane shell where the walls of the shell are incapable of maintaining compression, internal pressurization will allow the pressurized membrane to support a column load by virtue of a prestressing effect in the membrane. A good example is a cylindrical rubber balloon, which has column strength only if inflated.

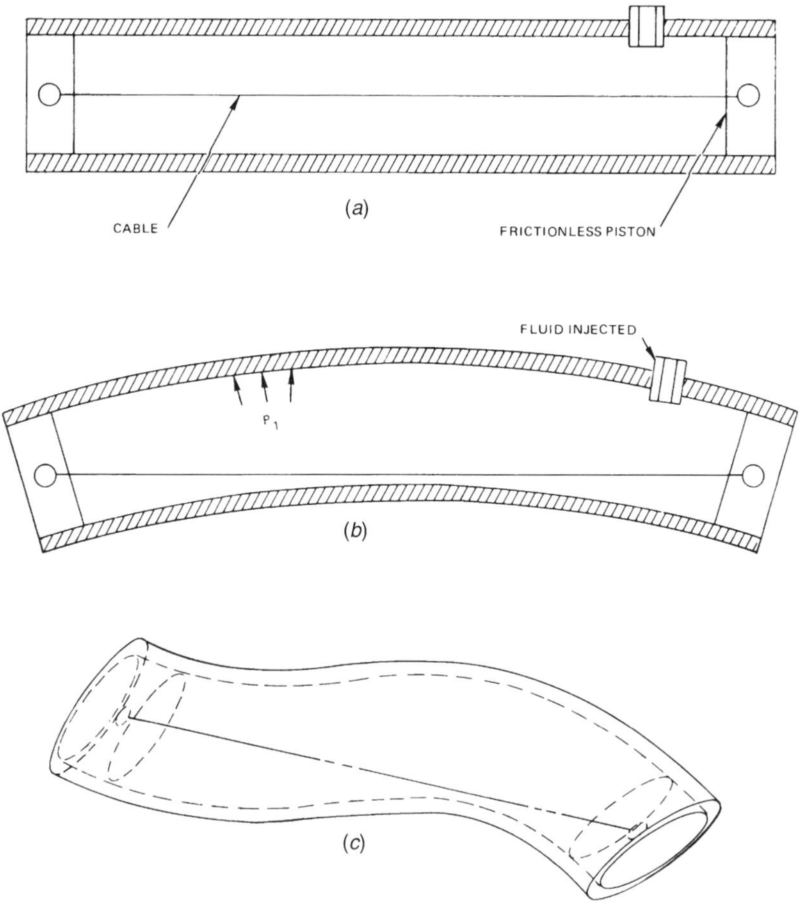

FIGURE 14.12 Buckling under internal pressure.

The axial tension force p_2A_2 and other effects resulting from external pressure acting alone can be deduced from the free-body diagram of Fig. 14.11. Hence in the presence of external and internal pressures, the overall stability of the column will depend on an effective compressive force given by

$$P_e = P + p_1A_1 - p_2A_2 \tag{14.56}$$

where P_e = effective-buckling force
P = any preexisting axial compression in the column
p_1 = internal pressure
p_2 = external pressure
A_1 = internal area of the cylinder
A_2 = external area of the cylinder

14.7 CYLINDERS SUBJECTED TO COMBINED LOADINGS

14.7.1 Combined Nonpressure Loadings

Gerard and Becker (1957) present interaction formulas together with experimental verification for several conditions of combined nonpressure loadings. Included among these are (1) axial compression and bending; (2) axial compression and torsion; (3) axial compression, bending, and torsion; and (4) bending and torsion. Although a slightly different interaction equation is proposed for each case, Schilling (1965) suggests that the following formula can be conservatively applied to all these combinations:

$$\frac{\sigma}{\sigma_{xc}} + \left(\frac{\tau}{\tau_c}\right)^2 = 1 \tag{14.57}$$

in which σ and τ are the normal and shear stresses, respectively, that must be applied simultaneously to cause failure; σ_{xc} is the critical stress for axial compression; and τ_c is the critical stress for torsion or transverse shear applied alone. If both axial compression and bending are present, it is conservative to consider σ as the sum of the two normal stresses. Schilling implies that this procedure is sufficiently accurate for use in both the elastic and inelastic stress ranges.

14.7.2 Axial Stress in Combination with Internal or External Pressure

Because radial stresses may be neglected in structural cylinders of usual proportions (D/t exceeding about 12), the following discussion will be limited to the interaction of hoop stresses caused by internal (or external) pressure and axial stresses resulting from a combination of axial load, bending, and hydrostatic pressure. In the general case, shear stresses may also be present. The overall problem is illustrated in Fig. 14.13, where the general yield failure of thick-walled cylinders based on the maximum-strain-energy theory is shown. Ellipses have been plotted for Poisson's ratio of 0.3 and 0.5.

The applicability of this criterion to the general yielding situation has been demonstrated analytically and experimentally (Holmquist and Nadai, 1939) and has been further validated by some 200 tests with oil well casings having yield strengths from 30 to 80 ksi (Edwards and Miller, 1939). It was found that Poisson's ratio varied between 0.3 and 0.5 during the inelastic behavior, and as a result, an average value of 0.4 was recommended. A somewhat simpler method of evaluating yield, however, is to determine a stress intensity

$$\sigma_i = (\sigma_\phi^2 + \sigma_x^2 - \sigma_\phi\sigma_x + 3\tau^2)^{0.5} \tag{14.58}$$

where σ_ϕ is the circumferential stress; σ_x is the total axial stress resulting from any combination of load, bending, and hdyrostatic pressure; and τ is the shear stress at the same location as the maximum σ_x. The yield criterion obtained by equating

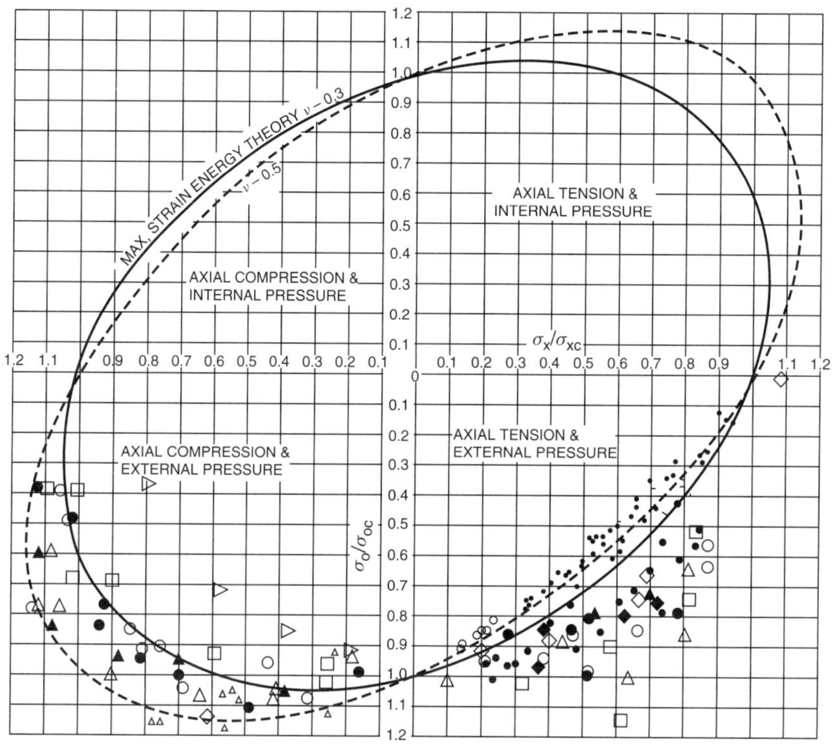

NOTE: σ_{xc} and $\sigma_{\phi c}$ are yield stresses or collapse stresses from uniaxial stress tests in the group with corresponding D/t and σ_y

Armco, 1966 (Data from Edwards and Miller, 1939, and Holmquist, 1938)
+ stretch failures
○ D/t = 16.5
• D/t = 21.7

3rd Ed. SSRC "Guide"
△

Miller, 1982: Kiziltug et al., 1985; Vojta, 1983
(includes unstiffened and ring stiffened cylinders)
◇ ◆ D/t = 33
○ ● D/t = 47
△ ▲ D/t = 62 solid symbols is are $\sigma_y \geq 50$ ksi
□ D/t = 96
▷ D/t = 1000

FIGURE 14.13 Normalized ellipse of biaxial yield stresses and data for combined pressure and axial load.

σ_i to the uniaxial yield stress is the Hencky–von Mises distortion energy criterion that corresponds to the ellipse in Fig. 14.13 for Poisson's ratio equal to 0.5.

The four quadrants of Fig. 14.13 represent the types of interaction between axial stress and pressure. In addition to general yielding, instability interactions must also be considered. An overview of the general problem is obtained by recognizing that the positive intercepts of the ellipse in Fig. 14.13 are not subject to reductions due to instability (pure tension for the abscissa intercept and pure internal pressure for the ordinate). The negative intercepts, however, will most likely not be achieved in thin cylinders due to buckling under axial compression or external pressure. The instability interaction in the four regions is discussed in the following.

Axial Tension in Combination with Internal Pressure Because there is no instability mode in this region, only the yield criterion need be considered. The general criterion of Eq. 14.58 may be used or it would be reasonable and only slightly conservative to neglect the interaction and consider the effects separately.

Axial Tension in Combination with External Pressure This combination of loading can lead to a phenomenon referred to as a propagating buckle. An indentation develops at a point around the circumference and spreads along the length of the cylinder.

Because the tension component does not lead directly to a stability failure, the results of the interaction are usually expressed in terms of a reduced critical external pressure stress, $\sigma'_{\phi c}$. Two approaches are used to obtain $\sigma'_{\phi c}$, one being a standard interaction equation and the other a reduced yield strength in the critical pressure (or stress) equations.

The interaction equation approach assumes an elliptic relation similar to the form for general yielding. The normalization of the hoop stress, however, is based on the critical stress for pressure only, rather than on yield. The resulting interaction equation is

$$\left(\frac{\sigma_x}{\sigma_y}\right)^2 + \left(\frac{\sigma'_{\phi c}}{\sigma_{\phi c}}\right)^2 + \frac{\sigma_x \sigma'_{\phi c}}{\sigma_y \sigma_{\phi c}} = 1 \qquad (14.59)$$

where σ_x is the imposed axial stress, $\sigma'_{\phi c} = p'_c D/2t$ is the reduced critical hoop stress, σ_y the yield stress, and $\sigma_{\phi c}$ the critical hoop stress when no axial loading is imposed. This normalization has been used to plot the data in Fig. 14.13, where it can be observed that Eq. 14.59 is conservative for the collapse failures.

The API (1989) procedure uses an effective hoop yield stress σ_{yr}, obtained by solving

$$\left(\frac{\sigma_{yr}}{\sigma_y}\right)^2 + \frac{\sigma_{yr}}{\sigma_y}\frac{\sigma_x}{\sigma_y} + \left(\frac{\sigma_x}{\sigma_y}\right)^2 = 1 \qquad (14.60)$$

The critical pressure can be obtained as the value of p_y in Eq. 14.44 using σ_{yr} in place of σ_y. It is also possible to use in empirical design curves that include inelastic and out-of-roundness effects in place of Eq. 14.44.

A variation on the reduced-yield approach is to base the inelastic properties on the stress intensity

$$\sigma_i = [(\sigma'_{\phi c})^2 + \sigma_x^2 - \sigma_x \sigma'_{\phi c}]^{0.5} \qquad (14.61)$$

Again the collapse pressure is based on Eq. 14.44 and the p_c term is based on the equations of Table 14.2 using a reduced modulus of elasticity.

The approaches used with Eqs. 14.60 and 14.61 are not as conservative as the interaction equation given by Eq. 14.59. It is also important to note that all of the above methods are for ultimate conditions. Actual design is more complicated in that different factors of safety are desirable for tension yielding and pressure collapse. Iterative solutions are typically required to obtain optimal designs from given tension and hydrostatic design loads.

Compression in Combination with External Pressure For very stocky cylinders for which failure occurs by general yielding, Fig. 14.13 indicates that it would be conservative and reasonable to ignore the interaction between axial compression and external pressure. For cylinders with larger D/t ratios, however, in which both external pressure and axial loading may result in a collapse failure mode, the two stresses acting on a preexisting imperfection often have an additive effect. In this case, Johns et al. (1975, 1976) suggest the construction of a composite failure envelope, as shown in Fig. 14.14. This method is conservative when compared to available bending and collapse test data (Johns et al., 1975, 1976) and with axial compression and external pressure tests (Miller and Vojta, 1984).

Compression in Combination with Internal Pressure The exact design of cylinders subjected to a combination of axial compression and internal pressure is very complicated. The following offers an approximate procedure that should be sufficiently accurate for most design purposes.

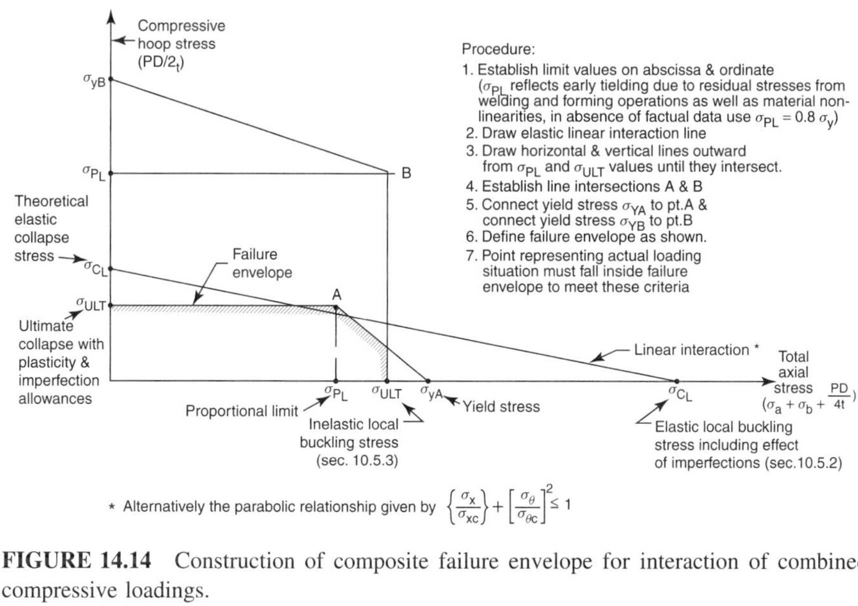

FIGURE 14.14 Construction of composite failure envelope for interaction of combined compressive loadings.

Many tests have been conducted (Lo et al., 1951; Fung and Sechler, 1957; Gerard, 1957; Seide, 1960; Baker et al., 1968; Mungan, 1974) that show that internal pressure increases the axial buckling stress as long as the failure stress remains elastic. A graphical method has been developed by Baker for determining this increase. The following equation given by Baker et al. (1968), when used in conjunction with Fig. 14.15, can be used for cylinders when the effective stress σ_i (calculated by Eq. 14.61) is less than the proportional limit,

$$\sigma_x < \sigma_{xc} + 2\,\overline{\Delta C}\,\frac{Et}{D} \tag{14.62}$$

with σ_{xc} determined from Section 14.3.1 and $\overline{\Delta C}$ from Fig. 14.15. When σ_i is greater than the proportional limit, the second term is no longer valid and therefore

$$\sigma_x \leq \sigma_{xc} \tag{14.63}$$

with σ_x determined from Section 14.3.1.

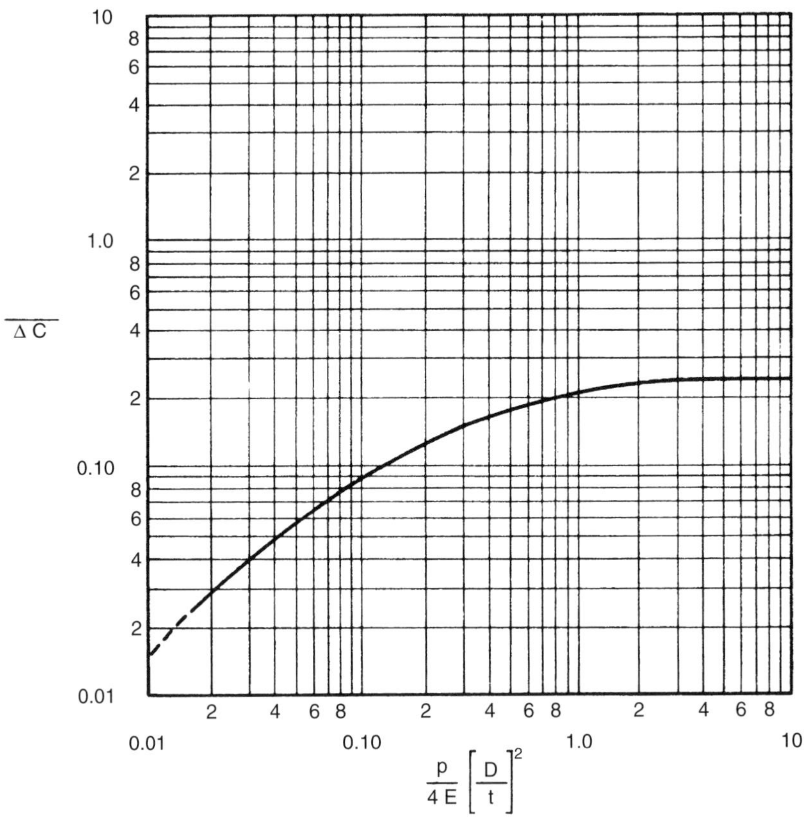

FIGURE 14.15 Increase in axial compressive elastic buckling stress due to internal pressure (Baker et al., 1968).

14.8 STRENGTH AND BEHAVIOR OF DAMAGED AND REPAIRED TUBULAR COLUMNS

There are several thousand major offshore fixed platforms devoted to the mining of crude oil in U.S. waters alone (Ricles et al., 1995). These structures are constructed from steel tubular members and they exist in a hazardous environment, being constantly in danger of collision with marine vessels and dropped objects, which can cause damage consisting of *dents* and *out-of-straightness*. Some of these dents may be quite large, resulting in significant changes of the original cross section. The marine environment also exposes the tubes to corrosion, requiring the use of counteractive corrosion measures, such as cathodic protection systems or protective coatings. Despite these precautions there are numerous cases reported of platform members that have suffered corrosion damage.

Because it is essential that the structural integrity of the platforms be maintained during their intended lifetime, it is necessary to be able to assess the remaining strength of a damaged tubular member and to provide a satisfactory repair procedure if that strength is inadequate. The strength of members with dents, both with and without out-of-straightness, has been a subject of extensive research since the late 1970s (Smith et al., 1981; Padula and Ostapenko, 1989; MacIntyre, 1991; Landet and Lotsburg, 1992; Duan et al., 1993; Salman, 1994). A tabular historic description and evaluation of this research has been prepared by Salman (1994). The strength of severely corroded tubular columns was extensively studied by Padula and Ostapenko (1989), Ostapenko et al. (1993), and Hebor and Ricles (1994). The significant corrosion damage occurs in patches anywhere on the surface of the tube. Damaged tubes are strengthened when necessary by internal or external grouting in the vicinity of the damaged portions of the member. Research on such tubes is reported in Parsanejad (1987), Parsanejad and Gusheh (1992), Ricles et al. (1994, 1995), and Ricles (1995). Computer methods for the analysis of the strength of internally grout repaired damaged members are available in Ricles (1995), and formulas for ultimate strength have been developed by Parsanejad (1987), Parsanejad and Gusheh (1992), and Ricles et al. (1994) for damaged and repaired tubular steel members.

REFERENCES

AA (2005), *Specification for Aluminum Structures*, Aluminum Association, Arlington, VA.

Ades, C. F. (1957), "Bending Strength of Tubing in Plastic Range," *J. Aeronaut. Sci.*, Vol. 24, pp. 605–610.

AISC (2005), *Specification for Structural Steel Buildings*, American Institute of Steel Construction, Chicago, IL.

AISI (2007), *North American Specification for the Design of Cold-Formed Steel Structural Members*, American Iron and Steel Institute, Washington, DC.

Almroth, B. O. (1966), "Influence of Edge Conditions on the Stability of Axially Compressed Cylindrical Shells," *AIAA J.*, Vol. 4, No. 1, pp. 134–140.

API (1977), *API Specification for Fabricated Structural Steel Pipe*, API Spec. 2B, 3rd., American Petroleum Institute, Division of Production, Washington, DC, Nov.

API (1989), *API Recommended Practice for the Planning, Design, and Construction of Fixed Offshore Platforms*, RP2A, 18th ed., American Petroleum Institute, Division of Production, Washington, DC, September.

ASME (1980), *Boiler and Pressure Vessel Code*, Section VIII, *Pressure Vessels*, Div. 1, American Society of Mechanical Engineers, Washington, DC, New York.

AWWA (1967), *AWWA Standard for Steel Tanks—Standpipes, Reservoirs, and Elevated Tanks—for Water Storage, AWWA D100-67*, American Water Works Association, Denver, CO.

Baker, E. H., et al. (1968), *Shell Analysis Manual*, NASA CR-912, Apr.

Ball, W. E. (1962), "Formulas and Curves for Determining the Elastic General-Instability Pressures of Ring-Stiffened Cylinders," David Taylor Model Basin Rep. No. 1570, Jan., Office of Naval Research, Bethesda, MD.

Basdekas, N. L. (1966), "A Survey of Analytical Techniques for Determining the Static and Dynamic Strength of Pressure-Hull Shell Structure," David Taylor Model Basin Rep. No. 2208, Sept., Office of Naval Research, Bethesda, MD.

Batdorf, S. B. (1947), "A Simplified Method of Elastic Stability Analysis for Thin Cylindrical Shells," NACA Rep. No. 874.

Batdorf, S. B., Schildcrout, M., and Stein, M. (1947a), "Critical Stress of Thin-Walled Cylinders in Axial Compression," NACA Tech. Note No. 1343.

Batdorf, S. B., Schildcrout, M., and Stein, M. (1947b), "Critical Stress of Thin-Walled Cylinders in Torsion," NACA Tech. Note No. 1344, July.

Becker, H. (1958), "General Instability of Stiffened Cylinders," NACA Tech. Note No. 4327, July.

Bijlaard, P. P. (1949), "Theory and Tests on the Plastic Stability of Plates and Shells," *J. Aeronaut. Sci.*, Vol. 16, No. 9, pp. 529–541.

Birkemoe, P. C., Prion, H. G. L., and Sato, J. A. (1983), "Compression Behavior of Unstiffened Fabricated Steel Tubes," paper presented at the ASCE Annu. Conv. Struct. Cong., Oct., Houston, TX.

Blumenberg, W. F. (1965), "The Effect of Intermediate Heavy Frames on the Elastic General-Instability Strength of Ring-Stiffened Cylinders Under External Hydrostatic Pressure," David Taylor Model Basin Rep. No. 1844, Feb., Office of Naval Research, Bethesda, MD.

Bodner, S. R. (1957), "General Instability and Ring-Stiffened Circular Cylindrical Shell Under Hydrostatic Pressure," *J. Appl. Mech.*, Vol. 24, No. 2, pp. 269–277.

Bodner, S. R., and Berks, W. (1952), "The Effect of Imperfections on the Stresses in a Circular Cylindrical Shell Under Hydrostatic Pressure," PIRAL Rep. No. 210, Polytechnic Institute of Brooklyn, New York, Dec.

Boichot, L., and Reynolds, T. E. (1964), "Inelastic Buckling Tests of Ring-Stiffened Cylinders Under Hydrostatic Pressure," David Taylor Model Basin Rep. No. 1992, May, Office of Naval Research, Bethesda, MD.

Bradford, M. A., Loh, H. Y., and Uy, B. (2001), "Slenderness Limits for CHS Sections," *Proceedings of the Ninth International Symposium and Euroconference on Tubular Structures*, Dusseldorf, Germany, Apr.

Brazier, L. G. (1927), "On the Flexure of Thin Cylindrical Shells and Other Thin Sections," *Proc. R. Soc. London*, Vol. A116, pp. 104–114.

Bryant, A. R. (1954), "Hydrostatic Pressure Buckling of a Ring-Stiffened Tube," Rep. No. R-306, Naval Construction Research Establishment, Washington, DC.

Chen, W. F., and Ross, D. A. (1978), "Tests of Fabricated Tubular Columns," Fritz Eng. Lab., Rep. No 393.8, Lehigh University, Bethlehem, PA, Sept.

Clark, J. W., and Rolf, R. L. (1964), "Design of Aluminum Tubular Members," *ASCE J. Struct. Div.*, Vol. 90, No. ST6, p. 259.

DeHeart, R. C., and Basdekas, N. L. (1960), "Yield Collapse of Stiffened Circular Cylindrical Shells," Report under NOMR Contract NR 2650(00), Proj. NR 064435, Southwest Reseach Institute, San Antonio, TX, Sept.

Denton, A. A., and Alexander, J. M. (1963a), "On the Determination of Residual Stresses in Tubes," *J. Mech. Eng. Sci.*, Vol. 5, No. 1, pp. 75–88.

Denton, A. A., and Alexander, J. M. (1963b), "New Method for Measurement of Local Residual Stresses in Tubes," *J. Mech. Eng. Sci.*, Vol. 5, No. 1, pp. 89–90.

Donnell, L. H. (1934), "A New Theory for the Buckling of Thin Cylinders Under Axial Compression and Bending," *Trans. ASME*, Vol. 56, pp. 795–806.

Donnell, L. H. (1956), "Effect of Imperfections on Buckling of Thin Cylinders Under External Pressure," *J. Appl. Mech.*, Vol. 23, No. 4, p. 569.

Donnell, L. H., and Wan, C. C. (1950), "Effect of Imperfections on Buckling of Thin Cylinders and Columns Under Axial Compression," *J. Appl. Mech.*, Vol. 17, No. 1, pp. 73–83.

Duan, L., Chen, W. F., and Loh, J. T. (1993), "Moment-Curvature Relationships for Dented Tubular Sections," *ASCE J. Struct. Eng.*, Vol. 119, No. ST3, pp. 809–831.

Eder, M. F., Grove, R. B., Peters, S. W., and Miller, C. D. (1984), "Collapse Tests of Fabricated Cylinders Under Combined Axial Compression and External Pressure," Final Rep., American Petroleum Institute Proj. 83-46, CBI Industries, Research Laboratory, Plainfield, IL, Feb.

Edwards, S. H., and Miller, C. P. (1939), "Discussion on the Effect of Combined Longitudinal Loading and External Pressure on the Strength of Oil-Well Casing," in *Drilling and Production Practice*, American Petroleum Institute, Washington, DC, pp. 483–502.

Farmer, L. E. (1966), "Method for Determining Buckling Loads of Circular Stiffening Rings Including Consideration of Torsional Displacement," AEDC-TR-66-188, Engineering Support Facility, Washington, DC, Sept.

Flügge, W. (1932), "Die Stabilität der Kreiszylinderschale," *Ing. Arch.*, Vol. 3.

Flügge, W. (1973), *Stresses in Shells*, 2nd ed., Springer-Verlag, New York, pp. 459–463.

Fung, Y. C., and Sechler, E. E. (1957), "Buckling of Thin Walled Circular Cylinders Under Axial Compression and Internal Pressure," *J. Aeronaut. Sci.*, Vol. 24, pp. 351–356.

Galletly, G. D., and Bart, R. (1957), "Effect of Boundary Conditions and Initial Out-of-Roundness on the Strength of Thin-Walled Cylinders Subjected to External Hydrostatic Pressure," David Taylor Model Basin Rep. No. 1066, Nov., Office of Naval Research, Bethesda, MD.

Gellin, S. (1980), "The Plastic Buckling of Long Cylindrical Shells Under Pure Bending," *Int. J. Solids Struct.*, Vol. 16, pp. 397–407.

Gerard, G. (1956), "Compressive and Torsional Buckling of Thin-Walled Cylinders in Yield Region," NACA Tech. Note No. 3726.

Gerard, G. (1957), "Plastic Stability Theory of Thin Shells," *J. Aeronaut. Sci.*, Vol. 24, No. 4, p. 269.

Gerard, G., and Becker, H. (1957), "Handbook of Structural Stability: Part 3. Buckling of Curved Plates and Shells," NACA Tech. Note No. 3783, Aug.

Graham, R. R. (1965), "Manufacture and Use of Structural Tubing," *J. Met.*, Vol. 17, No. 9, pp. 975–980.

Harari, A., and Baron, M. L. (1970, 1971), "Buckling of Vessels Composed of Combinations of Cylindrical and Spherical Shells," *J. Appl. Mech.*, July 1970, pp. 393–398; July 1971, pp. 571–572.

Harris, L. A., et al. (1957), "The Stability of Thin-Walled Unstiffened Circular Cylinders Under Axial Compression Including the Effects of Internal Pressure," *J. Aeronaut. Sci.*, Vol. 24, No. 8, pp. 587–596.

Hebor, M., and Ricles, J. M. (1994), "Residual Strength and Repair of Corroded Marine Steel Tubulars," ATLSS Rep. No. 94-10, Lehigh University, Bethlehem, PA.

Heise, O., and Esztergar, E. P. (1970), "Elastoplastic Collapse of Tubes Under External Pressure," *J. Eng. Ind.*, Vol. 92, pp. 735–742.

Hoff, N. J. (1966), "The Perplexing Behavior of Thin Cylindrical Shells in Axial Compression," *Isr. J. Technol.*, Vol. 4, No. 1.

Holmquist, J. L., and Nadai, A. (1939), "A Theoretical and Experimental Approach to the Problem of Collapse of Deep-Well Casing," in *Drilling and Production Practice*, American Petroleum Institute, Washington, DC, pp. 392–420.

Holt, M. (1952), "A Procedure for Determining the Allowable Out-of-Roundness for Vessels Under External Pressure," *ASME Trans.*, Vol. 74, p. 1225.

Hom, K. (1962), "Elastic Stress in Ring-Frames of Imperfectly Circular Cylindrical Shells Under External Pressure Loading," David Taylor Model Basin Rep. No. 1505, May, Office of Naval Research, Bethesda, MD.

Hom, K., and Couch, W. P. (1961), "Hydrostatic Tests of Inelastic and Elastic Stability of Ring-Stiffened Cylindrical Shells Machined from Strain-Hardening Material," David Taylor Model Basin Rep. No. 1501, Dec.

Hutchinson, J. W., and Amazigo, J. C. (1967), "Imperfection-Sensitivity of Eccentrically Stiffened Cylindrical Shells," *AIAA J.*, Vol. 5, No. 3, pp. 392–401.

Jirsa, J. O., et al. (1972), "Ovaling of Pipelines Under Pure Bending," Pap. 1569, presented at the 4th Annu. Offshore Technol. Conf., Houston, TX, May.

Johns, T. G., Mesloh, R. E., Winegardner, R., and Sorenson, J. E. (1975), "Inelastic Buckling of Cylinders Under Combined Loads," Pap. 2209, presented at the 7th Annu. Offshore Technol. Conf., Houston, TX, May.

Johns, T. G., Sorenson, J. E., Mesloh, R. E., and Attenbury, T. J. (1976), "Buckling Strength of Offshore Platforms Program: Phase I," Battelle Institute, Columbus, OH.

Johnston, B. G. (1976), *Guide to Stability Design Criteria for Metal Structures*, Structural Stability Research Council, Wiley, New York.

Kendrick, S. (1953a), "The Deformation Under External Pressure of Circular Cylindrical Shells with Evenly Spaced Equal Strength Nearly Circular Ring Frames," Rep. No. R-259, Naval Construction Research Establishment, Washington, DC, Oct.

Kendrick, S. (1953b), "The Buckling Under External Pressure of Circular Cylindrical Shells with Evenly Spaced Equal Strength Circular Ring Frames: Part III," Rep. No. R-259, Naval Construction Research Establishment, Washington, DC, Oct.

Kiziltug, A. Y., Grove, R. B., Peters, S. W., and Miller, C. D. (1985), "Collapse Tests of Short Tubular Columns Subjected to Combined Loads," Final Rep. Contract No. UI-851604, CBI Industries, Chicago, IL, Dec.

Korol, R. M. (1978), "Inelastic Buckling of Circular Tubes in Bending," Pap. 13944, *ASCE J. Eng. Mech. Div.*, Vol. 104, No. EM4, pp. 939–952.

Krenzke, M. A., and Kiernan, T. J. (1963), "Structural Development of a Titanium Oceanographic Vehicle for Operating Depths of 15,000 to 20,000 feet," David Taylor Model Basin Rep. No. 1677, Sept., Office of Naval Research, Bethesda, MD.

Krenzke, M. A., and Short, R. D. (1959), "Graphical Method for Determining Maximum Stresses in Ring-Stiffened Cylinder Under External Hydrostatic Pressure," David Taylor Model Basin Rep. No. 1348, Oct., Office of Naval Research, Bethesda, MD.

Kulak, G. L. (1994), "Tubular Members: Large and Small," *Proc. 50th Anniv. Conf.*, Structural Stability Research Council, Bethlehem, PA.

Kuper, E. J., and Macadam, J. N. (1969), "Quantitative Effects of Strain Hardening on the Mechanical Properties of Cold-Formed Electric Welded Steel Tubing," *Proc. 10th Mech. Working Steel Process. Conf.*, American Institute of Mining, Metallurgical, and Petroleum Engineers, New York.

Landet, E., and Lotsburg, I. (1992), "Laboratory Testing of Ultimate Capacity of Dented Tubular Members," *ASCE J. Struct. Eng.*, Vol. 118, No. 4, pp. 1071–1089.

Lo, H., Crate, H., and Schwartz, E. B. (1951), "Buckling of Thin Walled Cylinders Under Axial Compression and Internal Pressure," NASA Rep. No. 1027 (formerly 2021).

Lorenz, R. (1908), "Achsensymmetrische Verzerrungen im dünnwandigen Hohlzylinder," *Z. Ver. Dtsch. Ing.*, Vol. 52, No. 43, pp. 1706–1713.

Lubinski, A. (1951), "Buckling of Rotary Drilling Strings: Part III," *World Oil*, May, pp. 122–132.

Lubinski, A., Althouse, W. S., and Logan, J. L. (1952), "Helical Buckling of Tubing Sealed in Packers," *J. Petrol. Technol.*, June, pp. 655–670.

Lunchick, M. E. (1959), "Yield Failure of Stiffened Cylinders Under Hydrostatic Pressure," David Taylor Model Basin Rep. No. 1291, Jan., Office of Naval Research, Bethesda, MD.

Lunchick, M. E. (1961a), "Plastic Axisymmetric Buckling of Ring-Stiffened Cylindrical Shells Fabricated from Strain-Hardening Materials and Subjected to External Hydrostatic Pressure," David Taylor Model Basin Rep. No. 1392, Jan., Office of Naval Research, Bethesda, MD.

Lunchick, M. E. (1961b), "Graphical Methods for Determining the Plastic Shell-Buckling Pressures of Ring-Stiffened Cylinders Subjected to External Hydrostatic Pressure," David Taylor Model Basin Rep. No. 1437, Mar., Office of Naval Research, Bethesda, MD.

Lunchick, M. E. (1963), "Plastic General-Instability Pressure of Ring-Stiffened Cylindrical Shells," David Taylor Model Basin Rep. No. 1587, Sept., Office of Naval Research, Bethesda, MD.

Lunchick, M. E., and Short, R. D. (1957), "Behavior of Cylinders with Initial Shell Deflection," David Taylor Model Basin Rep. No. 1130, July, Office of Naval Research, Bethesda, MD.

Lundquist, E. E. (1933), "Strength Tests of Thin-Walled Duraluminum Cylinders in Compression," NACA Rep. No. 473.

MacIntyre, J. (1991), "An Analytical Study of Damaged Tubular Member Behavior," Ph.D. thesis, University of Toronto, Toronto, Ontario, Canada.

Marshall, P. W. (1971), "Design Criteria for Structural Steel Pipe," paper presented at the *Column Res. Conc. Proc.* Pittsburgh, PA.

Marzullo, M. A., and Ostapenko, A. (1978), "Tests on Two High-Strength Steel Large-Diameter Tubular Columns," Pap. 3086, *Proc. 10th Offshore Technol. Conf.*, Houston, TX, May.

Meck, H. R. (1965), "A Survey of Methods of Stability Analysis of Ring-Stiffened Cylinders Under Hydrostatic Pressure," *J. Eng. Ind.*, Aug., p. 385.

Miller, C. D. (1982), "Summary of Buckling Tests on Fabricated Steel Cylindrical Shells in USA," in *Buckling of Shells in Offshore Structures* (eds. J. E. Harding, P. J. Dowling, and C. M. Angelidis), Granada, London, pp. 429–472.

Miller, C. D. (1983), "Research Related to Buckling of Nuclear Containment," paper presented at the 7th Int. Conf. Struct. Mech. Reactor Technol., Chicago, Aug.

Miller, C. D. (1984), "API Bulletin on Stability Design of Shells" (draft), *Am. Petrol. Inst. Bull.*, Vol. 2u, Sept.

Miller, C. D., Grove, R. B., and Vojta, J. F. (1983), "Design of Stiffened Cylinders for Offshore Structures," paper presented at the AWS Weld. Offshore Struct. Conf., New Orleans, LA, Dec.

Miller, C. D., Kinra, R. K., and Marlow, R. S. (1982), "Tension and Collapse Tests of Fabricated Steel Cylinders," Pap. 4218, *Proc. 14th Offshore Technol. Conf.*, Houston, TX, May.

Miller, C. D., and Vojta, J. F. (1984), "Strength of Stiffened Cylinders Subjected to Combinations of Axial Compression and External Pressure," *Proc. SSRC Annu. Tech. Session*, San Francisco, CA.

Mungan, I. (1974), "Buckling Stress States of Cylindrical Shells," *ASCE Struct. Div.*, Vol. 100, No. ST11, pp. 2289–2306.

Nash, W. A. (1957), "Buckling of Initially Imperfect Shells Subject to Torsion," *J. Appl. Mech.*, Vol. 24, No. 1, pp. 125–130.

O'Shea, M. D., and Bridge, R. Q. (1997), "Local Buckling of Thin-Walled Circular Steel Sections with or without Internal Restraint," University of Sydney, Dept. of Civil Engineering Report R740, Apr.

Ostapenko, A., and Grimm, D. F. (1980), "Local Buckling of Cylindrical Tubular Columns Made of A-36 Steel," Fritz Eng. Lab. Rep. No. 450-7, Lehigh University, Bethelem, PA, Feb.

Ostapenko, A., and Gunzelman, S. X. (1976), "Local Buckling of Tubular Steel Columns," *Proc. Methods Struct. Analysis*," Vol. 2, American Society of Civil Engineers, New York, p. 549.

Ostapenko, A., and Gunzelman, S. X. (1978), "Local Buckling Tests on Three Steel Large-Diameter Tubular Columns," *Proc. 4th Int. Conf. Cold-Formed Steel Struct.*, St. Louis, MO, June, p. 409.

Ostapenko, A., Wood, B., Chowdhury, A., and Hebor, M. (1993), "Residual Strength of Damaged and Deteriorated Tubular Members in Offshore Structures," ATLSS Rep. No. 93-03, Lehigh University, Bethlehem, PA.

Padula, J. A., and Ostapenko, A. (1989), "Axial Behavior of Damaged Tubular Columns," Fritz Eng. Lab. Rep. No. 408. 11, Lehigh University, Bethlehem, PA.

Palmer, A. C., and Baldry, J. A. S. (1974), "Lateral Buckling of Axially Constrained Pipelines," *J. Petrol. Technol.*, Nov., pp. 1283–1284.

Parsanejad, S. (1987), "Strength of Grout-Filled Damaged Tubular Members," *ASCE J. Struct. Eng.*, Vol. 113, No. 3, pp. 590–603.

Parsanejad, S., and Gusheh, P. (1992), "Behavior of Partially Grout-Filled Damaged Tubular Members," *ASCE J. Struct. Eng.*, Vol. 118, No. 11, pp. 3055–3066.

Plantema, F. J. (1946), "Collapsing Stress of Circular Cylinders and Round Tubes," Rep. No. S.280, National Luchtfaart Laboratorium, Amsterdam.

Prescott, J. (1946), *Applied Elasticity*, 1st American ed., Dover, New York, pp. 552–554.

Prion, H. G. L., and Birkemoe, P. C. (1988), "Experimental Behaviour of Unstiffened Fabricated Tubular Steel Beam-Columns," Pub. No. 88-3, Department of Civil Engineering, University of Toronto, Toronto, Ontario, Canada.

Pulos, J. G. (1963), "Structural Analysis and Design Considerations for Cylindrical Pressure Hulls," David Taylor Model Basin Report No. 1639, Apr., Office of Naval Research, Bethesda, MD.

Pulos, J. G., and Salerno, V. L. (1961), "Axisymmetric Elastic Deformations and Stresses in a Ring-Stiffened, Perfectly Circular Cylindrical Shell Under External Hydrostatic Pressure," David Taylor Model Basin Rep. No. 1497, Structural Mechanics Laboratory, Sept., Office of Naval Research, Bethesda, MD.

Reynolds, T. E. (1957), "A Graphical Method for Determining the General Instability Strength of Stiffened Cylindrical Shells," David Taylor Model Basin Rep. No. 1106, Sept., Office of Naval Research, Bethesda, MD.

Reynolds, T. E. (1960), "Inelastic Local Buckling of Cylindrical Shells Under External Hydrostatic Pressure," David Taylor Model Basin Rep. No. 1392, Aug., Office of Naval Research, Bethesda, MD.

Reynolds, T. E. (1962), "Elastic Local Buckling of Ring-Supported Cylindrical Shells Under Hydrostatic Pressure," David Taylor Model Basin Rep. No. 1614, Sept., Office of Naval Research, Bethesda, MD.

Reynolds, T. E. (1971), "Stability of Cylindrical Shells," Graduate Course on Analysis and Design of Cylindrical Shells Under Pressure, Catholic University of America, Washington, DC, June.

Reynolds, T. E., and Blumenberg, W. F. (1959), "General Instability of Ring-Stiffened Cylindrical Shells Subjected to External Hydrostatic Pressure," David Taylor Model Basin Rep. No. 1324, June, Office of Naval Research, Bethesda, MD.

Ricles, J. M. (1995), "Analysis of Internally Grout Repaired Damaged Members," in *Analysis and Software of Cylindrical Members* (eds. W. F. Chen and S. Toma), CRC Press, Boca Raton, FL, Chap. 7.

Ricles, J. M., Bruin, W. M., Sooi, T. K., Hebor, M. F., and Schönwetter, P. C. (1995), "Residual Strength Assessment and Repair of Damaged Offshore Tubulars," OTC Pap. 7807, *Proc. 27th Offshore Technol. Conf.*, May 1–4, Houston, TX, pp. 41–59.

Ricles, J. M., Gillum, T., and Lamport, W. (1994), "Grout Repair of Dented Offshore Tubular Bracing: Experimental Behavior," *ASCE J. Struct. Eng.*, Vol. 120, No. 7, pp. 2086–2107.

Salman, W. A. (1994), "Damage Assessment of Tubular Members in Offshore Structures," Ph.D. thesis, University of Toronto, Toronto, Ontario, Canada.

Schilling, C. G. (1965), "Buckling Strength of Circular Tubes," *ASCE J. Struct. Div.*, Vol. 91, No. ST5, p. 325.

Seide, P. (1960), "The Effect of Pressure on the Bending Characteristics of an Actuator System," *J. Appl. Mech.*, Sept., p. 429.

Sherman, D. R. (1969), "Residual Stress Measurement in Tubular Members," Pap. 6502, *ASCE J. Struct. Div.*, Vol. 95, No. ST4, pp. 635–637.

Sherman, D. R. (1976), "Tests of Circular Steel Tubes in Bending," Pap. 12568, *ASCE J. Struct. Div.*, Vol. 102, No. ST11, pp. 218–2195.

Sherman, D. R. (1986), "Inelastic Flexural Buckling of Cylinders," in *Steel Structures: Recent Research Advances and Their Application to Design*, M. N. Pavlovic (eds.), Elsevier Applied Science, New York, Sept.

Sherman, D. R. (1992), "Tubular Members," in *Constructional Steel Design: An International Guide*, Dowling, P. J., Hanoing, J. E., and Bjorhovde, R. (eds.), Elsevier Applied Science, New York, Dec.

Singer, J. (1967), "The Influence of Stiffener Geometry and Spacing on the Buckling of Axially Compressed Cylindrical and Conical Shells," Prelim. Prepr. Pap., paper presented at the 2nd IUTAM Symp. Theory Thin Shells, Copenhagen, Sept.

Smith, C. S., Somerville, J. W., and Swan, J. W. (1981), "Residual Strength and Stiffness of Damaged Steel Bracing Members," OTC Paper 3981, *Proc. 13th Offshore Technol. Conf.*, Houston, TX.

Southwell, R. V. (1914), "On General Theory of Elastic Stability," *Philos. Trans. R. Soc. London*, Ser. A, Vol. 213A, pp. 187–244.

Southwell, R. V. (1915), "On the Collapse of Tubes by External Pressure," *Philos. Mag.*, Part 1, Vol. 25, May 1913, pp. 687–698; Part 2, Vol. 26, Sept. 1913, pp. 502–511; Part 3, Vol. 29, Jan. 1915, pp. 66–67.

Stein, M. (1968), "Some Recent Advances in the Investigation of Shell Buckling," *AIAA J.*, Vol. 6, No. 12, pp. 2339–2345.

Stephens, M. J., Kulak, G., and Montgomery, C. J. (1982), "Local Buckling of Thin-Walled Tubular Steel Members," Struct. Eng. Dept. Rep. No. 103, University of Alberta, Edmonton, Alberta, Canada, Feb.

Stephens, M. J., Kulak, G., and Montgomery, C. J. (1983), "Local Buckling of Thin-Walled Tubular Steel Members," *Proc. 3rd Int. Colloq. Stab. Met. Struct.* (George Mem. Session), Toronto, Ontario, Canada, May.

Sturm, R. G. (1941), "A Study of the Collapsing Pressure of Thin-Walled Cylinders," Univ. Ill. Eng. Exp. Sta. Bull., No. 329.

Timoshenko, S. P. (1910), "Einige Stabilitäts-Probleme der Elastizitäts-Theorie," *Z. Math. Phys.*, Vol. 58, No. 4.

Timoshenko, S. P., and Gere, J. M. (1961), *Theory of Elastic Stability*, McGraw-Hill, New York.

Tokugawa, T. (1929), "Model Experiments on the Elastic Stability of Closed and Cross-Stiffened Circular Cylinders Under Uniform External Pressure," *Proc. World Eng. Congr., Tokyo*, Vol. 29, pp. 249–279.

U.S. Steel (1964), *The Manufacture of Steel Tubular Products*, U.S. Steel Corporation, Pittsburgh, PA.

Vojta, J. F., and Miller, C. D. (1983), "Buckling Tests on Ring and Stringer Stiffened Cylindrical Models Subject to Combined Loads," Final Rep., Contract No. 11896, Vol. 1, Main Report, 3 Volume Appendix, CBI Industries, Chicago, IL, Apr.

von Kármán, T., and Tsien, H. S. (1941), "The Buckling of Thin Cylindrical Shells Under Axial Compression," *J. Aeronaut. Sci.*, Vol. 8, No. 8, pp. 303–312.

von Mises, R. (1931), "The Critical External Pressure of Cylindrical Tubes" (Der kritische Aussendruck zylindrischer Röhre), *VDI-Z.*, Vol. 58, No. 19, pp. 750–755; *David Taylor Model Basin Transl. 5*, Aug. 1931.

von Mises, R. (1933), "The Critical External Pressure of Cylindrical Tubes Under Uniform Radial and Axial Load" (Der kritische Aussendruck für allseits belastete zylindrische Röhre), Stodolas Festschrift, Zurich, 1929, pp. 418–430; *David Taylor Basin Transl. 5*, Aug. 1933.

von Sanden, K., and Tolke, F. (1949), "On Stability Problems in Thin Cylindrical Shells" (Über Stabilitätsprobleme dünner, kreiszylindrischer Schalen), *Ing. Arch.*, Vol. 3, 1932, pp. 24–66; *David Taylor Model Basin Transl. 33*, Dec. 1949.

Wah, T. (1967), "Buckling of Thin Circular Rings Under Uniform Pressure," *Int. J. Solids Struct.*, Vol. 3, pp. 967–974.

Weingarten, V. I., and Seide, P. (1961), "On the Buckling of Circular Cylindrical Shells Under Pure Bending," *Trans. ASME, J. Appl. Mech.*, Vol. 28, No. 1, pp. 112–116.

Weingarten, V. I., et al. (1965), "Elastic Stability of Thin-Walled Cylindrical and Conical Shells Under Combined Internal Pressure and Axial Compression," *AIAA J.*, Vol. 3, No. 6, pp. 118–125.

Weingarten, V. I., et al. (1968), "Buckling of Thin-Walled Circular Cylinders," NASA SP-8007, Aug.

Wenk, E., and Kennard, E. H. (1956), "The Weakening Effect of Initial Tilt and Lateral Buckling of Ring Stiffeners on Cylindrical Pressure Vessles," David Taylor Model Basin Rep. No. 1073, Dec., Office of Naval Research, Bethesda, MD.

Wilson, W. M. (1937), "Tests of Steel Columns," Univ. Ill. Eng. Exp. Sta. Bull. No. 292.

Wilson, W. M., and Newmark, N. M. (1933), "The Strength of Thin Cylindrical Shells as Columns," Univ. Ill. Eng. Exp. Sta. Bull. No. 255.

Wilson, W. M., and Olson, F. D. (1941), "Tests on Cylindrical Shells," Univ. Ill. Eng. Exp. Sta. Bull. No. 331, Sept.

Windenburg, D. F. (1960), "Vessels Under External Pressure," in *Pressure Vessel and Piping Design*, American Society of Mechanical Engineers, New York, pp. 625–632.

Windenburg, D. F., and Trilling, C. (1934), "Collapse by Instability of Thin Cylindrical Shells Under External Pressure," *Trans. ASME*, Vol. 56, No. 11, p. 819.

Wu, T. S., Goodman, L. E., and Newmark, N. M. (1953), "Effect of Small Initial Irregularities on the Stresses in Cylindrical Shells," Univ. Ill. Struct. Res. Ser. No. 50, Apr.

Yao, J. C. (1962), "Large Deflection Analysis of Buckling of a Cylinder Under Bending," *Trans. ASME J. Appl. Mech.*, Vol. 29, No. 4, pp. 708–714.

CHAPTER 15

MEMBERS WITH ELASTIC LATERAL RESTRAINTS

15.1 INTRODUCTION

Compression members are sometimes restrained laterally between their ends by intermittent elastic lateral supports. Typical examples include (1) the unbraced compression flange of a girder whose tension flange is laterally restrained by a bridge–floor or building–roof system and (2) the top chord of a pony truss, for which vertical clearance requirements prohibit direct lateral bracing. The methods of design presented in this chapter can also be used for guyed towers.

In the case of the girder, the web prevents the top flange from buckling in the vertical plane, and intermittent elastic lateral restraints may be provided in the plane normal to the web by means of vertical stiffeners in combination with the contiguous framing elements in the floor or roof system adjacent to the tension flange. In a truss supporting a floor or roof system in the plane of the tension chord members, the panel points provide vertical support and elastic restraint for the compression chords as a whole.

Although not as widely used as in the past, the pony truss has served as the prototype through the years for the development of theory and design procedures that currently may find applications in other similar situations. The behavior during buckling of a member with intermittent elastic lateral supports lies between two limiting extremes. If the elastic restraints are very stiff, nodal points can be induced at each restraint location; if the restraints are very flexible, and buckling can be in the shape of a single half-wave over the full member length as long as the ends of the member are laterally restrained. The actual buckled shape consists of a number of half-waves not greater than the total number of spaces between supports.

The design of a compression member with intermittent elastic lateral supports may be based on the computed critical load; or because of initial crookedness and moments introduced by bending of the floor beams, it could be based on a rigorous second-order analysis of a structural system. The latter approach is a rational one and is only now beginning to appear as a practical design procedure. It should be noted that a critical-load analysis gives only an upper bound to the actual strength of the member unless a reduction for the initial out-of-straightness is included. Current design procedures require evaluation of the stiffness supplied by the compression chord lateral support system. This stiffness forms the basis for evaluating the integrity of the compression chord. The critical load of the chord with elastic lateral supports at the panel points is then determined, and the design load is found by dividing this critical load by a suitable factor of safety.

Using the pony truss as the prototype, a procedure of analysis is developed to determine the critical load of a member with discrete elastic lateral restraints. To a lesser extent the rigorous second-order elastic analysis (combined-stress) procedure is reviewed in this chapter. The design of pony-truss transverse frames (floor beams, truss verticals, and connecting knee braces) has a direct bearing on both procedures. Proper design of transverse frames is essential to the safety of the pony-truss bridge. Toward the end of the nineteenth century the failure of several pony-truss bridges focused attention on the top-chord buckling problem. Engesser (1885, 1892) was the first to present a simple, rational, and approximate formula for the required stiffness, C_{req}, of elastic supports equally spaced between the ends of a hinge-ended column of constant section. An equivalent uniform elastic support was assumed in the Engesser analysis.

Early developments are reviewed by Bleich (1952). Using the energy method, Hu (1952) has studied the problem of elastically supported chords. He considered nonuniform axial forces, variable chord cross sections, and spring stiffness for both simple and continuous pony-truss bridges. Holt (1951, 1952, 1956, 1957), in work sponsored by the CRC, presented a method of analysis for determining the critical load of a pony-truss top chord that is essentially "exact" in that it includes most of the secondary effects that influence the behavior of the pony truss. In a similar manner, Lee and Clough (1958) and Elgaaly and Khalifa (1970) studied the stability of pony-truss bridges.

The effect of the floor-system deflections on the top-chord stresses was studied in another CRC-sponsored project by Barnoff and Mooney (1957). Tests on models of pony-truss bridges have been conducted by Holt (1957). Oliveto (1980) described a computer program and considered both elastic and inelastic buckling of columns partially restrained at intervals. Medland and Segedin (1979) evaluated brace forces for an initially crooked member.

15.2 BUCKLING OF THE COMPRESSION CHORD

The buckling problem of the compression chord of a pony truss can be reduced to that of a column braced at intervals by elastic springs whose spring constants corresponded to the stiffness of the truss transverse frames. The top-chord axial compression and the top-chord stiffness will vary from panel to panel, and the stiffness of the transverse frames may also vary from panel point to panel point, thus complicating the theoretical problem. In addition, there are secondary factors:

1. The stiffening effect of the truss diagonals
2. The torsional stiffness of the chord and web members
3. The initial crookedness of the chord and the eccentricity of the axial load
4. The effect of chord curvature, for non-parallel-chord trusses.

Engesser's (1885, 1892) solution is based on the following simplifying assumptions:

1. The top chord, including the end posts, is straight and of uniform cross section.
2. Its ends are taken as pin connected and rigidly supported.
3. The equally spaced elastic supports have the same stiffness and can be replaced by a continuous elastic medium.
4. The axial compressive force is constant through the chord length.

Engesser's analysis can be applied with reasonable accuracy to the case where the lateral support is supplied by equally spaced springs provided that the half-wavelength of the buckled shape of the continuously supported member is at least 1.8 times the spring spacing; this will be true if the bar is stable as a two-hinged column carrying the same axial load and having a length no less than 1.3 times the spring spacing.

Engesser's solution for the required stiffness of a pony-truss transverse frame is

$$C_{\text{req}} = \frac{P_{cr}^2 l}{4EI} \tag{15.1}$$

where C_{req} is the elastic transverse frame stiffness at a panel point that is required to ensure that the overall chord having panel lengths l and flexural rigidity EI will attain buckling load P_{cr}. If the proportional limit of the column material is exceeded, E should be replaced by the tangent modulus E_t.

The Euler equation for critical stress with E_t replacing E can be written as follows for a column of length l:

$$E_t I = \frac{P_{cr}(Kl)^2}{\pi^2} \tag{15.2}$$

Taking l in this equation as the panel length of the pony-truss compression chord, the following required spring constant can be obtained by substituting Eq. 15.2 into Eq. 15.1:

$$C_{\text{req}} = \frac{\pi^2 P_{cr}}{4K^2 l} \tag{15.3}$$

This equation has been shown by Hu (1952, p. 275) to be adequate when the half-wavelength of the buckled chord is no less than $1.8l$; this limiting value corresponds to a K factor of 1.3. It is not applicable to short bridges with a small number of panels.

The load P_{cr} can be considered as both of the following:

1. The buckling load of the entire compression chord laterally supported by the transverse frames and pinned at the ends
2. The buckling load of the portion of the compression chord *between* the transverse frames with end restraints producing effective-length factor K

According to Engesser's formulation, the maximum compression-chord buckling load and the corresponding required spring constant of each support can be determined as follows for a member of a given cross-sectional area A:

1. Determine the critical load P_{cr} for the member between spring supports using the expression

$$P_{cr} = A\sigma_{cr} \tag{15.4}$$

   Obtain σ_{cr} from an appropriate column strength curve, taking the equivalent column slenderness ratio as Kl/r, with $K = 1.3$ and r estimated on the basis of probable shape and size of member.
2. Determine the spring constant C_{req} such that the buckling load of the chord member as a whole is equal to P_{cr}:

$$C_{\text{req}} = 1.46\frac{P_{cr}}{l} \tag{15.5}$$

   It may be noted that Eq. 15.5 follows from Eq. 15.3, taking

$$\frac{\pi^2}{4K^2} = \frac{\pi^2}{4(1.3)^2} = 1.46$$

The Engesser simplifying assumption of taking the chord ends as pin connected may result in significantly unsafe errors in C_{req}, particularly in the case of short pony trusses. Holt (1952, 1956) provides an alternative design procedure that does not require this simplifying assumption. Holt's solution for the buckling load of the compression chord of a pony truss is based on the following assumptions (see Fig. 15.1):

1. The transverse frames at all panel points have identical stiffness.
2. The radii of gyration r of all top-chord members and end posts are identical.
3. The top-chord members are all designed for the same allowable unit stress; hence, their areas and (from step 2) their moments of inertia are proportional to the compression forces.
4. The connections between the top chord and the end posts are assumed pinned.

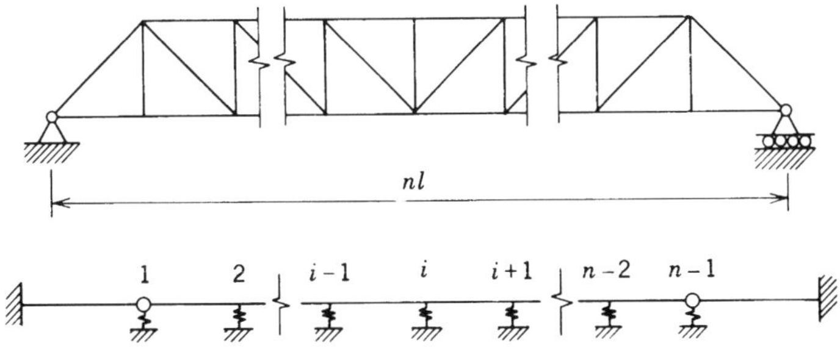

FIGURE 15.1 Pony truss and analogous top chord.

5. The end posts act as cantilever springs supporting the ends of the top chords.
6. The bridge carries a uniformly distributed load.

The results of Holt's studies are presented in Table 15.1, which gives the reciprocal of the effective-length factor K as a function of n (the numbers of panels) and Cl/P_{cr} (in which C is the furnished stiffness at the top of the least-stiff transverse frame). Where applicable, Table 15.1 provides a rapid design aid in checking the stability of a pony-truss compression chord. The procedure is as follows:

1. Design the floor beams and web members for their specified loads.
2. Calculate the spring constant C furnished at the upper end of the cross frame having the least transverse stiffness.
3. Calculate the value of parameter Cl/P_{cr}, where P_{cr} is the maximum design chord load multiplied by the desired factor of safety.
4. Enter the table with n and Cl/P_{cr} and find the corresponding value of $1/K$ for a compression-chord panel, interpolating as necessary.
5. Determine the value of Kl/r for the compression-chord panel (noting that this value of Kl/r is to be applied to all panels).
6. Determine the allowable top-chord compressive unit stress corresponding to this value of Kl/r, using the appropriate column curve or table.

Values of $1/K$ of less than 0.5 (i.e., $K > 2$) are only of academic interest, because usual bridge proportions and transverse-frame stiffnesses lead to values of $1/K$ greater than 0.5.

Hu (1952) developed the curves shown in Fig. 15.2. These curves give the stiffness of the compression-chord transverse supports that is required to make each panel of the chord buckle as one half-wave. Hu's results for a chord of constant section [$(EI)_{\text{end}}/(EI)_{\text{middle}} = 1.0$ in Fig. 15.2] can be compared with Holt's work for $1/K = 1$ (first line of Table 15.1) for the cases $n = 4.6, 8$, which were considered by both investigators. Hu's results give stiffness requirements approximately 7% less than those of Holt for $n = 4$ and 5% greater for $n = 6, 8$. Thus the results are in reasonable agreement, even though the procedures are somewhat different.

TABLE 15.1 1/K for Various Values of Cl/P_c and n

	n							
$1/K$	4	6	8	10	12	14	16	Lutz–Fisher (1985, Eq. 15-8)
1.000	3.686	3.616	3.660	3.714	3.754	3.785	3.809	4.000
0.980		3.284	2.944	2.806	2.787	2.771	2.774	3.730
0.960		3.000	2.665	2.542	2.456	2.454	2.479	3.478
0.950			2.595					
0.940		2.754		2.303	2.252	2.254	2.282	3.244
0.920		2.643		2.146	2.094	2.101	2.121	3.026
0.900	3.352	2.593	2.263	2.045	1.951	1.968	1.981	2.822
0.850		2.460	2.013	1.794	1.709	1.681	1.694	2.372
0.800	2.961	2.313	1.889	1.629	1.480	1.456	1.465	1.993
0.750		2.147	1.750	1.501	1.344	1.273	1.262	1.673
0.700	2.448	1.955	1.595	1.359	1.200	1.111	1.088	1.401
0.650		1.739	1.442	1.236	1.087	0.988	0.940	1.169
0.600	2.035	1.639	1.338	1.133	0.985	0.878	0.808	0.970
0.550		1.517	1.211	1.007	0.860	0.768	0.708	0.798
0.500	1.750	1.362	1.047	0.847	0.750	0.668	0.600	0.648
0.450		1.158	0.829	0.714	0.624	0.537	0.500	0.519
0.400	1.232	0.886	0.627	0.555	0.454	0.428	0.383	0.406
0.350		0.530	0.434	0.352	0.323	0.292	0.280	0.309
0.300	0.121	0.187	0.249	0.170	0.203	0.183	0.187	0.226
0.293	0							
0.259		0						
0.250			0.135	0.107	0.103	0.121	0.112	0.157
0.200			0.045	0.068	0.055	0.053	0.070	0.100
0.180			0					
0.150				0.017	0.031	0.029	0.025	0.056
0.139				0				
0.114					0			
0.100						0.003	0.010	0.025
0.097						0		
0.085							0	
0								0

Hu (1952) also studied the effects caused by parabolic variation of the length of the pony-truss verticals and the effect of parabolic variation of C. In both cases, the value of C_{req} will be less than that for the case where C has the same value at each transverse frame.

Because of the uncertainties involved in the analysis of pony-truss top chords, it is reasonable to require a factor of safety for overall top-chord buckling somewhat greater than that used for designing hinged-end columns. The transverse-frame

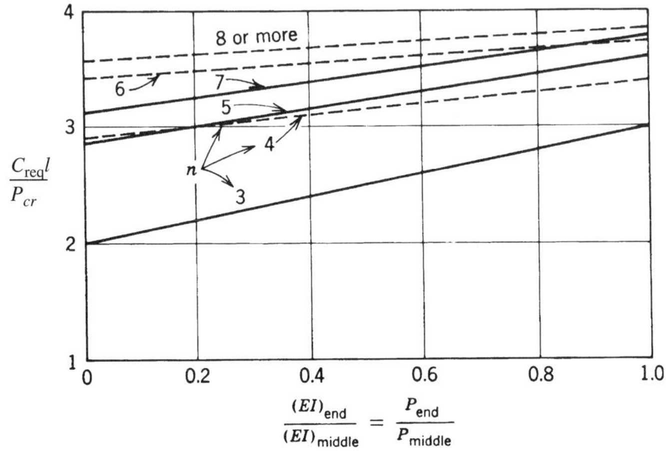

FIGURE 15.2 Effect of variation in compression chord on transverse stiffness requirements.

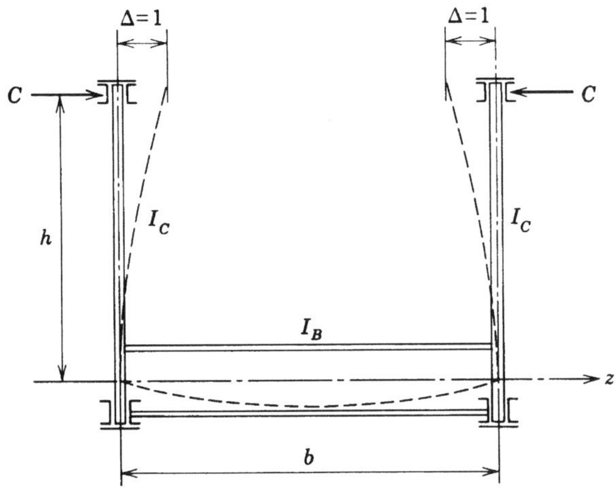

FIGURE 15.3 Frame loading for determining transverse-frame spring constant C.

spring constant C that is actually furnished can be determined for the frame loaded as shown in Fig. 15.3 by means of the following equation:

$$C = \frac{E}{h^2[(h/3I_C) + (b/2I_B)]} \tag{15.6a}$$

The first term within the denominator brackets represents the contribution of the truss vertical, and the second term represents the contribution of the floor beam. Thus, the contributions of the top-chord torsional strength and the web-diagonal bending strength to the frame stiffness are neglected in this equation. It is evident that if the floor beam is very stiff in comparison with the truss vertical, the frame stiffness is approximately

$$C = \frac{3EI_C}{h^3} \tag{15.6b}$$

When the two chords tend to move in the same direction, the floor beam will be in double curvature and the stiffness C will be greater than that given by Eq. 15.6a; therefore, C as found from Eq. 15.6a is always the lower bound.

If the diagonals of the truss system are effectively fixed at their base, their contribution to C in Eq. 15.6a may be included by introducing the additional term $L_d^3/3I_d$ into the denominator. The values L_d and I_d represent the length and moment of inertia, respectively, of the diagonal members. In this case C becomes

$$C = \frac{E}{h^2\{h/[3I_C + 3I_d(h/L_d)^3] + (b/2I_B)\}} \tag{15.6c}$$

15.3 EFFECT OF SECONDARY FACTORS ON BUCKLING LOAD

The consideration of secondary factors involves procedures that require significantly more computation. Most of these procedures use the usual methods of indeterminate structural analysis to set up a system of simultaneous, linear, homogeneous equations. The stability criterion is that the determinant of the coefficients of this system of equations must vanish (i.e., the eigenvalue problem given by Eq. 16.4 is solved).

Holt (1952) considered the following secondary factors:

1. Torsional stiffness of the chord and the web members
2. Lateral support given to the chord by the diagonals
3. Effect of web member axial stresses on the restraint provided by them
4. Effect of non-parallel-chord trusses
5. Error introduced by considering the chord and end posts to be a single straight member

Holt's analysis shows that the error in the critical load introduced by neglecting all of these factors is quite small and that satisfactory results in calculating the compression-chord buckling load can be obtained by assuming that the chord is a straight elasticity braced column with length equalling the total length of the chord and end posts. These conclusions are in agreement with those reached by Schibler (1946), who finds that the torsional stiffness of the top chord and the support furnished by the web diagonals increase the chord-buckling strength only slightly.

15.4 TOP-CHORD STRESSES DUE TO BENDING OF FLOOR BEAMS AND TO INITIAL CHORD ECCENTRICITIES

The compression chord of a pony truss is displaced laterally at some panel points as a result of live load on the bridge and because of initial crookedness and unintentional eccentricities of the chord. Such lateral deflections will, of course, reduce the maximum load capacity of the chord (and of the bridge), just as end eccentricity and initial curvature will reduce the compression strength of any column.

Lutz and Fisher (1985) did consider initial out-of-straightness for elastically restrained members in the same manner that Winter (1960) did for fully braced members, namely by requiring the actual stiffness supplied to be twice the ideal required stiffness. They followed the simplifying assumptions made by Engesser discussed earlier and rewrote Eq. 15.3 in the form

$$C_{\text{req}} = 2.5\frac{P_{cr}l}{L_e^2} \tag{15.7}$$

where $L_e = Kl$ and $\pi^2/4 \simeq 2.5$. They then extended the applicability to K-factors of less than 1.3 and as low as 1.0. Their empirical required stiffness expression is

$$C_i = \left[2.5 + 1.5\left(\frac{l}{L_e}\right)^4\right]\frac{P_{cr}l}{L_e^2} \tag{15.8}$$

At $K = 1$ each spring would provide full bracing to the compression chord, which requires a stiffness of $4P_{cr}/l$. They proposed that due to initial out-of-straightness, the actual stiffness should be twice the required stiffness given above [see Winter (1960) and accompanying discussion by Green].

Design procedures that take account theoretically of such imperfections are not presently available, although the basic concepts for the *direct analysis method* for buildings (Section 16.5.1) could be an excellent starting point. It is difficult to take initial imperfections into account because of both the complexity of the necessary calculations and the lack of knowledge with regard to probable initial imperfections. In any case, the analysis would need to be three dimensional and include the top-chord stiffness in both bending and torsion.

Holt (1957) has developed an empirical procedure for estimating bending moments in the top chord and end posts that is in agreement with his test results. He recommends that the end post be designed as a simple cantilever beam to carry the axial load combined with a transverse load of 0.5% of the axial load, applied at the upper end. Tentatively, a value of 1% should probably be used.

15.5 DESIGN EXAMPLE

Consider a pony truss that has 12 panels of 13 ft, 4 in. for a span of 160 ft. The transverse frame is shown in Fig. 15.4. The W27×84 floor beams are required

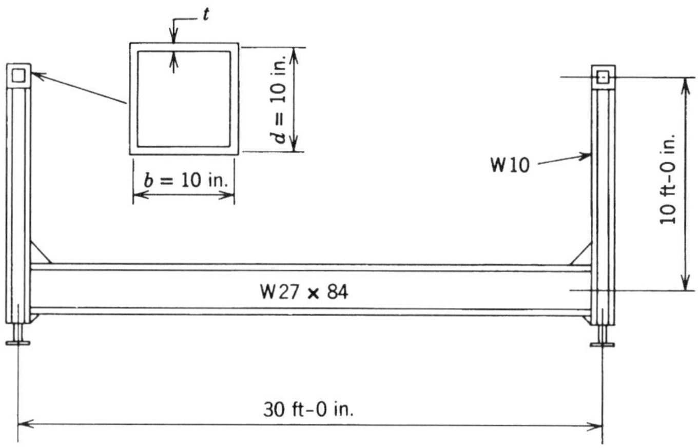

FIGURE 15.4 Transverse frame in pony truss example.

by bridge deck loads. The top chord is a box section, 10 in. by 10 in. with wall thickness to be determined by design requirements for a maximum compressive force of 360 kips (which includes unfactored dead load, live load, and impact). The verticals are W10 rolled sections.

Assuming that the vertical truss members, floor beams, and their connections need to be proportioned to resist a lateral force of not less than 300 pounds per linear foot of span applied at the top-chord panel points of each truss, the lateral force to be resisted is

$$0.3\,\text{kip/ft} \times 13.33\,\text{ft} = 4\,\text{kips}$$

The maximum moment in the transverse frame will be at the joint of the web vertical and the floor beam and is

$$M = 4\,\text{kips} \times 120\,\text{in.} = 480\,\text{kip-in.}$$

Using this moment and assuming that the maximum tension in the web vertical due to the bridge load is 24 kips in the region where the compression chord is most highly stressed, a W10×33 vertical with $I = 170\,\text{in.}^4$ is selected.

The I of the floor beam is $2850\,\text{in.}^4$, and from Eq. 15.6a the transverse-frame spring constant is

$$C = \frac{29{,}000}{120^2[120/(3 \times 170) + 360/(2 \times 2850)]} = 6.75\,\text{kips/in.}$$

Assuming a factor of safety against buckling of 2.12 (previous editions of AASHTO specifications used a factor of safety of 2.12 for the allowable stress design of concentrically loaded columns), the compression chord must be designed for a buckling strength of $P = 2.12 \times 360\,\text{kips}$. Because a square built-up tubular section is considered for the compression chord, its properties can be approximated as

$$\text{Area}: A = 4td$$

$$\text{Moment of inertia: } I = \frac{Ad^2}{6}$$

$$\text{Radius of gyration: } r = \frac{d}{\sqrt{6}}$$

If the distance between wall centerlines d is taken as 10 in., $r = 10/\sqrt{6} = 4.08$ in.

Using Holt's procedure, Kl/r is determined from Table 15.1. Using the actual supplied C and $P_{cr} = 763$ kips, we obtain

$$\frac{Cl}{P_{cr}} = \frac{6.75(160)}{763} = 1.415$$

Entering Table 15.1 in the column for $n = 12$ panels and $Cl/P_{cr} = 1.415$, $1/K$ falls between 0.750 and 0.800. Interpolation reveals $1/K = 0.776$ or $K = 1.288$. Considering the built-up 10-in. tube yields

$$\frac{Kl}{r} = \frac{1.288(160)}{4.08} = 50.5$$

$$F_a = 16,930 - 0.53(50.5)^2 = 15,580 \text{ psi} = 15.58 \text{ ksi}$$

The required $A = 360/15.58 = 23.1$ in.^{2} and $t = 0.577$ in. with $d = 10$ in. Had the number of panels been 8 instead of 12, the thickness t required would have increased to 0.603 in. using Holt's procedure.

Using the Lutz–Fisher expression gives

$$C_{\text{req}} = 2C_i = \frac{[5 + 3(l/L_e)^4]P_{cr}l}{L_e^2}$$

or with $l/L_e = 1/K$,

$$\frac{C_i l}{P_{cr}} = \left[2.5 + 1.5\left(\frac{1}{K}\right)\right]^4 \left(\frac{1}{K}\right)^2$$

which is included as the last column in Table 15.1. With $C_{\text{req}} = 6.75$,

$$\frac{C_i l}{P_{cr}} = \frac{(6.75/2)(160)}{763} = 0.708$$

From Table 15.1, $1/K = 0.52$ or $K = 1.923$. Thus $Kl/r = 1.923(160)/4.08 = 75.4$ for the 10-in. tube and $F_a = 16{,}930 - 0.53(75.4)^2 = 13{,}916$ psi.

Required $A = 360/13.92 = 25.86\,\text{in.}^2$ and $t = 25.86/(4 \times 10) = 0.646\,\text{in.}$ Using the Holt procedure one could end up with a design with $d = 9.5\,\text{in.}$ using a $\frac{5}{8}$-in. plate. With the Lutz–Fisher procedure, which considers out-of-straightness, one would need a $\frac{5}{8}$-in. plate with $d = 10.25\,\text{in.}$

15.6 PLATE GIRDER WITH ELASTICALLY BRACED COMPRESSION FLANGE

Although most of the research and references presented herein concern the pony-truss bridge, the design recommendations that have been given are also applicable to the design of plate girders with elastically braced compression flanges. Such girders will customarily have full-depth vertical stiffeners serving the dual function of web stiffener and top-flange transverse support. The lack of girder diagonal members is of no concern because Eq. 15.6a does not include the contribution of truss diagonals. In applying the design procedure to girders, one-third of the compression area of the web should be added to the area of the compression flange and introduced as the equivalent area of the top chord in the formulas as developed for the pony truss. This compression area should be reduced by multiplying by the plate-girder bending strength reduction factor (if less than 1) for slender webs. The same section may be used in calculating I_C. If girders are used in an arrangement differing from the section shown in Fig. 15.3, an independent calculation of C must be made appropriate to the framing arrangement that is used.

Additional information on beam bracing is provided in Section 12.10, which includes discussion and examples on lateral bracing and torsional bracing.

15.7 GUYED TOWERS

In determining the effective-length factor K for use in the design of a guyed tower, Springfield[1] has used Table 15.1 together with an adaptation of the pony-truss analysis. He reports on the application to a "mast" with guys at nine levels, as follows:

> Stiffnesses (of restraints) are not constants, but vary according to the magnitude of lateral displacement. The procedure we have adopted is as follows:
>
> 1. Compute the spring constants C at each guy level. This is done by dividing the horizontal guy reaction by the displacement under maximum wind load. The average spring stiffness is chosen rather than the instantaneous stiffness because the buckling mode could reduce the displacements at guy points, thus reducing the spring stiffness.
> 2. Calculate the parameter Cl/P_c for each column segment between guys. The smaller value of C is chosen from the top or bottom of the segment. l is the length of the

[1] June 9, 1971, letter to the editor of the fifth edition of this guide, from J. Springfield, formally of C.D. Carruthers & Wallace Consultants Ltd., Toronto, Canada.

column segment and P_c is the maximum column segment design load multiplied by the appropriate factor-of-safety.

3. The value of K for each segment is determined in turn from Table 15.1 (a total of nine K values). In its application, we have assumed a continuous compression strut with nine spans. Each span has identical spring supports of magnitude equal to the C value for the column segment being considered. The spacing of these supports is uniform, this distance being the length of column segment under consideration.

Holt assumes that the strut is fixed at each end and pinned at the first interior supports. Our approximation would be fairly good for the interior spans, but could be in error for the discontinuous ends, especially at the top end, where both joint displacement and rotation are possible.

The range of Cl/P_c for our mast is 1.3 to 2.3 and the corresponding range of K is 1.7 to 1.1. We have also applied Engesser's analysis for comparison, which shows the close agreement of C required for K values between 1.7 and 1.3. For K values smaller than 1.3, a rapid divergence appears as expected.

It is, of course, recognized that the complete design of guyed towers is a very complex problem, involving effects that may include preloading of guys, vertical movement of guys, vortex shedding, and other vibration dynamic problems. The foregoing is only a brief summary of the column-buckling aspects of the design problem as approached in an approximate and simplified manner.

REFERENCES

American Association of State Highway and Transportation Officials (AASHTO) (1994), *AASHTO LRFD Bridge Design Specifications*, AASHTO, Washington, DC.

Barnoff, R. M., and Mooney, W. G. (1957), "The Effect of Floor System Participation on Top Chord Stresses in Single Span Pony Truss Bridges," in *Stability of Bridge Chords Without Lateral Bracing,* Rep. No. 5, Column Research Council, Rolla, MC. Bethlehem, PA.

Bleich, F. (1952), *Buckling Strength of Metal Structures*, McGraw-Hill, New York.

Elgaaly, M. A., and Khalifa, M. E., Jr. (1970), *Stability of Pony-Truss Bridges*, on file, Institution of Structural Engineers, London, June.

Engesser, F. (1885), "Die Sicherung offener Brücken gegen Ausknicken," *Centralbl. Bauverwaltung*, 1884, p. 415; 1885, p. 93.

Engesser, F. (1892) *Die Zusatzkräfte und Nebenspannungen eisemer Fachwerkbrücken*. Vol. I, Die Zusatzkrafte, 18982, Vol. II, Die Nebenspannungen, 1893, Julius Springer, Berlin, Germany.

Holt, E. C. (1951), "Buckling of a Continuous Beam-Column on Elastic Supports," in *Stability of Bridge Chords without Lateral Bracing*, Rep. No. 1, Column Research Council, Bethlehem, PA.

Holt, E. C. (1952), "Buckling of a Pony Truss Bridge," in *Stability of Bridge Chords without Lateral Bracing,* Rep. No. 2, Column Research Council, Bethlehem, PA.

Holt, E. C. (1956), "The Analysis and Design of Single Span Pony Truss Bridges," in *Stability of Bridge Chords without Lateral Bracing,* Rep. No. 3, Column Research Council, Bethlehem, PA.

Holt, E. C. (1957), "Tests on Pony Truss Models and Recommendations for Design," in *Stability of Bridge Chords without Lateral Bracing,* Rep. No. 4, Column Research Council, Bethlehem, PA.

Hu, L. S. (1952), "The Instability of Top Chords of Pony Trusses," dissertation, University of Michigan, Ann Arbor, MI.

Lee, S. L., and Clough, R. W. (1958), "Stability of Pony Truss Bridges," *Publ. Int. Assoc. Bridge Struct. Eng.*, Vol. 18, p. 91.

Lutz, L., and Fisher, J. M. (1985), "A Unified Approach for Stability Bracing Requirements," *AISC Eng. J.*, Vol. 22, No. 4, p. 163.

Medland, I. C., and Segedin, C. M. (1979), "Brace Forces in Interbraced Column Structures," *ASCE J. Struct. Div.*, Vol. 105, No. ST7, pp. 1543–1556.

Oliveto, G. (1980), "Inelastic Buckling of Columns Partially Restrained Against Sway and Rotation," *Eng. Struct.*, Vol. 2, pp. 97–102.

Schibler, W. (1946), "Das Tragvermögen der Druckgurte offener Fachwerkbrücken mit parallelen Gurten," *Institut fuer Baustatik, Eidgenoessische Technische Hochschule*, *Zurich Mitteilungen*, No. 19.

Winter, G. (1960), "Lateral Bracing of Columns and Beams" with discussions, *Trans. ASCE*, Vol. 125, pp. 807–845.

CHAPTER 16

FRAME STABILITY

16.1 INTRODUCTION

Frame stability requires that all structural members and connections of a frame have adequate strength to resist the applied loads with *equilibrium satisfied on the deformed geometry of the structure.* In practice, this rather straightforward requirement presents difficulties for the design engineer. To further complicate matters, conventional analysis methods are typically limited in their ability to accurately calculate the inelastic deformations and internal forces that develop as structures approach their strength limit state. Simple calculations that are based on first-order elastic analysis, in which equilibrium is satisfied only on the original configuration and the material is always assumed elastic, are therefore inadequate to meet this requirement. Even when the best of analysis models are available, the designer still must account for uncertainties introduced by variability in the magnitude and distribution of loads as well as factors affecting strength and stiffness, such as material properties, connections, fabrication and erection tolerances, and residual stresses.

The objectives of this chapter are to address frame stability in the context of computer analysis methods that can accurately model behavior and to discuss how such methods can be used in design. For in-plane behavior of rigidly connected frames, the theory and formulations for conducting accurate second-order inelastic analyses are fairly well established. For planar frames with members fully supported out-of-plane, it is feasible to account for frame stability directly in the analysis. Although research has been performed on modeling the more complex aspects of three-dimensional behavior (see Section 16.3.5), this research has not been implemented in analysis algorithms in a manner that has achieved widespread use, nor is it readily available to structural designers in commercial software. Therefore, most approaches for stability design still separate in-plane frame and member behavior from out-of-plane stability checks.

Overview of Chapter The chapter begins with a review of the various types of elastic and inelastic second-order analysis methods and provides the behavioral assumptions implied within them. Section 16.3 provides a review of basic aspects of frame behavior by means of an example followed by discussion of specific behavioral effects and the modeling issues related to frame stability. In Section 16.4, the application and limitations of various types of elastic and inelastic analysis methods for evaluating frame and member stability are discussed. These applications form the basis of many of the specification-based approaches discussed in Section 16.5, which includes a brief summary and comparison of frame and beam-column design provisions from four steel design standards. Section 16.6 presents an overview of structural integrity as it relates to progressive collapse. The chapter concludes with brief comments on the current status of and future developments in frame stability.

16.2 METHODS OF ANALYSIS

The analysis methods presented in this chapter can be distinguished by whether or not they include geometric and material nonlinear behavior or, more specifically, (1) whether equilibrium is satisfied on the undeformed or deformed geometry of the structure and (2) whether member yielding (spread of plasticity through the cross section and/or along the member length) is considered. An overview of analysis methods to calculate frame response is shown by the schematic load–deflection curves in Fig. 16.1. These plots relate the magnitude of applied gravity and lateral loads to the drift of the frame. Brief descriptions of analysis types represented include the following:

- *First-order elastic analysis* is the most basic method of analysis in which the material is modeled as linear-elastic and equilibrium is satisfied on the undeformed configuration of the structure. As a result, forces and deformations are directly proportional throughout the analysis. First-order elastic analysis provides no direct measure of frame stability.
- *Second-order elastic analysis* models the material as linear-elastic, but equilibrium is formulated on the deformed geometry of the structure. For most frames, the load–deflection response curve obtained from a rigorous second-order elastic analysis asymptotically approaches the load level referred to as the *elastic stability limit* of the structure. The *elastic stability limit* calculated by a second-order incremental analysis is similar to but distinct from the *elastic critical load* (λ_{cr}) calculated by a classical stability (or eigenvalue) analysis. Differences in the two limits arise because the elastic stability load corresponds to equilibrium in the deformed configuration; whereas the elastic critical load is calculated as a bifurcation from equilibrium from the undeformed geometry (i.e., prebuckling deformations are ignored in the eigenvalue analysis).
- *First-order inelastic analysis* models the effects of member yielding under incremental loading, and it is limited to first-order response, because equilibrium is satisfied only for the undeformed geometry of the structure. The

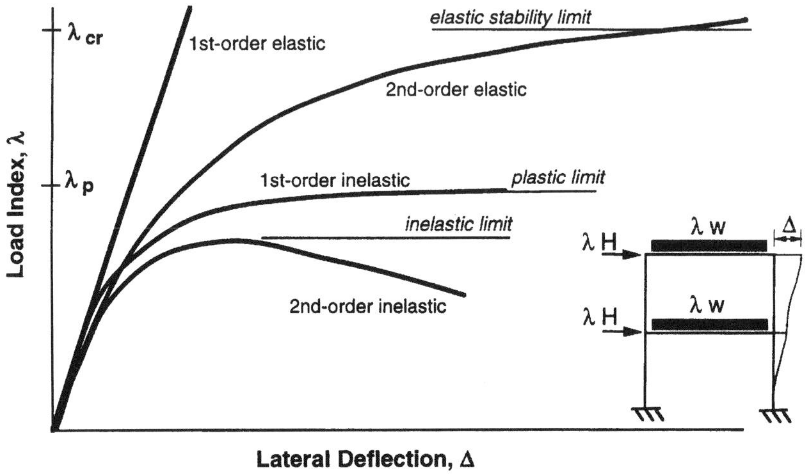

FIGURE 16.1 Schematic comparison of load–deflection behavior.

inelastic effects may be handled by techniques ranging from simple elastic–perfectly plastic hinge models to more detailed models that include spread of yielding or distributed plasticity effects. When the material representation is elastic–perfectly plastic (i.e., strain hardening is neglected), the load response curve from a first-order inelastic analysis asymptotically approaches the *plastic limit load*, which is identical to the load λ_p calculated by a plastic mechanism analysis.

- *Second-order inelastic analysis* includes both geometric and material nonlinearity. The analysis accounts for the decrease in stiffness due to both member yielding and large deflections. Subject to the limitations of behavioral effects considered in the analysis and accurate modeling of inelastic force redistribution, the *inelastic stability limit load* obtained by a second-order inelastic analysis is the most accurate representation of the true strength of the frame.

The following two sections describe the fundamental aspects of second-order (geometrically nonlinear) and inelastic (material nonlinear) analyses. In Section 16.2.3, a review of computer-based critical load (eigenvalue) methods is provided.

16.2.1 Second-Order Analysis

Referring to the simple frames shown in Fig. 16.2, the basic requirement of a second-order analysis is to ensure that equilibrium is satisfied for the structural system in the deformed condition. For the frame in Fig. 16.2*a*, a first-order analysis under gravity loads results in bending moments in the beam equal to those for a simple span condition with zero moments in the column. Equilibrium formulated on the deflected shape requires larger moments in the beam and nonzero

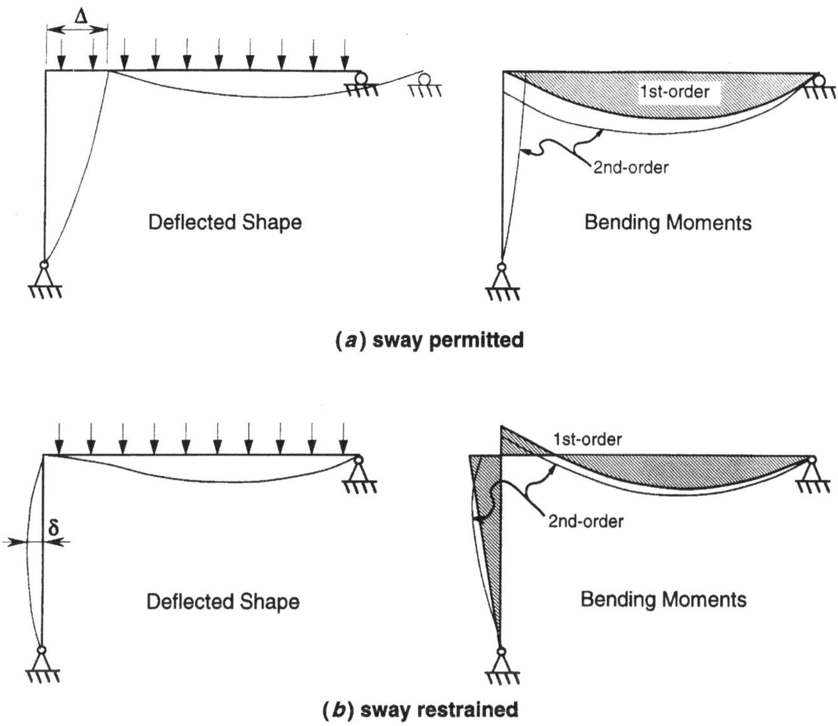

FIGURE 16.2 Second-order $P-\Delta$ and $P-\delta$ moments.

moments in the column. These additional moments are due primarily to the $P-\Delta$ *effect*, where Δ is the lateral drift of the frame and P is the total vertical load that translates through Δ. LeMessurier (1977) presented a similar example to this that demonstrates the extent of these second-order effects in a frame of realistic proportions.

For the frame in Fig. 16.2*b*, with sidesway prevented, second-order moments are created in the beam and column arising from the lateral deflection δ along the length of the column. The additional moment in the column is equal to the column load P times the deflection δ, hence the name $P-\delta$ *effect*. There is a reduction in the negative moment at the end of the beam due to the loss in rotational restraint of the column caused by the $P-\delta$ effect. This in turn increases the maximum positive moment within the beam span and may shift the location of maximum moment in the column. Further aspects of the second-order effects in a frame similar to this are discussed by Kanchanalai and Lu (1979).

Some points worth noting about these two examples, and about the calculation of frame second-order forces in general, are:

1. Second-order behavior affects moments in the beams, connections, and columns. Moreover, the axial forces and shears of all members also change

due to second-order effects. In other words, second-order behavior can affect all components and internal member forces within the structural system. In sway frames, however, the changes in axial forces and shears are usually of substantially lesser consequence than the changes in moments.

2. The second-order moments do not necessarily have the same distribution as the first-order moments, and hence the total moments are not simply a direct amplification of the first-order moments. As described in the literature (LeMessurier, 1977; Kanchanalai and Lu, 1979), and as incorporated in many design specifications, there are many practical cases where second-order moments may be calculated by amplifying the first-order moments. Using these methods for the calculation of the second-order forces, however, may become fairly involved. Furthermore, these amplification methods can be ineffective and cumbersome for calculating the maximum second-order moments in adjoining beams and connections.
3. All structures, whether braced or unbraced, will experience both $P-\Delta$ and $P-\delta$ effects (e.g., see LeMessurier, 1976, 1977; White et al., 2007a,b). While the examples in Fig. 16.2 are extreme cases dominated by one of the effects, both $P-\Delta$ and $P-\delta$ are present in each frame member to varying degrees. In the frame of Fig. 16.2*a*, second-order $P-\Delta$ effects cause flexure of the left column, which in turn generates $P-\delta$ moments. In Fig. 16.2*b* there is a slight joint translation due to axial shortening of the beam that causes small $P-\Delta$ effects. In many practical cases, one or the other of these effects might be safely ignored, but it is important to keep in mind that in most building frames the distinction between sway and nonsway moments is not as well defined as in these examples.
4. Linear superposition of effects cannot be applied with second-order analysis. Considering the frame in Fig. 16.2*a*, the deflections and internal forces from an analysis under lateral loads that caused a lateral deflection Δ cannot be linearly combined with gravity effects. Rather, the two loads (gravity plus lateral) must be applied simultaneously in a single analysis. Moreover, because second-order effects are generally nonlinear (i.e., second-order internal forces and deflections under factored loads do not scale linearly from a second-order analysis under service loads), all second-order analyses should be made under factored loads.

Theoretical Stiffness Formulation Various strategies to model geometric nonlinear effects can be developed based on conventional first- or second-order direct stiffness formulations and the use of incremental or incremental-iterative solution procedures. Formulations for second-order analysis can be generally expressed in the form

$$\{dF\} - \{dR\} = [K]\{d\Delta\} \tag{16.1}$$

where $\{dF\}$ is a vector of incremental applied nodal forces; $\{dR\}$ a vector of unbalanced nodal forces, calculated as the difference between the resultant of the current internal member forces and the applied loads; $[K]$ the stiffness matrix (typically a tangent stiffness, although a secant stiffness may be used), which expresses an approximate linearized relationship between the nodal forces and the nodal displacements; and $\{d\Delta\}$ a vector of incremental nodal displacements and rotations.

By defining $\{dR\}$ as the difference between the increment of applied loads $\{dF_{\text{applied}}\}$ and the counteracting internal member forces $\{dF_{\text{int}}\}$, Eq. 16.1 may be alternatively expressed in the form

$$\{dR\} = \{dF_{\text{applied}}\} - \{dF_{\text{int}}\} = [K]\{d\Delta\} \tag{16.2}$$

This form shows that $\{dR\}$ is an imbalance load vector that starts out as the magnitude of the incremental applied loads (because the resultant incremental internal member force vector is equal to $\{0\}$ in the first load step) and then becomes successively smaller as iteration is conducted. Chen and Lui (1987) illustrate how Eq. 16.2 can be applied if $[K]$ represents a secant rather than a tangent stiffness.

The features that distinguish the second-order analysis based on Eq. 16.1 from a first-order analysis are that the matrix equations are formulated in terms of incremental forces and displacements, and the unbalanced forces are calculated on the deformed geometry at the end of the linear loading step or iteration. As shown in Fig. 16.3, Eq. 16.1, which is generally nonlinear in the displacements through the term $\{dR\}$, can be solved in a piecewise linear fashion by applying successive increments of applied loads $\{dF_i\}$. For any force increment $\{dF_i\}$, a displacement increment $\{d\Delta_i^1\}$ can be computed based on the linear stiffness relationships, and then the actual internal nodal forces and the corresponding force imbalance $\{dR_i^1\}$ can be calculated. If the imbalance $\{dR_i^1\}$ is significant, Eq. 16.2 can be solved iteratively for successive values $\{dR_i^j\}$ and $\{d\Delta_i^j\}$, $j = 2, 3, 4, \ldots$, until the equilibrium error represented by the unbalanced force vector is negligible. If the force

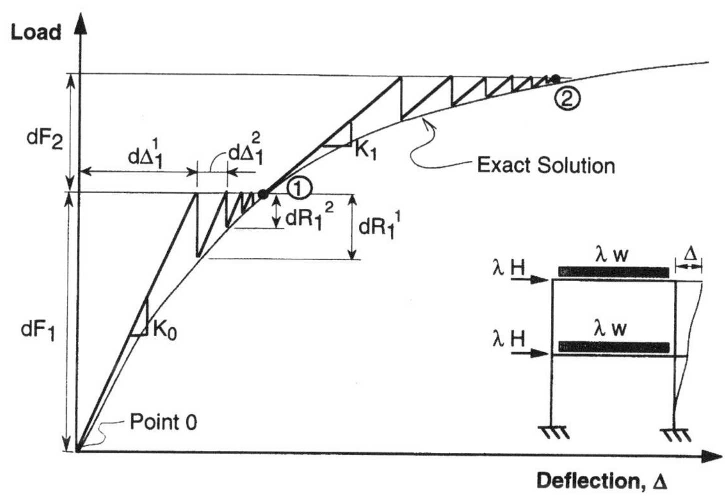

FIGURE 16.3 Schematic representation of incremental-iterative solution procedure.

increments $\{dF_i\}$ are small enough, iterations will not be necessary to obtain an accurate solution and the nonlinear solution becomes a pure or "simple" incremental method.

Figure 16.3 illustrates iteration at constant levels of load where a specified $\{dF_i\}$ is applied to the structure and iterations are conducted at a constant load until convergence is achieved. This type of procedure is commonly referred to as a Newton–Raphson incremental–iterative solution. Conventional Newton–Raphson procedures are not sufficient for continuing the solution through a stability limit load and into a postcollapse portion of the structural response. Therefore, alternative solution schemes such as arc length or displacement control methods can be useful to the engineer in certain situations. McGuire et al. (2000), Chen and Lui (1991), and Clarke (1994) provide useful overviews of several common solution approaches, and further details on solution methods are provided in the texts by Crisfield (1991), Yang and Quo (1994), and Bathe (1995).

Approximate Second-Order Methods With the capabilities of current computer hardware and software, rigorous second-order analysis is certainly feasible. Because superposition is not applicable when a second-order analysis is employed, many commercial design programs use approximate second-order, or $P-\Delta$, methods that allow for rapid linear combination of forces from multiple-load cases. For the design of a relatively simple building, there is the potential for a significantly large number of load combinations being generated based on the required application of wind and seismic forces such as those outlined in ASCE 7-05 (ASCE, 2005). While desktop computational capabilities continue to increase, it may not be generally practical to perform iterative analyses in all phases of the design process.

For elastic analysis of building frames in which deformations are usually not very large, it is common to utilize simplified methods to account for second-order effects. As examples, Rutenberg (1981, 1982) presents a computational method for approximate $P-\Delta$ analysis in computer applications, and Wilson and Habibullah (1987) present an approximate but computationally efficient method for modeling second-order effects in frames subjected to lateral loads. White et al. (2007a,b) present an approximate approach based on LeMessurier's (1976, 1977) second-order solutions. These methods, typical of many such approaches, apply fictitious lateral forces to capture interstory $P-\Delta$ effects that dominate the second-order sidesway response of most building frames. It is of particular importance that the engineer understands the assumptions built into any software package with respect to the second-order analysis, as different methods have different capabilities and limitations in the geometrically nonlinear behavior that they accurately capture. In addition, it is important to note that simplified $P-\Delta$ procedures do not typically include $P-\delta$ effects and, given the assumption of small deflections between member ends, they are limited in their ability to model slender structures that respond nonlinearly to second-order effects.

Amplification factors are often used in design to obtain approximate values of second-order moments by increasing first-order moments. For example, Chapter C of the AISC (2005b) specification includes separate amplification factors, B_1

and B_2, to account for $P-\delta$ and $P-\Delta$ effects, respectively. The application and development of these factors are discussed at length in the commentary of the AISC (2005b) specification. It is important to recognize that these factors are approximate and that in many situations they may not capture the full extent of second-order effects. For example, because the factors are developed for two-dimensional subassemblies with idealized boundary and loading conditions, they are cumbersome to apply in frames that have interdependent $P-\delta$ and $P-\Delta$ effects or for irregular and three-dimensional structures. In particular, the B_2 factor is story based, and it is therefore difficult to apply to frames with split story levels. Moreover, they often do not account directly for amplification of girder moments and/or axial forces in braces and columns. As a result, equilibrium may not exist at beam-to-column connections and in the adjoining beams as the result of amplifying only the column moments.

16.2.2 Inelastic Analysis

For the present discussion, inelastic behavior will be limited to consideration of rigidly connected steel frame structures in which the primary effect is yielding of members due to combined axial load and bending moments. Using an incremental stiffness analysis, such as that described above for second-order analysis (Eq. 16.1), inelastic effects may be included by modifying the incremental stiffness matrix $[K]$ to account for member plastification. Inelastic models are usually implemented in the same manner, whether the analysis is first- or second-order. Combined inelastic and second-order effects near the strength limit state generally require more robust solution methods. In addition, an incremental solution procedure is required to trace the applied load history because inelastic behavior is load path dependent.

Types of inelastic beam-column models can be broadly classified, depending on the degree to which they model the progression of yielding in the members due to the combined effects of axial load, moments, and residual stresses. Basic methods that do not model any spread of plasticity in the member are often termed *plastic hinge* or *concentrated plasticity models*, and those that do model gradual yielding through member cross sections and along their length are termed *spread-of-plasticity, distributed plasticity*, *plastic zone*, or *fiber models*. For evaluating frame stability, an important distinction between concentrated and distributed plasticity methods is that the latter capture reductions in member stiffness prior to full plastification, which can affect second-order deformations and internal forces and moments.

Basic aspects of inelastic behavior are illustrated by the frame in Fig. 16.4, which is subjected to proportional vertical and lateral loads. In this example, second-order effects are not considered. Response curves, shown in heavy lines in Figs. 16.4*c* and *d*, correspond to an elastic–perfectly plastic hinge analysis, while the light lines correspond to a distributed plasticity analysis. Assuming that the columns fully yield before the beam, the first hinge will form at the top of the right-hand column due to combined moment and axial compression. Upon further loading, the plastic limit load is reached with the formation of a second hinge in the left column that

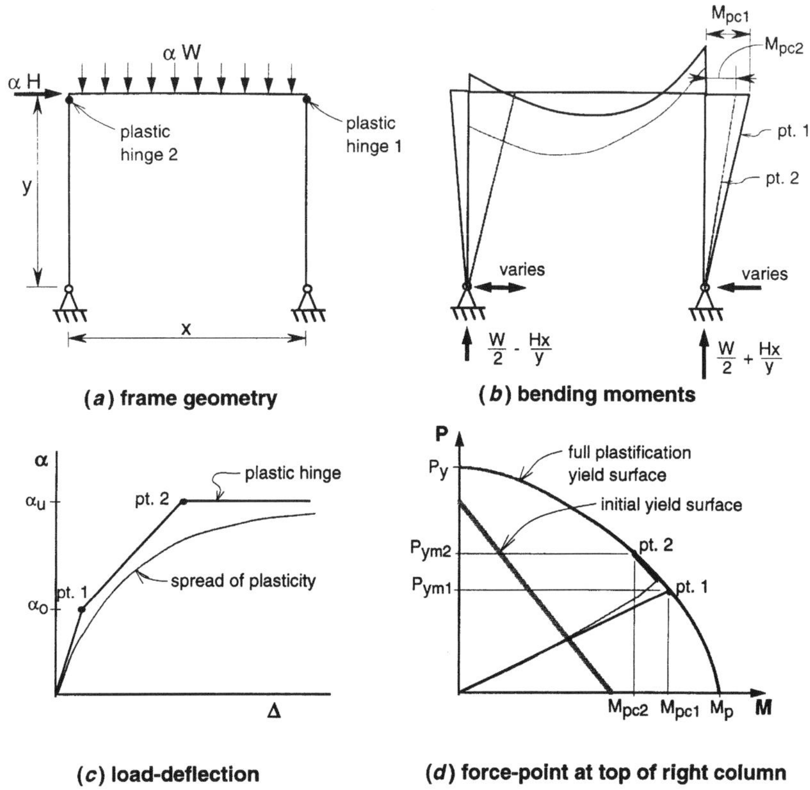

FIGURE 16.4 Example of elastic–plastic force redistribution.

creates a sidesway mechanism. Referring to Fig. 16.4*d*, the first hinge is detected when the combination of axial load and moment at the top of the right column (point 1) reaches the full-plastification yield criterion (surface) of the cross section. The internal bending moment diagram at this stage is shown by the heavy line in Fig. 16.4*b*. As shown in Fig. 16.4*c*, the structure continues to carry additional load until a second hinge forms at the top of the left column, indicated as point 2 in the figure. Referring to Fig. 16.4*b* and *d*, during loading from point 1 to 2, the right column redistributes some of its moment so that it can support additional axial load. The increased axial load is directly proportional to the vertical column base reactions, which are statically determinate as shown in Fig. 16.4*b*. By the time the second hinge forms in the left column the axial load resisted by the right column force increases from P_1 to P_2, and in adherence with the yield criteria the moment decreases from M_1 to M_2.

In accordance with the upper-bound plasticity theorem, the limit load obtained using first-order analysis, whether an elastic–perfectly plastic hinge or a distributed plasticity analysis, will be equal to that obtained by a plastic mechanism analysis (assuming that the analyses are all made using the same full-plastification criterion and that strain hardening is neglected). On the other hand, limit loads based on different types of second-order inelastic analyses are, in general, not equal to each other because of the interaction of spread-of-plasticity and second-order effects. In some frames this interaction can lead to large differences in second-order inelastic limit points obtained from plastic hinge versus distributed plasticity analyses.

Plastic Hinge Methods For framed structures, plastic hinge analyses are usually based on a yield criterion that considers longitudinal normal stresses due to axial loads and moments, often neglecting shear stresses due to shear forces and torsion. Although in theory it is possible to develop exact expressions for member cross-section yield criteria (see, e.g., Chen and Atsuta, 1976), most plastic hinge methods rely on simplified force interaction expressions to approximate the yield surface. For two-dimensional analyses either piecewise linear or continuous functions are commonly used, whereas for three-dimensional analyses, continuous functions are usually preferred over multifaceted surfaces. The following is an example of a yield surface presented in McGuire et al. (2000) for use in three-dimensional analyses:

$$p^2 + m_z^2 + m_y^4 + 3.5p^2 m_z^2 + 3p^6 m_y^2 + 4.5m_z^4 m_y^2 = 1.0 \tag{16.3}$$

where $p = P/P_y$, $m_z = M_z/M_{pz}$ (major axis bending), and $m_y = M_y/M_{py}$ (minor axis bending). As described in McGuire et al. (2000), features of this yield surface equation that make it amenable to computer-based inelastic analyses are that it is a smooth, continuous, and convex function. The above yield surface can accurately model the biaxial bending of light- to medium-weight wide-flange sections. The development of yield surface functions is provided by Orbison et al. (1982), Duan and Chen (1990), and Attalla et al. (1994).

The stiffness equation for a plastic-hinge analysis has the same form as Eq. 16.1, with the stiffness matrix $[K]$ updated during the analysis to reflect the formation of plastic hinges in the structure, and the unbalanced force vector $\{dR\}$ corrects for instances where the calculated element forces from the previous analysis increment violate the member yield criterion.

Ziemian and McGuire (2002) and Ziemian et al. (2008) developed a plastic hinge model that effectively captures the inelastic response of a beam-column element with partial yielding. In their *modified tangent modulus approach*, the base beam-column elastic–plastic hinge idealization is enhanced by using equivalent reduced member flexural stiffnesses EI_z and EI_y that vary as a function of the axial force and major and minor axis bending moments. The base beam-column elastic plastic hinge model is formulated using a fully plastic strength envelope that represents the cross-section strength interaction between the axial force and biaxial bending for doubly symmetric I-section members, assuming an elastic–perfectly plastic material. For cross sections where the strength envelope is reached, the plastic interaction between axial force and the bending moments is represented using plasticity concepts. This formulation can provide a more accurate estimate of member capacities in an inelastic analysis than a typical plastic hinge approach.

Distributed Plasticity Models At the member cross section, spread of plasticity can be idealized as a nonlinear moment–curvature response. Shown in Fig. 16.5*a* are two force interaction surfaces, the inner of which signifies the initiation of yielding calculated as a function of the cross-section area (A), elastic section moduli (S_y and S_z), residual stresses (F_{rc}), and material yield strength (F_y). The outer or full-plastification surface follows the same yield surface criteria as described previously for elastic–plastic hinge analyses. Considering a member under constant axial force and increasing moment (Fig. 16.5*b*), the resulting moment–curvature behavior is shown in Fig. 16.5*c*, where ϕ_e and ϕ_p refer to the elastic and plastic curvatures, and M_{rc} and M_{pc} refer to the initial and full-plastification moments in the presence of an axial force P. The nonlinear response shown in Fig. 16.5*c* reflects spread of plasticity through the member cross section for bending about

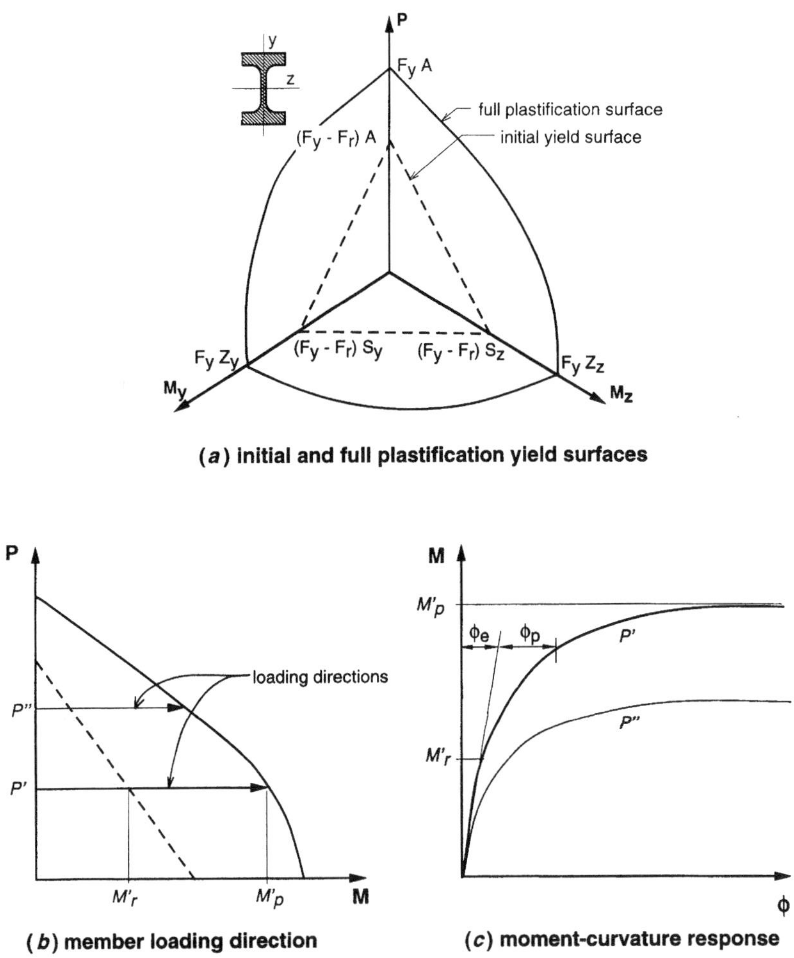

FIGURE 16.5 Inelastic cross-section behavior.

one axis under two levels of axial loads. Note that the departure from linearity begins sooner and is more significant for members with larger axial load. In general, the behavior shown in Figs. 16.5*b* and *c* can be extended to include nonlinear moment–curvature response about both bending axes under any arbitrary sequence of loading.

In addition to yielding through the cross section, spread of plasticity involves a progression of yielding along the length of the member. The length of the plastified regions are based on the initiation of yielding as calculated using an initial yield surface such as the one shown in Fig. 16.5*a*. Using force–deformation response from the cross sections (Fig. 16.5*c*), the stiffness of the member can be calculated by numerically integrating the inelastic cross-section properties along the length of the member. Many methods have been proposed for implementing spread-of-plasticity effects in frame analysis software, ranging from detailed fiber-type models to more computationally efficient procedures based on variants of the plastic hinge concept (i.e., by using stress-resultant or force–deformation models).

Fiber-type models involve discretization of the member cross sections into small areas that are considered as longitudinal fibers of the member. By applying the assumption that plane sections originally normal to the reference axis of the member remain plane and normal (with the exception of warping deformations due to nonuniform torsion), the strains in the longitudinal fibers can be calculated from the axial strain and curvatures of the section. Then, by relating the strains to stresses using an appropriate constitutive law and considering the residual stresses, the cross-section forces can be obtained by numerically integrating the stresses over the cross section (e.g., a simple numerical integration rule would be $P = \sum \sigma_i A_i$, $M_y = \sum \sigma_i A_i z_i$, and so on, where A_i is the area of fiber i, σ_i is the stress in each fiber, and z_i is the distance in the z-direction from the fiber to the centroidal bending axis). At any level of loading, inelastic cross-section stiffness can also be calculated by integration. For example, the tangent flexural stiffness can be calculated simply as $\sum E_i A_i z_i^2$, where E_i is determined from the stress–strain constitutive law for the material. Neglecting strain hardening by considering an elastic–perfectly plastic material behavior, E_i for any fiber will either be equal to the elastic modulus or zero. Axial and flexural stiffness of the cross section obtained in this way are then integrated along the length of the member to develop element tangent stiffness matrices.

Examples of fiber element implementations include Alvarez and Birnstiel (1969), El-Zanaty et al. (1980), Clarke (1994), and Alemdar and White (2005). While the basic formulations for fiber element methods are fairly straightforward, their direct implementation can require excessive computational requirements. In seeking an intermediate solution that has the computational efficiency of elastic–perfectly plastic hinge methods and the accuracy of spread-of-plasticity fiber methods, several researchers developed *quasi-hinge* or *stress-resultant constitutive models* (e.g., Liew et al., 1993; Attalla et al., 1994; El-Tawil and Deierlein, 1996). Although subject to some limitations of required calibration, such methods have been shown to make spread-of-plasticity analyses practical for large (realistic) building frames. Advances in these models extend the response to include three-dimensional stability behavior, such as the formulation by Nukala and White (2004). Additionally, fiber models utilizing flexibility (e.g., Neuenhofer and Filippou, 1997) or mixed element formulations (e.g., Spacone et al., 1996; Nukala and White, 2004), have been developed that better represent the nonlinear curvature fields in a yielded member and consequently more accurately capture the inelastic behavior than a displacement-based element (Alemdar and White, 2005).

Researchers have also examined alternate computational approaches designed to implement distributed plasticity-based solutions on larger-scale systems. Parallel and vector processing has been utilized to trace the complete load–deformation response of steel frames using the distributed plasticity methodology for large-scale fully and partially restrained frames (Foley, 1996, 2001; Sotelino, 2003). While vector computing has been relegated to obsolescence, parallel and distributed processing on networks of computers may enable the practical use of distributed plasticity methods.

16.2.3 Critical Load Analysis

Given that for most buildings the elastic critical loads are considerably higher than the inelastic limit point, critical load analyses are often of limited practical value for frame design. They have proved useful in determining amplification factors to approximate second-order effects; with modern second-order analysis software, however, such methods are of less importance. As described below and noting that methods have been proposed to improve critical load analyses to reflect inelastic behavior (Ziemian, 1999), most of these still provide only an estimate of the upper bound stability limit of a frame under a given loading. Methods of critical load analysis are reviewed here briefly given their traditional role in stability analysis and because they are still useful as screening tools for stability limits. Details for such an analysis are presented by McGuire et al. (2000) and in Chapter 20.

Using a stiffness matrix formulation, critical loads of frames and other structures may be calculated by solving the eigenvalue problem

$$[K + \mu_i K_g]\{\Delta_i\} = \{0\} \tag{16.4}$$

where $[K]$ is the structural stiffness matrix, $[K_g]$ the geometric stiffness matrix, μ_i the eigenvalue (critical load index), and $\{\Delta_i\}$ the corresponding eigenvector (buckled mode shape). The geometric stiffness matrix $[K_g]$ is typically of the same form as that employed in a second-order analysis. In this case, however, the member forces that comprise the geometric stiffness are calculated by a first-order analysis under a prescribed external load distribution $\{F\}$. Thus the resulting critical load is unique for the given loading condition and based on the distribution of internal forces that were calculated with the undeformed geometry of the structure. Note that in most cases the loading applied for critical load analysis is an idealized representation of the gravity loads that may not necessarily reflect the internal force distribution at the true limit point of the structure.

The critical load vector calculated by Eq. 16.4 is

$$\{F_{cr,i}\} = \mu_i\{F\} \tag{16.5}$$

where $\{F_{cr,i}\}$ is proportional to the prescribed load vector $\{F\}$. The subscript i in Eqs. 16.4 and 16.5 refers to the fact that there are as many buckling loads (eigensolutions) to the problem as there are number of degrees of freedom in Eq. 16.4. While the lowest (minimum) critical value is usually the one of interest, there may be instances where higher buckling modes provide useful information.

In addition to the rigorous solutions following the eigenvalue equations presented above, many simplified methods of determining critical loads of frames have been proposed (e.g., Stevens, 1967; Horne, 1979; Nair, 1987; ASCE Task Committee, 1997). Most of these are based on story-by-story checks with other simplifying behavioral assumptions.

Tangent-Modulus Adjustment In most applications, the structural stiffness used in Eq. 16.4 is the elastic stiffness, and the resulting critical load $\{F_{cr,i}\}$ corresponds to elastic buckling. By modifying $[K]$ using a tangent-modulus approach, however, one can obtain inelastic critical loads that more closely represent the actual limit state of the structure under pure axial loading. As described by Ziemian (1999), Eq. 16.4 can be adjusted to approximate the reduction in column stiffness due to axial compression and residual stresses in a manner similar to the method for adjusting effective column buckling lengths that was originally suggested by Yura (1971) and included in the AISC specification (AISC, 2005b). In this approach, the stiffness matrix $[K]$ in Eq. 16.4 is based on member stiffness that is calculated using a tangent modulus E_t for each member that is consistent with the axial force in that member at the critical load $\{F_{cr}\}$. Values of E_t can be calculated using either tangent-modulus formulas derived from the assumed residual stresses in the member or semiempirical formulas derived from inelastic column curves. The column stiffness reduction factors given in Part 4 of the AISC manual (AISC, 2005b) are one example of the latter type of adjustment. Because the terms in the stiffness matrix $[K]$ would be dependent on the magnitude of the applied load vector $\{F_{cr}\}$, the tangent-modulus adjustment necessitates an iterative solution procedure to solve Eq. 16.4. It is important to recognize that while this adjustment will result in critical loads closer to the inelastic limit point, the adjustment still does not take into account the loss in stiffness due to plastification of members under combined axial load and bending and the formation of plastic hinges.

16.3 FRAME BEHAVIOR

Technical Memorandum No. 5 of the SSRC (see Appendix B.5), which outlines general principles for stability design of metal structures, states:

Whenever possible, the procedure for the establishment of the load carrying capacity of frames, members, or elements on the basis of maximum strength should be based on a mathematical model which incorporates:

1. Experimentally determined physical characteristics, such as residual stresses, material nonlinearities, ... rationalized as may be appropriate.

2. A statistically appropriate combination of acceptance characteristics that are specified in supply, fabrication, and erection standards, such as out-of-straightness ... material properties, and erection tolerances.
3. Effect of boundary conditions, such as restraint applied to the end of members.

These fundamental guidelines form the starting point for assessing the most important attributes of frame stability and for evaluating how different methods of analysis and associated design procedures account for this behavior. In the following sections, considerations for analyzing frame behavior are described through a simple example, which is then followed by a more comprehensive summary of behavioral effects and modeling issues associated with frame stability assessment.

16.3.1 Planar Frame Example

Frame behavior and methods of stability analyses are reviewed by considering the frame shown in Fig. 16.6, which is based on studies by Iffland and Birnstiel (1982) and Ziemian et al. (1992a). The frame geometry and loading are representative of low-rise industrial structures, in which stability effects are accentuated by a high ratio of gravity to lateral load. Member sizes shown in Fig. 16.6 represent a least-weight design based on second-order inelastic analyses to satisfy minimum strength and service load criteria. Note that the leftmost W8 column of the frame is intentionally made smaller than the other columns and is stabilized by leaning on the stronger W14 columns.

The frame members are proportioned according to the following design criteria:

1. *Serviceability.* The roof and interstory drift indices under service wind loads are limited to 1/400 and 1/250, respectively, and the beam deflection index

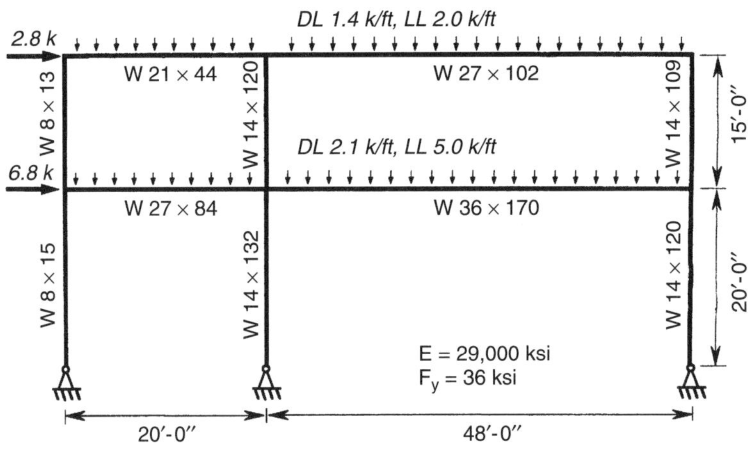

FIGURE 16.6 Example frame: geometry and loading.

under service live load is limited to 1/360. In addition, plastic hinges are not permitted to form under service loads.

2. *Strength.* A second-order inelastic analysis is used to check that the frame can safely resist the factored loads. Based on the AISC LRFD (1993) design criteria at the time of the frame study: load combinations of gravity and wind loads were investigated, including $1.2D + 1.6L$ and $1.2D + 0.5L + 1.3\,W$; the full-plastification strengths of member cross sections were reduced by resistance factors ($\phi = 0.9$ for axial tension and plastic moments, $\phi = 0.85$ for axial compression); and all members were required to meet section slenderness and material criteria for compact plastically designed structures.

Implied by these criteria is the assumption that the second-order inelastic structural analysis alone suffices for checking strength and stability of the frame and its members. Thus, subject to the assumptions that all members are rigidly connected and fully braced out of plane, the nonlinear analysis eliminates the need to check member strengths by beam-column interaction equations, such as in Chapter H of the AISC specification (2005b).

Response of the frame under gravity load ($1.2D + 1.6L$) is summarized in the plot of the applied load ratio versus displacement and the plastic-hinging sequence shown in Fig. 16.7*a*. Because of its asymmetric geometry, the frame begins to sway to the left until the first hinge forms at an applied load of 74% of the full factored load. Upon further loading, more hinges form and the frame stiffness changes dramatically, such that it begins to deflect to the right. Finally, with formation of the sixth hinge at a load ratio of 1.0 (100% of the factored load), the frame reaches its inelastic stability limit and fails by swaying back to the left. Results of this plastic hinge analysis are compared to those of a spread-of-plasticity (plastic zone) analysis in Fig. 16.7*b*. Also shown in Fig. 16.7*b* is a diagram indicating the percentage of cross-section yielding calculated at the inelastic limit point from the spread-of-plasticity analysis. Although the spread-of-plasticity solution indicates a less abrupt load–deformation response, both analyses give essentially the same stability limit strength. This is despite the fact that there is significant spread of plasticity in several columns and beams (Fig. 16.7*b*) that is not accounted for in the plastic hinge analysis (Fig. 16.7*a*).

The following observations are made from these analyses:

- At the inelastic limit point, a classical *plastic mechanism* has not formed. Rather, the instability is a function of the combined destabilizing effects of material and geometric nonlinear behavior.
- The distribution of internal moments at the limit point is quite different from the first-order elastic distribution. As described by Ziemian et al. (1992a), plastic hinge formations cause unloading of moments and/or axial forces in some members under increasing external load.
- Stability at the limit point is based solely on satisfying equilibrium on the deformed geometry with maximum internal forces limited to the fully plastified

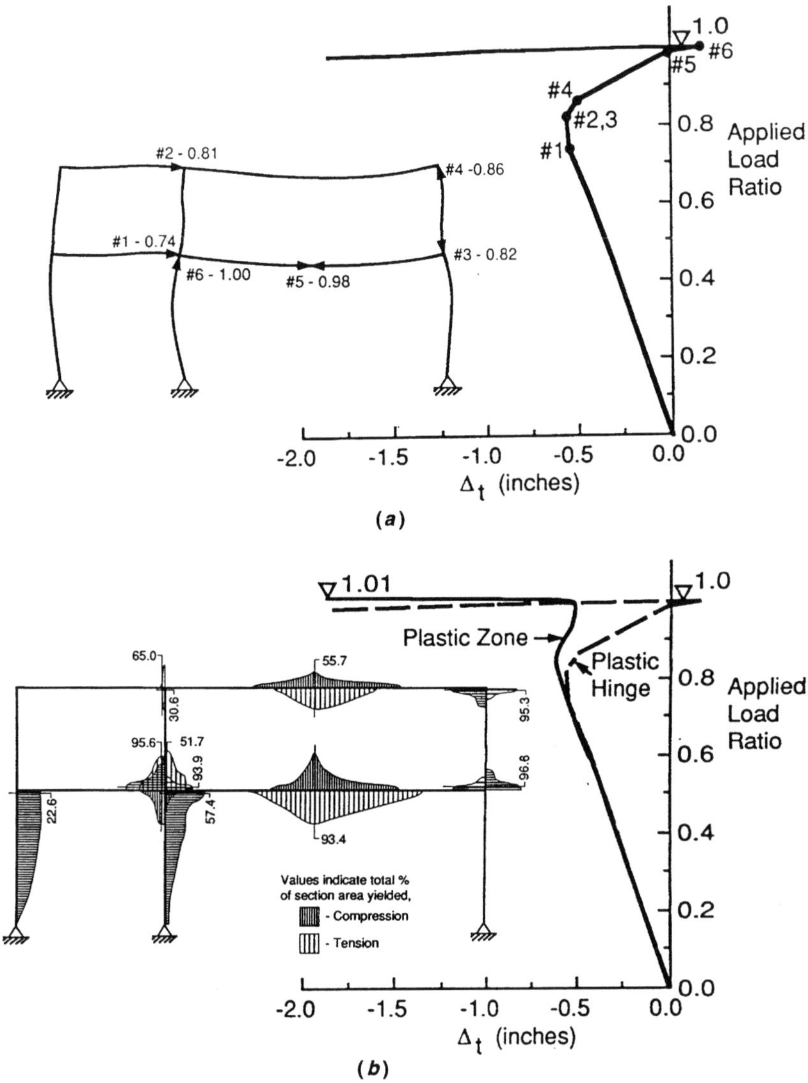

FIGURE 16.7 Example frame: (*a*) second-order elastic–plastic hinge analysis under factored gravity loads; (*b*) comparison of spread-of-plasticity (plastic zone) and plastic hinge results (Ziemian, 1991).

strength of the member cross sections. As noted above, the nominal yield strengths are reduced to design strengths by application of the AISC LRFD resistance factors to account for statistical variations in cross-section strengths.

Response curves for the frame under the factored gravity and wind loads using the four methods of analysis described previously are shown in Fig. 16.8. Under

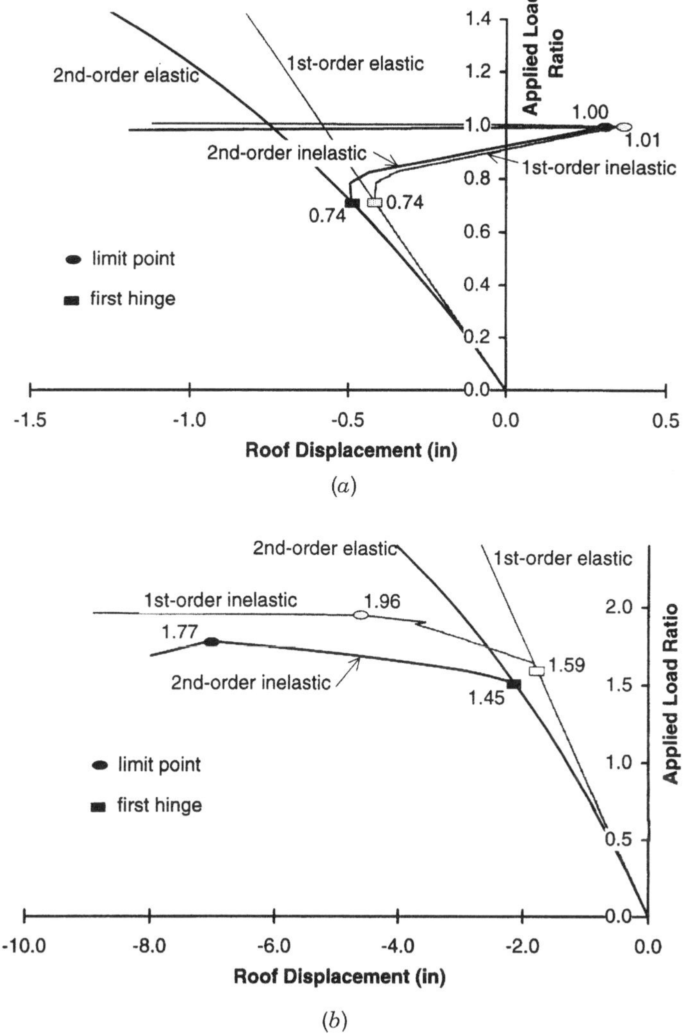

FIGURE 16.8 Comparison of analysis results for two-story frame: (*a*) gravity loading; (*b*) gravity plus wind loading.

gravity loading (Fig. 16.8*a*), there is only a slight (1%) difference in the limit point obtained using a first- versus second-order inelastic analysis. This is due to the fact that the net lateral displacement and resulting $P-\Delta$ effects are small at the limit load. There are significant inelastic $P-\delta$ effects in the light bottom-story column on the left-hand side, but these are not large enough to precipitate a "braced buckling" failure of that member prior to sway failure of the frame. As a result, lateral sway of the frame does not have a significant effect until the point when the abrupt sidesway collapse occurs. On the other hand, the second-order behavior reduces the inelastic limit strength by approximately 10% under combined gravity and lateral loading (Fig. 16.8*b*). These observations depend on the specific characteristics of this frame, and relative differences between first- and second-order response curves should not be generalized to other cases.

The analyses above do not include the effects of geometric imperfections. Based on the maximum allowable fabrication and erection tolerances as provided in the *Code of Standard Practice for Steel Buildings and Bridges* (AISC, 2005a), the geometric imperfection effects can be checked by imposing a uniform frame out-of-plumbness of $H/500$ on both stories of the building. To check for the significance of these imperfections, second-order plastic hinge analyses were rerun for cases where the initial geometry was modified to include an $H/500$ story imperfection. For the governing gravity load case, the sidesway imperfections do change the initial elastic sway of the frame but cause essentially no change in the inelastic strength limit for either the gravity or gravity plus wind load combinations (Ziemian, 1991).

Assuming that the second-order inelastic analyses provide an accurate representation of the stability limit strength of the frame, it is instructive to contrast these results with those from second-order elastic analyses. The second-order elastic response curves are superimposed on the inelastic curves in Fig. 16.8, although for presentation they have been truncated well below the elastic stability limit point. Had the full curves been plotted, they would asymptotically approach the elastic *stability limit loads* at an applied load ratio of 3.4 times the factored gravity loads (Fig. 16.8*a*) and at 6.6 times the factored gravity plus wind loads (Fig. 16.8*b*). Note that these limits, which are calculated on the deformed geometry, are close but not exactly equal to the *elastic critical load* ratios of 3.2 and 6.3 calculated by elastic eigenvalue analyses. The elastic limit load ratios of 3.4 and 6.6 are roughly three and one-half times larger than the corresponding second-order inelastic limit load ratios of 1.00 and 1.77. Differences between the elastic and inelastic limit points are considered to be one measure of the significance of geometric imperfection and spread-of-plasticity effects on frame behavior.

While elastic stability limits are often cited as a measure of frame stability, the internal force distribution and deformations for most practical frames at the elastic stability limit can be very different from those at the inelastic limit. Consider, for example, the two bending moment diagrams from the second-order elastic analysis of the frame under gravity loads shown in Figs. 16.9*a* and *b*. The moment distribution in Fig. 16.9*a* is at an applied load ratio of 1.0 (i.e., a load equal to the factored design load of $1.2D + 1.6L$) and in Fig. 16.9*b* is for a load ratio of 3.2 close to the elastic limit load. Comparing the two diagrams in Fig. 16.9, it is apparent that for some members the second-order behavior increases the first-order moments, whereas in other members the first- and second-order effects counteract one another. These counteracting effects are one reason why the frame was not sensitive to out-of-plumb imperfections. Finally, the highly nonlinear nature of the response near the stability limit load is why limits are often imposed on the use of approximate methods to calculate second-order forces by amplifying results from first-order analyses. For example, Eurocode 3 (CEN, 2005) limits the use of simplified second-order amplification factors to cases where the ratio of the critical

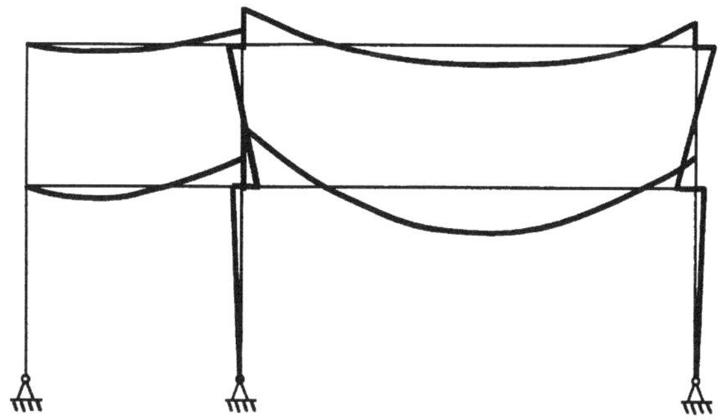

(*a*) factored design loads, applied load ratio = 1.0

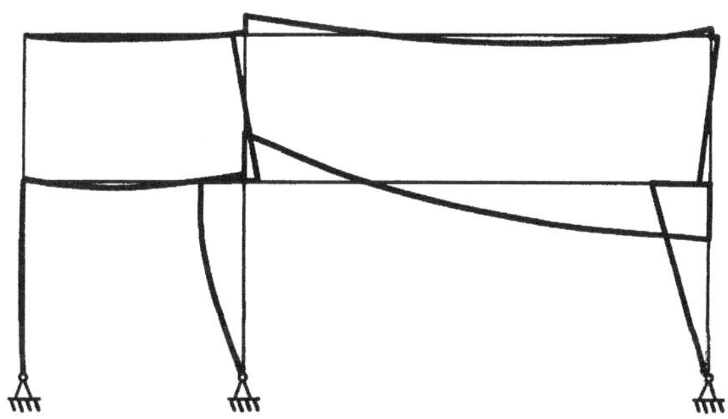

(*b*) near elastic limit load, applied load ratio = 3.2

FIGURE 16.9 Bending moments from second-order elastic analysis under gravity loads.

load of the frame to the factored design load, F_{cr}/F_u, is greater than or equal to 3. This translates to a limit on the second-order amplification factor of $AF \leq 1.5$, given that $\Delta F = 1/(1 - F_u/F_{cr})$.

The significance of second-order effects on the frame can be evaluated considering the data for the first-story drift and column moments calculated by first- and second-order elastic analyses summarized in Table 16.1. Under gravity load, the second-order story drift amplification is 44%, while the second-order column moments vary from +4 to −11% compared to the first-order values. Under gravity plus wind, the deflections increase +17%, while the change in moments range between +8 and −93%. This is an example in which distributed gravity loads, differences in the relative column stiffness, and gravity-induced sidesway make it cumbersome to calculate second-order forces accurately using approximate member-based amplification factors. Instead, it is clearly preferable to calculate these effects using a rigorous computer-based second-order analysis.

A final point regarding these analysis results is that a large amplification of deflections and moments due to elastic second-order effects does not necessarily result in a large change in the inelastic limit strength of the frame. This is an

TABLE 16.1 Comparison of Elastic First- and Second-Order Analysis Results at Factored Design Loads for Bottom Story

				Column Moments (kip-in.)								
	Story Drift (in.)			Left Column			Center Column			Right Column		
Loading	First Order	Second Order	Percent Change	First Order	Second Order	Percent Change	First Order	Second Order	Percent Change	First Order	Second Order	Percent Change
$1.2D + 1.6L$	0.36	0.52	44	28	29	3	3540	3690	4	3570	3180	−11
$1.2D + 0.5L + 1.3W$	0.93	1.09	17	70	72	3	3190	3450	8	330	22	−93

important point because most member-based design checks require an increase in member capacity to resist second-order moments that are often calculated in proportion to the drift amplification. Such procedures would be inconsistent for the previous frame, where despite rather large changes in the calculated first- and second-order *elastic* deflections and moments, the difference between the first- and second-order *inelastic* limit loads was much less. Referring back to Fig. 16.8, the second-order effects decrease the inelastic limit load by only 1% for the gravity load case and 10% for the gravity plus wind load case.

This example highlights some fundamental, yet sometimes subtle aspects of behavior that can affect the stability design of a relatively simple frame. The example is also intended to draw attention to aspects of behavior of indeterminate frames that are not necessarily captured in single-bay (often, single-story) frames that are the focus of much of the literature on frame stability. As this example shows, there are important aspects of behavior of practical frame systems that are not captured in simpler, less redundant structures. Finally, the example demonstrates some of the inherent limitations of applying elastic stability theory to structures in which the strength limit state is reached at inelastic limit loads that are much lower than elastic limit loads.

16.3.2 Factors Influencing Frame Stability

As presented in Tables 16.2A and 16.2B, there are various factors affecting frame stability that should be considered in design. The tables are adapted from similar summaries previously compiled by Birnstiel and Iffland (1980), McGuire (1992), and White and Chen (1993), and they follow the principles of Technical Memorandum No. 5. As indicated, the factors are distinguished between those related to the physical attributes of the structure and loading (Table 16.2A) and how the response phenomena are modeled in analysis and design (Table 16.2B). The categorization of the factors is imperfect because many of the effects are interrelated, but the listing is intended to provide some guidance on those effects that may influence behavior on typical building frames. The tables are by no means a comprehensive list of all possible effects that may arise in practice, and the degree to which various factors are significant for a given structure will vary from case to case. Whereas some factors are considered routinely in analysis, others are included in design implicitly through specification provisions and established practice.

The following is a summary of how the various factors from Tables 16.2A and 16.2B are addressed in the second-order elastic–plastic hinge analysis for the frame example presented in Section 16.3.1.

1. *Physical attributes and basic modeling.* The structure is modeled as a planar frame using centerline dimensions and ideal geometry. It is assumed that (1) the beam-column connections are fully restrained (rigid) moment connections, (2) the frame geometry and member proportions are such that the finite joint size and panel zone deformations are not significant, (3) column foundations provide negligible rotational restraint and any minor settlements

TABLE 16.2A Factors Affecting Steel Frame Stability: Physical Attributes of Structure and Loading

Frame geometry and configuration
Centerline framing dimensions
Member geometry and material
Connection details
Foundation and support conditions
Shear connections to slab
Infill walls or secondary structural elements
Finite member and joint size effects
Out-of-plane bracing elements
Material properties
Elastic moduli
Expected versus nominal strengths
Ductility and fracture toughness
Geometric imperfections
Erection out-of-plumbness
Member out-of-straightness
Incidental connection or loading eccentricities
Internal residual stresses
From manufacturing/fabrication processes
From erection fit-up
From construction sequencing
From incidental thermal loadings or support settlement
Loading
Magnitude and distribution
Loading rate and duration

will not affect frame behavior significantly, and (4) member properties are based on the bare steel members, assuming that they are designed neglecting composite action with the slab. Obviously, each of these assumptions is a simplification of the actual behavior and could be reconsidered in a more refined analysis. Residual stresses are not considered in the plastic hinge analysis; this assumption is justified by the close agreement with the more refined spread-of-plasticity analysis that did include residual stresses in the members (Fig. 16.7*b*). Erection out-of-plumbness was also shown to have a negligible effect on the inelastic limit point, and other geometric imperfections are assumed to be insignificant.

2. *Response phenomena.* The basic analysis model includes axial and flexural deformations in the members. Members are compact, and thus local buckling can be neglected. All members are assumed to be braced out-of-plane along their lengths, and thus torsional–flexural effects are not considered. Although this assumption is generally true for the beams, it is usually not true for columns and additional checks on the out-of-plane stability of the columns should be made. Loads are assumed to be static and are

TABLE 16.2B Factors Affecting Steel Frame Stability: Modeling Parameters and Behavioral Assumptions

- Linear elastic response
  - Flexural, axial, and shear deformations of members
  - Deformations of connections and beam-column panel zones
  - Uniform torsion and/or nonuniform warping torsion deformations in members
  - Foundation and support movement
  - Dynamic and inertial effects
- Geometric nonlinear response
  - $P-\delta$ effects: influence of axial force on stiffness and internal moments of beam-columns
  - $P-\Delta$: effects: influence of relative joint displacements on forces and displacements
  - Local buckling and cross-sectional distortion
  - Finite-rotation effects (three-dimensional structures)
- Material nonlinear response
  - Member plastification under the action of axial force and biaxial bending (spread of plasticity versus plastic hinge idealizations)
  - Member plastification due to shear forces, uniform torsion, and nonuniform warping torsion (bimoments)
  - Yielding in connection components and joint panels
  - Fracture of members and connections
  - Strain-hardening behavior
  - Cyclic plasticity effects
  - Load path effects, shakedown, and incremental collapse

applied proportionally. The geometric nonlinear response, modeled through an updated Lagrangian approach, includes both the $P-\Delta$ and $P-\delta$ effects (see McGuire et al., 2000). In this case, each column was discretized into two elements to capture the $P-\delta$ amplification of moments with reasonable accuracy. Material nonlinear effects are included through an elastic–perfectly plastic hinge model that is based on the full plastification cross-section strength of members considering the interaction of axial force and bending moments. Nominal yield strengths are used in the yield surface equations; strain hardening is neglected, and steel ductility and toughness are assumed sufficient to avoid fracture under the required inelastic deformation demands. Note that while in this case the accuracy of the plastic hinge analysis is verified by comparison to more detailed spread-of-plasticity analyses, it is not anticipated that such detailed verification would necessarily be applied in practice. Further discussion and guidance on gauging the significance of spread of plasticity are included in Section 16.4.

3. *Uncertainties.* Variability in the loads and member strengths is considered by incorporating load and resistance factors in the analysis. Resistance factors are included by reducing the yield surface criteria in the plastic hinge model. The potential for incremental collapse under repeated loading is assumed small, although this has not been verified.

Because it is impractical and often unnecessary to include all behavioral effects in the analysis, the engineer must judge the degree to which the analysis accurately models the important stability effects to ensure a safe design. The remainder of this chapter includes further discussion of a few of the more important factors.

16.3.3 Modeling Initial Geometric Imperfections

Geometric imperfections are often idealized as the combination of member out-of-straightness and frame out-of-plumbness. The true imperfections in actual structures, however, are less straightforward, often complicated by (1) actual column piece lengths and splice locations, (2) beam length and connection fit-up tolerances, (3) finite member size effects, (4) unavoidable eccentricities at foundations, (5) three-dimensional geometry, and (6) imperfect load placement, among other factors. With this in mind, frame imperfections cannot be idealized with the same precision that is done, for example, in modeling out-of-straightness in axially loaded columns. Fortunately, for most frames the applied lateral loads and resulting member forces are large enough that the effects of initial geometric imperfections are not critical; Surovek and Johnson (2008) provide several examples of this. Nevertheless, the engineer should consider the types and magnitude of imperfections that may occur and whether they are significant to stability of the structure.

Fabrication and Erection Tolerances In the absence of more accurate information, evaluation of imperfection effects should be based on the permissible fabrication and erection tolerances specified in the appropriate building code. In U.S. practice, for example, limits of member out-of-straightness and frame out-of-plumbness are prescribed in the *Code of Standard Practice for Steel Buildings and Bridges* (AISC, 2005a). The basic limits specified by AISC that should be fairly representative of good practice elsewhere are as follows:

- *Member out-of-straightness*: $\delta_0 < L/1000$, where L is the distance between brace points.
- *Interstory out-of-plumbness*: $\Delta_0 < h/500$, where h is the story height.
- *Maximum lack of verticality*: $\Delta_{0,\max} < 1$ in. (25 mm) below the 20th floor, $\Delta_{0,\max} < 2$ in. (50 mm) above the 20th floor.

In general, one could expect the actual imperfections to be less than the maximum permissible values given above. Note, for example, that the maximum member out-of-straightness of $L/1000$ is 50% larger than the out-of-straightness of $L/1500$ assumed in the development of SSRC column curve 2P, which as indicated in Chapter 3 is the basis of the column curve in the AISC specification (AISC, 2005b). Because a certain amount of imperfection is permitted, and the engineer has limited control over the final as-built imperfections in a structure beyond the specified tolerances, it is suggested that when initial imperfection effects are found significant, maximum erection tolerances be used as the basis of frame stability checks in design. In particular, checks for individual story instabilities should be checked using the maximum out-of-plumb tolerance.

Modeling of Imperfections with Notional Loads The most direct method of modeling geometric imperfections is by incorporating the imperfections into the geometrical description of the analytical model. Alternatively, the internal forces caused by initial imperfections can be calculated by applying equivalent lateral forces to the frame. These forces are often referred to as *notional loads* and are discussed further in Section 16.4.1. Consider the bar–spring structure shown in Fig. 16.10*a*, which includes an initial out-of-plumbness Δ_0. In Fig. 16.10*b* the initial imperfection is represented by the notional load of $P\Delta_0/h$, which produces internal first-order sway forces that are statically equivalent to those induced by vertical load P in Fig. 16.10*a*. By solving the equations of equilibrium on the deformed geometry for both cases, it can be shown that the resulting total second-order forces are equal, although the final deformed geometries differ by Δ_0. The resulting moment in the rotational spring is $M_s = 0.5P\Delta_0/(1 - 0.5Ph/k)$, where the denominator term represents the second-order amplification of the first-order spring moment, $M_{s1} = 0.5P\Delta_0$. These observations for this simple bar–spring structure generally hold true for elastic second-order moments created by initial out-of-plumbness in frames (ASCE Task Committee, 1997). Moreover, verification studies using spread-of-plasticity analyses have shown that notional loads can accurately represent initial out-of-plumbness effects for inelastic second-order analysis (Clarke and Bridge, 1992, 1996; Maleck and White, 1998).

Notional loads can be used to model story out-of-plumbness, as shown in Fig. 16.11*a*, where the story notional load F_{n1} is equal to

$$F_{n1} = \frac{\Delta_0}{h} \sum P_u \qquad (16.6)$$

and where $\sum P_u$ is equal to the summation of factored gravity loads added to the columns by the story in question, including all "gravity" or "leaning" columns

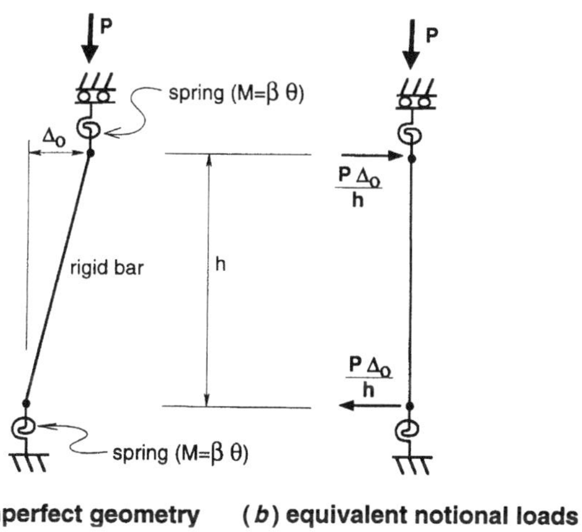

FIGURE 16.10 Modeling of out-of-plumb imperfections with notional loads.

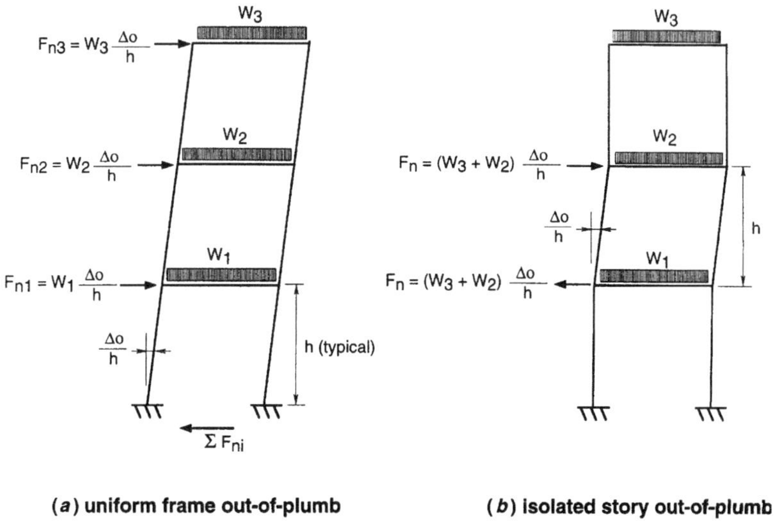

FIGURE 16.11 Notional loads to simulate out-of-plumb imperfections.

that may not be part of the lateral load resisting frame. As shown in Fig. 16.11*a*, where a uniform out-of-plumbness is assumed over the building height, the net notional loads are equivalent to applying forces based on the load at each floor that accumulate over the height of the frame.

In concept, member out-of-straightness can also be modeled by applying either concentrated or distributed notional loads along the member such that the first-order moments induced by the notional loads are equivalent to those induced by the member axial load P_u acting through the initial imperfection δ_0. For example, a transverse concentrated load of $4P_u\delta_o/L$ applied at midheight of a pin-ended column would be equivalent to the maximum moment caused by the out-of-straightness imperfection of δ_o.

When to Include Story Out-of-Plumbness in Analysis As previously noted, and as illustrated in the two-story frame example of Section 16.3.1, the effects of initial imperfections are often insignificant. Factors contributing to this behavior in the two-story frame include: (1) the unsymmetric geometry caused a significant first-order sway of $\Delta = h/670$ under gravity loads; (2) significant moments and shears are present in the columns due to gravity loads; and (3) the ratios of applied column axial loads to nominal compressive strengths were relatively small. In some instances, however, the effects of initial imperfections can be quite significant, for example, in symmetric frames with gravity loads that are significantly higher than the lateral loads, such as storage rack systems.

Regardless of whether a structure is sensitive to initial imperfections, for symmetric structures under gravity loads it is important to include some initial imperfections, particularly if the strength is to be directly assessed from the analysis independent of the specification interaction equation checks. This can be particularly important for second-order analyses being used to detect possible sway instabilities. For this reason, and because of the serious consequences of overlooking potentially large forces and moments induced by imperfections in frames governed by gravity loads, it is recommended that out-of-plumbness be included in the analysis of all frames under gravity loads. The imperfections can be included either by modifying the assumed frame geometry or by applying equivalent notional loads in combination with the factored gravity loads. Unfortunately, the sensitivity of frames to imperfections cannot easily be correlated to a single parameter, and it is therefore difficult to identify frames for which imperfections can be neglected (Surovek and Johnson, 2008).

Clarke and Bridge (ASCE Task Committee, 1997) note that it is generally conservative to specify a uniform out-of-plumbness equal to the maximum interstory values over the entire height of the building (e.g., Fig. 16.11*a*). For buildings up to seven stories tall, applying a uniform out-of-plumbness would be consistent with the maximum lack-of-verticality limit. For the sake of simplicity, this approach could be followed in taller structures. Studies by Clarke and Bridge (ASCE Task Committee, 1997) indicate that in multistory frames that are not extremely slender it is reasonable and not overtly conservative to assume a uniform out-of-plumbness of $h/500$ over the entire height of the structure. In slender frames, a more accurate strategy is typically warranted because a uniform out-of-plumbness applied throughout the height of the building may generate unrealistically large overturning effects near the base of the frame. Bridge (1998) suggests modeling a uniform nonverticality of e_{oh}/H over the full height H of the structure, where e_{oh} is the maximum permitted nonverticality. This approach suggests that the overall effect of imperfections in tall buildings can be captured as a cumulative effect over the height of the building, rather than being dominated by a critical story; this finding was also later supported by Maleck (2001). Alternatively, Maleck and White (1998) recommended modeling a "worst case" imperfection of $H/500$ over a defined portion of the building height up to the maximum out-of-plumb limit.

When to Include Member Out-of-Straightness The question of when to include member out-of-straightness in an analysis is related to two considerations: (1) whether or not separate beam-column stability checks are included in the design method and (2) whether the out-of-straightness has a significant effect on frame behavior. As will be described in Section 16.5, all current design specifications require separate member stability checks, which include allowances for member imperfections. In such cases, member out-of-straightness need not be considered in the analysis. When separate member stability checks are not made, however, member out-of-straightness should be included in the analysis, unless it can be shown to not have a significant effect on the behavior.

The significance of member out-of-straightness is related to (1) the relative magnitude of axial force to primary bending moments in the member, (2) whether the primary bending moments cause single or reverse curvature bending, and (3) the slenderness of the member. These factors can be evaluated by considering the relative magnitudes of second-order moments M_u induced by the applied factored loading to the second-order moment M^* produced by the imperfection, which can be approximated by $M^* = \delta_0 P_u/(1 - P_u/P_{cr})$, where $P_{cr} = \pi^2 EI/L^2$. When the ratio of M^*/M_u is on the order of 5% or less, the imperfection effects are considered small and the effects of out-of-straightness may be neglected. In addition to the simple ratio of moments M^*/M_u, the importance of imperfections will depend on whether the two moments are coincident along the member. Assuming that the maximum imperfection is at midlength of the member, the out-of-straightness imperfections will probably have minimal impact when the maximum applied moments occur at the member ends. Liew (1992) has proposed using the ratio of P_u/P_{cr} as an alternative measure of the effect of member imperfections. Liew's study, which is based on consideration of column strength relations from various standards and the results of distributed plasticity studies of isolated beam-columns and frames, suggests that out-of-straightness effects are small (on the order of 5%) when $P_u/P_{cr} \leq 0.2$. White and Nukala (1997) found similar results and suggest a limit of $P_u/P_{cr} < 1/7$.

16.3.4 Joint and Connection Effects

Referring to Fig. 16.12, the three basic aspects of beam-column joints that affect frame behavior are (1) the finite size of the joint, (2) shearing deformations of the joint panel, and (3) rotational deformations at the beam connections. Because the stiffening effect of the finite joint size may often offset the flexibility of the joint panel and connection deformations, common practice is to ignore all of these effects and model frames using centerline dimensions and ideally rigid joints. Alternatively, for frames with fully welded or bolted moment connections, where the relative rotation between the beams and columns is negligible, the combined effects of finite joint size and panel of deformations may be approximated by modeling some fraction of the finite joint as rigid, using member end offsets; for example, an effective rigid joint size is assumed as one-half the true size of the joint. As shown schematically in Fig. 16.13, force–deformation relationships for both joint panels (M_j versus γ_j) and connections (M_c versus θ_c) are generally nonlinear. To evaluate frame stability accurately, joints and connections should be modeled directly in the analysis whenever their effects may be significant. Fortunately, this is often feasible in analysis software by the combined use of rigid end links (member end offsets) and internal spring elements (McGuire et al., 2000).

Connections are typically idealized as either perfectly pinned or rigid. Classification schemes, such as those presented in Eurocode (CEN, 2005) and Bjorhovde et al. (1990), may be used to determine the cases for when a connection moment–rotation response may be idealized accordingly. Goto and Miyashita (1998) suggest minimum initial stiffnesses and moment capacities for connections in braced and unbraced frames if they are to be considered rigid in the context of using a second-order inelastic analysis to determine frame strength.

For partially restrained connections, both frame displacements and the distribution of forces within the structural system are affected by the connection

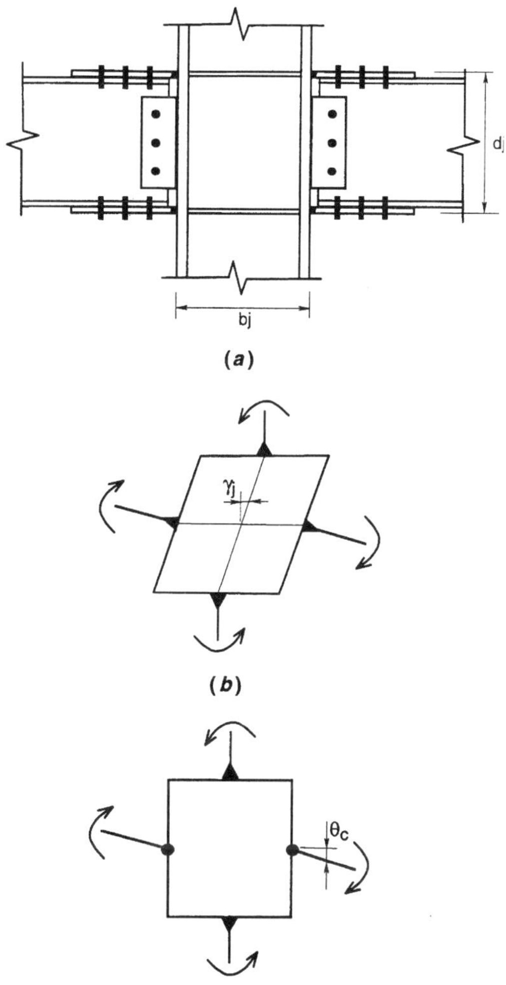

FIGURE 16.12 Schematic representation of *(a)* beam-to-column connection, *(b)* joint panel behavior, and *(c)* beam connection behavior.

stiffness and hence may need to be modeled explicitly. For design based on elastic analysis, the connection and joint response may be approximated through linear secant–stiffness relationships. The stiffness should be representative of conditions experienced under factored load combinations. Obviously, the selection of appropriate connection stiffness should be done with care and may require several iterations to be sure that the calculated response (joint/connection forces and deformations) correspond with values originally assumed. For a sampling of practical approaches of including partially restrained connection response in elastic design, the reader is referred to ASCE Task Committee (1998), Chen (1992), Chen et al (1996),

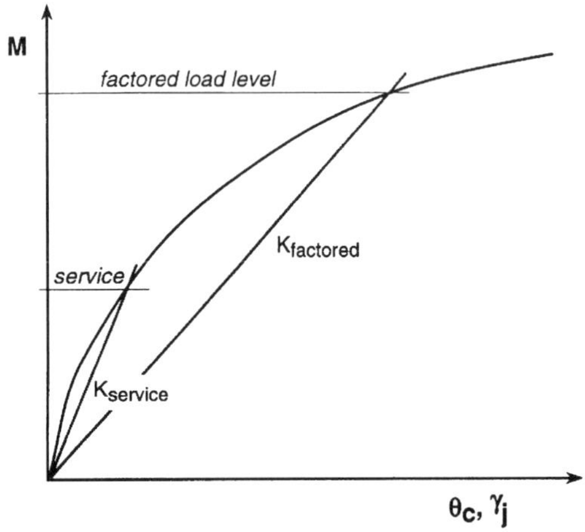

FIGURE 16.13 Idealized force–deformation response of connections and joint panels.

Christopher and Bjorhovde (1999), Gerstle and Ackroyd (1990), Leon and Hoffman (1995), Leon et al. (1996), and Surovek et al. (2005).

Joint panel flexibility is most significant under lateral loads and other situations when unbalanced beam and column forces result in a shearing distortion as shown in Fig. 16.12*b*. As described by Krawinkler (1978), behavior of the joint panel can be expressed in terms of the shear distortion and the net shear force in the joint. The joint shear is a function of the moments and shears in the beams and columns. Sometimes joint shear is expressed in terms of a joint moment so that the joint panel zone flexibility can be modeled using a rotational spring connecting the beams to the column. For further information and equations to calculate the joint stiffness, the reader is referred to Krawinkler (1978) and Krawinkler et al. (1995). As shown in Fig. 16.12*b*, connection deformation is highly localized and is caused by moment transferred from the beam to panel zone region of the column.

16.3.5 Three-Dimensional Frame Stability Behavior and Analysis

While discussion up to this point has dealt with two-dimensional frames, the influence of second-order effects induced by global twisting of the structure should be considered. Torsional stability may be particularly critical in mid- to high-rise buildings. To consider this behavior, assume the lateral response of a building with rigid floors (Fig. 16.14*a*) is characterized by two translational and one rotational degree of freedom (u_x, u_y, and θ_z) located at the center of stiffness of the framing system. In terms of second-order response, the translation $P-\Delta$ effects of individual columns (u_x and u_y) will create a torque defined by the distance between the column and the center of stiffness. Nair (1975) has investigated this behavior for the idealized structural configurations (Fig. 16.14*b*) with lateral systems that have varying ratios of lateral to torsional stiffness. His study shows that in structures such as configuration I, a global torsional instability would be the critical mode of failure. Of

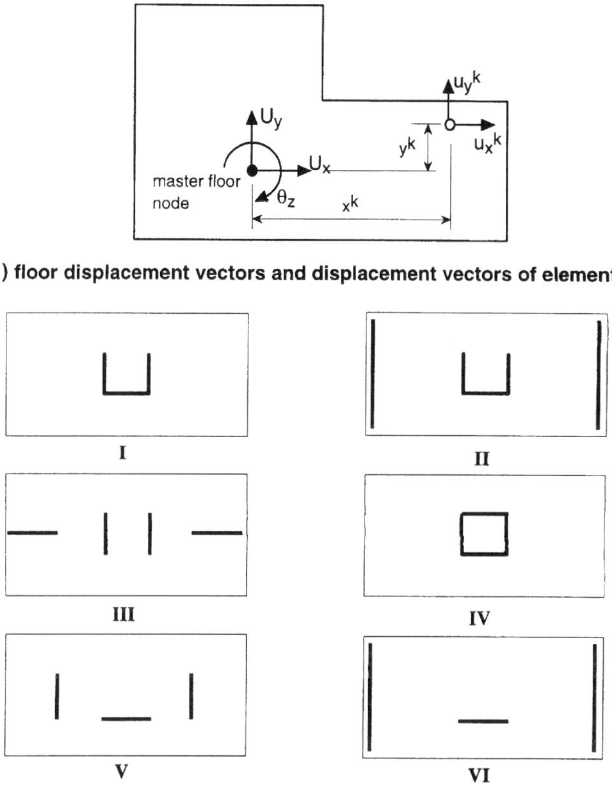

(*a*) floor displacement vectors and displacement vectors of element k

(*b*) alternate floor plan locations of lateral load frames

FIGURE 16.14 Three-dimensional second-order effects in buildings (adapted from Nair, 1975).

course, such instabilities are not captured by two-dimensional analyses or stability design methods that consider only behavior of the frame in two orthogonal directions. To investigate this mode of response accurately requires three-dimensional second-order analysis, including consideration of building out-of-plumbness and/or eccentricities of the lateral loading that will accentuate torsional response.

In general, the basic concepts of geometric nonlinear analysis described in Section 16.2.2 for two-dimensional structures can be extended into three dimensions. The rigorous development of appropriate geometric stiffness terms and transformation equations, however, requires consideration of some subtle but significant theoretical issues associated with the kinematics of finite three-dimensional rotations (McGuire et al., 2000). Specifically, these relate to the fact that finite rotations about more than one axis are noncommutative and do not follow conventional vector transformation equations. In building frames, special treatment of finite rotations are often ignored because the large second-order rotations are only about a single axis, the vertical axis of the building, and as result out-of-plane failure is not considered within the analysis. The method of modeling story $P-\Delta$ effects proposed by Wilson and Habibullah (1987) makes use of this assumption. When three-dimensional rotations about more than one axis are possible, such as the case where torsional–flexural member behavior is considered (Section 16.3.6), finite-rotation effects should be considered. For further information on this subject, the reader is referred to Argyris (1982), Yang and McGuire (1986a,b), Chen (1994), Yang and Quo (1994), Nukala (1997), Trahair and Chan (2003), and Chiorean and Barsan (2005).

16.3.6 Torsional–Flexural Member Response

Standard frame analysis software is usually incapable of modeling out-of-plane member instabilities (e.g., lateral–torsional beam buckling, torsional–flexural column buckling, etc.), and it is primarily for this reason that SSRC Technical Memorandum No. 5 recommends that separate checks be made for frame and member instabilities. Recent progress, however, has been made in research to develop frame analysis methods that include torsional–flexural member response and would thereby enable a single analysis that would capture both frame and member instabilities in a single analysis (McGuire et al., 2000; Ziemian et al., 2008). In addition to the basic attributes of an accurate second-order analysis, features that must be modeled to detect member instabilities include (1) an accurate representation of torsional member behavior, including the effects of warping and restraint of warping and twist; (2) member plastification under the combined effects of axial load, biaxial bending, and bimoments generated by warping restraint; and (3) second-order finite-rotation effects. A key aspect of analytical formulations to capture torsional–flexural member response are beam-column element stiffness with extra degrees of freedom related to warping and bimoment effects, such as those described by Barsoum and Gallagher (1970). For further information on developments related to this topic, the reader is referred to Yang and McGuire (1986a,b), Yang and Quo (1994), Pi and Trahair (1994), Attalla (1995), Attalla et al. (1996), Trahair and Chan (2003), and Nukala and White (2004).

16.4 FRAME STABILITY ASSESSMENT USING SECOND-ORDER ANALYSIS

It is useful to discuss second-order analysis in the context of limit-state design at factored load levels. Considering the format of the AISC LRFD approach (AISC, 2005b) as an example, strength and stability checks for design can be considered in terms of

$$\sum \gamma_i Q_i \leq \phi R_n \tag{16.7}$$

where the left-hand side consists of the effects of factored loads on a structural member or connection and the right-hand side represents the design resistance or design strength of that element. In Eq. 16.7, Q_i refers to the internal forces created by an applied load (e.g., dead load, live load, wind, etc.), R_n is the nominal member or connection strength, and γ_i and ϕ are factors to account for the variability in the loads and resistance. When stability effects are not involved, load effects calculated by analysis and represented on the left side of Eq. 16.7 are separated by the design resistance on the right side. The design check for frame stability, however, is complicated by splitting what is in essence an analysis problem between the two sides of Eq. 16.7. For example, member forces according to the AISC specification are calculated on the basis of an elastic analysis using factored loads (left-hand side of Eq. 16.7) and then compared to member limit states that are based on inelastic behavior. Therefore, the inelastic deflections implicit in the determination of the design resistances are not compatible with the elastic deflections calculated in the structural analysis. When inelastic stability effects are significant, this issue is generally addressed either through calibration of the strength formulas of the design standard (as is the case in the AISC specification) or by considering the inelastic effects on the design directly within the analysis. The latter approach, which is the goal behind more sophisticated second-order inelastic frame analysis methods, permits more direct checks for frame stability by accounting for the inelastic amplification of internal forces explicitly in the analysis (i.e., in the factored load terms on the left side of Eq. 16.7), as well as the member cross-section strengths associated with this inelastic behavior (i.e., the strength terms on the right side of Eq. 16.7).

In this section, guidance is provided for using different types of second-order elastic and inelastic analyses for frame stability checks. The basic requirement for second-order analyses is that equilibrium under the factored loads should be satisfied on the deformed configuration of the structure. The degree to which inelastic deformations influence the response at factored design loads is a central consideration, because most commercial programs prevalent in design practice consider only second-order *elastic* behavior. While second-order elastic analysis methods with simple member design checks are sufficient for the design of most structures, there are instances when inelastic effects can be significant.

In the following sections, three specific approaches to the direct assessment of frame stability are considered. Each requires as a basis a rigorous second-order analysis which includes geometric nonlinear effects arising from both interstory drifts ($P-\Delta$) and member curvature ($P-\delta$). Initial frame out-of-plumbness is also included when significant. The discussion is presented in a generic manner that is not linked to any specific design specification in order to focus on fundamental issues in design. The following three basic categories of elastic and inelastic analyses are studied:

1. *Elastic analysis with a first-hinge limit point.* Members and connections are modeled as elastic. Residual stresses and gradual plastification effects are not modeled, although approximate procedures for accounting for their effects are suggested. It should be noted that this approach will not take advantage of the redistribution of forces after the first plastic hinge forms and, therefore, can

be considered as a lower bound estimate to the true strength of the framing system. As described in Section 16.5, design standards are predominantly based on this approach.

2. *Elastic–perfectly plastic hinge analyses.* Members are modeled assuming elastic properties up to the point that they reach a fully plastic condition under combined axial loads and bending moments. Fully yielded sections are modeled using plastic hinges that enforce a yield surface criterion and allow for elastic unloading. This is essentially a continuation of procedures used in the elastic first-hinge limit-point approach that consider the redistribution of internal forces within the framing as yielding commences at various locations throughout the system.
3. *Inelastic distributed plasticity analyses.* Gradual yielding of members prior to full plastification is modeled through either an explicit fiber element analysis or a refined plastic-hinge model. This approach comes closest to modeling the true limit-state behavior of the structure. While full finite element analysis using shell or volume elements is sometimes employed for research purposes, it is currently considered impractical for design purposes. Refined plastic-hinge models provide for gradual yielding throughout the cross section prior to full plastification using initial yield and full-yield interaction surfaces or functions describing the transition from initial yield to full plastification.

16.4.1 Strength Assessment Based on Elastic Analysis with a First-Hinge Check

A rigorous second-order elastic analysis, including consideration of geometric imperfections, can be used to account for stability effects when a structure behaves elastically up to the formation of the first plastic hinge. One approach to design assessment using elastic analysis is to define the strength limit state as when either (1) the fully plastic yield condition is reached in one or more members of the structure or (2) the nominal strength of a connection, foundation, or other structural element is reached. Such an approach is referred to herein as a *first-hinge limit-state design.* In the context of an LRFD approach, this method implies that the left-hand side of design equation 16.7 is equal to the induced member forces calculated by a second-order elastic analysis under the factored loads, and the right side is equal to the nominal full-plastification strength of the member reduced by an appropriate strength reduction or ϕ factor. This approach includes both frame and in-plane member stability checks, provided that initial geometric imperfections are taken into account and inelastic effects (prior to the formation of the first plastic hinge) are not significant.

Referring back to the load–deformation response of the two-story frame shown in Fig. 16.8*a*, the first-hinge limit-state design would imply the frame strength under gravity load to be equal to 0.74 times the factored combination $1.2D + 1.6L$. Considering that the frame continues to carry load until reaching a factored gravity load ratio of 1.0, one obvious disadvantage of the first-hinge limit-point criterion is that the excess capacity afforded by inelastic force redistribution is neglected.

It is important, however, to recognize that the conservatism implied by this approach is inherent in all design specifications that are based on elastic analysis.

Based on the agreement between plastic-hinge analysis and the distributed plasticity analysis in Fig. 16.7*b*, the assumption of predominantly elastic response prior to formation of the first hinge is valid for this frame. The fact that spread-of-plasticity effects and initial imperfections are not significant in the frame is dependent in part on the fact that the column members are subject to major axis bending and the frame is nonsymmetrical. In a follow-up study, Ziemian and Miller (1997) investigated similar frames but with columns subjected to minor axis bending and found similar results suggesting that second-order elastic analysis with a first-hinge limit criterion would also be a sufficient check for in-plane stability of those frames.

Contrary to the good correlation reported above between elastic–plastic hinge and spread-of-plasticity analyses, there are cases where the simple second-order elastic analysis approach is unconservative, which is due to significant plastification prior to the formation of the first hinge. An example where such behavior is evident is shown in Fig. 16.15 for two portal frames that are similar to ones studied previously by El-Zanaty et al. (1980) and Attalla et al. (1994). The response curves shown in Figs. 16.15*b* and *c* contrast solutions obtained using plastic hinge versus distributed plasticity analyses for cases when the members are loaded in major or minor axis bending. These analyses were performed using elastic–plastic and quasi-plastic hinge software developed by Attala et al. (1994). As indicated in Fig. 16.15*a*, the applied loading consists of proportionally applied vertical loads (P) and a horizontal load of αP, where α values of 0.0001, 0.01, 0.05, and 0.15 were investigated. The case with $\alpha = 0.0001$ represents essentially pure gravity load where the small lateral load is applied to perturb the behavior of the perfectly symmetrical frame. Given the low degree of redundancy and the symmetry in these frames, the formation of the first plastic hinge essentially coincides with the inelastic limit point for the elastic–perfectly plastic hinge analyses. Of concern in Figs. 16.15*b* and *c* are the large differences between the maximum lateral loads calculated by the elastic–perfectly plastic analyses compared to those of the distributed plasticity analyses. As shown in the figures, the differences are largest for members subjected to minor axis bending with higher column loads. In the most extreme cases (minor axis bending with α values of 0.0001 and 0.01), the limit load is reached well before the columns reach their fully plastic condition. This is evident from the fact that the descending branches of the load–deformation response curves do not converge beyond the limit point.

The limit point of the curves in Figs. 16.15*b* and *c* are plotted in Figs. 16.16*a* and *b* in terms of normalized values of the average axial column force and maximum first-order moments in the columns (i.e., P/P_y and $\alpha PL/2M_p$, respectively). Results from similar analyses that included an initial out-of-plumb of $\Delta = h/500$ are also plotted in Fig. 16.16. Data from the distributed plasticity analyses with initial imperfections can be considered as representative of the true nominal in-plane strength of the frames. Thus comparisons with these data provide a measure of the

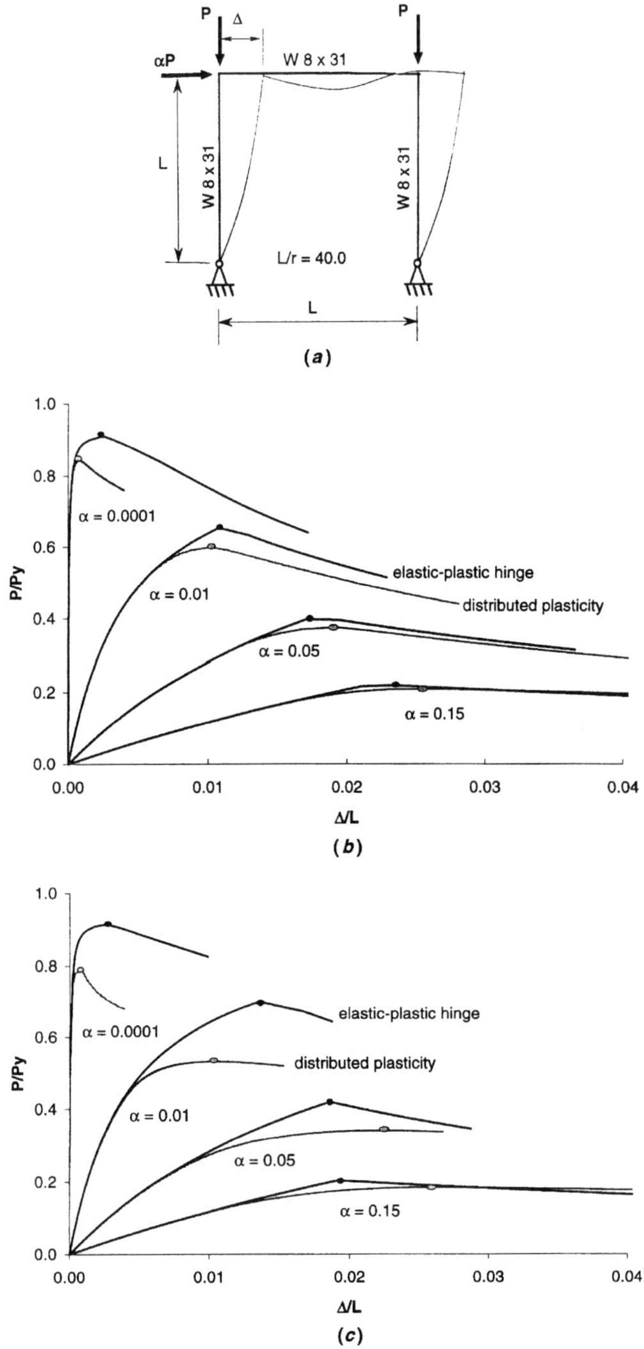

FIGURE 16.15 Load–deformation response from plastic hinge and distributed plasticity analyses: (*a*) portal frame; (*b*) major axis bending; (*c*) minor axis bending.

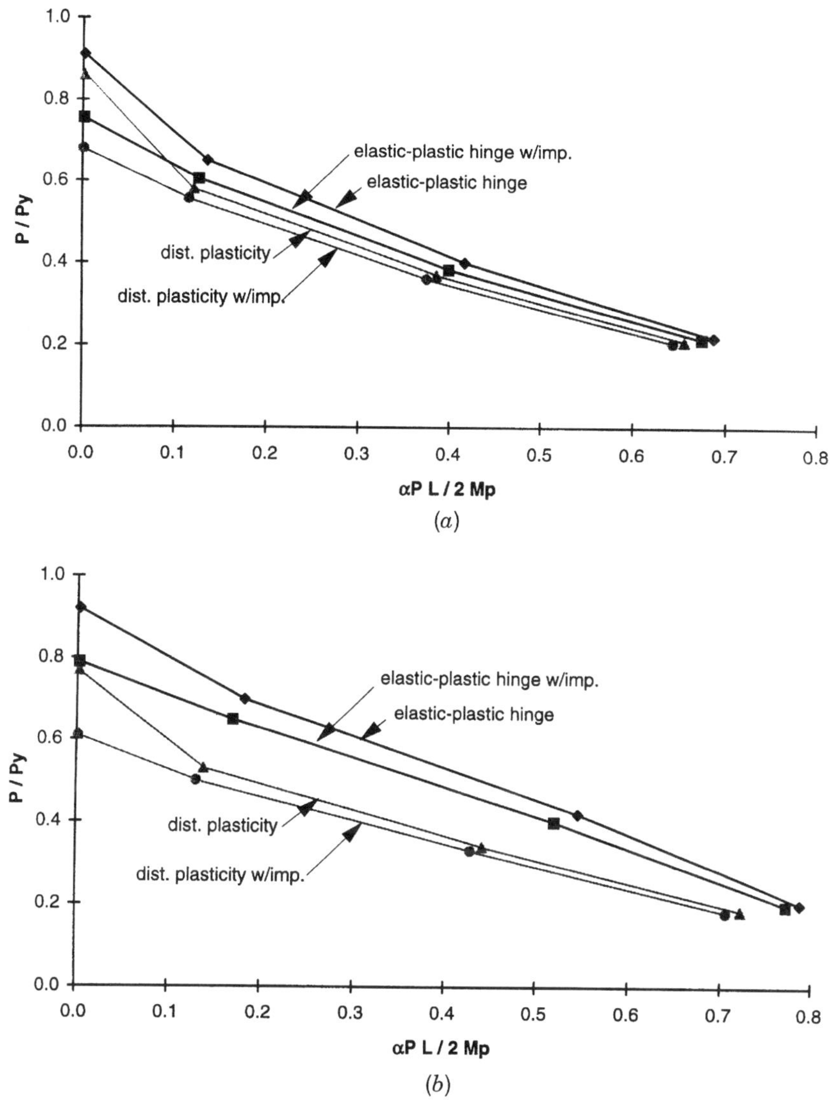

FIGURE 16.16 Comparison of maximum strengths for portal frame: (*a*) major axis bending; (*b*) minor axis bending.

errors introduced by neglecting spread-of-plasticity effects and/or initial imperfections. For example, comparing results for analyses with and without imperfections, the data show that imperfections alone are significant only for cases with small lateral loads (e.g., where $\alpha = 0.0001$). On the other hand spread-of-plasticity effects combined with initial imperfections play a more significant role. Tabulated values of the errors for the two sets of plastic hinge analyses (with and without

TABLE 16.3 Portal Frame Example: Error and Behavioral Indices[a]

	Major Axis Bending				Minor Axis Bending			
	0.0001P	0.01P	0.05P	0.15P	0.0001P	0.01P	0.05P	0.15P
% Error index 1	34	17	11	7	51	40	27	12
% Error index 2	11	9	6	5	30	30	21	9
Behavior index: P_u/P_y	0.91	0.65	0.40	0.22	0.92	0.70	0.42	0.20
Behavior index: $\Delta_{2nd}/\Delta_{1st}$	60	3.3	1.7	1.4	64	3.9	1.8	1.3
Behavior index: P_{cr}/P_u	1.0	1.4	2.3	4.2	1.0	1.3	2.2	4.6

[a]Error index 1 is the percentage difference between the plastic hinge analysis with ideal geometry compared to the distributed plasticity analysis with out-of-plumb imperfections. Error index 2 is the percentage difference between the plastic hinge analysis compared to the distributed plasticity analysis where both include out-of-plumb imperfections. Behavior indices are all based on loads and deformations calculated by the elastic–plastic hinge analyses.

imperfections) are summarized in Table 16.3. As indicated in this table, when both initial imperfections and spread-of-plasticity effects are neglected (% error index 1), differences from the exact solution vary from 7 to 34% for members under major axis bending and 12 to 51% for members under minor axis bending. Referring to % error index 2, inclusion of initial imperfections reduces the errors significantly for the cases with low lateral load, but even with this adjustment errors are still unacceptably large for most cases.

Differences of the type shown for the portal frame are important because they indicate the need for limitations on the application of second-order elastic analysis as the sole means of evaluating stability effects in design. It is vital to keep in mind that the portal frame example represents an extreme case that was devised to investigate the limitations of elastic and elastic–perfectly plastic hinge analyses. As listed in Table 16.3, behavioral indices that might be used to gage the sensitivity of frame stability to inelastic effects and initial imperfections include axial force levels in the columns (P/P_y), the ratio of elastic second- to first-order deflections ($\Delta_{2nd}/\Delta_{1st}$), and the ratio of the elastic critical load to the applied gravity load (P_{cr}/P_u). The data show that in the most sensitive cases, with $\alpha = 0.0001$, the gravity-induced column axial forces are simultaneously approaching both the full-yield (squash) load and elastic critical load of the columns, and in the absence of significant lateral loads, the $\Delta_{2nd}/\Delta_{1st}$ ratios are extremely high for these cases. On the other hand, for cases with high lateral loads ($\alpha = 0.15$), all three indices are in a more reasonable range when second-order effects are significant but not excessive. Note that none of the three indices, however, reflect the difference in errors associated with major axis versus minor axis bending conditions. Moreover, even for this simple frame, there is no clear relationship between the behavioral indices and the errors introduced by spread-of-plasticity and initial imperfections.

In considering what should be the limits on second-order elastic analysis, it is instructive to compare features of the two-story frame example when second-order elastic analysis is reasonably accurate (Figs. 16.6 to 16.9) to the portal frames (Figs. 16.5 and 16.16) in which it is not:

1. *Ratio of elastic critical load to applied loads.* For the two-story frame, the elastic critical load is equal to 4.7 times the factored gravity load at the first-hinge limit point, whereas for the portal frames (Table 16.3) the same ratios range from 4.6 down to 1.0.
2. *Ratio of second-order to first-order elastic response.* As reported in Table 16.1 for the gravity load case of the two-story frame, the difference between second- and first-order drift is 44% and the difference in moments is less than 11%. Recall that in the two-story frame the gravity loads contributed significantly to the first-order moments and drifts due to the nonsymmetrical geometry and the large gravity-induced moments in the beams. The second-order deflection increase of 44% in the two-story frame is close to the ratio of $\Delta_{2nd}/\Delta_{1st} = 1.4$ reported in Table 16.3 for the portal frames with $\alpha = 0.15$. The ratios of $\Delta_{2nd}/\Delta_{1st}$ are much higher for the portal frames with lower lateral loads because the vertical loads do not cause significant first-order moments in this frame.
3. *Ratio of applied column load to squash load.* For the two-story frame, the maximum column loads at the first-hinge limit are approximately $P_u/P_y =$ 0.2 to 0.4, compared to the ratios of 0.2 to 0.9 in the portal frames.
4. *Shape factor and inelastic bending stiffness of members.* Larger inelastic stiffness reductions occur in wide-flange columns subjected to minor versus major axis bending due to yielding at the flange tips, as a result of the applied and initial residual stresses. This is apparent in the portal frame example, where errors are much larger for the minor axis bending cases, even though the other behavioral indices (P_u/P_y, $\Delta_{2nd}/\Delta_{1st}$, and P_{cr}/P_u) are roughly equal for the two column orientations.
5. *Redundancy.* Due to redundancy in the two-story frame, there is still considerable elastic restraint after the first plastic hinge forms, and as a result up to 40% more gravity load can be carried before reaching the inelastic limit load. On the other hand, the inelastic limit strength of the frames in the portal frame examples essentially coincides with the formation of the first hinge. Thus the portal frame results basically reflect the behavior of individual beam-columns, as opposed to that of an indeterminate frame system that provides for inelastic force redistribution.
6. *Type of loading applied to the members.* In multistory frames, columns are often subjected to reverse-curvature bending, resulting in yielding under combined bending and axial loads tending to concentrate at the member ends. This is especially true for frames when sidesway instability governs. Thus the spread of yielding along members in many frames would be less severe than in the pinned-base portal frame, in which the columns bend in single curvature.

Given that many building frames will more closely resemble the two-story example than the simple portal frames, system stability checks based on second-order elastic analysis alone can often provide a reliable approach in engineering practice. There are limits on the accuracy of second-order elastic analysis, however, specifically in cases when inelastic effects become dominant in frames with (1) large ratios of second- to first-order effects, (2) columns with large axial compressive loads, (3) low redundancy, and (4) columns subjected to minor axis bending. White and Hajjar (1997) further show that large leaning column loads are another factor that can amplify inelastic effects. While some precise quantitative limits on the applicability of second-order elastic analysis have been proposed, none of these address all the factors noted above.

Although second-order amplification terms are useful guides for the safe limits of elastic analysis to evaluate frame stability, other parameters, such as those mentioned above, should be considered. A summary of these factors is presented in Table 16.4, which suggests, for example, that in redundant structures with lightly loaded columns subjected primarily to major axis bending it is likely that a second-order elastic analysis is sufficient to capture in-plane frame and member stability effects even for cases with second-order amplification higher than 1.4. Similarly, when a combination of the parameters from Table 16.4 suggests that the stability of the structure is susceptible to spread-of-plasticity effects, it is necessary to account for these effects directly within the frame analysis and/or through calibrated design procedures.

As described below, alternatives to a rigorous distributed plasticity analysis to account for spread-of-plasticity effects using elastic analysis are (1) the modified stiffness approach and (2) the notional load approach. Both methods are intended only to adjust for modest inelastic effects representative of the condition of frames under factored loads, and they are not meant to model inelastic force redistribution associated with yielding of the members. Neither of these approaches is intended to be applied to check forces or deformations under service loads.

Modified Stiffness Approach One approach to approximate modest spread-of-plasticity effects in an "elastic" second-order analysis is to reduce the stiffness of members in which gradual plastification could be significant. This would apply primarily to column members subjected to significant axial compression, but gradual plastification effects could also be significant in other highly stressed members with uniformly distributed (single-curvature) moments. As described by MacGregor (1993), such an approach has been adopted in the ACI-318 code (2008) provisions for slender columns. Similar ideas were previously proposed for steel structures by Cheong-Siat-Moy (1978).

In considering a modified stiffness approach for steel structures, it is instructive to review the ACI-318 (2008) provisions for frames with slender reinforced concrete columns which use essentially a first-hinge limit-point approach. In ACI-318, required member strengths under factored loads (P_u and M_u) are determined by a second-order elastic analysis and checked against the design strength of the member cross section. The design strength is essentially the axial force–moment interaction surface reduced by appropriate resistance for axial load and bending. A key aspect

TABLE 16.4 Suggested Guides to the Limits of Second-Order Elastic Analysis to Evaluate Frame Stability

- *Ratio of elastic first-order to second-order effects.* When the loading and/or structural configuration are such that first-order sidesway effects (moments, drifts, etc.) are significant, the ratio of second-order to first-order effects is one legitimate measure of the accuracy of elastic analysis to model stability limit states. Subject to adjustment by the other parameters noted below, maximum ratios of second- to first-order effects between 1.2 and 1.4 appear to be reasonably conservative limits of when distributed plasticity effects become significant. Limits such as these may not be appropriate for structures dominated by gravity loads, where first-order sidesway effects are small and economical column designs may require consideration of distributed plasticity effects.
- *Structural redundancy*. Highly redundant structures will, in general, have a larger margin between the onset of plastification and the inelastic limit than nonredundant structures or redundant structures where the stability is limited by a failure within a localized region of the structure. In many cases the onset of inelasticity in a few members will have a small influence on the overall response of the frame, and the overall frame behavior will remain predominantly elastic until larger load levels are reached.
- *Load magnitudes and moments in columns*. If $P/P_y \leq 0.2$, the existence of substantial distributed plasticity along the length of the column is not possible unless the column is loaded under large single-curvature bending. Moreover, in the absence of large moments, residual stress effects in combination with axial compression will not have a significant effect on column stiffness for $P/P_y \leq 0.6$.
- *Leaning columns*. If a large percentage of the gravity load is supported by leaning columns, reductions in the lateral stiffness of the structure due to distributed yielding are amplified. In the absence of extensive data on this, it is reasonable to assume that if loads supported by leaning columns are less than about twice the vertical loads supported by the lateral-load resisting columns, the amplification of distributed plasticity effects by the leaning columns should be negligible.
- *Major axis versus minor axis bending*. For H-shaped sections subjected to minor axis bending, the cross-section flexural rigidity decreases dramatically as yielding progresses inward from the flange tips. Furthermore, the general extent of the full-plastification surface is greater relative to the initial yield surface for minor axis bending of these shapes, as reflected by the shape factor for the case of bending alone. Due to this fact, the distribution of plasticity along the member length when the full-plastification strength is reached is generally greater for minor axis bending.

of the second-order analysis is that the flexural stiffnesses of beams and columns are chosen to give calculated deflections that are close to those at the strength limit of the structure under factored loads. This implies that member stiffnesses are defined to include the effects of concrete cracking and creep as affected by the distribution and duration of axial forces and moments.

In lieu of more precise calculations, ACI-318 (ACI, 2008) specifies values of stiffnesses for various types of members to use in the second-order analysis. For the overall frame analysis to calculate $P-\Delta$ effects, flexural stiffnesses of $0.35E_cI_g$ for beams and $0.7E_cI_g$ for columns are recommended, where E_c is the short-term elastic modulus of concrete and I_g is the gross moment of inertia. These values are intended to represent average properties throughout the structure. For calculating moments along the length of individual columns due to $P-\delta$ effects, a lower value of $0.4E_cI_g$ is suggested for the local moment amplification term. This is in recognition of the probabilistic nature of these estimated values, and the lower bound values should be used for checks of individual members, whereas for overall system behavior values closer to the means can be used. The stiffness values as given above are for use in checking stability under combined gravity plus lateral loads.

One method to approximate stiffness reductions in steel members with high axial load is through a tangent modulus derived from column curve formulas (Ziemian et al., 1992a, 2008; Ziemian and McGuire, 2002; White and Chen, 1993; Orbison et al., 1982). For example, the inelastic stiffness reduction implied by the AISC LRFD column curve is shown in Fig. 16.17*a*. Because this tangent-modulus adjustment does not affect members with $P/P_y < 0.5$, this adjustment alone would not reduce the errors shown in Figs. 16.15 and 16.16 significantly except for the cases with little bending. This is because the differences in these columns are due to the plastification under the combined action of axial compression and bending. For example, as shown in Fig. 16.17*b* for a column with $P/P_y = 0.5$, the flexural stiffness in minor axis bending drops off rapidly under the action of axial load and moment (Cheong-Siat-Moy, 1978). Thus for second-order analysis the modified—or reduced—stiffness must consider the combined effects of axial load, bending moments, residual stresses, and member shape factor (Ziemian et al., 2008; Ziemian and McGuire, 2002).

To investigate the feasibility of a modified stiffness approach, the portal frame examples from above (Fig. 16.15) were reanalyzed with a second-order plastic-hinge analysis using reduced column stiffnesses of $0.8EI$ for major axis bending and $0.6EI$ for minor axis bending. These values were chosen to be in line with data such as that shown in Fig. 16.17 and because they give good results in the portal frame example. First-hinge limit points from these analyses are shown in Fig. 16.18 that are analogous to the data shown in Fig. 16.16. Data from several analyses are provided, including:

- *DP:* distributed plasticity analysis
- *Elastic:* plastic hinge analysis with elastic properties
- *M-elastic:* plastic hinge analysis with modified column stiffness

All analyses are second order and include an initial frame out-of-plumb of $h/500$. Also plotted are results from an alternative notional load procedure (NL-elastic) that is discussed below. As shown in the figures, the modified stiffness approach (M-elastic) results for both major and minor axis bending are in nearly perfect agreement with the more accurate distributed plasticity (DP) solutions.

Although stiffness reduction ratios are input as constants that do not vary with axial load, the results are fairly accurate (Fig. 16.18) for axial loads ranging from about $P/P_y = 0.2$ to $P/P_y = 0.6$. The reason for this is that as shown in Fig. 16.17*a*, the drop in stiffness due to axial load alone is small up to about

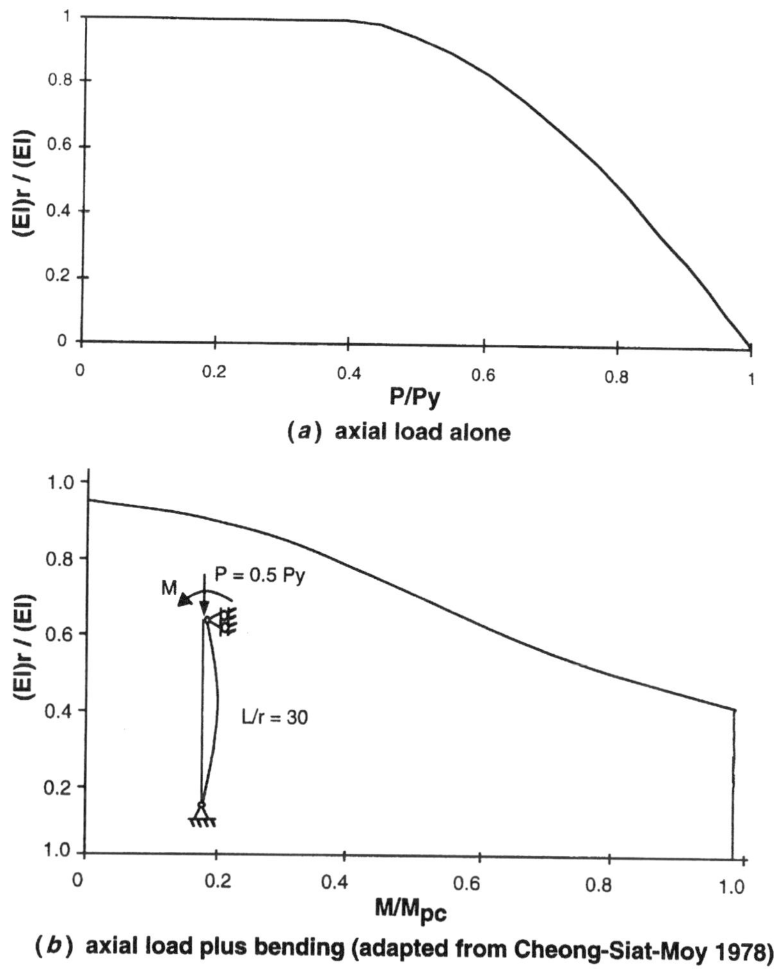

FIGURE 16.17 Effect of axial load and bending on flexural stiffness of beam-columns.

$P/P_y = 0.5$ or so. Presumably, the effective stiffness should be reduced further for cases with $P/P_y > 0.6$, perhaps by using a tangent-modulus adjustment (Fig. 16.17*a*) in combination with the 0.6 and 0.8 factors noted above.

As noted above, the portal frame examples are devised to accentuate a worst-case test for the limits of elastic analysis. Thus, the stiffness reductions used above are typically more severe than would be appropriate for more redundant frames. One virtue of the modified stiffness method, however, is that conservative errors due to underestimating the stiffness tend to be self-limiting in structures that are insensitive to stability effects. For this reason the relative change in second-order response due to a modest change in stiffness can be used as an indicator of a potential stability problem. Consider the data in Table 16.5, where amplification factors $\Delta_{2nd}/\Delta_{1st}$ are

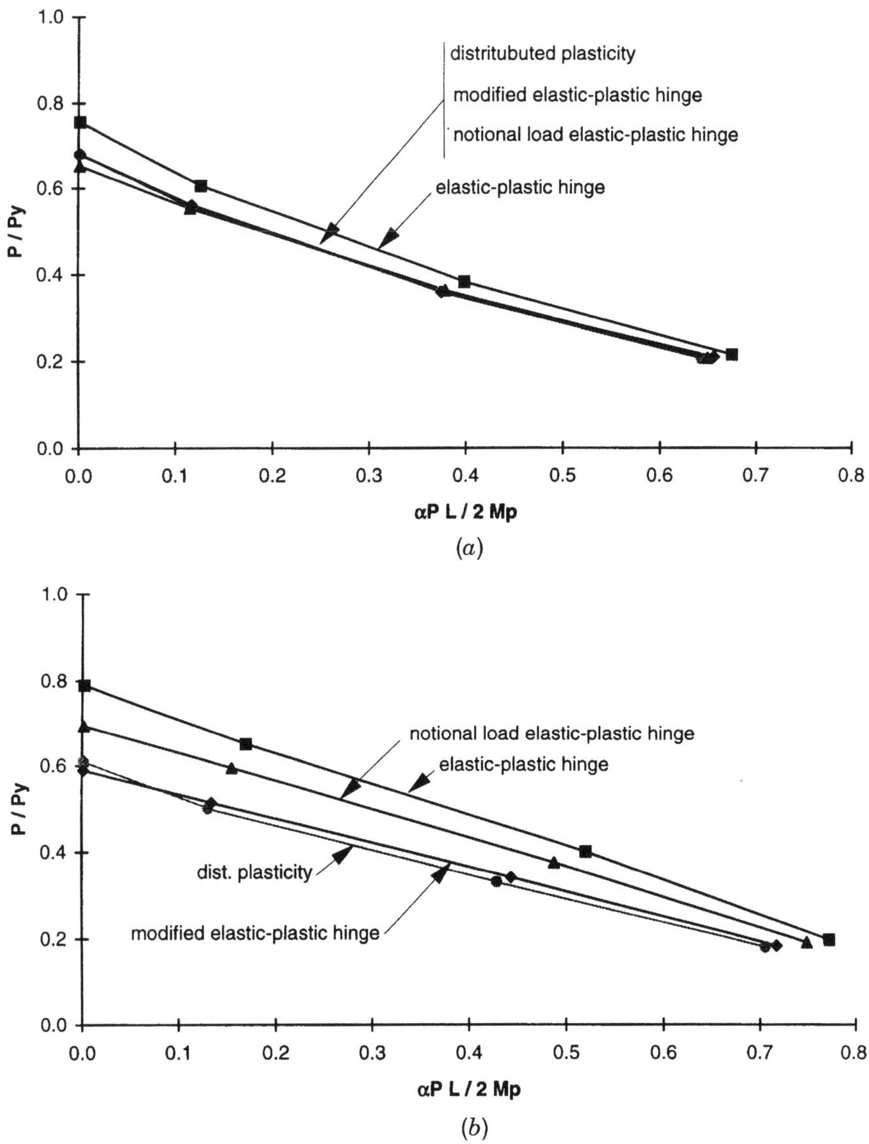

FIGURE 16.18 Comparison of strengths for portal frame with modified stiffness and notional load approaches: (*a*) major axis bending; (*b*) minor axis bending.

summarized for analyses of the portal frame using the full and modified stiffnesses (rows A and B). Row C lists ratios reflecting the increase in the second-order amplification factor due to the decrease in stiffness. Considering, for example, the major axis bending case with $\alpha = 0.15$, the fact that the second-order effects do not increase due to the stiffness reduction of -20% (i.e., the change ratio in row

TABLE 16.5 Comparison of $\Delta_{2nd}/\Delta_{1st}$ Index for Elastic and Reduced Stiffness Analyses

	Major Axis Bending ($\beta = 0.8$)				Minor Axis Bending ($\beta = 0.6$)			
	0.0001P	0.01P	0.05P	0.15P	0.0001P	0.01P	0.05P	0.15P
(A) Full elastic (EI)	5.3	2.8	1.7	1.4	6.3	3.3	1.7	1.3
(B) Modified (β EI)	7.4	3.4	1.8	1.4	14.6	5.2	2.1	1.4
(C) Change ratio = A/B	1.4	1.2	1.1	1.0	2.3	1.6	1.2	1.1

$C = 1.0$) indicates that the frame is not sensitive to stability effects. On the other hand, for frames with $\alpha = 0.0001$, the changes in the ratios for second-order effects (row C equals 1.4 and 2.3) are in excess of the percentage reduction in stiffness ($1/0.8 = 1.25$ and $1/0.6 = 1.67$), indicating that these structures are very sensitive to stability and spread-of-plasticity effects.

Notional Load Approach In Section 16.3.3, the notional load approach was introduced as a method of using fictional lateral loads to mimic the effects that initial out-of-plumbness will have on frame behavior. This concept has been expanded in several design specifications, including AISC (2005b), CSA-S16-01 (CSA, 2001), AS4100 (SAA, 1998), and Eurocode 3 (CEN, 2005), to include the effects of member inelasticity due to partial yielding and residual stresses (Clarke and Bridge, 1996; White and Clarke, 1997). In contrast to accounting for initial geometric imperfections, there is no theoretical basis for calculating notional loads appropriate for modeling spread-of-plasticity effects. Instead, equivalent notional loads are determined empirically through calibration to data from computational studies and experimental tests. An overview of this calibration can be found in Appendix F of the report by the ASCE Task Committee on Effective Length (1997).

16.4.2 Strength Assessment with Elastic–Perfectly Plastic Hinge Analysis

As demonstrated by the two-story-frame example (Fig. 16.6), the first-hinge limit criterion for indeterminate structures is a conservative estimate of the true inelastic limit strength of the frame. Often, the strength increase due to inelastic force redistribution beyond formation of the first hinge is on the order of 10 to 30%. Therefore, second-order elastic–plastic hinge analyses could enable the design of lighter and perhaps more cost-effective structures. More importantly, inelastic analyses provide insight on failure modes that ultimately control the strength limit-state behavior and thus improve the accuracy and reliability of the strength evaluation. When utilizing inelastic force redistribution beyond the first hinge, there are certain behavioral considerations that must be addressed in design. Beyond those described for the elastic first-hinge approach in Section 16.4.1 (including spread of plasticity, residual stress effects, initial imperfections, out-of-plane member failures), these additional considerations include (1) evaluating hinge rotation demands and capacities, (2) considering the increased potential for incremental collapse, and (3) proper modeling of strain hardening or softening.

Hinge Rotation Demands and Capacities When inelastic force redistribution is relied upon, a check should be made to verify that the inelastic rotation demands can be accommodated by the members and connections. In steel frames with rigid connections, the basic modes of failure that will limit member deformation capacities include local flange or web buckling, lateral–torsional member buckling, and fracture of the member or its connections. While each of these modes of failure is treated in design standards by independent slenderness limits, the modes of failure are often quite interrelated. For example, the rotation capacity of steel H-shaped members is highly dependent on the interactive effects of local flange and web buckling and lateral buckling (Kemp, 1986).

For the design of ordinary proportioned frames under gravity and wind loads, available evidence suggests that member compactness requirements, such as those contained in the plastic design provisions of the AISC specification (2005b), provide adequate deformation capacity required for the structure to reach the inelastic limit point (Ziemian et al., 1992a,b). Such provisions usually address section compactness criteria (b/t and d/t), lateral bracing criteria (L/r), and specified material requirements. In a study that included both low-rise planar structures and a midrise three-dimensional structure, Ziemian et al. (1992a,b) report that the maximum required plastic hinge rotations under gravity and wind loads were less than 0.01 rad, which were on the order of half of the inelastic deformation capacity implied in the AISC provisions for plastic design.

It is worth noting that the above study did not address the potential for fractures at beam-column connections, which, as experiences from the Northridge earthquake have shown, should not be discounted. There are, however, several reasons why fracture is less likely under gravity and wind loading compared to cases where seismic design governs. First, the maximum plastic rotation demands of 0.01 rad calculated by Ziemian et al. are considerably less than values of 0.02 to 0.04 rad that are considered as representative of the maximum hinge rotation demands for seismically designed frames. Moreover, for gravity and wind load design, repeated inelastic cyclic loading of the members and connections is not anticipated. Finally, the probable overstrength in the structure due to yield stresses that exceed the nominal values, strain hardening, composite beam action, and secondary structural elements will usually reduce plastic hinge rotation demands under the factored design loads compared to those predicted by an elastic–perfectly plastic analysis of the bare frame.

Consideration of Incremental Collapse Under repeated loading at levels that cause significant yielding, there is the potential for frames to fail due to incremental accumulation of lateral deflections at loads less than the inelastic monotonic limit load. This has implications for design by plastic hinge analysis, because the factored load combinations including live, wind, or other variable loads can approach the inelastic limit strength of the frame. Checks for incremental collapse are related to shakedown analysis, with the *shakedown load* being a value below which the structure will converge to a stable elastic state under repeated loading. Thus, for a given loading condition, the shakedown load lies between the load at which the first hinge forms and the inelastic limit point.

Previous studies dating back to research on plastic design in the 1950s and 1960s (Popov and McCarthy, 1960; WRC/ASCE, 1961; Neal, 1985; Horne, 1979) indicate that incremental collapse is probably not of concern for plastically designed building frames under variable gravity and wind loads. Since then, not much has been published related to the likelihood of incremental collapse. In considering the application of modern computer-based second-order plastic hinge analysis to evaluate the inelastic limit strength of frames, however, the potential for incremental collapse should not be discounted. Shakedown limit or incremental collapse analyses should be considered where there is the likelihood of repeated load excursions beyond the yield point (e.g., cases when the ratio of live to dead load is large and/or when the design strength is governed by lateral loads).

Classical methods for evaluating shakedown (e.g., Neal, 1985; Horne, 1979; Konig, 1987) are generally impractical for evaluating frames of realistic proportions and loading conditions, especially when second-order effects are significant. Guralnick et al. (1984, 1986, 1991) proposed an energy-based approach to evaluate incremental collapse that is more amenable to computer techniques than classical shakedown methods and can include second-order effects directly. The basic premise of this method is to evaluate shakedown through a convergence study of inelastic energy dissipation in a structure under repeated loading. A condition of shakedown is reached when there is a stabilization of the accumulation of plastic energy under repeated cycles of loading. While the method requires somewhat sophisticated computer analysis programs (e.g., plastic hinge analyses with features for cyclic loading), it does provide a systematic means for evaluating whether incremental collapse will occur under the applied loads.

Strain Hardening The question of whether to include strain hardening in an inelastic analysis is one that depends both on the basic design philosophy and circumstances specific to the structure being analyzed. For example, most steel design standards do not explicitly allow any increase in member strengths associated with strain hardening (i.e., member strengths are limited to the plastic strength of the cross section calculated with the minimum specified material yield strength). For these standards, plastic hinges should be modeled accordingly as elastic–perfectly plastic, thereby eliminating the possibility of member resistances exceeding the plastic strength of the cross section. This means that when numerical solution requirements necessitate that some small residual stiffness be included at the plastic hinges, such values should be kept to a minimum. For example, Eurocode 3 (CEN, 2005) suggests keeping strain-hardening values on the order of 0.01% of the elastic stiffness. Neglecting the beneficial effects of strain hardening tends to offset the destabilizing influence of local buckling and other effects that are not modeled in plastic hinge analyses.

16.4.3 Strength Assessment with Spread-of-Plasticity Analysis

Any analysis model that accurately represents the effects of distributed plasticity due to combined axial force, bending, and residual stresses is termed a *spread-of-plasticity* analysis method. The intent for applying spread-of-plasticity analysis methods in design is that the nominal residual stress effects and initial geometric imperfections can be modeled explicitly whenever they are important. This avoids the introduction of ad hoc modifications of the sort that might be used with elastic–plastic hinge methods, thereby relieving the engineer of the need to make assumptions that create uncertainty in the analysis. When it can be shown that one or both effects have a negligible influence on behavior, then as with elastic analysis they need not be included. The Australian steel design specification, AS4100 (SAA, 1998), includes provisions that permit the engineer to use what is termed an "advanced" spread-of-plasticity analysis as the sole basis for evaluating the in-plane stability of members and frames (i.e., in such cases member interaction equations for evaluating in-plane stability are not required).

Initial imperfections should at a minimum include frame out-of-plumbness, and member out-of-straightness should also be included in the analysis when significant. As described in Chapter 3, residual stresses in compression members can vary significantly depending on the member sizes and details of the particular manufacturing process. In the absence of more specific information, the residual stresses shown in Fig. 16.19 have been considered as representative of "typical" wide-flange shapes. This pattern has a precedent for use in design because it formed the basis of parametric studies used in the calibration and development of the AISC beam-column interaction equations (Kanchanalai, 1977; ASCE Task Committee, 1997). Of course, it can be left to the discretion of the designer to use residual stress patterns that are appropriate to the size and shape of the cross sections being modeled.

With spread-of-plasticity methods, it is important to check the inelastic deformation demands in plastified regions of the frame and to limit strain hardening to values that will not overestimate the nominal member strengths. Depending

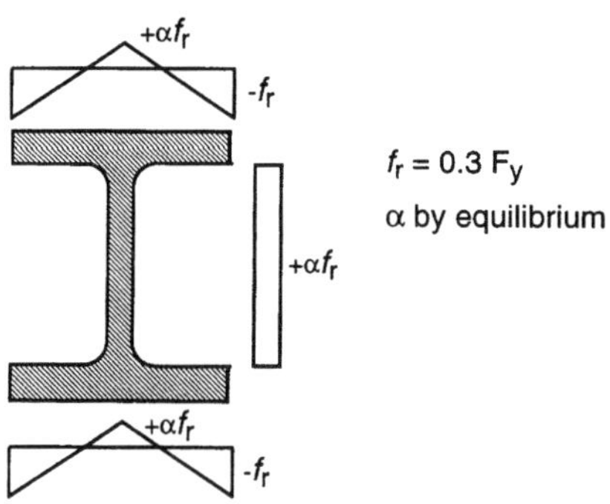

FIGURE 16.19 Idealized residual stresses for H-shaped members.

on the specific formulation employed in spread-of-plasticity analysis, the treatment of strain hardening and inelastic deformation demands will differ from those described above for plastic hinge methods. For example, in spread-of-plasticity methods that utilize a detailed fiber element approach, strain hardening is generally specified in terms of an assumed uniaxial stress–strain relationship. On the other hand, in quasi-hinge methods, strain hardening may be included indirectly in force–generalized strain relations of the cross sections, or in terms of some residual member stiffness. Whatever the method for specifying a "hardening" parameter, a check should be made at the end of the analysis to ensure that the member forces do not significantly exceed the full cross-section plastification criteria.

16.5 OVERVIEW OF CURRENT CODE PROVISIONS

This section compares several steel design standards, including AISC (AISC, 2005b), AS4100 (SAA, 1998), CSA-S16-01 (CSA, 2001), and Eurocode 3 (CEN, 2005). This overview is not intended to provide an in-depth explanation of the provisions appearing in each standard. Rather, the goals are to (1) further describe and contrast various methodologies for evaluating frame stability, (2) identify how the methods outlined in Section 16.4 have been implemented into the different approaches, and (3) consider the interdependence between design provisions and methods of analysis. Detailed comparisons of beam-column and frame stability are provided by White and Clarke (1997) and Hansoulle et al. (2007).

The design standards are similar in several key ways. They all use a limit-state design philosophy[1] that is comparable to the first-hinge design approach described in Section 16.4.1. They all require that (1) second-order effects be included in the elastic analysis, with options that include rigorous second-order analysis methods or approximate methods that are based on amplifying first-order analysis results; (2) the analysis account for member and frame geometric imperfections, either by explicitly including them in the analysis model or by the use of equivalent notional loads; and (3) the adequacy of the design be confirmed through the use of one or more interaction equations on a member-by-member basis.

Of course, there are also subtle and major differences in the specific approaches to frame stability assessment in each of these standards pertaining to (1) the adopted load and resistance factors, that is, partial safety factors for actions and resistances; (2) the nominal loads or actions selected for design within the jurisdiction of each of these standards; (3) column strength formulas, including the number of column curves provided; (4) beam strength formulas; (5) the number, format, and "shape" of curves used for calculating interaction of bending and axial compression; (6) use or nonuse of effective-length K factors; and (7) modification of moments to account for geometric imperfection and/or distributed plasticity effects through application of notional loads.

[1]The AISC specification (2005b) is the only standard compared that allows the implementation of allowable strength Design (ASD); the implementation of ASD in the Specification, however, utilizes the same strength equations as LRFD and differs primarily in how the factor of safety is applied.

16.5.1 Frame Stability in the AISC Specification

The AISC specification (2005b) permits any method for incorporating frame and member stability that considers the influence of second-order effects (including $P-\Delta$ and $P-\delta$); flexural, shear, and axial deformations; geometric imperfections; and member stiffness reduction due to partial yielding and residual stresses. Three approaches based on elastic analysis are presented in the specification, including the *effective length method*, the *direct analysis method*, and the *first-order elastic analysis method*. An excellent overview of these methods, including guidance and suggestions for modeling structural systems, examples, and detailed comparisons, is presented by Griffis and White (2010) and Nair (2009).

It should be noted that in several places within these methods the impact of second-order effects be assessed by the ratio of second-order drift to first-order drift. In most cases, it is intended that this ratio be represented by the story–stiffness-based moment amplification factor B_2 given by

$$B_2 = \frac{1}{1 - Y_i/(\beta_i L_i)} \tag{16.8}$$

where Y_i is the total factored gravity load at the ith level, L_i the story height, and the story stiffness $\beta_i = R_M \Sigma H/\Delta_H$ in which R_M accounts for $P-\delta$ effects and is taken as 1.0 for braced frames and 0.85 otherwise, Δ_H is the first-order interstory drift due to lateral forces, and ΣH is the story shear produced by the lateral forces used to calculate Δ_H.

Effective Length Method In this traditional AISC approach, the nominal column buckling resistance P_n is calculated using a single column curve that is a function of the in-plane or out-of-plane effective buckling lengths (KL). The effective lengths are usually calculated by use of alignment charts that may be adjusted to reflect differences between the elastic stiffness of the restraining beams and the potentially inelastic stiffness of the column. There are many complicating aspects to the proper calculation of effective lengths, and a thorough presentation of this topic is provided in an ASCE Task Committee on Effective Length report (ASCE Task Committee, 1997).

In several recent studies (Deierlein et al., 2002; Maleck and White, 2003; Surovek-Maleck and White, 2004a and 2004b), it was observed that the effective-length method could produce significantly unconservative results, especially in symmetric framing systems with low redundancy and high gravity-to-horizontal load ratios. Specifically, the second-order moment was being largely underpredicted as a result of not including an initial out-of-plumbness when modeling frames with nearly symmetrical geometry and loading patterns. Consequently, the 2005 AISC specification placed two additional requirements on the *effective length method:*

1. For load combinations that do not include a lateral load component (i.e., gravity loads only), a notional lateral load of $N_i = 0.002Y_i$ must be included, where Y_i is the total factored gravity load at the ith level.

2. The *effective length method* may be used only for frames with sufficient sidesway stiffness, which is defined by a ratio of second-order to first-order drift ($\Delta_{2nd}/\Delta_{1st}$) being less than 1.5.

Direct Analysis Method One of the most significant developments appearing in the 2005 AISC specification is that column strength may be based simply on the unbraced member length, thereby eliminating the need to calculate effective-length factors (i.e., the effective length factor may be taken as $K = 1.0$). As a result of calibration with sophisticated spread of plasticity analysis methods, the *direct analysis method* provides an improved representation of the internal force distribution at a structure's ultimate strength limit state. The method also has the advantages of being (1) applied in a logical and consistent fashion to all types of structural systems, including moment frames, braced frames, and combinations thereof, and (2) expanded in the future for use in conjunction with inelastic analysis (Lu et al., 2009; Surovek and Ziemen, 2005). The method has been further explored by the cold-formed steel industry (Sarawit and Peköz, 2006; Sarawit, 2003) and was recently adopted in the AISI (2007), *North American Specification for the Design of Cold-Formed Steel Structural Members*. It should also be noted that several aspects of the method are common to other major design specifications, including ACI 318, the Eurocodes, the Australian Standard, and the Canadian Standard.

With heavy reliance on an accurate second-order elastic analysis, the *direct analysis method* incorporates frame stability within the design process by employing several key features of the modified stiffness and notional load concepts presented in Section 16.4.1. At the heart of the method are adjustments made for geometric imperfections and the effects of partial yielding and residual stresses.

Geometric Imperfections. When their effects on stability are significant, initial geometric imperfections must be considered. In the development and calibration of the AISC *direct analysis method* provisions, such imperfections were assumed to equal the maximum fabrication and erection tolerances permitted by the AISC *Code of Standard Practice for Steel Buildings and Bridges* (AISC, 2005a), which for member out-of-straightness equals $L/1000$ and for frame out-of-plumbness $L/500$, where L is the member length or story height. By employing a column curve that accounts for member initial out-of-straightness, only the remaining initial out-of-plumbness needs to be incorporated within the analysis. This can be done by either explicitly including the imperfection within the geometry of the analysis model or using an equivalent notional lateral load of $N_i = 0.002Y_i$, where Y_i is the total factored gravity load applied at the ith level. For cases when the ratio of second-order drift to first-order drift $\Delta_{2nd}/\Delta_{1st}$ does not exceed 1.5, the imperfection or equivalent notional load need only be applied in the gravity-only load combinations and not in combination with other lateral loads.

Partial Yielding and Residual Stresses. The nominal stiffnesses of all components (including members, connections, etc.) contributing to the stability of the structural system are reduced by a uniform factor of 0.8. This factor accounts for the effects of partial yielding and uncertainties in the strength and stiffness of such components at the structure's strength limit states. For slender members, it is also noted that 0.8 is approximately the product of the resistance factor ϕ =0.9 and 0.877, the factor used within the AISC column curve (Eq. 3.16b) to modify the Euler buckling load to account for member out-of-straightness. Because the combination of high axial load and partial yielding (accentuated by the presence of residual stresses) has a greater effect on bending stiffness, particularly for cases of minor axis flexure of I-shapes, the bending stiffness of members with an axial load P_u in excess of $0.5P_y$ is further reduced by the stiffness reduction factor $\tau = 4\left(P_u/P_y\right)\left(1 - P_u/P_y\right)$, where the axial yield load $P_y = AF_y$. In lieu of applying the τ-factor, an additional notional lateral load of $N_i = 0.001Y_i$ may be added to the factor load combinations being investigated, where Y_i is the total factored gravity load applied at the ith level.

First-Order Elastic Analysis Method The *first-order elastic analysis method* is a simplified version of the *direct analysis method*, with additional notional lateral loads being applied to simulate second-order and partial yielding effects. To avoid the detrimental combination of large compressive forces and partial yielding, the axial force P_u in all members that contribute to the lateral stability of the structural system is limited to $P_u < 0.5\,P_y$. The notional lateral loads are back-calculated from a moment amplification factor B_2 that assumes (1) the ratio of second-order drift to first-order drift $\Delta_{2nd}/\Delta_{1st}$ equals 1.5 and (2) the stiffness of the lateral system is modified by the 0.8 factor used in the *direct analysis method*. Of course, the first assumption further requires that the method can only be applied when $\Delta_{2nd}/\Delta_{1st} \leq 1.5$. The resulting notional lateral load N_i that must be applied at each level i is

$$N_i = 2.1\left(\frac{\Delta}{L}\right)Y_i \tag{16.9}$$

in which Δ is the first-order interstory drift, L the story height, and Y_i the total factored gravity load applied at the ith level. Using a minimum first-order drift ratio of $\Delta/L = 0.002$, the minimum required notional lateral load N_i that must be included in all load combinations is

$$N_i = 0.0042Y_i \tag{16.10}$$

Note that a target maximum drift ratio Δ/L is often defined as a constraint in the design process. When this is the case, the defined ratio may be used in Eq. 16.9 to calculate the notional load, and as long as this ratio is not exceeded at strength load level, the design will be conservative.

Obviously, the two big advantages of this method are that only a first-order elastic analysis is required and effective length factors need not be calculated. Of course, these come with costs, including limitations on both the axial force in compression members, ratios of first-order to second-order effects ($B_2 \leq 1.5$), and the potentially weighty requirement of applying a notional lateral load that can be 1.4 to 2.1 times that needed in the *direct analysis method*.

16.5.2 Basic Approaches of the AISC Specification

As indicated in Chapter 8, the AISC specification provides for general use a single interaction equation to assess both in-plane and out-of-plane member strength. For the special case of doubly symmetric members in flexure and compression with moments primarily in one plane, two interaction equations may be employed, with one to check for in-plane instabilities and the other for out-of-plane flexural or flexural–torsional buckling. For in-plane instabilities, the AISC beam-column interaction equation either (1) implicitly accounts for geometric imperfections, partial yielding, and residual stress effects by using a column strength curve that is based on effective lengths or (2) explicitly accounts for these effects by use of the *direct analysis method* presented above with column strength based on actual member length. The basic concepts behind these approaches is illustrated in Fig. 16.20, where it is shown that the AISC LRFD beam-column interaction curves are a fit between two sets of "anchor points" representing the nominal strength under pure compression (P_n) and bending (M_n). These points are determined as follows:

1. *Nominal compression strength*, P_n: This represents the nominal axial strength of the column as limited by the effects of lateral buckling either in or out of the plane of the frame. When the *effective length method* is employed, $P_n = P_{n(KL)}$, the nominal axial strength is determined considering the limit of stability of the framing system, or of a local subassembly within the framing system, where all the columns are concentrically loaded such that the distribution of column forces is comparable to that which would occur under the factored design load. The nominal resistance $P_{n(KL)}$ is calculated using a single column curve that is a function of the effective buckling length; in other words, the nominal member resistance represented by the interaction equation is reduced inward to account for frame stability effects. When the *direct analysis method* is used, the nominal axial strength $P_{n(L)}$ is based on the same column curve but in this case is a function of the actual member length. Hence, as a result of the analysis being modified to account for frame stability effects and in turn calculating larger moments and forces associated with inelastic deformations and column out-of-plumbness, the interaction equation remains fixed at defining individual member strength as a function of unbraced length.
2. *Nominal bending strength*, M_n: The nominal moment strength is determined for bending in the plane of the frame considering the potential for elastic or inelastic lateral–torsional buckling out of plane. The nominal strength equation for calculating lateral buckling is based on the distribution of bending moments in the member per a second-order elastic analysis and assuming that the ends of the member are fully restrained against translation and

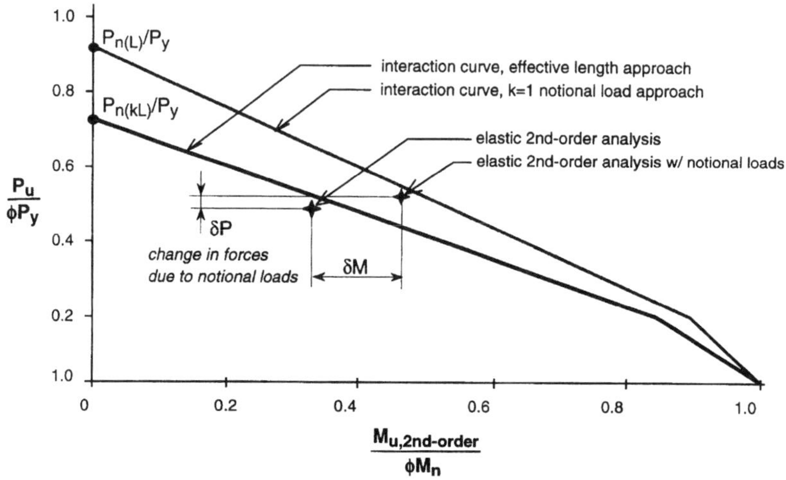

FIGURE 16.20 Distinction between effective length and notional load approaches (White and Clarke, 1997).

torsional rotation but free to warp and to rotate in or out of plane. In doing so, the beneficial effects of continuity with other framing members are neglected.

For cases governed by in-plane stability, the AISC beam-column design equations are calibrated to results from distributed plasticity analyses of small nonredundant frames, that include geometric imperfections and residual stresses (Yura, 1988; Yura et al., 1996; ASCE Task Committee, 1997). Calculated strengths for cases involving both in- and out-of-plane failures are also verified against test data (Kanchanalai, 1977; Yura et al., 1996).

16.5.3 Basic Approaches and Comparison of Other Design Standards

The stability provisions appearing in the AISC, Canadian (CSA-S16-01), European (Eurocode 3), and Australian (AS4100) steel design standards share several common features. A summary of these similarities along with subtle differences are provided below.

Member Length. All four of the standards permit the use of actual member length in frame and beam-column stability checks. Eurocode 3 is the only standard that allows the use of K factors that are less than 1 in the member stability checks for a sway frame.

Notional Loads. There are various considerations used in each standard for defining notional loads. AISC uses $0.002Y_i$ to account for initial frame out-of-plumbness and permits $0.001Y_i$ to represent inelastic effects at substantial axial loads in lieu of the more precise use of τ. CSA-S16-01 uses $0.005Y_i$ to account for both out-of-plumbness and inelastic effects, whereas AS4100 specifies a lower value of $0.002Y_i$, which reflects only the out-of-plumbness. Eurocode 3 employs a notional load of $\sqrt{0.5(1 + 1/m)}(2/\sqrt{h})0.005Y_i$, which in addition to geometric imperfections and partial yielding effects includes consideration of the number of columns m in a story and the height of the frame h (in meters). Eurocode 3 and CSA-S16-01 require notional loads be applied in combination with all loads, whereas AS4100 only requires that notional loads be used in the factored gravity load check and AISC requires notional loads for all gravity cases and to be additive with lateral loads for cases when the second-order to first-order drift ratio exceeds 1.5. Most of the standards are explicit in requiring that notional loads be applied in orthogonal directions to acknowledge the possibility of out-of-plumbness in both directions, but CSA-S16-01 requires that notional loads be applied only in the direction of sway.

Modified Stiffness. In allowing for the use of actual member length in design equations, the AISC specification is the only standard that uses a reduced lateral stiffness to account for inelastic effects.

Frame Flexibility. In all standards, the significance of second-order effects can be established by use of a sidesway moment amplification B_2 factor. The general method of using first-order elastic analysis supplemented by moment amplification factors to account for second-order effects is permitted when $B_2 \leq 1.5$ in the AISC and Eurocode 3 standards and $B_2 \leq 1.4$ in AS4100. Although previous editions of CSA-S16-01 required $B_2 \leq 1.5$, this restriction has been removed (Essa and Kennedy, 2000).

Interaction Equation. All of the standards rely on using a second-order elastic analysis in conjunction with interaction equations to confirm frame and beam-column stability. With the exception of AISC employing a single column curve, the nominal compressive strength term in the interaction equations, denoted by $P_{n(L)}$ in Fig. 16.20, is generally based on multiple column curves. The nominal moment strength, M_n, is by and large calculated in the same way by all of the standards, with AISC and CSA-S16-01 using essentially the same three-zone curve, AS4100 using a single continuous curve, and Eurocode 3 employing three curves. As indicated in Chapter 8, the Canadian and Australian standards include separate interaction equations to check (1) cross-section strength, (2) in-plane member strength, and (3) out-of-plane strength. Eurocode 3 uses two approaches to the coefficients appearing in the interaction equations: one with relatively complex coefficients that are intended to provide transparency to the physics of the problem and the other using much more compact coefficients to ease calculation.

Allowable Strength Design. Of all the standards, only the AISC specification recognizes both LRFD and ASD. Because the latter is based on service loads, the specification requires that these loads be multiplied by a factor of 1.6 when using a second-order elastic analysis to determine the distribution of internal member forces and moments. Frame and member stability is then assessed by dividing the resulting forces and moments by 1.6 and employing the ASD format of the interaction equation.

Beyond these conceptual differences, there are many additional differences in the provision formats and "shapes" of the interaction curves of the three standards. For example, AS4100 provide an "enhanced" strength interaction formula for members that have class 1 (i.e., plastic design type) cross sections, whereas the American, Canadian, and European standards do not distinguish between the design strengths for class 1 and class 2 type sections.

16.5.4 Summary

All four of the specifications described above share the common requirement for determining member forces using second-order elastic analysis. Most also permit the use of amplification factors to approximate second-order effects using first-order elastic analyses, although for the reasons noted earlier in this chapter it is preferable to perform a more rigorous second-order analysis, especially given the availability of modern computers and software. AISC, Eurocode 3, CSA-S16-01, and AS4100 are similar in that they rely solely on second-order elastic analyses coupled with the application of notional loads as the primary method for checking frame stability. In recognition that commonly available frame analysis programs do not model torsional–flexural member instabilities, all of the specifications maintain reliance on beam-column interaction equations for out-of-plane member instabilities.

16.6 STRUCTURAL INTEGRITY AND DISPROPORTIONATE COLLAPSE RESISTANCE

The built environment is subjected to many natural and man-made hazards. In fact, it has been suggested that the "risk of structural failure due to fire, vehicular impact, explosions, and other abnormal events in some instances may be comparable to the risks associated with hazards that traditionally have been addressed in structural design" [National Institute of Standards and Technology (NIST), 2005]. Effectively managing the risk associated with low-probability high-consequence hazards (e.g., terrorist-created explosion) is a substantial challenge facing owners and structural engineers; hence, it is one of the primary reasons for the development of performance-based and reliability-based engineering methodologies for design. At the heart of these design methods is the advanced structural analysis techniques discussed earlier in this chapter.

When an abnormal loading event occurs, it is likely that localized damage will be initiated somewhere within the structural framing system. If this relatively localized failure propagates into a failure that involves extensive damage, partial collapse, or full collapse of the building framing system, the collapse is termed *disproportionate* to the instigating event. This type of failure has also been coined *progressive collapse*. It is impossible to fully describe all aspects of design procedures for disproportionate collapse mitigation in this chapter. It is, however, desirable to provide an overview of the state-of-the-art in this subject area and, in doing so, outline how consideration of frame stability that is based on using sophisticated analysis procedures can become an integral tool for evaluating the potential for a disproportionate collapse event. Additional perspectives on this topic are provided by Nair (2004, 2006) and Hamburger and Whittaker (2004).

There are three excellent documents that provide design guidelines and procedures for mitigating the potential for disproportionate collapse [General Services Administration (GSA), 2003; DOD, 2005; and NIST, 2007]; note that national standards such as ASCE (2005) 7-05, *Minimum Design Loads for Buildings and Other Structures,* subscribe to the design philosophies included in these references. There are two fundamental approaches that can be taken to mitigate the risk of disproportionate collapse:

Direct Methods. These methods, which are becoming the basis for performance-based or reliability-based design methodologies, include two general approaches. The designer may quantify the ability of the structure to resist damage resulting from abnormal loading events and enhances this ability through engineering intervention, a procedure termed the *specific local resistance method*. As an alternative, the *alternate load path method* may be used, in which the designer admits to local damage occurring and then proceeds to ensure that the structural system is able to redistribute loading around the damaged area and remain stable while rescue operations and evacuation take place. This approach can also be thought of as means with which to ensure redundancy exists in the gravity load structural system. The specific local resistance method is a threat-dependent methodology and the alternate load path method is threat independent. However, one could argue that the alternate load path approach also requires that specific damage state scenarios be defined in order to allow the engineer to simulate the response of the structural system to them.

Indirect Methods. The methods are intended to provide a structural system with enhanced robustness by incorporating explicit measures to significantly increase the likelihood that the system will maintain general and inherent structural integrity. Although these measures can be considered prescriptive, they are relatively simple to implement by following a well-defined procedure that occurs during the standard design process. Means for improving structural integrity include integrating vertical and horizontal ties within the structural system; ensuring enhanced levels of ductility in the components of the system; and ensuring that connections within the system have demonstrable levels of strength and ductility. Details for these means, such as minimum recommended tie force magnitudes, are provided in many design guidelines (GSA, 2003; DOD, 2005; NIST, 2007). Indirect methods are typically provided in prescriptive-type design codes for buildings.

The following assumes that the engineer is considering a direct-design approach, specifically an alternate load path methodology. Because a comprehensive evaluation of the redundancy present in the structural system is needed, a structural analysis of the compromised system is essential. Several considerations must be made with respect to the analysis techniques employed, including:

- Extent to which nonlinear geometric effects are important (large vs. small displacements, large vs. small rotations, large vs. small strains)
- Material response (constitutive models, rate dependency)
- Importance of macro- versus microlevel behavior on response
- Overall stability of the structure and its components

Each of these issues will be discussed in the following sections. The methods of analysis presented in Section 16.2 are emphasized, noting a disproportionate collapse involves nonlinear behavior that is similar to that shown in Fig. 16.1, but most often in the absence of lateral loading.

16.6.1 Nonlinear Geometric Effects

As a structural system begins the process of reacting to damage, the deformations incurred will most likely be greater than those encountered in the evaluation of serviceability limit states. For example, damage or loss of a column within a moment-resisting frame will likely result in significant vertical deformation in the system as it redistributes loading and forms alternate load paths. While the geometric nonlinear behavior was not of significant concern in the original design of a girder or beam, it may become a predominant issue as this member compensates for column damage. Therefore, the need to conduct an accurate second-order (geometric nonlinear) analysis is likely to be very important. Further consideration of nonlinear kinematic effects may be needed to account for the resulting fiber strains (normal and shear) in the member. As a result, both large deformations and large strains may need to be incorporated into the analysis.

In most cases, the use of geometric stiffness matrices, such as those presented in McGuire et al. (2000), are appropriate for simulating the response of structural systems in damaged configurations. It should be noted that contributions to the geometric nonlinear behavior for all members, including columns and beams, should be included in system stiffness formulations. Furthermore, the location of damage within the structural system may result in "folding type" mechanisms in the column elements, thereby making P–δ effects significant. On the other hand, the gravity loading present at the occurrence of damage is often only a fraction of that used in the original design, thereby reducing P–δ effects. Without precisely knowing which controls, it is recommended that columns be discretized into at least two elements per member. As an alternative to the use of geometric stiffness matrices, approximate methods such as the use of moment amplification factors (as described in Section 16.2.1) can be employed.

It is highly likely that inelastic material behavior will occur during abnormal loading events. The use of second-order elastic analysis, however, can still be a very useful tool for identifying potentially important response characteristics and for preparing subsequent detailed studies using first-order or second-order materially nonlinear analyses.

16.6.2 Material Nonlinearity and Rate Dependency

It is often assumed that the response to abnormal loading events and development of alternate load paths in a damaged structural system involves nonlinear material behavior. As shown by Foley et al. (2007, 2008a, 2008b), this may or may not be the case. It is often good practice to study the response of the structural system with elastic material behavior and evaluate the extent to which nonlinear geometric effects are important. Ensuing analysis efforts can then be expanded to include the effects of material nonlinearity (Foley et al., 2007).

The strain rates observed or hypothesized in the creation of damage to a structural system (e.g., blast loading) are often very high and the material's response to strain rates of this magnitude requires special simulation procedures. If the investigation admits to damage in the structural system (i.e., subscribes to an alternate load path methodology), then the rate at which the damage is translated to response in the structural system is not as important. Foley et al. (2007) evaluated shear and normal strain rates after instantaneous column loss and determined that the strain rates were in the intermediate range. They concluded that elevated yield-stress effects and lowered toughness need not be included in the response simulation if localized connection details could be ignored (a condition which may or may not be realistic). Use of the specific local resistance method, however, may indeed require rate-dependent models for materials in order to simulate the extent of damage incurred by the system.

For steel structures, an evaluation of a system's ability to generate alternate load paths can be obtained by employing a computational model that is based on bare-frame behavior and employs a plasticity model defined by yield surfaces such as those shown in Figure 16.5*a*. It should be noted, however, that the material nonlinear behavior occurring at the ends of girders in a compromised system is a function of both the member and the girder-to-column connection properties. Therefore, yield surfaces used to model such behavior may need to be adjusted to include connection capacity; a process for doing this is described by Foley et al (2007, 2008a,b). In many cases, experimental testing of steel connections subjected to combined axial tension, shear, and bending moment demands is needed before these yield surfaces can be adequately calibrated and implemented with confidence. This is indeed a critical area of research that must be performed before the profession can comprehend and potentially take full advantage of the inherent resistance of steel framing systems to disproportionate collapse.

16.6.3 Macro- versus Microlevel Modeling of Behavior

Simulations of structural response can vary in complexity, depending on the level of detail investigated. As outlined earlier in this chapter, the most basic approach to modeling inelastic response is to employ a concentrated plasticity or plastic hinge model. This macrolevel modeling of behavior can be refined to a microlevel by using more advanced plastic zone or distributed plasticity analysis methods; in doing so, material-level response rather than overall cross-sectional response can be investigated.

When connection behavior must be included in predicting response, similar macro- and microlevel modeling approaches can be formulated. A macrolevel approach would simulate connection behavior with a linear or nonlinear spring, including hysteretic response or backbone modeling as appropriate. If fracture initiation in the localized details of the connection are considered exceedingly important to the simulation, then microlevel procedures consisting of detailed finite element modeling of the bolts, welds, connecting elements, and so on, should be included. Although the nonlinear behavior of the connection can be modeled to different degrees of complexity, the cost of preparing and running computational models at the microlevel may prohibit the analysis of realistic systems. Lim and Krauthammer (2006) provide a study of a wide range of connection models and their impact on the progressive collapse of steel frames.

Component modeling provides for the combined use of micro- and macrolevel modeling. In this approach, a macrolevel model is used to simulate the behavior of beams and columns (i.e. use of one-dimensional elements based on elastic, geometric, and plastic reduction matrices) and localized detailed-finite element models, similar to or perhaps more sophisticated than those presented in Section 16.3.4, are utilized to simulate microlevel connection behavior. In this manner, the stability analysis of damaged systems can be modeled with a resolution that is appropriate to the structural components, while at the same time at a computational efficiency that is needed to investigate complete systems.

16.6.4 System Stability and Linear Buckling Analysis

When considering the response of a steel framing system to damage, the overall stability of the structure must be ensured. Simple examples include the loss of girders providing end restraint to a column or the loss of story bracing, both tending to significantly increase the effective length of the column and thereby reduce its strength.

In general, a second-order inelastic analysis of the compromised system is the most appropriate method to assess strength limit states of damaged structures. Computing the buckling loads and corresponding modes using linearized eigenvalue analysis (Eq. 16.4) can be very useful for determining a variety of stable (in equilibrium) configurations that the system could shift to in a compromised state.

Buckling analysis for assessing stability should be used with caution, because the deformed shape associated with buckling of the compromised system will most likely be considerably different from that represented by the buckling mode of an elastic eigenvalue analysis. In this regard, a linear buckling analysis should only be used as a reference, providing insight for further detailed analyses. Foley et al. (2007) outline a methodology that incorporates linear eigenvalue analysis as an initial investigation into the response of a compromised frame.

16.6.5 General Commentary

This section has introduced basic concepts of analysis and design for disproportionate collapse. It should be noted that response to damage resulting in the loss of a column, girder, or portion of structural system is most accurately simulated by an inelastic transient analysis that accounts for geometric nonlinear effects. Although a distributed plasticity analysis can provide detailed representation of such behavior, the use of a yield surface approach is appropriate, especially when the yield surfaces are adjusted for elevated yield-stress effects resulting from dynamic strain rate.

16.7 CONCLUDING REMARKS

Although the second-order analysis methods prevalent in design practice go a long way toward addressing the problem of frame stability, they are still limited in their ability to model inelastic instability of frame and member behavior. This is particularly true for out-of-plane member response involving torsional–flexural behavior. Therefore, design standards continue to rely on methods that combine second-order elastic analysis with semiempirical member design equations and procedures to design for frame and member stability.

A key question in dealing with frame stability is the interplay between member and frame instability and whether they can be addressed together or separately. In this regard, it is important to distinguish between cases governed by in-plane versus out-of-plane stability effects. Cases that can be dealt with considering only in-plane behavior are planar frames with members that are either (1) fully braced in the out-of-plane direction or (2) oriented with their minor axes in the plane of bending such that out-of-plane failure is unlikely. In all other cases, out-of-plane or combined in- and out-of-plane behavior must be considered.

In-Plane behavior Available second-order analysis methods are capable of modeling in-plane member and frame instabilities in a single analysis. In particular, spread-of-plasticity analyses that include rigorous second-order routines to pick up both P–Δ and P–δ effects can be used as the sole basis for calculating strength and stability limit states of the frame and its members. Such analyses must, of course, include the effects of initial geometric imperfections, joints and connections, and so on, when they are significant. Short of performing detailed spread-of-plasticity analyses and, subject to certain limitations, elastic analyses can be used to calculate the first-hinge limit state for in-plane behavior. As with spread-of-plasticity analyses, any method that does not include separate member checks must take rigorous account of both P–Δ and P–δ effects, considering any significant initial geometric imperfections. In the absence of definitive guidelines to quantify the limitations and adjustments to elastic analyses, most current design standards require that member force interaction equations for in-plane behavior be used in conjunction with second-order elastic analysis methods.

Out-of-Plane Behavior Frames encountered in design practice may include members that can fail in an out-of-plane torsional–flexural mode, typically under combined axial load and major axis bending. Torsional–flexural behavior also arises in space frames where members are subjected to biaxial bending. At the present time, practical analysis techniques to account for this mode of behavior, which involves modeling of nonuniform (or warping) torsion, are not generally available. It is largely for this reason that current design standards require separate member-based strength and stability checks. It should be noted, however, that, especially in the case of planar frames, out-of-plane member behavior and design equations to check for this have much less interaction with overall frame behavior than do in-plane member checks. Similarly, notional story loads used, now common to many standards, only apply for in-plane behavior and do not affect the out-of-plane checks. Therefore, there is more justification for separately treating member and frame stability checks for out-of-plane behavior.

Final Remarks on Frame Stability With modern computer-based methods, the preference should be to conduct rigorous elastic second-order analyses that include as a minimum story $P-\Delta$ effects and, preferably, member $P-\delta$ effects. Such analyses should be three dimensional, taking into account the following factors to the extent that they are significant: (1) gravity loads carried by both the lateral resisting systems and "gravity" or "leaning" columns, (2) connection and joint flexibility and finite size, (3) initial geometric imperfections—both in elevation and plan, and (4) other modeling considerations. Without a clear consensus on the limits of elastic analysis to model inelastic stability effects, current codes and standards feature a variety of methods. It has been the intent of this chapter (particularly Section 16.4) to clarify the major issues and set a direction for the development of more consistent and rational approaches for frame stability that are more straightforward and take better advantage of existing and emerging analysis and limit-state design methods.

REFERENCES

ACI (2008), *Building Code Requirements for Structural Concrete and Commentary*, ACI 318–08, American Concrete Institute, Farmington Hills, MI.

AISC (1993) *Load and Resistance Factor Specification for Structural Steel Buildings,* American Institute of Steel Construction, Chicago, IL.

AISC (2005a), *Code of Standard Practice for Steel Buildings and Bridges*, American Institute of Steel Construction, Chicago, IL.

AISC (2005b), *Specification for Structural Steel Buildings*, American Institute of Steel Construction, Chicago, IL.

AISI (2007) *North American Specification for the Design of Cold-Formed Steel Structural Members,* AISI-S100-07, American Iron and Steel Institute Washington, DC.

Alemdar, B. N., and White, D. W. (2005), "Displacement, Flexibility and Mixed Beam-Column Finite Element Formulations for Distributed Plasticity Analysis," *ASCE J. Struct. Eng.*, Vol. 131, No. 12, pp. 1811–1819.

Alvarez, R. J., and Birnstiel, C. (1969), "Inelastic Analysis of Multistory Multibay Frames," *ASCE J. Struct. Div.*, Vol. 95, No. ST11, pp. 2477–2503.

Argyris, J. H. (1982), "An Excursion into Large Rotations," *Comput. Methods Appl. Mech. Eng.*, Vol. 32, pp. 82–155.

ASCE Task Committee on Effective Length (1997), *Effective Length and Notional Load Approaches for Assessing Frame Stability: Implications for American Steel Design*, American Society Civil Engineers, New York.

ASCE Task Committee on Design Criteria for Composite Structures in Steel and Concrete (1998), "Design Guide for Partially Restrained Composite Connections," *ASCE J. Struct. Eng.*, Vol. 125, No. 10, pp. 1099–1114.

ASCE (2005), *ASCE/SEI 7-05, Minimum Design Loads for Buildings and Other Structures,* American Society of Civil Engineers, New York.

Attalla, M. R. (1995), "Inelastic Torsional-Flexural Behavior and the Three-Dimensional Analysis of Steel Frames," Ph.D. thesis, Cornell University, Ithaca, NY.

Attalla, M. R., Deierlein, G. G., and McGuire, W. (1994), "Spread of Plasticity: A Quasi-Plastic Hinge Approach," *ASCE J. Struct. Eng.*, Vol. 120, No. 8, pp. 2451–2473.

Attalla, M. R., Deierlein, G. G., and McGuire, W. (1996), "Inelastic Torsional–Flexural Effects in Steel Structures: Theory and Applications," *Proc. 5th Int. Colloq. Stabil. Met. Struct.*, SSRC, Lehigh University, Bethlehem, PA, pp. 11–20.

Barsoum, R. S., and Gallagher, R. H. (1970), "Finite Element Analysis of Torsional–Flexural Stability problems," *Int. J. Numer. Meth. Engng.* Vol. 2, pp. 335–352.

Bathe, K. J. (1995), *Finite Element Procedures*, Prentice-Hall, Upper Saddle River, NJ.

Birnstiel, C., and Iffland, J. S. B. (1980), "Factors Influencing Frame Stability," *ASCE J. Struct. Div.*, Vol. 106, No. ST2, pp. 491–504.

Bjorhovde, R., Colson, A., and Brozzetti, J. (1990), "Classification System for Beam-to-Column Connections," *ASCE J. Struct. Eng.*, Vol. 116, No. 11, pp. 3059–3076.

Bridge, R. Q. (1998), "The Inclusion of Imperfections in Probability-Based Limit States Design," *Proceedings*, 1998 Structural Engineering World Congress, San Francisco, CA, July.

CEN (2005), Eurocode 3: *Design of Composite Steel and Concrete Structures, Part 1-1, General Rules and Rules for Buildings*, EN 1993-1-1, Comité Européen de Normalisation (CEN), European Committee for Standardization, Brussels, Belgium.

Chen, C. S. (1994), "Geometric Nonlinear Analysis of Three-Dimensional Structures," M.S. thesis, Cornell University, Ithaca, NY, pp. 197.

Chen, W. F., and Atsuta, T. (1976), *Theory of Beam-Columns: In-Plane Behavior and Design*, Vol. 1, McGraw-Hill, New York.

Chen, W. F., Goto, Y., and Liew, J. Y. (1996), *Stability Design of Semi-Rigid Frames*, Wiley, New York.

Chen, W. F., and Lui, E. M. (1991), *Stability Design of Steel Frames*, CRC Press, Boca Raton, FL.

Chen, W. F (1992), "Design Analysis of Semi-Rigid Frames with LRFD," in *Connections in Steel Structures II: Behavior Strength and Design* (Eds. R. Bjorhovde, A. Colson, G. Haaijer, and J. W. B. Stark) American Institute of Steel Construction, Chicago, IL, pp. 370–379.

Cheong-Siat-Moy, F. (1978), "Frame Design without Using Effective Column Length," *ASCE J. Struct. Div.*, Vol. 104, ST1, pp. 23–33.

Chiorean, C. G., and Barsan, G. M. (2005), "Large Deflection Distributed Plasticity Analysis of 3D Steel Frameworks," *Comput. Struct.*, Vol. 83, No. 19-20, pp. 1555–1571.

Christopher, J. E., and Bjorhovde, R. (1999), "Semi-Rigid Frame Design Methods for Practicing Engineers," *AISC Eng. J.*, Vol. 36, No. 1, pp. 12–28.

Clarke, M. (1994), "Plastic-Zone Analysis of Frames," in *Advanced Analysis of Steel Frames* (Eds. W. F. Chen and S. Toma), CRC Press, Boca Raton, FL.

Clarke, M., and Bridge, R. Q. (1992), "The Inclusion of Imperfections in the Design of Beam-Columns," *Proc. 1992 Annu. Tech. Session*, SSRC, Lehigh University, Bethelem, PA, pp. 327–346.

Clarke, M., and Bridge, R. Q. (1996), "The Design of Steel Frames Using the Notional Load Approach," *Proc. 5th Colloq. Stabil. Met. Struct.*, SSRC, Lehigh University, Bethlehem, PA, pp. 33–42.

Crisfield, M. A. (1991), *Nonlinear Finite Element Analysis of Solids and Structures*, Wiley, New York.

CSA (2001), *Limit States Design of Steel Structures*, CAN/CSA-S16-01, Canadian Standards Association, Rexdale, Ontario, Canada.

Deierlein, G. (2003), "Background and Illustrative Examples on Proposed Direct Analysis Method for Stability Design of Moment Frames," AISC TC 10 Report, July.

Deierlein, G. G., Hajjar, J. F., Yura, J. A., White, D. W., and Baker, W. F. (2002), "Proposed New Provisions for Frame Stability Using Second-Order Analysis." *Proc. 2002 Annu. Tech. Session*, SSRC, Seattle, Apr.

Department of Defense (DOD) (2005), *Unified Facilities Criteria: Design of Structures to Prevent Progressive Collapse*, Washington, DC.

Duan, L., and Chen, W. F. (1990), "A Yield Surface Equation for Doubly Symmetrical Sections," *Eng. Struct.*, Vol. 12, No. 2, pp. 114–119.

ECCS Technical Committee 8 (2006), *Rules for Member Stability in EN 1993–1–1, Background Documentation and Design Guidelines*, European Convention for Constructional Steelwork, Brussels, Belgium.

El-Tawil, S., and Deierlein, G. G. (1996), "Inelastic Analysis of Mixed Steel–Concrete Systems," *Proc. 5th Int. Colloq. Stab. Met. Struct.*, SSRC, Lehigh University, Bethelem, PA, pp. 53–62.

El-Zanaty, M. H., Murray, D. W., and Bjorhovde, R. (1980), "Inelastic Behavior of Multistory Steel Frames," Struct. Eng. Rep. No. 83, University of Alberta, Edmonton, Alberta, Canada.

Essa, H. S. and Kennedy, D. J. L. 2000, "Proposed Provisions for the Design of Steel Beam-columns in S16-2001," *Canadian J. Civil. Eng.*, Vol. 27, No. 4, pp. 610–619.

Foley, C. M. (1996), "Inelastic Behavior of Partially-Restrained Steel Frames Using Parallel Processing and Supercomputers," Ph.D. dissertation, Marquette University.

Foley, C. M. (2001), "Advanced Analysis of Steel Frames Using Parallel Processing and Vectorization," *Computer-Aided Civil and Infrastructure Engineering*, Blackwell Publishers, Vol. 16, No. 5, pp. 305–325.

Foley, C. M., Martin, K., and Schneeman, C. (2007), "Robustness in Structural Steel Framing Systems," Report to AISC, Chicago, IL.

Foley, C. M., Martin, K., and Schneeman, C. (2008a), "Quantifying and Enhancing Robustness in Steel Structures: Part 2—Floor Framing Systems," *AISC Eng. J.*, Vol. 45, No. 4, pp. 253–266.

Foley, C. M., Martin, K., and Schneeman, C. (2008b), "Quantifying and Enhancing Robustness in Steel Structures: Part 1—Moment-Resisting Frames," *AISC Eng. J.*, Vol. 45, No. 4. pp. 267–286.

General Services Administration (GSA) (2003), *Progressive Collapse Analysis and Design Guidelines for New Federal Office Buildings and Major Modernization Project*, GSA, Washington, DC.

Gerstle, K. H., and Ackroyd, M. H. (1990), "Behavior & Design of Flexibly-Connected Building Frames," *AISC Eng. J.*, Vol. 27, No. 1, pp. 22–29.

Goto, Y., and Miyashita, S. (1998). "Classification System for Rigid and Semi-Rigid Connections," *ASCE J. Struct. Eng.*, Vol. 124, No. 7, Sept., pp. 750–757.

Griffis, L. G., and White, D. W. (2010), *Stability Design of Steel Buildings—Design Guide*, American Institute of Steel Construction, Chicago, IL.

Guralnick, S. A., Singh, S., and Erber, T. (1984), "Plastic Collapse, Shakedown and Hysteresis," *ASCE J. Struct. Eng.,* Vol. 110, No. 9, pp. 2103–2117.

Guralnick, S. A., Erber, T., Stefanis, J., and Soudan, O. (1986), "Plastic Collapse, Shakedown, and Hysteresis of Multistory Steel Structures," *ASCE J. Struct. Eng.,* Vol. 112, No. 12, pp. 31–49.

Guralnick, S. A., Erber, T., Soudan, O., and Jixing He (1991). "Incremental Collapse of Framed Structures Subjected to Constant Plus Cyclically Varying Loads," *ASCE J. Struct. Eng.,* Vol. 117, No. 6, pp. 1815–1833.

Hamburger, R., and Whittaker, A. S. (2004), "Design of Steel Structures for Blast-Related Progressive Collapse Resistance," *AISC Modern Steel Const.*, Vol. 44, No. 3, pp. 45–51.

Hansoulle, T., Boissonnade, N., Goncalves, R., Camotim, D., and Jaspart, J. P. (2007), "Design of Beam-Columns in Steel Sway Frames: From the Actual Member to a Simply Supported Equivalent One," *Proc. 2007 Annu. Tech. Session*, SSRC, pp. 177–200.

Horne, M. R. (1975), "An Approximate Method for Calculating the Elastic Critical Loads of Multi-story Plane Frames," *Struct. Eng.*, Vol. 15, No. 6, pp. 242–248.

Horne, M. R. (1979), *Plastic Theory of Structures*, 2nd ed., Pergamon Press, Elmsford, NY.

Iffland, J. S. B., and Birnstiel, C. (1982), "Stability Design Procedures for Building Frameworks," AISC Proj. No. 21.62, AISC, Chicago, IL.

Kanchanalai, T. (1977), "The Design and Behavior of Beam-Columns in Unbraced Steel Frames," AISI Proj. No. 189, Rep. No. 2, Civil Engineering/Structures Research Laboratories, University of Texas, Austin, TX.

Kanchanalai, T., and Lu, L.-W. (1979), "Analysis and Design of Framed Columns Under Minor Axis Bending," *AISC Eng. J.*, No. 2, pp. 29–41.

Kemp, A. R. (1986), "Factors Affecting the Rotational Capacity of Plastically Designed Members," *Struct. Eng.*, Vol. 64B, No. 2, pp. 28–35.

Konig, J. A. (1987), *Shakedown of Elastic-Plastic Structures*, Elsevier, New York.

Krawinkler, H. (1978), "Shear in Beam-Column Joints in Seismic Design of Steel Frames," *AISC Eng. J.*, Vol. 15, No. 3, pp. 82–91.

Krawinkler, H., Alali, A., Thiel, C. C., and Dunlea, J. M. (1995), "Analysis of a Damaged Four-Story Building and an Undamaged Two-Story Building," in *SAC Rep. 95-04, Technical Report: Analytical and Field Investigations of Buildings Affected by the Northridge Earthquake*, SAC Joint Venture, Sacramento, CA, Chap. 3.

Leon, R. T. and Hoffman, J. J. (1995), "Plastic Design of Semi-Rigid Frames" *Connections in Steel Structures III, Proceedings of the Third International Workshop*, Trento, Italy, May.

Leon, R. T., Hoffman, J. J. and Staeger, T. (1996), *Partially Restrained Composite Connections*, Steel Design Guide Series 8, AISC, Chicago, IL.

LeMessurier, W. J. (1976), "A Practical Method of Second-Order Analysis: Part 1. Pin Jointed Systems," *AISC Eng. J.*, Vol. 13, No. 4, pp. 89–96.

LeMessurier, W. J. (1977), "A Practical Method of Second-Order Analysis: Part 2. Rigid Frames," *AISC Eng. J.*, Vol. 14, No. 2, pp. 49–67.

Liew, J. Y. R. (1992), "Advanced Analysis for Frame Design," Ph.D. dissertation, School of Civil Engineering, Purdue University, West Lafayette, IN.

Liew, J. Y. R., White, D. W., and Chen, W. F. (1993), "Second-Order Refined Plastic Hinge Analysis for Frame Design, Parts 1 and 2," *ASCE J. Struct. Eng.*, Vol. 119, No. 11, pp. 3196–3237.

Lim, J. and Krauthammer, T. (2006), "Progressive Collapse Analyses of 2D Steel-Framed Structures with Different Connection Models, *AISC Eng. J.*, Vol. 43, No. 3, pp. 201–215.

Lu, A. Y. C., MacRae, G. A., Peng, B. H. H., Ziemian, R. D., Hahn, C., Hyland, C., and Clifton, G. C. (2009). "Extended Direct Analysis of Steel Frames," *Struct. Eng. Soc. New. Zealand (SESOC) J.,* Vol. 22, No. 2.

MacGregor, J. G. (1993), "Design of Slender Concrete Columns, Revisited," *ACI Struct. J.*, Vol. 90, No. 3, pp. 302–309.

Maleck (Surovek), A. E. (2001), "Second-Order Inelastic and Modified Elastic Analysis and Design Assessment of Planar Steel Frames," Ph.D. dissertation, Georgia Institute of Technology, Atlanta GA.

Maleck (Surovek), A. E., and White, D. W. (1998), "Effects of Imperfections on Steel Framing Systems," *Proc. 1998 Annu. Tech. Session*, SSRC, Atlanta, Sept.

Maleck (Surovek), A. E., and White, D. W. (2003), "Direct Analysis Approach for the Assessment of Frame Stability: Verification Studies," *Proc. 2003 Annu. Tech. Session*, SSRC, Baltimore, MD, Apr.

McGuire, W. (1992), "Computer-Aided Analysis," in *Constructional Steel Design: An International Guide* (Eds. P. J. Dowling, J. E. Harding, and R. Bjorhovde), Elsevier Applied Science, New York, pp. 915–932.

McGuire, W., Gallagher, R. H., and Ziemian, R. D. (2000), *Matrix Structural Analysis,* Second Ed., John Wiley and Sons Publishers, Hoboken, NJ.

Nair, R. S. (1975), "Overall Elastic Stability of Multistory Buildings," *ASCE J. Struct. Div.*, Vol. 10, No. ST12, pp. 2487–2503.

Nair, R. S. (1987), "Tall Building Stability: Practical Considerations," *Proc. Struct. Congr. ASCE*, pp. 582–594.

Nair, R. S. (2004), "Progressive Collapse Basics," *AISC Modern Steel Constr.*, Vol. 44, No. 3, pp. 37–42.

Nair, R. (2006), "Preventing Disproportionate Collapse," ASCE *J. Perform. Constr. Facilities*, Vol. 20, No. 4, pp. 309–314.

Nair, R. (2009), "A Model Specification For Stability Design By Direct Analysis," *AISC Eng. J.*, Vol. 46, No. 1, pp. 29–37.

Neal, B. G. (1985), *The Plastic Methods of Structural Analysis,* 3rd ed., Chapman & Hall, London.

Neuenhofer, A., and Fillipou, F. (1997), "Evaluation of Nonlinear Frame Finite Element Models," *ASCE J. Struct. Div.*, Vol. 123, No. 7, pp. 958–966.

NIST (2005), *Final Report on the Collapse of the World Trade Center Towers, NIST NCSTAR 1—Federal Building and Fire Safety Investigation of the World Trade Center*, NIST, Technology Administration, U.S. Department of Commerce, Washington, DC.

National Institute of Standards and Technology (NIST) (2007), "Best Practices for Reducing the Potential for Progressive Collapse in Buildings," Report NISTIR 7396, NIST, U.S. Department of Commerce, Washington, DC.

Nukala, P. K. V. V. (1997), "Three-Dimensional Inelastic Analysis of Steel Frames," Ph.D. thesis, School of Civil Engineering, Purdue University, West Lafayette, IN.

Nukala, P. K. V. V., and White, D. W. (2004), "A Mixed Finite Element for Three-Dimensional Nonlinear Analysis of Steel Frames," *Comput. Methods Appl. Mech. Eng.*, Vol. 193, Nos. 23–26, pp. 2507–2545.

Orbison, J. G., McGuire, W., and Abel, J. F. (1982), "Yield Surface Applications in Nonlinear Steel Frame Analysis," *Comput. Methods Appl. Mech. Eng.*, Vol. 33, North-Holland, Amsterdam, pp. 557–573.

Pi, Y.-L., and Trahair, N. S. (1994), "Nonlinear Inelastic Analysis of Steel Beam-Columns: Theory and Applications," *ASCE J. Struct. Eng.*, Vol. 120, No. 7, pp. 2041–2085.

Popov, E. P., and McCarthy, R. E. (1960), "Deflection Stability of Frames Under Repeated Loads," *ASCE J. Eng. Mech. Div.*, Vol. 86, No. EM1, pp. 61–78.

Rutenberg, A. (1981), "A Direct P–Delta Analysis Using Standard Plane Frame Computer Programs," *Comput. Struct.*, Vol. 12, Nos. 1–2, pp. 97–102.

Rutenberg, A. (1982), "Simplified P–Delta Analysis for Asymmetric Structures," *ASCE J. Struct. Div.,* Vol. 108, No. ST9, pp. 1995–2013.

SAA (1998), *Steel Structures*, AS4100 Standards Association of Australia, Australian Institute of Steel Construction, Sydney, Australia.

Sarawit, A. T. (2003), Cold-Formed Steel Frame and Beam-Column Design, Ph.D. Dissertations, Research Report 03-03, Department of Civil and Environmental Engineering, Cornell University, Ithaca, NY.

Sarawit, A. T., and Peköz, T. (2006), "Notional Load Method for Industrial Steel Storage Racks," *Thin-Walled Struct.,* Vol. 44, No. 12, pp. 1280–1286.

Sotelino, E. D. (2003), "Parallel Processing Techniques in Structural Engineering Applications," *ASCE J. Struct. Eng.*, Vol. 129, No. 12, pp. 1698–1706.

Spacone, E., Ciampi, V., and Filippou, F. C. (1996), "Mixed Formulation of Nonlinear Beam Finite Element," *Comput. Struct.*, Vol. 58, No. 1, pp. 71–83.

Stevens, L. K. (1967), "Elastic Stability of Practical Multi-Story Frames," *Proc. Inst. Civ. Eng.*, Vol. 36, pp. 99–117.

Surovek, A. E., and Johnson, J. (2008), "Effects of Nonverticality on Steel Framing Systems—Implications for Design," *AISC Eng. J.*, Vol. 45, No. 1, pp. 73–85.

Surovek, A. E., White D. W., and Leon, R. T. (2005), "Direct Analysis for Design Evaluation and Design of Partially-Restrained Steel Framing Systems," *ASCE J. Struct. Eng.*, Vol. 131, No. 9, pp. 1376–1389.

Surovek, A. E., and Ziemian, R. D. (2005), "The Direct Analysis Method: Bridging the Gap from Linear Elastic Analysis to Advanced Analysis in Steel Frame Design," *Proc. Struct. Congr. ASCE,* New York, Apr.

Surovek-Maleck, A. E., and White, D. W. (2004a), "Alternative Approaches for Elastic Analysis and Design of Steel Frames. I: Overview," *ASCE J. Struct. Eng.*, Vol. 130, No. 8, pp. 1186–1196.

Surovek-Maleck, A. E., and White, D. W. (2004b), "Alternative Approaches for Elastic Analysis and Design of Steel Frames. II: Verification Studies," *ASCE J. Struct. Eng.*, Vol. 130, No. 8, pp. 1197–1205.

Trahair, N. S., and Chan, S.-L. (2003), "Out-of-Plane Advanced Analysis of Steel Structures," *Eng. Struct.*, Vol. 25, No. 13, pp. 1627–1637.

Welding Research Council/American Society of Civil Engineers (WRC/ASCE) (1961), *Commentary on Plastic Design in Steel,* WRC-ASCE Joint Committee Report, ASCE Manual No. 41, New York.

White, D. W., and Chen, W. F. (Eds.) (1993), *Plastic Hinge Based Methods for Advanced Analysis and Design of Steel Frames*, Structural Stability Research Council, Lehigh University, Bethlehem, PA.

White, D. W., and Clarke, M. (1997), "Design of Beam-Columns in Steel Frames: I. A Study of Philosophies and Procedures," *ASCE J. Struct. Eng.*, Vol. 123, No. 12, Dec. pp. 1556–1564.

White, D. W., and Hajjar, J. F. (1997), "Design of Frames without Consideration of Effective Length," *Eng. Struct.*, Vol. 19, No. 10, pp. 787–810.

White, D. W., and Nukala, P. K. V. V. (1997), "Recent Advances in Methods for Inelastic Frames Analysis: Implications for Design and a Look Toward the Future," *Proceedings, National Steel Construction Conference*, American Institute of Steel Construction, pp. 43-1–43-24.

White, D. W., Surovek, A. E. and Chang, C-J. (2007a), "Direct Analysis and Design Using Amplified First-Order Analysis. Part 2—Moment Frames and General Framing Systems," *AISC Eng. J.*, Vol. 44, No. 4, pp. 323–340.

White, D. W., Surovek, A. E. and Kin, S-C. (2007b), "Direct Analysis and Design Using Amplified First-Order Analysis: Part 1—Combined Braced and Gravity Framing Systems," *AISC Eng. J.*, Vol. 44, No. 4, pp. 305–322.

Wilson, E. L., and Habibullah, A. (1987), "Static and Dynamic Analysis of Multi-Story Buildings Including *P*–Delta Effects," *Earthquake Spectra*, Vol. 3, No. 2, pp. 289–298.

Yang, Y. B., and McGuire, W. (1986a), "Stiffness Matrix for Geometric Nonlinear Analysis," *ASCE J. Struct. Eng.*, Vol. 112, No. 4, pp. 853–877.

Yang, Y. B., and McGuire, W. (1986b), "Joint Rotation and Geometric Nonlinear Analysis," *ASCE J. Struct. Eng.*, Vol. 112, No. 4, pp. 879–905.

Yang, Y. B., and Quo, S. R. (1994), *Theory and Analysis of Nonlinear Framed Structures*, Prentice Hall, Upper Saddle River, NJ.

Yura, J. A. (1971–1972), "The Effective Length of Column in Unbraced Frames, *AISC Eng. J.*, Apr. 1971, pp. 37–42; "Closure to Discussion by Peter Adams," Jan. 1972, pp. 40–46.

Yura, J. A. (1988), *Elements for Teaching Load and Resistance Factor Design, Combined Bending and Axial Load*, American Institute of Steel Construction, Chicago, IL.

Yura, J. A., Kanchanalai, T., and Chotichanathawenwong, S. (1996), "Verification of Steel Beam-Column Design Based on the AISC–LRFD Method," *Proc. 5th Int. Colloq. Stab. Met. Struct.*, SSRC, Lehigh University, Bethelem, PA, pp. 21–30.

Ziemian, R. D. (1991), "Advanced Methods of Inelastic Analysis in the Limit States Design of Steel Structures," Ph.D. thesis, Cornell University, Ithaca, NY.

Ziemian, R. D. (1999), "Inelastic Critical Loads by Eigenvalue Analysis," *Proc. Struct. Congr. ASCE*, New Orleans, LA.

Ziemian, R. D., and McGuire, W. (2002), "Modified Tangent Modulus Approach: A Contribution to Plastic Hinge Analysis," *ASCE J. Struct. Eng.*, Vol. 128, No. 10, pp. 1301–1307.

Ziemian, R. D., McGuire, W., and Deierlein, G. G. (1992a), "Inelastic Limit States Design Part I: Planar Frame Studies," *ASCE J. Struct. Eng.*, Vol. 118, No. 9, pp. 2532–2549.

Ziemian, R. D., McGuire, W., and Deierlein, G. G. (1992b), "Inelastic Limit States Design Part II: Three-Dimensional Frame Study," *ASCE J. Struct. Eng.*, Vol. 118, No. 9, pp. 2550–2568.

Ziemian, R. D., McGuire, W., and Seo, D. W. (2008), "On the Inelastic Strength of Beam-Columns under Biaxial Bending," *Proc. SSRC Annual Stability Conference,* Nashville, TN.

Ziemian, R. D., and Miller, A. R. (1997), "Inelastic Analysis and Design: Frames with Members in Minor Axis Bending," *ASCE J. Struct. Eng.*, Vol. 123, No. 2, pp. 151–157.

CHAPTER 17

ARCHES

17.1 INTRODUCTION

An arch is a structure curved in elevation, loaded in its plane, with spreading of the supports prevented, and with its rib primarily in compression. The stability can be characterized by buckling in or out of the plane of the arch. In-plane buckling occurs when the arch is substantially braced against out-of-plane deformations, while out-of-plane buckling occurs for arches with significant free-standing portions. In-plane buckling is associated with combined compression and bending while out-of-plane buckling is associated with compression, biaxial bending, and torsion.

Arch-type structures are most efficient if they carry their load in such a way that the funicular curve coincides with the centroidal axis of the rib, which results in axial compression and no bending of the arch rib. Examples of arches under pure axial compression include circular arches subjected to uniform normal pressure, commonly called hydrostatic loading, parabolic arches subjected to uniform load on a horizontal projection, and catenary arches with load uniformly distributed along the arch axis. Arches can further be classified by their articulation: a three-hinged arch is statically determinate; two hinged and fixed arches are hyperstatic. In bridges, a distinction can be made between a through arch with the bridge deck between the supports and suspended from the arch rib and a deck arch with the bridge deck above the arch rib and supported by spandrel columns. The horizontal reaction forces (or thrusts) can be resisted in two ways: In the abutment arch, also called a true arch, the horizontal reactions are resisted by the abutments or supports, while in the tied arch, the reactions are carried by a tie between the two supports. The deck arch is an example of the former and the through arch the latter.

As shown in Fig. 17.1, arches are subject to two fundamentally different types of loading, including gravity-type loading and tilting loads, also referred

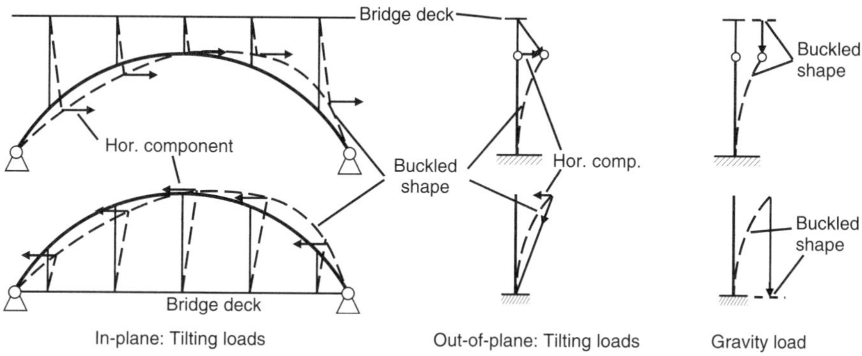

FIGURE 17.1 Gravity and tilting loads in arches.

to as conservative and nonconservative loading, respectively. In a through-arch bridge that is restrained to buckle in a sway mode, the hangers tilt and develop a horizontal-load component that supports the arch rib against further buckling both in and out of plane. Generally, the buckling loads associated with tilting loads are considerably higher than those of gravity loading cases. In a deck arch bridge, the in-plane horizontal-load component of the tilting spandrel columns tends to push the arch rib away from its funicular shape. The out-of-plane horizontal-load component drives the arch out of its plane and causes bending and torsion. Hence, the buckling loads for deck arch bridges are usually smaller than those of through arch bridges when considering gravity load conditions.

This chapter treats in-plane stability (Section 17.2) and out-of-plane stability (Section 17.3) for various shapes of arches and types of loading. In both cases, specific deign criteria are presented. Furthermore, bracing recommendations for arches are presented in Section 17.4 and stability design criteria for arch bridges are given in Section 17.5.

Because it is not feasible to give a complete overview of all research on the stability of arches within a single chapter, the more important and recent findings will be presented. There are, however, several excellent resources on arches. An extensive bibliography on the stability of arches prior to 1970 is given by DaDeppo and Schmidt (1970). The *Handbook of Structural Stability* (Hayashi, 1971) gives an overview of results of stability research of arches in which either the equations or graphs of the quoted literature are reproduced. Petersen (1982) dedicates a chapter to the in-plane and out-of-plane stability of arches and arch bridges. In *Stability of Metal Structures, a World View* (Beedle 1991), international building standards with provisions for in-plane and out-of-plane stability are compared with each other. An extensive state-of-the-art report on elastic and inelastic stability of arches is given in Fukumoto (1996). Singer et al. (1998) provides a chapter on experimental research that has been conducted on arches. King and Brown (2001) present a comprehensive study for the practical design of steel curved beams and arches, which includes several worked examples.

17.2 IN-PLANE STABILITY OF ARCHES

The analysis and design of arches may be broadly classified according to their response to load and their in-plane mode of failure. Based on these characteristics, the types of arches that will be treated in this chapter include the following:

- Slender arches, generally of solid-web rolled or built-up sections, subject primarily to axial force in the arch rib (analogous to axially loaded columns)
- Slender arches subject to significant bending and deformation, due primarily to asymmetric loading (analogous to beam-columns)
- Stocky arches, frequently of truss form, used because of high bending loads, in which strength limit states will be due primarily to chord or flange failure arising from axial load and excessive bending (similar to any truss subject to axial load and moment)
- Arches composed of arch ribs and deck-stiffening girders

When the loads acting on an arch are increased proportionally, it loses its stability as a certain critical value of the load is attained. In the case of elastic structures under conservative loads the critical load always corresponds to either a birfurcation point (see Section 2.2) or a stability limit point (Section 2.3). Various possible load–deformation plots for arches, called *equilibrium paths*, are shown in Fig. 17.2. Each point on a path represents an equilibrium configuration of the structure.

Figure 17.2*a* shows an equilibrium path for an unsymmetric problem in which the arch boundary conditions and/or the manner of loading may be unsymmetrical. The point on the equilibrium path at which the load is a relative maximum is called a limit point.

Figure 17.2*b* illustrates possible load–deformation relationships for a symmetrical and symmetrically loaded arch. Here the primary or fundamental equilibrium path is intersected by a secondary path. The primary branch represents a symmetric deflected mode and the secondary branch an antisymmetric mode, with bifurcation points defined where equilibrium paths intersect. If an antisymmetrical mode does not become dominant, the arch eventually will become unstable in a symmetrical mode with the load–deflection curve gradually reaching a limit point. On the other hand, the limit load may be significantly reduced if an antisymmetrical buckling mode dominates, as shown in Fig. 17.2*b*. Although the load capacity usually drops continuously beyond the bifurcation point, as is the behavior when the arch buckles in sidesway, it is possible for the load to increase slightly after bifurcation. In this case and as shown in Fig. 17.2*c*, the maximum load is attained at a limit point that corresponds to large in-plane displacements.

In the case of an unstable, rapidly descending postbuckling path, the arch is sensitive to geometric imperfections or load eccentricity. The associated drop in critical load is shown in Fig. 17.2*b* (dashed curve), and the buckling load at the bifurcation point is replaced by the snap load at the limit point on the dashed curve. The degeneration of a bifurcation point into a limit point due to the presence of imperfections suggests that bifurcation is an exception rather than the rule.

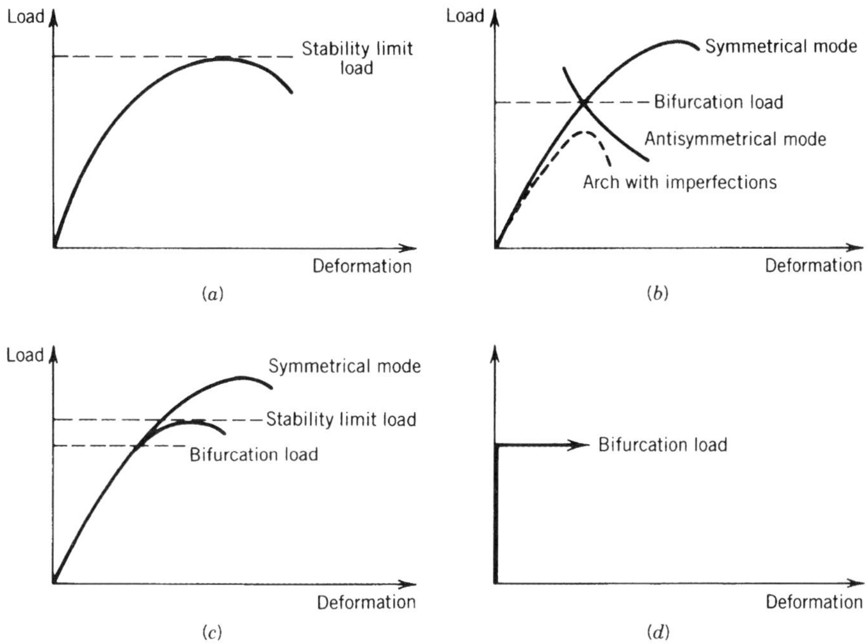

FIGURE 17.2 Some possible load–deformation paths for arches. (*a*) Nonlinear stability: unsymmetrical or unsymmetrically loaded arch. (*b*) Nonlinear stability: symmetrical and symmetrically loaded arch. (*c*) Nonlinear stability: symmetrical and symmetrically loaded arch. (*d*) Linear stability: negligible prebuckling deformation.

Arch ribs under pure compression experience only very small displacements before buckling. In the ideal situation such structures undergo birfurcation buckling. These bifurcation problems are far more amenable to analysis because they permit a linearization of the equilibrium equations for the prebuckling, or fundamental, state. As in small-deflection column theory, this linearization is based on the assumption of very small displacements and a horizontal equilibrium path (Fig. 17.2*d*).

For practical design loadings, however, the funicular curves normally do not coincide with the centroidal axis and in such cases the arch experiences substantial bending moments and displacements before buckling. Studies that take into account these prebuckling deformations are nonlinear stability problems, as illustrated by Figs. 17.2*a*–*c*.

17.2.1 Linear Stability

Early papers on arch stability were devoted to linear stability problems. This work by Gaber, Stüssi, Kollbrunner, Hilman, Dischinger, and Dinnik is summarized by Austin (1971) and Timoshenko and Gere (1961).

The buckling of arches of constant cross section (herein termed uniform arches) in which the arch is the funicular curve for the loading are first considered.

TABLE 17.1 Critical-Load Parameter and Critical Horizontal Reaction Parameter for Uniform Elastic Arches in Pure Compression[a]

$\frac{h}{L}$	Three-Hinged Arch		Two-Hinged Arch		Fixed Arch	
	qL^3/EI	HL^2/EI	qL^3/EI	HL^2/EI	qL^3/EI	HL^2/EI
Parabolic arches subjected to vertical load uniformly distributed on a horizontal projection						
0.10	22.5	28.1	29.1	36.3	60.9	76.2
0.15			39.5	32.9	85.1	70.9
0.20	39.6	24.8	46.1	28.8	103.1	64.5
0.25			49.2	24.6	114.6	57.3
0.30	49.5	20.6	49.5	20.6	120.1	50.0
0.35			47.8	17.1	120.6	43.1
0.40	45.0	14.1	45.0	14.1	117.5	36.7
0.50	38.2	9.6	38.2	9.6	105.3	26.3
Catenary arches subjected to vertical load uniformly distributed along the arch axis						
0.10			28.7	36.3	60.1	76.2
0.15			38.3	32.8	82.7	70.9
0.20			43.5	28.5	98.0	64.3
0.25			44.8	24.1	105.9	56.9
0.30			43.2	19.8	107.4	49.4
0.35			39.7	16.1	104.0	42.1
0.40			35.3	12.9	97.2	35.4
0.50			26.5	8.2	79.3	24.5
Circular arches subjected to normal load uniformly distributed along the arch axis						
0.10	22.2	26.7	28.4	34.1	58.9	70.7
0.20	33.5	17.6	39.3	20.6	90.4	47.5
0.30	34.9	9.3	40.9	10.9	93.4	24.9
0.40	30.2	3.4	32.8	3.7	80.7	9.1
0.50	24.0	0	24.0	0	64.0	0

[a] h, rise; L, span; q, critical intensity of distributed load; H, critical horizontal reaction at supports; E, Young's modulus of elasticity; I, moment of inertia of the cross section.

The critical values of distributed load and horizontal reaction of fixed two-hinged and three-hinged symmetric arches are summarized in Table 17.1 for the following cases: (1) parabolic arches subjected to vertical load uniformly distributed on a horizontal projection, (2) catenary arches under uniform vertical load along the arch axis, and (3) circular arches subjected to uniform normal pressure, commonly called hydrostatic loading. In each of these cases, the arch assumed is geometrically perfect and the loading results in pure axial compression (no bending) at every cross section of the arch. The arches are free to buckle in their plane without restraint. Axial compressive strain has been neglected in the analyses reported because it has been shown to have a very small effect on the behavior slender arches. In all cases, buckling is defined by bifurcation from the undeflected position, as repsented in

Fig. 17.2*d*. The critical values in Table 17.1 are given for a range of rise-to-span ratios from 0.10 to 0.50.

In Table 17.1, the critical values for two-hinged and fixed parabolic and catenary arches are taken from Austin and Ross (1976). The remaining values are from Timoshenko and Gere (1961). Kollbrunner (1936, 1942) showed that the elastic critical values obtained experimentally are in close agreement with the theoretical values reported in Table 17.1.

The fundamental modes that correspond to the critical values given in Table 17.1 are shown in Fig. 17.3. The fixed and two-hinged arches always buckle into an antisymmetrical mode in which the arch sways laterally, with the crown moving horizontally and becoming a point of contraflexure (Figs. 17.3*a* and *b*). In the case of a three-hinged arch, buckling is symmetric for low rise-to-span ratios, as shown in Fig. 17.3*c*, with the crown moving downward. For high rise-to-span ratios, the three-hinged arch buckles in an antisymmetrical mode similar to that of the two-hinged arch and at the same critical load.

The sidesway buckling behavior of uniform arches of low and moderate span-to-depth ratios under pure axial compression is similar to the buckling behavior of straight columns. For example, a fixed arch that buckles into two waves with a

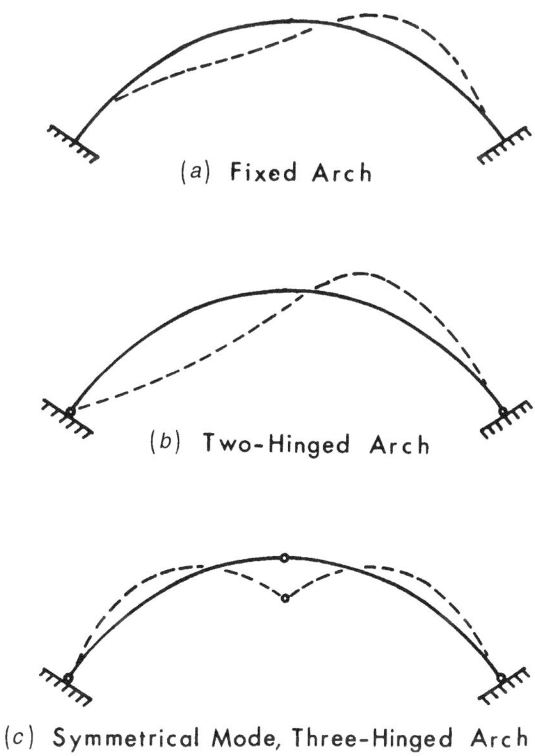

FIGURE 17.3 Fundamental buckling mode shapes.

point of contraflexure at the crown (Fig. 17.3*a*) has a mode shape from support to crown similar to that of a fixed-hinged column, and the critical axial thrust at the quarter-point of the arch agrees fairly well with the critical compressive force in the end-loaded, fixed-hinged column whose length is equal to the arc length of the arch from support to crown. This interpretation is developed fairly extensively in the third edition of this guide. It is not repeated here because it is not useful, except as a very broad concept, for studying the elastic buckling of slender arches under more general loadings that do not produce pure axial compression. It is used, however, in ultimate-strength studies and as a design tool, as will be described.

The elastic buckling of nonuniform parabolic arches subjected to vertical loading distributed uniformly on a horizontal projection has been studied for moment-of-inertia variations $I = I_c \sec \phi$ and $I = I_c \sec^3 \phi$, in which I_c is the moment of inertia at the crown and ϕ the angle between the tangent to the arch axis and the horizontal. These studies are reported in the third edition of the guide. A simple approximate procedure for applying uniform arch-buckling theory to cases of nonuniform arches with antisymmetrical buckling is to assume the critical load on a nonuniform arch equals that of a similar uniform arch with an "equivalent" moment of inertia. First, one-half of the nonuniform arch is straightened out to form a simply supported beam that is then loaded with a concentrated force at midspan. From the deflection at midspan, an equivalent uniform moment of inertia can be computed and used to represent a nonuniform arch. The result of such a computation is discussed in detail in the third edition of the guide. Similar procedures developed by Aas-Jakobsen are described by Forrester (1972) in a discussion of Austin's (1971) paper. More recent results on the stability of tapered arches are reported by Wolde-Tinsaie and Foadian (1989).

Stiffened arches are deck structures that consist of an arch connected to an overhead horizontal girder by closely spaced columns. As discussed in the third edition of this guide, the critical load and critical horizontal reaction for two- and three-hinged, low-rise parabolic arches subjected to uniform load on a horizontal projection can be closely approximated by use of the values in Table 17.1 with the moment of inertia I equal to the sum of the arch and girder moments of inertia. This simple concept recognizes that the arch and girder buckle together.

17.2.2 Nonlinear Elastic Stability

Section 17.2.1 addressed arches subjected to special loadings that produced only axial compression. Because the arches were considered to be inextensible, there were no prebuckling deflections. Actual design loadings on arches, however, usually produce both an axial compression and bending moment on a general cross section of the arch rib. These internal forces cause a change in shape of the arch before buckling occurs and results in the problem being nonlinear even though the material is deforming elastically.

Symmetrical Loadings on Symmetrical Arches The general behavior of symmetrical arches that are symmetrically loaded is shown in Fig. 17.2*b* and has

TABLE 17.2 Elastic Buckling Coefficients for Uniform Arches with Vertical Load Uniformly Distributed Along Arch Axis[a]

		Two-Hinged Arch			Fixed Arch		
θ (deg)	h_i/L	qL^3/EI	HL^2/EI	h_c/h_i	qL^3/EI	HL^2/EI	h_c/h_t
		Parabolic arches—antisymmetrical modes					
	0.10	28.6	36.3	1.002	60.4	76.1	1.003
	0.15	38.2	32.9	1.004	83.4	70.8	1.006
	0.20	43.4	28.8	1.006	99.3	64.3	1.010
	0.25	44.9	24.5	1.008	107.7	57.0	1.013
	0.30	43.5	20.5	1.009	109.6	49.5	1.015
	0.35	40.4	16.96	1.009	106.4	42.5	1.017
	0.40	36.5	13.98	1.009	100.0	36.2	1.017
	0.50	28.5	9.51	1.008	83.0	25.9	1.017
		Circular arches—antisymmetrical modes					
50	0.1109	31.2	35.6	0.994	64.8	75.2	0.992
70	0.1577	39.5	32.0	0.989	83.4	70.0	0.983
90	0.2071	44.0	27.4	0.981	95.5	63.3	0.969
106.26	0.2500	44.5	23.2	0.974	99.9	57.0	0.966
120	0.2887	42.8	19.34	0.968	99.8	51.2	0.938
140	0.3501	37.1	13.75	0.959	93.8	42.1	0.910
160	0.4196	28.9	8.72	0.953	81.9	32.6	0.880
180	0.5000	20.0	4.78	0.950	66.0	23.4	0.854
		Circular arches—symmetrical modes					
50	0.1109	63.0	74.6	0.88	90.9	110.6	0.94
70	0.1577	77.6	66.8	0.81	110.0	98.5	0.91
90	0.2071	85.6	57.9	0.75	119.4	85.7	0.88
106.26	0.2500	87.7	50.5	0.69	120.3	75.2	0.85
120	0.2887	86.7	44.3	0.64	117.2	66.7	0.82
140	0.3501	81.7	35.1	0.57	107.5	54.2	0.78
160	0.4196	73.8	27.1	0.45	93.5	43.1	0.73
180	0.5000	64.4	19.07	0.34	77.2	32.5	0.67

[a] h_i, initial rise of arch; h_c, height of arch at crown at instant of buckling; θ, angle of opening of the circular arch.

been described previously. Buckling data are given in Table 17.2 for two-hinged and fixed parabolic and circular arches subjected to vertical load uniformly distributed along the arch axis. These data are from Austin and Ross (1976). Similar data are also available in this paper for a single concentrated load at midspan. In these studies axial strains have been neglected because their influence is very small for slender arches.

The antisymmetrical buckling loads for circular arches are shown to be less than the symmetrical buckling loads in Table 17.2. The same is true for parabolic arches, although the symmetrical buckling critical loads are not available. In general, antisymmetrical buckling governs, although there are few cases in which symmetrical buckling controls, such as fixed circular arches with a single concentrated load at the crown and fixed parabolic arches with a rise-to-span ratio less than about 0.40 that are also subjected to a concentrated load at the crown. It can be observed from Table 17.2 that the arches tend to buckle in a sidesway mode when the ratio h_c/h_i is close to unity, which indicates that for these cases the profile at the instant of buckling is nearly the same as the unloaded profile.

In the fifth edition of the guide, it was shown that the critical conditions for antisymmetrical buckling can be expressed in terms of horizontal reactions. It was further shown that the critical horizontal reactions are insensitive to the arch shape and to a lesser extent to the loading; they vary primarily with the rise-to-span ratio. With this in mind, close estimates of the critical horizontal reaction are provided as a function of rise-to-span ratio.

Unsymmetrical Loading The general response behavior of unsymmetrically loaded arches is shown in Fig. 17.2*a* and has been described previously. Several studies have been made for the most important practical loading of parabolic arches which comprise a uniformly distributed dead load q (on a horizontal projection) combined with a uniformly distributed live load p extending a variable distance from one abutment, as shown in Fig. 17.4.

The first studies by Deutsch (1940), Harries (1970), and Kuranishi and Lu (1972) used half-span live load. Kuranishi and Lu noted that for elastic buckling the total dead plus live load intensity, $w = p + q$, at buckling seemed roughly equal to the buckling value for uniform load over the entire span. Studies have been made by Chang (1973) and Harrison (1982) in which the length of the live load s was varied. The results by Chang are given in Tables 17.3 and 17.4. Parabolic hinged and fixed arches are considered with span-to-depth ratios of 0.15 and 0.25 and live load–dead load ratios p/q of ∞, 0.50, 0.15, and 0, which should cover the range of practical values. Because a live load–dead load ratio $p/q = 0$ corresponds to uniform symmetrical load, the arch buckling is by bifurcation, and the values are the same as those reported in Table 17.1.

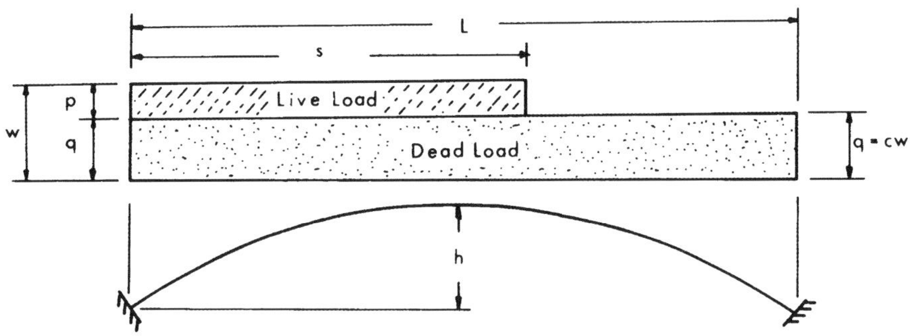

FIGURE 17.4 Unsymmetrical loading.

TABLE 17.3 Minimum Elastic Buckling Coefficients for Uniform Hinged- and Fixed-Ended Parabolic Arches Under Distributed Dead and Live Loads

		Hinged Ends			Fixed Ends		
h/L	p/q	s/L	wL^3/EI	HL^2/EI	s/L	wL^3/EI	HL^2/EI
0.15	∞	0.72	34.0	26.6	0.72	71.7	58.1
	0.50	0.67	35.5	29.4	0.66	75.9	62.2
	0.15	0.63	36.8	30.8	0.63	78.4	66.1
	0	—	39.5	32.9	—	85.1	70.9
0.25	∞	0.77	44.7	22.5	0.73	99.3	49.5
	0.50	0.68	46.0	23.3	0.66	104.0	51.9
	0.15	0.64	47.0	23.9	0.64	107.5	54.4
	0	—	49.2	24.6	—	114.6	57.3

TABLE 17.4 Elastic Buckling Coefficients for Uniform Hinged- and Fixed-Ended Parabolic Arches Under Distributed Dead and Half-Span Live Load ($s = 0.50L$)

		Hinged Ends		Fixed Ends	
h/L	p/q	wL^3/EI	HL^2/EI	wL^3/EI	HL^2/EI
0.15	∞	42.4	22.4	96.2	46.7
	0.50	37.1	28.4	80.2	60.4
	0.15	37.3	30.6	80.0	65.2
	0	39.5	32.9	85.1	70.9
0.25	∞	56.4	19.3	131.4	41.0
	0.50	48.3	22.8	109.8	50.4
	0.15	47.7	23.8	109.1	53.7
	0	49.2	24.6	114.6	57.3

In Table 17.3, the minimum total load intensity w is provided along with the corresponding horizontal reaction H and length of live load s. These results were obtained with a numerical solution using 24 equal divisions in the arch rib, with the load length being incremented one panel at a time. Although the values of w and H are accurate, the loaded length s is only approximate. Table 17.4 gives the values of w and H corresponding to the specific case of half-span live loading ($s = 0.50L$).

Several conclusions can be drawn for practical loading ranges. The minimum total buckling load is obtained when the live load acts over from 63 to 68% of the span. The buckling value of the total load s is minimized for high live load–dead load ratios but does not drop below about 89% of the full-span buckling value for the practical range, which is remarkably constant. Half-span loading is a good approximation for a partial loading case, as the corresponding elastic buckling load is not greater than about 6% more than the absolute minimum.

It should be noted that high live loads acting over partial spans produce very large moments and deformations. As a result, the elastic limit load is reached only after the arch is significantly distorted (Harrison, 1982; Yabuki and

Vinnakota, 1984). When inelastic behavior is considered, as will be discussed in the next section, the limit loads for these cases are much lower than the elastic buckling values reported above. One approach to greatly reducing the bending moments in the arch rib is to include hangers that are inclined in two directions in its plane; such a system is often referred to as a network arch (Tveit, 1966, 1999).

17.2.3 In-Plane Ultimate Load

Limit Analysis of Stocky Arches The limit analysis of arches was presented by Onat and Prager (1953) and Galli and Franciosi (1955). Following their investigations several researchers reported on the collapse load of arches (Stevens, 1957; Coronforth and Childs, 1967). In these early studies the collapse loads were obtained by the upper and lower bound methods of plasticity, which assumes rigid–plastic behavior. Based on a localized plastic hinge concept these methods cannot consider the impact of longitudinal spreading of yield zones or the second-order effects of deflections on moments. Hence, the studies are valid only for structures with very stocky cross sections subjected to predominant flexural moments. Stevens (1957) experimentally checked the validity of this plastic hinge approach on models of circular, elliptic, and parabolic arches with two-hinged and fixed support conditions. He found that a reasonably accurate estimate of the collapse load can be obtained by limit analysis for arches under a concentrated load. On the other hand, the experimental collapse load for an arch under half-span loading is considerably less than the calculated value. More recent experimental studies and comparisons with finite element analyses of the plastic collapse load of stocky arches are reported by Jukes et al. (1983), Mitri and Hassani (1990), and Khan et al. (1996).

Slender Arches in Pure Compression As indicated earlier, slender two-hinged and fixed arches with loadings that produce pure axial compression buckle in a sidesway mode with a point of contraflexure at the crown (Figs. 17.3*a* and *b*). Because the behavior is very much like that of a column, it is common to express the buckling strength of such arches in terms of the axial thrust at the quarter point of the arch P. The elastic critical value can be expressed as

$$P_c = \frac{\pi^2 EI}{(KS)^2} \tag{17.1}$$

where S is the length of the curved centroidal axis of the arch rib from support to crown, E the Young's modulus, and I the moment of inertia of the arch rib. The effective-length factor K depends on the support fixity condition, the rise-to-span ratio, and the shape of the arch profile. Values of K for arches are tabulated in the third edition of this guide and have values close to the well-known corresponding values for a column with a hinge at one end (crown) and a fixity condition at the other end that corresponds to the arch support condition.

Komatsu and Shinke (1977) have made inelastic ultimate-strength studies for two-hinged parabolic arches subjected to uniform load. Residual stress and initial crookedness were considered, and rise-to-span ratios of 0.10, 0.15, and 0.20 were used along with a wide range of other parameters. It was found that the ultimate value of the thrust at the quarter point of the arch was accurately predicted with the usual column curves adjusting for yield points, effective lengths, and so on. Thus, the standard relationships employed for columns relating the inelastic ultimate strength to the elastic critical load values can be used for arches.

Slender Arches under Symmetric and Asymmetric Load In this case, the loading produces combined axial compressive force and bending moment on a general cross section, which results in the arch behaving similar to a beam-column. It has been shown by Shinke et al. (1975) that the most demanding practical loading for low bridge arches of uniform cross section is the unsymmetrical loading discussed in the preceding section and shown in Fig. 17.4 with $s = 0.50$. The response behavior is similar to that shown in Fig. 17.2*a*, in which the buckling failure is a limit-point phenomenon. Arches of practical proportions develop extensive inelastic action before failure. To accurately predict the load-carrying capacity of steel arch structures, it is necessary to consider the effects of the spread of yielding zones in the cross section and along the longitudinal axial direction of the rib, the effects of initial residual stresses, and, of course, the amplification of the moments caused by the displacements. The following studies considered uniform arches subjected to uniform dead load over the entire span and half-span live load (as in Fig. 17.4, except as noted).

Harries (1970) reported analytical results on the ultimate strength of two-hinged parabolic steel arches in which the effects of prebuckling deformations and the spread of yielding zones were considered. Elastic–plastic deformations were calculated. Harries considered rectangular sections and circular tubes.

Kuranishi and Lu (1972) reported extensive parametric studies on the load-carrying capacity of two-hinged parabolic steel arches having either rectangular or idealized sandwich cross sections (approximating a wide-flange or box cross section). Residual stresses and strain hardening were taken into account. They found, as did Harries, that when the effects of partial yielding are considered, the strength of an arch can be drastically reduced under unsymmetrical loads; the buckling load is less when the live load acts over only one-half the span instead of over the entire span, and the larger the live load–dead load ratio, the smaller the buckling load becomes.

In a study that reevaluated the solutions of the previous paper, Kuranishi (1973) examined the ratio of the ultimate load to the elastic limit load found by a second-order analysis that uses a sandwich cross section to approximate I- and box-shaped cross sections. For a practical range of live load–dead load ratios, he showed that the ultimate load lies between that which produces initial yielding and about 93% of this value. He concluded that one could design efficiently by using second-order elastic analyses and keeping the maximum stress below 90% of the yield stress. The same could not be concluded for a solid rectangular section because the ultimate load always exceeded the load at first yielding.

Shinke et al. (1975) studied analytically the behavior of two-hinged and fixed parabolic arches of solid rectangular, pipe, and box cross sections. In this initial study residual stresses and strain hardening were not considered. Comparisons were made of the ultimate strength of several cases of uniform live load symmetrically distributed about the centerline and with arches having an unsymmetrically distributed load in the pattern of Fig. 17.4. For the same dead- and live-load intensities, it was found that unsymmetrically distributed load always governs. They also investigated the effect of length of the live load s and found that the buckling load is least when the length of live load s is roughly equal to one-half the span. With this loading configuration, a parametric study was also made of the ultimate strength of a box section. In a follow-up study, Shinke et al. (1977) considered the effects of residual stresses and initial crookedness of the arch rib. Residual stress effects were shown to be significant, while initial crookedness did not impact the results for the controlling unsymmetrical loading cases. An extensive series of experiments on two-hinged and fixed parabolic arches of solid rectangular cross section were also reported in this study, with experimental and analytical results agreeing well.

Kuranishi and Yabuki (1979) and Yabuki (1981) have published numerical studies on the load-carrying capacity of two-hinged parabolic steel arches with thin-walled box cross sections. These works have been summarized by Yabuki and Vinnakota (1984).

The strength of arches with stiffening girders has been studied by Shinke et al. (1980), Kuranishi et al. (1980), and Yabuki (1981). Within these papers formulas are proposed for estimating the ultimate loads.

Studies of the effects of the unsymmetrical distributed load of Fig. 17.4 supplemented by a single concentrated load at the quarter point of the span (a requirement of Japanese highway specifications) have been made by Yabuki and are summarized by Yabuki and Vinnakota (1984).

As indicated in the literature, the limit load depends on many factors, including rise-to-span ratio, live load–dead load ratio, slenderness ratio, yield stress, type of cross section (box section, wide-flange, rectangle, etc.), and residual-stress magnitudes and patterns. Only a brief review of the major points presented by Yabuki and Vinnakota (1984) will be given below.

The effects of several of these parameters on the ultimate load-carrying capacity of two-hinged, parabolic, uniform arches subject to uniform dead load plus half-space live load are shown in Figs. 17.5 through 17.7. In these figures, the ultimate or maximum load is expressed in terms of the ordinate $w_{\max}/w_y$, in which w_y is the magnitude of a uniformly distributed load ($p = 0$) that would cause the arch to yield at the springings (supports) under the direct axial thrust produced. Based on this definition, w_y can be calculated from

$$w_y = \frac{2A\sigma_y}{L\sqrt{\frac{1}{16}(L/h)^2 + 1}} \tag{17.2}$$

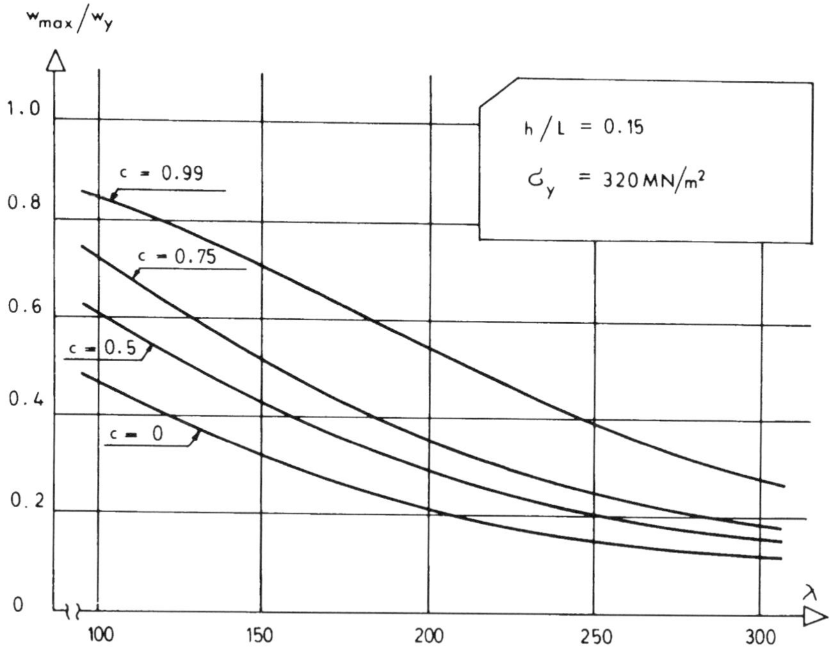

FIGURE 17.5 Variation of load-carrying capacity as a function of slenderness ratio.

where σ_y is the yield stress of the material. In these diagrams, $c = q/w$ (see Fig. 17.4) and the slenderness ratio $\lambda = L_s/r$, in which L_s is the curved length of the arch axis.

Figure 17.5 shows the variation of maximum strength $w_{\max}/w_y$ of steel arch ribs as a function of the slenderness ratio λ for different load ratios c. The load-carrying capacity decreases dramatically with the slenderness ratio. The capacity also decreases markedly as c decreases (i.e., as the live-load component increases) because the live load causes large bending moments and large displacements.

The influence of rise-to-span ratio on the load-carrying capacity of arch ribs with a slenderness ratio of 200 is shown in Fig. 17.6. It can be observed that the ultimate load, expressed in the nondimensional form with w_y given by Eq. 17.2, is not significantly affected by the rise-to-span ratio of the arch rib, especially for high values of c.

The reduction in the ultimate strength of arch ribs due to the effect of residual stresses can be as much as 20%. The reduction becomes especially noticeable when the arch is subjected to uniformly distributed load covering the entire span. The maximum variation of the ultimate strength using various distribution patterns for the residual stresses is within 10%. The reduction in ultimate strength becomes more significant as the level of compressive residual stress increases, and especially so when the compressive residual stress is greater than $0.4\sigma_y$.

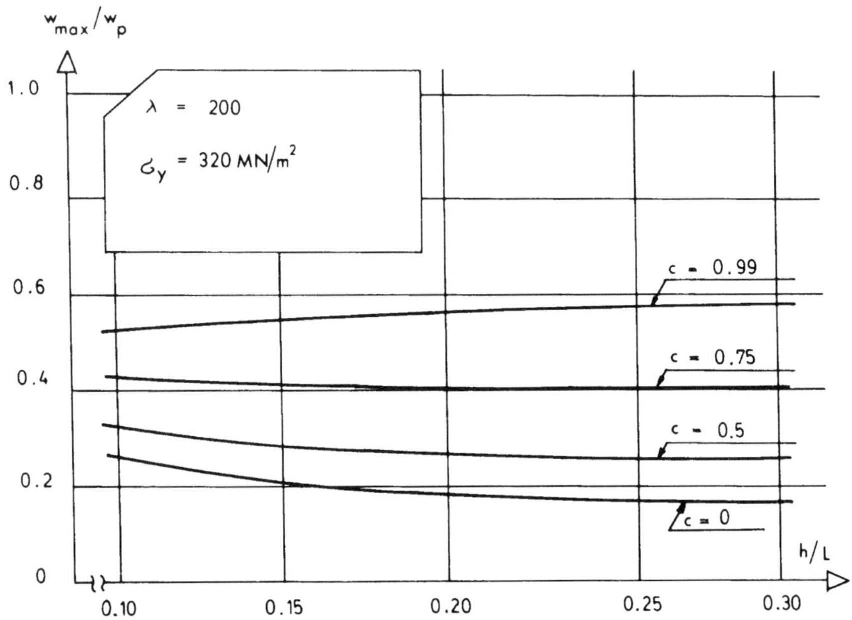

FIGURE 17.6 Variation of load-carrying capacity as a function of rise-to-span ratio.

The initial deformations of arch ribs, such as out-of-straightness, can result in a reduction of their ultimate strength. This reduction is important when the arch rib is subjected to uniformly distributed load covering the entire span. In the case of unsymmetrical loading, which is the critical loading condition for arch rib structures, the deformations due to the loading become large so that the effects of the initial deformations are not significant.

The nondimensional ultimate-load intensity $w_{\max}/w_y$ decreases in proportion to the square root of the yield-stress level of the material, as shown in Fig. 17.7.

The ultimate strength of stiffened two-hinged arch structures, with the arch and stiffening girder behaving as an entirely integrated structure, is analogous to that of two-hinged arches. The slenderness ratio of the stiffened arch structure can be defined by

$$\lambda_T = \frac{L_s}{\sqrt{(I + I_g)/A}} \tag{17.3}$$

where I_g is the moment of inertia of the stiffening girder. The ultimate strength of a stiffened arch structure with a slenderness ratio λ_T is always somewhat greater than that of a two-hinged arch using a slenderness ratio $\lambda = \lambda_T$. Judging from the analytical results, however, the two may be considered equal for all practical purposes. It is generally not required that attention be given to the *local failure* of arch rib members (buckling between the columns connecting the arch and the stiffening girder) for the unsymmetrical loading case, if a check is made for their

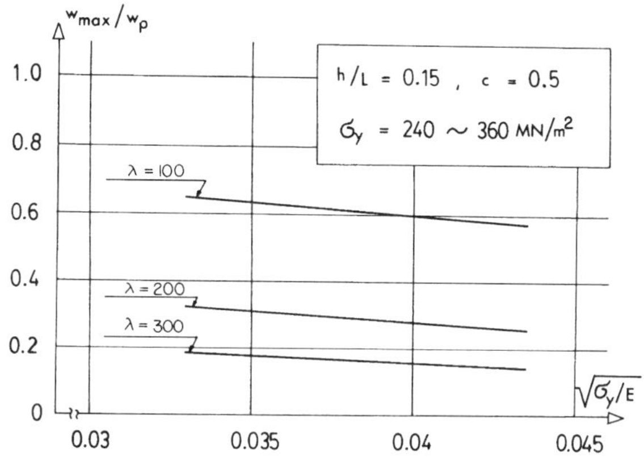

FIGURE 17.7 Influence of yield-stress level of material on the load-carrying capacity of arches.

strength for constant, uniformly distributed load. The *local failure strength* of arch rib members can be determined by the basic column strength curve when they have straight members between columns. For curved members, however, it is advisable to reduce this strength by 15%.

17.2.4 Design Criteria for In-Plane Stability

Arches of considerable span are normally not of uniform cross section but are composed of segments with different cross-sectional properties. Each of the segments must be designed for the applied loads that produce the maximum stresses in that segment. Many arch structures are complex structural systems such as deck bridge arches, in which the arch rib is connected to the roadway girder by rigidly attached columns. Studies of the behavior of uniform free-standing arches may be of limited usefulness in these cases. A general design approach is to factor the loads, use a second-order elastic analysis for the entire system, and keep the maximum combined stresses for each segment below some reference stress. Kuranishi (1973) recommends this procedure with the maximum stress less than 90% of the yield stress, as stated previously.

Design procedures for arches based on the ultimate inelastic strength studies reviewed in the previous section have been proposed by Kuranishi (1973), Komatsu and Shinke (1977), and Kuranishi and Yabuki (1984). Kuranishi proposed an interaction-type formula similar to beam-column formulas for two-hinged parabolic arches under unsymmetrical loading. Komatsu and Shinke presented practical formulas for the planar ultimate-load intensity of two-hinged and fixed parabolic steel arches that are a function of the normal thrust calculated at a quarter point of the arch rib by first-order elastic analysis. They also recommended that the ultimate-load capacity of arch ribs with varying and/or hybrid cross section can be evaluated accurately by using mean values of the cross-sectional area and/or yield-stress level of the material, which would be calculated by averaging along the entire axial length of the rib. Kuranishi and Yabuki also presented accurate practical formulas for the in-plane ultimate strength of parabolic two-hinged steel arch ribs and steel arch bridge structures with a stiffening girder. These formulas are expressed in terms of bending moment and axial thrust (or stresses provided by these cross-sectional forces) at a quarter point of the rib, with all calculations based on a first-order elastic analysis. More recently, Sakimoto and Watanabe (1995) proposed a design procedure based on nonlinear analysis that proportions each member automatically so as to meet the ultimate-strength requirements of the complete structural system. The procedure is based on the tangent-modulus method and eigenvalue analysis.

Circular Arches Loaded in Bending and Compression Pi and Trahair (1999) and Pi and Bradford (2004) studied the in-plane inelastic stability of hinged and fixed circular arches with I-shape cross sections. These studies included initial geometric imperfections and residual stresses. The load cases investigated included uniform compression, uniformly distributed load over half the span or over the full span, a concentrated load at the crown or at quarter span, and uniform bending (Table 17.5). The subtended angles studied ranged from 10° for shallow to 180° for deep arches. In Table 17.5, the slenderness of the arch is defined as $\lambda_s = S^2/(4r_x R)$, where S and R are the developed length and radius of the arch and r_x the radius of gyration of the cross section (Fig. 17.8). To check in-plane stability, the following interaction equation was proposed for a lower bound prediction (the notations are slightly changed from the cited papers to combine results for fixed and hinged arches):

$$\frac{N^*}{\alpha_{anx} \cdot N_{acx}} + \frac{M^*}{\alpha_{amx} \cdot M_p} \leq \phi \tag{17.4}$$

where N^* is the maximum axial compression based on first-order elastic analysis, M^* the maximum moment, α_{anx} and α_{amx} the axial compression and moment modification factors (Table 17.5), ϕ a capacity reduction factor with a value of 0.9 suggested for design, N_{acx} the axial compression capacity for uniform compression, and M_p the plastic section capacity about the major axis.

The axial compression capacity for uniform compression is defined as $N_{acx} = \alpha_{acx} N_y$ for hinged and $N_{acx} = \phi\alpha_{acx} N_y \leq N_y$ for fixed arches, with N_y equaling the yield (squash) load of the cross section and α_{acx} the arch slenderness reduction factor (see Eq. 17.5a). The maximum moment is defined as $M^* = \delta_b M$, where M is the maximum moment based on a first-order elastic analysis and $\delta_b = 1/(1 - N^*/N_A)$ the moment amplification factor.

The in-plane buckling load of an arch in uniform compression, N_A, is defined as $N_A = k_{ax} N_{Ex}$, in which k_{ax} is a factor for in-plane buckling and N_{Ex} is the elastic in-plane buckling load for uniform compression. For hinged arches $k_{ax} = 0.95$ for $50 \leq \lambda_s$; $k_{ax} = 0.7 + 0.1(\lambda_s - 9.9)^{1/4}$ for $10 < \lambda_s < 50$; $k_{ax} = 0.15 + 0.006\lambda_s^2$ for $\lambda_s \leq 10$, and for fixed arches $k_{ax} = 0.36 + 0.0011\lambda_s^2$ for $9.87 \leq \lambda_s \leq 18.6$; $k_{ax} = 0.6 + 0.4\sqrt{1 - 3.109\pi^2/\lambda_s^2}$ for $18.6 < \lambda_s < 50$; $k_{ax} = 1 + 0.049(\theta/\pi)^2$ for $\lambda_s > 50$. The elastic in-plane buckling load for uniform

TABLE 17.5 Factors for the In-Plane Design of Circular Arches

			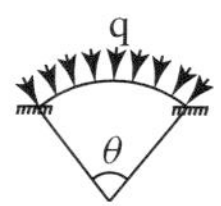	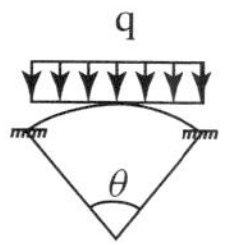	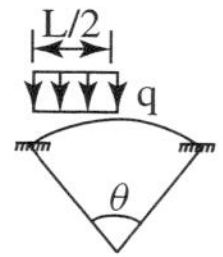	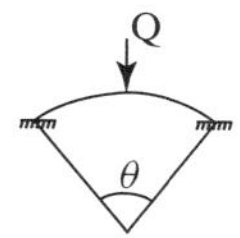	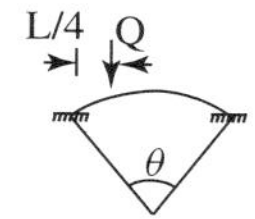	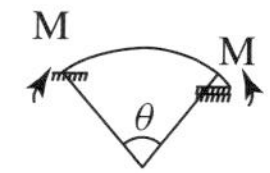
	α_{anx}		1.0	1.0	1.0	1.1	1.1	—
Fixed	α_{amx}	$\lambda_s < 18.6$	—	$2.0 + 0.1(18.6 - \lambda_s)\theta/\pi$	$1.2 + 0.5(18.6 - \lambda_s)\theta/\pi$	$1.5 + 0.05(18.6 - \lambda_s)\theta/\pi$	$1.8 + 0.05(18.6 - \lambda_s)\theta/\pi$	—
		$\lambda_s > 18.6$	—	2.0	1.2	1.5	1.8	—
Hinged	α_{amx}	$\lambda_s < 10$	—	$1.13 - 0.137\lambda_s$	$1 + 0.02\lambda_s$	$1.35 + 0.065\lambda_s$	$1.1 + 0.05\lambda_s$	1.0
		$\lambda_s > 10$	—	2.5	1.2	2.0	1.6	1.0

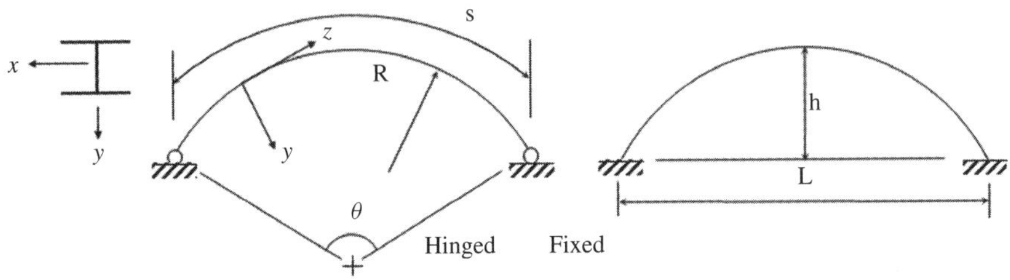

FIGURE 17.8 Geometry of arches studied by Pi and Trahair (1999).

compression is defined as $N_{Ex} = b\pi^2 EI_x/S^2$ with $b = 4$ for hinged and $b \approx 8.16$ for fixed arches. The remaining variables used are

$$\alpha_{acx} = \xi_{ax}\left[1 - \sqrt{1 - \left(\frac{90}{\xi_{ax}\cdot\lambda_{agx}}\right)^2}\right] \tag{17.5a}$$

$$\xi_{ax} = \frac{\left(\lambda_{agx}/90\right)^2 + 1 + \eta_{ax}}{2\left(\lambda_{agx}/90\right)^2} \tag{17.5b}$$

$$\lambda_{agx} = \frac{c\cdot S}{r_x}\sqrt{\frac{1}{k_{ax}}}\sqrt{\frac{\sigma_y}{250}} \tag{17.5c}$$

In Eq. 17.5b, $\eta_{ax} = 0.00326\left(\lambda_{agx} - 13.5\right) \geq 0$ is an imperfection parameter and in Eq. 17.5c, c equals 0.5 for hinged and 0.35 for fixed arches, and σ_y is the yield strength. Pi and Trahair (1999) and Pi and Bradford (2004) recommend that M^* and N^* should be based on a second-order elastic analysis if the moment amplification factor δ_b exceeds 1.4. In Fig. 17.9, the results from extensive finite-element analyses are compared to the interaction equation 17.4 for $\phi = 1.0$ (continuous line) and $\phi = 0.9$ (dashed line). The majority of the analysis results lie above the prediction of the interaction equation, especially when $\phi = 0.9$ is used.

Design Standards A comparison of design provisions for in-plane stability between international standards can be found in Beedle (1991). Some of the more recent provisions are listed below.

Eurocode 3, Part 2 (CEN, 2006a) provides charts with effective length factors for the elastic in-plane buckling of circular, parabolic, and caternary arches with unmovable supports and several articulations. For tied arches with vertical hangers, effective lengths are also given, as is a criterion which indicates if the arch is prone to snap-through buckling. Furthermore, the standard gives imperfections to be used for the in-plane analyses.

AASHTO (2004) provides effective-length factors for fixed, two-hinged and three-hinged arches with rise-to-span ratios of 0.1 to 0.4.

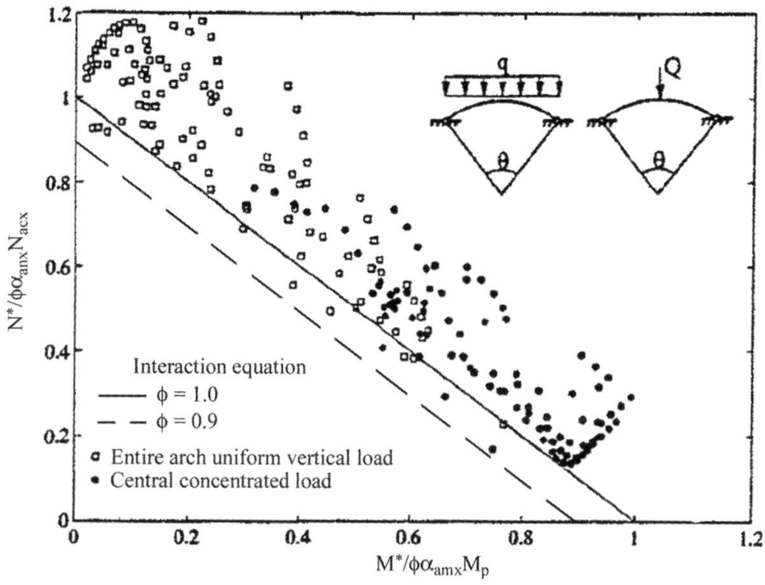

(*a*) Hinged arch, symmetric loading

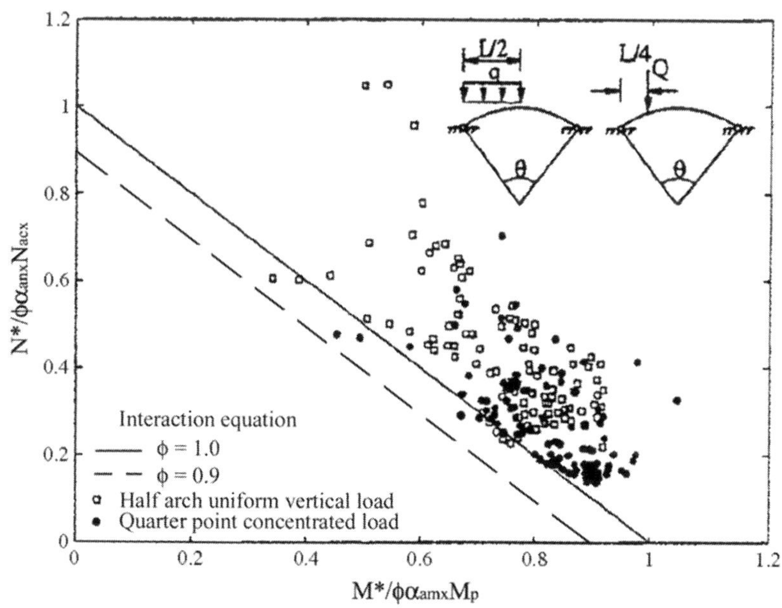

(*b*) Hinged arch, asymmetric loading

FIGURE 17.9 Interaction equation for in-plane stability compared to finite element analysis results (Pi and Trahair, 1999; Pi and Bradford, 2004).

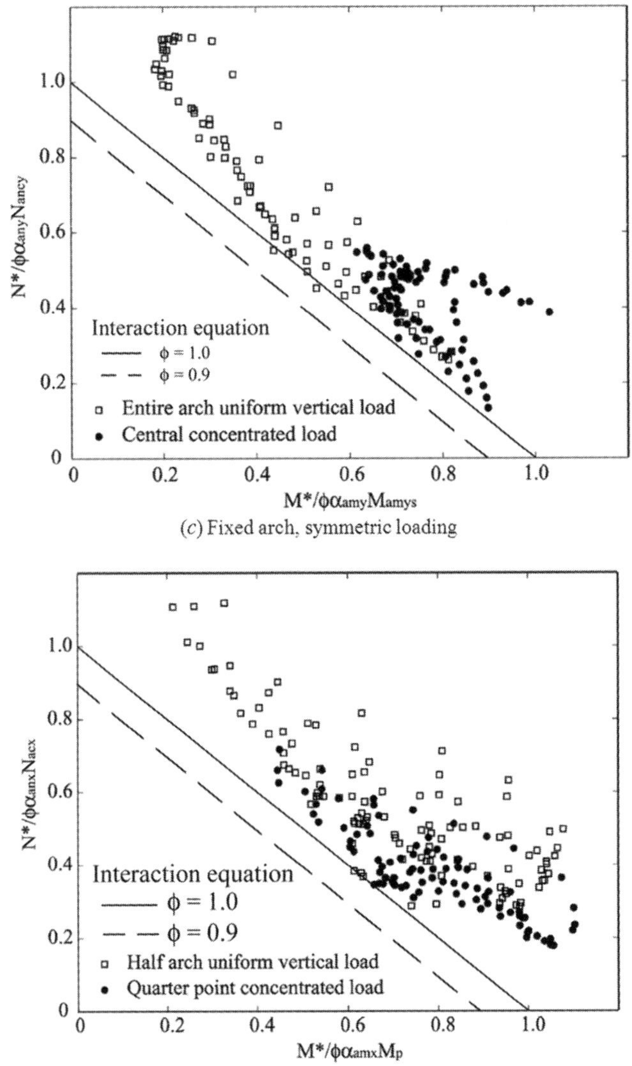

(*c*) Fixed arch, symmetric loading

(*d*) Fixed arch, symmetric loading

FIGURE 17.9 (*Continued*)

17.3 OUT-OF-PLANE STABILITY OF ARCHES

When applied forces acting in the plane of a curved member such as an arch reach a certain critical level, a combination of twisting and lateral bending will cause the member to deform out of its original plane. The critical load is influenced by the nature and distribution of the loads, the shape of the axis of the arch, the variation of the flexural and torsional stiffness of the cross sections along the length axis, and restraint available at the supports and other brace points. For a steel arch, residual stresses either resulting from nonuniform cooling of the material during fabrication or caused by welding or cold working (roller bending) the material into an arch are also important factors. The multiplicity of impacting parameters prevents the development of simple and widely applicable rules for the determination of buckling loads.

Subjects of research have progressed with time from elastic lincar buckling of a single arch to ultimate strength of total systems—that is, from simple and idealized theoretical models to more practical models to be encountered in actual structures. Some of the available solutions that may be useful as a guide to the design of practical structures are presented in formulas and tables. Results from elastic linear buckling theory are shown in the next sections while solutions for inelastic stability under general loading are presented in Section 17.3.3. Results from ultimate-strength analyses are presented in Section 17.5.

Figure 17.8 shows the notation and coordinate system that will be used in the following sections. Unless stated otherwise, the results presented apply to symmetric arches of constant cross section. The effect of warping torsional restraint is included in all results presented. Note that warping torsional resistance is negligible for closed-profile cross sections such as box sections, but it is of significant importance for open-profile cross sections such as I-shapes. In general, results obtained by ignoring warping torsional restraint are conservative for open-profile sections.

Isolated arches supported by ends that are free to rotate out of plane (y axis) have very little lateral stability and should be avoided. The practical significance of this should be emphasized, because in actual arches, completely rigid supports are difficult, if not impossible, to construct.

17.3.1 Elastic Buckling of Circular Arches

Uniform Compression An arch is loaded in uniform compression when uniformly distributed radial forces p are directed inward along the arch's centroidal axis and the ends are prevented from translation. Although this load case rarely exists in practice, it is useful as reference for design procedures.

For arches with rectangular cross sections and in- and out-of-plane hinged supports that prevent a twisting rotation about the length, or z, axis, Timoshenko and Gere (1961) obtained a closed-form solution for elastic out-of-plane buckling. Vlasov (1963) extended the solution to include warping. More refined solutions wcrc then obtained by Papangelis and Trahair (1987b), Rajasekaran and Padmanabhan (1989), and Yang and Kuo (1986, 1987, 1991). The most recent solutions for the elastic flexural–torsional buckling load of an arch in uniform compression have been obtained by Pi and Trahair (1998) for hinged arches and Pi and Bradford (2005) for fixed arches and are given by

$$N_{ays} = k_a N_{Ey} \tag{17.6a}$$

where, for hinged arches,

$$k_a = \frac{1}{2}\left[-\left(1+\frac{a^2}{b^2}\right)\frac{N_T}{N_{Ey}} - \left(1-a^2\right)^2 + \sqrt{\left(1+\frac{a^2}{b^2}\right)^2\frac{N_T^2}{N_{Ey}^2} + 2\left(\frac{a^2}{b^2}-1\right)\left(1-a^2\right)^2\frac{N_T}{N_{Ey}} + \left(1-a^2\right)^4}\,\right] \tag{17.6b}$$

and, for fixed arches,

$$k_a = \frac{1}{2}\left[\left(1+\frac{a^2}{b^2}\right)\frac{N_T}{N_{Ey}} + \left(1-a^2\right)^2 - \sqrt{\left(1+\frac{a^2}{b^2}\right)^2\frac{N_T^2}{N_{Ey}^2} + \left(\frac{a^2}{b^2}-1\right)\left(1-a^2\right)^2\frac{N_T}{N_{Ey}} + \left(1-a^2\right)^4}\,\right] \tag{17.6c}$$

with N_{Ey} equaling the minor axis buckling load of a column of length S, $N_{Ey} = \pi^2 EI_y/(S/c)^2$ where $c = 1$ for hinged and $c = 2$ for fixed arches, $a = (S/c)/(\pi R)$, and $b = \pi M_{yz}/\left[N_{Ey}\,(S/c)\right]$. The torsional buckling load N_T of a column and the lateral–torsional buckling moment M_{yz} for a beam in uniform bending are given by

$$N_T = \frac{A}{I_x + I_y}\left(GJ + \frac{\pi^2 EI_w}{(S/c)^2}\right) \tag{17.7}$$

$$M_{yz} = \sqrt{N_{Ey}\left(GJ + \frac{\pi^2 EI_w}{(S/c)^2}\right)} \tag{17.8}$$

Uniform Moment An arch is loaded in uniform bending when equal couples are applied about the x axis and the ends are fixed against translation in the x direction but are free to rotate about the x and y axes. Although this load case rarely exists in practice, it is useful as a reference load. Uniform bending can cause buckling out of the y–z plane if the flexural stiffness about the x axis is large in comparison to both the flexural stiffness about the y axis $B(= EI_y)$ and the St.-Venant torsional rigidity of the cross section about the z axis $C(= GJ)$. Solutions for the critical moment were obtained by Timoshenko and Gere (1961) and were extended by Vlasov (1963) and Kollár and Iványi (1966) to include the warping effects. More refined solutions were obtained by Papangelis and Trahair (1987b), Rajasekaran and Padmanabhan (1989), and Yang and Kuo (1986, 1987, 1991). The most recent solutions are given by

$$M_{ays} = k_{as} M_{yz} \tag{17.9a}$$

where, for hinged arches (Pi and Trahair 1998),

$$k_{as} = -ab - \frac{a}{2b} + \frac{a^3 b}{2} + \sqrt{\left(-ab - \frac{a}{2b} + \frac{a^3 b}{2}\right)^2 + \left(1 - a^2\right)^2} \tag{17.9b}$$

and, for fixed arches (Pi and Bradford 2005),

$$k_{as} = -\frac{ab}{2} - \frac{a}{2b} \pm \sqrt{\left(\frac{ab}{2} + \frac{a}{2b}\right)^2 + \left(1 - a^2\right)} \tag{17.9c}$$

The parameters a and b are the same as those defined for Eq. 17.6.

End Forces Directed along the Chord Two collinear forces applied to the ends of circular arches can cause lateral buckling (Klöppel and Protte, 1961; Ojalvo et al., 1969). Forces directed away from each other (pull loads) will generally cause antisymmetric buckling about the center of the arch, whereas forces directed toward each other (push loads) cause a symmetric buckling. The buckling load for ends hinged about the x axis may be expressed in the form (Klöppel and Protte, 1961)

$$F_c = m \frac{B}{(R\alpha/2)^2} \tag{17.10}$$

When the central angle α is less than π and the warping torsional stiffness is negligible, the value m (negative for pull loads) may be approximated by the following equation: For pull loads

$$m = a_0 + \frac{a_1}{\alpha} + \frac{a_2}{\alpha^2} \tag{17.11a}$$

and for thrust loads

$$m = 2.47 + a_3\alpha + a_4\alpha^2 \tag{17.11b}$$

Coefficients a_0 through a_4 are given in Table 17.6 for two C/B ratios. The effect of the warping torsional stiffness was also studied by Klöppel and Protte (1961).

TABLE 17.6 Coefficients for Use in Eq. 17.11

C/B	a_0	a_1	a_2	a_3	a_4
0.01	0.052	−0.217	−0.36	−0.24	0.09
0.5	0.815	−6.313	−7.411	−0.034	0.069

Uniformly Distributed Vertical Forces When the arch is fixed against rotation about the y axis at its ends, the uniform load w_c at buckling may be expressed in the form (Demuts, 1969)

$$w_c = m\frac{B}{L^3} \tag{17.12}$$

with several values of m provided in Table 17.7 for circular arches.

The usual closed-profile cross section used in arches has a C/B ratio in the range 0.5 to 1.5 and for the open-profile cross section C/B is in the range 0.01 to 0.001. If the arch cross section has a thin-walled open profile, the warping torsional rigidity, C_1, also has a primary effect on the buckling loads, as shown in Fig. 17.10 (Fukasawa, 1963; Namita, 1968). In this figure the buckling coefficient $m = w_c\alpha^2R^3/B$ for uniform vertical load w is shown by solid lines and the buckling coefficient $m_r = p_c\alpha_2R^3/B$ for uniform radial load p is shown by dashed lines. When the height-to-span ratio h/L is not large, $h/L \leq 0.2$, the values m and m_r are approximately related by $m_r = 2m \sin(\alpha/2)/\alpha$.

TABLE 17.7 Approximate Values of *m* for Use in Eq. 17.12

	Circular Arch				Parabolic Arch			
	Vertical Loads		Vertical Loads		Hanger Loads (Symmetric)		Column Loads (Antisymmetric)	
H/L \ *C/B*	0.01	0.5	0.01	0.5	0.01	0.5	0.01	0.5
0.1	18	28	16	28	39	70	12	18
0.2	17	39	15	39	35	110	12	29
0.3	13	38	13	37	28	116	10	32
0.4	9	30	11	30	24	104	8	31
0.5	5	20	9	24	20	87	6	28

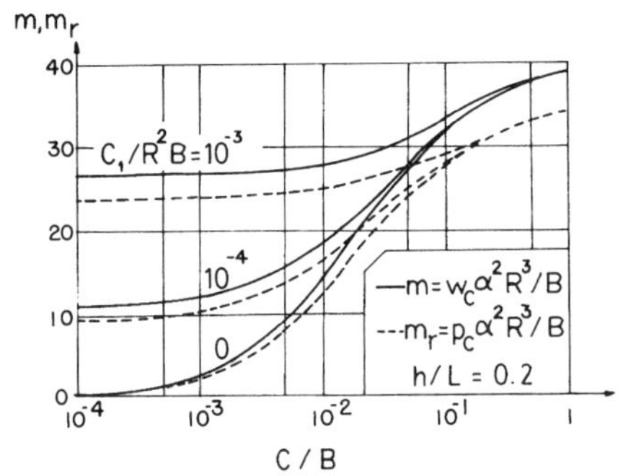

FIGURE 17.10 Effects of the warping rigidity C_1.

17.3.2 Buckling of Parabolic Arches

Uniformly Distributed Vertical Loads The uniformly distributed buckling load w_c for parabolic arches may also be expressed in the form of Eq. 17.12. Representative values of m are given in Table 17.7 (Tokarz and Sandhu, 1972). The ratio $C/B = 0.01$ is typical of thin-walled open-profile cross sections for which warping torsional restraint is significant. The effect of the warping torsional rigidity for these sections is similar to what is observed for circular arches (Fig. 17.10). Additional information on this case is presented by Stüssi (1943), Tokarz (1971), Kee (1961), and Papangelis and Trahair (1987a,b).

Parabolic Arches with Tilting Loads It has been recognized by Stüssi (1943) and others that the buckling load of an arch is increased if the arch is loaded by a system of vertical hangers connected to a laterally stiff girder at the elevation of the chord, as in the case of the through-type arch bridge shown in Fig. 17.1. If the loads are applied by means of columns connected to a laterally stiff girder above the arch and the arch is connected directly to the girder at the crown or by very short columns, the lateral deformations of the buckled arch will be antisymmetric about the crown. Both hanger- and column-loaded parabolic arches were studied theoretically by Östlund (1954), Godden (1954), and Shukla and Ojalvo (1971). The buckling load w_c may be expressed by Eq. 17.12 using appropriate values of m from Table 17.7. Again, it is assumed that the ends of the arch are fixed against rotation about the x axis. Östlund (1954) and Almeida (1970) reported that certain braced arches subjected to hanger loads are also controlled by antisymmetric buckling.

17.3.3 Design Criteria for Out-of-Plane Stability

Gravity Loading of Circular Arches Pi and Trahair (1998, 2000) and Pi and Bradford (2005) studied the out-of-plane stability of freestanding circular arches of I-sections including imperfections, residual stresses, and inelasticity. The support conditions studied included in- and out-of-plane hinged arches, which are prevented to twist about the longitudinal axis at the supports, and in- and out-of-plane fixed arches. The load cases investigated were the same as those for the in-plane stability, which is discussed in Section 17.2.4, and are included in Table 17.8. To check the out-of-plane stability, the following interaction equation, similar to the one for in-plane stability (Eq. 17.4), was proposed (to combine both fixed and hinged arches, notations are slightly changed from the original papers):

$$\frac{N^*}{\alpha_{any} \cdot N_{acys}} + \frac{M^*}{\alpha_{amy} \cdot M_{amys}} \leq \phi \tag{17.13}$$

where N^* and M^* are the maximum axial compression moments based on a first-order elastic analysis, α_{any} and α_{amy} axial compression and moment modification factors (Table 17.8), ϕ the capacity reduction factor with a value of 0.9

TABLE 17.8 Factors for the Out-of-Plane Design of Circular Arches

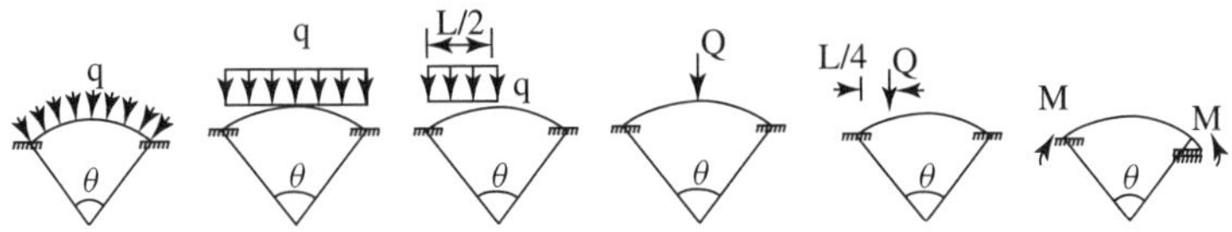

Fixed	α_{any}	—	1.2	1.6	1.4	2.2	—
	α_{amy}	—	2.2	2.2	1.5	1.8	—
Hinged	α_{any}	1.0	1.1	1.5	1.4	2.7	—
	α_{amy}	—	1.1	1.1	1.2	1.2	1.0

suggested for design, and N_{acys} and M_{amys} the out-of-plane strengths for uniform compression and uniform bending.

The out-of-plane strength for uniform compression is defined as $N_{acys} = \alpha_{acys} N_y$ for hinged arches and $N_{acys} = \phi\alpha_{acys} N_y \leq N_y$ for fixed arches, where N_y is the axial yield (squash) load of the cross section and α_{acys} the arch slenderness reduction factor for uniform compression (Eq. 17.14). Here, M^* is the nominal maximum moment, where for hinged arches $M^* = M$ and for fixed arches $M^* = \delta_{by} M$, in which M is the maximum moment based on a first-order elastic analysis and the moment amplification is $\delta_{by} = 1/\left(1 - N^*/N_{ays}\right)$ with N_{ays} given by Eq. 17.6a. Remaining variables used are

$$\alpha_{acys} = \xi_{ays}\left[1 - \sqrt{1 - \left(\frac{90}{\xi_{ays} \cdot \lambda_{agys}}\right)^2}\right] \tag{17.14a}$$

$$\xi_{ays} = \frac{\left(\lambda_{agys}/90\right)^2 + 1 + \eta_{ays}}{2\left(\lambda_{agys}/90\right)^2} \tag{17.14b}$$

$$\lambda_{agys} = \frac{c \cdot S}{r_y}\sqrt{\frac{1}{k_{acys}}}\sqrt{\frac{\sigma_y}{250}} \tag{17.14c}$$

In Eq. 17.14b, $\eta_{ays} = 0.00326\left(\lambda_{agys} - 13.5\right) \geq 0$ is an imperfection parameter, and in Eq. 17.14c, c equals 1.0 for hinged arches and 0.5 for fixed arches, σ_y is the yield strength, and r_y is the radius of gyration about the minor axis.

The out-of-plane strength of an arch under uniform bending is defined as $M_{amys} = \alpha_{says} M_{px}$ for hinged arches and $M_{amys} = \phi\alpha_{says} M_{px} \leq M_{px}$ for fixed arches, with M_{px} the full major axis plastic moment capacity and $\alpha_{says} = 0.6\left(\sqrt{\lambda_{amys}^4 + 3} - \lambda_{amys}^2\right)$ the arch slenderness reduction factor for uniform bending. The modified slenderness for uniform bending is defined as $\lambda_{amys} = \sqrt{M_{px}/M_{ays}}$ in which M_{ays} is the elastic flexural–torsional buckling moment for an arch in uniform bending (Eq. 17.9a). Pi and Trahair (1998, 2000) and Pi and Bradford (2005) both recommend that M^* and N^* should be based on a second-order elastic analysis when δ_{by} exceeds 1.4.

Figure 17.11 shows comparison of the interaction equation 17.13 and the finite-element analysis results from Pi and Trahair (2000) and Pi and Bradford (2005). In general, the interaction equation is accurate, noting that it seems to be conservative for fixed arches. Finally, it should be noted that the in- and out-of-plane load-bearing capacity for hinged arches vanishes when the subtended angle approaches 180° because the y axes of the hinges line up. Naturally, this is not the case for the fixed arches.

Circular Arches with Tilting Loads La Poutré (2005) studied both experimentally and numerically the inelastic out-of-plane stability of deep circular steel arches, with subtended angles ranging from 90° to 180° loaded by a single tilting load at the crown. The arches were in-plane hinged and out-of-plane fixed. The cross sections used were standard European wide-flange beams HEA 100 and HEB 600, with the latter tested to scale. The arches were produced by roller bending, which cold worked the material and changed the stress–strain characteristic of the steel and the residual-stress pattern.

To design for out-of-plane stability, the general method of Eurocode 3 (CEN, 2006b) was used. The plastic strength of the arch was determined by a material nonlinear analysis (MNA) and the elastic critical load by an eigenvalue analysis, also termed a linear buckling analysis (LBA). With these two parameters, the nondimensional slenderness of the arch was determined by

$$\bar{\lambda}_{GM} = \sqrt{\frac{F_{MNA}}{F_{LBA}}} \tag{17.15}$$

The actual buckling load can then be determined by either an experimental F_{EXP} or a finite element analysis F_{GMNIA} that include large displacements, material nonlinearity, imperfections, and residual stresses. The nondimensional reduction factor is determined by dividing the actual buckling load by the plastic strength

$$\chi_{EXP} = \frac{F_{EXP}}{F_{MNA}} \tag{17.16a}$$

$$\chi_{GM} = \frac{F_{GMNIA}}{F_{MNA}} \tag{17.16b}$$

Figures 17.12*a* and 17.12*b* show that the experimental reduction factors almost all lie above the most favorable buckling curve "*a*" of Eurocode 3 and all of the numerical results are above the same curve. Differences can be attributed to the actual imperfections in the tests being larger than the imperfections given by Eurocode 3, Part 2 (CEN, 2006a) for use with the numerical simulations. For design purposes, however, buckling curve "*b*" seems to be more appropriate.

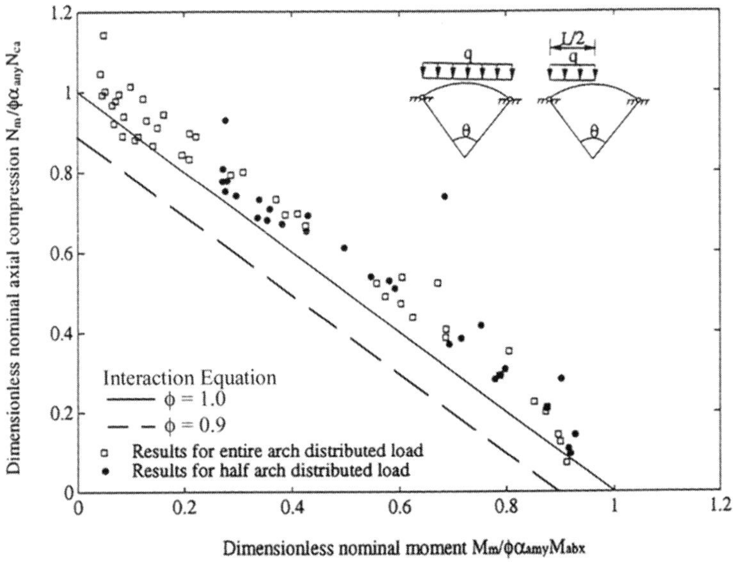

(*a*) Hinged arches with distributed loading

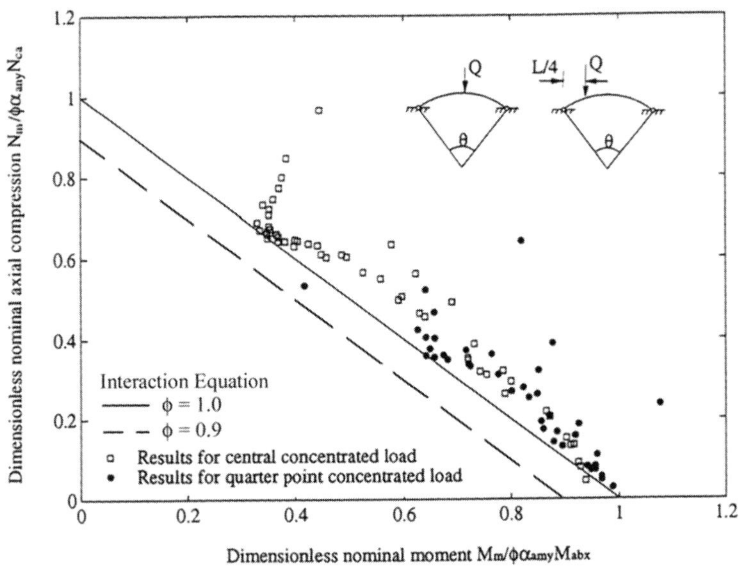

(*b*) Hinged arches with concentrated loading

FIGURE 17.11 Interaction equation for out-of-plane stability compared to finite-element analysis results (Pi and Trahair, 2000; Pi and Bradford, 2005).

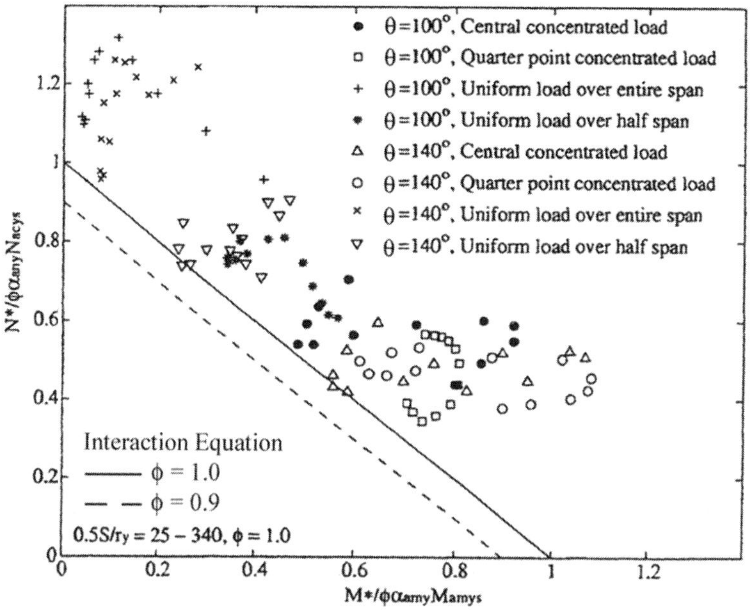

(*c*) Fixed deep arches with various loadings

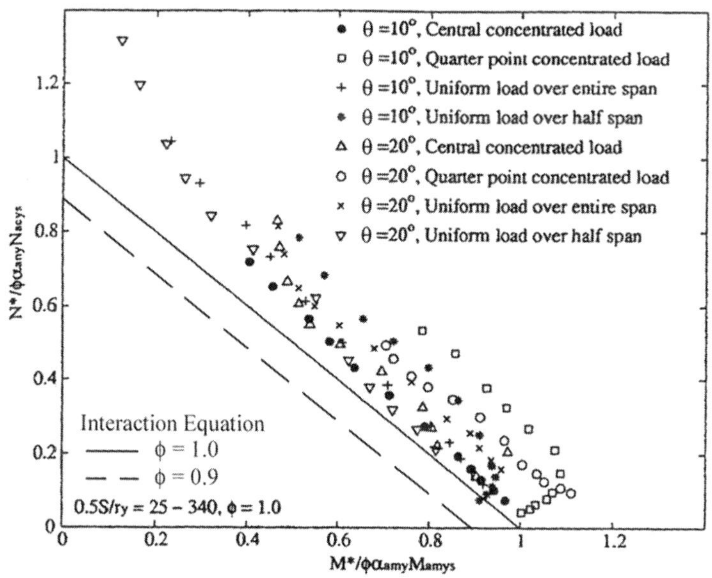

(*d*) Fixed shallow arches with various loadings

FIGURE 17.11 (*Continued*)

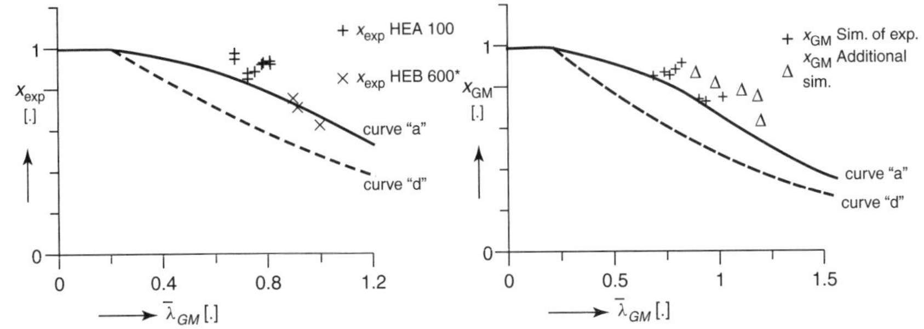

(*a*) Experimental reduction factors (*b*) Numerical reduction factors

FIGURE 17.12 Design for out-of-plane stability by the general method.

The experiments and numerical simulations by la Poutré (2005) have shown that the general method of Eurocode 3 is an accurate method for assessing the inelastic stability of arches under general loading, a condition for which no closed-form buckling equations exist.

Design Standards For the elastic out-of-plane buckling of free-standing arches of circular, parabolic, or caternary shaped arches with unmovable supports and different articulation, Eurocode 3, Part 2 (CEN, 2006a) provides charts with effective-length factors. Furthermore, the standard defines imperfections that should be used for out-of-plane buckling analyses. For through-arch bridges with wind bracing, only the stability of the end portal needs to be checked. Beedle (1991) provides a comparison of design methods for out-of-plane stability found in several international standards.

17.4 BRACED ARCHES AND REQUIREMENTS FOR BRACING SYSTEMS

17.4.1 Bracing by Transverse Bars

Most arches used in practice are braced against lateral movement either continuously or at regularly spaced intervals. An arch rib that is always connected to a curved roof is an example of continuous bracing. Unless such arch ribs are unusually deep, lateral buckling will not be a problem.

Twin-arch ribs with intermittent lateral bracing often appears in arch bridges. The elastic lateral buckling of twin-arch ribs braced with transverse bars normal to the plane of the ribs has been studied by several investigators, including Östlund (1954), Kuranishi (1961a), Almeida (1970), Tokarz (1971), and Sakimoto and Namita (1971). The following properties of the bracing system are important in preventing lateral buckling from occurring: (1) the location and spacing of the transverse bars; (2) the distance between the arch ribs b; (3) the flexural stiffness of the bars about the x axis, D_x; and (4) the flexural stiffness of the bars about the z axis, D_z.

It has been found that increasing the flexural stiffness D_x is a more effective way of suppressing lateral buckling than increasing the flexural stiffness D_z. The flexural stiffness D_z is less important except when the arch ribs have an open-profile cross section and the stiffness D_x cannot be provided. When either of the flexural stiffnesses D_x or D_z is increased independently, the buckling load of twin arches increases and tends to attain an asymptotic value. While the magnitude of the asymptotic value depends on the properties of the bracing system listed above, the asymptotic value of the buckling load for a closely braced arch system can be obtained when $D_x \geq 10b(B/R)$ or $D_z \geq b(B/R)$, respectively (Östlund, 1954; Sakimoto and Namita, 1971).

When pairs of arch ribs are braced closely with such stiff transverse bars, the elastic critical load for one arch rib of a set of braced arches can be increased up to 250% or more of that for the identical isolated single arch. For actual bridge arches, however, this increase in buckling strength will be limited by yielding of the material (as demonstrated later in Fig. 17.16).

Wästlund (1960) and others have suggested a simple approximate method for the determination of the lateral buckling load of braced arches by utilizing a planar system that is obtained by straightening out the arch in a horizontal plane. The compressive force required to buckle the longitudinal ribs of the assumed planar system would be computed as for a column with batten lacing, and these compressive forces would be the approximations to the rib forces required to buckle the actual arch structure. The approximate method can be accurate for the arches of a torsionally stiff (closed) cross section, but it is unconservative for the arches of an open-profile cross section (Östlund, 1954).

The strength of braced arches in the plane of bracing is similar to the buckling strength of columns with lacing bars and battens, which is given by Timoshenko and Gere (1961) as

$$\frac{1}{P_{cr}} = \frac{1}{P_E} + \frac{1}{S} \tag{17.17}$$

where P_{cr} is the critical load of the column, P_E the critical load of the column with stiff lacing, and S the shear stiffness provided by the bracing bars or battens. While increasing the shear stiffness of the arch in the lateral plane increases the critical load, the increase is limited by the flexural buckling strength in the vertical plane. Therefore, there is a critical value for the shear stiffness above which the buckling capacity is determined by the flexural strength. Based on extensive parametric studies using finite-element analysis, Kuranishi (1993) presented the following relationship for the critical value of the shear stiffness:

$$\mu_{S,cr} = \frac{\pi^2}{48} \frac{1}{(K_i \rho_i / K_o \rho_o)^2 - 1} \tag{17.18}$$

where $\mu_{S,cr}$ is the critical value for the ratio of the shear rigidity of the rib and bracing system in the plane containing the two ribs to the flexural rigidity of the two arch ribs, ρ_o the out-of-plane slenderness ratio of the rib, K_o the out-of-plane effective length factor, ρ_i the in-plane slenderness ratio of an arch rib, and K_i the in-plane effective length factor.

Although twin arches braced with transverse bars might be a favorable structure from an aesthetic point of view, a more effective bracing system for suppressing lateral buckling is to use diagonal members (Sakimoto, 1979; Sakimoto and Komatsu, 1982). If the lateral bracing consists of diagonals, diagonals in combination with transverse bars, or a K system, out-of-plane buckling of the arch tends not to be a problem except for through-type arch bridges, which cannot have lateral bracing over the entire arch length. Through-type steel arches with a double-diagonal bracing system will be discussed in the next section.

The stiffness of diagonal bracing members required to provide lateral stability of an arch system was studied by Sakimoto and Komatsu (1977a,b) and Kuranishi and Yabuki (1981), both focusing on the ultimate strength of braced steel arches subjected to the combination of vertical and horizontal uniform loads. More recent results on the effects of discrete elastic restraints on the out-of-plane buckling loads of circular arches loaded in uniform compression or uniform bending are given by Bradford and Pi (2002). Threshold values are given for the stiffness of the bracing at which the out-of-plane buckling mode changes from symmetrical to asymmetrical.

17.4.2 Ultimate Strength of Steel Arches Braced by Double-Diagonal Bracing System

Since the arch is basically a compressive member, initial out-of-plane deflections and residual stresses resulting from welding can have a significant influence on the lateral stability of steel arches. Numerical results obtained for typical theoretical models of square-box cross sections show that (1) the residual stress may cause at most 20% reduction in strength of mild-steel arches and 10% for high-strength steel arches and (2) the initial out-of-plane deflections may reduce the strength of a perfectly plane arch by a maximum of 15% (Sakimoto and Komatsu, 1977b; Komatsu and Sakimoto, 1977). These results are illustrated in Figs. 17.13 and 17.14, with the vertical axis representing the ultimate unit strength σ_u normalized by the yield stress σ_y and the horizontal axis defined by the slenderness ratio $\lambda = l_s/r_x$, in which l_s is the total curved length of the arch rib and r_x is the radius of gyration of the rib section about the x axis. The term σ_{rc} denotes the maximum compressive residual stress assumed in a trapezoidal distribution pattern and $\overline{v}_0$ denotes the amplitude of initial out-of-plane deflection that is assumed in the shape of a half-sine wave.

As shown in Fig. 17.15 and described in the introduction to this chapter (Fig. 17.1), the effect of load direction is also significant for steel arch bridges. In these figures, elastic buckling curves of the identical arches are also shown (Shukla and Ojalvo, 1971; Tokarz and Sandhu, 1972).

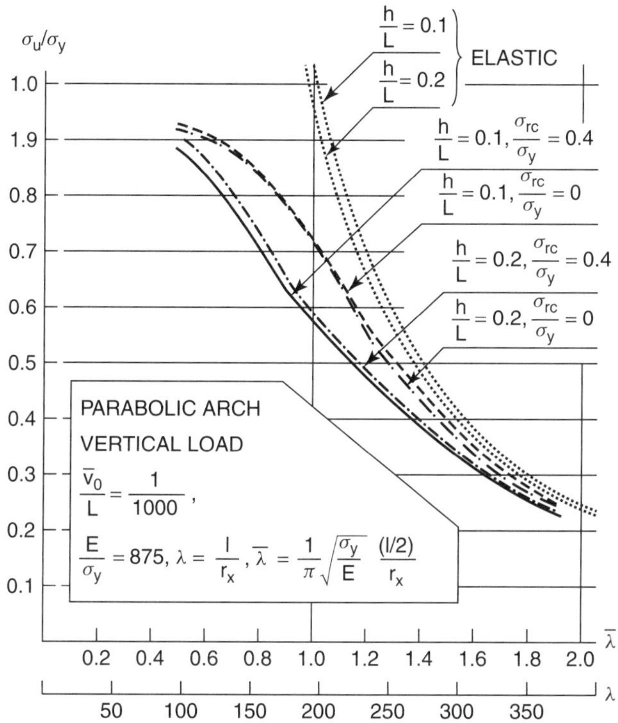

FIGURE 17.13 Effects of residual stresses.

Extensive parametric studies (Sakimoto, 1979; Sakimoto and Komatsu, 1982) have resulted in simple approximate methods for determining the strength of braced or unbraced steel arches that fail by lateral instability. For example, Sakimoto and Komatsu (1983a) use an analogy between an arch and a column to obtain the following equivalent slenderness parameter λ_a, which can be used to determine of the ultimate strength of through-type steel arches of box cross sections:

$$\lambda_a = \frac{1}{\pi}\sqrt{\frac{\sigma_y}{E}}\frac{l}{r_x}K_e K_l K_\beta \tag{17.19}$$

where σ_y is the lowest yield stress of the different steel grades used in the arch rib and K_e, K_l, and K_β are effective-length factors. The coefficient K_e relates to the rotational fixity of an arch rib at its ends with respect to the centroidal x axis. For the clamped condition K_e is 0.5 and for the hinged condition it is 1.0. The coefficient K_l corresponds to the direction of the loads and is taken as 0.65 for the tilting hanger case and 1.0 for nontilting hangers (i.e., for vertical loads). The coefficient K_β is related to the lateral restraint supplied by the bracing system and given by $K_\beta = 1 - \beta + (2r_x\beta/K_e b)$, where the term β is the ratio of the length of

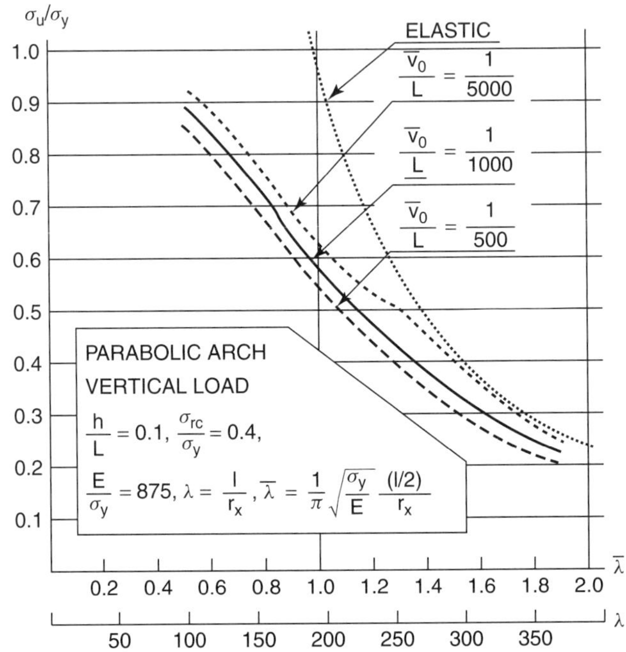

FIGURE 17.14 Effects of initial lateral deflections.

the braced portion to the total length of the arch rib. Since β is equal to zero for arches without bracing, K_β equals 1.0 for an isolated single arch.

The ultimate unit strength σ_u is computed for the column with the equivalent slenderness, and this ultimate unit strength would be an approximation of the ultimate stress required to buckle the actual arch structure. The ultimate unit strength σ_u for arches is defined as the tangential thrust at the support N_u divided by the cross-section area A; that is, $N_u = A \cdot \sigma_u$. The thrust N_u is determined from a linear analysis of the loaded arch. For a parabolic arch, for example, the uniformly distributed load per unit length of the arch span at the ultimate state w_u can be computed from Eq. 17.2, where $w_y = w_u$ and A is the cross-sectional area of an individual arch rib. When the rib cross section varies, A and r_x are weighted average values for the entire curved length l.

Although the Japanese column curve was used in this investigation (Sakimoto and Komatsu, 1983a), SSRC curve 2 (Eq. 3.8), curve C of the ECCS multiple column curves, or similar column curves are suggested as counterparts. Applicability of this approximate method has been examined by use of computer simulations for parabolic and circular steel arches composed of box section arch members with h/L ratios of 0.1 and 0.2. The results of one of these comparisons are shown in Fig. 17.16, noting that SSRC curve 2 was used in the approximate method. Practical applications for actual steel bridges are presented in Sakimoto and Komatsu

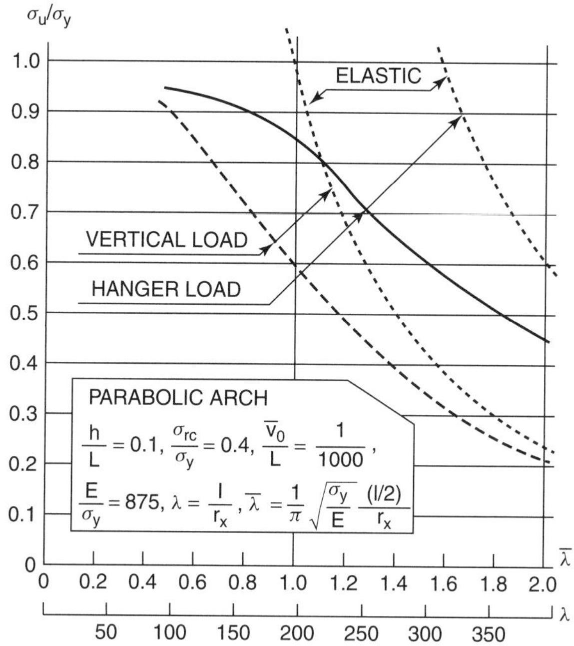

FIGURE 17.15 Effects of load directions.

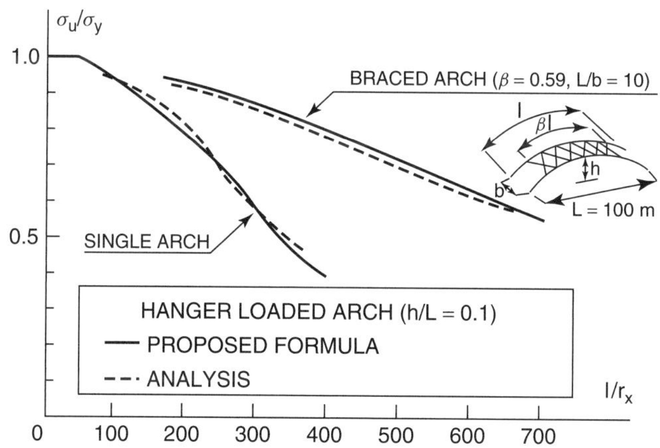

FIGURE 17.16 Accuracy of the approximation method.

(1983b), Sakimoto and Yamao (1983), Sakimoto et al. (1989), and Sakimoto and Sakata (1990).

When overall lateral buckling is suppressed, only buckling of rib segments between bracing points need be considered. The slenderness limit for which local instability of the arch segment may occur prior to overall lateral instability can be determined approximately from Eq. 17.19 as follows (Sakimoto and Komatsu, 1983a):

$$\frac{\sqrt{\sigma_y}}{E}\left(\frac{l_p}{r_x}\right)_{\max} \geq \frac{\sqrt{\sigma_y}}{E}\frac{l}{r_x}K_e K_l K_\beta \tag{17.20}$$

where $(l_p/r_x)_{\max}$ denotes the slenderness ratio for the most slender arch segment having an unbraced length l_p and the term σ_y in the left side is the yield stress of the arch segment material where $(l_p/r_x)_{\max}$ occurs.

Nazmy (1997) studied long-span half-through and deck arch bridges with K-bracing in vertical and inclined arches and found that there is a threshold for the stiffness of the lateral bracing system at which the critical buckling load is governed by the stiffness of the arch rib itself and not by the stiffness of the bracing elements. Nazmy also showed that the critical buckling load of through-arch bridges can be increased by changing the connections between the arch rib and the deck from pinned to fully restrained (rigid).

17.5 ULTIMATE STRENGTH OF STEEL ARCH BRIDGES

17.5.1 Bridges Subjected to Uniform Vertical Loads

In the previous section, Eq. 17.19 defines slenderness as a function of effective length L_e, with $L_e = \left(K_e K_l K_\beta\right) l$. The following procedure, suggested by Sakimoto et al. (1992), is based on using spatial eigenvalue (three-dimensional buckling) analyses, available in many finite element software packages, to determine effective lengths and in turn slenderness. Assuming linearity of the prebuckling state (i.e., the relative internal force distribution remains constant when scaling the applied load), the critical axial force of a member N_{cr} can be expressed as

$$N_{cr} = \alpha N_o = \frac{\pi^2 EI}{L_e^2} \tag{17.21}$$

where α is the smallest eigenvalue (buckling load factor), N_o the axial force of the member corresponding to the reference applied vertical load P_o, and L_e the effective length of the member. From Eq. 17.19, the slenderness can be defined as

$$\lambda_a = \frac{1}{\pi}\sqrt{\frac{\sigma_y}{E}}\frac{L_e}{r_x} = \sqrt{\frac{A\sigma_y L_e^2}{\pi^2 EI}} = \sqrt{\frac{A\sigma_y}{\alpha N_o}} \tag{17.22}$$

If the reference load is taken as the axial yield (squash) load of the arch rib, $N_o = A\sigma_y$, where A is the cross-sectional area, the slenderness becomes a function of the smallest eigenvalue for the system:

$$\lambda_a = \sqrt{\frac{1}{\alpha}} \tag{17.23}$$

By further substituting this slenderness into a column strength formula, as discussed in Section 17.4.2, an estimate of the ultimate strength of the arch bridge can be obtained.

To verify the validity of this procedure, the ultimate strengths of the arch bridge models shown in Fig. 17.17, with properties described in Table 17.9, were computed by an elastoplastic finite displacement analysis. The ultimate normal stress σ_{N_u} can be given as $\sigma_{N_u} = N_u/A$, where N_u is the axial force at the springing (support) calculated by linear theory for the ultimate load P_u.

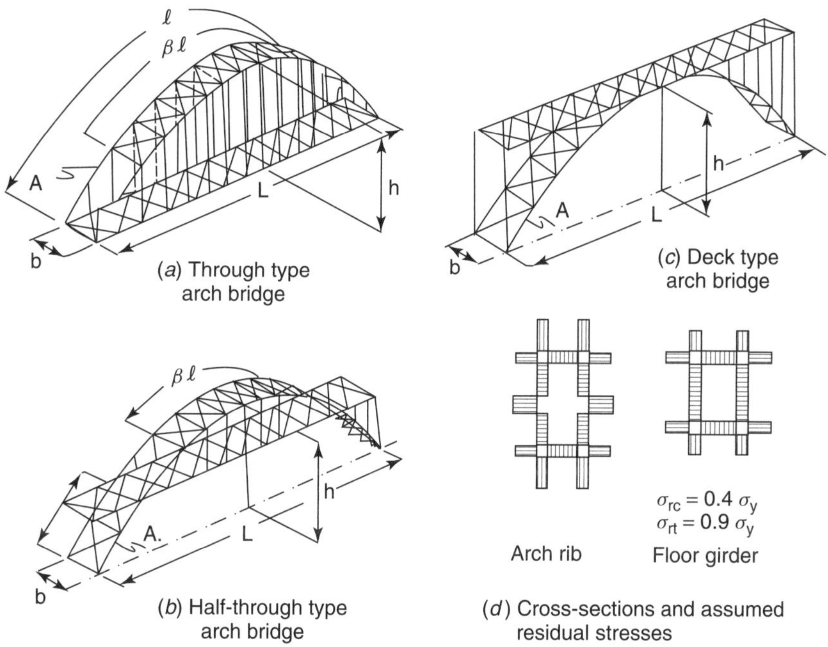

FIGURE 17.17 Numerical models of arch bridges.

TABLE 17.9 Structural Properties of the Numerical Models

	Through-Type Arch Bridge	Half-Through-Type Arch Bridge	Deck-Type Arch Bridge
Span length, L (m)	150	150	100, 200, 300
Rise, h (m)	22.5	22.5	20, 40, 60
Rise-to-span ratio, h/l	0.15	0.25	0.20
Arch rib distance, a (m)	20, 10, 5	20, 10, 5	5.2
Bending stiffness ratio, $\overline{I}$	0.1–3.0	0.1–3.0	0.1–1.0
Bracing ratio, β	0.48–0.86	0.64–0.88	1.0
Cross-sectional area, A (cm^2)	448	448	298.8–903.1

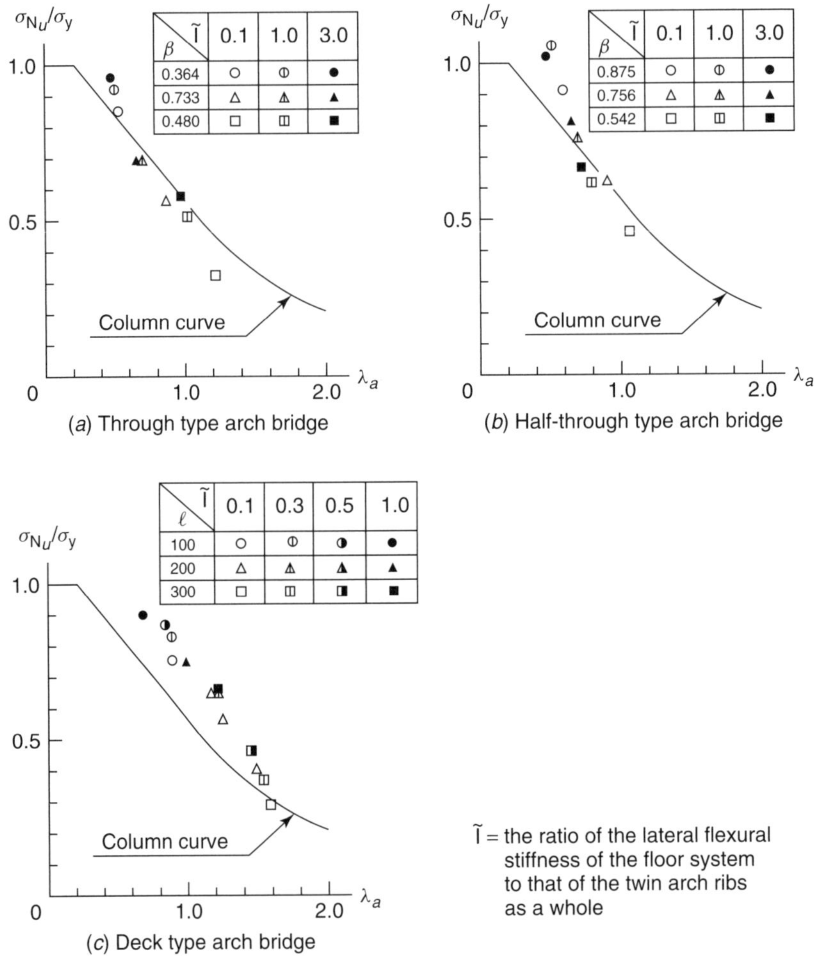

FIGURE 17.18 Validity of the slenderness procedure using the eigenvalue.

In Fig. 17.18, the σ_{N_u} values obtained are plotted as a function of the slenderness λ_a ,which is given by Eq. 17.22 for the computational models. The close agreement between the calculated values and the column strength curves demonstrates the validity of this procedure.

17.5.2 Bridges Subjected to Vertical and Lateral Uniform Loads

The spatial elastic–plastic behavior and the ultimate-load capacity of the through-type braced arch bridges of box cross sections was studied for a combination of the vertical and horizontal uniform loads by Sakimoto (Sakimoto and Komatsu, 1977a, 1979; Sakimoto et al., 1979). Based on computer analyses for various theoretical models, a simple approximate method for determining the ultimate lateral strength of bridge arches braced partly over the central portion was developed. By utilizing an analogy between a laterally loaded arch and a beam-column, the following interaction formula is proposed

$$\frac{N}{N_u} + \frac{M}{M_y[1 - (N/N_e)]} \leq 1.0 \tag{17.24}$$

Although Eq. 17.23 has a form similar to an interaction equation for a beam-column, attention should be placed on the precise definition of each of the terms, which include N being the tangential end thrust computed by linear theory for the uniformly distributed design loads; N_u the tangential end thrust corresponding to the uniformly distributed vertical loading that produces inelastic lateral buckling of the arch (which can be determined from $N_u = A\sigma_u$ as demonstrated in the previous section); M the lateral end moment of the individual arch rib computed by linear theory for the uniformly distributed horizontal loads [this can be approximated by the value computed for the planar system which is obtained by straightening out the arch in a horizontal plane; a simple approximation formula for the determination of this moment is also given by Sakimoto and Komatsu (1979)]; M_y the yield moment of the arch rib at the ends with respect to the x axis; and N_e the Euler buckling load computed for a hinged column of which the length and cross section are identical to those of the arch rib for the unbraced segment.

The validity of this interaction equation has been verified by extensive computer simulations for various braced bridge arches of practical proportions. If the twin ribs are braced closely over the entire length of the arch, the moment ratio term of Eq. 17.23 becomes negligible, which indicates that typical lateral loads are not significant for such braced arches.

Deflections and stresses due to forces normal to the plane of the arch are increased by the second-order effects associated with the combination of large axial loads and lateral deflections. Stüssi (1943), Östlund (1954), and Wästlund (1960) suggest that such increases may be estimated by multiplying the first-order deflections and stresses due to transverse loads by the amplification factor $1/(1 - w/w_c)$, in which w is the intensity of the uniform load in the plane of the arch and w_c is the intensity of the in-plane load that would cause lateral buckling. While this amplification factor is at best only a rough approximation of the second-order effects (Layrangues, 1959; Donald and Godden, 1961a,b), it is convenient for use in a preliminary design.

The effect of lateral horizontal force on the in-plane strength of arch bridges was studied by Kuranishi (1961b), Yabuki and Kuranishi (1973), and Kuranishi and Yabuki (1977). They determined that the in-plane strength of arch bridges with a close brace over the entire length of the arch is not significantly affected by lateral loads typically encountered in actual structures. Hence, it was concluded that through-type steel arch bridges can be designed with a lateral bracing system (between arch ribs) of sufficient out-of-plane stiffness so that the arch rib design will be determined primarily by the in-plane loads. With this in mind, the effect of lateral loads can be taken into account for practical purposes as a set of additional in-plane vertical loads. Thus, the arch rib design can be made on the basis of a quasi-planar model subjected to the principal vertical loads slightly modified to include the effects of wind loads. Kuranishi and Yabuki (1981, 1984) and Yabuki et al. (1983) present the lateral bracing stiffness required to ensure that two-hinged arch bridges behave basically as in-plane structures. Ju (2003) studied parabolic steel arch deck and through bridges with vertical and inclined ribs braced by transverse bars. Using a statistical analysis of a large set of computational results, he derived effective lengths for these bridges and found that the effective lengths used by AASHTO (2004) are conservative.

REFERENCES

AASHTO (2004). *LRFD Bridge Design Specifications*, American Association of State Highway and Transportation Officials, Washington, DC.

Almeida, P. N. (1970), "Lateral Buckling of Twin Arch Ribs with Transverse Bars," dissertation, Ohio State University, Columbus, OH.

Austin, W. J. (1971), "In-Plane Bending and Buckling of Arches," *ASCE J. Struct. Div.*, Vol. 97, No. ST5, pp. 1575–1592.

Austin, W. J., and Ross, T. J. (1976), "Elastic Buckling of Arches Under Symmetrical Loading," *ASCE J. Struct. Div.*, Vol. 102, No. ST5, pp. 1085–1095.

Beedle, L. S. (Ed.) (1991), *Stability of Metal Structures, a World View*, 2nd ed., Structural Stability Research Council, Bethlehem, PA.

Bradford, M. A., and Pi, Y.-L. (2002), "Elastic Flexural-Torsional Buckling of Discretely Restrained Arches," *ASCE J. Struct. Eng.*, Vol. 128, No. 6, pp. 719–727.

CEN (2006a) Eurocode 3: Design of Steel Structures, Part 2: Steel Bridges, EN 1993-2, Comité Européen de Normalisation (CEN), European Committee for Standardization Brussels, Belgium.

CEN (2006b), Eurocode 3: Design of Steel Structures, Part 1-1: General Rules and Rules for Buildings, EN 1993-1-1, Comité Européen de Normalisation (CEN), European Committee for Standardization Brussels, Belgium.

Chang, C. K. (1973), "Effect of Loaded Length on the Buckling Strength of Slender Arches," thesis, Rice University, Houston, TX.

Coronforth, R. C., and Childs, S. B. (1967), "Computer Analysis of Two Hinged Circular Arches," *ASCE J. Struct. Div.*, Vol. 93, No. ST2, pp. 319–338.

DaDeppo, D. A., and Schmidt, R. (1970). "Stability of an Arch Under Combined Loads; Bibliography on Stability of Arches," *Ind. Math.*, Vol. 20, Part 2, pp. 71–89.

Demuts, E. (1969), "Lateral Buckling of Circular Arches Subjected to Uniform Gravity Type Loading," dissertation, Ohio State University, Columbus, OH.

Deutsch, E. (1940), "Das Knicken von Bogenträgern bei unsymmetrischer Belastung," *Bauingenieur*, Dec., pp. 353–360.

Donald, P. T. A., and Godden, W. G. (1961a), "A Numerical Solution to the Curved Beam Problem," *Struct. Eng.*, Vol. 41, pp. 179–186.

Donald, P. T. A., and Godden, W. G. (1961b), "The Transverse Behavior of Laterally Unsupported Parabolic Arches," *Struct. Eng.*, Vol. 41, pp. 187–191.

Forrester, K. (1972), Discussion of "In-Plane Bending and Buckling of Arches" by W. J. Austin, *ASCE J. Struct. Div*., Vol. 98, No. ST1, pp. 378–379.

Fukasawa, Y. (1963), "Buckling of Circular Arches by Lateral Flexure and Torsion Under Axial Thrust," *Trans. Jpn. Soc. Civ. Eng.*, No. 96, pp. 29–47 (in Japanese).

Fukumoto, Y. (1996), *Structural Stability Design, Steel and Composite Structures*, Pergamon, Amsterdam.

Galli, A., and Franciosi, G. (1955), "Limit Analysis of Thin Arch Bridges with Stiffening Girders," *G. Genio Civ.,* Vol. 93, No. 11.

Godden, W. G. (1954), "The Lateral Buckling of Tied Arches," *Proc. Inst. Civ. Eng.*, Vol. 3, Part III, pp. 496–514.

Harries, H. (1970), "Traglast stahlerner Zweigelenkbogen mit ausgebreiteten Fliesszonen," *Stahlbau*, Vol. 6, pp. 170–177; Vol. 8, pp. 248–257.

Harrison, H. B. (1982), "In-Plane Stability of Parabolic Arches," *ASCE J. Struct. Div.*, Vol. 108, No. ST1, pp. 195–205.

Hayashi, T. (Ed.) (1971), *Handbook of Structural Stability*, Corona Publishing, Tokyo.

Ju, S.-H. (2003), "Statistical Analyses of Effective Lengths in Steel Arch Bridges," *Comput. Struct.*, Vol. 81, pp. 1487–1497.

Jukes, S. G., Hassani, F. P., and Whittaker, B. N. (1983), "Characteristics of Steel Arch Support Systems for the Mine Roadway. Part I, Modelling Theory, Instrumentation and Preliminary Results," *Mining Sci. Technol.*, Vol. 1, pp. 43–58.

Kee, C. F. (1961), "Lateral Inelastic Buckling of Tied Arches," *ASCE J. Struct. Div.*, Vol. 87, No. ST1, pp. 23–39.

Khan, U. H., Mitri, H. S., and Jones, D. (1996), "Full Scale Testing of Steel Arch Tunnel Supports," *Int. J. Rock Mech. Min.*, Vol. 33, No. 3, pp. 219–232.

King, C., and Brown, D. (2001), *Design of Curved Steel*, Steel Construction Institute, Ascot Berks UK.

Klöppel, K., and Protte, W. (1961), "A Contribution to the Buckling Problem of Circular Curved Bars," *Stahlbau*, No. 30, pp. 1–15 (in German).

Kollár, L. U., and Iványi, G. (1966), "Buckling of Shell Arches by the Energy Method," *Bautech. Arch.*, pp. 1–23 (in German).

Kollbrunner, C. F. (1936), "Versuche über die Knicksicherheit und die Grundschwingungzahl vollwandiger Bogen," *Bautechnik*, Mar., pp. 186–188.

Kollbrunner, C. F. (1942), "Versuche über die Knicksicherheit und die Grundschwingungzahl vollwandiger Dreigelenkbogen, *Schweiz. Bauzg.*, Vol. 120, No. 10, pp. 113–115.

Komatsu, S., and Sakimoto, T. (1977), "Ultimate Load Carrying Capacity of Steel Arches," *ASCE J. Struct. Div.*, Vol. 103, No. ST12, pp. 2323–2336.

Komatsu, S., and Shinke, T. (1977), "Practical Formulas for In-Plane Load Carrying Capacity of Arches," *Proc. Jpn. Soc. Civ. Eng.*, No. 267, pp. 39–51 (in Japanese).

Kuranishi, S. (1961a), "The Torsional Buckling Strength of Solid Rib Arch Bridge," *Trans. Jpn. Soc. Civ. Eng.*, No. 75, pp. 59–67 (in Japanese).

Kuranishi, S. (1961b), "Analysis of Arch Bridge Under Certain Lateral Forces," *Trans. Jpn. Soc. Civ. Eng.*, No. 73, pp. 1–6 (in Japanese).

Kuranishi, S. (1973), "Allowable Stress for Two-Hinged Steel Arch," *Proc. Jpn. Soc. Civ. Eng.*, No. 213, pp. 71–75.

Kuranishi, S. (1993), “Stiffening Effect of Lateral Bracing of Steel Arch Bridges on Their In-Plane Strength,” *Proc. SSRC Ann. Techn. Session*, Milwaukee, WI.

Kuranishi, S., and Lu, L. W. (1972), “Load Carrying Capacity of Two-Hinged Steel Arches,” *Proc. Jpn. Soc. Civ. Eng.*, No. 204, pp. 129–140 (in English).

Kuranishi, S., Sato, T., and Otsuki, M. (1980), “Load Carrying Capacity of Two-Hinged Steel Arch Bridges with Stiffening Deck,” *Proc. Jpn. Soc. Civ. Eng.*, No. 300, pp. 121–130.

Kuranishi, S., and Yabuki, T. (1977), “In-Plane Strength of Arch Bridges Subjected to Vertical and Lateral Loads,” paper presented at the 2nd Int. Colloq. Stabl. Steel Struct., Prelim. Rep., Liege, Belgium, Apr., pp. 551–556.

Kuranishi, S., and Yabuki, T. (1979), “Some Numerical Estimations of Ultimate In-Plane Strength of Two-Hinged Steel Arches,” *Proc. Jpn. Soc. Civ. Eng.*, No. 287, pp. 155–158.

Kuranishi, S., and Yabuki, T. (1981), “Required Out-of-Plane Rigidities of Steel Arch Bridges with Two Main Ribs Subjected to Vertical and Lateral Loads,” Techn. Rep. Tohoku Univ., Sendai, Japan, Vol. 46, No. 1.

Kuranishi, S., and Yabuki, T. (1984), “Lateral Load Effect on Arch Bridge Design,” *ASCE J. Struct. Eng.*, Vol. 110, No. 9, pp. 2263–2274.

la Poutré, D. B. (2005), "Inelastic Spatial Stability of Circular Steel Wide Flange Arches," Ph.D. thesis, Technische Universiteit Eindhoven, Eindhoven, The Netherlands.

Layrangues, P. (1959), “Elastic Deformations and Forces in a Fixed End Symmetrical Circular Arch Subjected to a Constant Normal Force and a Uniformly Distributed Lateral Force,” *Ann. Ponts Chaussees*, Vol. 129, pp. 323–343 (in French).

Mitri, H. S., and Hassani, F. P. (1990), "Structural Characteristics of Coal Mine Steel Arch Supports," *Int. J. Rock Mech. Min.*, Vol. 27, No. 2, pp. 121–127.

Namita, Y. (1968), “Second Order Theory of Curved Bars and Its Use in the Buckling Problem of Arches,” *Trans. Jpn. Soc. Civ. Eng.,* No. 155, pp. 32–41 (in German).

Nazmy, A. S. (1997), "Stability and Load-Carrying Capacity of Three-Dimensional Long-Span Steel Arch Bridges," *Comput. Struct.*, Vol. 65, No. 6, pp. 857–868.

Ojalvo, M., Demuts, E., and Tokarz, F. J. (1969), “Out-of-Plane Buckling of Curved Members,” *ASCE J. Struct. Div.*, Vol. 95, No. ST10, pp. 2305–2316.

Onat, E. T., and Prager, W. (1953), “Limit Analysis of Arches,” *J. Mech. Phys. Solids*, Vol. 1, pp. 71–89.

Östlund, L. (1954), “Lateral Stability of Bridge Arches Braced with Transverse Bars,” *Trans. R. Inst. Technol.*, Stockholm, No. 84.

Papangelis, J. P., and Trahair, N. S. (1987a), “Flexural–Torsional Buckling Tests on Arches,” *ASCE J. Struct. Eng.*, Vol. 113, No. ST7, pp. 1433–1443.

Papangelis, J. P., and Trahair, N. S. (1987b), “Flexural–Torsional Buckling of Arches,” *ASCE J. Struct. Eng.*, Vol. 113, No. ST4, pp. 889–906.

Petersen, C. (1982), *Statik und Stabilität der Baukonstruktionen*, 2nd ed. Vieweg & Sohn, Braunschweig (in German).

Pi, Y.-L., and Bradford, M. A. (2004), "In-Plane Strength and Design of Fixed Steel I-Section Arches," *Eng. Struct.*, Vol. 26 . No. 3, pp. 291–301.

Pi, Y.-L., and Bradford, M. A. (2005), "Out-of-Plane Strength Design of Fixed Steel I-Section Arches," *ASCE J. Struct. Eng.*, Vol. 131, No. 4, pp. 560–568.

Pi, Y.-L., and Trahair, N. S. (1998), "Out-of-Plane Inelastic Buckling and Strength of Steel Arches," *ASCE J. Struct. Eng.*, Vol. 124, No. 2, pp. 174–183.

Pi, Y.-L., and Trahair, N. S. (1999), "In-Plane Buckling and Design of Steel Arches," *ASCE J. Struct. Eng.*, Vol. 125, No. 11, pp. 1291–1298.

Pi, Y.-L., and Trahair, N. S. (2000), "Inelastic Lateral Buckling Strength and Design of Steel Arches," *Eng. Struct.*, Vol. 22, pp. 993–1005.

Rajasekaran, S., and Padmanabhan, S. (1989), "Equations of Curved Beams," *ASCE J. Eng. Mech.*, Vol. 115, No. 5, pp. 1094–1110.

Sakimoto, T. (1979), "Elasto-Plastic Finite Displacement Analysis of Three Dimensional Structures and Its Application to Design of Steel Arch Bridges," dissertation, Osaka University, Osaka, Japan.

Sakimoto, T., and Komatsu, S. (1977a), "A Possibility of Total Breakdown of Bridge Arches Due to Buckling of Lateral Bracings," paper presented at the 2nd Int. Colloq. Stab. Steel Struct., Final Rep., Liege, Belgium, Apr., pp. 299–301.

Sakimoto, T., and Komatsu, S. (1977b), "Ultimate Load Carrying Capacity of Steel Arches with Initial Imperfections," paper presented at the 2nd Int. Colloq. Stabil. Steel Struct., Prelim. Rep., Liege, Belgium, Apr., pp. 545–550.

Sakimoto, T., and Komatsu, S. (1979), "Ultimate Strength of Steel Arches Under Lateral Loads," *Proc. Jpn. Soc. Civ. Eng.*, No. 292, pp. 83–94.

Sakimoto, T., and Komatsu, S. (1982), "Ultimate Strength of Arches with Bracing Systems," *ASCE J. Struct. Div.*, Vol. 108, No. ST5, pp. 1064–1076.

Sakimoto, T., and Komatsu, S. (1983a), "Ultimate Strength Formula for Steel Arches," *ASCE J. Struct. Div.*, Vol. 109, No. 3, pp. 613–627.

Sakimoto, T., and Komatsu, S. (1983b), "Ultimate Strength Formula for Central-Arch-Girder Bridges," *Proc. Jpn. Soc. Civ. Eng.*, No. 33, pp. 183–186.

Sakimoto, T., and Namita, Y. (1971), "Out-of-Plane Buckling of Solid Rib Arches Braced with Transverse Bars," *Proc. Jpn. Soc. Civ. Eng.*, No. 191, pp. 109–116.

Sakimoto, T., and Sakata, T. (1990), "The Out-of-Plane Buckling Strength of Through-Type Arch Bridges," *J. Constr. Steel Res.*, Vol. 16, pp. 307–318.

Sakimoto, T., Sakata, T., and Tsuruta, E. (1989), "Elasto-Plastic Out-of-Plane Buckling Strength of Through Type and Half-Through Type Arch Bridges," *Struct. Eng., Earthquake Eng.*, Vol. 6, No. 2, pp. 307s–318s (*Proc. Jpn. Soc. Civ. Eng. No. 410/I-12*).

Sakimoto, T., Sakata, T., and Kobori, T. (1992), "An Effective Length Procedure for Out-of-Plane Buckling Strength Estimation of Steel Arch Bridges," *Proc. 3rd Pac. Struct. Steel Conf.*, Tokyo, pp. 251–258.

Sakimoto, T., and Watanabe, H. (1995), "A New Procedure for Frame Design," *Proc. SSRC Annual Tech. Session*, Pittsburgh, PA, pp. 23–31.

Sakimoto, T., and Yamao, T. (1983), "Ultimate Strength of Deck-Type Steel Arch Bridges," paper presented at the 3rd Int. Colloq. Stab. Met. Struct., Prelim. Rep., Paris, Nov.

Sakimoto, T., Yamao, T., and Komatsu, S. (1979), "Experimental Study on the Ultimate Strength of Steel Arches," *Proc. Jpn. Soc. Civ. Eng.*, No. 286, pp. 139–149.

Shinke, T., Zui, H., and Nakagawa, T. (1980), "In-Plane Load Carrying Capacity of Two-Hinged Arches with Stiffening Girder," *Trans. Jpn. Soc. Civ. Eng.*, No. 301, pp. 47–59.

Shinke, T., Zui, H., and Namita, Y. (1975), "Analysis of In-Plane Elasto-Plastic Buckling and Load Carrying Capacity of Arches," *Proc. Jpn. Soc. Civ. Eng.*, No. 244, pp. 57–69 (in Japanese).

Shinke, T., Zui, H., and Namita, Y. (1977), "Analysis of Experiment on In-Plane Load Carrying Capacity of Arches," *Proc. Jpn. Soc. Civ. Eng.*, No. 244, pp. 11–23 (in Japanese).

Shukla, S. N., and Ojalvo, M. (1971), "Lateral Buckling of Parabolic Arches with Tilting Loads," *ASCE J. Struct. Div.*, Vol. 97, No. ST6, pp. 1763–1733.

Singer, J., Arbocz, J., and Weller, T. (1998), *Buckling Experiments: Experimental Methods in Buckling of Thin-Walled Structures*, J Wiley, New York.

Stevens, L. K. (1957), "Carrying Capacity of Mild Steel Arches," *Proc. Inst. Civ. Eng.*, Vol. 6, pp. 493–514.

Stüssi, F. (1943), "Lateral-Buckling and Vibration of Arches," *Proc. Int. Assoc. Bridge Struct. Eng. Publ.*, Vol. 7, pp. 327–343 (in German).

Timoshenko, S. P., and Gere, J. M. (1961), *Theory of Elastic Stability*, 2nd ed., McGraw-Hill, New York.

Tokarz, F. J. (1971), "Experimental Study of Lateral Buckling of Arches," *ASCE J. Struct. Div.,* Vol. 97, No. ST2, pp. 545–559.

Tokarz, F. J., and Sandhu, R. S. (1972), "Lateral–Tosional Buckling of Parabolic Arches," *ASCE J. Struct. Div.*, Vol. 98, No. ST5, pp. 1161–1179.

Tveit, P. (1966), "Desing of Network Arches," *Struct. Eng.*, Vol. 44, No. 7, pp. 247–259.

Tveit, P. (1999), "Comparison of Steel Weights in Narrow Arch Bridges with Medium Spans," *Stahlbau*, Vol. 68, No. 9, pp. 753–757.

Vlasov, V. Z. (1963), *Thin Walled Elastic Beams*, 2nd ed., Israel Program for Scientific Translations, Jerusalem.

Wästlund, G. (1960), "Stability Problems of Compressed Steel Members and Arch Bridges," *ASCE J. Struct. Div.*, Vol. 86, No. ST6, pp. 47–71.

Wolde-Tinsaie, A. M., and Foadian, H. (1989), "Asymmetrical Buckling of Prestressed Tapered Arches," *ASCE J. Eng. Mech.*, Vol. 115, No. 9, pp 2020–2034.

Yabuki, T. (1981), "Study on Ultimate Strength Design of Steel Arch Bridge Structures," dissertation, Tokoku University, Sendai, Japan (in Japanese).

Yabuki, T., and Kuranishi, S. (1973), "Out-of-Plane Behavior of Circular Arches Under Side Loadings," *Proc. Jpn. Soc. Civ. Eng.*, No. 214, pp. 71–82.

Yabuki, T., and Vinnakota, S. (1984), "Stability of Steel Arch-Bridges, A State-of-the-Art Report," *Solid Mech. Arch.*, Vol. 9, No. 2, Martinus Nijhoff Publishers, The Hague, The Netherlands.

Yabuki, T., Vinnakota, S., and Kuranishi, S. (1983), "Lateral Load Effect on Load Carrying Capacity of Steel Arch Bridge Structures," *ASCE J. Struct. Div.*, Vol. 109, No. 10, pp. 2434–2449.

Yang, Y. B., and Kuo, S. R. (1986), "Static Stability of Curved Thin-Walled Beams," *ASCE J. Eng. Mech*, Vol. 112, No. 8, pp. 821–841.

Yang, Y. B., and Kuo, S. R. (1987), "Effect of Curvature on Stability of Curved Beams," *ASCE J. Struct. Eng.*, Vol. 113. No. 6, pp. 1185–1202.

Yang, Y. B., and Kuo, S. R. (1991), "Use of Straight-Beam Approach to Study Buckling of Curved Beams," *ASCE J. Struct. Eng.*, Vol. 117, No. 7, pp. 1963–1978.

CHAPTER 18

DOUBLY CURVED SHELLS AND SHELL-LIKE STRUCTURES

18.1 INTRODUCTION

A *shell-like metal structure* is one that supports loads in a manner similar to that of a thin shell. That is, the majority of its load-bearing behavior is provided by membrane actions in which forces are carried from point to point by biaxial tension or compression and by shear in the plane of the shell. In addition to the membrane action, shell-like structures can include some degree of bending resistance. Shell-like structures are often divided into two classes, although the distinction between them is often not obvious, continuous, and discrete. Continuous shell structures are basically curved plates, which may be stiffened in orthogonal or skew directions. Discrete shells are reticulated structures that are curved in space. Continuous shell structures are considered first.

A typical stiffened shell is shown in Fig. 18.1. Stiffening elements in the meridional direction are called *stringers.* They are usually small and spaced close together. Stiffening elements in the circumferential direction are called *ring stiffeners.* They are often large rings spaced fairly far apart. There are four modes in which buckling can occur for this structure. First, there is *local buckling*; the plate spanning between the stringers and stiffeners can buckle. Second, the shell plate and stringer combination spanning between the stiffeners can buckle; this is called *bay instability.* In the third form of instability, *general instability*, a large portion of the plate, stringers, and stiffeners buckles. In addition, the stringers or stiffeners may undergo *torsional buckling* as curved columns. This form of instability is not unique to shell structures and is considered in other chapters of this guide.

Shell design codes usually address the first three buckling cases. There are two ways in which stiffened shells can be described. The first approach is to "smear" stiffeners and stringers across the shell plate and analyze the system as a continuous orthotropic plate. The second approach is to use the split-rigidity concept. This approach is described here in conjunction with discrete shell structures.

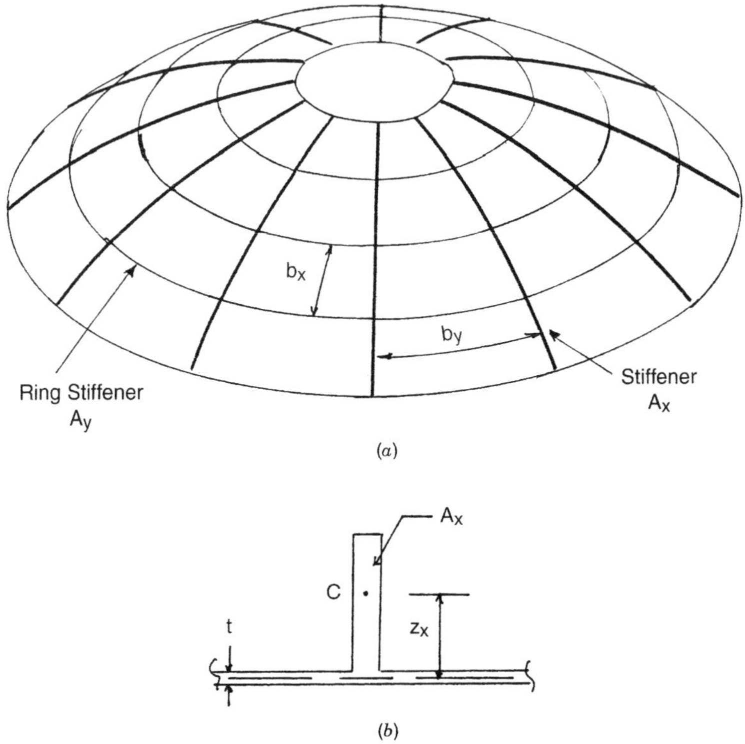

FIGURE 18.1 (*a*) Geometry of a typical stiffened shell; (*b*) geometry of stiffener.

The representation of shell stiffeners as orthotropic plates is presented in Flügge's (1962) text, among others. The constitutive equations for the equivalent orthotropic shell, modified slightly, are

$$N_x = A_{11}\epsilon_x + A_{12}\epsilon_y - B_{11}\kappa_x \tag{18.1a}$$

$$N_y = A_{12}\epsilon_x + A_{22}\epsilon_y - B_{22}\kappa_y \tag{18.1b}$$

$$N_{xy} = A_{66}\epsilon_{xy} \tag{18.1c}$$

$$M_x = B_{11}\epsilon_x - D_{11}\kappa_x - D_{12}\kappa_y \tag{18.2a}$$

$$M_y = B_{22}\epsilon_y - D_{12}\kappa_x - D_{22}\kappa_y \tag{18.2b}$$

$$M_{xy} = D_{66}\kappa_{xy} \tag{18.2c}$$

where ϵ_x denotes the middle-surface strain in the x direction and κ_x denotes the middle-surface curvature. The other ϵ and κ values are defined similarly.

The constants appearing in Eqs. (18.1) and (18.2) are

$$A_{11} = \frac{E_s t}{1-\nu^2}\frac{\overline{b}_y}{b_y} + E_x A_x \tag{18.3a}$$

$$A_{22} = \frac{E_s t}{1-\nu^2}\frac{\overline{b}_x}{b_x} + E_y A_y \tag{18.3b}$$

$$A_{12} = \frac{\nu E_s t}{1-\nu^2} \tag{18.3c}$$

$$A_{66} = \frac{G_s t}{2}\left(\frac{\overline{b}}{b_x} + \frac{\overline{b}_y}{b_y}\right) \tag{18.3d}$$

$$B_{11} = \frac{E_x A_x z_x}{b_y} \tag{18.3e}$$

$$B_{22} = \frac{E_y A_y z_y}{b_x} \tag{18.3f}$$

$$D_{11} = \frac{E_s t^3}{12(1-\nu^2)}\frac{\overline{b}_y}{b_y} + \frac{E_x I_x}{b_y} \tag{18.3g}$$

$$D_{22} = \frac{E_s t^3}{12(1-\nu^2)}\frac{\overline{b}_x}{b_x} + \frac{E_y I_y}{b_x} \tag{18.3h}$$

$$D_{12} = \frac{\nu E_s t^3}{12(1-\nu^2)} \tag{18.3i}$$

$$D_{66} = \frac{G_s t^3}{6}\left(\frac{\overline{b}_x}{b_x} + \frac{\overline{b}_y}{b_y}\right) + \frac{G_x J_x}{b_y} + \frac{G_y J_y}{b_x} \tag{18.3j}$$

where t = shell thickness
E_s = modulus of elasticity of the shell plate
E_x = modulus of elasticity of the x stiffener
G_s = shear modulus of elasticity of the shell plate
G_x = shear modulus of elasticity of the x stiffener
ν = Poisson's ratio
$\overline{b}_i$ = effective length in the i direction
z_i = distance from centroid of stiffening element to the center of the plate, outward positive

The effective lengths, $\overline{b}_i$, are introduced into these equations by some codes to take into account the fact that not all of the shell can be effective in interacting with stiffeners or stringers if these elements are spaced far apart (ASME, 2007). Because stringers are usually smaller members placed close together, representing them by an orthotropic shell is accurate. This, however, is not true for large stiffeners placed far apart. In this case, the stiffeners are separate units and, ideally, should be considered as such. Nonetheless, if one is attempting to solve the shell-buckling problem with a closed-form solution, as needed in a design code, there is no alternative to the use of the smeared stiffness approach for all shell properties.

18.2 THE BASIC PROBLEM

The basic problem of shell stability can be seen from an investigation of the buckling of a spherical shell under external pressure. This problem will be examined in some detail using a simplified analysis because many shell codes use equations similar to those presented.

A doubly curved shallow shell segment is shown in Fig. 18.2. An attempt will be made to compute the bifurcation load for this segment under constant membrane stress resultants. The membrane forces, $\overline{N}_x$ and $\overline{N}_y$, are found by a linear elastic analysis of the shell. The membrane shear stress resultants are neglected; their inclusion complicates the mathematics, but not the basic theory (Anderson and Bennet, 1981). The bifurcation equations describing buckling of the shell are three simultaneous equations defined in terms of derivatives of the buckled displacement functions, $u(x,y)$, $v(x,y)$, and $w(x,y)$:

$$\begin{aligned}\sum F_x &= \left(A_{11} - \frac{B_{11}}{R_x} - \frac{C_{11}}{R_x} + \frac{D_{11}}{R_x^2}\right)\frac{\partial^2 u}{\partial x^2} + \left(A_{66} + \frac{D_{66}}{R_x^2}\right)\frac{\partial^2 u}{\partial y^2} \\ &\quad + \left(A_{12} + A_{66} + \frac{D_{12}}{R_x R_y} + \frac{D_{66}}{R_x R_y}\right)\frac{\partial^2 v}{\partial x \partial y} + \left(\frac{D_{11}}{R_x} - B_{11}\right)\frac{\partial^3 w}{\partial x^3} \\ &\quad + \left(\frac{D_{12}}{R_x} + 2\frac{D_{66}}{R_x}\right)\frac{\partial^3 w}{\partial x \partial y^2} + \left(\frac{C_{11}}{R_x^2} - \frac{A_{11}}{R_x} - \frac{A_{12}}{R_y}\right)\frac{\partial w}{\partial x} = 0 \qquad (18.4)\end{aligned}$$

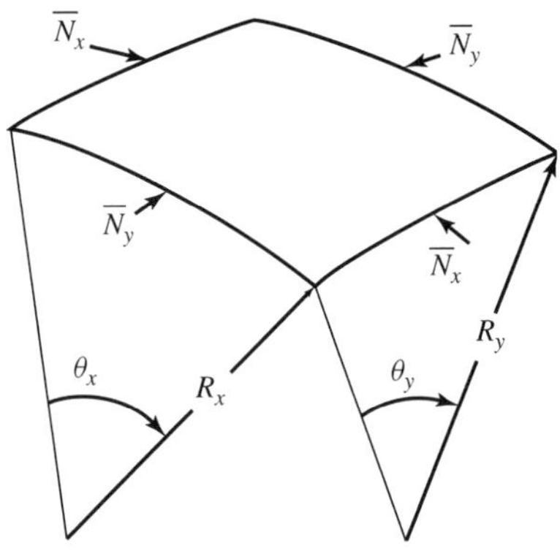

FIGURE 18.2 Shallow shell segment.

$$\sum F_y = \left(A_{22} - \frac{B_{22}}{R_y} - \frac{C_{22}}{R_y} + \frac{D_{22}}{R_y^2}\right)\frac{\partial^2 v}{\partial y^2} + \left(A_{66} + \frac{D_{66}}{R_y^2}\right)\frac{\partial^2 v}{\partial x^2}$$
$$+ \left(A_{12} + A_{66} + \frac{D_{12}}{R_x R_y} + \frac{D_{66}}{R_x R_y}\right)\frac{\partial^2 u}{\partial x \partial y} + \left(\frac{D_{22}}{R_y} - B_{22}\right)\frac{\partial^3 w}{\partial y^3}$$
$$+ \left(\frac{D_{12}}{R_y} + 2\frac{D_{66}}{R_y}\right)\frac{\partial^3 w}{\partial x^2 \partial y} + \left(\frac{C_{22}}{R_y^2} - \frac{A_{12}}{R_x} - \frac{A_{22}}{R_y}\right)\frac{\partial w}{\partial y} = 0 \tag{18.5}$$

$$\sum F_z = -D_{11}\frac{\partial^4 w}{\partial x^4} - (4D_{66} + 2D_{12})\frac{\partial^4 w}{\partial x^2 \partial y^2} - D_{22}\frac{\partial^4 w}{\partial y^4}$$
$$- \left(\frac{C_{11}}{R_x} + \frac{B_{11}}{R_x} - \overline{N}_x\right)\frac{\partial^2 w}{\partial x^2} - \left(\frac{C_{22}}{R_y} + \frac{B_{22}}{R_y} - \overline{N}_y\right)\frac{\partial^2 w}{\partial y^2}$$
$$- \left(\frac{A_{11}}{R_x^2} + 2\frac{A_{12}}{R_x R_y} + \frac{A_{22}}{R_y^2}\right) w$$
$$+ \left(C_{11} - \frac{D_{11}}{R_x}\right)\frac{\partial^3 u}{\partial x^3} - \left(2\frac{D_{66}}{R_x} + \frac{D_{12}}{R_x}\right)\frac{\partial^3 u}{\partial x\, \partial y^2}$$
$$+ \left(C_{22} - \frac{D_{11}}{R_x}\right)\frac{\partial^3 v}{\partial x^3} - \left(2\frac{D_{66}}{R_y} + \frac{D_{12}}{R_y}\right)\frac{\partial^3 v}{\partial x^2\, \partial y}$$
$$+ \left(\frac{A_{12}}{R_x} + \frac{A_{22}}{R_y} - \frac{B_{22}}{R_y^2} + \frac{\overline{N}_y}{R_y}\right)\frac{\partial v}{\partial y}$$
$$+ \left(\frac{A_{11}}{R_x} + \frac{A_{12}}{R_y} - \frac{B_{11}}{R_x^2} + \frac{\overline{N}_x}{R_x}\right)\frac{\partial u}{\partial x} = 0 \tag{18.6}$$

These equations can be solved by Galerkin's method. The shell segment is assumed to be simply supported on all sides. The displacement functions u, v, and w can then be approximated by a double Fourier series:

$$u = \sum_{m=M_L}^{M_u} \sum_{n=N_L}^{N_u} \overline{B}_{mn} \cos\frac{m\pi x}{\theta_x R_x} \sin\frac{n\pi y}{\theta_y R_y} \tag{18.7a}$$

$$v = \sum_{m=M_L}^{M_u} \sum_{n=N_L}^{N_u} \overline{C}_{mn} \sin\frac{m\pi x}{\theta_x R_x} \cos\frac{n\pi y}{\theta_y R_y} \tag{18.7b}$$

$$w = \sum_{m=M_L}^{M_u} \sum_{n=N_L}^{N_u} \overline{A}_{mn} \sin\frac{m\pi x}{\theta_x R_x} \sin\frac{n\pi y}{\theta_y R_y} \tag{18.7c}$$

in which $\overline{A}_{mn}$, $\overline{B}_{mn}$, and $\overline{C}_{mn}$ are unknown amplitudes and m and n are the meridional and circumferential half-wave numbers; M_u is the largest meridional half-wave number and M_L is the smallest; N_u and N_L have similar definitions.

Substitution of Eqs. (18.7) into Eqs. (18.4) and (18.5) permits a solution for $\overline{B}_{mn}$ and $\overline{C}_{mn}$ in terms of $\overline{A}_{mn}$. This requires the solution of two simultaneous algebraic equations for each choice of m and n. The results can be expressed as follows:

$$\overline{B}_{mn} = k_1 \overline{A}_{mn} \tag{18.8a}$$

$$\overline{C}_{mn} = k_2 \overline{A}_{mn} \tag{18.8b}$$

where k_1 and k_2 vary with the values of m and n as well as the physical constants describing the orthotropic shell.

Substitution of Eqs. (18.8) into Eq. (18.6), along with the requirement of nonvanishing $\overline{A}_{mn}$, results in an equation that is in terms of the critical values of the stress resultants $\overline{N}_x$ and $\overline{N}_y$:

$$\begin{aligned}
&-D_{11}\lambda_1^4 - (4D_{66} + 2D_{12})\lambda_1^2\lambda_2^2 - D_{22}\lambda_2^4 + \frac{2B_{11}}{R_x}\lambda_1^2 + \frac{2B_{22}}{R_y}\lambda_2^2 \\
&+ \left(B_{11} - \frac{D_{11}}{R_x}\right) k_1\lambda_1^3 + \left(B_{22} - \frac{D_{22}}{R_y}\right) k_2\lambda_2^3 - \left(\frac{A_{11}}{R_x^2} + 2\frac{A_{12}}{R_x R_y} + \frac{A_{22}}{R_y^2}\right) \\
&- (2D_{66} + D_{12})\left(\frac{k_1\lambda_1\lambda_2^2}{R_x} + \frac{k_2\lambda_1^2\lambda_2}{R_y}\right) - \left(\frac{A_{11}}{R_x} + \frac{A_{12}}{R_y} - \frac{B_{11}}{R_x^2}\right) k_1\lambda_1 \\
&- \left(\frac{A_{12}}{R_x} + \frac{A_{22}}{R_y} - \frac{B_{22}}{R_y^2}\right) k_2\lambda_2 - \left[\overline{N}_x\left(\lambda_1^2 + \frac{k_1\lambda_1}{R_x}\right) + \overline{N}_y\left(\lambda_2^2 + \frac{k_2\lambda_2}{R_y}\right)\right] = 0
\end{aligned} \tag{18.9}$$

where

$$\lambda_1 = \frac{m\pi}{\theta_x R_x} \quad \text{and} \quad \lambda_2 = \frac{n\pi}{\theta_y R_y} \tag{18.10}$$

Since the ratio of $\overline{N}_x$ to $\overline{N}_y$ is known, Eq. (18.9) can be used to solve for the critical stress resultant, $\overline{N}_{cr}$. Note, however, that $\overline{N}_{cr}$ depends on the choice of m and n, with different values of m and n yielding different buckling loads. The critical buckling stress resultant corresponds to the choice of m and n that minimizes $\overline{N}_{cr}$. Unfortunately, this computation requires a great deal of trial and error, thereby requiring a computer to complete the analysis.

It should be noted that with minor modifications, the same equations can be applied to cylindrical shells. Equations similar to Eq. 18.9 applied to cylinders are employed in Code Case N284 (ASME, 2007) and European Recommendations (ECCS, 1988).

For the case of a spherical shell under uniform pressure, $R_x = R_y = R$ and $\overline{N}_x = \overline{N}_y = pR/2$. The critical pressure found by using Eq. 18.9 corresponds closely to the classical buckling pressure for a complete sphere:

$$p_{cr} = \frac{E}{\sqrt{3(1-\nu^2)}}\left(\frac{t}{R}\right)^2 \tag{18.11}$$

At first glance, this result is surprising. It is due, however, to the fact that spherical shells buckle into a large number of waves in both directions. The value of n (or m) given by Timoshenko and Gere (1961) is

$$n = 3.46\sqrt{\frac{R}{t}} \tag{18.12}$$

Therefore, if the spherical segment is chosen long enough to develop the waves corresponding to Eq. 18.12, the buckling pressure will be the same as that for the complete sphere. The corresponding segment length must be greater than $1.82\sqrt{Rt}$. If the length of the spherical shell segment is less than $1.82\sqrt{Rt}$, the support conditions will affect the buckling pressure. Although Eq. 18.9 can still describe simply supported segments, it can no longer be applied independently of the boundary conditions.

Equation 18.9 is based on the assumptions of constant-stress resultants, as well as simple support on all four sides of the shell segment. Very few loadings produce a constant membrane state over a significant portion of a shell. Equation 18.9, however, can be used if the membrane state can be taken as constant over the buckling wave defined by Eq. 18.12. In addition, the assumption of simple support will be satisfactory for the buckling load if the shell segment is long enough for the buckling wave to develop.

As shown in Fig. 18.3, the comparison of theory with experiment for clamped spherical shells under uniform pressure is very disappointing. One might expect that the testing of any finite shell will differ somewhat from the results for a complete sphere, but the experiments show such scatter and low buckling pressures that the entire theory is suspect. Early researchers attributed the poor comparison between theory and experiment to the neglect of nonlinear terms in the shell equations (Kaplan, 1974). The results found by Budiansky (1959) and Huang (1964) when including these nonlinear terms are also shown in Fig. 18.3. Although these analytical results are more accurate than the linear bifurcation theory and agree with the experiments of Krenzke and Kiernan (1963), they still do not account for the large difference between theory and experiment.

The first satisfactory analysis of the problem, at least in a qualitative sense, was provided by Koiter (1977), who showed that the buckling load for spherical shells under uniform pressure is sensitive to small imperfections in the geometry of the shell. Koiter developed a perturbation analysis that helps the analyst to determine whether or not a structure is imperfection sensitive. If a shell is imperfection sensitive, its buckling load varies significantly with small imperfections. Koiter's theory

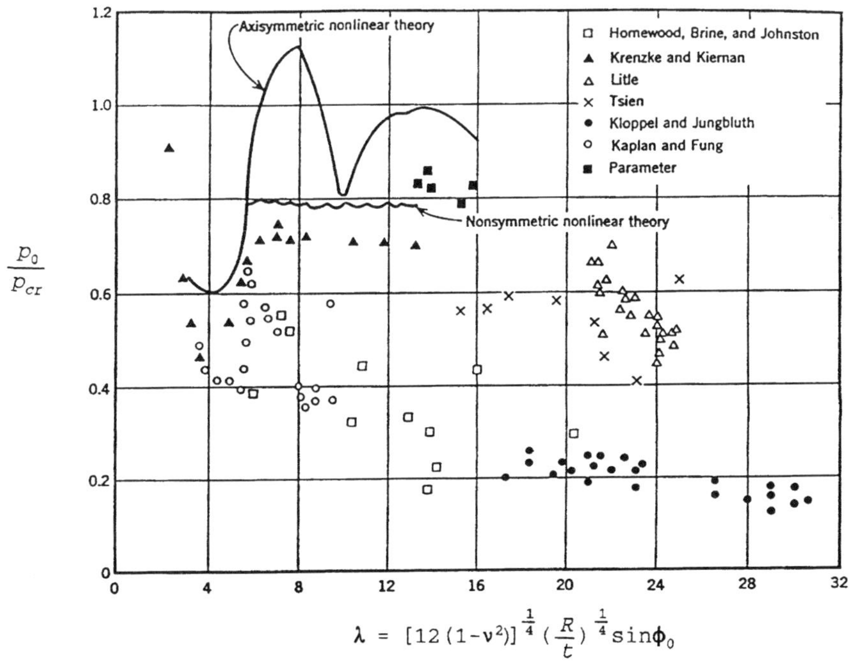

FIGURE 18.3 Results of experiments of spherical shells.

has been developed extensively, and some form of it is necessary for any application of shell stability (Hutchinson and Budiansky, 1970; Fitch and Budiansky, 1970). It should be noted, however, that imperfection sensitivity depends on the shell geometry, the geometry of the imperfections, and the type of loading. For example, a spherical shell is imperfection sensitive under uniform pressure, but it is not as imperfection sensitive to a concentrated load at the apex (Kaplan, 1974).

The information presented in Figure 18.3 is significant to the designer and may lead to several important questions, including: Can the shell being designed have less imperfection than the shells tested in various research laboratories? If the answer is no, what value should be chosen for the buckling pressure? Even if the shell can be fabricated with a great deal of precision, how close can the actual buckling pressure be to the theoretical value? This is the difficulty with design problems related to shell stability; the application of shell stability is at best based partly on theory and partly on experiment.

18.3 FINITE ELEMENT METHOD

One of the problems that plague the analytical investigation of shell structures is the description of boundary conditions. Shell equations are so complex that some form of Fourier series or other polynomial series must be employed in their solution. When this is done, all boundary conditions on the shell are determined by the assumed series. Usually, one or two of the conditions may not be physically possible on a constructed shell. It is hoped, however, that their effect will be small. Unfortunately, the fact that all actual boundary conditions cannot be satisfied mathematically makes it impossible to verify the critical loads by comparison with experiments. For those cases where the boundary conditions will influence the shell-buckling values, this makes it impossible to correlate analytical and experimental values. The finite element method allows one to avoid this problem because boundary conditions can be described correctly. (This is also a great advantage of the finite element method over finite difference analyses.)

It should be noted that a linear bifurcation analysis for the buckling load usually requires two separate analyses. A stress analysis must be made to determine the membrane state due to the applied load. This state must then be used as input to the bifurcation analysis.

Two distinct forms of finite element analysis can be applied to shell structures. If the structure is a shell that is axisymmetric, a special shell-of-revolution analysis can be applied to both the stress and eigenvalue computations. Such analyses are limited in that the stringers in the meridional direction must be modeled using the orthotropic shell properties described earlier. Stiffeners in the circumferential direction, however, can be modeled directly and need not be described by orthotropic shell properties. Shell-of-revolution algorithms must be employed with care because the smearing of the stringers is not valid when the shell is stiffened by relatively large stiffeners spaced far apart in both directions (Bushnell, 1977).

While stress analysis based on a shell-of-revolution algorithm is fairly straightforward, the corresponding eigenvalue computation can be complicated because a circumferential wave number N corresponding to the buckling mode must be chosen. Trial and error might be employed, in which first trial runs are made with large jumps in the N values used, but where N is allowed to vary over a large range. After the range of N values where buckling can occur is decided upon, a more refined analysis can be carried out with N values in that range. Investigators often perform a small study based on the analyses described in the preceding section to estimate the circumferential wave number prior to running the finite element code.

General-purpose three-dimensional finite element analysis software can, of course, also be applied to shell structures. They have the obvious advantage that both stiffeners and stringers can be modeled as discrete elements if so desired (Seide et al., 1979). A general-purpose finite element program would appear to be easier to use, in that the same model can be applied to both the stress and eigenvalue analyses. This advantage, however, may be more apparent than real. The node spacing required in the circumferential direction must be fine enough to permit modeling of a wave in that direction. In addition to requiring models with a large number of degrees of freedom, this requires some estimate of N prior to setting up the finite element mesh. Of course, the mesh must also be fine in the meridional direction in order to detect the rapid change in stress resultants—a problem which also exists when using shell-of-revolution routines.

A complete discussion of computer analyses of shell structures that has been prepared for the practicing engineer is provided by Bushnell (1985). An overview of the current state of the-art of nonlinear shell analysis is provided in a collection of articles edited by Kratzig and Onate (1990). An extensive bibliography on the subject is given by Noor (1989).

18.4 DESIGN CODES

In very few cases do existing design codes explicitly aid the designer of doubly curved shells (Beedle, 1991). Most codes give design equations for spherical shells under internal pressure.

A representative code for shell-related problems is ASME Code Case N284-2 (ASME, 2007). The code case defines the local buckling stress for spherical shells as

$$\sigma_{xe} = \eta\alpha_x\sigma_{cr} = \eta\alpha_x C_x \frac{Et}{R} \tag{18.13}$$

where

$$C_x = \begin{cases} 0.630 & \text{for } M_x \leq 1.5 \\ \dfrac{0.904}{M_x^2} + 0.1013M_x^2 & \text{for } 1.5 < M_x < 1.73 \\ 0.605 & \text{for } M_x \geq 1.73 \end{cases}$$

$$\alpha_x = \begin{cases} 0.627 & \text{for } M_x < 1.5 \\ 0.837 - 0.14M_x & \text{for } 1.5 \leq M_x < 1.73 \\ \dfrac{0.826}{M_x^{0.6}} & \text{for } 1.73 \leq M_x < 23 \\ 0.124 & \text{for } M_x \geq 23.6 \end{cases}$$

$$M_x = \frac{L_y}{\sqrt{Rt}}$$

L_y = arc length of support in the circumferential direction

In Eq. 18.13, σ_{cr} is the linear elastic buckling stress; α_x a capacity reduction factor, which takes into account the effect of initial imperfections; and C_x a coefficient describing the linear elastic buckling load. (Note that when $M_x \geq 1.73$, the value of σ_{cr} is the same as that found for the complete sphere, Eq. 18.11.) The plasticity reduction factor η is given by

$$\eta = \begin{cases} 1 & \text{for } \Delta \leq 0.55 \\ \dfrac{0.45}{\Delta} + 0.18 & \text{for } 0.55 < \Delta \leq 1.6 \\ \dfrac{1.31}{1 + 1.15\Delta} & \text{for } 1.6 < \Delta < 6.25 \\ \dfrac{1}{\Delta} & \text{for } \Delta \geq 6.25 \end{cases}$$

where

$$\Delta = \frac{\alpha_x \sigma_{cr}}{\sigma_y} \qquad \sigma_y = \text{yield stress}$$

ASME Code Case N284-2 also gives σ_{cr} values for the case in which one- or two-way orthogonal stiffeners are employed. This defines a general shell instability with the relevant equation

$$\sigma_{eG} = \frac{2.00Et_1^{0.25}}{Rt_2^{0.75}} \left(\frac{I_{E1}}{\overline{b}_2}\right)^{1/3} \left(\frac{I_{E2}}{\overline{b}_1}\right)^{1/6} \tag{18.14}$$

where $t_1 \geq t_2$ and $I_{E1} \geq I_{E2}$. The terms t_1 and t_2 are the equivalent orthotropic thicknesses and I_{E1} and I_{E2} are the equivalent moments of inertia of the orthotropic shell. If stiffeners act in only one direction, I_{E2} is taken as $\overline{b}_1(t^3/12)$.

Capacity reduction factors for general instability and stiffener instability are assumed to be the same as the values used for the local buckling stress. This does not seem to be reasonable, because it is doubtful whether stiffened shells buckling in either stiffener or general modes are as imperfection sensitive as unstiffened shells. The choice of the capacity reduction factor is significant because Eq. 18.13 is an ultimate-strength formula, which must be reduced by an appropriate factor of safety. Hence, the use of capacity reduction factors intended for unstiffened shells when designing stiffened shells can greatly reduce the design capacity of the stiffened shell.

Other double-curved shells such as toroidal or elliptical shells may be analyzed in the code case as equivalent spherical shells using the meridional radius as the sphere radius. The ECCS Eurocode does consider torispherical end closures subject to uniform pressure. Conical shells are designed using the buckling equations for an equivalent cylindrical shell (Weingarten et al., 1960). There does not appear to be any code that discusses stability problems of hyperbolic shells.

Although the ASME code case could provide more explicit help to the designer, it does present a rational design procedure that can be applied for any shell. Although it is basically an elastic procedure, it can be modified into a limit-state design form. The recommended steps are:

1. Carry out stress analyses of the shell for all factored load combinations deemed important. Only the membrane stress resultants are used in linear stability analyses. Complete collections of membrane solutions for common shell forms under various loadings are given in Pfluger (1961) and Baker et al. (1972). If a computational analysis (e.g., finite element analysis) is used at this stage, discontinuity stresses near abrupt changes of cross section or fixed supports should be noted. Only membrane stress resultant values at a distance greater than $0.50\sqrt{Rt}$ from the point of discontinuity need to be used.
2. Compute the capacity reduction factors and plasticity factors for the stress states computed in step 1. Note that the plasticity reduction factor depends on the membrane stress resultants causing buckling.

3. Carry out a linear bifurcation analysis for the membrane states found by dividing the membrane stress resultants computed in step 1 by the factors computed in step 2. The bifurcation analysis can be done by formula, where appropriate, using equations similar to Eq. 18.9 or rigorous finite element analyses. If an analysis using equations similar to Eq. 18.9 is employed, several different states found from a single load combination may have to be evaluated since it is not obvious at which position on the shell the critical stress combination will occur, particularly if the shell thickness varies or the shell is made of several different shell types.

Use of a shell-of-revolution algorithm for the linear bifurcation analysis may restrict the membrane states under investigation to axisymmetrical states. On the other hand, many important loads, both static and dynamic, yield asymmetric stress states. It has been found that it is conservative to use the largest asymmetric membrane stress resultants as axisymmetric stress states in the buckling equations (Klöppel and Roos, 1956). Note, however, that it is not a simple task to choose which asymmetric stress state is critical, so several may have to be checked. As noted in step 3, if the shell thickness varies or the shell is assembled from a combination of shapes, a large number of asymmetric stress states may have to be checked. A discussion of practical problems in applying the foregoing procedure is presented in Harstead et al. (1983).

While the foregoing procedure provides a reasonable design methodology, it is significantly limited by the lack of data for determining capacity reduction factors that would need to be applied to the membrane stress states for various shell configurations or loading conditions. As stated earlier, it would be expected that stiffened spherical shells would be less imperfection sensitive than unstiffened shells. Therefore, their capacity reduction factor should be larger for these structures. Since no codes provide these factors, their defined values would be up to the designer. Of course, one can be conservative and use factors that are based on the case of a sphere under external pressure, but these values will most likely lead to an extremely conservative design.

18.5 DESIGN AIDS

The dearth of information in design codes must be supplemented by the literature available on shell stability. Three excellent texts on the buckling of shells that may be of immediate and practical use to the designer are Buchert (1985), Kollár and Dulácska (1984), and Samuelson and Eggwertz (1992). The buckling of spheres and cylinders is covered in detail in all three texts as are the common problems related to construction tolerances for these shells. In addition to these texts, Yamada (1983) provides a complete study of spherical domes which includes a presentation and extensive comparison of formulas for stiffened domes under external pressure. A review of torispherical shells is presented in Galletly (1985). An interesting aspect of the design of these shells is that the use of linear analysis procedures results in very conservative designs.

A major shell type for which there is little design information is shells of revolution with different radii of curvature in each direction, ellipsoids, and hyperboloids. Simple buckling values can be computed using Eq. 18.9, but this assumes that the shell is similar to spherical shells in that buckling occurs in waves of small wavelength. It appears, however, from Kollár and Dulácska (1984) and Buchert (1985) that behavior is similar to spherical shells, at least if the principal radii are similar. Buchert recommends the same capacity reduction factor as that for a sphere under external pressure.

Another practical design problem for metal shells is buckling due to temperature effects. While computation of the buckling stress can be carried out easily using either Eq. 18.9 or finite element analysis, the important choice of an appropriate capacity reduction factor is not clear. Use of the spherical shell values may be too conservative in many problems. Intuitively, it would be expected that spherical shells under thermal buckling would not be imperfection sensitive because buckling is not due to gravity load but to restraint of deformation. Some discussion of this problem is presented in Samuelson and Eggwertz (1992) and in Bushnell (1985). Both agree that the capacity reduction factors need not be chosen as conservatively as for spherical shells under external pressure. Explicit values, however, are not given.

It would be ideal if data existed on initial imperfections so that computational analyses could be used to accurately determine capacity reduction factors (Arbocz, 1990). Without these data, imperfection sensitivity can only at best be estimateed. Koiter's method has been incorporated in some computer analysis algorithms (Riks et al., 1990). This might be of some assistance, but there still exists the open question of whether the Koiter approach can be used for quantitative analysis (Seide, 1974). An approach to imperfection sensitivity that may achieve quantitative results is that of Batista and Gonçalves (1994). They assume that the actual shell instability is due to the loss of membrane stiffness in any classical buckling mode. Therefore, one can obtain a lower bound to the actual buckling load by taking the ratio of reduced potential energy in any mode to the total potential energy in the mode. This reduced potential energy depends on the shell geometry and loading and is achieved by neglecting those stabilizing terms in the total potential energy that are severely undermined by the combined effects of initial imperfections and nonlinear mode coupling developed during the buckling process. The classical buckling load multiplied by this ratio is the lower bound to the failure load. Results using this approach are very promising.

18.6 RETICULATED SHELLS

Reticulated shells are space trusses or frames that are curved in space to behave as shell structures. Obviously, the first step in designing these structures is the design of individual members and joints. The stress resultants in these members are usually found from a linear frame analysis, although one could also use a shell analogy (Benjamin, 1963) if the number of elements and joints is extremely large. In many cases, these structures should be considered as shells, and therefore, the problem of overall instability must be investigated.

A good review of the current state of the art in reticulated shell analysis is presented in Gioncu (1992). It is possible to carry out rigorous analyses for reticulated shells, but the designer is still faced with the problem presented earlier in Section 18.2, the fairly unreliable relationship between the theoretical buckling loads found by analysis, linear or nonlinear, and the actual buckling loads determined by experimentally testing the structure.

Buchert (1973, 1985), has addressed this problem with the aid of the split-rigidity concept, which is described in detail in the fourth edition of this guide (Galambos, 1988). In general, the shell properties are averaged in such a way that the moments of inertia of the discrete members in each direction are described by a bending thickness, t_B, while the discrete areas are described by a membrane thickness, t_M. Note that it is common for there to be different thicknesses in each direction. A typical formula obtained from assumptions of this type is the following, valid when the same properties exist in both orthogonal directions:

$$p_{cr} = 2.67\eta E\left(\frac{t_M}{R}\right)^2\left\{\left[0.210\left(\frac{\delta}{t_M}\right)^2 + 0.0715\left(\frac{t_B}{t_M}\right)^3\right]^{1/2} - 0.459\frac{\delta}{t_M}\right\} \tag{18.15}$$

where δ is the combination of the imperfection in the dome segment and the deflection due to the applied loads. To compute the latter, one must be able to solve the shell equations for shells with the different bending and membrane thicknesses. Some solutions are provided by Buchert (1985) along with equations describing more complicated buckling cases.

It should be noted that the split-rigidity method may be used instead of the orthotropic shell approach to describe stiffened shells as well as lattice domes. The latter approach seems, however, to be more amenable to computational analyses, while the use of split rigidities is more useful for hand calculations. The split-rigidity concept has been criticized because some ingenuity is required to estimate the effective thicknesses when the stiffening elements are not orthogonally positioned. This is true, but, of course, a similar problem arises in replacing stiffener properties by orthotropic shell properties whenever the stiffeners are positioned arbitrarily. Another discrete approach to the reticulated shell problem is presented in Wright (1965). Yamada (1983) also provides equations for buckling of reticulated spherical shells using orthotropic shell properties.

A split-rigidity or orthotropic shell approach is useful as long as the reticulated shell is imperfection sensitive. There are few published experimental data on this problem, and the few computer studies that address the imperfection sensitivity problem are inconclusive (Batista et al., 1991; Morris, 1992; Batista and Alves, 1993; Kashani and Croll, 1994). It appears that some reticulated shells are imperfection sensitive, while others are not. Buchert gives the following criterion: *a reticulated shell is imperfection sensitive when the critical buckling wavelength includes two or more members in every direction.*

18.7 DESIGN TRENDS AND RESEARCH NEEDS

Although design codes and commercially available computer software are available to aid one in the design of shell-like structures, the designs actually arrived at might be either too conservative or not conservative enough. Use of the capacity reduction factors for spherical shells under external pressure for other load conditions or shell geometries seems to be very conservative. On the other hand, design of reticulated shells without considering the problem of geometric imperfections errs in the other direction. Most of the research work done on buckling of shell structures has been carried out in the aerospace industry. Shell structures in this field are very thin and made of high-strength alloys. Civil engineering shells are thicker and made of steel. Relatively little work has been carried out on shells of this type. The effect of geometric imperfections may not be as great, but the effect of inelastic behavior and residual stresses will become more pronounced. There currently exist very good computational software for the analysis of shell structures, but for them to be useful they must be used in conjunction with experimental data. Unfortunately, such data are rare and often simply do not exist.

REFERENCES

Anderson, C. A., and Bennet, J. G. (1981), "Containment Buckling Program," paper presented at the 9th Water Reactor Safety Research Information Meeting, Washington, DC, Oct., Los Alamos Scientific Laboratory, Los Alamos, NM.

Arbocz, J. (1990), "About the State of the Art of Shell Design," Final Rep., Int. Colloq. Stabil. Steel Struct., Budapest, Hungary, pp. 213–226.

ASME (2007), *ASME Boiler and Pressure Vessel Code*, ANSI/ASME BPV-III-1-NE, Section III, Code Case N-284-2, American Society of Mechanical Engineers, New York.

Baker, E. H., Kovalevsky, L., and Rish, F. L. (1972), *Structural Analysis of Shells*, McGraw-Hill, New York.

Batista, R. C., and Alves, R. V. (1993), "Asymptotic Modal Analysis of Lattice Domes," *Proc. SSRC Ann. Techn. Session*, pp. 211–221.

Batista, R. C, Antonini, R. C., and Alves, R. V. (1991), "An Asymptotic Modal Approach to Nonlinear Elastic Stability," *J. Comput. Struct.*, Vol. 38, No. 4, pp. 475–484.

Batista, R. C., and Gonçalves, P. B. (1994), "Nonlinear Lower Bounds for Shell Buckling Design," *J. Constr. Steel Res.*, Vol. 28, No. 28, pp. 101–120.

Beedle, L. S. (editor-in-chief) (1991), *Stability of Metal Structures: A World View*, 2nd ed., Structural Stability Research Council, Lehigh University, Bethlehem, PA.

Benjamin, B. S. (1963), *The Analysis of Braced Domes*, Asia Publishing House, New York.

Buchert, K. P. (1973), *Buckling of Shell and Shell-Like Structures*, K. P. Buchert and Associates, Columbia, MO.

Buchert, K. P. (1985), *Split Rigidity Theory of Plates, Shells, and Stability*, 2nd ed., K. P. Buchert and Associates, Columbia, MO.

Budiansky, B. (1959), "Buckling of Clamped Spherical Shells," *Proc. Symp. Theory Thin Elastic Shells*, Delft, The Netherlands, pp. 64–94.

Bushnell, D. (1977), "BSOR4: Program for Stress, Buckling and Vibration of Complex Shells of Revolution," in *Structural Mechanics Software Series*, Vol. 1 (Eds. N. Perrone and W. Pilkey), University of Virginia Press, Charlottesville, VA.

Bushnell, D. (1985), *Computerized Buckling Analysis of Shells*, Martinus Nijhoff, Dordrecht, The Netherlands.

European Convention for Constructional Steelwork (ECCS) (1988), *European Recommendations for Steel Construction: Buckling of Shells*, 3rd ed., Technical Committee 8, Structural Stability, ECCS, Brussels, Belgium.

Fitch, J. R., and Budiansky, B. (1970), "Buckling and Postbuckling Behavior of Spherical Caps Under Axisymmetric Load," *J. Am. Inst. Aeronaut. Astronaut.*, Vol. 8, pp. 686–693.

Flügge, W. (1962), *Stresses in Shells*, Springer-Verlag, Berlin.

Galambos, T. V. (Ed.) (1988), *Guide to Stability Design Criteria for Metal Structures*, 4th ed., Wiley, New York, Chap. 18.

Galletly, G. D. (1985), "Torispherical Shells," in *Shell Structures: Stability and Strength* (Ed. R. Narayan), Elsevier, New York.

Gioncu, V. (Ed.) (1992), *Stability of Space Structures*, special issue, *Int. J. Space Struct.*, Vol. 7, No. 4.

Harstead, G. A., Morris, N. F., and Unsal, A. I. (1983), "Containment Vessel Stability Analysis," *Nucl. Eng. Des.*, May, pp. 303–318.

Huang, N. C. (1964), "Unsymmetrical Buckling of Thin Shallow Spherical Shells," *ASME J. Appl. Mech.*, Vol. 31, pp. 447–457.

Hutchinson, J. W., and Budiansky, B. (1970), "Postbuckling Theory," *Appl. Mech. Rev.*, Vol. 23, pp. 1353–1366.

Kaplan, A. (1974), "Buckling of Spherical Shells," in *Thin Shell Structures: Theory, Experiment, and Design* (Eds. Y. C. Fung and E. E. Sechler), Prentice Hall, Upper Saddle River, NJ, pp. 247–288.

Kashani, M., and Croll, J. G. A. (1994), "Lower Bounds for Overall Buckling of Spherical Space Domes," *ASCE J. Eng. Mech.*, Vol. 120, No. 5, pp. 949–970.

Klöppel, K., and Roos, E. (1956), "Contribution to the Buckling Problem of Thin-Walled Stiffened and Unstiffened Spherical Shells Under Full and Half-Sided Loading," *Stahlbau*, Vol. 25, Mar., pp. 49–60 (in German).

Koiter, W. T. (1977), "The Stability of Elastic Equilibrium," Tech. Rep. No. AFFDL TR-70-25, Air Force Flight Dynamics Laboratory, Wright Patterson AFB, OH (translation of 1946 thesis).

Kollár, L., and Dulácksa, E. (1984), *Buckling of Shells for Engineers*, Wiley, London.

Kratzig, W. B., and Onate, E. (Eds.) (1990), *Computational Mechanics of Nonlinear Response of Shells*, Springer-Verlag, New York.

Krenzke, M. A., and Kiernan, T. J. (1963), "Elastic Stability of Near-Perfect Shallow Spherical Shells," *J. Am. Inst. Aeronaut. Astronaut.*, Vol. 1, pp. 2855–2857.

Litle, W. A. (1964), *Reliability of Shell Buckling Predictions*, Res. Monogr. No. 25, MIT Press, Cambridge, MA.

Morris, N. F. (1992), "Application of a Koiter-Type Theory to Buckling of Lattice Domes," *Int. J. Space Struct.*, Vol. 7, No. 4, pp. 335–343.

Noor, A. K. (1989), "List of Books, Monographs, Conference Proceedings and Survey Papers on Shells," in *Analytical and Computational Models of Shells* (Eds. A. K. Noor et al.), American Society of Mechanical Engineers, New York, pp. vii–xxxviii.

Pfluger, A. (1961), *Elementary Statics of Shells*, 2nd ed., F. W. Dodge Corporation, New York.

Riks, E., Brogan, F. A., and Rankin, C. C. (1990), "Numerical Aspects of Shell Stability Analysis," in *Computational Mechanics of Nonlinear Response of Shells* (Eds. W. B. Kratzig and E. Onate), Springer-Verlag, New York, pp. 125–151.

Samuelson, L. A., and Eggwertz, S. (1992), *Shell Stability Handbook*, Elsevier, New York.

Seide, P. (1974), "A Reexamination of Koiter's Theory of Initial Postbuckling Behavior and Imperfection Sensitivity of Structures," in *Thin Shell Structures: Theory, Experiment, and Design* (Eds. Y. C. Fung and E. E. Sechler), Prentice Hall, Upper Saddle River, NJ, pp. 247–288.

Seide, P., Weingarten, V. I., and Masri, S. (1979), "Buckling Criterion and Application of Criteria to Design of Steel Containment Shell," NUREG/CR-0793, International Structural Engineers, Glenndale, CA May.

Timoshenko, S. P., and Gere, J. (1961), *Theory of Elastic Stability*, McGraw-Hill, New York.

Weingarten, V. I., Morgan, E., and Seide, P. (1960), "Final Report on Development of Design Criteria for Elastic Stability of Thin Metal Shells," STL/TR-60–0000-19425, Space Technology Laboratories, El Segundo, CA.

Wright, D. T. (1965), "Membrane Forces and Buckling in Reticulated Shells," *ASCE J. Struct. Div.*, Vol. 91, pp. 173–202.

Yamada, M. (1983), "An Approximation on the Buckling Analysis of Orthogonally Stiffened Framed Spherical Shells Under External Pressure," in *Shell and Spatial Structures Engineering: Proc. Int. Symp. Shell Spat. Struct. COPPE, Federal University of Rio de Janeiro, Brazil, Sept*. (Eds. F. L. L. B. Carneiro et al.), Pentech Press, Plymouth, Devonshire, England, pp. 177–193.

CHAPTER 19

STABILITY UNDER SEISMIC LOADING

19.1 INTRODUCTION

19.1.1 Earthquake Effects

Earthquake ground motions at a given location can be characterized by three components, two horizontal and one vertical, of displacement, velocity, or acceleration that are distributed in a more-or-less random manner in the time domain, albeit with certain dynamic characteristics. Ground motions induce inertia forces and deformations in a structure, and the structure's response depends on the dynamic properties of the ground motions and the structure.

For design purposes, peak forces and deformations are needed and these maximum values can be obtained conveniently using response spectra to represent ground motion effects. For instance, an absolute acceleration response spectrum, S_a, gives the peak absolute acceleration that will be experienced by single oscillators having different periods T under a given earthquake ground motion. Figure 19.1*a* illustrates the development of such a spectrum for oscillators having periods ranging from 0 to 3 s and 5% of critical damping for the horizontal component of a ground acceleration recorded during the 1994 Northridge earthquake. In the figure, the peak acceleration is computed for three different oscillators ($T = 0.2$, 1.0, and 2.0 s), and the peak values are reported in the response spectrum. Other analyses are carried out at other periods to develop the entire spectrum. Knowing the period of vibration of a structure, the peak force can be obtained by multiplying the spectral acceleration by the structure mass.

Similar spectra can be developed to predict the peak displacement demand. The calculations can also be performed for the vertical component of ground motions. In practice, however, most structures possess inherent reserve strength in the vertical direction and the design for the vertical component of seismic ground motions is often ignored or treated in a simplified manner. Most of the attention in building or bridge design codes is directed toward the design against the effects of the

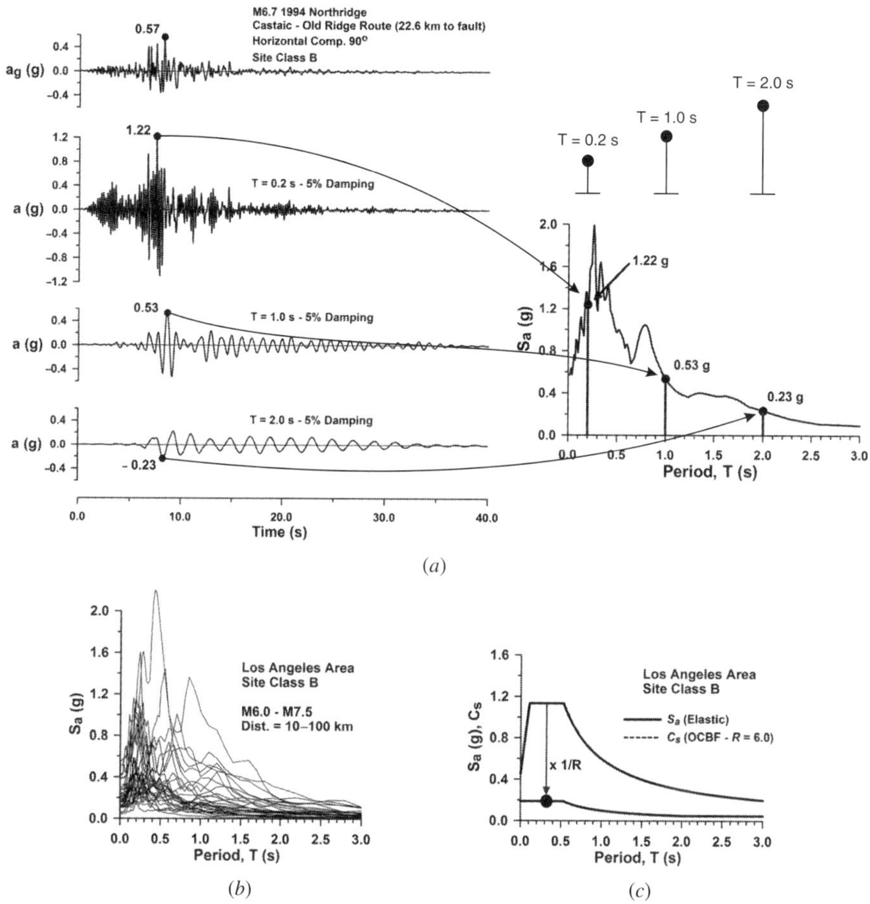

FIGURE 19.1 Acceleration spectra for the Los Angeles area: (*a*) development of a 5% damped absolute acceleration spectrum for a particular ground motion; (*b*) acceleration spectra of different earthquakes for a class B site; and (*c*) smoothed design spectrum for class B site.

horizontal ground motion components. The response spectrum in Fig. 19.1*a* was obtained for one particular ground motion. The seismic hazard at a given site arises from a wide variety of potential earthquakes that can occur in the vicinity. Ground motions at a site will also be affected by the local soil conditions. Figure 19.1*b* shows a group of response spectra for a rock site (site class B) in the Los Angeles area as obtained from potential earthquakes having different magnitudes and occurring at different distances. The large variation in the seismic demand is also due to differences in the faulting mechanism that would generate the potential earthquakes and in the geological conditions surrounding the site, including source directivity effects (Bozorgnia and Campbell, 2004). This variability is accounted for in design by using smoothed design spectra. Knowing the seismicity at the site, such as earthquake occurrence rates and attenuation of ground motion effects with distance and so on, probabilistic estimates of the spectral ordinates can be determined for different periods for a target probability of exceedance. Site-specific design spectra can then be built by anchoring straight or curved line segments to these spectral ordinates. Figure 19.1*c* shows a design spectrum for the Los Angeles area, as specified in ASCE 7-05 (ASCE, 2005) for a site class B.

Although response spectra are developed for simple single-degree-of-freedom systems, they can still be used to determine the forces in complex multi-degree-of-freedom structures by superposition of the response obtained in each of the individual vibration modes of the structure. This procedure is described in Chopra (2001). For regular structures, simplified methods are also proposed in most building and bridge codes to assess the demand on multi-degree-of-freedom systems: the demand is evaluated assuming that the structure responds essentially in its fundamental mode of vibration and correction factors are applied to account for the effects of the higher vibration modes.

19.1.2 Design for Ductile Earthquake Resistance

As shown in Fig. 19.1, forces induced by strong earthquake ground motions can be very large, with spectral accelerations in excess of 1.0 g for short period structures for this particular case. Clearly, designing common structures to remain elastic for such large lateral forces would be uneconomical. Therefore, code-specified lateral design forces are reduced from these elastic levels. In the United States, this reduction is achieved by dividing the required elastic force demand by a seismic response modification coefficient, R, to obtain the seismic design coefficient C_s, as illustrated in Fig. 19.1*c*. The design earthquake load is then equal to $C_s \cdot W$, where W is the effective seismic weight of the structure. Such reduced seismic design forces imply that a typical code-designed structure would remain serviceable (essentially elastic) after smaller frequent earthquakes, and safety and survival of the structure during a major earthquake will depend on the ability of the structure to withstand several reversed cycles of inelastic deformations.

The response to a severe ground motion is illustrated in Fig. 19.2 for a simple single-story moment-resisting frame. Under this particular ground motion, the analysis shows that the horizontal inertia loads would reach $1.28W$ if the frame was designed to remain elastic (Fig. 19.2*b*). If the beam is sized for a reduced horizontal load $V = 0.25W$ (Fig. 19.2*c*), the bending moments at the beam ends will quickly reach the beam flexural strength and cyclic plastic rotation will occur. The response of the frame will then be essentially governed by the hysteretic moment–rotation response of the beam. For instance, part of the energy fed in the structure produced by the ground motion is dissipated through cyclic plastic hinging response in the beam. This energy dissipation capacity has a beneficial dampening effect on the seismic response of the structure, which contributes to keeping the structure's lateral deformations within acceptable limits. In addition, the maximum horizontal load ($0.33W$ in this example) and, thereby, the forces in the structure

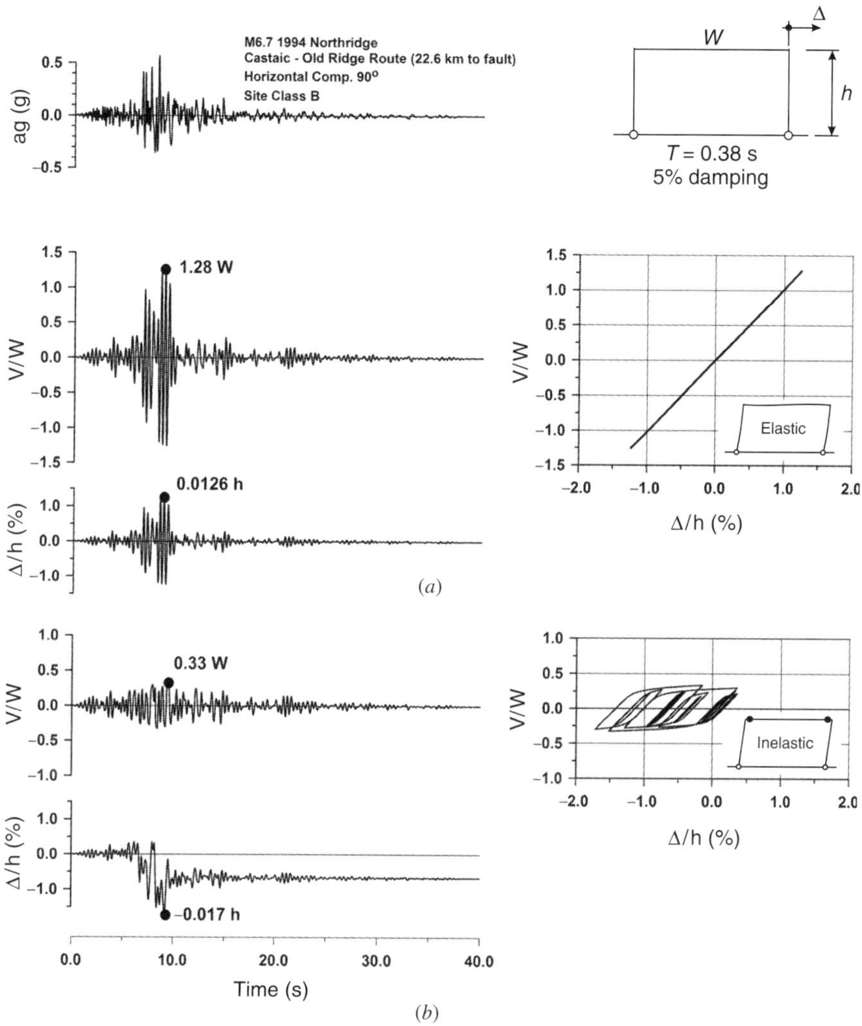

FIGURE 19.2 Seismic response of a single-story steel frame: (*a*) frame designed for elastic response ($V_{max} = 1.28W$); (*b*) frame designed for $V_y = 0.25W$ (inelastic response).

and its foundation are bounded by the actual flexural resistance that develops in the beam plastic hinges. This represents another advantage of this design approach as the inherent uncertainty attributed to the prediction of ground motion amplitude no longer translates into uncertainty in force demand. The variability in ground motion demand, however, translates into variation in the inelastic deformation demand that must be accommodated through adequate reserve plastic deformation capacity.

Adequate seismic inelastic response can only be achieved if the following two basic conditions are satisfied: (1) proper detailing and bracing are provided to ensure that the intended yielding components can sustain the anticipated cyclic inelastic deformation demand, without strength degradation or fracture, so that the lateral strength and energy dissipation capacity of the structure is maintained during the earthquake, and (2) proper strength and yielding hierarchy is implemented in the structure to ensure that the intended yielding mechanism actually forms and the structural integrity is preserved under strong ground shaking.

Steel represents an excellent choice for the energy dissipation mechanism in view of its inherently high ductility and its capacity to withstand reversed cyclic plastic deformations. The selection of a ductile material, however, does not suffice to meet the first condition and achieve ductile inelastic response, as schematically represented in Fig. 19.3*a*. For instance, local buckling of the yielding components can result in significant strength degradation or create highly localized strain demands that can lead to fracture of the steel and premature failure of the yielding mechanism. Figure 19.3*b* illustrates such undesirable behavior for a bracing member during the 1995 Kobe earthquake. Braces in concentrically braced steel frames

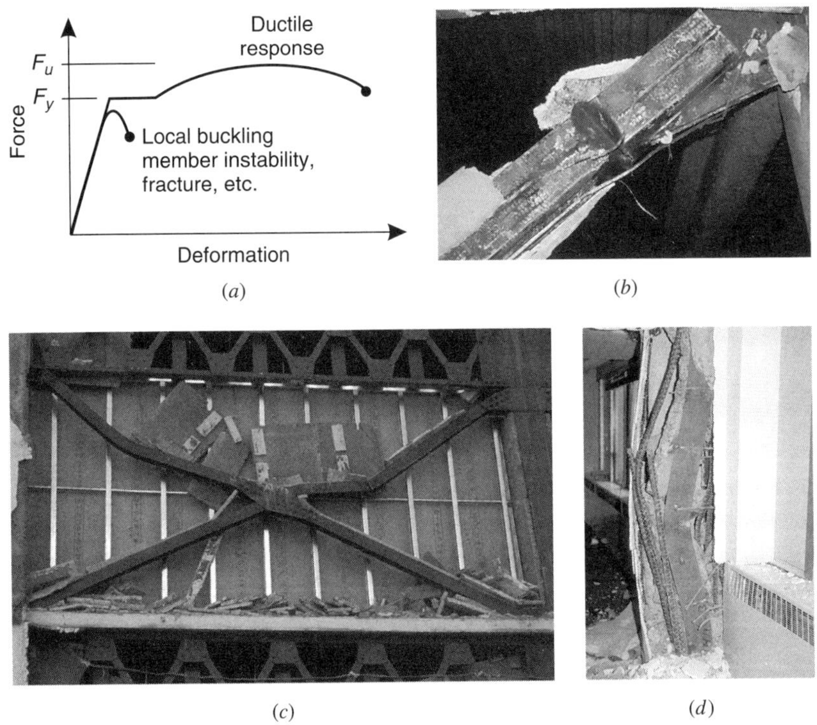

FIGURE 19.3 Failure of steel structures. (*a*) Possible load–deformation responses of steel components. (*b*) Local buckling of a tubular brace in the 1995 Kobe earthquake (Tremblay et al., 1996). (*c*) Brace overall buckling and fracture in the 1995 Kobe earthquake. (*d*) Buckling of a column during the 1964 Alaska earthquake. Courtesy of the National Information Service for Earthquake Engineering, EERC, University of California, Berkeley.

must be able to withstand several cycles of buckling in compression and yielding in tension. Material fracture due to local buckling (Fig. 19.3*c*) or net section fracture at the brace connections precludes the development of the desired inelastic response.

Failure of the yielding components by instability is also a condition that can have a negative impact on the inelastic performance of the system. For instance, sufficient bracing must be provided to prevent lateral–torsional buckling of the beam in the frame of Fig. 19.2. This is needed in order to rely on the ductile cyclic flexural yielding of the beam to maintain lateral strength and energy dissipation capacity for the duration of the strong ground motion. Mitigation of all potentially undesirable failure modes through appropriate detailing and bracing is a requirement to achieve good seismic performance. It may not be practical to prevent local and member buckling failure modes entirely; instead, the initiation and development of buckling can be controlled so that adequate ductility and inelastic deformation are achieved before buckling adversely affects the performance of the structure.

With regard to the structural system, it is equally important that all nonyielding components along the lateral load path of the structure possess sufficient capacity to ensure yielding at the desired locations prior to the occurrence of other failure modes. When carrying out this verification, particular attention must be devoted to brittle failure modes that can have severe consequences on the structure's response. For instance, column buckling must be prevented as it may endanger the gravity load-carrying capacity of the structure and, in turn, the safety of the building's occupants (Fig. 19.3*d*). The collapse of one of the three 21-story tower structures of the Pino Suarez complex in Mexico City during the 1985 Mexico earthquake is an example of such undesirable response. The failure was attributed to buckling of the columns at the fourth floor that became overloaded due to the large overturning moments that developed in the structure during the earthquake (Osteraas and Krawinkler, 1989, 1990; Ger et al., 1993).

Design for reduced seismic loads has gained general acceptance worldwide and several structural systems have been developed over the last decades to resist seismic ground motion effects through controlled yielding, without structural collapse. The most popular systems for building structures are given in Fig. 19.4, with the intended inelastic deformation modes for these systems illustrated. These will be discussed in more detail later in this chapter. Different R values have been assigned in codes for each of these systems to reflect their inelastic performance. Codes also provide detailing rules to ensure adequate deformation capacity in excess of the inelastic demand consistent with the R value assumed in design.

As shown in Fig. 19.5*a*, structures can experience large lateral deformations during severe earthquakes. These can lead to significant additional overturning moments due to the gravity loads acting on the laterally deformed structure. Such global second-order effects, or P–Δ effects, can amplify further the drifts and, thereby, the inelastic demand on the yielding components. The P–Δ effects can even lead to total structural collapse (Fig. 19.5*b*). This would be the case for structures supporting large gravity loads and that do not possess sufficient lateral stiffness and strength to control the deformation demand from earthquakes.

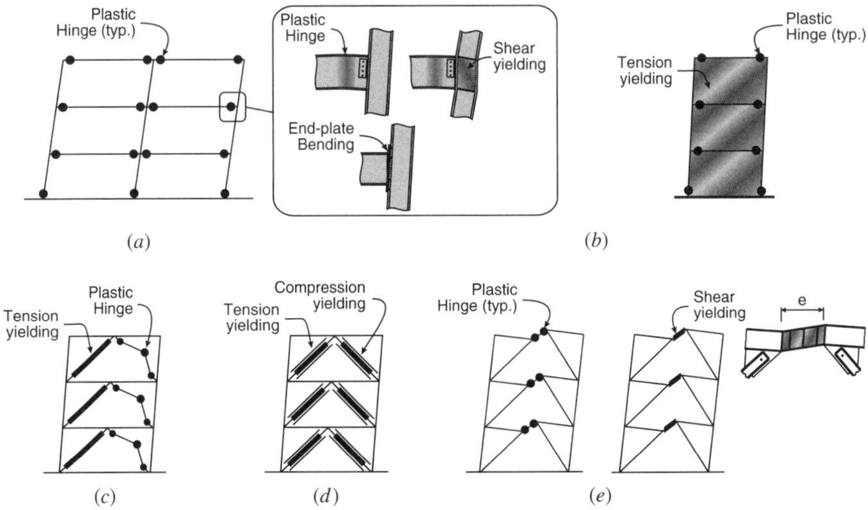

FIGURE 19.4 Intended inelastic mechanisms for (*a*) moment-resisting steel frames, (*b*) steel plate shear walls, (*c*) concentrically braced steel frames, (*d*) buckling-restrained steel frames, and (*e*) eccentrically braced steel frames.

(*a*) (*b*)

FIGURE 19.5 Global stability response of steel structures in the 1999 Chi Chi earthquake: (*a*) large story drift due to weak axis inelastic bending of the columns: (*b*) total building collapse due to a lack of lateral resistance and P–Δ effects.

Degradation of the lateral capacity of the structure due to nonductile inelastic behavior or accumulation of damage under repeated loading can also contribute to a global collapse response under strong ground shaking. Such an instability limit state must also be checked at the design stage.

This chapter addresses the main stability issues that need to be considered to achieve proper inelastic seismic performance for steel structures used for buildings. Local buckling and member instability effects are first discussed for each of the systems presented in Fig. 19.4. The moment-resisting frame system is presented first. It is reviewed in great detail in view of the significant research effort that has been devoted to this system in the last decade. This section also introduces concepts that apply to the other systems that will be reviewed. Seismic design provisions that have been developed in different countries to address stability issues are presented and compared. The global stability of steel seismic force resisting systems is described in the last section of this chapter.

It must be recognized that there exist several other limit states that may need to be considered to ensure adequate seismic performance, but these are not addressed in this chapter. It is also noted that earthquake engineering has drastically evolved in the last few decades and a considerable amount of new knowledge and information have been generated and published on a large variety of structural systems, techniques, and design methodologies to improve stability response under earthquakes. Due to space constraints, important topics such as bridge structures or structures constructed with materials other than steel could not be included and only part of the literature reviewed could be reported in the chapter. Additional information on the stability and ductility of steel structures under seismic loading can be found in the previous two editions of this guide, Fukumoto and Lee (1992), and Bruneau et al. (1998).

19.2 DESIGN FOR LOCAL AND MEMBER STABILITY

19.2.1 Moment-Resisting Frames (MRFs)

In multistory ductile MRFs, the global beam plastic hinging mechanism in Fig. 19.4*a* is preferred to a story (column) sway mode because energy dissipation can be distributed over several floors. In a story mechanism, plastic hinges only form at the ends of the columns of a single floor, resulting in large plastic rotations as energy dissipation is concentrated in fewer hinges (Roeder et al., 1993). This can result in excessive inelastic deformation demands in the columns as well as large story drifts that can lead to frame instability. Hence, the frame is typically designed such that inelastic response essentially develops through plastic hinging at the beam ends. Beams, which are typically made of I-shapes with wide flanges, must then be detailed and braced to sustain the rotation demands expected during earthquakes. Columns are also commonly made of I-sections, usually oriented to resist flexure about their major axis, but structural tubing is also used, especially when part of orthogonal MRFs. Composite columns, either concrete encased I-shapes or concrete-filled tubes, also represent effective design solutions in view of their inherent higher flexural stiffness and superior fire resistance. The strength of the columns must be adjusted relative to that of the beams to achieve the intended yielding mechanism. As shown in Fig. 19.4*a*, plastic hinging is expected at the base of columns that are rigidly connected to the foundation. In this case, attention must be given during the design stage to ensure that the columns can accommodate the plastic rotation demands.

This section will focus on local and member stability issues for MRFs that employ I-shaped beams and columns with fully restrained beam-to-column connections that lead to the formation of plastic hinges in the beams, away from the column faces. Emphasis is put on plastic rotation capacity of beam and column members, as limited by stability considerations. It must be noted that a number of alternative design strategies can be used to reach adequate seismic performance. In particular, solutions where inelastic rotational demand on beams is replaced by inelastic deformations of connection components can represent valuable alternatives from a member stability point of view. Shear yielding in the web panel zone of I-shaped columns and flexural yielding of beam end plates (Ghobarah et al., 1992; Murray and Sumner, 2003; Guo et al., 2006) are illustrated in Fig. 19.4*a*. Other similar schemes have been proposed that also have demonstrated high energy dissipation (e.g., Swanson and Leon, 2000; Choi et al., 2003; Garlock et al., 2003; Inoue et al., 2006; Sato and Uang, 2009). These could be considered when more traditional approaches to beam stability issues cannot be implemented in practice.

Beam Buckling Early cyclic tests on beam-to-column joints built with small- to medium-scale specimens, such as those by Popov and Pinkney (1967) and Popov and Stephen (1970), revealed that I-shaped beams can dissipate seismic input energy through stable inelastic cyclic flexural response. Although local buckling of the beam flanges was observed in these tests, the phenomenon did not result in sudden loss in capacity and hysteretic energy dissipation could be maintained when applying further cycles with increasing amplitudes. Flange buckling was perceived as beneficial in that it distributes damage in the members and extends fatigue life, but high strains concentrating at the flange buckles can eventually lead to cracking of the flange material. Flange local buckling (FLB) in inelastic cyclic tests is generally accompanied by web local buckling (WLB). Early test programs also indicated that plastic rotation capacity can also be influenced by lateral–torsional buckling (LTB) response of the beam (AIJ, 1975). These three interacting buckling modes are still observed today in cyclic tests of large-size ductile beam-to-column connections designed according to modern seismic provisions, as illustrated in Fig. 19.6*a* for a cyclic experiment on a haunch connection. Assessment of beam rotation capacity must therefore account for this buckling response and its impact on strength degradation and strain demand. Local buckling is influenced by the width-to-thickness ratio of the beam flanges ($b_f/2t_f$) and web (h/t_w), where h is the beam web clear depth. Lateral–torsional buckling is essentially governed by the lateral slenderness of the beam, L_b/r_y, where L_b is the unbraced length of the beam and r_y is the beam radius of gyration about its minor axis. Background information on these buckling modes is presented in Bruneau et al. (1998). Material properties, distribution of bending moment, and loading history are other parameters that can affect the inelastic rotation capacity of beams.

Plastic rotation response of beams was first examined for members subjected to monotonic loading, typically for cases where bending moments vary linearly from zero to a maximum value, as can be obtained from simple centrally loaded or cantilever beam tests (Fig. 19.6*b*). These conditions are representative of the conditions that prevail for beams and columns in moment-resisting frames subjected to lateral loads, as both members resist bending moments that increase linearly from

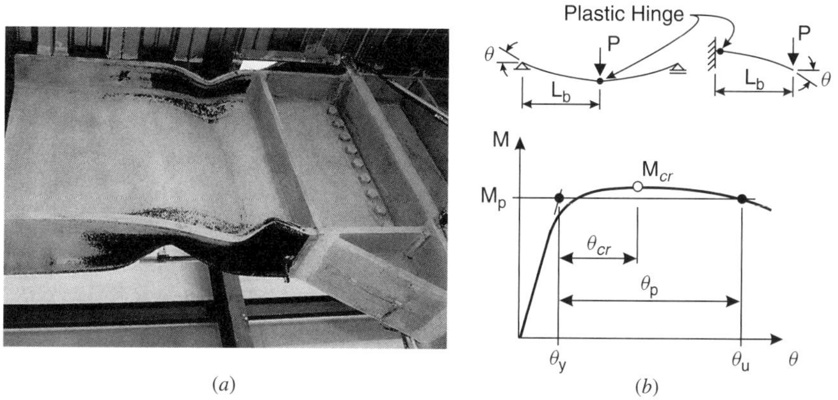

FIGURE 19.6 (*a*) Interaction of beam FLB, WLB, and LTB under reversed cyclic loading in a retrofitted beam-to-column connection with welded haunches. Courtesy of C. M. Uang (USDC). (*b*) Elastic and inelastic (plastic) beam rotation under monotonic loading.

the inflexion points to the beam-to-column joints. The rotation θ_y is the rotation reached over the length L_b at attainment of the beam plastic moment M_p at the point of maximum moment. A plastic hinge then forms at that location and the flexural resistance continues to increase due to strain hardening of the yielding material. The capacity eventually reaches a maximum at the onset of local buckling M_{cr} and then gradually diminishes as cross-section distortion takes place. Tension flange fracture may occur first in the case of a very compact section, but this scenario is very unlikely with common I-shapes.

The monotonic plastic rotation up to local buckling can be determined using classical beam rotation theory. Kato (1989) applied this technique for laterally braced doubly symmetric I-shaped beams. He adopted a simplified equivalent two-flange beam cross-section model and assumed rigid-plastic strain-hardening material behavior. With this method, the plastic rotation capacity at M_{cr} can be approximated by

$$R_{cr} = \frac{\theta_{cr}}{\theta_y} = \frac{1}{s}\left\{\frac{E}{E_{st}}(s-1)^2 + 2\frac{\epsilon_{st}}{\epsilon_y}(s-1)\right\} \tag{19.1}$$

in which E is the elastic modulus of steel, E_{st} is the strain-hardening modulus, ϵ_{st} is the strain at the onset of strain hardening, ϵ_y is the strain at yield ($= F_y/E$, where F_y is the yield strength, see Fig. 19.7*a*), and s is the critical stress ratio, that is, the ratio of the buckling stress of the compressed flange to the yield strength. The critical stress ratio s was obtained from regression analysis of stub-column tests. Accounting for interaction between FLB and WLB, it may be represented as

$$s = \frac{1}{0.6003 + 1.6\lambda_f^2 + 0.1535\lambda_w^2} \tag{19.2}$$

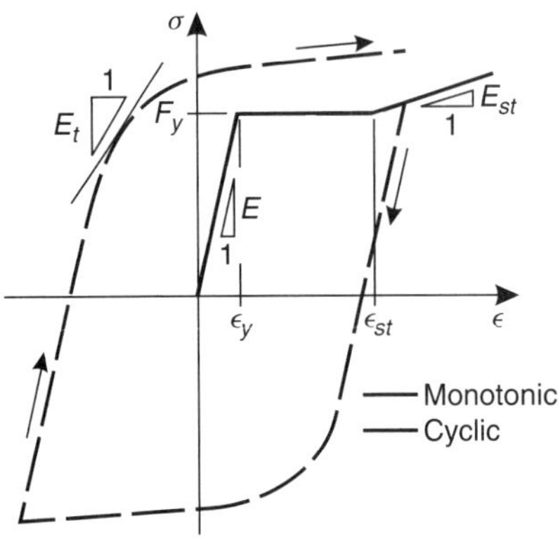

(a)

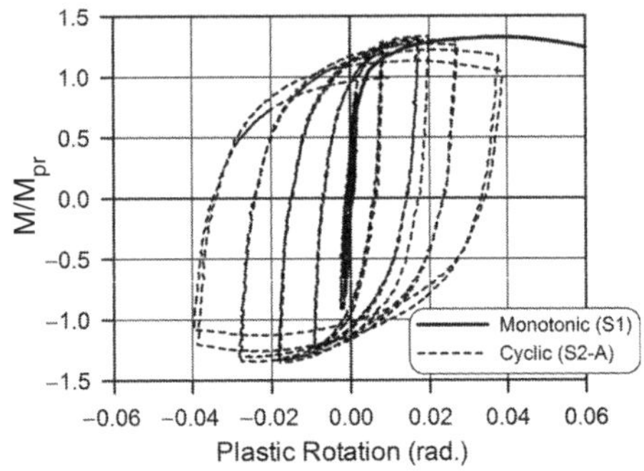

(b)

FIGURE 19.7 (*a*) Monotonic and cyclic stress–strain curves for steel and (*b*) moment–rotation responses of identical RBS specimens subjected to monotonic and cyclic loadings (Tremblay et al., 1997).

where λ_f and λ_w are the flange and web slenderness ratios normalized with respect to material properties ($\lambda_f = b_f/2t_f\sqrt{F_y/E}$, $\lambda_w = h_c/t_w\sqrt{F_y/E}$), and h_c is the clear depth of the web under compression ($= h/2$ for pure bending). An s-factor less than 1.0 indicates that local buckling occurs prior to yielding and $R_{cr} = 0$.

Using these expressions, interaction equations can be derived for limits on beam and web slenderness ratios that must be met to achieve a target plastic rotation capacity. Adopting a similar approach, Mazzolani and Piluso (1996) proposed the following equations when considering elastic–plastic with strain-hardening material response and including beam lateral–torsional buckling effects (Formisano et al., 2006),

$$R_{cr} = \frac{\theta_{cr}}{\theta_y} = \frac{1}{s}\left\{1 + \frac{E}{E_{st}}(s-1)^2 + 2\frac{\epsilon_{st}}{\epsilon_y}(s-1)\right\} - 1 \tag{19.3}$$

with

$$s = \frac{1}{0.695 + 1.632\lambda_f^2 + 0.062\lambda_w^2 - 0.602b_f/L_b} \leq \min\left\{\frac{F_u}{F_y}; 1.25\right\} \tag{19.4}$$

in which F_u is the ultimate tensile stress and the ratio b_f/L_b accounts for LTB and the stress gradient along the beam length. Kato, as well as Mazzolani and Piluso, also developed expressions for R_{cr} that account for the presence of axial compression loads. In such cases, a greater portion of the web is under compression ($h_c > h/2$) and the axial stress in the compressed flange reaches the buckling stress more rapidly, and both factors thereby result in reduced values of θ_{cr}.

Plastic rotation capacity at M_{cr}, however, generally represents a conservative estimate in the context of seismic design because most steel beams can sustain additional inelastic rotation while maintaining substantial postbuckling resistance to continue dissipating seismic input energy. The plastic rotation when the moment reduces to M_p (θ_p in Fig. 19.6*b*) or even below M_p (0.8 or 0.9 M_p) is typically recognized in codes as a better indicator of the available rotational ductility. Beam rotation at a given force limit criteria in the postbuckling range can be predicted using empirical expressions derived from physical or numerical test data or by conducting specific experimental or analytical studies. Kemp (1996) examined experimental results from tests on beams subjected to monotonic loading to establish a relationship between FLB, WLB, and LTB slenderness parameters and the plastic rotation θ_p over which the moment exceeds M_p. Poor correlation was found between $R_p = \theta_p/\theta_y$ with individual flange and web slenderness ratios or the lateral beam slenderness ratio. Instead, excellent correlation was found between R_p and an effective lateral slenderness ratio λ_e that accounts for the three instability modes. For noncomposite steel I-shaped beams, this relationship is given by

$$R_p = \frac{\theta_p}{\theta_y} = \frac{1400}{\lambda_e^{1.5}} \text{ with } \lambda_e = \left(\frac{b_f}{2t_f}\right)\frac{\gamma_f}{9}\left(\frac{h}{t_w}\right)\frac{\gamma_w}{70}\left(\frac{L_b}{r_y}\right)\gamma_f \tag{19.5}$$

in which γ_f and γ_w are normalizing yield strength factors for the flange and the web ($\gamma = \sqrt{F_y/250}$, with F_y in MPa), the terms $(b_f/2t_f)\gamma_f/9$ and $(h/t_w)\gamma_w/70$ are in the range between 0.7 and 1.5, and λ_e is in the range $25 < \lambda_e < 140$. If beam lateral restraint is provided between the maximum and zero moment, Eq. 19.5 is still applicable except that λ_e is divided by $1 + 0.61\beta$, where β is the ratio of the smaller moment to the maximum moment between the restraints. Results from additional tests when the beams were subjected to coincident axial compression load indicated a loss in rotational ductility. For this case, Kemp proposed a reduced rotation capacity $R'_p = R_p/\alpha$, where $\alpha = 2h_c/h$ (α is 1.0 for pure flexure and 2.0 when the axial load is sufficient to yield the web over its full depth). For the 44 reference specimens in flexure only, Eq. 19.5 results in test-to-predicted ratios with a mean value of 1.07 and a coefficient of variation (COV) equaling 0.30.

Gioncu and Petcu (1997) developed a yield-line model of the beam plastic hinge region to capture FLB and WLB effects on inelastic rotations of monotonically loaded beams. Beam flanges and web are defined using areas that yield axially and yield lines along which plastic rotation occurs. Those components are configured to reproduce potential local buckling patterns as observed in tests and the minimum M–θ solution among those obtained by applying the virtual work principle is retained. This approach represents an effective means of obtaining approximate yet realistic estimates of postbuckling beam response, for both research and practical applications. The method was implemented in a dedicated computer program to examine the influence of various parameters on beam rotation and assess code provisions. Their study indicated that beam rotation capacity reduces when increasing the steel yield strength or the bending moment gradient (shorter beams) and when applying axial loads. They also proposed that members be classified for ductility based on available plastic rotation capacity rather than cross-slenderness ratios. A similar yield-line model was developed by Lee and Stojadinović (2003) to assess the monotonic rotation capacity of beams and predict the forces imposed on beam lateral bracing when asymmetric buckled modes develop in plastic hinges.

When subjected to cyclic inelastic loading, steel develops an isotropic strain-hardening response that results in increasing yield strength with additional cycles (Fig. 19.7*a*). Upon reloading after a previous yielding excursion, however, the tangent modulus of steel gradually decreases before yielding occurs in the opposite direction ($E_t < E$ in Fig. 19.7*a*), which is a phenomenon typically referred to as the Baushinger effect (Bruneau et al., 1998). In members or member components subjected to successive inelastic buckling, that reduced modulus, combined with the fact that the section has residual deformations from previous inelastic buckling events, translates into progressive degradation of the compressive strength with additional cycles of loading. Several experimental programs included cyclic and monotonic tests on identical specimens (e.g., Lee and Lee, 1994; Green et al., 2002). As illustrated in Fig. 19.7*b*, the cyclic response leads to more pronounced postbuckling flexural strength degradation and, thereby, reduces available rotation capacity.

Numerous cyclic beam-to-column test programs have been carried out since the 1980s. Early studies on the influence of buckling on the cyclic rotation capacity of small I-shaped beams are reported in the previous edition to this guide and Bruneau et al. (1998). It was noted that the postbuckling strength degradation is more significant when LTB dominates the response instead of local buckling and that strength deterioration is severe when FLB is accompanied with WLB or LTB. Roeder (2000, 2002) summarized the main findings from recent tests conducted in the United States before and after the 1994 Northridge and 1995 Kobe earthquakes. These tests generally involved large- or full-scale specimens more representative of actual design and fabrication practice. A large portion of the factors affecting the connection response that are reported are related to detailing and fabrication practice (e.g., welding procedures). In terms of beam stability, it was concluded that beams with thicker webs and flanges exhibit higher plastic rotation than more slender sections and that connections with deeper beams and shorter beam spans have smaller plastic rotational capacity due to relatively high axial strain demands. The presence of a composite slab provides restraint against LTB. Although it also results in higher strain demand in the bottom beam flange, it was not accompanied by a greater tendency for bottom-flange buckling. FLB, WLB, and LTB were found to exhibit inelastic seismic performance sufficient for ductile moment-resisting frames ($R = 8.0$) provided that their slenderness ratios were limited to $0.30\sqrt{E/F_y}$, $2.45\sqrt{E/F_y}$, and $0.086E/F_y$, respectively. The data supporting the L_b/r_y bound, however, were limited and further study on LTB bracing requirements are needed.

Parametric experimental studies with cyclic tests confirmed the influence and interaction between FLB and WLB. Tests by Schneider et al. (1993) on specimens with nearly identical strength and stiffness but different flange and web slenderness ratios clearly indicated that more rapid deterioration occurs when local buckling dominates the inelastic behavior. From a similar test program, Castiglioni (2005) concluded that thin webs are less desirable as they cannot provide adequate support to thin beam flanges, which leads to premature distortion of the cross section with rapid loss of bending strength. Strain-hardening response could only be achieved, together with higher ductility and progressive loss in resistance and stiffness, by reducing the web slenderness. His tests indicated that strain hardening could be improved further when using both compact flanges and webs, but the rotation capacity was reduced due to cracks developing at the flange-to-web intersection. This latter observation suggests that completely preventing beam local buckling from occurring may not represent the most appropriate strategy for providing a ductile response. Gioncu and Petcu (1997) had already indicated that it could be preferable to select beams that exhibit local buckling response rather than sudden low-cycle fatigue failures at welds or other highly stressed locations. Tests performed on identical welded connections under different rotation histories also showed that beam local buckling can affect the connection response and failure mode; brittle failure close to the flange welds, without beam buckling or strength degradation, occurs when repeated low-amplitude inelastic cycles are applied, whereas tests with large amplitudes generated beam local buckling, with progressive strength degradation and failure in the buckles (Plumier, 2000; Castiglioni, 2005). Tests by Castiglioni (2005) also showed that successive large-amplitude cycles can cause a large reduction in strength but that applying only one large-amplitude cycle at an early stage in tests where specimens are subjected to small-amplitude cycles could change the failure mode from brittle to ductile and extend the endurance of the specimen. Formation of a plastic hinge in conjunction with local buckling is therefore seen as acting as a damper, preventing the accumulation of high inelastic strains in critical regions. Nevertheless, the amplitude of local buckling must be limited to avoid pronounced strength degradation or a premature low-cycle fatigue failure due to cracks that form in the buckles. Progressive development of FLB can also result in uneven force or strain demand in beam-to-column connections, with the potential for stress or strain concentrations and, hence, localized fracture. For instance, cracks observed at the ends of shear tabs connecting beam webs to columns were attributed to FLB (Lee et al., 2005). Therefore, slenderness limits must be established with the objective of controlling the extent of buckling and ensure good inelastic behavior, and not simply to prevent buckling from occurring.

Cyclic tests have also been instrumental for the development and calibration of numerical simulation tools that can now be used to help shed light on this problem. Finite element analyses that account for material and geometric nonlinearities have now become a reliable and useful research tool to complement physical test programs and carry out parametric studies on the inelastic buckling response of members, including fracture due to low-cycle fatigue of the material (e.g., Castiglioni, 2005; Chi et al., 2006; Castiglioni et al., 2007). Simpler finite element models with fiber dicretization of the cross section have been developed to predict the cyclic moment–rotation response of plastic hinges including local buckling effects of cross sections composed of flat elements (e.g., Sakimoto and Watanabe, 1997; Varma et al., 2002). Uniaxial fiber elements are assigned a stress–strain relationship that emulates the inelastic cyclic behavior of steel, which includes a buckling response that depends on the position of the fiber and the width-to-thickness ratio of the cross-section flat elements. Petcu and Gioncu (2003) introduced correction factors in their yield-line predictive model to account for the influence of cyclic loading as well as strain rate effects on beam rotation capacity. Lee and Stojadinović (2007) proposed a yield-line plastic hinge model with fiber discretization of the beam cross section that can predict the inelastic cyclic flexural response as well as flange fracture due to low-cycle fatigue in the yield lines. This model was used to examine the influence of beam flange and web slenderness on the cyclic fatigue life of different connections.

Nakashima et al. (2002) examined the LTB response of large I-shaped beams subjected to inelastic cyclic loading. Finite elements were used along the beam length with fiber discretization of the cross section assuming a plane section response. Hence, FLB and WLB were not included in the analysis. For beams that are laterally braced at the point of inflection and at the connection, the flexural strength is maintained above M_p up to a total rotation of 0.045 rad when the beam slenderness L_b/r_y is limited to $0.10E/F_y$. This rotation corresponds approximately to the demand anticipated for ductile moment-resisting frames in high-seismic regions in the United States. The same performance can be achieved for beams with n intermediate braces equally spaced at $L_{pd} = L_b/(n+1)$ when L_{pd}/r_y is limited to $[0.10 + 0.0188(n-1)/(n+1)]E/F_y$. Nakashima et al. (2003) carried out a detailed finite element analysis of beams laterally braced to meet the same performance criteria. Strong interaction was observed between LTB, FLB, and WLB for members with intermediate flange and web width–thickness ratios. Detailed finite element analysis by Green et al. (2002) showed that lateral braces that are too flexible may lead to a reduction in plastic rotation capacity but that little is gained by increasing the brace stiffness beyond the minimum required stiffness. Okazaki et al. (2006) investigated further the buckling response of cyclically loaded beams with various sizes through finite element analysis. A steel yield strength of 350 MPa was assumed in the calculations and three different unbraced lengths were considered, including L_b/r_y values of 60, 80 and 100 (equal to 0.11, 0.14, and 0.18 E/F_y). From a regression analysis of the results, the following expression could be obtained for determining the total rotation capacity ($\theta = \theta_y + \theta_p$) when the beam flexural strength is equal to M_p:

$$\theta = \left(95 - 21\frac{h}{b_f}\right)\left(\frac{b_f}{2t_f}\right)^{-1.6}\left(\frac{h}{t_w}\right)^{-1.1} \tag{19.6}$$

The slenderness parameter L_b/r_y influenced the rotation capacity only for shallow beams with a target rotation of 0.02 rad and for deep beams with a target rotation of 0.04 rad. In all other cases studied, the rotation was insensitive to LTB slenderness and this parameter was therefore omitted for simplicity in the regression study. Figure 19.8*a* shows the resulting limit curves for three different rotation values (0.02, 0.03, and 0.04 rad) and two different h/b_f ratios for 350 MPa yield strength steel. The $b/2t_f$ and h/t_w ratios of beam shapes available in North America are plotted on the graph for comparison purposes.

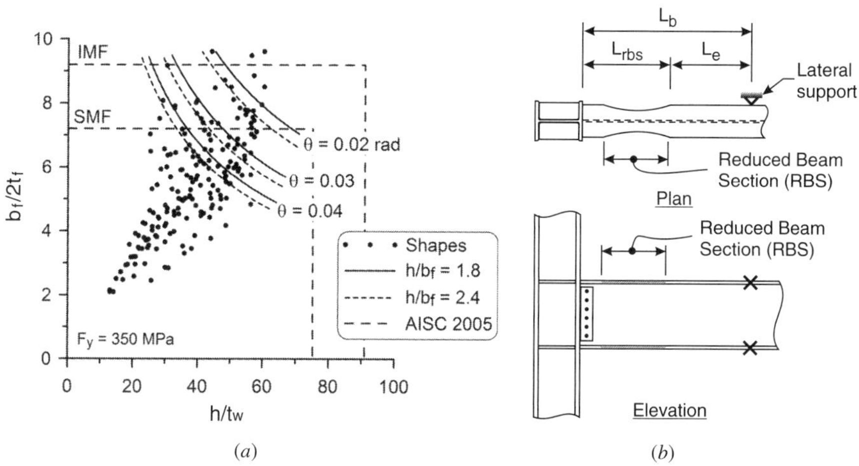

FIGURE 19.8 (*a*) Limit curves for cyclic rotation capacity; (*b*) reduced beam section (RBS) connection.

In the aftermath of the 1994 Northridge earthquake, alternative beam-to-column connection designs were investigated to develop the required inelastic rotation demand capacity for highly ductile steel moment-resisting frames. One such new design, the RBS concept (Plumier, 1997) with the circular cut detail (Fig. 19.8*b*), has become very popular in the United States. The width of the beam flanges is intentionally reduced near the connection to locally diminish the beam flexural strength and, thereby, reduce the demand on the beam-to-column welded connection. Plastic hinging develops in the reduced beam segment, where the flange slenderness is a minimum, which helps in controlling FLB. Extensive testing was performed on this type of connection (e.g., Engelhardt et al., 1998; Jones et al., 2002) and design guidance can be found in (AISC, 2005b). Uang and Fan (2001) examined the performance of 55 connection test specimens. Regression analysis of the data revealed the interaction between the slenderness ratios associated to FLB, WLB, and LTB modes on the plastic rotation that can be sustained before significant deterioration occurs. Web-buckling slenderness was found to have the greatest impact on the rotation capacity of this connection and beam lateral–torsional buckling slenderness (L_b/r_y) has the least effect. Neglecting the LTB mode, the authors proposed the following simplified empirical expression to predict the plastic rotation before the strength of the connection reduces to 80% of the peak strength:

$$\theta_p = 5.8\left(\frac{b_f}{2t_f}\right)^{-0.12}\left(\frac{h}{t_w}\right)^{-0.5} F_{yf}^{-0.5} \tag{19.7}$$

where F_{yf} is the yield strength of the beam flanges (in MPa) and b_f is the unreduced beam flange width. Similar correlation was observed when using the reduced beam sectional properties. In Eq. 19.7, web slenderness has a relatively greater impact on beam rotation when compared to the performance of prismatic beams. This is in agreement with test results by Jones et al. (2002) that suggest that RBS segments may be prone to earlier web local buckling compared to beams without an RBS. This is also consistent with the conclusion by Castiglioni (2005) on the key importance of web slenderness on beam ductility. To control WLB, Uang and Fan (2001) also recommended placing an upper limit on web slenderness of $h/t_w \leq 418/v\sqrt{F_y}$, with F_y in megapascals.

Moslehi Tabar and Deylamin (2006) performed numerical simulations of RBS connections including column web panel zone deformations. When examining numerical results and past test data, they found that the degradation of the flexural resistance due to lateral–torsional buckling correlated well with an equivalent beam slenderness parameter that accounts for the length of the reduced beam segment and the flexibility of the column web panel zone. Interaction between lateral–torsional and web local buckling effects was included by means of a combined equivalent slenderness parameter, λ_e. This parameter could then be used to predict the ratio of the moment capacity corresponding to a plastic rotation of 0.03 rad, $M_{0.03}$, and the maximum moment imposed at the column face, $M_{\max}$, according to

$$\frac{M_{0.03}}{M_{\max}} = \begin{cases} 1.0 & \text{for } \lambda_e \leq 1.5 \\ 0.5 + \dfrac{0.85}{\lambda_e} - \dfrac{0.15}{\lambda_e^2} & \text{for } \lambda_e > 1.5 \end{cases} \tag{19.8}$$

$$\text{with } \lambda_e = \left[\left(\frac{h}{t_w}\right)\frac{\sqrt{F_y}}{1365}\right]\left[\frac{1}{\sqrt{\kappa}}\frac{\sqrt{L_{rbs}L_e}}{2r_{y,rbs}}\sqrt{\frac{F_y}{C_b E_t}}\right]$$

in which κ accounts for the rotational flexibility of the beam-to-column connection contributed by the shear deformation in the column web panel zone (which equals the ratio of the moment at the column face induced by imposing a unit deflection at the tip of a beam of length L_b with and without panel zone deformations), L_{rbs} and L_e are as defined in Fig. 19.8*b*, $r_{y,rbs}$ is the minor axis radius of gyration of the beam at the reduced cross section, C_b is the equivalent uniform moment factor (see Chapter 5), and E_t is the tangent modulus of steel taken equal to 2100 MPa.

This new knowledge and the accompanying numerical simulation tools are now available to better assess the capacity of steel beams to accommodate the cyclic inelastic flexural demand from earthquakes. In view of the variability on both the capacity and demand sides, extensive probabilistic studies incorporating the knowledge gained on beam stability response are needed to develop guidelines that will provide the desired level of safety against connection failure and, more importantly, global collapse of structural systems. Numerous research projects that aim at accounting for the postbuckling strength degradation of beams on overall moment-resisting frame response have been undertaken to achieve this objective (e.g., Della Corte et al., 2002; Huang et al., 2004; Ibarra et al., 2005; Liao et al., 2007; Zareian and Krawinkler, 2007; Kazantzi et al., 2008).

Meanwhile, research results are progressively being incorporated into code documents as conclusive trends become available. For MRFs, current seismic design codes typically specify the target beam rotation capacity and the requirements that must be met to achieve this demand. These values vary according to the ductility category of the framing system and the seismic loads, or force modification factor, used in design. For example, per AISC seismic provisions (AISC, 2005a), the flexural resistance under cyclic rotations of connections in *special moment frame* (SMF, $R = 8.0$) and *intermediate moment frame* (IMF, $R = 4.5$) systems must not degrade below $0.8M_{pb}$ at interstory drift angles of 0.04 and 0.02 rad, respectively. In addition, the rotation capacity of the joint must be demonstrated by physical qualifying testing, unless a prequalified connection is used as described in AISC (2005b). Codes also typically prescribe limitations on cross-sectional geometry to achieve the intended rotation demand. For simplicity, FLB, WLB, and LTB are also generally treated as independent phenomena with slenderness limits being prescribed individually for each of the three instability modes. For instance, the $b_f/2t_f$ and h/t_w limits in AISC seismic provisions for SMFs and IMFs are plotted in Fig. 19.8*a* for $F_y = 350\,\text{MPa}$. It is noted that the AISC strength criteria of $0.8M_{pb}$ is less severe than the $1.0M_{pb}$ used by Okazaki et al. to generate their limit curves. For the SMF system, lateral bracing must be provided at least at the top and bottom flanges and along the beam length such that the brace spacing does not exceed $0.086E/F_y$. Beam lateral bracing can be designed according to the procedure described in Chapter 12, assuming that the beam develops its expected plastic bending moment resistance, but a minimum resistance corresponding to 6% of the expected yield resistance of the beam flange is prescribed for braces located adjacent to plastic hinges.

Similar requirements with some degree of variation exist in other seismic design codes. For instance, in CSA-S16 (CSA, 2001), the connection must maintain the full strength of the beam, M_{pb}, at the specified rotations (not $0.8M_{pb}$) but more stringent beam web slenderness limits apply (e.g., $h/t_w \leq 59$ for the most ductile system compared to the AISC limit of 75 illustrated in Fig. 19.8*a*). In Eurocode 8 (EC8) (CEN, 2004), plastic rotations of 0.035 and 0.025 rad must be achieved without degradation of strength and stiffness greater than 20% under cyclic loading in moment-resisting frames of the high- and medium-ductility classes, respectively. In EC8, upper limits on the axial loads and shear forces are prescribed to ensure that the rotation capacity implicitly assumed in design will be available. An upper limit on beam shear forces is also specified for the same reason in the NZS3404 standard (NZS, 2007). In this standard, interaction between LTB and the other buckling modes is accounted for as the limit on the slenderness L_b/r_y of I-shaped beams depends on the cross-section slenderness ratios. In Japan, interaction equations are prescribed for maximum b/t_f and h/t_w ratios (Fukumoto and Itoh, 1992). Interaction between all three buckling modes is considered in the 2005 Italian seismic provisions (Formisano et al., 2006) and beam member classification based on the parameter s (Eq. 19.3) with ductile members defined by $s > 1.2$, plastic members by $1.0 < s \leq 1.2$, and slender members by $s \leq 1.0$. The ductility-related seismic force modification factor used to determine the design seismic loads is also based on the s-factors determined for the members.

In general, less stringent seismic code provisions are prescribed when relatively high seismic loads are used in design and lower inelastic demands are expected. Similarly, higher seismic loads are specified in codes for less ductile systems. For instance, frames made of noncompact built-up or cold-formed steel members are common for low-rise applications due to their low weight-to-strength ratios. These members typically reach their maximum strength at small deformation levels, generally in the elastic range, and exhibit limited plastic rotation capacity with sharp postbuckling degradation in stiffness and strength. For this reason, these light framing systems have traditionally been designed based on elastic seismic demand, that is, using seismic design loads without ductility-related reductions (Dubina, 2004). Numerical and experimental studies of member response (Moldovan et al., 2000; Calderoni et al., 2006, 2009) indicate that noncompact thin wall beams with suitable proportions can exhibit repeatable hysteretic behavior and energy dissipation capacity at a postbuckling residual strength level before failure occurs by cracking at buckles. Proper consideration of this member response in analytical studies of frame responses could eventually lead to a reduction in design seismic loads for adequately detailed thin-walled steel structures.

Steels with F_y ranging between 235 and 350 MPa have been generally used for the fabrication of hot-rolled I-shaped beams for ductile moment-resisting frame applications and most of the available test data are limited to this range of steel grades. Past test studies also indicated that inelastic rotational capacity decreases as F_y is increased and the ductility of steel (strain at failure to strain at yield) typically also decreases with an increase in yield strength. As a result, seismic design codes generally prescribe an upper limit on F_y for the steel used in the components of the seismic load-resisting systems that are expected to undergo cyclic inelastic deformations. For instance, beams fabricated with steels with F_y exceeding 345 MPa are not permitted in SMFs and IMFs according to the AISC seismic provisions. Similar upper limits apply in CSA-S16 ($F_y \leq 350$ MPa) and in NZS3404 ($F_y \leq 360$ MPa).

In the few last years, rolled I-shapes with a yield strength of 345 MPa have become standard practice in North America. When strength controls the design, higher steel grades could represent a more cost-effective solution, but only limited data exist on the ductility of beams made of such materials. Takanashi (1973) compared the cyclic plastic rotation capacity from tests on small I-shaped beams with two different yield strength values, including 441 and 282 MPa. Inferior response was observed for the stronger specimen due to earlier LTB (Bruneau et al., 1998). Chen and Tu (2004) carried out tests on compact RBS moment connections with built-up 700-mm-deep beams with 50-mm-thick flanges made of ASTM A572 steel plates with $F_y = 414$ MPa. Failure occurred by cracking in the vicinity of the beam flange welds, prior to local buckling, at plastic rotations ranging between 0.028 and 0.043 rad. Absence of local buckling in that particular study may have contributed to the lower rotation values, as was noted in other studies.

In addition to lower deformability, higher strength steels typically also exhibit a higher yield strength ratio (F_y/F_u), which results in shorter plastic hinge lengths and, consequently, higher strain demand in plastic hinges. Kuwamura (1992) showed that increasing F_y/F_u contributes to a reduction in the plastic rotation capacity of beams. Test programs and finite element simulations were performed by Ricles et al. (1998) and Green et al. (2002) to assess the inelastic flexural deformation capacity of built-up I-shaped beams made of high-strength HSLA-80 steel ($F_y = 550\,\text{MPa}$, $F_u = 610\,\text{MPa}$ to $F_u = 690\,\text{MPa}$). This type of steel was perceived as a good candidate for ductile seismic applications as it features superior toughness and weldability properties. When compared to identical specimens made of ASTM A36 steel ($F_y = 248\,\text{MPa}$), the beams with HSLA-80 steel exhibited significantly less rotation capacity for a given normalized (with respect to $\sqrt{F_y}$) web or flange slenderness. It was also concluded that upper limits on the yield strength ratio (F_y/F_u) should be prescribed to ensure adequate rotation capacity and energy dissipation capacity. This ratio is limited to 0.80 in NZS3404 and to 0.85 in CSA-S16. The latter also applies to ASTM A992 steel, which is now commonly used in the United States for rolled shapes.

Column Buckling The desired global beam plastic hinge mechanism of Fig. 19.4*a* can be achieved by adopting a strong column–weak beam strategy in design (Roeder et al., 1993). The beams are proportioned first to resist the bending moments for load combinations that include seismic loads. Column sizes are then selected such that the story shear capacity associated to the story or column sway mode exceeds the story shear required to form the beam-hinging mechanism (Fig. 19.9*a*), which can be achieved by satisfying

$$\sum_{i=1}^{n+1} 2M_{pc}^{*}h \geq \sum_{i=1}^{n} 2M_{pb,e}h \tag{19.9}$$

where M_{pc}^{*} is the nominal flexural strength of the columns (reduced for the presence of axial force), $M_{pb,e}$ is the expected plastic moment capacity of the beams determined using expected beam yield strength and accounting for strain-hardening effects consistent with anticipated plastic deformations, h is the story height, and n is the number of bays. Alternatively, column sizes can be chosen such that the columns have a total flexural resistance larger than that of the beams at each beam-to-column joint,

$$\sum M_{pc}^{*} \geq \sum M_{pb,e} \tag{19.10}$$

Most design codes only require such verification at the joints, but variations exist on how the codes account for material variability and strain-hardening effects in the member capacities. In these equations, the flexural strength of the columns can be reduced further if the columns are also part of moment-resting frames acting in the orthogonal direction (NZS, 2004).

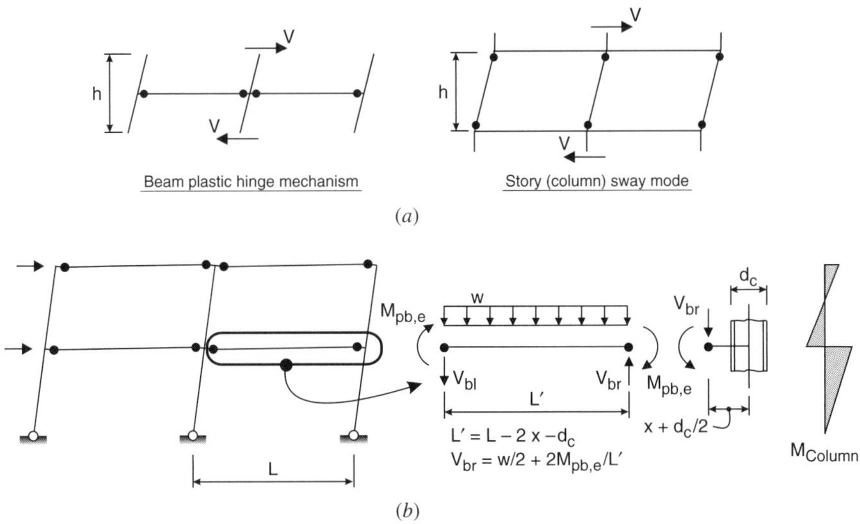

FIGURE 19.9 Overall behavior of moment-resisting frames: (*a*) simplified representation of plastic beam mechanism and column sway mode in MRFs; (*b*) beam forces imposed on columns.

Columns in moment-resisting frames, which are generally stocky and, therefore, satisfying the cross-sectional strength requirements of Eq. 19.9 or 19.10, as applicable, will generally suffice to achieve good performance. Slender columns, however, should be checked for in-plane instability and flexural–torsional buckling under anticipated combined axial load and bending moment. One possible loading scenario is illustrated in Fig. 19.9*b*, where beam plastic hinging is assumed to occur simultaneously at every level. Beam end shears due to gravity loads and attainment of $M_{pb,e}$ in the beam hinges are used to determine column axial load and bending moments (Leelataviwat et al., 1999; Goel and Chao, 2008). Gravity loads and bending moments from beams framing in the transverse direction must also be included. For columns that are part of two orthogonal MRFs, MacRae and Tagawa (2001) proposed a design procedure to control the amount of inelastic response due to bidirectional earthquake motions. In the column stability check, an effective-length factor $K = 1.0$ can be used to assess the column axial strength. Except at the first story where pinned base connections are used, it should be noted that the loading scenario of Fig. 19.9*b* leads to reversed column curvature, which may not represent the most critical condition for column stability. Multistory structures can undergo complex deformation patterns when responding dynamically to strong ground motions in the inelastic range, leading to various bending moment distributions in columns that could potentially represent more adverse stability conditions (MacRae, 1990). Current code provisions only provide limited guidance or simplified approaches to address this aspect and caution should be exercised when selecting slender columns in moment-resisting frames. In particular, deep and narrow columns bent about their major axis may represent a cost-effective solution for drift control, but this type of column can be prone to flexural–torsional buckling.

When used with RBS connections, deep narrow columns can also be subjected to twisting moments due to the eccentric compression force from the beam bottom flange upon lateral–torsional buckling of the beams. Based on tests conducted on RBS specimens without a concrete floor slab, Chi and Uang (2002) proposed a procedure to verify the column stresses at large story drifts, including torsional effects, and determine if additional beam bracing is needed. An analytical study by Shen et al. (2002) and analysis and experimental work by Zhang and Ricles (2006a,b) indicated that twisting of deep columns in RBS connections can be reduced by the presence of a floor slab or a supplemental brace located at the end of the RBS segment. Zhang and Ricles provide guidelines to determine the column torque at 4% story drift for the cases with and without a supplemental lateral brace.

Plastic hinging is expected at the lower end of fixed-base columns. In single-story moment-resisting frames, inelastic rotation is also permitted at the top of the columns, below the roof beams or trusses. In these cases, special attention must be given to control local buckling, as cross-section instability and subsequent flexural strength degradation can be aggravated by the presence of axial compression loads. It must also be recognized that even if designed in accordance with a strong column–weak beam approach (Eq. 19.9 or 19.10), columns in multistory MRFs are likely to experience inelastic rotation excursions under strong earthquakes because of the high variability in the amplitude and distribution of the bending moments under dynamic inelastic seismic response (MacRae and Tagawa, 2001). The expected flexural demand on columns can also be exceeded due to (1) the simultaneous bidirectional earthquake motions not accounted for in design, (2) lower than expected column-to-beam strength ratios resulting from actual versus nominal yield strength ratios being higher for the beams compared to the columns, or (3) the pronounced strain-hardening and composite slab effects on beam flexural capacity.

The consequences of column local buckling and subsequent loss in axial or flexural capacity can be dramatic, as was the case for the structural collapse in the Pino Suarez complex during the 1985 Mexico earthquake (Ger et al., 1993). The effects of column local buckling were recently examined in a three-dimensional shake table test performed on a full-scale four-story MRF with square hollow tubular columns fixed at their bases (Tada et al., 2007; Suita et al., 2008). Beams and columns were selected in accordance with the strong column–weak beam strategy, but the walls of the selected tubes did not meet the most stringent limits for slenderness to prevent local buckling from occurring; this limit is defined by $b - 3t = 30$, which approximately corresponds to the AISC (2005c) specification limit of 29.2 for compact cross sections assuming a nominal yield strength of 295 MPa. The axial loads in the columns were relatively low, 0.075 and 0.15 times the column yield load $P_y (= AF_y)$ for the exterior and interior columns, respectively. Local buckling developed at the base and upper ends of the first-story columns under 150% of the design ground motion level. When subjected to higher shaking level, the structure developed a column sway mechanism at the first story, even if a strong column–weak beam strategy had been followed. This was attributed to the deterioration of the column bending strength caused by local buckling. Similar behavior can occur if the bending capacity of the column-to-foundation attachment of fixed-base columns deteriorates during an earthquake. This could be the case if plastic elongation of the anchor rods connecting the base plates to the foundation or crushing of the concrete under the base plates occurs under reversed cyclic loading. This situation was observed during a two-dimensional quasi-static cyclic test of a full-scale three-story MRF specimen (Nakashima et al., 2006). The decrease in bending resistance of the base plate resulted in an increased flexural demand at the upper end of the first-story column segment. Plastic hinging and local buckling formed at that location, which then led to a story mechanism with degrading lateral capacity.

Only a few cyclic test programs have been carried out to study the inelastic flexural response of axially loaded steel columns. These tests confirmed the detrimental influence of axial loading but also indicated that adequate performance could still be achieved by properly controlling the cross-section slenderness. Popov et al. (1975) conducted cyclic tests on interior cruciform beam-column subassemblages with I-shaped (W200) columns. The amount of axial loads in the column and the direction of bending were varied in the tests. The column cross sections met the current 2005 AISC seismic compactness requirements with $b_f/2t_f$ and h/t_w ratios ranging from 0.66 to 0.85 and from 0.26 to 0.36 times the respective limits. The specimens were designed using a strong beam design approach to trigger yielding in the columns. In the tests, plastic rotation developed unevenly in the top and bottom column segments, indicating the propensity for a story mechanism to form when adopting a weak-column design strategy. The phenomenon was more pronounced for the columns supporting the highest axial loads ($0.6P_y$ and $0.8P_y$). The contribution of the column deformation to total horizontal displacements was also more important for such columns, with individual column segments experiencing drifts of 0.06 to 0.08 rad, but the response of these specimens was characterized by a sharp drop in strength resulting from flange buckling. Conversely, columns supporting an axial load of $0.30P_y$ could sustain cyclic flexural demand about minor and major axes corresponding to an interstory drift of 0.03 rad without significant (less than 10%) strength degradation. Similar behavior was observed in tests with $P/P_y = 0.5$, except that the drop in lateral resistance was more severe when the column was bent about its minor axis. A strain-hardening response was observed in all columns, which had a positive impact on the column flexural resistance. Popov et al. concluded that inelastic rotations should be avoided in columns with P/P_y greater than $0.5P_y$ and that moderate inelastic demand is permissible if the load ratio is less than 0.5.

Nakashima and Liu (2005) performed cyclic tests on reduced-scale fixed-end columns made of square tubing, which are classified as compact sections according to the AISC 2005 Specification (AISC, 2005c). The columns were subjected to cyclic loading with stepwise increasing lateral displacements and three different axial load levels were considered, including 0.0, $0.3P_y$, and $0.6P_y$. Flexural strength degradation due to column local buckling started in the form of a slight reduction in the two cycles at 0.06 rad, followed by a significant drop in the second cycle at 0.04 rad, and then a significant drop in the second cycle at 0.02 rad, respectively, for the three axial load levels. For $P/P_y = 0.6$, the specimens could not maintain the axial load in the cycles at 0.04 rad.

Bidirectional cyclic tests were recently conducted by Newell and Uang (2008) on full-scale heavy W360 I-shaped columns with cross sections complying with the 2005 AISC seismic compactness requirements. The columns were fixed at their two ends and were subjected to bidirectional transverse displacement histories while carrying axial loads of $0.35P_y$, $0.55P_y$, and $0.75P_y$. The specimens could sustain interstory drifts of 0.07 to 0.09 rad before a 10% reduction from the maximum moment was observed. Because all of the specimens had stocky webs, FLB dominated the buckling response in all tests. For the lightest column supporting $0.75P_y$, the $b_f/2t_f$ and h/t_w ratios corresponded to 1.0 and 0.43 times the 2005 AISC seismic compactness limits. Limited FLB was observed at 0.04 rad and the specimen could achieve a 0.07-rad-drift capacity.

In AISC seismic provisions, all columns in SMFs must meet the stringent seismic compactness requirements, even if a strong column design is enforced. The less stringent limits for compact sections from the AISC Specification (AISC 2005c) apply to the IMF system. In CSA-S16, the most severe cross-sectional slenderness limits apply for both type D (ductile) and type MD (moderately ductile) MRFs, but only to columns where plastic hinging is expected. Axial loads are also restricted in these columns to $0.3P_y$ and $0.5P_y$ for types D and MD frames, respectively. Less severe compacteness limits can be used for columns designed to remain elastic. In NZS3404, two limits are imposed on the design column axial loads in addition to cross-section slenderness requirements. One is related to the anticipated inelastic demand (ductility category), defined by $0.5P_y$ to $1.0P_y$ for the most ductile to the less ductile systems. The second limit depends on the column slenderness and aims at avoiding plastic hinges from forming between column ends. NZS3404 also limits the column axial loads due to gravity loads acting alone, and this limit is expressed as a function of the column web slenderness ratio. For fixed-base columns, strength degradation at the connection with the foundations can be prevented by designing column base plates for the full flexural strength of the columns. This is explicitly required in AISC and CSA-S16 seismic provisions and details to achieve this behavior can be found in Arlekar and Murty (2002) and Fahmy et al. (2000).

Alternatively, concrete-encased or concrete-filled tube composite columns can be used to enhance column local buckling behavior and flexural ductility (e.g., Ricles and Paboojian, 1994; El-Tawil and Deierlein, 1999; Inai et al., 2004; Varma et al., 2004). These columns also possess higher flexural stiffness (for drift control) and high torsional stiffness (in fact, flexural–torsional buckling need not be checked). In bridge structures, plastic hinging at the column base generally forms the main energy dissipation mechanism and findings from research in that field can be useful for high-loading applications. Large steel box or circular tubular columns are often used in bridges and the required rotational ductility is obtained by using hollow or concrete-filled rectangular or circular tubing columns (e.g., Kawashima et al.,

1992; Usami et al., 2000; Zheng et al., 2000; Susantha et al., 2002; Marson and Bruneau, 2004; Ishizawa and Iura, 2005; Mamaghani, 2008).

Depending on the beam-to-column connection typology, additional local stability verifications may be needed in the joint region of the columns. For instance, the column web must be checked for web crippling, web buckling in compression, and web buckling in shear for joints where the beams are directly welded to I-shaped columns bent about their major axis. Design guidance for these cases can be found in Eurocode 3 (CEN, 2002) and Carter (2003). Web crippling and compression web buckling are verified for unstiffened connections (without continuity plates or transverse stiffeners). Limits are prescribed in steel design codes to prevent stability failure of webs acting in shear. These limits are usually satisfied for the vast majority of hot-rolled I-shaped profiles used for columns, but they could control in cases of built-up columns with relatively thin webs and/or higher steel grades were used.

Past investigations (e.g., Krawinkler and Mohasseb, 1987; El-Tawil et al., 1999; Roeder, 2002; Lee et al., 2005) suggest that adequate seismic performance can be achieved when adopting an alternative design approach in which ductility is supplied partly by plastic rotation in the beams and partly by shear yielding in the column joint panel zones (Fig. 19.4*a*). This approach has now gained general acceptance in seismic design codes. Although the required shear resistance for the column web panel is reduced, more stringent web slenderness limits must be met to ensure that the shear yield resistance of the web panel can be maintained under cyclic inelastic loading. In AISC seismic provisions, this web slenderness limit is given by $t > 90\ (d_z + w_z)$, where t is the thickness of the web panel, d_z is the panel zone depth between continuity plates, and w_z is the panel zone width between the column flanges. If the columns do not satisfy the different web stability criteria, the designer can select columns with thicker webs on add web doubler plates or transverse stiffeners. The latter resolves issues of web crippling or compression buckling. Diagonal stiffeners can also be used to resist web shear in the elastic range. Experimental studies on various web-reinforcing details are presented by Lee et al. (2005) and by Ciutina and Dubina (2008). Economical aspects of column stiffening are discussed by Carter (2003).

19.2.2 Steel Plate Shear Walls (SPSWs)

In the last decade, seismic design provisions for plate shear walls with unstiffened infill plates have been introduced in codes in North America. At every floor, the infill plate is connected (typically by welding) to the surrounding beam- and column-framing members. The plates are usually relatively thin, with limited buckling capacity when subjected to shear. Lateral resistance is therefore primary obtained through diagonal tension-field action developing in the infill plates after buckling has occurred. Large interstory drifts anticipated resulting from strong earthquakes induce tension yielding of the plates along inclined diagonals (Fig. 19.4*b*). Under successive cycles of inelastic loading, progressive stretching of the infill plates takes place in both directions, resulting in pinched hysteretic behavior with limited lateral resistance around the initial position. In order to provide a minimum energy dissipation capacity under cyclic loading, seismic design provisions in the United States and Canada require that the framing members be rigidly connected to form a ductile moment-resisting frame with plastic hinges developing at the beam ends and at the column bases. Several inelastic cyclic test programs performed on full- and reduced-scale (1:2 to 1:4) single- to four-story specimens showed that well-proportioned SPSWs with slender unstiffened plates can exhibit stable inelastic cyclic performance up to interstory drifts of 0.03 to 0.05 rad (Caccese et al., 1993; Driver et al., 1998; Lubell et al., 2000; Astaneh-Asl, 2001; Berman and Bruneau, 2005; Qu et al., 2008; Sabouri-Ghomi and Gholhaki, 2008). Failure eventually takes place by tearing of the plates or at beam-to-column joints at large interstory drifts. Design guidance to achieve this seismic performance can be found in Sabelli and Bruneau (2007). Alternatively, the infill plates can be stiffened or laterally supported to delay or prevent out-of-plane buckling. Both approaches have been retained in early applications in the United States and Japan (Thorburn et al., 1983; Astaneh-Asl, 2001). Although the results are higher initial stiffness, increased story shear resistance, and reduced pinching in the hysteretic response, these approaches generally incur additional fabrication or construction costs. In both unstiffened and stiffened designs, infill plate yielding is the main source of energy dissipation, and the framing members must be sized to remain essentially elastic except at intended plastic hinge locations within the design. Stability issues include infill plate buckling as well as buckling of beams and columns.

Infill Plate Buckling Steel infill plates spanning across typical bay widths can develop high story shear resistance, and the design generally employs thin plates with high width-to-thickness slenderness ratios, typically between 400 and 2000. For unstiffened infill plates, buckling in shear then occurs in the elastic range at very low stress levels. The lateral resistance associated with this buckling mode is small and the contribution of the infill plates to story shear strength is solely based on the postbuckling yield capacity of the plates (Sabouri-Ghomi and Roberts, 1991; Berman and Bruneau, 2003). This postbuckling resistance may be expressed as

$$V_y \geq 0.5 F_y L t_w \sin(2\alpha) \tag{19.11}$$

where L and t_w are the width and the thickness of the infill plates and α is the angle of inclination of the tension field with respect to the vertical (Fig. 19.10*c*). Timler and Kulak (1983) developed the following equation for α that accounts for the panel width L and height h, the cross-section areas of the beams and columns, A_b and A_c, respectively, and the flexural stiffness of the columns I_c:

$$\alpha = \frac{1 + t_w L/2A_c}{1 + t_w h\left(1/A_b + h^3/360 I_c L\right)} \tag{19.12}$$

The angle α typically ranges between 38° and 45°. The horizontal shear capacity given by Eq. 19.11 assumes complete tension-field action and, hence, exceeds

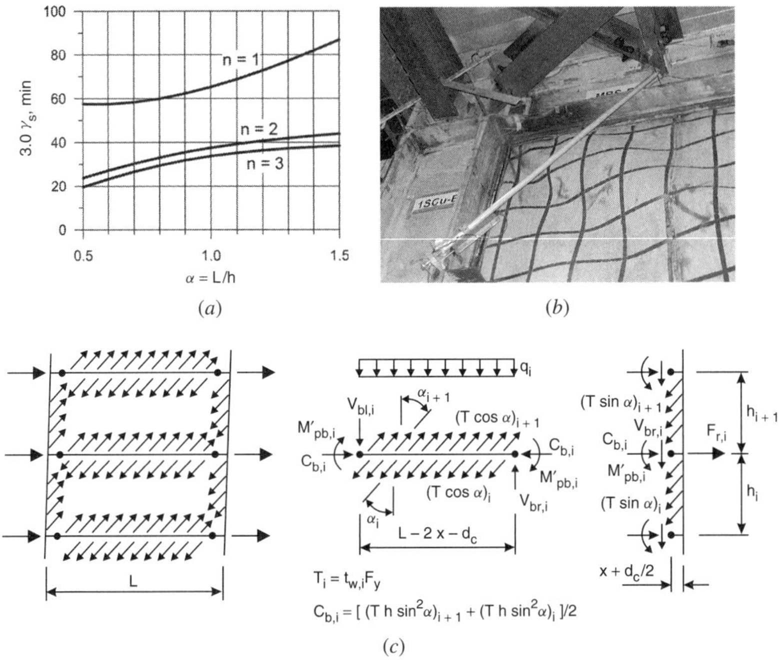

FIGURE 19.10 (*a*) Minimum rigidity requirement for regularly spaced vertical and horizontal infill plate stiffeners (Eq. 20.13). (*b*) Diagonal compression buckling and tension field in the infill plate and plastic rotation in the RBS connection during a cyclic inelastic SPSW test. Courtesy of B. Qu (CalPoly), M. Bruneau (Univ. at Buffalo), K. C. Tsai (NCREE), and C. H. Lin (NCREE). (*c*) Infill plate yield tension loads and beam plastic hinge moments imposed on beams and columns.

significantly the resistance provided by partial tension-field action as described for steel plate girders as in Chapter 6 (Berman and Bruneau, 2004). Complete tension-field action can only be achieved if the infill plates are well anchored at the foundation and the top beam and columns have sufficient stiffness to ensure uniform tension in the plates. From early work on aluminum beams with stiffeners and observations of SPSW behavior in previous tests, Montgomery and Medhekar (2001) proposed that I_c equal or exceed $0.00307 t_w h^4/L$ in order to keep the tensile stresses near the horizontal beams at the top and bottom of a wall panel from varying by more than 20% from the uniform-stress value. This requirement was adopted in the U.S. and Canadian seismic provisions. A limit of 20% is also specified on the variation in tensile stress along the top beam in CSA-S16. No such requirement is prescribed in AISC seismic provisions, but Sabelli and Bruneau (2007) recommend that the same column stiffness criterion also be applied to beams by inverting L and h in the equation and replacing t_w by the difference in plate thickness below and above the beams. In AISC provisions, the panel aspect ratio L/h must be defined between 0.80 and 2.5, which approximately corresponds to the range covered in past test programs. Boundary elements must also possess sufficient strength to resist the yield tension forces generated by the infill plates, as will be discussed in more detail below.

For seismic resistance, initial elastic buckling of the infill plates does not represent a limit state and is ignored in U.S. and Canadian codes. Initial out-of-straightness of the plates due to construction tolerances and potential buckling due to vertical compression stresses induced by gravity loads, however, may delay the development of the tension field and affect the elastic lateral stiffness of the wall. These effects can be minimized by adopting a proper construction technique and sequence. For instance, the plates can be field welded to the boundary members (with welding shrinkage reducing or eliminating the panel out-of-plane deflections) after most of the gravity loads are applied. The extent of buckling can also be reduced by selecting a low-yield steel material for the plates (Vian and Bruneau, 2004). Stiffeners or concrete panel encasement can also be used to improve both the initial stiffness and resistance to buckling. Under cyclic inelastic loading, these strategies also result in reduced pinching and enhanced energy dissipation capacity. Although past test programs conducted in Japan showed that good inelastic cyclic performance could be achieved with stiffened infill plates (Thorburn et al., 1983; Astaneh-Asl, 2001), no design guidance has been introduced in seismic code provisions for this concept. Using classical elastic stability theory, Chusilp and Usami (2002) derived the following expression for the stiffener rigidity ratio γ required to constrain shear buckling between regularly spaced horizontal and vertical stiffeners and prevent overall buckling of the panel and the stiffeners together:

$$\gamma_{s,\min} = \left(\frac{23.1}{n^{2.5}} - \frac{1.35}{n^{0.5}}\right) \frac{\left(1 + \alpha^{3/n-0.3}\right)^{2n-1}}{1 + \alpha^{5.3-0.6n-3/n}} \tag{19.13}$$

in which $\gamma_s = EI_s/Dh$, where EI_s is the flexural stiffness of the stiffeners and D is the plate flexural stiffness, $D = Et_w^3/[12(1 - v^2)]$, with v the Poisson's ratio of steel ($= 0.3$), $\alpha = L/h$ (different than the angle of inclination of the tension field, as previously defined), and n the number of stiffeners. Equation 19.13 applies to plates with one, two, or three stiffeners in both directions (n horizontal and n vertical stiffeners) and aspect ratios α defined between 0.5 and 2.0. A similar expression has also been derived for plates with stiffeners in one direction only. Finite element simulations of small panels subjected to inelastic cyclic loading showed that stiffened infill plates can develop and maintain the full shear yield resistance of the plates, without strength degradation up to a ductility of 20, provided that $\gamma_s \geq 3.0\ \gamma_{s,\min}$ from Eq. 19.13, $\alpha \leq 1.5$, and $h/t_w \leq 0.624\,(n + 1)\sqrt{k_v E/F_y}$ (Chen et al., 2006). In the last criterion, k_v is the plate-buckling coefficient for shear for the unstiffened infill plate (k_v equals $4.0 + 5.34/\alpha^2$ for $\alpha \leq 1$ and to $5.34 + 4.0/\alpha^2$ for $\alpha > 1$; see Chapter 4). Values of $3.0\ \gamma_{s,\min}$ from Eq. 19.13 are plotted in Fig. 19.10*a*. When these criteria are met, the authors showed that the cyclic inelastic response of stiffened walls can be accurately predicted with a simple bilinear model for nonlinear dynamic analysis.

Sabouri-Ghomi et al. (2008) also used elastic buckling theory to propose minimum stiffener rigidity values to prohibit overall buckling of infill plates. Finite element elastic buckling analysis revealed that the theoretical predictions better match the analysis results when double-sided, more closely spaced stiffeners are used. Stiffness values higher than those predicted are needed to prevent elastic overall buckling for one-sided or widely spaced stiffeners. Alinia and Dastfan (2007) performed numerical simulations of the inelastic cyclic response of infill plates stiffened with three and five stiffeners in both directions and aspect ratios equalling 1.0 and 1.5. Low-yield material was assumed for the plates ($F_y = 240\,\text{MPa}$). They found that shear-dominated response without pinching could be obtained when γ_s was approximately equal to 30, which is in agreement with the recommendations by Chen et al. (2006) (see Fig. 19.10*a*). Alinia and Dastfan (2007), however, indicated that fully constraining buckling to subpanels could reduce the wall deformability. It is noted that all the stiffener requirements presented herein have been essentially derived from analytical models and as of present (2009) have not been experimentally validated in the context of inelastic seismic wall response.

Berman and Bruneau (2005) examined the possibility of using corrugated infill plates to improve buckling characteristics. Cyclic testing was carried out on a half-scale specimen built with 38-mm-deep steel deck sheets with trapezoidal flutes. The sheets were inclined at 45° and were made of 0.76-mm-thick steel. Poor performance was observed as the specimen exhibited unsymmetrical hysteretic response (tension-field action could not develop perpendicular to the flutes) and fracture of the sheets occurred early, at an interstory drift of 0.014 rad, at the location of the repeated local buckling that developed in the deck sheets upon overall plate buckling. In a full-scale test on a two-story wall, Qu et al. (2008) showed that the amplitude of the out-of-plane buckling deformations of the infill plates under large interstory drifts could be controlled by horizontal tubular shapes installed at quarter points of the panels on both sides of the walls.

Encasement of the infill plates with reinforced concrete panels placed on one or both sides of the plates can be used to prevent shear buckling. The panels can be cast-in-place or precast elements, and uniformly spaced stud or bolt attachments are used between the infill plates and the concrete panels. Test data on this composite construction system are scarce and only limited seismic design guidance is available. Both the AISC seismic provisions and EC8 contain identical rules, in which the infill plates are assumed to be laterally braced and will develop their full shear yield capacity as long as the concrete panels include minimum vertical and horizontal reinforcement (0.25%) and have a minimum thickness of 100 mm (panels on both sides) or 200 mm (panels on one side). Mechanical attachments must be provided between the steel plates and the concrete panels to prevent separation and plate buckling. The same rules apply, regardless of the thickness and dimensions of the infill plates. In the commentary to the AISC seismic provisions, it is suggested that the shear buckling capacity of the composite wall be evaluated using elastic buckling theory and transformed section properties. When concrete panels are provided on one side of the infill plates, it is also recommended that the plate slenderness between mechanical attachment points meet the slenderness limits that are specified in the AISC specifications to prohibit shear buckling of beam webs, that is, $h/t_w \leq 1.1\sqrt{k_v E/F_y}$.

Cyclic testing performed by Zhao and Astaneh-Asl (2004) on half-scale two-story composite SPSW specimens showed that tension-field behavior can still dominate the response at large deformations when the infill plates are braced with concrete panels located on one side only. The test walls were built with 4.8-mm-thick, 1.8-m × 1.8-m infill plates made of steel with nominal yield strength of 248 MPa. Precast reinforced concrete panels, 76 mm thick, were attached on one side with structural bolts spaced 228 mm apart in both directions. The plate slenderness between the fasteners was therefore equal to 47.5 (228/4.8), which is less than the proposed h/t_w limit (70 when assuming $k_v = 5$). Diagonal compression buckling of the plates developed between the bolts in the 1.2-to-1.8% interstory drift range and punching of the bolts through the infill plates was observed at 0.03 rad drift, which is indicative of a significant out-of-plane buckling response. Hence, the panels could delay overall buckling of the infill plates until shear yielding initiated, but the wall eventually acted as an unstiffened wall primarily resisting shear through tension-field action. Sun et al. (2008) reported similar observations made in other test programs, which suggests that tension yielding of the infill plates should be considered in design, even when buckling restraining mechanisms are used.

Buckling of Beams and Columns Beams and columns acting as boundary elements must have not only the capacity to resist gravity loads but also the forces induced by tension yielding of the infill plates and plastic hinging at the beam ends. These yielding mechanisms can be observed in Fig. 19.9*b* and the forces imposed on the beams and columns are illustrated in Figure 19.9*c*. These forces can be determined from an incremental static (pushover) analysis of the wall with the infill plates being modeled with a series of inclined parallel tension-only strips exhibiting bilinear response (Thorburn et al., 1983; Shishkin et al., 2005). The calculations can also be made by hand assuming that a full plastic mechanism is reached and that the infill plates induce uniformly distributed tension-yield loads along the boundary elements, as shown in Fig. 19.10*c*. Details of these analysis methods can be found in Berman and Bruneau (2008a). As with any other capacity design check, the expected yield strength and strain-hardening effects must be considered when determining the forces imposed by the yielding components.

In-plane and out-of-plane stability of the columns must be verified against the combined effects of axial loads and bending moments. This is similar to the situation in moment-resisting frames except that (i) larger bending moments are expected due to the horizontal component of the tension field acting along the column height and (ii) axial loads resulting from the vertical component of the infill plate tension must be added to the beam-end shear forces. Occurrence of column buckling was observed in several past test programs when the column strength did not match the demand from the yielding elements (e.g., Caccese et al., 1993; Lubell et al., 2000; Park et al., 2007), and such undesirable behavior must be prevented to maintain structural integrity.

Attention must also be directed to the shear capacity of the columns as a result of the large distributed horizontal forces that are applied to the columns. Shear yielding should be prohibited because it can produce additional in-plane column deformations that may adversely affect column stability. As was discussed for MRFs, the amplitude and distribution of the bending moments in the structure, which respond nonlinearly to the complex dynamic input from earthquake ground motions, may deviate from that obtained from the simplified static analysis used in design. Plastic hinging in the columns therefore still remains a possibility and minimum section compactness requirements must be applied to delay local buckling. Tests by Park et al. (2007) showed the detrimental effect of local buckling along the column height on the system's overall response. Column hinging is also anticipated at the base of the columns (Driver et al., 1998; Park et al., 2007) and, hence, local buckling must also be controlled at that location. Both the U.S. and Canadian seismic codes include capacity design provisions as well as stringent width-to-thickness ratio requirements for the columns. In addition, CSA-S16 requires that column bases be reinforced so that plastic hinging forms at some distance from the base connection.

As would be expected, the choice of the infill plate thickness has a considerable impact on the design of the columns. Hence, a careful plate selection that closely matches the required strength along the structure height can help minimize the force demand on the columns. Such a tailored plate design will also contribute to the goal of achieving uniform inelastic demand over the height of the wall. When the plate thickness is governed by handling or fabrication or construction processes, the designer may control the force demand on the columns by adopting a perforated plate design. In this approach, circular perforations are introduced in the infill plates along the inclined tension-field pattern such that diagonal rectangular straps working in tension are formed between the inclined rows of holes (Roberts and Sabouri-Ghomi, 1992; Vian and Bruneau, 2004). Composite columns that inherently possess greater flexural stiffness can also represent a suitable solution for heavily loaded columns (Astaneh-Asl, 2001).

The horizontal component of the infill plate tension field acting on the columns is resisted in compression by the beams, which essentially act as struts (force $C_{b,i}$ in Fig. 19.10*c*). When different plate thicknesses are specified below and above the beams, additional axial loads and vertical forces are imposed on the beams by this diagonal tension. As a result of this compression, the stability of the beams must be checked together with the bending moments developing in the plastic hinges. It should be noted that the anticipated plastic hinge moments can be reduced by the axial compression present in the beams. Although floor diaphragms can provide out-of-plane lateral restraint to beams, in-plane flexural buckling of beams must be verified, as well as the cross-section strength outside of the plastic hinge regions. This check is more critical for top and bottom beams where infill plates are connected only from one side. At the base, the beams can be vertically anchored to the foundations to reduce the bending moment demand. For long spans, Sabelli and Bruneau (2007) suggest that vertical struts could be added at the center of the wall to provide a vertical support to the top beam.

Adopting an RBS beam-to-column connection design represents a good strategy as the force demand on the beams is reduced between the beam plastic hinges as well as at the face of the columns. Column forces are also reduced. As discussed in Section 19.2.1, well-proportioned RBS connections exhibit good plastic rotation capacity and several successful tests have been completed on SPSWs built with this connection (e.g., Vian and Bruneau, 2004; Qu et al., 2008). In view of the large axial compression loads acting in SPSW beams, it is expected that RBS connections in SPSWs may not develop the plastic rotation capacities observed in MRF applications (Eqs. 19.7 and 19.8) and further research is needed to develop design information on this aspect.

19.2.3 Concentrically Braced Steel Frames (CBFs)

Concentrically braced steel frames dissipate energy essentially through tensile yielding and inelastic buckling of the diagonal bracing members (Fig. 19.4*c*). Bracing members are expected to buckle at small interstory drifts, typically on the order of 0.25 to 0.50%. Upon buckling, plastic hinges form at the brace midlengths and at the brace ends. Plastic rotation at the ends can develop in the brace or in the connection, depending on the connection design and detail. Maximum inelastic rotation generally takes place at the brace midspan. Local buckling can form at that location under severe large cyclic loading and the resulting high localized strain demand can eventually lead to brace fracture. Global and local buckling of braces are therefore two aspects that must be accounted for in seismic design of CBFs. Other stability issues in concentrically braced steel frames include buckling of the brace gusset plates acting in compression and the stability of beams and columns resisting gravity loads together with lateral load effects corresponding to attainment of brace expected capacities. These aspects are discussed herein. Information on other seismic design requirements for CBFs can be found in Cochran and Honeck (2004) and AISC (2006).

Overall Brace Buckling Figure 19.11*a* shows the axial load–axial deformation hysteretic response of a typical steel bracing member with intermediate slenderness when subjected to cyclic inelastic displacements with stepwise incremented amplitudes. In this figure, the loading starts in compression and the brace buckles at a load C_u. If further negative axial deformation is applied, a plastic hinge forms at the brace midlength and the brace compressive resistance gradually decreases to maintain equilibrium as the brace lateral deformation increases. This behavior is illustrated in Fig. 19.11*b* for a simple brace with pinned end connections. Upon load reversal, the brace is straightened and the tension load increases up to the brace tensile-yield resistance $P_y = AF_y$, beyond which additional deformation results in plastic elongation. When reloading in compression, buckling occurs at a smaller load due to a combination of the Baushinger effect and the residual lateral deformation (kink) from previous buckling. Buckling also occurs earlier in terms of axial displacement because the brace has been lengthened in tension. Applying further deformation in compression produces additional hinge rotation with

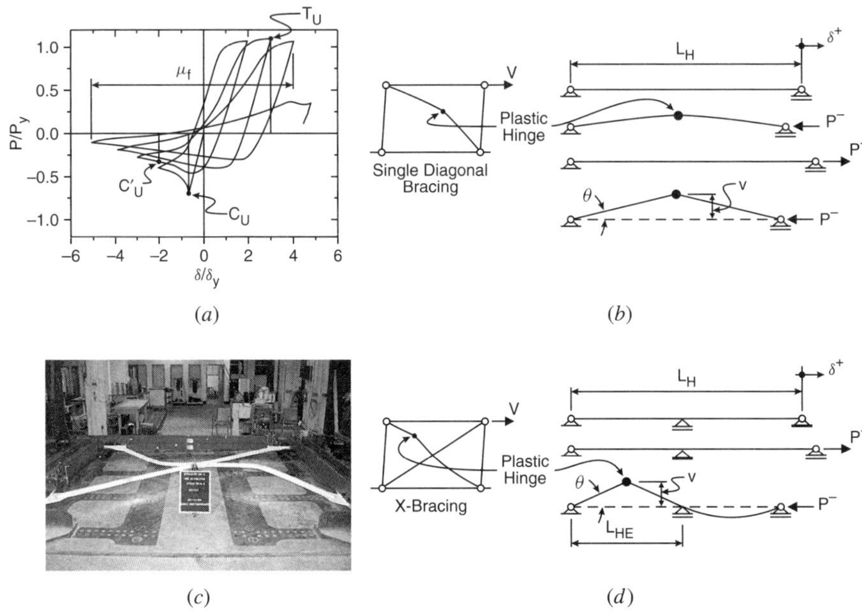

FIGURE 19.11 (*a*) Brace hysteretic response; (*b*) brace tension yielding and inelastic buckling in single diagonal brace; (*c*) photo of and (*d*) schematic of brace tension yielding and inelastic buckling in X-bracing configuration.

decreasing compressive resistance C_u'. Hence, previous brace plastic elongation is not recovered in compression; when the brace is pulled back in tension, the tension load will reach P_y and additional yielding in tension will occur only if the brace is stretched beyond the deformation attained in the previous tension loading cycle. Under large tension yielding excursions, strain hardening can develop and result in a maximum brace tension load T_u which exceeds P_y.

In design, the compressive strength of bracing members is based on an effective length KL equal to the centerline brace length, defined between working points. In X-bracing with braces connected at the intersection point, analytical and experimental studies have shown that the tension brace is capable of providing effective in-plane and out-of-plane support at midlength of the compression brace. For instance, an effective length factor $K = 0.5$ can be used to assess the brace-buckling strength in symmetrical bracing with continuous and identical pin-ended tension and compression braces carrying similar axial loads (e.g., Kitipornchai and Finch, 1986; Picard and Beaulieu, 1987, 1988). In X-bracing with identical braces, this response was found to carry over in the postbuckling range under cyclic inelastic loading in past earthquakes (Fig. 19.3*c*, in-plane buckling) and past cyclic inelastic test programs (e.g., Ballio and Perotti, 1987; Nakashima and Wakabayashi, 1992; Tremblay et al., 2003); once it is straightened, the tension-acting brace can still enforce second-mode buckling of the compression brace, as shown in Fig. 19.11*c* for out-of-plane buckling. Plastic rotation in X-braces, however, tends to concentrate in one of the two brace halves, as can be observed in Fig. 19.11*c* and schematically in Fig. 19.11*d*.

Research conducted since the 1970s on braced steel frames clearly demonstrates that the effective slenderness KL/r of a brace is the single most important parameter influencing the inelastic cyclic response of steel bracing members, including buckling and postbuckling compressive strengths, residual elongation, energy dissipation capacity, and brace fracture life (e.g., Jain et al., 1978, 1980; Popov and Black, 1981; Nakashima and Wakabayashi, 1992). This marked influence of brace slenderness can be observed for rectangular hollow section (RHS) bracing members in Fig. 19.12. In particular, the compression resistance and energy dissipation capacity (area enclosed by the hysteretic response curve) both decrease when brace slenderness is increased whereas more slender braces can withstand larger axial deformations before fracture occurs. In Fig. 19.12, λ is the nondimensional slenderness parameter ($\lambda = KL/r\sqrt{F_y/\pi^2 E}$) and δ_y is the brace axial deformation at yield. Limits on brace slenderness are imposed in modern seismic design provisions to ensure minimum brace compressive strength and energy dissipation capacity. For example, KL/r must not exceed 200 in CSA-S16 and AISC seismic provisions. In EC8, the limit is $2\pi\sqrt{E/F_y}$ ($= 150$ for $F_y = 350\,\text{MPa}$) for three-story and taller structures. In NZS3404, the limit is equal to $4.25\sqrt{E/F_y}$ ($= 102$ for $F_y = 350\,\text{MPa}$) for braces effective in tension and compression. Brace effective slenderness also has direct impact on the building height limit as well as on the amplitude of the design seismic loads. In design, realistic prediction of the brace compressive resistances C_u and C_u' are needed to assess the force demand on the other structural components along the lateral load path of the structure. Lastly, brace effective slenderness ratios dictate the plane along which a brace will buckle. This information is required to properly detail the brace and its connection such that a ductile response can be achieved under repeated inelastic buckling in that plane. For all these reasons, a realistic estimate of the brace critical KL/r is necessary for satisfactory CBF seismic design.

Data from past experimental studies can be used as guidance for this purpose. For single diagonal members, the effective length essentially depends on the stiffness and strength of the end connections. Early cyclic inelastic frame tests conducted in Japan confirmed that H-shaped braces with fixed ends behave as equivalent pin-ended braces having half the length (Nakashima and Wakabayashi, 1992). Both

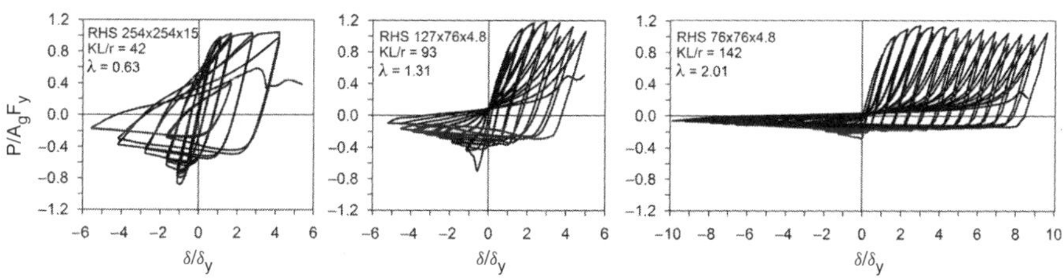

FIGURE 19.12 Influence of brace effective slenderness ratio on hysteretic response.

in-plane and out-of-plane buckling modes were observed in a full-scale test of a six-story CBF with inverted-V tubular braces (Roeder, 1989). At their top ends, the braces in the specimen were fully welded under the beams and at their lower ends they were embedded into the concrete slab. The estimated K-factor based on the clear length between the top of the floor slab and the bottom of the beams ranged between 0.63 and 0.77 for in-plane buckling and between 0.72 and 0.80 for out-of-plane buckling.

From cyclic tests on double-angle braces buckling out-of-plane and connected with single gusset plates, Astaneh-Asl et al. (1985) showed that good inelastic rotational response at the brace ends can be obtained, provided that a free length, equal to twice the gusset plate thickness t_g perpendicular to the brace axis, is left beyond the angles for the free formation of plastic hinges in the gusset plates. This detail is illustrated in Fig. 19.13*a* for a tubular brace. In their tests, the braces buckled as a pin-ended member between the gusset hinges, suggesting an effective length $KL = L_H$. This detail has gained wide acceptance in practice and is recommended in the AISC provisions as an acceptable means of accommodating inelastic rotation upon brace buckling. From stability analysis including gusset plate flexural stiffness, Tremblay et al. (2003) found effective lengths KL varying between $0.88L_H$ and $0.96L_H$ for test specimens built with this configuration. Lehman et al. (2008) recently proposed and experimentally validated the elliptical single gusset plate clearance design with an $8t_g$ offset (Fig. 19.13*b*). This alternative detail has been developed to provide for ductile brace-end rotations with more compact and flexible gusset plates and the authors suggest an effective length equal to the actual length of the bracing member.

Astaneh-Asl and Goel (1984) tested double-angle braces buckling in the plane of the frame. Plastic hinges formed in the braces next to the gusset plates and it was concluded that the K-factor could be approximated by 0.5 (Fig. 19.13*c*). If such a detail is adopted, CSA-S16 and AISC seismic provisions require that the connection has sufficient flexural strength to constrain plastic hinging in the brace. Alternatively, end rotation from in-plane buckling can be accommodated using a knife plate between the brace and the gusset plate (Fig. 19.13*d*). An effective length of $0.9L_H$ was computed for brace specimens of this type in a recent test program (Tremblay et al., 2008a).

For the X-bracing configuration, tests in Japan described by Nakashima and Wakabayashi (1992) revealed that the measured response of fix-ended H-shaped braces could be well reproduced using K-factors of 0.60 and 0.70 (of the half-diagonal length) for in-plane and out-of-plane buckling, respectively. Tests by El-Tayem and Goel (1986) showed that an effective length of 0.85 times the half-diagonal length was appropriate for X-bracing made of single equal-leg angles weld connected to single gusset plates. Using elastic stability analysis, Tremblay et al. (2003) determined the effective length of X-bracing specimens built with rectangular RHS braces buckling out-of-plane and terminating $2t_g$ before the line of restraint in the gusset plates. The computed KL values range between $0.83L_{HE}$ and $0.90L_{HE}$, where L_{HE} is the longer of the two brace segments measured between the gusset hinges and the brace intersection point (see Fig. 19.11*d*).

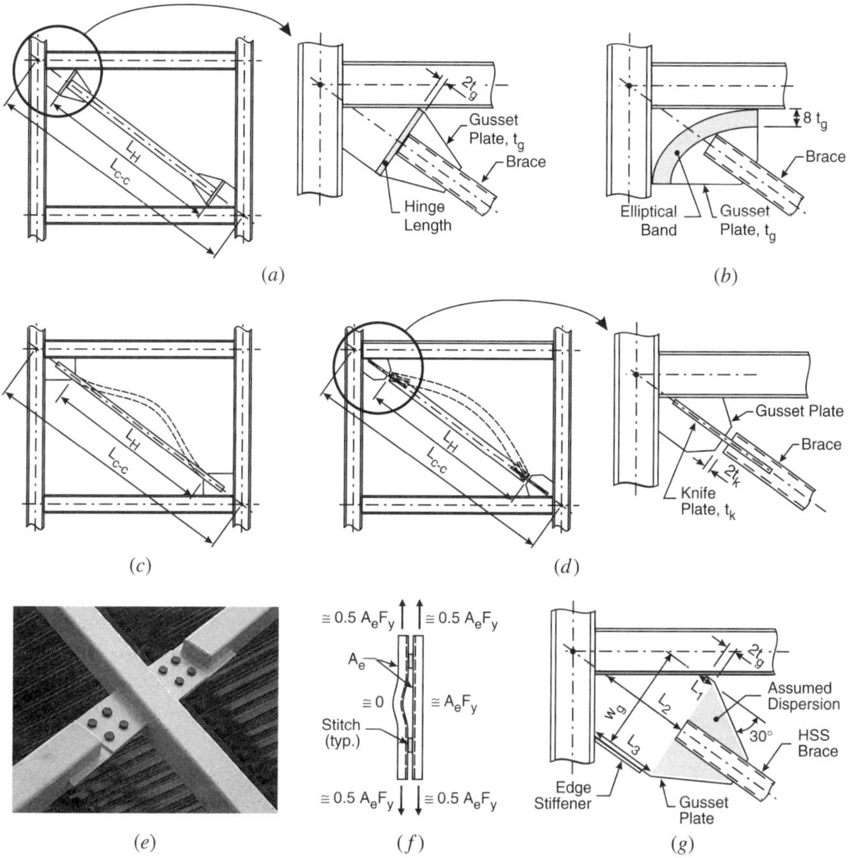

FIGURE 19.13 (*a*) Gusset hinge detail for out-of-plane buckling braces; (*b*) elliptical-clearance detail for out-of-plane buckling braces; (*c*) in-plane buckling of a fixed-end brace; (*d*) knife plate hinge detail for in-plate buckling braces; (*e*) discontinuous brace at the brace intersection point for X-bracing; (*f*) shear transfer in stitch connectors in a double-angle brace when reloading in tension after inelastic buckling in compression; and (*g*) equivalent column parameters for the assessment of the buckling capacity of gusset plates.

The braces were rigidly connected at the midpoint and the flexural stiffness and torsional stiffness of both the gusset plates and the tension brace contributed to the buckling strength of the compression brace. Accounting for these retraining effects, Sabelli and Hohbach (1999) analytically determined lower and upper bound brace effective-length values for symmetrical X-bracing such that the buckling mode and brace actual strength can be predicted while accounting for uncertainty in brace-end conditions. For in-plane buckling, they suggested that the *KL* of pin- and fix-ended braces could be safely taken equal to 0.84 and 0.59 times the half-diagonal length for the design of the brace in compression. These values increase to 1.0 and 0.70 for out-of-plane buckling and similar support conditions. Lower values are proposed to predict the actual brace strength and buckling axis. Guidance on brace effective length for X-bracing with unsymmetrical brace geometries, unequal brace forces, or semirigid brace-end connections can be found in Stoman (1989), Wang and Boresi (1992), Thevendran and Wang (1993), and Segal et al. (1994). Situations exist where effective lengths are in excess of the brace segment between the end connection and the brace intersection point.

In the above-mentioned tests and analytical studies on X-bracing, the two braces were continuous at their connecting point. In practice, one brace is often interrupted at the intersection and, thereby, has reduced out-of-plane flexural and torsional stiffness (Fig. 19.13*e*). When acting in tension, the discontinuous brace is less effective in supporting the compression brace and its buckling strength is reduced when loaded in compression (Davaran, 2001; Davaran and Hoveidae, 2009; Moon et al., 2008). These effects can be minimized by reducing the length of the connection or by ensuring minimum flexural continuity at the brace intersection. The latter can be achieved by cutting only one of the two angles in case of intersecting double-angle braces (Davaran, 2001) or by adding cover plates connecting the two segments of rectangular HSS braces (Tremblay et al., 2003; Davaran and Hoveidae, 2009). Similarly, Kim and Goel (1996) reported a local failure of the intermediate brace-to-beam connection in a two-story X-bracing during the 1989 Loma Prieta earthquake. Failure was attributed to a lack of adequate beam flange bracing. finite element analyses and cyclic tests showed that the addition of a diagonal beam web stiffener could prevent local beam twisting and restore the expected bracing response.

Brace buckling is accompanied with significant translational out-of-plane or in-plane deformations (v in Figs. 19.11*b* and 19.11*d*). The buckling behavior of braces can therefore be altered by the presence of secondary structural elements or nonstructural components (e.g., partition wall precluding in-plane brace buckling) and care should be exercised to ensure that the predicted buckling behavior can be achieved (Sabelli and Hohbach, 1999). Damage to partitions or wall cladding can also be caused by brace buckling out-of-plane deformations (Tremblay et al., 1996). These problems can be avoided by providing allowance for the expected brace translational deformations. Assuming the deformed shapes illustrated in Figs. 19.11*b* and 19.11*d*, Tremblay et al. (2003) suggest that the brace lateral deformation v and the plastic hinge rotation θ can be estimated from

$$\frac{v}{L_H} = \frac{1}{\sqrt{1 + L_H/L_{HE}}}\sqrt{\frac{(\mu_c + \mu_t - 1)}{L_H/L_{HE}}\left(\frac{F_y}{E}\right)} \tag{19.14}$$

$$\theta = 2\sqrt{\frac{(\mu_c + \mu_t - 1)}{1 + L_{HE}/L_E}\left(\frac{F_y}{E}\right)} \tag{19.15}$$

where L_H is the length between brace-end hinges, L_{HE} is as defined above, and μ is the brace ductility defined by $\mu = \delta/\delta_y$ (δ_y can be taken equal to $L_H F_y/E$).

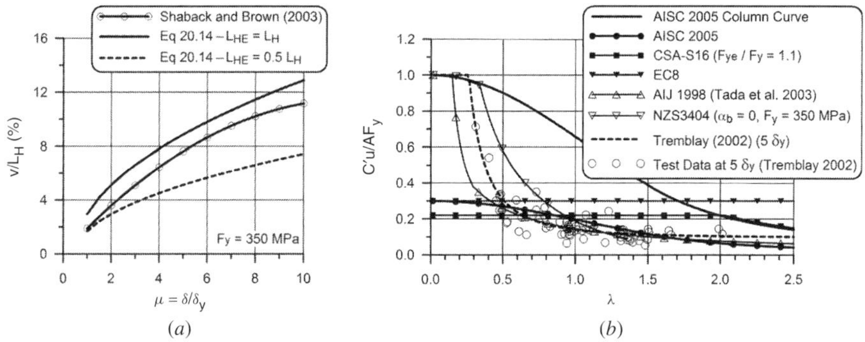

FIGURE 19.14 (*a*) Out-of-plane brace deformations; (*b*) postbuckling brace compression strength.

Equations 19.14 and 19.15 give values of v and θ at a ductility μ_c in compression after reaching a ductility μ_t in tension. They hold for single-diagonal braces (with L_{HE} set equal to L_H) and X-braces. From tests on single-diagonal RHS braces, Shaback and Brown (2003) proposed an empirical expression for brace out-of-plane deformation of $v/L_H = 0.02\mu - \mu^{1.9}/900$. Both expressions are compared in Fig. 19.14*a*.

For capacity design purposes, the expected brace-buckling strength C_u can be estimated using code column design curves together with representative slenderness ratios and expected yield strength values. Past test results indicated that the resistance so obtained can be exceeded in actual braces (Ballio and Perotti, 1987; Tremblay, 2002; Fell et al., 2009). Such resistance is thus amplified further by 1.1 and 1.2, respectively, in the U.S. and Canadian seismic provisions, to avoid underestimating brace compression forces used to design the surrounding components. Goggins et al. (2006) recently observed a similar difference for RHS braces and they recommended using the applicable column curve with the overall section yield strength, including cold forming effects, rather than the material yield strength from flat coupons.

In contrast, the brace postbuckling compressive resistance is generally considered in situations where the compression force acting in the brace reduces the force imposed on adjacent members. This is the case, for instance, for beams in chevron or X-bracing systems as well as for intermediate columns, as will be discussed later in this section. A lower bound estimate of the brace compressive strength, as observed at large inelastic deformation levels, is therefore needed. Degradation of the brace compression capacity under cyclic loading has been examined in past research (e.g., Jain et al., 1980; Lee and Bruneau, 2005; Goggins et al., 2006). In Fig. 19.14*b*, empirical expressions proposed in codes are compared with the AISC Specification (AISC 2005c) column design curve and brace compressive strength measurements at $5\delta_y$ reported by Tremblay (2002). The AIJ equation (Architectural Institute of Japan guidelines for limit-state design of steel structures; see Nakashima and Wakabayashi, 1992; Tada et al., 2003) is based on compression capacities observed at an axial compressive strain of 0.01 ($5.7\delta_y$ for 350-MPa steel), after which it remains nearly constant. The reduction factor proposed by Remennikov and Walpole (1998a) to approximate brace compression capacities at a ductility of 5.0 has been adopted in NZS3404. Significant loss in strength is observed in the common slenderness range $\lambda = 0.5$ to $\lambda = 1.5$. In that interval, all predictive curves are in good agreement. For stocky braces, the simplified expressions in AISC and CSA underestimate the test data. In CSA-S16, C_u' does not vary with slenderness and therefore exceeds the other values for slender braces. In EC8, the recommended value is $0.3AF_y$, which is appropriate for low slenderness but overestimates C_u' for intermediate and high brace slenderness.

Built-up members are commonly used for bracing members. The interaction between global and individual component-buckling modes must be considered when determining the compressive strength associated with overall buckling, which develops shear forces in the stitch connectors. In cyclic tests on double-channel and double-angle braces buckling in this mode (Astaneh-Asl et al., 1985; Goel and Xu, 1989; Aslani and Goel 1991, 1992a), individual components were found to locally bend between the connectors in the plastic hinge region that forms in the brace after global buckling. The severe local curvature demand on the individual components resulted in local buckling and premature fracture of the braces.

This individual inelastic behavior could be minimized by using stitch spacing much closer than the value required to prevent single-component buckling. Aslani and Goel (1992b) developed equations to predict the shear forces in the connectors at first buckling and when the brace carries compression in the postbuckling range. The cyclic tests also showed that the pronounced curvature of the brace in the buckled region could induce tension yielding of the component located on the convex (exterior) side of the brace. When reloading the brace in tension, the elongated segment of that component could not carry tension load and, as illustrated in Fig. 19.13*f*, a shear force approximately equal to half the yield tensile strength of the individual components had to be transferred by the stitches located on either side of the brace midpoint. Conversely, tests by Astaneh-Asl and Goel (1984) on in-plane buckling of double-angle braces with single gusset plate end connections showed no sign of significant shear forces in the stitches, suggesting that nominal stitches spaced such that single-angle buckling is prevented would be sufficient (Astaneh-Asl et al., 1986).

These observations served as a basis for current design and detailing rules for built-up braces. For instance, AISC seismic provisions limit the stitch spacing such that the slenderness of the individual element does not exceed 40% of the brace slenderness associated with the brace governing buckling mode. If that mode does not induce shear forces in the connectors, the limit is relaxed to 75%. If brace buckling causes shear in the stitch connectors, the total shear capacity of the stitches over the brace length should be sufficient to develop the yield tensile strength of one of the components being joined.

Recent cyclic tests by Lee and Bruneau (2008) on built-up members made of four angles connected with lacing and tie plates revealed a more complex response that included global, local (angles), and lacing buckling. Local and lacing buckling occurred at the brace hinge location as well as at other unexpected locations along the brace length. Lacing buckling had a detrimental impact on the brace compressive strength due to the reduction of the brace cross-section moment of inertia and flexural strength, and fracture of the angles was precipitated by local and lacing buckling. The influence of overall and cross-section slenderness on the member inelastic response was examined by the authors, but additional detailing requirements are needed before such members can be used safely for inelastic seismic response.

Good performance of the brace connections is critical for a satisfactory inelastic response behavior of the brace. As was previously discussed, attention must be given to the potential for local instability in brace connections at the intersection of X-braces. End connections must be designed to carry the expected brace axial strength in tension T_u and compression C_u. In addition, they must also have sufficient flexural capacity to constrain plastic hinging in the braces, next to the connections. That flexural strength verification can be performed without consideration of axial load because the maximum rotational demand occurs at maximum brace negative axial deformations, when the axial forces are small and hence can be neglected (see Figs. 19.11*a* and 19.12).

Due to their simplicity, single gusset plate connections have been used extensively in practice. Stability under compression is an issue for such gusset plates, especially when the buckling length is increased to create a gusset hinge detail to accommodate the end rotation associated with out-of-plane buckling of the braces in the gusset rather than forcing plastic hinging in the braces (Figs. 19.13*a* and 19.13*b*). Information on the seismic design and detailing of gusset plates can be found in Astaneh-Asl (1998) and Astaneh-Asl et al. (2006). Chambers and Ernst (2004) also reviewed past research on braced-frame gusset plates. They concluded that a comprehensive method is still needed to adequately predict the complex buckling response of gusset plates, which should include an interaction with framing members.

Until such a method becomes available, Chambers and Ernst (2004) recommended an approach proposed by Thornton (1984). In this method, the gusset plate is modeled as an equivalent column having a width equal to the Whitmore section assuming 30° stress dispersion on each side of the outermost bolt lines or welds (w_g in Fig. 19.11*g*) and an effective buckling length equal to 0.65 times the average of the three lengths l_1, l_2 and l_3 shown in Fig. 19.13*g*. Chambers and Ernst indicated that the approach had been shown to provide a wide variation of results, but all are consistently conservative when compared to test results. Nast et al. (1999) found that the method could predict well the gusset-buckling strength obtained in their tests. Astaneh-Asl (1998) and Cochran and Honeck (2004) recommended that the effective-length factor of the equivalent column be conservatively taken equal to 1.2 instead of 0.65 to account for the possibility of out-of-plane movement of the bracing member ends. They also recommended that edge buckling under cyclic loading be prevented by adding a stiffener when the length of the free edge (l_3 in Fig. 19.11*g*) exceeds $0.75t_g\sqrt{E/F_y}$. Dowswell (2006) compared the predictions from Thornton's model to the values obtained from tests or finite element analysis for various gusset plate configurations. For extended gusset plates connected to the beams and columns at beam-column joints (Fig. 19.13*a*), Dowswell proposed an effective length equal to $0.6l_2$.

Large-scale cyclic frame test programs have been recently initiated to examine the overall response of braced steel frames. In tests on a two-story frame, Uriz (2005) observed that gusset plates welded to the beam and the column at beam-to-column joints are subjected to diagonal transverse compression and tension due to the deformations of the frame subjected to lateral displacements. The compression due to frame action occurs when the connected brace is in tension, but it was sufficient in the test to induce out-of-plane buckling of the gusset. The presence of the gusset plates can also have a significant impact on the behavior of the beam-column joints. Although the test frame was designed assuming pin-ended members, the moment demand in the connection as built resulted in premature fracture of the column in one of the tests. Lehman et al. (2008) carried out a series of tests on a single-story diagonal braced frame with different gusset plate designs including those illustrated in Figs. 19.13*a* and 19.13*b*. When using the elliptical end offset detail with thinner gusset plates with greater flexibility, deformability and yielding resulted in an increased lateral deformation capacity of the frame. Both details were also examined in cyclic tests conducted on a two-story X-braced frame (Lin et al., 2009). Observations during these experiments and preliminary results indicate that both details performed satisfactorily. More detailed information on this test program should be forthcoming in the literature.

Local Brace Buckling Past tests have shown that the curvature experienced in brace plastic hinges upon buckling can be sufficient to trigger local buckling of the cross section, which contributes further to the reduction in the brace compressive strength at large negative axial deformations. More importantly, large strains develop in the local buckles that can eventually initiate cracks in the steel material and, upon reloading in tension, brace fracture.

Experiments revealed that this behavior could be controlled by limiting the brace cross-section slenderness. In cyclic tests on angle bracing, Goel and El-Tayem (1986) observed that the braces with legs not meeting the width-to-thickness requirement for plastic design, that is, $b/t \leq 0.30\sqrt{E/F_y}$, suffered early severe local buckling and fracture. They also noted that members with a smaller global slenderness ratio experienced more severe local buckling than the more slender ones. From cyclic tests on back-to-back double-angle bracing members, Astaneh et al. (1986) showed that the greatest demand is imposed on the outstanding legs when brace buckling occurs out-of-plane and on the back-to-back legs for in-plane buckling braces. They recommended that the b/t ratio of the critical legs meet plastic design requirements and suggested that the limit could be extended to $0.44\sqrt{E/F_y}$ for the other legs. I-shaped braces generally buckle about their minor axis, which leads to local buckling and subsequent fracture developing in the flanges (Black et al., 1980; Fell et al., 2009; Lin et al., 2009). Codes in the United States, Canada, and New Zealand have adopted the plastic limit $0.30\sqrt{E/F_y}$ for legs of angles and flanges of I-shaped brace members when high-ductility capacity is required. In CSA-S16, the limit linearly increases up to $0.38\sqrt{E/F_y}$ when the brace KL/r is between 100 and 200, recognizing the lesser inelastic strain demand imposed on more slender braces when buckling, and a limit of $0.44\sqrt{E/F_y}$ can be used for the back-to-back legs of double-angle bracing buckling out of the plane of symmetry, as suggested by Astaneh-Asl et al. (1986). Aslani and Goel (1992a) found that the energy dissipation capacity and fracture life response of angle bracing could be enhanced by forming box sections connected to single-gusset plates, recognizing that such braces exhibit higher compressive strength and are less prone to local buckling.

In view of their efficiency in carrying compressive loads and aesthetic appearance, structural tubing has become a very popular choice for bracing members. Fracture of RHS has been observed, however, in past earthquakes (e.g., Bertero et al., 1994; Tremblay et al., 1995) as well as in several past cyclic test programs (Gugerli, 1982; Foutch et al., 1987; Lee and Goel, 1987; Liu and Goel, 1988; Bertero et al., 1989; Fukuta et al., 1989; Walpole, 1996; Pons, 1997; Zhao et al., 2002; Shaback and Brown, 2003; Tremblay et al., 2003; Uriz, 2005; Goggins et al., 2006; Han et al., 2007; Lehman et al., 2008; Tremblay et al., 2008a; Fell et al., 2009; Lin et al., 2009). Local buckling develops inward on the concave side of the buckled brace, inducing severe localized curvature and strain demand in the corner regions of the cross section (see Fig. 19.15*a*). The key parameters influencing the occurrence of local buckling and subsequent fracture are the cross-section slenderness and brace effective slenderness. The effects of these two parameters can be seen in Fig. 19.15*b*, where the more slender and compact RHS braces generally exhibit a longer fracture life (μ_f is the sum of the maximum ductility in tension and compression, $\mu_c + \mu_t$, reached before occurrence of fracture; see Fig. 19.11*a*). The steel yield strength and the characteristics of the loading history are other influential parameters on brace local buckling.

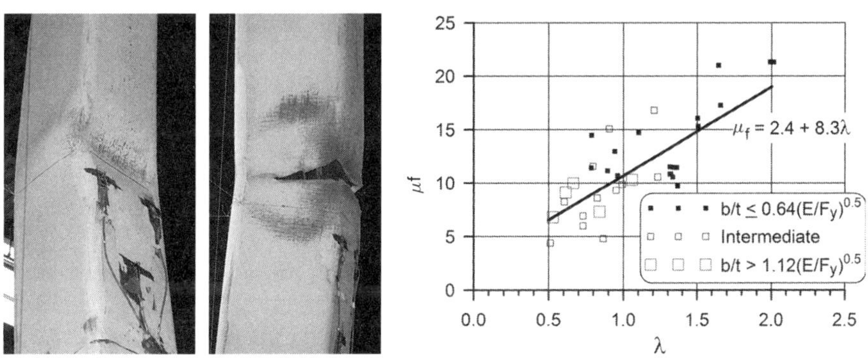

(*a*) (*b*)

FIGURE 19.15 (*a*) Local buckling and fracture of an RHS brace; (*b*) influence of brace slenderness on ductility at fracture (adapted from Tremblay, 2002).

Quantification of these effects represents a formidable task as local buckling and consequent fracture are complex phenomena resulting from the interaction of several factors. One possible approach to the prediction of RHS brace fracture life therefore consists in relying on empirical expressions developed on the basis of test results. From cyclic tests on tube members made of ASTM-A500 grade B steel, Tang and Goel (1989) proposed an empirical criterion to predict the fracture life of rectangular bracing members subjected to earthquake loading:

$$N_f = 262 \frac{(b/d)\,(KL/r)}{\left[(b-2t)/t\right]^2} \geq 262 \frac{(b/d)\,60}{\left[(b-2t)/t\right]^2} \tag{19.16}$$

where N_f is the number of equivalent cycles up to fracture, b and d are the widths of the flange and web of the tube cross section ($b \geq d$), and t is the wall thickness. When using this criterion, the brace axial deformation time history that is obtained from the analysis of the structure must be converted into numbers of equivalent cycles, as described in the references. This fracture life parameter decreases linearly with brace overall slenderness and to the square of the flange wall slenderness. For brace slenderness smaller than 60, only the width-to-thickness ratio has an impact on the fracture life. Good agreement was obtained by Tang and Goel when using Eq. 19.16 to predict the brace fracture response of a six-story braced test frame. Other similar models have been developed in subsequent research. For instance, based on the work by Lee and Goel (1987) and Hassan and Goel (1991), Shaback and Brown (2003) proposed the following predictive model:

$$\Delta_f = 0.065 \frac{\left(350/F_y\right)^{-3.5}\,(KL/r)^2}{\left[(b-2t)/t\right]^{1.2}} \left(\frac{4\,(b/d)-0.5}{5}\right)^{0.55} \quad \text{with } F_y \text{ in MPa} \tag{19.17}$$

In this expression, Δ_f is the cumulative axial deformation in tension and compression up to brace fracture. In calculating Δ_f demand from a time history analysis, more weight is given to the tensile deformations as more damage was believed to be created when straightening and stretching the brace. Compared to Eq. 19.16, brace slenderness in this model has a greater impact on brace fracture. It should be noted that *KL*/*r* is to be taken equal to 70 when it is less than 70. In the equation, the steel yield strength has a negative impact (reduces) brace fracture, reflecting the fact that flat elements made of higher yield strength material are more prone to local buckling. Goggins et al. (2006) also observed that the brace ductility at fracture decreased when increasing F_y.

The anticipated brace deformation history must be known in advance when using Eqs. 19.16 or 19.17, which may be problematic in design. Other empirical expressions have therefore been developed to predict the brace deformation capacity before fracture, irrespective of the loading history. Simple linear relationships, such as the one given in Fig. 19.15*b*, can be used to estimate the brace ductility capacity (Tremblay, 2002; Goggins et al., 2006). Adopting an approach similar to the one that led to the development of Eq. 19.7, Tremblay et al. (2003) proposed an expression to determine the brace plastic hinge rotation causing fracture θ_f that is a function of cross section and global slenderness ratios:

$$\theta_f = 0.091 \left\{ \frac{(b-4t)}{t} \frac{(d-4t)}{t} \right\}^{-0.1} \left(\frac{KL}{r} \right)^{0.3} \tag{19.18}$$

This rotation is intended to be compared to the plastic rotation associated with a given brace ductility, which can be determined from Eq. 19.15.

Fewer cyclic test programs have been conducted on circular hollow sections (CHS) used as bracing members. Local buckling of CHS also occurs inward, on the side subjected to the highest combined axial and flexural compressive stresses upon brace buckling. The shape of the cross section then gradually changes to an oval and eventually locally flattens to form a pronounced kink where high strains develop and tend to lead to the formation of cracks. Fracture is observed when reloading the brace in tension (Tremblay et al., 2008a). In a cyclic test program by Elchalakani et al. (2003), the braces were relatively stocky (λ equals 0.34 to 0.57) and the diameter-to-thickness ratio, D/t, varied from $0.044E/F_y$ to $0.11E/F_y$. The results indicated that the stockier braces generally had superior ductility capacity, including those with larger cross-section slenderness. This can be attributed to the difference observed in the local buckling mode and its possible effects on strain level: smooth kink for the more slender specimens compared to a multilobe diamond shape for the less slender ones. The influence of the loading sequence was also investigated in the tests. Stepwise incremented loading protocols were used with one and three cycles per displacement increment for two pairs of braces. In both cases, the ductility of the braces subjected to three cycles per increment was lower than that observed for an identical brace subjected to only one cycle per displacement increment. Single large inelastic cycles, as can be expected in structures subjected to near-field earthquakes, were also found to be capable of producing brace fracture.

Current seismic code provisions only prescribe cross-section slenderness limits to achieve the ductility intended in design. For RHS braces, the limit in the AISC (2005a) seismic provisions is $0.64\sqrt{E/F_y}$, which corresponds to two-thirds of the AISC plastic compactness limit. In CSA-S16, the limit is equal to $0.74\sqrt{E/F_y}$ for braces with $KL/r \leq 100$ but can be linearly increased with slenderness up to $0.93\sqrt{E/F_y}$ for $KL/r = 200$. Higher plastic design limits apply in NZS3404 with $0.95\sqrt{E/F_y}$, and EC8 at $1.13\sqrt{E/F_y}$. Variations also exist in D/t limits imposed for circular tubes: $0.044E/F_y$ in the United States and NZS (corresponding to the AISC plastic compactness requirement), $0.050E/F_y$ in CSA-S16, and $0.059E/F_y$ in EC8. In CSA-S16, the limit can be increased by up to 30% when slenderness is increased between 100 and 200.

Recent advances in numerical simulation techniques have expanded the available database on brace local buckling and fracture to examine the consequences of brace fracture on building seismic performance. In cyclic tests on angles, Park et al. (1996) noted that the maximum strains measured at crack initiation at critical locations in the plastic hinge varied between 0.25 and 0.40, irrespective of the specimen overall slenderness or b/t ratio. The experimental strain demand could be accurately reproduced using detailed nonlinear finite element analysis models accounting for local buckling behavior; this suggests that tracking the strain history could be used to predict the fracture life of bracing members. Using this technique, Haddad (2004) concluded that first cracking occurs in RHS braces at a cumulative plastic strain of 7.5. Yoo et al. (2008) also used the cumulative plastic strain from finite element analysis as an indicator of fracture to compare the influence of different brace connection designs on brace fracture life. Kanvinde and Deierlein (2007) developed a methodology to predict ductile crack initiation in structural steel members based on a micromechanics-based fracture model. Good agreement was found by Fell et al. (2006) when applying that model for the prediction of RHS brace fracture through refined finite element techniques using continuum three-dimensional brick elements and multiaxial plasticity. Huang and Mahin (2008) implemented a cyclic damage plasticity model based on simplified continuum damage mechanisms to predict crack initiation and propagation in a finite element analysis simulation tool. Brace local buckling and subsequent fracture as observed in a two-story concentrically braced steel frame test could be accurately reproduced using that methodology.

Simpler numerical tools have also been developed that could be more efficiently used in the analysis of complete structures. For instance, cycle counting techniques have been implemented in nonlinear beam-column elements with distributed plasticity and fiber discretization of the cross section to predict brace fracture (Uriz and Mahin, 2004; Uriz et al., 2008). Daravan and Far (2009) proposed a simple beam-column element with concentrated plastic hinge with linear cumulative damage theory. Although local buckling effects cannot be explicitly reproduced in such fiber or plastic-hinge-based representations, properly calibrated models can predict well the measured fracture response and represent effective tools when examining the response of building structures. Sabelli (2001) and Goel and Chao (2008) used the fracture model of Eq. 19.16 to evaluate the seismic performance of CBF structures with RHS braces designed according to current U.S. practice for high-seismic regions. Similar studies were performed using the fiber-based model and a probabilistic assessment approach (Uriz and Mahin, 2004; Salmanpour and Arbabi, 2008). These numerical studies revealed that brace fracture can occur in code-compliant braced steel frames, suggesting that further improvement may be needed to achieve adequate seismic performance.

Experience gained in past test programs showed that the seismic behavior of RHS brace members can be enhanced by further controlling cross-section slenderness ratios or by encouraging the use of more slender braces. Further performance assessment studies are needed before more definite limits can be proposed for these two parameters. Meanwhile, when large ductility demand is expected, designers may consider using double smaller tube sections stitched together and connected to single-plate gussets, as suggested in the commentary to the AISC seismic provisions. Such members would typically buckle in-plane, allowing the use of a reduced K-factor in design, and the compactness limits are more easily met when using smaller sections.

RHS braces can also be filled with concrete to delay local buckling or force local buckling to occur outward rather than inward. In either case, the severity of strain concentration at the corners of the cross section is reduced and the brace fracture life is increased (Lee and Goel, 1987; Liu and Goel, 1988; Zhao et al., 2002; Broderick et al., 2005; Fell et al., 2009). Filled tubes also maintain a higher plastic moment capacity under cyclic loading and, thereby, offer higher compressive strength in the postbuckling range and greater energy dissipation capacity. The contribution of the concrete fill to the brace-buckling strength can be accounted for in brace design, as proposed by Liu and Goel (1988), resulting in smaller steel tubes and reduced brace tensile yield strength for consideration in capacity design verifications. It is noted, however, that brace tensile capacity T_u of composite brace specimens in past tests was larger than that of their hollow counterparts (Lee and Goel, 1987; Broderick et al., 2005).

Locally reinforcing tube walls in the hinge region is another possible solution to enhance the seismic performance of RHS braces. Past tests (Liu and Goel, 1987; Fell et al., 2009) showed, however, that the plastic hinge in such braces tends to form at one end of the reinforcement, in the unreinforced cross section, which can cause more pronounced localized strain demand. Caution should therefore be exercised if this technique is used before detailing rules are proposed from experiments showing evidence of improvement.

Research has been initiated on tubular braces made of stainless steels as this material exhibits much higher strain-hardening response than carbon steels and can therefore contribute to controlling local buckling (e.g., DiSarno et al., 2008; Nip et al., 2008).

Another option for the designer is to use an alternative cross-section type that can provide for a greater resistance to fracture. Although less effective in compression, I-shaped brace members could represent a good choice in this context as they generally have demonstrated better ductility capacity than RHS members.

Column and Beam Buckling Beams and columns must be designed to carry gravity loads together with lateral load effects corresponding to the development of brace buckling in compression and brace yielding in tension. For the columns, this means that the vertical components of the brace forces acting at levels above the level under consideration must be accounted for, as illustrated in Fig. 19.16*a* with the tensile yield strength T_u and buckling strength of the braces C_u considered. Figure 19.16*b* illustrates the situation of braces connecting on both sides of an interior column. In this case, the braces acting in compression reduce the compression load in the column and the critical condition exists after the compression-acting braces have experienced significant inelastic deformations and exhibit their reduced postbuckling strength C'_u. The addition to the gravity loads of the net vertical component of the forces T_u and C'_u along the column height and will result in the maximum compression load in the column. In such situations, the compression braces could be omitted for simplicity, which would result in conservative column design. Similar scenarios for the determination of maximum anticipated axial compression loads in beams are shown in Figs. 19.16*c* and 19.16*d*. For V or

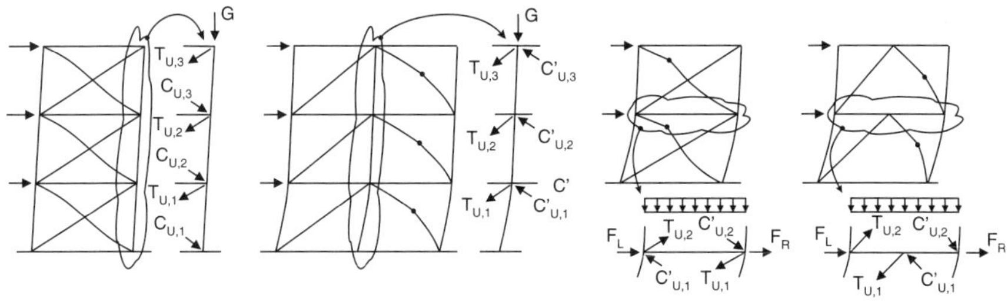

(*a*) (*b*) (*c*) (*d*)

FIGURE 19.16 Determination of the force demand on (*a*) columns in X-bracing; (*b*) interior columns; (*c*) beams in X-bracing; and (*d*) beams in chevron bracing.

inverted-V (chevron) bracing, the net unbalanced vertical load resulting from the brace loads T_u and C'_u at the beam midspan also induce vertical shear forces and bending moments in the beams (Khatib et al., 1988; Remennikov and Walpole, 1998b; Tremblay and Robert, 2000; Uriz, 2005).

When performing the calculation of the column axial loads, it is recognized that the probability that all braces reach their capacity at the same time reduces as the number of levels considered in the summation increases. Based on the work by Khatib et al. (1988) for zipper columns in chevron bracing, Redwood and Channagiri (1991) suggested that the axial loads in exterior columns of X-bracing or two-story X-bracing be determined by considering the loads fed by the first two braces connected immediately above the level under consideration plus the square root of the sum of the squares (SRSS) of the loads fed by the braces in the levels above. This method also accounts for the fact that the period of time during which the tension- and the compression-acting braces at a given level reach their peak strengths of C_u and T_u is very short as the compression brace loses its strength after buckling. A recent study by Lacerte and Tremblay (2006) indicated that the SRSS method may underestimate the column force demand in structures of up to 12 stories designed in accordance with the latest code provisions. In NZS3404, the braces at the level under consideration are considered to reach their actual capacities, whereas all braces above are assumed to reach their factored design strength. Further research is needed to develop design guidelines on this aspect. In chevron bracing, only the braces reaching their buckling strength induce compression in the exterior columns. For this case, it has been suggested to sum all brace forces C_u, without statistical reductions (Khatib et al., 1988; Tremblay and Robert, 2001). For this configuration, vertical shears induced by the unbalanced brace forces at the beam midspan (Fig. 19.16d) must be added to the contribution of the braces along the exterior columns.

In design, it may be convenient to multiply the seismic load effects obtained from elastic analysis by a single factor to obtain the maximum anticipated force demand on members and connections including system overstrength. This approach is permitted in current U.S. codes where amplified seismic loads defined as the horizontal seismic load effects multiplied by the system overstrength factor Ω_0 can be used for the design of columns. As described by Richards (2009), this approach fails in capturing local effects such as high individual brace overstrength or force redistribution after brace buckling in CBFs, which can lead to unconservative column loads in low-rise frames or in the upper part of taller structures. The same situation applies in beam design. For instance, seismic loads in elastic analysis would give negligible axial loads in beams of X-bracing systems. Amplifying this result will not produce anything close to the expected load resulting from the loading condition in Fig. 19.16*c*. Proper column and beam design therefore requires that all applicable brace force scenarios be examined to ensure that structural integrity is maintained while inelastic response develops in the braces. It must also be noted that column gravity loads from elastic analysis may not be appropriate for this calculation as the portion of the gravity loads that may be resisted by the braces as per elastic analysis will be transferred to the columns upon brace yielding and buckling.

As illustrated in Fig. 19.16b, even if columns of multistory CBFs are assumed to resist axial loads only, bending moments will develop during an earthquake due to differences between interstory drifts in adjacent levels and moment frame action at large deformations (Tremblay, 2000; Sabelli, 2001; MacRae et al., 2004; Uriz, 2005; Richards, 2009). This flexural demand cannot be obtained from elastic analysis and its magnitude can vary significantly depending on the amplitude and the vertical distribution of the inelastic response in the structure. In high-seismic regions, studies of code-designed CBFs by Sabelli (2001) and Richards (2009) showed differences in interstory drifts of 0.005 to 0.02 rad. This demand may lead to inelastic response in the columns, depending on their deformed shapes, lengths, and cross-section properties. It may also coexist with maximum compression forces when the frame is subjected to large deflections. Tests by Newell and Uang (2008) reported earlier showed that this combined demand can be accommodated by heavy I-shaped columns satisfying the AISC seismic cross-sectional compactness requirements. These limits apply to CBF columns in AISC seismic provisions and, hence, columns are permitted to be designed without consideration of the expected flexural demand. In this context, it may be advisable not to select columns that would be prone to flexural–torsional buckling. A different approach is used in CSA-S16 in which columns must be designed as beam-columns with an additional bending moment equal to $0.20M_p$. This bending moment corresponds to the average demand expected in seismic regions of Canada (Tremblay, 2000). Columns that are part of the gravity system undergo the same lateral deformations as the bracing bent columns and, thereby, the same bending demand due to variation in interstory drifts. These columns are designed for gravity axial loads that are larger than those considered present during the design earthquake, and this difference is deemed sufficient to accommodate the flexural demand without additional verifications. It is noted that column continuity in multistory braced frames has been identified to contribute to more uniform distribution of the inelastic demand over the height of braced frames, and this aspect is discussed in a later section on global stability. In K-bracing with compression and tension braces connecting to columns between floors, the braces will impose an unbalanced lateral load after buckling has occurred, similar to the situation for beams in chevron bracing (Fig. 19.16*d*). This represents an undesirable stability condition and such bracing configuration is not permitted when high-ductility demand is expected.

19.2.4 Buckling Restrained Braced Frames (BRBFs)

Buckling restrained braced (BRB) frames are concentrically braced frames that are built with braces that are specially designed to yield in both tension and compression without buckling. A typical BRB member is illustrated in Fig. 19.17. A steel core plate with a reduced cross section is inserted in a steel tube. The volume between the tube and the core is filled with mortar or concrete. At both ends, the core plate extends outside of the tube and is connected to the framework. The core is covered with an unbonding material prior to filling the tube such that the core can axially deform freely and, hence, axial loading on the brace is essentially resisted by the core. The unsupported core projections at both ends of the brace are typically made larger in size and/or stiffened to avoid buckling in compression. Under seismic loading, yielding is expected to take place in both compression and tension in the reduced segment of the core, L_c in Fig. 19.17. A small gap must be left between the concrete fill and the core to allow for transverse expansion of the core upon compression yielding.

Main stability issues for BRB frames include local buckling (ripping) of the brace core, overall brace buckling, buckling at the end connections, and buckling of the beams and columns. Design for stability of gusset plates, beams, and columns is very much the same as that for CBFs and only the main differences are discussed herein. In most countries, BRB members are proprietary products that are developed and supplied by specialized manufacturers to comply with project-specific performance requirements determined by the design engineer (e.g., minimum and maximum axial strength, cyclic deformation capacities, minimum axial stiffness, etc.). Local buckling and overall brace stability are therefore aspects that are generally addressed by the suppliers using techniques and methods specially developed for their products. Because a large variety of BRB systems have been proposed and used in the last two decades, only the main concepts and recent research findings that are common to most systems will be introduced using the simple configuration shown in Fig. 19.17. Additional information on the system and on seismic design

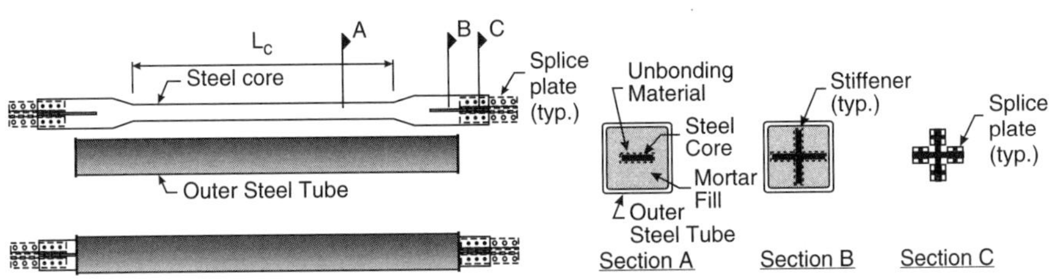

FIGURE 19.17 Buckling restrained bracing member.

procedures can be found in Uang and Nakashima (2004), Sabelli (2004), López and Sabelli (2004), and Xie (2005). Among the codes examined, buckling restrained braced frames have been introduced only in the AISC seismic provisions and only these provisions are referenced in the discussion.

Brace Local and Global Buckling The axial load deformation P–δ hysteretic response of a buckling restrained brace is illustrated in Fig. 19.18*a*. Figure 19.18*b* shows a core of rectangular cross section of a BRB member after testing. Evidence of local buckling and friction response between the core and the buckling restraining mechanism can be observed. Local buckling of the core as constrained by the buckling restraining mechanism is schematically illustrated in Fig. 19.18*c*. The inelastic cyclic response is stable and nearly symmetrical in tension and compression and shows marked strain-hardening behavior. The peak tension and compression forces are measured in cyclic tests at maximum anticipated deformations. The force $T_{\max}$ is defined as ωP_{yc}, where P_{yc} is the yield load of the core ($= A_c F_{yc}$) and ω is a factor that essentially accounts for strain hardening ($\omega \approx 1.2$ to 1.4, depending on the steel material and the imposed deformation history). The load $C_{\max}$ equals $\beta\omega P_{yc}$, where β mainly accounts for the longitudinal friction that develops between the core and the restraining system when pushing the locally buckled core into the brace ($\beta \approx 1.05$ to 1.2).

The minimum stiffness per unit length $\gamma_{\min}$ of the encasing material can be obtained by comparing the maximum anticipated compression force to the buckling load of a beam on an elastic foundation: $P_{cr} = 2\sqrt{\gamma_{\min} E_t I_c}$, where E_t is the

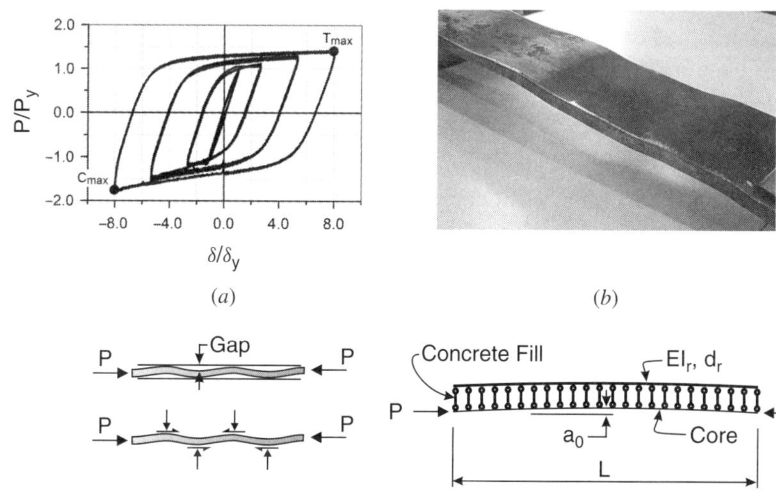

FIGURE 19.18 Buckling restrained bracing member: (*a*) hysteretic response; (*b*) evidence of core local buckling after cyclic testing; (*c*) core local buckling; and (*d*) overall brace buckling.

tangent modulus of steel ($E_t \approx 0.02E$) and I_c is the moment of inertia of the core in the direction of interest (Black et al., 2004). The stiffness of the encasing material must be reduced to account for the local transverse flexibility of the member containing the material (e.g., walls of tubes) and of the unbonding material used between the core and the restraining system, if any. The number of contact points between the core and the restraining system increases (wavelength reduces) as the axial compression load or deformation is increased. The magnitude of the contact forces or pressures between the core and the restraining system, normal to the longitudinal brace axis, depends on a number of factors including the brace axial load (or axial deformation), the spacing of the local buckles (local mode shape), the initial gap left between the core and the restraining mechanism, and the actual local stiffness of the restraining system. The amplitude of the friction forces depends on the contact forces and conditions at the interface. Low-friction, unbonding material can be used at the interface to ease sliding of the core. Only limited information has been published on this complex interaction between the brace core and the restraining mechanism. Furthermore, the available data often pertain to particular BRB systems or details and the findings may or may not apply directly to other designs. Iwata et al. (2000) conducted tests on BRB members with the same elastic buckling-to-yield-strength ratio, P_{er}/P_{yc} (see definition below) but different buckling restraining systems. The absence of an unbonding material led to less uniform strain demand and tensile fracture of the core. Failure of the restraining mechanism due to the outward forces induced by core local buckling was observed. Usami et al. (2003) developed a model to predict the buckling response of a steel core incorporating the flexibility of the unbonding material under monotonic and cyclic loading conditions. Iwata and Murai (2006) examined the influence of the core thickness and core width-to-thickness ratio together with the P_{er}/P_{yc} ratio on the local buckling response of the core and the cumulative ductility and the normalized plastic strain energy capacity of the brace. No clear trend could be found for the influence of the core geometry except that greater deformation capacity was achieved by specimens having the thickest core or the core with the smallest width-to-thickness ratio when compared to specimens with otherwise identical parameters. The positive impact of increasing the thickness of square restrainer tubes on the buckling behavior of the core was studied analytically and experimentally by Matsui et al. (2008) and through testing by Ju et al. (2009). Tests by Ding et al. (2009) showed that the normal forces are reduced when keeping the gap thickness to a minimum. Wei and Tsai (2008) carried out tests and developed equations to characterize the interaction between strength, stiffness, and thickness of the concrete fill material and the width-to-thickness ratio of the core plate in concrete-filled BRB members. It was concluded that the concrete thickness must be increased if the strength of the concrete is reduced or the width-to-thickness ratio of the core is increased.

The contribution of the concrete fill is typically ignored when examining the overall buckling response of BRB members. Assuming that the axial load in the restraining system is low enough to avoid inelastic response, the overall buckling strength of the system, P_{er}, is given by

$$P_{er} = \frac{\pi^2 E I_r}{L^2} \tag{19.19}$$

in which I_r is the moment of inertia of the restraining system and L is the length of the brace. In past studies, the latter has generally been taken equal to the length of the restraining mechanism (Fig. 19.18*d*) or the overall brace dimensions including end connections. As a minimum, the load P_{er} must be greater than the load $C_{\max}$; in addition, imperfections and friction response will induce axial stresses in the restraining system that must be limited to prevent yielding. Neglecting the contribution of the concrete fill, the maximum bending stress σ_{br} due to an initial out-of-straightness a_o under the load $C_{\max}$ in a restraining system with symmetrical cross-section depth d_r and moment of inertia I_r can be obtained from (Inoue et al., 2004; Wada and Nakashima, 2004)

$$\sigma_{br} = M_r \frac{d_r}{2I_r} \qquad \text{with} \quad M_r = \frac{C_{\max} a_o}{1 - C_{\max}/P_{er}} \tag{19.20}$$

With $C_{\max} = \beta \omega P_{yc}$, Eq. 19.20 may be rewritten as

$$\frac{P_{er}}{P_{yc}} \geq \beta \omega \left[1 + \frac{\pi^2 E}{2\sigma_{br}} \left(\frac{a_o}{L} \right) \left(\frac{d_r}{L} \right) \right] \tag{19.21}$$

If the axial stress in the restraining system needs to be limited to ϕF_{yr}, the stress σ_{br} in Eq. 19.21 could be taken equal to $F_{yr}[\phi - \omega(\beta - 1)P_{yc}/P_{yr}]$, where P_{yr} is the yield axial load of the restraining system. The second term in the brackets accounts for the additional axial stress induced by frictional response. For $F_{yr} =$ 300 MPa, $\phi = 0.9$, $\omega = 1.2$, $\beta = 1.1$, $a_o/L = 0.001$, $d_r/L = 0.03$, $P_{yc}/P_{yr} = 0.75$, and $\sigma_{br} = 0.81 F_{yr}$, a minimum ratio $P_{er}/P_{yc} = 1.48$ is required to achieve the desired behavior. Usami et al. (2008) suggested that the influence of the thickness of the gap and axial load eccentricity could also be considered by replacing a_o in Eq. 19.21 by the sum of these geometrical dimensions. Early tests by Watanabe et al. (1988) showed the influence of the P_{er}/P_{yc} ratio on the performance of BRB members and they suggested that a minimum ratio of 1.5 should be used in design.

Iwata and Murai (2006) found that the cumulative ductility and the normalized plastic strain energy capacity of BRB members up to the point where their compression resistance started to drop both increased linearly with the P_{er}/P_{yc} ratio. They suggested that these relationships could be used in design by selecting a restraining mechanism capable of achieving the expected demand from earthquakes. In their test program, one of the specimens had $P_{er}/P_{yc} < 1.0$ (0.9) and that specimen failed by global buckling early in the test. Four other specimens had the same core plate thickness but with a higher P_{er}/P_{yc} ratio (1.6, 2.3, 3.2, and 4.5) obtained by increasing the depth of the restraining mechanism. These specimens failed by local buckling of the core plate after several inelastic cycles at large axial deformations. The increase in local stiffness of the restraining system resulting from the increase in depth may also have contributed to the observed improvement in inelastic performance with P_{er}/P_{yc}.

Buckling at End Connections and of Beams and Columns In typical BRB design, the portion of the core protruding at the end of the restraining member has a larger cross section compared to that of the core. Axial yielding is then confined to the reduced segment of the core. When flat core plates are used, stiffeners can be added to prevent buckling of the protruding core portion. The portion of a core with a cruciform cross section that extends outside of the restraining mechanism may buckle in torsion when yielding in compression. Based on incremental theory of plasticity, Black et al. (2004) developed an equation for the critical stress,

$$\sigma_{cr} = \frac{E_t}{3}\left[\frac{\pi^2}{3}\frac{b_c^2}{l^2} + 1 + \frac{3F_{yc}}{E_t}\right]\frac{t_c^2}{b_c^2} \tag{19.22}$$

where t_c and b_c are the thickness and width, respectively, of each of the four flanges of the core and l is the maximum anticipated protruding length of the steel core. Unless specifically detailed to accommodate rotations, the connections of BRB members to gusset plates are generally fixed and flexural demand is expected on the bracing members when the frame deforms laterally (López et al., 2004; Tsai and Hsiao, 2008). This may cause inelastic rotation demand to concentrate in the steel core extension, just outside of the restraining system, and BRB members must be detailed to prevent this type of instability. In AISC seismic provisions, qualifying cyclic testing is required for BRB members and subassemblage tests on specimens with connections to the framing members are specified to investigate the response of BRB members including this rotation demand.

Out-of-plane instability has been observed at the gusset-to-brace connections in past BRB tests by Tsai and Hsiao (2008) and Ma et al. (2008). In the former study, gusset plates buckled in the early phase of the test program. The test was halted to allow the installation of stiffeners at the free edges of the gusset plates so that the tests could then be completed without further connection instability. The gusset plates had been initially designed for compression using the method proposed by Thornton that was described earlier for CBFs, which assumed an effective length equal to 0.65 times the maximum of the lengths l_1, l_2, l_3 (see Fig. 19.13*g*). Experimental observations (buckled shapes and buckling loads) indicated that an effective-length factor of 2.0 would have been more appropriate in the absence of free edge stiffeners. Koetaka et al. (2008) developed design criteria to prevent out-of-plane buckling of BRB members at the central connection of V and inverted-V bracing. The method accounts for the lateral stiffness of the connection as well as for the beam torsional stiffness. Cyclic tests were performed to validate the proposed procedure. For the same connection, Chou and Chen (2009) recommended that the BRB core terminations be extended beyond the gusset bending line and that stiffeners be used along gusset free edges. Chou and Chen further provided guidelines for the design of the stiffeners.

Similar to CBFs, beams and columns must be designed to carry the tributary gravity loads together with lateral load effects corresponding to the development of $C_{\max}$ and $T_{\max}$ in the braces. In V or inverted-V bracing, the beams must be designed such that the full brace capacities can be reached at small deflections. As was the case for CBFs, inelastic flexural demand is expected in multistory BRB frame columns (Sabelli et al., 2003) and it is required in the AISC seismic provisions that the stringent seismic compactness requirements be satisfied for these components. In the study by Richards (2009), the axial load demand from the amplified seismic loads, as prescribed in U.S. codes, underestimates the expected seismic demand in the uppermost floors of BRB frames. In this case, Richards indicates that this may have a limited impact on response as top-story columns typically possess excess strength capacity.

19.2.5 Eccentrically Braced Steel Frames (EBFs)

Inelastic response in EBFs is constrained to ductile link beam segments created at the beam ends, next to the columns when using single-diagonal members, or at the beam midspans when a chevron bracing configuration is employed (Fig. 19.19*a*). Yielding in the beam link may develop in the form of inelastic shear deformations in the web of the beam, plastic rotation at the ends of the links, or a combination thereof, depending on the length of the link e relative to the ratio of the link plastic moment capacity M_p to the link plastic shear capacity V_p. The first two inelastic deformation modes are illustrated in Fig. 19.4*e*. The plastic deformation imposed on link beam segments is defined by the plastic rotation γ shown in Fig. 19.19*a*. This plastic rotation depends on the link length relative to the bay width. Therefore, by selecting the link length e and a beam section with capacities M_p and V_p, the designer can control both the type of yielding mechanism (shear or flexure) and the amount of plastic deformation imposed on the links for a given inelastic story drift θ_p. Buckling of the web is more critical in shorter, shear-dominated links whereas flange buckling is more an issue in longer, flexure-critical links. Both phenomena are accentuated when γ is increased. Local buckling of the link beam segments is discussed in the following section, together with lateral stability requirements for the beam segments. Upon yielding, the link segments impose forces to the beam segments outside of the links, the bracing members, and the columns. These structural elements must be designed to withstand these forces combined with gravity loading without instability failure modes, as will also be discussed in the subsequent section. For simplicity, the description herein is limited to the chevron bracing configuration with the link at the beam midspan. Additional information on the seismic design of EBFs can be found in AISC (2006).

Buckling of Link Beam Segments In EBFs, the eccentricity introduced at the brace-to-beam connections causes the axial loads from the braces to induce high shear forces and reversed curvature bending moments in the link beam segments.

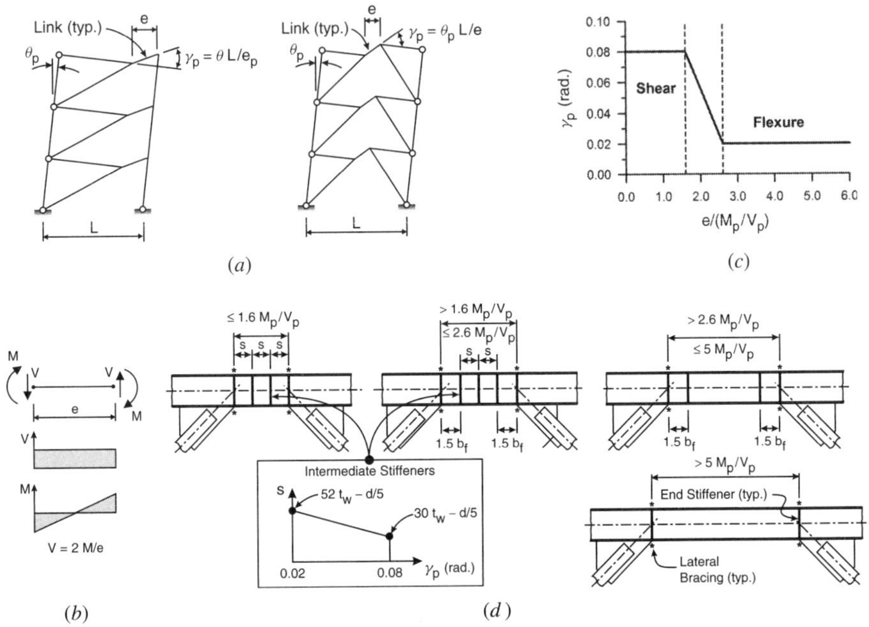

FIGURE 19.19 (*a*) Inelastic response of single-diagonal and chevron-type EBFs; (*b*) seismic-induced shear forces and bending moments in links of symmetrical chevron EBFs; (*c*) link plastic rotation capacity (AISC, 2005a); and (*d*) link stiffeners and lateral bracing requirements (AISC, 2005a).

An illustration of this loading condition imposed on a link specimen is shown in Fig. 19.20*a* (Okazaki and Engelhardt, 2007), where a short link is yielding essentially in shear. Buckling in shear is visible as this test specimen violated the minimum stiffener spacing prescribed in current code provisions, which will described later. Nevertheless, the performance of this link in Fig. 19.20*b* is very stable, with slight strength degradation after reaching a total rotation of approximately 0.10 rad. Links deforming in shear can show significant strain hardening with shear forces reaching $1.4V_p$ to $1.5V_p$ at large deformations. Longer links with beams yielding in flexure behave in a manner similar to beams in moment-resisting frame beam-to-column tests, with bending moments reaching $1.1M_p$ to $1.2M_p$ at plastic hinges.

Extensive test programs and analytical studies were performed in the 1980s and early 1990s to characterize the inelastic behavior of link beams and develop seismic design methodologies for EBF systems (e.g., Hjelmstad and Popov, 1983; Malley and Popov, 1984; Kasai and Popov, 1986a,b; Engelhardt and Popov, 1992). The links were defined as short or long, depending on the ratio M_p/V_p. For links in symmetrical chevron bracing, a point of zero moment exists at the link midlength under seismic loads and $M = V(e/2)$, as illustrated in Fig. 19.19*b*. The limiting case where shear and bending moment reach simultaneously their respective plastic

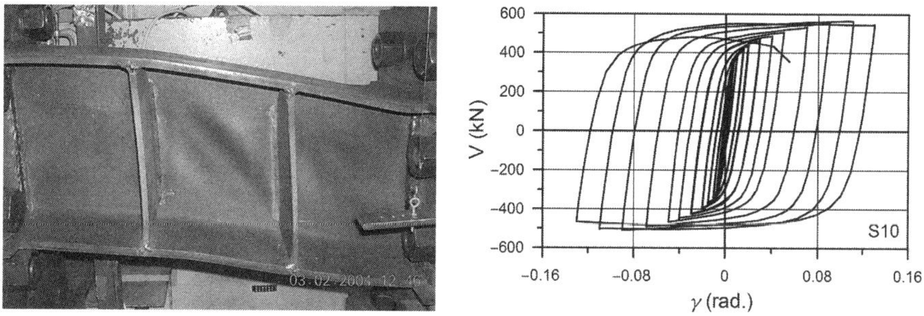

(*a*) (*b*)

FIGURE 19.20 EBF link response. (*a*) Shear yielding and buckling in EBF link. (*b*) Hysteretic response. Courtesy of T. Okazaki (University of Minnesota).

resistances V_p and M_p corresponds to $e = 2M_p/V_p$. Using the hardened capacities, short links are defined as those with $e \leq 2\ (1.2\ M_p)/(1.5\ V_p) = 1.6M_p/V_p$, whereas long links have been defined as links by a length exceeding $2.6M_p/V_p$ (AISC, CSA-S16) or $3.0\ M_p/V_p$ (EC8, NZS3404). Short links subjected to inelastic cyclic loading with increasing amplitude eventually develop web buckling, which in turn results in strength degradation. Test programs, however, demonstrated that this buckling response could be delayed by adding regularly spaced web stiffeners and this technique has been adopted in link design: in most codes, the plastic rotation capacity of short links is set equal to 0.08 rad, provided that web stiffeners are used. The spacing of these intermediate stiffeners, s in Fig. 19.19*d*, depends on the plastic rotation demand γ_p obtained from the anticipated plastic interstory drifts θ_p (γ_p and θ_p are shown Fig. 19.19*a*). For instance, in the AISC, CSA-S16, and EC8 seismic provisions, the stiffener spacing varies linearly from to $52t_w - 0.2d$ to $30t_w - 0.2d$ when γ_p increases from 0.02 to 0.08 rad, as illustrated in Fig. 19.19*d* (t_w and d are the link beam web thickness and depth, respectively). Slightly larger spacing is permitted in NZS3404. Detailing requirements for the stiffeners are also given in codes. Longer links yielding in flexure have lower rotation capacity and the limit has been set to 0.02 rad in AISC, CSA-S16, and EC8 seismic codes and to 0.03 rad in NZS3404. The rotation capacity varies linearly for intermediate links, as shown in Fig. 19.19*c*. As the shear inelastic deformation demand decreases when increasing eV_p/M_p, the need for web stiffeners becomes less critical and these stiffeners are no longer required for long links. As shown in Fig. 19.19*d*, web stiffeners, however, are still required at a distance of 1.5 times the beam flange width from the link ends for intermediate and long links, with the requirement $1.6M_p/V_p < e \leq 5.0M_p/V_p$. These stiffeners control flange buckling in the region where high rotational demand is anticipated. Flange buckling is also controlled by limiting the beam flange slenderness $b_f/2t_f$ to $0.30\sqrt{E/F_y}$.

As would be expected, the presence of an axial compressive load P in link beams reduces their shear and moment resistances. It also represents a more critical stability condition and the length of the links must be limited to prevent instability if P exceeds the link beam axial yield capacity AF_y. The limit depends on the ratio of the axial stress (P/A) to the web shear stress V/A_w in the link (with $A =$ link beam cross-sectional area and $A_w = t_w(d - 2t_f)$. When this ratio is less than 0.3, the length is limited to the short link limit $e \leq 1.6M_p/V_p$. If the ratio exceeds 0.3, this limit is reduced further by multiplying it by $[(1.15 - 0.5(P/A)/(V/A_w)]$. These limits have been adopted in the 2005 AISC seismic provisions. As in beam-columns, axial loading must be taken into account when verifying the link web slenderness against compactness requirements.

Most of the experimental research supporting the code design requirements for proper inelastic link response was carried out on I-shaped links made of ASTM A36 steel with $F_y = 248$ MPa. Additional series of tests recently completed by Okazaki et al. (2005) and Okazaki and Engelhardt (2007) showed that these code rules still apply for links made of the higher strength steel ASTM A992 ($F_y =$ 345 MPa) now commonly used in the United States for seismic applications. It is recommended to terminate the stiffener to beam web welds at some distance below the k-area of the beams to delay link web fracture in this region where steel exhibits higher strength and lower toughness due to shape-straightening processes typically employed during manufacturing. Chao et al. (2006) proposed to improve the resistance against web fracture by replacing the intermediate web stiffeners by a single horizontal stiffener placed at the link midheight between the end stiffeners. This scheme has been verified through finite element analysis simulations that include ductile fracture initiation modeling, but experimental validation has not yet been carried out to check the ability of this stiffener to control web buckling. For short links, the test programs by Okazaki et al. (2005) and Okazaki and Engelhardt (2007), together with the analytical work by Richards and Uang (2005), suggested that the limit on $b_f/2t_f$ could be extended to $0.38\sqrt{E/F_y}$ for short links and this change was included in the 2005 AISC seismic provisions.

Pairs of web stiffeners are required at both ends of the link beams. Lateral beam bracing is also required at both top and bottom flanges at the link ends. Experimental work by Hjelmstad and Lee (1989) emphasized the necessity for robust lateral bracing at the link ends. In the AISC seismic provisions, the bracing must be designed for a force of 0.06 times the link beam expected plastic moment divided by the distance between flange centroids. Minimum stiffness must also be provided in order to limit the bracing deformations to 0.45% of e under the bracing design loads. A similar minimum bracing strength level is prescribed in CSA-S16 and EC8 (0.06 times the expected flange yield strength). In NZS3404, top and bottom lateral bracing must be provided at the ends of the link, as well as along the link, at a spacing not exceeding $1.06r_y\sqrt{E/F_y}$. Such bracing must be designed for 2.5% of the axial force in the link flange at the point of restraint and minimum bracing stiffness must also be provided to limit the lateral beam flange displacement to 4 mm under the design bracing force. Tests by Ricles and Popov (1989) showed that lateral bracing of the link top flange can be provided by a composite floor, but the slab itself is not sufficient to prevent lateral–torsional buckling of the beam and independent bracing must be provided at the beam bottom flange at the link ends. Transverse beams framing from one side at both ends of the link have been found to provide effective lateral bracing. Berman and Bruneau (2008b) recently proposed and experimentally validated link beams made of built-up rectangular hollow cross sections that do not require lateral bracing. This could represent a very attractive alternative for cases when lateral bracing is difficult to provide, as would be the case along an exterior column line, or adjacent to an elevator or stairway. Built-up links also have the advantage that links can be sized to exactly match the design force demand and, hence, minimize the force demand when carrying out the capacity design verifications.

Buckling of Beams and Columns In order to achieve the intended yielding mechanism, the beams outside of the links, the braces, and the columns in EBF systems must be designed for the gravity loads plus the axial forces and bending moments induced by the links reaching their expected strain-hardened yield strength. Applications of such a capacity design procedure are given in AISC (2006) and Chao and Goel (2006). In this design phase, codes generally assume that the links will develop forces corresponding to approximately 1.25 to 1.30 times their nominal resistance determined with the expected yield strength. Although this overstrength factor is lower than the strain-hardening response observed for short links, the difference is assumed to be compensated by factors such as the fact that factored resistances are used in design. For the beam segment outside of the link, the overstrength factor in AISC seismic provisions is reduced to 1.1, which assumes that the beam strength will be increased by the presence of a floor slab and that limited yielding of that beam segment would not impair the global performance of the system. Beam segments outside of the links can be subjected to high combined axial load and bending moment demand and the verification against instability failure modes may pose difficulties in practice. The flexural demand can be reduced if short links are used instead of long links. Brace members can also be rigidly connected to the beams such that part of the bending moment induced by the links is resisted by the braces (Koboevic and Redwood, 1997). Another possible solution consists of using different shapes for the beam links and the beam segments outside of the links, as proposed by Stratan and Dubina et al. (2004) and Mansour et al. (2006).

For column design, the reduced expected link strength assuming an overstrength factor of 1.1 is also specified in AISC seismic provisions to account for the fact that several links may contribute to the column axial load at the level under consideration. As described earlier for CBFs, this approach may underestimate the demand on columns in low-rise structures and at the uppermost floors in tall multistory frames (Richards, 2009). Koboevic and Redwood (1997) proposed to use higher link overstrength for the design of the upper tier columns and this approach has been adopted in CSA-S16. Past dynamic time history analyses of multistory EBFs showed that the columns are also subjected to flexural demand, as is also observed in CBFs and BRB frames. Kasai and Han (1997) proposed that this demand be accommodated in design-bending by reducing the axial column strength by 15%. Koboevic and Redwood (1997) suggested increasing this allowance to 40% in the top two stories of multistory buildings due to higher localized story drifts resulting from higher mode response. These recommendations have been included in CSA-S16 by limiting the interaction equation for columns to 0.85 at every level except in the top two levels where the limit is set equal to 0.65.

19.3 GLOBAL SYSTEM STABILITY (P–Δ EFFECTS)

Structures must be capable of safely carrying the gravity loads they support while undergoing large inelastic lateral deformations resulting from strong earthquake ground motions. The design objective essentially consists in controlling the structure lateral displacements such that the inelastic demand on individual yielding components remains within acceptable limits and global instability of the entire structure or any part thereof is prevented. Achieving this performance objective in day-to-day practice still represents a formidable challenge in view of the complex and random nature of the seismic excitation and the difficulties in adequately predicting the dynamic nonlinear response of structures with proper consideration of all sources of geometric and material nonlinearities, damage accumulation, and consequent strength degradation and so on. In this regard, current code provisions for P–Δ effects in seismic design are generally based on simplistic models that may not properly address the actual inelastic seismic stability issues observed in building structures. Significant research efforts have been devoted in the last two decades to better understand dynamic seismic instability and the prediction of structural collapse. A vast portion of this work has been carried out within the wider context of the development of methodologies for probabilistic assessment of the seismic performance of structures and performance-based design strategies for buildings. While remarkable advances have been accomplished on various aspects related to global structural seismic response no simple and reliable design methods have been developed yet to ensure stable seismic inelastic response.

This section focuses on dynamic seismic stability. Basic concepts are first reviewed by examining P–Δ effects on the response of simple single-degree-of-freedom (SDOF) models subjected to static and dynamic seismic horizontal loading. Elastic and inelastic responses are examined. SDOF systems are useful to understand the effects of gravity on seismic response and assess the influence of key parameters, such as the hysteretic behavior of the structural system. Findings from studies on SDOF systems are obviously applicable to single-story structures, but they can also be extended to multistory buildings. For instance, most steel seismic force-resisting systems exhibit shear-dominated global inelastic response; that is, the lateral load response at a given level essentially depends on the properties of the structural components at that level. This situation favors the application of SDOF model results on a per-story basis in multistory structures, as is currently done in seismic code provisions for P–Δ effects. Results from SDOF systems can also be used, with modifications, when adopting design strategies where the global structure is reduced to an equivalent SDOF system. This is the case in the study of individual collapse mechanisms or when adopting displacement-based design approaches.

After reviewing the SDOF system, past studies on seismic stability effects on multistory structures are presented with emphasis on analysis methods. Building code provisions for seismic P–Δ effects are reviewed and discussed, and strategies that have been proposed to enhance the seismic stability of steel seismic force-resisting systems are presented. Given that this section cannot provide detailed coverage of this topic, additional information on the design for global seismic stability can be found in Gioncu (2000), Krawinkler (2006), Villaverde (2007), Zareian and Krawinkler (2007), and ATC (2008). Only the effects of horizontal ground motions are discussed in this section. Spears and Charney (2006) reviewed past studies on and examined the influence of vertical ground motions on seismic structural stability.

19.3.1 *P*–Δ Effects under Static Loading

Figure 19.21*a* shows a SDOF system subjected to a static horizontal load V while supporting a gravity load P. The seismic force-resisting system is represented by the horizontal spring with an elastic stiffness k and the force resisted by the system is $F_s = k\Delta$, where Δ is the horizontal displacement at the top of the structure. If the gravity load is omitted, $F_s^{\text{o}} = V$ and the deflection is $\Delta^{\text{o}} = F_s^{\text{o}}/k = V/k$. When P is included, equilibrium in the deformed configuration leads to

$$V = \left(k - \frac{P}{h}\right)\Delta = k\left(1 - \frac{P}{kh}\right)\Delta = k\,(1-\theta)\,\Delta \qquad \text{with} \quad \theta = \frac{P}{k\,h} = \frac{P\,\Delta^{o}}{V\,h} \tag{19.23}$$

In this expression, θ is referred to as the stability coefficient, a parameter that relates the geometric stiffness (P/h) to the system elastic stiffness k. The presence of gravity loads results in a reduced effective lateral stiffness of $k(1-\theta)$ and both the deflection Δ^{o} and the force F_s can be determined by amplifying the results from a first-order analysis without P–Δ effects by $1/(1-\theta)$ (Fig. 19.21*b*) to obtain

$$\Delta = \frac{V}{k\,(1-\theta)} = \frac{\Delta^{o}}{1-\theta} \tag{19.24}$$

$$F_s = k\,\Delta = \frac{V}{1-\theta} \tag{19.25}$$

The elastic buckling load of the structure, $P_{cr,}$ is the load that reduces the effective lateral stiffness to zero, that is, when $P_{cr} = kh$, $\theta = 1.0$. As will be discussed later, other forms of the stability coefficient are reported in the literature or used in building codes. The same concepts also apply to more complex multi-degree-of-freedom systems, except that the above are expressed in a matrix format as described in Chapter 16 and in more detail by McGuire et al. (2000). In seismic design, however, the SDOF representation is commonly applied on a per-story basis, with V being the story shear, Δ the interstory drift, P the total gravity loads carried by the columns at the story, and h (or h_s) the story height.

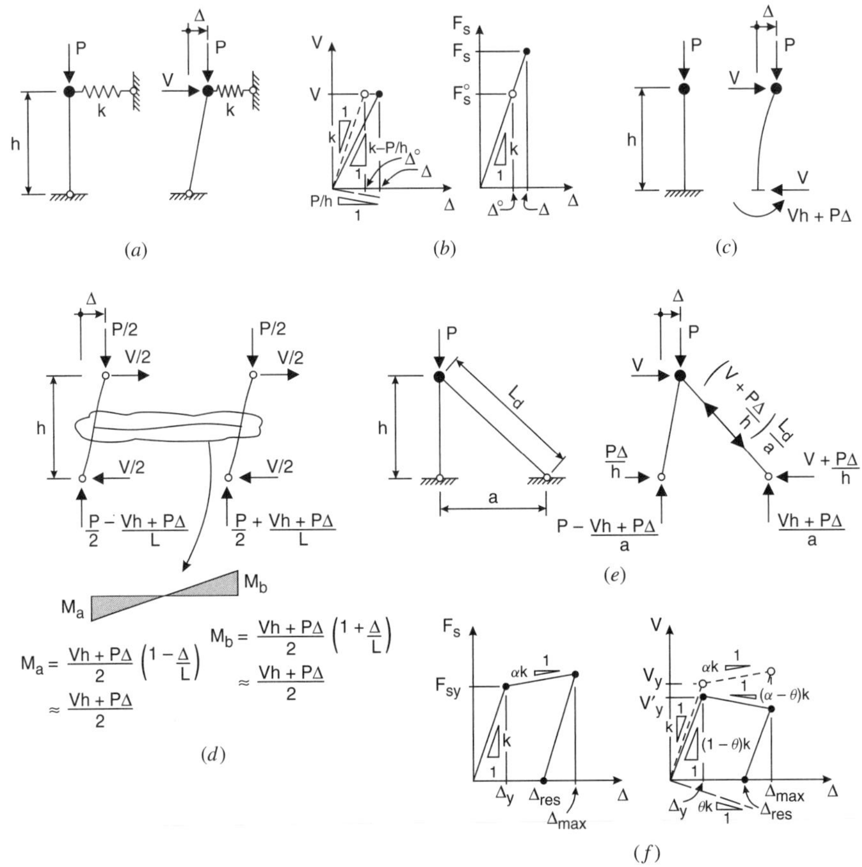

FIGURE 19.21 *P–Δ* effects under static loading: (*a*) SDOF model with lateral load-resisting system; (*b*) elastic response under lateral loading; (*c*) SDOF pole-type model; (*d*) moment-resisting frame; (*e*) Braced frame; and (*f*) inelastic response under lateral loading.

Gravity loads acting on a laterally deformed structure produce an additional overturning moment *P–Δ* that must be resisted by the lateral force-resisting system. This is illustrated in Figs. 19.21*c* to 19.21*e* for three different systems: a cantilevered column system, a single-bay moment-resisting frame at a given level in a structure, and a braced frame system. In the first case, resistance to lateral load is provided by bending of the column, and the base moment is amplified by the *P–Δ* moment. The beam moments in the moment-resisting frame are increased by the presence of the gravity loads. In the braced frame, the additional overturning moment induces additional axial load in the diagonal bracing member resisting lateral loads. In all three cases shown, the story shear and the total base horizontal reaction remain equal to the external lateral load V. Hence, the effects of the lateral loads, that is, member forces in the seismic force-resisting system and lateral deflections, are amplified by the P–Δ moments, not the lateral loads themselves or the story shears.

In Fig. 19.21f, the system has a defined yield strength F_{sy} and yield deformation Δ_y. Upon increasing the lateral load V in the elastic range, P–Δ effects are as above with a reduced elastic lateral stiffness $k(1-\theta)$ and a force F_s larger than the applied lateral load V. Once Δ_y is reached, yielding is initiated at a lateral load V_y', that is, at a reduced lateral load compared to the yield resistance F_{sy} or the load V_y the system would resist in the absence of gravity loads. Therefore, P–Δ causes a reduction of the effective resistance offered by the system to lateral loading. If the cantilever column in Fig. 19.21c has a flexural resistance of $V_y h$, yielding will occur under a lower lateral load in the presence of gravity loads. Similarly, if the beam in Fig. 19.21d or the brace in Fig. 19.21e is sized to yield under a lateral load V_y without gravity loads, the effective lateral yield strength of these systems will be reduced by $P\Delta_y/h$ if the structure supports a load P. On this basis, a logical design solution would consist of increasing the yield strength F_{sy} such that the system has sufficient capacity to resist the anticipated lateral load V, which can be done using Eq. 19.25 with $F_{sy} = V/(1-\theta)$.

While this solution is appropriate when yielding is prevented under defined static loading, it does not guarantee adequate performance when the structure is expected to deform in the inelastic range under seismic-induced lateral loading. If the system has an elastic–perfectly plastic F_s–Δ behavior ($\alpha = 0$ in Fig. 19.21f), the effective lateral stiffness becomes negative upon yielding, equal to $-\theta k$ (or $-P/h$), when including gravity load effects. This represents an unstable condition under static loading because the lateral load must be reduced to maintain equilibrium as the effective lateral resistance decreases when the lateral displacement is increased, from $V = F_{sy} - P\Delta/h$. Lateral loads due to seismic ground motions are dynamic inertia forces, not static loads, and their amplitude and direction change continually as the structure oscillates. In addition, damping forces are also induced in dynamic response and these forces also contribute to equilibrium. Therefore, stable inelastic response can still be achieved under an earthquake for this system, although a negative-yielding stiffness generally causes the structure to progressively drift toward one direction—a behavior that can eventually lead to large displacements and structural collapse, as will be discussed in the next section. In Fig. 19.21f, the postyield negative stiffness beyond Δ_y can be reduced and even cancelled out if the seismic-resisting system exhibits a positive-yielding stiffness αk. This is another possible design option to counteract P–Δ effects and it is examined in the next section. The figure also shows that residual permanent drifts (Δ_{res}) are likely after the structure has been laterally deformed in the inelastic range. This may represent a stability issue in the case of aftershocks and, as such, requires examination.

19.3.2 *P*–Δ Effects on Dynamic Seismic Response

Single-Degree-of-Freedom Systems The reduction in lateral elastic stiffness due to gravity loads lengthens the period of vibration of a structure and, thereby, its response to dynamic loading. For the system shown in Fig. 19.22*a* with a seismic weight W and lateral stiffness k, the period T' including P–Δ is

$$T' = 2\pi\sqrt{\frac{W}{g\,k\,(1-\theta)}} = \frac{T}{\sqrt{1-\theta}} \qquad \text{where} \quad T = 2\pi\sqrt{\frac{W}{g\,k}} \tag{19.26}$$

where g is the acceleration due to gravity ($g = 9.81\,\text{m/s}^2$) and T is the period of vibration of the system without consideration of P–Δ effects. This change in period,

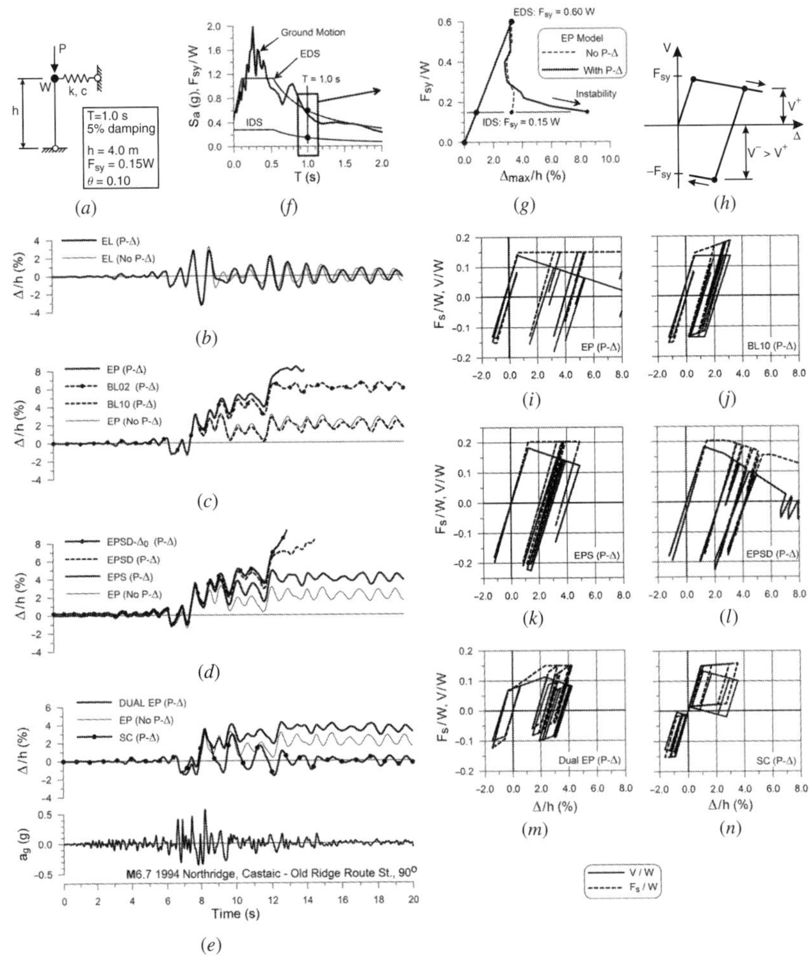

FIGURE 19.22 (*a*) SDOF model for dynamic analysis; (*b–e*) effects of gravity and hysteretic behavior on time history drift response; (*f, g*) design and ground motion response spectra; (*h*) influence of lateral strength on maximum displacement demand on the EP model; (*I–n*) Lateral force–deformation responses of SDOF systems with different hysteretic behavior.

however, is generally small and its effect on seismic response is not significant because the stability coefficient θ is generally less than 0.10 and very rarely exceeds 0.20 in actual structures.

Figures 19.22*b* to 19.22*e* show the normalized computed drift (Δ/h) time history response of SDOF systems with a period T of 1.0 sec, viscous damping of 5% of the critical damping, and $\theta = 0.10$ under the M6.7-1994 Northridge earthquake record. The response assuming elastic behavior (EL model) is given in Fig. 19.22*b* for the cases where P–Δ effects are omitted and included in the analysis. The difference is very small and is mainly characterized by a time shift in the decaying (free-vibration) response that follows the strong ground motion part of the earthquake record. Analytical studies on similar systems (e.g., Bernal, 1987) and on more complex structures (e.g., Cheng and Tseng, 1973) showed that period changes due to the P–Δ effects and cause the lateral displacements and forces to increase or decrease, but generally not by substantial amounts.

Most studies on the seismic stability of yielding structures have been performed using elastic–perfectly plastic (EP) and bilinear (BL) hysteretic models. The two models are identical except that the latter exhibits a positive load–deformation slope upon yielding ($\alpha > 0$ in Fig. 19.21*f*). Strain hardening of steel can contribute to this increase in strength with increasing lateral deformations. For complex structural systems, α can also represent progressive yielding taking place in the structure as the lateral displacement is increased (the components that are still elastic attract additional loads before they eventually yield). In the example of Fig. 19.22, the lateral strength F_{sy} of the EP and BL systems was determined using the elastic and inelastic design spectra (EDS and IDS) in Fig. 19.22*f*. The structure is assumed to have a global ductility of 4.0 and the EDS and IDS ordinates at $T = 1.0$ sec are $0.6W$ and $0.15W$, respectively. The 5% damped acceleration response spectrum of the ground motion is shown in the figure and the demand from the seismic record matches the EDS well at the model period of 1.0 sec. Inelastic time history dynamic analyses of the EP model are first carried out for values of F_{sy} varying stepwise from $0.6W$ (elastic) to $0.15W$ (design). In Fig. 19.22*g*, the peak lateral displacements Δ/h computed for the different resistance levels are plotted for the cases with and without P–Δ. Nearly identical results were obtained for the two cases until the lateral strength F_{sy} approaches $0.25W$ to $0.30W$. Above that point, the response is in accordance with the equal-displacement principle, that is, the peak horizontal displacements are not influenced much by the lateral strength of the system. When further reducing F_{sy}, the response with consideration of gravity loads starts to deviate significantly from the case without P–Δ, with excessive drifts and total collapse when approaching the design level. Montgomery (1981) observed this behavior for several structures and concluded that P–Δ has negligible effects on displacements until the lateral resistance is decreased to a critical level below which displacements increase very rapidly.

The time history response of the EP model with $F_{sy} = 0.15W$ is presented in Fig. 19.22*c* for the two cases with and without P–Δ. Figure 19.22*i* shows the hysteretic lateral load–lateral displacement (V-vs.-Δ/h) and lateral force–lateral displacement (F_s-vs.-Δ/h) response of the model with gravity loads. The response of the EP model supporting gravity loads is characterized by gradual drifting toward the positive-displacement side after a first large-amplitude inelastic excursion occurred in that direction. The reduced resistance at yield and the negative effective postyield stiffness of the system can be observed in the hysteretic (V-vs.-Δ/h) curve of Fig. 19.22i, together with the one-sided progressive shifting of the structure. Once the structure has yielded in one direction, it becomes unsymmetrical as the lateral load required to further yield the structure in the same direction is gradually decreased by the term $P\Delta/h$ (noting $V = F_{sy} - P\Delta/h$) whereas the horizontal load that must be applied to initiate yielding in the opposite direction, toward the initial position, is augmented by $P\Delta/h$. This bias in lateral strength is schematically illustrated in Fig. 19.22h and it is clear that such a difference in yield strength encourages subsequent yielding excursions to develop in the same direction, further increasing the displacements reached in previous yielding events. In Fig. 19.22i, yielding only occurred in the positive direction after the first large yielding event in that direction. This behavior can also be examined from the point of view of loading: Once the structure has displaced by yielding to a deformation of Δ, the overturning $P\Delta$ moment adds to subsequent seismic loads acting in the same direction but reduces seismic load effects acting in the opposite direction. The phenomenon becomes more pronounced as the displacement increases and it accelerates to cause structural collapse as the displacement approaches $\Delta_c = F_{sy}h/P$ (or Δ_y/θ), the point beyond which the system cannot support the static gravity loads acting alone. The P–Δ effects are more important when the strength of the structure relative to the force demand from the ground motion is decreased or when θ is increased. The detrimental effect of increasing the stability coefficient on the lateral displacements and spectral accelerations that can be resisted before collapse was confirmed by shake table tests on reduced-scale SDOF steel frames by Vian and Bruneau (2003). Structural collapse was also recently reproduced using hybrid simulation techniques (Schellenberg et al., 2008). It is however difficult to generalize on the magnitude of the P–Δ displacement amplification as it heavily depends on the characteristics of the ground motion (duration, frequency content, presence of pulses, etc.) and how the structure responds to that motion (Araki and Hjelmstad, 2000; Williamson, 2003; Bernal et al., 2006).

To illustrate the benefits of selecting a system with strain-hardening response upon yielding, plots of the time history responses of BL systems with α values of 0.02 and 0.10 (BL02 and BL10) are provided in Fig. 19.22c. In this case, P–Δ effects gradually reduce as α is increased and are eventually completely mitigated when $\alpha = \theta$. Husid (1967) and Jennings and Husid (1968) carried out a pioneering study in this area by developing expressions to predict the time to collapse for EP and BL systems. For EP systems, the time to failure was found to increase linearly with h and to be proportional to the square of the strength of the structure relative to the ground motion intensity. It also lengthens when increasing the ratio α/θ and no collapse was observed when that ratio was equal to 1.0. Several subsequent studies confirmed the stability of BL models with a positive effective yield slope

($\alpha - \theta > 0$). MacRae (1994) extended this concept and proposed a criterion to verify the stability of any hysteretic shape.

Providing sufficient strain-hardening response to achieve a stable response may not be practical in all cases, and an alternative approach consists of increasing the system lateral strength to compensate for P–Δ effects. In Fig. 19.22*c*, the peak displacement reached by the EP model without P–Δ effects is $\Delta_{\max} = 3.3\%$ of h. The strength of the model is therefore increased to $F'_{sy} = 1.35F_{sy}$ ($= 0.203W$) such that the new strengthened system (EPS) possesses an effective lateral resistance of $0.15W$ when $\Delta_{\max}$ is reached ($F'_{sy} - P\Delta_{\max}/h = 0.15W$). It is assumed that the drift of the EPS model will remain less than or equal to the drift of the EP model without P–Δ effects. In Fig. 19.22*e*, the EPS model underwent a peak drift of 4.9% of h, much larger (48%) than the target value of 3.3% of h. This result is typical and shows that compensating for dynamic P–Δ effects using methods based on static equilibrium, even when considering the inelastically deformed configuration, may not be sufficient to control the structure deformation or ductility (Humar et al., 2006). Bernal (1987) performed a parametric study on EP SDOF models to determine the strength amplification factor, $\varphi = F'_{sy}/F_{sy}$, that is needed to reach the same ductility μ with a resistance F'_{sy} when supporting the gravity load as that obtained with a resistance F_{sy} when P–Δ effects are neglected. Both F_{sy} and F'_{sy} are determined by successive iterations until the same target μ is attained. The parameters μ, θ, and T were varied. Limited correlation was found between the period and φ for the range studied ($T = 0.2$ sec to $T = 2.0$ sec) and, from regression analysis, an empirical expression was proposed for φ as a function of μ and θ

$$\varphi = \frac{1 + \psi\,(\mu - 1)\,\theta}{1 - \theta} \tag{19.27}$$

where ψ is a constant (1.87 and 2.69 for mean and mean plus one standard deviation results, respectively). The equation satisfies the limiting conditions $\varphi = 1.0$ when $\theta = 0$, $\varphi = 1/(1 - \theta)$ when $\mu = 1.0$ (elastic response, Eq. 19.25), and $\varphi = \infty$ when $\theta = 1.0$ (elastic critical load is reached). This study showed that large amplification can be obtained for ductile structures. For instance, for $\theta = 0.10$, the mean value of φ varies from 1.53 to 2.15 for μ comprised between 3.0 and 6.0. Mazzolani and Piluso (1993) reported on similar investigations performed in Italy that led to the same expression for φ except that an exponent ψ_2 is applied to the term ($\mu - 1$). For $T > 0.5$ sec, the mean amplification using this expression can be obtained with $\psi = 0.62$ and $\psi_2 = 1.45$, showing a greater influence of the ductility. Mazzolani and Piluso (1996) developed more refined expressions for φ that can be used to control ductility parameters that account for damage accumulation. Fenwick et al. (1992) proposed expressions for a P–Δ amplification factor β that must be applied to the lateral strength increase required at a ductility of μ, as determined from static analysis, to obtain the dynamic strength amplification φ:

$$\varphi = 1 + \beta\left(\frac{P\mu\Delta_y}{F_{sy}h}\right) = 1 + \beta\,(\mu\theta) \tag{19.28}$$

The factor β depends on the building period, the soil type (as it affects the ground motion signature), and the ductility. This approach has been adopted in the New Zealand code (Davidson and Fenwick, 2004) and is discussed later. For $\mu \geq 3.5$, a value of $\beta = 2.0$ is proposed for periods up to 2.0 sec (2.5 sec for structures on soft soils), confirming again the amplification of P–Δ effects under dynamic seismic loading. Yamazaki and Endo (2004) developed strength amplification factors based on the ratio of the seismic energy input required to reach a target ductility with and without consideration of P–Δ effects. This approach was found to give amplification levels nearly identical to those obtained from Eq. 19.27. Using the above expressions, values of θ can be determined to maintain the amplification small enough to be neglected. For instance, as suggested by Bernal (1987), if $\varphi < 1.1$ is deemed appropriate for no amplification, θ must be limited to 0.015 for a ductility of 4.0 when using Eq. 19.28 with $\psi = 1.87$, which is very restrictive.

The yielding components in a seismic structural system can experience strength degradation beyond a given deformation level. Such degradation may result in a negative effective lateral stiffness or a reduction in strength with cycles. Both of these situations can have detrimental impacts when combined with P–Δ effects. This is illustrated in Figs. 19.22*d* and 19.22*l* where strength degradation was implemented in the EPS model to form the EPSD model. In this case, the strength degradation was set to initiate at a drift of $0.027h$ and produce a slow linearly decreasing resistance such that the strength reduces by 10% at $0.04h$. This simple modification to the hysteretic response provoked collapse of the structure, emphasizing the importance of including this behavior in collapse analysis.

The interaction between strength degradation and gravity load effects has been studied analytically in past research (e.g., Della Corte et al., 2002; Williamson, 2003; Ibarra et al., 2005) and its impact on collapse response was demonstrated in shake table tests on steel MRFs (Rodgers and Mahin, 2006, 2008; Lignos et al., 2008; Suita et al., 2008). The deformation at which strength degradation begins, the slope of the descending (softening) branch, and the rate of cyclic degradation are key parameters influencing the response of structures. These aspects are discussed further in the next section.

Initial out-of-plumb of structures may also affect their seismic stability, as illustrated in Fig. 19.22*e* with the EPSD$-\Delta_0$ case. This model is identical to the EPSD model except that an initial deflection of $h/500$ corresponding to typical erection tolerances for steel structures was specified. This initial displacement creates a biased situation similar to the condition that exists after a first yielding excursion has occurred in one direction. Hence, seismic-induced drifts will tend to develop in the direction of Δ_0. An even more critical situation exists for structures that have been permanently deformed after an earthquake, as past studies showed that residual story drifts can be much larger than erection tolerances (e.g., Kawashima et al., 1998; Ruiz-García and Miranda, 2006). Limited studies have been carried out to predict the response of such initially deformed structures. In one study, Bernal (1987) derived from static equilibrium an upper limit on the product of μ and θ for EP systems that must be met if the structure needs to carry a gravity load P_u after the earthquake. The limit is given by $\mu\theta = P_s/P_u$. It assumes that the structure reached a displacement equal to $\mu\Delta_y$ while supporting a gravity load P_s during the earthquake. According to this criterion, the stability coefficient would need to be limited to 0.10 if, for example, P_u/P_s was equal to 2.5 and the anticipated ductility was 4.0.

Figures 19.22*e* and 19.22*m* show the response of a SDOF model representing a dual seismic force-resisting system composed of two framing systems exhibiting elastic–perfectly plastic behavior. In this example, the two frames have the same lateral strength $F_{sy} = 0.075W$, but the stiffness of one frame is four times the stiffness of the other frame. These stiffnesses have been adjusted such that $T = 1.0$ sec when considering the total initial lateral stiffness, which is identical to the original EP system. Yielding of the stiff frames occurs at a displacement of 0.58% of h whereas the more flexible frame remains elastic up to 2.33% of h. Hence, the flexible frame effectively provides $\alpha = 0.20$, larger than θ in the 0.58% to 2.33% of h displacement range. As shown, the seismic response of that dual-EP system is stable, even if its total lateral strength is the same ($0.15W$) as the EP model, which collapsed in Figs. 19.22*d* and 19.22*i*. This is attributed to the contribution of the flexible frame that provides the system with positive stiffness up to a drift of 2.33% of h, thereby reducing the duration and the number of yielding excursions during which an effective negative postyield stiffness prevails. Compared to the EPS model (EP model with resistance increased to $0.203W$), this design would represent an attractive solution because it would likely be lighter and impose lower forces on the other structural components located along the seismic load path while experiencing comparable lateral deformations. This example shows the benefits of designing for progressive yielding in the seismic load-resisting system or adding a back-up elastic system designed to yield at large deformations. The former application has been investigated by Tagawa et al. (2004) for steel MRFs, as will be presented below.

The response of a system with self-centering or recentering hysteretic properties is shown in Figs. 19.22*e* and 19.22*n*. These systems are forced to return to the undeformed position after each yielding excursion, which translates into a flag-shaped hysteretic curve. This behavior can be achieved by different means, such as using specialized materials displaying this particular hysteretic response, allowing the structure to rock, or introducing components connected by or built with pretensioned tendons. Compared to EP models, these systems lack energy dissipation capacity, but it has been demonstrated that this shortcoming can be compensated for by their ability to recenter (Christopoulos et al., 2002a). Of more significance for seismic stability, the increased displacement due to P–Δ effects during each individual yielding excursion is eradicated when the deformation is reduced to zero upon load reversal. Thus, the drifting or ratcheting behavior responsible for collapse is prevented, resulting in a more symmetrical response without residual drifts, as can be observed in Figs. 19.22*e* and 19.22*n*. Further discussion is presented at the end of this chapter on possible applications of this system to achieve stable seismic behavior.

Realizing that the influence of gravity loads remains small until the lateral strength of a structure is significantly reduced and dynamic instability is likely

(Fig. 19.22d), Bernal (1992) proposed an alternative treatment of P–Δ effects oriented toward collapse prevention instead of ductility control. In this case, the minimum resistance level that is needed for stable inelastic seismic response is determined from analysis, rather than the strength amplification factor required to meet a given target ductility. From parametric studies on EP and degrading models, expressions were proposed for collapse spectral ordinates (or a minimum F_{sy}/W ratio required to prevent collapse) as a function of the peak ground displacement and velocity, θ and T. For the EP model, the duration of strong ground shaking was also found to influence the strength at onset of instability. These ordinates increase with ground motion amplitude (and duration), as well as with θ, but decrease with the period. Asari et al. (2000) developed similar collapse spectra that also provide minimum strength values to achieve target ductility levels. In design, these spectra can be used to set the minimum lateral resistance necessary to achieve a desired safety margin against collapse. Alternatively, maximum values of the seismic response modification coefficient R can be determined from the ratios of the elastic seismic force demand from earthquakes to the lateral strength at which collapse occurs (Rutenberg and De Stefano, 2000; Miranda and Akkar, 2003). These critical values of R decrease when θ is increased and generally increase with T. For bilinear systems, the strength causing collapse generally decreases (R increases) when increasing the postyield stiffness ratio α. It must be noted that θ typically increases with T (Bernal, 1992), which reduces the apparent positive influence of the period on minimum strength requirements that is observed when θ and T are considered separately.

P–Δ Effects on Multistory Structures Experimental investigations and numerical simulations of the inelastic seismic response of multistory steel seismic force-resisting systems showed that the peak interstory drift demand generally varies along the height of buildings and the difference between peak roof drifts and peak interstory drifts accentuates when increasing the intensity of the ground motion relative to the lateral capacity of the structure (e.g., Roeder et al., 1987; Whittaker et al., 1987; Gupta and Krawinkler, 2000a; Marino and Nakashima, 2006; Chen et al., 2008). Figure 19.23*a* illustrates this phenomenon for an eight-story BRB frame subjected to two different ground motion levels. In the figure, the time histories of the horizontal displacement at each level, the vertical profile of the horizontal displacements at three points in time when the roof displacement reaches a maximum value, and the first-story (base) shear versus interstory drift hysteretic response are given for two ground motion intensities. As shown, the fairly uniform demand pattern observed under the lower shaking level changes to a one-sided response with more pronounced interstory drifts and residual deformations in the bottom levels when increasing the amplitude of earthquake motion.

A number of reasons contribute to nonuniform interstory drift demand along the height of structures. One reason is that inelastic deformations of the yielding components in steel frame systems (plastic hinging in beams in MRFs, plate yielding in SPSWs, brace yielding or buckling in CBFs or BRBFs, or yielding in EBF links)

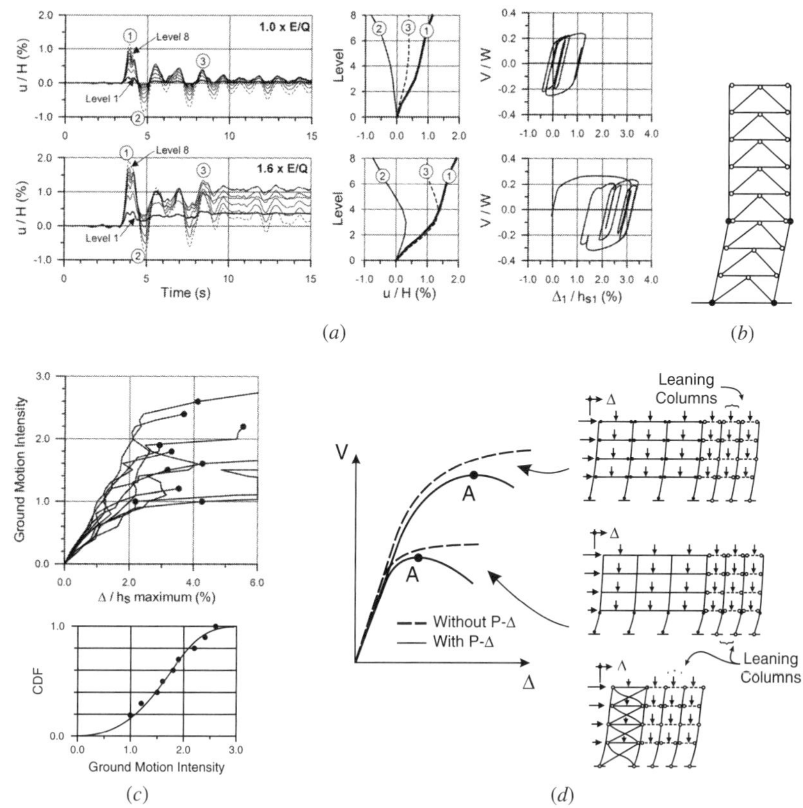

FIGURE 19.23 (*a*) Response of an eight-story BRB frame under increasing ground motion levels (time histories of story horizontal displacements, horizontal displacement profiles, and story shear vs. story drift response at level 1). (*b*) three-story collapse mechanism in an eight-story BRB frame; (*c*) incremental dynamic analysis results; and (*d*) non-linear static incremental analysis.

essentially translate into story shear deformations that can develop nearly independently of the story shear deformations taking place in adjacent stories. Columns of both the seismic and gravity load resisting systems are the only structural components that can contribute, through bending, to minimizing differences between stories, but they are generally not specifically designed for this particular purpose (Medina and Krawinkler, 2005). Another reason is that the vertical distribution of the horizontal demand (force or deformation) at any time during a particular earthquake can differ significantly from the lateral force pattern assumed in design when assigning story shear resistance to the structure. The design pattern represents an envelope, not the specific loading patterns that will be applied at various times during an earthquake. These actual time varying patterns can create large deformations at specific locations along the building height and there is no guarantee that the sequence and characteristics of such loading events will lead to the same inelastic deformation demand at every story at the end of a particular ground motion (e.g., Tjondro et al., 1992).

Amplification of lateral deformations and progressive drifting due to P–Δ may take place at stories where large interstory drifts have developed and a negative effective story shear stiffness exists, accentuating further the variations in interstory drifts between stories. This can lead to dynamic instability response characterized by the formation of a collapse mechanism with a vertical profile defined by the incremental inelastic story shear deformations due to P–Δ effects, thereby involving only one or a few adjacent stories in the region where large inelastic deformations had already developed (Bernal, 1992, 1998; Gupta and Krawinkler, 2000b; Tremblay and Poncet, 2005). The collapse induces plastic hinging of the continuous columns at the top and bottom ends of the mechanism, as illustrated in Fig. 19.23*b* for a three-story mechanism in a BRB frame assuming pinned brace-to-beam and beam-to-column connections. This behavior also occurs in steel MRFs, even if strong column–weak beam design was applied, as was recently demonstrated in the shake table test program by Lignos et al. (2008).

Occurrence of dynamic seismic instability depends on the numerous factors that dictate the overall inelastic response of the structure to the ground motion, including dynamic properties of the structure (periods and mode shapes, damping), structure lateral strength and vertical distribution of the story shear strength, hysteretic behavior of the yielding components, amplitude and vertical distribution of the gravity loads, continuity conditions and flexural properties of the columns, characteristics of the ground motion, and the local inelastic response to the large deformations developing at the levels involved in the collapse mechanism. Because inelastic structural deformations before and during collapse are essentially in the form of story shear deformations and the hysteretic behavior of the yielding components can be transposed into the story shear–story shear deformation response, the key parameters influencing frame stability can be related to each other on a story basis (e.g., stability coefficient, gravity load vs. story shear strength, etc.) and interstory drifts can be used to characterize both the structural response and the demand on the yielding components.

In past studies, inelastic deformation demand tends to concentrate at the building base, where the ground motion is input into the structure and the gravity loads are maximum, or in the upper levels as a result of higher vibration mode response inducing large story shear demand (e.g., Marino and Nakashima, 2006). Furthermore, comparative studies on identical buildings having different heights consistently showed that the deviation between peak interstory drifts and peak roof drifts increases when the height of the building is increased (e.g., Gupta and Krawinkler, 2000a; Sabelli et al., 2003; MacRae et al., 2004; Karavasilis et al., 2007; Tremblay et al., 2008b). There are many reasons that may explain this trend. As the structure height is increased, the relative contribution of the higher vibration modes augments, increasing uneven demand along the frame height. Gravity loads in the lower levels also increase with the building height whereas the design seismic load to seismic weight ratio diminishes due to the shape of current code design spectra. This creates a more critical situation for P–Δ effects with higher values of the term $P\Delta/h$ relative to F_{sy} for a given deformation Δ or smaller values of the interstory drift Δ_c where the effective lateral stiffness reduces to zero and instability is likely ($= F_{sy}h/P$ for EP models). A recent probabilistic study of the collapse response of 4- to 16-story steel MRFs by Krawinkler and Zareian (2007) seemed to confirm this tendency. It was found that the risk of collapse was much higher for the taller (longer period) frames and the authors suggested that a minimum lateral resistance be prescribed in codes such that sufficient margin exists against seismic instability. In this regard, a minimum design base shear ratio V/W expressed as a function of the seismic ground motion intensity at the site, has been reintroduced in ASCE 7 code to address this issue (ATC, 2008; ASCE, 2008).

As was observed for SDOF systems, comparative simulations performed on multistory building models with and without P–Δ effects showed that gravity loads have negligible effects when the structure experiences limited inelastic deformations (e.g., Tjondro et al., 1992; Jin and El-Tawil, 2005; Bhowmick et al., 2009). Montgomery (1981) concluded from his study on MRFs of up to 10 stories and with elastic–plastic hysteretic behavior that P–Δ effects were significant only when V/W was less than 0.10 or the interstory drifts were greater than 2.0 times the yield interstory drift. Fenwick et al. (1992) found that the ductility demand on ductile reinforced concrete frames could be well controlled by applying the P–Δ amplification factor developed for SDOF systems (Eq. 19.28). For structures near the collapse state, Tremblay et al. (1999) noted that increasing the lateral capacity using strength amplification factors such as those developed for SDOF systems or suggested in codes (see below) could reduce the interstory drifts to the values obtained when neglecting P–Δ effects in the analysis. These observations suggest that amplifying the seismic load effects in design is necessary when approaching instability or limiting the ductility demand when large inelastic deformations occur but no augmentation would be needed when sufficient lateral strength is already present.

When properly detailed, yielding components of most steel seismic force-resisting systems can develop substantial strain-hardening cyclic response up to large interstory drift deformations. In particular, shear yielding in short EBF links, axial yielding in BRB members, and plastic rotation in beams can be very pronounced. As described earlier for SDOF systems, such strain-hardening response contributes to counter dynamic P–Δ effects (e.g., Uetani and Tagawa, 1998) and should be modeled in stability analyses (e.g., Ricles and Popov, 1994; Byfield et al., 2005). Strain hardening may also improve the uniformity in the yielding demand over adjacent floors because a hardening response at a floor undergoing large inelastic deformations requires gradually increasing lateral loads which may eventually trigger yielding in adjacent floors and, thereby, lead to more uniform inelastic demand. In contrast, deteriorating hysteretic response is likely to aggravate the concentration of story shear deformations and exacerbate collapse response when approaching instability. This may have a major impact in collapse prediction analysis, that is, when pushing the structure to its limit, either statically or dynamically.

In steel MRFs, monotonic and cyclic strength degradation occurs as a result of local buckling or ductile fracture (see Fig. 19.7*b*) and several models have been developed and used to simulate this effect in performance assessment studies including stability effects (e.g., Mehanny and Deierlein, 2001; Della Corte et al., 2002; Ibarra et al., 2005; Kazantzi et al., 2008; Krawinkler and Zareian, 2007; Liao et al., 2007; Rodgers and Mahin, 2008). Degradation in well-detailed joints typically begins at 0.02 to 0.03 rad of plastic rotation and progressively takes place as the number of cycles and amplitude of deformation are increased. Ibarra and Krawinkler (2005) noted that the deformation where deterioration begins and the rate of degradation beyond that point had the largest effect on the collapse capacity of structures. Strength degradation is generally less pronounced for short links of eccentrically braced steel frames or buckling restrained braces. Conversely, buckling and fracture of steel bracing members can result in a marked reduction in story shear resistance and have detrimental impact on the seismic stability of braced frame structures. Brace buckling initiates at smaller interstory drifts (less than $0.005h_s$) compared to beam-to-column joints in MRFs and, thereby, has an earlier effect on frame response during an earthquake. The drop in brace compressive strength after buckling varies with the brace overall slenderness, with greater reductions for intermediate slenderness (see Fig. 19.14*b*). Because brace slenderness varies along the structure height, the variation in strength degradation may lead to greater concentration of inelastic damage.

Negative effects of brace buckling are minimized when slender braces sized not to buckle under the design seismic story shear are used in tension–compression pairs at every level (Tremblay, 2000, Karavasilis et al., 2007); in each direction, the slender tension-acting diagonal possesses sufficient reserve strength in tension to compensate for the strength reduction experienced by the buckled compression brace. This is not the case in chevron bracing when the beams are not specifically designed to resist the unbalanced brace loads illustrated in Fig. 19.16*d*. In this case, the postbuckling deterioration in brace compressive strength translates nearly entirely into story shear response degradation because the force developing in the tension-acting brace is limited by the capacity of the beams. The impact of this behavior on chevron-framing stability has been demonstrated experimentally (e.g., Fukuta et al., 1989) as well as in several numerical studies (e.g., Khatib et al., 1988; Remennikov and Walpole, 1998b; Tremblay and Robert, 2001). If braces are prone to low-cycle fatigue fracture (Eqs. 19.16 to 19.18), fracture is likely to occur at levels where large story deformations have already developed, creating more favorable conditions for the formation of a collapse mechanism. As described earlier, numerical brace models that could reproduce global buckling and fracture have been developed and can be used in braced frame seismic stability analyses (e.g., Uriz and Mahin, 2004; Daravan and Far, 2009).

Steel moment-resisting frames generally possess high lateral overstrength due to design requirements other than those related to the resistance of seismic loads; such requirements may include drift limits, meeting a weak beam–strong column design philosophy, gravity loads, and so on. Such extra-resistance reduces the extent of yielding, the likelihood of large interstory drifts, and initiation of a degradation situation and thus contributes to preventing dynamic instability. Similar overstrength exists in steel plate shear walls when the infill plate thickness is often governed by handling or welding processes rather than lateral strength requirements. In both systems, the columns are inherently strong and stiff in bending, which also helps in mitigating drift concentrations.

Braced steel frames, CBFs, BRBFs, or EBFs generally possess lower overstrength and their columns are of smaller sizes, resulting in greater likelihood for story response. For CBFs, this weakness contributes to a relatively limited strain-hardening response and strength degradation due to brace buckling, and this is compensated for by most building codes specifying lower seismic force modification factors. Column continuity can also have marked positive effects for braced steel frames, as was observed in several past studies (e.g., Tremblay and Stiemer, 1994; Martinelli et al., 2000; MacRae et al., 2004; Kimura and MacRae, 2006). Similarly, a combination with moment-resisting steel frames has been found to better the distribution of inelastic demand in multistory bracing systems (e.g., Whittaker et al., 1987; Shibata, 1988; Bertero et al., 1989; Hassan and Goel, 1991; Martenelli et al., 1996; Kiggins and Uang, 2006). Moment-resisting frames can be created by using rigid beam-to-column connections within the bracing bents and/or the gravity load supporting structure.

In view of the significant influence of the ground motion signature on stability response and the high variability in ground motion characteristics, randomness in ground motion must be accounted for in stability analysis. Similarly, the inherent variation in structural properties and the sensitivity of the response to modeling assumptions should also be included in the analysis. For instance, correlative time history analyses of shake table tests showed that asymmetric drift accumulations in CBFs were very sensitive to slight variations in modeling parameters (Broderick et al., 2008) or that collapse results were affected by damping assumptions (Vian and Bruneau, 2003). Gupta and Krawinkler (2000b) observed that refining an analytical steel MRF model by adding other dependable components (gravity columns, orthogonal frame columns, shear connections, and floor slabs) to a basic centerline representation can favourably impact the seismic stability assessment.

Methodologies have been recently developed to include these effects and obtain probabilistic estimates of the risk of collapse. In these methods, the ground motion intensity and the deformation at collapse are first obtained through incremental dynamic analysis (IDA) of the structure when subjected to a suite of ground motion records (Vamvatsikos and Cornell, 2002). For each record, nonlinear dynamic time history analysis is performed at stepwise increasing ground motion amplitudes until structure collapse occurs, with the maximum peak interstory drift value along the building height plotted as a function of the ground motion intensity measure. Figure 19.23*c* shows an example of IDA curves obtained for 10 ground motion records. Each dot in the graph corresponds to the point where collapse occurred under each record, that is, when interstory drifts become excessive. The structure model should include the frame studied together with the tributary leaning gravity columns supporting their gravity loads, as illustrated in Fig. 19.23*d*. Three-dimensional representation of the entire structure can also be used to better capture the contribution of all components as well as possible in-plane torsional response. Of course, geometric nonlinearities and gravity loads expected to be present during earthquakes should be included. Hysteretic behavior of the key yielding components should be specified, based on expected material properties and with proper consideration of strain hardening and deteriorating responses, together with a realistic representation of damping.

In the procedure proposed in FEMA 350 (FEMA, 2000a; Lee and Foutch, 2002), the level of confidence against collapse can be determined by comparing the interstory drifts at collapse (capacity) to those at the ultimate limit-state (collapse prevention) ground motion level (demand) from the IDA curves. The methodology accounts for the slope of the seismic hazard at the site, variations in capacity and demand, as well as uncertainties in the prediction of the demand (damping, material properties, etc.). From the IDA results, it is also possible to develop collapse fragility curves that correspond to the cumulative distribution function (CDF) or probability of collapse as a function of the intensity of ground motion, as illustrated in the bottom part of Fig. 19.23*c*. These curves are used in the ATC-63 methodology for the assessment of the seismic performance of structural systems (ATC, 2008). A collapse margin ratio is determined by comparing the ground motion intensity corresponding to a 50% probability of collapse to the ultimate limit-state (maximum considered earthquake) ground motion level. The probability of collapse at the ultimate limit-state ground motion level is also verified. In the procedure, the collapse margin ratio is modified to account for the spectral shape of rare extreme earthquakes and the accepted collapse margin ratio is defined with consideration of the variability in ground motions and uncertainties and the quality of the data and models used in the analysis. Selection and scaling of ground motions and accounting for modeling uncertainties are of key importance in these methodologies. Additional information on these two aspects can be found respectively in Haselton and Baker (2006) and Liel et al. (2009).

The above procedure can be used in design to determine a minimum lateral strength level required to achieve sufficient margin against collapse. These techniques, however, require intensive analysis and are mainly intended for the development and validation of design provisions for structural systems and may not be currently (2009) suitable for day-to-day design. Alternatively, incremental static (pushover) analysis can be used to identify potential collapse mechanisms and the lateral capacity of multistory structures. In such an analysis, the structure is subjected to a pattern of monotonically increasing horizontal loads, and the total lateral load–roof displacement response is plotted (Fig. 19.23*d*). Guidance on such analysis can be found in FEMA 356 (FEMA, 2000b). It is recommended in FEMA 356 and EC8 that at least two different load patterns be considered as an attempt to bound the solution; these include an inverted triangular pattern (according to the design seismic load pattern) and a uniform pattern (loads proportional to floor seismic weight). The analysis in each case should be run until a complete mechanism has formed.

A pushover analysis is particularly useful to determine the displacement and lateral capacity at the point where the effective lateral stiffness becomes negative and drifting of the response is likely (point A in Fig. 19.23*d*). The roof displacement at point A can be compared to the anticipated roof displacement demand established from the design spectrum and an equivalent SDOF system, as described in FEMA 356. Global instability is unlikely if the stiffness does not become negative under the anticipated deformation demand (Gupta and Krawinkler, 2000b; Humar et al., 2006). Greater lateral strength and deformation capacity before collapse can generally be achieved when the collapse mechanism mobilizes a larger number of stories and yielding components (Bernal, 1998); in general, the structure can be corrected as necessary to meet the anticipated demand. Compared to localized collapse mechanisms, a global mechanism is also characterized by a less pronounced negative P–Δ stiffness beyond point A, which should lessen seismic P–Δ effects in case of large deformations (Mazzolani and Piluso, 1996).

Other methods also based on the reduction of the structure to an equivalent SDOF system have been proposed for design purposes. Bernal (1992, 1998) suggested a method to identify the critical collapse mechanism for a steel MRF and evaluate the lateral strength and stability coefficient associated with that mechanism. The mechanism profile is assumed to correspond to the incremental lateral deformations between elastic and near-collapse conditions from a pushover analysis of the structure under a uniform load pattern. Safety against collapse is then verified by comparing the lateral strength capacity of the mechanism to SDOF collapse spectral values, as described earlier. Mazzolani and Piluso (1996) proposed to use a kinematic plastic analysis method to evaluate the lateral resistance of potential collapse mechanisms in steel MRFs. The method can then be used to control the governing mechanism and assess local ductility demand. Adam et al. (2004) proposed using a nonlinear dynamic time history analysis to examine the stability of an equivalent SDOF system having properties derived from a pushover analysis of the entire structure. The conditions for stability found from the equivalent SDOF system can then be transformed back into the structure domain to assess the stability of the structure.

It must be recognized that methods based on static analysis procedures have limitations (Krawinkler and Seneviratna, 1998; Krawinkler, 2006) and may not predict the collapse mechanism that would actually form under dynamically applied ground motions. In particular, static analysis cannot reproduce the dynamic drifting or ratcheting response due to gravity loads under cyclic loading, the strength deterioration or fracture of yielding components due to cyclic loading, and, more importantly, the continuously varying horizontal load patterns, including higher mode effects, that contribute to drift concentrations. Although enhanced static analysis methods, such as the modal incremental analysis that accounts for higher mode response, may improve the prediction compared to a single load pattern analysis, the technique may still fail in predicting accurately the demand on structures that deform significantly in the inelastic range with marked negative stiffness, as is the case when investigating dynamic P–Δ instability (Goel and Chopra, 2004).

19.3.3 Design for Seismic Stability

Building Code Requirements for Global Stability Recent research on global seismic stability of steel structures has not yet resulted in simple and generally applicable methods that could be implemented in codes to account for P–Δ effects in seismic design. Current methods proposed in codes essentially require that seismic-induced member forces be amplified based on the value of stability coefficients. The calculations are performed on a per-story basis and the amplification of seismic effects is generally based on static equilibrium. In all codes reviewed, P–Δ effects can be ignored if the stability coefficient is below a certain value, and the design must be corrected if the stability coefficient exceeds a maximum value. Although similarities exist, there are also marked differences in the treatment of the topic within many building codes. This is probably a consequence of the rapid evolution of knowledge in this field that has taken place in the last two decades, and it is expected that more consistent modifications will be implemented in the near future to better reflect these recent developments.

In the United States, the ASCE 7-05 standard (ASCE, 2005) is generally the reference document for loads and general design requirements for building structures. In ASCE 7, a stability coefficient is evaluated at every story,

$$\theta_{\mathrm{ASCE}} = \frac{P\,(C_d\,\Delta)}{Vh_s C_d} \tag{19.29}$$

where P is the total gravity load above the level under consideration, C_d is an amplification factor to account for inelastic deformations, Δ is the elastic deflection under the design seismic loads, V is the story shear due to the design seismic loads, and h_s is the story height. The C_d-factor varies depending upon the seismic force-resisting system. For most systems, it is approximately equal to 0.7 times the force modification factor R. It is important to note that the C_d-factor is contained in both the numerator and denominator in Eq. 19.29, which means that the stability coefficient is in fact based on elastic response. According to ASCE 7, P–Δ effects can be ignored in design if θ is less than 0.10. If the stability coefficient exceeds 0.10, amplification of displacements and member forces due to P–Δ effects must be determined using a rational analysis. No such method is given in ASCE 7, but it is permitted to determine P–Δ amplification by multiplying the seismic effects by $1/(1-\theta)$, thus assuming elastic response amplification. This approach is believed to give a reasonable estimate provided that the negative stiffness region is avoided at the target displacement (Krawinkler, 2000). It is noted here that strength amplification only applies to the design of the yielding components of seismic force-resisting systems. The design loads for the capacity-protected components are driven and limited by the capacity of the yielding components and need not be amplified for P–Δ effects.

An upper limit for θ is specified in ASCE 7,

$$\theta_{\max} = \frac{0.5}{\beta\, C_d} \leq 0.25 \tag{19.30}$$

In this equation, β is the story shear demand-to-capacity ratio. This ratio reflects the overstrength present in the structure and is typically less than 1.0. If θ exceeds $\theta_{\max}$, ASCE indicates that the structure may become unstable during strong ground motions and must be redesigned. By setting $V = \beta V_y$ in Eqs. 19.29 and 19.30, this verification becomes

$$\frac{P\,(C_d \Delta)}{V_y h_s} \le \min\,(0.5;\ 0.25\beta C_d) \tag{19.31}$$

In this equation, the left-hand side corresponds to the ratio of the story moment due to gravity loads acting on the inelastically deformed structure to the overturning moment capacity. This ratio also represents the loss in lateral strength caused by P–Δ (which equals $PC_d\Delta/h_s$) relative to the structure actual story shear resistance V_y. For most ductile systems, the limit will be equal to or slightly less than 0.5 because the factor C_d in ASCE 7 varies between 4.0 and 5.5 and system overstrength typically varies between 1.5 and 2.5. If the limit is exceeded and because P and h_s are parameters that cannot be modified, the designer has the option of increasing either the stiffness (reducing Δ) or the lateral strength V_y.

A similar upper limit exists in FEMA-350 (FEMA, 2000a), in which the parameter Ψ is used to characterize P–Δ effects at every level,

$$\Psi = \frac{P\,(R\Delta)}{V_y h_s} \tag{19.32}$$

where the product $R\Delta$ represents the expected story drift in the yielded condition. In FEMA-350, it is suggested to ignore P–Δ effects if $\Psi \le 0.1$ and to redesign the structure if the factor exceeds 0.30. No amplification is required for intermediate values of Ψ. Taking $C_d = 0.70R$, the limit of $\Psi = 0.3$ corresponds to $PC_d\Delta V_y h_s = 0.21$, which is more restrictive than the ASCE 7 limit. The FEMA approach is recommended in the commentary to the 2005 AISC seismic provisions.

The format in EC8 is similar to ASCE 7, except that inelastic deformations are included in the interstory drift values used in the calculation of the stability coefficient,

$$\theta_{\text{EC8}} = \frac{P\,(q\Delta)}{Vh_s} \tag{19.33}$$

where q is the behavior (ductility) factor. For the high-ductility-class systems, it takes a value of 6.5 for redundant moment-resisting frames, 6.0 for eccentrically braced frames, 4.0 for concentrically braced frames, and 2.5 for frames with chevron bracing. The P–Δ effects need not be taken into account if θ is less than 0.10. If θ is between 0.1 and 0.2, second-order effects must be included and this can be done by multiplying lateral load effects by $1/(1-\theta)$, that is, based on elastic response, but using a reduced stiffness accounting for inelastic deformations. If θ exceeds 0.3, the structure must be stiffened to meet this upper limit. Compared to ASCE 7, the EC8 provisions for P–Δ effects are more severe due to the q-factor included in θ, and the difference is more pronounced for the more ductile systems. Moreover, EC8 also requires that the formation of a soft story plastic mechanism be prevented to avoid excessive demand on the columns. It is suggested that this be achieved by performing a static incremental analysis as discussed earlier.

In the National Building Code of Canada (NRCC, 2005), the design seismic loads are determined by dividing the elastic earthquake forces by the product $R_o R_d$, where R_o and R_d are the dependable overtrength and ductility force modification factors, respectively. The R_d-factor varies from 1.0 for the most brittle systems to 5.0 for the most ductile ones such as type D (ductile) moment-resisting frames. For this system, the product $R_o R_d = 7.5$ (Mitchell et al., 2003). The anticipated interstory drifts including inelastic deformations are obtained by multiplying the story drifts under the design seismic loads by $R_o R_d$. As in FEMA 350, the stability coefficient corresponds to the story shear due to the gravity loads acting on the deformed structure divided by the lateral resistance V_y taken equal to $R_o V$, which results in

$$\theta_{\text{NBCC}} = \frac{P\,(R_o R_d \Delta)}{R_o V\, h_s} = \frac{P\,(R_d \Delta)}{V\, h_s} \tag{19.34}$$

In NBCC, P–Δ effects need not be considered if θ is less than 0.10. The structures must be stiffened or strengthened if the stability coefficient exceeds 0.40, which is similar to the requirements in EC8. For intermediate θ values, however, the seismic-induced forces at every level must be multiplied by $1 + \theta$, which means that additional lateral resistance must be supplied to fully compensate for P–Δ effects (the effective lateral strength is equal to $R_o V$ at the anticipated deformed interstory drift).

In the New Zealand Standard NZ 1170 (NZS, 2004), the stability coefficient at each level is obtained from

$$\theta_{\text{NZS}} = \frac{P\,(\mu \Delta)}{V_y h_s} \tag{19.35}$$

where μ is the structural ductility factor and, similar to FEMA 350 and NBCC, the anticipated deflections including inelastic response $\mu\Delta$ are used together with the story shear strength V_y. The ductility factor varies from 1.0 to 6.0. The latter value applies for the more ductile categories of MRFs, CBFs, or EBFs. In the determination of θ, the potential for a concentration of inelastic demand over the building height must be considered as the inelastic interstory drift $\mu\Delta$ must not be less than the value obtained from any possible mechanism with an amplitude at a roof level equal to μ times the elastic roof displacement. For concentrically braced steel frames, the so-computed interstory drifts must be increased further to account for the greater tendency of the system to form a sway mechanism, including 1.5 times for tension–compression X-bracing and 2.0 times for tension-only and chevron-bracing systems. In regular structures, however, when a design procedure is used to prevent story mechanisms, as is the case when adopting a strong column–weak beam design approach for ductile moment-resisting frames, the stability coefficient need not be calculated in the upper half of the structure.

In NZ 1170, the structure must be redesigned if the stability coefficient at any level, where applicable, exceeds 0.3. For lower values of θ, P–Δ effects must be considered in the calculation of the seismic-induced forces except if the fundamental period is less than 0.4 sec (0.6 sec if the structure height is less than 15 m) or the computed values of θ do not exceed 0.10 over the building height. Two methods are provided to account for P–Δ effects.

In the first method, the seismic effects for the entire structure are amplified by the single factor $1 + k_p W/V$, where W is the total seismic weight, V is the design base shear, and k_p is given by

$$k_p = 0.015 + 0.0075\,(\mu - 1) \quad \text{for } 0.015 < k_p < 0.03 \tag{19.36}$$

In this method, amplification is independent of the anticipated deflections. For ductile structures, it only varies according to the amplitude of the seismic load ratio V/W ($k_p = 0.03$ for $\mu \geq 3.0$). For instance, amplification would range between 1.15 and 1.6 for V/W between 0.20 and 0.05, respectively. This method thus yields greater amplification when higher inelastic response is expected (μ large) and for structures designed for lower seismic loads.

In the alternative method, a story shear force equal to $\beta P(\mu\Delta)/h_s$ is added to compensate for P–Δ effects, where β is the P–Δ amplification factor that accounts for P–Δ effects on dynamic response (as described earlier with Eq. 19.28). This factor is plotted in Fig. 19.24*a* and its value depends on the period of the structure, the type of soil (site class), and the structural ductility factor. For ductile systems, the amplification factor equals 2.0 for systems in the low- and intermediate-period range. This method is identical to the approach used in Canada, except that the dynamic nature of P–Δ effects on inelastic structures responding to seismic motions is included through the β-factor and the inelastic interstory drifts $\mu\Delta$ must account for the possibility of concentration of inelastic deformations along the building height. For high-ductility systems, the NZ provisions are significantly more severe

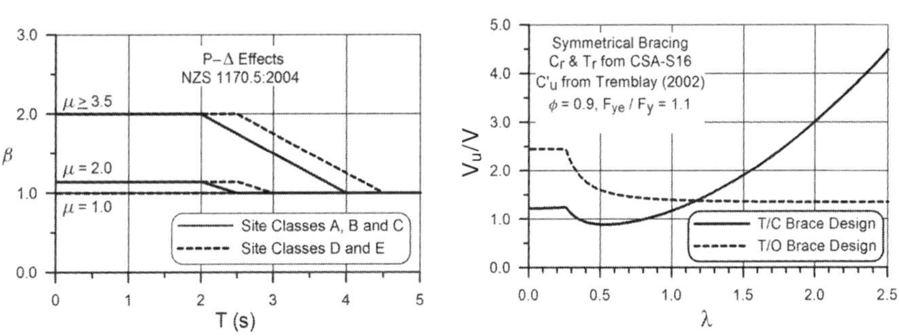

(*a*) (*b*)

FIGURE 19.24 (*a*) P–Δ amplification factor in NZS1170 Standard (NZS, 2004); (*b*) story shear overstrength from tension–compression (T/C) and tension-only (T/O) brace designs.

than those included in ASCE 7. Additional background information on the method can be found in Davidson and Fenwick (2004).

Design Strategies for Enhanced Seismic Stability As discussed in previous sections, P–Δ effects on the seismic response are generally small and can be neglected when the structure response is nearly elastic and/or far from dynamic instability conditions. Verifying the safety margin against collapse after a preliminary design has been performed could be one of the first steps in design for seismic stability. If P–Δ is a problem, increasing the lateral strength is a solution, but other options are reviewed herein that may be considered.

The potential for collapse by instability can be verified by performing a static incremental (pushover) analysis (thereby verifying that the anticipated target displacement is less than the displacement where negative stiffness starts) or by adopting approaches such as those described earlier and proposed by Bernal (1992), Mazzolani and Piluso (1996), or Adam et al. (2004). When assessing the capacity of steel MRFs, consideration should be given to the strength of the joint panel zones as these components could govern the frame lateral capacity (Krawinkler and Mohasseb, 1987; Tsai et al., 1995). Using the proposed methods, the lateral strength of the structure can be increased as necessary to achieve the required margin against possible instability. This can be done by trial and error or by using strength amplification factors from SDOF systems using equivalent SDOF properties reflecting specific structural behavior. For this purpose, one may also consider the use of the amplification approach proposed for the displacement-based design method (e.g., Asimakopoulos et al., 2007; Pettinga and Priestley, 2007). As described earlier, all these methods may not capture potential inelastic demand concentrations due to dynamic effects and caution should be exercised for cases of tall structures and structures exhibiting strength and stiffness irregularities or strength degradation or when the structure is subjected to large inelastic deformations and/or is close to the point of dynamic instability.

For multistory structures, many of the solutions proposed to mitigate P–Δ effects share the same objective of distributing more uniformly the inelastic demand over the structure height and, thereby, prevent the formation of localized mechanisms. Achieving this goal offers several advantages. In particular, a greater number of yielding components will be mobilized in the structure, leading to a greater energy dissipation capacity of the structure being exploited without imposing excessive demand on individual components and, thereby, better control of lateral displacements. Uniform drift demand also results in a more uniform ratio of the P–Δ overturning moment to the story overturning moment capacity ($= P\Delta/Vh_s$) over the building height as P and V increase toward the base. Other options can be considered to limit P–Δ effects are (i) providing the structure with larger lateral displacement capacity before the development of a negative effective lateral stiffness, (ii) using back-up systems, or (iii) taking advantage of newly developed self-centering technologies. Those are also briefly introduced below. It must be noted that most of these methods or structural systems may have not been fully validated and may have not been approved by regulating authorities, and, of course, it is the responsibility of the design engineer to ensure the method or system selected will provide for at least the level of protection implicitly assumed or expected in the applicable building code.

Inelastic interstory drifts are closely related to the supplied story shear resistance, and drift concentrations could be minimized by closely matching the story shear demand without abrupt changes in capacity along the structure height. For instance, Rossi and Lombardo (2007) observed for EBF structures that determining the story shear demand from a spectrum analysis, rather than from an equivalent static force procedure, and providing uniform link overstrength along the building height led to better collapse capacities. In EC8, a global yielding mechanism is encouraged by requiring that the actual strength of the links including strain hardening at any floor not exceed the minimum specified value by more than 25%. Consideration may be given to vertical story shear distributions that have been recently proposed to achieve more uniform yielding along the height of structures (Chao and Goel, 2006; Park and Medina, 2007).

Taking advantage of the flexural stiffness and strength of continuous columns is another means of reducing variations in interstory drifts in multistory structures. The columns must be designed, however, to fulfill this function. Even for steel MRFs designed according to the weak beam–strong column code requirements for preventing story sway mechanisms, additional column capacity must be provided to prevent plastic hinging in columns resulting from differences in interstory drifts developing between adjacent floors during an earthquake (Krawinkler, 2006). Evaluating this flexural demand on columns is not straightforward because the bending moments have to be determined from a nonlinear dynamic time history analysis and they depend on the deformation profile of the structure, which in turn depends on the flexural stiffness of the columns. A simpler approach has been proposed that consists of sizing the columns to achieve a target collapse mechanism. For steel MRFs, Mazzolani and Piluso (1996) recommend that a global collapse mechanism could be achieved by proportioning the beams for gravity loads and, subsequently, designing the columns such that the global mechanism exhibits the lowest lateral capacity among all other possible collapse mechanisms. The calculations in their procedure are based on the plastic analysis method described earlier. The approach resulted in an enhanced inelastic demand distribution compared to a current code design procedure for a five-story building (Mazzolani and Piluso, 1997). To enhance this method, Neri (2000) developed a solution for the base columns such that iterations are avoided when including P–Δ in the calculations. Mastrandrea et al. (2003) applied the method to eccentrically braced steel frames. Ghersi et al. (2006) combined this global collapse strategy to the capacity spectrum (N2) method (Fajfar, 1999) to ensure that the structure has a minimum lateral strength consistent with the expected seismic demand. The approach has been verified by comparing the roof displacements from the method to those obtained from nonlinear dynamic analyses for a number of frames having different properties. Although not reported in the literature, a similar approach (adjusting column sizes to achieve a desired collapse mechanism) could probably be adopted using the stability check procedure proposed by Bernal (1992, 1998). Application of the method to tall structures, when the response is dominated by higher vibration modes, has not been reported in the literature. More recently, Ohsaki et al. (2007) used a multiobjective optimization technique for steel MRFs to simultaneously minimize steel tonnage and maximize the dissipated energy such that a uniform interstory drift is obtained at collapse of the structure. In this procedure, collapse is determined through an incremental dynamic analysis, thus accounting for structural dynamic response. Validation of the method was carried out for a five-story building.

Continuity of the columns that are not part of the moment-resisting frames (gravity columns or columns part of perpendicular frames) can also improve the drift response of steel MRFs (Gupta and Krawinkler, 2000b; Tagawa et al., 2004), and the above methods should be extended to include these columns because they will experience the same displacement demand during an earthquake.

The stability of braced steel frames can also be enhanced by exploiting column continuity. For these structures, MacRae et al. (2004) and Kimura and MacRae (2006) developed equations to predict the drift concentration factor (ratio of maximum interstory drift to roof story drift) and the bending moment demand on continuous columns as a function of the flexural stiffness of the columns, the ductility of the structure, the number of stories, and the vertical distribution of the story shears. These expressions have been validated through nonlinear dynamic analysis for frames up to 20 stories. In view of limitations imposed by assumptions used in their development (e.g., uniform column size over the building height), further refinement may be needed, however, before they can be used efficiently in practice. In CSA-S16, minimum column continuity is required for concentrically braced steel frames, both for the gravity columns and the bracing bent columns, and minimum flexural strength is required for the bracing bent columns of braced frame systems to delay plastic hinging and maintain the capacity to distribute vertically the inelastic demand.

For steel MRFs, Gupta and Krawinkler (2000b) propose providing a flexible back-up system with sufficient story stiffness to overcome the negative P–Δ effects on lateral stiffness (i.e., provide $\alpha \geq \theta$ in Fig. 19.21*f*) and suggest that the gravity load system could be used for this function. This solution should be applicable in practice and could be extended to other systems as well. For instance, the benefits of adding an MRF back-up system to braced steel frames have long been recognized in codes by means of relaxations (extended height limits, higher force modification factors, etc.). In order to be effective, such a parallel system should be designed to remain essentially elastic over the anticipated deflection range. Gupta and Krawinkler also proposed providing MRFs with more redundancy and adjusting the strength of members such that plastic hinging occurs at widely spaced interstory drifts, and the displacement when a full mechanism forms increases. The benefit of this concept was illustrated in Fig. 19.22*e* for a simple SDOF dual system exhibiting trilinear response.

In order to develop this behavior, Tagawa et al. (2004) inserted pinned beam-to-column connections at one end of all beams located in the exterior bays of a nine-story MRF. When column continuity was ignored in the analysis, the peak interstory drifts were generally reduced when compared to the response of the reference frame with rigid connections that had higher lateral stiffness and strength. For the frame studied, the benefits of the multilinear response were surpassed by those resulting from providing column continuity. Similarly, Charney and Atlayan (2008) proposed an MRF design using members and connections with a variety of detailing rules, including those associated with ordinary (OMF), intermediate (IMF), and special moment frames (SMF). The more ductile components were designed to yield at force levels well below the design basis earthquake, whereas the less ductile ones could remain elastic under this seismic demand. The cost of this design should be less than the conventional SMF solution because only a limited number of elements and connections have special detailing. Preliminary studies of a nine-story MRF structure showed that the margin against collapse was improved when compared to employing a conventional SMF system. The authors did indicate that this strategy resulted in a larger ductility demand on the ductile components and that this aspect would require further examination.

As discussed earlier, concentrically braced steel frames may be more vulnerable to a sway mechanism due to story strength degradation associated with brace buckling, and several provisions have been implemented in codes to compensate for this weakness. For instance, minimum symmetry requirements are provided to avoid one-sided response; in this regard, story shear is shared between tension- and compression-acting braces or tension-acting braces with comparable resistances supplied in each direction. More stringent height limits apply to the braced frames designed assuming higher ductility (or using higher R_d- or μ-factors) in CSA-S16 and NZS3404, which is consistent with the fact that P–Δ effects are more pronounced in taller braced frames or when greater inelastic demand is expected. Lower height limits are also specified in both codes when T/O bracing design is used, that is, the braces are designed assuming that the total story shear is entirely resisted by the tension-acting braces. This design allows for slender braces that have lower energy dissipation capacity compared to tension–compression braces, which causes larger interstory drifts. In NZS3404, the prescribed height limits also reduce if the brace slenderness is increased, and the design seismic loads must be amplified if the building height or the brace slenderness is increased, reflecting the influence of brace energy dissipation on braced frame response (Remennikov and Walpole, 1997). Furthermore, 50% amplification (braces effective in tension and compression) and 100% amplification (tension-only and chevron bracing) must be applied to interstory drift predictions from code seismic loads in view of the tendency for soft-story response in braced steel frames.

Tension–compression bracing in which braces are designed not to buckle under the design seismic loads may possess story shear overstrength due to the reserve tension capacity available in the tension-acting braces. This overstrength can be expressed as the ratio of the ultimate story shear capacity V_u contributed by the brace postbuckling compression strength C'_u and the tension brace tensile strength T_u to the design story shear V. This ratio is plotted against brace slenderness in Fig. 19.24*b* assuming that the brace design strength exactly matches the demand V. Abrupt variations in V_u/V along the building height are likely to result in drift concentrations, and selecting braces to minimize such variations should be considered in design. In Japan, the braces are designed such that V_u equals or exceeds the story shear demand V (Tada et al., 2003; Marino et al., 2006). With this approach, it is therefore easier to achieve uniformity in the V_u/V ratio. This method, however, will generally lead to lower lateral overstrength compared to cases where brace buckling is prevented under the design story shear, and as a result, increased seismic design loads (lower force modification factors) may be needed to achieve similar seismic performance.

In Fig. 19.24*b*, the variation of V_u/V with brace slenderness for T/C bracing shows that higher story shear capacity after brace buckling, that is, strain-hardening response, can be obtained if more slender braces are selected in design. This strategy can have positive impacts on braced frame stability by limiting interstory drifts and triggering brace buckling and yielding in several adjacent stories (Lacerte and Tremblay, 2006). In addition, longer brace fracture life would also be obtained by increasing λ. Brace slenderness, however, should be kept within a reasonable range because slender braces will lead to the use of additional bracing steel tonnage as well as higher design loads for brace connections and adjacent beam and column members. In EC8, the contribution of the compression-acting braces is ignored in brace design (T/O design). In this case, beneficial story shear overstrength is obtained by selecting less slender braces (Fig. 19.24*b*). The resulting extra-capacity provided by the compression braces, however, must be considered in the capacity design check.

Several innovative bracing configurations have been proposed to achieve more uniform interstory drift demand in braced steel frames. Some of them are illustrated in Fig. 19.25. For chevron bracing, Khatib et al. (1988) suggested adding a zipper column to tie together the beams and prevent a story mechanism (Fig. 19.25*a*). The further addition of a hat truss as shown in Fig. 19.25*b* allows for the full development of the tension brace capacity along the building height and further enhances the robustness of the system (Yang et al., 2008a,b). Rai and Goel (2003) found that the seismic stability of chevron bracing could be improved significantly by using a two-story X (split-X) configuration (Fig. 19.25*c*). This configuration has become very popular for CBF applications in North America. Mazzolani et al. (1994) proposed to connect two CBF bracing bents by weak beams rigidly connected to the inner columns (Fig. 19.25*d*). The CBFs would be designed to remain elastic such that a global collapse mechanism can be achieved with the energy dissipation concentrated in plastic hinges in the beam links. In Fig. 19.25*e*, vertical ties are added between the link ends of an EBF to form two braced substructures coupled by the links (Ghersi et al., 2000, 2003; Rossi, 2007). With this configuration, more uniform and effective exploitation of the overall energy dissipation capacity of the links can be achieved while mitigating the potential for story mechanisms.

Figure 19.25*f* shows a similar concept for buckling-restrained braced frames. The frame is divided into two parts, including an energy-dissipating frame and an elastic braced frame to prevent damage concentration (Tremblay, 2003; Merzouq and Tremblay, 2006). Mega braced frames built with conventional or buckling-restrained braces spanning over several levels along the exterior walls of

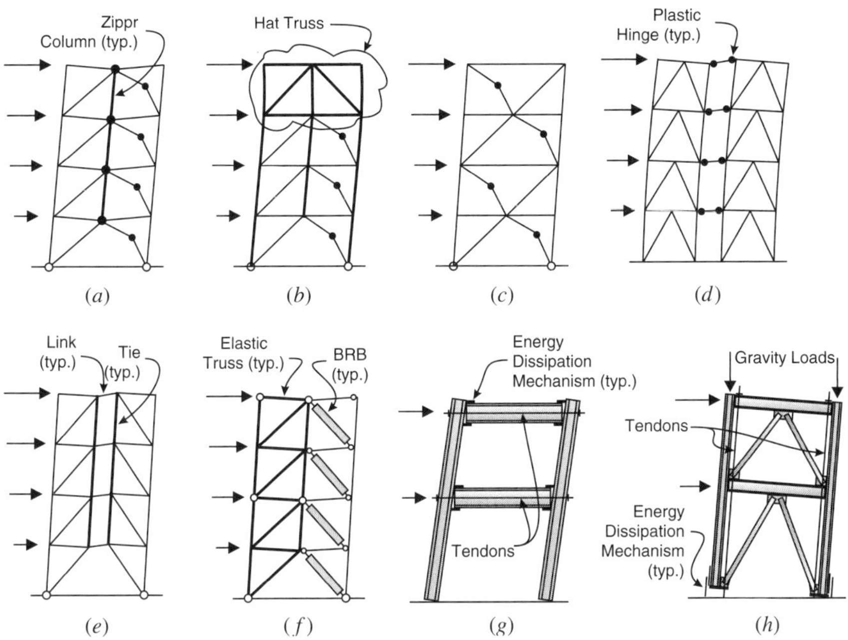

FIGURE 19.25 (*a*) Zipper column in chevron bracing; (*b*) suspended zipper system in chevron bracing; (*c*) two-story X-bracing; (*d*) coupled CBFs; (*e*) tied EBF; (*f*) dual BRB frame; (*g*) posttensioned MRF; and (*h*) controlled rocking CBF.

buildings have also been shown effective in developing uniform drift demand (Di Sarno and Elnashai, 2009).

By virtue of their capacity to mitigate residual deformations and permanent damage, systems exhibiting self-centering hysteretic behavior have attracted much attention in the last decade by the design profession. As illustrated in Fig. 19.22*e*, recentering properties may also be used to counter P–Δ effects. In the systems studied to date, the required hysteretic response has been essentially obtained by connecting framing components with either (i) tendons or bars made of superelastic shape memory (SMA) alloys that exhibit flag shape hysteretic behavior or (ii) tensioned steel tendons or bars acting in combination with a dedicated energy dissipation mechanism (yielding, friction etc.). Bracing members built with sliding components and tendons have also been designed to produce the desired self-centering behavior in compression and tension. SMA tendons have been used in beam-to-column moment connections (e.g., Graesser and Cozzarelli, 1991; Ocel et al., 2004; Sepúlveda et al., 2008) or in the fabrication of self-centering braces (e.g., Dolce et al., 2000; Auricchio et al., 2006; McCormick et al., 2007; Zhang and Zhu, 2008; Zhu and Zhang, 2008). Christopoulos et al. (2008) also proposed a bracing member with pretensioned tendons and friction energy dissipation capacity. Moment-resisting frames built with horizontal posttensioned steel tendons that run parallel to the beams have been examined in several past research programs (e.g.,

Ricles et al., 2001, 2002; Christopoulos et al., 2002b; Garlock et al., 2005; Tsai et al., 2007). As shown in Fig. 19.25*g*, the beams rotate relative to the face of the columns, which creates a gap and imposes elongation to the tendons. Energy is dissipated through yielding or friction in components connecting the beam to the columns. Rocking of the structure with respect to the foundations represents another means of developing self-centering capability (Fig. 19.25*h*). Recentering is provided by gravity loads acting alone or in combination with supplemental downward forces using posttensioned steel tendons. Energy dissipation can be introduced to control the motion and reduce impact forces when the columns land back on the foundations. Analytical studies and test programs have been conducted on several rocking systems for steel structures (Midorikawa et al., 2006, 2008; Azuhata et al., 2006; Ikenaga et al., 2006; Ishihara et al., 2006; Sause et al., 2006; Pollino and Bruneau, 2007, 2008; Eatherton et al., 2008; Tremblay et al., 2008c).

REFERENCES

Adam, C., Ibarra, L., and Krawinkler, H. (2004), "Evaluation of P-Delta Effects in Non-Deteriorating MDOF Structures from Equivalent SDOF Systems," Paper No. 3407, *Proc. 13th World Conf. on Earthquake Eng.*, Vancouver, Canada.

AIJ (1975), *Guide to the Plastic Design of Steel Structures*, Architectural Institute of Japan, Tokyo, Japan.

AISC (2005a), *Seismic Provisions for Structural Steel Buildings, Including Supplement No. 1*, ANSI/AISC 341-05, American Institute of Steel Construction, Chicago, IL.

AISC (2005b), *Prequalified Connections for Special and Intermediate Steel Moment Frames for Seismic Applications*, ANSI/AISC 358-05, American Institute of Steel Construction, Chicago, IL.

AISC (2005c), *Specification for Structural Steel Buildings*, American Institute of Steel Construction, ANSI/AISC 360-05, Chicago, IL.

AISC (2006), *Seismic Design Manual*, American Institute of Steel Construction and Structural Educational Council, Chicago, IL.

Alinia, M.M., and Dastfan, M. (2007), "Cyclic Behaviour, Deformability and Rigidity of Stiffened Steel Shear Panels," *J. Constr. Steel Res.*, Vol. 63, No. 4, pp. 554–563.

Araki, Y., and Hjelmstad, K. D. (2000), "Criteria for Assessing Dynamic Collapse of Elastoplastic Structural Systems," *Earthquake Eng. Struct. Dyn.*, Vol. 29, No. 8, pp. 1177–1198.

Arlekar, J.N., and Murty, C. V. R. (2002), "P-V-M Interaction Curves for Seismic Design of Steel Column Base Connections," *AISC Eng. J.*, Vol. 39, No. 3, pp. 154–165.

ASCE (2005), *Minimum Design Loads for Buildings and Other Structures, Includes Supplement No. 1*, ASCE/SEI 7-05, ASCE, Reston, VA.

ASCE (2008), *Supplement No. 2 to ASCE 7-05*, ASCE, Reston, VA.

Asari, T., Inoue, K., and Ishiyama, Y. (2000), "Evaluation of Destructiveness of Earthquake Motions by Collapse Base Shear Coefficient Spectra," Paper No. 1537, *Proc. 12th World Conf. Earthquake Eng.*, Auckland, NZ.

Asimakopoulos, A. V., Karabalis, D. L., and Beskos, D. E. (2007), "Inclusion of P–Δ Effect in Displacement-Based Seismic Design of Steel Moment Resisting Frames," *Earthquake Eng. Struct. Dyn.*, Vol. 36, No. 14, pp. 2171–2188.

Aslani, F., and Goel, S. C. (1991), "Stitch Spacing and Local Buckling in Seismic-Resistant Double-Angle Braces." *ASCE J. Struct. Eng.*, Vol. 117, No. 8, pp. 2442–2463.

Aslani, F., and Goel, S. C. (1992a), "Stitch Spacing and End Fixity in Seismic Resistant Boxed Angle Braces," *ASCE J. Struct. Eng.*, Vol. 118, No. 10, pp. 2872–2889.

Aslani, F., and Goel, S. C. (1992b), "Analytical Criteria for Stitch Strength of Built-up Compression Members," *AISC Eng. J.*, Vol. 29, No. 3, pp. 102–110.

Astaneh-Asl, A. (1998), *Seismic Behavior and Design of Gusset Plates*, Steel Tips, Structural Steel Educational Council, Moraga, CA.

Astaneh-Asl, A. (2001), *Seismic Behavior and Design of Steel Shear Walls*, Steel Tips, Structural Steel Educational Council, Moraga, CA.

Astaneh-Asl, A., Cochran, M. L., and Sabelli, R. (2006), *Seismic Detailing of Gusset Plates for Special Concentrically Braced Frames*, Steel Tips, Structural Steel Education Council, Moraga, CA.

Astaneh-Asl, A., and Goel, S. C. (1984), "Cyclic In-Plane Buckling of Double Angle Bracing," *ASCE J. Struct. Eng.*, Vol. 110, No. 9, pp. 2036–2055.

Astaneh-Asl, A., Goel, S. C., and Hanson, R. D. (1985), "Cyclic Out-of-Plane Buckling of Double-Angle Bracing," *ASCE J. Struct. Eng.*, Vol. 111, No. 5, pp. 1135–1153.

Astaneh-Asl, A., Goel, S. C., and Hanson, R. D. (1986), "Earthquake-Resistant Design of Double-Angle Bracing," *AISC Eng. J.*, Vol. 23, No. 4, pp. 133–147.

ATC (2008), *ATC-63, Recommended Methodology for Quantification of Building System Performance and Response Parameters, 90 % Interim Draft Report*, Applied Technology Council, Redwood City, CA.

Auricchio, F., Fugazza, D., and Desroches, R. (2006), "Earthquake Performance of Steel Frames with Nitinol Braces," *J. Earthquake Eng.*, Vol. 10, No. 1, pp. 45–66.

Azuhata, T., Ishihara, T., Midorikawa, M., and Wada, M. (2006), "Seismic Response Reduction of Steel Frames with Multi-Spans by Applying Rocking Structural System," paper presented at the Proc. STESSA 2006 Conf., Yokohama, Japan, pp. 659–664.

Ballio, G., and Perotti, F. (1987), "Cyclic Behaviour of Axially Loaded Members: Numerical Simulation and Experimental Verification," *J. Constr. Steel Res.*, Vol. 7, No. 1, pp. 3–41.

Berman, J. W., and Bruneau, M. (2003), "Plastic Analysis and Design of Steel Plate Shear Walls," *ASCE J. Struct. Eng.*, Vol. 129, No. 11, pp. 1448–1456.

Berman, J. W., and Bruneau, M. (2004), "Steel Plate Shear Walls Are Not Plate Girders," *AISC Eng. J.*, Vol. 41, No. 3, pp. 95–106.

Berman, J. W., and Bruneau, M. (2005), "Experimental Investigation of Light-Gauge Steel Plate Shear Walls," *ASCE J. Struct. Eng.*, Vol. 131, No. 2, pp. 259–267.

Berman, J. W., and Bruneau, M. (2008a), "Capacity Design of Vertical Boundary Elements in Steel Plate Shear Walls," *AISC Eng. J.*, Vol. 45, No. 3, pp. 57–71.

Berman, J. W., and Bruneau, M. (2008b), "Tubular Links for Eccentrically Braced Frames I: Finite Element Parametric Study; II: Experimental Verification," *ASCE J. Struct. Eng.*, Vol. 134, No. 5, pp. 692–712.

Bernal, D. (1987), "Amplification Factors for Inelastic Dynamics P–Δ Effects in Earthquake Analysis," *Earthquake Eng. Struct. Dynamics*, Vol. 15, No. 5, pp. 635–651.

Bernal, D. (1992), "Instability of Buildings Subjected to Earthquakes," *ASCE J. Struct. Eng.*, Vol. 118, No. 8, pp. 2239–2260.

Bernal, D. (1998), "Instability of Building During Seismic Response," *Eng. Struct.*, Vol. 20, Nos. 4–6, pp. 496–502.

Bernal, D., Nasseri, A., and Bulut, Y. (2006), "Instability Inducing Potential of Near Fault Ground Motions," in *Proc. of the SMIP06 Seminar on Utilization of Strong-Motion Data* (Ed. M. Huang), California Strong Motion Instrumentation Program, Oakland, CA, pp. 41–62.

Bertero, V. V., Anderson, J., and Krawinkler, H. (1994), "Performance of Steel Building Structures During the January 17, 1994 Northridge Earthquake," Report No. UBC/EERC-94/09, Earthquake Engineering Research Center, Univ. of California, Berkeley, CA.

Bertero, V. V., Uang, C.-M., Llopiz, C. R., and Igarashi, K. (1989), "Earthquake Simulator Testing of Concentric Braced Dual System," *ASCE J. Struct. Eng.*, Vol. 115, No. 8, pp. 1877–1894.

Bhowmick, A. K., Driver, R. G., and Grondin, G. Y. (2009), "Seismic Analysis of Steel Plate Shear Walls Considering Strain Rate and P–Delta Effects," *J. Constr. Steel Res.*, Vol. 65, No. 5, pp. 1149–1159.

Black, C. J., Makris, N., and Aiken, I. D. (2004), "Component Testing, Seismic Evaluation and Characterization of Buckling-Restrained Braces," *ASCE J. Struct. Eng.*, Vol. 130, No. 6, pp. 880–894.

Black, R. G., Wenger, W. A. B., and Popov, E. P. (1980), "Inelastic Buckling of Steel Struts under Cyclic Load Reversals", Report UCB/EERC-80/40, Earthquake Eng. Research Center, Univ. of California, Berkeley, CA.

Bozorgnia, Y., and Campbell, K. W. (2004), "Engineering Characterization of Ground Motions," in *Earthquake Engineering—From Engineering Seismology to Performance-Based Engineering* (Eds. Y. Bozorgnia and V. V. Bertero), CRC Press, Boca Raton, FL, Chap. 5.

Broderick, B. M., Elghazouli, A. Y., and Goggins, J. (2008), "Earthquake Testing and Response Analysis of Concentrically-Braced Sub-Frames," *J. Constr. Steel Res.*, Vol. 64, No. 9, pp. 997–1007.

Broderick, B. M., Goggins, J. M., and Elghazouli, A. Y. (2005), "Cyclic Performance of Steel and Composite Bracing Members," *J. Constr. Steel Res.*, Vol. 61, No. 4, pp. 493–514.

Bruneau, M., Uang, C.-M., and Whittaker, A. S. (1998), *Ductile Design of Steel Structures*, McGraw-Hill, New York.

Byfield, M. P., Davies, J. M., and Dhanalakshmi, M. (2005), "Calculation of the Strain Hardening Behaviour of Steel Structures Based on Mill Tests," *J. Constr. Steel Res.*, Vol. 61, No. 2, pp. 133–150.

Caccese, V., Elgaaly, M., and Chen, R. (1993), "Experimental Study of Thin Steel-Plate Shear Walls Under Cyclic Load," *ASCE J. Struct. Eng.*., Vol. 119, No. 2, pp. 573–587.

Calderoni, B., Giubileo, C., and De Martino, A. (2006), "Assessment of Hysteretic Cyclic Behaviour of Plastic Hinge in Cold-Formed Steel Beams," *Proc. STESSA 2006*, Yokohama, Japan, pp. 185–190.

Calderoni, B., De Martino, A. Formisano, A., and Fiorino, L. (2009), "Cold-Formed Steel Beams Under Monotonic and Cyclic Loading: Experimental Investigation," *J. Constr. Steel Res.*, Vol. 65 No. 1, pp. 219–227.

Carter, C. J. (2003), *Stiffening of Wide-Flange Columns at Moment Connections: Wind and Seismic Applications*, Steel Design Guide 13, 2nd printing, American Institute of Steel Construction, Chicago, IL.

Castiglioni, C. A., (2005), "Effects of the Loading History on the Local Buckling Behavior and Failure Mode of Welded Beam-to-Column Joints in Moment-Resisting Steel Frames," *ASCE J. Eng. Mech.*, Vol. 6, No. 1, pp. 568–585.

Castiglioni, C. A., Mouzakis, H. P., and Carydis, P. G. (2007), "Constant and Variable Amplitude Cyclic Behavior of Welded Steel Beam-to-Column Connections," *J. Earthquake Eng.*, Vol. 11, No. 6, pp. 876–902.

CEN. (2002), prEN 1993-1-8:E, *Eurocode 3: Design of Steel Structures—Part 1.8 : Design of Joints*, Comité Européen de Normalisation, (CEN), European Committee for Standardization, Brussels, Belgium.

CEN. (2004), EN 1998-1:2004:E, *Eurocode 8: Design of Structures for Earthquake Resistance—Part 1: General Rules, Seismic Actions and Rules for Buildings*, Comité Européen de Normalisation, (CEN), European Committee for Standardization, Brussels, Belgium.

Chambers, J. J., and Ernst, C. J. (2004), *Brace Frame Gusset Plate Research—Phase I Literature Review*, Report, Dept. of Civil and Environmental Eng., University of Utah, UT.

Chao, S. H., and Goel, S. C. (2006), "Performance-Based Seismic Design of Eccentrically Braced Frames Using Target Drift and Yield Mechanism as Performance Criteria," *AISC Eng. J.*, Vol. 43, No. 3, pp. 173–200.

Chao, S.-H., Khandelwal, K., and El-Tawil, S. (2006), "Ductile Web Fracture Initiation in Steel Shear Links," *ASCE J. Struct. Eng.*, Vol. 132, No. 8, pp. 1192–1200.

Charney, F. A., and Atlayan, O. (2008), "Hybrid Moment Resisting Steel Frames," Paper No. 05-06-0164, *Proc. 14th World Conf. on Earthquake Eng.*, Beijing, China.

Chen, C.-H., Lai, J. W., and Mahin, S. A., (2008), "Seismic Performance Assessment of Concentrically Braced Steel Frame Buildings," Paper No. 05-05-0158, *Proc. 14th World Conf. on Earthquake Eng.*, Beijing, China.

Chen, S.-J., and Tu, C.-T. (2004), "Experimental Study of Jumbo Size Reduced Beam Section Connections Using High-Strength Steel," *ASCE J. Struct. Eng.*, Vol. 130, No. 4, pp. 582–587.

Chen, Z., Ge, H., and Usami, T. (2006), "Hysteretic Model of Stiffened Shear Panel Dampers," *ASCE J. Struct. Eng.*, Vol. 132, No. 3, pp. 478–483.

Cheng, F. Y., and Tseng, W. H. (1973), "Dynamic Instability and Ultimate Capacity of Inelastic Systems Parametrically Excited by Earthquakes: Part 1," Tech. Report, National Science Foundation, August (available from U.S. Department of Commerce, National Technical Information Service, Springfield, VA, NTIS Access No. PB261096jAS).

Chi, B., and Uang, C.-M. (2002), "Cyclic Response and Design Recommendations of Reduced Beam Section Moment Connections with Deep Columns," *ASCE J. Struct. Eng.*, Vol. 128, No. 4, pp. 464–473.

Chi, W.-M., Kanvinde, A. M., and Deierlein, G. G. (2006), "Prediction of Ductile Fracture in Steel Connections Using SMCS Criterion," *ASCE J. Struct. Eng.*, Vol. 132, No. 2, pp. 171–181.

Choi, J., Stojadinovic, B., and Goel, S. C. (2003), "Design of Free Flange Connections," *AISC Eng. J.*, Vol. 40, No. 1, pp. 24–41.

Chopra, A. K. (2001), *Dynamics of Structures—Theory and Applications to Earthquake Engineering*, 2nd ed., Prentice Hall, Upper Saddle River, NJ.

Chou, C.-C., and Chen, P.-J. (2009), "Compressive Behavior of Central Gusset Plate Connections for a Buckling-Restrained Braced Frame," *J. Constr. Steel Res.*, Vol. 65, No. 5, pp. 1138–1148.

Christopoulos, C., Filiatrault, A., and Folz, B. (2002a), "Seismic Response of Self-Centering Hysteretic SDOF Systems," *Earthquake Eng. Struct. Dyn.*, Vol. 31, No. 5, pp. 1131–1150.

Christopoulos, C., Filiatrault, A., Uang, C. M., and Folz, B. (2002b), "Post-Tensioned Energy Dissipating Connections for Moment-Resisting Steel Frames," *ASCE J. Struct. Eng.*, Vol. 128, No. 9, pp. 1111–1120.

Christopoulos, C., Tremblay, R., Kim, H.-J., and Lacerte, M. (2008), "Development and Validation of a Self-Centering Energy Dissipative (SCED) Bracing System for the Seismic Resistance of Structures," *ASCE J. Struct. Eng.*, Vol. 134, No. 1, pp. 96–107.

Chusilp, P., and Usami, T. (2002), "New Elastic Stability Formulas for Multiple-Stiffened Shear Panels," *ASCE J. Struct. Eng.*, Vol. 128, No. 6, pp. 833–836.

Ciutina, A. L., and Dubina, D. (2008), "Column Web Stiffening of Steel Beam-to-Column Joints Subjected to Seismic Actions," *ASCE J. Struct. Eng.*, Vol. 134, No. 3, pp. 505–510.

Cochran, M. L., and Honeck, W. C. (2004), *Design of Special Concentric Braced Frames (with Comments on Ordinary Concentric Braced Frames)*, Steel Tips, Structural Steel Education Council, Moraga, CA.

CSA. (2001), "CAN/CSA-S16.1 Limits States Design of Steel Construction, Including CSA-S16S1-05 Supplement No. 1," Canadian Standards Association, Willowdale, ON.

Davaran, A. (2001), "Effective Length Factor for Discontinuous X-Bracing Systems," *ASCE J. Struct. Eng.*, Vol. 127, No. 2, pp. 106–112.

Davaran, A., and Far, N. E. (2009), "An Inelastic Model for Low Cycle Fatigue Prediction in Steel Braces," *J. Constr. Steel Res.*, Vol. 65, No. 3, pp. 523–530.

Davaran, A., and Hoveidae, N. (2009), "Effect of Mid-Connection Detail on the Behavior of X-Bracing Systems," *J. Constr. Steel Res.*, Vol. 65, No. 2, pp. 290–298.

Davidson, B. J., and Fenwick, R. C. (2004), "Proposed P-Delta Design for the Australian and New Zealand Loadings Standard," *Proc. 2004 Annual Stability Conf.*, Structural Stability Research Council, Long Beach, CA, pp. 581–602.

Della Corte, G., De Matteis, G., Landolfo, R., and Mazzolani, F. M. (2002), "Seismic Analysis of MR Steel Frames Based on Refined Hysteretic Models of Connections," *J. Constr. Steel Res.*, Vol. 58, No. 10, pp. 1331–1345.

Ding, Y., Zhang, Y., and Zhao, J. (2009), "Tests of Hysteretic Behavior for Unbonded Steel Plate Brace Encased in Reinforced Concrete Panel," *J. Constr. Steel Res.*, Vol. 65, No. 5, pp. 1160–1170.

Di Sarno, L., and Elnashai, A. S. (2009), "Bracing Systems for Seismic Retrofitting of Steel Frames," *J. Constr. Steel Res.*, Vol. 65, No. 2, pp. 452–465.

Di Sarno, L., Elnashai, A. S., Nethercot, D. A. (2008), "Seismic Response of Stainless Steel Braced Frames," *J. Constr. Steel Res.*, Vol. 64, Nos. 7–8, pp. 914–925.

Dolce, M., Cardone, D., and Marnetto, R. (2000), "Implementation and Testing of Passive Control Devices Based on Shape Memory Alloys," *Earthquake Eng. Struct. Dyn.*, Vol. 29, No. 7, pp. 945–968.

Dowswell, B. (2006), "Effective Length Factors for Gusset Plate Buckling," *AISC Eng. J.*, Vol. 43, No. 2, pp. 91–101.

Driver, R. G., Kulak, G. L., Kennedy, D. J. L., and Elwi, A. E. (1998), "Cyclic Test of Four-Story Steel Plate Shear Wall," *ASCE J. Struct. Eng.*, Vol. 124, No. 2, pp. 112–120.

Dubina, D. (2004), "Ductility and Seismic Performance of Thin-Walled Cold-Formed Steel Structures," *Int. J. Steel Struct.*, Vol. 4, No. 4, pp. 209–222.

Eatherton, M. R., Hajjar, J. F., Deierlein, G. G., Krawinkler, H., Billington, S., and Ma, X. (2008), "Controlled Rocking of Steel-Framed Buildings with Replaceable Energy-Dissipating Fuses," Paper No. 05-06-0026, *Proc. 14th World Conf. on Earthquake Eng.*, Beijing, China.

Elchalakani, M., Zhao, X.-L., and Grzebieta, R. (2003), "Tests of Cold-Formed Circular Tubular Braces under Cyclic Axial Loading," *ASCE J. Struct. Eng.*, Vol. 129, No. 4, pp. 507–514.

El-Tawil, S., and Deierlein, G. G. (1999), "Strength and Ductility of Concrete Encased Composite Columns," *ASCE J. Struct. Eng.*, Vol. 125, No. 9, pp. 1009–1019.

El-Tawil, S., Vidarsson, E., Mikesell, T., and Kunnath, S. K. (1999), "Inelastic Behavior and Design of Steel Panel Zones," *ASCE J. Struct. Eng.*, Vol. 125, No. 2, pp. 183–193.

El-Tayem, A. A., and Goel, S. C. (1986), "Effective Length Factor for the Design of X-Bracing Systems," *AISC Eng. J.*, Vol. 23, No. 1, pp. 41–45.

Engelhardt, M. D., and Popov, E. P. (1992), "Experimental Performance of Long Links in Eccentrically Braced Steel Frames," *ASCE J. Struct. Eng.*, Vol. 118, No. 11, pp. 3067–3088.

Engelhardt, M. D., Winneberger, T., Zekany, J., and Potyraj, T. J. (1998), "Eexperimental Investigation of Dog Bone Moment Connections," *AISC En. J.*, Vol. 35, No. 4, pp. 128–139.

Fahmy, M., Goel, S. C., and Stojadinovic, B. (2000), "Hysteretic Behavior of Steel Moment Resisting Column Bases," *Proc. STESSA 2000 Conf.*, Montreal, Canada, pp. 191–197.

Fajfar, P. (1999), "Capacity Spectrum Method Based on Inelastic Demand Spectra," *Earthquake Eng. Struct. Dyn.*, Vol. 28, No. 9, pp. 979–993.

FEMA. (2000a), *FEMA-350, Recommended Seismic Design Criteria for New Steel Moment-Frame Buildings*, SAC Joint Venture for the Federal Emergency Management Agency, Washington, DC.

FEMA. (2000b), *FEMA 356 Prestandard and Commentary for the Seismic Rehabilitation of Buildings*, ASCE, Federal Emergency Management Agency, Washington, DC.

Fell, B. V., Kanwinde, A. M., Deierlein, G. G., and Myers, A. T. (2009), "Experimental Investigation of Inelastic Cyclic Buckling and Fracture of Steel Braces," *ASCE J. Struct. Eng.*, Vol. 135, No. 1, pp. 1–19.

Fell, B. V., Kanwinde, A. M., Deierlein, G. G., Myers, A. T., and Fu, X. (2006), *Buckling and Fracture of Concentric Braces under Inelastic Cyclic Loading,* Steel Tips, Structural Steel Education Council, Moraga, CA.

Fenwick, R. C., Davidson, B. J., and Chung, B. T. (1992), "P-Delta Actions in Seismic Resistant Structures," *Bull. New Zeal. Nat. Soc. Earthquake Eng.*, Vol. 25, No. 1, pp. 56–69.

Formisano, A., Faggiano, B., Landolfo, R., and Mazzolani, F. M. (2006), "Ductile Behavioural Classes of Steel Members for Seismic Design," *Proc. STESSA 2006 Conf.*, Yokohama, Japan, pp. 225–232.

Foutch, D. A., Goel, S. C., and Roeder, C. W. (1987), "Seismic Testing of Full-Scale Steel Building—Part I," *ASCE J. Struct. Eng.*, Vol. 113, No. 11, pp. 2111–2129.

Fukumoto, Y., and Itoh, Y. (1992), "Width-to-Thickness Ratios for Plate Elements in Earthquake Engineering Design of Steel Structures," in *Stability and Ductility of Steel Structures Under Cyclic Loading* (Ed., Y. Fukumoto and G. Lee), CRC Press, Boca Raton, FL, pp. 285–296.

Fukumoto, Y., and Lee, G. (Eds.) (1992), *Stability and Ductility of Steel Structures Under Cyclic Loading*, CRC Press, Boca Raton, FL.

Fukuta, T., Nishiyama, I., Yamanouchi, H., and Kato, B. (1989), "Seismic Performance of Steel Frames with Inverted V Braces," *ASCE J. Struct. Eng.*, Vol. 115, No. 8, pp. 2016–2028.

Garlock, M. M., Ricles, J. M., Sause, R., and Zhao, C. (2005), "Experimental Studies of Full-Scale Posttensioned Steel Connections," *ASCE J. Struct. Eng.*, Vol. 131, No. 3, pp. 438–448.

Garlock, M. M., Ricles, J. M., and Sause, R. (2003), "Cyclic Load Tests and Analysis of Bolted Top-and-Seat Angle Connections," *ASCE J. Struct. Eng.*, Vol. 129, No. 12, pp. 1615–1625.

Ger, J.-F., Cheng., F. Y., and Lu, L.-W. (1993), "Collapse Behavior of Pino Suarez Building During 1985 Mexico City Earthquake," *ASCE J. Struct. Eng.*, Vol. 119, No. 3, pp. 852–870.

Ghersi, A., Marino, E. M., and Neri, F. (2006), "A Seismic Design Method for High Ductility Steel Frames," *Proc. STESSA 2006 Conf.*, Yokohama, Japan, pp. 33–38.

Ghersi, A., Neri, F., Rossi, P. P., and Perretti, A. (2000), "Seismic Response of Tied and Trussed Eccentrically Braced Frames," *Proc. STESSA 2000 Conf.*, Montreal, Canada, pp. 495–502.

Ghersi, A., Pantano, S., and Rossi, P. P. (2003), "On the Design of Tied Braced Frames," *Proc. STESSA 2003 Conf.*, Naples, Italy, pp. 413–419.

Ghobarah, A., Korol, R. M., and Osman, A. (1992), "Cyclic Behavior of Extended End-Plate Joints," *ASCE J. Struct. Eng.*, Vol. 118, No. 5, pp. 1333–1353.

Gioncu, V. (2000), "Framed Structures. Ductility and Seismic Response. General Report," *J. Constr. Steel Res.*, Vol. 55, Nos. 1–3, pp. 125–154.

Gioncu, V., and Petcu, D. (1997), "Available Rotation Capacity of Wide-Flange Beams and Beam-Columns: Part 1. Theoretical Approaches and Part 2. Experimental and Numerical Tests," *J. Constr. Steel Res.*, Vol. 43, Nos. 1–3, pp. 161–244.

Goel, S. C., and Chao, S.-H. (2008), *Performance-Based Plastic Design: Earthquake-Resistant Steel Structures*, ICC Publications, Country Club Hills, IL.

Goel, R. K., and Chopra, A. K. (2004), "Evaluation of Modal and FEMA Pushover Analyses: SAC Buildings," *Earthquake Spectra*, Vol. 20, No. 1, pp. 225–254.

Goel, S. C., and El-Tayem, A. A. (1986), "Cyclic Load Behavior of Angle X-Bracing," *ASCE J. Struct. Eng. J.*, Vol. 112, No. 11, pp. 2528–2539.

Goel, S. C., and Xu, P. (1989), "Brace Connections for Ductility," *Proc. ASCE Structures Congress*, San Francisco, CA, pp. 559–568.

Goggins, J. M., Broderick, B. M., Elghazouli, A. Y., and Lucas, A. S. (2006), "Behaviour of Tubular Steel Members Under Cyclic Axial Loading," *J. Constr. Steel Res.*, Vol. 62, Nos. 1–2, pp. 121–131.

Graesser, E. J., and Cozzarelli, F. A. (1991), "Shape-Memory Alloys as New Materials for Aseismic Isolation," *ASCE J. Eng. Mech.*, Vol. 117, No. 11, pp. 2590–2608.

Green, P. S., Sause, R., and Ricles, J. M. (2002), "Strength and Ductility of HPS Flexural Members," *J. Constr. Steel Res.*, Vol. 58, Nos. 5–8, pp. 907–941.

Gugerli, H. (1982), "Inelastic Cyclic Behavior of Steel Bracing Members", Ph.D. thesis, University of Michigan, Ann Arbor, MI.

Guo, B., Gu, Q., and Liu, F. (2006), "Experimental Behavior of Stiffened and Unstiffened End-Plate Connections under Cyclic Loading," *ASCE J. Struct. Eng.*, Vol. 132, No. 9, pp. 1352–1357.

Gupta, A., and Krawinkler, H. (2000a). "Estimation of Seismic Drift Demands for Frame Structures," *Earthquake Eng. Struct. Dyn.*, Vol. 29, No. 9, pp. 1287–1305.

Gupta, A., and Krawinkler, H. (2000b). "Dynamic P–Δ Effects for Flexible Inelastic Steel Structures," *ASCE J. Struct. Eng.*, Vol. 126, No. 1, pp. 145–154.

Haddad, M. A. (2004), "Design of Concentrically Braced Steel Frames for Earthquakes," Ph.D. thesis, Dept. of Civil Engineering, University of Calgary, Calgary, Alberta, Canada.

Han, S.-W., Kim, W. T., and Foutch, D. A. (2007), "Seismic Behavior of HSS Bracing Members according to Width–Thickness Ratio under Symmetric Cyclic Loading," *ASCE J. Struct. Eng.*, Vol. 133, No. 2, pp. 264–273.

Haselton, C. B., and Baker, J. W. (2006), "Ground Motion Intensity Measures for Collapse Capacity Predictions: Choice of Optimal Spectral Period and Effect of Spectral Shape," Paper No. 461, *Proc. 8th U.S. Nat. Conf. Earthquake Eng.*, San Francisco, CA.

Hassan, O. F., and Goel, S. C. (1991), "Modeling of Bracing Members and Seismic Behavior of Concentrically Braced Steel Structures," Research Report UMCE 91-1, Dept. of Civil Engineering, University of Michigan, Ann Arbor, MI.

Hjelmstad, K. D., and Popov, E. P. (1983), "Cyclic Behavior and Design of Link beams," *ASCE J. Struct. Eng.*, Vol. 109, No. 10, pp. 2387–2403.

Hjelmstad, K. S., and Lee, S.-G. (1989), "Lateral Buckling of Beams in Eccentrically-Braced Frames," *J. Constr. Steel Res.*, Vol. 14, No. 4, pp. 251–272.

Huang, Y., and Mahin, S. A. (2008), "Evaluation of Steel Structure Deterioration with Cyclic Damaged Plasticity," Paper No. 14-0240, *Proc. 14th World Conf. on Earthquake Eng.*, Beijing, China.

Huang, Z., Varmat, A. H., Arichadran, R. S., and Cordero, A. (2004), "Effects of Inelastic Local Buckling on the Seismic Behavior of Members and Frames," *Proc. 2004 Annual Stability Conference*, Structural Stability Research Council, Long Beach, CA, pp. 329–351.

Humar, J., Mahgoub, M., and Ghorbanie-Asl, M. (2006), "Effect of Second-Order Forces on Seismic Response," *Can. J. Civ. Eng.*, Vol. 33, No. 6, pp. 692–706.

Husid, R. (1967), "Gravity Effects on the Earthquake Response of Yielding Structures," Ph.D. thesis, California Institute of Technology, Pasadena, CA.

Ibarra, L. F., and Krawinkler, H. (2005), "Global Collapse on Frame Structures Under Seismic Excitations," Report No. 152, John A. Blume Earthquake Engineering Research Center, Stanford University, San Francisco, CA.

Ibarra, L. F., Medina, R. A., and Krawinkler, H. (2005), "Hysteretic Models That Incorporate Strength and Stiffness Deterioration," *Earthquake Eng. Struct. Dyn.*, Vol. 34, No. 12, pp. 1489–1511.

Ikenaga, M., Nagae, T., Nakashima, M., and Suita, K. (2006), "Development of Column Bases Having Self-Centering and Damping Capability," *Proc. STESSA 2006 Conf.*, Yokohama, Japan, pp. 703–709.

Inai, E., Mukai, A., Kai, M., Tokinoya, H., Fukumoto, T., and Mori, K. (2004), "Behavior of Concrete-Filled Steel Tube Beam Columns," *ASCE J. Struct. Eng.*, Vol. 130, No. 2, pp. 189–202.

Inoue, K., Sawaizumi, S., and Higashibata, Y. (2004), "Stiffening Requirements for Unbonded Braces Encased in Concrete Panels," *ASCE J. Struct. Eng.*, Vol. 127, No. 6, pp. 712–719.

Inoue, K., Suita, K., Takeuchi, I., Chusilp, P., Nakashima, M., and Zhou, F. (2006), "Seismic-Resistant Weld-Free Steel Frame Buildings with Mechanical Joints and Hysteretic Dampers," *ASCE J. Struct. Eng.*, Vol. 132, No. 6, pp. 864–872.

Ishihara, T., Azuhata, T., and M. Midorikawa, M. (2006), "Effect of Dampers on Dynamic Behavior of Structures Allowed to Uplift," *Proc. STESSA 2006 Conf.*, Yokohama, Japan, pp. 709–715.

Ishizawa, T., and Iura, M. (2005), "Analysis of Tubular Steel Bridge Piers," *Earthquake Eng. Struct. Dyn.*, Vol. 34, No. 8, pp. 985–1004.

Iwata, M., Kato, T., and Wada, A. (2000), "Buckling Restrained Braces as Hysteretic Dampers," *Proc. STESSA 2000,* Montreal, Canada, pp. 33–38.

Iwata, M., and Murai, M. (2006), "Buckling-Restrained Brace Using Steel Mortar Planks; Performance Evaluation as a Hysteretic Damper," *Earthquake Eng. Struct. Dyn.*, Vol. 35, No. 14, pp. 1807–1826.

Jain, A. K., Goel, S. C., and Hanson, R. D. (1978), "Inelastic Response of Restrained Steel Tubes," *ASCE J. Struct. Div.*, Vol. 104, No. ST6, pp. 897–910.

Jain, A. K., Goel, S. C., and Hanson, R. D. (1980), "Hysteretic Cycles of Axially Loaded Steel Members," *ASCE J. Struct. Div.*, Vol. 106, No. ST8, pp. 1777–1795.

Jennings, P. C., and Husid, R. (1968), "Collapse of Yielding Structures During Earthquakes," *ASCE J. Eng. Mech. Div.*, Vol. 94, No. 5, pp. 1045–1065.

Jin, J., and El-Tawil, S. (2005), "Seismic Performance of Steel Frames with Reduced Beam Section Connections," *J. Constr. Steel Res.*, Vol. 61, No. 4, pp. 453–471.

Jones, S. L., Fry, G. T., and Engelhardt, M. D. (2002), "Experimental Evaluation of Cyclically Loaded Reduced Beam Section Moment Connections," *ASCE J. Struct. Eng.*, Vol. 128, No. 4, pp. 441–451.

Ju, Y. K., Kim, M.-H., Kim, J., and Kim, S.-D. (2009), "Component Tests of Buckling-Restrained Braces with Unconstrained Length," *Eng. Struct.*, Vol. 31, No. 2, pp. 507–516.

Kanvinde, A. M., and Deierlein, G. G. (2007), "A Cyclic Void Growth Model to Assess Ductile Fracture in Structural Steels Due to Ultra Low Cycle Fatigue," *ASCE J. Eng. Mech.*, Vol. 133, No. 6, pp. 701–712.

Karavasilis, T. L., Bazeos, N., and Beskos, D. E. (2007), "Estimation of Seismic Drift and Ductility Demands in Planar Regular X-Braced Steel Frames," *Earthquake Eng. Struct. Dyn.*, Vol. 36, No. 15, pp. 2273–2289.

Kasai, K., and Han, X. (1997), "New EBF Design Method and Application: Redesign and Analysis of US-Japan EBF," *Proc. STESSA 1997 Conf.*, Kyoto, Japan, pp. 242–249.

Kasai, K., and Popov, E. P. (1986a), "General Behavior of WF Steel Shear Link Beams," *ASCE J. Struct. Eng.*, Vol. 112, No. 2, pp. 362–382.

Kasai, K., and Popov, E. P. (1986b), "Cyclic Web Buckling Control for Shear Link Beams," *ASCE J. Struct. Eng.*, Vol. 112, No. 3, pp. 505–523.

Kato, B. (1989), "Rotation Capacity of H-Section Members as Determined by Local Buckling," *J. Constr. Steel Res.*, Vol. 13, Nos. 2-3, pp. 95–109.

Kawashima, K., Hasegawa, K., MacRae, G. A. and Ikeuchi, T. (1992), "Ductility of Steel Bridge Piers from Dynamic Loading Tests," in *Stability of Steel Structures Under Cyclic Loading* (Eds. T. Fukumoto and G. C. Lee), CRC Press, Boca Raton, FL.

Kawashima, K., MacRae, G. A., Hoshikuma, J. I., and Nagaya, K. (1998), "Residual Displacement Response Spectrum," *ASCE J. Struct. Eng.*, Vol. 124, No. 6, pp. 523–530.

Kazantzi, A. K., Righiniotis, T. D., and Chryssanthopoulos, M. K. (2008), "The Effect of Joint Ductility on the Seismic Fragility of a Regular Moment Resisting Steel Frame Designed to EC8 Provisions," *J. Constr. Steel Res.*, Vol. 64, No. 9, pp. 987–996.

Kemp, A. R. (1996), "Inelastic Local and Lateral Buckling in Design Codes," *ASCE J. Struct. Eng.*, Vol. 122, No. 4, pp. 374–382.

Khatib, I. F., Mahin, S. A., and Pister, K. S. (1988), "Seismic Behavior of Concentrically Braced Steel Frames." Report UCB/EERC-88/01, Earthquake Engineering Research Center, University of California, Berkeley, CA.

Kiggins, S., and Uang, C.-M. (2006), "Reducing Residual Drift of Buckling-Restrained Braced Frames as a Dual System," *Eng. Struct.*, Vol. 28, No. 11, pp. 1525–1532.

Kim, H. I., and Goel, S. C. (1996). "Upgrading of Braced Frames for Potential Local Failures," *ASCE J. Struct. Eng.*, Vol. 122, No. 5, pp. 470–475.

Kimura, Y., and MacRae, G. A. (2006), "Effect of Column Flexural Characteristic on Seismic Behavior of Braced Frame with Fixed Column Base," *Proc. STESSA 2006 Conf.*, Yokohama, Japan, pp. 437–443.

Kitipornchai, S., and Finch, D. L. (1986), "Stiffness Requirements for Cross Bracing," *ASCE J. Struct. Eng.*, Vol. 112, No. 12, pp. 2702–2707.

Koboevic, S., and Redwood, R. (1997), "Design and Seismic Response of Shear Critical Eccentrically Braced Frames," *Can. J. Civ. Eng.*, Vol. 24, No. 5, pp. 761–777.

Koetaka, Y., Kinoshita, T., Inoue, K., and Iitani, K. (2008), "Criteria of Buckling-Restrained Braces to Prevent Out-of-Plane Buckling," Paper no. 05-05-0121, *Proc. 14th World Conf. Earthquake Eng.*, Beijing, China.

Krawinkler, H. (2000), "State of the Art Report on Systems Performance of Steel Moment Frames Subject to Earthquake Ground Shaking," Report No. FEMA-355C, SAC Joint Venture for the Federal Emergency Management Agency, Washington, DC.

Krawinkler, H. (2006), "Importance of Good Nonlinear Analysis," *Struct. Design Tall Spec. Build.*, Vol. 15, pp. 515–531.

Krawinkler, H., and Mohasseb, S. (1987), "Effects of Panel Zone Deformations on Seismic Response," *J. Constr. Steel Res.*, Vol. 8, 1987, pp. 233–250

Krawinkler, H., and Seneviratna, G. D. P. K. (1998), "Pros and Cons of a Pushover Analysis for Seismic Performance Evaluation," *J. Eng. Struct.*, Vol. 20, Nos. 4–6, pp. 452–464.

Krawinkler, H., and Zareian, F. (2007), "Prediction of Collapse—How Realistic and Practical Is It, and What Can We Learn From It?" *Struct. Design Tall Spec. Build.*, Vol. 16, pp. 633–653.

Kuwamura, H. (1992), "High Performance Steels for Earthquake Resistant Building Structures in Japan," In: *Stability and Ductility of Steel Structures Under Cyclic Loading* (Eds. Y. Fukumoto and G. Lee), CRC Press, Boca Raton, FL, pp. 49–60.

Lacerte, M., and Tremblay, R. (2006), "Making Use of Brace Overstrength to Improve the Seismic Response of Multi-Storey Split-X Concentrically Braced Steel Frames," *Can. J. Civ. Eng.*, Vol. 33, No. 8, pp. 1005–1021.

Lee, D., Cotton, S. C., Hajjar, J. F., Dexter, R. J., and Ye, Y. (2005), "Cyclic Behavior of Steel Moment-Resisting Connections Reinforced by Alternative Column Stiffener Details. I. Connection Performance and Continuity Plate Detailing," *AISC Eng. J.*, Vol. 42, No. 4, pp. 189–214.

Lee, G. C., and Lee, E. T. (1994), "Local Buckling of Steel Sections Under Cyclic Loading," *J. Constr. Steel Res.*, Vol. 29, Nos. 1–3, pp. 55–70.

Lee, K., and Bruneau, M. (2005), "Energy Dissipation of Compression Members in Concentrically Braced Frames," *ASCE, J. Struct. Eng.*, Vol. 131, No. 4, pp. 552–559.

Lee, K., and Bruneau, M. (2008), "Seismic Vulnerability Evaluation of Axially Loaded Steel Built-Up Laced Members I: Experimental Results & II: Evaluation," *Earthquake Eng. Eng. Vib.*, Vol. 7, No. 2, pp.113–136.

Lee, K., and Foutch, D. A. (2002), "Performance Evaluation of New Steel Frame Buildings for Seismic Loads," *Earthquake Eng. Struct. Dyn.*, Vol. 31, No. 3, pp. 653–670.

Lee, K., and Stojadinović, B. (2003), "Seismic Bracing Requirements for US Steel Moment Connections," *Proc. 2003 Annual Technical Session & Meeting*, Structural Stability Research Council, Baltimore, MD, pp. 85–104.

Lee, K., and Stojadinović, B. (2007), "Rotation Capacity of Pre-Qualified Moment Connections: A Yield Line Approach," Paper No. 34, *Proc. ASCE Structures Congress 2007,* Long Beach, CA.

Lee, S., and Goel, S. C. (1987), "Seismic Behavior of Hollow and Concrete-Filled Square Tubular Bracing Members," Research Report UMCE 87-11, Dept. of Civil Engineering, University of Michigan, Ann Arbor, MI.

Leelataviwat, S., Goel, S. C., and Stojadinovic, B. (1999), "Toward Performance-Based Seismic Design of Structures," *Earthquake Spectra*, Vol. 15, No. 3, pp. 435–461.

Lehman, D. E., Roeder, C. W., Herman, D., Johnson, S., and Kotulka, B. (2008), "Improved Seismic Performance of Gusset Plate Connections," *ASCE J. Struct. Eng.*, Vol. 134, No. 6, pp. 890–901.

Liao, K. W., Wen, Y.-K., and Foutch, D. A. (2007), "Evaluation of 3D Steel Moment Frames under Earthquake Excitations. I: Modeling," *ASCE J. Struct. Eng.*, Vol. 133, No. 3, pp. 462–470.

Liel, A. B., Haselton, C. B., Deierlein, G. G., and Baker, J. W. (2009), "Incorporating Modeling Uncertainties in the Assessment of Seismic Collapse Risk of Buildings," *Struct. Safety*, Vol. 31, No. 2, pp. 197–211.

Lignos, D. G., Krawinkler, H., and Whittaker, A. S. (2008), "Shaking Table Collapse Tests of Two Scale Models of a 4-Story Moment Resisting Steel Frame," Paper No. 12-01-0186, *Proc. 14th World Conf. Earthquake Eng.*, Beijing, China.

Lin, C.-H., Tsai, K.-C., Wei, C.-Y., Wang, K.-J., Tsai, C.-Y., Wu, A.-C., Powell, J., Clark, K., and Roeder, C. (2009), "Cyclic Tests of a Full-Scale 2-Story Steel Concentrically Braced Frame," *Proc. 5th Int. Symposium on Steel Structures*, Seoul, Korea (in press).

Liu, Z., and Goel, S. C. (1987), "Investigation of Concrete-Filled Steel Tubes Under Cyclic Bending and Buckling," Research Report No. 87-3, Department of Civil Engineering, University of Michigan, Ann Arbor, MI.

Liu, Z., and Goel, S. C. (1988), "Cyclic Load Behavior of Concrete-Filled Tubular Braces," *ASCE J. Struct. Eng.*, Vol. 114, No. 7, pp. 1488–1506.

López, W. A., Gwie, D. S, Lauck, T. W., and Saunders, M. (2004), "Structural Design and Experimental Verification of a Buckling-Restrained Braced Frame System," *AISC Eng. J.*, Vol. 41, No. 4, pp. 177–186.

López, W. A., and Sabelli, R. (2004), *Seismic Design of Buckling-Restrained Braced Frames*, Steel Tips, Structural Steel Education Council, Moraga, CA.

Lubell, A. S., Prion, H. G., Ventura, C., and Rezai, M. (2000), "Unstiffened Steel Plate Shear Wall Performance under Cyclic Loading," *ASCE, J. Struct. Eng.*, Vol. 126, No. 4, pp. 453–460.

Ma, N., Wu, B., Zhao, J., Li, H., Ou, J., and Yang, W. (2008), "Full Scale Test of All-Steel Buckling Restrained Braces," Paper No. 11-0208, *Proc. 14th World Conf. Earthquake Eng.*, Beijing, China.

MacRae, G. A. (1990), "The Seismic Response of Steel Frames," Research Report No. 90-6, University of Canterbury, Christchurch, New Zealand

MacRae, G. A. (1994), "P-Delta Effects on Single-Degree-of-Freedom Structures in Earthquakes," *Earthquake Spectra*, Vol. 10, No. 3, pp. 539–568.

MacRae, G. A., Kimura, Y., and Roeder, C. (2004), "Effect of Column Stiffness on Braced Frame Seismic Behavior," *ASCE J. Struct. Eng.*, Vol. 130, No. 3, pp. 381–391.

MacRae, G. A., and Tagawa, H. (2001), "Seismic Behavior of 3D Steel Moment Frame with Biaxial Columns," *ASCE J. Struct. Eng.*, Vol. 127, No. 5, pp. 490–497.

Malley, J., and Popov, E. P. (1984), "Shear Links in Eccentrically Braced Steel Frames," *ASCE J. Struct. Eng.*, Vol. 110, No. 9, pp. 2275–2295.

Mamaghani, I. H. P. (2008), "Seismic Design and Ductility Evaluation of Thin-Walled Steel Bridge Piers of Box Sections," *Transportation Res. Record: J. Transportation Res. Board*, No. 2050, pp. 137–142.

Mansour, N., Christopoulos, C., and Tremblay, R. (2006), "Seismic Design of EBF Steel Frames Using Replaceable Nonlinear Links," *Proc. STESSA 2006 Conf.*, Yokohama, Japan, pp. 745–750.

Marino, E. M., Ghersi, A., and Nakashima, M. (2006), "Design of Chevron Braced Frames Based on Ultimate Lateral Strength I: Theory; II Evaluation of the Behavior Factor," *Proc. STESSA 2006 Conf.*, Yokohama, Japan, pp. 51–65.

Marino, E. M., and Nakashima, M. (2006), "Seismic Performance and New Design Procedure for Chevron-Braced Frames," *Earthquake Eng. Struct. Dyn.*, Vol. 35, No. 4, pp. 433–452.

Marson, J., and Bruneau, M. (2004), "Cyclic Testing of Concrete-Filled Circular Steel Bridge Piers Having Encased Fixed-Based Detail," *ASCE J. Bridge Eng.*, Vol. 9, No. 1, pp. 14–23.

Martinelli, L., Mulas, M. G., and Perotti, F. (1996), "The Seismic Response of Concentrically Braced Moment-Resisting Steel Frames," *Earthquake Eng. Struct. Dyn.*, Vol. 25, No. 11, pp. 1275–1299.

Martinelli, L., Perotti, F., and Bozzi, A. (2000), "Seismic Demand and Response of a 14-Story Concentrically Braced Steel Buildings," *Proc. STESSA 2000 Conf.*, Montreal, Canada, pp. 327–334.

Mastrandrea, L., Montuori, R., and Piluso, V. (2003), "Failure Mode Control of Seismic Resistant EB-Frames," *Proc. STESSA 2003 Conf.*, Naples, Italy, pp. 435–441.

Matsui, R., Takeuchi, T., Hajjar, J. F., Nishimoto, K., and Aiken, I. (2008), "Local Buckling Restraint Condition for Core Plates in Buckling Restrained Braces," Paper No. 05-05-0055, *Proc. 14th World Conf. On Earthquake Eng.*, Beijing, China.

Mazzolani, F. M., Georgescu, D., and Astaneh-Asl, A. (1994), "Remarks on Behaviour of Concentrically and Eccentrically Braced Steel Frames," *Proc. STESSA 1994 Conf.*, Timisoara, Romania, pp. 310–322.

Mazzolani, F. M., and Piluso, V. (1993), "*P*–Δ Effect on Seismic Resistant Steel Structures," *Proc. 1993 Annual Technical Session*, Structural Stability Research Council, Milwaukee, WI, pp. 271–282.

Mazzolani, F. M., and Piluso, V. (1996), *Theory and Design of Seismic Resistant Steel Frames*, E & FN Spon, London.

Mazzolani, F. M., and Piluso, V. (1997), "Platic Design of Seismic Resistant Steel Frames," *Earthquake Eng. Struct. Dyn.*, Vol. 26, No. 2, pp. 167–191.

McCormick, J., DesRoches, R., Fugazza, D., and Auricchio, F. (2007), "Making Seismic Assessment of Concentrically Braced Steel Frames with Shape Memory Alloy Braces," *ASCE J. Struct. Eng.*, Vol. 133, No. 6, pp. 862–870.

McGuire, W., Gallagher, R. H., and Ziemian, R. D. (2000), *Matrix Structural Analysis*, 2nd ed., John Wiley and Sons, Inc., Hoboken, NJ.

Medina, R. A., and Krawinkler, H. (2005) "Strength Demand Issues Relevant for the Seismic Design of Moment-Resisting Frames," *Earthquake Spectra*, Vol. 21, No. 2, pp. 415–439.

Mehanny, S. S. F., and Deierlein, G. G. (2001), "Seismic Damage and Collapse Assessment of Composite Moment Frames, *ASCE J Struct. Eng.*, Vol. 127, No. 9, pp. 1045–1053.

Merzouq, S., and Tremblay, R. (2006), "Seismic Design of Dual Concentrically Braced Steel Frames for Stable Seismic Performance for Multi-Storey Buildings," Paper No. 1909, *Proc. 8th U.S. Nat. Conf. on Earthquake Eng.*, San Francisco, CA.

Midorikawa, M., Azuhata, T., Ishira, T., and Wada, A. (2006), "Shaking Table Tests on Seismic Response of Steel Braced Frames with Column Uplift," *Earthquake Eng. Struct. Dyn.*, Vol. 35, No. 14, pp. 1767–1785.

Midorikawa, M., Toyomaki, S., Hori, H., Asari, T., Azuhata, T., and Ishihara, T. (2008), "Seismic Response of Six-Story Eccentrically Braced Steel Frames with Column Partially Allowed to Uplift," Paper No. ST17-03-006, *Proc. 14th World Conf. on Earthquake Eng.*

Miranda, E., and Akkar, S. D. (2003), "Dynamic Instability of Simple Structural Systems," *ASCE J. Struct. Eng.*, Vol. 129, 1, pp. 1722–1726.

Mitchell, D., Tremblay, R., Karacabeyli, E., Paultre, P., Saatcioglu, M., and Anderson, D. (2003), "Seismic Force Modification Factors for the Proposed 2005 NBCC," *Can. J. Civ. Eng.*, Vol. 30, No. 2, pp. 308–327.

Moldovan, A., Petcu, D., and Gioncu, V. (2000), "Calibration of Thin-Walled Members Ductility," *Proc. STESSA 2000 Conf.*, Montreal, Canada, pp. 63–71.

Montgomery, C. J. (1981), "Influence of P-Delta Effects on Seismic Design," *Can. J. Civ. Eng.*, Vol. 8, No. 1, pp. 31–43.

Montgomery, J., and Medhekar, M. (2001), "Unstiffened Steel Plate Shear Wall Performance under Cyclic Loading : Discussion," *ASCE J. Struct. Eng.*, Vol. 127, No. 8, p. 973.

Moon, J., Yoon, K.-Y., Han, T.-S., and Lee, H.-E., (2008), "Out-of-Plane Buckling and Design of X-Bracing Systems with Discontinuous Diagonals," *J. Constr. Steel Res.*, Vol. 64, No. 3, pp. 285–294.

Moslehi Tabar, A., and Deylam, A. (2006), "Investigation of Major Parameters Affecting Instability of Steel Beams with RBS Moment Connections," *Steel and Composite Structures*, Vol. 6, No. 3, pp. 203–219.

Murray, T. M., and Sumner, E. A. (2003), *Extended End-Plate Moment Connections—Seismic and Wind Applications*, 2nd ed., AISC Design Guide No. 4, American Institute of Steel Construction, Chicago, IL.

Nakashima, M., Kanao, I., and Liu, D. (2002), "Lateral Instability and Lateral Bracing of Steel Beams Subjected to Cyclic Loading," *ASCE J. Struct. Eng.*, Vol. 128, No. 10, pp. 1308–1316.

Nakashima, M., and Liu, D. (2005), "Instability and Complete Failure of Steel Columns Subjected to Cyclic Loading," *ASCE J. Eng. Mech.*, Vol. 131, No. 6, pp. 559–567.

Nakashima, M., Liu, D., and Kanao, I. (2003), "Lateral–Torsional and Local Instability of Steel Beams Subjected to Large Cyclic Loading," *Int. J. Steel Struct.*, Vol. 3, No. 3, pp. 179–189.

Nakashima, M., Matsumiya, T., Suita, K., and Liu, D. (2006), "Test on Full-Scale Three-Storey Steel Moment Frame and Assessment of Ability of Numerical Simulation to Trace Cyclic Inelastic Behavior," *Earthquake Eng. Struct. Dyn.*, Vol. 35, No. 1, pp. 3–19.

Nakashima, M., and Wakabayashi, M. (1992), "Analysis and Design of Steel Braces and Braced Frames in Building Structures," in *Stability and Ductility of Steel Structures under Cyclic Loading* (Eds.) Y. Fukumoto and G. C. Lee), CRC Press, Boca Raton, FL, pp. 309–321.

Nast, T. E., Grondin, G. Y., and Cheng, R. J. J. (1999), *Cyclic Behaviour of Stiffened Gusset Plate-Brace Member Assemblies*, Struct. Eng. Report No. 229, Dept. of Civil Engineering, University of Alberta, Edmonton, Alberta, Canada.

Neri, F. (2000), "Influence of P–Δ Effect on Proposed Procedure for Seismic Design of Steel Frames," *Proc. STESSA 2000 Conf.*, Montreal, Canada, pp. 511–518.

Neuss, C. F., and Maison, B. F. (1984), "Analysis for P–Δ Effects in Seismic Response of Buildings," *Comput. Struct.*, Vol. 19, pp. 369–380.

Newell, J. D., and Uang, C.-M. (2008), "Cyclic Behavior of Steel Wide-Flange Columns Subjected to Large Drift," *ASCE J. Struct. Eng.*, Vol. 134, No. 8, pp. 1334–1342.

Nip, K. H., Gardner, L., and Elghazouli, A. Y. (2008), "Comparative Study of Stainless Steel and Carbon Steel Tubular Members Subjected to Cyclic Loading," *Proc. XII Int. Symposium on Tubular Structures*, Shanghai, pp. 219–226.

NRCC (2005), *National Building Code of Canada*, 12th ed., National Research Council of Canada Ottawa, Canada.

NZS (2007), NZ 3404:Parts 1 and 2:1997, *Steel Structures Standard, Incorporating Amendment No. 1 and Amendment No. 2(2007)*, New Zealand Standards, Wellington, New Zealand.

NZS (2004), NZ 1170.5:2004, *Structural Design Actions, Part 5: Earthquake Actions—New Zealand*, New Zealand Standards, Wellington, New Zealand.

Ocel, J. DesRoches, R., Leon, R. T., Hess, W. G., Krumme, R., Hayes, J. R., and Sweeney, S. (2004), "Steel Beam-Column Connections Using Shape Memory Alloys," *ASCE J. Struct. Eng.*, Vol. 130, No. 5, pp. 732–740.

Ohsaki, M., Kinoshita, T., and Pan, P. (2007), "Multiobjective Heuristic Approaches to Seismic Design of Steel Frames with Standard Sections," *Earthquake Eng. Struct. Dyn.*, Vol. 36, No., pp. 1481–1495.

Okazaki, T., Arce, G., Ryu, C., and Engelhardt, M. D. (2005), "Experimental Study of Local Buckling, Overstrength, and Fracture of Links in Eccentrically Braced Frames," *ASCE J. Struct. Eng.*, Vol. 131, No. 10, pp. 1526–535.

Okazaki, T., and Engelhardt, M. D. (2007), "Cyclic Loading Behavior of EBF Links Constructed of ASTM A992 Steel," *J. Constr. Steel Res.*, Vol. 63, No. 6, pp. 751–765.

Okazaki, T., Liu, D., Nakashima, M., and Engelhardt, M. (2006), "Stability Requirements for Beams in Seismic Steel Moment Frames," *ASCE J. Struct. Eng.*, Vol. 132, No. 9, pp. 1334–1342.

Osteraas, J., and Krawinkler, H. (1989), "The Mexico Earthquake of September 19, 1985—Behavior of Steel Buildings," *Earthquake Spectra*, Vol. 5, No. 1, pp. 51–88.

Osteraas, J., and Krawinkler, H. (1990), "Strength and Ductility Considerations in Seismic Design", Rep. No. 90, The John A. Blume Earthquake Engineering Research Center, Dept. of Civil and Environmental Engineering, Stanford University, Stanford, CA.

Park, H.-G., Kwack, J.-H., Jeon, S.-W., Kim, W.-K., and Choi, I.-R. (2007), "Framed Steel Plate Wall Behavior under Cyclic Lateral Loading," *ASCE J. Struct. Eng.*, Vol. 133, No. 3, pp. 378–388.

Park, K., and Medina, R. A. (2007), "Conceptual Seismic Design of Regular Frames Based on the Concept of Uniform Damage," *ASCE J. Struct. Eng.*, Vol. 133, No. 7, pp. 945–955.

Park, Y. S., Iwai, S., Kameda, H., and Nonaka, T. (1996), "Very Low Cycle Failure Process of Steel Angle Members," *ASCE J. Struct. Eng.*, Vol. 122, No. 2, pp. 133–141.

Petcu, D., and Gioncu, V. (2003), "Computer Program for Available Ductility Analysis of Steel Structures," *Comput. Struct.*, Vol. 81, Nos. 22–23, pp. 2149–2164.

Pettinga, J. D., and Priestley, M. J. N. (2007), *Accounting for P-Delta Effects in Structures When Using Direct Displacement-Based Design*, Research Report No. ROSE -2007/02, IUSS Press, Pavia, Italy.

Picard, A., and Beaulieu, D. (1987), "Design of Diagonal Cross Bracings Part 1: Theoretical Study," *AISC Eng. J.*, Vol. 24, No. 3, pp. 122–126.

Picard, A., and Beaulieu, D. (1988), "Design of Diagonal Cross-Bracings Part 2: Experimental Study," *AISC Eng. J.*, Vol. 25, No. 4, pp. 156–160.

Plumier, A. (1997), "The Dogbone: Back to the Future," *AISC Eng. J.*, Vol. 34, No. 2, pp. 61–67.

Plumier, A. (2000), "General Report on Local Ductility," *J. Constr. Steel Res.*, Vol. 55, Nos. 1–3, pp. 91–107.

Pollino, M., and Bruneau, M. (2007), "Experimental Study of the Controlled Rocking Response of Steel Braced Frames," Paper No. 46, *Proc. ASCE Struct. Congress,* Long Beach, CA.

Pollino, M., and Bruneau, M. (2008), "Dynamic Seismic Response of Controlled Rocking Bridge Steel-Truss Piers," *Eng. Struct.*, Vol. 30, No. 6, pp. 1667–1676.

Pons, H. F. (1997), "Dynamic Behavior of Tubular Bracing Members with Single Plate Concentric Connections," M.Sc. Thesis, University of Toronto, Toronto, Ontario Canada.

Popov, E. P., Bertero, V. V., and Chandramouli, S. (1975), "Hysteretic Behavior of Steel Columns," Report No. UCB/EERC-75/11, Earthquake Engineering Research Center, University of California, Berkeley, CA.

Popov, E. P., and Black, R. G. (1981), "Steel Struts Under Severe Cyclic Loadings," *ASCE J. Struct. Eng. Div.*, Vol. 107, No. ST9, pp. 1857–1881.

Popov, E. P., and Pinkney, R. B. (1967), "Behavior of Steel Building Connections Subjected to Repeated Inelastic Strain Reversal, Experimental Data," Reports

UCB/SESM-1967/30&31, Dept. of Civil Engineering, University of California, Berkeley, CA.

Popov, E. P., and Stephen, R. M., (1970), "Cyclic Loading of Full Size Steel Connections," UCB/EERC Report 70-3, University of California, Berkeley, CA.

Porter, F. L., and Powell, G. H. (1971), "Static and Dynamic Analysis of Inelastic Structures," Report No. EERC 71-3, Earthquake Engineering Research Center, University of California, Berkeley, CA.

Qu, B., Bruneau, M., Lin, C.-H., and Tsai, K.-C. (2008), "Testing of Full-Scale Two-Story Steel Plate Shear Wall with Reduced Beam Section Connections and Composite Floors," *ASCE J. Struct. Eng.*, Vol. 134, No. 3, pp. 364–373.

Rai, D. C., and Goel, S. C. (2003), "Seismic Evaluation and Upgrading of Chevron Braced Frames," *J. Constr. Steel Res.*, Vol. 59, No. 8, pp. 971–994.

Redwood, R. G., and Channagiri, V. S. (1991), "Earthquake Resistant Design of Concentrically Braced Steel Frames," *Can. J. Civ. Eng.*, Vol. 18, No. 5, pp. 839–850.

Remennikov, A. M., and Walpole, W. R. (1997), "Analytical Prediction of Seismic Behaviour for Concentrically-Braced Steel Systems," *Earthquake Eng. Struct. Dyn.*, Vol. 26, No. 8, pp. 859–874.

Remennikov, A. M., and Walpole, W. R. (1998a), "A Note on Compression Strength Reduction Factor for a Buckled Strut in Seismic-Resisting Braced System," *Eng. Struct.*, Vol. 20, No. 8, pp. 779–782.

Remennikov, A. M., and Walpole, W. R. (1998b), "Seismic Behavior and Deterministic Design Procedures for Steel V-Braced Frames," *Earthquake Spectra*, Vol. 14, No. 2, pp. 335–355.

Richards, P., Okazaki, T., Engelhardt, M., and Uang, C.-M. (2007), "Impact of Recent Research Findings on Eccentrically Braced Frame Design," *AISC Eng. J.*, Vol. 44, No. 1, pp. 41–53.

Richards, P., and Uang, C.-M. (2005), "Effect of Flange Width-Thickness Ratio on Eccentrically Braced Frames Link Cyclic Rotation Capacity," *ASCE J. Struct. Eng.*, Vol. 131, No. 10, pp. 1546–1552.

Richards, P. W. (2009), "Seismic Column Demands in Ductile Braced Frames," *ASCE J. Struct. Eng.*, Vol. 135, No. 1, pp. 33–41.

Ricles, J. M., Sause, R., Peng, S. W., and Lu, L. W. (2002), "Experimental Evaluation of Earthquake Resistant Posttensioned Steel Connections," *ASCE J. Struct. Eng.*, Vol. 128, No. 7, pp. 850–859.

Ricles, J. M., and Paboojian, S. D. (1994), "Seismic Performance of Steel-Encased Composite Columns," *ASCE J. Struct. Eng.*, Vol. 120, No. 8, pp. 2474–2494.

Ricles, J. M., and Popov, E. P. (1989), "Composite Action in Eccentrically Braced Frames," *ASCE J. Struct. Eng.*, Vol. 115, No. 8, pp. 2046–2066.

Ricles, J. M., and Popov, E. P. (1994), "Inelastic Link Element for EBF Seismic Analysis," *ASCE J. Struct. Eng.*, Vol. 120, No. 2, pp. 441–463.

Ricles, J. M., Sause, R., Garlock, M. M., and Zhao, C. (2001), "Post-Tensioned Seismic-Resistant Connections for Steel Frames," *ASCE J. Struct. Eng.*, Vol. 127, No. 2, pp. 113–121.

Ricles, J. M., Sause, R., and Green, P. S. (1998), "High Strength Steel: Implications of Material and Geometric Characteristics on Inelastic Flexural Behavior," *Eng. Struct.*, Vol. 95, Nos. 4–6, pp. 323–335.

Roberts, T. M., and Sabouri-Ghomi, S. (1992), "Hysteretic Characteristics of Unstiffened Perforated Steel Plate Shear Panels," *Thin Walled Struct.*, Vol. 14, No. 2, pp. 139–151.

Rodgers, J. E., and Mahin, S. A. (2006), "Effects of Connection Fractures on Global Behavior of Steel Moment Frames Subjected to Earthquakes," *ASCE J Struct. Eng.*, Vol. 132, No. 1, pp. 78–88.

Rodgers, J. E., and Mahin, S. A. (2008), "Effects of Deformation Softening on Seismic Behavior of Steel Moment Frames," Paper No. 12-01-0093, *Proc. 14th World Conf. Earthquake Eng.*, Beijing, China.

Roeder, C. W. (1989), "Seismic Behavior of Concentrically Braced Frame," *ASCE J. Struct. Eng.*, Vol. 115, No. 8, pp. 1837–1857.

Roeder, C. W. (2000), *FEMA-355D State of the Art Report on Connection Performance*, prepared for the SAC Joint Venture, SAC Joint Venture, Federal Emergency Management Agency, Washington, DC.

Roeder, C. W. (2002), "General Issues Influencing Connection Performance," *ASCE J. Struct. Eng.*, Vol. 128, No. 4, pp. 420–428.

Roeder, C. W., Foutch, D. A., and Goel, S. C. (1987), "Seismic Testing of Full-Scale Steel Building—Part II," *ASCE J. Struct. Eng.*, Vol. 113, No. 11, pp. 2130–2145.

Roeder, C. W., Schneider, S. P., and Carpenter, J. E. (1993), "Seismic Behavior of Moment-Resisting Steel Frames: Analytical Study," *ASCE J. Struct. Eng.*, Vol. 119, No. 6, pp. 1866–1884.

Rossi, P. P. (2007), "A Design Procedure for Tied Braced Frames," *Earthquake Eng. Struct. Dyn.*, Vol. 36, No. 14, pp. 2227–2248.

Rossi, P. P., and Lombardo, A. (2007), "Influence of the Link Overstrength Factor on the Seismic Behavior of Eccentrically Braced Frames," *J. Constr. Steel Res.*, Vol. 63, No. 11, pp. 1529–1545.

Ruiz-García, J., and Miranda, E. (2006), "Evaluation of Residual Drift Demands in Regular Multi-Storey Frames for Performance-Based Seismic Assessment," *Earthquake Eng. Struct. Dyn.*, Vol. 35, No. 13, pp. 1609–1629.

Rutenberg, A., and De Stefano, M. (2000), "Approximate Stability Bounds on the Seismic Force Reduction Factor," Paper No. 1635, *Proc. 12th World Conf. Earthquake Eng.*, Auckland, New Zealand.

Sabelli, R. (2001), "Research on Improving the Design and Analysis of Earthquake Resistant Braced Frames," 2000 NEHRP Fellowship in Earthquake Hazard Reduction, Final Report to FEMA and EERI Earthquake Eng. Res. Institute, Oahland, CA.

Sabelli, R. (2004), "Recommended Provisions for Buckling-Restrained Braced Frames," *AISC Eng. J.*, Vol. 41, No. 4, pp. 155–175.

Sabelli, R., and Bruneau, M. (2007), *Steel Plate Shear Walls,* Steel Design Guide 20, American Institute of Steel Construction, Chicago, IL.

Sabelli, R., and Hohbach, D. (1999), "Design of Cross-Braced Frames for Predictable Buckling Behavior," *ASCE J. Struct. Eng.*, Vol. 125, No. 2, pp. 163–168.

Sabelli, R., Mahin, S., and Chang, C. (2003), "Seismic Demands on Steel Braced Frame Buildings with Buckling Restrained Braces," *Eng. Struct.*, Vol. 25, No. 5, pp. 656–666.

Sabouri-Ghomi, A., and Gholhaki, M. (2008), "Ductility of Thin Steel Plate Shear Walls," *Asian J. Civ. Eng. (Building and Housing)*, Vol. 9, No. 2, pp. 153–166.

Sabouri-Ghomi, S., Kharrazi M. H. K., Mam-Azizi, S.-E-D., and Sajadi, R. A. (2008), "Buckling Behavior Improvement of Steel Plate Shear Wall Systems," *Struct. Design Tall Special Buildings,* Vol. 17, No. 4, pp. 823–837.

Sabouri-Ghomi, S., and Roberts, T. M. (1991), "Nonlinear Dynamic Analysis of Thin Steel Plate Shear Walls," *Comput. Struct.*, Vol. 39, Nos. 1–2, pp. 121–127.

Sakimoto, T., and Watanabe, H. (1997), "Simplified Methods of Analysis for Cyclic Behavior of Steel Box Beam-Columns," *Proc. 1997 Annual Technical Meeting and Session*, Structural Stability Research Council, Toronto, Ontario, Canada, pp. 265–276.

Salmanpour, A. H., and Arbabi, F. (2008), "Seismic Reliability of Concentrically Braced Steel Frame," Paper No. 05-05-0062, *Proc. 14th World Conf. on Earthquake Eng.*, Beijing, China.

Sato, A., and Uang, C.-M. (2009), "Seismic Design Procedure Development for Cold-Formed Steel—Special Bolted Moment Frames," *J. Constr. Steel Res.*, Vol. 65, No. 4, pp. 860–868.

Sause, R., Ricles, J. M., Roke, D., Seo, C.-Y., and Lee, K.-S. (2006), "Self-Centering Seismic Resistant Steel Concentrically-Braced Frames," *Proc. STESSA 2006 Conf.*, Yokohama, Japan, pp. 85–90.

Schellenberg, A., Yang, T. Y., Mahin, S. A., and Stojadinovic, B. (2008), "Hybrid Simulation of Structural Collapse," Paper No. S16-02-004, *Proc. 14 World Conf. Earthauke Eng.*, Beijing, China.

Schneider, S. P., Roeder, C. W., and Carpenter, J. E. (1993), "Seismic Behavior of Moment-Resisting Steel Frames: Experimental Investigation," *ASCE J. Struct. Eng.*, Vol. 119, No. 6, pp. 1885–1902.

Segal, F., Levy, R., and Rutenberg, A. (1994), "Design of Imperfect Cross-Bracings," *ASCE J. Eng. Mech.*, Vol. 120, No. 5, pp. 1057–1075.

Sepúlveda, J., Boroschek, R., Herrera, R., Moroni, O., and Sarrazin, M. (2008), "Steel Beam–Column Connection Using Copper-Based Shape Memory Alloy Dampers," *J. Constr. Steel Res.*, Vol. 64, No. 4, pp. 429–435.

Shaback, J. B., and Brown, T. (2003), "Behaviour of Square Hollow Structural Steel Braces with End Connections Under Reversed Cyclic Axial Loading," *Can. J. Civ. Eng.*, Vol. 30, No. 4, pp. 745–753.

Shen, J.-H. J., Astaneh-Asl, A., and McCallen, D. B. (2002), "Use of Deep Columns in Special Steel Moment Frames," Steel Tip Report No. 24, Structural Steel Educational Council, Moraga, CA.

Shibata, M. (1988), "Hysteretic Behavior of Multi-Story Braced Frames," *Proc. 9th World Conf. Earthquake Eng.*, Vol. IV, Tokyo, Japan, pp. 249–254.

Shishkin, J. J., Driver, R. G., and Grondin, G. Y. (2005), "Analysis of Steel Plate Shear Walls Using the Modified Strip Model," Struct. Eng. Report No. 261, Dept. of Civil Engineering, University of Alberta, Edmonton, Alberta, Canada.

Spears, P. W., and Charney, F. A. (2006), "The Effect of Vertical Ground Accelerations on the Dynamic Stability of Simple Structures," *Proc. 2006 Annual Stability Conference*, Structural Stability Research Council, San Antonio, TX, pp. 443–461.

Stoman, S. H. (1989), "Effective Length Spectra for Cross Bracings," *ASCE J. Struct. Eng.*, Vol. 115, No. 12, pp. 3112–3122.

Stratan, A., and Dubina, D. (2004), "Bolted Links for Eccentrically Braced Steel Frames," *Proc. of Fifth Int. Workshop on Connections in Steel Structures V*, Amsterdam, The Netherlands, pp. 223–232.

Suita, K., Yamada, S., Tada, M., Kasai, K., Matsuoka, Y., and Sato, E. (2008), "Results of Recent E-Defense Tests on Full-Scale Steel Buildings: Part 1—Collapse Experiments on 4-Story Moment Frames," Paper No. 266, *Proc. ASCE Structures Congress 2008*, Vancouver, BC.

Sun, F. F., Liu, G. R., and Li, G. Q. (2008), "An Analytical Model for Composite Steel Plate Wall," Paper No. 05-05-0097. *Proc. 14th World Conf. on Earthquake Conf.*, Beijing, China.

Susantha, K. A. S., Ge, H., and Usami, T. (2002), "Cyclic Analysis and Capacity Prediction of Concrete-Filled Steel Box Columns," *Earthquake Eng. Struct. Dyn.*, Vol. 31, No. 2, pp. 195–216.

Swanson, J. A., and Leon, R. T. (2000), "Bolted Steel Connections: Tests on T-Stub Components," *ASCE J. Struct. Eng.*, Vol. 126, No. 1, pp. 50–56.

Tada, M., Fukui, T., Nakashima, M., and Roeder, C. W. (2003), "Comparison of Strength Capacity for Steel Building Structures in the United States and Japan," *Earthquake Eng. Eng. Seismol.*, Vol. 4, No. 1, pp. 1–13.

Tada, M., Ohsaki, M., Yamada, S., Motoyui, S., and Kasai, K., (2007), "E-Defense Tests on Full-Scale Steel Buildings Part 3—Analytical Simulation of Collapse," Paper No. 19, *Proc. ASCE Structures Congress 2007,* Long Beach, CA.

Tagawa, H., MacRae, G. A., and Lowes, L. (2004), "Evaluation and Mitigation of P–Δ Effects on 2D MDOF Frame Behavior," *Proc. 2004 Annual Stability Conference*, Structural Stability Research Council, Long Beach, CA, pp. 499–518.

Takanashi, K. (1973), "Inelastic Lateral Buckling of Steel Beams Subjected to Repeated and Reversed Loadings," *Proc. 5th World Conf. on Earthquake Eng.*, Vol. 1, Rome, Italy, pp. 795–798.

Tang, X., and Goel, S. C. (1989), "Brace Fractures and Analysis of Phase I Structure," *ASCE J. Struct. Eng.*, Vol. 115, No. 8, pp. 1960–1976.

Thevendran, V., and Wang, C. M. (1993), "Stability of Nonsymmetric Cross-Bracing Systems," *ASCE J. Struct. Eng.*, Vol. 119, No. 1, pp. 169–180.

Thorburn, L. J., Kulak, G. L., and Montgomery, C. J. (1983), "Analysis of Steel Plate Shear Walls," Struct. Eng. Report No. 107, Dept. of Civil Engineering, University of Alberta, Edmonton, Alberta, Canada.

Thornton, W. A. (1984), "Bracing Connections for Heavy Construction," *AISC Eng. J.*, Vol. 21, No. 3, pp. 139–148.

Timler, P. A., and Kulak, G. L. (1983), "Experimental Study of Steel Plate Shear Walls," Struct. Eng. Report No. 114, Dept. of Civil Engineering, University of Alberta, Edmonton, Alberta, Canada.

Tjondro, J. A., Moss, P. J., and Carr, A. J. (1992), "Seismic P–δ Effects in Medium Height Moment Resisting Steel Frames," *Eng. Struct.*, Vol. 14, No. 2, pp. 75–90.

Tremblay, R. (2000), "Influence of Brace Slenderness on the Seismic Response of Concentrically Braced Steel Frames," *Proc. STESSA 2000 Conf.*, Montréal, Quebec, Canada, pp. 527–534.

Tremblay, R. (2002), "Inelastic Seismic Response of Steel Bracing Members," *J. Constr. Steel Res.*, Vol. 58, Nos. 5–8, pp. 665–701.

Tremblay, R. (2003), "Achieving a Stable Inelastic Seismic Response for Concentrically Braced Steel Frames," *AISC Eng. J.*, Vol. 40, No. 2, pp. 111–129.

Tremblay, R., Archambault, M. H., and Filiatrault, A. (2003), "Seismic Performance of Concentrically Braced Steel Frames Made with Rectangular Hollow Bracing Members," *ASCE J. Struct. Eng.*, Vol. 129, No. 12, pp. 1626–1636.

Tremblay, R., Bruneau, M., Nakashima, M., Prion, H. G. L., Filiatrault, A., and DeVall, R. (1996), "Seismic Design of Steel Buildings: Lessons from the 1995 Hyogoken-Nanbu Earthquake," *Can. J. Civ. Eng.*, Vol. 23, No. 3, pp. 727–756.

Tremblay, R., Côté, B., and Léger, P. (1999), "An Evaluation of P–Δ Amplification Factors in Multistorey Steel Moment Resisting Frames," *Can. J. Civ. Eng.*, Vol. 26, No. 5, pp. 535–548.

Tremblay, R., Haddad, M., Martinez, G., Richard, J., and Moffatt, K. (2008a), "Inelastic Cyclic Testing of Large Size Steel Bracing Members,", Paper No. 05-05-0071, *Proc. 14th World Conf. Earthquake Eng.*, Beijing, China.

Tremblay, R., Lacerte, M., and Christopoulos, C. (2008b), "Seismic Response of Multi-Storey Buildings with Self-Centering Energy Dissipative Steel Braces," *ASCE J. Struct. Eng.*, Vol. 134, No. 1, pp. 108–120.

Tremblay, R., Poirier, L.-P., Bouaanani, N., Leclerc, M., René, V., Fronteddu, L., and Rivest, S. (2008c), "Innovative Viscously Damped Rocking Braced Steel Frames," Paper No. 05-01-0527, *Proc. 14th World Conf. on Earthquake Eng.*, Beijing, China.

Tremblay, R., and Poncet, L. (2005), "Seismic Performance of Concentrically Braced Steel Frames in Multi-Storey Buildings with Mass Irregularity," *ASCE J. Struct. Eng.*, Vol. 131, No. 9, pp. 1363–1375.

Tremblay, R., and Robert, N. (2000), "Design of Low- and Medium-Rise Chevron Braced Steel Frames," *Can. J. Civ. Eng.*, Vol. 27, No. 6, pp. 1192–1206.

Tremblay, R. and Robert, N. (2001), "Seismic Performance of Low- and Medium-Rise Chevron Braced Steel Frames," *Can. J. Civ. Eng.*, Vol. 28, No. 4, pp. 699–714.

Tremblay, R., and Stiemer, S. F. (1994), "Back-Up Stiffness for Improving the Stability of Multi-Storey Braced Frames Under Seismic Loading," *Proc. 1994 SSRC Annual Task Group Technical Session*, Structural Stability Research Council, Bethlehem, PA, pp. 311–325.

Tremblay, R., Tchebotarev, N., and Filiatrault, A. (1997), "Seismic Performance of RBS Connections for Steel Moment Resisting Frames: Influence of Loading Rate and Floor Slab," *Proc. STESSA '97 Conf.*, Kyoto, Japan, pp. 664–671.

Tremblay, R., Timler, P., Bruneau, M., and Filiatrault, A. (1995), "Performance of Steel Structures During the January 17, 1994 Northridge Earthquake," *Can. J. Civ. Eng.*, Vol. 22, No. 2, pp. 338–360.

Tsai, K.-C., Chou, C.-C., Lin, C.-L., Chen, P.-C., and Jhang, S.-J. (2007), "Seismic Self-Centering Steel Beam-to-Column Moment Connections Using Bolted Friction Devices," *Earthquake Eng. Struct. Dyn.*, Vol. 37, No. 4, pp. 624–645.

Tsai, K.-C., and Hsiao, P.-C. (2008), "Pseudo-Dynamic Test of a Full-Scale CFT/BRB Frame—Part II: Seismic Performance of Buckling-Restrained Braces and Connections," *Earthquake Eng. Struct. Dyn.*, Vol. 37, No. 7, pp. 1099–1115.

Tsai, K. C., Wu, S., and Popov, E. P. (1995), "Experimental Performance of Seismic Steel Beam-Column Moment Joints," *ASCE J. Struct. Eng.*, Vol. 121, No. 6, pp. 925–931.

Uang, C.-M., and Fan, C.-C. (2001), "Cyclic Stability Criteria for Steel Moment Connections with Reduced Beam Section," *ASCE J. Struct. Eng.*, Vol. 127, No. 9, pp. 1021–1027.

Uang, C.-M., and Nakashima, M. (2004), "Steel Buckling-Restrained Braced Frames," In *Earthquake Engineering—From Engineering Seismology to Performance-Based Engineering* (Eds. Y. Bozorgnia and V. V. Bertero), CRC Press, Boca Raton, FL, Chap. 16.

Uetani, K., and Tagawa, H. (1998), "Criteria for Suppression of Deformation Concentration of Building Frames Under Severe Earthquakes," *Eng. Struct.*, Vol. 20, Nos. 4–6, pp. 372–383.

Uriz, P. (2005), "Towards Earthquake Resistant Design of Concentrically Braced Steel Structures," Ph.D. thesis. Dept. of Civil Engineering, University of California, Berkeley, CA.

Uriz, P., Filippou, F. C., and Mahin, S. A. (2008), "Model for Cyclic Inelastic Buckling of Steel Braces," *ASCE Struct. Eng.*, Vol. 134, No. 4, pp. 619–628.

Uriz, P., and Mahin, S. A. (2004), "Seismic Vulnerability Assessment of Concentrically Braced Steel Frames," *J. Int. Steel Struct.*, Vol. 4, No. 4, pp. 240–248.

Usami, T., Gao, S., and Ge, H. (2000), "Stiffened Steel Box Columns. Part 2: Ductility Evaluation," *Earthquake Eng. Struct. Dyn.*, Vol. 29, No. 11, pp. 1707–1722.

Usami, T., Ge, H. B., and Kasai, A. (2008), "Overall Buckling Prevention Condition of Buckling Restrained Braces as a Structural Control Damper," Paper No. 05-05-0128, *Proc. 14th World Conf. on Earthquake Eng.*, Beijing, China.

Usami, T., Kasai, A., and Kato, M. (2003), "Behavior of Buckling-Restrained Brace Members," *Proc. STESSA 2003*, Naples, Italy, pp. 211–216.

Vamvatsikos, D., and Cornell, C. A. (2002), "Incremental Dynamic Analysis," *Earthquake Eng. Struct. Dyn.*, Vol. 31, No. 3, pp. 491–514.

Varma, A. H., Ricles, J. M., Sause, R., and Lu, L-W. (2002), "Seismic Behavior and Modeling of High-Strength Composite Concrete-Filled Steel Tube (CFT) Beam-Columns," *J. Constr. Steel Res.*, Vol. 58, Nos. 5–8, pp. 725–758.

Varma, A. H., Ricles, J. M. Sause, R., and Lu, L.-W. (2004), "Seismic Behavior and Design of High-Strength Square Concrete-Filled Steel Tube Beam-Columns," *ASCE J. Struct. Eng.*, Vol. 130, No. 2, pp. 169–179.

Vian, D., and Bruneau, M. (2003), "Tests to Structural Collapse of Single Degree of Freedom Frames Subjected to Earthquake Excitations," *ASCE J. Struct. Eng.*, Vol. 129, No. 12, pp. 1676–1685.

Vian, D., and Bruneau, M. (2004), "Testing of Special LYS Steel Plate Shear Walls," Paper No. 978, *Proc. 13th World Conf. on Earthquake Eng.*, Vancouver, Canada.

Villaverde, R. (2007), "Methods to Assess the Seismic Collapse Capacity of Building Structures: State of the Art," *ASCE J. Struct. Eng.*, Vol. 133, 1, pp. 57–66.

Wada, A., and Nakashima, M. (2004), "From Infancy to Maturity of Buckling Restrained Braces Research," Paper No. 1732, *Proc. 13th World Conf. on Earthquake Eng.*, Vancouver, Canada.

Walpole, W. R. (1996), "Behaviour of Cold-Formed Steel RHS Members Under Cyclic Loading," Research Rep. No. 96-4, University of Canterbury, Christchurch, New Zealand.

Wang, D. Q., and Boresi, A. P. (1992), "Theoretical Study of Stability Criteria for X-Bracing Systems," *ASCE J. Eng. Mech.*, Vol. 118, No. 7, pp. 1357–1364.

Watanabe, A., Hitomi, Y., Saeki, E., Wada, A., and Fujimoto, M. (1988), "Properties of Brace Encased in Buckling-Restraining Concrete and Steel Tube," *Proc. 9th World Conf. on Earthquake Eng.*, Vol. IV, Tokyo, Japan, pp. 719–724.

Wei, C.-Y., and Tsai, K.-C. (2008), "Local Buckling of Buckling Restrained Braces," Paper No. 05-05-0139, *Proc. 14th World Conf. on Earthquake Eng.*, Beijing, China.

Whittaker, A. S., Uang, C.-M., and Bertero, V. V. (1987), "Earthquake Simulation Tests and Associated Studies of a 0.3-Scale Model of a Six-Story Eccentrically Braced Steel Structure," Report No. UCB/EERC-87/02, Earthquake Eng. Research Center, University of California, Berkeley, CA.

Williamson, E. B. (2003), "Evaluation of Damage and P–Δ Effects for Systems Under Earthquake Excitation," *ASCE J. Struct. Eng.*, Vol. 129, No. 8, pp. 1036–1046.

Xie, Q. (2005), "State of the Art of Buckling-Restrained Braces in Asia," *J. Constr. Steel Res.*, Vol. 61, No. 6, pp. 727–748.

Yamazaki, S., and Endo, K. (2004), "Stability Ratio and Dynamic P-D Effects in Inelastic Earthquake Responses," Paper No. 466, *Proc. 13th World Conf. Earthquake Eng.*, Vancouver, British Columbia.

Yang, C. S., Leon, R. T., and Desroches, R. (2008a), "Pushover Response of a Braced Frame with Suspended Zipper Struts," *ASCE J. Struct. Eng.*, Vol. 134, No. 10, pp. 1619–1626.

Yang, C. S., Leon, R. T., and Desroches, R. (2008b), "Design and Behavior of Zipper-Braced Frames," *Eng. Struct.*, Vol. 30, No. 4, pp. 1092–1100.

Yoo, J.-H., Roeder, C. W., and Lehman, D. E. (2008), "Analytical Performance Simulation of Special Concentrically Braced Frames," *ASCE J. Struct. Eng.*, Vol. 134, No. 6, pp. 881–889.

Zareian, F., and Krawinkler, H. (2007), "Assessment of Probability of Collapse and Design for Collapse Safety," *Earthquake Eng. Struct. Dyn.*, Vol. 36, No. 13, pp. 1901–1914.

Zhang, X., and Ricles, J. M. (2006a), "Experimental Evaluation of Reduced Beam Section Connections to Deep Columns," *ASCE J. Struct. Eng.*, Vol. 132, No. 3, pp. 346–357.

Zhang, X., and Ricles, J. M. (2006b), "Seismic Behavior of Reduced Beam Section Moment Connections to Deep Columns," *ASCE J. Struct. Eng.*, Vol. 132, No. 3, pp. 358–367.

Zhang, Y., and Zhu, S. (2008), "Seismic Response Control of Building Structures with Superelastic Shape Memory Alloy Wire Dampers," *ASCE J. Struct. Eng.*, Vol. 134, No. 3, pp. 240–251.

Zhao, Q., and Astaneh-Asl, A. (2004), "Cyclic Behavior of Traditional and Innovative Composite Shear Walls," *ASCE J. Struct. Eng.*, Vol. 130, No. 2, pp. 271–284.

Zhao, X.-L., Grzebieta, R., and Lee, C. (2002), "Void-Filled Cold-Formed Rectangular Hollow Section Braces Subjected to Large Deformation Cyclic Axial Loading," *ASCE J. Struct. Eng.*, Vol. 128, No. 6, pp. 746–753.

Zheng, Y., Usami, T., and Ge. H. (2000), "Ductility of Thin-Walled Steel Box Stub-Columns," *ASCE J. Struct. Eng.*, Vol. 126, No. 11, pp. 1304–1311.

Zhu, S., and Zhang, Y. (2008), "Seismic Analysis of Concentrically Braced Frame Systems with Self-Centering Friction Damping Braces," *ASCE J. Struct. Eng.*, Vol. 134, No. 1, pp. 121–131.

CHAPTER 20

STABILITY ANALYSIS BY THE FINITE ELEMENT METHOD

This chapter provides a methodology and corresponding nomenclature for the development of the finite element method as a means for solving the governing equations of structural systems that are susceptible to instabilities. This development, presented in Sections 20.1 and 20.2, uses principles and notation that are not typically encountered in elementary finite element courses but are nonetheless of great importance as an introduction to the conventions that are typically encountered within modern literature on the subject.

With this overview, Section 20.3 presents more practical aspects of stability analysis by the finite element method. After providing a survey of the dominant formulations behind the finite element buckling procedures employed in commercial software, comparisons in predicted capacities are made and differences discussed. Additionally, observations regarding the nature of the first eigenmodes (as obtained from finite element buckling analyses) are made and their usefulness as devices for defining seed (initial geometric) imperfections when using incremental nonlinear finite element analysis methods to study structural stability problems is assessed.

20.1 INTRODUCTION

This discussion on the finite element method, as it relates to the evaluation of stability of equilibrium, takes as its point of departure the integral statement for the governing equations of motion. While, historically, it has been more common to employ the principles of *virtual work*, or equivalently *stationarity in the total potential*, when developing the finite element equations for solids and structures (Bathe, 1996; Cook et al., 2002; McGuire et al., 2000; Zienkiewicz and Taylor 2000a), this treatment focuses on the *Galerkin method*, as it applies to the weak form of the governing equations in a structural problem (Reddy, 1993). This approach is taken in order to equip the reader with concepts that will be useful in understanding modern research advances concerning the finite element method (Ciarlet and Lions, 1991; Chapelle and Bathe, 2003; Duarte and Oden, 1996;

Babuska and Melenk, 1997; Melenk and Babuska, 1996; Belytschko and Black, 1997; Farhat et al., 2003) as well as enhancing an understanding of the method's convergence properties (Reddy, 1993; Langtangen, 2000).

It is ultimately the goal of the finite element method to idealize a continuous system in a discrete form that preserves salient physical responses while at the same time admitting insight into convergence properties and accuracy. With this in mind, consider the statement of some homogeneous and continuous model problem of the form

$$\mathcal{L}\left(u(x)\right) = 0 \quad \forall x \in \Omega \tag{20.1}$$

where $\mathcal{L}$ is a general linear differential operator, $u\left(x\right)$ is the solution of this model problem, and Ω is some open set defining the problem domain. In a discrete numerical solution to the problem posed in Eq. 20.1, an approximate solution of the form $\hat{u}\left(x\right)$ will be obtained with[1]

$$u \approx \hat{u} = u_i N_i\left(x\right) \tag{20.2}$$

subject to the Fourier requirement (Reddy, 1993; Kreyszig, 1978) that $\|u - \hat{u}_i\| \rightarrow 0$ as $i \rightarrow \infty$. Given that $\hat{u}$ is an approximation, it is not reasonable to expect that Eq. 20.1 will be satisfied by Eq. 20.2 In general $\mathcal{L}\left(\hat{u}\left(x\right)\right) \neq 0$, and thus it is appropriate to define a residual for the model problem as

$$\mathcal{R} = \mathcal{L}\left(\hat{u}\left(x\right)\right) \tag{20.3}$$

In any discrete solution of the model problem, it can only be ensured that the residual, $\mathcal{R}$, is arbitrarily small, and it is desired that $\mathcal{R} \rightarrow 0 \Rightarrow \|u - \hat{u}_i\| \rightarrow 0$. Considering now exclusively the *residual*, the goal is to minimize approximation error. With Eq. 20.4 below as an alternate form of Eq. 20.3, an important observation can be made: because $\mathcal{R}$ varies in space, the requirement of a minimum can only be imposed in an average or integral sense:

$$\mathcal{R}\left(u_i, x\right) = \mathcal{L}\left(\hat{u}\left(x\right)\right) \tag{20.4}$$

One approach to providing such a minimization involves the l*east squares method*, wherein the objective is to minimize the average square of the residual, which can be expressed by

$$\min\left(\int_\Omega \mathcal{R}^2 d\Omega\right) \text{ w.r.t. } u_i \tag{20.5}$$

Because Eq. 20.5 describes a discrete system, the minimization may be obtained using partial derivatives:

$$\frac{\partial}{\partial u_i}\int_\Omega \mathcal{R}^2 d\Omega = \int_\Omega 2\mathcal{R}\frac{\partial \mathcal{R}}{\partial u_i} d\Omega = 0 \tag{20.6}$$

[1]Einstein's summation convention is used throughout this discussion.

It is actually instructive to recast the results of Eq. 20.6 into the slightly different form

$$\int_{\Omega}\left(2\frac{\partial \mathcal{R}}{\partial u_i}\right)\mathcal{R}\ d\Omega = 0 \qquad (20.7)$$

From Eq. 20.7, it is now more obvious that the least squares method is a type of *weighted residual method* with a canonical form expressed as

$$\int_{\Omega} w_i \mathcal{R}\ d\Omega = 0 \qquad (20.8)$$

Thus, in the case of the least squares method, Eq. 20.7 implies that $w_i = 2\left(\partial\mathcal{R}/\partial u_i\right)$. In general, the *weighting functions*, w_i, are somewhat arbitrary, although they do need to satisfy certain requirements, including:

- They are linearly independent functions.
- They vanish on the part of the boundary, $\partial_E\Omega$, where *essential* (*Dirichlet*) boundary conditions are specified.

These requirements may be stated more formally as

$$w_i \in H_o^m\left(\Omega\right) \qquad (20.9)$$

where H_o^m is the *Sobolev space* of order m, where $2m$ refers to the order of the highest derivative present in the linear differential operator $\mathcal{L}$.

Example For illustrative purposes, these somewhat formal concepts may be applied to a simple problem. Figure 20.1 depicts a bar of length L that is loaded by end forces and a distributed body force that takes the form of a traction over the entire length of the member. Considering now the equilibrium of an isolated rod segment of differential length, forces may be summed as (consistent with Fig. 20.2)

$$\rightharpoonup \sum F_x = 0 = -\int_A \sigma\ dA + \int_A \left(\sigma + d\sigma\right)\ dA + \beta\ dx \qquad (20.10)$$

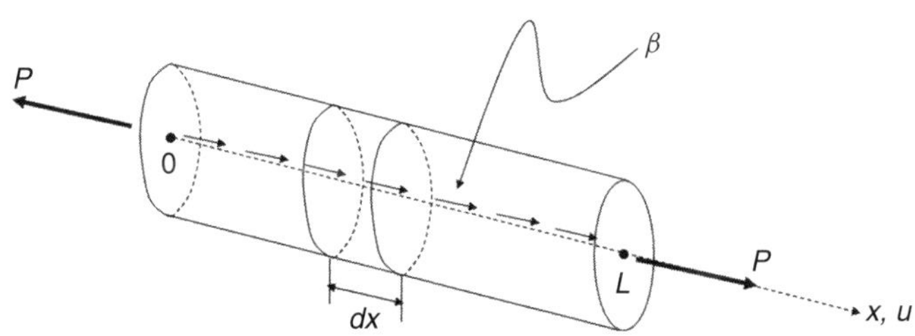

FIGURE 20.1 Rod of length L loaded by end forces P and a distributed body force β.

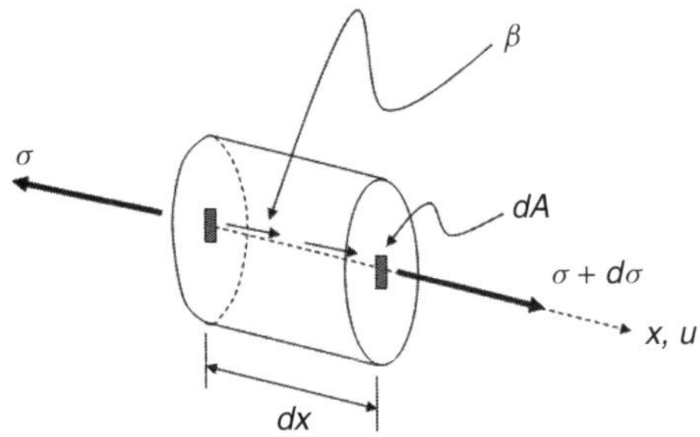

FIGURE 20.2 Free-body diagram of differential length of rod.

Equation 20.10 simplifies to

$$A\frac{d\sigma}{dx} + \beta = 0 \tag{20.11}$$

Subsequent consideration of linear kinematics and constitutive relations, $\varepsilon = du/dx$ and $\sigma = E\varepsilon$, respectively, yields the governing differential equation (strong-form statement) of the problem,

$$\frac{d}{dx}\left(EA\frac{du}{dx}\right) + \beta = 0 \tag{20.12}$$

and thus provides

$$\mathcal{L}\left(u\left(x\right)\right) = \frac{d}{dx}\left(EA\frac{du}{dx}\right) + \beta = 0 \tag{20.13}$$

over the domain $\Omega = \{x\,|x \in (0, L)\}$

This continuous problem may be discretized by replacing the unknown solution, $u\left(x\right)$, with an approximation in the form of that given by Eq. 20.2. The quantities N_i are frequently referred to as *shape functions* in the literature on the finite element method or as *trial functions* within the context of the classical applications of the *Rayleigh–Ritz method*. Given that these individual functions, N_i, satisfy *compact support* (Reddy, 1993) with respect to individual finite elements, their superposition leads to the creation of a *Schauder basis* (Kreyszig, 1978) that spans the Sobolev space of dimension i. As $i \to \infty$, the approximation space approaches the space of continuous functions where $u\left(x\right)$ occurs.

Using $\hat{u}\left(x\right)$ in $\mathcal{L}$ yields the residual per-unit length of prismatic bar of homogeneous material,

$$\mathcal{L}\left(\hat{u}\left(x\right)\right) = EA\frac{d^2\left(\hat{u}\left(x\right)\right)}{dx^2} + \beta = \mathcal{R} \tag{20.14}$$

Upon substituting Eq. 20.2 into Eq. 20.14, the following expression for the residual per unit length is obtained:

$$\mathcal{R} = (EA)\, u_i N_i''(x) + \beta \tag{20.15}$$

where the primes signify spatial differentiation.

The least squares method may now be used to identify a suitable weighting function for use in a *weighted residual statement* (i.e., *weak form*) of this example problem:

$$w_i = 2\frac{\partial \mathcal{R}}{\partial u_j} = (2EA)\,\frac{\partial}{\partial u_j} u_i N_i''(x) = (2EA)\, N_j''(x) \tag{20.16}$$

Note that the multiplier of $2EA$ may be dropped, as it is merely a constant and thus does not impact the linear independence or satisfaction of essential (Dirichlet) boundary conditions.

Continuing with the discussion of the least squares method as a *weighted integral form*, the goal is to minimize the residual, $\mathcal{R}$, as

$$\int_\Omega w_i \mathcal{R}\, d\Omega = \int_0^L \big((EA)\, u_i N_i''(x) + \beta\big) N_j''(x)\, dx = 0 \tag{20.17}$$

The contents of Eq. 20.17 may be placed in slightly more familiar form as

$$-\left(EA \int_0^L N_i''(x)\, N_j''(x)\, dx\right) u_i = \int_0^L \beta N_j''(x)\, dx \tag{20.18}$$

thus leading to a matrix representation of

$$[A]\,\{u\} = \{b\} \tag{20.19}$$

where $[A]$ is a stiffness matrix having entries

$$A_{ij} = -EA \int_0^L N_i''(x)\, N_j''(x)\, dx \tag{20.20}$$

and the force vector $\{b\}$ is of the form

$$\int_0^L \beta N_j''(x)\, dx \tag{20.21}$$

Galerkin Method A similar approach to the foregoing replaces the least squares method with an approach known as the *Galerkin method*. This latter approach is more common in the finite element analysis of solids and structures as a result of the usual *positive definiteness* and *strong ellipticity* (Naylor and Sell, 1982) of the governing functionals in structural mechanics. One important benefit in using the Galerkin method is that the required order for continuity in the shape functions, N_i, is reduced, as will be seen below.

In the Galerkin method, the weighting functions, w_i, are selected to be identical to the shape functions, N_i, underpinning the approximate solution $\hat{u}_i$, as outlined in Eq. 20.2. It should be noted that the foregoing approach is sometimes referred to as the *Bubanov–Galerkin method*, in contrast to the *Petrov–Galerkin method*, in which the weighting functions and shape functions are not of the same form. This distinction is pointed out to avoid confusion when working with the finite element literature. For the purposes of the current discussion, reference to the Bubanov–Galerkin method will be made simply as the *Galerkin method*.

Reconsidering the above example problem in terms of the Galerkin method, the following weighted integral statement (i.e., weak-form statement) is obtained:

$$\int_{\Omega} w_i \mathcal{R}\, d\Omega = \int_0^L N_j \left(EA \frac{d^2u}{dx^2} + \beta \right) dx = 0 \tag{20.22}$$

In addition, it is instructive to include a set of practically useful boundary conditions:

- End fixity, $u\,(0) = 0$
- Loading applied to the free end, $EA\;\,[du\,(L)/dx] = P$

In this example, the linear differential operator, $\mathcal{L}$, is of order 2, and thus $m = 1$ (see Eq. 20.9), and the weighting functions w_i in general, and N_i in particular, must be admissible, stated formally as

$$N_i \in H_0^L\,(0, 1) \tag{20.23}$$

Standard integration-by-parts approaches from elementary differential calculus may be applied to Eq. 20.22 using the following integration and substitution formulas:

$$\int u\; dv = uv - \int v\; du \qquad u = N_j \quad du = \frac{dN_j}{dx} \quad dv = EA\frac{d^2u}{dx^2} \quad v = EA\frac{du}{dx} \tag{20.24}$$

leading to

$$uv - \int v\; du = N_j \left(EA \frac{du}{dx} \right)\Bigg|_0^L - EA \int_0^L \frac{dN_j}{dx}\frac{du}{dx} dx + \int_0^L N_j \beta\; dx = 0 \tag{20.25}$$

Recognizing that $N_j(0) = 0$ (because $N_j \in H_0^m$), $N_j(L) = 1$, and $EA\big[du(L)/dx\big] = P$, it is seen that the *natural* (*Neumann*) boundary conditions are "automatically" recovered and Eq. 20.25 simplifies to

$$P - EA\int_0^L \frac{dN_j}{dx}\frac{du}{dx}dx + \int_0^L N_j\beta\, dx = 0 \qquad (20.26)$$

The approximation described by Eq. 20.2 may now be employed within Eq. 20.26 by replacing u by $\hat{u}$, and rearranging terms yields

$$EA\int_0^L \frac{dN_j}{dx}\frac{dN_i}{dx}u_i\, dx = P + \int_0^L N_j\beta\, dx \qquad (20.27)$$

The left-hand terms in Eq. 20.27 lead to the stiffness matrix, expressing the parameters associated with the response of the structural system under external actions. The terms on the right lead directly to the generalized force vector, expressing the external actions on a structural system.

The benefit of using the Galerkin method is now apparent when comparing Eqs. 20.18 and 20.27. In Eq 20.18, the approach based on the least squares method yielded an integral statement wherein the shape functions of the approximate solution needed to be twice differentiable, and able to satisfy *both* the essential (Dirichlet) and natural (Neumann) boundary conditions. In contrast, the outcome of the Galerkin method is that the application of integration by parts results in a shifting of the derivative from one term to another in Eq. 20.22, thus weakening the continuity requirements on the shape functions used in forming the approximate solution. Furthermore, it is observed (see Eq. 20.26) that the natural (Neumann) boundary conditions are *implied* within the integral statement (Eq. 20.22) of the example problem. As a result of their implied sense, the natural boundary conditions did not require overt consideration during the selection of the shape functions used in the approximate solution, $\hat{u}$.

The foregoing consideration of the weighted integral approach to developing a finite element representation of a structural problem may be brought into consonance with other common formulation approaches (Bathe, 1996; Cook et al., 2002; Zienkiewicz and Taylor, 2000a) through further consideration of the example depicted in Fig. 20.1. When considering Eq. 20.27, the form of the so-called strain–displacement matrix, commonly given as $[B]$ in the finite element literature (Bathe, 1996; Belytschko et al., 2000; Cook et al., 2002; Crisfield, 1991a,b; Hughes, 2000; Zienkiewicz and Taylor, 2000a), may be developed as

$$\varepsilon = \frac{du}{dx} \approx \frac{d\hat{u}_i}{dx} = \frac{dN_i}{dx}u_i \rightarrow [B]\{u\} \qquad (20.28)$$

Substituting the foregoing result into Eq. 20.27 yields (in matrix form)

$$\left(EA\int_0^L [B]^{\mathrm{T}}[B]\, dx\right)\{u\} = P + \int_0^L \{N\}^{\mathrm{T}}\beta\, dx \qquad (20.29)$$

Equation 20.29 lends itself to being cast in the familiar general form of the equilibrium statement

$$[K]\{u\} = \{F\} \tag{20.30}$$

where $[K]$ is the stiffness matrix formed through integration of the product of the strain–displacements terms and $\{F\}$ the force vector representation (in this case, the aggregate effect of concentrated loads and distributed body forces).

20.2 NONLINEAR ANALYSIS

Nonlinearities in structural analysis may arise from three primary sources: boundary conditions that change during the problem solution (e.g., hard contact between bodies); nonlinear material response (e.g., yielding of steel in a structural member); and geometric nonlinearity (e.g., elastic column buckling). It is the last of these, geometric nonlinearity, which is most closely aligned with the notion of quality of equilibrium and thus is the focus of the present discussion. This is not to say that the other two sources do not pertain to stability problems; each of these can lead to manifestations of instability in their own right. The effects of nonlinear material and boundary conditions, however, will not be treated herein due to space limitations and instead reference is given to Simo and Hughes (1991), Wriggers (2002), Bathe (1996), and Belytschko et al. (2000).

A convenient starting point to a discussion on ways of admitting geometric nonlinearity into the finite element formulation is through a consideration of large displacements as well as large strains. Since instability problems invariably require the formulation of equilibrium on some deformed configuration of the structural system being investigated, it appears as a truism that, at the very least, large displacements must be admitted. In the case of the present discussion, large strains are also included for completeness. With this in mind, a new strain measure that is suitable for large displacement and large strain response is now introduced.

The *Green–Lagrange strain* is commonly encountered in treatments of the nonlinear finite element method (Bathe, 1996; Belytschko et al., 2000; Crisfield, 1991; Zienkiewicz and Taylor, 2000b), but important foundation theory can be found within relevant texts on continuum mechanics (Malvern, 1969; Fung, 1994; Ogden, 1984; Novozhilov, 1953).

In general, the Green–Lagrange strain is a second-order tensor describing the deformation of a body and whose tensorial components are

$$E_{ij} = \frac{1}{2}\left(\frac{du_i}{dx_j} + \frac{du_j}{dx_i} + \frac{du_k}{dx_i}\frac{du_k}{dx_j}\right) \tag{20.31}$$

One very important property of the Green–Lagrange strain, as compared with the usual linear engineering strain, is its invariance under rigid-body rotations, a critically important feature with respect to stability-related finite element analyses that involve large displacements (rotations).

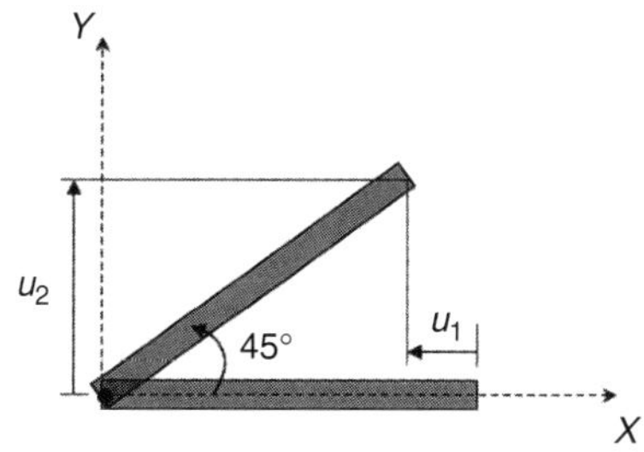

FIGURE 20.3 Large rotation of rigid rod.

Consider a simple unit-dimensional example problem depicted in Fig. 20.3. This example will highlight two important ideas that are at the heart of the present discussion:

- The unsuitability of the engineering strain in cases of large rotation
- The invariance of the Green–Lagrange strain in such cases

Clearly, a strain measure that yields nonzero terms when a body merely rotates as a rigid body (i.e., no deformation) would prove disastrous for applications involving geometric nonlinearity. Such a rigid-body rotation condition is depicted in Fig. 20.3, wherein a bar that is initially aligned with the x axis of a two-dimensional Cartesian reference frame rotates 45° about its left end. The displacement components of the right end may then be given as

$$u_1 = -\left(L - \frac{\sqrt{2}}{2}L\right) \quad \text{and} \quad u_2 = \frac{\sqrt{2}}{2}L$$

Considering these displacements, the response of the element according to engineering strain is given by

$$\varepsilon = \frac{du_1}{dx} = \frac{(\sqrt{2}/2)L - L}{L} = \frac{\sqrt{2}}{2} - 1 \neq 0 \tag{20.32}$$

Clearly the strains obtained according to linear engineering strain theory (Eq. 20.32) are incorrect because there can be no deformation of the structural element under a pure rigid-body rotation.

Consider now the Green–Lagrange strain of this problem. The components of the tensor that measure deformation along the x axis yield

$$\begin{aligned} E_{11} &= \frac{1}{2}\left(\frac{\partial u_1}{\partial x_1} + \frac{\partial u_1}{\partial x_1} + \frac{\partial u_1}{\partial x_1}\frac{\partial u_1}{\partial x_1} + \frac{\partial u_2}{\partial x_1}\frac{\partial u_2}{\partial x_1}\right) \\ &= \frac{\partial u_1}{\partial x_1} + \frac{1}{2}\left(\frac{\partial u_1}{\partial x_1}\right)^2 + \frac{1}{2}\left(\frac{\partial u_2}{\partial x_1}\right)^2 \end{aligned}$$

$$= \left(\frac{\sqrt{2}}{2} - 1\right) + \frac{1}{2}\left(\frac{\sqrt{2}}{2} - 1\right)^2 + \left(\frac{\sqrt{2}}{2}\right)^2$$

$$= \left(\frac{\sqrt{2}}{2} - 1\right) + \frac{1}{2}\left(\frac{1}{2} - \sqrt{2} + 1\right) + \frac{1}{4} = 0 \tag{20.33}$$

Thus, the Green–Lagrange strain remains invariant (i.e., null) during the rigid-body rotation depicted in Fig. 20.3.

In light of the foregoing discussion on the Green–Lagrange strain, an extension of the equilibrium relation of Eq. 20.29 to include large deformations may now be defined. Such an extension will permit the consideration of *geometric nonlinearity*, a condition that is at the heart of essentially all structural stability problems.

Toward this end, consider an extension of the notion of a strain–displacement matrix $[B]$ to account for geometrically nonlinear effects. Recognizing that the first two terms on the right-hand side of Eq. 20.31 correspond to the usual, linear engineering strain, a matrix designation of $[B_L]$ can be applied to this portion of the strain tensor. The remaining terms, which are responsible for the consideration of large deformations, can be included in a second strain–displacement relation, $[B_{NL}]$. Using this approach, the Green–Lagrange strain tensor now may be represented (in matrix form) as

$$[E] = [B_L]\{u\} + [B_{NL}]\{u\} = [B_{GNL}]\{u\} \tag{20.34}$$

This result may now be used to extend the equilibrium relation (Eq. 20.29), for the example problem of Fig 20.2, for applications that include geometrically nonlinear behavior. This extension takes the straightforward form

$$\left(EA\int_0^L [B_{GNL}]^{\mathrm{T}}[B_{GNL}]\,dx\right)\{u\} = P + \int_0^L \{N\}^{\mathrm{T}}\beta\,dx \tag{20.35}$$

Evaluation of the left-hand side of Eq. 20.35 results in the *tangent stiffness matrix*, a linear expansion of the equilibrium path about a particular point in the configuration space of the particular problem. The right-hand side, in this case, embodies the loading effects related to the specific example problem. In the incremental nonlinear solution of Eq. 20.35, within a Lagrangian reference frame, the treatment of the solution of an unknown equilibrium configuration, using a previously solved-for configuration, leads naturally to the identification of the system internal force vector that arises in order to balance applied loadings from the right-hand side. A more detailed treatment of these concepts is provided in Chapter 6 of Bathe (1996).

To facilitate the discussion to follow, the tangent stiffness matrix is represented as

$$[K_T] = [K_o] + [K_\sigma] \tag{20.36}$$

in which the linear elastic stiffness matrix is

$$[K_o] = EA\left(\int_0^L [B_L]^{\mathrm{T}}[B_L]\,dx\right) \tag{20.37}$$

and the initial stress matrix is

$$[K_\sigma] = EA\left(\int_0^L [B_{NL}]^{\mathrm{T}}[B_{NL}]\,dx\right) \tag{20.38}$$

In all subsequent discussions, a *total Lagrangian* reference frame is assumed, and thus all formulations take the initially undeformed configuration of the structure as the reference frame from which to measure kinematic relations (Bathe, 1996). It is pointed out that the initial state from the total Lagrangian perspective is used to formulate nonlinear finite element equations of motion. This initial state has no bearing on the baseline and characteristic equilibrium states that are subsequently introduced in the discussions pertaining to linearized eigenvalue buckling approaches.

20.3 LINEARIZED EIGENVALUE BUCKLING ANALYSIS

There is an increasing availability of commercial finite element software that permits the consideration of the effects of geometric nonlinearity in structural analysis. Oftentimes these software systems will have the capability to treat stability problems through eigenvalue extraction routines applied to the global system stiffness matrix, an approach often referred to as *buckling*, *eigenvalue buckling*, or *linearized eigenvalue buckling* analysis.

This type of a stability analysis is attractive from the standpoint that it is not computationally expensive. As compared with a more general incremental analysis that traces the entire nonlinear equilibrium path of the structural system, the eigenvalue buckling approach concerns itself with only one or two points on the equilibrium path. In addition, results obtained from eigenvalue buckling analyses, when applied to stability problems exhibiting bifurcation instability, are usually quite accurate (Earls, 2007), and this accuracy is obtained without much intervention on the part of the software user. Care must be taken, however, when applying this technique to problem types exhibiting other manifestations of instability (e.g., limit point instability).

In practice, situations may arise in the design office where the application of eigenvalue buckling may seem attractive for problems involving elastic beam buckling (e.g., lateral–torsional buckling of a beam or planar truss) or elastic snap-through buckling (e.g., lattice dome, arch, or shallow truss assembly). In the domain of stability research, linearized eigenvlaue buckling is also oftentimes attractive as a means for identifying an initial or *seed imperfection* for application within a more detailed incremental nonlinear finite element analysis of a beam or framework, for instance. In all of the foregoing, there are finite structural deformations prior to the onset of instability that are additive to the governing buckling mode (as compared to a bifurcation instability where the pre- and postbuckling deformations may be thought of as being orthogonal to one another). This fact creates an inconsistency with regard to assumptions made in the formulation of the linearized eigenvalue buckling procedure itself.

The present discussion examines the underlying assumptions within the formulation of the eigenvalue buckling method in order to highlight the problem types that most readily lend themselves to solution by this method. In addition, problems presenting responses that violate these fundamental assumptions are also examined. In this latter case, it becomes very important to understand the nature of the implementation of eigenvalue buckling in the given software system (example problems, solved by MASTAN2, ADINA, ANSYS, and ABAQUS, are included in this discussion); certain implementations will make application of eigenvalue buckling to other than bifurcation problems extremely problematic. The present discussion related to the various finite element buckling formulations employs a single, standardized notation to allow for a transparent comparison of underlying assumptions.

20.3.1 Overview of Finite Element Buckling Analysis Approaches

The literature adopts the term buckling analysis when referring to a family of finite element techniques applied to structural systems for the identification of critical load levels through the solution of an *eigenproblem* arising out of assumptions made relative to changes in structural stiffness and associated applied loadings. While the technique is applicable to structures that exhibit critical responses arising from the *limit point* as a well as *bifurcation of equilibrium*, the term buckling is nonetheless universally applied. While this may seem inconsistent, since buckling is normally associated with the condition of bifurcation in the equilibrium path only, the nomenclature is defensible nonetheless as a result of the fact that the eigenproblem posed within the finite element context resembles the case where the vanishing of the determinant of the stiffness matrix is associated with a form of the Sturm–Liouville problem (Reddy, 1993; Boyce and DiPrima, 1986). All formulations within the present discussion will be presented in a standardized notation (to facilitate comparison) that is defined subsequently.

Because the analyses considered here are strictly static, time will be used to signify an equilibrium point within the configuration space of a given structure corresponding to certain load level. Definitions for several left subscripts are now introduced to facilitate subsequent discussions. A generic place holder term, Ω, will be used to show the relative locations of the subscripts:

$_{o}\Omega \equiv$ generic quantity, Ω, evaluated at the equilibrium configuration associated with the trivial case of no external actions.

$_{t}\Omega \equiv$ generic quantity, Ω, evaluated at an intermediate equilibrium configuration occurring between the unloaded and critical configurations.

$_{\Delta t}\Omega \equiv$ denotes the change in a generic quantity, Ω, occurring as a result of movement from location t to $t + \Delta t$ on the equilibrium path.

$_{t+\Delta t}\Omega \equiv$ generic quantity, Ω, evaluated at an intermediate equilibrium configuration occurring between the unloaded and critical configurations, that is, arbitrarily close to $_t\Omega$.

In subsequent discussions, it will also be helpful to define three applied loading conditions that are used to surmise an assumed characteristic change in the system stiffness. In general, the applied loading will be denoted with P.

$\{P_{\text{baseline}}\} \equiv$ loading condition used to bring the structure to a point in configuration space associated with the left subscript t.

$\{P_{\text{characteristic}}\} \equiv$ loading condition used to bring the structure to a point in configuration space associated with the left subscript $t + \Delta t$.

$\{P_{cr}\} \equiv$ the critical load associated with the equilibrium configuration at incipient instability.

Structural stiffness will be denoted by the usual quantity $[K]$, more specifically defined by:

$[_oK_o] \equiv$ linear elastic stiffness matrix whose elements are independent of the current structural configuration.

$[_\tau K_\sigma] \equiv$ *initial stress matrix* dependent on the state of stress at an arbitrary time, t. This matrix is populated with terms that include both linear and quadratic dependencies on the current displacement field.

The sum of the foregoing two stiffness matrices is typically what is referred to as the *tangent stiffness matrix*, and associated with a specific equilibrium point in configuration space. Some readers may be more familiar with the notion of the tangent stiffness being associated with the linear terms of a Taylor series expansion of the internal force vector about the current configuration during the solution; while others may recognize it as emanating from the stationarity of the total potential functional with an internal energy term that includes the influence of finite strains. While other options exist for the population of the tangent stiffness matrix (Wood and Schrefler, 1978; Holzer et al., 1990; Chang and Chen, 1986), the former definition has emerged as the most popular to date. In particular, the instantaneous stiffness of the structure arrived at by retaining only the linear terms in a Taylor series expansion of the load–deflection response of the structure about the point in configuration space may be defined according to the applied loading as follows:

$[K_{\text{baseline}}] \equiv$ instantaneous stiffness corresponding to the applied loading $\{P_{\text{baseline}}\}$

$[K_{\text{characteristic}}] \equiv$ instantaneous stiffness corresponding to the applied loading $\{P_{\text{characteristic}}\}$

Classical Formulation The initial treatment of the finite element buckling analysis appeared in the literature prior to the formal naming of the finite element method (Gallagher et al., 1967); this earliest reference identified the approach as being based on the *discrete element procedure*. In light of the foregoing, and based on a survey of the literature, it appears that in the most commonly held definition of the classical formulation for finite element buckling analysis, the following problem is solved (Cook et al., 2002; Holzer et al., 1990; Chang and Chen, 1986; Brendel and Ramm, 1980):

$$\det\left([{}_oK_o] + [{}_tK_\sigma]\right) = 0 \tag{20.39}$$

It is frequently assumed in the literature that the equilibrium point at time t is very close to the initial configuration at time 0, but this is not a strict requirement. The subsequent buckling load is computed as

$$\{P_{cr}\} = \lambda \{P_{\text{baseline}}\} \tag{20.40}$$

Secant Formulation The present discussion adopts the name *secant formulation* to describe the variation of the finite element buckling problem that is referred to variously as the secant formulation (Bathe and Dvorkin, 1983; ADINA, 2007) and the *linear and nonlinear analysis* (Holzer et al., 1990). This problem is posed as

$$\det\left([K_{\text{baseline}}] + \lambda\left([K_{\text{characteristic}}] - [K_{\text{baseline}}]\right)\right) = 0 \tag{20.41}$$

and the subsequent buckling load is computed as

$$\{P_{cr}\} = \{P_{\text{baseline}}\} + \lambda\left(\{P_{\text{characteristic}}\} - \{P_{\text{baseline}}\}\right) \tag{20.42}$$

20.3.2 Example Problems

In the following examples, comparisons of results from the two approaches to finite element buckling analyses are considered (i.e., those involving software packages employing buckling approaches characterized by Eqs. 20.39 and 20.40 and 20.41and 20.42, respectively). In some instances the results from closed-form solutions are also presented. In addition, several cases also include results from manual implementation of the finite element buckling approaches as encapsulated in Eqs. 20.39 and 20.40.

This discussion begins by distinguishing between bifurcation and limit point instability in a formal way. Consider the classical form of the finite element statement of the incremental equilibrium equations:

$$\left([{}_oK_o] + [{}_tK_\sigma]\right)\{{}_{t+\Delta t}\Delta u\} = \{{}_{t+\Delta t}R\} \tag{20.43}$$

where $\{\Delta u\}$ are the incremental nodal displacements and $\{R\}$ is the residual vector representing the imbalance between the internal forces at time t and the desired load levels associated with some set of external forces, $\{P\}$. It is pointed out that this residual differs in context from the quantity sharing the same name and defined in Eq. 20.3. The standard form of the eigenvalue problem will be used to compute eigenvalues, ω_i, and eigenvectors, $\{\phi\}_i$, for the tangent stiffness matrix according to

$$([_oK_o] + [_tK_\sigma])\{\phi\}_i = \omega_i\{\phi\}_i \tag{20.44}$$

These eigenpairs lead to the spectral representation of the tangent stiffness matrix

$$[K_T] = \sum_i \omega_i \{\phi\}_i \{\phi\}_i^{\mathrm{T}} \tag{20.45}$$

These same eigenvectors may also be used to define a projection operator that takes the original displacement and load vectors and projects them onto a new vector space spanned by the eigenvectors, $\{\phi\}_i$, such that

$$\{_{t+\Delta t}\Delta u\} = \sum_i \alpha_i \{\phi\}_i \tag{20.46a}$$

$$\{_{t+\Delta t}P\} = \sum_i \rho_i \{\phi\}_i \tag{20.46b}$$

where $\{P\}$ is the externally applied load vector on the structure and $\rho_i = \{\phi\}_i^{\mathrm{T}}\{P\}$. The transformations embodied in Eqs. 20.45 and 20.46 effectively diagonalize Eq. 20.43 and result in the following transformation (Pecknold et al. 1985):

$$\omega_i\alpha_i = \lambda\rho_i \tag{20.47}$$

In an elastic structure, the first critical point occurs when the equilibrium equations become singular or, in other words, when the tangent stiffness matrix is no longer positive definite. Well-ordered eigenpairs result from this being one case of the *Sturm–Liouville problem*, and thus it can be recognized that loss of positive definiteness of the stiffness occurs when $\omega_1 = 0$, at which point the condition $\lambda\rho_1 = 0$ occurs. Based on this fact, it can be recognized that a distinction between bifurcation and limit point instability exists (Pecknold et al., 1985). If $\rho_1 \neq 0$, then it is clear that the eigenvector is not orthogonal to the externally applied loading vector $\{P\}$, and thus λ will have to be zero in order for $\lambda\rho_1 = 0$ to be true. In this case, a limit point instability condition exists (this point will become more clear subsequently). Conversely, $\rho_1 = 0$ when the loading is orthogonal to the eigenmode and buckling is occurring. This may be summarized as:

- Limit point instability: $\omega_1 = 0$ and $\{\phi\}_1^{\mathrm{T}}\{P\} \neq 0$
- Bifurcation instability: $\omega_1 = 0$ and $\{\phi\}_1^{\mathrm{T}}\{P\} = 0$

These results lead to the following interpretation. In the neighborhood of limit points on the equilibrium path in configuration space, there is no increasing load, and the eigenvector is not orthogonal to the external load vector. Conversely, if an increase in loading is possible in the neighborhood of the critical point, and the eigenvector is orthogonal to the load vector, then bifurcation instability is present.

Bifurcation Instability Two classical examples of bifurcation instability are considered next. In these problems, prebuckling deformations are small, and thus the assumptions in this regard, which underlie the linearized eigenvalue buckling approach, are preserved. The first problem considers a one-dimensional structural idealization of a three-dimensional problem, while the second example treats a two-dimensional idealization of a three-dimensional case.

A classical Euler column example is depicted in Fig. 20.4. This figure displays the problem geometry, along with boundary condition, cross section, and material response information. The well-known solution of the example problem is given by

$$P_{cr} = \frac{\pi^2 EI}{L^2} \tag{20.48}$$

which for the given problem results in a theoretically "exact" answer of $P_{cr} = 238$ kips. Clearly this agrees well with both linear eigenvalue buckling solutions (also shown in Fig. 20.4), irrespective of baseline loading, P_{baseline}.

The second bifurcation instability example is given in Fig. 20.5. This plate-buckling problem also has a well-known solution, which is given by

$$\sigma_{cr} = k \frac{\pi^2 E \sqrt{n}}{12\left(1 - \nu^2\right)(b/t)^2} \tag{20.49}$$

Substitution of the specific properties and boundary conditions described in Fig. 20.5 yields a critical elastic stress $\sigma_{cr} = 1.114$ ksi. As might be expected, finite element results for both buckling approaches under discussion are in close agreement with this result.

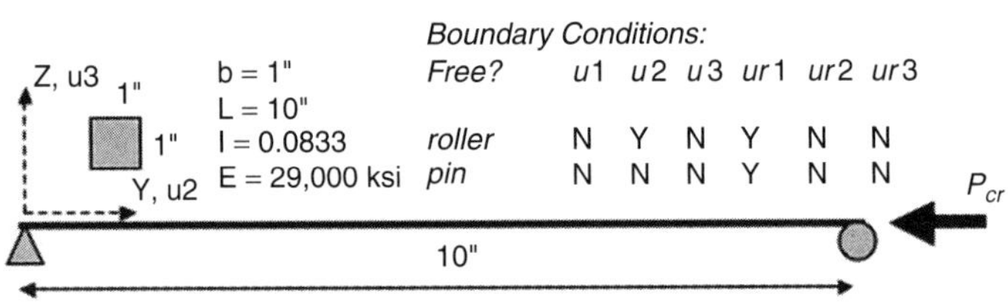

Linearized Eigenvalue Buckling Results:
(5 ADINA Hermitian beam elements)

P_{baseline}	P_{cr}	Method
200 k	235	*Classical*
100 k	235.8	*Classical*
10 k	236.5	*Classical*
1 k	236.6	*Classical*
200 k	235	*Secant*
100 k	235.8	*Secant*
10 k	236.5	*Secant*
1 k	236.6	*Secant*

FIGURE 20.4 Euler column example.

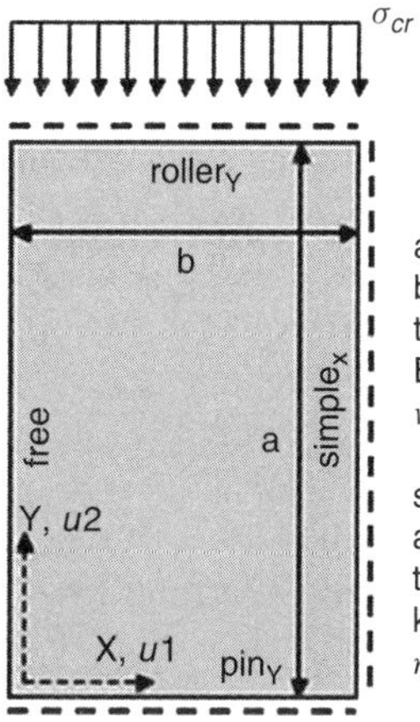

$a = 50''$
$b = 10''$
$t = 0.1''$
$E = 29{,}000$ ksi
$\nu = 0.3$

since $a > b$
and $b \gg t$
then
$k = 0.425$
$\eta = 1$

Boundary Conditions:

Free?	*u*1	*u*2	*u*3	*ur*1	*ur*2	*ur*3
roller$_Y$	Y	Y	N	Y	N	N
pin$_Y$	Y	N	N	Y	N	N
simple$_x$	N	Y	N	N	Y	N
free	Y	Y	Y	Y	Y	N

Linearized Eigenvalue Buckling Results:
(10 ADINA MITC4 four node shell elements across width)

$P_{baseline}$	σ_{cr}	Method
0.001 k/in	1.222 ksi	*Classical*
0.01 k/in	1.217 ksi	*Classical*
0.04 k/in	1.217 ksi	*Classical*
0.05 k/in	1.217 ksi	*Classical*
0.12 k/in	NA	*Classical*
0.001 k/in	NA	*Secant*
0.01 k/in	1.217 ksi	*Secant*
0.04 k/in	1.217 ksi	*Secant*
0.05 k/in	1.217 ksi	*Secant*
0.12 k/in	NA	*Secant*

FIGURE 20.5 Plate-buckling example.

Based on the two simple bifurcation buckling examples presented, it may seem as though linearized eigenvalue buckling approaches satisfactorily predict critical loads in this type of problem. This, however, is not always the case (Earls, 2007) and caution should be taken against merely accepting the buckling loads from finite element software, even under the favorable conditions of bifurcation buckling. For example, it is noted here that in the case of the ADINA two-point buckling formulation (known as the secant formulation in the ADINA literature), it is not possible to know what the load level of $P_{characteristic}$ is because the information contained in the output files are incomplete in this regard. This is considered to be a critical shortcoming in ADINA in terms of reliance on the secant formulation for finite element buckling analysis. As will be seen subsequently, it is not possible to take full advantage of the secant method when using ADINA, because sufficient information regarding the underlying solution process is not available to the user, and thus critical judgment related to the specification of $P_{baseline}$, or even simply the interpretation of results, is compromised.

Limit Point Instability The remaining cases for discussion all exhibit *limit point* (*snap-through*) instability, and thus prebuckling deformations tend to be finite. Strictly speaking, this may be viewed as a violation of the underlying assumptions used in the formulations of Eqs. 20.39 and 20.42. Thus, it might be reasonably concluded that finite element buckling analyses are, technically, not applicable to such instances. Engineers, however, do employ this type of approach to cases that are not strictly in consonance with the underlying assumptions of the formulation, and thus it is important to consider this class of problems within the present discussion. In addition, it is not possible to escape the fact that, under certain circumstances, accurate solutions are obtained when comparing finite element buckling results with the results of more exact methods of analysis.

In the subsequent discussion, it will be useful to refer to a class of diagrams known as eigenvalue plots (Brendel and Ramm, 1980; Holzer et al., 1990). As shown in Fig. 20.6, such a plot depicts a graph of P_{critical} versus P_{baseline} (both normalized by dividing by the "exact" critical load). The depiction of the eigenvlaue plot in Fig. 20.6 is useful to consider in the case of limit point instabilities, because it highlights the dependence of the finite element buckling solution on the selection of a reasonable baseline loading, P_{baseline}. The predicted critical loading from the finite element buckling solution is arrived at by multiplying the eigenvalues by the baseline loads, as described in Eqs. 20.40 and 20.42, respectively, for the classical and secant approaches. The results presented in Fig. 20.6 are consistent with what is expected, from the standpoint that the approximate finite element buckling loads improve in accuracy as the magnitude of the baseline load increases.

Next, consider the truss arch shown in Fig. 20.7, with several height-to-span ratios studied. In all cases, the span length is held constant at 20 in., and the height is varied from 2 to 17 in. The truss arch is a particularly useful example, because it affords the opportunity to easily obtain results using hand calculations. In the case of the truss arch shown Fig. 20.7, with a height of 2 in. (i.e., shallow case), an energy formulation involving the stationarity of the total potential yields a critical load of 85.4 kips. For this particular truss arch geometry, finite element buckling results were obtained using MASTAN2, ANSYS, and ADINA. In the case of the first two software packages, only a very small axial force is considered in the

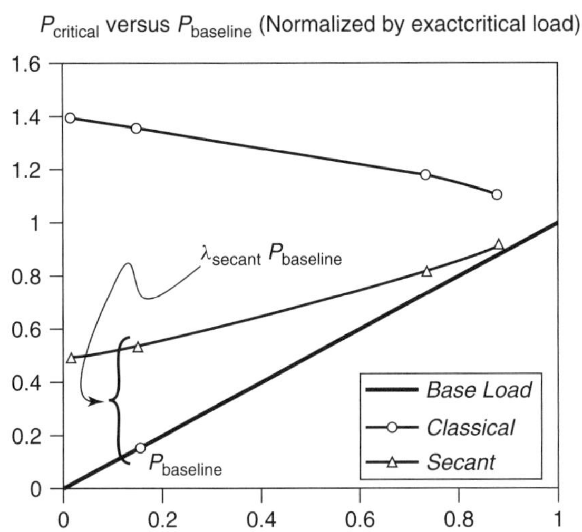

FIGURE 20.6 Representative eigenvalue plot.

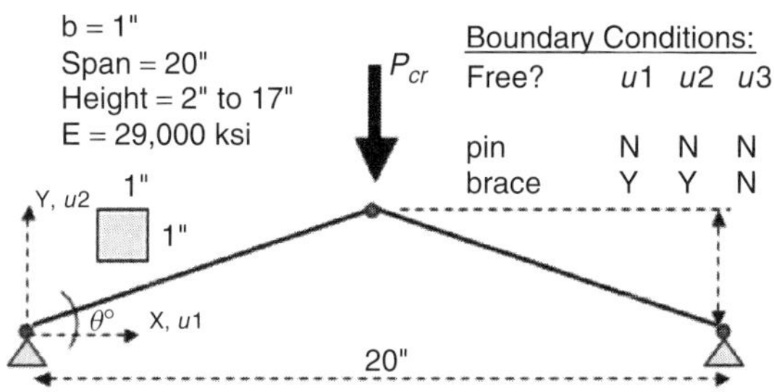

FIGURE 20.7 Truss arch example.

formulation of the tangent stiffness matrix used in the classical approach from Eq. 20.39. These results, along with a hand calculation meant to parallel the classical formulation, as presented in Eqs. 20.39 and 20.41, appear in Fig. 20.8. It is clear that MASTAN2, ADINA, ANSYS, and the hand calculation all agree reasonably well for small values of P_{baseline}. As P_{baseline} is increased, MASTAN2 and ANSYS remain constant in their predictions, because their implementation of the classical method does not admit the possibility of a varying P_{baseline}. In addition, while the hand calculations and ADINA both permit a variation in P_{baseline}, their agreement at high levels is less favorable than at low values of P_{baseline}. This may be a result of subtle differences in the way the classical formulation is implemented in ADINA; but as an unfortunate byproduct of a lack of inclusion (within the output files of ADINA) of intermediate values in the solution process, it is very difficult to test any theories aimed at understanding the nature of the observed differences.

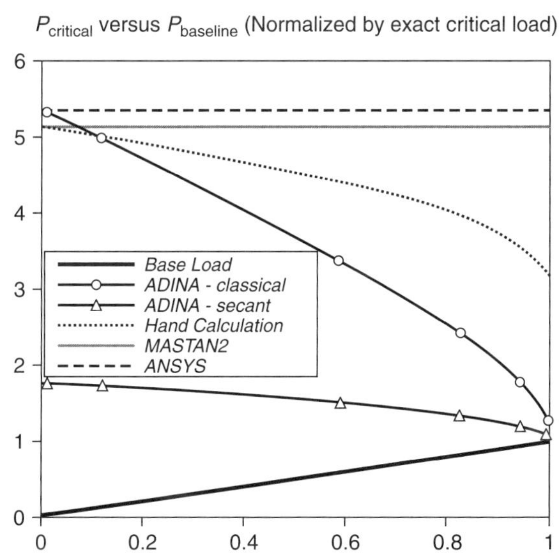

FIGURE 20.8 Truss arch example results for shallow case.

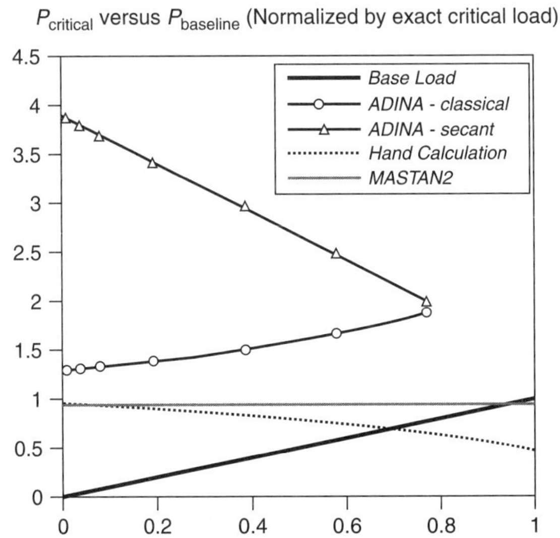

FIGURE 20.9 Truss arch example results for deep case.

In the case of the 17-in. high truss arch (i.e., deep case), we see a difference in the trend of the response observed in the eigenvalue plots of Fig. 20.9. While the MASTAN2 and hand calculations agree well with each other at low loads and produce a reasonable estimate for the critical load, the same is not true for the ADINA results. The ADINA secant results are clearly diverging from the correct solution (13,000 kips; obtained from an incremental nonlinear finite element analysis) while the ADINA classical result begins at a point inexplicably far away from the other two classical implementations.

20.3.3 Linearized Eigenvalue Buckling for Imperfection Seeding in Incremental Nonlinear Finite Element Analysis

Linearized eigenvalue buckling analysis techniques are sometimes employed as a means for obtaining a seed (initial) imperfection to the structural geometry in order to permit a realistic incremental nonlinear finite element analysis (e.g., to facilitate equilibrium branch switching in the study of a bifurcation problem). In such an approach, it is the components of the eigenvector (which represent the relative displacement values at each degree of freedom associated with all nodes) that are scaled to define an initial displacement field that is assigned (as a perturbation) to the ideal node locations of a mesh employed in an incremental nonlinear analysis. Such an approach is predicated on the notion of repeatability across commercial codes (i.e., commercial codes should all yield similar mode shapes from models possessing identical parameters). It has been clearly illustrated, however, that each of the dominant formulations for eigenvalue analysis solves a slightly different eigenproblem, as seen in Eqs. 40.39 through 40.42, and thus complete uniformity in predictions across commercial codes is unlikley. In order to further analyze the significance of this variation in formulation, eigenvector results are quantified to facilitate comparison.

The eigenvectors computed with the secant and classical formulations for a given model may be very similar or, unfortunately, drastically different. This difference or similarity in results can accompany minor changes in geometry and/or loading conditions as well as changes in eigenproblem formulation. Finite element postprocessing software most often scales any reported eigenvectors in order to facilitate visualization of the mode shapes. This scaling can cause problems when comparing results from different commercial programs. In order to more precisely compare the eigenvectors from different formulations (applied to models with identical geometries and material parameters), a sampling of eigenvector values along the centerline of two example beams (Fig. 20.10) is taken. A discrete Fourier analysis is then conducted with these data using the position along the beam as the independent variable.

Nominal dimensions for the two cases depicted in Fig. 20.10 are presented in Table 20.1. Twist is prevented at the ends of the beams by guiding the rectangular stiffener plates to resist out-of-plane deformation. Both beams are modeled in ADINA and ABAQUS using shell elements. Although the shell element formulations are not the same for ADINA and ABAQUS, any differences in formulation are not significant to this comparison because the observed behavior is within the elastic, small-strain-response region (all element meshes were of equivalent density using linear, four node shells; the meshes were also seen to be converged). Two ADINA analyses were performed: one using the classical formulation, and the other with the secant formulation. The ABAQUS analyses only employed a classical type of formulation because secant buckling analysis can only be approximated through the use of a "restart" option.

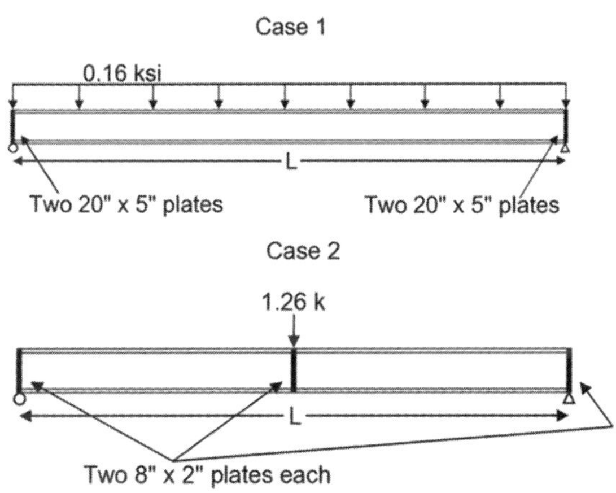

FIGURE 20.10 Beam imperfection example cases.

TABLE 20.1 Nominal Dimensions of Example Beams

	Case 1		Case 2	
Measure	in.	mm	in.	mm
Length, L	250	6350	29	736
Depth, d	19.00	483	8.25	210
Flange width, b_f	6	152.4	4	101.6
Web thickness, t_w	1.00	25.4	0.20	5.1
Flange thickness, t_f	1.00	25.4	0.25	6.4

The first example (case 1) is a steel I-beam that is loaded with a uniformly distributed pressure on the top flange. Figure 20.11 shows that the resulting mode shapes are very similar for both programs. The eigenvectors obtained from ADINA and ABAQUS postprocessing have been normalized (scaled), to enable comparison, using the relation

$$\|\phi_i\| = \frac{\phi_i}{\phi_{\max}} \tag{20.50}$$

where ϕ represents the eigenvector of interest, with the subscripts corresponding to the discrete nodal degree of freedom displacement components of the vector.

When using eigenvector data corresponding to the minor axis (x-direction) lateral displacement component, from nodes located along the centerline of the web, it becomes apparent (Fig. 20.12) that the first-mode eigenvectors for the different formulations employed are only slightly different. This difference in mode shape becomes more definable once a discrete Fourier transform of the minor axis direction eigenvector displacement data is carried out. Figure 20.12 displays the results from this approach, highlighting that the first mode shape resulting from the ADINA classical formulation, the ADINA secant formulation, and ABAQUS formulation are related. They are obviously all harmonic, and their peaks occur at the same frequency. The frequency of the Fourier transform corresponds to the number of cycles of the harmonic function, which for this particular beam example describes the deformation per length of the member. The peak amplitude of the ABAQUS Fourier analysis is 8% less than that obtained using both ADINA formulations; the

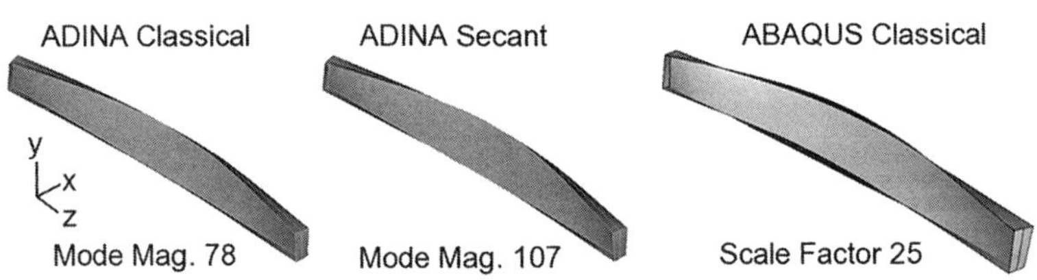

FIGURE 20.11 Case 1 buckling mode predictions.

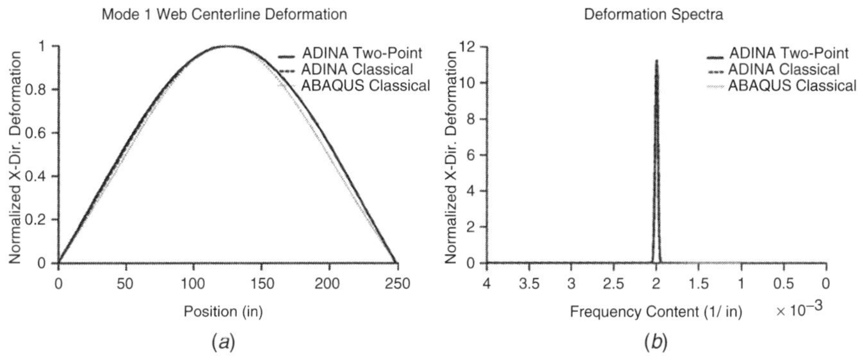

FIGURE 20.12 Case 1 results: (*a*) eigenvector plots along web centerline; (*b*) discrete Fourier transformation of these eigenvectors.

difference in peak frequency is only 1% between the ADINA classical and secant formulation results. In the second mode the difference in Fourier peaks between ABAQUS and ADINA results is 30%.

The second beam example (case 2, Fig. 20.10, with nominal dimensions listed in Table 20.1) incorporates an extra stiffener plate at the center of the steel I-beam and includes a midspan concentrated load, which is distributed linearly along the upper edge of the center stiffener plates. The first buckling modes for this model, obtained using ADINA and ABAQUS, are shown in Fig. 20.13. The results vary to such an extent that the discrepancies are clearly not an artifact of magnification or scaling. Normalized eigenvector values along the centerline nodes are plotted in Fig. 20.14. The discrete Fourier transform of the minor axis (x-direction) component of the eigenvector data, which is also presented in Fig. 20.14, indicates not only that there is a difference in the peak height of the results but also that the peaks are also located at different points along the member longitudinal axis. The secant formulation solution has a much more pronounced harmonic displacement field when compared to the classical formulation. In this case, the ABAQUS results are nearly equivalent to those of the ADINA classical formulation. ADINA classical and secant formulations differ significantly in that the major peaks do not coincide in height or location.

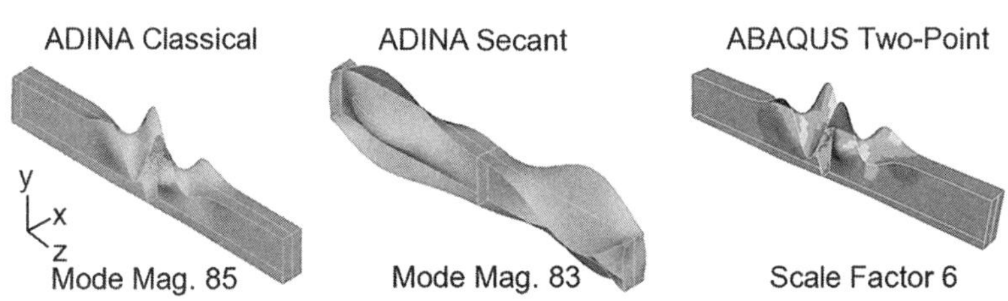

FIGURE 20.13 Case 2 buckling mode predictions.

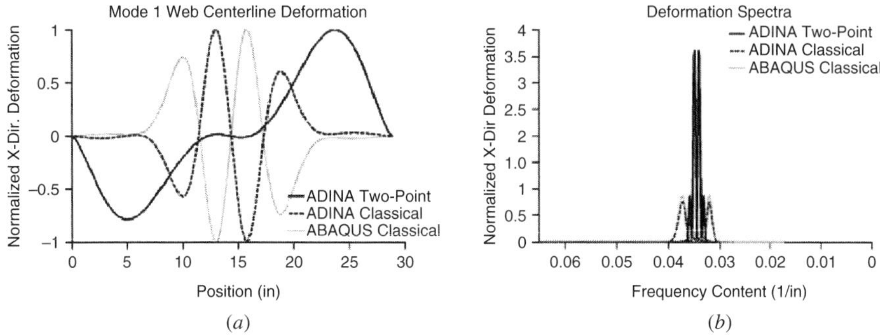

FIGURE 20.14 Case 2 results: (*a*) eigenvector plots along web centerline; (*b*) discrete Fourier transformation of these eigenvectors.

20.3.4 Observations on Linearized Eigenvalue Buckling

Finite element buckling analysis results should be interpreted with great care. The results of ostensibly identical formulations (e.g., as described in theory manuals) within various software packages frequently lead to estimates of critical loads that vary significantly for identical structural configurations.

It seems reasonable to avoid using such finite element buckling approaches for all but the simplest cases of bifurcation buckling, but even then care must be taken when considering the validity of the results.

It also appears that the approach of using eignenmodes, obtained from linearized eignevalue buckling analyses, as an imperfection seed within a more general incremental nonlinear finite element analysis should be approached with caution; as models with identical geometries and materials properties may result in very different imperfection fields, depending on the finite element software that is employed and with what formulation options the analysis is carried out.

REFERENCES

ADINA (2007), *Theory Manual*, ADINA R and D, Watertown, MA.

Babuska, I., and Melenk, J. M. (1997), "The Partition of Unity Method," *Int. J. Numer. Methods Eng.*, Vol. 40, pp. 727–758.

Bathe, K. J. (1996), *Finite Element Procedures*, Prentice Hall, Englewood Cliffs, NJ.

Bathe, K. J., and Dvorkin, E. N. (1983), "On the Automatic Solution of Nonlinear Finite Element Equations," *Comput. Struct.*, Vol. 17, Nos. 5–6, pp. 871–879.

Belytschko, T., and Black, T. (1997), "Elastic Crack Growth in Finite Elements with Minimal Remeshing," *Int. J. Numer. Methods Eng.*, Vol. 45, pp. 601–620.

Belytschko, T., Liu, W. K., and Moran, B. (2000), *Nonlinear Finite Elements for Continua and Structures*, Wiley, New York.

Boyce, W. E., and DiPrima, R. R. (1986), *Elementary Differential Equations and Boundary Value Problems*, 4th ed., Wiley, New York.

Brendel, B., and Ramm, E. (1980), "Linear and Nonlinear Stability Analysis of Cylindrical Shells," *Comput. Struct.*, Vol. 12, pp. 549–558

Chang, S. C., and Chen, J. J. (1986), "Effectiveness of Linear Bifurcation Analysis for Predicting the Nonlinear Stability Limits of Structures," *Int. J. Numer. Methods Eng.*, Vol. 23, pp. 831–846.

Chapelle, D., and Bathe, K. J. (2003), *The Finite Element Analysis of Shells—Fundamentals*, Springer, New York.

Ciarlet, P. G., and Lions, J. L. (1991), *Handbook of Numerical Analysis*, Vol. II: *Finite Element Methods (Part 1)*, North-Holland, Amsterdam.

Cook, R. D., Malkus, D. S., Plesha, M. E., and Witt, R. J. (2002), *Concepts and Applications of Finite Element Analysis*, 4th ed., Wiley, Hoboken, NJ.

Crisfield, M. A. (1991a), *Non-Linear Finite Element Analysis of Solids and Structures*, Vol. 1, Wiley, New York.

Crisfield, M.A. (1991b), *Non-Linear Finite Element Analysis of Solids and Structures*, Vol. 2, Wiley, New York.

Duarte, C. A., and Oden, J. T. (1996), "An H-P Adaptive Method Using Clouds," *Comput. Methods Appl. Mech. Eng.*, Vol. 139, pp. 237–262.

Earls, C. J. (2007), "Observations on Eigenvalue Buckling Analysis Within a Finite Element Context," *Proceedings of the Structural Stability Research Council, Annual Stability Conference*, New Orleans, LA.

Farhat, C., Harari, I., and Hetmaniuk, U. (2003), "The Discontinuous Enrichment Method for Multiscale Analysis," *Comput. Methods Appl. Mech. Eng.*, Vol. 192, pp. 3195–3209.

Fung, Y. C. (1994), *A First Course in Continuum Mechanics*, Prentice Hall, Englewood Cliffs, NJ.

Gallagher, R. H., Gellatly, R. A., Padlog, J., and Mallett, R. H. (1967), "A Discrete Element Procedure for Thin-Shell Instability Analysis," *AIAA J.*, Vol. 5, No. 1, pp. 138–145.

Holzer, S. M., Davalos, J. F., and Huang, C. Y. (1990), "A Review of Finite Element Stability Investigations of Spatial Wood Structures," *Bull. Int. Assoc*. Shell Spatial Struct., Vol. 31, No. 2, pp. 161–171.

Hughes, T. J. R. (2000), *The Finite Element Method: Linear Static and Dynamic Finite Element Analysis*, Dover, New York.

Kreyszig, E. (1978), *Introductory Functional Analysis with Applications,* Wiley, New York.

Langtangen, H. P. (2000), *Computational Partial Differential Equations*, Springer, New York.

Malvern, L. E. (1969), *Introduction to the Mechanics of a Continuous Medium*, Prentice-Hall, Englewood Cliffs, NJ, 1969.

McGuire, W., Gallagher, R. H., and Ziemian, R. D. (2000), *Matrix Structural Analysis*, 2nd ed., Wiley, New York.

Melenk, J. M., and Babuska, I. (1996) "The Partition of Unity Finite Element Method: Basic Theory and Applications," *Comput. Methods Appl. Mech., Eng.*, Vol. 139, pp. 289–314.

Naylor, A. W., and Sell, G. R. (1982), *Linear Operator Theory in Engineering and Science*, Springer. New York.

Novozhilov, V. V. (1953), *Foundations of the Nonlinear Theory of Elasticity*, Dover, New York.

Ogden, R. W. (1984), *Non-Linear Elastic Deformations*, Dover, New York.

Pecknold, D. A., Ghaboussi, J., and Healey, T. J. (1985), "Snap-Through and Bifurcation in a Simple Structure,". *J Eng. Mech.*, Vol. 111, No. 7, pp. 909–922.

Reddy, B. D. (1998), *Introductory Functional Analysis*, Springer, New York.

Reddy, J. N. (1993), *An Introduction to the Finite Element Method*, McGraw-Hill, New York.

Simo, J. C., and Hughes, T. J. R. (1991), *Computational Inelasticity*, Springer, New York.

Wood, R. D., and Schrefler, B. (1978), "Geometrically Non-linear Analysis—A Correlation of Finite Element Notations," *Int. J. Numer. Methods Eng.*, Vol. 12, pp. 635–642.

Wriggers, P. (2002), *Computational Contact Mechanics*, Wiley, Hoboken, NJ.

Zienkiewicz, O. C., and Taylor, R. L. (2000a) *The Finite Element Method*, Vol. 1: *The Basics,* Butterworth-Heinemann, Oxford, UK

Zienkiewicz, O. C., and Taylor, R. L. (2000b), *The Finite Element Method*, Vol. 2: *Solid Mechanics,* Butterworth-Heinemann, Oxford, UK.

APPENDIX A

GENERAL REFERENCES ON STRUCTURAL STABILITY

Aghayere, A. O., and Vigil, J. (2008), *Structural Steel Design: A Practice Oriented Approach*, Prentice Hall, Upper Saddle River, NJ.

Allen, H. G., and Bulson, P. S. (1980), *Background to Buckling*, McGraw-Hill, Maidenhead, Berkshire, England.

ASCE (1979), *Plastic Design in Steel: A Guide and a Commentary*, American Society of Civil Engineers, New York.

ASCE (1997), *Effective Length and Notional Load Approaches for Assessing Frame Stability: Implications for American Steel Design*, American Society of Civil Engineers, New York.

ASTM (1967), "Test Methods for Compression Members,'' ASTM Spec. Publ. No. 419, American Society for Testing and Materials, Philadelphia, PA.

Baker, J. F., Horne, M. R., and Heyman, J. (1956, 1961), *The Steel Skeleton*, Vol. II: *Plastic Behavior and Design*, Cambridge University Press, Cambridge.

Ballio, G., and Mazzolani, F. M. (2007), *Theory and Design of Steel Structures*, Spon Press, Taylor and Francis, London.

Bath, K. J. (1996), *Finite Element Procedures*, Prentice Hall, Upper Saddle River, NJ.

Bažant, Z. P., and Cedolin, L. (1991), *Stability of Structures: Elastic, Inelastic, Fracture, and Damage Theories*, Oxford University Press, New York.

Beedle, L. S. (Ed.) (1991), *Stability of Metal Structures: A World View*, 2nd ed., Structural Stability Research Council, Bethlehem, PA.

Bleich, F. (1952), *Buckling Strength of Metal Structures*, McGraw-Hill, New York.

Bruneau, M., Uang, C.-M., and Whittaker, A. S. (1997), *Ductile Design of Steel Structures*, McGraw-Hill, New York.

Brush, D. O., and Almroth, B. O. (1975), *Buckling of Bars, Plates, and Shells*, McGraw-Hill, New York.

Chajes, A. (1974), *Principles of Structural Stability Theory*, Prentice Hall, Upper Saddle River, NJ.

Chen, W. F., and Atsuta, T. (1977), *Theory of Beam—Columns, Vols. 1 and 2*, McGraw-Hill, New York.

Chen, W. F., Goto, Y., and Liew, J. Y. R. (1996), *Stability Design of Semi-Rigid Frames*, Wiley, New York.

Chen, W. F., and Han, D. J. (1985), *Tubular Members in Offshore Structures*, Pitman Advanced Publishing Program, Boston, MA.

Chen, W. F., and Kim, S.-E. (1997), *LRFD Steel Design Using Advanced Analysis*, CRC Press, Boca Raton, FL.

Chen, W. F., and Lui, E. M. (1987), *Structural Stability: Theory and Implementation*, Elsevier, New York.

Chen, W. F., and Lui, E. M. (1991), *Stability Design of Steel Frames*, CRC Press, Boca Raton, FL.

Chen, W. F., and Sohal, I. (1995), *Plastic Design and Second-Order Analysis of Steel Frames*, Springer-Verlag, New York.

Chen, W. F., and Toma, S. (Eds.) (1993), *Advanced Analysis of Steel Frames*, CRC Press, Boca Raton, FL.

Chilver, A. H. (Ed.) (1967), *Thin-Walled Structures*, Wiley, New York.

Column Research Committee of Japan (1971), *Handbook of Structural Stability*, Corona Publishing Co., Tokyo (in English).

Column Research Council (1962), "Column Research Council Symposium on Metal Compression Members,'' *ASCE Trans*., Vol. 127.

Cook, R. D., Malkus, D. S., Plesha, M. E., and Witt, R. J. (2001), *Concepts and Applications of Finite Element Analysis*, Wiley, New York.

Crawley, S. W., and Dillon, R. M. (1993), *Steel Buildings: Analysis and Design,* Wiley, New York.

Crisfield, M. A. (1996), *Non-Linear Finite Element Analysis of Solids and Structures*, Vol. 1 and Vol. 2, Wiley, New York.

Doyle, J. F. (2001), *Nonlinear Analysis of Thin-Walled Structures: Statics, Dynamics, and Stability*, Springer-Verlag, New York.

Eckhaus, W. (1965), *Studies in Nonlinear Stability Theory*, Springer-Verlag, New York.

Englekirk, R. E. (1994), *Steel Structures: Controlling Behavior through Design*, Wiley, New York.

Falzon, B. G., and Aliabadi, M. H. (2008), *Buckling and Postbuckling Structures: Experimental, Analytical and Numerical Studies*, Imperial College Press, London.

Fukumoto, Y., and Lee, G. C. (1992), *Stability and Ductility of Steel Structures Under Cyclic Loading*, CRC Press, Boca Raton, FL.

Galambos, T. V. (1968), *Structural Members and Frames*, Prentice-Hall, Upper Saddle River, NJ.

Galambos, T. V., and Surovek, A. E. (2008), *Structural Stability of Steel, Concepts and Applications for Structural Engineers,* Wiley, Hoboken, NJ.

Gambhir, M. L. (2004), *Stability Analysis and Design of Structures*, Springer, New York.

Gerard, G. (1962), *Introduction to Structural Stability Theory*, McGraw-Hill, New York.

Ghersi, A., Mazzolani, F. M., and Landolfo, R. (2001), *Design of Metallic Cold-Formed Thin-Walled Members,* Routledge-Taylor & Francis, New York.

Gjelswik, A. (1981), *Theory of Thin-Walled Bars*, Wiley, New York.

Godoy, L. A. (1996), *Thin-Walled Structures with Structural Imperfections*, Pergamon Press–Elsevier, New York.

Godoy, L. A. (1999), *Theory of Elastic Stability: Analysis and Sensitivity*, CRC Press, Boca Raton, FL.

Gregory, M., (1967), *Elastic Instability*, E & FN Spon, London.

Hancock, G. J., Murray, T., and Ellifrit, D. S. (2001), *Cold-Formed Steel Structures to the AISI Specification*, CRC Press, Boca Raton, FL.

Johnston, B. G. (1981), "Selected Papers,'' FERS and SSRC, Bethlehem, PA.

Jones, R. M. (2006), *Buckling of Bars, Plates, and Shells*, Bull Ridge Publishing, Blacksburg, VA.

Kirby, P. A., and Nethercot, D. A. (1979), *Design for Structural Stability*, Wiley, New York.

Kissell, J. R., and Ferry, R. L. (2002), *Aluminum Structures: A Guide to Their Specifications and Design,* Wiley, Hoboken, NJ.

Knowles, P. R. (1977), *Design of Structural Steelwork*, Trans-Atlantic, Philadelphia, PA.

Kumar, A. (1998), *Stability of Structures,* Allied Publishers, Mumbai, India.

Lam, D., Ang, T.-C., and Chiew, S.-P. (2004), *Structural Steelwork: Design to Limit State Theory*, Elsevier/Butterworth-Heinemann, Boston, MA.

Li, G.-Q., and Li, J.-J. (2007), *Advanced Analysis and Design of Steel Frames*, Wiley, Hoboken, NJ.

Liew, J. Y. R., Thevendran, V., and Shanmugam, N. E. (1998), *Thin-Walled Structures: Research and Development*, Elsevier Science, Oxford.

Little, W. A., (1964), *Reliability of Shell Buckling Predictions*, M.I.T. Press, Cambridge, MA.

Mazzolani, F. M. (1985), *Aluminum Alloy Structures*, Pitman Advanced Publishing Program, Boston, MA.

McGuire, W. (1968), *Steel Structures*, Prentice Hall, Upper Saddle River, NJ.

McGuire, W., Gallagher, R. H., and Ziemian, R. D. (2000), *Matrix Structural Analysis*, Wiley, Hoboken, NJ.

Narayanan, R. (Ed.) (1982), *Axially Compressed Structures*, Applied Science Publishers, Barking, Essex, England.

Narayanan, R. (Ed.) (1983a), *Beams and Beam-Columns*, Applied Science Publishers, Barking Essex, England.

Narayanan, R. (Ed.) (1983b), *Plated Structures*, Elsevier Applied Science Publishers, London.

Narayanan, R. (Ed.) (1985a), *Steel Framed Structures*, Elsevier Applied Science Publishers, London.

Narayanan, R. (Ed.) (1985b), *Shell Structures*, Elsevier Applied Science Publishers, Barking Essex, England.

Nethercot, D. (2001), *Limit States Design of Structural Steelwork*, Taylor & Francis, Oxford.

Roorda, J. (1980), *Buckling of Elastic Structures*, University of Waterloo, Waterloo, Ontario, Canada.

Salmon, C. G., Johnson, J. E., and Malhas, F. A. (2008), *Steel Structures: Design and Behavior*, Prentice Hall, Upper Saddle River, NJ.

Seyranian, A. P, and Elishakoff, I. (2004), *Modern Problems of Structural Stability,* Springer-Verlag, New York.

Seyranian, A. P., and Mailybaev, A. A. (2003), *Multiparameter Stability Theory with Mechanical Applications*, World Scientific, Hackensack, NJ.

Shanley, F. R., (1960), *Weight-Strength Analysis of Aircraft Structures*, Dover Publications, New York.

Sharp, M. L. (1992), *Behavior and Design of Aluminum Structures*, McGraw-Hill, New York.

Simitses, G. J. (1976), *An Introduction to the Elastic Stability of Structures*, Prentice Hall, Upper Saddle River, NJ.

Simitses, G. J., and Hodges, D. H. (2006), *Fundamentals of Structural Stability*, Elsevier/Butterworth-Heinemann, Boston, MA.

Singer, J., Arbocz, J., and Weller, T. (1998), *Buckling Experiments*, Vol. 1 and Vol. 2, Wiley, New York.

Sputo, T., and Turner, J. (2005), *Bracing Cold-Formed Steel Structures*, American Society of Civil Engineers, Reston, VA.

SSRC (1993), *Is Your Structure Suitably Braced*, Structural Stability Research Council, Bethlehem, PA.

Supple, W. J. (1973), *Structural Instability: Fundamentals of Post-Buckling Behavior of Structures*, IPC Science and Technology Press, Guildford, England.

Tall, L. (Ed.) (1974), *Structural Steel Design*, Ronald Press, New York.

Thompson, J. M. T., and Hunt, G. W. (1984), *Elastic Instability Phenomena*, Wiley, New York.

Timoshenko, S. P., and Gere, J. M. (1961), *Theory of Elastic Stability*, McGraw-Hill, New York.

Trahair, N. S. (1993), *Flexural-Torsional Buckling of Structures*, E & FN Spon, London.

Trahair, N. S., and Bradford, M. A. (1998), *The Behaviour and Design of Steel Structures to AS4100*, E & FN Spon, London.

Trahair, N. S., Bradford, M. A., and Nethercot, D. A. (2001), *The Behaviour and Design of Steel Structures to BS 5950*, E & FN Spon, London.

Trahair, N. S., Bradford, M. A., Nethercot, D. A., and Gardner, L. (2008), *The Behaviour and Design of Steel Structures to EC3*, E & FN Spon, London.

Vinnakota, S. (2005), *Steel Structures: Behavior and LRFD,* McGraw-Hill, New York.

Vlasov, V. Z. (1961), *Thin-Walled Elastic Beams*, Israel Program for Scientific Translation, for NSF, Jerusalem (in English; original Russian edition, 1959).

Wang, C. M., Wang, C. Y., and Reddy, J. N. (2004), *Exact Solutions for Buckling of Structural Members*, CRC Press, Boca Raton, FL.

White, D. W., and Chen, W. F. (1993), *Plastic Hinge Based Methods for Advanced Analysis and Design of Steel Frames*, Structural Stability Research Council, Bethlehem, PA.

Williams, D. G., and Aalami, B. (1979), *Thin Plate Design for In-Plane Loading*, Wiley, New York.

Winter, G. (1975), *The Collected Papers of George Winter*, Cornell University Press, Ithaca, NY.

Wong, M. B. (2008), *Plastic Analysis and Design of Steel Structures*, Elsevier/Butterworth-Heinemann, Boston, MA.

Yang, Y. B., and Kuo, S. R. (1994), *Theory and Analysis of Nonlinear Framed Structures*, Prentice-Hall, Englewood Cliffs, NJ.

Yu, W. W. (2000), *Cold-Formed Steel Design*, Wiley, New York.

Ziegler, H. (1968), *Principles of Structural Stability*, Blaisdell, Waltham, MA.

Zienkiewicz, O. C., and Taylor, R. L. (2000a), *The Finite Element Method*, Vol. 1: *The Basics*, Elsevier/Butterworth-Heinemann, Boston, MA.

Zienkiewicz, O. C., and Taylor, R. L. (2000b), *The Finite Element Method,* Vol. 2: *Solid Mechanics*, Elsevier/Butterworth-Heinemann, Boston, MA.

APPENDIX B

TECHNICAL MEMORANDA OF STRUCTURAL STABILITY RESEARCH COUNCIL

Technical memoranda have been periodically issued by the Structural Stability Research Council (formerly the Column Research Council) to establish recommended procedures and recording requirements for studying specific issues related to structural stability and thereby provide a means for researchers to obtain useful and consistent results. Technical Memorandum No. 1 (TM No. 1), issued in 1952, states the method most appropriate to that time for determining the axial strength of centrally loaded columns. Subsequent research has increased the knowledge of column behavior and although TM No. 1 was replaced by Technical Memorandum No. 4, it still appears in this guide to provide a historical perspective. Researchers are strongly encouraged to perform tests in accordance with these technical memoranda so that their results can be compared with previous and future experimental and/or analytical data.

B.1 TECHNICAL MEMORANDUM NO. 1: THE BASIC COLUMN FORMULA*

The Column Research Council has brought out that it would be desirable to reach agreement among engineers as to the best method for predicting the ultimate load-capacity in compression of straight, prismatic, axially loaded, compact members of structural metals. It was proposed that Research Committee A of the Council be assigned the problem of reporting on the correctness and desirability of the tangent-modulus column formula. This formula involves simply the substitution of the tangent modulus, E_t, for E in the Euler formula. This formula may be written

$$\frac{P}{A} = \frac{\pi^2 E_t}{(KL/r)^2} \tag{B.1}$$

*Issued May 19, 1952. Technical Memorandum No. 5 reflects the current position of the SSRC and replaces Technical Memorandum No. 1.

where P = the ultimate load
A = the cross-sectional area
E_t = the compressive tangent modulus (slope of the compressive stress–strain curve) of the material in the column at the stress P/A
r = least radius of gyration of cross section
L = the length of the column
K = a constant depending on end conditions
K = 2 for one end fixed and the other end free
K = 1 for both ends simply-supported
K = 0.7 for one end fixed and the other end simply-supported
K = 0.5 for both ends fixed

For materials which exhibit upper and lower yield points in compression, the lower yield point is to be considered as the limiting value of P/A.

Information and reference to literature supporting the foregoing statement will be made available on request to the Secretary of the Column Research Council.

It is the considered opinion of the Column Research Council that the tangent-modulus formula for the buckling strength affords a proper basis for the establishment of working-load formulas.

The column formula presented here differs in form from the familiar Euler formula only in that the tangent modulus-of-elasticity is substituted for the ordinary modulus-of-elasticity. There is, however, a great practical difference between the two formulas, for whereas the Euler formula can be solved directly for the average stress corresponding to any given slenderness ratio, the tangent-modulus formula cannot. It is not the intention to advocate the use of the tangent-modulus formula in design, but rather to propose it as the basis for relating the compressive stress–strain properties of the material to the column strength of the material. The formula furnishes the information for approximating to the average stress in terms of the ratio of slenderness, for any type of centrally loaded column under consideration, by making suitable assumptions with respect to such items as accidental eccentricity, initial curvature of member, residual stresses, and variation in properties of the material.

ADVISORY PREFACE TO TECHNICAL MEMORANDA NOS. 2, 3, AND 4

The reader is advised that Technical Memoranda Nos. 2, 3, and 4, although accurate, need care and interpretation when used in conjunction with modern testing and data acquisition equipment. Thus, it may be inappropriate to adhere to a specific sensitivity of measured strain in Technical Memorandum No. 2 when the strain-measuring device referred to is no longer in general use. Similarly for Technical Memoranda Nos. 3 and 4, it may not be appropriate to specify crosshead speeds in the testing machine when the testing machine has stroke or strain controlled by a servomechanism. In Technical Memorandum No. 4, the first alignment method of preparing the column for testing is no longer in general use—rather, most tests use the second alignment method, by which columns are tested "as is" with exact measurements of initial out-of-straightness, which are used with analytical strength predictions. Although these technical memoranda remain correct, it has not been possible to update them to reflect today's practice in this edition of the Guide, because the methods in use today have not yet been standardized. These technical memoranda have been retained in this edition so as to provide a basis of comparison for the researcher.

Three types of testing machines can be used to perform tension and compression tests: mechanical (screw driven), hydraulic, and servo-controlled. When Technical Memoranda Nos. 2, 3, and 4 were written, servo-controlled machines were not in general use as they were when Technical Memoranda Nos. 7 and 8 were written. Some servo-controlled machines have self-aligning heads and grips which reduce the tedious alignment issues discussed in the technical memoranda. The technical memoranda emphasize the use of static values of the yield stress, which theoretically is obtained at a zero strain rate. Because strain is difficult to control and can cause problems as the test continues to high strains, zero crosshead motion or stroke is generally used to obtain static yield strengths. In mechanical testing machines, zero crosshead motion is easily obtained. A skilled operator is needed to sustain zero crosshead motion with a hydraulic machine. Servo-controlled machines can be set to a zero stroke rate, but the machine must be finely tuned to minimize flutter caused by the difference between the control and feedback signals.

The manual measurement of strain with extensometers that use dial gages has largely been replaced with electronic extensometers or strain gages that can give continuous indications of strain. Displacements can also be measured manually with dial gages or with ruled scales read through a transit. Potentiometers, LVDT's (linear voltage displacement transducers), and non-contact optical devices, however, can be used to produce signals that can be recorded with multi-channel digital acquisition recorders. Care must be taken to calibrate these devices to ensure accurate measurements.

Test specimens must be cold cut and machined in order to avoid altering residual stresses or material properties. Traditionally this has involved sawing or machining with cooling fluid. Modern techniques, such as water jet cutting, can also be used as long as avoiding excessive heating of the material is ensured. The electric discharge eroding/machining technique is even more advanced. It introduces virtually no heat into the specimen and is extremely precise.

B.2 TECHNICAL MEMORANDUM NO. 2: NOTES ON THE COMPRESSION TESTING OF METALS

For predicting column strength it is necessary to have compressive and tensile stress–strain curves of the column material. In order that the tension and compression specimens be as nearly as possible equally representative of the material, the thickness of compression specimens should approximate that of the tension specimens. Preferably each pair of tensile and compressive specimens should be

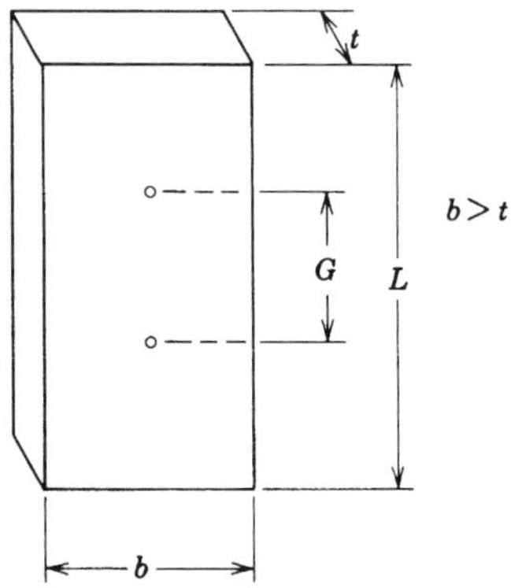

FIGURE B.1 Compression test specimen.

cut from the same coupon (conveniently, the section or slice remaining from the residual strain measurements).

Specimens taken from a flange or web of a rolled shape, or from a plate, should be rectangular in cross section. They should be machined on all four sides with grinding as the final machining operation (on a magnetic grinder for steel). The ends of the specimens should be ground plane and normal to the longitudinal axis of the specimen. The ends should be parallel within close limits.

In general, compression specimens (Fig. B.1) should be no longer than necessary to accommodate a compressometer or resistance strain gages and have, between each end of the specimen and the adjacent end of the gage length, a distance, $\frac{1}{2}$ $(L - G)$, at least equal to the greatest cross-sectional dimension, b. The compressometer should meet the specifications for Class A Extensometers (ASTM E83-06 *Standard Practice for Verification and Classification of Extensometer Systems*) that limit the error of indicated strain to 0.00001. The length should not be less than the greatest cross-sectional dimension; and in order to keep the specimen short, the gage length should not exceed twice this dimension. If the length of a rectangular specimen is more than about 4.5 times the length of the shorter side of the rectangle or, for a circular specimen, more than four times the diameter, difficulty may be expected in avoiding premature bending (column action), and special precautions must be taken to prevent excessive bending.*

Referring to Fig. B.1, the following relationships summarize the preceding requirements for a member of rectangular cross section:

$$G = b \tag{B.2a}$$

$$G = t \tag{B.2b}$$

$$4.5t = L = (G + 2b) \tag{B.2c}$$

*For rectangular specimens see ASTM E9-89a (2000) *Standard Test Methods of Compression Testing of Metallic Materials at Room Temperature*.

in which G = gage length; b = specimen width; L = specimen length; and t = specimen thickness. Specimens should be measured with a micrometer reading to 0.001 in. (0.025 mm) and the dimensions recorded. Nominal dimensions should not be used in computations.

Both ends of a compression specimen should bear on smoothly finished plane surfaces. The bearing blocks should be made of, or faced with, a suitably hard material so that the faces of the blocks will not suffer permanent deformation during the test. The blocks should be at least as thick as the smallest cross-sectional dimension of the specimen and should project beyond the contact area a distance at least half as great as the smallest cross-sectional dimension.

Precautions should be taken to ensure uniform distribution of strain over the cross section and to prevent relative rotation of upper and lower bearing surfaces during testing. The following are three suggested procedures for attaining these goals:

1. Use of a subpress loaded through a push rod acting at the lower end of the hollow plunger.
2. Use of bearing blocks that will permit initial adjustment for parallelism of bearing surfaces.†
3. Use of a thin capping layer of Hydrostone between the upper bearing block and the testing machine crosshead. While the Hydrostone is setting, a small load should be maintained.

If tilting or lateral displacement of one testing machine crosshead relative to the other during loading is a suspected possibility, it is suggested that a subpress be used to load the compression specimen so as to reduce the probability of bending during loading. The hollow plunger should fit closely within the annulus of the subpress frame (but vertical motion of the plunger should not be restricted).

Adjustable bearing blocks cannot be depended upon to compensate for tilting of the testing machine heads during loading and should be used only if appreciable relative tilting of the heads does not occur. If a spherical bearing block is used, it should be at the upper end of the specimen (for specimens tested with the longitudinal axis vertical). It is desirable that the center of the spherical surface lie within the flat surface on which the specimen bears. Also, it is essential that the longitudinal axis of the specimen pass, closely through the center of the spherical surface, so that the eccentricity of loading may not be great enough to overcome the friction necessary to rotate the block.

The compression specimen should be aligned so that the deviation in strain indicated by any gage is less than 5% of the average of all gages when the specimen is subjected to a stress of about one-half the yield strength of the material. At least three strain gages need to be mounted on the specimen and monitored during the aligning operation to meet this requirement.

†See Fig. 1 in ASTM Standard E9-70, *Methods of Compression Testing of Metallic Materials*.

If the length of the specimen does not exceed the maximum recommended length (4 or 4.5 times a cross section dimension for a circular or rectangular specimen, respectively), strains should be measured during the test with an averaging compressometer or with two strain gages mounted opposite each other. In the case of longer specimens tested without lateral support, reasonable certainty of uniform strain distribution can be obtained only with the use of not fewer than two strain gages on the wide sides of thin rectangular specimens, or three gages on thick rectangular or square specimens (one at the center of each of three sides), or three gages on circular specimens. Strain measurement with only one gage is unreliable. The stress–strain curve should extend from zero stress, and strain, to values for which the ratio of total stress to total strain is less than $0.7E$, or to a total strain of at least 0.01, whichever results in the larger strain.

Material properties will be a function of the loading rate and therefore the rate should be recorded. Dynamic loading should be interrupted in the yielding region so as to obtain at least three values of the static yield stress (yield stress at essentially zero strain rate) for metals with a yield plateau.

Compressive stress–strain curves should be plotted with stress as ordinate and strain as abscissa to as large a scale as the quality of the data justifies. The individual values of stress and strain should also be reported. When applying the procedures above to material that is suspected of showing a considerable variation in properties, the specimens should be taken from a sufficient number of locations to define the extent of the variation in properties.

B.2.1 Determination of Typical Stress–Strain Curve from a Number of Stress–Strain Curves

It is assumed that the compressive stress–strain relationships of enough specimens will be determined so that all variations of the material likely to be submitted under a given specification will be represented. The yield strength values determined from the individual tests should be presented in the form of a distribution plot in which the percentage of the total number of tests for which the yield strength is within a certain range is plotted against the average of the range. The standard deviation should be indicated if the number of specimens is significant.

Several methods for constructing a typical stress–strain curve from a number of individual curves have been proposed. One simple method, which has had considerable use, is described as follows:

1. Record the strain departures from the modulus line for various fixed percentages of the particular individual yield strength value. These percentages should cover stresses from the proportional limit to above the yield strength.
2. Average all offset values for each of the fixed percentages. (For steel shapes it is recommended that the offsets be weighted in proportion to relative flange and web areas.) A curve may be plotted in which the ordinate is the percent of yield strength and the abscissa is the average strain offset from the initial modulus line.

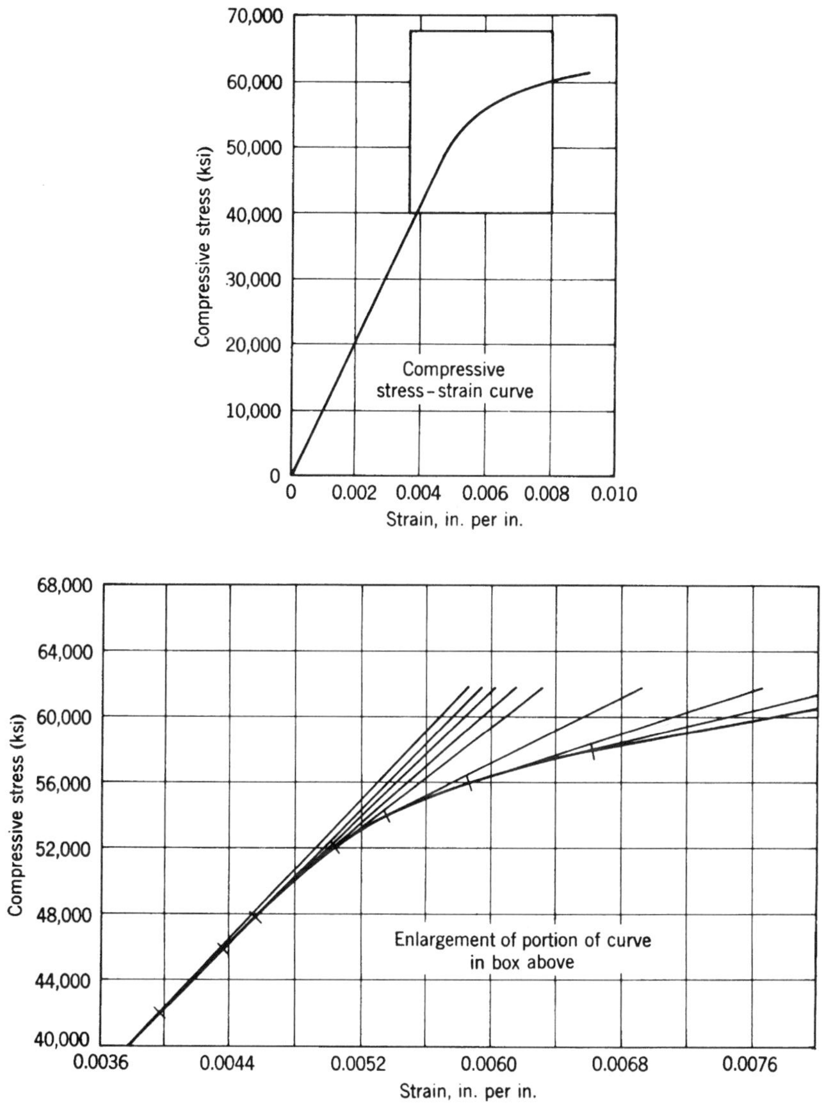

FIGURE B.2 Typical compressive stress-strain curve of a high-strength aluminum alloy. 1 ksi = 6.9 MPa.

3. For any appropriate yield strength value, a typical curve can then be plotted by adding the offset values to the strain consistent with the elastic-modulus values.

Figure B.2 shows a typical compressive stress–strain curve of a high-strength aluminum alloy. Lines have been drawn tangent to this curve at different values of stress, P/A. The slopes of these lines define the corresponding tangent modulus, E_t, essential to the determination of the basic column strength.[*†]

B.3 TECHNICAL MEMORANDUM NO. 3: STUB-COLUMN TEST PROCEDURE[‡§]

A stub column is a member sufficiently short so as to preclude member buckling when compressed, but sufficiently long to contain the same initial residual stress pattern as a much longer member cut from the same stock. For cold-formed steel sections, which generally have thin-walled plate elements, the stub-column test is aimed at determining the effect of local buckling as well as the effect of cold-forming on the column performance. For these sections, the stub-column length should be sufficiently long to exhibit such behavior. Because column strength may be expressed as a function of the tangent modulus determined from the stress–strain relationship of the stub-column test,[§§] this test is an important tool for investigating column strength.

The difference between Young's modulus and the tangent modulus at any load level, determined from a compression test on the complete cross section, essentially reflects the effect of residual stresses. This may be realized when one considers that the cross section, hitherto completely elastic under load, becomes elastic-plastic at the proportional limit as the member is loaded further. The presence of residual stresses in the cross section implies that some fibers are in a state of residual compression. The fibers in a state of residual compression are the first to reach the yield point under load.

The difference between the behavior of a column free of residual stresses and one containing residual stresses lies in the fact that, beyond the proportional limit for the latter, both the tangent modulus and the effective moment of inertia are

*Convenient and accurate techniques are available for determining the tangent modulus; one such technique is described in NACA TN, No. 2640, "Interaction of Column and Local Buckling in Compression Members," by P. P. Bijlaard and G. P. Fisher.

†See "The Basic Column Formula," Technical Memorandum No. 1, Column Research Council, May 19, 1952 (presented previously in this appendix).

‡This document was originally prepared by L. Tall under the technical guidance of Task Group 1 of the Column Research Council as Lehigh University Fritz Eng. Lab. Rep., No. 220A.36 (February 1961), and was revised by an International Institute of Welding Working Group consisting of H. Louis (Belgium) Chair, M. Marincek (Yugoslavia), and L. Tall (U.S). It was approved by the IIW at the Annual Conference, Oslo, 1962, as Class C Document No. X-282-61. Task Group 6 of the Column Research Council further revised the document in 1974.

§See Advisory Preface to Technical Memoranda Nos. 2, 3, and 4.

§§Column strength is not always a direct function of the tangent modulus, for example, for an H-shape bent about the major axis, irrespective of the stress–strain relationship and the pattern of residual stress, the function is direct. For an H-shape bent about the minor axis (only for rolled or welded built-up shapes with universal mill plates whose stress–strain curve can be considered as elastic–perfectly plastic), however, the strength is approximately a function of the cube of the tangent modulus. Because there is no direct or simple relationship for other cases, care must be taken in applying the stub-column test results to the prediction of column strength.

greater for the column free of residual stresses. (The behavior of a stub column, however, because of its shorter length, reflects only the effect of residual stresses on the tangent modulus; the reduction in the effective moment-of-inertia due to plastification has no effect on its behavior.) Under load, some parts of the column cross section will yield before others, leading to a decrease of the effective moment of inertia and hence in the strength of the column, as those proportions of the cross section which have yielded support no additional load if strain-hardening is neglected. The residual stress distribution over the cross section, through its influence on the effective moment-of-inertia, is the connecting link between column strength and the tangent modulus of the stress–strain relationship of the stub column. That residual stresses are, indeed, a major factor affecting the strength of axially loaded, initially straight columns, and that a conservative value for this strength may be specified in terms of the tangent modulus determined from the results of a stub-column test, have been documented extensively.

B.3.1 Stub-Column Test Procedure

1. *Object.* To determine the average stress–strain relationship over the complete cross section by means of a stub column test.
2. *Specimen.*

   a. The stub column should be cold-sawed from the stock at a distance at least equal to the shape depth away from a flame-cut end.
   b. The length of hot-rolled stub columns should not be less than $2d + 10$ in. ($2d + 250$ mm) or $3d$, whichever is smaller, and not greater than $20r_y$ or $5d$, whichever is larger, in which d = depth of the shape and r_y = radius-of-gyration about the minor axis. For cold-formed shapes the length of the stub column should not be less than three times the largest dimension of the cross section and no more than 20 times the least radius-of-gyration.
   c. The ends of the columns should be milled plane and perpendicular to the longitudinal axis of the column.* This operation may be omitted for light gage members that are difficult to mill, if their ends are welded to base plates.
   d. The thickness of the flanges and webs and the length and cross-sectional area of the stub column should be measured and recorded.

3. *Instrumentation.* Mechanical dial indicator gages or electrical resistance gages may be used to determine the strains during testing. The use of dial gages over a comparatively large gage length is to be preferred as they provide a better average strain indication. The dial gages should read to 0.0001 in. (0.0025 mm) when read over a 10-in. (250-mm) gage length, or to 0.001 in. (0.025 mm) when installed between base plates over the whole length of the

*A tolerance across the milled surface of ±0.001 in. (±0.025 mm) is usually satisfactory.

stub column. Where it can be demonstrated that electrical resistance gages give the same or better results, they may be used instead of dial gages. The gage length should be placed symmetrically about the mid-height of the stub column. At least two gages in opposite positions should be used and the average of the readings taken. Corner gages over the complete column length are used for alignment; mid-height gages are used for determining the stress–strain relationship. When four mid-height gages are used instead of two, the corner gages may be omitted. (This is possible with the flange tips of an H-shape.) Figure B.3 depicts typical gage arrangements for structural shapes.

For uniformity in stub-column testing, the following instrumentation is recommended for H-shapes:

a. Four 0.001-in. (0.025-mm) dial gages over the complete length of the stub column, at the four corners; to be used during alignment.
b. Two 0.0001-in. (0.0025-mm) dial gages on opposite flanges over a 10-in. (250-mm) gage length at mid-height; to be used to determine the stress–strain relationship. The points of attachment for the gage length are to be at the junction of the flange and web, to avoid the influence of local flange crippling on the readings. When early local flange crippling is unlikely, four 0.0001-in. (0.0025-in) dial gages over a 10-in. (250-mm) gage length may be clamped at mid-height to each flange tip. Corner alignment gages are then not needed.
c. As-rolled steel specimens should be whitewashed before testing. Flaking of the mill scale during testing gives a general area of the progress of yielding during the test.

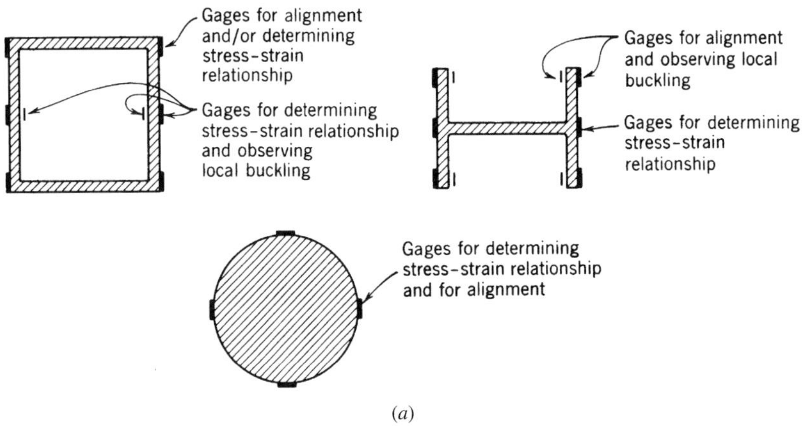

FIGURE B.3 Position of gages for alignment and testing: (*a*) location of electrical resistance gages; (*b*), (*c*), (*d*) location of dial indicators.

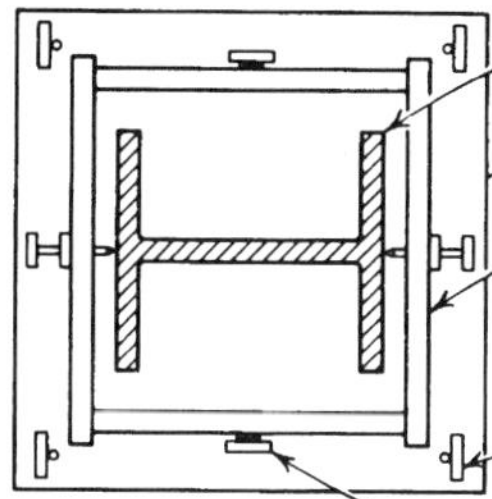

(*b*)

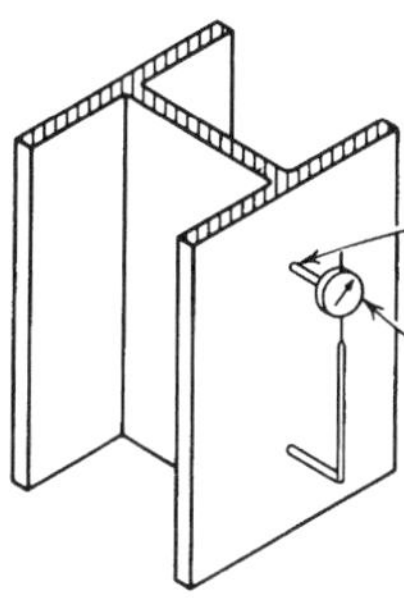

(*c*)

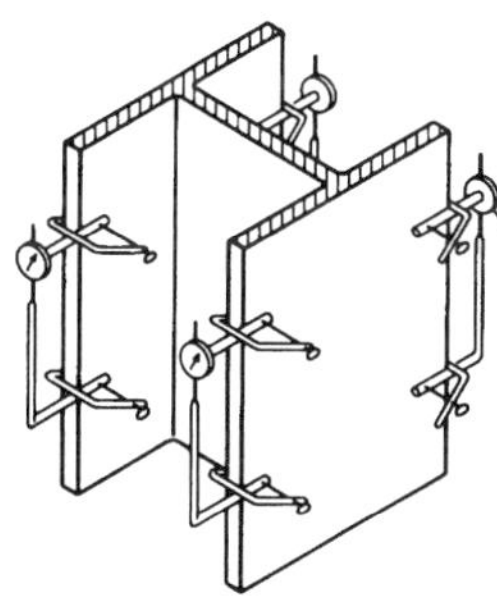

(*d*)

FIGURE B.3 (*Continued*)

4. *Test set-up.* The specimen should be set in the testing machine between flat bearing plates. These plates should be thick enough to ensure a uniform distribution of load through the specimen. The test set-up is shown in Fig. B.4.

   Alignment may be achieved with the use of special beveled bearing plates, or else by the use of spherical bearing blocks that are fixed by wedges after alignment to prevent rotation. Hydrostone bedding for the column ends has been used successfully as an aid to alignment, especially for light-gage members.

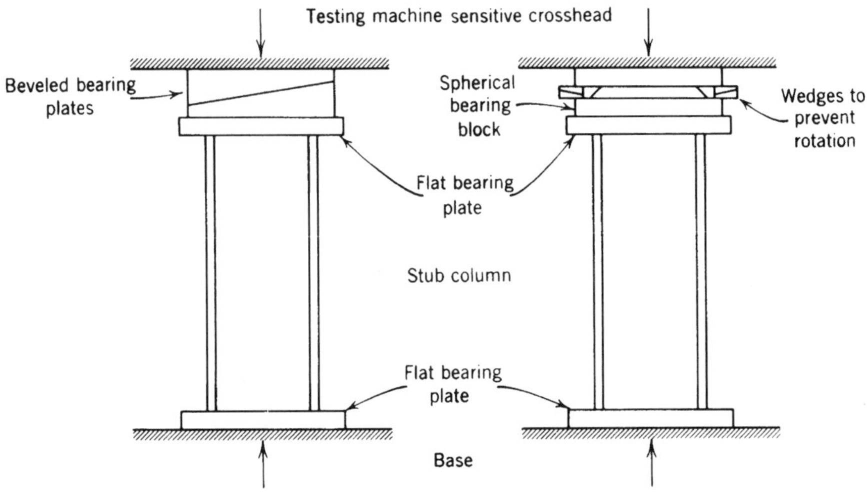

FIGURE B.4 Subcolumn set-up for testing.

5. *Alignment.* The specimen should be aligned at loads less than that corresponding to the proportional limit stress. For rolled H-shapes of mild structural steel this limit is about one-half of the predicted yield level load; for welded shapes the limit may be as low as one-quarter of the yield level load.

   The alignment is performed by noting the variation of strain at the four corners of the specimen. The variation between the strains at any corner from the average strain should be less than 5% at the maximum alignment load. Alignment at low loads is unsatisfactory. The alignment loading should consist of several increments up to the maximum alignment load.

   To check that the load is below the proportional limit, the stress–strain relationship may be plotted during the test and its linearity observed. It is inadvisable to exercise this control by observing the whitewash for flaking of the mill scale, because flaking begins at a load greater than that corresponding to the proportional limit stress as indicated by the plotted stress–strain relationship.

6. *Testing.* The stress–strain curve should be constructed from as many experimental data points as possible. To this end, the load increments in the elastic region should be less than 10% of the expected yield load. After the proportional limit has been reached, the load increments should be reduced so that there are sufficient data points to delineate the "knee" of the stress–strain curve. Strain increments may be more convenient than load increments to delineate the "knee" in the inelastic region. The proportional limit* will be

*It is assumed that the residual stresses are symmetrical with respect to the principal axes of the cross section and constant in the longitudinal direction, so that the proportional limit does not indicate localized yielding.

marked by the beginning of the deviation of the stress–strain relationship from linear behavior. Yield lines (made clearly visible by the whitewash as the mill scale flakes off) will indicate the progress of yielding. This matter is covered further in Item 10.

After the onset of yielding, readings should be recorded when both load and strain have stabilized. The criteria used to specify when data may be recorded depend on the type of machine used for testing. This is explained further in Item 7.

To ensure correct evaluation of the yield level and other material properties, the test should be continued until one of the following conditions is satisfied:

a. After an immediate drop in load due to local plate buckling, the test should be continued until the load had dropped to about half the predicted yield level load.
b. For a specimen that exhibits a plastic region of considerable extent, the test should be continued until the load has dropped to about 80% the predicted yield level load.
c. For a specimen that strain hardens without apparent buckling, or which strain hardening occurs without a plastic range, the test should be continued until the load is about 25% above that corresponding to the load computed from the yield strength based on the 0.2% strain offset criterion.

The load and strain at all initial load levels should be recorded. This is further outlined in Item 9. It may be necessary to remove some of the dial gages before the test is completed to avoid damage due to local buckling.

7. *Criteria for stabilization of load.* Standard criteria should be followed for recording of test data when the load is greater than that at the proportional limit. The criterion depends on the type of testing machine used, whether hydraulic or mechanical.

For mechanical testing machines (screw type) the criterion is as follows: no relative crosshead movement with both the loading and bypass valves closed. For hydraulic systems that leak, the criterion is a simulation of that for the screw-type machine. This is accomplished by balancing the load and bypass flows so that no relative motion of the crosshead occurs, and then by waiting for the load to stabilize.

These criteria are best applied by plotting the load change, or crosshead movement versus time, and noting the value corresponding to the asymptote (see Fig. B.5). The test data are recorded when:

a. The asymptotic load is approached when using the load criterion.
b. The asymptotic crosshead movement is approached when using the crosshead movement criterion.

Readings should not be recorded until the asymptote is definite. Experience will indicate the time intervals required, but 3 minute intervals are usually

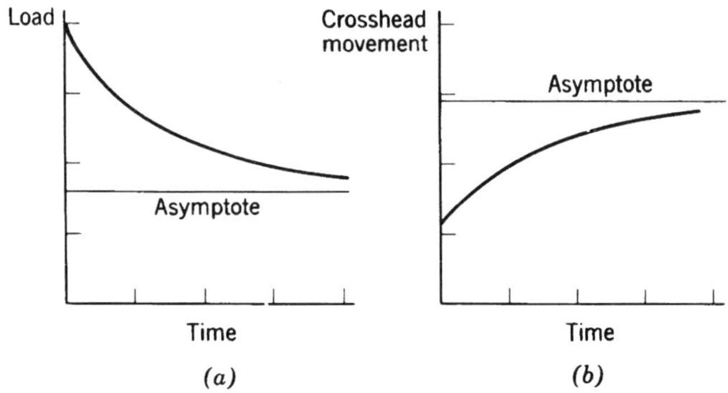

FIGURE B.5 Criteria for load stabilization.

satisfactory. The crosshead movement should be measured with a 0.001-in. (0.0025-mm) dial gage.

8. *Evaluation of data.* The test data should be evaluated in the following manner:

   a. Plot the data during the test to detect any inconsistencies.
   b. Translate the test data to stress versus strain (based on the actual cross-sectional area) and plot the stress–strain relationship to a large scale. A typical stress–strain diagram is shown in Fig. B.6. A Ramberg-Osgood type curve is often fitted to the test data.
   c. Determine the tangent-modulus curve from the stress–strain relationship. This may be done by using a strip of mirror. The mirror is held normal to the curve, the normal being determined from the continuity of the stress–strain curve and its mirror image at the tangent point considered. Then a line is drawn along the mirror edge.

9. *Data to be reported.* In addition to presenting the stress–strain and stress-tangent modulus curves, the following information, obtained from the stress–strain relationship given by the stub-column test, should be reported:

   1. Young's modulus of elasticity
   2. Proportional limit stress
   3. Yield strength
   4. Yield stress level
   5. Elastic range
   6. Elastic–plastic range
   7. Plastic range
   8. Onset of strain hardening
   9. Strain-hardening range
   10. Strain-hardening modulus

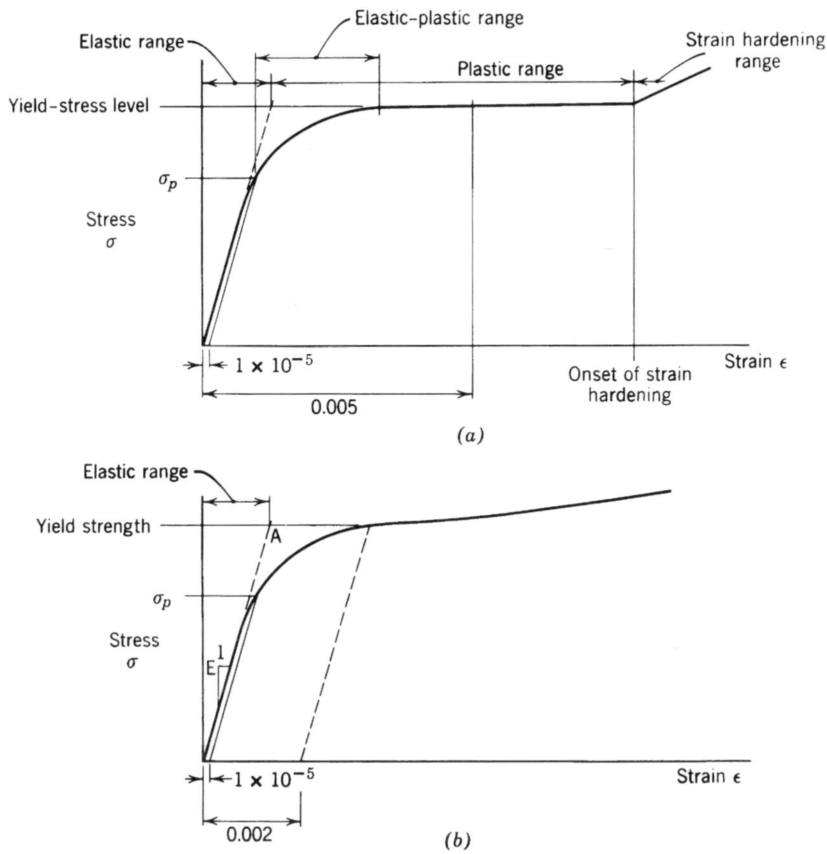

FIGURE B.6 The stress-strain diagram.

The occurrence of local buckling and any unusual phenomena during the test should be recorded.

10. *Definition of terms.* The terms just listed in Item 9 should be defined and measured as follows:

    1. *Young's modulus, E*, is the ratio of stress to strain in the elastic range. (The method of measuring is defined by ASTM E111-04 *Standard Test Method for Young's Modulus, Tangent Modulus, and Chord Modulus.*)
    2. *Proportional limit stress*, σ_p, is the stress corresponding to the strain above which the stress is no longer proportional to strain. It is best determined by the use of a strain offset of 1×10^{-5}.
    3. *Yield strength* is the "stress corresponding to the load which produces in a material, under specific conditions of the test, a specified limiting plastic strain." This is the definition from ASTM E370-08a *Standard Test Methods and Definitions for Mechanical Testing of Steel Products*,

and a strain offset of 0.002 is suggested. (The yield strength criterion is normally used when there is gradual yielding without a yield plateau. For stub-column stress–strain curves, the yield stress level is mainly used, and as it is an average value in the plastic range, it is more representative.)

4. *Yield stress level* is the stress corresponding to a strain of 0.005. This stress usually corresponds to the constant stress under yield when the stress–strain relationship is such as that shown in Fig. B.6*a*.
5. *Elastic range* is defined as the increment of strain between zero strain and the strain at Point A in Fig. B.6*b*.
6. *Elastic-plastic range* is the increment of strain between the strain at the proportional limit stress and the strain at which the stress first reaches the yield stress level.
7. *Plastic range* is defined as the increment of strain between the elastic range and the onset of strain hardening.
8. *Onset of strain hardening* may be defined as the strain corresponding to the intersection on the stress–strain curve of the yield stress level in the plastic range with the tangent to the curve in the strain-hardening range. This tangent is drawn as the average value in the strain increment of 0.002 after the apparent onset of strain hardening.
9. *Strain-hardening range* is the range of strain after the plastic range in which the material no longer strains at a constant or near-constant stress.
10. *Strain-hardening modulus* is the ratio of stress to strain in the initial strain-hardening range. It is taken as the average value in the strain increment of 0.005 after the onset of strain hardening.

B.4 TECHNICAL MEMORANDUM NO. 4: PROCEDURE FOR TESTING CENTRALLY LOADED COLUMNS*†

A column may be defined as a member whose length is considerably larger than any of its cross-sectional dimensions and which is subjected to compression in the longitudinal direction. If the resultant compressive force is approximately coincident with the longitudinal centroidal axis of the member, the column is said to be centrally loaded. Although columns have been extensively studied for more than two centuries, both analytically and experimentally, technological developments may necessitate further testing of centrally loaded columns. The purpose of this memorandum is to set forth a suggested procedure for conducting such experiments.

The experimentally determined values of column strength form a wide scatter band when plotted versus the effective slenderness ratio, KL/r, in which KL

*This document was prepared by Task Group 6 of the Column Research Council based on Lehigh University Fritz Eng. Lab. Rep. No. 351.6, authored by N. Tebedge and L. Tall (1970).
†See Advisory Preface to Technical Memoranda Nos. 2, 3, and 4.

denotes the effective column length and r the appropriate radius-of-gyration of the cross section. The scatter is due to geometrical imperfections of the column specimens, eccentric application of load, non-homogeneity of the column material, residual stresses from the rolling and fabricating processes, variations in the action of loading machines, imperfections in the end fixtures, and other factors. The major sources of scatter are briefly discussed subsequently.

A geometrically perfect centrally loaded column would not deflect laterally at loads less than the critical load. All column specimens, however, deflect from the beginning of loading because of bending that results from the initial curvature and twist of the specimen and the unavoidable eccentricity of load application. Non-homogeneity of the column material results in bending of columns that are stressed above the proportional limit because the pattern of yielded zones of the cross section is not perfectly symmetrical with respect to the principal axes of the cross section.

Residual stresses from the rolling and fabricating processes, present in the column specimen prior to testing, cause scatter in the observed column strength because of patterns of the residual stresses among different specimens of the same size and shape cause variations in the load at the onset of yielding and, as residual stress patterns are generally not symmetric about the principal axes of the cross section, cause variations in the amount of column bending and twisting which, in turn, affects the column strength.

In column tests, as in stability tests of other structural elements, the response of the column is influenced by the action of the loading device. Loading devices may be categorized as gravity, deformation, and pressure types. The force-deflection characteristics of these types differ. The oldest form of testing device used for columns is the gravity-type. For such a system, the load-deflection characteristic is simple and can be graphically represented by a series of straight lines parallel to the deflection axis. Later, the screw-type testing machine became a common laboratory apparatus. These machines have the advantage of a well-defined load–deflection characteristic, the slope of which depends, essentially, on the elastic response of the loading system. As higher capacity loading machines were needed, the hydraulic testing machine was developed. This system, however, does not have an easily defined load-deflection characteristic as it depends on the properties of the hydraulic system, temperature, and other factors. Loading of a column in a testing machine is always conducted under some finite loading rate and the experimental results are influenced by this rate.

Centrally loaded columns may have different end conditions, ranging theoretically, from full restraint (fixed) to zero restraint (pinned), with respect to end rotation and warping. Most investigators have used the pinned-end condition for column testing for a number of reasons. Under the pinned-end condition the critical cross section is located near the mid-height of the column, thus making the cross section of interest remote from the boundary and, therefore, little influenced by end effects. For the same effective slenderness ratio, the pinned-end condition requires the use of only half the column length used for the fixed-end condition. With the pinned-end condition, however, it is necessary to provide end fixtures that

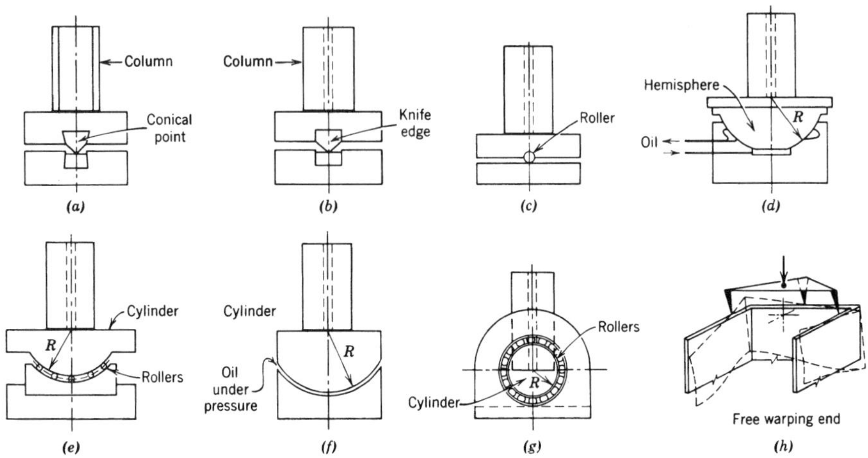

FIGURE B.7 End fixtures for pin-ended columns.

offer virtually no restraint to column-end rotation. Rotational restraint influences the effective slenderness ratio of the column and thereby contributes to scatter of experimental results.

Several schemes have been utilized to provide the pinned-end condition, some of which are shown in Fig. B.7, which is reproduced from Estuar and Tall (1967). The fixtures differ in that they are either "position-fixed" or "direction-fixed" (Salmon, 1931).

Probably the best way to reduce end restraint is by means of a relatively large hardened cylindrical surface bearing on a hard flat surface. Rotation will be virtually frictionless, even with some indentation under load. Another interesting feature of cylindrical fixtures is that the effective column length can be made equal to the actual length of the column by designing the fixtures so that the center of the cylinder coincides with the centroidal axis of the cross section at the column end. With a cylindrical fixture, the column is essentially pin-ended about one axis (usually the minor principal axis) and is essentially fixed-ended about the other.

A schematic diagram of the end fixtures used at Lehigh University's Fritz Engineering Laboratory is shown in Fig. B.8. A description of the fixtures, and their performance as "pins," is given by Huber (1958).

In testing columns under the fixed-end condition, the full restraint may not be provided over the entire range of the test loads; thus the effective length of the column is not a constant but a function of the applied load. This may be due partly to the fact that the rigidity of the testing machine varies with the applied load and partly to the indeterminate nature of the stress distribution at the column end, particularly in the load range in which the material yields. These problems are eliminated by using pinned-end conditions because the critical condition exists at about the mid-height cross section.

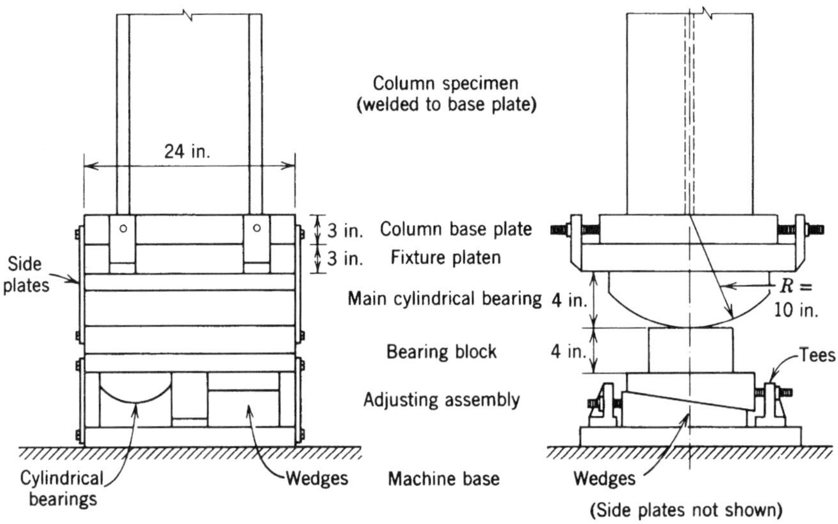

FIGURE B.8 Standard column end fixture at Fritz Engineering Laboratory (capacity = 2.5 million pounds).

B.4.1 Column Test Procedure

Preparation of Specimens To minimize initial geometrical imperfections in the specimen, the column specimen is cut from a straight portion of the stock. Both ends of the specimen are milled. Columns may be tested with the ends bearing directly on the loading fixtures, provided the material of which the loading fixtures are made is sufficiently harder than that of the column to avoid damaging the fixtures. Otherwise, base plates should be welded to the specimen ends, matching the geometric center of the specimen to the center of the base plate. The welding procedure should be such that compressive residual stresses at the flange tips caused by the welding are minimized. For columns initially curved, the milled surfaces may not be parallel to each other, but will be perpendicular to the centerline at the ends because milling is usually performed with reference to the end portions of the columns. For relatively small column specimens, it is possible to machine the ends flat and parallel to each other by mounting the specimens on an arbor in a lathe. For small deviations in parallelism, the leveling plates at the sensitive crosshead of the testing machine may be adjusted to improve alignment. The tolerance in deviation must not exceed the range of adjustment of the leveling plates of the particular testing machine.

Initial Dimensions The variation in cross-sectional area and shape, and the initial curvature and twist, will affect the column strength. Therefore, initial measurement of the specimen is an important step in column testing.

The cross section is measured to determine the variation between the actual dimensions of the section and the nominal catalog dimensions and to enable the

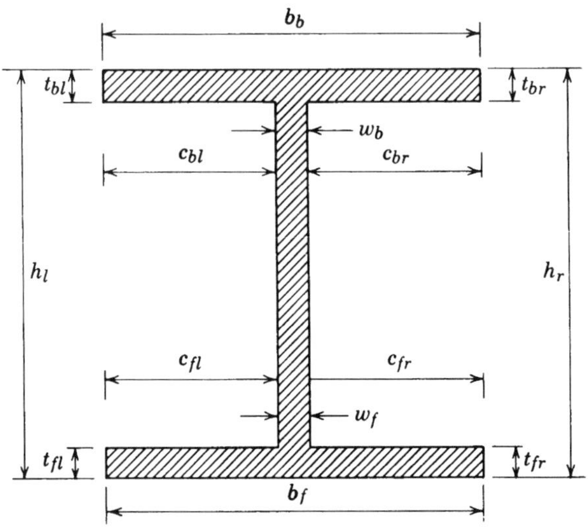

FIGURE B.9 Measurements required to determine cross-sectional properties of H-shaped column.

computation of the required geometrical cross-sectional parameters. The dimensions shown in Fig. B.9 are measured at a number of stations along the column (the quarter points of the specimen are recommended, as a minimum).

The initial camber, sweep, and twist of the specimen are measured at intervals. Nine stations, spaced at one-eighth of the column length, are suggested. A method of determining the initial out-of-straightness and twist is shown in Fig. B.10. Readings are taken with a theodolite (stationed in line with the column and near one of the ends) on a strip scale mounted onto a movable carpenter's square. The sweep offset is determined from four readings—each referenced to a flange tip. The average of the four readings is considered as the sweep offset. The camber offset is obtained from two readings, one referenced to each flange surface at the intersection of web and flange. The initial angle of twist is computed from the sweep offset readings and the cross-sectional dimensions. Values of initial out-of-straightness and twist are used in evaluating test results.

Aligning the Column Specimen Aligning the specimen within the testing machine is the most important step in the column testing procedure prior to loading. Two approaches have been used to align centrally loaded columns. In the first approach, the column is aligned under load such that the axial stresses are essentially uniform over the mid-height and the quarter-point cross sections. The objective in this alignment method is to maximize the column load by minimizing the bending stresses caused by geometrical imperfections of the specimen.

In the second alignment method, the column is carefully aligned geometrically, but no special effort is made to secure a uniform stress distribution over the critical cross section. Geometric alignment is performed with respect to a specific reference

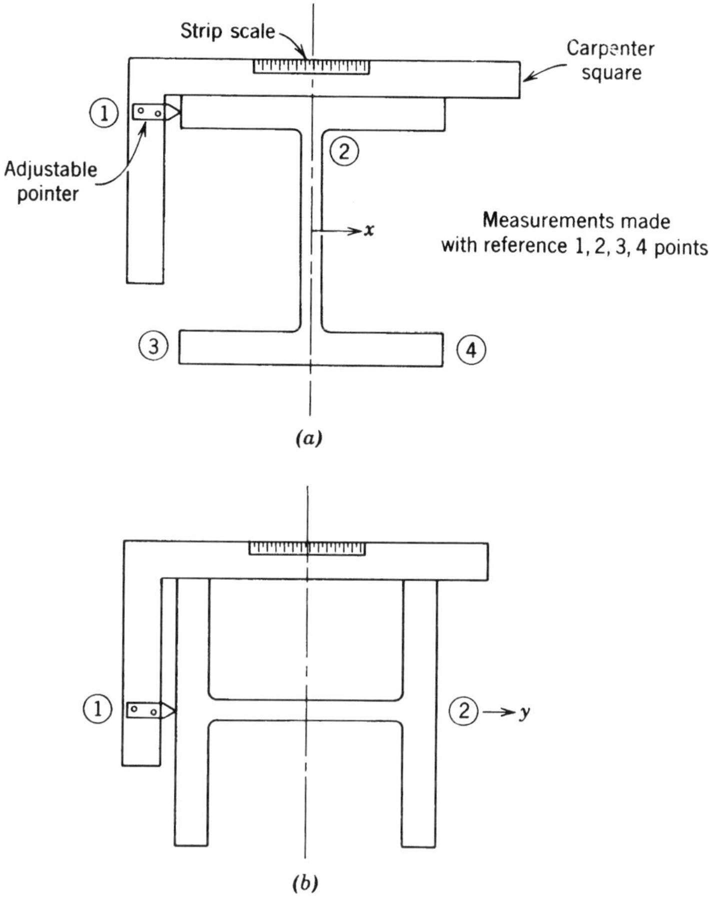

FIGURE B.10 Method for measuring initial camber and sweep: (*a*) measurement normal to minor axis (sweep); (*b*) measurement normal to major axis (camber).

point on the cross section (the specific reference point will be defined later). The method of geometric alignment is recommended as it is, generally, simpler and quicker. The end plates can easily be centered with reference to the centerline of the testing machine (Tebedge et al., 1971).

The specific reference point on the cross section utilized in geometrical alignment depends on the cross-sectional shape. For H-shaped cross sections the best centering point is the center of flanges because the web has little effect on buckling about the minor axis. This reference point may be located at the midpoint of the line connecting the two centers of the flanges (Tebedge et al., 1971).

Instrumentation In some column investigations, only the ultimate load is measured during the test. It is, however, usually desirable to measure the more important deflections and twists to compare the behavior of the column specimen under load with theoretical predictions of behavior. The instrumentation for column tests has changed markedly in the past few years due to progress made in measuring techniques and data acquisition systems, and it is now possible to obtain automatic recordings and plotting of the measurements. Such recordings have been found to be more convenient and more precise than manual readings.

The most important records needed in column testing are applied load and corresponding lateral displacements, twist, and overall column shortening. A typical column set-up and instrumentation are shown in Fig. B.11.

Lateral deflections normal to both principal cross-sectional axes may be automatically recorded by means of potentiometers attached at quarter points of the column (more points may be used for larger columns). Lateral deflections may also be measured from strip scales attached to the column and read with the aid of a theodolite.

Strains are measured using electric-resistance strain gages. For ordinary pinned-end column tests, it is sufficient to mount eight strain gages at each end and at the mid-height level. For long columns, it may be necessary to mount eight more strain gages at the quarter- and three-quarter points. As shown in Section *A-A* of Fig. B.11, the gages should be mounted in pairs "back-to-back" to enable the local flange bending effects to be cancelled by averaging the readings of each pair of "back-to-back" gages.

In the fixed-end test condition more strain gages are mounted below and above the quarter- and three-quarter levels. This is done to determine the actual effective length of the column by locating the inflection points using the strain gage measurements.

End rotations are measured by mechanical or electrical rotation gages. Mechanical rotation gages (Johnston and Mount, 1939) are assembled by mounting level bars on support brackets welded to the base plate and the top plate of the column as shown in Fig. B.12*a*. Angle changes due to column-end rotation are measured by centering the level bubble with the micrometer screw adjustment. A dial gage attached to the end of the level bar gives an indication of the rotation of the bar over a gage length of 20 in. (508 mm). In the electrical rotation gage, rotations are determined from bending strains induced in a thin metallic strip from which a heavy pendulum is suspended as depicted in Fig. B.12*b*. It has been shown that the strain at any location of the strip is proportional to the end rotation (Yarimci et al., 1968).

The angles of twist are determined at mid-height and at the two ends by measuring at each level the differences in lateral deflections of the two flanges. For better accuracy, the measurements may be taken at points located at the ends of two rods attached transversely on the adjacent sides of the column, as shown in Section *B-B* of Fig. B.11.

The overall shortening is determined by measuring the movement of the sensitive crosshead relative to the fixed crosshead using a dial gage or potentiometer.

Steel column specimens are whitewashed with hydrated lime. During testing, the whitewash cracking pattern indicates the progression of yielding in the column (the cracking reflects the flaking of the mill-scale at yielding zones).

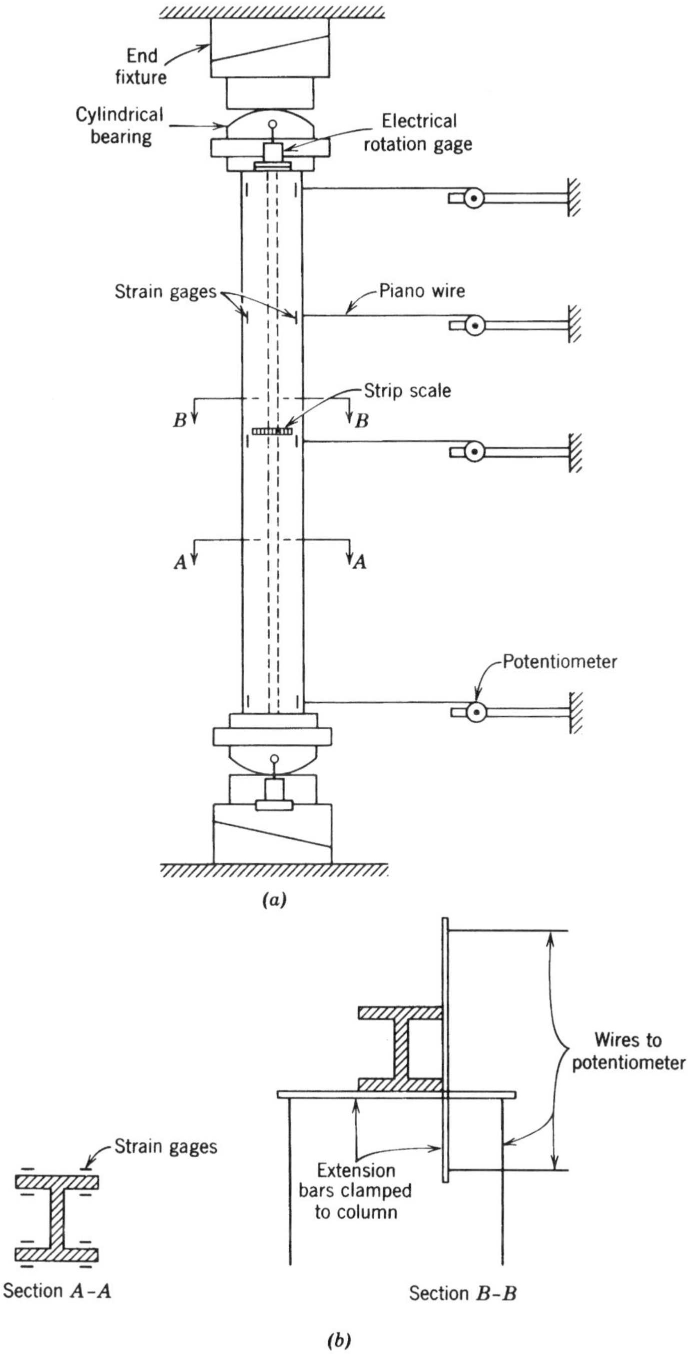

FIGURE B.11 Column test set-up.

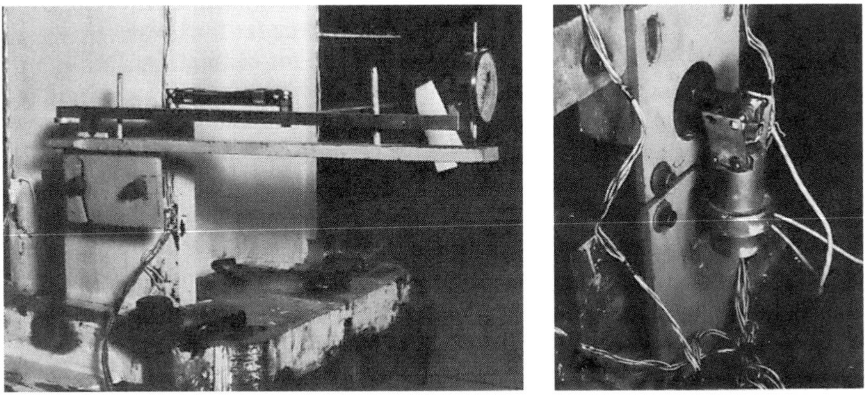

(*a*) (*b*)

FIGURE B.12 Rotation gages: (*a*) mechanical; (*b*) electrical.

Testing Procedure After the specimen is aligned in the testing machine, the test is started with an initial load of $\frac{1}{20}$ to $\frac{1}{15}$ of the estimated ultimate load capacity of the column. This is done to preserve the alignment established at the beginning of the test. At this load all measuring devices are adjusted for initial readings.

Further load is applied at a rate of 1 ksi/min (6.9 MPa/min), and the corresponding deflections are recorded instantly. This rate is established when the column is still elastic. The dynamic curve is plotted until the ultimate load is reached, immediately after which the "maximum static" load is recorded. (The procedure for determining a "static" load is described subsequently.) After the maximum static load is recorded, compression of the specimen is resumed at the "strain rate" which was utilized for the elastic range. In hydraulic testing machines this may be accomplished, approximately, by using the same bypass valve and load valve settings as had been used in the elastic range. The specimen is compressed in the "unloading range" until the desired load-displacement curve is attained. An example of such a curve is shown in Fig. B.13.

A static condition, as is needed to obtain the "maximum static" load, is when the column shape is unchanged under a constant load for a period of time. This means that the chord length of the column must remain constant, or practically, the distance between the crossheads must remain constant during the period. For screw-type testing machines the criteria can normally be satisfied by maintaining the crossheads in a stationary position. It is difficult, however, to maintain the distance between crossheads in hydraulic machines because of oil leakage and changes in oil properties due to the temperature changes that accompany pumping and throttling. To attain the static condition in the hydraulic machine, from the dynamic state, the bypass valve is further opened slowly until further lateral deflection of the column at mid-height ceases. The cessation of lateral deflection amounts to the condition of constant chord length. Alternatively, the relative crosshead displacements may be monitored, but this parameter is usually not as sensitive as lateral displacement.

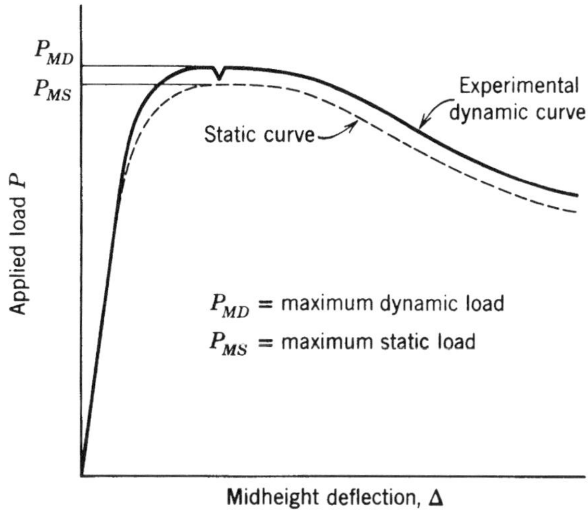

FIGURE B.13 Typical load–deflection curve of column.

B.4.2 Test Results

Presentation of the Data The behavior of the test specimen under load is determined with the assistance of measurements of lateral deflections at various levels along the two principal directions, rotations at the ends, strains at selected cross sections, angles of twist, and the column shortening. These measurements are compared to theoretical predictions. The results of the test are best presented in diagrammatic form. Such plots are shown in Figs. B.14 through B.19.

In Fig. B.14, typical plots of initial camber and sweep for a column are shown. They are used to determine the reduction in column strength due to initial out-of-straightness.

Figure B.15*a* shows the mid-height load-deflection curve of the column along the minor axis, and Fig. B.15*b* along the major axis. The load-deflection curves give the most significant data of the column test.

Plots of the measured strains at mid-height of the column are shown in Fig. B.16. This plot may be compared with the stub column test results to detect any unusual behavior of the column.

End rotations of the column measured using both mechanical and electrical rotation gages are shown in Fig. B.17. The results may be checked by comparing them with the lateral displacements along the length of the column (see Fig. B.15).

The angles of twist at mid-height and at the two ends are shown in Fig. B.18. The values are determined as discussed in the section on instrumentation.

Figure B.19 shows a typical plot of the load versus the overall shortening of the column.

For hot-rolled steel columns the progression of yielding is detected from the cracking and flaking of the whitewash. The sequential development of the whitewash cracks may be recorded to indicate the yielding pattern during loading.

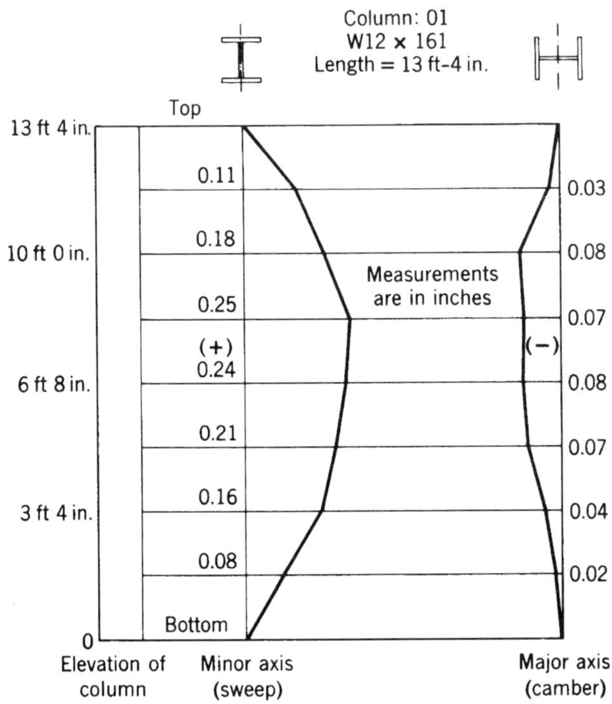

FIGURE B.14 Initial camber and sweep of a column specimen.

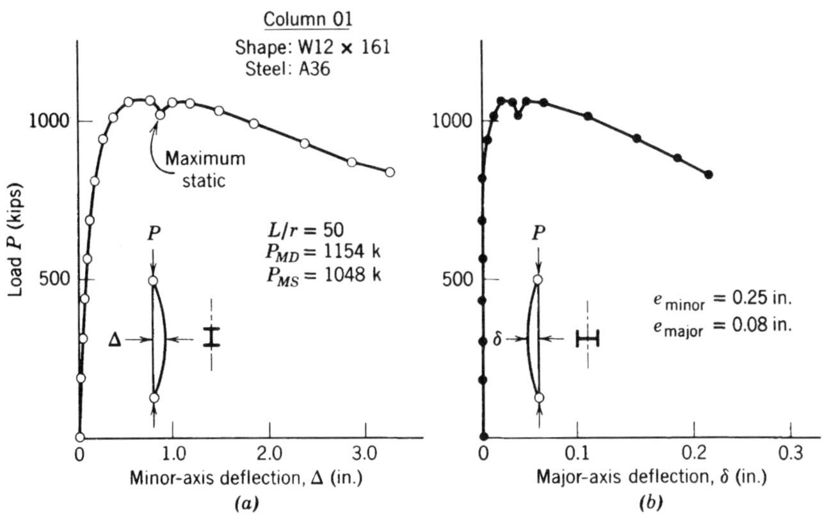

FIGURE B.15 Load–deflection curves.

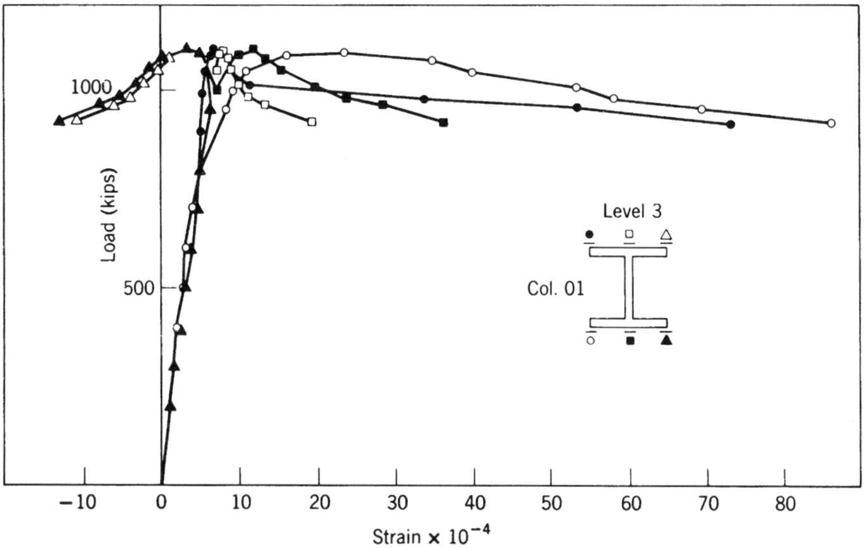

FIGURE B.16 Strain measurements at mid-height section using strain gages.

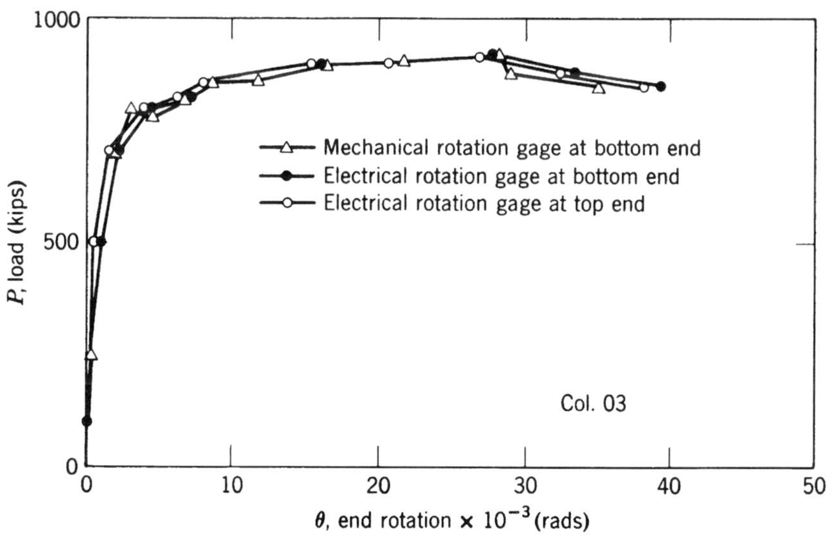

FIGURE B.17 End rotations of column 03 determined from mechanical and electrical rotation gages.

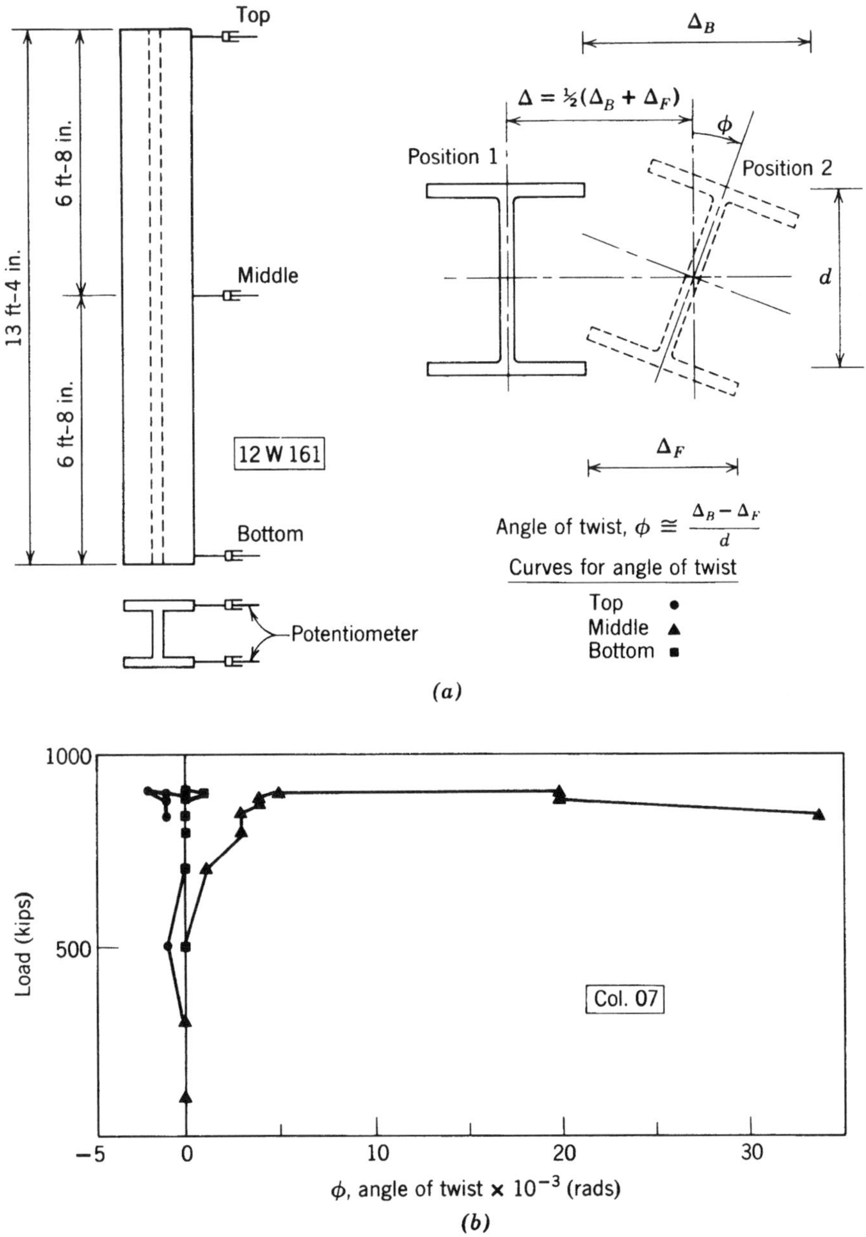

FIGURE B.18 Angles of twist at three levels.

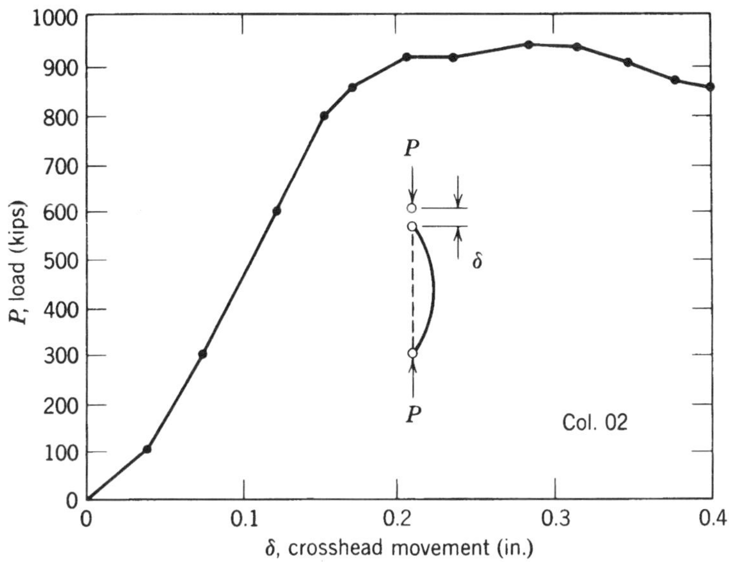

FIGURE B.19 Load versus overall shortening curve.

Whenever local buckling or any other phenomena occur during the test they should be recorded.

Evaluation of Test Results The test results may be evaluated by comparing the experimental load–deflection behavior and the theoretical prediction. A preliminary theoretical prediction can be made based on simplified assumptions of material properties, residual stresses, and measured initial out-of-straightness. The prediction may be improved if the actual residual stresses and the variations in material properties are used in the analysis. These properties should be determined from preliminary tests of specimens obtained from the original source stock. The preliminary test specimens should be cut immediately adjacent to that portion of the stock utilized for the column specimen.

B.5 TECHNICAL MEMORANDUM NO. 5: GENERAL PRINCIPLES FOR THE STABILITY DESIGN OF METAL STRUCTURES*

Research has shown that while the maximum strength of some types of centrally loaded columns is predicted by the tangent modulus buckling concept, this concept alone is not adequate for all types of structures and structural elements (for example, most hot-rolled and welded built-up columns, cylindrical shells and stiffened panels). To account for some of these inadequacies, the CRC Column Strength

*Published in February 1981, *Civil Engineering* (ASCE).

Formula, when adopted in 1960 as the basis for design by the American Institute of Steel Construction, included a variable factor of safety that, within the inelastic buckling range, increases as a function of column slenderness. In addition to the material non-linearities and residual stresses, which can be incorporated in the tangent modulus buckling concept, geometric imperfections (such as out-of-straightness), loading history, large deflections, post-buckling strength and behavior and connection response may affect significantly the limit of structural usefulness.

The maximum load resisting capacity of a member or frame determined inelastically by the inclusion of the effects mentioned above, has been termed its "maximum strength." Although the strength of an element may be thought of as a uniquely defined quantity, "maximum strength" is definitive in contrast with other concepts of strength such as tangent modulus, elastic modulus or first yield strength. For members, as distinct from frames, the dominant factor in the determination of the maximum strength, in addition to residual stress and material non-linearity, is member out-of-straightness.

In elements for which the effect of imperfections must be assessed in the determination of strength, end restraint may be significant, as it is in the tangent-modulus approach, and likewise, should be considered in the formulation of design criteria.

In accordance with the foregoing discussion, the following principle represents the currently (1979) held position of the Structural Stability Research Council with respect to stability in the design of metal structures:

> Maximum strength, determined by evaluation of those effects that influence significantly the maximum load-resisting capacity of a frame, member or element, is the proper basis for the establishment of strength design criteria.

This philosophy underlies the research effort and the *Guide* of SSRC. Also, it is the stated philosophy of the *Manual of Stability of Steel Structures* of the European Convention for Constructional Steelwork. It incorporates the tangent modulus buckling approach insofar as it provides the proper basis for defining the maximum strength of certain types of structures and elements, but it encompasses the members of frames in which initial geometric imperfections, large deflections, post-buckling strength and behavior, residual stresses, material non-linearities, load eccentricities, and end restraint must be considered.

Implicit in the tangent modulus approach recommended in Technical Memorandum No. 1 was a procedure to establish the column buckling load/slenderness curve, determined theoretically from the tangent-modulus curve for the specific column section under consideration. Whenever possible, the procedure for the establishment of the load carrying capacity of frames, members or elements on the basis of maximum strength should be based on a mathematical model which incorporates:

1. Experimentally determined physical characteristics, such as residual stresses, material non-linearities, and cross-sectional variations in yield strength, rationalized as may be appropriate.

2. A statistically appropriate combination of acceptance characteristics that are specified in supply, fabrication and erection standards, such as out-of-straightness, underrun of cross section, cross section dimensional variations, material properties and erection tolerances.
3. Effect of boundary conditions, such as restraint applied to the end of members.

When it is not possible to determine maximum strength theoretically, experimentally determined values of maximum strength may be accepted provided that the tests have been controlled and the results adjusted to compensate for the inclusion of the most adverse combination of unfavorable factors, contributing to a reduction in strength below the experimentally determined values, which has an acceptable probability of occurrence.

Although the maximum strength of frames and the maximum strength of the component members are interdependent (but not necessarily coexistent), it is recognized that in many structures it is not practical to take this interdependence into account rigorously. At the same time, it is known that difficulties are encountered in complex frameworks when attempting to compensate automatically in column design for the instability of the entire frame (for example, by adjustment of column effective length). Therefore, SSRC recommends that, in design practice, the two aspects, including stability of individual members and elements of the structure and stability of the structure as a whole, be considered independently. The proper basis for member design is the maximum strength of the restrained imperfect member. Where appropriate, second-order effects (such as $P-\Delta$ effects in frames) determined with due regard for non-linear and non-coexistent response, should be included with the first-order effects among the actions for which the member is to be designed.

B.6 TECHNICAL MEMORANDUM NO. 6: DETERMINATION OF RESIDUAL STRESSES*

Abstract This Technical Memorandum explains the origin of residual stresses as well as the techniques and procedures for measurement. Residual stresses may be induced during manufacture as a result of non-uniform cooling of the metal and/or from cold working. The magnitude and pattern of residual stress can have a pronounced influence on the behavior of structural members. Until recently, the most widely used technique of determining residual stresses was the method of sectioning and is fully described in this Technical Memorandum. Other methods that are commonly being used today are the center hole drilling method, x-ray diffraction, and ultrasonic techniques.

*Prepared by an SSRC Task Group. Chair: T. Peköz; Members: R. Bjorhovde, S. J. Errera, B. G. Johnston, D. R. Sherman, and L. Tall. Published in *Exp. Tech.*, Vol. 5, No. 3, September 1981.

B.6.1 List of Symbols

d = diameter of the contact edge in the gage hole
t = thickness of strip
L_g = gage length
N_i = initial middle ordinate
N_f = final middle ordinate
S_i = correction for initial middle ordinate
S_f = correction for final middle ordinate
a = internal angle of the extensometer gage point
d = difference between initial and final middle ordinates
$\lambda_h, \lambda_n, \lambda_x, \lambda_t$ = correction factor

B.6.2 Introduction

Most metal products contain residual stresses induced during manufacture. One source of these stresses in hot-rolled steel shapes is the non-uniform cooling of the metal after it leaves the rolls. Steel shapes fabricated from hot-rolled products by means of welding have additional residual stresses due to non-uniform heating of the base metal as the weld metal is deposited, and the subsequent non-uniform cooling of the weldment.

In addition to these thermal residual stresses, steel products may possess residual stresses as a result of cold working. For example, rolled structural shapes and welded built-up shapes are often bent by gagging to remove camber or sweep, or to introduce camber or sweep. In some mills hot-rolled shapes are rotary straightened, a process in which the shape is successively bent in opposite curvature several times as it passes through rolls at ambient temperature, rendering the shape essentially straight. The metal undergoes plastic deformation during rotary straightening as evidenced by flaking of the mill scale on the product.

The magnitude of the residual thermal and cold-work stresses resulting from the processes described above is far greater in the longitudinal direction of the shape than in any transverse direction except for surface effects. Longitudinal residual stresses can have a pronounced influence on the behavior of structural members, especially in the case of columns and plate structures built up by welding, and it is with the determination of such stresses that this memorandum is primarily concerned. Longitudinal residual stresses vary with respect to the width and thickness of each plate-element comprising the shape. Except for very thick elements and for walls of cold-formed tubular members, however, the variation of residual stresses through the thickness is usually not important.

Residual stresses due to cold work are present in all products whose cross section is cold-formed from sheets, either by press-braking or roll-forming. The magnitude of residual stresses is greatest in the direction transverse to the bent line, and the variation through the thickness is pronounced. Longitudinal stresses, however, are created by the transverse stresses because of the Poisson effect and these longitudinal stresses impact the behavior of members subjected to longitudinal compression, bending, and twist.

Aluminum structural shapes are usually extruded. Extruded shapes are straightened by stretching so that the finished product is virtually free of residual stresses. Aluminum sheets and plates may contain residual stresses from rolling, or from quenching if subjected to heat treatment after rolling. The flattening operation, which may consist of roller leveling or stretching, or both, does not always produce a stress-free product. These stresses may cause inconvenience in machining operations, but are rarely of structural significance. Shapes built-up by welding contain residual stresses due to non-uniform heating and cooling during the welding process. The magnitude of the longitudinal stresses from this source may reach the yield strength of the aluminum in its annealed state.

B.6.3 Techniques for Determining Residual Stress Magnitude

The techniques for determining the magnitude and distribution of residual stresses may be classified as nondestructive, semi-destructive, and destructive. X-ray and ultrasonic methods are termed as nondestructive. Until recently, these methods were not practical for determining residual stresses in structural members, but research, such as that reported by Castex and Maso (1975), Niku-Lari (1986), Noyan and Cohen (1987), Hoffmann (1987), and Drescher-Krasicka and Ostertag (1999), show significant potential benefits. The ultrasonic technique provides information on the difference between the principal (in the geometric sense) stresses only, and it is not practicable to interpret the residual stress in a specific direction.

In the semi-destructive and destructive techniques the residual stresses are determined from distortions caused by the removal of material. Because the residual stresses in the body are in equilibrium, removal of stressed material by cutting, planing, drilling, grooving, or etching causes a relaxation in stress and the corresponding strain. The strain is measured, and the relaxation of stress is obtained by using Hooke's Law. The testing technique is said to be semi-destructive if the amount of material removed is small compared to the initial volume of the specimen and if the specimen can be made whole again, as by welding. The technique is termed destructive if so much material is removed that the specimen is virtually destroyed.

Semi-destructive techniques usually involve drilling holes in the specimen. The action of drilling alters the internal stress distribution resulting in deformation at the surface of the specimen. This deformation is interpreted as caused by residual stress. The hole-drilling method of residual stress determination as developed by Mathar, and that of Soete and Vancrombrugge were studied by Tebedge et al. (1972, 1973). In Mathar's method two gage points are installed diametrically opposite and equidistant from the center of the hole to be drilled, to suit a sensitive mechanical extensometer. The axis of the gage points and hole should be in the direction of the stress to be determined. The initial distance between the gage points is measured, the hole drilled, and the final spacing of the gage points measured. From the displacement of the gage points, caused by the deformation of the specimen during drilling, the relaxation in stress may be determined theoretically or by utilizing a calibration test. Mathar's method using an extensometer is no longer widely used.

In the other semi-destructive technique, developed by Soete and Vancrombrugge (1955) and Randle and Vigness (1966) and now standardized in ASTM E837-08 *Standard Test Method for Determining Residual Stresses by the Hole-Drilling Strain-Gage Method*, the strains induced in the specimen by drilling are measured with the aid of electrical strain gages. These commercially available strain gages are more convenient and reliable for one not experienced in the use of mechanical extensometers. They are available in rosette form and in small size. The advantage of small rosettes is that the drilled hole can be of correspondingly small diameter, thus minimizing damage to the specimen and allowing the work to be done with smaller tools. The direction of the principal strains can then be determined. Although Soete and Vancrombrugge's technique seems to possess many advantages compared with destructive techniques, until recently it has not been utilized extensively in the United States. The foregoing semi-destructive techniques are described in more detail in the papers by Tebedge, Alpsten, and Tall, which also includes an extensive bibliography. ASTM E837-08 provides a practical method for determining residual stresses utilizing a hole-drilling procedure.

Most studies of residual stresses in structural shapes have been performed using the destructive technique. A portion of the specimen (the test piece), located at a suitable distance from the ends of the specimen, is marked into strips as shown in Fig. B.20*a*. Gage holes are drilled near each end at the mid-width of each strip, and their longitudinal spacing is measured. The test piece is then cold-sawed from the specimen, appearing as in Fig. B.20*b* after this operation, and the strips are cut from the test piece often by means of a thin milling cutter or band-saw. An alternative to cold-sawing is the electric discharge eroding/machining technique. They have come to be termed *sections* and such a section is depicted in Fig. B.20*c*. The distance between gage holes of each section is measured and the change in length of a section is interpreted with Hooke's Law as the average value of the residual stress present in that section prior to its removal from the specimen. It is believed the method of sectioning was first used in Lehigh University's Fritz Laboratory in the late 1940's by Luxion and Johnston (1948).

If it is desired to determine the variation of residual stress through the thickness of the section, the section is marked into strips as shown in Fig. B.20*c*, gage holes are drilled at the ends of each strip, and the longitudinal distance between these is measured. The strips are then cut from the section to obtain what are termed *slices*. The distance between the gage holes of each slice is measured again, and the change in length of a slice is a measure of the average residual stress in the section from which it was cut. This destructive technique has been named *the method of sectioning* and will be described in more detail subsequently.

Electrical strain gages can be used to measure the strains in the method of sectioning. They have the distinct advantage that smaller sections may be used and fewer corrections are necessary. Very careful and time-consuming techniques, however, are required to protect the gages from damage during the cutting operation and to ensure the same zero base if the gage wires have to be re-attached to the instrumentation for the final readings.

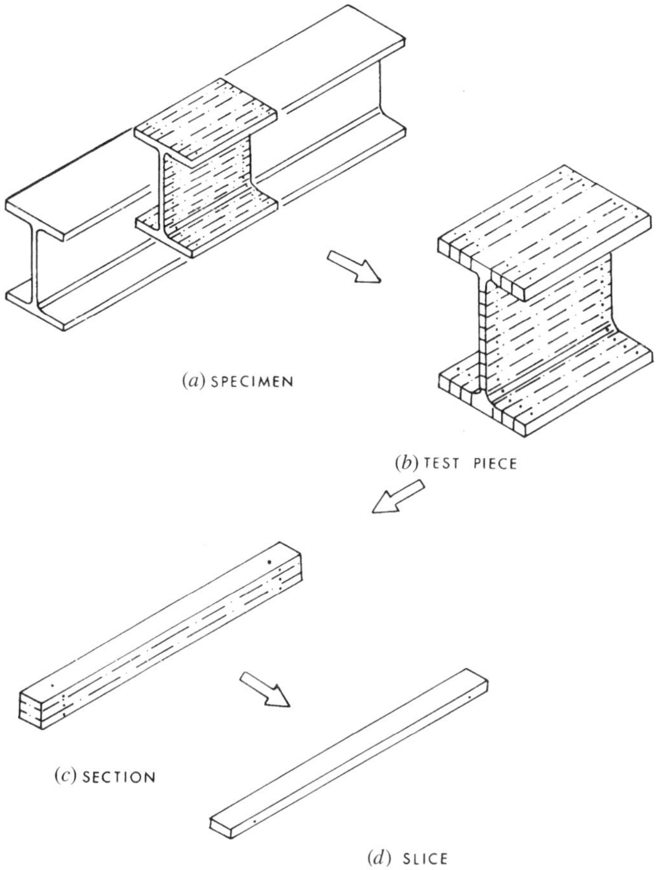

FIGURE B.20 Method of sectioning.

B.6.4 Method of Sectioning

Preparation of Test Piece The stock should be carefully examined for evidence of cold work prior to selecting the portion to be used as a specimen. If it is desired to determine the residual stresses due to thermal effects only, then the specimen should be cut from the portion of the stock that does not have transverse cracks in the mill scale if any are present. Such cracks are usually evidence that the member has been cold-straightened, either by gagging or rotary straightening. On the other hand, if it is desired to evaluate the residual stresses caused by both cold work and thermal effects (the usual situation), then the specimen should have a representative pattern of mill scale cracks. In either case, the specimen should be cut to a minimum length equal to three times the largest transverse dimension, plus the gage length, plus 2 in. (50 mm), if it is to be cold-sawed from the stock. If the specimen is removed by flame cutting, it should be longer by an amount equal to the largest transverse dimension in order to minimize the possibility that this operation will disturb the residual stress pattern in the central portion of the specimen. The specimen should be wire brushed and washed with a solvent to remove traces of cutting oil and its ends should be deburred. The specimen is then ready for marking of the test piece.

The central portion of the specimen is usually coated with machinist's blueing or a similar compound for a distance equal to at least the gage length plus 2 in. (50 mm). Next, the lines defining the sections are scribed on both sides of all elements of the cross section, where possible. The width of the strips should be based on the expected residual stress gradient and the needs of the investigation.

When the sections have been scribed on the test piece, the gage holes for the extensometer are laid out. A gage length of 10 in. (254 mm) or 100 mm is recommended, partly because these are the lengths of the standard Whittemore and Pfender Gages, respectively. The centers of the holes are punched on the centerline of each strip. These punch marks serve to guide the drill bit during drilling. It is convenient to use the punch fixture in the Whittemore kit, as the gage hole centers will then be located within the measuring range of the gage. If the relaxation of strain is to be measured on one side of the section only (either as a matter of convenience or because the second side is not accessible) the gage holes need be located only on the first side. If strains are to be measured on both sides, but only one side is accessible for drilling and the thickness of the section is less than about 1 in. (25 mm), gage holes centers need only be punched on the side accessible to the drill because those on the opposite side can be located by drilling through the section. In the case of thick sections, the strains should be measured on both sides. It is preferable to punch the gage hole centers on both sides provided that drilling on both sides is possible.

The importance of careful preparation of the gage holes to receive the gage points of the Whittemore Gage cannot be overemphasized. The diameter of the gage holes should be as small as practicable. Where shallow holes are drilled into the sections, or through thin sections, drill bits of about 0.03 in. (0.75 mm) diameter are suitable. For holes through thick material it is advisable to double the drill bit diameter so as to avoid excessive "wander" as the bit advances. It is desirable to utilize a drill press or, for large shapes, a magnetic base drill stand, to ensure that the holes will be as close to normal to the surface as possible. The holes must be chamfered to remove burrs from the drilling operation and to bring the contact surfaces between the test piece and the gage points of the instrument below the surface of the test piece, in order to protect them from damage during subsequent handling and machining. The depth of chamfer should be on the order of 0.02 in. (0.50 mm), and it is conventional practice to accomplish this with the reamer provided in the Whittemore kit. Because the reamer is hand-held, however, some researchers have found it desirable to use a guide block to aid in maintaining the reamer normal to the test piece. Such a guide block can be readily made by drilling an appropriately sized hole in a steel bar of about 1 in. (25 mm) thickness. Some investigators have reported success with a special bit with which the holes can be drilled and chamfered in a single operation (Tebedge et al., 1972, 1973).

Measuring Technique For measuring the relaxation of strain resulting from removal of material, the Whittemore Gage, or other comparable quality mechanical gage, with a nominal gage length of 10 in. (254 mm) is recommended. This is a portable mechanical extensometer with a dial gage graduated to 0.0001 in. (0.0025 mm) that is clamped to one of two steel coaxial tubes and bears on the other. To minimize temperature changes of the tubes due to handling, they are attached (internally) to a three-sided metal housing by means of flexible links. The observer holds the housing to operate the gage.

Measurements should be made with respect to the steel reference bar that is part of the Whittemore kit so that, if the temperatures of the bar and the coaxial tubes are essentially equal, no correction need be made to account for instrument errors due to ambient temperature variations during the test. Furthermore, if the test piece is steel and precautions are taken to maintain the reference bar, Whittemore Gage, and test piece at the same temperature, then all the errors due to temperature change will be minimal and no temperature corrections are necessary. The process can be made easier when all the measurements are made in a temperature controlled environmental chamber (when available).

In addition to measuring the longitudinal gage lengths it is also necessary to measure the curvature in planes normal and parallel to the strip surface for both the "initial" and "final" states when through the thickness residual stresses exist. Because the equilibrium of the strip is disturbed during the cutting operation, these curvatures change. This phenomenon is especially evident in strips cut from structural tubes (Sherman, 1969). It is necessary to determine these curvature changes in order to correct for the following errors:

1. Error due to "final" gage length measurement along chord between gage holes, rather than along the arc lying in the surface plane. Correction denoted as λ_s.
2. Error due to "final" gage length measurement along chord, rather than along arc in plane normal to strip surface. Correction denoted as λ_n.
3. Error due to misalignment of instrument gage point and gage hole axes due to change in curvature in plane normal to strip surface. Correction denoted as λ_h.

In order to compute corrections for the first of these errors (due to the change in curvature in the surface planes), the middle ordinates in the "initial" and "final" states, respectively S_i and S_f, should be measured. Normally, $S_i = 0$. The corrections for the last two errors are functions of the change in curvature in a plane normal to the strip and can be computed from the middle ordinates N_i and N_f.

The suggested procedures for taking the "initial" and "final" measurements are essentially alike and may be outlined as follows:

1. Thoroughly clean all gage holes utilizing cotton swabs, solvent, and compressed air blasts. If the holes were covered with adhesive tape for protection during machining, they may be coated with a gummy residue because some

adhesives are soluble in some cutting oils. Such deposits must be removed in order to secure repeatability of measurements.

2. Place the test piece, sections, or slices upon a sturdy table or other support so as to bring the work to a convenient height. The table should be located away from windows, radiators, and fans, and out of the path of drafts. The material should preferably not rest directly upon the table but rather on strips of wood, to reduce heat conduction. The parts should be arranged so that minimal handling is required after testing is commenced. Place the Whittemore Gage and the reference bar directly upon the material to be measured, cover with a clean cloth, and wait (usually overnight) for the specimens, reference bar, and instrument to reach the same temperature.
3. When testing is to begin, remove the cloth cover and proceed with the measurements as expeditiously as possible. The data required are indicated on the suggested data sheet shown as Fig. B.21. For each side of a strip take a reference bar reading before (columns 2 and 8) and after (columns 4 and 10) then three readings for the gage length (record average in columns 3 and 9). If the reference bar readings differ by more than one or two dial indicator units, discard the set of measurements. Such a difference indicates excessive temperature change, slipping of the dial indicator in its clamp, other instrument malfunction, or improper operation of the gages. If the largest difference between the three "initial" gage length readings exceeds three units, discard the set and redress the gage holes with the reamer. (After the "initial" readings have been completed, the gage holes must not be touched, except with a cotton swab.) The uncorrected gage lengths (columns 5 and 11) are determined by subtracting the average reference bar readings from the average gage-length readings.
4. The values of the middle ordinates, S (columns 6 and 12) and N (columns 7 and 13), are often measured with a dial indicator gage and fixture of the type described by Sherman (1969). If the strips are cut from the test piece by a milling machine it is reasonable to take $S_i = 0$. When using the fixture, precautions should be taken to avoid contact with the gage holes.

B.6.5 Calculation of Curvature Corrections

The bending deformations accompanying removal of a section from the test piece, or a slice from a section, may have to be accounted for in determining the final gage distances. The three errors are listed previously. The first two errors, due to "final" measurement along the chord instead of the arc, may be corrected adequately by adding the following quantity to the final measured values:

$$\lambda_n = \lambda_s = \frac{8\delta^2}{3L_g} \tag{B.3}$$

where L_g is the gage length, and $\delta = (S_f - S_i)$ or $(N_f - N_i)$ depending on the change of curvature under consideration. To account for misalignment of the conical

DATA SHEET—RESIDUAL STRAIN MEASUREMENTS

Date ____________________ Test piece indent. ____________________

Time start ____________ Temp. start ____________ Initial/final measure. (indicate which).

Time finish ____________ Temp. finish ____________

Observer ____________________ Recorder ____________________

	Side A					
(1)	(2)	(3)	(4)	(5)	(6)	(7)
				Uncorrected Length with Respect to Ref. Bar	Middle Ordinates	
Section/Slice Identification	Ref. Bar Reading	Gage Length Avg. of 3	Ref. Bar Reading		S	N

	Side B					
	(8)	(9)	(10)	(11)	(12)	(13)
				Uncorrected Length with Respect to Ref. Bar	Middle Ordinates	
	Ref. Bar Reading	Gage Length Avg. of 3	Ref. Bar Reading		S	N

FIGURE B.21 Headings for residual strain data sheet.

extensometer point and gage hole axes, the correction to the observed final gage length is

$$\lambda_h = \frac{4d\delta}{L_g} \tan \frac{\alpha}{2} \tag{B.4}$$

where d is the diameter of the contact edge in the gage hole and α is the internal angle of the extensometer gage point.

If measurements are made on both surfaces of a section or slice, the middle thickness relaxation of strain is usually obtained from the average of the gage length changes on both surfaces, neglecting the corrections λ_n and λ_h. On the other hand, if measurements are made only on one side, then all corrections λ_n, λ_s, and λ_h may be required. Furthermore, because curvature corrections λ_n, λ_s, and λ_h refer to the surface of the strip, it may be necessary to apply a correction for curvature to obtain the final length at mid-thickness, if measurements are made on one surface only. This correction is

$$\lambda_t = \frac{5t\delta}{L_g} \tag{B.5}$$

where t is the thickness of the strip. A further discussion of the above corrections is given by Sherman (1969).

B.6.6 Acknowledgment

The Council wishes to thank the former members of Task Group 6 who were active during the time of preparation of earlier drafts of the report. They were: L. S. Beedle, C. Birnstiel, J. W. Clark, E. W. Gradt, R. A. Hechtman, T. R. Higgins, and B. M. McNamee.

B.7 TECHNICAL MEMORANDUM NO. 7: TENSION TESTING*

B.7.1 Introduction

The tension yield strength is the key mechanical property required by most material specifications and design practice. Because of its standard usage, it is the most accepted value for analyzing and comparing test data. Usually, the comparison is performed by "normalization," that is, the test results are non-dimensionalized with respect to the yield strength (stress). Thus, the tension test becomes a most important aspect of a "test of stability" in which all or some portion of a structural shape is tested in compression. Because the tensile yield is sensitive to the rate of straining and the location where the specimen is taken, normalized stability test data can easily be shifted by more than 20% if care is not exercised in conducting the tension test and in reporting the test method employed and its results.

Yield strength is not the only parameter that is important in evaluating tests of stability and theory, as it is often desirable to know other material properties such as the proportional limit and the strain hardening characteristics, which can be obtained from a tension test. At the same time, the tension test method must conform as closely as possible to those of standard quality control tests so that the stability test results can be interpreted with respect to design standards.

Ideally, the strain rate in the tension test and the stability test should be the same. Due to the difficulty in conducting a stability test at a constant, known strain rate, the SSRC advocates the use of the static yield strength in stub column tests and the static load in tests of stability. Static values are obtained by loading the specimen with a load or deflection increment and then holding a constant distortion until the load is stabilized. This stable load is the static value. A static tensile yield strength is, therefore, the most appropriate value to be used in normalizing test data.

B.7.2 Purpose

This technical memorandum is intended to provide guidelines for conducting a tension test so that consistent, uniform values are obtained and reported for the static yield strength, dynamic yield strength, proportional limit, strain value at the initiation of strain hardening, strain hardening modulus, ultimate strength, percent elongation, and percent area reduction.

*Prepared by an SSRC Task Group; Chair: L. Tall; Members: P. C. Birkemoe, R. Bjorhovde, S. J. Errera, K. H. Klippstein, R. A. LaBoube, T. Peköz, D. R. Sherman, and R. B. Testa. Approved by SSRC Executive Committee: October 23, 1986.

B.7.3 Equipment

Tension tests are performed on different types of testing machines in different laboratories. These machines can be grouped as screw-driven machines, manually controlled hydraulic machines, and servo-controlled closed loop systems of the screw or hydraulic type. One common aspect of all these machines is that the specimen is loaded by the motion of a crosshead, although a feedback mechanism can be used to relate the crosshead motion to load or strain in the specimen. In manually operated machines, the feedback mechanism is the operator who watches a load dial and the output of a strain extensometer, whereas a closed loop machine uses a servo controller which provides the appropriate feedback to drive the crosshead and thus maintain the desired rate of load, stroke, or strain. These systems control strain, load, or stroke more precisely and with faster response than a manually operated machine.

Another difference in testing machines is the manner of gripping the specimen. Threads, button heads, wedge grips or hydraulic grips are the most common methods. The characteristics of the grips have a distinct influence on the relationship between crosshead motion and strain in the specimen. Due to inelastic creep in the gripping, the strain in the specimen may change even though the crosshead motion is completely stopped.

A Class 2B extensometer (strain error < 0.0002) will normally be satisfactory for monitoring strain. If accurate values of the modulus of elasticity are required in addition to the yield information, a more accurate extensometer or strain gage should be used. As stated in the Purpose this technical memorandum is to be used for conducting a tension test and not a test to determine the modulus of elasticity. ASTM E111-04 *Standard Test Method for Young's Modulus, Tangent Modulus and Chord Modulus* is to be used to obtain this material property.

B.7.4 Specimen

The location of each tension test specimen should be selected in accordance with ASTM A6/A6M-08 *Standard Specification for General Requirements for Rolled Structural Steel Bars, Plates, Shapes, and Sheet Piling* or EN 10025 (European Committee for Standardization CEN, 2004). When these prescribed locations cannot be met, it should be recorded where each specimen has been taken to ensure that comparisons can be made with other tests. It should also be recorded if the specimen was full thickness or milled out of a plate or part of a structural shape. When steel is being ordered, a certificate from the steel mill or steel supplier should be requested that will state the yield point, ultimate tensile strength and percent elongation. This steel heat production information should be close to the values obtained by this technical memorandum, but will not include a static yield strength value.

The tension test specimen should be prepared in accordance with ASTM E8/E8M-08 *Standard Test Methods for Tension Testing of Metallic Materials*, EN 10002 (European Committee for Standardization CEN, 2001b), ISO 377 (1997) and any applicable product specification for the specimen in the test for stability.

The end section for gripping, and in some cases the size of the tension test specimen, will be dictated by the testing machine. If the test for stability involves a rolled or standard shape, the tension test specimen should be taken from the piece as required by ASTM A6/A6M, EN 10025, or other product specification. For fabricated test pieces, two tension test specimens are desirable. One should be taken from the plate or sheet material prior to fabrication. This provides a correlation with the material requirements. The other tension test specimen should be taken from the fabricated piece at a location that represents average properties resulting from strain hardening or work hardening and residual stresses. The tensile properties of this specimen are required for comparison with other data and for correlation with theory.

B.7.5 Procedure

Although it is desirable to operate the testing machine in a strain control mode, this may not always be practical. Even servo-controlled machines may go out of control if the extensometer slips, or it may be difficult to switch to a crosshead or displacement control mode during loading. Consequently, operation of the testing machine in a crosshead or stroke control mode is acceptable. In this event, it will be necessary to determine the rate of crosshead motion that will produce the desired strain rate by loading the specimen to less than 50% of the anticipated yield and making adjustments in the rate of crosshead motion as required. When yielding occurs, it is usually necessary to reduce the rate of crosshead motion to obtain approximately the same rate of straining.

The rate of strain should be approximately the same as that obtained during loading in the stability test, but still within ASTM or EN limits. A graphic plot of stress (or load) vs. strain is desirable but the data may be taken manually in sufficient increments to produce a well defined curve. When the strain reaches a value corresponding to approximately 0.2% offset $(0.002 + \sigma_y E)$,* the test should be interrupted by holding a constant strain or stopping the crosshead motion. This condition should be maintained for at most five minutes or until the load stabilizes. The lowest value of the load and the corresponding strain should be recorded. Straining is then resumed at the original strain rate. The test should be interrupted with static load values recorded at least two more times before strain hardening begins or at 0.005 increments of strain. Straining at the original rate should continue until the initial strain hardening characteristics of the material are evident, at which time the rate can be increased according to ASTM or EN procedures until failure occurs.

B.7.6 Results

The stress–strain information should form one of the two curves shown in Fig. B.22. If the resulting curve is the flat yielding Type A, the static yield should

*If the applicable material or product specification defines yield as 0.005 total strain, that value should be used in place of 0.2% offset throughout this technical memorandum.

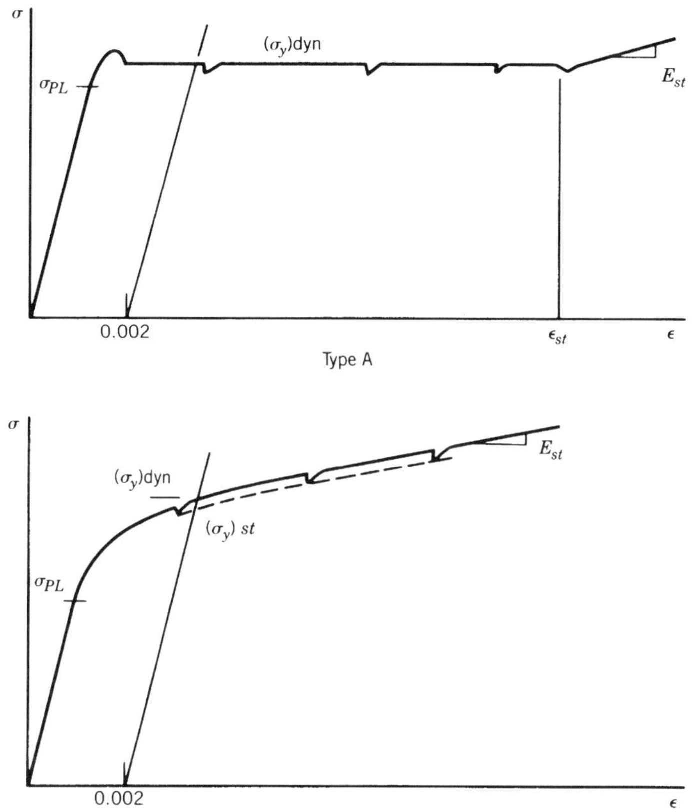

FIGURE B.22 Typical tension test stress-strain curves.

be reported as the average of the three low values obtained. For a rounded Type *B* stress–strain curve, a line should be drawn through the low points of the three interruptions and the static yield determined by the 0.2% offset intercept.

Because the yield strength is not the only parameter important in stability theory, the following information obtained in the tension test should be reported:

1. Static yield strength (indicate if obtained by 0.2% offset method or at 0.005 strain).
2. Dynamic yield strength and strain rate.
3. Proportional limit.
4. Strain at initiation of strain hardening.
5. Strain hardening modulus. Refer to ASTM E111-04.
6. Modulus of elasticity. Refer to ASTM E111-04.

Three additional values that are normally part of a tension test should be reported for quality control purposes, although they are not a factor in stability.

7. Ultimate strength.
8. Percent elongation with statement of gage length.
9. Percent area reduction.

B.8 TECHNICAL MEMORANDUM NO. 8: STANDARD METHODS AND DEFINITIONS FOR TESTS FOR STATIC YIELD STRESS*

B.8.1 Introduction

Static yield stress is a reliable and consistent measure of the value at which steel yields and is independent of testing procedures and testing machine behavior. It is the yield stress corresponding to a strain rate of zero (or to a testing speed of zero). It is applicable to both tension and compression tests and to tests on both a coupon and the full cross section. Because the static yield stress is a consistent and reliable measure, it serves as a valid basis of comparison between different steels compared to other measures of yield. For this reason it is also used to normalize† data from column tests or from other tests of stability so that comparisons with other data or with theory can be made without the influence of variation of yield. The Structural Stability Research Council has advocated the use of static yield stress in normalizing data from tests of stability.

The yield point and yield strength of steel are affected directly by the rate of straining. Generally, the higher the rate, the higher the yield tends to become, until the ultimate load is reached without clearly defined yielding (Nadai, 1950).

Because a particular type of steel could have many yield values depending on the strain rate, the speed of testing is of the utmost importance when defining yield stress. Actually, there are a number of definitions for the yield stress (Tall, 1974), and justification exists, to a greater or lesser degree, for using various values for design. Specifications do not take account of the size effect in the specimens and the differences in testing machines. Although ASTM A370-08a *Standard Test Methods and Definitions for Mechanical Testing of Steel Products* limits the maximum testing speed for structural steel, some investigators use lower speeds than others with the result that discrepancies as high as 20% may exist in the measured value for yield stress (Beedle and Tall, 1960). In addition, the testing speed and strain rate are two different quantities without a defined relationship. Hence, the use of the term "yield stress" has limited value, unless it is qualified by a strain rate. Strain rate, however, does not account for all the variation between tests; it cannot account for material differences or manufacturing methods. The difference due to chemistry and manufacturing procedures can be evaluated more clearly if the superimposed effects of strain rate are removed.

*Prepared by SSRC Task Group 6, *Test Methods.* Chair: L. Tall; Members: P. C. Birkemoe, R. Bjorhovde, S. J. Errera, K. Klippstein, R. A. LaBoube, T. Peköz, D. R. Sherman, R. B. Testa. (Adopted Nov. 1987).

†*Normalization* is the nondimensionalizing of test results with respect to the yield stress.

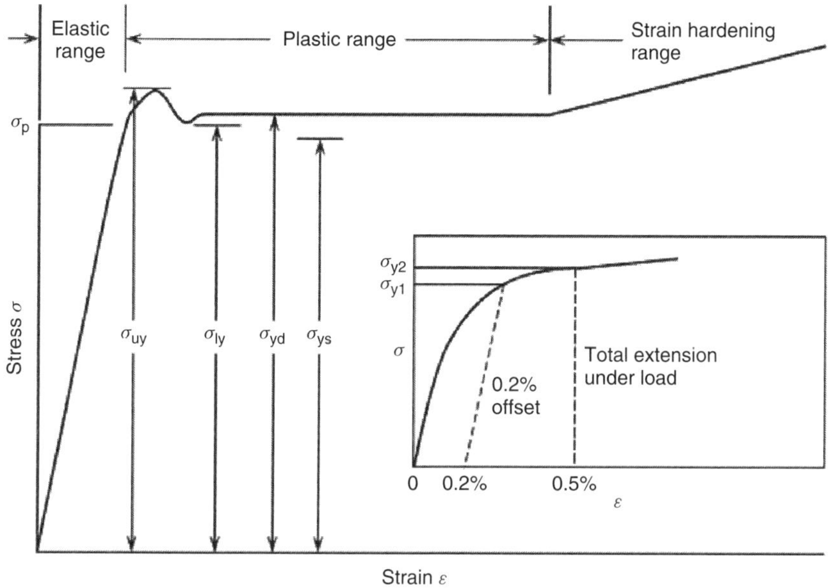

FIGURE B.23 Stress-strain curve and definition of terms.

B.8.2 Definitions

Symbols used in this technical memorandum are σ, average stress, and ε, strain. Two of the terms used to define the strength of steel are yield point and yield strength as defined in ASTM E6-08 *Standard Terminology Relating to Methods of Mechanical Testing*. Various methods to determine the quantities are described in ASTM A370-08a *Standard Test Methods and Definitions for Mechanical Testing of Steel Products*. In this technical memorandum, the following terms are used: upper yield point (σ_{uy}), lower yield point (σ_{ly}), dynamic yield stress (σ_{yd}), and static yield stress (σ_{ys}). These terms are shown in Fig. B.23, and are defined as:

Upper Yield Point, σ_{uy}: same as ASTM E6-08 definition for yield point: "the first stress in a material, less than the maximum attainable stress, at which an increase in strain occurs without an increase in stress."*

Lower Yield Point, σ_{ly}: lowest level of yield stress immediately following the upper yield point while maintaining a constant strain rate.

Dynamic Yield Stress, σ_{yd}: average stress during actual yielding in the plastic range. It remains fairly constant, provided that the strain rate remains constant, and is also referred to as the "dynamic yield stress level."

Static Yield Stress, σ_{ys}: average stress during actual yielding in the plastic range at zero strain rate; this stress remains fairly constant, and is also referred to as the "static yield stress level." When the static yield stress is not a plateau

**ASTM Note*: It should be noted that only materials that exhibit the unique phenomenon of yielding have a "yield point."

(that is, constant with respect to strain), it is taken as the stress at zero strain rate corresponding to a total strain of 0.005 or to 0.2% offset, depending on the product specification.

Yield Stress, σ_y***:*** general term that includes all the definitions for the yield value.

Plastic Range of Strain, ϵ_y ***to*** ϵ_{st}***:*** range between the yield strain and the strain at the onset of strain hardening.

Yield Strain, ϵ_y***:*** strain at which the yield stress first occurs. Its value depends on the definition of the yield stress.

Speed of Testing: rate of crosshead motion or rate of stress increase.

Maximum Speed: limiting speed in ASTM A370-08a *Standard Test Methods and Definitions for Mechanical Testing of Steel Products*.

It is impossible to test a specimen at zero strain rate and obtain a stress–strain diagram. A method, however, for obtaining the stress corresponding to zero strain rate is described below. Assuming that the strain rate is the only factor that influences the dynamic yield stress, a relationship between the strain rate and the ratio of dynamic yield stress to static yield stress has been established (Nagaraja et al., 1966). For convenience, this ratio is termed the dynamic yield stress ratio, and it always will be greater than unity.

The term "yield strength" is defined in ASTM E6-08 *Standard Terminology Relating to Methods of Mechanical Testing* as: "the stress at which a material exhibits a specified limiting deviation from the proportionality of stress to strain. The deviation is expressed in terms of strain."*

B.8.3 Method of Testing

The basic method of testing conforms to the usual standards, that is, to ASTM E8/E8M-08 *Standard Test Methods for Tension Testing of Metallic Materials* and ASTM E9-89a (2000) *Standard Test Methods for Compression Testing of Metallic Materials at Room Temperature* for tension testing and compression testing, respectively (ASTM, 1982), and to SSRC TM 2, TM 3, and TM 7, for the testing of tension and compression specimens and stub columns. The usual test procedure and testing speed is followed up to the yield point, or until the plot shows that the material is starting to strain in the plastic range. At this point, the crosshead motion of the machine is stopped to record the static yield stress. At this stage the strain is between two to five times the yield strain—and the control mode is crosshead motion. (Although the modern servo-controlled testing machines enable control by strain, it is generally more convenient to control crosshead motion in order to avoid changing the control mode when the extensometer must be removed.) At this point, the plot of the load–strain curve drops from its original course indicating a decrease in load—this is accompanied by a slight increase in strain. About 5 minutes are needed to stabilize the load at zero crosshead motion. At the end of the 5 minutes,

**ASTM Note*: Whenever yield strength is specified, the method of test must be stated along with the percent offset or the total strain under load. The values obtained by the two methods may differ.

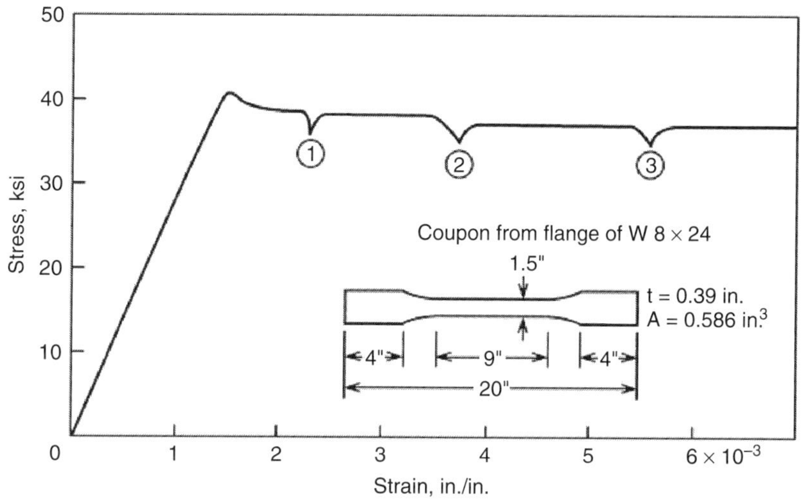

FIGURE B.24 Typical stress-strain diagram.

or when the load has stabilized, the magnitude of the load is recorded, either automatically on the running plot, or else visually. (Five minutes is suggested also as a maximum time limit because, in the case of hydraulic machines, strain reversal and a lower equilibrium load may be recorded due to leakage of oil in the system.)

The test is then continued by returning to the standard testing speed for a brief interval. This interval is approximately 5 seconds for a tension specimen being tested at the ASTM maximum speed of $\frac{1}{16}$ in./min per inch of gage length; the interval will be longer for slower test speeds. The foregoing is then repeated by stopping the crosshead motion of the machine again. The minimum value of load corresponding to the zero crosshead motion indicates the yield stress at a strain rate of zero (Fig. B.24). Depending on the extent of the plastic range of strain, two or three values of the static yield stress are recorded. Thus, a line is drawn between points 1, 2, and 3 on Fig. B.24. If the line is horizontal, the value indicates σ_{ys}. If the line is inclined to the horizontal, then the ordinate of the line at the 0.005 strain, or at the 0.2% offset, is taken as σ_{ys}.

B.8.4 Commentary

Static Yield Stress Normalization of test data is usually made with respect to the static yield stress. Ideally, the strain rate in tests used to obtain the yield stress for normalizing data and the strain rate in the stability test in question should be the same. Due to the difficulty of conducting stability tests at a constant known strain rate, the Structural Stability Research Council advocates the use of the "static yield stress" in the stub column test and in the tension test, and the use of the static load in a stability test. These static values are obtained by loading the specimen with a load or deflection increment and then holding a constant distortion until the load has stabilized. This stable load or stress is the static value.

Both the upper and lower yield points have been used in the past as the basis for designating the yield stress. It is common practice in the testing of tensile specimens to record the "yield" as the highest reading indicated by the free follower pointer on the load indicator dial, the actual load having dropped somewhat. In this technical memorandum, emphasis is given to the static yield stress and the dynamic yield stress. By its definition, the static yield stress is independent of the speed of testing. It is not affected by size of specimen or testing machine. Moreover, most loads on structures such as buildings are static. Therefore, the static yield stress level is a more uniform standard for comparison, and it is of direct application to the testing of structures and structural components and to research into their strength.

Relationship between Static and Dynamic Yield Stress The yield point and yield strength of steel are affected directly by the rate of straining. Generally, the higher the rate, the higher the yield stress tends to become until the limit when the ultimate load is reached without yielding. The dynamic yield stress is influenced by the speed of testing, size of specimen, and testing machine. Thus, for the same testing machine and size of specimen, the dynamic yield stress is a function of the speed of testing. The speed of testing may be defined by either the rate of separation of the crosshead, or by the strain rate—these have entirely different meanings.

The current method of designating testing speed introduces a problem. The ASTM specifies the rate of separation of the crossheads, because the specification of strain rate may not be a practicable method of controlling machines currently used in production testing. In Section 11.4.1 of ASTM A370-08a *Standard Test Methods and Definitions for Mechanical Testing of Steel Products*, the rate of separation of crosshead is limited to $\frac{1}{16}$ inch (1.6 mm) per minute per inch (25.4 mm) of gage length for tension specimens with reduced cross sections, hereafter called the maximum speed. In Section 11.4.3 of ASTM A370-08a, the speed of testing is limited by the rate of stressing of 100,000 psi (689.47 MPa) per minute.

Suppose the rate of separation of crossheads of $\frac{1}{16}$ in./min per inch of gage length is used. For a 0.505-in.-diameter tension specimen of 2 in. gage length, the crosshead separation speed is $\frac{1}{8}$ in./min, and for a plate-type tension specimen of 8 in. gage length, the crosshead separation speed is $\frac{1}{2}$ in./min. The two specimens theoretically will have approximately the same strain rate of $\frac{1}{16}$ in./in./min or 1042 μin./in./sec, although the actual measured strain rates in the gage length will be significantly less. If the rate of stressing of 100,000 psi/min is used as the criterion, the corresponding calculated strain rate would be 55 μin./in./sec. (In this calculation, it is assumed that all of the deformation goes into the specimen in the indicated gage length.) Thus, the two criteria are too far apart for valid comparison of the results.

The strain induced in a specimen in the elastic range has been shown to be influenced by the response of the testing machine (Nagaraja et al., 1966; Gozum and Huber, 1955). It can be assumed that in the plastic range all the extension is absorbed by the specimen, and the testing machine does not undergo further deformation. This would indicate that the crosshead speed may not be linearly proportional to the strain rate. Thus, the same crosshead speed would produce a lower strain rate in the elastic range and a relatively higher strain rate in the plastic range.

The results of a study (Nagaraja et al., 1966) into tension tests of 0.50 in. (13 mm) round specimens with 2 in. (51 mm) gage length and plate-type specimens with 8 in. (203 mm) gage length have shown that the dynamic yield stress, the static yield stress, and the strain rate are related by

$$\frac{\sigma_{yd}}{\sigma_{ys}} = 1 + k\dot{\varepsilon}^n \tag{B.6}$$

where σ_{yd}/σ_{ys} = dynamic yield stress ratio

$\dot{\varepsilon}$ = strain rate, μin./in./sec.
k = a constant, 0.021 for A36 steel
0.020 for A441 steel
0.023 for A514 steel
n = a constant, 0.26 for A36 steel
0.18 for A441 steel
0.08 for A514 steel

An alternative form of this relationship is

$$\sigma_{yd} - \sigma_{ys} = C\dot{\varepsilon}^m \tag{B.7}$$

where

C = a constant, 0.87 for A36
1.06 for A441 steel
2.57 for A514 steel
m = a constant, 0.24 for A36
0.18 for A441 steel
0.08 for A415 steel

Because the difference $(\sigma_{yd} - \sigma_{ys})$ is essentially the same for A36, A441, and A514 steels (Fig. B.25), an average curve was proposed (Nagaraja et al., 1966)

$$\sigma_{yd} - \sigma_{ys} = 3.2 + 0.001\dot{\varepsilon} \tag{B.8}$$

that is valid for the range of strain rate $200 < \dot{\varepsilon} < 1600$. (Because the strain rate does not greatly affect the difference between static and dynamic yield stress for practical values of strain rate, the crosshead speed per inch of gage length per second may be used in place of strain rate in Eq. B.8.)

The relationship of Eqs. B.6, B.7, and B.8 are shown in Fig. B.25 for the three grades of steel considered.

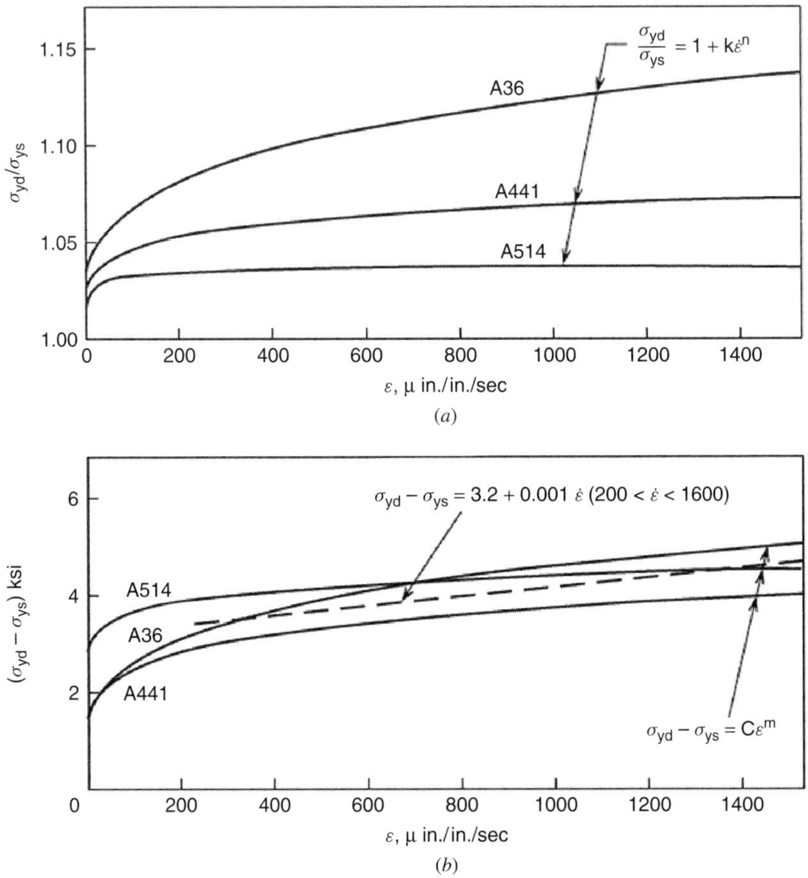

FIGURE B.25 Estimated curves relating dynamic yield stress, static yield stress, and strain rate—A36, A441, and A514 steels.

Tension tests of A36 and A633C steels using standard plate-type coupons with 2 in. (51 mm) and 8 in. (203 mm) gage lengths were also conducted (Roloff, 1984). A low elastic strain rate of 10 μin./in./sec was obtained in a servo-controlled testing machine using the stroke-control mode. The comparison of average dynamic and static yield stress values is summarized in Table B.1. The ratios and differences for the 8 in. (203 mm) specimens are similar to those in Fig. B.25 for A36 and A441 steels. The results are slightly higher for the 2 in. (51 mm) specimens.

A strain rate of about 100 μin./in./sec corresponds to the testing rate normally used, and at such a strain rate (Fig. B.25*a*) the average dynamic yield stress ratios are 1.126, 1.070, and 1.040 for A36, A441, and A514 steels, respectively (Nagaraja et al., 1966). Thus if no static yield stress has been recorded, as in the case of a mill test, the static yield stress may be inferred approximately from the recorded dynamic yield stress value by reducing it by 13, 7, and 4%, respectively, for those

TABLE B.1 Comparison of Dynamic σ_{yd} and Static σ_{ys} Yield Stress Values

	Number of Tests	Average Values [ksi (MPa)]			
		σ_{yd}	σ_{ys}	$\sigma_{yd}/_{ys}$	$\sigma_{yd} - \sigma_{ys}$
A36, 2 in.	16	42.4 (292)	39.2 (270)	1.08	3.2 (22)
A36, 8 in.	8	40.8 (281)	38.6 (266)	1.06	2.2 (15)
A633C, 2 in.	13	61.1 (421)	58.0 (400)	1.05	3.1 (21)
A633C, 8 in.	7	60.2 (415)	58.1 (401)	1.04	2.1 (14)

grades of steel—this approximation assumes that the mill test result, the upper yield stress σ_{uy}, is the same as the dynamic yield stress, σ_{yd}.

Further Considerations There is a relationship between the yield stress obtained from a stub column test and from tension specimens from its cross section. One of the results of a stub column test is the static yield stress for the complete cross section (Technical Memorandum No. 3). It has been shown (Tall and Alpsten, 1969) that the σ_{ys} of the stub column corresponds to the weighted average of the σ_{ys} from tension specimens from the flange and web (weighted according to the respective areas).

B.9 TECHNICAL MEMORANDUM NO. 9: FLEXURAL TESTING

B.9.1 Introduction

The design of beams and beam-columns depends on knowledge of the flexural behavior of the member as established theoretically and confirmed by test. The modes of flexural behavior recognized in limit states design standards include attainment of cross-sectional strength, and elastic or inelastic lateral-torsional buckling. The fundamental purpose of flexural testing is to develop or verify design equations, i.e., to establish how well proposed equations predict the behavior as given by the mean value of the ratios of the test-to-predicted moments and the associated coefficient of variation for a series of tests. In establishing these statistical parameters, both the material and geometric properties of the test beams must be determined in order that the test-to-predicted ratios reflect only the variability of the test strength relative to the design equation. Ancillary tests and measurements on representative samples are therefore conducted to determine the relevant geometric and material properties. See also Technical Memorandum No. 10 for statistical evaluation of test data.

B.9.2 Specimen

Provide a sufficient additional length of material from each different heat for the ancillary tests. The main flexural specimens vary in length from whatever length is required for lateral buckling tests to that for those tests with a constant moment region of length sufficient for all the instrumentation but not greater than $1.2r_y/\sqrt{F_Y/E}$, for example, for tests to determine M_p. For such strength tests, a length of two times the beam depth is recommended for the shear span.

B.9.3 Geometric Properties

Establish the width, depth, flange thickness, and web thickness from a sufficient number of measurements of each taken along the length of the beam to be statistically representative of the beam. Thirty measurements of each are recommended. The contribution of fillets to the cross-sectional properties, A, S, Z, I, and J, should be included as it may be significant. For lateral-torsional buckling tests especially, establish the sweep, camber, out-of-parallel of flanges, and out-of-straightness of web and compare to the manufacturing or rolling tolerances. The latter two are especially important for welded three-plate shapes.

B.9.4 Material Properties

Conduct residual stress measurements and tension coupon tests on sections taken longitudinally from the entire cross section in accordance with Technical Memoranda No. 6 and No. 7, respectively. A stub-column test curve, Technical Memorandum No. 3, can provide supplementary information on the overall behavior in compression, the maximum compressive residual stress, and the local buckling behavior.

B.9.5 Boundary Conditions

Because any and all unforeseen restraints increase both the stiffness and flexural strength of a beam, it is absolutely imperative that no restraints be introduced inadvertently. Boundary conditions must be known. Friction must be reduced to an absolute minimum. The devices shown here for attaining certain boundary and loading conditions are examples of what is considered to be good practice. They are in no way meant to preclude other designs that achieve comparable results. The objective is to have highly predictable conditions so that the loads and reactions, as vector quantities, are known both in magnitude and direction. It cannot be overemphasized that unwanted frictional forces leave the value of test results in serious doubt.

Figure B.26 shows the set-up for a typical flexural test, with two-point loading and simple boundary conditions at all load points and reactions, to determine the cross-sectional strength of a beam. For simple supports, when the coordinate axes are: x out-of-plane, y in the vertical direction, and z along the length of the member, translation in the z-direction and rotation about the x-axis must be allowed. The knife edge assemblies, placed adjacent to the specimen, allow the specimen to rotate about the x-axis as it deflects under load. The transverse rollers allow translation in the z-direction and, reacting against horizontal surfaces, ensure that the reactions

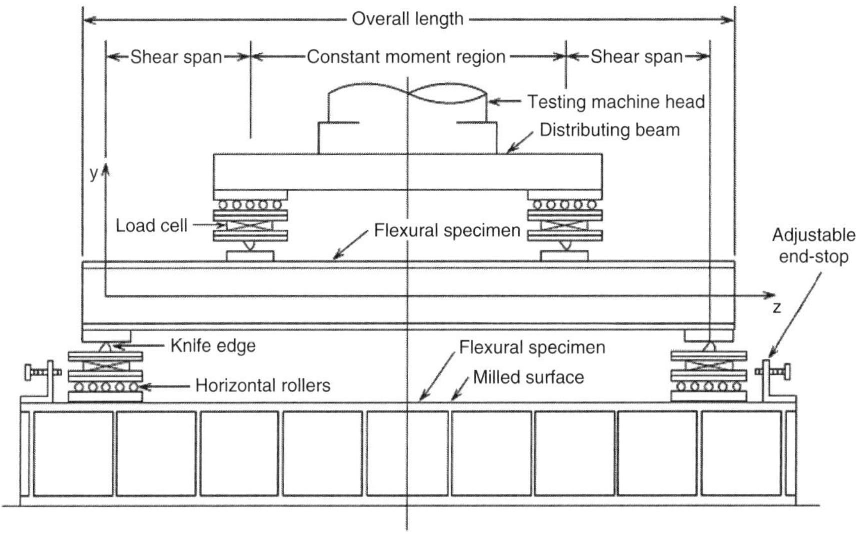

FIGURE B.26 Flexural test set-up to determine cross-sectional strength.

and loads are vertical. The four sets of rollers are needed to keep the loading symmetric as the specimen deforms. Notice, that with no longitudinal restraint, the specimen is in neutral equilibrium. Adjustable stops should therefore be provided to keep the specimen centered but not restrained longitudinally. These are adjusted during the course of the test. As the specimen deflects, the load points move inward and the reactions outward changing the length of the shear span, which therefore must be measured.

Figure B.27 is a schematic diagram of a knife edge-roller assembly with a capacity of 3000 kN (675 kips) that is designed for large rotations. The design of knife edges and rollers depends on the magnitude of the contact stresses (Boresi and Schmidt, 2002). Air-hardening chromoloy A2 steel is excellent for these because it distorts little on hardening and provides surface yield strengths in excess of 1800 MPa (260 ksi). Serrated knife edges and matching sloped serrated plates, as shown in Fig. B.27 (Brattland and Kennedy, 1992), acting like a rack and pinion system, can accommodate rotations of 30° or more. The slope is set at half the maximum anticipated rotation. For relatively small rotations, horizontal plate surfaces without serrations may be used, but, without serrations, the angle of rotation is limited by the frictional force that can be developed between the knife edge and the adjacent plate.

In lateral-torsional buckling tests, the displacement in the vertical or y direction, which is the displacement of the load or reaction, is imposed. Any of the other 5 degrees of freedom, Δx, Δz, θx, θy, or θz, may or may not be constrained as desired. In these tests, translations in the z and x directions are allowed by two perpendicular sets of rollers as illustrated subsequently, as are the other devices. Rotations about x and z axes are provided for by two orthogonal knife edges or

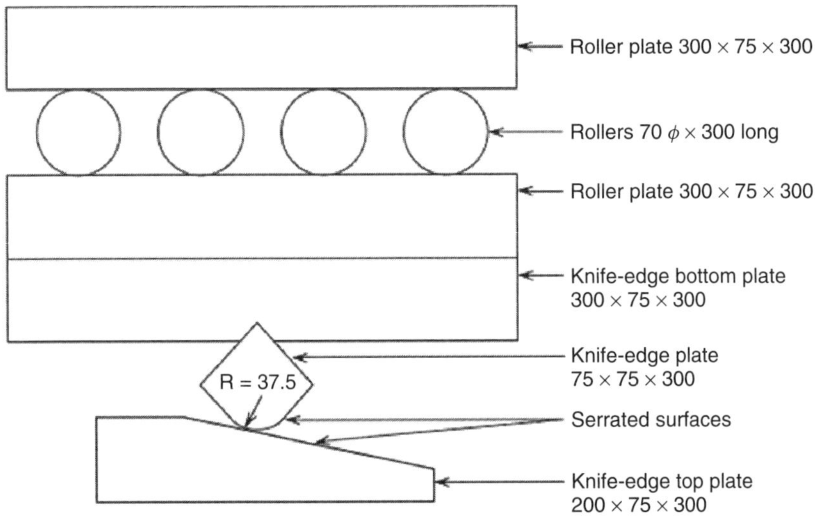

Notes: 1. Steel is air-hardening chromoloy A2. 2. Dimensions in millimeters

FIGURE B.27 Knife-edge roller assembly.

semi cylindrical rockers. A hemispherical rocker can be considered to provide three rotational degrees of freedom as the torsional restraint about a vertical axis due to friction is limited by the size of the contact area. The small bearing area, however, limits the loads that can be carried. Hence, when rotation about the vertical or y-axis is to be allowed, a thrust bearing should preferably be used. Thrust bearings are commerically available. Depending on their capacity, they are about 9 to 60 mm high with a circular race of ball bearings at mid-height that can transfer compressive forces but allow rotation about the y-axis.

Figure B.28 shows an example of a test set-up for a lateral-torsional buckling test of a beam with a single overhang and with loads applied at the cantilever tip and at five locations on the main or back span.

Figure B.29 shows a loading frame that ensures that the load is applied vertically even though the beam specimen displaces laterally during the test. The inner frame is constrained to move vertically, even when it is loaded eccentrically by the displaced beam, by the rollers fastened to it at its corners and reacting against the outer frame. The horizontal forces of the rollers at opposite corners form a couple acting on the inner frame equal and opposite to that formed by the vertical load of the loading rods and the reaction of the displaced beam acting on the inner frame. Only vertical loads can be applied.

Figure B.30 shows different combinations of load devices and the corresponding cross-sectional displacements. The surfaces of roller and knife-edge assemblies must be kept clean as any dirt may introduce unwanted frictional forces. Typical reaction systems and the corresponding cross-sectional displacements or distortions are shown in Fig. B.31. The four rounded noses in Fig. B.31 have a radius of about

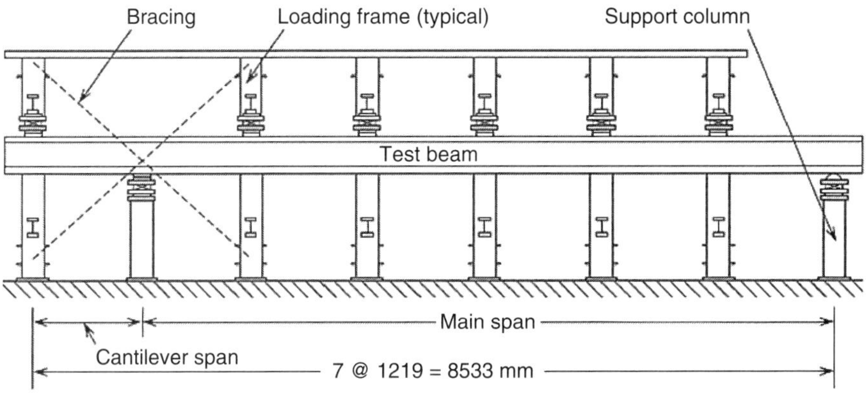

FIGURE B.28 Set-up for lateral-torsional buckling tests (section on longitudinal centerline) (B.25).

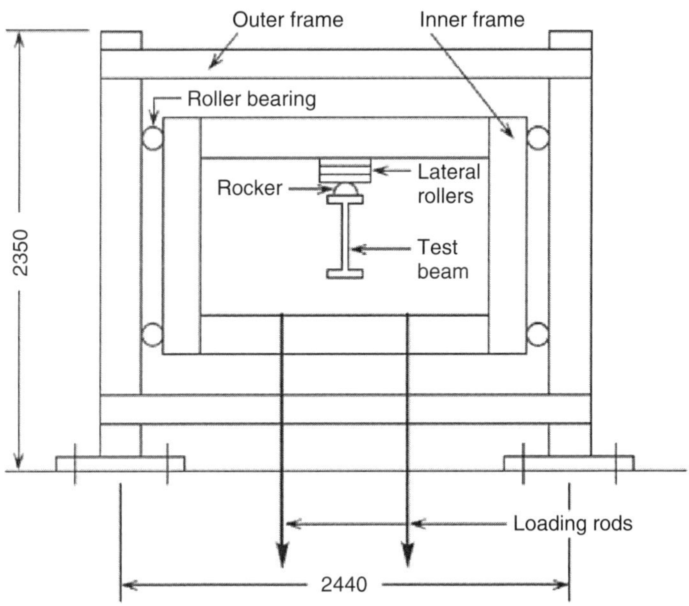

FIGURE B.29 Loading frame to apply vertical loads to beams that displace laterally.

20 mm and simply bear lightly against the web. They therefore prevent the web from moving laterally and provide a forked support. Any tangential frictional forces developed between the web and noses, should relative longitudinal movement be impending, are only a fraction of the bearing forces and are considered to be negligibly small. In Fig. B.31*b*, the width of the bearing, load cells and roller plates must be sufficient to prevent uplift as the beam cross section distorts.

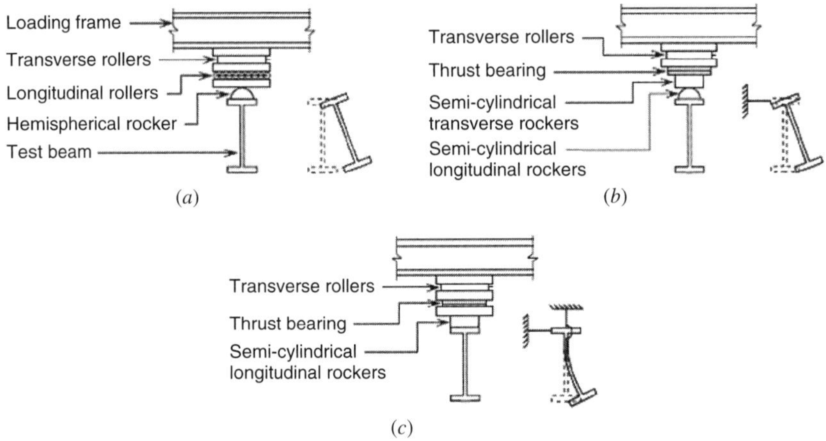

FIGURE B.30 Loaded flange restraint conditions: (*a*) free to rotate and displace laterally; (*b*) free to rotate; (*c*) fixed against rotation about the longitudinal axis and lateral translation.

B.9.6 Test Measurements

Loads and Reactions By measuring all loads and reactions using calibrated load cells, as shown in Fig. B.26, statics is then available to confirm the calibrations, to check the progress of the test, and to monitor any frictional losses. In no case should only the loads or the reactions be measured, as errors in calibration or frictional losses would then remain undetected, and increase the experimental error. In Fig. B.27, the load applied to the loading frame and hence to the test beam is measured by a load cell. By also measuring the two reactions (see Fig. B.28) any losses in the roller bearings can be deduced. Note as well, by statics, that the magnitudes of the reactions can be checked against the loads during the course of the test and problems detected before it is too late.

Displacements and Distortions Consideration should be given to making some redundant measurements. Cable, linearly variable differential transformers, LVDTs (Sabnis et al., 1985) or potentiometers are ideal for measuring displacements and can be used to measure large displacements with excellent sensitivity. Readings can be recorded electronically and very quickly. The number of LVDTs required depends on the displacements to be determined and whether the section distorts or not. For example, two LVDTs attached horizontally to the flange tips give the horizontal centroidal displacement provided the section does not distort. With six LVDTs at one cross section, two attached horizontally and four attached vertically to the flange tips, web distortions can be calculated. Three dimensional displacement can also be measured by triangulation using two or more high precision electronic theodolites that sight on a mylar target mounted on the test specimen. A precision of 0.05 mm (0.002 in.) can be obtained regardless of the magnitude of the

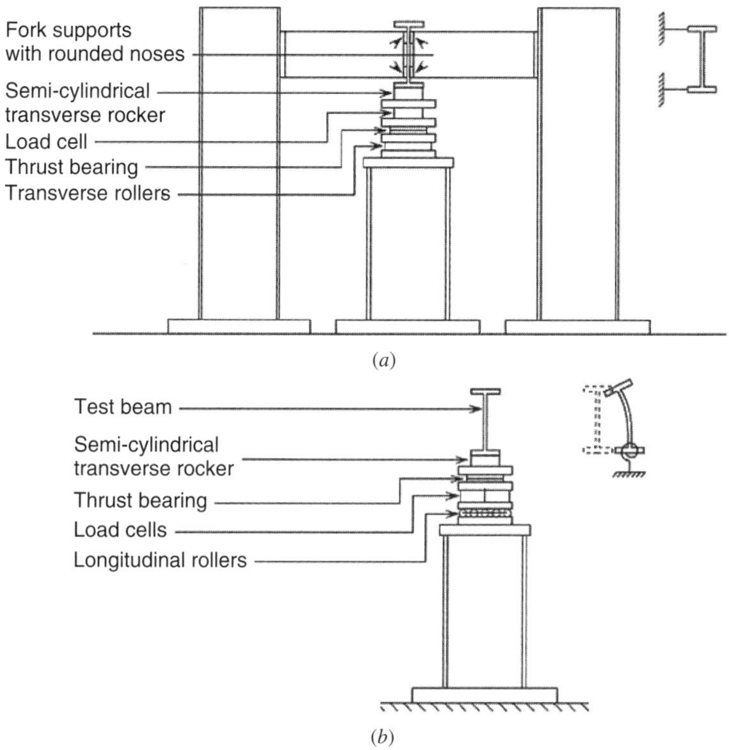

FIGURE B.31 Reaction restraint conditions: (*a*) beam fixed against lateral displacement and rotation; (*b*) bottom flange fixed against rotation, beam free to displace laterally.

displacement. This system is particularly useful for tests with large displacements and rotations. Dial gages are also useful for monitoring deflections during tests.

Strains Electrical resistance strain gages may be used to establish:

1. Strain distributions across the depth, and hence, the curvature
2. Strain distributions across the flange due to warping
3. Lateral bending strains in the web due to web distortion
4. Variations in strains along the beam

B.9.7 Test Procedure

Tests not taken to failure are of little or no value in establishing the behavior or strength of flexural members. A general procedure follows:

1. Set up reaction devices, load cells, and align the flexural specimen in the test frame.

2. Position loading devices with load cells.
3. Install and test instrumentation such as LVDTs, potentiometers, strain gages, curvature meters, and monitoring instrumentation, take measurements to establish position of loads, reactions, and any restraining devices.
4. "Seat" the test specimen by applying about 20% of the anticipated maximum load while staying in the elastic range, perform statics checks, unload, and set instruments to zero. Seating of the specimen needs to be done with great care and it may be necessary to repeat this step before proceeding with the test. Also, after unloading, some minimal load needs to be maintained, i.e. less than 0.1% of the anticipated maximum load. This will keep the specimen from inadvertently moving longitudinally or laterally on the support reactions.
5. Load monotonically to failure with increments determined from a load deflection curve plotted as the test progresses. Use displacement control for load increments as failure is approached when the stiffness of the system may be diminishing. Keep a test log book to record general observations in addition to the data recorded electronically. The book is invaluable later in establishing what happened and when. At each increment: (a) record measurements, (b) perform statics checks, (c) make visual observations, (d) take photographs as necessary, and (e) record general behavior and any unusual occurrences. Make sketches in the log book.
6. Take residual measurements, notes and photographs as appropriate at the end of the test.

B.9.8 Test Report and Analyses

The test report should describe the test set-up, procedures and specimens in sufficient detail, so that other researchers can check the results independently. All relevant material properties, geometric properties, boundary conditions, loading, and deflections must therefore be presented. The analysis of the results should include a complete analysis of the geometric and material data to establish predicted moment resistances.

For tests to determine cross-sectional strengths, as a minimum, the moment-curvature relationship should be reported with an explanation of the effect of residual stresses, a description of overall behavior, the mode of failure, error analyses, and the ratio of test flexural strength to that predicted for the cross section. After local buckling, curvatures are difficult to define and angle changes are considered to be a more appropriate measure.

For lateral-torsional buckling tests, the analysis may be very extensive but should include a description of the overall behavior, test-to-predicted moment ratios, modes of failure, buckling shapes, cross-sectional distortions, error analyses, strain variations, and residual deformations.

The most important factor to establish in any test is the ratio of the test-to-predicted moment, the rationale, if it is the case, why it differs substantially from 1.0, and for a series of tests, the coefficient of variation of the test-to-predicted moment ratio for the series.

B.10 TECHNICAL MEMORANDUM NO. 10: STATISTICAL EVALUATION OF TEST DATA FOR LIMIT STATES DESIGN

B.10.1 Introduction

The philosophy of limit states design (load and resistance factor design) is represented by the equation

$$\phi R \geq \sum \alpha_i S_i \tag{B.9}$$

that states the factored resistance (design strength) shall be equal to or greater than the effect of the factored loads. Resistance factors, ϕ, and load factors, α_i, are chosen to provide selected small probabilities of failure. Although Eq. B.9 indicates that the resistance factors and the load factors are related to each other, Galambos and Ravindra (1973, 1977, 1978) show that the resistance factor can be approximately expressed as

$$\phi = \rho_R e^{-\beta \alpha_R V_R} \tag{B.10}$$

where ρ_R = the bias coefficient for the resistance, i.e., the mean value of the ratio of the measured resistance to the nominal resistance

= the product of $\rho_G \rho_M \rho_P$, where in turn

ρ_G = the bias coefficient for the cross-sectional geometry, i.e., the mean value of the ratio of the relevant geometric property, such as *A, b, d, t, I, S, Z,* C_w, to the nominal value as given in steel design handbooks

ρ_M = the bias coefficient for the material property, i.e., the mean value of the ratio of the relevant material property, such as F_y, F_u, E to the nominal value as given in published steel material specifications

ρ_P = the bias coefficient for the design equation, i.e., the mean value of the ratio of the measured or test strength to that predicted by a design equation such as $M_p = F_y Z$

V_R = the coefficient of variation associated with ρ_R

= $\sqrt{V_G^2 + V_M^2 + V_P^2}$, when the quantities are not correlated, and V_G, V_M, and V_P are the coefficients of variation of the geometric properties, material properties, and test-to-predicted strength ratios

β = reliability index, currently taken to have a target value of about 3.0 for structural steel members in buildings and 4.5 for structural connections

α_R = the coefficient of separation taken as 0.55 (Galambos and Ravindra, 1973) (not to be confused with the load factors α_i)

Because of the interdependence of the resistance factor, ϕ, and the load factors, α_i, when load factors are established for building structures as a whole based on a target value of 3.0 for β, an adjustment factor less than one must be applied to Eq. B.10 when calculating resistance factors for a target value for β of 4.5 for connections. Fisher et al. (1978) show that this factor varied only from 0.86 to 0.90

for a broad range of live/dead load ratios; therefore a value of 0.88 is considered appropriate. Hence for members, from Eq. B.10 with $\beta = 3.0$ and $\alpha_R = 0.55$,

$$\phi = \rho_R e^{-1.65V_R} \tag{B.11}$$

and for connections and their connectors from Eq. B.10 with $\beta = 4.5$, $\alpha_R = 0.55$ and the adjustment factor of 0.88,

$$\phi = 0.88\rho_R e^{-2.47V_R} \tag{B.12}$$

Therefore, for test data to be used statistically to determine resistance factors for limit states design, all the statistical parameters, ρ_G, ρ_M, and ρ_p must be determined or be available. It follows that, in tests, to determine how well an equation predicts the behavior of a component, i.e. the fit of the equation to the test results or the test-to-predicted ratio, the results must be based on measured geometric and material properties and not on nominal values. Thus the test-to-predicted ratio does not include any of the variation in the geometric and material properties. Test strengths predicted using nominal properties are of no value.

Also, incomplete reporting of the statistical parameters renders the test data virtually useless. Furthermore, as the sample size must be statistically significant, the work of several researchers will likely need to be combined. This emphasizes the importance of complete and accurate reporting of the test data. Raw data in the form of measurements of geometric and material properties and test strengths should be available so that statistical analyses can be carried out independently. The test must be described in detail so that, for example, the effects of boundary conditions on the behavior can be assessed.

When Eurocode EN 1990 was being developed guidelines were provided in Section 5.2 and Annex D for design assisted by testing. Each additional part of Eurocode (e.g. Eurocode 3 for steel structures, EN 1993-1-1) has a chapter on design assisted by testing. Therefore, standardized in this code is a methodology to perform statistical analysis.

B.10.2 Reporting Test Data

In addition to the general data reported such as the objectives and scope of the experimental program, the data should include:

1. Description of the test set-up and instrumentation (sensitivity and range).
2. Load-deflection curves and displaced shapes that illustrate the behavior.
3. Description of the failure and the failure criteria used.
4. Sufficient data so that calculations and statistical analyses can be reproduced. Thus the basic geometry, initial imperfections (if relevant), nominal and measured geometric and material properties and test-to-predicted ratios are required, and an error analysis.

For unique tests, possibly with elaborate specimens, boundary and load conditions, some simpler tests on specimens of known or readily predictable behavior need to be made to verify test procedures and the performance of the test equipment.

B.10.3 Geometric and Material Properties

Many data on the statistical variation of geometric and material properties of steel members are available in the literature, Galambos and Ravindra (1978), Kennedy and Gad Aly (1980), Baker and Kennedy (1984), Kennedy and Baker (1984), Kennedy et al. (1992), and Rang et al. (1978). Care must be taken that the data are appropriate to the circumstances at hand. Various national standards may have quite different requirements. Changes in manufacturing and rolling practices and in national standards may mean that some data are no longer applicable.

The bias coefficient, the coefficient of variation, and the type of distribution are required for both the geometric and material properties.

Chernenko and Kennedy (1991) give procedures for calculating statistical data for cross-sectional properties, such as for the area and moment of inertia when the variations in the plate width and thickness are known. Kennedy and Gad Aly (1980) introduce participation factors to handle the case when, for example, in determining resistance factors for columns, the strength is governed by one material parameter (the yield strength) at low slenderness ratios and another (the modulus of elasticity) governs at high slenderness ratios.

B.10.4 Ratio of Test-to-Predicted Strengths

Physical Tests To compute the test-to-predicted ratio for any test, the measured geometric and material properties of the test specimen are used in the design equation without a resistance factor to predict the strength. The measure of the fit of the design equation to the test results is given by the distribution of the test-to-predicted ratio as characterized by the bias coefficient, ρ_P, and the coefficient of variation, V_P.

Computer-Simulated Tests Computer simulated tests, using finite element models that can model both material and geometric nonlinearities, offer several advantages. They are quick, inexpensive, and there are no physical limitations on the size of the model that can be "tested." They are, however, only as good as the assumptions inherent in the models and, therefore, they themselves must be verified by actual physical tests on real specimens. This comparison should demonstrate correspondence not only of maximum loads but also of the load-deflection behavior. Computation of the test-to-predicted bias coefficient is a two-step multiplicative process. First, physical tests are used to verify the computer model, and then the computer model is used to verify the design equation. This results in one bias coefficient for the ratio of the physical test strength to the computer simulated strength, which is multiplied by the ratio of the computer simulated strength to the strength predicted by the design equation. The coefficients of variation are combined as the square root of the sum of the squares. In some instances, when the computer simulation itself is used to predict the design strength, as no relatively simple model exists, the ratio of the physical test strength to the computer simulated strength gives the test-to-predicted ratio directly.

When dealing with computer simulated tests, Chernenko and Kennedy (1991) proposed a sequential method for considering the effects of out-of-straightness and residual stresses on column strengths that avoids the conservative approach of combining the worst residual stress effects with the maximum out-of-straightness effects. Test data are required on the distribution of out-of-straightness and some measure of residual stress effects. The bias coefficient of the test-to-predicted ratio is determined at the mean value of each of these two effects with a coefficient of variation for each effect in turn. The overall bias coefficient is then the product of all the contributing coefficients while the coefficients of variation are combined as the square root of the sum of the squares.

B.10.5 Other Factors

Discretization Factor, D Steel shapes and plates are rolled in discrete sizes. The structural engineer tends to, or should, select sizes with a resistance equal to or greater than that required. (Not doing so encroaches upon the margin of safety.) Thus the value of the bias coefficient of the resistance supplied to that required, ρ_D, is greater than 1.0. The corresponding coefficient of variation for the data set, V_D, can also be calculated. This allows the effect of discretization to be included in the calculation of the resistance factor as another distribution.

Errors in Measurement, E_m Errors in measurement may be random, systematic or blunders. Blunders are generally obvious and do not require statistical examination to be rejected. Systematic errors can be averted through proper calibration, redundant measurements, checks and due care in testing. For example, when the equations of statics are satisfied we are confident that the loads have been determined without systematic error. Chauvenet's criterion may be applied for the rejection of outliers of a data set that may have resulted from errors in measurement. The random errors that remain are as likely to be positive as negative, resulting in a mean value of the bias coefficient, ρ_{Em}, of 1.00. The coefficient of variation of the errors in measurement, V_{Em}, will not be zero, and will have typical values of about 0.02 to 0.04. This inflates the coefficient of variation of the measured quantity and may be accounted for in the calculations if an error analysis is performed.

Monte Carlo Simulation For cases when the probability distributions to be combined are of different types or for cases when the data are limited, and provided that the general types of distribution and statistical parameters, ρ and V are known, a Monte Carlo simulation may be used to create the combined distribution for the resistance of the component. This will allow for the calculation of ρ_R and V_R. Care should be taken to ensure that the simulation process can be repeated. Note that the chi-square test or the Kolmogorov–Smirnov method

TABLE B.2 Statistical Data for Computing Resistance Factors

Parameter	Number of Tests or Simulations	Bias Coefficient, ρ	Coefficient of Variation, V	Ref.
Geometric factor				
Plastic section modulus	>352	0.99	0.038	(B.30)
Elastic section modulus	>352	0.99	0.021	
Material factor				
Elastic modulus	341	1.036	0.045	(B.28)
Yield strength	4796	1.060	0.051	(B.30)
Test/predicted ratio				
Finite element analysis	31	0.990	0.060	(B.38)
Design charts	257	1.016	0.013	(B.37)

can be used to verify the fit of a given distribution to the data set (Kennedy and Neville, 1986; Ang and Tang, 1975).

Example B.1 This example of the calculation of resistance factors is based on the experimental data given in Essa and Kennedy (1995) for 31 tests to determine the distortional buckling strength of steel beams under a broad range of loading and restraint conditions. The results of two other tests have been excluded because unwanted frictional restraints caused higher energy buckling modes and therefore test strengths greater than expected. The relevant statistical data are given in Table B.2, noting that data for the modulus of elasticity is a re-evaluation of the data reported by Galambos and Ravindra (1978).

Because the failure strengths for distortional buckling may range all the way from the fully plastic cross-sectional strength to an elastic buckling strength, material data are given for both the yield strength and the modulus of elasticity. Statistical data for two different test-to-predicted ratios are also given, including (a) the data determined from an inelastic finite element analysis that considers the effect of residual stresses and the distortion of the web (Albert et al., 1992) and (b) as an extension of the finite element analysis, the data from design charts proposed by Essa and Kennedy (1995). The coefficient of variation for the test-to-predicted ratio for the 31 tests has been adjusted using a coefficient of variation for errors in measurement, V_{Em}, equal to the conservative value of 0.02. It is noted in passing that the statistical data for these tests reflect all the variations arising from the tests themselves as well as the simplifications in the finite element model. The resistance factors are calculated for the four combinations of failure modes and methods of predicting the strength. In the interest of simplicity, the discretization factor has not been considered.

1. *Fully plastic strength and finite element analysis.* Employing Eq. B.11, $\phi = \rho_R e^{-1.65 V_R}$, with

$$\rho_R = \rho_G \rho_M \rho_P = \rho_Z \rho_{F_y} \rho_{FEM} = 0.99 \times 1.060 \times 0.99 = 1.039$$

and

$$V_R^2 = V_Z^2 + V_{Fy}^2 + V_{FEM}^2 = 0.038^2 + 0.051^2 + 0.060^2 = 0.088^2$$

results in

$$\phi = 1.039e^{-1.65\times 0.088} = 0.899, \text{ say } \phi = 0.90$$

2. *Fully plastic strength and design charts.* Here, there is an additional term for the test-to-predicted ratio as the procedure to get the test-to-predicted ratio has two steps. For the series of 31 physical tests, the ratio of each test strength to that determined from the finite element analysis was found and hence, the mean value and the coefficient of variation of this ratio are known. Next, many finite element analysis results for situations that had not been physically tested were compared to the design charts. The latter, as derived, contain some approximations to make them generally applicable. This introduces a second set of variations. Thus,

$$\rho_R = \rho_G \rho_M \rho_{P1} \rho_{P2} = \rho_Z \rho_{F_y} \rho_{FEM} \rho_{DC} = 0.99 \times 1.060 \times 0.99 \times 1.016$$
$$= 1.056$$

and

$$V_R^2 = V_Z^2 + V_{Fy}^2 + V_{FEM}^2 + V_{DC}^2 = 0.038^2 + 0.051^2 + 0.060^2 + 0.013^2$$
$$= 0.089^2$$

which results in

$$\phi = 1.056e^{-1.65\times 0.089} = 0.912, \text{ say } \phi = 0.90$$

3. *Elastic buckling and finite element analysis.* This differs from the first calculation because the geometric property is taken to be the elastic section modulus and the material factor is taken to be the modulus of elasticity. There are other geometric factors that are also relevant such as moments of inertia, but it is tacitly assumed that their statistical distributions are not dissimilar to that of the elastic section modulus. Thus,

$$\rho_R = \rho_G \rho_M \rho_P = \rho_S \rho_E \rho_{FEM} = 0.99 \times 1.036 \times 0.99 = 1.015$$

and

$$V_R^2 = V_S^2 + V_E^2 + V_{FEM}^2 = 0.021^2 + 0.045^2 + 0.060^2 = 0.078^2$$

providing

$$\phi = 1.015e^{-1.65\times 0.078} = 0.892, \text{ say } \phi = 0.90$$

4. *Elastic buckling and design charts.* As in the second calculation, there is an additional term for the test-to-predicted ratio. Thus,

$$\rho_R = \rho_G \rho_M \rho_{P1} \rho_{P2} = \rho_S \rho_E \rho_{FEM} \rho_{DC} = 0.99 \times 1.036 \times 0.99 \times 1.016$$
$$= 1.031$$

and

$$V_R^2 = V_S^2 + V_E^2 + V_{FEM}^2 + V_{DC}^2 = 0.038^2 + 0.045^2 + 0.060^2 + 0.013^2$$
$$= 0.085^2$$

resulting in

$$\phi = 1.013e^{-1.65 \times 0.085} = 0.895, \text{ say } \phi = 0.90$$

A resistance factor of 0.90 for the four cases studied appears to be appropriate, especially, if it was considered desirable in the interests of simplicity to have only one value. Using this value and by rearranging Eq. B.10, a reliability index, β, can be obtained

$$\beta = \frac{-\log_e(\phi/\rho_R)}{\alpha_R V_R} = \frac{-\log_e(0.99/\rho_R)}{0.55 V_R} \tag{B.13}$$

The values of β for the four cases are computed to be 2.97, 3.26, 2.80, and 2.91, respectively, which are grouped relatively closely to the target value of 3.0.

REFERENCES

Albert, C., Essa, H. S., and Kennedy, D. J. L. (1992), "Distortional Buckling of Steel Beams in Cantilever Suspended Span Construction," *Can. J. Civ. Eng.s*, Vol. 19, pp. 767–780.

Ang, A., and Tang, W. H. (1975), *Probability Concepts of Engineering Planning and Design*, Vol. I: *Basic Principles*, Wiley, New York.

ASTM (1982), *Annual Book of ASTM Standards: Part 10: Metals*, American Society for Testing and Materials, Philadelphia, PA.

Baker, K. A., and Kennedy, D. J. L. (1984), "Resistance Factors for Laterally Unsupported Steel Beams and Biaxially Loaded Steel Beam-Columns," *Can. J. Civ. Eng.*, Vol. 11, pp. 1008–1019.

Beedle, L. S., and Tall, L. (1960), "Basic Column Strength," *ASCE J. Struct. Div.*, Vol. 86, No. ST7, pp. 138–178.

Boresi, A. P., and Schmidt, R. J. (2002), *Advanced Mechanics of Materials*, 6th ed., Wiley, New York.

Brattland, A., and Kennedy, D. J. L. (1992), "Flexural Tests of Two Full-Scale Composite Trusses," *Can. J. Civ. Eng.*, Vol. 19, pp. 279–295.

Castex, L., and Maso, J. C. (1975), "Études des contraites résiduelles, diffracto-metriex, dans des profilés laminés d'acier doux," *Constr. Met.* (CTICM, France), No. 2.

CEN (2001a), Eurocode–Basis of Structural Design, EN 1990, Comite Europeen de Normalisation (CEN) European Committee for Standardization, Brussels, Belgium.

CEN (2001b), Metallic Materials–Tensile Testing–Part 1: Method of Test at Ambient Temperature, EN 10002-1 Comite Europeen de Normalisation (CEN), European Committee for Standardization, Brussels, Belgium.

CEN (2004), Hot Rolled Products of Structural Steels– Part 1:General Technical Delivery Donditions; Part 2: Technical Delivery Conditions for Non-Alloy Structural Steels; Part 3: Technical Delivery Conditions for Normalized/Normalized Rolled Weldable Fine Grain Structural Steels; Part 4: Technical Delivery Conditions for Thermomechanical Rolled Weldable Fine Grain Structural Steels; Part 5: Technical Delivery Conditions for Structural Steels with Improved Atmospheric Corrosion Resistance; Part 6: Technical Delivery Conditions for Flat Products of High Yield Strength Structural Steels in the Quenched and Tempered Condition, EN 10025-1 thru 6, respectively, Comite Europeen de Normalisation (CEN), European Committee for Standardization, Brussels, Belgium.

Chernenko, D. E., and Kennedy, D. J. L. (1991), "An Analysis of the Performance of Welded Wide Flange Columns," *Can. J. Civ. Eng.*, Vol. 18, pp. 537–555.

Drescher-Krasicka, E., and Ostertag, C. P. (1999), "Residual Stress Measurements in Welded Steel Beam Column Connections by Scanning Acoustic Microscopy," *J. Mater. Sci.*, Vol. 34, pp. 4173–4179.

Essa, H. S., and Kennedy, D. J. L. (1995), "Design of Steel Beams in Cantilever-Suspended Span Construction," *ASCE J. Struct. Eng.*, Vol. 121, No. 11. pp. 1667–1673.

Estuar, F. R., and Tall, L. (1967), "Testing Pinned-End Steel Columns," in *Test Methods for Compression Members*, ASTM STP419, American Society for Testing and Materials, Philadelphia, PA.

Fisher, J. W., Galambos, T. V., Kulak, G. L., and Ravindra, M. K. (1978), "Load and Resistance Factor Design Criteria for Connectors," *ASCE J. Struct. Eng.*, Vol. 104, No. ST9, pp. 1427–1441.

Galambos, T. V., and Ravindra, M. K. (1973), "Tentative Load and Resistance Factor Design Criteria for Steel Buildings," Res. Rep. 18, Civil Engineering Department, Washington University, St. Louis, MO.

Galambos, T. V., and Ravindra, M. K. (1977), "The Basis for Load and Resistance Factor Design Criteria for Steel Building Structures," *Can. J. Civ. Eng.*, Vol. 4. pp. 178–189.

Galambos, T. V., and Ravindra, M. K. (1978), "Properties of Steel for Use in LRFD," *ASCE J. Struct. Eng.*, Vol. 104, No. ST9, pp. 1459–1468.

Gozum, A. T., and Huber, A. W. (1955), "Material Properties, Residual Stresses and Column Strength," Fritz Eng. Lab. Rep. No. 220.14, Lehigh University, Bethlehem, PA.

Hoffmann F. (1987), "Eine Einführung in die Technik des Messens mit Dehnungsmeβstreifen," HBM GmbH, Darmstadt, Germany.

Huber, A. W. (1958), "Fixtures for Testing Pin-Ended Columns", *ASTM Bull.*, No. 234, Dec.

International Organization for Standardization (ISO) (1997), *Steel and Steel Products—Location and Preparation of Samples and Test Pieces for Mechanical Testing*, ISO 377, ISO, Geneva.

Johnston, B. G., and Mount, E. H. (1939), "Designing Welded Frames for Continuity," *Weld. Res. J.*, Vol. 18, pp. 355–374.

Kennedy, D. J. L., and Baker, K. A. (1984), "Resistance Factors for Steel Highway Bridges," *Can. J. Civ. Eng.*, Vol. 11, pp. 324–334.

Kennedy, D. J. L., and Gad Aly, M. M. (1980), "Limit States Design of Steel Structures: Performance Factors," *Can. J. Civ. Eng.*, Vol. 7, pp. 45–77.

Kennedy, D. J. L., Gagnon, D. P., Allen, D. E., and MacGregor, J. G. (1992), "Canadian Highway Bridge Evaluation: Load and Resistance Factors," *Can. J. Civ. Eng.*, Vol. 19, pp. 992–1006.

Kennedy, J. B., and Neville, A. M. (1986), *Basic Statistical Methods for Engineers and Scientists*, 3rd ed., Harper & Row, New York.

Luxion, W., and Johnson, B. G. (1948), "Plastic Behavior of Wide Flange Beams," *Weld. J.*, Vol. 27, p. 538s.

Nadai, A. (1950), *Theory of Flow and Fracture of Solids*, McGraw-Hill, New York.

Nagaraja Rao, N. R., Lohrmann, M., and Tall, L. (1966), "Effects of Strain Rate on the Yield Stress of Structural Steels," *ASTM J. Mater.*, Vol. 1, No. 1, pp. 241–262.

Niku-Lari, A. (1986), *Advances in Surface Treatments*, Vol. 4: *International Guidebook on Residual Stresses*, Pergamon, Oxford, U.K..

Noyan, I. C., and Cohen, J. B. (1987), *Residual Stress, Measurement by Diffraction and Interpretation*, Springer-Verlag, New York.

Randle, N. J., and Vigness, F. (1966), "Hole-Drilling Strain Gage Method of Measuring Residual Stresses," *Exp. Mech.*, Vol. 6, pp. 577–586.

Rang, T. N., Galambos, T. V., Yu, W.-W., and Ravindra, M. K. (1978), "Load and Resistance Factor Design of Cold-formed Steel Structural Members," *Proc. 4th Int. Spec. Conf. Cold-formed Steel Struct.*, University of Missouri–Rolla, Rolla, MO, June.

Roloff, S. J. (1984), "Stress–Strain Properties of Fabricated Steel Pipe," *Proc. Annu. Techn. Session SSRC*, Bethlehem, PA.

Sabnis, G., Harris, H. G., White, R. N., and Mirza, M. S. (1985), *Structural Modeling and Experimental Techniques*, Prentice Hall, Upper Saddle River, NJ.

Salmon, E. H. (1931), *Materials and Structures*, Longmans, London, England.

Sherman, D. R. (1969), "Residual Stress Measurement in Tubular Members," *ASCE J. Struct. Div.*, Vol. 95, No. ST4, pp. 635–648.

Soete, W., and Vancrombrugge, A. (1955), "An Industrial Method for the Determination of Residual Stresses," *Proc. Soc. Exp. Stress Anal.*, Vol. 8, No. 1, pp. 17–26.

Tall, L. (Ed.) (1974), *Structural Steel Design*, Ronald Press, New York.

Tall, L., and Alpsten, G. A. (1969), "On the Scatter in Yield Strength and Residual Stresses in Steel Members," *Proc. IABSE Symp. Concepts Safety Struct. Methods Des.*, London, England.

Tebedge, N., Alpsten, G., and Tall, L. (1972), "Measurement of Residual Stresses: A Comparative Study of Methods," *Proc. Conf. Rec. Interpret. Eng. Meas.*, Institute of Marine Engineers, London, England, Apr. 5 (also available as Fritz Eng. Lab. Rep. No. 337.8, Lehigh University, Bethlehem, PA, Feb. 1971).

Tebedge, N., Alpsten, G., and Tall, L. (1973), "Residual Stress Measurement by the Sectioning Method," *Exp. Mech.*, Vol. 13, pp. 88–96.

Tebedge, N., Marek, P., and Tall, L. (1971), "On Testing Methods for Heavy Columns," Fritz Eng. Lab. Rep. No. 351.4, Lehigh University, Bethlehem, PA, Mar.

Yarimci, E., Yura, J. A., and Lu, L. W. (1968). "Rotation Gages for Structural Research," *Exp. Mech.*, Vol. 8, No. 11, pp. 525–526.

Zhou, F., and Young, B. (2007), "Experimental and Numerical Investigations of Cold-Formed Stainless Steel Tubular Sections Subjected to Concentrated Loading," *J. Constr. Steel Res.*, Vol. 63, pp. 1452–1466.

APPENDIX C

STRUCTURAL STABILITY RESEARCH COUNCIL

The following is a summary of pertinent information about the SSRC.

PURPOSE

The general purposes of the Structural Stability Research Council are:

1. To maintain a forum where the structural stability aspects of metal and composite metal-and-concrete structures and their components can be presented for evaluation and pertinent structural research problems proposed for investigation.
2. To review the world's literature on structural stability of metal and composite metal-and-concrete structures and study the properties of materials available for their construction and to make the results widely available to the engineering profession.
3. To organize, administer, and guide cooperative research projects in the field of structural stability and to solicit financial support for such projects.
4. To promote publication and dissemination of research projects in the field of structural stability.
5. To study the application of the results of research to stability design of metal and composite metal-and-concrete structures and to develop comprehensive and consistent strength and performance critera and encourage consideration thereof by specification-writing bodies.

MEMBERSHIP

Organizations or firms concerned with investigation and design of metal and composite structures may be invited by the Council to become donor societies.

Euler Society: minimum fee of $2400 per year; organization and firms may appoint up to five representatives.

Johnston Society: minimum fee of $900 to $2399 per year; organization and firms may appoint up to three representatives.

Shanley Society: minimum fee of $600 to $899 per year; organization and firms may appoint up to three representatives.

Beedle Society: minimum fee of $300 to $599 per year; organization and firms may appoint up to three representatives.

Winter Society: minimum fee of $120 to $299 per year.

The voting membership of the Council consists of representatives of donor societies, members-at-large, contact members, honorary members, and student members.

Representatives: Individuals appointed by donor societies subject to the approval of the executive committee.

Members-at-large: Individuals who have expressed interest in the work of the Council and who have done or are doing work germane to its interest may be elected by the Council, following nomination by the executive committee. Fee $60 per year.

Contact members: Individuals appointed by the executive committee to maintain contact with organizations in other countries that are active in areas of interest to the Council. Fee voluntary.

Honorary members: Active Council members of appropriate age and service to SSRC may be elected to the Council, following nomination by the executive committee. Fee voluntary.

Student members: Full-time graduate students who have expressed interest in the work of the Council. Fee $20 per year.

SPONSORS (2009)

Euler Society

American Institute of Steel Construction
American Iron & Steel Institute
American Petroleum Institute
HNTB Corporation
Nucor Research & Development

Johnston Society

Chicago Bridge & Iron Company

Shanley Society

Leslie E. Robertson Assoc., RLLP
Loomis and Loomis, Inc.

Metal Building Manufactures Association
Steel Deck Institute
Walter P. Moore & Assoc., Inc.

Beedle Society

Aluminum Association
Canadian Institute of Steel Construction
Cantor Seinuk Group, PC
Columbia Engineering, Inc.
Degenkolb Engineers
Easterling, W. Samuel
Galambos, Theodore
Graef, Anhalt, Schloemer & Associates, Inc.
HDR Engineering, Inc.
Modjeski & Masters, Inc.
Sherman, Donald
Skidmore, Owings & Merrill, LLP
Steel Construction Institute
Steel Institute of New York
Steel Joist Institute
T.Y. Lin International
Thornton-Tomasetti, Engineers
Whitney, Bailey, Cox & Magnani
Wiss, Janney, Elstner Assoc., Inc.
Yolles Partnership, Inc.

ACTIVITIES (2009)

No.	Task Group	Chairman
1	Centrally Loaded Columns	R. G. Driver
4	Frame Analysis	A. A. Surovek
6	Test Methods	P. S. Green
11	International Copperation on Stability	D. R. Z. Camotim and M. Ivanyi
13	Thin-Walled Metal Construction	B. W. Schafer
14	Horizontally Curved Girders	D. G. Linzell
15	Beams	D. W. White
20	Composite Members and Systems	A. H. Varma
24	Stability under Seismic Loading	R. Tremblay
26	Stability of Angle Members	I. H. P. Mamaghani
27	Plate and Box Girders	J. S. Davidson
30	Bracing Members	J. A. Yura

OFFICERS (2009)

Chair: Ronald D. Ziemian
Vice Chair: Benjamin W. Schafer
Treasurer: Donald R. Sherman
Immediate Past Chair: W. Samuel Easterling
Faculty Liaison: Roger A. LaBoube
Administrative Secretary: Christina Stratman

EXECUTIVE COMMITTEE (2009)

Peter C. Birkemoe, University of Toronto
Dinar R. Z. Camotim, Technical University of Lisbon
W. Samuel Easterling, Virginia Tech
Perry S. Green, Steel Joist Institute
Todd A. Helwig, The University of Texas at Austin
LeRoy A. Lutz, Computerized Structural Design (Retired)
Clarence D. Miller, Consulting Structural Engineer
Benjamin W. Schafer, The John Hopkins University
Donald R. Sherman, University of Wisconsin—Milwaukee (Emeritus)
Andrea Surovek, South Dakota School of Minos
Donald W. White, Georgia Institute of Technology
Ronald D. Ziemian, Bucknell University

SSRC CONTACT INFORMATION

http://stabilitycouncil.org

NAME INDEX

B

D

E

H

I

M

N

Q

R

S

X

Y

Z

SUBJECT INDEX

B

D

E

G

J

L

M